FOUNDATIONS *of* ECOLOGY II

FOUNDATIONS

Classic Papers with

OF ECOLOGY II

Commentaries

EDITED BY
*Thomas E. Miller
and Joseph Travis*

THE UNIVERSITY OF CHICAGO PRESS
Chicago and London

The University of Chicago Press, Chicago 60637
The University of Chicago Press, Ltd., London
© 2022 by The University of Chicago
Published 2022
Printed in the United States of America

31 30 29 28 27 26 25 24 23 22   1 2 3 4 5

ISBN-13: 978-0-226-12536-7 (paper)
ISBN-13: 978-0-226-12553-4 (e-book)
DOI: https://doi.org/10.7208/chicago/9780226125534.001.0001

Library of Congress Cataloging-in-Publication Data

Names: Miller, Thomas E. (Ecologist), editor. | Travis, Joseph, 1953– editor.
Title: Foundations of ecology II : classic papers with commentaries / [compiled by]
    Thomas E. Miller and Joseph Travis.
Other titles: Foundations of ecology 2
Description: Chicago : University of Chicago Press, 2022. | Includes bibliographical
    references and index.
Identifiers: LCCN 2021055341 | ISBN 9780226125367 (paperback) | ISBN
    9780226125534 (ebook)
Subjects: LCSH: Ecology. | Plant ecology. | Animal ecology.
Classification: LCC QH541.145.F683 2022 | DDC 577—dc23/eng/20211123
LC record available at https://lccn.loc.gov/2021055341

♾ This paper meets the requirements of ANSI/NISO Z39.48-1992
(Permanence of Paper).

Our collection includes papers by leaders in the field who have passed away. The losses of Joseph Connell, Robert May, and Robert Paine, especially, remind us that this volume represents a crossroads in the history of ecology. From this crossroads, we look to a future in which ecology is led by a diverse population committed to using science to improve the environment and human welfare. We dedicate this volume to our predecessors, who have brought us to this point, and to the ecologists of the future who will lead us beyond.

Contents

PART THREE

Productivity and Resources

PART SIX

Evolutionary and Behavioral Ecology

General Introduction

*Joseph Travis, Thomas E. Miller,
and F. Helen Rodd*

       The original *Foundations of Ecology* (edited by Leslie A. Real and James H. Brown) was published in 1991 and has served as an introduction for new ecologists and a touchstone for established ecologists. A thoughtful reader would be forgiven were they to wonder just what, precisely, is a second set of foundational papers in ecology? In the strictest sense, a "foundation," once laid down, is the underlying support for all that is built upon it. Can one really recognize more than one foundation?

Clearly, we think the answer is "yes." Science does not just grow linearly, it expands and shifts. If a science is like a building with a foundation, it is a building that is constantly being enlarged. Each era of a science is a new wing added onto the existing structure, with its own foundation, with more sophisticated features, built perhaps in a different style, but always inspired by and connected to what preceded it.

The papers collected here represent the foundation of what became new halls of ecology and come from a quarter-century era from the late 1960s to the mid-1990s that encompassed many critical developments in the science of ecology. The end of the era covered by the previous volume of *Foundations* saw the onset of some of these changes, including the embrace of mathematical theory, greater rigor in experimental design, and the integration of evolutionary biology and systems science into ecology. The era reflected in the papers included here saw the dramatic growth of those trends

as well as a number of other changes, scientific and demographic, that would define ecology for years to come.

A major challenge of the first *Foundations of Ecology* was to reach back to the very beginnings, from Stephen Forbes's "The Lake as a Microcosm" in 1887 and on through to the early 1970s. Our challenge in this volume has been different: surveying a rapidly expanding field and an exponentially increasing literature from a much shorter period. The number of scientific journals publishing original ecological research grew over twice as fast from 1960 to 2000 than from 1920, when *Ecology* was first published, to 1960 (Graham and Dayton 2002). In addition, the number of pages published per year in *Ecology, Journal of Ecology,* and *Journal of Animal Ecology* increased at an increasing rate in the forty years after 1960 (Graham and Dayton 2002, fig. 3) and continues to accelerate (Gorham and Kelly 2014).

There were also significant changes in this period in how ecologists published papers and in the demographics of ecologists themselves. As we noted above, the size of the ecological literature began a rapid expansion in this period. So did the number of authors per paper. From 1970 to 1995, the proportion of papers published in *Ecology* with only one author declined from about 55% to 30%, a decline that has continued (Gorham and Kelly 2014). This pattern reflects the rise of team science, which is evident in a simple comparison of this volume with the first volume. In the first *Foundations of Ecology*, three-fourths (30 of 40) of the papers had a single author; in this collection, the proportion shrunk to below one half (22 of 54). In addition, only 2 of the 40 (5%) papers in the first volume had three or more authors; in this collection, 12 of the 54 (22%) papers had three or more authors.

There was a broadening of the types of people doing and publishing ecology, including women, racialized minorities, and ecologists from developing countries. For example, during the period covered in this volume, there was a striking rise in the proportion of women authors in ecology. We sampled 781 papers from

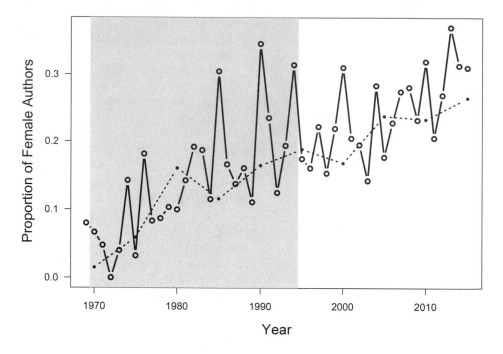

FIGURE 1. *The proportions of female authorship in the journal* Ecology *(solid line) and* American Natu-*ralist (dashed line) have increased over the last forty five years, including the period covered by this volume (grey zone). These estimates come from random subsamples of fifteen to twenty* Ecology *papers from each year and from surveying all* American Naturalist *papers in specific years at five-year intervals. Where gen-der could not be determined, the paper was not included in the presented data*

the pages of *Ecology* from the years 1969–2015 (randomly choosing 15–20 papers per year), finding that the proportion of women authors increased during the time covered in this volume, from less than 5% in the early 1970s to around 20% in 1996 and upwards of 30% by 2015 (see figure). We will discuss underrepresented groups and the process of putting together this volume further below.

Selection of Papers for This Volume

Just as was the case for the first *Foundations of Ecology* (1991), it was not easy to select the papers to include in this volume. We invited a diverse group of established colleagues to join us as a board of editors whose first task was identifying candidate papers. All of us solicited ideas from others so that, ultimately, we had suggestions from a broad swath of ecologists

from a diversity of backgrounds and disciplines. From these suggestions, we created a "super-list" of almost 200 papers and then solicited comments about each paper on the list.

This board then discussed and argued about each paper, while making new discoveries and painful cuts as the process played out. We specifically chose to rely on consensus and not select papers through balloting. We made the decision not to include review papers, book chapters, and papers that were largely statistical, but to instead focus on peer-reviewed papers that introduced influential concepts through theory and/or experiments. While continuing to consult with others, the board winnowed the superlist until we settled on the papers included in this volume. A subset of the advisers then agreed to write commentaries on sets of related papers, as was done with the first volume of *Foundations of Ecology*.

The process has not been easy and has taken over twelve years to complete. The limitations of publishing a book required us to make a hard cut at around 50 papers, selected from the tens of thousands of papers published during this time. But it was in many ways a labor of love, allowing us to reread and think hard about papers that had influenced the earlier stages of our careers, often starting from the list of papers that many of us keep on hand for our graduate students.

The decision to exclude review papers was especially difficult. To include these generally longer papers, we would need to decrease the total number of papers substantially. This decision forced us to leave out such influential papers as Paine's (1980) delineation of types of food webs, Peterson and Fry's (1987) paper on stable isotopes, Turner's (1989) and Levin's (1992) seminal works on spatial ecology, Harrison's (1991) critique of metapopulation theory, Ludwig et al.'s (1993) linking of exploitation and conservation, and Aber et al.'s (1989) description of nitrogen in forests, among many others.

This was an imperfect process. Leslie Real and James Brown discussed the issue of what makes a foundational paper very well in the first volume and we cannot improve on that discussion. We emphasize three points. First, as McIntosh (1989) discussed, influential papers do not necessarily become "citation classics." To be sure, some of the papers in this collection have been cited thousands of times, but others have been cited much less than one might expect. The real measure of a paper, as Real and Brown noted, is that it made a substantial contribution to our thinking about ecological processes. While often the paper itself is that contribution, sometimes a paper makes its contribution by inspiring its successors. Second, it is important to note that, in some cases, the individual papers selected here represent a larger body of work by an individual or a group that, in its entirety, proved enormously influential. For example, Polis's (1991) critique of food web theory has not drawn thousands of citations, but his subsequent work with Holt on food

web theory (Polis and Holt 1992) and his work on omnivory and intraguild predation, which grew out of this paper, have changed the way ecologists view food webs. Third, we would be disappointed if readers did not find fault with our selections. One of the goals of a volume such as this is to encourage discussion and debate about not only which papers we included but the many others we left out. We encourage scientists at all levels to create and share their own personal list of "foundation" papers.

Statistical Methods and the Foundations of Ecology

Our second difficult decision was to exclude papers that were focused primarily on statistical methods in ecology. The years between 1969 and 1995 saw a substantial increase in the statistical sophistication with which ecologists analyzed their data. Alongside this growth came controversies about the appropriate use of statistical inference in ecology. To have included these papers would have made this large volume much larger. However, the controversies of this period over best practices in statistical inference and the innovations that followed have exercised a profound influence on the practice of modern ecological research. An outline of some of these issues can help students and younger ecologists appreciate how ecology became a hard science and perhaps inspire some of them to assemble a list of candidate papers for a future *Foundations of Statistical Ecology*.

Ecologists have long been interested in species diversity, but quantifying that diversity has always been a challenge. In the 1950s and 1960s, ecologists drew concepts and methods from information theory to measure diversity. The stinging critique of the use of these methods by Hurlbert (1971) led to a major reassessment of just what ecologists thought they were measuring. Subsequent papers like those of Simberloff (1972), Heck et al. (1975), and Smith and van Belle (1984) offered improved methods for estimating diversity via the design of sampling programs and use of nonparametric methods for assessing sampling error and confidence intervals.

In 1984, Hurlburt offered another set of stinging criticisms, this time of the design and analysis of ecological experiments (Hurlbert 1984). This paper introduced ecologists to the term "pseudoreplication," referring to several different mistakes in study design and analysis that stem from incorrectly identifying the unit of observation. While there had been other critiques of similar practices in ecology before Hurlbert's paper, this paper became famous for the clarity of its exposition and notorious for its tone and the explicit citation of papers considered to be egregious offenders.

Younger ecologists today may be surprised at some of the elementary errors highlighted by Hurlbert. If so, this is a measure of how far ecology has advanced and the role this paper played in accelerating that advancement. Papers describing statistical methods for addressing the problems that Hurlbert diagnosed appeared soon thereafter, including the use of multivariate analysis for handling certain types of data structures (Simms and Burdick 1988) and employing so-called BACI and BACIP designs for large-scale studies (Osenberg et al. 1994; Underwood 1994). The eventual development of the computing power required to employ generalized mixed models (Bolker et al. 2009) enabled ecologists to handle many of the other data structures for which analyses using traditional regression or analyses of variance were flawed.

Other papers were also critical of how ecologists were using statistical methods. In 1990, James and McCulloch (1990) reviewed the use of multivariate statistics in ecology and taxonomy and reported high frequencies of mistakes in the use of each of the methods they reviewed. However, in this case the authors offered examples of well-done analyses for each of the techniques that they reviewed, thereby pointing their readers toward models of good practice.

Among the multivariate techniques that James and McCulloch (1990) reviewed were those used in ecological gradient analyses.

Ecological gradient analyses describe how assemblages of species change as generally abiotic habitat factors change. Ter Braak (1987) changed how ecologists approach these analyses. Prior to Ter Braak's paper, these analyses were indirect: the multivariate dataset of species occurrences or densities across locations was subjected to a dimension-reduction technique like Bray-Curtis ordination (Bray and Curtis 1957) or a principal components analysis. The ecologist then examined how well the values of a given environmental variable at a site explained the variation of species' scores along the first or second axes of the ordination.

There is no single best way to perform indirect gradient analyses, and different schools of thought advocated different approaches (see Legendre and Legendre 1998 for a full review of these methods). Ter Braak's paper championed a method called canonical correspondence analysis, which finds the vector of environmental factors that maximizes the dispersion of species scores among locations. This is a direct gradient analysis, a single analysis with a more straightforward approach and less ambiguity in interpretation.

The largest controversy over statistical inference in ecology in the period revolved around the analysis of binary matrices in biogeography. These binary matrices consist of rows that correspond to different locations and columns that correspond to different species. The entries in the matrix are "1" if a species occurs in that location and "0" if it does not. There is a statistical question and an ecological question about such matrices. The statistical question is whether the patterns of ones and zeroes are nonrandom. If the answer is "yes," then the ecological question emerges: does the pattern indicate that certain pairs of species tend to have similar occurrence patterns or that certain pairs of species have mutually exclusive occurrence patterns? These questions are at the heart of whether one can use biogeographic patterns to make inferences about the roles of phylogenetic relatedness in determining distributions or interspecific competition in doing so.

The statistical question is very difficult to

answer. One reason is that there is no single null hypothesis for a random pattern; "random" can be unconditional, conditional on the row totals, the column totals, or both. For another, there is no analytic solution for a statistical test of a pattern against any null hypothesis; randomization methods are necessary. Further, there are many reasonable but different randomization methods, with different computational requirements and different balances of statistical power and type I error (Strona et al. 2018).

The ecological question is difficult to answer because the different statistical null hypotheses suggest different mechanisms. For example, fixed row totals can imply a limit on how many species can occur in a location, which, in turn, assumes that species will compete for occupancy.

Analyzing and interpreting binary matrices of this type is an old challenge in ecology, dating back to at least Williams (1951). It became a very contentious issue in ecology when Connor and Simberloff (1979) criticized how Diamond (1975) had drawn inferences about competitive exclusion within pairs of species from such binary matrices. Different approaches to handling such matrices appeared throughout the period of this volume (e.g., Gilpin and Diamond 1982; Wilson 1987), with subsequent papers extending to the present (e.g., Wright et al. 1997; Gotelli 2000; Almeida-Neto et al. 2008; Strona et al. 2018).

Foundational Papers Not Reprinted in This Volume

Our third difficult decision was to forgo reprinting seven papers that we had chosen for this volume. The fees required by their publishers to allow us to reprint those papers were prohibitively expensive. These papers are Brander (1981), Greenwood (1980), Holt (1977), Martin and Fitzwater (1988), May (1972), Paine (1992), and Tilman et al. (1996). We hope that our readers will be able to read these papers through their institutional libraries.

Underrepresented Groups in *Foundations* Papers

We recognize that papers selected for both volumes of *Foundations of Ecology* have a strong male bias; this has caused significant discussion and concern amongst the organizers of this volume. We made every attempt to consider papers from the widest possible diversity of authors, recognizing the limits of our abilities to identify members of most minority groups based on authors' names. Despite this, from our very first list, we were aware of a dearth of authors from underrepresented groups in science, including women, racialized minorities, ecologists from non-Western countries and nonnative English speakers.

Underrepresented groups have faced, and still face, significant challenges to participation and recognition in ecology. Compiling this volume gave us the opportunity to discuss those challenges and highlight some of the initiatives for removing them. Our goal is to stimulate interest in the history of underrepresented groups in ecology and provoke discussion about how to further improve equity, diversity, and inclusion. If you are using this book in a course, we encourage you to specifically discuss and use this section to make students aware of both the past and current status of these groups, as well as encourage everyone to take proactive steps for the future.

Few data are available about the rates of participation by most underrepresented groups in ecology. In surveys by the Ecological Society of America, data for American racialized minorities show that there was not an increase in their membership between 1992 and 2005 and numbers remained well below US population estimates (Perkins 2006). However, participation by non-US-born Latinx did increase over that time period. Recent membership surveys of the Animal Behavior Society are more encouraging—over the last 10 years, gender and racial diversity have increased in the society and in the field as a whole (Beltran et al. 2020; Lee 2020).

We do have more information about the historical participation of women than for other groups in ecology. For example, the original *Foundations of Ecology* had one woman among the 54 authors (Margaret Davis, who had a sole-authored paper in 1969). Our volume has 8 women authors, about 9% of the total number of authors, and 6 of the 54 papers have women as the first authors. Our data from subsamples of the journals *Ecology* and *American Naturalist* (see figure), as well as data used in Bronstein and Bolnick (2018) and Frances et al. (2020), suggest that the proportion of women authors during this time ranged from less than 5% in the early 1970s to around 20% by 1995, so the representation of women in this book is within the range, but slightly lower than the average proportion of women publishing in ecology over this period.

The low number of women authors in this volume may be influenced by variety of conscious and unconscious obstacles that reduce the recognition of the work by underrepresented groups (Langenheim 1988, 1996; Martin 2012). Academic hierarchies are well known to reproduce and reinforce the societal norms (Bendels et al. 2018). Although the proportion of ecologists drawn from underrepresented groups is growing (for women, see figure and Perkins 2006, fig. 11), it still does not come close to reflecting society as a whole. Why, and what should be done about this?

Despite the discrimination that women have experienced historically, the proportion of women among authors in the field of ecology increased steadily in the period covered by this collection of papers (grey zone in the figure). Women's membership in the Ecological Society of America (ESA) has also increased continuously from 1970, with levels at 18% in 1987, 23% in 1992, and 40% in 2005 (Perkins 2006). In 2017, more than 50% of the Ph.D.s in the biological sciences awarded in the US were awarded to females (Frances et al. 2020). In this same period, overt sexism has been recognized in many professional contexts and in many (but not all) parts of society is less tolerated. For example, scientific societies have acknowledged explicitly the problem of overt sexism and have developed policies to fight it (e.g., ESA's Code of Conduct, https://www.esa.org/saltlake/about/code-of-conduct/). However, there is still inequity in inclusiveness for women across subdisciplines, and regions, especially in positions of power (Jones and Solomon 2019; Eaton et al. 2020). Even now, women scientists are still advancing more slowly through academic ranks (Larivière et al. 2013), publishing fewer papers during their career, receiving lower citation rates (Holman et al. 2018; Huang et al. 2020), being appointed to fewer key leadership roles, and receiving fewer top awards (National Academy of Science 2007; Ma et al. 2019). There is obviously more work to be done.

While there has been progress in the participation of and recognition for females, this is not so for other groups. For those groups, which include racialized minorities, LGBTQ2+, the economically disadvantaged, first in family to attend university, and those with visible and invisible mental and physical disabilities, progress has been glacial in pace (Perkins 2006; Gewin 2011; National Center for Science and Engineering Statistics 2018, table 19; Beltran et al. 2020).

There are many ways to improve the situation for members of all underrepresented groups. Significant problems that must be tackled are systemic discrimination, harassment, and unconscious bias—attitudes and stereotypes that impede the progress of all underrepresented groups (Vijayaraghavan et al. 2017; Gewin et al. 2020). More education is necessary about all of these problems, their prevalence, their effects, and how individuals can overcome those biases. For example, unconscious gendered bias in wording used to describe individuals in reference letters affects acceptance to graduate school and hiring into academic positions (Madera et al. 2019). Bias can be reduced by minimizing the opportunities for it to occur. For example, double-blind reviewing is being used to address concerns about conscious and unconscious bias in peer review (Fox and Paine 2019; reviewed in Fox et al. 2019).

A number of other barriers slow the inclusion of members of underrepresented groups, including inadequate advising and insufficient funds, as well as the lack of emotional and appropriate mental health support (Perkins 2006). Over the last few decades, organizations at a number of levels have started to document the inclusion and exclusion of all underrepresented groups and to implement progressive initiatives to reduce these barriers. Local and national programs have helped to stem the loss of girls and women at every stage from grade school to upper administration positions at universities (Beltran et al. 2020). Some professional societies, recognizing the value of diversity and inclusion of all types, have worked to include more individuals from minorities and other marginalized groups, including those from disadvantaged economic or educational backgrounds (Perkins 2006; Lee 2020). Lee (2020) highlights the immense value of role models for young underrepresented minority scientists and how they, in turn, go on to act as mentors and role models for new cohorts of underrepresented minorities (URMs). Many students emerge from their primary and secondary educations without confidence in their abilities (Beltran et al. 2020) and feel that they don't belong at university ("impostor syndrome"). These feelings are enhanced in URMs when they experience microaggressions (e.g., Where are you really from?), where fewer role models are available, and when they experience overt racism (Schinske et al. 2016; Jones and Solomon 2019; TEAM-UP 2020; Gewin et al. 2020).

For many scientists, their interest in science was sparked before they were eleven years old (Perkins 2006; Maltese and Tai 2010) and most were inspired by family and/or teachers. Enrichment activities in the classroom are especially important where opportunities to engage with science in the home are limited. Therefore, programs focused on early education and mentoring are critical (Maltese and Tai 2010; Rippon 2019). Career advice, mentoring, training, opportunities to engage with role models, and encouragement to take on leadership roles should then continue at all levels to help develop skill

sets, resilience, and confidence (Powell 2013; Walker 2014; Schinske et al. 2016; Jones and Solomon 2019; Gewin 2020; TEAM-UP 2020).

In some situations, affirmative action initiatives (e.g., targeted scholarships and internships, URM-specific faculty positions) have been successfully implemented, but they can be controversial. However, affirmative action may be necessary when there are few or no role models in an organization (e.g., Jones and Solomon 2019) and when the trajectory for increasing diversity is unacceptably slow. A number of factors can increase the efficacy of affirmative action hires, including ensuring that such positions are merit-based from a broad pool of applicants, hiring several new colleagues at the same time to provide critical mass, working with members of the department to change the departmental culture to be supportive of URM faculty, and providing adequate institutional support networks (Gewin 2020). URM faculty hires are like "compound interest"—they then go on themselves to make greater contributions to diversity-enhancing activities than many of their peers (Jimenez et al. 2019).

Universities can improve outlooks for undergraduates from underrepresented minority groups in several ways. Inquiry-based learning in courses, including research and field courses, increases confidence in scientific skills and builds networks, but departments must be vigilant for potential barriers to enrollment, such as fees or field requirements (Beltran et al. 2020). Paid internships and summer research jobs are also effective because they help to reduce financial roadblocks, provide experience and connections, and enhance self-confidence and a sense of belonging (TEAM-UP 2020); more of these paid positions should be established. Advising systems that can help detect students encountering challenges and that direct them to assistance, including mental health services, are key for vulnerable individuals (TEAM-UP 2020). All of these things are known to enhance retention in STEM programs and staying on a STEM career path (Beltran et al. 2020). Universities should also recognize and reward faculty and departments engaged in these activities and in outreach to members of URMs (Jimenez et al. 2019; TEAM-UP 2020).

Universities, academic societies, and funding agencies must have effective policies on equity and harassment. However, having policies on equity and harassment in place is not sufficient if those policies are not endorsed and implemented, with mechanisms to protect those reporting issues and with consequences for harassers and abusers (Vijayaraghavan et al. 2017). Communicating the importance of these policies can help clarify which behaviors are inappropriate as well as increase feelings of belonging in members of underrepresented groups (e.g., Jones and Solomon 2019). Funding agencies should continue to require and reward innovative/successful diversity programs including outreach to URMs (Gewin 2011, 2018; Jimenez et al. 2019). Organizations should regularly collect data on participation of URMs at all levels so that they can evaluate the successes of their programs and identify areas for improvement (e.g., Jones and Solomon 2019).

The proven strategies that can improve the situation for members of underrepresented groups will not make a difference unless they are implemented broadly. We look forward to a time when all underrepresented groups are appropriately represented at all levels in scientific communities. Without this, we are losing bright minds and their ability to bring new perspectives to ecology, ask new questions, and find new answers to some of its most stubborn, unanswered questions (Gewin 2018; Jimenez et al. 2019; Hofstra et al. 2020; Lee 2020).

Acknowledgments

We are grateful to Judie Bronstein, Catherine Scott, Tania Barrera, Anna Miller, Amber Gigi Hoi, Pasan Samarasin, Nicole Mideo, John Stinchcombe, Jenny Rodd, Alice Winn, Anna Li, Yun Cheng, Alex De Serrano, Ben Downer-Bartholomew, Ingrid Parker, and Chantelle Lam for extremely helpful discussions, suggestions, and feedback.

Literature Cited

* indicates a paper that we were unable to include in this volume.

Aber, J. D., K. J. Nadelhoffer, P. Steudler, and J. M. Melillo. 1989. Nitrogen saturation in northern forest ecosystems. BioScience 39: 378–386.

Almeida-Neto, M., P. Guimarães, P. R. Guimarães Jr., R. D. Loyola, and W. Ulrich. 2008. A consistent metric for nestedness analysis in ecological systems: reconciling concept and measurement. Oikos 117: 1227–1239.

Beltran, R. S., E. Marnocha, A. Race, D. A. Croll, G. H. Dayton, and E. S. Zavaleta. 2020. Field courses narrow demographic achievement gaps in ecology and evolutionary biology. Ecology and Evolution 10: 5184–5196.

Bendels, M. H. K., R. Müller, D. Brueggmann, and D. A. Groneberg. 2018. Gender disparities in high-quality research revealed by Nature Index journals. PLOS ONE 13:e0189136.

Bolker, B. M., M. E. Brooks, C. J. Clark, S. W. Geange, J. R. Poulsen, M. H. H. Stevens, and J.-S. S. White. 2009. Generalized linear mixed models: a practical guide for ecology and evolution. Trends in Ecology & Evolution 24: 127–135.

*Brander, K. 1981. Disappearance of common skate *Raia batis* from Irish sea. Nature 290 (5801): 48–49.

Bray, J. R., and J. T. Curtis. 1957. An ordination of the upland forest communities of southern Wisconsin. Ecological Monographs 27: 325–349.

Bronstein, J. L., and D. I. Bolnick. 2018. "Her joyous enthusiasm for her life-work . . .": early women authors in *The American Naturalist*. American Naturalist 192: 655–663.

Connor, E. F., and D. Simberloff. 1979. The assembly of species communities: chance or competition? Ecology 60: 1132.

Davis, M. B. 1969. Climatic changes in southern Connecticut recorded by pollen deposition at Rogers Lake. Ecology 50: 409–422.

Diamond, J. M. 1975. Assembly of species communities. Pages 342–444 in M. L. Cody and J. M. Diamond, eds. Ecology and Evolution of Communities. Belknap Press.

Eaton, A. A., J. F. Saunders, R. K. Jacobson, and K. West.

2020. How gender and race stereotypes impact the advancement of scholars in STEM: professors' biased evaluations of physics and biology post-doctoral candidates. Sex Roles 82: 127–141.

Fox, C. W., M. A. Duffy, D. J. Fairbairn, and J. A. Meyer. 2019. Gender diversity of editorial boards and gender differences in the peer review process at six journals of ecology and evolution. Ecology and Evolution 9: 13636–13649.

Fox, C. W., and C. E. T. Paine. 2019. Gender differences in peer review outcomes and manuscript impact at six journals of ecology and evolution. Ecology and Evolution 9: 3599–3619.

Frances, D. N., C. R. Fitzpatrick, J. Koprivnikar, and S. J. McCauley. 2020. Effects of inferred gender on patterns of co-authorship in ecology and evolutionary biology publications. Bulletin of the Ecological Society of America 101: e01705.

Gewin, V. 2011. Equality: the fight for access. Nature 469: 255–257.

Gewin, V. 2018. What does it take to make an institution more diverse? Nature 558: 149–151.

Gewin, V. 2020. What Black scientists want from colleagues and their institutions. Nature 583: 319–322.

Gilpin, M. E., and J. M. Diamond. 1982. Factors contributing to non-randomness in species co-occurrences on islands. Oecologia 52: 75–84.

Gorham, E., and J. Kelly. 2014. Multiauthorship, an indicator of the trend toward team research in ecology. Bulletin of the Ecological Society of America 95: 243–249.

Gotelli, N. J. 2000. Null model analysis of species co-occurrence patterns. Ecology 81: 2606–2621.

Graham, M. H., and P. K. Dayton. 2002. On the evolution of ecological ideas: paradigms and scientific progress. Ecology 83: 1481–1489.

*Greenwood, P. J. 1980. Mating systems, philopatry and dispersal in birds and mammals. Animal Behaviour 28 (Nov.): 1140–1162.

Harrison, S. 1991. Local extinction in a metapopulation context: an empirical evaluation. Biological Journal of the Linnean Society 42: 73–88.

Heck, K. L., G. van Belle, and D. Simberloff. 1975. Explicit calculation of the rarefaction diversity measurement and the determination of sufficient sample size. Ecology 56: 1459–1461.

Hofstra, B., V. V. Kulkarni, S. Munoz-Najar Galvez, B. He, D. Jurafsky, and D. A. McFarland. 2020. The Diversity–Innovation Paradox in Science. Proceedings of the National Academy of Sciences 117: 9284–9291.

Holman, L., D. Stuart-Fox, and C. E. Hauser. 2018. The gender gap in science: how long until women are equally represented? PLOS Biology 16:e2004956.

*Holt, R. D. 1977. Predation, apparent competition, and structure of prey communities. Theoretical Population Biology 12 (2): 197–229.

Huang, J., A. J. Gates, R. Sinatra, and A.-L. Barabási. 2020.

Historical comparison of gender inequality in scientific careers across countries and disciplines. Proceedings of the National Academy of Sciences 117: 4609–4616.

Hurlbert, S. H. 1971. The nonconcept of species diversity: a critique and alternative parameters. Ecology 52: 577–586.

Hurlbert, S. H. 1984. Pseudoreplication and the design of ecological field experiments. Ecological Monographs 54: 187–211.

James, F. C., and C. E. McCulloch. 1990. Multivariate analysis in ecology and systematics: panacea or Pandora's box? Annual Review of Ecology and Systematics 21: 129–166.

Jimenez, M. F., T. M. Laverty, S. P. Bombaci, K. Wilkins, D. E. Bennett, and L. Pejchar. 2019. Underrepresented faculty play a disproportionate role in advancing diversity and inclusion. Nature Ecology & Evolution 3: 1030–1033.

Jones, M. S., and J. Solomon. 2019. Challenges and supports for women conservation leaders. Conservation Science and Practice 1: e36.

Langenheim, J. H. 1988. Address of the past president: Davis, California, August 1988: The path and progress of American women ecologists. Bulletin of the Ecological Society of America 69: 184–197.

Langenheim, J. H. 1996. Early history and progress of women ecologists: emphasis upon research contributions. Annual Review of Ecology and Systematics 27: 1–53.

Larivière, V., C. Ni, Y. Gingras, B. Cronin, and C. R. Sugimoto. 2013. Bibliometrics: global gender disparities in science. Nature 504: 211–213.

Lee, D. N. 2020. Diversity and inclusion activisms in animal behaviour and the ABS: a historical view from the U.S.A. Animal Behaviour 164: 273–280.

Legendre, P., and L. Legendre. 1998. Numerical Ecology. Elsevier.

Levin, S. A. 1992. The problem of pattern and scale in ecology: the Robert H. MacArthur award lecture. Ecology 73: 1943–1967.

Ludwig, D., R. Hilborn, and C. Walters. 1993. Uncertainty, resource exploitation, and conservation: lessons from history. Science 260: 17–36.

Ma, Y., D. F. M. Oliveira, T. K. Woodruff, and B. Uzzi. 2019. Women who win prizes get less money and prestige. Nature 565: 287–288.

Madera, J. M., M. R. Hebl, H. Dial, R. Martin, and V. Valian. 2019. Raising doubt in letters of recommendation for academia: gender differences and their impact. Journal of Business and Psychology 34: 287–303.

Maltese, A. V., and R. H. Tai. 2010. Eyeballs in the fridge: sources of early interest in science. International Journal of Science Education 32: 669–685.

*Martin, J. H., and S. E. Fitzwater. 1988. Iron deficiency limits phytoplankton growth in the north-east Pacific subarctic. Nature 331 (6154): 341–343.

Martin, L. J. 2012. Where are the women in ecology?

Frontiers in Ecology and the Environment 10: 177–178.

*May, R. M. 1972. Will a large complex system be stable? Nature 238 (5364): 413–414.

McIntosh, R. P. 1989. Citation classics of ecology. Quarterly Review of Biology 64: 31–49.

National Academy of Science. 2007. Beyond Bias and Barriers: Fulfilling the Potential of Women in Academic Science and Engineering. National Academies Press.

National Center for Science and Engineering Statistics. 2018. Doctorate Recipients from U.S. Universities: 2017. Special Report NSF 19-301. National Science Foundation, Alexandria, VA.

Osenberg, C. W., R. J. Schmitt, S. J. Holbrook, K. E. Abu-Saba, and A. R. Flegal. 1994. Detection of environmental impacts: natural variability, effect size, and power analysis. Ecological Applications 4: 16–30.

Paine, R. T. 1980. Food webs: linkage, interaction strength and community infrastructure. Journal of Animal Ecology 49: 666.

*Paine, R. T. 1992. Food-web analysis through field measurement of per capita interaction strength. Nature 355 (6355): 73–75.

Perkins, A. 2006. Profile of Ecologists: Results of a Survey of the Membership of the Ecological Society of America. Ecological Society of America.

Peterson, B. J., and B. Fry. 1987. Stable isotopes in ecosystem studies. Annual Review of Ecology and Systematics 18: 293–320.

Polis, G. A. 1991. Complex trophic interactions in deserts: an empirical critique of food-web theory. American Naturalist 138: 123–155.

Polis, G. A., and R. D. Holt. 1992. Intraguild predation: the dynamics of complex trophic interactions. Trends in Ecology & Evolution 7: 151–154.

Powell, K. 2013. Higher education: on the lookout for true grit. Nature 504: 471–473.

Real, L. A., and J. H. Brown, eds. 1991. Foundations of Ecology: Classic Papers with Commentaries. University of Chicago Press.

Rippon, G. 2019. Gender and Our Brains: How New Neuroscience Explodes the Myths of the Male and Female Minds. 1st US ed. Pantheon.

Schinske, J. N., H. Perkins, A. Snyder, and M. Wyer. 2016. Scientist Spotlight Homework Assignments Shift Students' Stereotypes of Scientists and Enhance Science Identity in a Diverse Introductory Science Class. CBE—Life Sciences Education 15: ar47.

Simberloff, D. 1972. Properties of the rarefaction diversity measurement. American Naturalist 106: 414–418.

Simms, E. L., and D. S. Burdick. 1988. Profile analysis of variance as a tool for analyzing correlated responses in experimental ecology. Biometrical Journal 30: 229–242.

Smith, E. P., and G. van Belle. 1984. Nonparametric estimation of species richness. Biometrics 40: 119.

Strona, G., W. Ulrich, and N. J. Gotelli. 2018. Bi-dimensional null model analysis of presence-absence binary matrices. Ecology 99: 103–115.

TEAM-UP. 2020. The Time Is Now: Systemic Changes to Increase African Americans with Bachelor's Degrees in Physics and Astronomy. (The AIP National Task Force to Elevate African American Representation in Undergraduate Physics & Astronomy). American Institute of Physics.

Ter Braak, C. J. F. 1987. The analysis of vegetation-environment relationships by canonical correspondence analysis. Vegetatio 69: 69–77.

*Tilman, D., D. Wedin, and J. Knops. 1996. Productivity and sustainability influenced by biodiversity in grassland ecosystems. Nature 379 (6567): 718–720.

Turner, M. G. 1989. Landscape ecology: the effect of pattern on process. Annual Review of Ecology and Systematics 20: 171–197.

Underwood, A. J. 1994. On beyond BACI: sampling designs that might reliably detect environmental disturbances. Ecological Applications 4: 3–15.

Vijayaraghavan, R., K. L. Duran, K. Ramirez, J. Zelikova, E. Lescak, and 500 women scientists. 2017. It's time for science and academia to address sexual misconduct. Scientific American Blog.

Walker, C. 2014. Equality: standing out. Nature 505: 249–251.

Williams, C. B. 1951. Intra-generic competition as illustrated by Moreau's records of East African bird communities. Journal of Animal Ecology 20: 246.

Wilson, J. B. 1987. Methods for detecting non-randomness in species co-occurrences: a contribution. Oecologia 73: 579–582.

Wright, D. H., B. D. Patterson, G. M. Mikkelson, A. Cutler, and W. Atmar. 1997. A comparative analysis of nested subset patterns of species composition. Oecologia 113: 1–20.

1 Diversity and Predation

Mark McPeek and Joseph Travis

Introduction

The study of predators and prey is among ecology's oldest topics. More specifically, the study of how predator and prey affected each other's abundance, and whether they can coexist, is perhaps the oldest challenge in theoretical ecology (Kingsland 2005). For much of its early history, the scrutiny of predator-prey systems was predominantly a search for ways in which a fundamentally unstable pairwise interaction might persist (Lotka 1920; Rosenzweig and MacArthur 1963; Murdoch and Oaten 1975).

Two lines of research broadened this focus. First, laboratory studies were showing that predator-prey systems could persist longer in a highly structured spatial environment in which prey could find temporary refuges from predators or in which prey could recolonize patches from which predators had previously extirpated them (Huffaker 1958; Huffaker et al. 1963; Pimentel et al. 1963). Second, field studies were showing that selective predators promote the coexistence of competing prey species (Brooks and Dodson 1965; Connell 1970; Dodson 1974; Harper 1969; Janzen 1970; Hall et al. 1970; Paine 1966). Both lines of research forced ecologists to step back from looking at pairwise species interactions in a uniform environment

and consider selective predators and multiple prey interacting across larger, heterogeneous landscapes.

The papers in this section brought these lines of inquiry together. They carried ecology from debating dichotomies—biotic versus abiotic forces, competition versus predation—toward examining how these ecological agents acted in combination. They were also the vanguard of papers that moved ecologists from focusing on how a predator and prey or two competitors might coexist to developing a general paradigm for the maintenance of species diversity.

These papers are important for a second reason: they accelerated the integration of theoretical and empirical work. To be sure, theory and observation had been informing each other since William Thompson's models of insect host-parasite systems in the 1920s (Kingsland 1995). The papers collected here were instrumental in making ecology a fully integrated science that blended theory, observation, and experiment. The conceptually theoretical papers (Connell 1978; Menge and Sutherland 1976) drew on a substantial empirical literature in developing their sweeping hypotheses. The mathematically theoretical papers (Caswell 1978; Holt 1977) used empirical observations to motivate their questions as well as the specific mathematical approaches they deployed. The authors of the empirical studies (Dayton 1971; Lubchenco 1978; Wilbur 1972) placed their work explicitly in the context of current mathematical theory, while Menge and Sutherland (1976) combined new conceptual theory and new experimental results in a single paper.

These papers also promoted three new themes that would emerge in the ensuing years as critical facets of a unified ecology. First, spatial structure is often critical for maintaining diversity (Dayton 1971; Connell 1978; Caswell 1978; Lubchenco 1978) (see also the papers in part 4 of this volume). Second, the indirect effects of interactions between two species are strong and cannot be easily predicted from pairwise species interactions (Holt 1977; Wilbur 1972) (see also the papers in part 2 of this volume). Third, ecological systems may never achieve a locally stable equilibrium in the conventional sense (Caswell 1978; Connell 1978; Menge and Sutherland 1976) (see also the papers in part 5 of this volume).

Some of the papers also exerted a profound influence on the future conduct of ecological work. Dayton (1971), Lubchenco (1978), and Wilbur (1972) demonstrated that sophisticated ecological experiments, which included multiple trophic levels, were possible in field settings. Holt's (1977) elegant theoretical paper on apparent competition illustrated the danger of inferring ecological process from observations on patterns alone; his work demonstrated the need for empirical work for rigorously testing hypotheses about alternative processes that could generate the same patterns in nature. This theme is explored in several of the papers in part 2 of this volume.

Competition, Predation, and Indirect Effects

Four of our papers illustrate the myriad ways in which predation can combine with competition and other forces to foster diversity and organize communities. To appreciate their impact on ecology, one has to appreciate the context in which they appeared. The rise of niche theory in the 1950s and 1960s had led to an intense focus on how interspecific competition could structure communities (see also Miller and Cooper's introduction to part 2 in this volume). Specifically, how could competing species coexist while feeding on shared resources (see also Tilman 1977 [reprinted in part 2 of this volume])? Predation was an alternative structuring force because predators would keep prey densities too low for competition among them to be effective.

Viewing competition and predation as alternative forces for structuring communities may seem naïve to the modern reader. If this seems so, it reflects how thoroughly ecologists' ways of thinking have changed. What had been missing from the paradigm of "alternative forces" was an appreciation of indirect effects propagated across multiple trophic levels. In extrapolating from simple two-species models of competitive or predator-prey dynamics to effects on an entire community, ecologists focused on the direct interactions among species. These four papers, each in a different manner, demonstrated the importance of strong indirect effects of species interactions in structuring communities.

Henry Wilbur's (1972) paper "Competition, predation, and the structure of the *Ambystoma–Rana sylvatica* community" reports three separate complicated experiments performed in enclosures set into a natural pond. In one experiment, Wilbur showed that the outcome of larval competition among three closely related species of salamanders differed from what one would expect from knowing only the results of direct pairwise species interactions. This result implicated an important role for indirect effects in a relatively simple strictly competitive situation. In another experiment, an invading species could encroach on the native community but was at a competitive disadvantage that would make invasion unlikely.

In the final, critical experiment, Wilbur explored the interactions among three different types of species—an herbivore (tadpoles of *Rana sylvatica*), a secondary consumer (larval salamanders in the *Ambystoma maculatum* group), and an omnivore (larval *Ambystoma tigrinum*) that could feed on both the herbivore and the consumer. The results revealed strong indirect effects. For example, when the omnivore was absent, the survival of the herbivore was higher in the presence of the secondary consumers. Wilbur pointed out that this was likely because the secondary consumers, the salamander larvae, preyed upon the invertebrate herbivores that would otherwise compete with the tadpoles for algae and detrital food resources. When the omnivore was present, tadpole (the herbivore) survival was dramatically lower in the presence of the secondary consumer. This result was driven by predation by the omnivore on the herbivore and a dramatic reduction in the survival of the secondary consumers (the other salamanders, which were also being preyed upon by the omnivore), which in turn reduced net predation on the invertebrates that were competing with the tadpoles.

Wilbur's paper is notable for two other reasons. First, the discussion section is devoted to exploring what his experimental results meant for prevailing mathematical theories of community ecology. This was a novel feature for an entirely empirical paper and heralded the metamorphosis of ecology into a synthetic discipline. Second, like Vandermeer (1969) (see discussion in part 2 of this volume), Wilbur used the term "higher order interactions" to denote the failure of pairwise interactions to account for all of the results of his experiments. The more descriptive term, "indirect effects," would eventually replace "higher order interactions."

Robert Holt's (1977; not included in this volume) analysis of "apparent competition" in the paper "Predation, apparent competition, and structure of prey communities" brought indirect effects to the forefront of ecologists' discussions (see Holt 1977, fig. 5). This paper appeared as ecologists were reevaluating two ideas: (1) the general importance of resource competition in structuring communities in the face of increasing evidence of the importance of predation; and (2) the validity of using empirical observations as tests of theoretical predictions. This

approach had long been controversial in ecology (e.g., making inferences about competition solely from species abundance distributions: see Cohen 1968), but this paper brought the practice into sharp and critical relief.

Holt began the paper by reviewing the use of static patterns of abundance, habitat use, or resource partitioning as evidence for the importance of resource competition. He then developed a model of one predator feeding on two prey that do not compete with each other. This model showed that increasing the abundance of one prey species would decrease the abundance of the other prey species, just as if the two species were competing for resources. Further, if the predator and each prey species alone could stably coexist, then each prey species will show a reduced density when the prey species co-occur than when only of them occurs with the predator. The reduced density of the two prey species when co-occurring, compared to when one was absent, was precisely the type of observation that was being used to infer that interspecific competition was governing both species' abundances. In Holt's words (1997, 201):

> Were one unaware of the importance of the shared predator, it would be easy—but wrong—to ascribe this pattern of reduced densities to direct competitive interactions between the two species.

The remainder of the paper, indeed its larger portion, is an analysis of a multispecies assemblage in which several prey species, which do not directly compete with one another, are preyed upon by a shared predator species. Each prey species is limited in its population growth by its ability to withstand predation and transform resources into its own population growth.

The indirect connections among prey species through their shared predator produced a number of results that were counterintuitive to the ecological wisdom of the time. For example, at any moment in time the per capita growth rate of an individual prey species is limited only by the predator density and its own density relative to the resources available to it; yet that species' equilibrium density in the community depends upon the densities and growth parameters of all other prey species. As another example, two prey species can coexist if their ability to withstand the common predator is sufficiently different, a result analogous to the coexistence of competitors by reducing their overlap in the food they consume. However, two prey species that could coexist as a pair might not be able to coexist in the presence of additional prey species. Finally, a decrease in "harshness," that is, a decrease in density-independent mortality, could actually decrease, not increase, species diversity.

Holt also examined the effects of evolution in a prey species upon the community, an early example of "eco-evo feedbacks" (Travis et al. 2014; see also Greenwood 1980; Lande and Arnold 1983; Felsenstein 1985; and Berenbaum et al. 1986 [the last three are reprinted in part 6 of this volume]). He showed that if there were no trade-offs among the trio of prey growth rate, the intraspecific density-dependence of prey growth rate, and ability to withstand predation, then adaptive evolution in one prey species would increase the apparent competition among the prey and make coexistence of all prey species more difficult to attain.

Holt's paper demonstrated the enormous complexity created by indirect effects, even in a single community. The full model had implications for a variety of issues. For example, in the multispecies model, the predator limits the entire prey trophic level; however, as long as there are at least two prey species present, the growth rate of each individual species will be limited by each one's food resources. This result touched upon the older debate among Hairston et al. (1960), Murdoch (1966), and Ehrlich and Birch (1967), who argued about which factors were limiting on different trophic levels; Holt's results showed that, in some sense, all of the disputants were correct. In addition, these results gave support in mathematical theory for the idea that prey species would partition "escape space," which is a niche for resisting predators (Ricklefs and O'Rourke 1975), an idea born of empirical observations that did not, at that point, have a strong underlying theoretical foundation.

The remaining two of the four papers on the interaction of predation and competition extended the reach of experiments and hypotheses to more than one community or habitat.

Jane Lubchenco's (1978) paper "Plant species diversity in a marine intertidal community: importance of herbivore food preference and algal competitive abilities" is a defining example of how ecological context shapes a predator's role in a larger community. In this case, the complexity emerges because the preferred algal food of a consumer (the periwinkle *Littorina littorea*) is a superior competitor in one habitat and an inferior competitor in another. This was one of the first papers to show such contingent effects of a consumer on the assemblage structure of the species that it consumes.

The paper is a tour de force of careful empirical observation and experimentation, drawing on new and previously published results. Lubchenco discerned the rank order of the snails' preference for algal food among approximately three dozen species of green, brown, and red algae. She then reported the results of a series of experimental removals and additions of snails to tidal pools (where the snail's preferred food is competitively dominant) and on emergent substrata (where the snail's preferred food is competitively inferior).

The results of these elegant experiments were striking. In pools, algal diversity was highest at intermediate grazing levels, much like the pattern described by Connell (1978; see below) for frequency of disturbance. Very low grazing levels allowed the competitive dominant to prevail, and very high grazing levels eliminated most species. The effects of competition and predation balanced each other at intermediate grazing levels because the predator was disproportionately removing the competitive dominant. In contrast, on the exposed substratum, the snail was removing inferior competitors, and algal diversity was a steadily decreasing function of grazing pressure.

In the discussion, Lubchenco sets her results

against the paradoxical results of many other studies of how grazers affect the diversity of their resources and suggests that understanding contingency, specifically the relationships among consumer preference, competitive ability of resource species, and background habitat, might allow a unified view of consumer-resource relationships to emerge. The power of this approach is evident in the longevity and diversity of this paper's influence, from further studies of marine algal systems (Guerry and Menge 2017) to studies of freshwater phytoplankton (Leibold et al. 2017) and experimental metacommunities of protists (Johnston et al. 2016).

An integrated view of multiple ecological forces is the theme of Bruce Menge and John Sutherland's (1976) paper "Species diversity gradients: synthesis of the roles of predation, competition, and temporal heterogeneity," not only of conceptual theory and data in one paper, but of the roles of predation and competition in creating repeatable patterns of species richness across the different trophic levels of food webs. The paper appeared as ecologists were debating whether competition or predation would prove the more important process governing species diversity. In contrast, Menge and Sutherland argued that competition and predation were complementary forces, each being more important at different trophic levels and in communities with different levels of trophic complexity.

The hypothesis that different factors limit different trophic levels can be traced to Hairston et al. (1960), as Menge and Sutherland acknowledged (see also the discussion above of Holt 1977). Whereas Hairston et al. offered a succinct verbal argument, Menge and Sutherland offered a far-reaching argument with graphical models and extensive empirical studies of communities on two coasts. They concluded that interspecific competition would be more important at higher trophic levels and in communities of simpler trophic structure, whereas predation would be the stronger force at lower trophic levels and in communities with a more complex trophic structure. In short, where predation is stronger, competition

is weaker, hence the complementarity of the two forces. Their synthesis of natural history, rigorous experimentation and conceptual theory stands as a great advance in the growing sophistication of ecological science.

Disturbance and Nonequilibrial Perspectives

Three of our papers represent pioneering studies of the idea that ecological systems might not be at a stable equilibrium on the local scale (see also Grime 1977 [reprinted in part 6 of this volume]). Instead, ecological systems might consist of sets of species that interact on a local scale, with other forces disrupting those interactions from time to time and place to place. On the local scale, the system never settles to a stable equilibrium but instead displays nonequilibrial dynamics. In these systems, the maintenance of diversity occurs through the dynamics of the entire system as it unfolds over a large spatial area and across a long period.

The same idea about local dynamics, equilibria, and spatial scale was emerging in the study of population dynamics. The papers in part 5 of this volume describe theoretical, methodological, and empirical efforts to describe population dynamics in terms more nuanced than *density-independent* and *density-dependent* and to liberate ecological thinking from a focus on point equilibria toward a focus on steady states that might themselves be dynamic ones.

At first glance, Paul Dayton's (1971) meticulous study of an intertidal community reported in "Competition, disturbance, and community organization: the provision and subsequent utilization of space in a rocky intertidal community" might appear to be a simple analogue to studying how predation affects community structure. On deeper reading, this paper breaks completely new conceptual ground; not only does it address the role of density-independent disturbance (as opposed to selective, density-dependent predation), it is one of the first papers to demonstrate that such abiotic disturbance can have profound consequences for local species interactions and diversity.

Dayton compared species assemblages of intertidal organisms in areas with different wave exposures and combined these observations with experimental manipulations of predation and competition. Wave exposure brings the disruptive energy of the waves along with impacts from large physical structures like drifting logs. These two forces have complementary effects. Log damage dislodges mussels, the competitive dominant, and creates a patch of open space. Wave action enlarges that space by removing individuals at the periphery of the disturbed patch. The open space is then available for colonization by and subsequent competition among algae, barnacles, and mussels.

These interactions unfold against a background of predation on the encrusting animals from guilds of limpets and gastropods and the sea star *Pisaster ochraceus*. The interactions among the predators are synergistic; the effects of limpets change with the presence or absence of gastropods. The competitive dominants, a mussel and a barnacle, can grow too large for the gastropod to prey upon them. They cannot outgrow *Pisaster*. The effects of predators and log damage combine to enhance the diversity of the community, which would otherwise decrease as the competitive dominants displaced the other species.

Dayton's paper was a pioneering effort in many ways. First, it was one of the first demonstrations in nature of synergistic effects of different predators on a prey community. Second, it was the first paper to explore how the intensity and frequency of abiotic disturbances would disrupt local biotic interactions, thus forcing a more expansive view of the processes regulating diversity. In a very short time, this idea bore fruit in new mathematical theory (Levin and Paine 1974). Third, the work was an observational and experimental tour de force in studying multiple factors across multiple environmental conditions (variation in intertidal height and variation in wave exposure).

The paper has had a lasting influence on ecology. Not only is the work a seminal demon-

stration of the role of disturbance in governing species diversity, but it also offers classic demonstrations of predator-mediated coexistence, the complexity introduced when multiple predators exploit a common resource base, and the role of size-limited predation, an interaction that would be called "trait-mediated."

The role of disturbance in preventing communities from reaching a steady state is the explicit subject of Joseph Connell's (1978) essay "Diversity in tropical rain forests and coral reefs: high diversity of trees and corals is maintained only in a nonequilibrium state." In this influential paper, Connell argued that repeated disturbance disrupts the biotic interactions among species and creates patches in which different subsets of the community interact. The repeated disturbances, combined with the different sets of species with which each individual species finds itself interacting in different patches, create conditions that will increase species diversity.

Connell's ideas emerged from his empirical work in tropical forests and coral reefs. Both communities are composed of sessile organisms for which space in which an individual can settle and grow is at a premium. The paradox of these systems is that they have high species richness despite the fact that a small number of dominant species appear capable of outcompeting the rest of the species for space.

To resolve this paradox, Connell proposed the "intermediate disturbance hypothesis." In this hypothesis, physical disturbances, such as treefalls, which open space in the forest, or storm waves, which destroy segments of coral reefs, disrupt ongoing interspecific competition and "reset" the system for a new start in small patches. In each newly opened patch, a subset of species arrive, settle, and begin interacting. Which species make up this subset is determined by two factors: which species happen to be producing propagules when the disturbance occurs and which propagules are in close proximity to the newly opened patch.

Connell proposed that species diversity will be maximized at intermediate levels of disturbance. If disturbances occurred infrequently, diversity would decrease because the superior competitors would prevail. If species were equal in competitive ability, then at low frequencies of disturbance only those species most resistant to disturbance would persist. If disturbances occurred frequently, diversity would also decrease but for a different reason. At high disturbance frequencies, species with lower dispersal abilities and slower growth rates would be unable to reach enough new patches to sustain populations.

Connell's hypothesis drew on a vast literature on seemingly disparate ecosystems, coral reefs and tropical forests. It integrated several ecological agents into a synthetic hypothesis for species diversity. The intermediate disturbance hypothesis has inspired studies of communities from very small spatial scales (Santillan et al. 2019) to disturbances over long temporal scales (Kuneš et al. 2019). Subsequent theory, for example Miller et al. (2011), has defined how rates and magnitudes of disturbance can affect species diversity and the circumstances in which intermediate rates of disturbance will have the largest effects on species diversity.

In the same year that Connell offered his conceptual theory of disturbance, Hal Caswell (1978) offered a groundbreaking mathematical analysis of the interaction of a predator and its prey in an explicitly spatial context in his paper "Predator-mediated coexistence: nonequilibrium model." This paper reflected the experimental laboratory studies of predator-prey systems that had shown the importance of spatial structure in the environment for those systems to persist (Huffaker 1958; Pimentel et al. 1963). There was an emerging mathematical theory for interspecific competition in a patchy environment that examined how tradeoffs between dispersal ability and competitive ability might facilitate coexistence (Levins and Culver 1971; Horn and MacArthur 1972; Levin and Paine 1974). While Levin (1974) included a section on predation, Caswell was the first to examine predator-prey interactions explicitly in a much broader context.

Caswell's treatment stood apart from previous work because he built a general theory of predator-prey interactions that could be set in the context of either open or closed systems and characterized by either equilibrial or nonequilibrial theory. While a modern reader may take these distinctions for granted, it is important to realize that all of these ideas were relatively new and percolating among ecologists in various ways at the time. Caswell's paper was the first substantial review of these ideas that clarified their distinctions, set them in context, and showed how important those distinctions were.

Caswell's investigation of these systems revealed striking results. He showed that a general property of open systems was that transient dynamics would persist for extremely long periods. This property in turn could permit predator-mediated coexistence on the broad spatial scale that would be easier to achieve than in models of only local interactions. He offered predictions about observing nonequilibrial coexistence in nature, discussed approaches that could, in principle, falsify other theoretical possibilities for coexistence, and reviewed the empirical literature that did and did not appear congruent with nonequilibrial coexistence. The paper is remarkable for the depth to which Caswell used empirical information not only to inform theory, but to examine the theory's importance.

These papers span a narrow period, 1972 to 1978. Yet they contributed substantially to changing the focus and scope of ecological inquiry. From their syntheses of theory and data to their exploration of nonequilibrial community properties, they enlarged the vision of ecologists and helped define what modern ecology has become. If some of their insights seem intuitively obvious now, that is itself a tribute to their lasting contribution to our thinking.

Literature Cited

Berenbaum, M. R., A. R. Zangerl, and J. K. Nitao. 1986. Constraints on chemical coevolution: wild parsnips and the parsnip webworm. Evolution 40 (6): 1215–1228.

Brooks, J. L., and S. I. Dodson. 1965. Predation, body size, and composition of plankton. Science 150 (3692): 28–35.

Caswell, H. 1978. Predator-mediated coexistence: a nonequilibrium model. American Naturalist 112 (983): 127–154.

Cohen, J. E. 1968. Alternative derivations of a species-abundance relation. American Naturalist 102:165–172.

Connell, J. H. 1970. A predator-prey system in the marine intertidal region. I. *Balanus glandula* and several predatory species of *Thais*. Ecological Monographs 40 (1): 49–78.

Connell, J. H. 1978. Diversity in tropical rain forests and coral reefs. Science 199 (4335): 1302–1310.

Dayton, P. K. 1971. Competition, disturbance, and community organization: the provision and subsequent utilization of space in a rocky intertidal community. Ecological Monographs 41 (4): 351–389.

Dodson, S. I. 1974. Adaptive change in plankton morphology in response to size-selective predation: a new hypothesis of cyclomorphosis: predation and cyclomorphosis. Limnology and Oceanography 19 (5): 721–729.

Ehrlich, P. R., and L. C. Birch. 1967. The "balance of nature" and "population control." American Naturalist 101 (918): 97–107.

Felsenstein, J. 1985. Phylogenies and the comparative method. American Naturalist 125 (1): 1–15.

Greenwood, P. J. 1980. Mating systems, philopatry and dispersal in birds and mammals. Animal Behaviour 28 (4): 1140–1162.

Grime, J. P. 1977. Evidence for the existence of three primary strategies in plants and its relevance to ecological and evolutionary theory. American Naturalist 111 (982): 1169–1194.

Guerry, A. D., and B. A. Menge. 2017. Grazer impacts on algal community structure vary with the coastal upwelling regime. Journal of Experimental Marine Biology and Ecology 488 (March): 10–23.

Hairston, N. G., F. E. Smith, and L. B. Slobodkin. 1960. Community structure, population control, and competition. American Naturalist 94 (879): 421–425.

Hall, D. J., W. E. Cooper, and E. E. Werner. 1970. An experimental approach to the production dynamics and structure of freshwater animal communities. Limnology and Oceanography 15 (6): 839–928.

Harper, J. L. 1969. The role of predation in vegetational diversity. Brookhaven Symposia in Biology 22: 48–62.

Holt, R. D. 1977. Predation, apparent competition, and the structure of prey communities. Theoretical Population Biology 12 (2): 197–229.

Horn, H. S., and R. H. MacArthur. 1972. Competition among fugitive species in a harlequin environment. Ecology 53 (4): 749–752.

Huffaker, C. B. 1958. Experimental studies on predation: dispersion factors and predator-prey oscillations. Hilgardia 27 (14): 343–383.

Huffaker, C. B., K. P. Shea, and S. G. Herman. 1963. Experimental studies on predation: complex dispersion and levels of food in an acarine predator-rey Interaction. Hilgardia 34 (9): 305–330.

Janzen, D. H. 1970. Herbivores and the number of tree species in tropical forests. American Naturalist 104 (940): 501–528.

Johnston, N. K., Z. Pu, and L. Jiang. 2016. Predator identity influences metacommunity assembly. Journal of Animal Ecology 85 (5): 1161–1170.

Kingsland, S. E. 2005. The Evolution of American Ecology, 1890–2000. Johns Hopkins University Press.

Kingsland, S. E. 1995. Modeling Nature: Episodes in the History of Population Ecology. 2nd ed. University of Chicago Press.

Kuneš, P., V. Abraham, and T. Herben. 2019. Changing disturbance-diversity relationships in temperate ecosystems over the past 12000 years. Journal of Ecology 107 (4): 1678–1688.

Lande, R., and S. J. Arnold. 1983. The measurement of selection on correlated characters. Evolution 37 (6): 1210–1226.

Leibold, M. A., S. R. Hall, V. H. Smith, and D. A. Lytle. 2017. Herbivory enhances the diversity of primary producers in pond ecosystems. Ecology 98 (1): 48–56.

Levin, S. A., and R. T. Paine. 1974. Disturbance, patch formation, and community structure. Proceedings of the National Academy of Sciences 71: 2744–2747.

Levins, R., and D. Culver. 1971. Regional coexistence of species and competition between rare species. Proceedings of the National Academy of Sciences 68 (6): 1246–1248.

Lubchenco, J. 1978. Plant species diversity in a marine

intertidal community: importance of herbivore food preference and algal competitive abilities. American Naturalist 112 (983): 23–39.

Menge, B. A., and J. P. Sutherland. 1976. Species diversity gradients: synthesis of the roles of predation, competition, and temporal heterogeneity. American Naturalist 110 (973): 351–369.

Miller, A. D., S. H. Roxburgh, and K. Shea. 2011. How frequency and intensity shape diversity-disturbance relationships. Proceedings of the National Academy of Sciences 108 (14): 5643–5648.

Murdoch, W. W. 1966. "Community structure, population control, and competition": a critique. American Naturalist 100 (912): 219–226.

Paine, R. T. 1966. Food web complexity and species diversity. American Naturalist 100: 65–75.

Pimentel, D., W. P. Nagel, and J. L. Madden. 1963. Space-time structure of the environment and the survival of parasite-host systems. American Naturalist 97 (894): 141–167.

Ricklefs, R. E., and K. O'Rourke. 1975. Aspect diversity in moths: a temperate-tropical comparison. Evolution 29 (2): 313–324.

Santillan, E., H. Seshan, F. Constancias, D. I. Drautz-Moses, and S. Wuertz. 2019. Frequency of disturbance alters diversity, function, and underlying assembly mechanisms of complex bacterial communities. npj Biofilms and Microbiomes 5 (1): 8.

Tilman, D. 1977. Resource competition between planktonic algae: an experimental and theoretical approach. Ecology 58 (2): 338–348.

Travis, J., D. Reznick, R. D. Bassar, A. López-Sepulcre, R. Ferriere, and T. Coulson. 2014. Do eco-evo feedbacks help us understand nature? Answers from studies of the Trinidadian guppy. Advances in Ecological Research 50: 1–40.

Vandermeer, J. H. 1969. The competitive structure of communities: an experimental approach with protozoa. Ecology 50 (3): 362–371.

Wilbur, H. M. 1972. Competition, predation, and the structure of the *Ambystoma–Rana sylvatica* community. Ecology 53 (1): 3–21.

COMPETITION, PREDATION, AND THE STRUCTURE OF THE *AMBYSTOMA–RANA SYLVATICA* COMMUNITY[1]

HENRY M. WILBUR

Society of Fellows and Museum of Zoology, University of Michigan, Ann Arbor 48104

Abstract. Populations of six species of amphibians were manipulated in field enclosures to study the biological tractability of current concepts of the organization of natural communities. Experimental communities with a known composition of mature eggs were introduced into screen enclosures in a pond to assay the importance of competition and predation to the ecology of amphibian larvae in temporary ponds. The competitive ability of each population was measured by its survivorship, mean length of its larval period, and mean weight at metamorphosis. Three simultaneous experiments (requiring 70 enclosures and 137 populations) were replicated in a randomized complete-block design for variance analysis.

The assumptions of the classical Lotka-Volterra model of competition were tested by raising *Ambystoma laterale, Ambystoma tremblayi,* and *Ambystoma maculatum* in all combinations of three initial densities (0, 32, and 64). All three measures of competitive ability were affected by competition with other species. Higher-order interactions decreased the variance of the outcomes of the experiments as species were added to the communities. The statistical effects of these higher-order interactions between the densities of competing species often exceeded the simple effects of competition. The increase in community stability with the addition of species to the community is not predicted by the classical models of community ecology.

The second experiment tested the effects of adjacent trophic levels on the structure of the three-species community. Eggs of *Ambystoma tigrinum,* a predator, and *Rana sylvatica,* an alternate prey of *Ambystoma tigrinum,* were added singly and together into systems with 16 eggs of species in the Maculatum species-group. *Ambystoma tigrinum* was a predator if it acquired an initial size advantage by preying on *Rana sylvatica* tadpoles; otherwise it was principally a competitor. *Rana sylvatica* adversely affected the Maculatum group by competing with invertebrate prey for periphyton and photoplankton. The three species in the Maculatum group had nearly the same response to the addition of both *A. tigrinum* and *R. sylvatica.*

Ambystoma texanum, which occurs sporadically in southern Michigan at the northern limit of its range but not on the study area, was introduced as a test for community saturation. *Ambystoma texanum* was successfully raised alone. When mixed with the Maculatum group, *Ambystoma texanum* had a low survivorship, a small body size, and a long larval period. The native species were affected equally by the introduction of *Ambystoma texanum,* demonstrating the complexity of the food web and the ecological pliability of salamander larvae.

The uncertainty of the temporary pond environment precludes extreme ecological specialization among these species of salamanders. Coexistence is a consequence of the relative advantages of the species in different years and the long adult life spans. The complexity of the food web and "predator switching" are probably important elements of the density-dependent interactions that contribute to the stability of communities within seasons.

The concept of the community has its historical roots in the thought of the seventeenth-century naturalists who recognized that species of plants and animals occurred in natural assemblages. The implications of this observation were slowly realized by natural historians. By the middle of this century, ecologists were discussing a theory of the community. Allee et al. (1949:436) defined the community as "a natural assemblage of organisms which, together with its habitat, has reached a survival level such that it is relatively independent of adjacent assemblages of equal rank; to this extent, given radiant energy, it is self-sustaining." Dice (1952) added that the community is the highest level at which life can be organized. Hence, it is in the context of the community that evolution occurs.

Recently ecologists have departed from the func-

tional definition of the community to a rather arbitrary concept that defines the community as the group of organisms being studied. Ecologists now study "grassland bird communities" (Cody 1968), "Drosophila communities" (Levins 1968), and "diatom communities" (Patrick 1968). The concept of the "community matrix" (Levins 1968, Vandermeer 1970) originally was restricted to groups of species that compete with one another.

My thesis is that these natural groupings of species are organized by interspecific interactions that can be discovered and evaluated by experimentally dissecting the community into smaller components. There is a large corpus of laboratory studies, beginning with those of Chapman (1928) and Gause (1934), that involved simple ecological communities with few interacting species. These simple communities are described satisfactorily by elementary

[1] Received April 9, 1971; accepted November 18, 1971.

4   HENRY M. WILBUR   Ecology, Vol. 53, No. 1

mathematical models such as the logistic equation and its extensions. The logistic model proved inadequate as laboratory ecologists advanced from the protozoan systems of Gause (1934) to more-complex systems (Slobodkin 1953). For example, studies of flour beetles (*Tribolium*) required the addition of age structure (Park et al. 1965, Taylor 1968, Mertz 1969), genetic components (Lerner and Ho 1961; Dawson and Lerner 1962; Lerner and Dempster 1962; Park, Leslie, and Mertz 1964), and stochastic processes (Park 1954; Leslie and Gower 1958; Barnett 1962; Sokal and Huber 1963; Bartlett, Gower, and Leslie 1960; Leslic 1962; Niven 1967; Mertz and Davies 1968).

Most laboratory studies have considered only unstable associations in which one species always "wins" by driving the other to extinction. This tradition produced the "competitive exclusion principle," which was first suggested by Gause (1934:19, 45 f) and has even risen to the status of being called "Gause's axiom" (Slobodkin 1962). The period of coexistence can be prolonged by increasing the complexity of the environment (Huffaker 1958, predator-prey systems), or may be stabilized if each species has a refugium in which it has a competitive advantage (Frank 1957).

The observation that natural communities are diverse and stable for long periods of time poses the question of what limits the similarity of coexisting species. Many field ecologists have studied a sympatric pair of species with superficially similar ecologies, frequently congeners, by observing their feeding and microhabitat requirements until they discover "the differences that permit coexistence."

Ecologists have long realized the value of experimentation in the field (Clements and Weaver 1924, Gause 1936, Nicholson 1954, Murdoch 1970, St. Amant 1970), but there have been few well-designed experiments that used the theory and techniques of classical laboratory ecologists to assay the importance of competition and predation in the organization of natural communities. Connell (1961a), in a pioneering study, concluded that upper limits of intertidal invertebrates are set by physical factors, but that lower limits are set by biotic ones, such as competition for space. Eisenberg (1966, 1970) demonstrated that the availability of high-quality food may regulate the size of populations of the pond snail *Lymnaea elodes*. Paine (1966, 1969) demonstrated that a top predator (*Pisaster*) mediated competition for space among marine invertebrates. Wilbur (1971a) discussed the role of competition in maintaining the coexistence of the salamander *Ambystoma laterale* and its sexual parasite, *A. tremblayi*.

My study uses an experimental approach to assay the importance of competition among three species of aquatic salamander larvae in the organization of temporary pond communities. The effect of a predator, and an alternate food source for the predator, on the structure of the community of smaller salamanders is examined. The community is tested for ecological saturation by adding a foreign species, which occurs south of the study area.

Most ecologists require a demonstration that a resource is in short supply before an interaction between species can be termed competitive (Milne 1961). The assumption that aquatic salamander populations are limited by their food supply has not been directly tested by manipulating the densities of prey populations. The examination of stomach contents (unpublished data) and the effect of population density on body size at metamorphosis support this assumption. A more operational description of the relationship between two species is to define the relationship as competition if the two species have adverse effects on one another, mutualism if both species have a positive effect on one another, and predation or parasitism (depending on the biology of the mechanism) if one species benefits from the association at the expense of the other species. This approach defines the relationship between two species by the sign of their interaction coefficients, which permits simple statistical tests for identifying relationships between species in communities.

The amphibian community of temporary ponds in southeastern Michigan consists of four salamanders of the genus *Ambystoma* and the frog *Rana sylvatica*. The genus *Ambystoma* contains four species groups. Three of the species used in the study: *A. laterale*, *A. tremblayi*, and *A. maculatum*, are in the Maculatum group, primarily an eastern group of woodland forms. The fourth species in the study area, *A. tigrinum*, is in the Tigrinum group. This group evolved in the American Southwest and invaded the northern part of its present range in the early Pliocene (Tihen 1958). The present distributions of the four species include an area of sympatry in northern Ohio and southern Michigan. The present amphibian fauna of Michigan has probably been stable for the last 4,000–5,000 years (Smith 1957, Zumberg and Potzer 1955). Within the area of sympatry, *A. maculatum*, *A. laterale*, and *A. tremblayi* and the frog *Rana sylvatica* frequently occur in the same ponds. *A. tigrinum* is often found alone. The literature of the natural histories of these species is reviewed in Wilbur (1971a,b).

METHODS AND MATERIALS

The study was conducted on the University of Michigan's Edwin S. George Reserve, a 513-ha field station 7.2 km west of Pinckney in south-central Livingston County, Michigan. Extensive descriptions of the history and ecology of the Reserve can be found in Rogers (1942), Cantrall (1943), Cooper (1958), and Wilbur (1971*b*).

During 1970 field experiments were used to investigate the structure of the amphibian community in woodland ponds. All experiments were conducted in Burt Pond, a permanent pond near the center of the Reserve, which has been described by Young and Zimmerman (1956), Mitchell (1964), Botzler (1967), Brockelman (1969), and Wilbur (1971*b*). Burt Pond is surrounded by an oak-hickory woodlot except for the southeastern end, which opens into a tamarack bog. The pond receives water from a seepage area along its northern margin and from George Pond, a smaller pond about 100 m to the north. Burt Pond was dredged and dammed in 1942. The maximum depth is 1.25 m, the surface area is 2,655 m², and the volume is about 2,112 m³. The pond overflows in the early summer, but by late August the water has receded to expose a strip of rich muck about 3 m wide on the southern side and 1–2 m wide on the northern side. The pH is slightly acidic (6.4–6.8) in early May, but it becomes neutral by late June (Botzler 1967). Since 1963 the bottom of the permanent portions of the pond has been covered by a solid bed of *Chara* (Mitchell 1964). Predators that ambush prey (odonate naiads, *Ranatra, Lethocerus,* etc.) hunt in the *Chara* so that few areas of the pond are accessible only to swimming predators. The littoral areas have a mixture of *Chara, Ceratophyllum,* and *Potamogeton berchtoldii* and emergent vegetation of *Leersia oryzoides, Sagittaria latifolia,* and *Eleocharis erythropoda.* By mid-June about half of the surface is covered by *Lemna minor* and filamentous algae. This mat is shifted by the wind so that only the cove at the western end of the pond is constantly covered. The mat retains heat in the afternoon by buffering the effect of cooling winds (Young and Zimmerman 1956). Burt Pond supported populations of *Lepomis macrochirus, Micropterus salmoides,* and *Pimephales notatus* (R. M. Bailey, personal communication) until a winter kill in 1961–62 (Mitchell 1964). The highest trophic level is now composed of reptiles, amphibians, large insects, and leeches. During the summer of 1970 a large adult (carapace length of 30 cm) and several juvenile *Chelydra serpentina,* an adult *Natrix sipedon,* and about 30 *Chrysemys picta* were resident. Carnivorous insects large enough to capture young *Ambystoma* larvae included *Lethocerous americanus, Belostoma* sp., *Notonecta undulata,* and *Ranatra fusca.* The leech *Batrachbdella picta* was a common parasite on both adults and larval *Ambystoma.* Other leeches in the pond include *Macrobdella decora, Erpobdella punctata, Placobdella hollensis, P. parasitica, P. pappillifera,* and *Helobdella papillata* (Sawyer 1968). The newt *Notophthalmus viridescens* is also common. The anuran fauna consists of large populations of *Bufo americanus, Rana clamitans,* and *Hyla versicolor. Rana catesbiana* and *R. pipiens* are present in low densities. There is a large population of *Ambystoma tigrinum* in the pond, which was estimated in June 1970 by Bailey's (1952) triple-catch method as 1,759 ± 1,500 (SE) larvae. Larval *A. laterale* and *A. tremblayi* were occasionally taken. There are no records of *A. maculatum* in the pond.

The experimental populations were retained in "pens" with external dimensions of 60 by 60 by 240 cm constructed with 5 by 5 cm fir frames and fiberglass window screening (7 meshes per cm). Lids were made of the same screening to exclude aerial predators. Kingfishers *Megaceryle alcon* and blackbirds, *Quiscalus quiscula* and *Agelaius phoeniceus,* forage along the pond margins, and raccoons *Procycon lotor* nightly patrol the edge of the pond. The pen bottoms were made of steel-reinforced plastic sheeting to provide a solid substrate. Some of these pens were also used by Brockelman (1969), Wilbur (1971*a*), and DeBenedictis (1970:11, see photograph). The screening excluded large invertebrate predators but permitted circulation of water and the entry of food species.

On May 2 and 3, 1970, all eggs of one species were mixed in two pans to assure a heterogeneous foundation for each experimental population. This mixing randomized both genetic effects and the dates of laying of the different clutches. The experimental populations were drawn from these mixtures and recounted before introduction into the pens. Each experimental treatment was represented once in each of two "blocks" of pens. Treatments were assigned to pens by a separate random procedure for each block. Each of the source ponds from which eggs were obtained contained all four species of *Ambystoma* and *Rana sylvatica.* The developmental stage of the stock at the time of introduction on May 4 varied from advanced tail-bud embryos to day-old hatchlings. All experiments were run simultaneously.

Each pen was examined for transformed juveniles every 2 days from 50 days after introduction of the populations (June 23) until day 70 (July 13), then every third day until day 173 (October 24). The pens were checked between 0600 and 0900 EST in order to capture the salamanders before they retreated to the water as the sun warmed boards, which had been placed at the inshore end of the pens on day 40 (June 13). All amphibians under the boards were collected. This ecological definition of metamorphosis is unambiguous and uses the same criterion for both frogs and salamanders. At metamorphosis salamanders have reduced gill stubs and the rudiments of adult pigmentation patterns; frogs have reduced tails. Both forms have suspended feeding.

Animals were identified and anesthetized in chloro-tone within an hour of capture. Salamanders from pens containing both *A. laterale* and *A. tremblayi* were identified by blood samples as described in Wilbur (1971*b*). The wet weight and body length of each animal were measured before it was fixed, individually tagged, and stored in formalin. The wet weights of towel-dried specimens were determined to 0.001-g precision. Subsamples of each species were dried over silica gel at 35°–45° C for 24 hours and then stored for an additional 24 hours over calcium chloride before weighing. This treatment assured a constant dry weight for even the largest specimens of *A. tigrinum*. All specimens are deposited in the amphibian collections of the University of Michigan Museum of Zoology. Listings of the composition of each population, the number of individuals that transformed, their body lengths, weights, and larval periods, and copies of all statistical analyses are in the library of the Division of Reptiles and Amphibians, Museum of Zoology.

Analysis of three simultaneous measures of competitive ability in systems with as many as five species is complex. One measure, survivorship, is a characteristic of the population. The other two measures, body weight and larval period, are characteristics of individuals. A reductionist approach was used in the analysis. First, single-species populations were examined, then all pairs of species, and finally the three-species communities were analyzed. Insight into the biological mechanisms regulating the community was obtained at each step of this progressive increase in the complexity of the analysis. Lewontin (1968) commented that this reductionist approach has been of paramount importance to molecular biologists but is virtually untouched by evolutionary ecologists. The assumptions of my approach are that the larval stage is the period in which the diversity of salamanders is regulated and that the most complex level of the analysis is a reasonable approximation of nature, which is tantamount to saying that the field enclosures and randomized design represent an adequate sample of the pond. However, as Gause (1934:120) warned, "we may be told that after we have 'snatched' two components out of a complex natural community and placed them under 'artificial' conditions we shall certainly not obtain anything valuable and shall come to absurd conclusions."

Each of the measures of competitive ability is assumed to be associated with Darwinian fitness, the expectation of the contribution of an individual to future generations. The demonstration of the relationship between these measures of competitive ability and the actual fitness would require following individual salamanders from the time of metamorphosis until at least their first breeding.

Most amphibians breed at a certain size rather than at a specific age (Tilley 1968). A large body size,

relative to conspecifics, is assumed to be advantageous because it would increase the probability of breeding at an early age. In temporary pond habitats, an early metamorphosis is an escape from the invertebrate predators and parasites in the ponds. An early metamorphosis is also an escape from the increasing food shortage when prey populations crash during the summer as ponds shrink. In exceptionally wet years the ponds may retain water and support large populations of invertebrates throughout the summer. However, the annual rainfall in southern Michigan is virtually never greater than 125% of the average (Visher 1954). In the wet years it would be advantageous for salamanders to remain in the ponds if they are growing at a faster rate than they would in a terrestrial habitat (C. R. Shoop, personal communication). Survivorship *per se* is always advantageous, but it is a measure of the probability that a given individual in a population will live to metamorphosis.

The distributions of the lengths of the larval period and the body weights at metamorphosis were examined in each population. Tests of skewness and kurtosis were made on the larger samples. These two measures did not deviate significantly from normal distributions within populations. Within species the variances of body weights and larval periods were not correlated with the population means. Survivorship was always computed as the percentage of the individuals introduced that survived. This percentage survivorship was transformed to the arcsine of its square root for all parametric analyses (Bartlett 1947).

Competition among the Maculatum Species-Group

The three species of the Maculatum group: *A. laterale*, *A. tremblayi*, and *A. maculatum*, were raised at all combinations of three initial densities of each species: 0, 32, and 64. This $3 \times 3 \times 3$ factorial design for variance analysis has 26 treatments per replicate (the treatment of $0 + 0 + 0$ was not used). The treatments were randomly assigned to pens separately within each block. The full design included two replications of three species, each at three densities. The outcome was judged by three measures of competitive ability. The $3 \times 3 \times 3 \times 3 \times 2$ dimensionality of the data matrix is difficult to interpret. Therefore, a stepwise approach was used in the analysis. First, all pens that contained only one species were analyzed; within this analysis each measure of competitive ability was studied. Then all two-species systems were analyzed. Finally the pens with all three species were studied. Each of these analyses used separate portions of the data for independent comparisons of the species to determine the effect of increasing the complexity of the community. The re-

TABLE 1. Comparison of the species in the Maculatum group in single-species systems

	A. laterale		A. tremblayi		A. maculatum	
Initial density per pen	32	64	32	64	32	64
Number survivors (both blocks)	21	55	39	80	40	29
Percentage survivorship	33%	42%	62%	63%	63%	22%
Mean body weight (g)	0.940	0.639	0.792	0.774	0.989	1.054
Mean larval period (days from hatching)	87	94	81	92	95	91

TABLE 2. Analysis of variance in body weight of Maculatum group in single-species systems

Source	df	M.S.[a]	F	P
Species differences	2	1.335	15.837	<0.001
Density of conspecifics	1	0.389	4.620	<0.050
Species X conspecifics	2	0.595	7.061	<0.001
Error	258	0.084		

[a]Mean square.

sults from the two blocks were always pooled before the analysis to avoid missing values in four of the three-species treatments, which had no survivors of one of the species in one replicate.

Single-species systems

Four pens for each species were used at the first level of the analysis. Within each of the two blocks the species were raised alone at initial densities of 32 and 64. The data for all three species (12 pens) were used in the analysis of each measure of competitive ability. Table 1 compares the survivorship, mean body weight at metamorphosis, and the mean length of the larval period of each species. This analysis used a 3 × 2 factorial design with unequal subclass numbers to test the effects of species differences, the density of conspecifics, and the interaction between these two main effects.

The greatest difference between the species is the body weight at metamorphosis (Table 2). The order of the species by increasing body sizes at the low density was A. tremblayi, A. laterale, and A. maculatum, but at the high density the order was A. laterale, A. tremblayi, then A. maculatum. This resulted from the strong response of A. laterale to the density of conspecifics.

The length of the larval period was analyzed with the same design as for the body weight analysis. The only significant term was the interaction between the species difference and the density of conspecifics ($F_{2, 258} = 3.652$, $P < 0.05$). The order of the species, with respect to increasing length of larval period, was A. tremblayi, A. laterale, and A. maculatum at the low density, and A. maculatum, A. tremblayi, then A. laterale at the high density. A. tremblayi had the strongest response to density—

the first suggestion of a difference in adaptations for competition among the species.

The analysis of variance of the survivorship data used the transformed survivorship of each population from the date of introduction until metamorphosis. The observations from each block were treated at replicate observations in a 3 × 3 factorial design for variance analysis. There were no significant effects in this analysis.

In summary, the three species reacted differently to an increase in the density of conspecifics. Ambystoma laterale increased both the number of survivors and the percentage survivorship when the density of conspecifics was increased. Presumably, this was accomplished by dividing the available food among smaller larvae and by increasing the mean larval period. The response of A. tremblayi was similar. An increase in density resulted in a higher number of survivors (but not a significantly higher survivorship), but each animal was smaller and required a longer time to reach metamorphosis. The plastic growth rate and variable size at metamorphosis of these species are adaptations to the uncertain environment of temporary ponds. Ambystoma maculatum is less plastic. At the high density only a few larvae survived, but they were large and had a relatively rapid larval period. This suggests that the effect of larval density occurred early in the season. The few larvae that survived this competition were able to exploit the food supply and rapidly grow to a large size.

Two-species systems

A second set of pens contained all possible pairs of species, each at an initial density of 32 or 64. These populations were analyzed by a 2 × 2 × 2 factorial analysis of variance in which the factors and levels were: species (e.g., A. laterale and A. tremblayi), density of species A (32 or 64), and density of species B (32 or 64). The eight treatment combinations for each pair of species were represented once in each block. The results from the two blocks were pooled before the analysis.

A. laterale and A. tremblayi

In the absence of A. maculatum, the eight systems with A. laterale and A. tremblayi had a final mean composition of 21 A. laterale and 17 A. tremblayi. The variance (the mean square of the Euclidean distances from the final compositions to the mean final composition) around this mean was 76. Lewontin (1969) has borrowed the term "temperature" from statistical mechanics to describe this variance. This term will only cause confusion if it is used in a biological context—"the temperature of the system varied with the temperature of the environment"— so I will use "mean square" or "variance" to de-

TABLE 3. Two-species systems of *Ambystoma laterale* and *A. tremblayi*: data summary

Species	Initial density per pen *A. laterale*	*A. tremblayi*	Number of survivors (both blocks)	Percentage survivorship	Mean larval period (days)	Mean body weight (g)
A. laterale	32	32	37	58%	98	0.608
		64	60	97%	98	0.556
	64	32	46	28%	88	0.518
		64	31	24%	100	0.573
A. tremblayi	32	32	32	48%	91	0.724
		64	61	48%	84	0.612
	64	32	22	34%	82	0.732
		64	31	24%	99	0.630

scribe the scatter in the final outcomes. The significance of this outcome is tested by the analysis of variance in the survivorships. The variance between replicates was large enough to obscure differences between all treatment effects; an exception was the effect of the density of *A. laterale*, which caused a strong decrease in the survivorship of both species ($F_{1,8} = 8.900$, $P < 0.05$).

The length of the larval period (Table 3) of *A. laterale* was significantly longer than that of *A. tremblayi* in all treatment combinations ($F_{1,311} = 7.148$, $P < 0.01$). The highly significant interaction between the densities of the two species ($F_{1,311} = 10.660$, $P < 0.005$) resulted from a much stronger response to increases in the density of competitors when the density of conspecifics was also high. This nonlinear response to density will be discussed in a later section.

The mean body weight at metamorphosis was a sensitive measure of the effect of competition (Table 3). The difference between the species was highly significant (Table 4). Both species reduced their body size when in competition with either conspecifics or with the other species. At all treatment levels *A. tremblayi* had a larger body weight at metamorphosis. At a low density of conspecifics *A. laterale* was strongly affected by *A. tremblayi*, whereas the effect of *A. laterale* on *A. tremblayi* was about the same at both densities of conspecifics.

In summary, the three measures of competition reflect the species differences that were evident in the single-species systems. In general, these differences were intensified by competition. The divergence was most evident with the addition of competitors at the lower densities of conspecifics.

A. laterale and A. maculatum

The eight systems of *A. laterale* and *A. maculatum* tended toward a final composition of 21 *A.*

TABLE 4. Analysis of variance of body weight of *Ambystoma laterale* and *A. tremblayi* in two-species systems

Source	df	M.S.	F	P
Species differences	1	0.882	38.121	<0.005
Density of *A. laterale*	1	0.010	0.445	
Density of *A. tremblayi*	1	0.198	8.567	<0.005
Species X *A. laterale*	1	0.044	1.915	
Species X *A. tremblayi*	1	0.210	9.069	<0.005
A. laterale X *A. tremblayi*	1	0.061	2.633	
3-way interaction	1	0.042	1.798	
Error	311	0.023		

laterale and 16 *A. maculatum*. This poorly defined cluster had a mean square of 69. The analysis of variance of survivorship had no significant terms, a reflection of the high variance between replicates.

In contrast to survivorship, the length of the larval period and the body weight at metamorphosis are sensitive measures of competition (Table 5). *Ambystoma maculatum* had a longer larval period, especially at the low density of conspecifics. The mean square of the interaction between the species differences and the density of *A. maculatum* was greater than the mean square of either of the main effects (Table 6). This result is evidence for the importance of density-dependent competition coefficients in the system. An increase in the density of conspecifics increased the length of the larval period of *A. laterale*, but it had a mixed effect on *A. maculatum*. *Ambystoma maculatum* was more sensitive to competition with *A. laterale* at the low density of conspecifics. *Ambystoma laterale* was affected more strongly by *A. maculatum* at the high density of conspecifics.

These results are confirmed by the analysis of body weight (Tables 5 and 6). *Ambystoma maculatum* has a larger body size. Both species have a strong effect on the other species, but there is not usually a strong divergence in the characteristics of *A. late-*

TABLE 5. Two-species systems of *Ambystoma laterale* and *A. maculatum:* data summary

Species	Initial density per pen A. laterale	A. maculatum	Number of survivors (both blocks)	Percentage survivorship	Mean larval period (days)	Mean body weight (g)
A. laterale	32	32	40	62%	91	0.589
		64	26	39%	88	0.766
	64	32	67	52%	102	0.484
		64	33	26%	117	0.473
A. maculatum	32	32	15	24%	125	0.614
		64	40	31%	105	1.119
	64	32	34	53%	91	0.734
		64	41	29%	112	0.681

TABLE 6. Analyses of variance of larval period and body weight of *Ambystoma laterale* and *A. maculatum* in two-species systems

Source	Larval period (days) df	M.S.	F	P	Body weight (g) df	M.S.	F	P
Species differences	1	4605	9.614	<0.005	1	2.753	46.140	<0.005
Density of *A. laterale*	1	774	1.615		1	1.506	25.245	<0.005
Density of *A. maculatum*	1	652	1.361		1	2.014	33.744	<0.005
Species X *A. laterale*	1	430	0.899		1	0.320	5.371	<0.050
Species X *A. maculatum*	1	17767	37.093	<0.005	1	0.024	0.409	
A. laterale X *A. maculatum*	1	12979	27.096	<0.005	1	2.192	36.740	<0.005
3-way interaction	1	2304	4.810	<0.050	1	0.541	9.074	<0.005
Error	288	479			288	0.050		

rale and *A. maculatum* when in competition, compared to when they are in single-species populations. As in the analysis of larval period, there is a highly significant interaction between the densities of the two species.

Ambystoma tremblayi and A. maculatum

The eight systems with only *A. tremblayi* and *A. maculatum* had a mean outcome of 18 *A. tremblayi* and 20 *A. maculatum*. The mean square was 70, which was due as much to differences within treatments as between treatments.

Ambystoma maculatum had a significantly longer larval period than *A. tremblayi* (Tables 7 and 8). Both species increased the length of their larval period at high densities of either conspecifics or competitors. The two species reacted differently to increases in the density of *A. maculatum*. *Ambystoma tremblayi* was less affected than *A. maculatum*.

The results of the analysis of body weight are similar to those for the analysis of the length of the larval period (Tables 7 and 8). *Ambystoma maculatum* had a significantly larger body size and a stronger response to an increase in the density of conspecifics. Both species decreased their body size

in the presence of competition, but the significant interactions between the densities of the two species and the differences between species confuses the analysis.

Ambystoma tremblayi and *A. maculatum* are significantly different with respect to their body weights and the lengths of their larval periods, but they respond to density in similar ways. There is no evidence for divergence in the characteristics of the two species together compared to when they are in single-species populations.

In summary, the analyses of each pair of species have shown that differences between species have been maintained or even magnified by competition. Each species competes significantly with all other species and, in most cases, the first-order interaction coefficients are not constants but functions of density.

Three-species systems

The third set of pens initially contained either 32 or 64 individuals of each species. In several of the pens either one or two of the species were eliminated; this result was considered a valid experimental outcome. The eight treatment combinations were repli-

TABLE 7. Two-species systems of *Ambystoma tremblayi* and *A. maculatum*: data summary

Species	Initial density per pen A. tremblayi	Initial density per pen A. maculatum	Number of survivors (both blocks)	Percentage survivorship	Mean larval period (days)	Mean body weight (g)
A. tremblayi	32	32	37	58%	81	0.768
	32	64	52	81%	94	0.683
	64	32	28	22%	97	0.635
	64	64	31	24%	99	0.586
A. maculatum	32	32	30	47%	94	1.001
	32	64	57	45%	112	0.709
	64	32	25	39%	97	0.774
	64	64	51	40%	122	0.866

TABLE 8. Analyses of variance of larval period and body weight of *Ambystoma tremblayi* and *A. maculatum* in two-species systems

Source	Larval period (days) df	Larval period (days) M.S.	Larval period (days) F	Larval period (days) P	Body weight (g) df	Body weight (g) M.S.	Body weight (g) F	Body weight (g) P
Species differences	1	13549	19.071	<0.005	1	2.048	54.683	<0.005
Density of *A. tremblayi*	1	5409	7.613	<0.010	1	0.403	10.764	<0.005
Density of *A. maculatum*	1	15132	21.298	<0.005	1	0.496	13.237	<0.005
Species X *A. tremblayi*	1	297	0.418		1	0.111	2.968	
Species X *A. maculatum*	1	3737	5.261	<0.050	1	0.019	0.519	
A. tremblayi X *A. maculatum*	1	33	0.047		1	0.789	21.068	<0.005
3-way interaction	1	1641	2.310		1	0.536	14.309	<0.005
Error	303	710			303	0.037		

cated in each block, but the results were pooled before the analysis. Since each species was represented at least once for each treatment combination, the pooling of replicates circumvented the need for specific techniques to evaluate the designs with missing cells and permitted a direct comparison between the analyses of all levels of complexity. Again, the following analyses are based only on the pens that initially contained all three species; therefore, the results are statistically independent of the single-species and two-species systems. But the experimental error and environmental component of variation are the same for all levels of the analysis because all pens were randomized together within the blocks.

The final outcomes of the 16 three-species systems have a center at 12 *A. laterale,* 17 *A. tremblayi,* and 11 *A. maculatum.* The mean square of this cluster is 36. As in the two-species systems, the variance between replicates was high. The sole significant term in the analysis of variance of survivorship is the interaction between the density of *A. laterale* and the density of *A. maculatum* ($F_{1,24} = 3.604, P < 0.05$), which is due to the high survivorship of *A. laterale* and *A. tremblayi* in the pens with 32 *A. laterale* and 32 *A. maculatum* regardless of the density of *A.*

tremblayi. The mechanism of this result is not obvious; probably little biological significance should be attached to this inconsistent result among the 45 comparisons that were made.

The mean lengths of the larval periods of the three species (Table 9) were significantly different ($F_{2,524} = 12.755, P < 0.001$), as expected from the analyses of the two-species systems. The order of the species according to the lengths of their larval periods is also predicted: *A. tremblayi, A. laterale,* and *A. maculatum.* As the density of *A. maculatum* increased, *A. laterale* and *A. tremblayi* lengthened their larval periods ($F_{1,524} = 4.970, P < 0.05$). But *A. maculatum* was not significantly affected by its own density.

The body weight at metamorphosis was a more sensitive measure of the effect of competition (Table 9). Again species were significantly different (Table 10); *A. laterale* was smaller than *A. tremblayi,* and both were smaller than *A. maculatum.* The densities of *A. laterale* and *A. tremblayi* significantly affected the system both as main effects and in their interaction at the low density of *A. laterale. Ambystoma maculatum* was important in the system only through its interaction with *A. laterale* and *A. tremblayi.* The

TABLE 9. Three-species systems of the Maculatum group: data summary

Initial density of competitors			Percentage survivorship			Mean larval period (days)			Mean body weight (g)		
A. laterale	A. tremblayi	A. maculatum	A.l.	A.t.	A.m.	A.l.	A.t.	A.m.	A.l.	A.t.	A.m.
		32	34%	42%	14%	97	78	98	0.686	0.652	0.972
	32	64	17%	21%	9%	92	90	98	0.817	0.866	0.972
32		32	44%	59%	24%	97	78	110	0.575	0.596	0.854
	64	64	7%	14%	8%	95	90	110	0.411	0.428	0.968
	32	32	41%	19%	1%	89	83	108	0.494	0.685	0.807
		64	42%	7%	51%	104	105	107	0.606	0.634	0.762
64		32	34%	21%	6%	99	88	100	0.697	0.647	0.821
	64	64	6%	10%	13%	111	95	109	0.415	0.479	0.658
Grand means			26	22	13	98	88	105	0.587	0.623	0.851

TABLE 10. Analysis of variance of body weight of the Maculatum group in three-species systems

Source	df	M.S.	F	P
Species differences	2	1.112	26.742	<0.001
Density of A. laterale	1	0.497	11.953	<0.001
Density of A. tremblayi	1	0.820	19.731	<0.001
Density of A. maculatum	1	0.091	2.191	
A. laterale X A. tremblayi	1	0.309	7.436	<0.010
A. laterale X A. maculatum	1	0.219	5.257	<0.050
A. tremblayi X A. maculatum	1	0.591	14.227	<0.001
Species X A. laterale	2	0.100	2.414	
Species X A. tremblayi	2	0.063	1.518	
Species X A. maculatum	2	0.003	0.071	
Sp. X A. lat. X A. trem.	2	0.072	1.737	
Sp. X A. lat. X A. mac.	2	0.014	0.338	
Sp. X A. trem. X A. mac.	2	0.117	2.822	
A. lat. X A. trem. X A. mac.	1	0.002	0.045	
4-way interaction	2	0.114	2.735	
Error	524	0.042		

insignificance of the interactions involving species differences implies that there are no strong differences in the effect of one species on another. The significant two-way interactions between each pair of species in the system implies that competition is a function of the total number of salamanders in the system, and is not greatly influenced by the initial configuration of the system (e.g., 32-32-64, 32-64-32, or 64-32-32). This result is consistent with the strong similarity of the ecologies of the three species.

Community complexity and stability

The value of the stepwise approach is two-fold: by proceeding from the simple to the complex, intuition is accumulated that can be used to suggest mechanisms that integrate the whole community. We can further ask if the three-species system has emergent properties that could not be predicted from separate analyses of the subsystems. For example, in proceeding from the one-species to the two-species systems, the effects of species were not additive. The

impact of a competing species depended on the density of conspecifics. The importance of this nonlinearity was measured by the interaction between the density of conspecifics and the density of competitors in the analyses of variance (Tables 2, 4, 6, 8, 10). Response surfaces were construced from the data for the complete design (Fig. 1). These surfaces display the interactions between the densities of competing species. The systems without significant interaction terms have smooth responses to density. For example, A. tremblayi has an even response in body weight to increasing densities of both A. laterale and A. maculatum. The response to A. maculatum is greater as shown by the relative slope of the surface. But A. laterale has an uneven response to the density of the other species. As the densities are increased from 0 to 32 there is a sharp decrease in the body weight of A. laterale in the low-density pens. An addition of 32 more larvae of either species does not have a striking effect; but, if the densities of both species are increased to 64, there is another sharp reduction in the body size of the 32 A. laterale. This is the kind of interaction that contributed to the A. tremblayi X A. maculaum interaction in Table 5. The response of the low-density population of A. laterale was different from the response of the high-density population. This is the kind of difference that contributed to the A. laterale X A. tremblayi and A. laterale X A. maculatum components of Table 5. The species X A. maculatum interaction resulted from differences such as the contrast between the response of A. laterale and the response of A. tremblayi to the same increases in the density of A. maculatum. The highly significant interactions cause twisted and folded surfaces, such as the A. maculatum body-weight diagrams.

These nonlinearities can also be examined by analyzing the trajectories of the species compositions

12 HENRY M. WILBUR Ecology, Vol. 53, No. 1

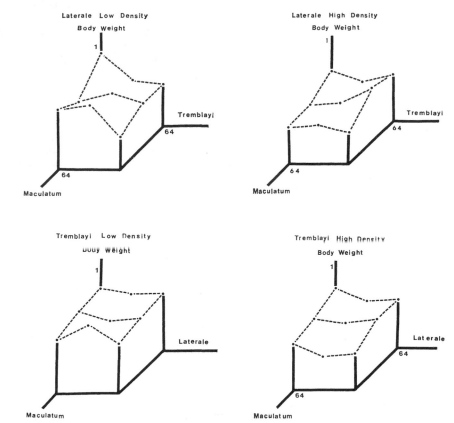

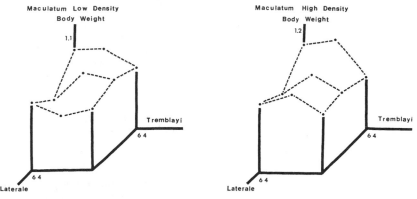

FIG. 1. Body-weight response surface for full design. Vertical axes represent mean body weights (grams) of all survivors in both replicate populations of species with low (32) or high (64) initial densities. The horizontal axes are the densities of competing species (0, 32, or 64). Smooth surfaces indicate additive interaction coefficients; complex surfaces indicate density-dependent responses. The relative amount of competition between species is measured by the symmetry of the surfaces.

of the systems from the start to the conclusion of the experiment. The trajectories of the systems demonstrated a tendency to converge on a mean composition. A linear relationship between the mean squares of the final states of the systems and the number of species in the systems is expected if the effects of species on each other are additive. Because the mean square is the sum of the Euclidean distances of each pen's composition to the mean composition, by definition the mean square v^2 is

$$v^2 = (1/[N-1]) \sum_{j=1}^{S} \sum_{i=1}^{N} (\bar{x}_j - X_{ij})^2$$

in which $\bar{x}_j$ is the mean size of the population of species j, S is the number of species in the community, and N is the number of experimental systems. In Fig. 2 the mean squares of all systems are plotted against the number of species in the communities. There is a point for each species raised alone ($N = 4$), a point for each pair of species ($N = 8$), and a point for all three-species systems ($N = 16$). By inspection the curve is decreasing. The three-species systems have a lower observed mean square than would be expected from the lower-order systems. This conclusion is robust due to the increased degrees of freedom associated with each point, a consequence of the factorial property of the experimental design.

Frequently the higher-order interactions are as important as the main effects. This result means that competition among salamander larvae is not a simple additive process that is a function of the total number of larvae in the community, but it is a complex

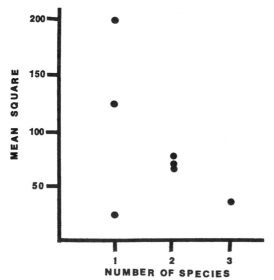

FIG. 2. Relationship between the mean square of the final compositions and the number of species in the systems.

interaction between the proportions as well as the abundances and identity of the species.

The relationship between complexity and stability has been discussed but rarely demonstrated in ecology (MacArthur 1955; Odum 1959, 1962; Leigh 1965). Hairston et al. (1969) found no relationship between diversity and stability within trophic levels of bacteria-protozoa systems. The relationship between climatic stability and tropical diversity has been widely discussed (see review by Pianka 1966), but the evidence for a return to a steady state after perturbations is scant. The salamander community is highly interactive, making the outcome of complex communities less variable than the outcome of simple communities.

THE ROLE OF PREDATION

Predation is less frequently studied than competition as an organizing mechanism in communities. The power of predation to organize communities has been demonstrated by Paine (1966, 1969, and see Spight 1967), Murdoch (1969), and Connell (1961b) in marine communities; by Hutchinson (1961), Brooks and Dodson (1965), Brooks (1968), and Maguire, Belk, and Wells (1968) in freshwater communities; and suggested by Hairston, Smith, and Slobodkin (1960) in terrestrial communities.

Methods

The second experiment manipulated the trophic levels adjacent to the secondary carnivore level occupied by the Maculatum group. In some years *Ambystoma tigrinum* larvae feed almost exclusively on smaller amphibian larvae. On June 20, 1968, 16 *A. tigrinum* larvae were caught in the Southwest Woods Pond on the George Reserve. Their snout-vent lengths ranged from 39 to 57 mm. A total of nine *Rana sylvatica* and seven *Ambystoma* tadpoles were found in 13 of the stomachs. The other three stomachs contained odonate nymphs, culicid larvae and pupae, *Chaoborus* larvae, and dytiscid larvae. There is no a priori reason to predict ecological interactions between *R. sylvatica* tadpoles and the Maculatum group (*A. laterale*, *A. tremblayi*, and *A. maculatum*) except indirectly through the complex food web (Fig. 3).

A system of 16 eggs of each species in the Maculatum group was the object of a 2×2 factorial experiment. The four combinations of *A. tigrinum* (0 or 5) and *R. sylvatica* (0 or 300) were introduced on May 2–3, 1970, into pens that were randomized within the blocks of the competition experiment. This randomization permitted the generalizations of one experiment to apply to the other. All eggs were from matings between animals from the Large West Woods Pond on the George Reserve, which contains populations of all five species. All of the *A. tigrinum*

TABLE 11. Effects of *Ambystoma tigrinum* and *Rana sylvatica* on the Maculatum group: data summary

Species	Initial density per pen *Rana sylvatica*	Percentage survivorship		Mean larval period (days)[a]		Mean body weight (g)	
		Density of 0	*A. tigrinum* 5	Density of 0	*A. tigrinum* 5	Density of 0	*A. tigrinum* 5
A. laterale	0	69%	59%	94 (22)	76 (19)	0.814	0.661
	300	59%	3%	98 (19)	126 (1)	0.746	0.793
A. tremblayi	0	56%	38%	74 (18)	74 (12)	0.826	0.771
	300	38%	3%	93 (12)	92 (1)	0.773	1.403
A. maculatum	0	50%	28%	77 (16)	86 (9)	1.086	1.061
	300	72%	22%	110 (23)	107 (7)	0.940	1.328

[a] The number of survivors from both blocks is in parentheses; all species of the Maculatum group had an initial density of 16 hatchlings per pen.

TEMPORARY POND FOOD WEB

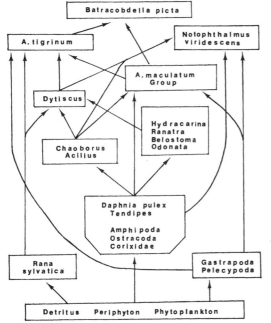

FIG. 3. Temporary pond food web. The feeding relationships are based on observations and collections in the Southwest Woods Pond, E. S. George Reserve.

eggs were from one female to avoid introducing an age structure into the small population of predators.

Results

All three species of the Maculatum group had a reduced survivorship ($F_{1,12} = 6.066, P < 0.05$) when raised with *A. tigrinum* (Table 11). The presence of *A. tigrinum* did not change the length of the larval period of the Maculatum group, except as a com-

TABLE 12. Analysis of variance of body weight of the Maculatum group in communities with *Ambystoma tigrinum* and *Rana sylvatica*

Source	df	M.S.	F	P
Species differences	2	0.681	18.775	<0.001
Density of *A. tigrinum*	1	0.253	6.985	<0.010
Density of *R. sylvatica*	1	0.214	5.900	<0.050
Species X *A. tigrinum*	2	0.180	2.966	
Species X *R. sylvatica*	2	0.075	2.075	
A. tigrinum X *R. sylvatica*	1	0.616	16.993	<0.001
Species X *A. tigrinum* X *R. sylvatica*	2	0.050	1.379	
Error	147	0.036		

ponent of the three-way interactions ($F_{2,147} = 3.235$, $P < 0.05$). In the absence of *R. sylvatica*, *A. tigrinum* reduced the body size, but in the presence of *R. sylvatica*, *A. tigrinum* increased the body size of the Maculatum group (Tables 11 and 12). *R. sylvatica* significantly increased the larval period by as much as 50 days ($F_{1,147} = 22.128, P < 0.05$).

A critical question is whether *A. tigrinum* acted as a predator or a competitor and if *R. sylvatica* had an effect on the Maculatum group in the absence of *A. tigrinum*. The reduced survivorship and the increased body size of the Maculatum group support the hypothesis of the predatory action of *A. tigrinum*. The analysis of body size suggests that *R. sylvatica* reduced the amount of food available to the Maculatum group by cropping primary production, which is the food base of the invertebrates that are eaten by the salamanders (Fig. 3). *R. sylvatica* reduced the effect of predation by *A. tigrinum*. This result confirms the observation that *R. sylvatica* is the preferred food of *A. tigrinum* in natural ponds.

The causal mechanisms regulating the amphibian community are more evident in the analysis of the effects of the experiment on the *R. sylvatica* and *A. tigrinum* populations. The small sample sizes of *A. tigrinum* preclude statistical verification of conclu-

TABLE 13. Effects of the amphibian community on *Ambystoma tigrinum*

Initial density per pen Maculatum group *Rana sylvatica*		*A. tigrinum*		
		Percentage survivorship	Mean larval period (days)[a]	Mean body weight (g)
0	0	70%	95 (7)	3.727
	300	20%	102 (2)	11.606
16 + 16 + 16	0	11%	97 (3)	5.783
	300	40%	77 (4)	5.982

[a]Sample size from both blocks is in parentheses.

TABLE 14. Effects of the amphibian community on *Rana sylvatica*

Initial density per pen Maculatum group *A. tigrinum*		*Rana sylvatica*		
		Percentage survivorship[a]	Mean larval period (days)	Mean body weight (g)
0	0	11.2% (67)	53	0.194
	5	9.0% (54)	51	0.186
16 + 16 + 16	0	15.8% (95)	53	0.174
	5	0.3% (3)	54	0.245

[a]Sample size for both blocks is in parentheses.

sions; therefore, only trends in the data will be mentioned. All *A. tigrinum* larvae were from the same clutch so there was no age structure in the populations. In the absence of vertebrate prey, many small *A. tigrinum* evenly divided the invertebrate resources (Table 13). If *R. sylvatica* was available, *A. tigrinum* had a low survivorship, long larval period, and a large body size. A few larvae probably got an initial growth advantage and were able to exploit the *R. sylvatica* population. *A. tigrinum* had a similar response to the Maculatum group. When both prey types were available, *A. tigrinum* had a very short larval period and a relatively high survivorship.

Neither *A. tigrinum* nor the Maculatum group had a significant effect on the survivorship or the length of the larval period of *R. sylvatica* tadpoles because of the great variation between replicates (Table 14). The mean body size of *R. sylvatica* was significantly affected by *A. tigrinum*, both as a main effect ($F_{1,215} = 4.763$, $P < 0.05$) and in its interaction with the Maculatum group ($F_{1,215} = 7.523$, $P < 0.01$). The mean body size of *R. sylvatica* was reduced if either *A. tigrinum* or the Maculatum group was added to the system. However, *R. sylvatica* increased in body size and reduced its survivorship if both were added. In the absence of *A. tigrinum,* the effect of the Maculatum group was to increase the survivorship, but decrease the body size, of *R. sylvatica*. The total biomass of frogs increased from 13 to 16.5 g wet-weight. This suggests that the salaman-

ders were preying on invertebrate herbivores (corixids, cladocerans, gastropods, etc.; see Fig. 3), which compete with *R. sylvatica* for food. *Ambystoma tigrinum* reduced the total biomass of *R. sylvatica,* probably by predation. In the presence of the Maculatum group, *A. tigrinum* increased its predation on *R. sylvatica* because of competition for invertebrate prey with the smaller species of *Ambystoma*. In conclusion, the five species of temporary pond amphibians have a complex relationship, which involves a mixture of competition and predation within and between preconceived "trophic levels." The very similar larvae of the Maculatum group have complex competitive interactions. There is no convincing evidence for predation among this group. *Ambystoma tigrinum* is a predator on all members of the Maculatum group if there is an opportunity to acquire a size advantage early in the summer, perhaps by eating *R. sylvatica* tadpoles. *Ambystoma tigrinum* preys on invertebrates and is a competitor of the Maculatum group if the opportunities to gain a size advantage do not occur. The role of *R. sylvatica* in the community as a herbivore and a food source for *A. tigrinum* also reduces the food available for the Maculatum complex.

This experiment suggests a possible network of biotic relationships that may regulate the entire community in temporary ponds. Briefly, primary productivity is consumed by either invertebrates or frog tadpoles. If predation is low, frogs may be food limited. Otherwise they are limited by *A. tigrinum*. The invertebrate consumers may be regulated by both salamander larvae and invertebrate predators. Beetle larvae (*Dytiscus, Acilius,* etc.) are abundant and may be as important as salamanders. The role of leech parasitism on amphibians is difficult to evaluate. Adult leeches (*Batracobdella picta*) feed on breeding salamanders in April. By late May leech broods are leaving their mother's ventral surface and attacking amphibian larvae. Very few dead larvae are found in the ponds, which suggests that leech parasitism may retard growth but is not fatal. This aspect of the larval ecology needs further study.

COMMUNITY SATURATION

The evidence of the ecology of colonization suggests that communities are ecologically saturated. Ecological saturation is operationally defined by introducing a foreign species into a community and asking if any of the native species are displaced by the invader. If the introduced species does not survive, saturation has not been demonstrated. The species may not have the physiological tolerances demanded by the physical environment. Thus, a test for ecological saturation must be accompanied by a demonstration of the ability of the species to survive in the new environment under reduced competition.

TABLE 15. Competition between *Ambystoma texanum* and the Maculatum group

Native species	Number of survivors[a]		Percentage survivorship		Mean larval period (days)		Mean body weight (g)	
	Density 0	*A. texanum*[b] 16	Density 0	*A. texanum* 16	Density 0	*A. texanum* 16	Density 0	*A. texanum* 16
A. laterale	22	20	69%	63%	94	100	0.814	0.516
A. tremblayi	18	12	56%	38%	74	85	0.826	0.595
A. maculatum	16	15	50%	47%	77	88	1.086	0.937

[a]Sample size from both blocks.
[b]Two *A. texanum* survived with a mean larval period of 101 days and a mean body weight of 0.491 g.

There are numerous examples of competitive displacements both planned and accidental. Elton (1958), Baker and Stebbins (1965), DeBach and Sundby (1963), and MacArthur and Wilson (1967) review this literature. Theoretical arguments are presented by MacArthur and Wilson (1967) and MacArthur and Levins (1967) and are applied by Orians and Horn (1969).

Methods

The final experiment added 16 hatchling *Ambystoma texanum* to a community of 16 hatchlings of each species in the Maculatum species-group (*A. laterale, A. tremblayi,* and *A. maculatum*). *A. texanum* was also raised alone at an initial density of 16 to demonstrate its ability to survive in Burt Pond. The three treatments (Maculatum group alone, *A. texanum* alone, and *A. texanum* with the Maculatum group) were replicated and randomized with the other experiments. The four species are remarkably similar in size (Table 15).

Ambystoma texanum occurs locally in extreme southern Michigan (University of Michigan Museum of Zoology [UMMZ] collections) at the northern limit of its geographical range. The materials for this study were eggs from two females from a population about 32 km south of the Reserve in Scio Township, Washtenaw County. *Ambystoma texanum* have been collected from this locality since 1942 (UMMZ 95415). *Ambystoma maculatum, A. laterale,* and *A. tremblayi* also breed in this pond (Stevan Arnold, personal communication), but they have low densities.

Results

One of the pens with a population of 16 *A. texanum* was damaged. No salamanders transformed in this pen. It is evident that *A. texanum* is physiologically able to live in Burt Pond because 7 of 16 larvae in the replicate pen survived to metamorphosis. Only two *A. texanum* survived when they were raised with the Maculatum group; both were in the same pen. The *A. texanum* from the mixed population had a smaller body size (0.492 vs. 0.610 g, not significantly different, $P > 0.05$) and a longer larval

period (101 vs. 78 days, not significantly different, $P > 0.05$) than the seven survivors from the single-species population.

In the presence of *A. texanum* each species in the Maculatum group had a reduced survivorship, but the variance between replicates was too great for the effect to be significant (Table 15). The presence of *A. texanum* caused a highly significant increase in the length of the larval period ($F_{2,96} = 7.707$, $P < 0.001$) and a highly significant reduction in the body weight at metamorphosis of the species in the Maculatum group (Table 16). There were no significant interactions between the species differences and the density of *A. texanum*, indicating that the addition of *A. texanum* to the community equally affected *A. laterale, A. tremblayi,* and *A. maculatum*.

In conclusion, *A. texanum* was able to encroach into the community, but it had a low performance with respect to all measures of competitive ability. If the initial densities of the native species had been higher, the introduction of *A. texanum* may not have been successful. This experiment emphasizes the extreme plasticity of the ecology of salamander larvae. The addition of a foreign species affected the whole community because of the complex interactions between the species and their differential ability to modify growth rates. The concept of a trophic level as a resource line that is divided among species into discrete "food niches" is too simplistic (MacArthur and Levins 1967, Colwell 1969). The resources are part of a complex food web with direct and indirect connections between species in the same trophic level (Fig. 3).

The ecological plasticity of species and the flexibility of the food web have been noted by field

TABLE 16. Analysis of variance of body weights of the Maculatum group in competition with *Ambystoma texanum*

Source	df	M.S.	F	P
Species differences	2	1.175	40.930	<0.001
Density of *A. texanum*	1	1.262	43.963	<0.001
Species X *A. texanum*	2	0.049	1.709	
Error	96	0.029		

biologists, but not laboratory and theoretical ecologists. Crowell (1962) studied the niche expansion of three species of birds on Bermuda, where there is reduced competition from other species. Lack and Southern (1949) found a similar niche expansion in the birds of Tenerife. Lack (1969) has argued that the process of ecological segregation of species may be a lengthy process. He compared the tits of Europe and North America and concluded that the European fauna is more complex because of a longer period of evolution. The salamanders are a counter-example. Lack attributes complexity of the fauna to perfection of niche segregation. The salamander fauna is complex because of the uncertainty of the pond environment, which precludes extreme specialization. The annual variation and the transience of ponds in evolutionary time do not permit the full course of competition. The complexity of the present fauna is due to the presence of good years and good ponds for each species, rather than a highly refined partitioning of a constant set of resources.

Conclusions

The classical models of laboratory ecologists are inadequate for interpreting natural communities. The inadequacies of these models were discovered by evaluating their assumptions in a complex community of predaceous vertebrates living in a changing environment. The evaluation of several assumptions of the classical models will be summarized and guidelines for constructing suitable models will be outlined.

Models of ecological communities usually are systems of differential equations. Typically each equation describes the rate change of one species through time as a function of the densities of all the species in the community. In its most general form, the model of a three-species community would be a system of three simultaneous differential equations:

$$dN_1/dt = N_1 f_1 (N_1, N_2, N_3)$$
$$dN_2/dt = N_2 f_2 (N_1, N_2, N_3)$$
$$dN_3/dt = N_3 f_3 (N_1, N_2, N_3)$$

in which f_1, f_2, and f_3 are unspecified functions of the initial densities of each species in the community. The empirical definition of these functions has been the *raison d'etre* of most laboratory studies. Charles Elton, hardly a laboratory ecologist, proposed in a presidential address to the British Ecological Society (1949) that the specification of these functions should be the goal of all ecology: "The ultimate goal of an ecological survey I would suggest is: 'an attempt to discover the main dynamic relations between populations living on an area.'" Mathematically oriented workers have specified the form of the functions by intuitively reasonable arguments (Rescigno and Richardson 1965, Kerner 1957, Rescigno 1968, Coutlee

and Jennrich 1968); other ecologists have used more empirical arguments (Lotka 1932, Volterra 1928, Gause and Witt 1935). The Lotka-Volterra formulation has been a useful model for laboratory ecologists (Gause 1934; Miller 1964, 1967; Vandermeer 1969), and has been used as a model for evolutionary arguments (MacArthur and Wilson 1967; Hairston, Tinkle, and Wilbur 1970). The form of the system of equations invented by Lotka (1932) in the form of the logistic equation

$$dN_1/dt = [r_1 N_1 (K_1 - N_1 - \alpha_{12} N_2 - \alpha_{13} N_3)]/K_1$$
$$dN_2/dt = [r_2 N_2 (K_2 - N_2 - \alpha_{21} N_1 - \alpha_{23} N_3)]/K_2$$
$$dN_3/dt = [r_3 N_3 (K_3 - N_3 - \alpha_{31} N_1 - \alpha_{32} N_2)]/K_3$$

has been a popular and intuitively acceptable model (Gause 1934, Slobodkin 1962, Levins 1968).

This model has several implicit assumptions. All individuals in the population are considered ecologically equivalent to other individuals from the moment they enter the population until they leave by death or emigration (Slobodkin 1953). The omission of age structure with its differences in mortality, fecundity, and competitive ability has not been serious in applications of the model to laboratory systems of protozoa (Gause 1934, Vandermeer 1969). More complex systems such as flour beetle communities require more sophisticated models with components for age differences and stochastic processes. The Lotka-Volterra model also assumes that the effect of one individual on another is independent of density (Smith 1952, Slobodkin 1955). If the system of equations describing the behavior of the system near its equilibrium point contains density-dependent terms (α_{ij} becomes some function of N_i), then the locus of points of no change in the rate of change of species i in the N_1, N_2, N_3 space would not be a system of three planes, but a complex of curved surfaces perhaps with several mutual intersections, each defining possible equilibria of the system. The initial abundances of the species would determine the final outcome of the community.

Another assumption of the model is that the competing species have a linear, additive effect on the capacity of a species to increase. The form of the "drag" terms $(K - N)/K$ in the Lotka-Volterra equations assumes that the rate of increase, or decrease, of species i is proportional to the remaining resources in the habitat as determined by the number of conspecifics N_i and the number of each competing species, the N_j's, which are appropriately transformed by the set of coefficients, the α_{ij}. The effect of each species is additive if each individual removes a constant share of the resources, thereby increasing the "environmental resistance" $(K - N)/K$ by an amount that is independent of the other individuals in the community. The preceding assumption of linearity requires that the share is independent

18   HENRY M. WILBUR   Ecology, Vol. 53, No. 1

of the density of species i; the additivity assumption requires that the share is independent of the species composition of the community. Mathematically, the additivity assumption neglects higher-order interactions, such as the β's in the following equations. A failure of the assumption means that the drag terms of the model must be modified to the form

$$(K_1 - N_1 - \alpha_{12}N_2 - \alpha_{13}N_3 - \beta_{1\cdot23}N_2N_3)/K_1$$
$$(K_2 - N_2 - \alpha_{21}N_1 - \alpha_{23}N_3 - \beta_{2\cdot13}N_1N_3)/K_2$$
$$(K_3 - N_3 - \alpha_{31}N_1 - \alpha_{32}N_2 - \beta_{3\cdot12}N_1N_2)/K_3$$

which is a nonlinear system of equations when it is set to zero to solve for the equilibrium densities. Such a nonlinear system does not have the analytical power of the first-order system derived from the Lotka-Volterra model (Vandermeer 1969).

This additivity assumption was originally a simplification, which has been supported by the one resource–one species law that (unlike the one gene–one enzyme law of genetics) remains a tenet of ecological theory (Levin 1970). My study does not deny the competitive exclusion principle, which is the result of laboratory studies and field observations that two species cannot coexist indefinitely on the same resource. This study does emphasize the complexity of the food web, the role of environmental uncertainty, and the extreme plasticity of the species in the structure of communities. The *A. texanum* experiment demonstrated that a community with a high density of salamanders could accept another species, not by eliminating one of the native species, but by changing the growth curves of all the species. The laboratory ecologists would argue that species could be added to the system until each species specialized on a single prey species. At this point the addition of a new species would require the extinction of another. Soon this reasoning becomes a semantic argument over the definition of the concept of a resource in ecology. Is *Daphnia* a resource? *Daphnia pulex*? Penultimate instars of *Daphnia pulex*? Penultimate instars of *Daphnia pulex* in the top 10 cm of the pond?

If the effects of species are not additive, the analytical tool of adding species to the community until the determinant and all subdeterminants of the community matrix pass through zero is also a gross oversimplification, in spite of its aura of elegance (Levins 1968, Vandermeer 1970). Nonadditivity was certainly the rule in the salamander community, both within and between species. The concept of "predator switching" (Murdoch 1969) is a step in the direction of an operational concept of community saturation. This theory tries to predict when a predator will "switch" from one resource to another as the composition of the community changes.

One assumption of the classical models withstood examination. Environmental heterogeneity did not obscure the effects of competition or the differences between species. This was true temporally (1968 [Wilbur 1971a] and 1970) and spatially (the analysis of blocks). The endurance of this important assumption is assurance that natural communities are organized by biotic mechanisms and that field experimentation can be used to elucidate the principles of community structure.

The final assumption that will be considered is common to all community models. It requires that the rules of community organization are constant throughout the study period. For example, the carrying capacity K must be constant with respect to the generation time of the species. Andrewartha and Birch (1954:661) claim that climatic fluctuations prevent insects from reaching steady-state populations that would be predicted by a model based on competitive interactions. The mosaic pattern of the environment may change the rules of community structure in space. This concept has not been exploited by ecologists except for the first steps of Levene (1953), Levins (1968), and Colwell (1969).

The validity of the Lotka-Volterra model as a description of salamander communities has several aspects. The experimental method used an input-output approach rather than the collection of time-series data. The differential growth of the species, the succession of prey species during the summer, and the variance in the length of the larval period suggest that interaction coefficients will not be constant throughout the experiment. The prolonged period of metamorphosis complicates the analysis of time-series data because the effective density is continually changing and the importance of time lags are unknown. My earlier study (Wilbur 1971a) also indicated that competition coefficients were density-dependent. The failure of an empirical model, which was based on a single species, suggested that the effect of competition cannot be modeled by a linear combination of first-order coefficients. For this reason, a statistical approach was used to identify the important components that affect the system. The factorial design tested for the effect of the density of each species but did not assume an ordinal relationship between density levels, as in the design of the empirical, regression model. If precise models of salamander communities are desired, simple, linear, additive effects of competing species must be replaced by density-dependent functions, and it must be recognized that coalitions between species may result in either increased or decreased amounts of overlap of the niches of competitors. Also, the interaction between measures of competitive ability must be considered as part of an integrated strategy for competition. If only qualitative statements about the community are desired some of these components may be omitted (Vandermeer 1969), but the price of such simplification can be determined only by knowing the behavior of the full model. The significance of interaction components when none of the main effects are significant warns against a simplistic approach to complex communities.

Most community models have described the rate of change of the number of individuals of each species through time. A few ecologists have substituted biomass for the number of individuals as a description of the state of a species. Each approach uses a single measure of the quantity of each species. A useful evolutionary description of the state of an amphibian species requires two variables: the number of individuals and their mean body weight. For example, compare the outcomes of *A. laterale* populations raised from initial densities of 32 and 64 without competition with other species. At the high density there are more survivors (mean 27.5 vs. 10.5) and more biomass (mean 35 vs. 20 g wet-weight) than from low-density populations, but the mean body weight was much less (0.63 vs. 0.94 g wet-weight). The cohort of juveniles from the low-density population would probably have a higher survivorship and an earlier age of maturity, and therefore would contribute more offspring to future generations than the cohort from the high-density population. If this is so, the traditional measures of population fitness, population size, or biomass would be erroneous. A bivariate representation of the population size and mean body weight is a more cumbersome index of population fitness, but it is a correct measure because it is a summation of the fitness of individuals. This argument refutes the indices chosen by many population geneticists (Carson 1957, 1958, biomass; Ayala 1965, 1966, 1968, 1970, 1971, population size). This problem has been recognized, in another context, by demographers. Models of populations with age structure represent the state of the population by an age vector, which has the size of each age class as elements (Leslie 1945, 1948). A model of the dynamics of an amphibian population might take the form of operations on a vector of size distributions.

Although I have played the role of the devil's advocate, I strongly feel that the classical models should not be disregarded. They are a valuable heuristic construct, and in their failure they point to the important questions of ecology. Without the simplified models, field ecologists could do little more than continue to gather data that "might be interesting." Usually they are not.

ACKNOWLEDGMENTS

I am grateful to my doctoral committee (Charles F. Walker, Nelson G. Hairston, Donald W. Tinkle, and Frank B. Livingstone) for their critical discussion and encouragement of my dissertation, which is the source of this paper. Daniel A. Livingstone, John H. Vandermeer, and the ecology group at the University of Michigan contributed ideas and techniques.

I inherited some field enclosures from Warren Y. Brockelman and Paul A. DeBenedictis and profited from their experiences. Francis C. Evans permitted the full use of the E. S. George Reserve facilities and gave me encouragement and insight. Dorothy S. Wilbur helped with the field work and typing. Kenneth Guire and the staff of the University of Michigan Statistical Research Laboratory offered advice and support, but any errors in judgment and computation are my responsibility.

I was supported by: 1967–68, E. S. George Scholarships; 1967–68, Research Assistantship, Museum of Zoology; 1968–71, National Science Foundation Graduate Fellowships; and 1970–71, Junior Fellowship, University of Michigan Society of Fellows. The research assistantship and materials were funded by a National Science Foundation grant for research in systematic and evolutionary biology (GB 8212, N. G. Hairston, principal investigator) and by the E. S. George Reserve Equipment Fund.

LITERATURE CITED

Allee, W. C., A. E. Emerson, O. Park, and K. P. Schmidt. 1949. Principles of animal ecology. Saunders, Philadelphia. 837 p.

Andrewartha, H. G., and L. C. Birch. 1954. The distribution and abundance of animals. Univ. Chicago Press, Chicago. 782 p.

Ayala, F. J. 1965. Relative fitness of populations of *Drosophila serrata* and *Drosophila birchi*. Genetics **51**: 527–544.

———. 1966. Dynamics of populations. I. Factors controlling population growth and population size in *Drosophila serrata*. Amer. Natur. **100**: 333–344.

———. 1968. Genotype, environment, and population numbers. Science **162**: 1453–1459.

———. 1970. Population fitness of geographic strains of *Drosophila serrata* as measures by interspecific competition. Evolution **24**: 483–494.

———. 1971. Competition between species: frequency dependence. Science **171**: 820–823.

Bailey, N. T. J. 1952. Improvements in the interpretation of recapture data. J. Anim. Ecol. **21**: 120–127.

Baker, H. G., and G. L. Stebbins. 1965. The genetics of colonizing species. Academic Press, New York. 588 p.

Barnett, V. D. 1962. The Monte Carlo solution of a competing species problem. Biometrics **49**: 76–103.

Bartlett, M. S. 1947. The use of transformations. Biometrics **3**: 39–52.

Bartlett, M. S., C. Gower, and P. H. Leslie. 1960. A comparison of theoretical and empirical results for some stochastic population models. Biometrika **47**: 1–11.

Botzler, R. G. 1967. The incidence of *Listeria monocytogenes, Pasteurella pseudoturberculosis,* and *Yersinia enterocolitica* in frogs and turtles of the Edwin S. George Reserve. Masters Wildl. Manage. Thesis. Univ. Mich. 70 p.

Brockelman, W. Y. 1969. An analysis of density effects and predation in *Bufo americanus* tadpoles. Ecology **50**: 632–644.

Brooks, J. L. 1968. The effect of prey size selection by lake planktivores. Syst. Zool. **17**: 272–291.

Brooks, J. L., and S. I. Dodson. 1965. Predation, body size, and composition of plankton. Science **150**: 28–35.

Cantrall, I. J. 1943. The ecology of the Orthoptera and the Dermaptera of the George Reserve, Michigan. Misc. Pub. Mus. Zool. Univ. Mich. No. 54. 182 p.

Carson, H. L. 1957. Production of biomass as a measure of fitness of experimental populations of *Drosophila*. Genetics **42**: 363–364.

——→. 1958. Increase in fitness in experimental populations resulting from heterosis. Nat. Acad. Sci. (U.S.), Proc. **44**: 1136–1141.

Chapman, R. N. 1928. The quantitative analysis of environmental factors. Ecology **9**: 111–122.

Clements, F. E., and J. E. Weaver. 1924. Experimental vegetation. Carnegie Inst., Washington, D.C. 335 p.

Cody, M. L. 1968. On the methods of resource division in grassland bird communities. Amer. Natur. **102**: 107–147.

Colwell, R. K. 1969. Ecological specialization and species diversity of tropical and temperate arthropods. Ph.D. Thesis. Univ. Mich. 79 p.

Connell, J. H. 1961a. The influence of interspecific competition and other factors on the distribution of the barnacle *Chthamalus stellatus*. Ecology **42**: 710–723

——. 1961b. Effect of competition, predation by *Thias lapillus,* and other factors on natural populations of the barnacle *Balanus balanoides*. Ecol. Monogr. **31**: 61–104.

Cooper, A. W. 1958. Plant life-forms as indicators of microclimate. Ph.D. Thesis. Univ. Mich. 387 p.

Coutlee, E. L., and R. I. Jennrich. 1968. The relevance of logarithmic models for population interaction. Amer. Natur. **102**: 307–321.

Crowell, K. L. 1962. Reduced interspecific competition among the birds of Bermuda. Ecology **43**: 75–88.

Dawson, P. S., and I. M. Lerner. 1962. Genetic variation and indetermination in interspecific competition. Amer. Natur. **96**: 379–380.

DeBach, P., and R. A. Sundby. 1963. Competitive displacement between ecological homologues. Hilgardia **35**: 145–183.

DeBenedictis, P. A. 1970. Interspecific competition between tadpoles of *Rana pipiens* and *Rana sylvatica:* an experimental field study. Ph.D. Thesis. Univ. Mich. 72 p.

Dice, L. R. 1952. Natural communities. Univ. Mich. Press, Ann Arbor. 547 p.

Eisenberg, R. M. 1966. The regulation of density in a natural population of the pond snail, *Lymnaea elodes*. Ecology **47**: 889–906.

——. 1970. The role of food in the regulation of the pond snail, *Lymnaea elodes*. Ecology **51**: 680–684.

Elton, C. 1949. Population interspersion: an essay on animal community studies. J. Ecol. **37**: 1–23.

——. 1958. The ecology of invasions by animals and plants. Methuen, London. 181 p.

Frank, P. W. 1957. Coactions in laboratory populations of two species of *Daphnia*. Ecology **38**: 510–519.

Gause, G. F. 1934. The struggle for existence. Williams and Wilkens, Baltimore. 163 p.

——. 1936. The principles of biocoenology. Quart. Rev. Biol. **11**: 320–336.

Gause, G. F., and A. A. Witt. 1935. Behavior of mixed populations and the problem of natural selection. Amer. Natur. **69**: 596–609.

Hairston, N. G., J. D. Allen, R. K. Colwell, D. J. Futuyma, J. Howell, M. D. Lubin, J. Mathias, and J. H. Vandermeer. 1969. The relationship between species diversity and stability: an experimental approach with protozoa and bacteria. Ecology **49**: 1091–1101.

Hairston, N. G., F. E. Smith, and L. B. Slobodkin. 1960. Community structure, population control, and competition. Amer. Natur. **94**: 421–425.

Hairston, N. G., D. W. Tinkle, and H. M. Wilbur. 1970. Natural selection and the parameters of population growth. J. Wildl. Manage. **34**: 681–690.

Huffaker, C. R. 1958. Experimental studies on predation: dispersion factors and predator-prey oscillations. Hilgardia **27**: 343–383.

Hutchinson, G. E. 1961. The paradox of the plankton. Amer. Natur. **95**: 137–145.

Kerner, E. H. A. 1957. A statistical mechanics of interacting biological species. Bull. Math. Biophys. **19**: 121–146.

Lack, D. 1969. Tit niches in two worlds: or homage to Evelyn Hutchinson. Amer. Natur. **103**: 43–49.

Lack, D., and H. N. Southern. 1949. Birds on Tenerife. Ibis **91**: 607–626.

Leigh, E. 1965. On the relation between productivity, biomass, diversity, and stability of a community. Nat. Acad. Sci., Proc. **53**: 777–783.

Lerner, I. M., and E. R. Demster. 1962. Indeterminism in interspecific competition. Nat. Acad. Sci. (U.S.), Proc. **48**: 821–826.

Lerner, I. M., and F. K. Ho. 1961. Genotype and competitive ability of *Tribolium* species. Amer. Natur. **95**: 329–343.

Leslie, P. H. 1945. On the use of matrices in certain population mathematics. Biometrika **33**: 183–212.

——. 1948. Some further notes on the use of matrices in population mathematics. Biometrika **35**: 213–245.

——. 1962. A stochastic model for two competing species of *Tribolium* and its application to some experimental data. Biometrika **49**: 1–25.

Leslie, P. H., and J. C. Gower. 1958. The properties of a stochastic model for two competing species. Biometrica **45**: 316–330.

Levene, H. 1953. Genetic equilibrium where more than one ecological niche is available. Amer. Natur. **87**: 331–333.

Levin, S. A. 1970. Community equilibria and stability, and an extension of the competitive exclusion principle. Amer. Natur. **104**: 413–423.

Levins, R. 1968. Evolution in changing environments. Princeton Univ. Press, Princeton. 120 p.

Lewontin, R. C. 1968. Introduction to the symposium, p. 1–4. *In* R. C. Lewontin [ed.] Population biology and evolution. Syracuse Univ. Press, Syracuse.

——. 1969. The meaning of stability, p. 13–24. *In* G. M. Woodwell and H. H. Smith [ed.] Diversity and stability in ecological systems. Brookhaven Symp. Biol. No. 22.

Lotka, A. J. 1932. The growth of mixed populations: two species competing for a common food supply. J. Washington Acad. Sci. **22**: 461–469.

MacArthur, R. H. 1955. Fluctuations of animal populations and a measure of community stability. Ecology **35**: 533–536.

MacArthur, R. H., and R. Levins. 1967. The limiting similarity, convergence, and divergence of coexisting species. Amer. Natur. **101**: 377–385.

MacArthur, R. H., and E. O. Wilson. 1967. The theory of island biogeography. Princeton Univ. Press, Princeton. 203 p.

Maguire, B., Jr., D. Belk, and G. Wells. 1968. Control of community structure by mosquito larvae. Ecology **49**: 207–210.

Mertz, D. B. 1969. Age-distribution and abundance in

populations of flour beetles. I. Experimental studies. Ecol. Monogr. **39**: 1–31.

Mertz, D. B., and R. B. Davies. 1968. Cannibalism of the pupal stage by adult flour beetles: an experiment and stochastic model. Biometrics **24**: 247–275.

Miller, R. S. 1964. Larval competition in *Drosophila melanogaster* and *D. simulans*. Ecology **45**: 132–148.

——. 1967. Pattern and process in competition. Adv. Ecol. Res. **4**: 1–74.

Milne, A. 1961. Definition of competition among animals. Symp. Exp. Biol. **15**: 44–61.

Mitchell, R. 1964. A study of sympatry in the water mite genus *Arrenurus* (Family Arrenuridae). Ecology **45**: 546–558.

Murdoch, W. W. 1969. Switching in general predators: experiments on predator specificity and stability of prey populations. Ecol. Monogr. **39**: 335–354.

——. 1970. Population regulation and population inertia. Ecology **51**: 497–502.

Nicholson, A. J. 1954. An outline of the dynamics of animal populations. Aust. J. Zool. **2**: 9–65.

Nivin, B. S. 1967. The stochastic simulation of *Tribolium* populations. Physiol. Zool. **40**: 67–82.

Odum, E. P. 1959. Fundamentals of ecology. 2nd ed. Saunders, Philadelphia. 546. p.

——. 1962. Relationship between structure and function in the ecosystem. Jap. J. Ecol. **12**: 108–118.

Orians, G. H., and H. S. Horn. 1969. Overlap in foods and foraging of four species of blackbirds in the Potholes of central Washington. Ecology **50**: 930–938.

Paine, R. T. 1966. Food web complexity and species diversity. Amer. Natur. **100**: 65–75.

——. 1969. The *Pisaster-Tegula* interaction: prey patches, predator food preference, and intertidal community structure. Ecology **50**: 950–961.

Park, T. 1954. Competition: an experimental and statistical study, p. 175–195. *In* O. Kempthorne, T. A. Bancroft, J. L. Gowen, and J. L. Lush [ed.] Statistics and mathematics in biology. Univ. Iowa Press, Ames.

Park, T., P. H. Leslie, and D. B. Mertz. 1964. Genetic strains and competition in populations of *Tribolium*. Physiol. Zool. **37**: 97–162.

Park, T., D. B. Mertz, W. Grodzinski, and T. Prus. 1965. Cannibalistic predation in populations of flour beetles. Physiol. Zool. **38**: 289–321.

Patrick, R. 1968. The structure of diatom communities in similar ecological conditions. Amer. Natur. **102**: 173–183.

Pianka, E. R. 1966. Latitudinal gradients in species diversity: a review of concepts. Amer. Natur. **100**: 33–46.

Rescigno, A. 1968. The struggle for life: II. Three competitors. Bull. Math. Biophys. **30**: 291–298.

Rescigno, A., and I. W. Richardson. 1965. On the competitive exclusion principle. Bull. Math. Biophys. **27**: 85–89.

Rogers, J. S. 1942. The crane flies (Tipulidae) of the George Reserve, Michigan. Misc. Pub. Mus. Zool. Univ. Mich. No. 53. 128 p.

St. Amant, J. 1970. The detection of regulation in animal populations. Ecology **51**: 823–828.

Sawyer, R. T. 1968. Notes on the natural history of the leeches (Hirundinea) on the George Reserve, Michigan. Ohio J. Sci. **68**: 226–228.

Slobodkin, L. B. 1953. An algebra of population growth. Ecology **34**: 513–519.

——. 1955. Conditions for population equilibrium. Ecology **36**: 530–533.

——. 1962. Growth and regulation of animal populations. Holt, Rinehart and Winston, New York. 184 p.

Smith, F. E. 1952. Experimental methods in population dynamics: a critique. Ecology **33**: 441–450.

Smith, P. W. 1957. An analysis of post-Wisconsin biogeography of the Prairie Peninsula Region based on distribution phenomena among terrestrial vertebrate populations. Ecology **38**: 205–218.

Sokal, R. R., and I. Huber. 1963. Competition among genotypes in *Tribolium casteneum* at varying densities and gene frequencies (the *sooty* locus). Amer. Natur. **97**: 169–184.

Spight, T. M. 1967. Species diversity: a comment on the role of the predator. Amer. Natur. **101**: 467–474.

Taylor, N. W. 1968. A mathematical model for two *Tribolium* populations in competition. Ecology **49**: 843–848.

Tihen, J. A. 1958. Comments on the osteology and phylogeny of ambystomatid salamanders. Bull. Florida State Mus. **3**: 1–50.

Tilley, S. G. 1968. Size-fecundity relationships and their evolutionary implications in five desmognathine salamanders. Evolution **22**: 806–816.

Vandermeer, J. H. 1969. The competitive structure of communities: an experimental approach with protozoa. Ecology **50**: 362–371.

——. 1970. The community matrix and the number of species in a community. Amer. Natur. **104**: 73–83.

Visher, S. S. 1954. Climatic atlas of the United States. Harvard Univ. Press, Cambridge.

Volterra, V. 1928. Variations and fluctuations of the number of individuals in animal species living together. *Trans.* by M. E. Wells, p. 409–448. *In* R. N. Chapman. 1931. Animal ecology. McGraw-Hill, New York.

Wilbur, H. M. 1971*a*. The ecological relationship of the salamander *Abystoma laterale* to its all-female, gynogenetic associate. Evolution **25**: 168–179.

——. 1971*b*. Competition, predation and the structure of the *Ambystoma–Rana sylvatica* community. Ph.D. Thesis. Univ. Mich. 142 p.

Young, F. N., and J. R. Zimmerman. 1956. Variations in temperature in small aquatic situations. Ecology **37**: 609–611.

Zumberg, J. H., and J. E. Potzer. 1955. Pollen profiles, radiocarbon dating, and geological chronology of the Lake Michigan Basin. Science **121**: 309–311.

Vol. 112, No. 983   The American Naturalist   January–February 1978

PLANT SPECIES DIVERSITY IN A MARINE INTERTIDAL
COMMUNITY: IMPORTANCE OF HERBIVORE
FOOD PREFERENCE AND ALGAL COMPETITIVE ABILITIES

JANE LUBCHENCO*

Biological Laboratories, Harvard University, Cambridge, Massachusetts 02138

Since Hutchinson (1959) drew attention to the question "Why are there so many kinds of animals?" investigation of the causes of species diversity has proven to be a fertile area of ecological endeavor. A major current emphasis is on mechanisms creating and/or maintaining diversity (Connell 1971, 1975; Dayton 1971, 1975; Dayton et al. 1974; Jackson and Buss 1975; Janzen 1970; MacArthur 1972; B. Menge and Sutherland 1976; Paine 1966, 1971, 1974; Pianka 1967, 1969; Ricklefs 1973). One such mechanism, the predation hypothesis, suggests that predators, by keeping the abundance of their prey in check, prevent competitive exclusion and thus permit or maintain a higher species richness than would occur in their absence (Paine 1966, 1971).

Experimental removals or additions of aquatic carnivores have resulted in changes (decreases or increases, respectively) in the local species diversity of lower trophic levels over ecological time (Paine 1966, 1971, 1974; Hall et al. 1970; B. Menge 1976). In contrast, the effect of herbivores on local species diversity patterns is confusing, in part because few experimental studies have been done. In some instances, herbivores appear to increase plant diversity (Harper 1969; Paine and Vadas 1969), decrease plant diversity (Harper 1969), or both (Harper 1969; Paine and Vadas 1969; Vadas 1968). The key to understanding such variable results may reside in understanding consumer prey preferences and competitive abilities of the food species. A number of authors have suggested that only when a consumer (predator or herbivore) preferentially feeds on the competitively dominant prey can the consumer increase diversity (Hall et al. 1970; Harper 1969; MacArthur 1972; Paine 1971; Patrick 1970; Van Valen 1974). In this paper I present results of an experimental evaluation of the effect of generalized herbivores on plant diversity in a rocky intertidal community. In this system, knowledge of (1) food preferences of the herbivores, (2) competitive relationships between the plants, and (3) how these relationships change according to physical regimes in microhabitats permits an analysis of the importance of the relationship between herbivore food preference and competitive ability of the plants.

* Present address: Department of Zoology, Oregon State University, Corvallis, Oregon 97331.

Amer. Natur. 1978. Vol. 112, pp. 23–39.

TABLE 1

FOOD PREFERENCES OF *Littorina littorea**

Preference Ranking	Chlorophyceae (Greens)	Phaeophyceae (Browns)	Rhodophyceae (Reds)
High	*Cladophora*	*Ectocarpus-Pylaiella*	*Ceramium*
	Enteromorpha	*Elachistea*	*Porphyra*
	Monostroma	*Petalonia*	...
	Spongomorpha	*Scytosiphon*	...
	Ulva	...	...
	Ulothrix-Urospora	...	...
Medium	*Rhizoclonium*	*Dictyosiphon*	*Asparagopsis*
	...	...	*Cystoclonium*
	...	...	*Dumontia*
	...	...	*Halosaccion*
	...	...	*Phycodrys*
	...	...	*Polysiphonia lanosa*
	...	...	*P. flexicaulis*
Low	*Chaetomorpha*	*Agarum*	*Ahnfeltia*
	Codium	*Ascophyllum*	*Chondrus*
	...	*Chorda*	*Euthora*
	...	*Chordaria*	*Gigartina*
	...	*Desmarestia*	*Polyides*
	...	*Fucus*	*Rhodymenia*
	...	*Laminaria*	...
	...	*Ralfsia*	...
	...	*Saccorhiza*	...

* Preferences were determined by laboratory two-way choice experiments. Only large individuals of any algal species were used. A group of 20–40 snails was placed in the middle of the bottom of a filled 20-gal aquarium (standing new seawater) and surrounded by equal amounts of two species of algae, with the same species on opposing sides. The probability of any snail's contacting species 1 was equal to that of its contacting species 2. These periwinkles did not appear to detect food at a distance, but relied on tactile-chemical methods once plants were contacted. Once a snail contacted a piece of alga it would either move away or remain there and feed. The numbers of snails on the two species of algae were compared using χ^2 after 30–90 min. All large algae had had micro- and macroscopic epiphytic algae removed from them. Results were usually clear-cut and are arranged here in three preference categories. Most experiments were repeated at least once, rotating positions of algae and using a new group of snails. Further details and discussion of these experiments will be presented in a later paper.

HERBIVORE FOOD PREFERENCES

Along the rocky shores of New England, the most abundant and important herbivore in the mid and low intertidal zones is the periwinkle snail *Littorina littorea* (J. Menge 1975). Often attaining a length of 2–3 cm, this snail forages primarily when under water or during cool, humid low tides. It is a generalist with respect to both size and species of food, consuming most local species of microscopic and macroscopic algae (J. Menge 1975). Laboratory choice experiments indicate that *L. littorea* has strong food preferences (table 1). In general, the preferred algae are primarily ephemeral small and tender species (like the green *Enteromorpha* spp.), which appear to lack either structural or chemical means of deterring herbivorous snails. Algae in the lowest preference category are either never eaten by *L. littorea* or are eaten only if no other food has been available for a considerable length of time. These plants (like the perennial red *Chondrus crispus* [Irish moss]) are all tough compared to those in the high category. More detailed information on these preferences and how they correlate to potential antiherbivore mechanisms of the algae will be published later. The snails and many of the algae occur both in tide pools and on emergent substrata, i.e., rock exposed to air at low tide. Since different mechanisms and relationships exist in these two different habitats, the effects of the herbivores on the algae will be considered separately for each habitat.

EFFECT OF *Littorina littorea* ON COMPOSITION
OF TIDE POOL ALGAE

Normally there is considerable variation in the algal composition of tide pools in the upper half of the rocky intertidal region ($+5.9-+12.0$ ft or $+1.8-+3.7$ m). Comparable variation exists for European pools where a classification scheme of tide pools based solely on the dominant type of algae present has been suggested (Gustavsson 1972). Algal composition of high tide pools (hereafter called pools) in New England ranges from the extremes of almost pure stands of the opportunistic green alga *Enteromorpha intestinalis* or of the perennial red alga *Chondrus crispus* with a variety of intermediate situations (i.e., pools inhabited by many different types and species of algae). *Littorina littorea* appears to colonize these pools primarily by settlement from the plankton as newly metamorphosed snails (≤ 0.2 cm long). Experimental manipulations and subsequent monitoring of snail density (described below) indicate that immigration and emigration of adult *L. littorea* ($\geq 1.2-1.5$ cm long) are rare despite the pools being inundated approximately every 6 h. There is wide variation in *L. littorea* density between but not within pools. In any single pool, the snail density remains relatively constant over time. To examine the role of *L. littorea* in controlling the macroscopic algal composition of these pools, periwinkle densities were experimentally altered. In September 1973, three pools of similar height, salinity, size, depth, and exposure to light at the Marine Science Institute, Nahant, Massachusetts, were selected which subjectively appeared to represent the two extremes of the continuum in types of algae present. One pool was dominated by an almost pure stand of *Enteromorpha* (97% cover, see initial point in fig. 1*B*) and had a low density of *L. littorea* (four per m²). The other two pools were dominated by *Chondrus* (85% and 40% cover; see initial points in fig. 1*A* and 1*C*, respectively) and had high densities of *L. littorea* (233 and 267/m²).

Because *Enteromorpha* is one of *L. littorea*'s preferred food species (table 1) and *Chondrus* is not eaten by the snails, I hypothesized that the observed correlation between littorine abundance and algal composition was causal. It appeared that intense snail grazing may be eliminating ephemeral algae such as *Enteromorpha* and allowing inedible *Chondrus* to persist. To test whether *L. littorea* was responsible for the algal differences between pools, I removed all *L. littorea* from one *Chondrus* pool, added them to the *Enteromorpha* pool, and left the second *Chondrus* pool undisturbed as a control. These experiments were

EFFECT OF <u>LITTORINA</u> <u>LITTOREA</u> ON

ALGAL COMPOSITION OF TIDE POOLS

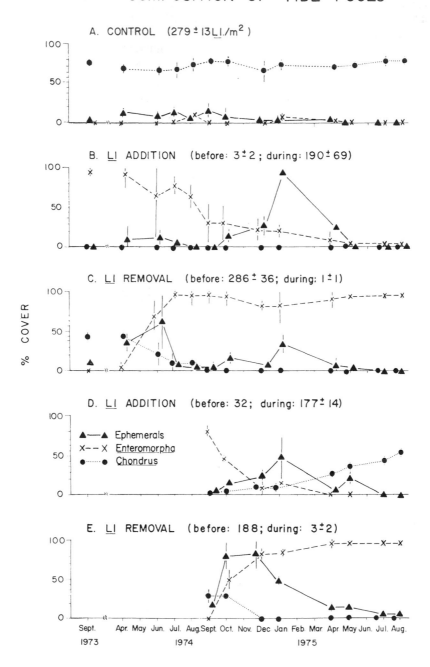

EFFECT OF HERBIVORES ON PLANT DIVERSITY　　27

initiated in April 1974 after the samples shown for that date had been taken (fig. 1*A–C*). The percentage of cover of algae and density of herbivores were monitored throughout the following $1\frac{1}{2}$ yr. The few species of encrusting algae (*Ralfsia, Hildenbrandia,* and encrusting corallines) are not included here because of sampling and/or field identification problems. Their distribution in the various pools appears uniform. Percentage of cover of upright algae was estimated by observing what species was under each of 100 dots on a 0.25-m², $\frac{1}{2}$-inch Plexiglas quadrat placed over the area. The coordinates of the dots were obtained using a table of random numbers. As many quadrats as could be fit onto the bottom of each pool were sampled. Three quadrats fit in pools A, D, and E and four quadrats in pools B and C.

In the control pool, *Chondrus* abundance remained high throughout this monitoring period (fig. 1*A*); *Enteromorpha* and other ephemeral algae were present but never abundant. In *Chondrus* pools, the periwinkles feed on microscopic plants and sporelings and germlings of many ephemeral algae that settle on *Chondrus*. In the *L. littorea* addition pool (formerly *Enteromorpha* dominated, fig 1*B*), *Enteromorpha* gradually declined in abundance to $< 5\%$ cover by April 1975. Snails could be seen actively ingesting *Enteromorpha* throughout this time period. Comparison of this experimental pool to the control (compare fig. 1*B* with 1*A*) supports the hypothesis that *L. littorea* is the cause of this decline in *Enteromorpha*. Note that ephemeral algal species (e.g., *Ectocarpus confervoides, Petalonia fascia,* and *Scytosiphon lomentaria*) became seasonally abundant even in the presence of *L. littorea*. Periwinkles are less active during the winter (J. Menge 1975), and as a result, ephemeral algae can temporarily swamp them (e.g., in January 1975). From January to April, periwinkles became increasingly active and eliminated nearly all edible algae from the pool. (Inedible algae include *Chondrus* and crusts.) No *Chondrus* has yet appeared in this pool. However, my observations in the low intertidal suggest that *Chondrus* recruits slowly (J. Menge 1975). I therefore predict that this alga will eventually settle and become abundant in the *L. littorea* addition pool. In the *L. littorea* removal pool (formerly with 40% cover of *Chondrus*, fig. 1*C*), *Enteromorpha* and several seasonal, ephemeral species immediately settled or grew from microscopic sporelings or germlings and became abundant. These include *Cladophora* sp., *Rhizoclonium tortuosum, Spongomorpha lanosa, Ulva lactuca* (all green algae), *Chordaria flagelliformis, Petalonia, Scytosiphon* (browns), *Ceramium* spp., *Dumontia incrassata* (reds), and filamentous diatoms. In spite of the presence of these species, *Enteromorpha* quickly became

---

Fig. 1.—Effect of *Littorina littorea* on algal composition of high tide pools at Nahant, Massachusetts. Means $\pm 95\%$ confidence intervals of angularly transformed (Sokal and Rohlf 1969) percentage of cover data are indicated. All pools are 1–2 m² in surface area and 10–15 cm deep. Three to four permanent quadrats (0.25 m²) were sampled per pool. The mean density of *L. littorea* (1974) is indicated after each caption. "Before" percentage of cover and density data were taken in September and April for *A, B,* and *C* and in September for *D* and *E.* Removals or additions were begun immediately after April (*A, B, C*) or September (*D, E*) sampling. See legend in *D. Chondrus* is deemed "present" only when upright thalli occur.

the most abundant alga in this pool. Although individuals of *Enteromorpha* are ephemeral, the species appears able continually to monopolize space by reproducing and recruiting throughout the year. Hence, individuals of *Enteromorpha* initially colonizing this pool are probably not still present a year later, but have been replaced by other individuals. I have often observed *Enteromorpha* in tide pools releasing swarmers (spores or gametes) during low tide which may increase the probability that offspring will recruit near their parents (e.g., Dayton 1973).

Careful examination of the *L. littorea* removal pool revealed that the disappearance of upright thalli of *Chondrus* was not simply the result of its being hidden by the canopy of *Enteromorpha*. *Enteromorpha* settled on *Chondrus* and on primary substratum and appears to have outcompeted the long-lived *Chondrus*. Following settlement of *Enteromorpha* on *Chondrus*, the thalli (upright portions) of the latter became bleached and then disappeared. However the encrusting holdfasts of *Chondrus* remain.

A second set of *L. littorea* addition and removal experiments were initiated in September 1974 to determine the effect of seasonal differences on these results, since many ephemeral algal species in tide pools are different in spring-summer and fall-winter (J. Menge 1975). These experiments demonstrated that upon removal of *L. littorea* (fig. 1*E*), ephemerals (primarily the brown algae *Ectocarpus*, *Scytosiphon*, and *Petalonia*) are initially more abundant. However, *Enteromorpha* eventually prevails, as in the removal experiments initiated in April. Notes (but no data) taken on other pools lacking *L. littorea* indicated that *Enteromorpha* had continually covered about 80%–100% of the pool substratum for at least 3 yr.

In the second set of experiments, addition of *L. littorea* resulted in an immediate increase in ephemeral algal abundance (fig. 1*D*). These algae were apparently able to settle because the *Enteromorpha* abundance had been reduced by grazers. However, both the ephemerals and *Enteromorpha* were eventually eaten. Contrary to the first addition experiment, encrusting holdfasts of *Chondrus* were present in this pool. Upright *Chondrus* thalli began to appear after grazers removed the ephemerals and *Enteromorpha*. Thus *Chondrus* increased in abundance in pool D via vegetative growth whereas it had not yet appeared in pool B, probably because of slow recruitment. Excepting these variations in *Chondrus* abundance, the outcome of both sets of experiments was the same. Removal of *L. littorea* resulted in a near pure stand of *Enteromorpha* (fig. 1*C* and *E*), while addition of this snail resulted in elimination of *Enteromorpha* and, in one case, eventual dominance by *Chondrus* (fig. 1*D*).

These experiments suggest that in tide pools *Enteromorpha* is the dominant competitor for space. However, because it and most of the ephemeral seasonal algae are preferred species of *L. littorea* (table 1), their abundance decreases when this grazer is common and active. *Chondrus* persists in tide pools where *L. littorea* is dense because it is not eaten. In such pools, the periwinkles feed on microscopic plants, sporelings, and germlings of many ephemeral algae that

EFFECT OF HERBIVORES ON PLANT DIVERSITY 29

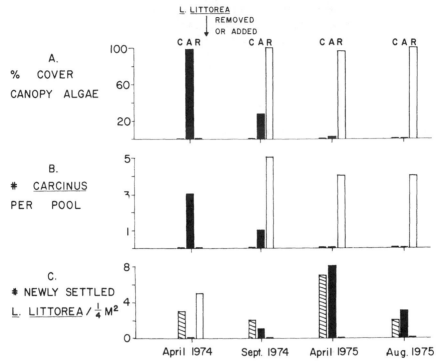

FIG. 2.—Changes in canopy algae, abundance of green crabs, and abundance of newly settled *Littorina littorea* following periwinkle manipulations in tide pools. *L. littorea* settles primarily during the spring and early summer. C (control, cross-hatched bars), A (*L. littorea* addition, solid bars), and R (*L. littorea* removal, blank bars) pools are the same as in figure 1*A*, *B*, and *C*, respectively. The other two experimental pools show similar changes. Canopy algae ($\geq$ 10 cm tall) is primarily *Enteromorpha* but may also include up to 5% total of *Scytosiphon* and/or *Rhizoclonium*. Percentage of canopy algae (X) and no. *Carcinus*/pool (Y) are positively correlated (linear regression: $Y = -0.02 + 0.04X$, $r^2 = .96$). No. *Carcinus*/pool (X) and no. newly settled *L. littorea* are negatively correlated ($Y = 5.69 - 1.59X$, $r^2 = .73$ for spring data).

settle on *Chondrus*. Thus *L. littorea* exerts a controlling influence on the algal composition of these pools.

In both of the *L. littorea* removal experiments (pools C and E), the green crab *Carcinus maenas* became abundant after a canopy of *Enteromorpha* was established (fig. 2*A* and *B*). Thereafter, very few tiny, newly metamorphosed *L. littorea* (0.2–0.3 cm long) were counted in the pools, in contrast to an abundance of them in the *L. littorea* addition and control pools (fig. 2*C*). Examination of numerous nonexperimental pools confirms that *Enteromorpha*-

dominated pools harbor many *Carcinus*, while *Chondrus*-dominated pools lack this crab. A dense canopy probably provides protection for this crab from sea gull (*Larus argentatus* and *L. marinus*) predators. Laboratory experiments demonstrate *Carcinus* readily preys upon small (but not medium or large) *L. littorea*. Thus, once *L. littorea* is absent from a pool long enough for *Enteromorpha* to become abundant, *Carcinus* may invade and prevent young *L. littorea* from recruiting from the plankton into the pools. Such a mechanism would explain the continued existence of low periwinkle density pools filled with highly desirable food. Hence, gulls probably indirectly affect the type of algae in pools. The pools may represent two alternative stable nodes (Lewontin 1969): (1) pools dominated by *Chondrus* because a dense contingent of herbivores continually removes superior competitors, and (2) pools dominated by the competitively superior *Enteromorpha* because predators prevent herbivores from being established. In this situation such alternative stable points may exist because the pools are essentially islands for which immigration and emigration are limited.

EFFECT OF *L. littorea* ON DIVERSITY OF ALGAE

When Competitive Dominant Is Preferred

The relationship between *Littorina littorea* density and the diversity of algae in tide pools can be seen in figure 3*A* and *B*. These data are from September 1974 after *L. littorea* grazing and competition between algal species have eliminated heavy spring and early summer recruitment of many short-lived algal species (Lubchenco and Menge 1978). Because the food preferences of *L. littorea* are known (table 1), and because this herbivore has been demonstrated to have a controlling effect on algae in tide pools (fig. 1), the source of between-pool variations in algal compositions and diversity can be interpreted as follows: When *L. littorea* is absent or rare, *Enteromorpha* outcompetes other algal species in pools, reducing the diversity. When *L. littorea* is present in intermediate densities, the abundance of *Enteromorpha* and various ephemeral algal species is reduced, competitive exclusion is prevented, and many algal species (ephemerals and perennials) coexist. At very high densities of *L. littorea*, all edible macroscopic algal species are consumed and prevented from appearing, leaving an almost pure stand of the inedible *Chondrus*. (New microscopic algae probably continue to settle and provide food for *L. littorea*.) Both the number of species (S) and H', an index of diversity based on both species number and the relative abundances, reveal the same unimodal relationship of macroscopic algal diversity to *L. littorea* density. A similar relationship was found for sea urchins grazing algae (Vadas 1968; Paine and Vadas 1969) and is suggested by qualitative results in certain terrestrial systems (Jones, cited in Harper 1969). These results support the theoretical predictions of Emlen (1973).

When Competitive Subordinate Is Preferred

Other workers have suggested that the effect of consumers on prey species diversity depends on the relationship between food preferences of the poten-

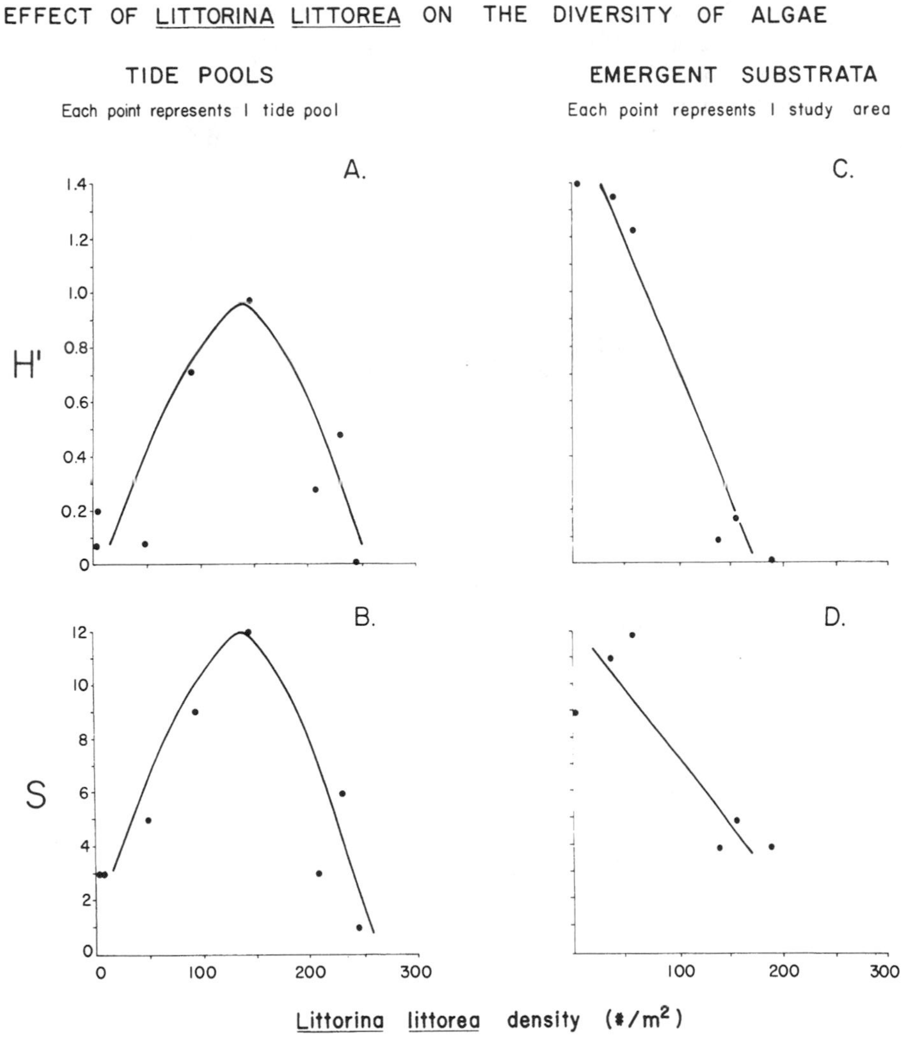

EFFECT OF HERBIVORES ON PLANT DIVERSITY   31

EFFECT OF LITTORINA LITTOREA ON THE DIVERSITY OF ALGAE

TIDE POOLS

Each point represents 1 tide pool

EMERGENT SUBSTRATA

Each point represents 1 study area

Fig. 3.—Effect of *Littorina littorea* density on the diversity of algae in high tide pools (*A, B*) and on emergent substrata in the low intertidal zone (*C, D*). *S* = no. species, *H'* is an index of diversity, here based on the percentage of cover of each species. Each point in *A* and *B* is from four (0.25 m^2) quadrats. Each point represents a different pool at Nahant, Massachusetts, September 1974. Each emergent substratum point was from 10 (0.25 m^2) quadrats in the low zone at six different areas in Massachusetts and Maine, June and July 1974 (see J. Menge [1975] for descriptions of areas). Regression equations: (*A*) tide pool $H' = -0.0409 + 0.01250X - 0.00005X^2$, $r^2 = .65$; (*B*) tide pool $S = 1.64 + 0.1357X - 0.00056X^2$, $r^2 = .73$; (*C*) low emergent substratum $H' = 1.58 - 0.0089X$, $r^2 = .94$; (*D*) low $S = 11.58 - 0.0415X$, $r^2 = .72$. For mid-zone regressions (not illustrated), density of *L. obtusata* was converted to units of *L. littorea* density where 1 g wet weight *L. obtusata* is presumed to be equal to 1 g wet weight *L. littorea*. *X*, then, = "units of *L. littorea*," i.e., actual density of that snail plus presumably equivalent units of *L. obtusata*. Mid $H' = 1.65 - 0.004X$, $r^2 = .91$; mid $S = 8.03 - 0.017X$, $r^2 = .81$.

tially controlling consumer and competitive hierarchies of the food species (Paine 1969; Harper 1969; Patrick 1970; MacArthur 1972). The importance of this relationship can be seen when the effect of *L. littorea* on diversity of algae in tide pools is compared to its effect on diversity of algae on emergent substrata. In the New England rocky intertidal zone the competitive dominance of the most abundant tide pool plants is actually reversed when they interact on emergent substrata. Perennial brown algae (*Fucus vesiculosus, F. distichus,* and *Ascophyllum nodosum*) are competitively superior to other algae in the mid zone, while the perennial red alga *Chondrus* is competitively dominant in the low zone (J. Menge 1975; Lubchenco and Menge 1978). *Littorina littorea's* preferences remain the same in and out of tide pools. Consequently there is an inverse correlation between periwinkle abundance and algal species diversity on emergent substrata (fig. 3*C* and *D*). Specifically where *L. littorea* is scarce or absent (e.g., at areas exposed to wave action, or in experimental removals at protected sites) at least 14 ephemeral species coexist with *Fucus, Ascophyllum,* and *Chondrus.* (These include *Enteromorpha, Spongomorpha spinescens, S. arcta, Rhizoclonium tortuosum, Ulva lactuca* [greens]; *Chordaria flagelliformis, Dictyosiphon foenicularis, Ectocarpus* spp., *Elachistea fucicola, Petalonia fascia, Pylaiella littoralis, Scytosiphon lomentaria* [browns]; and *Ceramium* spp. and *Dumontia incrassata* [reds].) Although the ephemeral species are all eventually outcompeted by the perennials on primary space, the former can coexist with the latter by occupying patches of primary space cleared by disturbances or by settling and growing epiphytically upon the perennials. If no such refuge were possible, there would be no relationship between *L. littorea* density and algal diversity, and the herbivores would simply increase the rate at which the eventual dominance by competitively superior plants was attained. On emergent substrata in New England, when *L. littorea* is abundant, it preferentially eats the ephemerals, leaving the more unpalatable fucoids and *Chondrus,* thus decreasing diversity. *Littorina obtusata* has the same effect as its congener in the mid zone (J. Menge 1975). Similar results have been suggested by Patrick (1970) for freshwater snails grazing diatoms (based on unpublished laboratory data of K. Roop) and by Harper (1969) for Milton's data on sheep grazing pastures.

DISCUSSION

Space has been shown to be a primary limiting resource in many rocky intertidal communities (Connell 1961, 1971, 1972; Dayton 1971, 1975; Lubchenco and Menge 1978; B. Menge 1976; J. Menge 1975; Paine 1966, 1974). This resource is modified by physical conditions that may determine the outcome of competitive interactions. Thus in tide pools in New England, ephemeral algae like *Enteromorpha* are competitively superior while on emergent substrata the more hardy perennials dominate primary space. Spatial and temporal heterogeneity in physical conditions are undoubtedly important in maintaining the coexistence of these competing species.

The result that in some habitats an opportunistic species can continually outcompete perennials is worth emphasizing. Evidently in the absence of herbivores, ephemerals such as *Enteromorpha* can outcompete the perennial fucoids and Irish moss in tide pools but probably not on emergent substrata. In the latter habitat, ephemerals like *Enteromorpha* can temporarily "outcompete" perennials on newly opened primary substrata by growing faster (J. Menge 1975). However, these ephemerals are eventually replaced by perennial species (e.g., *Fucus* in the mid zone), which can recruit slowly and take over when the ephemerals die. The key difference between the ephemerals' performance in the two habitats appears to be that they can continually recruit and replace themselves in tide pools but not on emergent substrata. The brown alga *Postelsia* is another short-lived species which appears to outcompete later successional species, barnacles and mussels, by virtue of its ability to recruit offspring near parent plants (Dayton 1973). Thus in microhabitats where ephemerals outcompete perennials, the successional sequence does not progress except where herbivores remove the ephemerals.

Concomitant with the reversal of competitive dominance in different microhabitats is the alteration of effects of herbivores. The plants that seem best suited to cope with the physical rigors of the intertidal zone, the perennials, are also least attractive to periwinkle herbivores. Thus these herbivores have a negative or small effect on algal species diversity on emergent substrata. In tide pools, however, because the preferred algae are also competitively dominant, herbivores determine the algal composition and species diversity in this microhabitat. This shift in competitive dominance in different microhabitats and consequent alteration of effects of consumers may be widespread, but it needs additional documentation.

The effects of *Littorina littorea* on plant species diversity in New England are complex. The key to understanding these effects lies in knowledge of the food preferences of the herbivore. These results may typify the effects of many generalized consumers: When the competitively dominant species is preferred by the consumer, there is a unimodal relationship between prey diversity and consumer density, with the highest diversities at intermediate consumer densities. When the competitively inferior species are preferred, there is an inverse correlation between prey diversity and consumer density (e.g., fig. 3). Under the former conditions, effects of consumers counteract competitive dominance. In the latter situation, feeding reinforces effects of competitive dominance.

As indicated earlier, the general effects of herbivores on plant species diversity have not been clear. After reviewing a number of case studies of effects of terrestrial herbivores on plant species diversity, Harper (1969) concluded that the results were too variable and inconsistent to warrant any generalizations. In view of the results presented here, it is possible to reinterpret the studies cited by Harper. I believe knowledge of three critical factors is necessary to obtain the proper insight into these studies. These factors are (1) the relationship between herbivore preferences and competitive ability of the plants, (2) the length of time the experiment was monitored after the manipulation, and (3) the initial relative abundance of the herbivores. The importance of the second and third factors can be seen in the following example. In figure 3, if the initial density of *L. littorea* were $250/m^2$, its removal would result in an initial increase in species richness followed by a decrease as competitive interactions occurred. In contrast, if the initial density were only 100/m², snail removal would result in a decrease in algal diversity. Thus the initial relative intensity of grazing or predation and the amount of time the experiment is monitored will determine what effect removal of the consumer has.

The importance of all three of the above factors is indicated by numerous studies. First, experimental manipulations of predators or herbivores that selected competitively superior food species conform to either the left half of the or the whole unimodal curves in figure 3A and B, depending on the range in abundance of consumers (Hall et al. 1970; Harper 1969; Paine 1966, 1971, 1974; Paine and Vadas 1969; Vadas 1968). Second, studies in which consumers preferred competitively inferior species comply with results in figure 3C and D (Harper 1969; Patrick 1970). Third, examples of overgrazing (i.e., the right half of fig. 3A and B curves) caused by a high density of consumers abound (Bartholomew 1970; Dayton 1971, 1975; Earle 1972; Harper 1969; Kitching and Ebling 1961, 1967; Leighton et al. 1966; Lodge 1948; Ogden et al. 1973; Paine and Vadas 1969; Randall 1965; Southward 1964; Vadas 1968). Taken together, these results suggest that generalized herbivores can have the same effect on species diversity as do generalized predators. This effect appears to depend primarily on the relationship between the consumer's food preferences and the competitive interactions of the food species.

The effects of periwinkles on algae presented here are the result of local manipulations of herbivores over ecological time. To what degree consumers affect broadscale biogeographic patterns of species diversity of lower trophic levels over evolutionary time remains to be seen. However, I believe the effects may be comparable. In the following example, regional differences in plant species diversity perhaps caused by the effects of an herbivore may be intermediate between local and broadscale patterns.

There is a striking difference in the low intertidal algal species diversity of New England rocky coasts and that of the rocky shores in the Bay of Fundy, which may be a function of the abundance of the herbivorous sea urchin *Strongylocentrotus droebachiensis*. Unlike *L. littorea*, this urchin readily eats *Chondrus*, and several lines of evidence suggest it has an effect on low-zone algal diversity comparable to the periwinkle's effect in tide pools.

In the low zone of New England, this urchin is rare and *Chondrus* outcompetes most other algae and dominates the zone (cols. 1, 2 in table 2; Lubchenco and Menge 1978). Changes in algal species diversity at different New England areas are caused by varying densities of periwinkles affecting algal epiphytes on *Chondrus* (fig. 3). Sea urchins are common in the low zone at rocky areas in the Bay of Fundy. Field experiments done with this urchin indicate it readily eats *Chondrus* and can prevent this plant from monopolizing space in the low zone (fig. 4; Lubchenco and Menge 1978). Thus sea urchins may prevent *Chondrus* from dominating the low zone in the Bay of Fundy and allow many other algae to coexist there (col. 3, table 2). This example would parallel the results of Dayton (1975) and Vadas (1968) for the Pacific northwest

EFFECT OF HERBIVORES ON PLANT DIVERSITY 35

TABLE 2

EFFECT OF SEA URCHINS ON ALGAL SPECIES DIVERSITY IN THE LOW
INTERTIDAL ZONE IN NEW ENGLAND AND THE BAY OF FUNDY

	NEW ENGLAND		BAY OF FUNDY	
	Chamberlain, Maine (1)	Canoe Beach, Cove, Nahant, Mass. (2)	Cape Forchu, Yarmouth, Nova Scotia (3)	Quoddy Head, Maine (4)
Herbivore densities:*				
Strongylocentrotus	0	0	4.2 ± 2.6	26.4 ± 13.8
Acmaea	0	$.1 \pm .2$	$.5 \pm .8$	21.2 ± 7.1
L. littorea	0	126.8 ± 60.0	0	0
Algal diversity and percentage of cover:				
No. species†	8	3	27	6
H'_n	1.20	.23	2.03	1.14
$\bar{X}\%$ cover canopy	0	0	78.6 ± 19.7	$.4 \pm .7$
$\bar{X}\%$ cover understory ..	80.8 ± 17.0	89.9 ± 10.2	$125.6 \pm 20.7‡$	2.6 ± 2.1
$\bar{X}\%$ cover *Chondrus* ...				
(= in understory) ...	74.5 ± 18.2	83.3 ± 11.4	14.6 ± 10.2	0

NOTE.—Data are from June–July 1975–1976.

* Densities are $\bar{X} \pm 95\%$ confidence intervals/$0.25 m^2$ for 10 quadrats at each area; see Lubchenco and Menge (1978) or J. Menge (1975) for methods.

† Includes both canopy and understory species. In the low zone, *L. littorea* grazes and affects only epiphytic algae on *Chondrus*; sea urchins graze and affect *Chondrus* and most other understory and canopy species.

‡ Percentage of cover > 100% reflects the dense multilayer arrangement of the understory at Cape Forchu.

coast of North America. In New England, urchins occur only in the subtidal region; in the Bay of Fundy they are both subtidal and intertidal. It is not clear why urchins are more abundant in the low intertidal zone in the Bay of Fundy. Possible factors are (1) a lower level of gull predation, (2) less desiccation stress, and (3) less wave action in Fundy. When urchins are exceedingly abundant, e.g., at Quoddy Head, Maine (on the Bay of Fundy), algal species diversity and abundance are very low (col. 4, table 2). In such areas, only encrusting coralline algae, a few unpalatable species (*Desmarestia* and *Agarum*; Vadas 1968), and some recently settled ephemeral species are present. Thus there is evidently a unimodal relationship between low intertidal algal species diversity and sea urchin density similar to that in figure 3A and B (see also Vadas 1968). That this correlation is causal can be demonstrated only through experimental manipulations.

Throughout this paper I have emphasized the importance of preferential predation on the competitively dominant prey. However, it may be possible for nonselective predation to have comparable effects on prey species diversity. This could be accomplished with differential recruitment and/or growth rates of the prey such that with nonselective predation the competitively dominant prey are prevented from outcompeting other prey. For example, *Chondrus* may

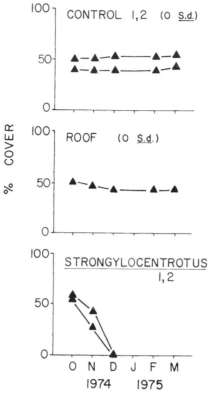

FIG. 4.—Effect of sea urchins on *Chondrus* in the low zone at Grindstone Neck, Maine, where urchins are normally absent. Two 10 × 10 × 4-cm stainless steel mesh cages containing one (3.5 cm test diameter) and two (both 3.2 cm) urchins were placed over *Chondrus*. Two unmanipulated controls and one stainless steel mesh roof indicate the normal percentage of cover of *Chondrus* without urchins. In the urchin enclosures, urchins completely removed *Chondrus*, including the encrusting holdfast.

be prevented from dominating the low zone by urchins in the following manner. If urchins graze most plants present in proportion to their abundance (except for certain unpalatable algae), and many of these plants have higher recruitment and/or growth rates than does *Chondrus*, *Chondrus* could not take over. In other words, the critical factor is the effect the consumer has on the competitive dominant (here, prevention of competitive exclusion), not the means (selective or nonselective grazing) by which it is accomplished.

From the above discussion, it is obvious that not all herbivores have similar effects on plant species diversity. If the sea urchins are removing *Chondrus* and preventing it from outcompeting other low zone algae in the Bay of Fundy, they have a very different effect on the emergent substratum community than do the periwinkle snails that do not graze *Chondrus*, but only its epiphytes. The size, manner of feeding, degree of food specialization, and mobility of herbivores may be important determinants of their effects on vegetation. For example, spatial escapes of seeds and young trees from their specialized and less mobile herbivores may provide an important mechanism of maintaining tropical rain forest tree diversity (Connell 1971, 1975; Janzen 1970). Nonetheless, it appears that a critical determinant of the effect many generalized consumers have on their food resources is the relationship between consumer food preferences and food competitive ability.

SUMMARY

Field experiments demonstrate that the herbivorous marine snail *Littorina littorea* controls the abundance and type of algae in high intertidal tide pools in New England. Here the highest species diversity of algae occurs at intermediate *Littorina* densities. This unimodal relationship between algal species diversity and herbivore density occurs because the snail's preferred food is competitively dominant in tide pool habitats. Moderate grazing allows inferior algal species to persist and intense grazing eliminates most individuals and species. In contrast to pools, on emergent substrata where the preferred food is competitively inferior, this herbivore decreases algal diversity. Thus, the effect of this consumer on plant species diversity depends on the relationship between herbivore food preference and competitive abilities of the plants. These results may apply to most generalized consumers and provide a framework within which previously confusing results can be understood.

Thus predators or herbivores do not simply increase or decrease species diversity of their food, but can potentially do both. The precise effect a consumer has probably depends both on the relationship between its preferences and the food's competitive abilities and on the intensity of the grazing or predation pressure.

ACKNOWLEDGMENTS

I gratefully acknowledge P. K. Dayton, B. A. Menge, R. T. Paine, T. W. Schoener, F. E. Smith, J. R. Young, and an anonymous reviewer for discussions and comments on this manuscript. This paper is contribution no. 42 from the Marine Science Institute, Northeastern University, Nahant, Massachusetts, where facilities were kindly made available by N. W. Riser and M. P. Morse. This paper represents part of a Ph.D. thesis submitted to Harvard University. The research was supported in part by National Science Foundation grants to J. Lubchenco Menge (GA-40003) and to B. A. Menge (GA35617 and DES72-01578 A01).

LITERATURE CITED

Bartholomew, B. 1970. Bare zone between California shrub and grassland communities: the role of animals. Science 170:1210–1212.
Connell, J. H. 1961. The influence of interspecific competition and other factors on the distribution of the barnacle *Chthamalus stellatus*. Ecology 42:710–723.

————. 1971. On the role of natural enemies in preventing competitive exclusion in some marine animals and in rain forest trees. Pages 298–312 *in* P. J. den Boer and G. R. Gradwell, eds. Dynamics of populations. Proceedings of the Advanced Study Institute on Dynamics of Numbers in Populations, Oosterbeek, 1970. Centre for Agricultural Publishing and Documentation, Wageningen.

————. 1972. Community interactions on marine rocky intertidal shores. Annu. Rev. Ecol. Syst. 3:169–192.

————. 1975. Some mechanisms producing structure in natural communities: a model and evidence from field experiments. Pages 460–490 *in* M. L. Cody and J. Diamond, eds. Ecology and evolution of communities. Belknap, Cambridge, Mass.

Dayton, P. K. 1971. Competition, disturbance, and community organization: the provision and subsequent utilization of space in a rocky intertidal community. Ecol. Monogr. 41:351–389.

————. 1973. Dispersion, dispersal and persistence of the annual intertidal alga *Postelsia palmaeformis* Ruprect. Ecology 54:433–438.

————. 1975. Experimental evaluation of ecological dominance in a rocky intertidal algal community. Ecol. Monogr. 45:137–159.

Dayton, P. K., G. A. Robilliard, R. T. Paine, and L. B. Dayton. 1974. Biological accommodation in the benthic community at McMurdo Sound, Antarctica. Ecol. Monogr. 44:105–128.

Earle, S. A. 1972. The influence of herbivores on the marine plants of Great Lameshure Bay, with an annotated list of plants. Sci. Bull. Los Angeles County Natur. Hist. Mus. 14:17–44.

Emlen, J. M. 1973. Ecology: an evolutionary approach. Addison-Wesley, Reading, Mass. 493 pp.

Gustavsson, Ulla. 1972. A proposal for a classification of Marine rockpools on the Swedish West Coast. Bot. Marina 15:210–214.

Hall, D. J., W. E. Cooper, and E. E. Werner. 1970. An experimental approach to the production dynamics and structure of freshwater animal communities. Limnol. Oceanogr. 15:839–928.

Harper, J. L. 1969. The role of predation in vegetational diversity. Brookhaven Symp. Biol. no. 22, pp. 48–62.

Hutchinson, G. E. 1959. Homage to Santa Rosalia, or why are there so many kinds of animals? Amer. Natur. 93:145–159.

Jackson, J. B. C., and L. W. Buss. 1975. Allelopathy and spatial competition among coral reef invertebrates. Proc. Nat. Acad. Sci. USA 72:5160–5163.

Janzen, D. H. 1970. Herbivores and the number of tree species in tropical forests. Amer. Natur. 104:501–528.

Kitching, J. A., and F. J. Ebling. 1961. The ecology of Lough Ine. XI. The control of algae by *Paracentrotus lividus* (Echinoidea). J. Anim. Ecol. 30:373–383.

————. 1967. Ecological studies at Lough Ine. Advance. Ecol. Res. 4:198–291.

Leighton, D. L., L. G. Jones, and W. J. North. 1966. Ecological relationships between giant kelp and sea urchins in Southern California. Pages 141–153 *in* E. G. Young and J. L. McLachlan, eds. Proceedings of the Fifth International Seaweed Symposium. Pergamon, New York.

Lewontin, R. C. 1969. The meaning of stability. Brookhaven Symp. Biol. no. 22, pp. 13–23.

Lodge, S. M. 1948. Algal growth in the absence of *Patella* on an experimental strip of foreshore, Port St. Mary, Isle of Man. Proc. Trans. Liverpool Biol. Soc. 56:78–85.

Lubchenco, J., and B. A. Menge. 1978. Community development and persistence in a low rocky intertidal zone. Ecol. Monogr. (in press).

MacArthur, R. H. 1972. Geographical ecology. Harper & Row, New York. 269 pp.

Menge, B. A. 1976. Organization of the New England rocky intertidal community: role of predation, competition and environmental heterogeneity. Ecol. Monogr. 46:355–369.

Menge, B. A., and J. P. Sutherland. 1976. Species diversity gradients: synthesis of the roles of predation, competition and temporal heterogeneity. Amer. Natur. 110:351–369.

Menge, J. L. 1975. Effect of herbivores on community structure of the New England rocky intertidal region: distribution, abundance and diversity of algae. Ph.D. diss. Harvard University.

EFFECT OF HERBIVORES ON PLANT DIVERSITY   39

Ogden, J. C., R. A. Brown, and N. Salesky. 1973. Grazing by the echinoid *Diadema antillarum* Philippi: formation of halos around West Indian patch reefs. Science 182:715–717.

Paine, R. T. 1966. Food web complexity and species diversity. Amer. Natur. 100:65–75.

———. 1969. The *Pisaster-Tegula* interaction: prey patches, predator food preference and intertidal community structure. Ecology 50:950–961.

———. 1971. A short-term experimental investigation of resource partitioning in a New Zealand rocky intertidal habitat. Ecology 52:1096–1106.

———. 1974. Intertidal community structure: experimental studies on the relationship between a dominant competitor and its principal predator. Oecologia 15:93–120.

Paine, R. T., and R. L. Vadas. 1969. The effects of grazing by sea urchins, *Strongylocentrotus* spp., on benthic algal populations. Limnol. Oceanogr. 14:710–719.

Patrick, R. 1970. Benthic stream communities. Amer. Sci. 58:546–549.

Pianka, E. R. 1967. On lizard species diversity: North American flatland deserts. Ecology 48:333–351.

———. 1969. Habitat specificity, speciation, and species density in Australian desert lizards. Ecology 50:498–502.

Randall, J. E. 1965. Grazing effect of sea grasses by herbivorous reef fishes in the West Indies. Ecology 46:255–260.

Ricklefs, R. E. 1973. Ecology. Chiron, Newton, Mass. 861 pp.

Sokal, R. R., and F. J. Rohlf. 1969. Biometry. Freeman, San Francisco. 776 pp.

Southward, A. J. 1964. Limpet grazing and the control of vegetation on rocky shores. Pages 265–273 *in* D. J. Crisp, ed. Grazing in terrestrial and marine environments. Blackwell, Oxford.

Vadas, R. L. 1968. The ecology of *Agarum* and the kelp bed community. Ph.D. diss. University of Washington. 280 pp.

Van Valen, L. 1974. Predation and species diversity. J. Theoret. Biol. 44:19–21.

Vol. 110, No. 973   The American Naturalist   May–June 1976

SPECIES DIVERSITY GRADIENTS:
SYNTHESIS OF THE ROLES OF PREDATION,
COMPETITION, AND TEMPORAL HETEROGENEITY

Bruce A. Menge and John P. Sutherland

Department of Biology, University of Massachusetts, Boston, Massachusetts 02125; and
Duke University Marine Laboratory, Beaufort, North Carolina 28516

A major goal of many ecologists working toward the development of a broad theory of community organization is to understand the causes of patterns of species diversity (Hutchinson 1959). Factors potentially affecting species diversity have been reviewed several times (e.g., Pianka 1966, 1967, 1974a; Ricklefs 1973) and include (1) time, (2) spatial heterogeneity, (3) competition, (4) predation, (5) climatic stability, and (6) productivity, plus several combinations of these (e.g., Pianka 1974a). Various attempts have been made to synthesize several of these hypotheses into a broad theory accounting for certain sets of observations or experiments. Among these are the predation hypothesis (Paine 1966, 1971), the stability-time hypothesis (Sanders 1968, 1969), and, most recently, a synthesis offered by MacArthur (1972). These syntheses focus mainly on theories (3)–(6) above but differ in their emphasis on the relative importance of each factor. This difference in emphasis revolves around the seemingly contradictory roles of competition and predation in the determination of community structure. Below we suggest that these roles are in fact complementary in their effects, the relative importance of each depending on the trophic level being considered and the overall trophic complexity in a community. We present data from the New England rocky intertidal zone supporting our contention and view this synthesis against patterns of environmental behavior and structural complexity. Our synthesis is not particularly new or novel. Hairston et al.'s classic paper (1960) makes essentially the same point, though it is concerned with regulation of populations and whole trophic levels and not species diversity. However, a realization of the fundamental lesson of their paper seems peculiarly absent from the controversy considered here.

Species diversity is here the number of species present in a community. Following MacArthur (1965), we distinguish between within-habitat comparisons (e.g., comparisons within rocky intertidal, grassland, or forest habitats) and between-habitat comparisons (e.g., comparisons between grasslands and forests, or rocky intertidal and kelp communities). In our view, terms such as lizard, bird, or plant "communities" are misleading and lead to confusion. We refer to such associations as "guilds," assemblages of species utilizing a specific type of resource (Root 1967). We reserve the term "community" for collections of interacting organisms of all trophic positions occurring in a given habitat.

Amer. Natur. 1976. Vol. 110, pp. 351–369.

We further define environmental "stability" as statistical variation in an environmental parameter (e.g., Margalef 1969; Holling 1973). Environmental "predictability" refers to the level of serial autocorrelation (Levins 1965; Lewontin 1969). Finally, environmental "stress" refers to the frequency that physical environmental conditions approach or exceed the physiological tolerance limits of an organism. These terms all have at least three temporal components: short term, or time scales of days or weeks; seasonal, or time scales ranging from months to 1 yr; and long term, or time scales greater than 1 yr. The former two scales generally correspond to ecological time (at least to macroscopic organisms), while the latter ranges from ecological to evolutionary time, depending on the organism. All of these factors may or may not be strongly correlated. That is, a predictable environment may be stable or unstable, a stable environment may be stressful or benign, etc. We combine these terms with levels and patterns of production and refer to them as "temporal heterogeneity" (e.g., Pianka 1974a).

THEORY

Role of Predation and Competition

The predation hypothesis (Paine 1966, 1971) suggests that selective predation on dominant competitors can maintain a relatively high local species diversity over ecological time by preventing the dominant competitors from monopolizing the major resource (food or space). By keeping species in lower trophic levels below their carrying capacities (K), competition is alleviated and additional species can invade the system. Diversity may be further increased by a positive feedback mechanism whereby new invading predators can be supported by the new invading prey species (e.g., Dodson 1970). This process is presumably limited by the stability and level of primary production (Paine 1966). Severe predation, by creating genetically disconnected allopatric populations of a species, can presumably lead eventually to speciation, thus increasing diversity over evolutionary time (e.g., Stanley 1973). This theory is supported by several experimental studies (Paine 1966, 1971, 1974; Hall et al. 1970; Harper 1969) and a variety of comparative studies (Porter 1972a, 1974; Janzen 1970; Connell 1970b).

The competition hypothesis argues that highly diverse communities arise in environments which are stable over long periods of time as a result of competition-maintained niche diversification (Dobzhansky 1950; Pianka 1966, 1974a). Specifically, the argument maintains that interspecific competition is more intense in stable environments because such environments allow most species to reach carrying capacity. Theoretically, increased interspecific competition usually acts to reduce the array of habitats or patches used by a species (the "compression hypothesis" [MacArthur and Wilson 1967; MacArthur 1972; Schoener 1974b]). Interspecific competition thus selects for increased specialization, which serves to reduce competition intensity. Species diversity is then increased via successful invasion of additional species. The limit to this

SPECIES DIVERSITY GRADIENTS   353

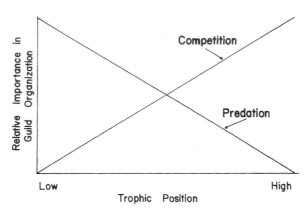

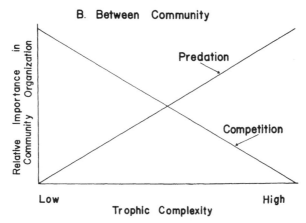

FIG. 1.—Qualitative models of the relative within-community (*A*) and between-community and habitat (*B*) importance of interspecific competition and predation in organizing and maintenance of diversity in communities. Trophic complexity is a function of several factors including at least the number of trophic levels, the number of species at each level, and the abundance and foraging strategy of each species. The relationships are linear for convenience; their precise shapes are not obvious and are unimportant for the purposes of this paper.

mechanism is theoretically a function of the number of discrete resources present (MacArthur 1965) or the maximum tolerable niche overlap on continuous resources (MacArthur and Levins 1967; MacArthur 1972), or both. Although maximum niche overlap is supposedly not directly sensitive to environmental variation (May and MacArthur 1972; Pianka 1974*b*), it should decrease as competition intensity increases (Pianka 1974*b*). Since increased competition intensity is theoretically a function of decreased environmental variation, maximum niche overlap is indirectly affected by environmental behavior. The ways in which the opposing effects of intra- and interspecific competition interact to produce broad or narrow niches are currently a matter of debate (Van Valen 1965; Soulé and Stewart 1970; Roughgarden 1972). Indirect evidence that competition regulates diversity is available for birds (e.g., Orians and Horn 1969; Karr 1971), lizards (e.g., Schoener 1968; Pianka 1967, 1969, 1971), gastropods (e.g., Kohn 1967, 1968, 1971), fish (e.g., Zaret and Rand 1971), crustacea (e.g., Culver 1970; Vance 1972), and corals (e.g., Lang 1973; Connell 1973; Porter 1972b, 1974).

As indicated by Pianka (1966), these theories appear contradictory. Thus, in the predation hypothesis, predation is believed to allow high diversity by reducing competition intensity. In contrast, the competition hypothesis argues that diverse communities can only arise if differential specialization alleviates competition. This contradiction seems partly responsible for a recent exchange in the literature concerning the causes of diversity patterns in the deep sea (Sanders 1968, 1969; Slobodkin and Sanders 1969; Dayton and Hessler 1972; Grassle and Sanders 1973).

In fact, these theories are complementary rather than contradictory. The essential points of our hypothesis are (1) competition regulates the number of species in a guild only when the members of that guild actually compete, i.e., when they are at or near carrying capacity. This is usually true at relatively *higher* trophic levels because of the absence of other controlling factors, e.g., predation (fig. 1A). Conversely, (2) predation characteristically regulates the number of species present in guilds at relatively *lower* trophic levels (fig. 1A). Extending this hypothesis to between-community and between-habitat comparisons, we predict that in communities with few trophic levels competition will be relatively more important than predation as an overall organizing factor. As the number of trophic levels and the number of species per level increase, predation will become relatively more important as an organizing factor (fig. 1B).

ROCKY INTERTIDAL COMMUNITY ORGANIZATION

Species occupying the hard substrata of the rocky intertidal are either sessile (e.g., mussels, barnacles, algae) or slow moving (e.g., snails, starfish, limpets). The major limiting resource for sessile species is usually primary space on the rock substratum (e.g., Paine 1966, 1971, 1974; Connell 1961a, 1961b, 1970a; Dayton 1971) or, for algae, space in the light (Dayton 1975). Food, space, or both may limit more mobile species (Stimson 1970, 1973; Haven 1973; Menge 1972b; Sutherland 1970). However, because distribution and abundance patterns of sessile species are the dominant features of rocky intertidal community structure, we will focus below on patterns of utilization of primary space.

New England

The rocky mid-intertidal zone of New England harbors a very simple community, whose trophic organization is given in figure 2. Headlands exposed to the full force of storm-generated waves are usually typified by sharp zonation

SPECIES DIVERSITY GRADIENTS   355

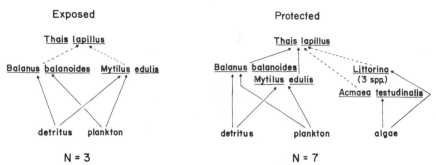

Fig. 2.—Generalized food webs at typical exposed and protected rocky intertidal communities in New England. The number of animal species occurring in
each community is given below each web. The dashed lines indicate that *Thais*
rarely preys upon the prey (protected) or has little effect (exposed).

and dense coverage of primary space by barnacles (*Balanus balanoides*, 69%–
89%) in the high intertidal and mussels (*Mytilus edulis*, 64%–89%) in the
mid intertidal (table 1). Moreover, exposed areas have relatively little free
space (ranging from 2% to 25% in the mid intertidal [table 1]). At protected
areas, the high intertidal also has a barnacle zone (table 1), but neither barnacles
(0%–2% cover) nor mussels (25%–39% cover) occupy nearly as much space
in the mid intertidal at protected areas as they do at exposed areas (table 1).
Availability of free space at protected areas is generally high (e.g., 47%–72%
at Grindstone Neck [table 1]).

The only mid-intertidal invertebrate predator in New England is the gastropod *Thais lapillus* (fig. 2). At all exposed areas, this species is relatively scarce
(table 2) and seems confined to crevices (B. Menge, unpublished data). In
contrast, mid-intertidal areas at more protected areas are characterized by
higher predator densities (table 2), and from April to October these snails are
dispersed widely throughout the mid intertidal. Other kinds of invertebrate
predators observed in the mid intertidal on other coasts (e.g., starfish, crabs)
are confined to the low (between $+2.0$ and -2.0 ft, or $+0.61$ and -0.61 m)
intertidal and subtidal habitats in New England (B. Menge, unpublished data).
Vertebrate predators (e.g., fish, gulls) evidently prey infrequently on the species
in this food web (personal observations).

To determine the relative intensity of predation and competition in these
communities, experiments similar in technique and design to those of Connell
(1961a, 1961b, 1970a) and Dayton (1971) were performed at several areas in
New England (B. A. Menge 1976). Briefly, each experiment, replicated four to
eight times over 1–3 yr from 1972 to 1974, consisted of a stainless-steel mesh
predator exclusion cage ($10 \times 10 \times 3$ cm), a mesh roof (10×10 cm), and a
control (also 10×10 cm). The cage tests the effect of predation by excluding
Thais, the roof tests the effect of shading by the mesh on survival of species
settling under it, and the control provided the natural situation. These experiments were established on patches of substratum cleared of all sessile
species in March of each year. In 1973 and 1974, a second cage from which

TABLE 1

UTILIZATION OF PRIMARY SPACE AT AN EXPOSED AREA (PEMAQUID POINT) AND A RELATIVELY PROTECTED AREA (GRINDSTONE NECK)[a]

INTERTIDAL LEVEL AND DATE	% COVER Balanus balanoides			% COVER Mytilus edulis			% FREE SPACE[b]		
	Pemaquid Point	Grindstone Neck	t	Pemaquid Point	Grindstone Neck	t	Pemaquid Point	Grindstone Neck	t
High:									
August 1973	83 ± 6	68 ± 15	2.04	15 ± 6	0.1 ± 0.2	8.72**	1 ± 1	32 ± 15	6.46**
October 1973	88 ± 5	67 ± 12	3.89**	10 ± 5	12 ± 9	0.23	2 ± 2	19 ± 12	4.81**
December 1973	89 ± 5	54 ± 15	5.31**	9 ± 5	12 ± 15	0.51	2 ± 2	34 ± 16	5.69**
March 1974	69 ± 10	48 ± 18	2.42*	10 ± 6	9 ± 16	1.22	20 ± 6	41 ± 19	1.99
May 1974	...[c]	62 ± 20	...	...	3 ± 7	...	...	28 ± 19	...
July 1974	85 ± 7	64 ± 17	2.51*	5 ± 3	9 ± 12	0.14	10 ± 7	28 ± 15	2.28*
Mid:									
August 1973	11 ± 8	2 ± 2	2.51*	81 ± 22	42 ± 18	2.83*	8 ± 17	47 ± 12	4.61**
October 1973	7 ± 7	0	2.72*	85 ± 21	39 ± 27	3.37**	8 ± 14	47 ± 24	4.06**
December 1973	8 ± 7	0.1 ± 0.2	2.86*	89 ± 8	37 ± 23	4.28**	2 ± 2	51 ± 22	5.03**
March 1974	5 ± 6	0	2.65*	79 ± 18	25 ± 18	4.60**	16 ± 14	72 ± 16	5.76**
May 1974	...	0	...	...	28 ± 22	...	...	59 ± 24	...
July 1974	8 ± 4	2 ± 2	3.30**	64 ± 13	26 ± 19	3.69**	25 ± 16	53 ± 20	2.28*

[a] Mean percentage cover and 95% confidence limits of $\frac{1}{4}$-m² quadrats ($N = 10$) located randomly in the high and mid-intertidal zones of the study areas. Angular transformations (Sokal and Rohlf 1969) were performed on the data for t tests; degrees of freedom = 18 in all cases.

[b] "Free" space is primary space on the rock which is either completely bare or covered with encrusting algae or lichens upon which mussels and barnacles will settle.

[c] No data for May 1974 at Pemaquid Point.

* $P < .05$.

** $P < .01$.

TABLE 2

DENSITY* OF *Thais lapillus* AT AN EXPOSED (PEMAQUID POINT) AND A PROTECTED (GRINDSTONE NECK) AREA

	INTERTIDAL ZONE	
AREA AND YEAR†	High	Mid
Pemaquid Point:		
1973	0.1 ± 0.2	4 ± 5
	(10)	(10)
1974	0	13 ± 8
	(10)	(10)
Grindstone Neck:		
1973	130 ± 48	95 ± 21
	(15)	(54)
1974	76 ± 41	60 ± 20
	(26)	(40)

* Mean number/$\frac{1}{4}$-m² quadrat (± 95% confidence limits). Number of quadrats in parentheses.

† Data are from June–August, since *Thais* are most active during the summer months.

Thais was excluded and mussels were removed was added to each experiment. This experiment tested the hypothesis that mussels would outcompete barnacles for space. The experiments were monitored every 2–4 wk. Abundance (percentage of primary substratum occupied) of mussels and barnacles in each treatment was estimated using the random-dot technique (e.g., Connell 1970*a*, Dayton 1971) either in the laboratory from photographs of the treatments or directly in the field.

Figures 3 and 4 show typical data from the mid intertidal at two exposed and two protected areas. At exposed areas such as Pemaquid Point and Little Brewster Point, barnacles settled in May, followed by mussels in July and August (fig. 3). Initially the barnacles rapidly occupied virtually all available space. However, mussels settled on and outcompeted the barnacles by late summer or early autumn. Since there was no important difference between predator exclusions and controls, predation seems to be an unimportant organizing agent at exposed areas. Here, competition for space between barnacles and mussels is apparently the dominant organizing biological interaction. Qualitative observations at seven other exposed headlands suggest these results are general for such shorelines.

At more protected areas, predation is more intense in the mid intertidal. Barnacles again settled before mussels, and in the exclusion cages *Balanus* were again overtaken and outcompeted by *Mytilus* by late summer to early autumn (fig. 4). Barnacles covered by mussels may survive for varying periods of time, but most are dead after 2 mo of complete coverage (B. Menge unpublished). In contrast, barnacles persist in the mussel removal cages, which supports the hypothesis that barnacles are outcompeted for space by mussels when predators are effectively absent (exposed areas) or excluded (protected areas). At protected areas, few barnacles or mussels survive through summer in the roof and

358   THE AMERICAN NATURALIST

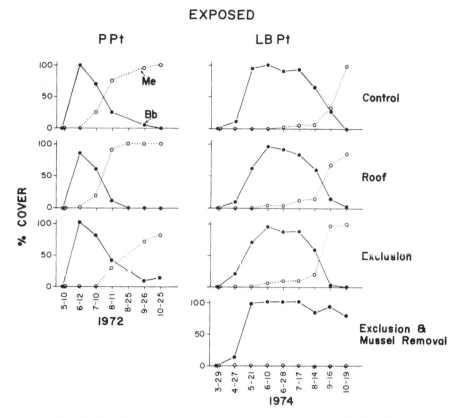

FIG. 3.—Results of exclusion experiments at two exposed areas in New England; *PPt* = Pemaquid Point, Maine; and *LBPt* = Little Brewster Point, Outer Boston Harbor, Massachusetts; *Bb* = *Balanus balanoides*; *Me* = *Mytilus edulis*. See text for explanation of experimental design.

control. At these areas eventually all mussels and barnacles are eaten by *Thais*. Hence, space is rarely limiting and competition for space is relatively unimportant in structuring these areas (e.g., Grindstone Neck and Little Brewster Cove). Rather, predation seems largely responsible for the major observed patterns of community structure. However, observations and experiments at five areas for up to 3 yr indicate that even at protected areas, mussels occasionally escape predation and may occupy substantial amounts of space (e.g., 26%–42% cover at Grindstone Neck [table 1]). We conclude that in the trophically simple New England rocky intertidal, competition for space generally ranges from chronic (at exposed areas) to intermittent (at protected areas) and is a relatively important structuring agent. Further, the importance of competition in structuring New England communities evidently varies inversely with that of predation.

SPECIES DIVERSITY GRADIENTS   359

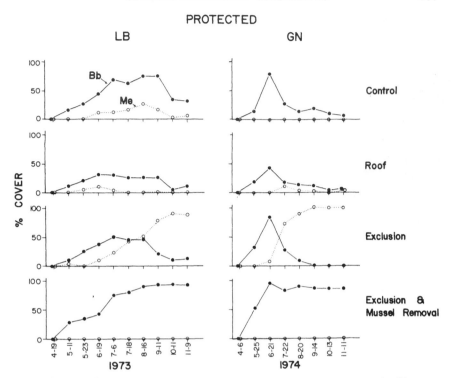

FIG. 4.—Results of exclusion experiments at two protected areas in New England; *LB* = Little Brewster Cove, Outer Boston Harbor, Massachusetts; *GN* = Grindstone Neck, Maine. See caption of fig. 3 for species code.

West Coast

The mid intertidal of the northwest coast of North America is considerably more diverse than that of New England (fig. 5). The major functional differences between the West and East Coast communities are that the former has a secondary carnivore trophic level (starfish) and a much larger herbivore guild, including limpets, chitons, and gastropods. This community is structured largely by predation. Paine (1966, 1974) found that the top carnivore, *Pisaster ochraceus* (a starfish), increases the species richness in this community by preventing the mussel *Mytilus californianus* from monopolizing space in the mid and low intertidal. Dayton (1971) and Connell (1970a) have shown that this phenomenon is widespread on the West Coast and that primary carnivores (*Thais* spp.) also play an important, but subordinate, role in clearing space. Finally, Menge (1972b) has indicated that the small starfish *Leptasterias hexactis* may exert an important effect on intertidal community structure, especially when *Pisaster* is scarce. Hence, in the West Coast rocky intertidal, predation predominates in determining the distribution, abundance, and

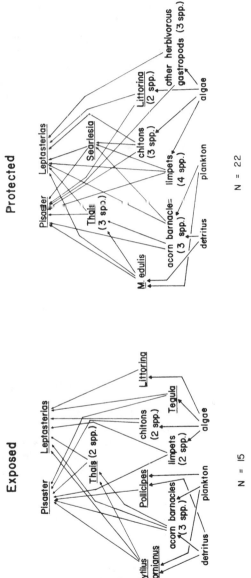

Fig. 5.—Generalized food webs at exposed and protected communities in Washington state. The left side represents open coast communities and is partly adapted from Paine (1966). The right side represents San Juan Island communities and is adapted from Menge (1972a, 1972b, unpublished data), Louda (unpublished data), and Lloyd (1971). The number of animal species in each community is given below each web.

diversity of species occupying lower trophic levels (e.g., mussels, barnacles, limpets, herbivorous snails, etc.). Competition at lower trophic levels, at least among space users, appears to occur only sporadically.

At the top trophic level, Menge (1972b, 1974) and Menge and Menge (1974) have shown that *Pisaster* competes with *Leptasterias* and that, as a result, these cooccurring predators partition their food resources. Further, Connell (1970a) has suggested that three primary carnivores, *Thais* spp., also partition resources by habitat, presumably as a result of competition. We conclude that, on the West Coast, competition seems to be important only at high trophic levels and is generally unimportant overall in structuring the community in the mid-intertidal region.

Within- and Between-Coast Comparisons

The above evidence provides several insights concerning the relative intensity of competition and predation along gradients of temporal heterogeneity. If we compare local communities on the East Coast, the less stable (more variable over short-term and seasonal time scales), exposed headlands tend to have less dense predator populations (table 2) and hence functionally fewer trophic levels (fig. 2) than protected areas. As a result, competition for space between the suspension feeders (*Mytilus* and *Balanus*) occupying the "top" trophic level is largely responsible for the observed mid-intertidal community organization (fig. 3). At more protected areas, the predator *Thais* is more abundant and greatly reduces the occurrence of competition between barnacles and mussels (fig. 4). In other words, moving from exposed to more protected areas in New England effectively adds a trophic level (fig. 2) which results in a change from a competition-dominated to a predation-dominated community.

If we compare local communities along a comparable exposure gradient on the West Coast, predators (especially *Pisaster*) are almost always abundant enough to prevent competition for space in the mid intertidal regardless of local stability (fig. 5; see also Dayton 1971). Hence, predation is the predominant organizing factor along most of this intracoastal gradient.

If we contrast the two coasts, the East Coast rocky intertidal community appears to experience a more rigorous (less stable, more stressful, and, in some respects at least, less predictable) environment over at least short-term and seasonal time scales than does the West Coast rocky intertidal. Variability in wave shock, temperature, and precipitation is noticeably greater on the East Coast than on the West Coast (B. Menge, personal observations). Moreover, deaths of mid-intertidal organisms from freezing or overheating and desiccation seem to occur more frequently on the East Coast than on the West Coast (F. E. Smith, personal communication; B. Menge, personal observations). Finally, such events as severe storms and short-term weather variations appear less predictable on the East Coast than on the West. (These qualitative patterns will be considered more quantitatively in a future publication [B. Menge, in preparation].) Correlated with these patterns of environmental variation is a difference in the relative importance of predation and competition as structuring agents in the rocky intertidal community. On the relatively more rigorous, trophically simpler East Coast, competition seems to be the dominant organizing factor for extensive segments of the shore (exposed areas), is responsible for the maintenance of the low diversity of these regions (via competitive exclusion), and hence may be at least partly responsible for the relatively low overall diversity in this community. On the relatively less rigorous, trophically complex West Coast, predation dominates community organization over most of the shore and seems mostly responsible for maintaining the relatively high diversity of this community.

DISCUSSION

An earlier synthesis, Sanders's "stability-time" hypothesis (1968; we use quotation marks to distinguish Sanders's terminology from ours), was specifically proposed to account for changes in diversity of marine soft-sediment infaunal communities arranged over gradients of decreasing environmental "stress" (shallow to deep sea). Sanders suggested that, where the environment is "severe" and "unpredictable," adaptations are primarily to the physical environment and communities are "physically controlled." Biological interactions between these eurytopic species are "unrefined," and their effects "are often drastic or catastrophic" (Sanders 1969). In "benign" and "predictable" environments, adaptations are primarily to other organisms and the community is "biologically accommodated." These adaptations are thought to ameliorate the intensity of biological interactions, e.g., by the production of narrow, nonoverlapping niches. He offers the hypothesis that "the resulting stable, complex, and buffered assemblages are always characterized by a large number of stenotopic species" (Sanders 1968).

In our view, the usefulness of this synthesis is impaired by the failure to recognize that (1) animals are usually physiologically adapted to their environment such that they are rarely stressed by it (e.g., Wolcott 1973) and (2) all communities have trophic organization, i.e., interactions occur between (predation) as well as within (competition) trophic levels.

First, those species which do not possess the appropriate physiological adaptations are probably simply absent from the habitat. However, there is no reason to believe that the biological interactions between the remaining species are any less "refined" or that their effects are any more "drastic or catastrophic" (Sanders 1969). Thus, there is considerable evidence for "biological accommodation" in the rocky intertidal (Connell 1961a, 1961b, 1970a, 1970b; Paine 1966, 1969, 1974; Dayton 1971, 1975; Menge 1972b, 1974; Menge and Menge 1974) and some for the muddy intertidal as well (Woodin 1974; Fenchel 1975a, 1975b), communities which are termed "physically controlled" by Sanders (1968) and Grassle and Sanders (1973).

It seems more appropriate to label a community "physically controlled" only if environmental catastrophes are a direct, and primary, cause of distribution and abundance patterns. In these habitats, biological interactions are likely to be of lower intensity or absent rather than "unrefined." Catastrophic mortality is probably a typical result of physiological stress primarily in such areas as the upper intertidal zone, the upper reaches of estuaries, ice-scoured benthic habitats, mountaintops, etc. (see Paine 1974). For example, Sutherland (1970) demonstrated a reduced level of food competition and an increase in the effects of physical factors with increasing tidal height in the limpet, *Acmaea scabra*.

The second consideration affecting the value of Sanders's synthesis is the failure to recognize that all communities have trophic organization. In Sanders's view, the increase in infaunal diversity he observed (1968, 1969) in sampling increasingly deeper benthic habitats results from an increase in "biological accommodation" as environmental "stress" is reduced. Here "biological accommodation" evidently means tighter species packing and finer resource subdivision via competition. We reinterpret Sanders's data to be a reflection of (1) a probable increased number of predator species, total predator abundance, and perhaps trophic levels with decreasing temporal heterogeneity and, consequently, (2) fewer occurrences of competitive exclusion as a result of the increase in predation intensity (see also Dayton and Hessler 1972). In addition, there are probably an unknown number of species which disappear with increasing temporal heterogeneity (i.e., in shallow habitats) because they do not possess the physiological adaptations necessary to survive. Support for this interpretation comes from the work of Jackson (1972), who found an increased predator effect on bivalves with depth, and of M. Rex (in preparation), who has found that the fraction of gastropod species which are apparently predators in samples of deep-sea gastropods (i.e., H' gastropod predators/H' total gastropod species) increases with depth, at least down to the abyssal plain. These patterns are consistent with the above reinterpretation of causes of variations in patterns of deep-sea diversity.

Role of Temporal Heterogeneity

Temporal heterogeneity may affect community organization both directly and indirectly. Some direct effects have been suggested by other workers. For example, with trophically simple communities consisting of either primary producers (e.g., grasslands) or primary space occupiers (e.g., rocky intertidal), local physical disturbance (a local stress) creating patchiness may increase species richness (e.g., Loucks 1970; Taylor 1973; Levin and Paine 1974). In these examples, physical disturbances (fires, wave-borne logs) occur frequently enough so that competitively inferior opportunists are maintained in the system. Hence, physical disturbance may "switch" some systems from ones in which competitive exclusion would lead to reduced richness to ones where disturbance-mediated competitive coexistence occurs. However, if the disturbance is more frequent and widespread, diversity may be reduced (e.g., Dayton 1971) and, at the extreme, the community may be "physically controlled" or essentially nonexistent.

As trophic complexity increases, the ways in which temporal heterogeneity can affect community organization also increase. First, trophic complexity itself is undoubtedly a function of temporal heterogeneity. The precise number of trophic levels in a community is probably dependent in part on the rate and predictability of primary and secondary production (Hutchinson 1959; Paine 1966; Connell 1970b), which in turn is probably influenced by environmental stability and predictability (Connell and Orias 1964). Several studies indicate there may be fewer trophic levels with increasing temporal heterogeneity (Pianka 1967; Jackson 1972; MacArthur 1972; Tinkle and Ballinger 1972). One effect of increased temporal heterogeneity is undoubtedly the elimination of more specialized consumers of high trophic status, since the array of resources becomes less predictably available (e.g., Menge 1972a). For example, Connell (1970a) suggested there are three species of *Thais* on the West Coast of North America and only one (*T. lapillus*) in England (and New England) because food resources are spatially and temporally more predictable on the West Coast. A second potentially important effect of temporal heterogeneity on trophically higher species is to make the environment unfavorable for certain periods of time (e.g., winter in temperate and boreal regions, gales, summer in deserts and certain other habitats). This would reduce the foraging period of consumers and might permit prey species to escape predation. For example, this effect seems partly responsible for the ineffectiveness of *Thais* at exposed areas in New England (table 2 and fig. 3 herein; and B. Menge, in preparation).

In general, then, we suggest that commonly observed reductions in within-habitat species richness along gradients of temporal heterogeneity (Sanders 1968; Johnson 1970; Jackson 1972) are due in large part to the increased incidence of competitive exclusion as trophic levels are lost or become in-effective. However, if highly localized physical disturbances or stresses occur with sufficient frequency, the diversity of such simple systems may be increased. Local physical disturbance apparently enhances the effects of predators in trophically complex systems (Dayton 1971).

Role of Spatial Heterogeneity

The effect of competition or predation on diversity is strongly influenced by the structural complexity or spatial heterogeneity (number of habitats) of a given locality. Thus, in the absence of regular local disturbance in structurally simple environments (i.e., two-dimensional habitats like the rocky intertidal or grasslands), competition can reduce diversity through the process of competitive exclusion (e.g., fig. 4). On the other hand, predation evidently increases and then decreases diversity as it increases in intensity in structurally simple environments (e.g., Paine 1966, 1971, 1974; Paine and Vadas 1969; Emlen 1973; Harper 1969; J. Menge, in preparation). Thus, when predation intensity is low, competition will be intense and competitive exclusion will reduce species diversity (figs. 1A and 4). As predation intensity increases, diversity increases (fig. 1B) until a point is reached where predation becomes so intense it reduces diversity (e.g., Emlen 1973).

Structurally complex localities with many habitats may allow the co-existence of many more species through competition-induced habitat specialization or through moderation of predation. Both theory (the "niche compression hypothesis" [see MacArthur and Wilson 1967; Schoener 1974*b*]) and available evidence (e.g., MacArthur 1964; Schoener 1974*a*) implicate habitat segregation as the major means of coexistence between competing species. Hence, at high trophic levels, competition in structurally complex environments can maintain high diversity through the production of narrow, nonoverlapping niches. At lower trophic levels, structural complexity may moderate predation and increase diversity in two ways. First, such localities might provide many more refuges for prey species than might structurally simple localities. Both in theory (Rosenzweig and MacArthur 1963) and in nature (Paine 1969; Dayton et al. 1974; Connell 1972), refuges stabilize otherwise unstable predator-prey interactions. Hence, more refuges should result in a higher diversity of prey. Second, structurally complex environments might decrease the foraging efficiency of the predator, which is another stabilizing influence on predator-prey interactions (Rosenzweig and MacArthur 1963). For example, Ware (1972) demonstrated that trout forage less efficiently in more complex habitats. Thus, the overall effect of structural complexity should be to increase diversity at all trophic levels.

Finally, a refuge of potentially great significance is that of size. This may be an especially important refuge at the primary producer level. In the low intertidal on the West Coast, Dayton (1975) has shown that interspecific competition between kelp species is an important structuring agent in this habitat. Dayton suggests these algae essentially "escape" control by herbivores, i.e., many individuals survive and grow large because the herbivores are decimated by predators. Terrestrial forests may represent an analogous situation; adult trees surely must compete for light (e.g., Horn 1971), and few consumers are capable of eating whole trees. In other words, trees have evidently also escaped control from herbivores. However, (1) predation by higher trophic levels may be indirectly responsible for these kinds of escapes, and (2) herbivores may still regulate diversity of primary producers by preying selectively on juvenile stages of the plants (e.g., Janzen 1970; Connell 1970*b*). Hence, predation may have important direct and indirect effects on overall community structure. Obviously, a key refinement of our hypothesis will be the examination of the regulation of the structure of such primary·producer associations.

SUMMARY

We suggest that the "predation" and "competition" hypotheses of community organization and species diversity are complementary. Maintenance of high diversity by competition appears to be relatively more important at higher trophic levels, while maintenance of high diversity by predation seems relatively more important at lower trophic levels. Further, predation is probably the dominant organizing interaction in trophically complex communities, while competition is probably the dominant organizing interaction in trophically simple communities. These hypotheses are supported on a local scale by experimental studies in the rocky intertidal communities of New England and the West Coast. A probable consequence of its greater temporal heterogeneity (i.e., a less stable, less predictable, and more stressful environment) is that the East Coast is trophically more simple and has an increased incidence of competitive exclusion. As a result, diversity is lower on the East Coast compared with the West Coast. A similar interpretation is possible for differences in diversity along other gradients of temporal heterogeneity such as the shallow to deep-sea soft-sediment communities.

In structurally simple environments, competition reduces diversity through competitive exclusion. On the other hand, predation first increases and then decreases diversity in spatially simple environments, presumably because refuges are few and hence overexploitation of a resource is more probable. In structurally complex environments, competition may increase diversity through increased habitat specialization. Such environments undoubtedly have more refuges and reduce predator foraging efficiency, both of which may allow the coexistence of more species. Predator-mediated escapes by primary producers from herbivores may explain the apparent importance of interspecific competition in certain primary producer associations.

ACKNOWLEDGMENTS

J. Menge provided field assistance, thoughtful discussion, and shared ideas and made constructive comments while the fieldwork was in progress and on the manuscript. In addition, the genesis and development of the paper benefited from discussions with, and constructive criticism from, many colleagues, chief among whom are J. Commito, D. Gill, J. Hatch, R. Karlson, J. Jackson, A. Kohn, R. Paine, M. Rex, D. Rubinstein, T. Schoener, and S. Woodin. Field assistance was provided by M. Cohen, S. Garrity, M. Lubchenco, P. McKie, S. Riggs, D. Spero, and B. Walker, plus a cast of thousands. To all we are grateful. Research support was provided by NSF grants GA 35617 and DES72-01578 A01 and a Sigma Xi grant-in-aid (B. A. M.) and ONR Contract N00014-67-A-0251-0006 (J. P. S.).

LITERATURE CITED

Connell, J. H. 1961a. Effects of competition, predation by *Thais lapillus*, and other factors on natural populations of the barnacle *Balanus balanoides*. Ecol. Monogr. 31:61–104.

———. 1961b. The influence of interspecific competition and other factors on the distribution of the barnacle *Chthamalus stellatus*. Ecology 42:710–723.

———. 1970a. A predator-prey system in the marine intertidal region. I. *Balanus glandula* and several predatory species of *Thais*. Ecol. Monogr. 40:49–78.

———. 1970b. On the role of natural enemies in preventing competitive exclusion in some marine animals and in rain forest trees. Proc. Advance. Stud. Inst. Dynamics Numbers Pop. 1970:298–312.

———. 1972. Community interactions on marine rocky intertidal shores. Annu. Rev. Ecol. Syst. 3:169–192.

———. 1973. Population ecology of reef-building corals. *In* O. A. Jones and R. Endean, eds. Biology and geology of coral reefs. Vol. 2, pt. 1. Academic Press, New York. 480 pp.

Connell, J. H., and E. Orias. 1964. The ecological regulation of species diversity. Amer. Natur. 98:399–414.

Culver, D. 1970. Analysis of simple cave communities: niche separation and species packing. Ecology 51:949–958.

Dayton, P. K. 1971. Competition, disturbance, and community organization: the provision and subsequent utilization of space in a rocky intertidal community. Ecol. Monogr. 41:351–389.

———. 1975. Experimental evaluation of ecological dominance in a rocky intertidal algal community. Ecol. Monogr. 45:137–159.

Dayton, P. K., and R. R. Hessler. 1972. Role of biological disturbance in maintaining diversity in the deep-sea. Deep-Sea Res. 19:199–208.

Dayton, P. K., G. A. Robilliard, R. T. Paine, and L. B. Dayton. 1974. Biological accommodation in the benthic community at McMurdo Sound, Antarctica. Ecol. Monogr. 44:105–128.

Dobzhansky, Th. 1950. Evolution in the tropics. Amer. Sci. 38:208–221.

Dodson, S. I. 1970. Complementary feeding niches sustained by size-selective predation. Limnol. Oceanogr. 15:131–137.

Emlen, J. M. 1973. Ecology: an evolutionary approach. Addison-Wesley, Reading, Mass. 493 pp.

Fenchel, T. 1975a. Factors determining the distribution patterns of mud snails (Hydrobiidae). Oecologia 20:1–17.

———. 1975b. Character displacement and coexistence in mud snails (Hydrobiidae). Oecologia 20:19–32.

Grassle, J. F., and H. L. Sanders. 1973. Life histories and the role of disturbance. Deep-Sea Res. 20:643–659.

Hairston, N. G., F. E. Smith, and L. B. Slobodkin. 1960. Community structure, population control, and competition. Amer. Natur. 94:421–425.

Hall, D. J., W. E. Cooper, and E. E. Werner. 1970. An experimental approach to the production dynamics and structure of freshwater animal communities. Limnol. Oceanogr. 15:839–928.

Harper, J. L. 1969. The role of predation in vegetational diversity. Brookhaven Symp. Biol. 22:48–61.

Haven, S. B. 1973. Competition for food between the intertidal gastropods *Acmaea scabra* and *Acmaea digitalis*. Ecology 54:143–151.

Holling, C. S. 1973. Resilience and stability of ecological systems. Annu. Rev. Ecol. Syst. 4:1–23.

Horn, H. S. 1971. The adaptive geometry of trees. Monographs in Population Biology, vol. 3. Princeton University Press, Princeton, N.J. 144 pp.

Hutchinson, G. E. 1959. Homage to Santa Rosalia, or why are there so many kinds of animals? Amer. Natur. 93:145–159.

Jackson, J. B. C. 1972. The ecology of molluscs of *Thalassia* communities, Jamaica, West Indies. II. Molluscan population variability along an environmental stress gradient. Mar. Biol. 14:304–337.

Janzen, D. H. 1970. Herbivores and the number of tree species in tropical forests. Amer. Natur. 104:501–528.

Johnson, R. G. 1970. Variations in diversity within benthic marine communities. Amer. Natur. 104:285–300.

Karr, J. R. 1971. Structure of avian communities in selected Panama and Illinois habitats. Ecol. Monogr. 41:207–233.

Kohn, A. J. 1967. Environmental complexity and species diversity in the gastropod genus *Conus* on Indo–West Pacific reef platforms. Amer. Natur. 101:251–260.

———. 1968. Microhabitats, abundance, and food of *Conus* on atoll reefs in the Maldive and Chagos Islands. Ecology 49:1046–1062.

———. 1971. Diversity, utilization of resources, and adaptive radiation in shallow-water marine invertebrates of tropical oceanic islands. Limnol. Oceanogr. 16:332–348.

Lang, J. 1973. Interspecific aggression by scleractinian corals. II. Why the race is not only to the swift. Bull. Mar. Sci. 23:260–279.

Levin, S. A., and R. T. Paine. 1974. Disturbance, patch formation, and community structure. Proc. Nat. Acad. Sci. 71:2744–2747.

Levins, R. 1965. Theory of fitness in a heterogeneous environment. V. Optimal genetic systems. Genetics 52:891–904.

Lewontin, R. C. 1969. The meaning of stability. Brookhaven Symp. Biol. 22:13–24.

Lloyd, M. 1971. The biology of *Searlesia dira* (Mollusca: Gastropoda) with emphasis on feeding. Ph.D. diss. University of Michigan. 97 pp.

Loucks, O. L. 1970. Evolution of diversity, efficiency, and community stability. Amer. Zool. 10:17–25.

MacArthur, R. H. 1964. Environmental factors affecting bird species diversity. Amer. Natur. 98:387–397.

———. 1965. Patterns of species diversity. Biol. Rev. 40:510–533.

———. 1972. Geographical ecology. Harper & Row, New York. 269 pp.

MacArthur, R. H., and R. Levins. 1967. The limiting similarity, convergence, and divergence of coexisting species. Amer. Natur. 101:377–385.

MacArthur, R. H., and E. O. Wilson. 1967. The theory of island biogeography. Monographs in Population Biology, vol. 1. Princeton University Press, Princeton, N.J. 205 pp

Margalef, R. 1969. Diversity and stability: a practical proposal and a model of inter-dependence. Brookhaven Symp. Biol. 22:25–37.

May, R. M., and R. H. MacArthur. 1972. Niche overlap as a function of environmental variability. Proc. Nat. Acad. Sci. 69:1109–1113.

Menge, B. A. 1972a. Foraging strategy of a starfish in relation to actual prey availability and environmental predictability. Ecol. Monogr. 42:25–50.

———. 1972b. Competition for food between two intertidal starfish and its effect on body size and feeding. Ecology 53:635–644.

———. 1974. Effect of wave action and competition on brooding and reproductive effort in the seastar, *Leptasterias hexactis*. Ecology 55:84–93.

———. 1976. Organization of the New England rocky intertidal community: role of predation, competition, and environmental heterogeneity. Ecol. Monogr., vol. 46 (in press).

Menge, J. L. 1976. Algal species diversity: importance of herbivore food preference and plant competitive abilities. Unpublished manuscript.

Menge, J. L., and B. A. Menge. 1974. Role of resource allocation, aggression, and spatial heterogeneity in coexistence in two competing intertidal starfish. Ecol. Monogr. 44:189–209.

Orians, G. H., and H. S. Horn. 1969. Overlap in foods of four species of blackbirds in the potholes of central Washington. Ecology 50:930–938.

Paine, R. T. 1966. Food web complexity and species diversity. Amer. Natur. 100:65–75.

———. 1969. The *Pisaster-Tegula* interaction: prey patches, predator food preference and intertidal community structure. Ecology 50:950–961.

———. 1971. A short-term experimental investigation of resource partitioning in a New Zealand rocky intertidal habitat. Ecology 52:1096–1106.

———. 1974. Intertidal community structure: experimental studies on the relationship between a dominant competitor and its principal predator. Oecologia 15:93–120.

Paine, R. T., and R. L. Vadas. 1969. The effects of grazing by sea urchins, *Strongylocentrotus* spp., on benthic algal populations. Limnol. Oceanogr. 14:710–719.

Pianka, E. R. 1966. Latitudinal gradients in species diversity: a review of concepts. Amer. Natur. 100:33–46.

———. 1967. On lizard species diversity: North American flatland deserts. Ecology 48:333–351.

———. 1969. Sympatry of desert lizards (*Ctenotus*) in Western Australia. Ecology 50:1012–1030.

———. 1971. Lizard species density in the Kalahari Desert. Ecology 52:1024–1029.

———. 1974a. Evolutionary ecology. Harper & Row, New York. 356 pp.

———. 1974b. Niche overlap and diffuse competition. Proc. Nat. Acad. Sci. 71:2141–2145.

Porter, J. 1972a. Predation by *Acanthaster* and its effect on coral species diversity. Amer. Natur. 106:487–492.

———. 1972b. Patterns of species diversity in Caribbean reef corals. Ecology 53:745–748.

———. 1974. Community structure of coral reefs on opposite sides of the Isthmus of Panama. Science 186:543–545.

Ricklefs, R. 1973. Ecology. Chiron, Portland, Ore. 861 pp.

Root, R. B. 1967. The niche exploitation pattern of the blue-gray gnatcatcher. Ecol. Monogr. 37:317–350.

Rosenzweig, M. L., and R. H. MacArthur. 1963. Graphical representation and stability conditions of predator-prey interactions. Amer. Natur. 97:209–223.

Roughgarden, J. 1972. Evolution of niche width. Amer. Natur. 106:683–718.

Sanders, H. L. 1968. Marine benthic diversity: a comparative study. Amer. Natur. 102:243–282.

———. 1969. Benthic marine diversity and the stability-time hypothesis. Brookhaven Symp. Biol. 22:71–80.

Schoener, T. W. 1968. The *Anolis* lizards of Bimini: resource partitioning in a complex fauna. Ecology 49:704–726.

———. 1974a. Resource partitioning in ecological communities. Science 185:27–39.

———. 1974b. The compression hypothesis and temporal resource partitioning. Proc. Nat. Acad. Sci. 71:4169–4172.

Slobodkin, L. B., and H. L. Sanders. 1969. On the contribution of environmental predictability to species diversity. Brookhaven Symp. Biol. 22:82–93.

Sokal, R. R., and F. J. Rohlf. 1969. Biometry. Freeman, San Francisco. 776 pp.

Soulé, M., and B. R. Stewart. 1970. The "niche-variation" hypothesis: a test of alternatives. Amer. Natur. 104:85–97.

Stanley, S. 1973. An ecological theory for the sudden origin of multicellular life in the late Precambrian. Proc. Nat. Acad. Sci. 70:1486–1489.

Stimson, J. 1970. Territorial behavior of the owl limpet, *Lottia gigantea*. Ecology 51:113–118.

———. 1973. The role of the territory in the ecology of the intertidal limpet *Lottia gigantea* (Gray). Ecology 54:1020–1030.

Sutherland, J. P. 1970. Dynamics of high and low populations of the limpet, *Acmaea scabra* (Gould). Ecol. Monogr. 40:169–188.

Taylor, D. L. 1973. Some ecological implications of forest fire control in Yellowstone National Park, Wyoming. Ecology 54:1394–1396.

Tinkle, D. W., and R. E. Ballinger. 1972. *Sceloporus undulatus*: a study of the intraspecific comparative demography of a lizard. Ecology 53:570–584.

Vance, R. R. 1972. Competition and mechanism of coexistence in three sympatric species of intertidal hermit crabs. Ecology 53:1062–1074.

Van Valen, L. 1965. Morphological variation and width of ecological niche. Amer. Natur. 99:377–390.

Ware, D. M. 1972. Predation by rainbow trout (*Salmo gairdneri*): the influence of hunger, prey density, and prey size. J. Fisheries Res. Board Can. 29:1193–1201.

Wolcott, T. G. 1973. Physiological ecology and intertidal zonation in limpets (*Acmaea*): a critical look at "limiting factors." Biol. Bull. 145:389–422.

Woodin, S. A. 1974. Polychaete abundance patterns in a marine soft-sediment environment: the importance of biological interactions. Ecol. Monogr. 44:171–187.

Zaret, T., and A. S. Rand. 1971. Competition in tropical stream fishes: support for the competitive exclusion principle. Ecology 52:336–342.

COMPETITION, DISTURBANCE, AND COMMUNITY ORGANIZATION: THE PROVISION AND SUBSEQUENT UTILIZATION OF SPACE IN A ROCKY INTERTIDAL COMMUNITY[1]

Paul K. Dayton[2]

Department of Zoology, University of Washington, Seattle, Washington

Table of Contents

ABSTRACT

An understanding of community structure should be based on evidence that the growth and regulation of the component populations in the community are affected in a predictable manner by natural physical disturbances and by interactions with other species in the community. This study presents an experimental evaluation of the effects of such disturbances and competitive interactions on populations of sessile organisms in the rocky intertidal community, for which space can be demonstrated to be the most important limiting resource. This research was carried out at eight stations on the Washington coastline which have been ranked according to an exposure/desiccation gradient and subjected to comparable manipulation and observation.

Physical variables such as wave exposure, battering by drift logs, and desiccation have important effects on the distribution and abundance of many of the sessile species in the community. In particular, wave exposure and desiccation have a major influence on the distribution patterns of all the algae and of the anemone *Anthopleura elegantissima*. The probability of damage from drift logs is very high in areas where logs have accumulated along the intertidal. Log damage and wave exposure have complementary effects in the provision of free space in a mussel bed, as wave shock enlarges a patch created by log damage by wrenching the mussels from the substratum at the periphery of the bare patch.

Competition for primary space results in clear dominance hierarchies, in which barnacles are dominant over algae. Among the barnacles, *Balanus cariosus* is dominant over both *B. glandula* and *Chthamalus dalli*; *B. glandula* is dominant over *C. dalli*. The mussel *Mytilus californianus* requires secondary space (certain algae, barnacles, or byssal threads) for larval settlement, but is capable of growing over all other sessile species and potentially is the competitive dominant of space in the community.

[1] Received November 23, 1970; accepted June 2, 1971.
[2] Present address: University of California, San Diego; Scripps Institution of Oceanography, La Jolla, California 92037.

352   PAUL K. DAYTON   Ecological Monographs
                                                                Vol. 41, No. 4

Three general levels of biological disturbance result from a limpet guild, a carnivorous gastropod guild composed mainly of species of *Thais*, and the asteroid *Pisaster ochraceus*. Limpets have a negative effect on the recruitment of algae and barnacles. By the end of the summer, however, the limpet disturbance in the absence of *Thais* increases the survival of *Chthamalus* by reducing the survival of the competitively superior *Balanus* spp.; in the presence of *Thais*, all barnacle species are negatively affected by limpets. The effect of *Thais* in the absence of limpets is to increase *Chthamalus* survival by selectively eating the *Balanus* spp.; *Thais*, in the presence of limpets, have a negative effect on all the barnacle species. The competitive dominants, *Mytilus californianus* and *B. cariosus*, have an escape in growth from the gastropod predators and could eventually monopolize all the space. The combined effects of the predation of *Pisaster* and the log damage prevent this from happening.

In natural areas disturbances are sufficient to prevent the monopolization of space by any of the sessile organisms. In contrast to many communities which are thought to be structured around competitive interactions, this intertidal community is characterized by continuous physical and biological disturbances, an abundance of free space, and a large number of species which utilize this same potentially limiting resource.

INTRODUCTION

Many theories of community organization have grown from a descriptive basis heavily dependent upon sampling approaches which stress the structural aspects of the community. Very few biological communities can be manipulated so that there are comparable altered (experimental) and unaltered (control) situations. For this reason little emphasis has been placed on experimentally elucidating the functional roles of the major community components.

The concept of community organization suggests more than a description of the assemblage of populations. Such an organization, incorporating dynamic interactions, trophic structure, and patterns of distribution and abundance, should react predictably to physical and biological disturbances; the most convincing demonstration of community structure would be proof that the growth and regulation of the component populations are affected in a predictable manner by natural physical disturbances and by changes in abundances of other species in the community. Thus a dynamic theory of community structure should be built around an understanding both of the effects of the major community disturbances and of the interactions among the component populations. If such disturbances and interactions do contribute to a community structure, they can be evaluated by studying the factors influencing community succession. Such an examination is useful (1) for identifying the ecological conditions necessary for the establishment of a population and (2) for evaluating the competitive interactions among populations which share a potentially limiting resource. Once the patterns and rates of succession are determined, the effects of physical disturbances can be evaluated and the effects of the predator populations on the competitive patterns can be assessed by selectively removing the predators. The information gained by doing this should suggest the important parameters controlling the community structure and should explain and predict the observed distribution patterns and abundances of most of the species in the community.

The rocky intertidal community of the northeast Pacific offers many opportunities to assess the importance to its organization of various physical disturbances and biological interactions. It has the logistic advantage of being readily accessible, and it has geographic continuity, in that most of the species have widely overlapping geographic ranges—from Kodiak, Alaska, to Point Conception, California (reviewed by Glynn 1965, Paine 1969). Furthermore, most of the species are limited to the intertidal region and are confined to rocky substrata. The sluggish behavior of most of the invertebrate predators and the sessile habitus of most of the prey species make the community amenable to experimental manipulation. Most important, almost all of the prey species share an essentially two-dimensional space as their single potentially limiting resource. The successful utilization of this resource is easy to evaluate and quantify. Competition theory predicts that in the absence of disturbance a single or very restricted number of species should effectively occupy this simple resource. The fact that space is very rarely fully utilized by those species which are the competitive dominants, i.e., those that successfully outcompete others for space, in this community suggests that much of the community organization is a function of factors interfering with the recruitment and survival of these sessile species.

The main processes involved with the provision, procurement, and subsequent utilization of space are (1) physical stress, (2) natural death, especially of annual algal species, and the defoliation of the perennial algal species, (3) competition for a limiting resource, and (4) biological disturbances or predation-caused mortality.

The basic goal of this research has been to test the hypothesis that the growth and regulation of most of the populations of sessile species are predictably affected by physical disturbance and by the dominant competitive and predatory populations in the community. Since the competitive dominance is exerted via the common resource of space, one main theme has been the identification of patterns of provision and allocation of free primary space, here taken to mean unoccupied substratum available for colonization of sessile organisms. In situations where there is an algal cover over, but not attached to, available substratum, this substratum is considered "shaded free space." The total available primary substratum or "total free space" is the sum of the shaded free space and the "uncovered free space." In areas where there is no canopy, the total available primary substratum may be referred to as "free space."

Intertidal research has classically involved elaborate descriptions of distribution and abundance patterns of various algal and animal populations. These distribution patterns along the Pacific coast of the United States are described by Ricketts, Calvin, and Hedgpeth (1968). The suggested role of various physical stresses in determining the distribution patterns in the marine intertidal has been summarized by Lewis (1964). This emphasis was on the role of physical stresses and suggested that the populations were indeed random aggregations of species sharing similar physical tolerances. However, a number of other factors appear to influence the distribution patterns and spatial relations of many of the important species in the intertidal community.

Abrupt discontinuous distribution patterns of sessile populations are frequently considered to be the result of competitive interactions (Daubenmire 1966). Such patterns have been so explained for intertidal populations (Endean, Kenny, and Stephenson 1956), and Connell (1961a) has experimentally proven that competitive pressures limit the lower distribution of the barnacle *Chthamalus stellatus* at Millport, Scotland.

Important and often dramatic effects of predation on distribution patterns of marine organisms have been experimentally demonstrated by Stephenson and Searles (1960), Randall (1961, 1963), and Bakus (1964) on benthic communities in tropical waters and by Blegvad (1928) in the Baltic Sea. Connell (1961b) demonstrated the important effects of the predation of the gastropod *Thais lapillus* at Millport, and Paine (1966) has experimentally demonstrated that the asteroid *Pisaster ochraceus*, by selectively preying on the dominant competitor for space, can be responsible for maintaining much of the species richness in the middle intertidal community at Mukkaw Bay, Washington. An intensive study by Connell (1970) has shown that many facets of the distribution pattern of the barnacle *Balanus glandula*

on San Juan Island, Washington, are a result of predation by three species of *Thais*.

As the marine environment merges into the terrestrial environment above it, the intertidal organisms are subject to increasingly severe physiological stresses. Because all the intertidal species have different tolerances to these stresses, the different upper limits of their distribution patterns result in conspicuous zones. It is well known that temperate intertidal shores are divided into a lower algal zone and an upper barnacle-mussel zone (see Stephenson and Stephenson 1961, Lewis 1964, and Ricketts, Calvin, and Hedgpeth 1968, for reviews). This paper will be restricted to the discussion of biological interactions in the barnacle-mussel zone with emphasis on the functionally more important species.

Study Areas

Study areas that represented a gradient of physical exposures were chosen because one of the objectives of the work was to describe effects of physical conditions on the intertidal community (Fig. 1). Tatoosh Island (48°24′N, 124°44′W) is completely exposed to the oceanic swells and wave action of the eastern Pacific Ocean. The upper barnacle-mussel association here is dominated by an extremely dense bed of the mussel *Mytilus californianus*, which occupies from 65–97% of the total space. The barnacles *Balanus cariosus* and *Pollicipes polymerus* occur in scattered patches, and the upper level is dominated by *Balanus glandula*. *Postelsia palmaeformis*, the palm alga, is a conspicuous alga in the upper intertidal. No experimental work was done at Tatoosh Island because logistic difficulties inherent in getting to the island prohibited regular visits.

Waadah Island (48°23′N, 124°36′W) is slightly less exposed to wave shock than is Tatoosh Island. At this site the predation of *Pisaster ochraceus* on *Mytilus californianus* seems to be much more effective than at Tatoosh Island, and the *Mytilus* are found only in small patches, usually above the *Pisaster* foraging zone. There are scattered patches of *Balanus cariosus, Balanus glandula, Chthmalus dalli,* and *Pollicipes*. Experiments were done on Waadah Island at Postelsia Point (Rigg and Miller 1949) as well as on the protected side of the breakwater. Waadah Island is a particularly useful experimental area, both because of the many available habitats and because the U.S. Coast Guard protects it from human disturbance. Both Tatoosh and Waadah Islands represent the typical exposed outer coast communities described by Ricketts et al. (1968).

Shallow offshore water offers sufficient protection to Shi Shi reef (48°16′N, 124°41′W) to break much

PAUL K. DAYTON   Ecological Monographs
Vol. 41, No. 4

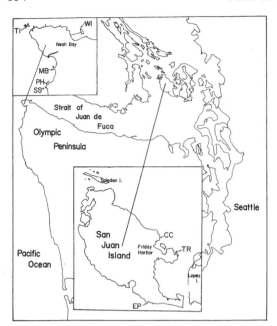

FIG. 1. Map of Olympic Peninsula region of Washington State showing locations of study sites. SS, Shi Shi; PH, Portage Head; MB, Mukkaw Bay; TI, Tatoosh Island; WI, Waadah Island; EP, Eagle Point; TR, Turn Rock; CC, Colin's Cove.

of the power of the deep sea swells so conspicuous at Tatoosh and Waadah Islands. Offshore reefs shelter Mukkaw Bay (48°19′N, 124°40′W), and shallow water and a complex series of offshore reefs considerably protect the Portage Head site (48°17′N, 124°41′W). The topography of the intertidal zone of these three sites is broken by large boulders, surge channels, and deep tidepools. None of these sites has *Postelsia*; all three sites would be classified as protected outer coast environments by Ricketts et al. (1968). Large *Balanus cariosus*, frequently under the *Mytilus*, are also found above the effective level of *Pisaster* predation. *Pollicipes polymerus*, a well-known component of this community (Ricketts et al. 1968), is rare on the rather flat study sites in the protected outer coast areas and on San Juan Island. Its local distribution pattern represents a specialized situation (Barnes and Reese 1960), and it will not be considered further. Large aggregations of the anemones *Anthopleura elegantissima* and *A. xanthogrammica* as well as occasional heavy sets of *Balanus glandula* and *Chthamalus dalli* occupy much of the intermediate levels between the algal association and the *Mytilus* refuge at all three protected outer coast areas.

The Shi Shi and Portage Head experimental sites were chosen because they offer protected outer coast environments and because of their relative inaccessibility to marauding human activity. Mukkaw Bay is subject to an increasingly intense human disturbance, and only a few experiments survived the 4 years there. The substratum at all of the outer coast sites is a soft siltstone.

Although the San Juan Islands lack the deep swells of the open seas, the Eagle Point site (48°27′N, 123°2′W) faces the prevailing winds and is frequently exposed to severe wave action. Two areas were studied at Eagle Point: (1) Eagle Point main area, where almost all the experimental work was done, and (2) Eagle Point log area, a spot where drift logs accumulate and are moved around during high tides and storms. The barnacle-mussel association at Eagle Point includes patchy but conspicuous aggregations of the anemone *Anthopleura elegantissima*, many large *Balanus cariosus*, a few large *Mytilus californianus*, a few small patches of *M. edulis*, and frequently heavy sets of *Balanus glandula* and *Chthamalus dalli*.

Turn Rock (48°32′N, 122°58′W) is in a protected channel and has very little wave action, but is exposed to strong tidal currents (3 knots). The barnacle-mussel association above the level of *Pisaster* predation is dominated by a complete cover of very large *Balanus cariosus* and *B. glandula*. Very few mussels were present during the course of this study. Large patches of the alga *Fucus distichus* have settled on the barnacles.

Colin's Cove (48°33′N, 123°0′W), approximately 200 m north of the Cantilever Pier of the Friday Harbor Laboratories, is the most protected study area. Occasional winter storms with winds from the northeast produce heavy wave action at this site, but usually the heaviest waves at Colin's Cove come from passing ferry boats and pleasure craft. The barnacle-mussel association here is essentially nonexistent; the upper levels are mostly bare rock with isolated individuals of *Balanus cariosus, B. glandula,* and *Chthamalus dalli*. The intertidal substratum of San Juan Island is a very hard metamorphosed chert (McLellan 1927).

Mean monthly temperatures and salinities have been described by Connell (1970) for the San Juan Islands and are similar on the outer coast. Although it is easy to measure and has received much attention, temperature per se probably does not limit the distribution of the various organisms among these study areas. As indirect evidence for this, the ranges of most of these organisms extend well into Alaska to the north, where temperatures are much lower, and to southern California and Mexico to the south, where the temperature extremes are much higher than those generally encountered in these study areas.

Desiccation (Kensler 1967) is probably the most important physical factor limiting the local distribution of most of these organisms within the sites studied. The San Juan sites are subjected to a much greater degree of desiccation than are the outer coast sites. The latter are buffered by the wet maritime weather brought on by offshore upwellings. In contrast, the San Juan Islands have more wind, about twice as many sunny days in the summer, and probably more extreme cold in the winter. The effect of the different climatic conditions on the marine intertidal organisms is exaggerated in the San Juan Islands because the summer spring low tides usually occur around the middle of the day, subjecting the exposed marine organisms to the extreme conditions of summer desiccation, and the winter spring low tides are in the middle of the night, subjecting them to the most extreme climatic conditions of that season. The spring low tides on the outer coast tend to be late in the afternoon in the winter and very early in the morning during the summer. Thus the outer coast intertidal is relatively well protected from desiccation and climatic extremes both by the moist weather and by the time of day of the tides.

Within each area there are obvious biological differences among habitats with differing degrees of exposure. Since one of the objectives was to evaluate the effects of wave action on communities in exposed environments, the study has been restricted to the exposed rocky intertidal localities in each of the study areas.

The height of a level along the shore is given as feet above or below the mean lower low water. The levels at each site were obtained by observing the low point of the low tide. This measure varied considerably from the predicted low of the tide tables because of wind, surf activity, and barometric pressure, but the reference point at each site is believed to be accurate to within ±5 cm. All the other levels at each site were measured in relation to the reference point with surveying equipment. Because of the limited number of visits, comparable data are not available for Tatoosh Island.

The relative utilization of substratum space by the competitively dominant space consumers in the barnacle-mussel association is presented in Table 1 in terms of percentage cover. The occupation of primary space was usually evaluated photographically by placing a piece of acetate with 100 randomly placed points over it or by the use of a planimeter. The former method is similar to the technique described by Connell (1970) and yields a measure of percentage cover which can be converted into square centimeters of cover. The planimeter yields a proportion cover which can be converted into either percentage cover or actual square centimeters of cover. Repeated measurements of large organisms on the same photograph with the planimeter vary

less than measurements with the random dot technique (1–2% as opposed to 1–4%), but with small scattered organisms, such as barnacles, it is essentially impossible to use the planimeter. The planimeter and the random dot method are interchangeable and repeatable with less than 5% difference.

In some situations primary substratum alone is not the most important limiting resource. The distribution patterns of the algae in particular are also critically influenced by light intensity and protection from desiccation. Clear evidence of this are the differences in vertical distribution of algal zones on north- and south-facing exposures. In the areas studied the south exposure is subjected to more desiccation; this probably is why the algal zones occur at lower depths on the south side than do the corresponding zones on the north side of the same rocks. The distribution patterns of the algae are described in Table 1 in terms of general categories such as percentage cover of any algal canopy and percentage cover of an obligate understory category, which includes algal species which either disappear when the canopy cover is removed or are found only in permanently damp spots. The fugitive species (Hutchinson 1951) in Table 1 are those algal species which appear when the canopy or barnacle-mussel cover is disturbed.

The encrusting coralline algae, such as various species of *Lithothamnion* and *Lithophyllum*, often cover large areas of primary substratum. Since these algae have proven to be satisfactory substrata for most of the other attached intertidal algae, the primary space utilized by them will not be discussed. The actual degree of interference via cellular sloughing or allelopathic agents of these encrusting algae with the recruitment of algal spores or animal larvae is unknown. No interference with other algae has been observed, and certainly barnacle cyprids readily settle on the encrusting corallines.

PHYSICAL EFFECTS

The most important physical factors correlated with differences in the relative distributions and abundances of the important sessile species in the intertidal are (1) wave exposure, (2) battering by drift logs, and (3) physiological stresses such as desiccation and heat.

Wave exposure

The importance of wave exposure or wave shock to the distribution and abundances of intertidal populations is well known (Ricketts et al. 1968). Bascom (1964) discusses the degrees of wave activity and the many factors complicating its measurement; Harger (1970) describes a simple device which estimates the force of wave impact directed vertically

356   Paul K. Dayton   Ecological Monographs
                                                                            Vol. 41, No. 4

Table 1. Mean percentage cover of major sessile organisms in upper intertidal zones. Data are from ten to thirty 0.25-m² quadrates taken randomly along horizontal transects. Variance is presented as 95% confidence limits. "Fugitive species" and "obligate understory" categories refer to algae and are described in the text. (* = *Postelsia palmaeformis*; ** = *Fucus distichus*; + = *Pelvetiopsis limitata*; ° = *Hedophyllum sessile*.) Blanks mean species were not seen in samples

Site	Level (ft)	Chthamalus dalli	Balanus glandula	Balanus cariosus	Mytilus californianus	Anthopleura elegantissima	Fugitive species
Tatoosh Island	2–4				94.4± 8.6		1.3± 0.9
	4–6			0.7± 0.1	82.1±27.3		7.5±14.3
	6–8			2.2± 0.1	53.7±20.7		20.6±13.9
	8–10			2.6 ±2.1	54.7±35.7		30.1±25.5
	>10				61.1±26.8		8.6± 6.7
Waadah Island	5–7				20.7±29.8		7.5± 5.6
	7–9			1.1± 1.8	76.3±15.5		1.6± 1.5
Shi Shi	2–4	7.5± 2.3	33.9±16.1		11.8± 9.6	4.0±3.6	7.0± 4.6
	4–6		72.1±16.7		19.0± 6.4	4.4±5.1	
	>6				74.0±24.5		10.3±22.2
Portage Head	2–4	5.7± 2.0	1.5± 1.0			4.1±2.6	72.9±13.1
	4–5	10.1± 3.5	16.5± 9.1			13.6±8.5	5.7± 6.7
	5–6	2.1± 3.1	10.5± 8.5	21.8±10.5	46.9±15.2		1.0± 4.2
	6–7						20.0± 9.5
	7–9	25.0± 6.7	16.0± 6.1				
Eagle Point	1–2	0.6± 0.6		4.3± 3.0			1.0± 4.3
	2–4			30.3± 7.0	3.3± 1.8	6.8± 3.7	8.2± 7.9
	4–6	5.9± 1.8	4.0± 5.8	63.2± 4.6			
	>6	3.0± 2.0	0.8± 0.8	39.7±12.4			
Turn Rock	1–2			2.4± 2.1			20.0±15.8
	2–3			100.0			9.2± 9.1
	3–4			98.3± 4.5			
	4–6		61.3± 5.8	38.9± 9.3			
Colin's Cove	2–4	0.1± 0.7					19.3± 5.8
	4–6	2.0± 3.0	1.7± 1.6	0.8± 1.1			7.5± 9.3
	>6	0.5± 0.9	2.0± 1.8				0.9± 0.7

Site	Level (ft)	Obligate understory	Canopy cover	Holdfast cover	Uncovered free space	Shaded free space
Tatoosh Island	2–4	3.7± 0.7				
	4–6	10.0±15.1				
	6–8	6.5± 4.6	*6.9± 8.1	4.7±2.5	9.0± 4.2	3.8± 3.6
	8–10	6.2± 5.6			8.4± 7.9	
	>10	23.3±16.8	*4.0± 6.4	3.1±0.9	3.6± 2.7	
Waadah Island	5–7	24.1±12.8	*76.8±24.1	31.8±5.8		17.9±11.3
	7–9				20.9±13.4	
Shi Shi	2–4				35.9±16.2	
	4–6				5.1± 3.7	
	>6				15.1± 5.5	
Portage Head	2–4		**4.5± 6.1		10.8± 5.7	4.5± 1.9
	4–5		**7.0± 7.9		51.1±11.8	3.7± 1.2
	5–6		**9.0± 4.2		19.3± 9.2	
	6–7	3.0± 5.4	**66.7± 22.0		14.0± 5.7	64.4± 8.9
	7–9		+25.0± 5.7		42.1±15.9	15.4± 6.7
Eagle Point	1–2	1.1± 1.3	°5.5± 6.4		92.2±11.7	
	2–4	0.9± 2.1	°1.0± 0.7		50.8± 8.1	
	4–6				26.5± 2.9	
	>6				57.1±11.2	
Turn Rock	1–2	7.8±7.2	**4.8± 5.0		67.0±18.0	4.8± 7.5
	2–3		**37.5±18.5			
	3–4		**83.7±11.4			1.6± 2.4
	4–6					
Colin's Cove	2–4		**53.5±10.7		27.2± 9.0	53.5±11.4
	4–6		**34.5±20.4		54.0±11.4	34.5±22.5
	>6		**6.5± 5.7		97.3± 2.4	

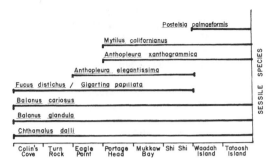

STUDY SITES

FIG. 2. Distributions between sites of conspicuous sessile organisms. Colin's Cove represents the most sheltered area; Tatoosh Island is the site of greatest wave exposure.

toward the substratum. The complications attendant to an adequate description of wave exposure along broken, heavily exposed shorelines were prohibitive for this study. Study sites were chosen which represent varying degrees of exposure to wave action and were subjectively ranked in decreasing order of the magnitude of wave shock. Tatoosh Island is most exposed, with Waadah Island, Shi Shi, Mukkaw Bay, Portage Head, Eagle Point, Turn Rock, and Colin's Cove following in decreasing order of exposure.

The distribution of the conspicuous sessile organisms in environments representing different degrees of wave-shock conditions is plotted in Fig. 2. Most of these species are found in a number of sites and apparently have wide tolerances to wave exposure.

Drift logs

Many areas along the shores of the San Juan Islands and the outer coasts of Washington and Vancouver Island have large accumulations of drift logs. The battering by these logs destroys the intertidal organisms and is an important factor in the provision of space. On San Juan Island 50% of the logs stranded on shore have been cut and appear to be there as a result of human activity; 15% of the logs still have their root systems intact and are therefore known to be natural drift, probably from erosion along the large coastal river systems or from the coastal shores. The remaining 35% of the San Juan Island drift logs are old and worn, and their source is obscure. The outer coast beaches frequently are covered with great piles of logs; in some cases, particularly where salvage is impossible, the log jam which stretches in a strip along the upper level of the beach is between 30 and 75 m wide. At these coastal sites fewer than 1% of the drift logs show signs of being cut by man, while 85% of them still have their root systems intact, indicating a natural source. The remaining 14–15% are too battered for

a positive evaluation, but also appear to represent natural drift. I have frequently observed natural erosion at the Shi Shi and Portage Head study sites causing large trees to slide onto the intertidal zone. Similar observations were made before logging became a major environmental problem in the Pacific Northwest (Swan 1857). Erosion along the shores of the sea and along the many river systems of the Pacific Northwest has probably always been a source of drift logs. Since these logs persist for many years, their impact on the intertidal community may have been even more profound before commercial interests made their salvage profitable. Certainly the disturbance from this battering is a natural and important ecological phenomenon in this community.

The probability of log disturbance at each study site was measured by embedding cohorts of nails haphazardly into the substratum at three different intertidal levels. The nails were embedded with a construction stud gun using .32 caliber blanks; each nail stood approximately 2 cm high. Survival curves of these nail cohorts show that within most of the study areas there is a 5–30% probability of any given spot being struck by a log within 3 years (Fig. 3). There was no consistent level effect on the "mortality" of the nails.

The exposed organisms are killed regularly in areas such as the Eagle Point log area, where drift logs concentrated by local current action make the site particularly susceptible to log battering. Here the only sessile organisms are found in deep crevices; all the exposed shore is devoid of long-lived sessile organisms. There are many such areas in the San Juan Islands and on the outer coast; however, I confined my experimental studies to sites which, unlike the Eagle Point log area, are struck only by logs carried by the currents, rather than by the shifting of an accumulated drift at high tide.

Turn Rock is an unusual site in the San Juan Islands, because a rapid current pattern carries the drift around the small island and protects it from being struck by this material. Postelsia Point on Waadah Island is apparently similarly protected by an offshore rip current.

Figure 3 also compares the percentage cover at the 4- to 5-ft level of three areas in the San Juan Islands which are similar in most physical parameters, but which have different probabilities of being struck by logs. The 4- to 5-ft level was chosen because the dominant occupant of space, the long-lived *Balanus cariosus*, is not subjected to *Pisaster* predation at this level. Thus the relative amount of available substratum is very likely a result of log damage.

No nails were placed at Tatoosh Island, but four sites were found on the island in 1968 and three in 1970, in an area of about 208 m², which had been

PAUL K. DAYTON   Ecological Monographs
Vol. 41, No. 4

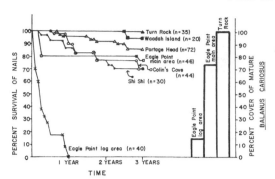

Fig. 3. Survival curves of nails placed at each study site. The percentage cover of mature *Balanus carious* at the 4- to 5-ft level at three San Juan Island sites is shown at right.

struck recently by logs, as evidenced by wood chips in crevices in the rocks.

In areas where *Mytilus californianus* dominates, the effects of log battering and wave shock complement each other. *Mytilus* larvae preferentially settle in filamentous algae and among the byssal threads of adult *Mytilus* (Bayne 1964, 1965; also personal observation). In areas where wave action is sufficient to provide the lower individuals with aerated water and adequate food, a mat of mussels up to 25 cm thick can develop. Usually the mat is anchored to the substratum by the byssal threads of relatively few individuals; the majority of individuals are anchored to each other's valves and byssal threads. A thick mussel bed such as this is partially dependent upon its spatial continuity to keep the mussels from working loose in the surf, because such a mat of mussels, developing on an exposed ridge or rocky promontory, can be ripped off by wave action. When log damage or *Pisaster* causes a clearing, the mussels around the edge of the clearing are susceptible to violent twisting by wave action and are often torn off the substratum. Even a small cleared area can be considerably enlarged. In this way wave shock can be responsible for the provision of substratum free from the *Mytilus*.

Much patchiness in the upper intertidal in exposed areas results from log damage creating an initial clearing in the *Mytilus* bed in which the space is then considerably enlarged by wave shock ripping the newly vulnerable mussels from the substratum. The original and ultimate sizes of such clear patches caused by log damage at the Shi Shi and Portage Head sites are shown in Table 2. The mean increases in available primary space, expressed as a percentage of the original clearing, are 3,897% and 24%, respectively. The original clearing was known to be made by a log because either the log was seen or bark and wood were jammed into crevices in the

rock. Seven patches originally created by log damage were observed at Tatoosh Island (Table 2). The percentage increase in these clearings was estimated by assuming that the original clearing was 0.09 m² (the size of the largest Shi Shi original clearing) and using the observed size of each clearing as an estimate of the ultimate size, even though in most cases the clearings were still being enlarged. The resultant conservative estimate of the percentage increase of a cleared patch at Tatoosh is 4,884%. Differences in percentage increase of patch size at the three sites are an indication of relative differences in wave exposure.

Predation by *Pisaster* along the lower edge of the *Mytilus* aggregation creates small clearings which are enlarged by wave action. Thus, clear patches at the lower level of the *Mytilus* bed can result either from *Pisaster* predation or log damage, and it is impossible a posteriori to differentiate between them. In 1970 the 208-m² area on the exposed point of Tatoosh Island had three such cleared patches in the lower intertidal totaling 10.0 m². Twenty-two patches in the upper intertidal, which were sufficiently high so that they were almost certainly originated by logs, totaled 63.8 m². Since the recovery of the *Mytilus* bed is much faster at the lower levels, the relative rates of the provision of space in these two levels cannot be compared. But 35% of the total area in this *Mytilus* bed is composed of free primary space. The physical force of log battering and of wave shock is largely responsible for the clearing of this space.

Physiological stress

The algae seem to be particularly sensitive indicators of physical conditions (Lewis 1964, Druehl 1967). The luxuriant intertidal algal growths of the outer coast seem to be prevented from being established in the San Juan Islands by the higher degree of desiccation there. Most of the outer coast species are found very occasionally in the San Juan Islands, but only in places extremely well protected from desiccation. In comparison to the algal species, the major space-consuming animal species have much wider distribution patterns across the desiccation gradient represented geographically by the study sites and locally by vertical and topographical exposure differences. This suggests that the distribution patterns of sessile animals are less likely to be determined by physiological stress than are those of the algal species. The major exception is *Anthopleura elegantissima*, a spatially important organism at certain study sites (Table 1), whose distribution seems to be limited partially by desiccatory stresses. It is not found at the two exposed outer coast sites, nor at the two most protected San Juan Island sites. However, large areas (50–100 m²), protected from extreme

Table 2. Increase of patch size cleared in *Mytilus californianus* beds by log damage and wave shock at three outer coast sites

Site	Date	First observed size (m²)	Ultimate observed size (m²)	Average percentage increase in size of patch due to wave shock
Tatoosh Island	1968	—[a]	0.5[b]	
	(June)	—	4.3[b]	
		—	6.2[b]	
		—	16.0	4,884%
	1970	—	3.1[b]	
	(June)	—	0.6[b]	
		—	0.7[b]	
Shi Shi	1967	0.062	0.950	
	(May)	0.050	0.500	
	1968	0.090	2.790	3,897%
	(June)	0.070	8.100	
	1969	0.050	1.390	
	(May)			
Portage Head	1967	0.035	0.085	
	(May)	0.030	0.030	
		0.045	0.045	24%
	1968	0.050	0.050	
	(May)	0.035	0.035	
	1969	0.040	0.040	
	(June)			

[a]For calculations, original size of the clearing was assumed to be 0.09 m².
[b]Patch still being enlarged at time of observation.

Table 3. Percentage cover of *Anthopleura elegantissima* on protected and exposed sides of the same surge channel at Shi Shi (means of 10 randomly sampled quadrates at each level; variance is presented at 95% confidence limit)

Height (ft)	Exposed slope	Protected slope
0–2	0	81.7±24.0
2–4	4.0±3.6	78.7±27.0
4–6	4.4±5.1	7.4± 4.7

surf pounding, at Tatoosh and Waadah Islands have up to 100% cover of this anemone. This correlation of *A. elegantissima* with protected outer coast sites is well demonstrated in a surge channel at Shi Shi, where the side exposed to wave shock always had less than 10% *A. elegantissima* cover, and the protected side had over 80% *A. elegantissima* cover at the 0–2-ft level (Table 3). The restriction of *A. elegantissima* to protected outer coast sites seems to be the result of both competition and physiological stress. Competition appears to reduce its abundance and distribution at the more exposed outer coast sites (see below), whereas physiological stress restricts its abundance and distribution at many of the more protected sites.

The distribution of *Anthopleura elegantissima* seems directly affected by desiccation. On San Juan Island the anemone distribution is essentially confined to isolated areas along the west side of the island

which are exposed to sufficient splash to reduce the effects of desiccation. The only natural population of *A. elegantissima* on the calmer east side of the island was found at Colin's Cove in a small crack between two rocks where it is permanently shaded and well protected from exposure to the wind. The survival of *A. elegantissima* seems to be affected more by desiccation than by temperature. To test this, a calibrated thermometer was placed inside the coelenterons of isolated anemones. The area where the anemones were located was marked with paint or with a screw, and the survival of the anemones was followed on successive tides. With no air movement, the dark green anemones absorb the radiant heat and can reach temperatures considerably above the ambient air temperature (Table 4). The highest *A. elegantissima* temperature recorded was 33.6°C, and the greatest temperature differential between the ambient air temperature and the anemone was 13.3°C. When there was no air movement, the anemones were healthy and suffered no mortality. The *A. elegantissima* seem to be protected against desiccation by a volume of sea water maintained in the coelenteron. Wind produces an evaporative effect which often lowers the anemone temperature below that of the surrounding air; however, this evaporation depletes the water in the coelenteron, and eventually the anemone is killed by desiccation. Although the relative humidity of the air is somewhat increased with the wind, this does not seem to help anemone survival.

360   Paul K. Dayton   Ecological Monographs
                                                                          Vol. 41, No. 4

Table 4. Temperature and percentage mortality of *Anthopleura elegantissima* (n = 50) under various conditions of ambient air temperature, relative humidity, and wind—June–July 1966

Air temperature (°C)	Relative humidity (%)	Wind (mph)	Temperature of *A. elegantissima* (°C)	Total percentage mortality
13.0	84	0	28.2±1.1[a]	None
21.5	75	0	31.1±2.4	None
17.6	95	0	27.6±1.9	None
19.5	98	5–15	19.0±1.7	35%
16.5	95	10–20	13.2±1.0	55%
13.0	90	15–20	9.0±1.8	45%

[a] Mean of 20 samples ± standard deviation.

To test whether Turn Rock and the study area at Colin's Cove will support *Anthopleura elegantissima*, patches of anemones were transferred to each of these sites during the autumn night tides of 1966. Two aggregates, each composed of between 100 and 150 anemones, became successfully established at each site. All the anemones that became attached survived the winter, but in March the tides moved into daylight hours, and the anemones began to turn brown and die from desiccation. All had died by the end of April.

Biological Interactions

Three different types of biological interactions affecting the distribution of the sessile barnacle-mussel association were considered: (1) a sweeping or barrier effect of *Fucus distichus* on barnacles; (2) interspecific competition; and (3) disturbance, especially the effects of predation.

Biological disturbances in the upper intertidal result from the activities of limpets, carnivorous gastropods, and asteroids. Limpets graze on settling algal spores and sporelings (Moore 1938, Southward 1953), and numerous workers have shown that the removal of limpets results in a heavy growth of algae (Jones 1948, Lodge 1948). The large *Patella* in Europe consumes and dislodges settling barnacle cyprids (Hatton 1938, Lewis 1954, Southward 1956, Connell 1961*b*); and Stimson (1970) has found that the large owl limpet, *Lottia gigantea*, dislodges small *Balanus glandula*. Limpet disturbance is, therefore, an important factor in reducing the settlement and establishment of both the algae and the barnacles.

The importance of carnivorous gastropods as a cause of barnacle mortality in Scotland and on San Juan Island has been experimentally demonstrated by Connell (1961*b*, 1970), who showed that with the exclusion of all predators, particularly *Thais* species, barnacle mortality is very much reduced.

Paine (1966) demonstrated that in 1963 the exclusion of *Pisaster ochraceus*, an asteroid which preferentially preys on *Mytilus californianus*, was sufficient for *M. californianus* to outcompete all the other upper intertidal species and monopolize the substratum in the outer coast community at Mukkaw Bay.

Each of these biological disturbances can have a different effect on the populations of sessile organisms. In the analysis of these effects, the limpets and *Thais* are considered as "guilds" such as defined by Root (1967). The limpet guild basically is composed of four species (Table 5); the *Thais* guild is less well defined, as it includes in various areas three species of *Thais*, *Ceratostoma foliatum*, and *Searlesia dira* (Tables 6 and 7).

Methods and field experiments

Experiments designed to measure the effects of competition and predation in the barnacle-mussel association are closely related, and their designs are discussed together. The precise comparisons and tests are discussed in the following sections; the general experimental scheme is diagrammed in Fig. 4. Barriers designed to restrict limpet movement were made by cutting the bottoms out of flexible plastic cereal bowls or dog dishes (Fig. 5). These have the advantages of being inexpensive and sufficiently pliable so that they can be bent over the natural contours of the rock substratum and attached with eye screws. Limpets can be excluded or included with few invasions or escapes; monitoring the dishes every 10–14 days was sufficient to maintain the desired limpet concentration. A dog dish with normal limpet density rather than a "control" area without a barrier around it was used to compare with the exclusion dish in order to keep constant any unknown side effects of the dishes. The limpet concentration to be included in each dish was determined by placing the barrier on the substratum haphazardly 10 times to determine the average densities of the limpet species at that site and level. The dish was then attached to the substratum and the average limpet density was maintained in the dish. Diving observations showed that *Thais* species crawl over these barriers as readily as over the many rock projections and discontinuities of their natural environment. The plastic dishes were

TABLE 5. Density of four species of limpets (*Acmaea*) at various levels at each study site. Data have been converted to number/m² and are derived from transects of 10–25 quadrates. Quadrate sizes ranged from 0.01 to 0.250 m². Blanks mean data unavailable

Site	Limpet	Level (ft)								
		−2–0	0–1	1–2	2–3	3–4	4–5	5–6	6–7	7–8
Waadah Island	*A. scutum*	0		0	0	2	96	485	48	0
	A. pelta	0		0	0	15	35	250	96	0
	A. digitalis	0		0	0	2	10	170	540	1150
	A. paradigitalis	0		0	0	0	80	310	110	25
Shi Shi	*A. scutum*		0	12	32	11	3	0		
	A. pelta		0	0	29	33	44	29		
	A. digitalis		0	0	105	110	183	272		
	A. paradigitalis		0	0	571	600	576	251		
Mukkaw Bay	*A. scutum*							0	0	
	A. pelta							15	0	
	A. digitalis							544	680	
	A. paradigitalis							240	208	
Portage Head	*A. scutum*			16		10				
	A. pelta			32		26	38	26		
	A. digitalis			0		101	215	355		
	A. paradigitalis			12		381	360	180		
Eagle Point	*A. scutum*	2	5	176	54	59	64	73	67	
	A. pelta	2	54	131	310	167	115	22	0	
	A. digitalis	0	0	0	565	592	675	503	378	
	A. paradigitalis	0	0	64	544	528	625	314	144	
Turn Rock	*A. scutum*	131	112	122	48	46	29	32		
	A. pelta	120	40	60	90	150	0	5		
	A. digitalis	0	0	20	90	83	80	153		
	A. paradigitalis	100	98	110	143	204	0	97		
Colin's Cove	*A. scutum*	120	132	122	94	29	36	0	0	
	A. pelta	160	9	9	75	16	119	128	110	
	A. digitalis	0	0	0	68	85	76	81	101	
	A. paradigitalis	180	0	0	110	170	310	280	20	

placed either in spots not accessible to *Pisaster ochraceus* or in the middle of *Pisaster* removal areas. *Pisaster* removal areas were maintained at Shi Shi, Portage Head, Eagle Point, and Turn Rock by manually removing *Pisaster*. The areas were policed regularly, but since *Pisaster* is a very motile predator, there is a large edge effect to an experiment of this type; therefore, the *Pisaster* removal areas were usually larger than 1,000 m².

Stainless steel mesh cages identical in design and materials to those described by Connell (1961b) were used to totally exclude or regulate the densities of both *Thais* and limpets (Fig. 6). These cages, covering approximately 100 cm², were attached to the substratum with stainless steel screws, which were inserted into plastic wall anchors embedded in holes drilled in the substratum. The objective of these cages was to exclude *Thais* while maintaining limpet density at zero or at the normal density for that site and level. The limpet density to be included in the normal density cage was sampled in the same manner as for the dog dish. Stainless steel mesh "roofs" of 100 cm² supported about 2 cm above the

substratum by plastic washers were also placed beside the cages. The stainless steel cage and roof enclosures were placed in both *Pisaster* removal and normal *Pisaster* areas. In the *Pisaster* removal areas, these roofs served as controls approximating the physical conditions under the cages while allowing free entry of limpets and *Thais*. The effects of *Pisaster* predation were evaluated by comparing roofs in normal *Pisaster* areas with adjacent control plots which allowed access to *Pisaster*, limpets, and *Thais*.

If the comparisons undertaken in the experimental design are legitimate, there should be no consistent differences between the results under the roofs and in the limpet inclusion dog dishes; nor should there be a consistent difference between those areas under the roofs and in the adjacent control plots in *Pisaster* removal areas. Wilcoxon matched-pair signed rank tests (Siegel 1956) on these pairs at Shi Shi and Eagle Point, chosen as areas representative of the outer coast and San Juan Island environments, respectively, showed no significant differences.

One potential artifact or indirect effect of the design was considered. The limpet exclusion dishes and

TABLE 6. Densities at various tidal levels of carnivorous gastropods at three outer coast sites. *Thais lamellosa* is rare at all these sites, as is *T. canaliculata* at Shi Shi and Portage Head. Tagging studies show that *T. emarginata* at Waadah Island do forage as low as the −1 ft level; therefore their mean density is calculated from their entire foraging range despite the fact that they were not found in the lower levels in this sample

Site	Species	Level (ft)	Date	Sample size (m²)	Sample mean	N	SD	Mean density (number/m²)	Mean density in total foraging area (number/m²)
	Thais canaliculata	−1 to +3	5/3/68	0.250	2.6	15	2.3	10.4	
		−1 to +3	6/2/69	0.250	0.9	10	1.3	3.6	
		+3 to +6	5/3/68	0.062	0.1	15	1.2	1.6	5.1
		+3 to +6	6/2/69	0.062	0.3	10	1.4	4.8	
Waadah Island	*Ceratostoma foliatum*	−1 to +3	5/3/68	0.250	2.7	15	3.8	10.8	
		−1 to +3	6/2/69	0.250	1.2	10	0.6	4.8	
		+3 to +6	5/3/68	0.062	0	15		0	6.8
		+3 to +6	6/2/69	0.062	0.1	10	1.4	1.6	
	Thais emarginata	+3 to +6	5/3/68	0.062	3.9	15	2.6	62.4	28.0
		+3 to +6	6/2/69	0.062	3.1	10	8.0	49.6	
Shi Shi	*Thais emarginata*	0 to +1	6/23/68	0.250	9.3	10	8.5	37.2	
		+1 to +3	6/23/68	0.250	14.9	20	12.0	59.6	
		+3 to +4	6/23/68	0.250	56.2	20	26.7	224.8	245.7
		+4 to +5	6/23/68	0.062	25.1	21	8.3	401.6	
		+4 to +5	6/23/68	0.062	26.6	21	7.8	425.6	
		+4 to +5	6/23/68	0.250	81.3	10	31.4	325.2	
Portage Head	*Thais emarginata*	+1 to +2.5	5/2/68	0.062	1.0	10	1.3	16.0	
		+2.5 to +3.5	5/2/68	0.062	1.7	10	2.1	27.2	
		+3.5 to +4.5	6/23/68	0.062	16.0	12	9.6	256.0	198.5
		+3.5 to +4.5	5/2/68	0.062	17.7	10	10.0	283.2	
		+3.5 to +4.5	6/24/68	0.250	70.1	10	32.3	280.4	
		+4.5 to +5.5	6/24/68	0.250	82.0	10	27.2	328.0	

cages, if left unattended, usually developed a heavy cover of *Porphyra* and ulvoids. This protection from desiccation sometimes enabled large populations of the nemertean *Emplectonema gracile* to congregate and eat the barnacles. Thinning this algal growth on the semimonthly visits to the experimental areas was sufficient to eliminate the *Emplectonema* predation.

Fucus *barrier and whiplash effects*

Hatton (1938) and Southward (1956) described a barrier effect in which three species of Fucaceae interfere with the recruitment of *Balanus balanoides* by preventing access of the cyprids to the substratum. Furthermore, a whiplash effect of fucoid fronds may dislodge cyprids and young barnacles (Lewis 1964). This interference effect was tested at Portage Head and Colin's Cove in areas where there was 100% canopies of *Fucus*. At each site two areas ranging from 0.5 m² to 1.5 m² were cleared of only *Fucus distichus*. The numbers of metamorphosed barnacles in these areas where there was no possible *Fucus* effect were compared with the number in control strips. The potential interference of limpets and predators was not controlled, but their densities were monitored, and they did not respond to the altered canopy and thus probably did not differentially affect the barnacle recruitment. Table 8 shows that there was no differ-

ence in barnacle density at Portage Head between the experimental and control strips, but that at Colin's Cove, an area of high desiccation, there was significantly higher barnacle recruitment and survival under the *Fucus*. The positive effect at Colin's Cove is probably due to protection from desiccation. Probably *Fucus* interference such as that found in Europe was not seen in this study because *Fucus distichus* is a smaller plant and does not form as thick or heavy a canopy as is formed by the European fucoids.

Density of limpets and predators

The density data for acmaeid limpets presented in Table 5 are oversimplifications of the natural situation. They were estimated from 0.062-m² or 0.25-m² quadrats placed randomly along a 30-m rope laid horizontally across the intertidal. The data are means of 10–40 such samples at each level. The reliability of these estimates is reduced by environmental heterogeneity, which makes proper sampling difficult (Frank 1965). Furthermore, the limpet larvae have a tendency to settle in the lower levels of the intertidal and move upward (Castenholz 1961, Frank 1965, and personal observation). Thus at times one can find enormous numbers (over 1,000/m²) of tiny (less than 4 mm in length) limpets, unidentifiable as to species, in the lower levels. These limpets as well as many of the adults are consumed

TABLE 7. Densities at various tidal levels of carnivorous gastropods at the San Juan Island sites. The foraging area is calculated from the observed mobility of marked animals

Site	Species	Level (ft)	Date	Sample size (m²)	Sample mean	N	SD	Mean density (number/m²)	Mean density in total foraging area (number/m²)
	Thais canaliculata	−2 to 0	6/20/67	0.062	0.3	10	1.2	4.8	
		0 to +2	6/20/67	0.062	4.3	10	3.9	68.8	
		+2 to 3	6/20/67	0.062	6.1	10	4.7	97.6	41.3
		3 to 4	6/20/67	0.062	4.0	10	3.0	64.0	
		4 to 5	6/20/67	0.062	0.4	10	0.6	6.4	
		5 to 6	6/20/67	0.062	0.4	10	1.3	6.4	
Eagle Point	Thais emarginata	0 to 2	6/20/67	0.062	0.4	10	0.7	6.4	
		2 to 3	6/20/67	0.062	0.2	10	0.5	3.2	
		3 to 4	6/20/67	0.062	8.0	10	2.9	128.0	56.3
		4 to 5	6/20/67	0.062	2.6	10	3.4	41.6	
		5 to 6	6/20/67	0.062	9.9	10	5.1	158.4	
	Searlesia dira	−2 to 0	6/20/67	0.062	0.3	10	0.1	4.8	4.0
		0 to +2	6/20/67	0.062	0.2	10	0.3	3.2	
	Thais lamellosa	0 to 1	5/17/67	0.250	0.6	10	1.2	2.4	
		1 to 2	5/17/67	0.250	0.1	10	0.5	0.4	
		2 to 3	7/10/68	0.062	1.2	20	1.1	19.2	8.7
		3 to 4	5/17/67	0.250	2.5	10	4.0	10.0	
		3 to 4	7/10/68	0.062	1.1	20	1.4	17.1	
		4 to 5	5/17/67	0.250	0.7	10	1.2	2.8	
Turn Rock	Thais canaliculata	0 to 1	5/17/67	0.250	0.4	10	0.9	1.6	
		1 to 2	5/17/67	0.250	2.2	10	3.5	8.8	
		2 to 3	7/10/68	0.062	5.6	20	4.0	89.2	42.8
		3 to 4	5/17/67	0.250	25.5	10	22.8	102.0	
		3 to 4	7/10/68	0.062	3.0	20	3.7	48.0	
		4 to 5	5/17/67	0.250	1.8	10	2.5	7.2	
	Thais emarginata	0 to 1	5/17/67	0.250	0.4	10	1.2	1.6	
		1 to 2	5/17/67	0.250	0.2	10	1.0	0.8	
		2 to 3	7/10/68	0.062	1.2	20	7.0	19.2	22.1
		3 to 4	5/17/67	0.250	13.4	10	12.3	53.6	
		3 to 4	7/10/68	0.062	3.0	20	3.4	48.0	
		4 to 5	5/17/67	0.250	2.3	10	2.7	9.2	
	Thais lamellosa	0 to 1	6/30/68	0.250	21.8	10	15.7	87.2	
		1 to 2	6/30/68	0.250	14.7	10	13.7	58.8	41.2
		2 to 3	6/30/68	0.250	3.9	10	6.2	15.6	
		3 to 4	6/30/68	0.250	0.8	10	0.7	3.2	
Colin's Cove	Thais emarginata	0 to 1	6/30/68	0.250	0.3	10	1.1	1.2	
		1 to 2	6/30/68	0.250	2.0	10	2.4	8.0	
		2 to 3	6/30/68	0.250	3.6	10	3.1	14.4	7.9
		3 to 4	6/30/68	0.250	0.3	10	1.2	1.2	
		4 to 5	6/30/68	0.250	3.4	10	3.3	13.6	
	Searlesia dira	−1 to 0	6/30/68	0.250	0.8	10	1.0	3.2	
		+1 to 2	6/30/68	0.250	2.1	10	2.8	8.4	7.2
		2 to 3	6/30/68	0.250	2.5	10	2.4	10.0	

by *Leptasterias hexactis* (Menge 1970), *Pisaster* (Paine, *personal communication*, and personal observation), and *Searlesia* (Louda, *personal communication*, and personal observation) as well as by the nemertean *Emplectonema gracile* and various polyclad worms (Frank 1965, and personal observation). The result of this predation is that much of the acmaeid population turns over each year, and their observed densities and size frequencies are strongly influenced by the time of year and the level at which they are sampled. The data in Table 5 represent limpets at least 4 mm long sampled in the spring and early summer, when their activity is most likely to affect the recruitment and survival of the algae and barnacles.

It is equally difficult to estimate the population densities of the species of *Thais* at each site because of bias from at least one of the following sources: (1) *Thais* have a marked tendency to retire into crevices, under algae, and among anemones or mussel byssal threads during low tide; (2) this tendency may be exaggerated during spring tides; (3) during the summer months, when these data were collected, an unknown proportion of the populations of *T.*

PAUL K. DAYTON
Ecological Monographs
Vol. 41, No. 4

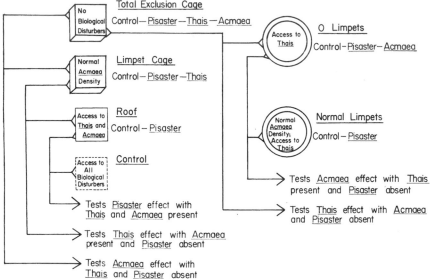

STAINLESS STEEL MESH COVERS
(approximately 100 cm²)

PLASTIC DOG DISH BARRIERS
IN PISASTER-EXCLUSION AREAS
(approximately 177 cm² – corrected to
100 cm² for comparison)

FIG. 4. A diagram of the experimental design used to test the effects of competition and biological disturbance in the barnacle-mussel association. Listed within each enclosure are those biological disturbers enclosed in or having access to the area within. Beside each drawing is the name (underlined) given to that particular type of enclosure and the situation that it represents relative to the control area (— means minus). Comparisons between cages and the effects that they are designed to test are indicated by arrows.

emarginata and *T. canaliculata* retreat into crevices and other protected spots to copulate and to lay eggs; and (4) individuals are highly motile and seem to be able to locate patches of food successfully, thus forming concentrated clumps of individuals. The *Thais* and *Ceratostoma* densities estimated in Table 6 were derived from transects like those for estimating limpet density. Means from these transects at different heights were pooled in a given area at the outer coast sites, where observations of tagged *Thais* have shown that even though the *Thais* are found concentrated at higher levels where there is more food, they definitely also forage in the lower levels.

The same procedure was followed in the San Juan Islands, but was complicated because there were two and occasionally three species of *Thais* involved, as well as another carnivorous gastropod, *Searlesia dira* (Table 7). In the San Juan Islands there is a tendency for *T. emarginata* to occur above *T. lamellosa* and *T. canaliculata* in the intertidal, but at any given site some *T. emarginata* can usually be found feeding and reproducing at the lower levels as well. Furthermore, in situ marking with fast-drying paint has shown that *T. emarginata* will also forage horizon-

tally as far as 15 m within 2 weeks. *Thais canaliculata* and *T. lamellosa* tend to be more sedentary in their foraging behavior and seem to be restricted to lower levels, although on occasion they may forage almost as high as does *T. emarginata*. Therefore, although there is an apparent vertical separation between *T. emarginata* and the other two *Thais* species, after a heavy barnacle set *T. emarginata* readily moves down to forage; *Thais canaliculata* and *T. lamellosa* are less able to exploit prey higher in the intertidal. At the sites where *Searlesia dira* is found, it occurs mainly in the lower levels. Thus lower levels are subjected to a much higher degree of snail predation than are the higher levels, and it is very difficult to predict accurately that a certain area of the shore will be exploited by any given density of predators. For these reasons population estimates from transects or from studying an isolated locality can be very misleading.

During the data analysis it became apparent that the 1969 and 1970 *Thais* densities were much below those of 1967 and 1968. This reduction is probably a result of an unusually cold period in December 1968 and January 1969 which coincided with very

TABLE 8. The mean number of *Balanus glandula* per 0.062 m² to settle and metamorphose in areas where 100% canopy cover of *Fucus distichus* had been removed, contrasted to control areas with 100% canopy cover. The experiment was begun in May 1967 and concluded in August 1967. The variance (based on 8 samples in areas of 0.5 m² and 10 samples in all other areas) is given as the 95% confidence limit. The data for Colin's Cove show significant differences between the *Fucus* removal and control areas

	Fucus removal area		Control (100% *Fucus* cover)	
Site	Area (m²)	*Balanus glandula*	Area (m²)	*Balanus glandula*
Portage Head	0.5	191.1±47.2	0.5	239.0±62.3
	1.5	267.4±38.4	1.5	217.1±44.4
	1.5	139.2±27.2	1.5	175.2±56.3
Colin's Cove	0.5	5.0± 6.2	0.5	39.5±10.5
	0.75	19.4± 5.7	0.5	41.3± 7.9
	0.6	32.1±10.1	0.5	65.9±10.5

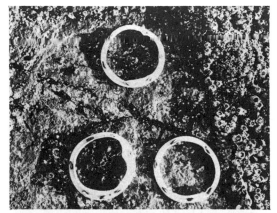
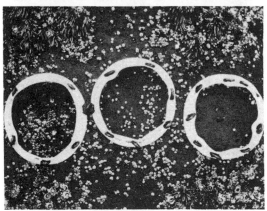

FIG. 5. Photograph of two sets of dog dishes at Mukkaw Bay. The top plate was taken 3 weeks after the experiment was begun and shows algal settlement in the absence of limpet grazing in the lower left dish. The upper dish had normal limpet density while the lower right dish had twice the normal density. The bottom plate shows the effects of limpet disturbance on barnacle recruitment in an experiment 7 weeks old. The left dish excluded limpets completely, the middle dish had normal limpet density, and the right dish had twice the normal limpet density.

low night tides. Since the effect of the cold on *Thais* populations was not evident at the time, no accurate assessment of the *Thais* populations was made, and unfortunately, direct censuses of *Thais* density in 1969 were limited. Control areas at Portage Head were extensively photographed during all the years of the study. The Portage Head *Thais* densities in Table 9 were obtained from the photographs of one such control area taken during the first day of a minus tide series and are realistic estimates of the *Thais* densities. The data for the San Juan Islands were obtained under less controlled conditions, but still suggest a significant decline in the *Thais* density.

The foraging behavior of *Pisaster ochraceus* involves a vertical movement in which it forages upward during high tide, captures its prey, and then returns back down into a tidepool or surge channel to digest the prey (Mauzey 1966, and personal observation). This retreat to areas protected from desiccation is more apparent during spring tides, which are associated with dry conditions. The *Pisaster* densities reported in Table 10 were derived by counting all the *Pisaster* in a particular locality and estimating the area over which they forage. At Shi Shi, Portage Head, and Waadah Island, this is relatively simple, as one can count all the *Pisaster* in a surge channel, and the upper limits of their foraging excursions are conspicuously marked by browse lines along the *Mytilus* and *Balanus* populations. Repeated counts in these areas show high consistency. The estimate of *Pisaster* density on the exposed point at Tatoosh Island is less reliable. Here the lower limit to the *Mytilus* bed was at the 0–1-ft level. I counted 16 *Pisaster* along a 3–4-m strip below the *Mytilus* on June 7, 1970. Due to a heavy cover of the kelp *Lessoniopsis littoralis*, this count is almost certainly an underestimate. The estimates of *Pisaster* densities in the San Juan Islands were obtained by diving and are in excellent agreement with those of Mauzey (1967) and Menge (1970).

FIG. 6. Photograph of a stainless steel cage experiment at Shi Shi. The center cage is a total exclusion cage, the left is a limpet cage excluding *Thais* and *Pisaster*. On the right is the "roof" which excludes only *Pisaster*. To the right of the roof is the control area. Both cages and the roof have been moved so that the experimental areas can be seen below them in the photo. A light set of *Chthamalus dalli* is not noticeably affected by the limpet cage or the roof, but a heavy settlement of *Balanus glandula* in the total exclusion cage occupies 100% of the space. The scale marker is 15 cm long.

TABLE 9. Densities of *Thais emarginata* at Portage Head determined from photographs taken along the same transect before and after the cold period in December 1968–January 1969; densities of *Thais canaliculata* at Eagle Point were taken from random quadrates along a 10-m line in 1967, 1968, and 1970 and from haphazardly tossed quadrates during high tide in 1969

Site	Date	Sample size (m²)	N	Sample mean	SD	Number/m²
Portage Head	June 8, 1968	0.250	10	70.1	21.4	280
(*T. emarginata*)	June 27, 1969	0.250	10	17.3	28.2	109
	June 8, 1967	0.250	10	30.2	25.3	121
Eagle Point	June 6, 1968	0.062	15	7.2	3.8	115
(*T. canaliculata*)	July 2, 1969	0.062	21	2.1	5.3	34
	July 2, 1970	0.062	20	2.4	6.9	38

Recruitment in the upper intertidal

The most consistent occupants of space in the barnacle-mussel association are the three barnacle species *Chthamalus dalli*, *Balanus glandula*, and *B. cariosus*, the mussel *Mytilus californianus*, and the anemone *Anthopleura elegantissima*. Due to settlement uneven in time and space, the recruitment patterns of these species are difficult to quantify in an experimental manner. However, inferences about the barnacle recruitment patterns can be drawn from the experimental data. The *Mytilus* and *Anthopleura* patterns discussed at the end of this section are based only on general observations from 1965 to 1970.

The percentage cover and numbers of individuals per 100 cm² for each barnacle species under each experimental condition are shown for selected years in Fig. 7–12. In order to present representative patterns for one year, the first year for which there were complete data at each site was used. Inferences about the potential recruitment (i.e., the availability of cyprids ready to settle) at a given level and time can be made by considering the number of barnacles present in each experimental situation in the following manner. Early in the season the number of metamorphosed and identifiable individuals of each species at each level could be observed directly in the "total exclusion cage," where there were no distur-

TABLE 10. Density of *Pisaster* in its total foraging area at each of the study sites. Samples collected on a particular date represent a single search of a given area. The means (given with N and standard deviation) are from a number of searches in a given area

Site	Level (ft)	Date	Foraging area (m²)	Mean	N	SD	Density (number/m²)
Tatoosh Island	? to +1	7/6/70	102	16			0.16
Waadah Island	−2 to +7	1967–69	75	28	7	0.9	0.46
Shi Shi Reef	−2 to +6	1967–69	52	81	14	6.2	1.56
	0 to +5	2/2/67	158	137			0.87
Portage Head	0 to +6	1966–69	442	310	22	7.9	0.70
Eagle Point	−3 to +4	6/22/67	193	17			0.09
Turn Rock	−2 to +3	1/6/67	1,258	78			0.06
Colin's Cove	−3 to +7	6/22/67	487	19			0.04

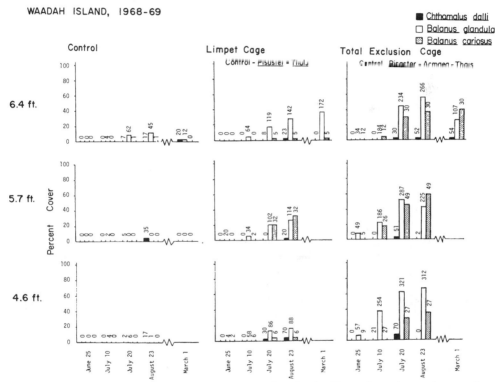

FIG. 7. Percentage cover (bars) and density/100 cm² (above bars) of barnacles under varying experimental conditions at Waadah Island from June 1968 through March 1969. A conservative estimate of the recruitment is shown by an increase in the density. A decrease in density in predator-free situations (limpet cage and total exclusion cage) indicates barnacles killed as a result of competition for space.

bances to prevent their settling or to kill or remove them. Substratum quickly became limiting in this cage at most levels, however, and after about 50% of the total space was occupied, competition for space became an important factor (see below), and the observed changes in barnacle numbers were no longer a good index of potential recruitment. After the early part of the season, changes in numbers of barnacles in the "0 limpet" and "limpet cage" experiments (Fig. 7–9, 12) were sometimes a better indication

of recruitment. In such experiments where competition or biological disturbances occur, an increase in barnacle number is definite evidence of recruitment. Such demonstrated increases are often underestimates of the recruitment which has occurred, since they are net increases beyond the settlement lost to competition and disturbance; these increases underestimate the potential recruitment even more than they do the the actual recruitment. A decrease in barnacle number cannot be interpreted as a lack of recruitment,

368   PAUL K. DAYTON   Ecological Monographs
                                                                            Vol. 41, No. 4

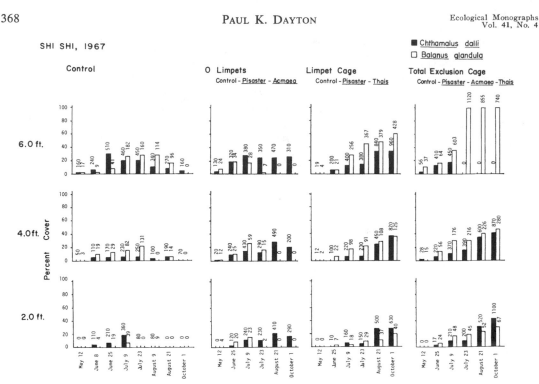

FIG. 8. Percentage cover (bars) and density/100 cm² (above bars) of barnacles under varying experimental conditions at Shi Shi in the summer of 1967. A conservative estimate of the recruitment is shown by an increase in the density. A decrease in density in predator-free situations (limpet cage and total exclusion cage) indicates barnacles killed as a result of competition for space.

as it is possible that new barnacles have settled and these, along with some previously established individuals, have been eliminated. Another complication relevant to all estimates of potential recruitment is that there is a 5- to 20-day lag between the time a cyprid settles and the time that it is identifiable at the next low tide series. Despite these complications, the following results are apparent from considerations of the experimental cages and dishes.

Chthamalus dalli recruitment was seen at all sites and was essentially the same at each level within each site (Fig. 7–10, 12), except at Turn Rock where there was a sporadic recruitment (Fig. 11). On the outer coast the settlement usually began in May (Fig. 8 and 9) and continued throughout the summer (Fig. 7–9, total exclusion and limpet cages). The single outstanding exception was an extremely heavy *Chthamalus* set in August 1968 at Portage Head. In the San Juan Islands *Chthamalus* has a lighter set which begins later in the summer (Fig. 10–12). *Chthamalus dalli*, like its European congener, *C. stellatus* (Connell 1961a), is a very small barnacle, usually less than 4 mm in basal diameter.

Balanus glandula was the most nearly ubiquitous barnacle species in the study sites; its settlement had

begun by late April and was usually heavy from late May through July (Fig. 7–12), continuing at a somewhat reduced rate until September. On the protected outer coast sites there is a strong tendency for *B. glandula* to settle relatively high in the intertidal (Fig. 8 and 9). The growth rates of *B. glandula* are discussed by Connell (1970), who found that they attain diameters of 4–5 mm by October and 9–15 mm by the end of 5 years.

Large *Balanus cariosus* occur in all of the study sites, but the species has a peculiar settlement pattern which is very patchy in time and space. During the years 1965 to 1969, small areas of shoreline, particularly along the west side of San Juan Island, received heavy but very localized recruitment. These areas of recruitment involved perhaps 50 m of shoreline and received 40–60% cover, while the shore 200–300 m away received almost no settlement. Of the study areas, Eagle Point received the most regular *B. cariosus* recruitment; Turn Rock and Colin's Cove generally received light settlements. Although they occur as large adults at Shi Shi, Portage Head, and Mukkaw Bay, I did not observe *B. cariosus* to settle in these areas during this study. Light settlements were observed at Waadah Island, while cleared patches in the exposed areas of Tatoosh

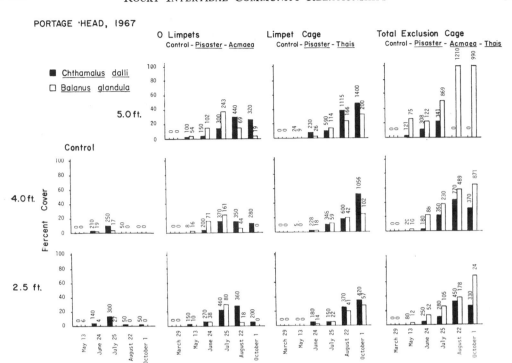

FIG. 9. Percentage cover (bars) and density/100 cm² (above bars) of barnacles under varying experimental conditions at Portage Head in the summer of 1967. See Fig. 8 for explanation.

Island (summer 1968 and 1970) received very heavy (approaching 100% cover) settlements. *Balanus cariosus* tends to occur at lower middle levels in the intertidal (Fig. 7, 10–12) and never is found as high as the upper limits of *B. glandula* or *Chthamalus dalli*. *Balanus cariosus* is the largest of the three species, attaining a basal diameter of at least 42 mm at one site; it also has the fastest growth rate of the local barnacles. Since accurate measurements of growth could not be obtained in the experiments because intraspecific competition led to hummocking and distortion of basal diameters, no detailed growth study was done. However, the presence of a few solitary *B. cariosus* in the controls allowed some observation of growth (Fig. 13). Although the growth rates shown are based on inadequate sample sizes, they do indicate that *B. cariosus* has a potentially fast growth rate and grows considerably larger than *B. glandula* and *Chthamalus dalli*.

Interspecific competition in the upper intertidal

With the exception of some evidence of algal competition for light, most of the interspecific interactions between the sessile organisms in this intertidal zone involve competition for primary space (bare, unoccupied rock substratum) as the potentially limiting resource. In the experimental situations when primary space is not abundant, most of the sessile species will utilize secondary space (another organism used as substratum) by settling on barnacles and mussels, thus becoming entirely dependent upon the well-being of their "hosts." Rare instances were observed in which barnacles are overgrown by the secondary settlement of other barnacles and algae and apparently eventually starved; the whole hummock then disappears, thereby freeing primary space. As primary space is rarely in limiting supply in natural situations in the upper intertidal (Table 1), the phenomenon of secondary space has been considered relatively unimportant and has not been studied.

As primary substratum becomes limiting, the various species of algae always lose in the upper intertidal to any of the three species of barnacles. The mechanisms are straightforward: as the barnacle grows in diameter, its plates dislodge or weaken the holdfasts of the nonencrusting algae, and the plants are carried away by wave action; the encrusting algae are simply overgrown by the barnacles. This interaction was observed frequently in upper intertidal situations where algae and barnacles occupied primary substratum immediately adjacent to each other. When the barnacles occupy all the primary space, many algae, particularly *Fucus*, *Endocladia*, and *Gigartina*, will settle and grow on the barnacles. The

370   PAUL K. DAYTON   Ecological Monographs
                                                                    Vol. 41, No. 4

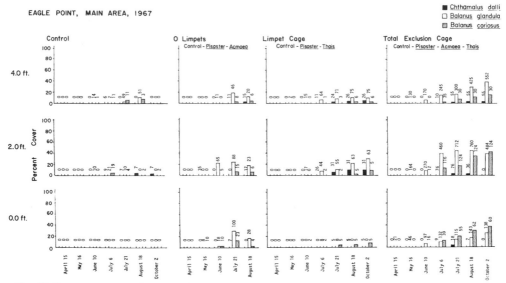

FIG. 10. Percentage cover (bars) and density/100 cm² (above bars) of barnacles under varying experimental conditions at Eagle Point in the summer of 1967. See Fig. 8 for explanation.

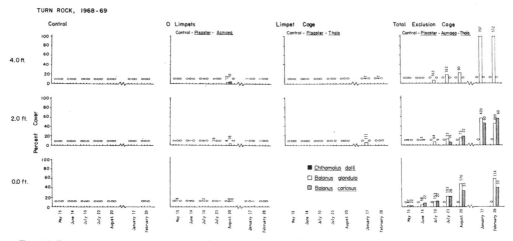

FIG. 11. Percentage cover (bars) and density/100 cm² (above bars) of barnacles under varying experimental conditions at Turn Rock from May 1968 through February 1969. See Fig. 8 for explanation.

articulated coralline algae have large encrusting holdfast systems which appear to offer barnacles an unstable substratum and might be an exception to the generalization that in the upper intertidal the barnacles outcompete the algae. No instance of competitive interaction between barnacles and articulated coralline algae was observed.

The different growth rates of the three barnacle species suggest differences in their ability to competitively dominate the primary space. Such trends were observed in most total exclusion cages with *B. cariosus* dominating in space competition with *B.*

glandula and *Chthamalus*, while *B. glandula* successfully dominates in the competition with *Chthamalus*. The competitive abilities are very clear; the mechanisms in this competition are identical to those described by Connell (1961a): the dominant individual completely grows over or squeezes the other individual.

The competitive trends under the experimental conditions at each site are presented in Fig. 7–12, in which the ordinates are given as percentage cover, an indication of the successful domination of primary space, the resource for which the barnacles

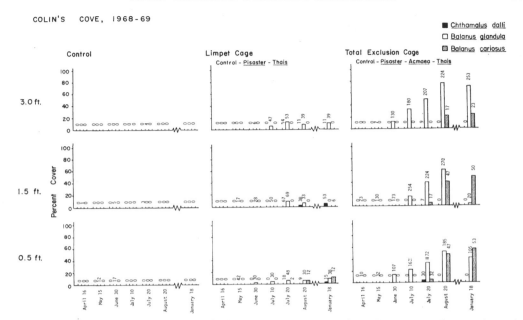

FIG. 12. Percentage cover (bars) and density/100 cm² (above bars) of barnacles under varying experimental conditions at Colin's Cove from May 1968 through February 1969. A conservative estimate of the recruitment is shown by an increase in the density. The loss of *Balanus glandula* in the middle level limpet cage and total exclusion cage is a result of a winter invasion of *Onchidoris* into the cages; in all other cases a decrease in density in those predator-free situations indicates barnacles killed as a result of competition for space.

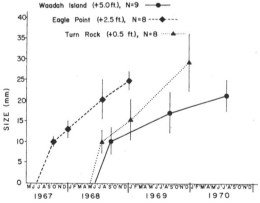

FIG. 13. Growth rates of uncrowded *Balanus cariosus* at three sites. Each point represents the mean basal diameter; *N* and standard deviation are shown.

compete. In every predator-free situation (total exclusion cage and limpet cage) in Fig. 7–12 in which either the number or the percentage cover of a species is reduced, the reduction resulted directly from barnacles being competitively overgrown or squeezed. The result of this competition for substratum in which *B. glandula* reduces or eliminates *Chthamalus dalli* can be seen in most of the total exclusion cages at Shi Shi and Portage Head (Fig. 8 and 9), while situations in which *B. cariosus* is dominant over both *B. glan-*

dula and *Chthamalus* can be seen in most total exclusion cages at Waadah Island, Eagle Point, Turn Rock, and Colin's Cove (Fig. 7, 10, 11, and 12). As is predictable from the recruitment data, *B. glandula* is strongest in the higher levels (Fig. 8–12), while *B. cariosus* is usually more common in the middle-lower levels (Fig. 7, 10–12).

Another spatially important organism is *Mytilus californianus*, whose larvae settle among filamentous algae and barnacles as well as among byssal threads of their own species (Glynn 1965, and personal observation). After initial settlement, the pediveligers and young mussels are capable of extensive crawling (Bayne 1965, Harger 1968), and frequently the *Mytilus* attach and grow on top of barnacles. Figures 14 and 15 show secondary space utilization during 3 years' competition in the total exclusion cages, which were maintained because many of the competitive situations were incomplete at the end of 1 year and because *Mytilus* rarely were observed in the first year. It was not determined whether the *Mytilus* in these total exclusion cages actually settled on the *Balanus* or settled elsewhere and crawled onto them. Twelve long-term total exclusion cages at Shi Shi and Portage Head showed that in the absence of any disturbance, *Balanus glandula* initially dominates the primary space, but by the end of the second summer, *Mytilus* completely covers the *Balanus*. Although during the time of this research the *Balanus* were generally not killed by this cover, Paine (1966) has shown that in the absence of starfish predation the *Mytilus* eventually completely dominate the space.

In the San Juan Islands where adult *Mytilus californianus* are rare, long-term total exclusion cages showed that *Balanus cariosus* completely dominates the primary space (Fig. 16). Most of the long-term cages in the San Juan Islands also developed a heavy cover of *Fucus*, which utilized *Balanus cariosus* as secondary space. In those cages from which *Fucus* was not removed, large numbers (50-150) of *Mytilus californianus* and *M. edulis* appeared. How-

ever, because the *Fucus* cover made photoanalysis impossible, the *Mytilus* cover in these cages is not precisely known and does not appear in Fig. 16. It is reasonable to suppose that with protection against desiccation and predation, the San Juan Islands could develop a situation analogous to that of the outer coast.

To test whether the abundance patterns of *Anthopleura elegantissima* described in Table 3 were affected by competition, five separate 0.062-m² patches were cleared of the anemones from the middle of a very large aggregation (50 m²) at Shi Shi in May 1966. In each case the cleared patch was quickly covered with diatoms and *Enteromorpha linza* and *E. intestinalis*. During the first summer there was limited anemone immigration into the clearings. By July 1966 there was a 25–35% cover of *Balanus glandula*. By August, *Thais emarginata* were killing the barnacles, and all the barnacles were dead by

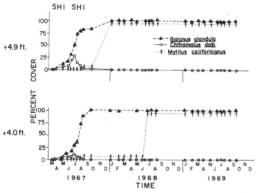

FIG. 14. Percentage cover of barnacles and *Mytilus californianus* in total exclusion cages in 3-year competition experiments at two levels at Shi Shi. The *Mytilus* are occupying secondary space.

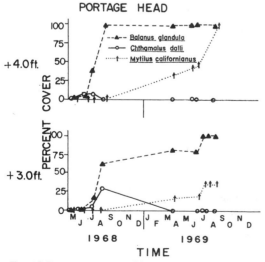

FIG. 15. Percentage cover of barnacles and *Mytilus californianus* in total exclusion cages in 2-year competition experiments at two levels at Portage Head. The *Mytilus* are occupying secondary space.

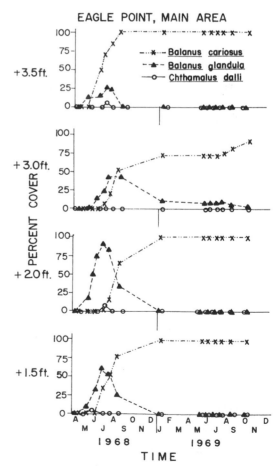

FIG. 16. Percentage cover of barnacles in total exclusion cages in 2-year competition experiments at four levels at Eagle Point.

October. Similar events occurred in 1967 and 1968. By 1969 a few anemones had moved into the cleared patches, but the patches remained distinct and again the barnacles were killed by *Thais*. The anemones offer *Thais* protection from desiccation and thereby increase their predatory efficiency by allowing them to remain close to their major prey patch. But the anemones do not effectively crowd or cover the barnacles, and thus are not efficient space competitors until they completely cover the substratum, at which point the anemones probably could successfully control the space by preventing larval recruitment of potentially competing species. However, such growth of an *Anthopleura elegantissima* population may take many years and was never seen during the period of this study.

Space competition with *Mytilus californianus* and *Balanus cariosus* may prevent the anemones from moving into more exposed conditions. Paine (1966) has shown that when *Pisaster* is removed, *M. californianus* grows over and kills *Anthopleura elegantissima*. However, the reciprocal experiment in 1966 of clearing *Balanus glandula* and *Mytilus californianus* from the exposed side of the surge channel at Shi Shi to see if *Anthopleura* utilized the space was not successful, as neither larval recruitment nor adult immigration into the clearings had occurred by August 1970.

Aggregations of *Anthopleura elegantissima* on the west side of San Juan Island are frequently found among clumps of large *Balanus cariosus*. To test possible effects of competition between these two sessile species, *A. elegantissima* was removed from three 0.062-m² areas in 1966. In contrast to the results on the outer coast, these Eagle Point cleared areas were completely reclaimed by immigrant anemones within 3 months. It is not clear why the San Juan Island *A. elegantissima* are so much more motile than those of the outer coast.

The reciprocal experiment of removing all the *Balanus cariosus* was performed, using two 1.0-m² areas. Figure 17 shows the results of the experimental removal of 45% and 55% covers of large *Balanus cariosus* in these two patches at Eagle Point. The anemone population immediately responded to the release in space competition with a sharp increase in their population as the result of both immigration and asexual fission. These populations were then severely reduced by desiccation during the summer months. Figure 17 also shows that the anemones in the controls, often clumped around the base of large *B. cariosus* and thus protected from the wind and hence from desiccation, suffered relatively little mortality. Thus the *A. elegantissima* populations in the controls appeared more stable in the presence of their space competitors than those experimental anemones with-

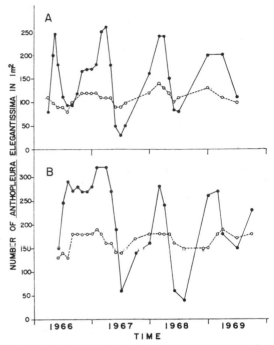

FIG. 17. Eagle Point *Anthopleura elegantissima* density in two 1-m² patches following removal of 55% (A) and 45% (B) covers of *Balanus cariosus*. The adjacent controls had 48% (A) and 55% (B) covers of *B. cariosus*. Solid circles represent *Balanus cariosus* cover removed, open circles the control.

out *B. cariosus*. By 1969 the *A. elegantissima* in cleared patches were beginning to form aggregations which offer some protection from wind desiccation and, possibly as a result, their population fluctuations seemed to be damping. It appears that the San Juan Island *A. elegantissima* populations are in a more marginal environment than are those on the outer coast; the result is that the San Juan Island populations are maintained in a continuous state of flux, as a result of both desiccation and competitive pressures. The only hypothesis which is forthcoming is that *A. elegantissima* behaviorally respond to desiccation by increasing their movement until they encounter a locality such as an *A. elegantissima* aggregation, a crevice, or a barnacle clump, any of which offer some protection.

Effects of limpets on algal recruitment and survival

Experimental results of limpet exclusions are given as means of all levels and over three summers (Table 11). The dishes were cleared at the end of each season. In all cases where limpets were excluded there was a heavy growth of algae which covered 100% of the surface. In each instance, the initial algal growth was composed of various species of diatoms;

374   PAUL K. DAYTON   Ecological Monographs
                                                                                  Vol. 41, No. 4

however, it was not possible to exclude the adventitious grazing of littorine gastropods from the cages or dishes. As Castenholz (1961) has shown, this grazing is sufficient to reduce or eliminate most of the diatom growth. The diatom growth quickly gave way to *Porphyra* and various species of *Enteromorpha* and *Ulva*. This is probably a result of both littorine grazing and natural succession, with the larger *Porphyra* and ulvoids outcompeting the diatoms. In the dog dishes, where *Thais* predation prevented the barnacles from outcompeting the algae, there was usually a modest tendency toward succession during the first summer from the diatom-*Porphyra*-ulvoid-group to *Gigartina papillata*, *Iridaea heterocarpa*, *Halosaccion glandiforme*, and *Microcladia borealis*.

Different dishes at Shi Shi, Portage Head, and Mukkaw Bay were left uncleared for 3 years to assess algal succession in the absence of limpets. The limpet exclusion dishes developed up to 100% canopy cover and averaged 21.2 *Fucus distichus* (Portage Head and Shi Shi; $N = 6$; SD $= 5.2$), 37.4 *Pelvetiopsis limitata* (Mukkaw Bay; $N = 3$; SD $= 10.1$), 17.8 *Gigartina papillata* (mean of three sites; $N = 9$; SD $= 3.0$), and 34.0 *Endocladia muricata* (mean of three sites; $N = 9$; SD $= 5.5$). This can be compared with the control means of 0.8 *Fucus* (Portage Head and Shi Shi; $N = 6$; SD $= 0.6$), 3 *Pelvetiopsis* (Mukkaw Bay; $N = 3$; SD $= 1.4$), and means of 3 *Gigartina* and 1 *Endocladia* ($N = 9$; SD $= 0.2$ and 0.5) for the three sites. This supports the conclusion of European workers that the normal limpet grazing retards the algal succession (see Lewis 1964 for review).

Effects of limpets on barnacle populations

Effects on recruitment.—The effects of limpets on the initial recruitment of barnacles were dramatic. The mechanisms by which limpets interfere with barnacle recruitment involve eating, pushing, and dislodging the cyprid or newly metamorphosed barnacle from the substratum. The quantitative effect of these procedures on barnacle recruitment was evaluated by using pairs of cages. Each pair consisted of a cage from which all *Pisaster*, *Acmaea*, and *Thais* had been excluded and an adjacent cage from which *Pisaster* and *Thais* had been excluded, but in which limpets occurred at approximately the same density as they did naturally outside the cages (Fig. 4). Periodic surveys were done by removing the cages, counting or photographing the organisms underneath, and then replacing the cages.

The dates for statistical comparison of barnacle recruitment in the pairs of cages were selected to allow sufficient settlement to have occurred to show whether limpets had an effect on recruitment and to avoid any effects of competition that might occur in the cages as available space became limiting. The pairs of cages were compared at the last survey before 50% or more of the primary space in the total exclusion cage was occupied. For example, if a pair was surveyed on April 15, May 16, June 10, and July 6, with 50% cover observed on June 10, the numbers of barnacles occurring in the cages on May 16 were used for the comparison. If 50% cover had not been reached by the final survey of a given year, the final survey date was used for the comparison. A cover of 50% was chosen because space competition should not yet have had an undue influence on the number of barnacles observed.

The pairs of cages were compared by using the randomization test for matched pairs (Siegel 1956) and computing the probability that, if limpets did not affect the barnacles, recruitment (the cyprids which settled, metamorphosed, and grew large enough to identify), the statistic obtained would be as large as or larger than that observed. The results are shown in Table 12. In almost every pair there was less recruitment of *Balanus* spp. in the dish with the natural density of limpets. The limpets generally had less effect on the recruitment of *Chthamalus*. This is probably because the metamorphosed *Chthamalus* is very much smaller than *Balanus* species, and once a small *Chthamalus* settles in a crack, it is much less likely to be bulldozed off than either of the other two barnacle species.

The influence of limpets on barnacle recruitment at each study area for a single year is shown in the limpet cage experiments (Fig. 8–11). In general the limpet interference is much more effective in the San Juan Island sites than in the outer coast sites. There are a number of reasons why this is true. First, the San Juan Island substratum is much smoother than the coastal siltstone. Therefore, the barnacle cyprids and recently metamorphosed barnacles are more likely to be dislodged by the bulldozing effect of the limpets than are the coastal ones, which are somewhat protected by the uneven substratum. Second, the populations of *Acmaea pelta* and *A. scutum* in the San Juan Islands are much denser than they are on the coast, and individuals of these two species, particularly *A. scutum*, are both larger and much more active than are those of *A. digitalis* and *A. paradigitalis*. Therefore there is proportionately more limpet movement in the San Juan sites. This factor is particularly important at Turn Rock and Colin's Cove. And third, at Eagle Point there is a much higher absolute density of all limpet species than at any other site.

Effects on survival.—To evaluate the significance of the net effects of limpet grazing on the populations of barnacles at each of three levels over 3 years, two sets

Table 11. Mean number of individuals of each algal species per 177-cm^2 dog dish observed before the end of one summer. Data are averaged for three summers, 1967–1969, and for all levels. Blank areas mean that algae were not seen

Species	Shi Shi (6)[a] O[b]	N[b]	Mukkaw Bay (5) O	N	Portage Head (6) O	N	Eagle Point (5) O	N	Turn Rock (4) O	N
Porphyra perforata	5.3	0.7	7.4	0.4	6.3	1.1	2.0		2.4	
Ulvoids	13.7	2.6	13.4	1.9	15.0	P[c]	3.2	0.6	3.9	
Spongomorpha coalita	P		5.6		2.4		P		P	
Gigartina papillata	7.4	1.0	9.0		1.3		2.0		1.2	
Polysiphonia hendryii	P		P		P		4.1		5.8	
Ceramium pacificum					P					
Heterochordaria abietina	1.4		2.2		P					
Colpomenia sinuosa	4.9	1.6	4.0		4.3	1.1	2.2		0.9	
Halosaccion glandiforme	6.0	1.2	2.1		1.8					
Microcladia borealis	1.4	P	1.3		0.8				P	
Cladophora trichotoma	P		P		P		P		P	
Iridaea spp.	2.3		P		P		P		P	
Ptilota spp.	P				P					
Ralfsia spp.	P		6.4		1.7		2.1		1.3	
Hildenbrandia rosea	p				P					
Membranoptera spp.	P									
Gelidium sp.	P				P					
Rhodomela larix	1.3				P		5.7		3.2	
Prionitis lyallii	P				P					
Fucus distichus	1.4				2.1		3.8		6.8	
Pelvetiopsis limitata			5.3		1.0					
Endocladia muricata	4.2		4.9		3.4		1.4		1.4	
Odonthalia floccosa							3.2		2.9	
Bangia fuscopurpurea							P		P	
Gloiopeltis furcata							2.4		3.0	

[a]Number of experiments in parentheses.
[b]The densities of limpets are represented as normal (N) or excluded (O) (see Table 7).
[c]Where mean was less than 0.5 or where numbers were not available, a species is represented as present (P).

of experiments were conducted at most sites between March and late August, 1967–69. In both sets *Pisaster*, due to its patchy but usually devastating foraging habits, was excluded from the experimental area. To evaluate the effects of limpets in the absence of *Thais*, the numbers of barnacles in total exclusion cages were compared to the numbers in the cages containing the normal density of *Acmaea* for each area and height. To find the effects of limpets in the presence of *Thais*, the numbers of barnacles in the plastic dog dishes with the normal densities of limpets were compared with those in which there were no limpets. The areas under the cages were approximately 100 cm^2; densities of barnacles have been corrected to this dimension for analysis. The area surrounded by the plastic barriers or dog dishes was about 177 cm^2; densities of barnacles were again corrected to 100 cm^2 for comparison. The densities were surveyed during the last August low tide series each year.

When possible, the means were compared with a 3-way analysis of variance considering treatment, height, and year effects, in order to identify significant interactions. Treatment and height effects were considered to be fixed, while year effects were considered to be random. The data were transformed using the log $(x + 1)$ transformation (Snedecor 1966) to provide for additivity, because height, year,

and the grazing habit of the limpets would logically be expected to have a proportional rather than strictly additive effect on the number of barnacles surviving. If no significant difference between treatments was discovered with the 3-way analysis, and interaction effects had been found at a level significant enough to prevent pooling of sums of squares (i.e., "H: no interaction effects" could not be accepted at the 25% significance level) and thus to limit severely degrees of freedom in testing for treatment effects, the untransformed data were subjected to one of two nonparametric tests for matched pairs. The randomization test for matched pairs (Siegel 1956) was used whenever the number of pairs showing some barnacle survival was eight or less, in order to use the most information possible. When $N \geqq 9$, the Wilcoxon signed-rank matched-pairs test (recommended by Siegel 1956) was used, due to the unwieldiness of the randomization test with a large N.

Effects in the absence of Thais.—Limpet activity in the absence of *Thais* caused significant reductions in the number of both *Balanus glandula* and *B. cariosus* (Table 13). The one exception was at Colin's Cove, where all of the total exclusion cages had more *B. cariosus* than did the limpet cages, but because there were only three pairs

376
PAUL K. DAYTON
Ecological Monographs
Vol. 41, No. 4

TABLE 12. Effects of *Acmaea* spp. on recruitment of barnacle species. Pairs of cages with and without limpets were compared (Fig. 4), before they had 50% cover of primary space, with the randomization test for matched pairs (Siegel 1956). The *P* values given are for one-tailed tests

Species	Area	Level (ft)	Years	Number of pairs started	Number with settlement	Number in which limpets reduced recruitment	Significance level
Balanus glandula	Waadah Island	−1 to +5	1968–1969	8	7	7	$P = 0.008$
	Shi Shi Reef	2.7 to 5.4	1967–1969	9	9	8	$P = 0.004$
						1: 58% cover by June 28, 1969	
	Portage Head	2.5 to 5.0	1967–1969	9	9	9	$P = 0.002$
	Eagle Point	0 to +1	1967–1969	6	6	5	$P = 0.031$
						1: 52% cover by May 29, 1969	
		1.5 to 2	1967–1969	9	9	7	$P = 0.008$
						2: 100% cover by May 29, 1969	
		2 to 3.5	1967–1969	11	11	10	$P = 0.001$
						1: 100% cover by May 29, 1969	
	Turn Rock	−1 to +5	1968–1969	8	8	7	$P = 0.008$
						1: 72% cover by May 30, 1969	
	Colin's Cove	0.5 to 3	1968–1969	4	4	4	$P = 0.062$
Balanus cariosus	Waadah Island	−1 to +5	1968–1969	8	4	4	$P = 0.062$
	Eagle Point	0 to 1	1967–1969	6	6	4	$P = 0.125$
						1: 52% cover by May 29, 1969	
		1.5 to 2	1967–1969	9	8	6	$P = 0.016$
						2: 100% cover by May 29, 1969	
		2 to 3.5	1967–1969	11	6	5	$P = 0.031$
						1: 100% cover by May 29, 1969	
	Turn Rock	−1 to +5	1968–1969	8	6	5	$P = 0.031$
						1: 72% cover by May 30, 1969	
	Colin's Cove	0.5 to 3	1968–1969	4	4	4	$P = 0.062$
Chthamalus dalli	Waadah Island	−1 to +5	1968–1969	8	5	5	$P = 0.031$
	Shi Shi Reef	2.7 to 5.4	1967–1969	9	9	7	$P = 0.008$
						1: 50% cover by June 28, 1969	
	Portage Head	2.5 to 5	1967–1969	9	9	5	$P > 0.05$
	Eagle Point	0.0 to 2	1967–1969	15	9	5	$0.047 \geq P \geq 0.031$
						3: 52% or 100% cover by May 29, 1969	
		2 to 3.5	1967–1969	11	7	3	$P = 0.172$
						1: 100% cover by May 29, 1969	
	Turn Rock	−1 to +5	1968–1969	8	4	1	$P = 0.88$
						1: 72% cover by May 30, 1969	
	Colin's Cove	0.5 to 3	1968–1969	4	3	2	$P = 0.375$

TABLE 13. The effect of *Acmaea* disturbance on barnacle density by the end of each summer in the absence of *Pisaster* and *Thais*. Significance levels are for one-tailed tests. In addition to the comparison of the means, interactions between treatment, year, and intertidal height were evaluated. Anova = analysis of variance

Barnacle species	Site	Tide height (feet)	Year	Mean number of barnacles/100cm² Limpets present (limpet cage)	Mean number of barnacles/100cm² Limpets absent (total excl. cage)	Type of analysis	Significance level	Significant interactions
Balanus glandula	Shi Shi	+2.7 – +5.4	1967–1969	88	257	3–Way Anova	0.05 > P > 0.025	None
	Portage Head	+2.5 – +5.0	1967–1969	76	419	3–Way Anova	0.025 > P > 0.01	*Treatment –Year
	Waadah Island	+4.6 – +7.7	1968–1969	55	222	3–Way Anova	P< 0.001	None
	Eagle Point	0.0 – +3.5	1967–1968	61	459	3–Way Anova	P< 0.001	None
	Colin's Cove	+0.5 – +3.0	1968	39	223	2–Way Anova	0.05 > P > 0.025	—
	Turn Rock	−1.0 – +4.2	1968	1.9	124	2–Way Anova	0.005 > P > 0.001	—
Balanus cariosus	Waadah Island	+4.6 – +7.7	1968–1969	3.0	10	3–Way Anova	0.05 > P > 0.025	None
	Eagle Point	0.0 – +3.5	1967–1968	19	84	Wilcoxon Signed rank	0.01 > P > 0.005	None
	Colin's Cove	+0.5 – +3.0	1968	1.3	34	2–Way Anova	0.10 > P > 0.05	—
	Turn Rock	−1.0 – +4.2	1968	0.7	14	2–Way Anova	0.05 > P > 0.025	—
Chthamalus dalli	Shi Shi	+2.7 – +5.4	1967–1969	529	61	3–Way Anova	P< 0.001	***Height –Treatment
	Portage Head	+2.5 – +5.0	1967–1969	795	73	3–Way Anova	P< 0.001	***Height –Treatment
	Waadah Island	+4.6 – +7.7	1968–1969	57	5.5	Randomization	P=0.28	*Height –Treatment
	Eagle Point	0.0 – +3.5	1967–1968	25	9.3	Wilcoxon Signed rank	P> 0.05	None
	Colin's Cove	+0.5 – +3.0	1968	21	0	2–Way Anova	0.025 > P > 0.01	None
	Turn Rock	−1.0 – +4.2	1968	0	0	All Zero	—	—

(*$0.05 \leq p > 0.01$; ***$p \leq 0.001$).

of cages (i.e., few degrees of freedom), P was between 0.10 and 0.05. Although there was a significant treatment-year interaction for *B. glandula* at Portage Head, the average number of *B. glandula* in the total exclusion cages was significantly higher than in the limpet cages.

In contrast to the two *Balanus* species, the survival of *Chthamalus* was significantly increased at all sites in the presence of limpet disturbance, because the limpets had a disproportionately strong negative effect on the populations of the competitively superior *Balanus glandula* and *B. cariosus*. This is true despite the fact that limpet activity reduces the success of initial *Chthamalus* recruitment (Table 12).

Thus, *Balanus* species very quickly dominated most of the available substratum in the absence of limpet and *Thais* disturbance, but the limpet disturbance alone maintained sufficient substratum for the settlement and survival of the very much smaller *Chthamalus*.

The highly significant height-treatment interactions for *Chthamalus* at Shi Shi and Portage Head result from the limpets reducing *Chthamalus* recruitment (see also Table 12) in the lower intertidal and having a positive influence at the higher levels by alleviating competition. Indeed, the upper level positive effect is sufficiently strong that the overall site effect is positive (Table 13), even though at both

TABLE 14. The effect of *Acmaea* disturbance on barnacle density by the end of each summer with *Thais* present at their normal densities for each level and area (Tables 6 and 7) but with *Pisaster* absent. Significance levels are for one-tailed tests. In addition to the comparison of the means, interactions between treatment, year, and intertidal height were evaluated. Anova = analysis of variance

Barnacle species	Site	Tide height (feet)	Year	Mean number of barnacles/100cm² Limpets present (normal limpet)	Mean number of barnacles/100cm² Limpets absent (0 limpet)	Type of analysis	Significance level	Significant interactions
Balanus glandula	Shi Shi	+2.7 − +5.4	1967–1968	0.8	0	Randomization	P=0.25	None
	Mukkaw Bay	+6.0 − +7.6	1968–1969	68	286	Randomization	P=0.062	None
	Portage Head	+2.5 − +5.0	1967–1969	3.2	9.2	3–Way Anova	0.005>P>0.001	*Height-year
	Eagle Point	+0.5 − +4.5	1967–1968	0.8	16	Randomization	P=0.016	None
	Turn Rock	0.0 − +4.0	1967–1968	0	6.8	Randomization	P=0.031	None
Balanus cariosus	Eagle Point	+0.5 − +4.5	1967–1968	0.3	4.0	3–Way Anova	0.005>P>0.001	None
	Turn Rock	0.0 − +4.0	1967–1968	0	2.0	3–Way Anova	0.025>P>0.01	None
Chthamalus dalli	Shi Shi	+2.7 − +5.4	1967–1968	200	386	3–Way Anova	0.025>P>0.01	None
	Mukkaw Bay	+6.0 − +7.6	1968–1969	158	129	3–Way Anova	P>0.25	None
	Portage Head	+2.5 − +5.0	1967–1968	235	323	3–Way Anova	P<0.001	None
	Eagle Point	+0.5 − +4.5	1967–1968	2.4	3.7	Randomization	P=0.094	None
	Turn Rock	0.0 − +4.0	1967–1968	0	4.0	3–Way Anova	0.005>P>0.001	None

(*0.05 ≥ p > 0.01).

Shi Shi and Portage Head in each of the years tested there were two pairs of cages at the lower levels and only one pair at the higher level. An explanation is that *Balanus glandula* has a strong tendency to settle higher in the intertidal (see above), whereas *Chthamalus* settles much more uniformly throughout (see above). At Waadah Island, all cages were at the higher levels, the highest being above the optimum settling height for *Balanus glandula*. Thus, the Waadah Island height-treatment interaction resulted from *Chthamalus* being positively affected by limpets at 4.6 ft and 5.7 ft, where *Balanus glandula* competition was a factor, and negatively affected at 7.7 ft, where few *B. glandula* settled. The net result, as indicated by the height-treatment interactions, at all these areas is that the disturbance activities of the *Acmaea*, by countering the strong competition effects, are important in providing adequate space for *Chthamalus*.

Effects in the presence of Thais.—In most cases limpet activity significantly reduces the net barnacle survival in the presence of normal *Thais* predation (Table 14). Thus, despite selective predation by *Thais* on the *Balanus* species (see below), the combined influence of *Acmaea* and *Thais* is to increase mortality for all barnacle species (Table 14). This situation contrasts with events in the absence of *Thais* (Table 13), when a significant increase in *Chthamalus* occurs in the presence of limpets.

The significant year-height interaction at Portage Head is the result of relatively small survival of *B. glandula* in the dog dishes in 1968, particularly at the higher levels, where *B. glandula* was relatively more numerous in other years.

Effects of Thais *predation on barnacle populations*

Two independent approaches were used to evaluate the effect of *Thais* predation on the barnacle populations. The first involved the removal of *Thais* from an area and the second was the more refined experimental design outlined in Fig. 4.

At representative outer coast (Portage Head) and San Juan Island (Eagle Point) sites in *Pisaster* removal areas, *Thais* were manually removed from areas of approximately 20 and 25 m² respectively. Clearly this procedure is not as effective as the cage technique, but by carefully removing the *Thais* every 2 weeks for 2 years and by comparing immigration rates and densities with the controls, effective *Thais* predation was reduced by 60–80%. Table 15 shows the degree to which *Thais* eats *Balanus* before eating *Chthamalus*. It can be seen that *Thais* selectively kill most of the *Balanus* by the end of the summer, at which time the *Chthamalus* density is actually greater than in the control, presumably as a result of the release from space competition with *Balanus*. These crude experiments show that *Thais* predation, in addition to other disturbance, can have an important effect on barnacle populations.

Continuing interference caused by the normal activity of limpets reduces all barnacle recruitment;

TABLE 15. *Thais* removal experiment at Portage Head and Eagle Point in 1967. Mean number ± standard deviation of living and dead barnacles from 25 randomly sampled quadrats of 16 cm² at Portage Head and 100 cm² at Eagle Point

Site	Dates		Chthamalus dalli		Balanus glandula		Balanus cariosus	
			Experimental	Control	Experimental	Control	Experimental	Control
Portage Head	May 14	Alive	9.2±7.0	8.5±6.2	9.7±5.2	10.5±2.6		
		Dead	0.4±1.2	0.4±1.0	3.2±2.5	3.4±0.8		
	June 9	Alive	15.9±12.2	14.6±13.5	13.1±12.3	10.9±8.6		
		Dead	0.9±1.7	2.1±3.0	4.0±4.1	12.2±7.3		
	July 10	Alive	12.5±2.5	17.4±14.1	15.3±8.3	4.1±4.4		
		Dead	1.0±2.2	1.7±2.7	3.0±3.3	15.0±9.8		
	August 21	Alive	19.8±12.8	34.6±17.2	19.6±7.5	2.0±2.6		
		Dead	1.7±2.7	1.9±3.0	4.1±5.0	23.0±8.9		
Eagle Point	June 11	Alive	4.0±5.3	3.7±3.7	6.5±8.2	6.7±12.8	6.0±4.3	5.1±8.0
		Dead	0.8±0.5	0	3.1±4.7	4.0±5.2	1.2±2.1	0.4±3.9
	July 21	Alive	4.3±3.8	4.7±5.0	7.2±6.5	2.1±2.9	7.7±3.2	2.2±3.4
		Dead	0.5±2.1	0.2±1.2	4.5±5.9	9.5±4.1	0.8±1.4	4.4±2.9
	August 18	Alive	3.9±4.7	5.0±4.2	8.0±6.5	0.2±2.1	8.4±3.8	1.9±4.4
		Dead	1.0±1.7	0.5±1.3	5.0±3.6	10.3±3.3	1.1±1.0	5.8±2.6
	October 2	Alive	4.4±4.6	5.4±7.7	7.1±8.9	0	9.2±4.0	0.2±3.7
		Dead	0.8±1.2	0.8±1.4	6.2±5.5	10.0±0.2	1.4±2.2	8.1±1.8

however, some barnacles do metamorphose and escape the limpet disturbance. Within 10–20 days after metamorphosis, the barnacles are sufficiently large that they are not killed by limpets. The escape in growth from limpet disturbance means that the net effect of the limpets is only to introduce a time lag into the eventual barnacle domination of all the available space resource in the upper intertidal. This time lag, due to a necessary interval for *Thais* to find and consume the barnacles, is longer at higher than at lower levels, since *Thais* tend to seek lower level refuge from desiccation during spring tides and must move up to forage during neap tides. During such time lags the continuous limpet pressure on the barnacle recruitment maintains a generally significant negative effect on each of the barnacle populations.

To measure the effect of *Thais* predation in the presence of limpets, comparisons were made of 100-cm² areas under cages containing normal limpet densities but excluding *Thais* and 100-cm² areas under "roofs" which allowed access to both *Thais* and limpets at their normal densities. The effect in the absence of limpets was evaluated by comparing the number of barnacles surviving in the limpet exclusion plastic dishes, corrected to 100 cm², to the number surviving in the total exclusion cages (Fig. 4). Statistical analyses were performed as for the limpet experiments.

In almost every case predation by *Thais* in the absence of other disturbance had a pronounced effect on the survival of all three barnacle species (Table 16). Again there was a trend for *Chthamalus* to have a much higher survival with disturbance because *Thais* selectively eats the competitively dom-

inant *Balanus* species (Table 15). The increased *Chthamalus* survival due to the presence of *Thais* is not as dramatic as was the increase due to limpets. This is because limpets reduce the *Balanus glandula* and *B. cariosus* settlements before they have a chance to outcompete *Chthamalus*, whereas in the absence of limpets, the *Balanus* species often have already grown over many *Chthamalus* by the time *Thais* find and eat the *Balanus*. The significant height-treatment interactions for *Chthamalus* at Portage Head and Shi Shi indirectly result from the strong tendency of *Balanus glandula* to settle above the +4-ft level. *Thais* removes essentially all the *Balanus* in the dog dishes at all the heights. *Chthamalus*, then, survives relatively much better at the higher levels in the presence of *Thais*, because of significant reduction in the intensity of interspecific competition.

Table 17 shows significant reduction of survival of all three barnacle species due to *Thais* predation. As the reduction is not significant at Turn Rock, this site presents an exception to this generalization for each barnacle species, as does Colin's Cove for *Balanus cariosus*. The reason is the extremely effective limpet interference at both sites; limpet activity both in the *Thais* exclusion cage and under the roof is sufficient to prevent the settlement of enough barnacles to allow a proper analysis of *Thais* predation.

Relative influence of limpets and Thais

When the relative influences of limpets and *Thais* on barnacle recruitment and survival over 1 year are examined (Fig. 7–12), the following patterns emerge: (1) the control areas at each level have low recruitment and essentially no survival in the outer coast sites and neither recruitment nor survival in the San Juan Island sites; (2) in the absence of limpets there tends to be a strong recruitment of all barnacle

380   PAUL K. DAYTON   Ecological Monographs
                                                                      Vol. 41, No. 4

TABLE 16. The effect of *Thais* predation on barnacle density by the end of each summer with *Acmaea* and *Pisaster* absent. Significance levels are for one-tailed tests. In addition to the comparison of the means, interactions between treatment, year, and intertidal height were evaluated

Barnacle species	Site	Tide height (feet)	Year	Mean number of barnacles/100cm² *Thais* present (0 limpet)	Mean number of barnacles/100cm² *Thais* absent (total excl. cage)	Type of analysis	Significance level	Significant interactions
Balanus glandula	Shi Shi	+2.7 − +5.4	1967–1968	0	253	3–Way Anova	$P<0.001$	*Height −Treatment
	Portage Head	+2.5 − +5.0	1967–1969	0	431	3–Way Anova	$0.05>P>0.025$	None
	Eagle Point	0.0 − +4.5	1967–1968	24	504	3–Way Anova	$0.005>P>0.001$	None
	Turn Rock	0.0 − +4.0	1968	25	128	2–Way Anova	$0.10>P>0.05$	—
Balanus cariosus	Eagle Point	0.0 − +4.5	1967–1968	4.0	49	3–Way Anova	$P<0.001$	None
	Turn Rock	0.0 − +4.0	1968	2.2	17	2–Way Anova	$0.025>P>0.01$	—
Chthamalus dalli	Shi Shi	+2.7 − +5.4	1967–1968	386	60	Randomization	$P=0.28$	**Height −Treatment
	Portage Head	+2.5 − +5.0	1967–1969	313	73	3–Way Anova	$P<0.001$	***Height −Treatment
	Eagle Point	0.0 − +4.5	1967–1968	3.7	11.2	Randomization	$P=0.047$	None
	Turn Rock	0.0 − +4.0	1968	3.0	0	2–Way Anova	$0.25>P>0.10$	—

(*$0.05 \geq p > 0.01$; **$0.01 \geq p > 0.001$; ***$p \leq 0.001$).

species, but *Thais* predation strongly reduces or eliminates these barnacle populations; (3) in the presence of limpets but in the absence of *Thais*, the initial recruitment is reduced, but eventually barnacles do escape the limpet disturbance and these barnacles persist in the absence of predation; and (4) strong interspecific competition is usually seen in the total exclusion cages within the first season in the absence of both limpet and *Thais* disturbance.

Escape by growth of Balanus cariosus and Mytilus californianus *from* Thais *predation*

Two cages at Eagle Point and one cage at Turn Rock came loose and were removed in June 1968. These cages had been in place since the summer of 1967 and had mixed size classes of *Balanus glandula* and *Balanus cariosus* under them. The fate of these barnacles was followed throughout the remainder of the period of this study. *Thais canaliculata* and *T. emarginata* killed all 320 of the *Balanus glandula* within 10 days of the removal of the cages, and by the end of July 1968 had killed all 47 of the *Balanus cariosus* with basal diameters less than 10 mm. However, a total of 27 *B. cariosus* at Eagle Point and 10 at Turn Rock with basal diameters between 13 and 19 mm survived. In January 1969 ten were accidentally killed at Turn Rock. The Eagle Point barnacles were still alive when the site was last observed in August 1970. Further evidence of *Balanus cariosus* reaching a refuge in size was observed at the Eagle Point *Thais* removal area. A patch of *Balanus cariosus*

settled in June 1967 and grew in hummocks at the 1- to 3-ft tide levels, with their basal diameters ranging from 7 to 14 mm by October 1967. At this time *Thais* removal was terminated in the area. Very few of the *Balanus cariosus* in this patch had been killed by *Thais* by August 1970; their most severe mortality seems to come from exfoliation of the hummocks. When a *B. cariosus* attains a basal diameter of 13–15 mm it seems quite safe from *Thais* predation. The optimal growth rates found at the lower levels (Fig. 13) suggest that a *B. cariosus* which settles in May or June may have reached this refuge in size by the end of the calendar year. Certainly most *B. cariosus* that escape predation attain this size refuge by the end of their second year.

The escape in growth of *Balanus cariosus* involves the time, energy, and risk parameters affecting *Thais* (Emlen 1966, Connell 1970). Unsuccessful attacks by each of the three species of *Thais* on large *B. cariosus* have been observed, and misplaced or incomplete drill marks are commonly seen on large *B. cariosus*. The actual size at which an individual *B. cariosus* becomes a poor risk for a *Thais* to attack depends on its level in the intertidal and the time of year of its settlement, that is, (1) the ability of *Thais* to drill and consume a barnacle is obviously related to submergence time, and (2) *Thais emarginata* and *T. canaliculata* tend to be relatively inactive during the winter months, and at this time *T. lamellosa* ceases feeding and forms breeding aggregations. Thus, if the *B. cariosus* settle in the fall (see Fig.

TABLE 17. The effect of *Thais* predation on barnacle density by the end of each summer with *Acmaea* present at normal densities (Table 5) but with *Pisaster* absent. Significance levels are for one-tailed tests. In addition to the comparison of the means, interactions between treatment, year, and intertidal height were evaluated, but none was found to be statistically significant (all $p > 0.05$)

Barnacle species	Site	Tide height (feet)	Year	Mean number of barnacles/100cm² *Thais* present (roof)	Mean number of barnacles/100cm² *Thais* absent (limpet cage)	Type of analysis	Significance level
Balanus glandula	Shi Shi	+2.7−+5.4	1967–1969	7.9	102	Wilcoxon Signed rank	$P<0.005$
	Portage Head	+2.5−+5.0	1967–1969	4.3	76	3–Way Anova	$P<0.001$
	Waadah Island	+4.6−+7.7	1968–1969	0.8	54	3–Way Anova	$P<0.001$
	Eagle Point	0.0−+3.5	1967–1968	4.2	61	3–Way Anova	$P<0.001$
	Colin's Cove	+0.5−+3.0	1968	0	43	2–Way Anova	$0.005>P>0.001$
	Turn Rock	−1.0−+4.5	1968	0.5	1.9	2–Way Anova	$P>0.25$
Balanus cariosus	Waadah Island	+4.6−+7.7	1968–1969	0.1	3.0	3–Way Anova	$0.005>P>0.001$
	Eagle Point	0.0−+3.5	1967–1968	2.1	19	3–Way Anova	$P<0.001$
	Colin's Cove	+0.5−+3.0	1968	0	0.8	2–Way Anova	$P>0.25$
	Turn Rock	−1.0−+4.2	1968	0.8	0.7	2–Way Anova	$P>0.25$
Chthamalus dalli	Shi Shi	+2.7−+5.4	1967–1969	312	529	3–Way Anova	$0.05>P>0.025$
	Portage Head	+2.5−+5.0	1967–1969	467	795	Wilcoxon Signed rank	$P<0.005$
	Waadah Isalnd	+4.6−+7.7	1968–1969	29	57	3–Way Anova	$0.25>P>0.10$
	Eagle Point	0.0−+3.5	1967–1968	4.9	25	3–Way Anova	$0.01>P>0.005$
	Colin's Cove	+0.5−+3.0	1968	0	21	2–Way Anova	$0.025>P>0.01$
	Turn Rock	−1.0−+4.2	1968	0.9	1.0	2–Way Anova	$0.25>P>0.10$

11—total exclusion and Fig. 12—total exclusion and limpet cage), they are subjected to reduced *Thais* pressure and have a good chance of attaining a refuge in size by the end of the following summer.

Mytilus californianus has a similar refuge in growth from *Thais*. The actual safe size depends on many factors, particularly the level in the intertidal and the size of the local *Thais*. No detailed study of this was made, but when two protective cages at about the 4-ft level in the *Pisaster* removal areas at Shi Shi were removed late in the second summer after their placement, the newly exposed *Mytilus* were subjected to severe *Thais* predation. No *Mytilus* under 3 cm in length survived, and the *Thais* did kill a number which were over 3 cm. In the San Juan Islands the larger *Thais* appear to be able to kill much larger *Mytilus*. A rock in a *Pisaster* removal area at Turn Rock, heavily covered with both *Mytilus californianus* and *M. edulis*, including individuals at least as large as 8 cm, was denuded in 1967. Many large valves ($\cong$8 cm) were found with drill holes. Thus the size refuge from *Thais* is quite variable, and when *Mytilus* do escape *Thais*, they do so initially by numerically swamping the local *Thais* populations.

Effect of Pisaster *predation on barnacles and mussels*

The importance of *Pisaster ochraceus* to the outer coast community has been stressed by Paine (1966, 1969); yet it is apparent from Tables 15, 16, and 17 that even by late August the *Thais* had killed many of the barnacles at each site. It is of interest, then, to test the effect of *Pisaster* beyond that of limpets and *Thais*. The experimental design allowed the effect of *Pisaster* to be assessed by comparing the number of barnacles remaining in a 100-cm² area under a "roof," which protects the barnacles from *Pisaster*, with the number of barnacles remaining on an adjacent 100-cm² control plot (Table 18). The analyses for Table 18 are similar to those for limpet and *Thais* effects.

On the outer coast the *Pisaster* effect is variable ($P \leq 0.001$ to $P = 0.35$). To a certain extent this is artifactual, because the terminal data each year were collected in late August, while all the predators remain active until at least November. But the most important explanation is related to the foraging behavior of *Pisaster*. Paine (1969) has observed that *Pisaster* tends to bypass *Balanus glandula* in search of the preferred *Mytilus californianus*. It is only late

382   PAUL K. DAYTON   Ecological Monographs
                                                                      Vol. 41, No. 4

TABLE 18. The effect of *Pisaster* predation on barnacle densities by the end of each summer in addition to *Acmaea* disturbance and *Thais* predation. Significance levels are for one-tailed tests. In addition to the comparison of the means, interactions between treatment, year, and intertidal height were evaluated, but none was found to be statistically significant (all $p > 0.05$)

Barnacle species	Site	Tide height (feet)	Year	Mean number of barnacles/100cm² *Pisaster* present (control)	Mean number of barnacles/100cm² *Pisaster* absent (roof)	Type of analysis	Significance level
	Shi Shi	+4.2−+5.4	1967–1969	3.1	10.1	Randomization	$P=0.22$
Balanus glandula	Portage Head	+3.0−+4.0	1967–1969	0	14	3–Way Anova	$P<0.001$
	Waadah Island	+4.6−+7.7	1968–1969	4.2	0.8	3–Way Anova	$0.10>P>0.05$
	Colin's Cove	0.0−+3.7	1968–1969	0	0	3–Way Anova	$P>0.25$
Balanus cariosus	Waadah Island	+4.6−+7.7	1968–1969	1.2	0.1	3–Way Anova	$0.25>P>0.10$
	Colin's Cove	0.0−+3.7	1968–1969	0	0	All zero	—
	Shi Shi	+4.2−+5.4	1967–1969	141	245	Randomization	$P>0.50$
Chthamalus dalli	Portage Head	+3.0−+4.0	1967–1969	5.1	546	Randomization	$P=0.016$
	Waadah Island	+4.6−+7.7	1968–1969	34	29	Randomization	$0.36>P>0.34$
	Colin's Cove	0.0−+3.7	1968–1969	0	0.6	3–Way Anova	$P>0.25$

in the summer or in the absence of the preferred prey that *Pisaster* will begin to consume appreciable quantities of *Balanus glandula*. Therefore, if a *Pisaster* had foraged over a control area in late August, before the last census, the area would be devoid of suitable *Pisaster* prey, but if it had not, a number of barnacles might be surviving.

With the exception of one insignificant difference in *Chthamalus* populations at Colin's Cove, no analysis of *Pisaster* effect was possible in the San Juan Islands, because 100% of all species of barnacles in the experimental areas were killed by the normal limpet and *Thais* activity before the late August census. The fact that there were no interaction effects (Table 18) suggests that the *Pisaster* pressure measured was essentially the same at all levels in all years.

Paine (1969) has discussed the implications of the fact that by consuming *Mytilus californianus*, *Pisaster* procures space for *Endocladia muricata*, the most important settling resource of the *Mytilus* larvae. In the absence of *Pisaster, Endocladia* is overgrown and eliminated from the community by *Mytilus*. A substantial proportion of the diet of *Pisaster* includes limpets (Menge 1970, Paine, *personal communication*). Therefore the theme of the indirect effect of *Pisaster* can be elaborated somewhat by the demonstration that limpets can severely reduce the recruitment of *Endocladia* (see Table 11). Thus, besides procuring space for *Endocladia* by eating *Mytilus*, *Pisaster* increases the chance of *Endocladia* recruitment by preying on limpets and reducing their grazing pressure. The implications are that by using limpets as a secondarily preferred prey, *Pisaster* in-

directly has a positive effect on the recruitment of its major prey.

Effect in absence of Thais.—Paine (1966) has claimed that even though Mukkaw Bay *Thais* may actually kill more barnacles than do *Pisaster*, the *Thais* effect is less important than that of *Pisaster*. We have not seen, however, an important *Pisaster* effect in addition to that of *Thais* and limpets in the experiments of this study. This appropriate controlled experiment, of measuring the *Pisaster* effect completely independently of the *Thais* effect, has not been done, as it would necessitate the removal of *Thais* while maintaining the natural *Pisaster* predation, and this would involve a prohibitive amount of time. To a limited extent, however, this experiment was done naturally by an unusually cold period in December 1968 and January 1969 coinciding with very low night tides, which resulted in the death of many *Thais* (Table 9).

An immediate result of this population crash of *Thais* on the outer coast was an increased survival of the 1969 spring *Balanus glandula* set, many of which became fertile by September 1969, and apparently succeeded in releasing into the plankton an unusually large number of cyprids in the fall of 1969. The late autumn plankton conditions are certainly very different from the conditions prevailing in the spring and summer, and normally the winter *Balanus glandula* settlement and metamorphosis is very rarely seen in my experimental areas. However, the survival of the 1969 winter set seemed to be extremely high, and in March 1970 neither *Thais* nor *Pisaster* had eaten much of the heavy set. By late July 1970 *Pisaster* was consuming large numbers of

Balanus, but by this time many of the barnacles had reproduced, and it appeared possible that there would be another successful set in the fall of 1970. In this case, however, most of the barnacles settling within the *Pisaster* levels will be consumed before reproduction. The general space allocation pattern, though immediately influenced by the cold kill of *Thais* will have no long-term effect, due to the activities of *Pisaster*, which is capable of consuming all sizes of barnacles within its foraging range.

Additional ramifications in the upper intertidal above the level of *Pisaster* predation could be predicted to result from the *Thais* cold kill. Here the predicted heavy set of *Balanus glandula* in the fall of 1970 should have survived and dominated much of the primary space. Thus, by the spring, 1971, there should have been a reduction in the number of *Chthamalus* and algae utilizing primary space. By February 1971 these predictions were verified at Mukkaw Bay (R. T. Paine, *personal communication*). In addition to the *Chthamalus* and algae which suffered in the space competition with *B. glandula*, Paine reports that the *B. glandula* have forced even the limpets off the primary space and that intraspecific competition among the *B. glandula* has resulted in severe hummocking and exfoliation of the barnacles.

The implications to patterns of space utilization of this cold kill are both more delayed and more important in the San Juan Islands. Here the immediate response of *Balanus glandula* is somewhat buffered, as there are additional predators, especially *Leptasterias* (Menge 1970, and personal observation), *Onchidoris bilamellata* (personal observation) and to a certain extent, *Searlesia dira* (Louda 1968, and personal observation), which may respond to an increase in *B. glandula* by increasing their predation on it. On the other hand, *B. cariosus*, the largest, most rapidly growing and hence competitively effective intertidal barnacle. is significantly influenced. The usual strategy of *B. cariosus* is to swamp *Thais* with a patchy summer settlement. Some *B. cariosus* also appear to release a late summer brood of cyprids which results in a light autumn set. Under normal conditions (here 1966–68) the autumn settlement is insignificant and the abundant *Thais* populations are capable of eliminating most of the *B. cariosus* which escape the limpets. With the *Thais* populations drastically reduced in the summer of 1969 (Table 9) many of the individuals of the spring and fall set survived in 1969 and released a very large number of cyprids in early 1970; the spring 1970 *B. cariosus* set was extremely large. In July 1970 *B. cariosus* covered 80–100% (mean: 87%; $SD = 5.1\%$; $N = 20$) of all the available surface between +1-ft and +3-ft levels at Eagle Point. The density of *Thais canaliculata* was still low (Table 9). It is clear that many of the *B. cariosus* will escape *Thais* predation during the summer of 1970; further, since many of the lower intertidal *B. cariosus*, which usually do not survive the summer, had well developed gonads in late July 1970 (S. A. Woodin, *personal communication*), there should have been a successful autumn set in 1970. The result of the cold kill will be a prolonged, in terms of years, occupancy of space, an event predictable from knowledge of the escape in size of *B. cariosus* from molluscan carnivores, its superior competitive ability, and the relatively low density of *Pisaster*.

Effects of other predators

The cages which have been analyzed as representing *Thais* exclusion effects actually also represent the exclusion of *Leptasterias hexactis*, a small carnivorous asteroid characterized by a generalized diet. Menge (1970) has found that *Leptasterias* derives most of its energy from various molluscan herbivores, but that when *Balanus glandula* is available, it consumes many of these small barnacles. However, Menge has also demonstrated that *Leptasterias* is very susceptible to desiccation, which limits its foraging effectiveness to low areas in the intertidal. Nevertheless, there is an overlap between the *Leptasterias* foraging range and the cage experiments at Eagle Point and Colin's Cove. At these sites it is possible that *Leptasterias* may have contributed to the barnacle mortality in the dog dishes and control plots, and thus may have been mistakenly interpreted as part of the *Thais* effect. But the relatively low densities of *Leptasterias* at these sites (Menge 1970) are probably not sufficient to alter appreciably the conclusions.

Onchidoris bilamellata, a dorid nudibranch, is another potentially important barnacle predator in this community. It is present in the winter months, from October to March. The cages were not entirely effective in excluding *Onchidoris*, but those that entered were quickly removed. Cages containing *Onchidoris* egg masses in association with dead barnacles were removed from the evaluation. *Onchidoris* is usually found in protected bays, on pilings, and on floats, where it has a better supply of *Balanus glandula* and *B. crenatus*. *Onchidoris* were never seen in controls or dog dishes and only very rarely along the rocky intertidal.

Colonization in the high intertidal

The most common means of experimentally denuding an intertidal substratum are scraping, burning, and saturating with formalin. None of these methods is absolutely certain to kill all of the organisms, and all of them leave various organic residues

384 PAUL K. DAYTON

Ecological Monographs
Vol. 41, No. 4

which may differentially affect bacterial growth and diatom recruitment and growth. To evaluate the effects of leaving organic and sometimes living residues, each month for a year virgin rock surfaces were exposed by blasting with dynamite, and the trends compared with those characterizing burned areas, at Portage Head.

A certain amount of leaching or weathering is necessary before any substantial colonization occurs (see also Moore 1938). All the rocks in the middle intertidal area blasted between August 1967 and March 1968 stayed completely bare of sessile organisms until August 1968, when a heavy *Chthamalus dalli* settlement almost covered them. Many *Littorina* had come onto the rocks soon after blasting and may have been responsible for the fact that a heavy diatom film never developed (see also Castenholz 1961). No limpets or *Thais* were seen on the blasted rocks during the first summer. One rock blasted in July, 1967, split along a crack where seawater had already penetrated (as evidenced by the presence in the crack of serpulid worm tubes). This rock was very quickly covered with *Balanus glandula* and *Chthamalus dalli* by August 1967.

Those rocks blasted between April and August 1968 remained bare of sessile organisms until April 1969, when *Balanus glandula* began to appear. In all cases, between 6 months and a year elapsed before there was appreciable colonization on any of the rocks. During and after colonization small pieces of rock flaked off, creating clear patches in the heavy barnacle cover and allowing further colonization to be observed. Approximately 12–18 months after the blast, the algae *Porphyra* spp. and *Gigartina papillata* as well as new *Chthamalus* and *Balanus glandula* appeared. By September 1969 *Gigartina* covered 45–60% of the blasted surfaces, and *Acmaea paradigitalis*, *A. pelta*, *A. digitalis*, and *Thais emarginata* had appeared and had begun to influence the barnacle population.

Of the more common means of denuding an intertidal surface, burning is probably the most satisfactory, as it is much more likely to kill the algal sporelings and gametophytes than is scraping, and it leaves less residue than does soaking in formalin. The surfaces of three large rocks were burned in the upper intertidal at Portage Head in November and December 1966 following a soaking with a mixture of white gas and diesel oil. A careful study of the basal systems of the coralline algae *Gigartina papillata* and *Ralfsia* spp. showed that it was not until after the third burn that all of the plants had been killed. Despite the fact that this was done in the winter and *Littorina* moved in, a very heavy diatom film developed over these rocks after each burning. This strongly contrasts with the situation on the blasted

rocks, which also had *Littorina*, but which had no growth of any sort for many months after blasting. By February 1967 the burned rocks had a very heavy cover of *Porphyra*, which lasted until late March, when the tides moved into daylight hours; at this time most of the *Porphyra* was killed by desiccation. *Balanus glandula* and *Chthamalus dalli* settled on the burned rocks during the summer of 1967, again in contrast to the blasted rocks, which had no barnacle settlement the first summer. As *Porphyra* died out, scattered individuals of *Ulva* spp., *Enteromorpha* spp., *Odonthalia floccosa*, *Halosaccion glandiforme*, and *Polysiphonia hendryii* appeared in 1967. In 1968 most of the same red algae individuals were recognized by their position. In addition, there were numerous cases of *Gigartina papillata* fronds growing up from what appeared to be an extensive rhizoidal growth of the holdfast. By 1969, *Gigartina* had taken over 65% of the surface area and 85% of the canopy, and all of the other algal species were gone. It was not until August 1969 that a few isolated individuals of *Fucus distichus* and *Endocladia muricata* appeared. By early 1970, *Fucus* seemed to be growing through the *Gigartina* canopy. It will certainly be at least 4 years after the burning before *Fucus* regains its original dominance (70%) of the canopy cover.

Patchiness and interactions among the dominant high intertidal species on the outer coast

Much of the flat upper intertidal community at Shi Shi and Portage Head is dominated by either *Mytilus californianus* clumps or *Fucus distichus* or *Gigartina papillata* canopy; there is often very little apparent overlap between these patches (Table 19). In order to test whether there were competitive interactions involved in these apparently nonoverlapping distribution patterns, patches of each species were cleared at Portage Head and Shi Shi in the spring of 1967. The five 0.5- to 1.5-m² *Mytilus*-removed patches all developed essentially as the burning experiment, except that the *Gigartina* appeared during the same summer that the *Mytilus* were removed. By July 1968 these five patches were almost completely dominated by *Gigartina*, which has persisted through 1969. By March 1970 a few *Fucus* individuals had appeared. The failure of *Mytilus* to reoccupy the area is best explained by their failure to recruit in sufficient numbers to swamp the predation of the resident *Thais emarginata*.

Fucus individuals were manually removed in the spring of 1967 from seven patches varying from 0.5 to 2.5 m². The encrusting holdfast system of *Gigartina* was observed to cover much of the substratum under the heavy *Fucus* canopy. The individual *Gigartina* plants had few fronds, but removal of the *Fucus* canopy initiated a large increase in the number of

TABLE 19. Summary of samples of conspicuous organisms in distinct patches of *Mytilus californianus*, *Gigartina papillata*, and *Fucus distichus* in the flat upper (+6 ft to 8 ft) intertidal at Portage Head. Variance is presented as 95% confidence limits. The total percentage cover sums to over 100%, as *Fucus* frequently covers *Mytilus* and *Gigartina*

Species	*Mytilus* patches ($N=17$)		*Gigartina* patches ($N=10$)		*Fucus* patches ($N=5$)	
	Mean % cover	Density[a]	Mean % cover	Density[a]	Mean % cover	Density[a]
Mytilus californianus	100	Not taken	0	0	0	0
Fucus distichus	5.0±7.1	0.1±0.3	12.2±9.3	1.7±0.7	100	6.0±2.6
Gigartina papillata	0	0	100	11.8±0.9	15.2±7.0	2.4±2.0

[a] Number /0.062 m².

fronds. A few of the *Fucus* grew back by October 1967; these individuals were probably from holdfasts which had not been properly removed at the initial clearing. By the end of 1969, *Fucus* had regained about 50% of the canopy in each of the clearings. Experimental removals of *Gigartina* were not effective.

The following pattern of succession for flat upper intertidal surfaces emerges from the blasting experiments, the burning experiments, and the selective removal experiments. *Gigartina* is the first persistent perennial to appear, and it seems to dominate the canopy for at least one summer. In the second season *Fucus* appears but does not completely dominate the canopy for at least 2 years, and probably much longer. Even when *Fucus* does overgrow *Gigartina* and dominate the canopy, the *Gigartina* persists below the *Fucus* as an encrusting rhizoidal system which tends to respond to a break in the *Fucus* canopy with a rapid growth of fronds. At this tidal level (+6 ft to +8 ft) in the protected outer coast habitats, it probably takes many years for a *Mytilus californianus* population to become established. Once *Mytilus* does become established and reaches a refuge in size from *Thais* predation, the patch has a refuge in space from *Pisaster* predation and may survive for a great many years, until the individuals are killed by such physical phenomena as log damage or unusual cold.

DISCUSSION

Sessile marine organisms have in common two resource pools: (1) the primary space, the substratum on which they attach, and (2) the aquatic milieu around them, which is the source of their physical and organic nutrients. There are two corresponding levels of interaction among these organisms in the competition for a potentially limiting resource; they can compete for the primary space and (or) they can grow above and then over their competitors and compete for the physical resources and organic nutrients. Adaptations to both of these levels of competition can be seen in intertidal organisms. Algae and barnacles compete for primary space. *Mytilus* recruitment, on the other hand,

usually requires a secondary space in the form of filamentous algae or barnacles for the initial larval settlement; then its byssal threads become attached to the primary substratum. Thus the *Mytilus* essentially concede the initial competition for primary space, in fact, needing the secondary space for their larval recruitment, and then dominate the competition in the space above the primary substratum. Eventually the *Mytilus* dominate the primary substratum also, as the underlying barnacles either starve or are smothered by sedimentation.

In each study area the barnacle-mussel association is characterized by one potential dominant capable of monopolizing the resource and completely excluding the other species; the identification and evaluation of the effects of the disturbance factors preventing this monopolization are crucial to any functional understanding of the organization of the community.

Three levels of biological disturbance which predictably affect the monopolization of space by the dominant sessile species include (1) grazing by limpets; (2) predation by carnivorous gastropods, particularly species of *Thais*; and (3) predation by the asteroid *Pisaster ochraceus*. At their normal densities, various species of limpets severely restrict the recruitment and survival of almost all the sessile species in the community. However, individuals in all the barnacle species and most of the algae may eventually attain a size at which they are immune to the bulldozing effect and the grazing of the limpets. Nevertheless, after 4 years, in none of the areas cleared of algae but which had normal limpet densities has the algal community returned to its natural state. In contrast to this slow natural succession rate, there was an algal bloom in the areas where limpets had been removed. This usually resulted in a cover of "climax" algal species which within 2 years equaled and usually far surpassed the cover in the undisturbed control areas. A similar algal response to the release from limpet pressure has been observed in Europe (Jones 1948, Southward 1956).

The grazing activity of the limpets significantly reduced the recruitment of all the barnacles in the presence of *Thais*, but the direct detrimental effects of the limpet grazing were much reduced on the smaller, competitively inferior *Chthamalus dalli*. By the end of the summer season, the *Chthamalus* population in the absence of *Thais* was indirectly but significantly increased through the limpet interference with the competitively superior *Balanus glandula* and *B. cariosus*.

The effects of predators on their prey are detrimental to the prey populations, but the question that concerns us here is how the various predators affect the community through the medium of the common space resource. Paine (1966) has argued that by killing large numbers of barnacles and small *Mytilus* at one time, the Mukkaw Bay *Pisaster* are more important providers of primary space than are *Thais*, which must kill these prey separately; he further points out that the *Pisaster* dislodge the dead barnacles, thereby immediately freeing the space, while the plates of the barnacles killed by *Thais* may occupy space for a number of weeks. However, this work has shown that in 1967–69 the continued limpet disturbance and the predation of *Thais* were usually sufficient to eliminate all the *Balanus glandula* and *B. cariosus* from the experimental areas. There was rarely a significant effect of *Pisaster* in this regard beyond the observed *Thais* effect. What contributions to community structure, then, are made by these two levels of predation?

Part of the answer is found in the growth refuge phenomenon of the prey. In both the outer coast system and the San Juan Islands, those species capable of monopolizing the space, *Mytilus californianus* and *B. cariosus*, respectively, have an escape in growth from *Thais* predation. For example, there were two 3- by 4-m patches in the Shi Shi *Pisaster* removal area where the *Mytilus* recruitment was sufficiently heavy to swamp the ability of *Thais* to kill them, and many of the *Mytilus* reached the length of approximately 3 cm necessary to escape this source of mortality. Likewise in the San Juan Islands, although it did not happen in my experimental areas, the extremely patchy settlement pattern of *Balanus cariosus* sometimes overwhelms the ability of the local *Thais* to kill them, and a few escape (personal observation). Thus, given enough time, one would expect all the space in each of the areas to be monopolized by one of these two long-lived, competitively superior species with the other species being eliminated or forced to subsist on secondary space. The reason that such monopolies fail to develop on the outer coast is that there is very rarely an escape in growth from *Pisaster*. In the study areas of the San Juan Islands, however, the *Pisaster* densities are very much lower, and desiccation apparently prevents *Pisaster* from foraging very high in the intertidal. In the lower areas

(below +2 ft) *Pisaster* eventually kills the *Balanus cariosus* that have escaped the limpet and *Thais* disturbance. Frequently a sharp line of large *B. cariosus* demarcates the upper limit of *Pisaster* predation, and in areas where there are few *Pisaster*, this line may be much lower (Mauzey 1967, Menge 1970, and personal observation).

In the San Juan Islands in the intertidal area immediately above the *Pisaster* foraging levels, *Balanus cariosus* occasionally escapes the limpets and *Thais* and would be expected eventually to occupy 100% of the space. The probability of log damage (Fig. 3) suggests the hypothesis that the frequency of logs striking an area is as high as or higher than the rate of the escape of *B. cariosus*. A test of this hypothesis would be to examine areas where logs do little or no damage; in such areas one would expect to find a heavy cover of escaped *B. cariosus*. This is indeed seen to be the case in cracks which are not accessible to logs. The best test of the hypothesis is Turn Rock, which was never struck by a log during the course of this study and, as expected, has 100% cover of *B. cariosus* above the level of *Pisaster* predation.

Competition, disturbance, and community organization

Proof that the co-occurring populations do form a community is prerequisite to any study of community structure. This community is open ended since much of the primary productivity and decomposition occur elsewhere; however, all communities are open ended in an energetic sense. Many of the species have planktonic larvae, but the relative abundances of most of the reproducing populations are most critically influenced by interactions among the component intertidal species rather than by plankton phenomena. Most of the changes in populations in this community, therefore, occur as predictable responses to other populations in the intertidal community.

This community permits experimental evaluation of the relative importance of competition and predation to the community structure. Much of community theory is classically based on competitive processes (Elton 1946, Elton and Miller 1954, Whittaker 1965, 1969, McNaughton and Wolf 1970). The identification and appraisal of the relative degrees of dominance has been a continuing descriptive problem (Whittaker 1965); some of the measures of a species' importance summarized by McNaughton and Wolf (1970) are (1) cover (2) biomass, and (3) relative abundance; Whittaker (1965) also suggests (4) productivity. With the exception of cover, none of these criteria directly measures competitive dominance of a potentially limiting resource. Most of the above criteria of "important species" have been considered and found to be of limited or conflicting application by Fager (1968) for a simple benthic marine community. McNaughton and Wolf (1970) proposed the following generalizations from the communities they summarized: (1) dominance was characteristic of the most abundant species; (2) dominant species had broader niches than subordinate species; (3) species are added to the system by compression of niches or expansion of "K," or both; and (4) community dominance is minimum on the most equitable sites.

In this intertidal community on the outer coast and in the San Juan Islands, where *Mytilus californianus* and *Balanus cariosus*, respectively, are the competitive dominants in the community, none of McNaughton and Wolf's generalizations seem to apply. For example, considering (1), the competitive dominants are numerically much less abundant than any number of fugitive species in this intertidal community. Furthermore, the number and distribution of the competitive dominants may be suppressed by disturbance to the point that at times they are even absent. In regard to (2), and considering "the breadth of niche" in the more general sense of Levins (1968), referring here to broadness of physical tolerances as indicated by vertical distribution in the intertidal, we find the competitively inferior *Balanus glandula* and *Chthamalus dalli* having a broader tolerance range, as they survive higher in the intertidal zone than do the dominant *Mytilus* and *Balanus cariosus*. Generalization (3) of McNaughton and Wolf can be evaluated only by examining the relative success of invasions of sessile species into rocky intertidal communities. The best documented invasion was by *Elminius modestus* into the European intertidal; *Elminius modestus* is an "r" species (Crisp 1964). In considering the exposed point of Tatoosh Island as most equitable (referring to the least amount of physiological stress), generalization (4) is contradicted because Tatoosh Island has the maximum dominance, as *Mytilus californianus* monopolizes most of the space.

What are the general implications of the fact that the experimentally derived conclusions of this study conflict with the carefully drawn conclusions of McNaughton and Wolf? The intertidal community may represent a special case, but since controlled experiments have been done in few other communities, it is too early to evaluate this possibility. Another explanation is that the descriptive criteria used to evaluate dominance in the studies discussed by McNaughton and Wolf are not sufficient to characterize functional or competitive dominance. In all of the communities considered by McNaughton and Wolf, the data were collected and analyzed with the assumption that the respective communities were structured solely around competitive niche differentiation and dominance. Another explanation for the differences between this study and the conclusions of McNaugton and Wolf might be that all the earlier community studies which they summarized were analyses of one trophic level, and it is risky to draw generalizations about community organization from one trophic level.

While competition seems to play a rather minor role in structuring this intertidal community, it is of interest to consider the circumstances in which competition does seem to have a significant role in influencing patterns of distribution and survival. The only species considered which is regularly exposed to strong space competition is *Chthamalus dalli*. When two of the three levels of disturbance (*Pisaster* and either *Thais* or the limpets) are removed, the *Chthamalus* population is significantly enhanced by the activities of either of the remaining disturbances (*Thais* or limpets), which exert a stronger negative effect against *Balanus*. However, this situation is artificial, as the community is naturally exposed to both the limpet and the *Thais* levels of disturbance and is usually exposed to *Pisaster* and logs. Thus, the only time that competition is naturally very important is when the dominants have attained a size refuge from limpets and *Thais* and have not been killed by *Pisaster* or logs. This has been seen to occur naturally above the level of *Pisaster* predation in areas not struck by logs, such as Turn Rock, and after the cold-induced *Thais* die-off in the winter of 1968.

The major conclusion of the present study is that although there are clear competitive dominants, this intertidal community is characterized by continuous physical and biological disturbance including the effects of carnivores and herbivores, an abundance of the potentially limiting spatial resource, and a large number of species which utilize this same resource.

ACKNOWLEDGMENTS

This paper is based on a dissertation submitted in partial fulfillment of the requirements for the Ph.D. in the Department of Zoology, University of Washington, Seattle. I thank my major advisor, Dr. Robert T. Paine, for his continued support. The comments and constructive criticism of my thesis committee, Drs. Paine, A. J. Kohn, and R. L. Fernald have helped me throughout my research program and have considerably improved drafts of this manuscript. Discussions with them and with J. Connell, L. Druehl, P. Illg, J. Kain, B. Menge, G. Orians, T. Spight, R. Vadas, and R. Vance have been particularly helpful.

Many people have helped with the field work; in particular I thank A. DeVries, J. Drescher, S. Fullilove, S. Heller, S. Jordan, N. Lellelid, S. Louda, M. McKey, J. P. Mauck, and B. Rabbit. M. Harlin, P. Lebednik, and R. Norris have kindly helped me with algal taxonomy.

I wish to further acknowledge the help of Dr. R. L. Fernald, who provided laboratory space and facilities

388   PAUL K. DAYTON   Ecological Monographs
                                                               Vol. 41, No. 4

at Friday Harbor Laboratories. Research support has been provided by an NSF Marine Science Training Grant to Friday Harbor Laboratories, an NSF Ecology Training Grant to the Department of Zoology, and two NSF grants to R. T. Paine. While preparing this manuscript I have received support from Sea Grant GH-112 awarded to University of California, San Diego.

The many talents and patient, conscientious help of my wife, Linnea, are apparent in every page of this paper. She helped with the field work, the statistics, and the figures; for all this and much more I am very grateful.

LITERATURE CITED

Bakus, G. J. 1964. The effects of fish-grazing on invertebrate evolution in shallow tropical waters. Allan Hancock Foundation. Occas. Pap. 27: 1–29.

Barnes, H., and E. S. Reese. 1960. The behaviour of the stalked intertidal barnacle Pollicipes polymerus J. B. Sowerby, with special reference to its ecology and distribution. J. Anim. Ecol. 29: 169–185.

Bascom, W. 1964. Waves and beaches, the dynamics of the ocean surface. Anchor Books, Doubleday & Co., Inc. Garden City, N. Y. 267 p.

Bayne, B. L. 1964. Primary and secondary settlement in Mytilus edulis L. (Mollusca). J. Anim. Ecol. 33: 513–523.

————. 1965. Growth and the delay of metamorphosis of the larvae of Mytilus edulis (L.). Ophelia 2: 1–47.

Blegvad, H. 1928. Quantitative investigation of bottom invertebrates in the Limfjord 1910-1927 with special reference to plaice food. Rep. Danish Biol. Sta. 34: 33–52.

Castenholz, R. W. 1961. The effect of grazing on marine littoral diatom populations. Ecology 42: 783–794.

Connell, J. H. 1961a. The influence of interspecific competition and other factors on the distribution of the barnacle Chthamalus stellatus. Ecology 42: 710–723.

————. 1961b. Effect of competition, predation by Thais lapillus, and other factors on natural populations of the barnacle Balanus balanoides. Ecol. Monogr. 31: 61–104.

————. 1970. A predator-prey system in the marine intertidal region. 1. Balanus glandula and several predatory species of Thais. Ecol. Monogr. 40: 49–78.

Crisp, D. J. 1964. An assessment of plankton grazing by barnacles. In D. J. Crisp [ed.] Grazing in terrestrial and marine environments. Brit. Ecol. Soc. Symposium 4, Blackwell, Oxford.

Daubenmire, R. 1966. Vegetation: identification of typal communities. Science 151: 291–298.

Druehl, L. D. 1967. Vertical distributions of some benthic algae in a British Columbia inlet as related to some environmental factors. J. Fish. Res. Bd. Can. 24: 33–46.

Elton, C. 1946. Competition and the structure of ecological communities. J. Anim. Ecol. 15: 54–68.

Elton, C., and R. S. Miller. 1954. The ecological survey of animal communities with a practical system of classifying habitats by structural characters. J. Ecol. 42: 460–496.

Emlen, J. M. 1966. The role of time and energy in food preference. Amer. Natur. 100: 611–617.

Endean, R., R. Kenny, and W. Stephenson. 1956. The ecology and distribution of intertidal organisms on rocky shores of the Queensland mainland. Australian J. Mar. Freshw. Res. 7: 88–146.

Fager, E. W. 1968. A sand-bottom epifaunal community of invertebrates in shallow water. Limnol. Oceanogr. 13: 448–464.

Frank, P. W. 1965. The biodemography of an intertidal snail population. Ecology 46: 831–844.

Glynn, P. W. 1965. Community composition, structure and interrelationship in the marine intertidal Endocladia muricata-Balanus glandula association in Monterey Bay, California. Beaufortia 12: 1-198.

Harger, J. R. E. 1968. The role of behavioral traits in influencing the distribution of two species of sea mussel, Mytilus edulis and Mytilus californianus. The Veliger 11: 45–49.

————. 1970. The effect of wave impact on some aspects of the biology of sea mussels. The Veliger 12: 401–414.

Hatton, H. 1938. Essais de bionomie explicative sur queleues especes intercotidales d'algues et d'animaux. Ann. Inst. Monaco 17: 241–348.

Hutchinson, G. E. 1951. Copepodology for the ornithologist. Ecology 32: 571–577.

Hutchinson, G. E., and E. S. Deevey. 1949. Survey of biological progress. Vol. 8. Academic Press, New York. 325 p.

Jones, N. S. 1948. Observations and experiments on the biology of Patella vulgata at Port St. Mary, Isle of Man. Proc. Liverpool Biol. Soc. 56: 60–77.

Kensler, C. B. 1967. Desiccation resistance of intertidal crevice species as a factor in their zonation. J. Anim. Ecol. 36: 391–406.

Levins, R. 1968. Evolution in changing environments. Monographs in Population Biology (2), Princeton Univ. Press, Princeton, N. J.

Lewis, J. R. 1954. Observations on a high-level population of limpets. J. Anim. Ecol. 23: 85-100.

————. 1964. The ecology of rocky shores. English Univ. Press Ltd., London. 323 p.

Lodge, S. M. 1948. Algal growth in the absence of Patella on an experimental strip of foreshore, Port St. Mary, Isle of Man. Proc. Liverpool Biol. Soc. 56: 78–83.

Louda, S. M. 1968. Characterization of field populations of Searlesia dira Reeve. Unpublished Zool. 533 Report, Friday Harbor Laboratories, Friday Harbor, Washington.

Mauzey, K. P. 1966. Feeding behavior and reproductive cycles in Pisaster ochraceus. Biol. Bull. 131: 127–144.

————. 1967. The interrelationship of feeding, reproduction, and habitat variability in Pisaster ochraceus. Ph.D. Thesis. Univ. Washington, Seattle, Wash.

McLellan, R. D. 1927. The geology of the San Juan Islands. Univ. Wash. Publ. Geol. 2: 1–185.

McNaughton, S. J., and L. L. Wolf. 1970. Dominance and the niche in ecological systems. Science 167: 131–142.

Menge, B. A. 1970. The population ecology and community role of the predaceous asteroid, Leptasterias hexactis (Stimpson). Ph.D. Thesis. Univ. Washington, Seattle, Wash.

Moore, H. B. 1938. Algal production and food requirements of a limpet. Proc. Malacol. Soc. London 23: 117-118.

————. 1939. The colonization of a new rocky shore at Plymouth. J. Anim. Ecol. 8: 29–38.

Paine, R. T. 1966. Food web complexity and species diversity. Amer. Natur. 100: 65-75.

————. 1969. The Pisaster-Tegula interaction: prey patches, predator food preference, and intertidal community structure. Ecology 50: 950–961.

Randall, J. E. 1961. Overgrazing of algae by herbivorous marine fishes. Ecology **42**: 812.

——. 1963. An analysis of the fish populations of artificial and natural reefs in the Virgin Islands. Carib. J. Sci. **3**: 1–16.

Ricketts, E. F., J. Calvin, and J. W. Hedgpeth. 1968. Between Pacific tides. 4th ed. Stanford Univ. Press, Stanford, Calif. 614 p.

Rigg, G. B., and R. C. Miller. 1949. Intertidal plant and animal zonation in the vicinity of Neah Bay, Washington. Proc. Calif. Acad. Sci. **26**: 323–351.

Root, R. B. 1967. The niche exploitation pattern of the blue-gray gnatcatcher. Ecol. Monogr. **37**: 317–350.

Siegel, S. 1956. Nonparametric statistics for the behavioral sciences. McGraw-Hill Book Co., Inc., New York. 312 p.

Snedecor, G. W. 1966. Statistical methods. Iowa State Coll. Press, Ames, Iowa. 534 p.

Southward, A. J. 1953. The ecology of some rocky shores in the south of the Isle of Man. Proc. Liverpool Biol. Soc. **59**: 1–30.

——. 1956. The population balance between limpets and seaweeds on wave-beaten rocky shores. Rep. Mar. Biol. Sta. Pt. Erin, No. **68**: 20–29.

Stephenson, T. A., and A. Stephenson. 1961. Life between tidemarks in North America, IVa: Vancouver Island, I. J. Ecol. **49**: 1–29.

Stephenson, W., and R. B. Searles. 1960. Experimental studies on the ecology of intertidal environments at Heron Island. 1. Exclusion of fish from beach rock. Australian J. Mar. Freshw. Res. **11**: 241–267.

Stimson, J. 1970. Territorial behavior of the owl limpet, *Lottia gigantea*. Ecology **51**: 113–118.

Swan, J. S. 1857. The Northwest Coast; or three years' residence in Washington Territory. Harper and Brothers, New York. 435 p.

Whittaker, R. H. 1965. Dominance and diversity in land plant communities. Science **147**: 250–260.

——. 1969. Evolution of diversity in plant communities. *In* Diversity and ability in ecological systems. Brookhaven Symposia in Biology, Number 22. Upton, N. Y.

ERRATUM

The following references were omitted from the end of the article "Succession after fire in the chaparral of southern California" by Ted L. Hanes, appearing in Ecological Monographs 41:27–52, Winter 1971:

Visher, S. S. 1966. Climatic atlas of the United States. Harvard Univ. Press, Cambridge, Mass. 403 p.

Vlamis, J., E. C. Stone, and C. L. Young. 1954. Nutrient status of brushland soils in southern California. Soil Sci. **78**: 51–55.

Wells, P. V. 1962. Vegetation in relation to geological substratum and fire in the San Luis Obispo Quadrangle, California. Ecol. Monogr. **32**: 79–103.

Went, F. W., G. Juhren, and M. C. Juhren. 1952. Fire and biotic factors affecting germination. Ecology **33**: 351–364.

Wieslander, A. E., and C. H. Gleason. 1954. Major brushland areas of the coast ranges and Sierra-Cascade foothills in California. U.S. Forest Serv., Calif. Forest and Range Exp. Sta. Misc. Paper 15. 9 p.

niques is taking place, and new information is already yielding fresh insights into chemical, physical, and biological systems.

References and Notes

1. B. M. Kincaid, *J. Appl. Phys.* **48**, 2684 (1977); J. P. Blewett and R. Chasman, *ibid.*, p. 2692.
2. D. A. G. Deacon, L. R. Elias, J. M. J. Madey, G. J. Ramian, H. A. Schwettman, T. I. Smith, *Phys. Rev. Lett.* **38**, 892 (1977).
3. D. Iwanenko and I. Pomeranchuk, *Phys. Rev.* **65**, 343 (1944).
4. J. P. Blewett, *ibid.* **69**, 87 (1946).
5. J. S. Schwinger, *ibid.* **70**, 798 (1946); *ibid.* **75**, 1912 (1949).
6. B. Alpert and R. López-Delgado, *Nature (London)* **263**, 445 (1976).
7. O. B. d'Azy, R. López-Delgado, A. Tramer, *Chem. Phys.* **9**, 327 (1975).
8. L. F. Wagner and W. E. Spicer, *Phys. Rev. Lett.* **28**, 1381 (1972).
9. D. E. Eastman and W. D. Grobman, *ibid.*, p. 1379.
10. I. Lindau, in *Synchrotron Radiation Research*, A. N. Mancini and I. F. Quercia, Eds. (International Colloquium on Applied Physics, Istituto Nazionale di Fisica Nucleare, Rome, in press).
11. R. J. Smith, J. Anderson, G. J. Lapeyre, *Phys. Rev. Lett.* **37**, 1081 (1976).
12. E. Spiller, R. Feder, J. Topalian, D. Eastman, W. Gudat, D. Sayre, *Science* **191**, 1172 (1976); R. Feder, E. Spiller, J. Topalian, A. N. Broers, W. Gudat, B. J. Panessa, Z. A. Zadunaisky, J. Sedat, *ibid.* **197**, 259 (1977); E. Spiller, D. E. Eastman, R. Feder, W. D. Grobman, W. Gudat, J. Topalian, *J. Appl. Phys.* **47**, 5450 (1976); E. Spiller and R. Feder, in *X-ray Optics*, H. J. Quesser, Ed. (Springer-Verlag, Berlin, in press); B. Fay, J. Trotel, Y. Petroff, R. Pinchaux, P. Thiry, *Appl. Phys. Lett.* **29**, 370 (1976). For a general review of x-ray lithography with conventional sources see S. E. Bernacki and H. I. Smith, *IEEE Trans. Electron Devices*, **ED-22**, 421 (1975).
13. P. Horowitz and J. A. Howell, *Science* **178**, 608 (1972).
14. B. M. Kincaid, P. Eisenberger, K. O. Hodgson, S. Doniach, *Proc. Natl. Acad. Sci. U.S.A.* **72**, 2340 (1975).
15. K. D. Watenpaugh, L. C. Sieker, J. R. Herriott, L. H. Jensen, *Acta Crystallogr. Sect. B* **29**, 943 (1973).
16. See, for example, R. G. Shulman, P. Eisenberger, W. E. Blumberg, N. A. Stombaugh, *Proc. Natl. Acad. Sci. U.S.A.* **72**, 4003 (1975).
17. R. S. Goody, K. C. Holmes, H. G. Mannherz, J. Barrington Leigh, G. Rosenbaum, *Biophys. J.* **15**, 687 (1975).
18. T. Tuomi, K. Naukkarian, P. Rabe, *Phys. Status Solidi A* **25**, 93 (1974); M. Hart, *J. Appl. Crystallogr.* **8**, 436 (1975); J. Bordas, S. M. Glazer, H. Hauser, *Philos. Mag.* **32**, 471 (1975); B. Buras and J. Staun Olsen, *Nuc. Instrum. Methods* **135**, 193 (1976).
19. Research was performed at Brookhaven National Laboratory under contract with ERDA.

Diversity in Tropical Rain Forests and Coral Reefs

High diversity of trees and corals is maintained only in a nonequilibrium state.

Joseph H. Connell

The great variety of species in local areas of tropical rain forests and coral reefs is legendary. Until recently, the usual explanation began with the assumption that the species composition of such assemblages is maintained near equilibrium (*1*). The question thus became: "how is high diversity maintained near equilibrium?" One recent answer

communities is a consequence of past and present interspecific competition, resulting in each species occupying the habitat or resource on which it is the most effective competitor. Without perturbation this species composition persists; after perturbation it is restored to the original state (*3*).

In recent years it has become clear

Summary. The commonly observed high diversity of trees in tropical rain forests and corals on tropical reefs is a nonequilibrium state which, if not disturbed further, will progress toward a low-diversity equilibrium community. This may not happen if gradual changes in climate favor different species. If equilibrium is reached, a lesser degree of diversity may be sustained by niche diversification or by a compensatory mortality that favors inferior competitors. However, tropical forests and reefs are subject to severe disturbances often enough that equilibrium may never be attained.

for tropical bird communities is given as follows: "The working hypothesis is that, through diffuse competition, the component species of a community are selected, and coadjusted in their niches and abundances, so as to fit with each other and to resist invaders" (*2*). In this view, the species composition of tropical

that the frequency of natural disturbance and the rate of environmental change are often much faster than the rates of recovery from perturbations. In particular, competitive elimination of the less efficient or less well adapted species is not the inexorable and predictable process we once thought it was. Instead, other

forces, often abrupt and unpredictable, set back, deflect, or slow the process of return to equilibrium (*4*). If such forces are the norm, we may question the usefulness of the application of equilibrium theory to much of community ecology.

In this article I examine several hypotheses concerning one aspect of community structure, that is, species richness or diversity (*5*). I first explore the view that communities seldom or never reach an equilibrium state, and that high diversity is a consequence of continually changing conditions. Then I discuss the opposing view that, once a community recovers from a severe perturbation, high diversity is maintained in the equilibrium state by various mechanisms.

Here I apply these hypotheses to organisms such as plants or sessile animals that occupy most of the surface of the land or the firm substrates in aquatic habitats. I consider two tropical communities, rain forests and coral reefs, concentrating on the organisms that determine much of the structure, in these cases, trees and corals. Whether my arguments apply to the mobile species, such as insects, birds, fish, and crabs, that use these structures as shelter or food, or to nontropical regions, remains to be seen. I deal only with variations in diversity within local areas, not with large-scale geographical gradients such as tropical to temperate differences. While the hypotheses I present may help explain them, such gradients are just as likely to be produced by mechanisms not covered in the present article (*6*).

Various hypotheses have been proposed to explain how local diversity is produced or maintained (or both). I have reduced the number to six, which fall into two general categories:

Joseph H. Connell is a professor of biology at the University of California, Santa Barbara 93106.

0036-8075/78/0324-1302$02.00/0 Copyright © 1978 AAAS

1) The species composition of communities is seldom in a state of equilibrium. High diversity is maintained only when the species composition is continually changing. (i) Diversity is higher when disturbances are intermediate on the scales of frequency and intensity (the "intermediate disturbance" hypothesis). (ii) Species are approximately equal in ability to colonize, exclude invaders, and resist environmental vicissitudes. Local diversity depends only on the number of species available in the geographical area and the local population density (the "equal chance" hypothesis). (iii) Gradual environmental changes, that alter the ranking of competitive abilities, occur at a rate high enough so that the process of competitive elimination is seldom if ever completed (the "gradual change" hypothesis).

2) The species composition of communities is usually in a state of equilibrium; after a disturbance it recovers to that state. High diversity is then maintained without continual changes in species composition. (iv) At equilibrium each species is competitively superior in exploiting a particular subdivision of the habitat. Diversity is a function of the total range of habitats and of the degree of specialization of the species to parts of that range (the "niche diversification" hypothesis). (v) At equilibrium, each species uses interference mechanisms which cause it to win over some competitors but lose to others (the "circular networks" hypothesis). (vi) Mortality from causes unrelated to the competitive interaction falls heaviest on whichever species ranks highest in competitive ability (the "compensatory mortality" hypothesis).

Nonequilibrium Hypotheses

The intermediate disturbance hypothesis. Organisms are killed or badly damaged in all communities by disturbances that happen at various scales of frequency and intensity. Trees are killed or broken in tropical rain forests by windstorms, landslips, lightning strikes, plagues of insects, and so on; corals are destroyed by agents such as storm waves, freshwater floods, sediments, or herds of predators. This hypothesis suggests that the highest diversity is maintained at intermediate scales of disturbance (Fig. 1).

The best evidence comes from studies of ecological succession. Soon after a severe disturbance, propagules (for example, seeds, spores, larvae) of a few species arrive in the open space. Diver-

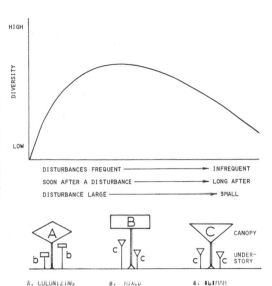

Fig. 1. The "intermediate disturbance" hypothesis. The patterns in species composition of adults and young proposed by Eggeling (8) for the different successional stages of the Budongo forest are shown diagrammatically at the bottom.

sity is low because the time for colonization is short; only those few species that both happen to be producing propagules and are within dispersal range will colonize. If disturbances continue to happen frequently, the community will consist of only those few species capable of quickly reaching maturity.

As the interval between disturbances increases, diversity will also increase, because more time is available for the invasion of more species. New species with lower powers of dispersal and slower growth, that were excluded by more frequent disturbances, can now reach maturity. As the frequency declines further and the interludes between catastrophes lengthen, diversity will decline, for one of two reasons. First, the competitor that is either the most efficient in exploiting limited resources or the most effective in interfering with other species (or both) will eliminate the rest. Second, even if all species were equal in competitive ability, the one that is the most resistant to damage or to death caused by physical extremes or natural enemies will eventually fill much of the space. This process rests on the assumption that once a site is held by any occupant, it blocks all further invasion until it is damaged or killed. Thus it competitively excludes all potential invaders, which are by assumption incapable of competitively eliminating it (7).

Thus, diversity will decline during long interludes between disturbances unless other mechanisms, such as those given in the other hypotheses below, intervene to maintain diversity. Disturbances interrupt and set back the pro-

cess of competitive elimination, or remove occupants that are competitively excluding further invaders. Thus, they keep local assemblages in a nonequilibrium state, although large geographic areas may be stable in the sense that species are gained or lost at an imperceptible rate.

Evidence that this model applies to tropical rain forests comes from several sources. Eggeling (8) classified different parts of the Budongo forest of Uganda into three stages: colonizing, mixed, and climax stands. Using observations made many years apart, he showed that the colonizing forest was spreading into neighboring grassland. In these colonizing stands the canopy was dominated by a few species (class A in Fig. 1), but the juveniles (class B in Fig. 1) were of entirely different species. Adults of class B species occurred elsewhere as canopy trees in mixed stands of many species. In these mixed stands, the juveniles were also mainly of different species (class C, Fig. 1), those with even greater shade tolerance. Adults of class C species occurred in the canopy of other climax stands where a few species dominated (mainly ironwood, *Cynometra alexandrei*, which comprised 75 to 90 percent of the canopy trees). However, in these stands, the understory was composed mainly of juveniles of the canopy species. Thus, an assemblage of self-replacing species (that is, a climax community of low diversity) had been achieved. This is not a special case; the Budongo forest is the largest rain forest in Uganda, and one-quarter of it is dominated by ironwood. Later and more ex-

Table 1. Mortality of young trees (between 0.2 and 6.1 meters tall) in relation to their abundance for two rain forests in Queensland. Not all species had enough young trees to analyze; only those whose adults were capable of reaching the canopy and that had at least six young trees are included. The mortality rate between 1965 and 1974 was plotted against the original numbers mapped in 1965; the least-squares regression slope and correlation coefficient are shown.

Site	Number of species	Regression of mortality (%) on abundance	
		Slope	r*
Tropical, North Queensland, 16°S	49	0.039	0.217
Subtropical, South Queensland, 26°S	46	0.002	0

*Neither correlation coefficient is significantly different from zero at $P < .05$

tensive surveys (9) showed that the proportion so dominated in other forests in Uganda is even higher and have confirmed that, where *Cynometra* dominates the canopy, its juveniles also dominate the understory.

Another excellent example is the work of Jones (10) in Nigeria. In this diverse tropical forest, many of the larger trees, aged about 200 years, were dying. They probably became established in abandoned fields in the first half of the 18th century, a time when the countryside was depopulated by the collapse of the Benin civilization. These trees had few offspring; most regeneration was by other species, shade-tolerant and of moderate stature. This mixed forest was in fact an "old secondary" forest that had invaded after agricultural disturbances. It was in about the same state as Eggeling's mixed forest in Uganda. In both Nigeria and Uganda, high diversity was found in a nonequilibrium intermediate stage in the forest succession.

In many studies of forest dynamics, the abundance of juvenile stages constitutes the evidence as to whether a species is expected to increase or to die out. Such inferences are of course open to the criticism that, if the mortality rate of juveniles increases with their abundance, it is not necessarily a good indicator of more successful recruitment. I tested this for young trees in two rain forest plots in Queensland that several colleagues and I have been studying since 1963 (11). Over a 9-year period, mortality showed no correlation with abundance (Table 1). Thus, it seems safe to assume that species which now have many offspring will be more abundant in the next generation of adult trees as compared to those species which now have few offspring.

In most of the mixed, highly diverse stands of tropical rain forests that have been studied, some species are represented by many large trees with few or no offspring, whereas others have a superabundance of offspring (8, 10–12). (Of course, many species are so rare as

adults that one would not expect to find many offspring.) My interpretation of this finding is that these mixed tropical forests represent a nonequilibrium intermediate stage in a succession after a disturbance, in which some species populations are decreasing whereas others are increasing. Since mixed rain forests are common in the tropics, this hypothesis suggests that disturbance is frequent enough to maintain much of the region in the nonequilibrium state.

If this is so, tropical forests dominated by a single canopy species that has abundant offspring in the understory must not have been disturbed for several generations. Such forests, similar to the ironwood climax of Eggeling (8), also occur commonly elsewhere in Africa as well as in tropical America and Southeast Asia (13). Two lines of evidence indicate that they have been less frequently disturbed than have mixed forests. First, the only papers that I have found in which the incidence of storms was described in relation to single-dominant forests state that destructive storms "never occur" in these regions (8, 14). Second, many of these forests are unlikely to have been disturbed by man, because they lie on poor soils, in swamps or along creek margins, on steep stony slopes, or on highly leached white sands (12, 15). All of these are soils that the farmers of shifting cultivation in forests avoid since they produce very poor crops (12). Such agriculture is confined mainly to the well-drained good soils, and these are the soils where the mixed diverse forests exist. Thus, mixed forests occur in the places most likely to have been disturbed by man, whereas single-dominant forests occur in those least likely to have been disturbed.

Since single-dominant forests often (though not always) lie on poor soils, it has usually been assumed that this is because only a few species have evolved adaptations to tolerate them (12, 15). However, the difference between forests on good soils and those on poor soils lies in the dominance of a single species in

the canopy rather than in the total number of species. Thus, in comparing plots in rain forests in Guyana, the commonest species constituted 16 percent of the large trees (more than 41 centimeters in diameter) on good soils and 67 percent on poor soils (leached white sands), yet the number of species of trees more than 20 centimeters in diameter was 55 and 49, respectively [table 27 in (12)]. Thus, a large number of moderate- to large-sized tree species occupy poor soils, even though only a few are common. The best evidence that single-species dominance is not necessarily due to poor soils is the example of the Budongo forest. Here, various forest stands, ranging from ones of mixed high diversity to those with single-species dominance, each occur on similar soils. Single-species dominance seems to be explained more satisfactorily by the absence of disturbance rather than by poor soil quality.

On coral reefs, the relation between disturbance and diversity is similar to that in tropical forests. At Heron Island, Queensland, the highest number of species of corals occurs on the crests and outer slopes that are exposed to damaging storms. Since I began studying this reef in 1962, two hurricanes have passed close to it, one in 1967 and one in 1972. Each destroyed much coral on the crest and outer slopes but failed to damage another slope protected by an adjacent reef. The disturbed areas have been recolonized by many species after each hurricane, but colonization has not been so dense that competitive exclusion has yet begun to reduce the diversity (Fig. 2A). Other workers on corals have witnessed the same phenomenon; disturbances caused both by the physical environment and by predation remove corals and then recolonization by many species follows (16, 17).

In contrast, in permanently marked quadrats observed over several years without disturbance at Heron Island, I found that competitive elimination of neighboring colonies was a regular feature, either by one colony overshadowing or overgrowing another, or by direct aggressive interactions (18). Here competition is by interference, rather than by more efficient exploitation of resources. On one region of the south outer slope, protected from storm disturbance by an adjacent reef, huge old colonies of a few species of "staghorn" corals occupy most of the surface (Fig. 2B). Since these are able to overshadow neighbors (18) at a height sufficient to be out of reach of the mesenteric filaments used as defenses (19), I infer that such staghorns have in fact competitively

1304

eliminated many neighbors during their growth. Here competitive elimination has apparently gone to completion, with a consequent reduction in local diversity. A similar situation has been described for Hawaii (16) and for the Pacific coast of Panama (17).

The discussion so far has concerned mainly the frequency of disturbance. However, the same reasoning applies to variations in intensity and area perturbed; diversity is highest when disturbances are intermediate in intensity or size, and lower when disturbances are at either extreme. For example, if a disturbance kills all organisms over a very large area, recolonization in the center comes only from propagules that can travel relatively great distances and that can then become established in open, exposed conditions. Species with such propagules are a small subset of the total pool of species, so diversity is low. In contrast, in very small openings, mobility is less advantageous: the ability to become established and grow in the presence of resident competitors and natural enemies is critical. In addition, recolonizing propagules are more likely to come from adults adjacent to the small opening. Therefore, colonizers will again be a small subset of the available pool of species, and diversity will tend to be low. When disturbances create intermediate-sized openings, both types of species can colonize and the diversity should be higher than at either extreme.

Not only the size, but also the intensity of disturbances makes a difference. If the disturbance was less intense so that some residents were damaged and not killed, in a large area recolonization would come both from propagules and from regeneration of survivors, so that diversity would be greater than was the case when all residents were killed and colonization came only from new propagules.

Direct evidence linking diversity with variations in intensity and total area of disturbance in tropical communities is meager. However, there is evidence that the processes described above do occur. For example, a 40-kilometer-wide swath of reef in Belize was heavily damaged by a hurricane in 1961, with lesser damage on both sides. Four years later, in the middle of the swath, new colonies of a few species were present, but the only significant frame-building corals, mainly *Acropora palmata*, were the survivors of the original storm (20). Ten years later many of the new colonies were of this species. In contrast, in the zone of lesser damage, colonies and broken fragments of many species had survived the storm

and had regenerated quickly so that recovery was complete.

Likewise, in rain forests, the size and intensity of a disturbance influences the process of recolonization. In a long-term study of a small experimental opening made in a Queensland rain forest, the most successful colonists after 12 years were either stump sprouts from survivors of the initial bulldozing or seedlings that came from adult trees at the edge of the clearing (21). Farther from the forest edge, in a much larger clearing, only species with great powers of seed dispersal had colonized (22).

It has recently been suggested (23) that in a nonequilibrium situation, any conditions that increase the population growth rates of a community of competitors should result in decreased diversity (since faster growth produces faster competitive displacement). In places with a lower rate of competitive elimination, there is also a greater chance for interruption by further disturbances. This "rate of competitive displacement" hypothesis is an extension of the intermediate disturbance hypothesis and should be true, other things being equal. How relevant it is for explaining differences in local diversity remains to be seen. However, present evidence from tropical communities does not support it. Forests on extreme soils (such as

leached white sands, heavy silt, or steep stony slopes) that have slower growth rates than those on less extreme soils have either few species or strong single-species dominance (24). Likewise, coral diversity shows little correlation with growth rates. Coral diversity varies with increasing depth, sometimes decreasing, sometimes increasing, or sometimes being greatest at intermediate depths. Coral growth rates tend to be faster at intermediate depths (24). Thus, among neither tropical rain forest trees nor corals is there a consistent correlation of diversity and growth rates, as predicted by the hypothesis.

In summary, variations in diversity between local stands of these tropical communities are more likely to be due to differences in the degree of past disturbances than to differences in the rate of competitive displacement during recovery from the disturbances. The high diversities observed in tropical rain forest trees and in corals on reefs appear to be a consequence of disturbances intermediate in the scales of frequency and intensity.

The equal chance hypothesis. In contrast to the previous model, let us assume that all species are equal in their abilities to colonize empty spaces, hold them against invaders, and survive the vicissitudes of physical extremes and natu-

Fig. 2. Species diversity of corals in the subtidal outer reef slopes at Heron Island, Queensland. (A) Changes over 11 years on one of the permanently marked plots on the north slope. The number at each point gives the years since the first census at year 0 (no censuses were made in years 3, 5, and 10). The dashed lines indicate changes caused by hurricanes in 1967 and 1972. (B) Results from line transects done 3 to 4 months after the 1972 hurricane. (△) Data from the heavily damaged north slopes; (○) data from the undamaged south slope; the line drawn by eye. Where disturbances had either great or little effect (very low or high percent cover, respectively) there were few species, with maximum numbers of species at intermediate levels of disturbance.

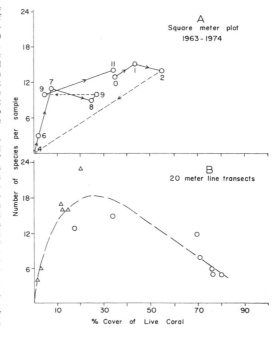

ral enemies. Then local diversity would simply be a function of the number of species available and the local population densities. The species composition at any site would be unpredictable, depending upon the history of chance colonization.

What conditions would produce this? First, for all species the number of young (such as larvae and seeds) invading empty places must be independent of the number produced by the parent population. Otherwise, any species that increased its production of offspring per parent would progressively increase at the expense of those with lesser production. Second, any occupant must be able to hold its place against invaders until it is damaged or killed. Otherwise, any species that evolved the ability to oust an occupant would also progressively increase. Last, all species must be equal in ability to resist physical extremes and natural enemies. Otherwise, the most resistant species will gradually increase, as was discussed in the previous hypothesis.

Do communities exist that satisfy these conditions? Sale (25) has proposed that certain guilds of coral reef fish do. He assumes that, as with some temperate fish, recruitment to newly vacated sites is independent of the stock of eggs released into the plankton. One must probably assume that the fecundity and mortality of all species are equal. The juveniles grow quickly after they colonize vacant places, and thus they are able to hold their territory against further invasion by smaller juveniles of any species from the plankton. Space is limiting, as judged from the rapid colonization of vacated sites. Since the juvenile fish seem to be generalists in the use of food and space, Sale suggests that local diversity would be a function of chance colonization from the available pool of species. Clearly, the initial assumption of independence of stock and recruitment is critical and needs to be tested for these tropical fish.

Likewise, for rain forest trees, Aubreville (26) has suggested that many species have such similar ecological requirements that it would be impossible to predict which subset would occur together on a site. He based this suggestion on the observation that some of the commoner large trees on his study plot in the Ivory Coast had few or no offspring on the plot. He inferred that their offspring must be elsewhere, so that the species composition of the forest would continually shift in space and time. While this might be so, his original observation of few offspring could be explained if the

1306

forest was an old secondary one, similar to Jones' (10) in Nigeria.

Other characteristics of trees and corals do not satisfy the requirements of the equal chance hypothesis. For example, dispersal of propagules of many trees and corals is quite restricted so that local recruitment of juveniles may not be as independent of local production of propagules as it apparently is in some fish populations. Likewise, species differ in fecundity, competitive ability, and resistance to environmental stresses, and the differences often result in predictable patterns of species distribution along environmental gradients (27). Therefore, it seems unlikely that either rain forest trees or corals conform to the equal chance hypothesis.

The gradual change hypothesis. This model was suggested by Hutchinson (28) to explain why many species coexist in phytoplankton assemblages. Seasonal changes in, for example, temperature and light, occur in a lake, and different species are assumed to be competitively superior at different times. It is postulated that no species has time to eliminate others before its ability to win in competition is reduced below that of another species by changes in the environment.

Climates change on all time scales from seasonal to annual to millennial and longer, and hence, this hypothesis may apply to organisms with any length of generation. With long-lived organisms such as trees or corals, gradual changes in climate over several hundred years represent the same scale as seasons do to a phytoplankton community. Drier periods producing a savanna vegetation in regions now covered with rain forest occurred about 3000 and 11,000 years ago in the Amazon basin; similar changes occurred in Africa and Australia (29). As Livingstone (30) pointed out, "Climates change and vegetational adjustments are not rare and isolated events, they are the norm." As climates changed, marine transgressions shifted and altered coral reef environments (31).

Whether such gradual transitions would also produce the highly intermingled diverse assemblages seen in present forests and reefs depends on the rate of competitive elimination compared to the rate of environmental change. If the time required for one tree species to eliminate another in competition is much shorter than the time taken for an environmental change that reversed their positions in the hierarchy, they would not coexist. Therefore, very slow changes would not maintain diversity, but higher rates might do so.

Equilibrium Hypotheses

The niche diversification hypothesis. The key point in this model is the degree of specialization to subdivisions of the habitat. For a given range of habitat variation, more species can be packed in the more they are specialized. The question is: Are the species so often observed living in diverse local assemblages sufficiently specialized to coexist at equilibrium? Some ecologists believe that motile animals have reached the required degree of specialization, particularly if different aspects of habitat subdivision are considered (32). The different aspects such as food, habitat space, and time of activity are called "niche axes."

Specialization along niche axes does not seem to have evolved to this extent in plants and in sessile aquatic animals such as corals. For long-lived organisms there exists no regular temporal variation to which they could have specialized. Plants in general have not specialized along the food niche axis. They all have similar basic resource requirements (such as light, water, carbon dioxide, and mineral nutrients). Niche subdivisions are made on degrees of tolerance to different quantities of these resources. As a consequence, plants subdivide space along gradients of quantitative variations in light, water, and nutrients. These variations are often associated with variations in elevation, slope, aspect, soil type, understory position, and so on. Exceptions to this idea are marine algae that have adapted to the qualitative changes in wavelength of light at different depths. General observations and some statistical analyses (12, 15, 33) have revealed associations between sets of species and certain subdivisions of the habitat in tropical rain forests, for example, to broad variations in soil properties (such as parent material, drainage, and tip-up mounds and hollows at the roots of fallen trees), and topography (ridges, steep slopes, creek margins, and so on). Other analyses have shown little association between species and local soil types (15, p. 188). As was discussed earlier, plants are also specialized according to differences in habitats caused by variations in the frequency and intensity of disturbance. The aftermath of a disturbance presents a new local environment in which species with different traits are at an advantage. It has been suggested that tropical trees may have subdivided this niche axis finely (34); at present there is little direct evidence to support this view. It seems unlikely that tropical trees are so highly specialized to such small differences in the local physical environment that more than 100 species of trees could coexist at equilibrium on a single hectare of rain forest. In fact, the forests closest to equilibrium are those dominated by a single tree species, as was discussed earlier.

Corals seem as general in their requirements as trees; for example, although some of their energy comes from feeding on zooplankton, much comes from photosynthesis by their symbiotic zooxanthellae, which consist of a single species in all coral species studied to date, although several different strains detected by electrophoresis show some degree of specificity (35). It has been suggested (36) that corals have differentiated along the food niche axis between the extremes of autotrophy and heterotrophy. However, in shallow water where both light and species diversity is high, this differentiation could promote the coexistence of several species on the same space only if they were stratified vertically, autotrophs above, heterotrophs in the understory. Yet the layering observed thus far has not revealed specialized "shade" species adapted for life in the understory. Corals have been seen beneath open-branched species such as the Caribbean *Acropora cervicornis* (37) but, to my knowledge, never beneath close-branched species. One might expect that hetertrophy would be advantageous where light is reduced by deeper water. Yet there is evidence that a predominantly autotrophic coral was able, over a day's time, to meet its energy requirements down to a water depth of 25 meters (38). Thus, the proposed niche differentiation along the food axis has apparently contributed little to coexistence, and corals seem very generalized in their use of resources. On the habitat niche axis, corals are also generalized. Although some species are confined to certain zones, most corals have broad ranges of distribution with respect to depth and location on reefs, which indicates little precise specialization in habitat (39). Thus, like rain forest trees, corals do not seem to have specialized to the degree required to maintain the observed high diversities at equilibrium.

The circular networks hypothesis. This model suggests that, instead of the linear and transitive hierarchy (A eliminates B, B eliminates C, implying A eliminates C) presumed in the other hypotheses, the competitive hierarchy is circular (A > B > C, but C eliminates A directly). This hypothesis was first applied to sessile invertebrates living beneath ledges on coral reefs (40). Since it seems unlikely that the same competitive

mechanism could apply throughout such circular interactions, the reverse pathway acting against the highest ranked species is likely to be a different mechanism. For example, if species A and B overshadow the species below them in the hierarchy, but C poisons A, the network is biologically more plausible. A difficulty arises if the interactions are not exactly balanced: if A eliminated B first, then C, no longer reduced by B, would quickly eliminate A. However, if the species in this network competed only in pairs, none would be eliminated.

I tested this hypothesis (18) for interactions between adjacent coral colonies on a permanently marked plot (12 species, 55 colonies, 82 interactions observed over 9 years) and found no circular pathways, even though two mechanisms, overshadowing and direct extracoelenteric digestion, were acting. It is more likely that these networks would operate between more distantly related organisms. The original observations involved different phyla of invertebrates (40).

Among trees of the rain forest this hypothesis has not been examined. Shading, root competition, and allelopathy are different mechanisms, so that some circular networks might be possible. However, trees may also be too similar for this to maintain diversity.

The compensatory mortality hypothesis. If mortality falls most heavily on whichever species is ranked highest in competitive ability, or, if they are all of approximately equal rank and it falls heaviest on whichever species is commonest (that is, mortality is frequency-dependent), competitive elimination may be prevented indefinitely. In tropical forests, if herbivores attack and kill seeds or seedlings of common species more frequently and to a greater extent than they attack those of less common or rare species competitive elimination could be prevented. For example, if herbivores attack the offspring of a species more heavily nearer than farther from the parent tree, that species would probably not be able to form a single-species grove (41). This possibility has been tested by either observations or field experiments and rejected for four out of five species of seeds of rain forest trees and vines, but not rejected for seedlings in two other species (41, 42). In the analysis reported in Table 1, the mortality of seedlings or saplings did not increase significantly with their abundance. Thus, mortality of young trees is not generally frequency-dependent. Destruction of trees by elephants also does not seem to be compensatory. In Ugandan rain forests it has

been observed (43) that elephants preferentially destroy young of the fast-growing early and middle succession trees, leaving the young of the late succession ironwood alone, thus hastening progression toward the low-diversity forest. Therefore, contrary to my own earlier work on this aspect (41), I feel that while compensatory mortality may occur in some instances it does not seem to be a generally important factor in maintaining the high diversity of mixed tropical rain forests.

Watt's (44) "cyclic succession" is probably an example of this mechanism. The dominant species does not replace itself; other species intervene before the dominant becomes reestablished. This process has never, to my knowledge, been demonstrated in the tropics, but there seems no reason not to expect it to happen there.

In coral reefs, some predation does not act in a frequency-dependent fashion. An earlier claim (45) that the starfish *Acanthaster planci* might act in this way has now been demonstrated to be in error. Studies in Hawaii and Panama (46) indicate that the starfish attacks rarer species preferentially, which would reduce diversity. Studies in a much more diverse coral community on Saipan (47) suggest that *Acanthaster* might eliminate certain preferred species, although no data were given which indicated whether these were the common or rare species. In Panama, evidence indicates that other types of predators may possibly act in a compensatory manner, increasing diversity (17, 46).

In my studies of corals I found that the physical environment can inflict mortality in a manner that compensates for the competitive advantage of branching species that overshadow others. I measured the mortality of corals over a 4-year period that included a hurricane at Heron Island in Queensland (18). As described above, I had ranked these species in competitive ability by observing dynamic interactions over a period of 9 years on permanent quadrats. On the part of the reef crest that was badly damaged by the hurricane, the mortality of those species that ranked high in the competitive hierarchy was much greater than those that ranked low. In contrast, the high-ranked species on an undamaged part of the reef crest had a lower mortality than low-ranked species, over the same period. The reason for the difference was that the high-ranked corals were branching species observed to grow above their neighbors, overshadowing and thus killing them. However, these branching species were more heavily damaged in the storms. Thus, species of corals that ordinarily win in competition suffer proportionately more from storm damage, compensating for their advantage.

In certain situations, diversity, instead of decreasing with high coral cover (Fig. 2), increases. This occurs on the very shallow reef crests at Heron Island and is due to compensatory mortality. The larger colonies that are spreading horizontally and eliminating their neighbors tend to die in the center, where they have grown up above the low tide level. This provides open spaces in which new species can colonize. Thus, on the reef crest no species is capable of monopolizing the space, in contrast to the slope situations shown in Fig. 2B.

High diversity at high cover has also been found in the Caribbean, and it was proposed that a balance in competitive abilities exists at equilibrium (48). Last, it occurred in the deepest samples at Eilat, Red Sea (49). Although no explanation was suggested for this last-mentioned instance, it and the Caribbean one could be explained by the intermediate disturbance hypothesis. In both cases, the slope is steep where diversity is highest. In such places, small-scale disturbances occur by slumping of coral blocks (17, 46). The deep corals at Eilat are very small (more than 100 colonies in some 10-meter line transects), which might indicate that they are recent colonists after local disturbances.

Tests of the Hypotheses

Hypotheses are made to be tested, and, in ecology, field experiments are often an excellent way to do so (50). The intermediate disturbance hypothesis can be tested in various ways. It will be necessary to verify that the sequence observed by Eggeling (8) also occurs in other rain forests. Probably the best way to do this would be to examine gaps of various sizes within forests dominated by a single species. In very small openings the shade-tolerant offspring of the dominant should grow and survive better than other species, whereas in larger openings, juveniles of less shade-tolerant species should perform better. To estimate the probability of replacement, one would need to measure the abundance and sizes of each species having juveniles in the light gap, and if possible their rates of growth and mortality.

Even better than such observations are experimental transplants into different-sized light gaps of seedlings of species whose adults live in mixed and in

single-dominant stands. These experiments would test the prediction that the species of the single-dominant stands will be more successful in small openings near the parent tree, whereas those of mixed stands will be more successful in larger openings. The alternative hypothesis, that single-dominant stands are due to poor soils, could be tested by experimentally improving soils (by draining, for example, or by fertilizing) and then planting seedlings of species that do or do not live in poor soils, in these plots and in unmodified control plots.

Tests of the equal chance hypothesis involve determining whether recruitment is (i) independent of adult stock, and (ii) equal among the different species. This is a difficult problem if propagules are distributed widely. In addition, equality in ability to resist invaders, extremes of the physical environment, and natural enemies must be established. Sale (25) has made a start on this in his experiments with coral reef fish.

The hypothesis of continual change is difficult to test because of the impracticality of determining the fate of organisms as long-lived as trees or corals. Pollen records in lake sediments are seldom precise enough to distinguish species, although genera are often identifiable.

Attempts to test the niche-diversification hypothesis are sometimes made by postulating how the different species could divide up resources and then seeing whether the coexisting species overlap significantly in their use of resources. The degree of overlap is sometimes judged indirectly by the range of variation in those aspects of morphology associated with resource use, such as root depth in plants, or degree of branching and polyp size in corals. However, these indirect measures are open to the criticism that the particular resource chosen (or the structure used to indicate it) may not be the one for which the species are competing. Another criticism is that competition may not be taking place through superior efficiency in exploiting resources, but by superior ability in interfering with competitors. Until a precise definition of the range of resources of each species is specified, this hypothesis will remain untestable.

The circular networks hypothesis might be tested either by observing as many interactions as possible, or better, by transplanting individuals into mixed and single-species groups. Since circular networks are apparently rare, many replicate observations must be made if such a network is found. For example, if a single set of observations indicated that

(A > B > C > A), further observations might uncover an instance where (A > C), indicating "equal chance," which I found in observing coral interactions (18).

The compensatory mortality hypothesis can be tested in various ways. Observations of density and mortality before and after storms or predator attacks would reveal whether highly ranked species suffered greater mortality (18). Experiments in which seeds were placed both near and far from adult trees, or in both dense clumps and sparsely, have been done with several species of tropical trees (41, 42). Observations on the mortality of naturally occurring seedlings and on experimental plantings both near and far from adults have also been made (11). In addition, I have used cages to exclude insects and larger herbivores from seeds and seedlings, using open-sided cages as controls. The purpose was to establish whether natural enemies act in a compensatory way. The experiments done so far should be regarded as pilot ones, since they were done with few replicates on a few species. More experiments need to be done before the role of compensatory mortality can be established.

Conclusion

This article discusses two opposing views of the organization of assemblages of competing species such as tropical trees or corals. One is that stability usually prevails, and, when a community is disturbed, it quickly returns to the original state. Natural selection fits and adjusts species into this ordered system. Therefore, ecological communities are highly organized, biologically accommodated, coevolved species assemblages in which efficiency is maximized, life history strategies are optimized, populations are regulated, and species composition is stabilized. Tropical rain forests and coral reefs are generally regarded as the epitomes of such ordered systems. The last three hypotheses presented in this article detail the mechanisms that may maintain these systems.

In the contrasting view, equilibrium is seldom attained: disruptions are so common that species assemblages seldom reach an ordered state. Communities of competing species are not highly organized by coevolution into systems in which optimal strategies produce highly efficient associations whose species composition is stabilized. The first three hypotheses represent this view.

My argument is that the assemblages

of those organisms which determine the basic physical structure of two tropical communities (rain forest trees and corals) conform more closely to the nonequilibrium model. For these organisms, resource requirements are very general: inorganic substances (water, carbon dioxide, minerals) plus light and space, and, for corals, some zooplankton. It is highly unlikely that these can be partitioned finely enough to allow 100 or more species of trees to be packed, at equilibrium, on a single hectare (12). Instead, if competition is allowed to proceed unchecked, a few species eliminate the rest. The existence of high local diversity in the face of such overlap in resource requirements is a problem only if one assumes equilibrium conditions. Discard the assumption and the problem vanishes.

Although I have presented these ideas as separate hypotheses, they are not mutually exclusive. Within a local area, there are usually enough variations in habitats and resources to enable several species to coexist at equilibrium as a result of niche differentiation. In addition, a certain amount of compensatory mortality probably occurs, as some evidence from rain forests indicates (41, 42). In special circumstances, circular networks might also increase diversity. Thus, a certain amount of local diversity would exist under equilibrium conditions.

However, climates do change gradually, which probably results in changes in the competitive hierarchy. On a shorter time scale, disturbances frequently interrupt the competitive process. These variations prevent most communities from ever reaching equilibrium. In certain special cases, species may be so alike in their competitive abilities and life history characteristics that diversity is maintained by chance replacements.

Thus, all six hypotheses may contribute to maintaining high diversity. My contention is that the relative importance of each is very different. Rather than staying at or near equilibrium, most local assemblages change, either as a result of frequent disturbances or as a result of more gradual climatic changes. The changes maintain diversity by preventing the elimination of inferior competitors. Without gradual climatic change or sudden disturbances, equilibrium may be reached; diversity will then be maintained by the processes described in the hypotheses of niche diversification, of circular networks, and of compensatory mortality, but at a much lower level than is usually observed in diverse tropical forests and coral reefs.

Although tropical rain forests and cor-

al reefs require disturbances to maintain high species diversity, it is important to emphasize that adaptation to these natural disturbances developed over a long evolutionary period. In contrast, some perturbations caused by man are of a qualitatively new sort to which these organisms are not necessarily adapted. In particular, the large-scale removal of tropical forests with consequent soil destruction (51), or massive pollution by biocides, heavy metals, or oil, are qualitatively new kinds of disturbances, against which organisms usually have not yet evolved defenses. Tropical communities are diverse, thus species populations are usually smaller than those in temperate latitudes, which increases the chances that such new disturbances will cause many species extinctions.

References and Notes

1. Equilibrium of species composition is usually defined as follows: (i) if perturbed away from the existing state (equilibrium point or stable limit cycle), the species composition would return to it; (ii) without further perturbations, it persists in the existing state. A perturbation is usually regarded as a marked change; death and replacement of single trees or coral colonies would not qualify.
2. J. Diamond, in Ecology and Evolution of Communities, M. L. Cody and J. Diamond, Eds. (Belknap, Cambridge, Mass., 1975), p. 343.
3. For discussions on ecological stability, natural balance, and related topics, see F. E. Clements, Carnegie Inst. Wash. Publ. 242, 1 (1916); A. J. Nicholson, J. Anim. Ecol. 2, 132 (1933); L. B. Slobodkin, Growth and Regulation of Animal Populations (Holt, Rinehart and Winston, New York, 1961), p. 46; R. M. May, Stability and Complexity in Model Ecosystems (Princeton Univ. Press, Princeton, N.J., 1973); R. M. May, Ed., Theoretical Ecology (Saunders, Philadelphia, 1976), pp. 158–162.
4. For discussions of nonequilibrium communities, see H. A. Gleason [Bull. Torrey Bot. Club 43, 463 (1917)] and H. G. Andrewartha and L. C. Birch [The Distribution and Abundance of Animals (Univ. of Chicago Press, Chicago, 1954), pp. 648–665]. The case for the importance of catastrophes in keeping forests away from an equilibrium state is discussed in several of the subsequent references. See also J. D. Henry and J. M. A. Swan, Ecology 55, 772 (1974); H. E. Wright and M. L. Heinselman, Quat. Res. (N.Y.) 3, 319 (1973). For corals, see D. W. Stoddart [in Applied Coastal Geomorphology, J. A. Steers, Ed. (Macmillan, New York, 1971), pp. 155–197; Nature (London) 239, 51 (1972)]. The case for catastrophes on a geological scale is convincingly presented by C. Vita-Finzi [Recent Earth History (Wiley, New York, 1973)] and D. V. Ager [The Nature of the Stratigraphical Record (Wiley, New York, 1973)].
5. It has been suggested that the term diversity be restricted to measures that include the relative abundance of species. However, since species number is certainly an indicator of diversity in the common usage of the word and since it is almost always closely correlated with indices based on relative abundance (16), I use the number of species as a measure of diversity.
6. R. W. Osman and R. B. Whitlatch, Paleobiology, in press.
7. A recent summary of various models of ecological succession is given by J. H. Connell and R. O. Slatyer [Am. Nat. 111, 1119 (1977)].
8. W. J. Eggeling, J. Ecol. 34, 20 (1947).
9. Later surveys are summarized by I. Langdale-Brown, H. A. Osmaston, and J. G. Wilson [The Vegetation of Uganda and Its Bearing on Land-Use (Government of Uganda, Kampala, 1964)]; they point out that in certain smaller forests in Uganda, the ironwood occurs in pure stands mainly on poorer soils. However, this is not the case in the Budongo forest, by far the most extensive in Uganda.
10. E. W. Jones, J. Ecol. 44, 83 (1956).
11. J. H. Connell, J. G. Tracey, L. J. Webb, in preparation.

12. P. W. Richards, The Tropical Rain Forest (Cambridge Univ. Press, Cambridge, 1952).
13. Forests dominated by single species are common in northern South America; examples are species of Mora, Eperua, Ocotea, Dicymbe, Dimorphandra, Aspidosperma, and Peltogyne. In Africa, the dominants are species of Macrolobium, Cynometra, Berlinia, Brachystegia, Tessmannia, and Parinari. In Southeast Asia, they are species of Eusideroxylon, Dryobalanops, Shorea, and Diospyros. It is important to distinguish climax stands from colonizing forests that are also often dominated by a single canopy species that, in contrast, has few or no offspring in the understory (Fig. 1).
14. T. A. W. Davis, J. Ecol. 29, 1 (1941).
15. T. C. Whitmore, Tropical Rain Forests of the Far East (Clarendon, Oxford, 1975).
16. R. W. Grigg and J. E. Maragos, Ecology 55, 387 (1974); Y. Loya, ibid. 57, 278 (1976).
17. P. W. Glynn, R. H. Stewart, J. E. McCosker, Geol. Rundsch. 61, 483 (1972).
18. J. H. Connell, in Coelenterate Ecology and Behavior, G. O. Mackie, Ed. (Plenum, New York, 1976), pp. 51–58.
19. J. Lang, Bull. Mar. Sci. 23, 260 (1973).
20. D. R. Stoddart, in Proceedings of the Second International Coral Reef Symposium, P. Mather, Ed. (Great Barrier Reef Committee, Brisbane, Australia, 1974), vol. 2, p. 473.
21. L. J. Webb, J. G. Tracey, W. T. Williams, J. Ecol. 60, 675 (1972).
22. M. Hopkins, thesis, University of Queensland, Brisbane, Australia (1976).
23. M. Huston, Am. Nat., in press.
24. For variation in tropical tree diversity and growth rates, see (12) and D. H. Janzen, Biotropica 6, 69 (1974). Coral diversity may either increase or decrease with depth or be highest at intermediate depths: see J. W. Wells, U.S. Geol. Surv. Prof. Pap. 260, 385 (1954); D. R. Stoddart, Biol. Rev. 44, 433 (1969); J. E. Maragos, Pac. Sci. 28, 257 (1974); T. F. Dana, thesis, University of California at San Diego (1975); Y. Loya, Mar. Biol. 13, 100 (1972); T. J. Done, in Proceedings of the Third International Coral Reef Symposium, D. L. Taylor, Ed. (Univ. of Miami, Fla., 1977), vol. 1, p. 9. Growth of corals has been found to be greater at intermediate depths: see P. H. Baker and J. N. Weber, Earth Planet. Sci. Lett. 27, 57 (1975); S. Neudecker, in Proceedings of the Third International Coral Reef Symposium, D. L. Taylor, Ed. (Univ. of Miami, Fla., 1977), vol. 1, p. 317.
25. P. Sale, Am. Nat. 111, 337 (1977).
26. A. Aubreville, Ann. Acad. Sci. Colon. Paris 9, 1 (1938).
27. R. H. Whittaker, Taxon 21, 213 (1972).
28. G. E. Hutchinson, Am. Nat. 75, 406 (1941); ibid. 95, 137 (1961).
29. J. Haffer, Science 165, 131 (1969); B. S. Vuilleumier, ibid. 173, 771 (1971); A. Kearst, R. L. Crocker, C. S. Christian, Biogeography and Ecology in Australia Monogr. Biol. 8 (1959).
30. D. A. Livingstone, Ann. Rev. Ecol. Syst. 6, 249 (1975).
31. D. R. Stoddart, Symp. Zool. Soc. London 28, 3 (1971); J. G. Tracey and H. S. Ladd, in Proceedings of the Second International Coral Reef Symposium, P. Mather, Ed. (Great Barrier Reef Committee, Brisbane, Australia, 1974), vol. 2, p. 537; D. Hopley, in ibid., p. 551.
32. T. W. Schoener, Science 185, 27 (1974).
33. W. T. Williams, G. N. Lance, L. J. Webb, J. G. Tracey, J. H. Connell, J. Ecol. 57, 635 (1969); M. P. Austin, P. S. Ashton, P. Grieg-Smith, ibid. 60, 305 (1972).
34. R. Ricklefs, Am. Nat. 111, 376 (1977).
35. D. A. Schoenberg and R. K. Trench, in Coelenterate Ecology and Behavior, G. O. Mackie, Ed. (Plenum, New York, 1976), pp. 423–432.
36. R. K. Trench, Helgol. Wiss. Meeres 26, 174 (1974); J. W. Porter, Am. Nat. 110, 731 (1976).
37. J. Lang, personal communication.
38. D. S. Wethey and J. W. Porter, in Coelenterate Ecology and Behavior, G. O. Mackie, Ed. (Plenum, New York, 1976), pp. 59–66.
39. T. F. Goreau, Ecology 40, 67 (1959). See also (17) and (24).
40. J. B. C. Jackson and L. Buss, Proc. Natl. Acad. Sci. U.S.A. 72, 5160 (1975); M. Gilpin, Am. Nat. 109, 51 (1975).
41. D. H. Janzen, Am. Nat. 104, 501 (1970); J. H. Connell, in Dynamics of Populations, P. J. den Boer and G. R. Gradwell, Eds. (PUDOC, Wageningen, 1970), pp. 298–312. This mechanism could also produce the mosaic pattern envisaged by Aubreville (26).
42. D. H. Janzen, Ecology 53, 258 (1972); D. E. Wilson and D. H. Janzen, ibid., p. 955; D. H. Janzen, ibid. p. 350.
43. R. M. Laws, I. S. C. Parker, R. O. B. John-

stone, *Elephants and Their Habitats* (Clarendon, Oxford, 1975).
44. A. S. Watt, *J. Ecol.* **35**, 1 (1947).
45. J. Porter, *Am. Nat.* **106**, 487 (1972).
46. J. M. Branham, S. A. Reed, J. H. Bailey, J.
Caperon, *Science* **172**, 1155 (1971); P. W. Glynn,
Environ. Conserv. **1**, 295 (1974); *Ecol. Monogr.*
46, 431 (1976).
47. T. E. Goreau, J. C. Lang, E. A. Graham, P. D.
Goreau, *Bull. Mar. Sci.* **22**, 113 (1972).
48. J. W. Porter, *Science* **186**, 543 (1974): "There
appears to be a balance of abilities divided
among the Caribbean corals such that no one

species is competitively superior in acquiring
and holding space. The effect of this balance of
competitive abilities is to retard, even in high-
density situations, the rapid competitive exclusion that takes place on undisturbed eastern
Pacific reefs" (p. 544); L. A. Maguire and J. W.
Porter, *Ecol. Modelling* **3**, 249 (1977).
49. Y. Loya, *Mar. Biol.* **13**, 100 (1972).
50. J. H. Connell, in *Experimental Marine Biology*,
R. Mariscal, Ed. (Academic Press, New York,
1974), pp. 21–54.
51. A. Gómez-Pompa, C. Vázquez-Yanes, S. Gue-
vara, *Science* **177**, 762 (1972).

52. I thank the following for critical discussions and
readings of earlier drafts: J. Chesson, P. Chesson, J. Dixon, M. Fawcett, L. Fox, S. Holbrook, J. Kastendiek, A. Kuris, D. Landenberger, B. Mahall, P. Mather, J. Melack, W.
Murdoch, A. Oaten, C. Onuf, R. Osman, D.
Potts, P. Regal, S. Rothstein, W. Schlesinger, S.
Schroeter, A. Sih, W. Sousa, R. Trench, R.
Warner, G. Wellington, and two anonymous reviewers. Supported by NSF grants GB-3667,
GB-6678, GB-23432, and DEB-73-01357, and by
fellowships from the J. S. Guggenheim Memo-
rial Foundation.

Reputational Ratings of Doctoral Programs

Rodney T. Hartnett, Mary Jo Clark, Leonard L. Baird

Undoubtedly the best-known efforts to assess the quality of doctoral programs in recent years have been the collection of prestige or reputational ratings by the American Council on Education (ACE) in 1964 (*1*) and 1969 (*2*). In those surveys the ACE obtained from samples of university faculty members ratings of the quality of graduate faculties in their own fields at other U.S. institutions. In addition to serving their primary purpose in the graduate education community, these surveys produced data that have been used to gain a better understanding of the meaning of reputational ratings, particularly how they are related to other characteristics of doctoral programs. As a result, we have learned that the reputational ratings—often called peer ratings—tend to be fairly highly related to program size (*3, 4*) and various indices of research productivity (*4, 5*), though the magnitude of these relationships varies considerably across disciplines. In particular, it appears that the ratings are more highly correlated with various traditional measurements (for example, number of Ph.D.'s produced, levels of funding) in the biological and physical sciences than in the social sciences or the humanities. One plausible explanation for this is that in the biological and physical sciences there tends to be greater consensus about accepted knowledge and standards (*6*).

There has been a good deal of concern

The authors are research psychologists at Educational Testing Service, Princeton, New Jersey 08540.

about the use of reputational ratings in making judgments about program quality. The chief objections have been (i) that the ratings are unfair to doctoral programs which do not place primary emphasis on doing research and preparing their students to do research; (ii) that there is a strong halo effect, the ratings of a department being unduly affected by the prestige or reputation of the university of which it is a part; (iii) that there is a time lag, that is, the ratings are usually based on impressions of what a department used to be like, not on knowledge of its current strengths and weaknesses; and (iv) that the rating information seldom makes for a better understanding of a specific program's strengths and weaknesses and therefore is not useful for program improvement.

It was largely in response to some of these dissatisfactions with reputational ratings that the Council of Graduate Schools (CGS) and Educational Testing Service (ETS), in 1975, conducted a multidimensional study of quality in doctoral programs in three disciplines (*7*). This project was designed primarily as a study of the feasibility of employing information about a wide variety of characteristics in making judgments about the quality of programs. An important feature of the project was the idea that a single ranking is too simplistic, that it does not allow for the possibility that doctoral programs relatively strong in one respect (such as publication rates of the faculty) might be less strong in another (such as the quality of their teaching).

A major procedural characteristic was that most of the information collected from respondents had to do with their own departments; for example, faculty members reported their own publication rates or journal-refereeing activities, students their opinions about the quality of teaching they received, alumni their dissertation experiences, and so on. These reports were obtained from students, faculty, and alumni by means of questionnaires. A general conclusion of the study was that such reports can be obtained without great difficulty, are usually reliable, and augment the description of characteristics relevant to appraisals of doctoral program quality.

Though they were not a crucial element in the CGS/ETS study, peer ratings were also obtained from the faculty respondents, each of whom was asked to rate the quality of the faculties of the other departments in his or her discipline which participated in the study. This aspect of the CGS/ETS study paralleled the two earlier ACE surveys, and it is this aspect of the CGS/ETS study that is the focus of this article.

The primary reason for obtaining the peer ratings was to examine their relationship to the broader array of program characteristics reported in the main part of the survey, a line of inquiry that was not possible with either of the earlier ACE studies. But interest in peer ratings per se remains strong. The Conference Board of the Associated Research Councils convened a planning conference, in the fall of 1976, to investigate issues involved in conducting another peer rating survey (*8*). In spite of ACE's announced intention of refraining from further efforts of this kind, it appears likely that some agency concerned with graduate education in the United States will conduct some kind of reputational rating survey in the not too distant future. Our interest in an improved understanding of the nature and meaning of peer ratings therefore goes beyond pure intellectual curiosity.

This article draws on the data gathered in the CGS/ETS study (*7*) and the two earlier ACE studies (*1, 2*) to address

Vol. 112, No. 983   The American Naturalist   January–February 1978

PREDATOR-MEDIATED COEXISTENCE:
A NONEQUILIBRIUM MODEL

HAL CASWELL

Biological Sciences Group, U-42, University of Connecticut, Storrs, Connecticut 06268

THE EMPIRICAL PROBLEM

One of ecology's central problems is the coexistence of species. Within a trophic level, explanations for coexistence are usually framed in terms of competition by asking what properties of the species, and of their environment, prevent competition from excluding some members of the assemblage. Predation, impinging on a group of (presumably) competing species, has been implicated as a potentially important factor which might allow coexistence.

Observations of such "predator-mediated coexistence" are by now commonplace, including Darwin (1859) on mowing of grassland plants, Summerhayes (1941) on voles and grassland plants, Paine (1966, 1971) and Dayton (1971) on starfish and intertidal invertebrates, Harper (1969) on grazers and plants, Slobodkin (1964) on harvesting of laboratory hydrids, Neill (1972) on fish and laboratory zooplankton, and Porter (1972) on starfish and corals. Brooks and Dodson (1965), Wells (1970), and Hall et al. (1970) showed that fish predation could mediate zooplankton coexistence, although it is not certain that the interaction being affected is purely competitive (Dodson 1974). Finally, predation has also been shown to contribute to the coexistence of different genetic morphs within single species, e.g., zooplankton (Zaret 1972) and moths (Kettlewell 1955; Lees and Creed 1975).

There are also studies that have failed to demonstrate predator-mediated coexistence or actually have shown a decrease in the number of coexisting species under the impact of predation (e.g., Harper 1969; Paine and Vadas 1969; Hurlbert et al. 1972; Adicott 1974; Janzen 1976). These studies are particularly useful as tests of any theory devised to explain predator-mediated coexistence. Not only must such a theory be able to generate the observed positive effect of predation on coexistence, but it should also explain the conditions under which the effect is not seen or is negative.

The empirical observations of predator-mediated coexistence have led naturally to the suggestion that it may play a major role in structuring communities (Paine 1966; Janzen 1970; Connell 1970) and thus to attempts to incorporate it into the mathematical framework of population biology. These attempts to date have been frustrating. While the *possibility* of predator-mediated coexistence has been demonstrated, most analyses make its occur-

Amer. Natur. 1978. Vol. 112, pp. 127–154.

rence seem very unlikely. The following brief review builds a framework for presenting a new model for predator-mediated coexistence.

THEORETICAL APPROACHES

Population models can be conveniently classified according to the kinds of systems they are designed to mimic. I distinguish first closed population systems from open population systems. A closed system is one in which the population exists in a single, closed, roughly homogeneous volume of habitat. There is no migration into or out of this single habitat cell. An open system consists, in its simplest form, of a set of habitat cells coupled by migration. The flux of population between cells must be small enough that the cells retain some measure of independence but large enough that the cells are not totally isolated. Notice that an open population system, considered as a whole, may itself be closed. In this sense the distinction between open and closed systems might be described equally well as between subdivided and nonsubdivided systems. Also note that the terms "open" and "closed" are not used here in their thermodynamic sense, in which an open system is one which exchanges energy with its environment. All populations are thermodynamically open.

The bottle ecosystem studies pioneered by Gause (1934) have provided an experimental paradigm for closed systems, and the vast majority of ecological theory is based on the properties of such systems. The most important of these properties is that zero is an absorbing state for a population in such a system. Extinction is final, permanent, irreversible.

For open systems, an experimental paradigm can be found in the multicelled systems of Huffaker (1958), mites on connected oranges; Pimentel et al. (1963), wasps and houseflies in connected cages; or Neill (1972), algae, zooplankton, and fish in connected aquaria. The most crucial distinction between closed and open systems is that in open systems local extinction is not an absorbing state. Recolonization of a local population that becomes extinct is now a possibility.

A second way to classify population models is as equilibrium or nonequilibrium theories. Equilibrium theories focus their attention on the properties of the system at an equilibrium point. Such theories abstract time out of the picture completely, since a system at equilibrium exhibits no temporal dynamics at all. A nonequilibrium theory, by contrast, would be concerned with the transient behavior of the system away from equilibrium and would be intimately concerned with time.

These two categories generate four classes of population systems within which the effect of predation on coexistence can be examined. The most straightforward way to do this is to construct a model of interacting prey species and then study the coexistence properties of the model with and without predators. Some results follow.

The Predation Effect: Closed, Equilibrium Systems

In a closed system, zero is an absorbing state for any population. This means that coexistence is assured only by a stable equilibrium point at which all species have positive values (or a limit cycle which does not get too close to axes). Moreover, laboratory studies suggest that unless there is a stable equilibrium extinction occurs relatively rapidly. From this the inference is commonly drawn that in nature one will see primarily or exclusively equilibrium situations. Most of theoretical community ecology is devoted to seeking conditions on the properties of the species and their environment that will assure a positive stable equilibrium. In fact, it is customary, when looking for predator-mediated coexistence, to refer to the problem as one of predator-generated stability. We will see later that the two cannot be equated.

Several attempts have been made to model predator-mediated coexistence in closed, equilibrium systems. The results depend on the form in which the predation process is modeled. Let N_1 and N_2 be the population sizes of two competing prey species, and N_3 that of the predator. Equations for the system dynamics can be written

$$\frac{dN_1}{dt} = r_1 N_1 \left(\frac{K_1 - N_1 - \alpha N_2}{K_1} \right) \quad f_1(N_1, N_2, N_3),$$

$$\frac{dN_2}{dt} = r_2 N_2 \left(\frac{K_2 - N_2 - \beta N_2}{K_2} \right) - f_2(N_1, N_2, N_3),$$

$$\frac{dN_3}{dt} = g(N_1, N_2, N_3),$$

where the function $f_i(N_1, N_2, N_3)$ expresses the losses to population i due to predation. These losses in general depend on the densities of all three species.

Constant dilution rate $[f_i(N_1, N_2, N_3) = k_i N_i]$.—This is the simplest possible model of "predation," in which a constant fraction of the prey population is continuously removed. It is more appropriately termed dilution than predation, since there is no feedback from the abundance of the prey to the predation process.

Slobodkin (1961) analyzed this case graphically and found that such predation easily could reverse the outcome of competition if the losing competitor had a higher growth rate at low densities. In order to maintain both species in stable coexistence, however, the coefficients in the model must be very precisely balanced indeed.

Yodzis (1976) has recently analyzed an even simpler analogue of predation: harvesting at a constant rate, where $f_i(N_1, N_2, N_3) = k_i$. He found that, in a deterministic environment, such harvesting can eliminate competitive dominance and result in stable coexistence. When the environment is subject to random fluctuations, however, harvesting at a constant rate increases the niche separation required for stable coexistence, leading to the loss of species from the community (Yodzis 1977).

Lotka-Volterra predation $[f_i(N_1, N_2, N_3) = k_i N_i N_3]$.—This description of predation adds a little more realism to the situation; the intensity of predation now depends on the densities of both the predator and the prey. May (1973) and Cramer and May (1971) have studied the coexistence properties of this model. As in the previous case, it turns out that predation can stabilize the system but that this requires a delicate balancing of the parameters. Over most of the parameter space, the stability properties of the model with and without predators are identical.

The results for both of these models can be understood qualitatively in terms of Ayala's (1972) analysis of frequency dependence in competition systems. In order for a competitive equilibrium to be stable, the fitnesses of the competing species must exhibit reverse frequency dependence near the equilibrium point. That is, a perturbation away from equilibrium which increases the frequency of N_1 relative to N_2 must decrease the fitness of N_1 relative to that of N_2. Only in this case will the relative frequencies move back in the direction of the equilibrium.

Neither the constant dilution rate nor the Lotka-Volterra predation terms are frequency dependent; the predation intensity on one prey species is independent of the abundance of the other. Thus predator-generated stability in these cases requires that the addition of a non-frequency-dependent predation pressure to a non-frequency-dependent situation results in frequency dependence. This is possible (since fitnesses depend on the absolute as well as the relative values of the N_i and because the system with predation is three rather than two dimensional), but it does not seem very likely.

Frequency-dependent predation.—A frequency-dependent form of predation, in which the $f_i(\cdot)$ would depend explicitly on all the N_i, might be more successful at stabilizing competitive interactions. What is needed is a form of predator switching in which any prey species that begins to dominate the system receives a disproportionate increase in predation pressure, while a species that begins to disappear gets a correspondingly disproportionate decrease. This process has not been extensively analyzed, but Roughgarden and Feldman (1975) have shown that if predator preference increases with the relative density of the prey then the permissible niche overlap between prey can be significantly increased.

The theories of Janzen (1970) and Connell (1970) for predator-mediated coexistence in tropical trees can be interpreted in frequency-dependent terms. The frequency dependence is generated by the dispersal and searching abilities of seed and/or seedling predators. A local high-density aggregation of seedlings near the parent tree will be decimated by predation; isolated seedlings have a much higher probability of escaping. Because of the spatial element involved, this hypothesis can also be interpreted in open-system terms (see below).

It is worth noting that reverse frequency-dependent predation can occur in which rare species are preyed upon especially heavily. This can happen any time that the competitive pressure in the community is lowered to the point that a very unpalatable (or inedible) species can persist. If such a species begins to become abundant, the predator has no choice but to continue to seek out the ever-rarer palatable forms. This has been found to happen in certain heavy grazing situations (Harper 1969), in certain cases of physical disturbance due to elephant populations in Africa (Laws 1970), and in a generalized form seems to be responsible for the "disclimax" situations caused by practicing swidden agriculture in areas of insufficient moisture (Geertz 1969).

The most encouraging of the attempts to explain predator-mediated coexistence in terms of equilibria in closed systems seem to be Yodzis's (1976) analysis of the reversal of competitive dominance and Roughgarden and Feldman's (1975) work on frequency-dependent predation. The major weakness of both of these attempts is the neglect of the predator population dynamics. In those models which have included the predator population, this form of predator-mediated coexistence seems to require a very strict balancing of parameters.

The Predation Effect: Open, Equilibrium Systems

Several authors recently have examined the properties of sets of competitive systems connected by migration (e.g., Levins 1970; Levins and Culver 1971; Cohen 1970; Horn and MacArthur 1972; Levin 1974; Slatkin 1974). This is usually approached by switching attention from N_1 and N_2, the abundances of the competing species, to the proportions of cells (P_1 and P_2) in which the two species occur. Local extinction, within a single cell, is no longer an absorbing state. Overall extinction ($P_i = 0$) is, of course, still absorbing and in these studies coexistence has been equated with a stable equilibrium with both P_1 and P_2 positive.

The effect of predation on such equilibria has not been extensively investigated, but some hints are forthcoming from the preceding analyses. The models of Levins and Culver (1971), and Horn and MacArthur (1972) turn out to be formally equivalent to the Lotka-Volterra competition equations. Frequencies of occurrence take the place of abundances, and immigration and extinction rates replace birth and death rates. This suggests that the effects of predation may be similar to those outlined above, depending on the form in which it is modeled.

Slatkin (1974) considered briefly the effect of predation on an open system at equilibrium. He found that if competition acted only through colonization probability or only through extinction probability then increased extinction rates (as by a predator) could stabilize an otherwise unstable system. Again, no predator dynamics were included in the model.

The Predation Effect: Closed, Nonequilibrium Systems

The possibility of nonequilibrial (and thus, of course, temporary) coexistence in closed systems has received little attention in ecology. The crucial quantity in a nonequilibrium theory of competition is the rate of competitive exclusion, or, equivalently, extinction time. This contrasts with the equilibrium theory in which temporal dynamics are abstracted out of the picture by taking limits as "$t \to \infty$."

The factors affecting the rate of competitive exclusion have not been thoroughly studied. Some simulation work (Caswell, unpublished) suggests, however, that the rate of exclusion decreases with increasing similarity of the competing species and is higher for interference than for exploitation competition (agreeing with data summarized by Miller [1969]). The possibility to be explored here is that predation might increase extinction time to the point that the competitors would appear to coexist indefinitely.

Preliminary analyses (Caswell, unpublished simulation data) suggest that predation may slow down competitive exclusion. Parrish and Saila (1970) reported that predation (Lotka-Volterra type) stabilized an otherwise unstable competitive system, but Cramer and May (1971) later showed analytically that their example was actually unstable even under the impact of predation. Predation had, however, prolonged coexistence of the prey dramatically enough that it was mistaken for stability.

Hutchinson (1953, 1961) proposed nonequilibrial coexistence as a possible explanation for the persistence of ecologically similar plankton species. In such a situation, Hutchinson distinguishes three cases: depending on whether the time required for competitive exclusion is much less than, approximately equal to, or much greater than the time scale on which the environment fluctuates. In the first case, a competitive equilibrium would be established, and there would be no coexistence. The third case would lead to an equilibrium obtained by time averaging of the rapidly varying environmental conditions, and again one species would be excluded. In the second case, however, nearly permanent nonequilibrium coexistence may result.

It seems entirely possible that a fluctuating predator population could provide a varying environment on precisely the right time scale to keep competition from reaching its equilibrium. In Parrish and Saila's (1970) simulation results, the decline in the prey species which is being excluded becomes apparent as soon as the predator's oscillations die out. On the other hand, Caswell (1972) examined the interaction of predation and competition in a complex model including time lags, predator satiation, and prey refuges. This system oscillated violently, and the competitors appeared to coexist indefinitely. How general such phenomena are in nature is unknown.

While these closed system, nonequilibrial results look promising, there are still problems with generalizing them to explain predator-mediated coexistence in nature. Local extinction is still an absorbing state, and it is not yet clear that predation can delay that extinction for ecologically relevant lengths of time. Nature, moreover, is not organized in closed system fashion.

The Effect of Predation: Open, Nonequilibrium Systems

Nonequilibrium coexistence in open systems can be phrased as a special case of a more general problem. Consider a system composed of a large number of similar components, each of which interacts with some of its fellows. How long does it take such a system to reach equilibrium (the usual nonequilibrium theory question) as a function of the number of components, the time required for each component to reach its local equilibrium, and the pattern of connection between components? This question has arisen in the study of adaptive behavior in the nervous system (Ashby 1960), in the study of general hierarchical system theory (Simon 1962), in design theory (Alexander 1964), and in the study of epigenetic control networks (Kauffman 1969). The answer in each case is the same: for even very simple components in reasonably small numbers, high levels of connectedness lead to astronomically long delays in reaching equilibrium. Consider this simple example from Ashby (1960). The system is a set of 100 light bulbs, each of which is in one of two states (on or off). For each bulb, in each second, the transition (on → off) occurs with probability .5. For each bulb, in each second, the reverse transition (off → on) occurs with probability .5 *if* at least one bulb in the "on" state is connected to the bulb in question, otherwise the probability is zero. The equilibrium for the system is the state with all bulbs off; this equilibrium is stable. If the system is started with all 100 lights on, how long will it take to reach equilibrium?

If there are no connections among the bulbs at all, the expected time required to reach equilibrium is on the order of 2^1 s = 2 s. At the other extreme, when the system is totally connected, with every bulb interacting with every other bulb, the time to equilibrium is on the order of 2^{100} s = 10^{22} yr. This is an immensely long time; the estimated age of the universe is only 10^{10} yr. For all practical purposes this system, containing a reasonable number of components, each of which reaches its own equilibrium on a time scale of seconds, will never reach equilibrium at all.

This suggests that in an open system of population cells joined by migration, nonequilibrium conditions might last long enough to be ecologically relevant—perhaps even more relevant than the equilibrium conditions.

In the following sections of this paper I will examine in some detail the effect of predation on nonequilibrium coexistence in open systems. To show the possibility of predator-mediated coexistence in such a system, I have built a simple but very general model from which the closed-system equilibrium, closed-system nonequilibrium, and open-system equilibrium effects of predation are eliminated. After demonstrating the possibility of predator-mediated coexistence in this system, I will try to make some deductions about its probability in nature and then examine some observed cases of predator-mediated coexistence in light of the model results.

A MODEL

The model consists of a set of cells (the number of cells is an independent variable whose effect is to be studied) connected stochastically by migration. A model of an entire ensemble of cells is generated from a description of the within-cell population interactions and between-cell migrations.

The description of population dynamics within a single cell is an abstraction of the results typically observed in closed, single-cell systems in the laboratory. In figure 1 the top panel shows the model for single-species population growth; the logistic curve has been abstracted into a discrete binary variable registering presence (1) or absence (0). The middle panel shows the model for the competitive interaction of two such populations. (The losing competitor, species *A*, coexists with the winning competitor, species *B*, for *TC* time units and then becomes extinct.) The bottom panel shows the model for predation. Over the course of *TP* time units the predator population increases, decimates its prey, and then crashes to extinction (*TC* is the competitive exclusion time and *TP* the predatory decimation time; these are also independent variables

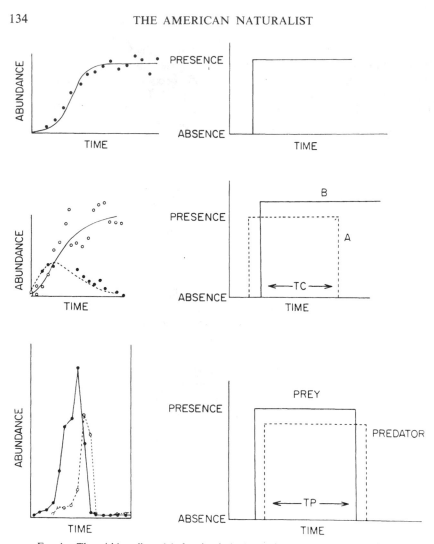

FIG. 1.—The within-cell models for (top) single species population growth (data on *Paramecium* in laboratory culture, after Gause [1934]), (middle) interspecific competition (two species of *Paramecium* in the laboratory, after Gause [1934]), and (bottom) predation (two species of mites, after Huffaker [1958]).

whose effect is of interest). Time units in the model could be measured either in generations or calendar time.

The instantaneous situation in each cell is described by a five-dimensional state variable:

$$XA = \text{abundance of the losing competitor } (0 \text{ or } 1),$$
$$XB = \text{abundance of the winning competitor } (0 \text{ or } 1),$$
$$XC = \text{abundance of the predator } (0 \text{ or } 1),$$
$$tc = \text{competitive "clock," } 0 \leq tc \leq TC, \text{ and}$$
$$tp = \text{predatory "clock," } 0 \leq tp \leq TP.$$

The competitive and predatory clocks keep track of how long competitive and predatory interactions have been going on. In a sense, they are alarm clocks; when $tc = TC$ or $tp = TP$ the losing competitor or the prey, respectively, goes extinct.

Migration between cells is modeled in similarly abstract fashion. The probability of a given species colonizing a cell from which it is absent and which is suitable for colonization (as specified below) is given by

$$P[\text{colonization}] = D_i \times \frac{\text{no. of cells occupied by species } i}{\text{total number of cells}} \ ; \qquad i = A, B, C.$$

Dispersal is thus directly proportional to the number of cells occupied by a given species. The proportionality constant (D_A, D_B, and D_C) measures the "per cell" dispersal ability of the species. This version of the model does not incorporate the spatial configuration of the cells, so there is no proximity effect on colonization. This will be added to future versions of the model.

Coupling the within-cell dynamics with the migration process, and adding a few additional details to complete the descriptions of predation and competition, results in an overall set of state-transition rules (table 1). Note particularly the following points: (i) Competitive exclusion of species A by species B is total. A cell can be invaded by species A only if species B is absent; the presence of the predator does not affect this restriction. (ii) The presence of the predator cannot result in permanent coexistence of the prey within a single cell (the closed system, equilibrium effect). (iii) The presence of the predator cannot even extend coexistence within a single cell (the closed system, nonequilibrium effect). In a cell containing all three species, species A becomes extinct when $tc = TC$, regardless of the presence of the predator. (iv) The system as a whole is a finite-state Markov process with three absorbing states: (1) species B present in every cell and both other species extinct, (2) species A present in every cell and both other species extinct, (3) all three species extinct. The system will eventually end up in one of these states with probability 1.0 (see, e.g., Feller 1968; Karlin 1966). Thus there is no possibility of overall equilibrium coexistence (or its stochastic analogue, a limiting probability distribution) with all species present (the open-system equilibrium effect).

In establishing these transition rules, I have purposely made it as difficult as possible for the presence of the predator to generate coexistence of the prey, by eliminating all modes of predator-mediated coexistence except the open system, nonequilibrium effect. This is not because I am convinced that the other effects never operate in nature (they probably do), but simply to maximize the power of any demonstration of this mode of coexistence.

RESULTS

The independent variables of interest in this model are the number of cells in the system (NC), the per cell dispersal rates of the three species (D_A, D_B, D_C), the competitive exclusion time (TC), and the predatory decimation time (TP). It is interesting to note that the first four of these are of interest because the model describes an open system, the last two because I am concerned with

TABLE 1

STATE-TRANSITION RULES

State Variable	Transition Rule
XA (the losing competitor):	
$1 \rightarrow 0$	If $XB = 1$ and $t_c = TC$ or $XC = 1$ and $t_p = TP$
$1 \rightarrow 1$	Otherwise
$0 \rightarrow 1$	With probability $D_A P_A$ if $XB = 0$ with probability 0 if $XB = 1$
$0 \rightarrow 0$	With probability $1 - D_A P_A$ if $XB = 0$ with probability 1 if $XB = 1$
XB (the winning competitor):	
$1 \rightarrow 0$	If $XC = 1$ and $t_p = TP$
$1 \rightarrow 1$	Otherwise
$0 \rightarrow 1$	With probability $D_B P_B$
$0 \rightarrow 0$	With probability $1 - D_B P_B$
XC (the predator):	
$1 \rightarrow 0$	If $XA = 0$ and $XB = 0$
$1 \rightarrow 1$	If $XA = 1$ or $XB = 1$
$0 \rightarrow 1$	With probability $D_C P_C$
$0 \rightarrow 0$	With probability $1 - D_C P_C$
t_c (competitive clock)	The competitive clock is started as soon as species A and B are both present and is incremented each time unit until it reaches TC
t_p (predatory clock)	The predatory clock is started as soon as species C and either A or B are both present and is incremented each time unit until it reaches TP

NOTE.—P_A = the proportion of cells containing species, A, similarly for P_B and P_C; D_A = the per colony dispersal ability of species A, similarly for D_B and D_C; TC = the time required for competitive exclusion of A by B; TP = the time required for predatory decimation of A and/or B by C.

nonequilibrium dynamics, in which the time scale cannot be abstracted out of the picture. A fully general treatment of this problem would also include the spatial configuration of the cells; the dispersal rates would then be defined in terms of dispersal over distance.

The mere possibility of predator-mediated coexistence is easy to demonstrate. Figure 2 shows the results of a single realization of the model, with predators present. All three species persisted until the simulation was terminated after 1,000 generations. In fact, all three species persisted for the full 1,000 generations in each of 10 replicates of this particular system. In contrast, in the absence of the predator the persistence time of species A averaged only 64 generations (range in 10 replicates 53–80). In fact, in 160 replications using 16 different parameter combinations (see fig. 3), the maximum persistence time for species A was 92 generations, an order of magnitude less than that demonstrated in figure 2 under the impact of predation.

Species A, the otherwise doomed competitor, is not necessarily being maintained as a rare species. In the 10 replicates using the parameter set of figure 2,

PREDATOR-MEDIATED COEXISTENCE 137

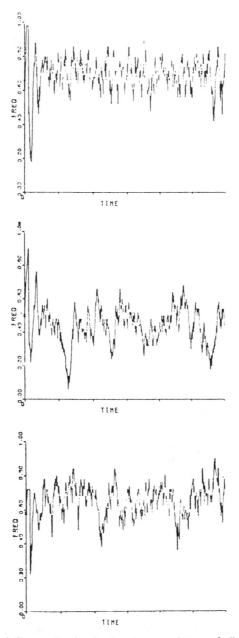

FIG. 2.—Simulation results showing long-term coexistence of all three species in a 50-cell system (species A–C from top to bottom). The simulation was terminated after 1,000 generations, although all three species were present and showed no signs of disappearing. The ordinate measures frequency of occurrence. Parameter values: $NC = 50$, $D_A = .25$, $D_B = .10$, $D_C = .25$, $TC = 20$, $TP = 20$. In the absence of predators, species A was eliminated in less than 80 generations under these conditions.

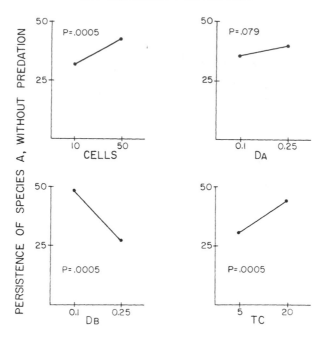

FIG. 3.—Main effects and significance levels for the persistence of species *A* in the absence of predation. Independent variables defined in text.

the frequencies (mean ± SE) for the three species at the end of 1,000 generations were *A*, 0.68 ± 0.034; *B*, 0.48 ± 0.13; *C*, 0.75 ± 0.24. Species *A* was actually more frequent than the superior competitor, *B*.

To explore the effect of the model parameters, a statistical examination of the parameter space was carried out. Each of the independent variables was examined at two values, generating a 2^6, fixed effect, factorial design. The results were examined by analysis of variance. This design is standard for exploring the effect of a large number of variables simultaneously (Kirk 1968); it sacrifices information on nonlinearities for the ability to examine and screen a large number of variables.

The dependent variable of interest in this analysis is extinction time. This creates some operational problems, since the only way to evaluate extinction time is to let the simulation run until extinction occurs. This can take a lot of computer time, since this system may persist for any time short of infinity. To get around this problem, I used the following procedure. For each combination of parameter values I ran 10 replicates of the system, terminating the simulation after 1,000 generations. From these replicates I calculated the median extinction time and used this as the response variable in the analysis of variance. Unlike the mean, the median can be unambiguously calculated if more than half of the extinction times are less than 1,000. The region of the parameter space to be explored was chosen, after some pilot runs, so that the median extinction time of species A was almost always less than 1,000 generations. In the few cases where it exceeded 1,000 generations, its value was set equal to 1,000.

Using the median as the response variable means that there is only a single value per cell in the analysis of variance. The four-, five-, and six-way interactions were pooled and used as an error term (Kirk 1968). This assumes that these interactions are negligible; if the assumption is violated, the significance tests are conservatively biased due to inflation of error mean squares.

Three response variables were measured and analyzed: the extinction time of species A without predators, the extinction time of species A with predators present, and the effect of predation on the extinction time of species A. The latter quantity was defined as the difference between the median extinction time of species A with and without predators. Positive values indicate that predation increased persistence.

Initial conditions for the simulations were arbitrarily defined by randomly and independently allocating species A, B, and C into 50% of the cells. A few spot checks revealed no great sensitivity to different initial conditions.

Figures 3–5 show the main effects and significance levels for the three dependent variables. The extinction of species A in the absence of predators is determined by how long it takes species B to disperse to all of the cells, since there is no mechanism to remove the winning competitor once it occupies a cell. Thus there are significant positive effects on persistence time of the number of cells and TC, and a negative effect of D_B. The dispersal rate of species A has a marginally nonsignificant ($P = .079$) positive effect; in this situation there is little advantage to a high dispersal rate for species A, since species B permanently occupies any cell it colonizes.

In the presence of predation (fig. 4), the persistence time of species A is increased significantly by increases in the number of cells, its own dispersal rate and that of the predator, and the time required for within-cell interaction (either predation or competition) to be completed. The dispersal rate of species B has, as would be expected, a significant negative effect.

The results for the measure of predation effect (fig. 5) parallel those for persistence under predation. The fact that this effect is significantly greater than zero is statistical proof of the possibility of open-system, nonequilibrium, predator-mediated coexistence. This demonstration is particularly powerful because the analysis is triply biased against demonstrating the effect. First, the parameter space investigated was chosen to result in median extinction times of less than 1,000 generations. Then those parameter combinations which resulted in more than 1,000 generations of coexistence were constrained to equal 1,000. Finally, there is the complication of predator extinction. Strictly speaking, a demonstration of predator-mediated coexistence could be made conditional on the continued presence of the predator, not merely its presence at the beginning of the simulation. In fully 40% of these runs, the predator went extinct before species A. These cases speak not of the failure of the predator to generate coexistence of the prey but rather of the inability of the predator to maintain a viable population, yet they were included in the analysis.

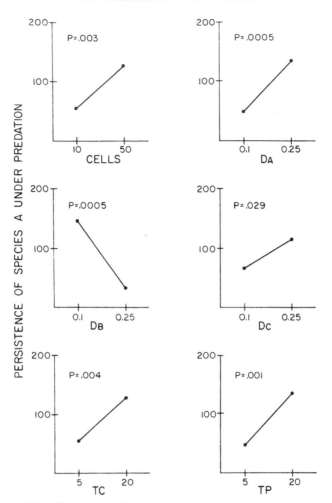

FIG. 4.—Main effects and significance levels for the persistence of species *A* under the impact of predation. Note change in scale from that of fig. 3.

The Likelihood of Predator-mediated Coexistence: A Crude Evolutionary Argument

The possibility of open-system, nonequilibrium, predator-mediated coexistence is established, but how likely is it to occur? The answer to this question requires a prediction of the course of evolution to predict where in the parameter space real systems are likely to be located. To make a crude attempt at answering this question, I will use the treatment effects shown in figures 3–5, as well as similar data on the extinction time of the predator which is not included here.

The only measure of fitness available in this model is persistence time. Since

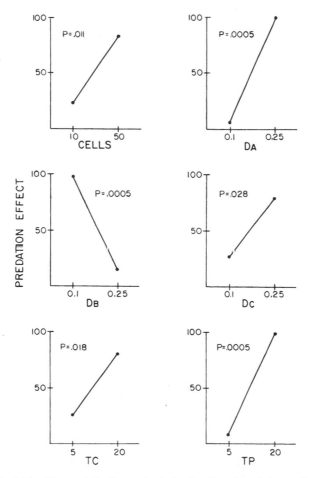

Fig. 5.—Main effects and significance levels for the effect of predation on the coexistence of species *A*. Note change in scale from those of figs. 3 and 4.

the system is open, it is possible to select (interdemic selection) for delayed extinction. Genotypes leading to shorter extinction times will disappear and be replaced by those with longer extinction times (see Gilpin [1975] for a detailed example). The evolution of decreased virulence by myxomatosis virus in Australian rabbits (Ratcliffe 1959) is a model case.

From the treatment effects in figures 3 and 4 and the data on species *C* it is possible to predict that those parameters likely to be under the genetic control of either species *A* or *C* (D_A, *TC*, *TP*, and D_C) should all be increased by selection for persistence. Predictions about species *B* cannot be made from this data; it seldom went extinct. The crucial point is that these increases all tend to increase the effect of predation. That is, this form of selection will act to increase the importance of predation in maintaining coexistence of the prey.

This first-order analysis can be extended somewhat by examining the two-way interactions. All the interactions involving combinations of D_A, TC, TP, and D_C are of the reinforcing type. That is, the effect of one of the variables is increased by an increase in the level of the other. Thus these second-order interactions will only act to increase selection pressure for an increased effect of predation.

This analysis is crude in several respects (ignoring species B, considering only extinction time as a measure of fitness, and particularly in ignoring possible nonlinear responses), but it does suggest that this type of coexistence can be selected for.

SOME GENERALIZATIONS

The model used here is highly abstracted—cells are discrete permanent fixtures, species are modeled by binary switches, etc. As I have repeatedly pointed out, however, the abstractions have been designed to stack the deck against predator-mediated coexistence whenever possible. In addition, the model can be generalized easily, suggesting that the results obtained could be of widespread occurrence and importance in natural communities.

What Is a Predator?

In this model, the predator opens up otherwise closed cells for colonization by the inferior competitor. There is no reason why this function must be performed by an actual predator. Physical disturbance would serve equally well. Forest fires have been implicated as having a major role in maintaining coexistence of trees in northern temperate forests (e.g., Loucks 1970; Taylor 1973; Wright and Heinselman 1973) and in maintaining the precarious existence of the Kirtlands warbler in Michigan (Mayfield 1960). Osman (1977) has documented the importance of physical disturbance by wave action on the maintenance of coexistence in a marine epifaunal community, and Abele (1976) has shown that physical disturbance generates an increase in diversity in the crustacean community inhabiting coral heads. The major difference between actual predation and abiotic physical disturbance is in the feedback between the abundance of the "prey" and the activity of the disturbing agent. This would lead to different dynamics of the system but would probably make the coexistence effect more dramatic because the "predator" could not become extinct.

The role of the predator can also be taken by nonpredatory biological disturbance. Werner (1972) introduced a rosette-forming biennial plant species (teasel, *Dipsacus sylvestris*) into several old fields and found an increase in the number of plant species in the plots to which the teasel had been added. Close investigation revealed that this increase was due to the colonization of openings in the field generated by the teasel rosettes. The colonizing species were clearly nonequilibrial, since they were found only in the teasel-generated openings and were unable to reestablish themselves on the site through reproduction. It is hard to guess how generally important this type of biological disturbance is, but it seems likely to occur in certain stages of succession, where one life form invades a community dominated by a few species of another life form. Platt (1975) has documented an even more dramatic case in which a well-defined assemblage of plants persists in the Iowa prairie only because of the generation of open space by badger mounds.

It is even possible for competitors to act as predators in this model. A nonlinear competitive hierarchy (e.g., *A* excludes *B*, *B* excludes *C*, *C* excludes *A*) would have the same basic properties as the predator-prey system modeled here: between-cell interactions would perturb the progress to local equilibrium. Jackson and Buss (1975) report such an interaction among coral reef invertebrates and suggest that it might be partly responsible for the high diversity of coral reefs.

What Is a Cell?

This model is applicable directly to situations with discrete, permanently defined, habitat patches (islands, lakes, mountaintops, host plants, etc.), but there are a variety of manifestations of this idea in different form.

In many cases, the cells are formed by the action of the "predator" itself (e.g., forest fires, badger mounds). Levin and Paine (1974) have modeled the generation of open cells in the rocky intertidal by starfish grazing. Coupling a model like theirs for patch formation with something like my model for species interaction within and between patches would generate a first approximation to a model of community dynamics in the intertidal.

An important generalization of the model would be to include cells of more than one type. Of particular importance is the division of a habitat into areas which the predator can reach and those which he cannot. The latter areas form a particular type of cell—a prey "refuge." Such refuges are known to be important in maintaining coexistence of predator-prey systems (e.g., Gause 1934; Caswell 1972).

The most important result of the division of an area into cells is the maintenance of nonuniformity from one spot to another. Since extinction must happen everywhere at once to be final, this nonuniformity acts to prolong coexistence. This effect also can occur in a homogeneous expanse of habitat simply because of the time-lag inherent in the propagation of ecological effects over distance. This continuous form of "cells" would stand in the same relation to my discrete model as the "isolation-by-distance" to the "stepping-stone" model in population genetics (see, e.g., Crow and Kimura 1970). Earlier (Caswell 1972) this phenomena was implicated as a way to obtain refuges for prey populations. I am inclined to view Luckenbill's (1973, 1974) experiments as examples of this, although he might not. Dubois (1975) and Levin and Segel (1976) have begun work on modeling the generation of nonuniform conditions in ecological systems in homogeneous media. The much more general work of Glansdorff and Prigogine (1971), Iberall (1972), and Thom (1975) may ultimately provide a framework for this sort of phenomenon.

If cell-like phenomena can be generated by temporal asynchrony over large expanses of uniform habitat, then the results of this analysis become extremely general. They suggest that nonequilibrium coexistence may occur not only in obviously patchy habitats, but in such apparently uniform areas as large lakes or oceans.

In the model presented here, cells are cells to all three species involved. This clearly need not be the case in nature (see, e.g., Janzen [1967] in a different context). A whole spectrum of locally effective population areas may exist with interactions occurring between species at all levels. This will make any direct application of the model to a real community very difficult but is no deterrent to applying the model's qualitative conclusions.

Hierarchical arrangements of cells are an interesting possibility. Such a hierarchy would consist of sets of clusters of cells, each cluster being described by a model like the one developed here. Nonequilibrium coexistence in a hierarchy will be extended beyond its value within a single cluster. At the second level of the hierarchy, TC and TP will be greatly lengthened over their values in a single cell at the first level. And, while the dispersal probability from any single cell of one cluster to any single cell of a second cluster must be small (or else the two clusters would not be divided), the probability of dispersal from some cell of one cluster to some cell of the second should be rather higher. The results in figure 5, applied to the second level of the hierarchy, suggest that predation in this case will extend coexistence beyond its already extended value at the first level.

Interaction of TP and TC

The closed-system, nonequilibrium effect of predation is essentially an interaction between the competitive and predatory exclusion times, TC and TP. If the predator is present in the cell, competitive exclusion is delayed for a greater or lesser period of time. In the most extreme case, TC would become equal to TP; that is, competitive exclusion would not occur at all while the predator was present.

This effect is not included in the model, although it may occur in nature. It would definitely occur at the cell-cluster level in the hierarchical systems discussed above. Its effect, again, will be to prolong predator-mediated coexistence. This effect could be approximated by an increase in TC to somewhere between TC (in the absence of predators) and TP, the exact location determined by the fraction of cells in which both predator and prey occur.

Continuous and Discrete Mathematics

How many of the results of this analysis are artifacts of the use of discrete (presence-absence) mathematics, as opposed to a continuous representation? This is a difficult question to answer, but some justifications for the use of discrete mathematics are available.

First, the logical structure underlying the modeling of dynamic systems applies equally to continuous and to finite-state systems (Arbib 1966). Presence and absence of a species are perfectly good ecological variables, and a finite-state model describing the dynamics of these variables is certainly of ecological relevance.

However, presence and absence can also be obtained from continuous mathematical models—abundances of zero and greater than zero, respectively. To evaluate the agreement between two different models describing the same quantities, we need more than just a formal equivalence between discrete and continuous modeling approaches; we need to look for properties of the particular system under consideration that will guarantee that we draw the same conclusions from the two analyses.

Rosen's (1968) analysis of this problem showed that a discrete representation is obtained by a decomposition of the state space and input space of the continuous system into equivalence classes. The equivalence classes of states correspond to the basins of the stable orbits in the continuous state space; the equivalence classes of inputs are groups of inputs which transfer the continuous system from one basin to another. The transition rules for my model correspond exactly to such a decomposition. Attaching single values to competitive exclusion and predatory decimation times tacitly assumes that the basins in the continuous state space are very steep sided, a condition which Rosen (1968) points out results in a system with strong digital characteristics.

My results thus will be most quantitatively accurate for systems in which competition and predation generate extinction abruptly and at rates relatively independent of initial conditions. When these conditions are not met, the time scale of the model will become blurred, but the qualitative results will be unaffected.

It is worth noting briefly here the relation between my model and the similar models of epigenetic control networks studied by Kauffman (1969, 1972). Modeling the switching of genetic control systems, he coupled randomly defined Boolean functions in a randomly selected pattern and followed the approach of the systems to stable cycles of behavior. In Kauffman's systems, the components of the network are specified at random, and the system then operates deterministically. The components of my networks, on the other hand, are specified deterministically but are operated stochastically. The lags involved in competitive and predatory extinction are another phenomenon which does not appear in Kauffman's nets, in which elements switch on and off instantaneously.

A LOOK AT THE REAL WORLD

This model can be tested against real-world observations in several ways. Since it predicts the existence of predator-mediated coexistence, all of the examples cited in the opening section, as well as the cases involving nonpredatory disturbance mentioned above, can be taken as corroborating the model. More interesting, however, are those predictions of the model that distinguish it from the other three types of predator-mediated coexistence.

The first of these is the prediction that the observed effect of predation will vary with the scale of observation. If an investigator happens to focus his attention within a single cell, predation will have a negative effect on coexistence. (Total extermination of all prey, in the limiting case described in the model.) On the other hand, if a study takes in an area large enough to contain a whole ensemble of cells, it should demonstrate a positive predation effect. Neither of the closed-system theories predicts an area effect of this sort.

In addition, the model relies on the temporary action of predation, opening up new cells for colonization and then releasing the predation pressure to allow nonequilibrium population growth. If the predation pressure does not disappear periodically, the nonequilibrium effect will not be seen. Instead, a new equilibrium will be established within the cell—with no more, perhaps fewer, species than in the absence of predation. Neither of the equilibrium models of predator-mediated coexistence makes this prediction.

A number of examples in the literature which have failed to show predator-mediated coexistence seem to be explained by these predictions. Adicott (1974) searched for predator-mediated coexistence in the protozoan communities in pitcher plant leaves under the impact of mosquito larva predation. The isolated water bodies in a single pitcher plant are clearly single cells for both the prey and the predator. Within these cells, Adicott demonstrated a significant negative relationship between predation intensity and the number of coexisting protozoans. Unfortunately, he presents no data on the effect of predation at the multicell level in these systems. However, it would not be surprising if it also were small or nonexistent, because the predation pressure does not seem to be intermittent. Mosquito larvae occupy a given cell more or less continuously throughout the summer. Abele (1976) studied a similar system: the crustacean community inhabiting heads of the coral *Pocillopora damicornis*. He contrasted the diversity of this community in two sites differing in the level of abiotic environmental variability and found that the disturbance generated no increase in diversity within a single coral head, but at the multicell level it generated a 50% increase in species numbers in a sample of 35 heads.

The rocky intertidal communities studied by Paine and his students also provides an interesting comparison. The work on the starfish/mollusc system (Paine 1966, 1971; Dayton 1971) clearly demonstrated a positive predation effect on a spatial scale of tens of meters of coastline. It would probably not occur to anyone to study this system within a single cell; this would involve following a starfish around and studying the patches he leaves behind him after feeding. Paine and Vadas (1969), however, seem to have found a within-cell effect in another intertidal system, this one involving grazing by sea urchins on tide-pool benthic algae. When the urchins were removed from individual tidal pools, the immediate response was an increase in the number of algal species, which eventually decreased as a few species became dominant. I suspect that a single tidal pool functions as a cell to a group of grazing sea urchins. If there is some mechanism available to occasionally remove urchins from pools, thus making predation intermittent, urchin grazing could be responsible for maintaining the diversity of algal species over stretches of shoreline. It turns out that there is such a mechanism. The starfish *Pycnopodia* occasionally enters pools, consuming some urchins and stampeding the rest out of the pool (see also Dayton 1972).

Fish predation has been clearly implicated in maintaining coexistence among zooplankton in ponds (Hall et al. 1970) and lakes (Brooks and Dodson 1965; Wells 1970). Such systems are large and spatially complex enough to contain many "cells" for zooplankton, and predation is made locally intermittent by the fact that the fish cannot be everywhere at once. In contrast to these systems, Hurlbert et al. (1972) reported on the effect of fish predation (by the mosquitofish, *Gambusia*) on zooplankton in small plastic wading pools. In this system, where the fish were almost certainly treating the pool as a single cell, and in which there was no fluctuation in predation intensity, the fish totally decimated their prey. They eliminated all aquatic insects and all zooplankton species larger than rotifers. Neill (1972) also studied the effect of fish predation on zooplankton in microcosms (1.5-liter aquaria in this case). He found a significant increase in the number of coexisting zooplankton under the impact of fish predation, but his system differed from Hurlbert's in two respects. Neill generated migration of prey species between aquaria, thus creating an open, multicelled system, and he made predation locally intermittent by periodically removing the fish from the aquaria. Thus Neill's system fits somewhere between Hurlbert's closed system and the complexity of a real lake.

When grazing herbivores are fenced into enclosed pastures, both the density of grazers and the temporal pattern of the grazing will affect the pattern of predator-mediated coexistence. At high densities, the entire pasture will be treated as a single cell, receiving uniformly high predation pressure. In this situation the nonequilibrium model predicts that predation will have a positive effect on coexistence only if it is interrupted by removal of the grazers. Harper (1969) cites several such cases. At much lower densities of grazers, or in situations where they are not enclosed, predator movement will generate intermittent predation pressure in a patchy fashion over the landscape and thus increase coexistence. Harper (1969) and Summerhayes (1941) discuss several examples; Nicholson et al. (1970) report some observations on the spatial pattern of grazing by livestock.

Janzen (1976) has proposed an interesting special case of this effect to explain the low density of snakes and lizards in Africa compared with Central America. He hypothesizes that the decline is due to a higher predation pressure in Africa caused by the availability of carrion as an alternative food source, which keeps the predation pressure constant rather than intermittent.

Cairns et al. (1971) studied the protozoan fauna colonizing artificial sponges suspended in a lake and imposed an intermittent, nonselective disturbance on the community by periodically squeezing the sponges and allowing them to refill with water. As predicted by the nonequilibrium model, they found a significant increase in the number of protozoan species in the squeezed sponges. It would be very interesting to repeat this experiment with continuous rather than intermittent "predation," perhaps by pumping a continuous flow of water through the center of the sponge. The model predicts that this manipulation should not increase the number of coexisting species.

Finally, at a very general level the nonequilibrium model predicts that as a community becomes more protected from the action of disturbance (biotic or abiotic) it will become less diverse. This prediction agrees with the observation that, in the absence of disturbance, population-level interactions result in decreases, not increases, in diversity (Caswell 1976).

The agreement between the open system, nonequilibrium model of predator-mediated coexistence and studies of real communities is impressive. Each of the four types of predator effect are no doubt important in some natural situations. The open-system, nonequilibrium model seems as robust as any of the others at this point since it includes the dynamics of the predator, is insensitive to environmental variation, and can make at least crude evolutionary predictions. Still, more information is needed on parameter effects than is included here, and distinguishing the different predator effects in nature should receive more attention.

EQUILIBRIUM AND NONEQUILIBRIUM CONCEPTS
IN ECOLOGICAL THEORY

The distinction between closed and open, equilibrium and nonequilibrium systems has implications extending beyond the problem of predator-mediated coexistence. The structure of ecological theory is heavily influenced by its spatial and temporal scales of reference.

A theory which focuses on a single closed cell, on a time scale corresponding to a single extinction, will view zero as an absorbing state for the system. Since extinction in such a system is permanent, this theory will devote considerable energy to the search for conditions which guarantee the existence of stable equilibria.

A theory encompassing many cells, over a time scale of many local extinctions, will have very different properties. It will differentiate, implicitly or explicitly, between the extinction of local populations and of entire species. Local extinctions will be regarded as relatively common, and, since such extinction is not an absorbing state, this theory will not be overly concerned with conditions leading to stable equilibria. Dispersal, migration, and colonization will be fundamental to such a theory, and spatial effects likely will be explicitly considered.

The arguments which raged in the 1950s over density dependence are one example of this. In addition to the empirical evidence mustered by each side, this dispute was characterized by a peculiar logical argument. The density-dependent theorists, largely inspired by closed laboratory systems (Gause 1934; Nicholson 1957), argued that density-dependent effects are necessary to generate stable equilibria and that stable equilibria are necessary to prevent extinction. The champions of density independence replied,

Nevertheless, the misconception prevails that the extinction of a population is a very rare event. This leads our colleagues who hold to the dogma of "density-dependent factors" to propound this riddle. On hearing us expound our views on ecology, they ask: "How is it, if there is no density-dependent factor in the environment, that the population does not become extinct?" This places us in a position like that of the man in the dock who was asked to answer Yes or No to the question: "Do you still beat your wife?" We cannot answer the question until we have cleared up the misconception in the mind of the questioner ... we can witness the extinction of local populations, of even the most abundant species, going on all around us all the time. [Andrewartha and Birch 1954, pp. 663 ff.]

That they were defending an open-system, nonequilibrium theory of population dynamics can be seen in their explicit distinction between extinction of local populations and of entire species, and in their insistence (reflected in the title of their book) that distribution and abundance be considered together.

The frequency of local extinction has since been documented in studies of island communities (Mayr 1965; MacArthur and Wilson 1967; Simberloff and Wilson 1969; Diamond 1969; Simberloff 1974) although the quality of some of these data is now in doubt (Lynch and Johnson 1974). The theory of island biogeography (MacArthur and Wilson 1967; Simberloff 1974) maps this open-system nonequilibrium view of populations upward to the community level. This theory focuses on an equilibrium in species number, generated by an implicit, open-system, nonequilibrium conception of the dynamics of the species. The unexpectedly high species turnover rates discovered in these studies are clearly a nonequilibrium phenomenon.

While the problem of succession is only beginning to be rigorously formulated in terms of population theory, it also seems to involve open-system, nonequilibrium phenomena. Successional development is strewn with the corpses of locally extinct populations (not species); this fact is so obvious that it excites little attention until a conceptual framework is available in which it is important. The process of dispersal, necessary to supply colonists to open systems, has always figured prominently in succession theory (Thoreau 1860; Keever 1950; Drury and Nisbet 1973; Horn 1975). There is recent evidence strongly suggesting that in the mature stages of succession diversity declines in the absence of disturbance (Loucks 1970; Auclair and Goff 1971; Wright and Heinselman 1973; Caswell 1976).

With the seeming defeat of density independence, equilibrium theory became the dominant mode of explanation in population biology in the late 1960s. Nonequilibrium ideas remained in evidence (Hutchinson [1951, 1961, 1965] in particular seems to have been well aware of the distinction), but until very recently the mainstream of ecological thought has regarded them as special cases and sought explanations for community structure primarily in terms of equilibria of species interactions.

The fact that nonequilibrium situations in multicell connected systems can persist for immense periods of time (Ashby 1960), and that such systems can be easily interpreted in terms of ecological communities, suggests the need to evaluate the importance of nonequilibrium coexistence. It certainly will not do to suggest that communities as a whole are entirely equilibrium or entirely nonequilibrium systems. It seems more reasonable to view the populations within a community as existing in a spectrum of cell sizes, reflecting within-cell equilibria and between-cell disequilibria, modified by biotic and abiotic disturbance on a variety of levels. Perhaps a community consists of a core of dominant species, which interact strongly enough among themselves to arrive at equilibrium, surrounded by a larger set of nonequilibrium species playing out their roles against the backdrop of the equilibrium species (see, e.g., Diamond 1975).

Certainly, to evaluate this requires looking at the whole spectrum of abundance in the community. The diversity trends analyzed by Loucks (1970), and Auclair and Goff (1971), and Caswell (1976) are a promising approach. The correlation of rarity with spatial clumping demonstrated in soil mites (Hairston 1959), periodical cicadas (Dybas and Davis 1962), and snails in Egyptian and Sudanese canals (Hairston, personal communication) is also suggestive. Developing these ideas rigorously will require much more work on the transient behavior of ecological systems, the topological properties of multispecies interactive models, and the properties of open systems.

SUMMARY

There is now convincing evidence, from a variety of ecological systems, that predation is capable of maintaining coexistence among a set of competing prey species, some of which would be excluded in its absence. This interaction has been suggested as a major factor determining the structure of some communities, but attempts to incorporate it into the mathematical framework of population theory have been frustrating. Although the possibility of predator-mediated coexistence is easily shown, parameter-space studies of simple three-species models suggest that it is an improbable occurrence, requiring a very delicate balancing of parameter values.

To attack this problem, I classify population systems as open or closed, and equilibrium or nonequilibrium. Closed systems consist of a single homogeneous patch of habitat; open systems, in their simplest form, are a collection of such patches (or cells) connected by migration. Equilibrium theories are restricted to behavior at or near an equilibrium point, while nonequilibrium theories explicitly consider the transient behavior of the system. Almost all of the work on predator-mediated coexistence has been limited to closed, equilibrium systems. In such studies conditions are sought which guarantee that predation results in a stable equilibrium at which all species are present.

A general property of open systems is that transient, nonequilibrium behaviors may persist for extremely long periods of time. A model is presented which uses this fact to generate long-term, but nonequilibrial, coexistence among competitors under the impact of predation. The model is a discrete, or logical, model, in which presence and absence of each of two competitors and a predator is followed in a set of stochastically connected local population cells. The predator acts to open up new cells for nonequilibrium growth of the prey species. All forms of predator-mediated coexistence other than the open-system, nonequilibrium effect were purposely eliminated from the model.

The results of the model clearly demonstrate the possibility of long-term predator-mediated coexistence in such a system. In spite of stacking the deck against it, the positive effect of the predator on coexistence is statistically highly significant. Moreover, a crude evolutionary analysis suggests that the effect is not only possible, but probable. These results can be generalized far beyond the highly abstracted framework of the model. Such generalizations suggest that this phenomenon may be of major importance in natural systems.

The predictions of the model can be tested against the real world in several ways. First, it predicts the possibility of predator-mediated coexistence, which agrees with the numerous observations of that phenomenon. More importantly, it makes predictions (concerning the spatial and temporal organization of the predation process) that distinguish it from the other modes of predator-mediated coexistence. These predictions are corroborated by observations of a number of cases where predation has failed to generate coexistence.

The distinction between closed and open, equilibrium and nonequilibrium, population systems has an important impact on the form and substance of ecological theory.

ACKNOWLEDGMENTS

During its long gestation, this paper benefited from discussion with the ecology group at Michigan State University, and from B. Maguire, Jr., G. Bush, L. Lawlor, D. Janzen, and S. Levin. The original impetus for the work was provided by Don Beaver. Support from NSF grants GI-20, GB-43378, and DEB76-19278 is gratefully acknowledged.

LITERATURE CITED

Abele, L. G. 1976. Comparative species richness in fluctuating and constant environments: coral-associated decapod crustaceans. Science 192:461–463.

Adicott, J. F. 1974. Predation and prey community structure: an experimental study of the effect of mosquito larvae on the protozoan communities of pitcher plants. Ecology 55:475–492.

Alexander, M. 1964. Notes on the synthesis of form. Harvard University Press, Cambridge, Mass.

Andrewartha, H. G., and L. C. Birch. 1954. The distribution and abundance of animals. University of Chicago Press, Chicago.

Arbib, M. A. 1966. Automata theory and control theory—a rapprochement. Automatica 3:161–189.

Ashby, W. R. 1960. Design for a brain. 2d ed. Chapman-Hall, London.

Auclair, A., and F. G. Goff. 1971. Diversity relations of upland forests in the western Great Lakes area. Amer. Natur. 105:499–528.

Ayala, F. J. 1972. Competition between species. Amer. Sci. 60:348–357.

Brooks, J. L., and S. Dodson. 1965. Predation, body size, and composition of plankton. Science 150:28–35.

Cairns, J., K. L. Dickson, and W. H. Yongue. 1971. The consequences of nonselective periodic removal of portions of fresh-water protozoan communities. Trans. Amer. Microscop. Soc. 90:71–80.

Caswell, H. 1972. A simulation study of a time-lag population model. J. Theoret. Biol. 34:419–439.

———. 1976. Community structure: a neutral model analysis. Ecol. Monogr. 46:327–354.

Cohen, J. E. 1970. A Markov contigency table model for replicated Lotka-Volterra systems near equilibrium. Amer. Natur. 104:547–559.

Connell, J. H. 1970. On the role of natural enemies in preventing competitive exclusion in some marine animals and in rain forest trees. Pages 298–312 in P. J. DenBoer and G. R. Gradwell, eds. Dynamics of populations. Pudoc, Netherlands.

Cramer, N. F., and R. M. May. 1971. Interspecific competition, predation, and species diversity: a comment. J. Theoret. Biol. 34:289–293.

Crow, J. F., and M. Kimura. 1970. An introduction to population genetics theory. Harper & Row, New York.

152   THE AMERICAN NATURALIST

Darwin, C. 1859. On the origin of species. Reprint ed. Harvard University Press, Cambridge, Mass.,
     1964.
Dayton, P. K. 1971. Competition, disturbance, and community organization: The provision and
     subsequent utilization of space in a rocky intertidal community. Ecol. Monogr.
     41:351–389.
————. 1972. Toward an understanding of community resilience and the potential effects of
     enrichment to the benthos at McMurdo Sound, Antarctica. Pages 81–95 in B. C. Parker,
     ed. Proceedings of the colloquium on conservation problems in Antarctica. Allen,
     Lawrence, Kans.
Diamond, J. M. 1969. Avifaunal equilibria and species turnover rates on the Channel Islands of
     California. Proc. Nat. Acad. Sci. USA 64:57–63.
————. 1975. Assembly of species communities. Pages 342–444 in M. L. Cody and J. M. Diamond,
     eds. Ecology and evolution of communities. Harvard University Press, Cambridge, Mass.
Dodson, S. I. 1974. Zooplankton competition and predation: an experimental test of the size-
     efficiency hypothesis. Ecology 55:605–613.
Drury, W. H., and I. C. T. Nisbet. 1973. Succession. J. Arnold Arboretum 54:331–368.
Dubois, D. M. 1975. A model of patchiness for prey-predator plankton populations. Ecol.
     Modelling 1:67–80.
Dybas, H. S., and D. D. Davis. 1962. A population census of seventeen-year periodical cicadas
     (Homoptera: Cicadidae; *Magicicada*). Ecology 43:432–444.
Feller, W. 1968. An introduction to probability theory and its applications. Vol. 1. Wiley, New
     York.
Gause, G. F. 1934. The struggle for existence. Reprint ed. Dover, New York, 1971.
Geertz, C. 1969. Two types of ecosystems. Pages 3–28 in A. P. Vayda, ed. Environment and cultural
     behavior: ecological studies in cultural anthropology. Natural History Press, New York.
Gilpin, M. E. 1975. Group selection in predator-prey communities. Princeton University Press,
     Princeton, N.J.
Glansdorff, P., and I. Prigogine. 1971. Thermodynamic theory of structure, stability and fluctua-
     tions. Wiley-Interscience, London.
Hairston, N. G. 1959. Species abundance and community organization. Ecology 40:404–416.
Hall, D. J., W. E. Cooper, and E. E. Werner. 1970. An experimental approach to the production
     dynamics and structure of freshwater animal communities. Limnol. Oceanogr.
     15:839–928.
Harper, J. L. 1969. The role of predation in vegetational diversity. Brookhaven Symp. Biol.
     22:48–62.
Horn, H. 1975. Markovian properties of forest succession. Pages 196–213 in M. L. Cody and J. M.
     Diamond, eds. Ecology and evolution of communities. Harvard University Press,
     Cambridge, Mass.
Horn, H., and R. H. MacArthur. 1972. Competition among fugitive species in a harlequin
     environment. Ecology 53:749–752.
Huffaker, C. B. 1958. Experimental studies on predation: dispersion factors and predator-prey
     oscillations. Hilgardia 27:343–383.
Hurlbert, S. H., J. Zedler, and D. Fairbanks. 1972. Ecosystem alteration by mosquitofish (*Gambusia
     affinis*) predation. Science 175:639–641.
Hutchinson, G. E. 1951. Copepodology for the ornithologist. Ecology 32:571–577.
————. 1953. The concept of pattern in ecology. Proc. Acad. Natur. Sci. Philadelphia 105:1–12.
————. 1961. The paradox of the plankton. Amer. Natur. 95:137–1945.
————. 1965. The ecological theater and the evolutionary play. Yale University Press, New Haven,
     Conn.
Iberall, A. S. 1972. Toward a general theory of viable systems. McGraw-Hill, New York.
Jackson, J. B. C., and L. Buss. 1975. Allelopathy and spatial competition among coral reef
     invertebrates. Proc. Nat. Acad. Sci. USA 72:5160–5163.
Janzen, D. H. 1967. Why mountain passes are higher in the tropics. Amer. Natur. 101:233–249.
————. 1970. Herbivores and the number of tree species in tropical forests. Amer. Natur.
     104:501–528.
————. 1976. The depression of reptile biomass by large herbivores. Amer. Natur. 110:371–400.

Karlin, S. 1966. A first course in stochastic processes. Academic Press, New York.

Kauffman, S. 1969. Metabolic stability and epigenesis in randomly constructed genetic nets. J. Theoret. Biol. 22:437–467.

———. 1971. The organization of cellular genetic control systems. Pages 65–116 *in* Lectures on mathematics in the life sciences. Vol. 3. American Mathematical Society, Providence, R.I.

Keever, C. 1950. Causes of succession on old fields of the Piedmont, N.C. Ecol. Monogr. 20:229–250.

Kettlewell, H. B. D. 1955. Selection experiments on industrial melanism in the Lepidoptera. Heredity 9:323–342.

Kirk, R. E. 1968. Experimental design procedures for the behavioral sciences. Brooks/Cole, Belmont, Calif.

Laws, R. M. 1970. Elephants as agents of habitat and landscape change in East Africa. Oikos 21:1–15.

Lees, D. R., and E. R. Creed. 1975. Industrial melanism in *Biston betularia*: the role of selective predation. J. Anim. Ecol. 44:67–84.

Levin, S. 1974. Dispersion and population interactions. Amer. Natur. 108:207–228.

Levin, S., and R. T. Paine. 1974. Disturbance, patch formation, and community structure. Proc. Nat. Acad. Sci. USA 71:2744–2747.

Levin, S. and L. A. Segel. 1976. An hypothesis for the origin of planktonic patchiness. Nature 259:659.

Levins, R. 1970. Extinction. Pages 77–107 *in* M. Gerstenhaber, ed. Lectures on mathematics in the life sciences. Vol. 2. American Mathematical Society, Providence, R.I.

Levins, R., and D. Culver. 1971. Regional coexistence of species and competition between rare species. Proc. Nat. Acad. Sci. USA 68:1246–1248.

Loucks, O. 1970. Evolution of diversity, efficiency and community stability. Amer. Zool. 10:17–25.

Luckenbill, L. S. 1973. Coexistence in laboratory populations of *Paramecium aurelia* and its predator, *Didinium nasutum*. Ecology 54:1320–1327.

———. 1974. The effects of space and enrichment on a predator-prey system. Ecology 55:1142–1147.

Lynch, J. F., and N. K. Johnson. 1974. Turnover and equilibria in insular avifaunas, with special reference to the California Channel Islands. Condor 76:370–384.

MacArthur, R. H., and E. O. Wilson. 1967. The theory of island biogeography. Princeton University Press, Princeton, N.J.

May, R. 1973. Stability and complexity in model ecosystems. Princeton University Press, Princeton, N.J.

Mayfield, H. 1960. The Kirtlands warbler. Cranbrook Institute of Science, Bloomfield Hills, Mich.

Mayr, E. 1965. Avifauna: turnover on islands. Science 150:1587–1588.

Miller, R. S. 1969. Competition and species diversity. Brookhaven Symp. Biol. 22:63–70.

Neill, W. 1972. Effects of size-selective predation on community structure in laboratory aquatic microcosms. Ph.D. thesis. University of Texas.

Nicholson, A. J. 1957. The self-adjustment of populations to change. Cold Springs Harbor Symp. Quant. Biol. 22:153–174.

Nicholson, I. A., I. S. Patterson, and A. Currie. 1970. A study of vegetational dynamics: selection by sheep and cattle in *Nardus* pasture. Pages 129–143 *in* A. Watson, ed. Animal populations in relation to their food resources. Blackwell Scientific, Oxford.

Osman, R. W. 1977. The establishment and development of a marine epifaunal community. Ecol. Monogr. 47:37–63.

Paine, R. T. 1966. Food web complexity and species diversity. Amer. Natur. 100:65–75.

———. 1971. A short-term experimental investigation of resource partitioning in a New Zealand rocky intertidal habitat. Ecology 52:1096–1106.

Paine, R. T., and R. L. Vadas. 1969. The effects of grazing by sea urchins, *Strongylocentrotus spp.* on benthic algal populations. Limnol. Oceanogr. 14:710–719.

Parrish, J. D., and S. B. Saila. 1970. Interspecific competition, predation and species diversity. J. Theoret. Biol. 27:207–220.

Pimentel, D., W. P. Nagel, and J. L. Madden. 1963. Space-time structure of the environment and the survival of parasite-host systems. Amer. Natur. 97:141–167.

Platt, W. J. 1975. The colonization and formation of equilibrium plant species associations on badger disturbances in a tall-grass prairie. Ecol. Monogr. 45:285–305.

Porter, J. W. 1972. Predation by *Acanthaster* and its effect on coral species diversity. Amer. Natur. 106:487.

Ratcliffe, F. N. 1959. The rabbit in Australia. Pages 545–564 *in* A. Keast, R. L. Crocker, and C. S. Christian, eds. Biogeography and ecology in Australia. Junk, The Hague.

Rosen, R. 1968. Discrete and continuous representations of metabolic models. Int. Rev. Cytol. 23:25–32.

Roughgarden, J., and M. Feldman. 1975. Species packing and predation pressure. Ecology 56:489–492.

Simberloff, D. 1974. Equilibrium theory of island biogeography and ecology. Annu. Rev. Ecol. and Syst. 5:161–182.

Simberloff, D., and E. O. Wilson. 1969. Experimental zoogeography of islands: the colonization of empty islands. Ecology 50:278–296.

Simon, H. A. 1962. The architecture of complexity. Proc. Amer. Phil. Soc. 106:467–482.

Slatkin, M. 1974. Competition and regional coexistence. Ecology 55:128–134.

Slobodkin, L. B. 1961. The growth and regulation of animal populations. Holt, Rinehart & Winston, New York.

———. 1964. Experimental populations of Hydrida. J. Anim. Ecol. 33:131–148.

Summerhayes, V. S. 1941. The effect of voles (*Microtus agrestis*) on vegetation. J. Ecol. 29:14–48.

Taylor, D. L. 1973. Some ecological implication of forest fire control in Yellowstone National Park, Wyoming. Ecology 54:1394–1396.

Thom, R. 1975. Structural stability and morphogenesis. Benjamin, Reading, Mass.

Thoreau, H. D. 1860. Succession of forest trees. Pages 11–23 *in* Eighth annual report of the secretary of the Massachusetts board of agriculture. (Boston, Mass.)

Wells, L. 1970. Effects of alewife predation on zooplankton populations in Lake Michigan. Limnol. Oceanogr. 15:556–565.

Werner, P. A. 1972. The effect of the invasion of *Dipsacus sylvestris* on plant communities in early old-field succession. Ph.D. thesis. Michigan State University.

Wright, H. E., and M. L. Heinselman, eds. 1973. The ecological role of fire in natural conifer forests of Western and Northern America. Quaternary Res. 3:317–513.

Yodzis, P. 1976. The effects of harvesting on competitive systems. Bull. Math. Biol. 38:97–109.

———. 1977. Harvesting and limiting similarity. Amer. Natur. 111:833–843.

Zaret, T. 1972. Predators, invisible prey, and the nature of polymorphism in the Cladocera (class Crustacea). Limnol. Oceanogr. 17:171–184.

2 Competition, Coexistence, and Extinction

Thomas E. Miller and Greg Cooper

Much of twentieth-century ecology was dominated by three fundamental questions. The first of these questions concerns population regulation. Why do populations in nature fluctuate within rather limited bounds in comparison to what might be possible? For some ecologists working in the first half of the century it was close to a conceptual truth that populations must be regulated by density-dependent factors, the most likely candidate being intraspecific competition (see Nicholson and Bailey 1935 [reprinted in Real and Brown 1991]; Lack 1954). Those arguing the other side emphasized the importance of stochastic abiotic factors (Andrewartha and Birch 1954). This debate has continued throughout the recent history of ecology (Strong et al. 1984; Strong 1986; see discussion in part 5 of this volume). The second fundamental question directly asked about community patterns. In his presidential address to the American Society of Naturalists, Hutchinson asked: Why are there so many kinds of animals? (see Hutchinson 1959 [reprinted in Real and Brown 1991]). The nature and causes of species diversity had been a background issue for ecology for a long time, however, Hutchinson's seminal paper brought it center stage, where it has remained ever since. The related third fundamental question was raised most clearly by Elton (1927): why do the organisms that comprise ecological communities consist of a limited subset of those that might occur together? Ecologists have been searching for the conditions that limit community membership ever since.

Thanks to another seminal paper by Hutchinson (1957 [reprinted in Real and Brown 1991]) as well as pioneering work by Robert MacArthur (e.g., MacArthur 1958 [reprinted in Real and Brown 1991]; MacArthur and Levins 1967), answers to these questions began to coalesce in the late 1950s and 1960s. In fact, the Hutchinson/MacArthur paradigm offered a univocal answer to all three questions: competition. Competition for resources was taken to be the central mechanism that regulated populations, generated the adaptive changes that lead to taxonomic diversity, and determined which species can coexist due to niche differences. This approach also made apparent the Darwinian roots of ecology, and offered the potential for the development of relevant theory. This capacity for unification goes a long way towards explaining the dominance of competition and niche concepts in ecology throughout the sixties and much of the seventies, though, as

151

it turns out, the connection with evolution did not depend on the prominence of competition (see part 6 of this volume).

However, while both theory and experiments supporting competition ruled the literature, competition was an uneasy crown to wear for many ecologists. The eight papers in part 2 are all, in one way or another, concerned with the limitations of the Hutchinson/MacArthur paradigm. The challenges came from a number of directions and with varying degrees of sympathy for the paradigm. Furthermore, these papers are concerned primarily with that aspect of the paradigm that addresses the Eltonian question of the conditions that limit community membership. The most sympathetic challenges do not call into question the soundness of resource partitioning as the key to understanding community structure; however, they do call for a more complete, and more mechanistic, understanding of the processes involved. Other papers in this volume are more of a direct attack on competition, suggesting that it may not be the only, or even the central, mechanism shaping patterns of coexistence. These papers call attention to alternative processes such as predation (see part 1 in this volume), parasitism, and environmental variability (see part 4 of this volume). Finally, there are challenges to the very reality of the phenomena that the Hutchinson/MacArthur paradigm sees itself as explaining—are the patterns of community membership really distinct from what we would expect by chance? We will discuss each of these challenges in turn.

Theory associated with competition and the niche was a key component of the Hutchinson/MacArthur paradigm. The predominant model of multispecies species competition was and arguably still is the Lotka-Volterra competition equation, where the density of a species is assumed to have negative linear effects on the growth rate of a competitor and the effects of multiple competitors are additive (see Volterra 1926 [reprinted in Real and Brown 1991]; MacArthur and Levins 1967). This model and its well-known graphical representations with isoclines have both tortured and enticed under-

graduate students in ecology for the last fifty years. It has also formed the backbone for the development of much of competition theory. For example, Robert May's (1972) use of community matrices to explore the relationships among interaction strength, diversity, and stability helped to frame many later questions about the determinants of species diversity (see part 4 of this volume for further discussion)

Sympathetic Challenges

There is a deceptive simplicity to these models that led to a number of predictions that are still debated today. These include the ideas that for coexistence, intraspecific competition must be greater than interspecific competition (e.g., Siepielski and McPeek 2010; Barabás et al. 2016) and that the number of coexisting species must be equal to or less than the number of limiting resources. The first three papers in part 2 challenged some of the assumptions and predictions of the Lotka-Volterra models and, in doing so, demonstrated that even the simplest competition theory can lead to complex outcomes.

Part 2 begins with one of the first papers published by John Vandermeer (1969) titled "The competitive structure of communities: an experimental approach with protozoa." He explored the concept of additivity among competitors and proposed that "higher-order" interactions among species are likely to occur. Vandermeer defined "higher-order" to be when the per-individual effect of one species on the growth rate of another is also the function of the abundance of a third species; he said that this would lead to "interactive" communities. He recognized that higher-order terms would lead to a much more diverse set of outcomes that would be hard to predict and "difficult to work with analytically" (Vandermeer 1969). In fact, it would require that ecologists consider much more complex mathematical models to understand species interactions.

Vandermeer's goal was to simply show that these effects qualitatively occurred. He was not abandoning the competition-based paradigm of the time, but simply arguing that the simplistic view of competition of that time might be insufficient. Interestingly, Vandermeer chose an experimental system that he felt was unlikely to have higher-order interactions, under the assumption that if he could demonstrate such interactions, then perhaps they were general properties of communities To this end, he chose to work with ciliates, whose monoculture growth in the laboratory was well known to conform to simple population models. Ultimately, he concluded that simple equations actually were sufficient to describe the protozoa system, although as he later noted, it really depends on the context of what they are asked to do (Vandermeer 1981). However, for protozoa, the additive linear equations were sufficient to predict which species would persist and the general patterns of population change.

Though his strategy for demonstrating the importance of higher-order interactions did not quite work out, his ideas took hold. Vandermeer's concept of higher-order interactions was very influential for both the experimental and theoretical work that followed over the next two decades. Following soon thereafter, Wilbur's 1972 work (reprinted in part 1 of this volume) on competition among larval amphibians, Ayala et al.'s 1973 studies with *Drosophila* (Ayala et al. 1973), and Neill's 1974 work with algae in lakes (Neill 1974) all suggested that linear models of species interactions often were insufficient to explain experimental results. Vandermeer's paper also stimulated thinking about a broader class of interactions, indirect effects. Indirect effects, defined as how "one species alters the effect that another species has on a third" (Abrams 1987), may include everything from trophic cascades (e.g., Hairston et al. 1960 [reprinted in Real and Brown 1991]) and apparent competition (Holt 1977; Holt and Bonsall 2017), to indirect mutualisms within trophic levels (e.g., Stone and Roberts 1991; Miller 1994). Vandermeer's study of protozoa stimulated ecologists to think about how indirect interactions occur, as well as how to mathematically describe such interactions (see Bender et al. 1984 [reprinted in part 4 of this volume];

Miller and Kerfoot 1987; Strauss 1991; Wootton 1992). The types of interactions originally discussed by Vandermeer are now generally divided into numerical or density-mediated indirect effects (e.g., Strauss 1991; Wootton 1994) and behavioral or "trait-mediated" indirect effects (e.g., Werner and Peacor 2003). Both the concepts and the terminology continue to bedevil community ecologists. Vandermeer continued in his career to ask questions about multispecies interactions in a variety of systems, including agricultural work on intercropping and pest management (e.g., Vandermeer 1981; Vandermeer and Perfecto 2019).

One especially important prediction from the Lotka-Volterra competition equations was that the number of "resources" determined the maximum number of coexisting species (e.g., MacArthur and Levins 1967; Levin 1970). This extended an idea from Gause (1932) that one species can dominate any single given resource or niche: this is commonly referred to as the "competitive exclusion principle" (Hardin 1960). This principle left ecologists asking whether there were really enough resources or niche space in natural systems to allow the observed diversity (e.g., Hutchinson 1961; Grubb 1977). Relaxing particular assumptions of the standard Lotka-Volterra theory can allow coexistence of more than one species on a single resource; however, most such models described ways in which one resource could act as two. For example, Haigh and Maynard Smith (1972) showed that two predators could coexist on a single prey if each specialized on different life stages of the prey. However, this general prediction that the number of coexisting species does not exceed the number of available resources has remained an important prediction from competition theory. Simple resource-ratio models (see discusson of Tilman 1977, below) result in a similar prediction.

Robert Armstrong and Richard McGehee published several papers on the coexistence of competitors, most notably their 1980 work simply titled "Competitive exclusion" (Armstrong and McGehee 1980). This paper reevaluated some of the major assumptions of Lotka-

Volterra theory and proposed that the theory may be in fact too limiting. They both summarized previous models and presented some new approaches to understanding how more than one species can coexist per limiting resource. In particular, they noted that there are two problematic assumptions of the standard Lotka-Volterra equations. First, many of these models assumed that growth rates will be linear functions of resource availability. However, other types of functional response curves are known and have been shown to allow greater coexistence. Second, most models investigated single-point equilibria where densities stabilize to some constant value. Previous models had not considered nonequilibrium conditions. Relaxing these two assumptions, linearity and point-stability, allows the number of coexisting species to be free of the number of resources. Armstrong and McGehee (1980) presented a thorough explanation of coexistence on a single resource resulting from different nonlinear responses to a shared and fluctuating resource and showed that coexistence could be generated by endogenous population cycles that create a trade-off allowing each species to dominate at different times in the cycle (e.g., Okuyama 2015). Armstrong and McGehee also took pains to note the conditions under which the number of species cannot exceed the number of resources.

An important legacy of Armstrong and McGehee (1980) was to move ecology beyond a strict and simple interpretation of the competitive exclusion principle. Their paper stimulated thinking about the linearity and stability assumptions in previous work (Chesson 2000). Assumptions of stability were already being challenged by work on disturbance theory and equivalence (see discussion of Hubbell 1979 below); however, this is very different from the coexistence with endogenously generated oscillations envisioned by Armstrong and McGehee. Subsequent theoretical work relaxed other assumptions of the Lotka-Volterra model, including assumptions about spatial distributions of individuals both within (e.g., metapopulations; Levins 1969) and among species

(e.g., metacommunities; Leibold et al. 2004) or assumptions about time and equilibria (Sale 1977 and below; Chesson and Warner 1981; Chesson 2000). These ideas were particularly important for the development of ideas in applied and conservation ecology (see part 5 of this volume).

The Lotka-Volterra approach to competition persists today because it is relatively simple, especially near equilibrium, and provides a good first-order description of observed patterns. However, while mathematically convenient, the terms in these models confound traits of the species with characteristics of the environment. For example, the carrying capacity reflects both the availability of resources and the efficiency with which a species uses these resources. More mechanistic models began to be developed in the early 1970s, in which the growth was described explicitly as a function of resource availability, which was in turn a function of species abundances and resource consumption (e.g., MacArthur 1972; Stewart and Levin 1973; Neill 1974). In these models, competition is explicitly modeled through the effects of species on the dynamics of shared resources.

One of the first people to successfully apply these models was David Tilman, who used algal systems for one of the first applications of MacArthur's original model. Tilman came out of a limnology tradition in Peter Kilham's lab at the University of Michigan. His 1977 paper "Resource competition between planktonic algae: an experimental and theoretical approach" describing competition between two species of freshwater algae (Tilman 1977) was the start of a body of work that forced ecology to rethink how we view competition and coexistence. Using what are referred to as "resource-ratio" models, Tilman extended MacArthur's (1972) original model of multispecies competition in which the dominant competitor is defined as the species that can persist at the lowest resource levels. Tilman went on to provide several clear descriptions of resource-ratio theory (Tilman 1980, 1982) and later conducted several experimental tests with terrestrial plants (Tilman 1984; Tilman and Wedin 1991; Wedin and Tilman 1993). Further, he explored a diversity of other major questions in ecology with this theory, including succession, invasion, diversity-productivity relationships, and even the extension of resource-ratio theory to other types of species interactions, such as predation (Tilman 1982).

Resource-ratio theory represented a paradigm shift in ecology for at least two reasons. First, the previous understanding of competition (i.e., Lotka-Volterra) was largely phenomenological, while with resource-ratio theory the mechanisms behind resource availability and use, coexistence, and diversity were explicit (see Chase and Leibold 2003). Resource competition was now explicitly an indirect effect through shared resources, and so recognized that the dynamics of the resources themselves are also important. Second, resource-ratio theory led to important arguments about the definition of competition itself. The Hutchinson-MacArthur paradigm generally assumed that the species that could take up the most resources was the better competitor. In resource-ratio theory, individual growth is based on Monod curves and the best competitor becomes the species that can survive at the lowest resource levels. This prediction did not conform to previous ideas about the nature of competition, competitive traits, and competitive outcomes, and so instigated an important argument about the nature of competition (Grace and Tilman 2012) that has continued ever since (e.g., Craine 2005). In fact, the role of resource-ratio theory in ecology is still evolving (e.g., Chase and Leibold 2003) and, almost 40 years after Tilman's original papers, there are remarkably few tests of the theory in natural systems (Miller et al. 2005). It is also telling that almost all introductory ecology texts still teach the simpler Lotka-Volterra equations and do not mention resource-ratio ideas or other explicit consumer-resource models. It may be that since the resource-ratio theory is more conceptually complex than the Lotka-Volterra models, it will always remain a lesser-used view of competition.

These three papers (Vandermeer 1969; Armstrong and McGehee 1980; Tilman 1977) fall in the category of sympathetic challenges to the Hutchinson/MacArthur paradigm and illustrate the pervasive influence of competition during this period. However, there were thoughtful criticisms that more directly challenged the preeminence of competition. For example, some authors were asking whether observed community patterns were really consistent with niche partitioning among competitors. One of the earliest of these papers, and included here, was "Taxonomic diversity of island biotas" by Daniel Simberloff in 1970, which quantified co-occurrence patterns among related species on islands (Simberloff 1970; see also Pielou and Pielou 1968). Niche theory predicted that islands would be colonized by species with low relatedness (MacArthur and Wilson 1967); instead, Simberloff found that the actual ratio of species to genera on islands was higher than that expected by chance. He interpreted this to mean that competition was not the primary force determining species assemblages and that instead species that have similar ecologies (niches) or dispersal abilities may actually be more likely to co-occur.

Simberloff's paper directly challenged earlier papers, specifically Moreau (1966) and Grant (1966), and was just the first of several influential papers published by Simberloff and colleagues that challenged the competition paradigm. Connor and Simberloff (1979) investigated patterns of species co-occurrence on islands, suggesting that the patterns were generally random, rather than supportive of competition. Strong et al. (1979) and Simberloff and Boecklen (1981) proposed statistical tests of character displacement patterns and also found that many well-known patterns of character differences from the literature (e.g., Darwin's finches) could not be differentiated from random patterns or were otherwise inconsistent with competition-driven mechanisms. There were numerous rebuttals and defenses associated with these papers (e.g., Grant and Abbott 1980; Diamond and Gilpin 1982); at the time, Simberloff and colleagues were thought by some to be deconstructionists and were referred to as the "Florida State mafia."

This literature was sometimes contentious, as it often necessarily described perceived flaws in the methods and/or conclusions of specific previous studies. It ultimately forced a major reevaluation of the existence of character displacement and competition and evidence for the Hutchinson/MacArthur paradigm. Importantly, it also forced a reevaluation of the methods used in ecology and evolutionary biology and, in particular, promoted the use of null models (see Gotelli and Graves 1996).

Is There More Than Competition?

While mechanisms promoting diversity were largely associated with competition, the importance of other types of species interactions were also being discussed. It had long been recognized that predation could limit population growth, but it was not generally thought to promote coexistence. However, three papers in particular described how predation could also be important for explaining diversity. The first was Robert Paine (1966 [reprinted in Real and Brown 1991]). In this paper, Paine suggested that predators could minimize the abundance of dominant prey species, allowing poorer competitors to persist and thereby increasing diversity. Paine (1969) later termed this "keystone predation" and in various papers provided several examples from marine systems. Keystone species have since been identified in a wide variety of ecosystems.

The second paper, included in this volume, was "Herbivores and the number of tree species in tropical forests" by Daniel Janzen (1970), whose work was inspired by the high diversity of tropical communities. Rain forests, with hundreds of species of trees per hectare, provide a particular challenge to niche theory, as the abiotic environment seems insufficiently diverse to allow resource partitioning. Janzen was a pioneer in ecological studies in tropical systems and is particularly well known for promoting tropical conservation biology, especially in Guanacaste, Costa Rica. He suggested that predation could operate in a density-dependent

fashion to promote diversity among tropical forest trees. His early work was well known for demonstrating an ant-plant mutualism, whereby ants protect some tropical plants from herbivory (Janzen 1966). Janzen noticed that few new adult trees generally became established around older adult trees of the same species. He suggested that this was due to localized predation or plant diseases being more prevalent near established adults (Janzen 1970). Seeds are more likely to fall near adults but, because of disease and predation or herbivory, more likely to establish far from adults. This dynamic prevents dominance by a single species in local patches. Equal credit should go to Joseph Connell, who proposed a similar idea (Connell 1970a), and so the concept has been called the Janzen-Connell hypothesis.

The Janzen-Connell hypothesis is unique in several ways. It does not require competition and so coexistence is not based on resource partitioning. In a sense, it is a niche-based model in that it requires species-specific predators. But what makes it most interesting as an explanation for tropical tree diversity is the density dependence that prevents monopoly by a single species. This is another nonequilibrium explanation for diversity, as species are not expected to replace themselves. By extension, it also suggests that the monoculture stands of trees more often found in temperate forests may be due to the ability of juveniles to become established or replace adults, perhaps because of a lack of species-specific predators or because extreme variability in abiotic conditions provides periods of escape from such predators.

There have been many tests of the Janzen-Connell hypothesis since 1970 (e.g., see Hubbell 1979, discussed below), including studies by Janzen and Connell themselves. Evidence could include distribution patterns or patterns of mortality in adults, seedlings, and juvenile trees. Stronger evidence comes from experimental studies of establishment and growth of propagules as a function of distance from the parents. Surprisingly, two meta-analyses of this growing body of tests of the Janzen-

Connell hypothesis come to diametrically opposite conclusions. Hyatt et al. (2003) found no consistent support and went so far as to state that the issue was resolved and should be pursued no further. However, Comita et al. (2014) found that natural enemies were frequently the cause of density-dependent mortality in plant communities worldwide, strongly supporting the hypothesis. Clearly the role of density-dependent predation in maintaining diversity remains a stimulating, if open, question.

Our third paper to call attention to the role of predation, while also reevaluating the role of competition, is the highly cited "Mechanisms of succession in natural communities and their role in community stability and organization" by Joseph Connell and Ralph Slatyer (1977). Many of the roots of community ecology can be traced back to arguments about plant succession, as was covered in papers by Henry Cowles (1899), Frederic Clements (1936), and Henry Gleason (1926 [reprinted in Real and Brown 1991]). The field was greatly influenced by Clements in the early 1900s, who viewed plant communities as predictable associations of codependent species (see Clements 1936 [reprinted in Real and Brown 1991]; Kingsland 2005). Opposing this view were Gleason (1926) and later Frank Egler (e.g., Egler 1954) and Robert Whittaker (e.g., Whittaker 1965), who all proposed that succession was more of a free market venture, where each species acts independently in its own best interests. The two camps had opposite views on the nature of the interactions among plants, with Clements's "superorganism" view based more on facilitation between species, while Gleason and colleagues were suggesting plant communities developed from suites of independently acting competitors. No single accepted theory of succession developed from this debate; it could be argued that none exists today (McIntosh 1999; Walker and del Moral 2003).

Through much of his career, Slatyer studied how plants differed in physiological traits (e.g., Noble and Slatyer 1980), while Connell was already well known for work in marine inter-

tidal communities (Connell 1961, 1970b), but both were interested in plant succession. Connell and Slatyer (1977) noted that explanations of succession were based on patterns that were poorly documented, often failed to account for factors such as predation and disease, and lacked testable hypotheses. In particular, they specifically challenged the competition-based theory prevalent at that time in plant ecology. They took on this challenge by clearly defining mechanisms by which species change occurs during plant succession. Their facilitation, tolerance, and inhibition models encompassed all the mechanisms of succession discussed in the earlier debate between the Clementsian and Gleasonian views (see part 1 in Real and Brown 1991). They directly associated these early concepts with specific testable mechanisms and evaluated their likelihood based on evidence available at that time, concluding that the concept of species evolving different resource-use strategies to reduce competition was not supported. And importantly, they recognized both facilitation and effects of other trophic levels as important forces modifying the role of direct competition. Like Janzen, they also recognized that localized factors such as herbivory and disease will influence whether species are self-replacing, which has implications for stability and diversity of communities.

It has been argued that the facilitation, tolerance, and inhibition mechanisms described in Connell and Slatyer had been well described earlier (e.g., Pickett et al. 1987; McIntosh 1999) and were too simplistic (e.g., Walker and Chapin 1987), and that Connell and Slatyer only contributed to the confusion of ideas and terms about succession (Pulsford et al. 2016). However, by providing a larger conceptual framework, Connell and Slatyer helped to clarify what we did know about succession and what remained to be investigated and in doing so stimulated later work. Perhaps more importantly, their work was an explicit attack on the Hutchinson/MacArthur competition paradigm, suggesting that "in addition to the competitive interactions between plants or sessile animals, interactions with herbivores, predators, and pathogens are of critical importance in the course of succession" (Connell and Slatyer 1977, 1120).

The penultimate paper in this section is "Maintenance of high diversity in coral reef fish communities" by Peter Sale (1977). Sale also took on accepted dogma while considering the high diversity in tropical coral reefs (Sale 1977). Tropical reefs are thought to have some of the highest diversity on earth, capable of having over 2000 species of fishes on the same reef, and sometimes over 100 species on a single coral head. Such high diversity was taken as evidence that there must be high specialization and reduced overlap in resource use by different species, minimizing competition. Sale noted that, as more information became available, individual species of fish were being shown to have relatively broad resource use with significant overlap among species. He used this to suggest that coral reefs were likely unstable and that diversity was maintained by stochastic migration following small-scale disturbances, which he described as a lottery process.

Sale's paper was very influential because it directly refuted the idea that niche specialization was the only process to explain diversity. Instead, he argued that the limiting factor for many reef species was finding open space during the colonization and establishment phase. Species were in fact all generalists, with successful species dispersing numerous and "small, frail" offspring. However, once established, these individuals could hold onto territory and not be easily displaced. Note that Sale felt competition was still important, both for establishment and for priority effects among established individuals. Yet this competition in a highly unpredictable environment selected for many species sharing the same niche. In an important later work, Chesson and Warner (1981) confirmed Sale's general conclusion that coexistence can occur in such lottery systems, although they make it clear that their "explanations and predictions differ in detail" (1981, 940) from those of Sale.

The Role of Chance in Nature

Sale's ideas were another attack on equilibrium concepts that were often a part of the Hutchinson/MacArthur paradigm. In the competition paradigm, communities were viewed as the endpoints of repeatable and predictable processes (e.g., succession) and were largely at equilibrium. This allowed ecologists to envision a time when we would be able to accurately predict the natural world around us. As Slobodkin stated:

We may reasonably expect to have eventually a complete theory for ecology that will not only provide a guide for the practical solution of land utilization, pest eradication, and exploitation problems but will also permit us to start with an initial set of conditions on the earth's surface (derived from geological data) and construct a model that will incorporate genetics and ecology in such a way as to explain the past and also predict the future of evolution on earth. (Slobodkin 1961, 72)

Sale instead was suggesting a very different worldview in which high variability in the assembly mechanisms, due to both stochastic migration and priority effects, caused assembly to vary unpredictably from site to site. The variation from site to site and across time in coral reef habitats allowed coexistence and diversity at larger spatial and temporal scales. Interestingly, part 1 of this volume describes how the works of Paul Dayton, Joseph Connell, and Hal Caswell around this same time also investigated nonequilibrium ideas.

The final paper in this part of the volume, "Tree dispersion, abundance, and diversity in a tropical dry forest" by Stephen Hubbell (1979), touches on many themes already discussed in this set of papers, including the role of competition, the Janzen-Connell hypothesis, succession, and the predictability of nature. There are really two different studies in this paper, each important in its own way. The primary goal was to analyze the distribution of woody plants within a tropical deciduous forest in Guanacaste, Costa Rica, for evidence of non-random spatial patterns of adults. In particular, overdispersion would support the negative density-dependence inherent in the Janzen-Connell hypothesis. Each individual plant was mapped to determine juvenile and adult densities at different distances from each adult tree for each species. Hubbell found largely clumped or underdispersed patterns, which is contrary to the predictions of Janzen and Connell. Hubbell went on to correlate different distribution patterns of adults and juveniles with specific traits of the plants, such as general abundance, pollination system, and seed size.

This portion of Hubbell's paper was a clever use of spatial patterns to infer factors that determine persistence and coexistence. Hubbell went on in later work with others to repeat this approach on a 50-hectare forest tract on Barro Colorado Island in Panama, in a long-term study where the woody species have been recensused at five-year intervals beginning in 1980. Both this initial study in Costa Rica and the later work on Barro Colorado have served as models for similar studies that have been established in both tropical and temperate forests around the world (Anderson-Texeira et al. 2014).

But the second portion of Hubbell's paper was equally insightful and presents more of a challenge to the Hutchinson/MacArthur paradigm. It was also an early first work on neutral theory, an important concept that wasn't generally recognized by ecologists until more than 20 years later (see Hubbell 2005). Hubbell investigated the patterns in relative abundance of different tree species in the community using rank-abundance curves. On a log scale of abundance, such curves often show a predictable progression from the most to least abundant species that has generally been thought to reflect the relative competitive abilities of the component species (see Whittaker 1965). Hubbell used a simple stochastic model to show that random disturbance can act as a nonequilibrium force to generate log-normal rank-abundance patterns, even if the species are ecologically "neutral" in that they have equal abilities for reproduction, dispersal, and competition. Furthermore, highly speciose communities with stochastic disturbance will lose species at a very slow rate, such that diversity can be maintained among neutral species for meaningful periods of time.

This represents a direct challenge to niche theory, as diversity is now associated with species being more similar to each other, rather than more different as presupposed by niche partitioning. Yet this portion of the paper was probably viewed more as support for Connell's and others' recent work on the importance of disturbance for maintaining diversity than for its implications for the Hutchinson/MacArthur paradigm. It wasn't until some 20 years later that Bell (2000) and Hubbell (2001) more explicitly proposed that diversity might be associated with ecological similarity rather than differences, in what is now recognized as "neutral theory."

The 1960s and 1970s were a time of a significant development of ideas based largely on competition and the niche. This was also a time of great social change, much of which was directed towards environmental issues. Rachel Carson's *Silent Spring* was published in 1962, Paul Ehrlich's *The Population Bomb* was published in 1968, and Norman Borlaug won the Nobel Peace Prize as the "father of the Green Revolution" in 1970. Various versions of the Endangered Species Act were passed in 1966, 1969, and 1973. The general public was shocked by events such as the fire on the polluted surface of the Cuyahoga River near Cleveland in 1969. Ecologists were beginning to be asked hard questions about the future of the environment, including the effects of air and water pollution, population growth, and species loss. Many ecologists therefore hoped that their field could develop into a more quantitative hard science and the Hutchinson/MacArthur paradigm seemed to hold much promise for doing so. Exciting papers on resource partitioning and character states, such as Brown and Davidson's work on desert granivores (e.g., Davidson et al. 1980), Pianka's studies of desert lizard communities (e.g., Pianka 1975), and Pulliam's (1975) work with sparrows, held promise that the strong reductionist and competition-based view of the niche would give us the tools to explain most patterns of diversity in nature.

These eight papers therefore exemplify a revolution of sorts during what is arguably one of the most dynamic times in the history of the young field of ecology. They represent the growing body of work in the 1970s that suggested that the world wasn't going to be as simple to explain as we had hoped. Ecologists were forced to recognize that the factors explaining local and regional patterns of diversity were much more complex and perhaps much more interesting than the world described by simple niche theory. Ecologists are still dealing with the problem not of replacing niche theory, but incorporating it into a larger, integrated understanding of the forces constraining populations and communities.

Literature Cited

Abrams, P. A. 1987. On classifying interactions between populations. Oecologia 73: 272–281.

Anderson-Teixeira, K. J., S. J. Davies, A. C. Bennett, E. B. Gonzalez-Akre, H. C. Muller-Landau, S. J. Wright, K. Abu Salim, et al. 2015. CTFS-ForestGEO: a worldwide network monitoring forests in an era of global change. Global Change Biology 21: 528–549.

Andrewartha, H. G., and C. Birch. 1954. The Distribution and Abundance of Animals. University of Chicago Press.

Armstrong, R. A., and R. McGehee. 1980. Competitive exclusion. American Naturalist 115: 151–170.

Ayala, F. J., M. E. Gilpin, and J. G. Ehrenfeld. 1973. Competition between species: theoretical models and experimental tests. Theoretical Population Biology 4: 331–356.

Barabás, G., M. J. Michalska-Smith, and S. Allesina. 2016. The effect of intra- and interspecific competition on coexistence in multispecies communities. American Naturalist 188: E1–E12.

Bell, G. 2000. The distribution of abundance in neutral communities. American Naturalist 155: 606–617.

Bender, E. A., T. J. Case, and M. E. Gilpin. 1984. Perturbation experiments in community ecology: theory and practice. Ecology 65: 1–13.

Chase, J. M., and M. A. Leibold. 2003. Ecological Niches: Linking Classical and Contemporary Approaches. University of Chicago Press.

Chesson, P. 2000. Mechanisms of maintenance of species diversity. Annual Review of Ecology and Systematics 31: 343–366.

Chesson, P. L., and R. R. Warner. 1981. Environmental variability promotes coexistence in lottery competitive systems. American Naturalist 117: 923–943.

Clements, F. E. 1936. Nature and structure of the climax. Journal of Ecology 24: 252–284.

Comita, L. S., S. A. Queenborough, S. J. Murphy, J. L. Eck, K. Xu, M. Krishnadas, N. Beckman, et al. 2014. Testing predictions of the Janzen-Connell hypothesis: a meta-analysis of experimental evidence for distance- and density-dependent seed and seedling survival. Journal of Ecology 102: 845–856.

Condit, R. 1998. Tropical Forest Census Plots: Methods and Results from Barro Colorado Island, Panama and a Comparison with Other Plots. Springer.

Connell, J. H. 1961. The influence of interspecific competition and other factors on the distribution of the barnacle Chthamalus stellatus. Ecology 42: 710–723.

Connell, J. H. 1970a. On the role of natural enemies in preventing competitive exclusion in some marine animals and in rain forest trees. Pages 298–312 in P. J. Den Boer and G. R. Gradwell, eds. Dynamics of Population. Pudoc.

Connell, J. H. 1970b. A predator-prey system in the marine intertidal region. I. Balanus glandula and several predatory species of Thais. Ecological Monographs 40: 49–78.

Connell, J. H., and R. O. Slatyer. 1977. Mechanisms of succession in natural communities and their role in community stability and organization. American Naturalist 111: 1119–1144.

Connor, E. F., and D. Simberloff. 1979. The assembly of species communities: chance or competition? Ecology 60: 1132.

Craine, J. M. 2005. Reconciling plant strategy theories of Grime and Tilman: reconciling plant strategy theories. Journal of Ecology 93: 1041–1052.

Davidson, D. W., J. H. Brown, and R. S. Inouye. 1980. Competition and the structure of granivore communities. BioScience 30: 233–238.

Diamond, J. M., and M. E. Gilpin. 1982. Examination of the "null" model of Connor and Simberloff for species co-occurrences on islands. Oecologia 52: 64–74.

Egler, F. E. 1954. Vegetation science concepts I. Initial floristic composition, a factor in old-field vegetation development with 2 figs. Vegetatio 4: 412–417.

Elton, C. S. 1927. Animal Ecology. Macmillan.

Gause, G. F. 1932. Experimental studies on the struggle for existence. Journal of Experimental Biology 9: 389.

Gleason, H. A. 1926. The individualistic concept of the plant association. Bulletin of the Torrey Botanical Club 53: 7.

Gotelli, N. J., and G. R. Graves. 1996. Null Models in Ecology. Smithsonian Institution Press.

Grace, J., and D. Tilman. 2012. Perspectives on Plant Competition. Elsevier Science.

Grant, P. R. 1966. Ecological compatibility of bird species on islands. American Naturalist 100: 451–462.

Grant, P. R., and I. Abbott. 1980. Interspecific competition, island biogeography and null hypotheses. Evolution 34: 332–341.

Grubb, P. J. 1977. The maintenance of species-richness in plant communities: the importance of the regeneration niche. Biological Reviews 52: 107–145.

Haigh, J., and J. Maynard Smith. 1972. Can there be more predators than prey? Theoretical Population Biology 3: 290–299.

Hairston, N. G., F. E. Smith, and L. B. Slobodkin. 1960. Community structure, population control, and competition. American Naturalist 94: 421–425.

Hardin, G. 1960. The competitive exclusion principle. Science 131: 1292–1297.

Holt, R. D. 1977. Predation, apparent competition, and the structure of prey communities. Theoretical Population Biology 12: 197–229.

Holt, R. D., and M. B. Bonsall. 2017. Apparent competition. Annual Review of Ecology, Evolution, and Systematics 48: 447–471.

Hubbell, S. P. 1979. Tree dispersion, abundance, and diversity in a tropical dry forest. Science 203: 1299–1309.

Hubbell, S. P. 2001. The Unified Neutral Theory of Biodiversity and Biogeography. Princeton University Press.

Hubbell, S. P. 2005. Neutral theory in community ecology and the hypothesis of functional equivalence. Functional Ecology 19: 166–172.

Hubbell, S. P., R. B. Foster, S. T. O'Brien, K. E. Harms, R. Condit, B. Wechsler, S. J. Wright, et al. 1999. Light-gap disturbances, recruitment limitation, and tree diversity in a neotropical forest. Science 283: 554.

Hutchinson, G. E. 1957. Concluding remarks. Cold Spring Harbor Symposium on Quantitative Biology 22: 415–427.

Hutchinson, G. E. 1959. Homage to Santa Rosalia or why are there so many kinds of animals? American Naturalist 93: 145–159.

Hutchinson, G. E. 1961. The paradox of the plankton. American Naturalist 137–145.

Hyatt, L. A., M. S. Rosenberg, T. G. Howard, G. Bole, W. Fang, J. Anastasia, K. Brown, et al. 2003. The distance dependence prediction of the Janzen-Connell hypothesis: a meta-analysis. Oikos 103: 590–602.

Janzen, D. H. 1966. Coevolution of mutualism between ants and acacias in Central America. Evolution 20: 249–275.

Janzen, D. H. 1970. Herbivores and the number of tree species in tropical forests. American Naturalist 104: 501–528.

Kingsland, S. E. 2005. The Evolution of American Ecology, 1890–2000. Johns Hopkins University Press.

Lack, D. 1954. Cyclic mortality. Journal of Wildlife Management 18: 25.

Leibold, M. A., M. Holyoak, N. Mouquet, P. Amarasekare, J. M. Chase, M. F. Hoopes, R. D. Holt, et al. 2004. The metacommunity concept: a framework for multi-scale community ecology. Ecology Letters 7: 601–613.

Levin, S. A. 1970. Community equilibria and stability, and an extension of the competitive exclusion principle. American Naturalist 104: 413–423.

Levins, R. 1969. Some demographic and genetic consequences of environmental heterogeneity for biological control. Bulletin of the Entomological Society of America 15: 237–240.

MacArthur, R. H. 1958. Population ecology of some warblers of northeastern coniferous forests. Ecology 39: 599–619.

MacArthur, R. H. 1972. Geographical Ecology: Patterns in the Distribution of Species. Princeton University Press.

MacArthur, R. H., and R. Levins. 1967. The limiting similarity, convergence, and divergence of coexisting species. American Naturalist 101: 377–385.

MacArthur, R. H., and E. O. Wilson. 1967. The Theory of Island Biogeography. Princeton University Press.

May, R. M. 1972. Will a large complex system be stable? Nature 238: 413–414.

McIntosh, R. 1999. The succession of succession: a lexical chronology. Bulletin of the Ecological Society of America 80: 256–265.

Miller, T. E. 1994. Direct and indirect species interactions in an early old-field plant community. American Naturalist 143: 1007–1025.

Miller, T. E., J. H. Burns, P. Munguia, E. L. Walters, J. M. Kneitel, P. M. Richards, N. Mouquet, et al. 2005. A critical review of twenty years' use of the resource-ratio theory. American Naturalist 165: 439–448.

Miller, T. E., and W. C. Kerfoot. 1987. Redefining indirect effects. Pages 33–37 in W. C. Kerfoot and A. Sih, eds., Predation: Direct and Indirect Impacts on Aquatic Communities. University Press of New England.

Moreau, R. E. 1966. The bird faunas of Africa and its islands. Academic Press.

Neill, W. E. 1974. The community matrix and interdependence of the competition coefficients. American Naturalist 108: 399–408.

Nicholson, A. J., and V. A. Bailey. 1935. The balance of animal populations. Part I. Proceedings of the Zoological Society of London 105: 551–598.

Noble, I. R., and R. O. Slatyer. 1980. The use of vital attributes to predict successional changes in plant communities subject to recurrent disturbances. Vegetatio 43: 5–21.

Okuyama, T. 2015. Demographic stochasticity alters the outcome of exploitation competition. Journal of Theoretical Biology 365: 347–351.

Paine, R. T. 1969. A note on trophic complexity and community stability. American Naturalist 103: 91–93.

Pianka, E. R. 1975. Niche relations of desert lizards. Pages 292–314 in M. L. Cody and J. M. Diamond, eds. Ecology and Evolution of Communities. Harvard University Press.

Pickett, S. T. A., S. L. Collins, and J. J. Armesto. 1987. A hierarchical consideration of causes and mechanisms of succession. Vegetatio 69: 109–114.

Pielou, D. P., and E. C. Pielou. 1968. Association among species of infrequent occurrence: The insect and spider fauna of Polyporus betulinus (Bulliard) Fries. Journal of Theoretical Biology 21: 202–216.

Pulliam, H. R. 1975. Coexistence of sparrows: a test of community theory. Science 189: 474–476.

Pulsford, S. A., D. B. Lindenmayer, and D. A. Driscoll. 2016. A succession of theories: purging redundancy from disturbance theory. Biological Reviews 91: 148–167.

Real, L. A., and J. H. Brown, eds. 1991. Foundations of Ecology: Classic Papers with Commentaries. University of Chicago Press.

Sale, P. F. 1977. Maintenance of high diversity in coral reef fish communities. American Naturalist 111: 337–359.

Siepielski, A. M., and M. A. McPeek. 2010. On the evidence for species coexistence: a critique of the coexistence program. Ecology 91: 3153–3164.

Simberloff, D., and W. Boecklen. 1981. Santa Rosalia reconsidered: size ratios and competition. Evolution 35: 1206.

Simberloff, D. S. 1970. Taxonomic diversity of island biotas. Evolution 24: 23–47.

Slobodkin, L. B. 1961. The Growth and Regulation of Animal Populations. Holt, Rinehart and Winston.

Stewart, F. M., and B. R. Levin. 1973. Partitioning of resources and the outcome of interspecific competition: A model and some general considerations. American Naturalist 107: 171–198.

Stone, L., and A. Roberts. 1991. Conditions for a species to gain advantage from the presence of competitors. Ecology 72: 1964–1972.

Strauss, S. Y. 1991. Indirect effects in community ecology: their definition, study and importance. Trends in Ecology & Evolution 6: 206–210.

Strong, D. R. 1986. Density-vague population change. Trends in Ecology & Evolution 1: 39–42.

Strong, D. R., D. S. Simberloff, L. G. Abele, and A. B.

Thistle. 1984. Ecological Communities: Conceptual Issues and the Evidence. Princeton University Press.

Strong, D. R., L. A. Szyska, and D. S. Simberloff. 1979. Test of community-wide character displacement against null hypotheses. Evolution 33: 897.

Tilman, D. 1977. Resource competition between planktonic algae: an experimental and theoretical approach. Ecology 58: 338–348.

Tilman, D. 1980. Resources: a graphical-mechanistic approach to competition and predation. American Naturalist 116: 362–393.

Tilman, D. 1982. Resource Competition and Community Structure. Princeton University Press.

Tilman, G. D. 1984. Plant dominance along an experimental nutrient gradient. Ecology 65: 1445–1453.

Tilman, D., and D. Wedin. 1991. Dynamics of nitrogen competition between successional grasses. Ecology 72: 1038–1049.

Vandermeer, J. H. 1969. The competitive structure of communities: an experimental approach with protozoa. Ecology 50: 362–371.

Vandermeer, J. 1981. A further note on community models. American Naturalist 117: 379–380.

Vandermeer, J., and I. Perfecto. 2019. Hysteresis and critical transitions in a coffee agroecosystem. Proceedings of the National Academy of Sciences 116: 15074.

Volterra, V. 1926. Fluctuations in the abundance of a species considered mathematically. Nature 118: 558–560.

Walker, L. R., and F. S. Chapin. 1987. Interactions among processes controlling successional change. Oikos 50: 131.

Walker, L. R., and R. del Moral. 2003. Primary Succession and Ecosystem Rehabilitation. Cambridge University Press.

Wedin, D., and D. Tilman. 1993. Competition among grasses along a nitrogen gradient: initial conditions and mechanisms of competition. Ecological Monographs 63: 199–229.

Werner, E. E., and S. D. Peacor. 2003. A review of trait-mediated indirect interactions in ecological communities. Ecology 84: 1083–1100.

Whittaker, R. H. 1965. Dominance and diversity in land plant communities. Science 147: 250–260.

Wilbur, H. M. 1972. Competition, predation, and the structure of the Ambystoma–Rana sylvatica community. Ecology 53: 3–21.

Wootton, J. 1992. Indirect effects, prey susceptibility, and habitat selection: impacts of birds on limpets and algae. Ecology 73: 981–991.

Wootton, J. T. 1994. The nature and consequences of indirect effects in ecological communities. Annual Review of Ecology and Systematics 25: 443–466.

JOHN H. VANDERMEER

Ecology, Vol. 50, No. 3

leaves in Illinois as influenced by soil type and soil composition. Soil Sci. **68:** 317–328.

Metz, L. J., G. G. Wells, and B. F. Swindel. 1966. Sampling soil and foliage in a pine plantation. Soil Sci. Soc. Amer. Proc. **30:** 397–399.

Moore, C. W. E. 1959. The interaction of species and soil in relation to the distribution of eucalypts. Ecology **40:** 734–735.

———. 1961. Competition between *Eucalyptus melliodora and Eucalyptus rossii* at varying levels of exchangeable calcium. Aust. J. Bot. **9:** 92–97.

Mulder, D. 1953. Nutritional studies on fruit trees. II. The relation between potassium, magnesium and phosphorus in apple trees. Plant and Soil **4:** 107–117.

Pirson, A. 1956. Functional aspects in mineral nutrition of green plants. Ann. Rev. Plant Physiol. **6:** 71–114.

Pryor, L. D. 1959. Species distribution and association in *Eucalyptus*. p. 461–471. *In* A. Keast, R. L. Crocker and C. S. Christian (ed.) Biogeography and ecology in Australia. Dr. W. Junk, The Hague, Netherlands. 640 p.

Reuther, W., F. E. Gardner, P. F. Smith, and W. R. Roy. 1949. Phosphate fertilization trials with oranges in Florida. I. Effect on yield, growth, and leaf and soil composition. Amer. Soc. Hort. Sci. Proc. **53:** 71–84.

Shomaker, C. E., and V. J. Rudolph. 1964. Nutritional relationships affecting height growth of planted yellow-poplar in southwestern Michigan. Forest Sci. **10:** 66–76.

Slatyer, R. P. 1967. Plant-water relationships. Academic Press, New York, 366 p.

Slatyer, R. P. and I. C. McIlroy. 1961. Practical microclimatology. CSIRO and UNESCO, Australia. 5.12–5.17.

Snedecor, G. W. 1956. Statistical methods. 5th ed. Iowa State Coll. Press, Ames, Iowa. 534 p.

Tanada, T. 1955. Effects of ultraviolet radiation and calcium and their interaction on salt absorption by excised Mung bean roots. Plant Physiol. **30:** 221–225.

Walker, L. C. 1956. Foliage symptoms as indicators of potassium-deficient soils. Forest Sci. **2:** 113–210.

Walker, R. B. 1954. The ecology of serpentine soils. II. Ecology **35:** 259–266.

Wiklander, L. 1958. The soil, p. 118–169. *In* W. Ruhland (ed.) Encycl. Pl. Physiol. IV, Springer Verlag, Berlin.

THE COMPETITIVE STRUCTURE OF COMMUNITIES: AN EXPERIMENTAL APPROACH WITH PROTOZOA

John H. Vandermeer[1]

Department of Zoology, University of Michigan, Ann Arbor, Michigan

(Received November 25, 1968; accepted for publication January 31, 1969)

Abstract. An empirical test of the existence of higher order interactions was carried out using four ciliate protozoans, *Paramecium caudatum, P. bursaria, P. aurelia,* and *Blepharisma* sp.

All four ciliates were cultured individually, and their population histories were described quite well by the simple logistic equation.

Attempts to explain minor deviations of the data from the logistic by use of the one or two time lag logistic failed. A more complicated time lag phenomenon must be operative.

Every possible pair of the four ciliates was cultured and a trial and error procedure was used to estimate α and β of the Gause equations. In all cases the simple Gause equations seemed adequately to describe the data.

All four ciliates were cultured together and compared to predictions made by use of the competition coefficients estimated from pair-wise competition and population parameters estimated from single species population growth. The correspondence between prediction and data suggests that the higher order interactions have slight or no effect on the dynamics of this artificial community.

The central goal of community ecology is to understand the dynamics of community organization. Most likely that goal would be best approached from the point of view of mechanism, beginning with basic principles and deducing how communities should behave. However, such attempts have in the past been stifled by certain methodological difficulties, diverting the attention of community ecologists toward more empirical observations. These empirical observations have usually taken the form of measuring the relative

abundance of species (number of species with a given number of individuals).

Most of the work concerned with the latter may be dichotomized as follows. One series of papers (Fisher, Corbet, and Williams 1943; Preston 1948, 1962a, b; Hairston and Beyers 1954; Hairston 1959) emphasizes underlying mathematical distributions and causes of deviance from these distributions. Another series, stimulated by Margalef (1957), and rigorously formalized by Pielou (1966a, b), is concerned with measuring the relative abundance of species, usually to compare communities to each other.

[1] Present address: Department of Biology, University of Chicago, 1103 E. 57th Street, Chicago, Illinois 60637.

There have also been some attempts at building either conceptual or mathematical models of communities using supposedly basic components. Kendall (1948) has shown how certain patterns of population growth will lead to Fisher's logarithmic series. MacArthur (1955) discussed several aspects of community stability as deduced from considerations of energy transfer among populations, and Watt (1964) later tested some of these ideas. MacArthur (1960) suggested mechanisms whereby lognormal distributions should arise. Hairston, Smith, and Slobodkin (1960) presented a general explanation for the structure of terrestrial communities which stimulated a recent controversy (Murdoch 1966, Ehrlich and Birch 1967, Slobodkin, Smith, and Hairston 1967). Garfinkel (1967) studied stability properties of theoretical systems using the Lotka-Volterra predator-prey equations.

These approaches to community ecology have established the basic attitude for the following empirical tests. Simple assumptions are made, namely that populations grow according to the logistic equation, and populations on the same trophic level obey the Gause equations of competition. I shall be concerned with communities which derive their structure through inter- and intraspecies competition.

The basic Gause competition equations may be extended to include m species. The differential equation for the ith species is,

$$\frac{dN_i}{dt} = \frac{r_i N_i}{K_i}\left\{K_i - N_i - \sum_{j=1}^{m}\alpha_{ij}N_j\right\}\ j \neq i \tag{1}$$

(MacArthur and Levins 1967), where there are m species in the community, α_{ij} is the effect of the jth species on the ith species, r is the intrinsic rate of natural increase, N is number of individuals and K is saturation density. Assumed in the above extension is that there are no higher order interaction terms. That is, we need not write,

$$\frac{dN_i}{dt} = \frac{r_i N_i}{K_i}\left\{K_i - N_i - \sum\alpha_{ij}N_j - \sum_{jk}\beta_{ijk}N_jN_k - \ldots - \sum_{jk\ldots m}\omega_{ijk\ldots m}N_jN_k\ldots N_m\right\} \tag{2}$$

where $j \neq i,\ k \neq i, \ldots, m \neq i$, and the Greek letters are interaction coefficients of increasingly higher order, e.g. β_{ijk} is the joint effect of species j and k on the ith species. Communities represented by equations (1) will be called noninteractive communities, and communities represented by equations (2) will be called interactive communities.

Equations (1) (noninteractive communities) are fairly tractable and have already led to some rather interesting conclusions about community structure (Levins 1968, Vandermeer 1968). On the other hand, equations (2) (interactive communities) are difficult to work with analytically.

The qualitative effects of higher order interactions are made clear in the following discussion. Consider the competitive effect, on one particular species (say species A), of adding further species, holding the average α constant and assuming the higher order terms to be negligible. The mean α is held constant in such a way that each new species overlaps with 25% of species A. If another species is added, species A only occupies 75% of its former niche without competitive pressure. If another species is added only 50% of its niche is free of competition. In general, as competitors accumulate in the system, if the average competition coefficient remains constant, the weight of competition eventually gets so strong that other species cannot make their way into the **community.**

Now suppose the three way interactions (the β's in equation (2)) are important. Considering again species A, after the addition of another species, as before only 75% of the former niche is occupied without competition. If an additional species is added, part of the niche overlap will be contained in a three way overlap. Thus, instead of having the niche reduced to 50% of its former size, something greater than 50% but less than 75% of the niche remains free of competition (since a portion of the potential competition is absorbed in the three way interaction).

For instance, consider a hypothetical organism living in a stream. Suppose species A occupies the entire length of the stream. Suppose further that experiments are undertaken in which species B and C are added one at a time. If B and C each have X effect on species A, we would expect that the combination of B and C together would have effect 2X. But suppose that the result of B and C competing with A is that in either case A is eliminated only from the upstream section of the stream. It does not matter to species A which of B or C does the eliminating; the fact is, it is being eliminated. Thus, the predicted effect of B and C together on A would be 2X but the observed effect would be X.

The type of higher order interaction discussed above implies $\beta, \gamma, \ldots, \omega$ of equation 4 are positive. If, on the other hand, the higher order terms were negative, it would imply a coalition formed by two or more species against some other species. The higher order interaction reported by Hairston, et al. (1969) is an example of a coalition type interaction.

Thus, it is not possible at the present time to predict the general effect of higher order interactions except qualitatively. However, it is clear that their effects may be highly significant. It, therefore, would be quite important to discover whether or not they commonly occur in nature.

The motivation for undertaking the experiments reported below derives from the foregoing discussion of interactive communities. Since several interesting consequences are deducible from the premises of noninteractive communities, it would be desirable to know either the importance of higher order interactions in real situations, or the theoretical effect which higher order interactions would have on the consequences derived from the noninteractive assumptions. As remarked earlier, it is virtually certain that higher order interactions will have an effect on the outcomes of the model, but the nature and extent of this effect are difficult to assess owing to the intractable nature of the model when applied to interactive communities. Thus, the simplest approach is to verify the presence or absence of higher order interactions in nature, and this is the approach of this paper.

Induction in biology—the implied approach suggested in the above discussion—has often been eased by first studying extremes. If the existence of higher order interactions can be demonstrated for a group of species which would be expected on an *a priori* basis not to exhibit them, it may be safely assumed that such interactions are more or less of universal occurrence. If that group of species does not exhibit higher order interactions, the broad implications for which the experiment was designed are lost.

Obtaining an extreme example and satisfying the basic postulates of the model have been the two guidelines in choosing experimental subjects and situations. Different species of free-living ciliates from different localities seem to satisfy both of these guidelines. Ciliates are simple organisms from the population point of view, and in several cases are known to conform to simple population mathematics, making it likely that most of the basic premises of the model are satisfied.

Therefore, four species of ciliates from four different localities were studied in a laboratory situation. Parameters of population growth were estimated for each species separately. Competition coefficients were estimated for each possible pairwise combination of species. Then all four species were cultured together and compared to the prediction made on the basis of the simple growth parameters and competition coefficients measured earlier.

METHODS AND MATERIALS

Three of the four ciliates, *Paramecium bursaria* (PB), *Paramecium aurelia* (PA), and *Paramecium caudatum* (PC) were obtained from three different localities in the vicinity of Ann Arbor, Michigan. The fourth species *Blepharisma* sp. (BL), was obtained from a biological supply house. The Varieties are unknown except for PA which is Variety 3.

Culture techniques are those of Sonneborn (1950), with various modifications after Hairston and Kellerman (1965). Cultures were maintained at 15°C and all experiments were done at 25°C. All cultures and experiments were kept in darkness—i.e., PB was without effective symbiotic algae.

Experiments were done in 10-ml test tubes, filled with 5 ml of culture. Each day a sample of 0.5 ml was rapidly removed with a coarse pipette after the test tube had been vigorously shaken. The tubes were shaken after a section of rubber glove had been sterilized in hot distilled water and positioned as a covering for the test tube. Verification of the test tube sampling procedure was accomplished by repeated sampling and subsequent replacement of samples in several groups of test tubes. The sampling procedure was extensively experimented with to obtain a reasonably close correspondence between the mean and variance of a set of repeated samples; i.e., to assure a random sampling technique.

Bacterized culture medium was added to each culture to replace exactly the 0.5 ml which had been removed as a sample. The culture medium had been aged for various lengths of time and was used in an orderly sequence based on this age. This was originally done to introduce a forced environmental oscillation on the experiments. Since absolutely no relationship between food age (food quality) and population growth or competition was observed anywhere in the study, the food source is considered as a constant in this paper.

POPULATION GROWTH

The mathematical model used for the simple population growth experiments was the familiar logistic equation, $dN/dt = (rN/K)(K - N)$, where N is the number of individuals, K is saturation population density, and r is the intrinsic rate of natural increase. The parameters K, r, and N_o were estimated exactly as in Gause (1934) excepting the case of PC in which several points were discarded as wild points for purposes of estimating the parameters—not in plotting the figures. In all cases the intercept of the equation $\ln[(K-N)/N] = a + rt$ was used to estimate N_o, which was needed to generate the expected curve. Expected curves were obtained from the recurrence relation given by Leslie (1957),

$$N_{t+1} = \lambda N_t/(1 - bN_t)$$
where $\lambda = e^r$ and $b = (\lambda - 1)/K$.

Statistical testing of the fit of the data to the model is at best difficult (Smith 1952). Leslie's (1957) procedure depends on the assumption of independence of observations from one time to the next. It was obvious by inspection that there was a rather high correlation between times within replicates—i.e., that replicate which had the highest value at time t was most likely to be the replicate with the highest value at time $t + 1$. Because of this statistical difficulty, and the difficulties raised by Smith (1952), the statistical testing of Leslie (1957) was not used.

Figure 1 presents the expected and observed values for population growth of BL. The fit is

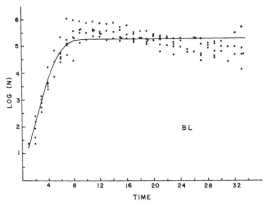

FIG. 1. Population growth of BL.

generally good, except that a systematic deviation of observed from expected is notable. This deviation is biased above the curve at early times, and below the curve at later times. The pattern is reminiscent of that obtained when time lags are added to the logistic equation (Cunningham 1954).

Using the equation of Wangersky and Cunningham (1956), the value of the reproductive time lag τ_1, and density time lag τ_2 were estimated. Systematic variation in the value of these two parameters provided no improvement of fit. If the deviations of fit are due to a time delay, the relationship is more complicated than that described by Wangersky and Cunningham's equation.

In Figure 2 are shown expected and observed values for PB. Of the four species this is the

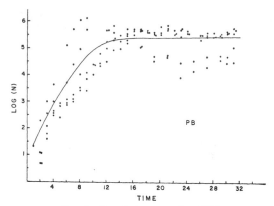

FIG. 2. Population growth of PB.

most highly variable in its population growth. The relatively large variance in observed values is understood by noting that two of the replicates are suggestive of a very large time lag and three are suggestive of a smaller, if at all existent, time lag. The curve predicted by the logistic (Fig. 2) seems to be an adequate representation of the basic trend of population growth in this species.

The expected and observed values for PC and PA are shown in Figures 3 and 4. The fits are obviously excellent and need no further discussion here.

The parameters for population growth for all four species are summarized in Table 1.

TWO-WAY COMPETITION

The mathematical methods used here are even more inexact than those used in single species population growth. A smooth curve was drawn by eye through the observed data points, and estimates of α and β were obtained for each time from 2 to 32, with the equations given by Gause (1934),

$$\alpha = \frac{1}{N_2}\left[K_1 - \frac{(dN_1/dt)K_1}{r_1N_1} - N_1\right]$$

$$\beta = \frac{1}{N_1}\left[K_2 - \frac{(dN_2/dt)K_2}{r_2N_2} - N_2\right]$$

The K's and r's were obtained from the single species experiments. The N's were taken off of the smooth curve, and the derivatives were estimated as

$$\frac{dN_t}{dt} = \frac{N_{t+1} - N_{t-1}}{2}$$

The resultant values of α and β—31 values of each for each of the six competition experiments—were highly variable within experiments. The values were exceptionally extreme at very early times and very late times. Intermediate values

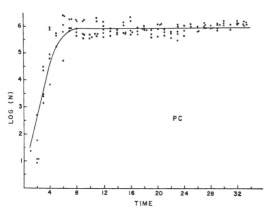

FIG. 3. Population growth of PC.

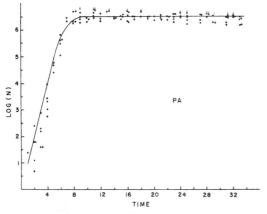

FIG. 4. Population growth of PA.

TABLE 1. Parameters of the logistic equation

Species	r	K	N_o
PA	1.05	671	2.5
PC	1.07	366	5.0
PB	0.47	230	5.0
BL	0.91	194	3.0

were somewhat less variable, but the variation seemed always to form some sort of trend. However, under the hypothesis of this paper, α and β are defined as constants so the objective is to obtain those values of α and β which produce the curve which best fits the observed data.

To obtain a reasonable first approximation to α and β I took the average of the intermediate values as described above—i.e., the average of those values that tended to form a reasonably invariant clump. With these values, integral curves were computed using the predictor corrector method of Hamming (Ralston and Wilf 1960; see also System/360 Scientific Subroutine Pack-

age 1968). Usually, these first approximations gave curves that rather poorly represented the observed data. A trial and error procedure was then pursued in which α and β were systematically varied and the behavior of the integral curves investigated under this variation. It eventually became apparent that the curves obtained were about the best obtainable under the present model. In all cases the boundary conditions were considered as random variables, and were thus relatively free to vary in the trial and error procedure.

The comments about statistical testing made in the previous section apply equally well here. However, a relative measure of the goodness of fit is needed for a later section, and is introduced here to facilitate easy comparison of the goodness of fit of one experiment to that of another. The

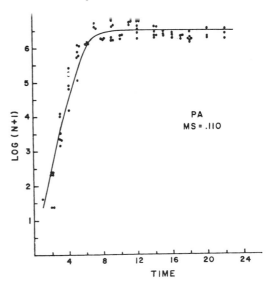

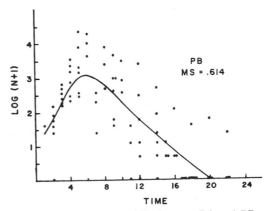

FIG. 5. Results of competition between PA and PB.

arbitrary criterion used is mean square deviation of logs of observed from log of expected.

The results of the six pairwise competition experiments are presented in Figures 5–10, in order of decreasing mean square deviation. The aforementioned variability of PB is reflected in Figure 5, but appears to be somewhat dampened by the presence of BL (Fig. 9). Figures 6 and 7 both show some indication of a time lag, but since the population growth data for these two species (PA and PC) alone did not suggest any frictional components, one must presume that the presence of a competitor is the factor which initiates a nonlinear response. However, the difference between the linear and nonlinear response in this case appears to be negligible.

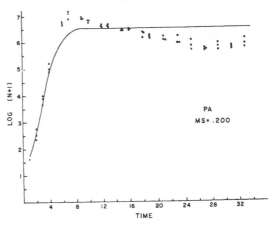

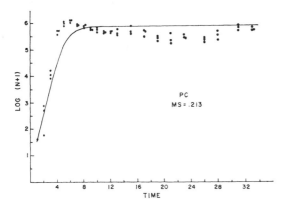

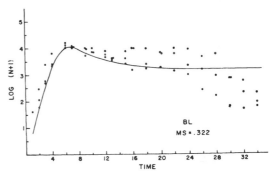

Fig. 7. Results of competition between PA and BL.

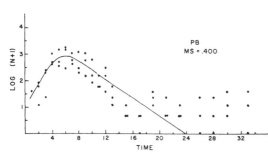

Fig. 6. Results of competition between PC and PB.

Thus, the six cases of pairwise competition are represented quite well by the basic theory, minor deviations being interpretable as a result of unknown but insignificant frictional components or excessive inherent variability. Since at least the general trend of the data is in agreement with that of the model, these small deviations most likely will not interfere with the basic hypothesis to be tested.

In Table 2 is presented the community matrix for this artificial community. The α_{ij}th entry refers to the effect of the jth species on the ith species.

Four-Way Competition

Having estimates of the intrinsic rate of natural increase and saturation density for each species, and estimates of all interaction coefficients on a pairwise basis, it is now necessary to substitute into equations (1)—four-way competition without higher order interactions—solve the equations, and see if the solution agrees with the data obtained from the laboratory experiments. It must be emphasized that up to this point all curves in this paper have been fit. That is, the expected curves are based on the data which they are describing and therefore cannot deviate too greatly. In the case of four-way competition, however, the expected curves are based on independent estimates of the parameters, estimates which were made from different experiments. As a consequence a reasonable expectation, if higher order

368   JOHN H. VANDERMEER   Ecology, Vol. 50, No. 3

TABLE 2. Community matrix for four species of protozoa in test tubes

*i*th Species	*j*th Species			
	PA	PC	PB	BL
PA	1.00	1.75	−2.00	−0.65
PC	0.30	1.00	0.50	0.60
PB	0.50	0.85	1.00	0.50
BL	0.25	0.60	−0.50	1.00

interactions are not important, is that the fit to the four-way situation is, at best, as good as the fits to the pair-wise competitions. Though it cannot be expected that the four-way competition experiments will fit the model as well as the two-way competition, it is not clear exactly how "good" a fit is necessary to judge a model as being valid. Such a judgment, of course, depends on the purposes for which the model is to be used. In the present case it seems that there are three vaguely distinquishable levels of purpose which might be used in helping to judge the validity of this model. First, if only the equilibrium conditions of the model are needed—say in predicting the maximum number of species in an equilibrium community—we might only require that the model predict accurately which species will persist and which will

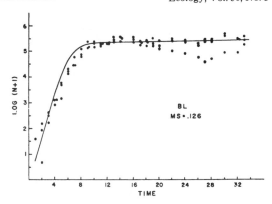

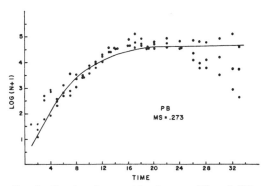

FIG. 9. Results of competition between BL and PB.

go to extinction. Second, the relative numbers of individuals at any particular point in time may be desired—say, in a theoretical study of the relative abundance of species—or, more or less equivalently, the general behavior of the populations within the framework of the larger system must be known—i.e., does species i oscillate, does it tend to be peaked and then drop off fast, and similar general questions. Under this requirement, we accept the model as valid if the general shape of the curves reflects the general trend of the data. Third, we may wish to make very precise predictions with the model such as exactly how much pesticide must be applied at time x to produce y per cent change in species i by time $x + a$. At this level we require the fit to four-way competition to be almost as good as the fit to the two-way competitions. Certainly these three levels are not discrete, but are merely useful points in a continuum, defined so as to be able to judge the relative validity of the model as applied to a simple community.

In Figures 11–14 are shown the results of simulating four-way competition with the community

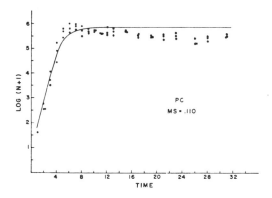

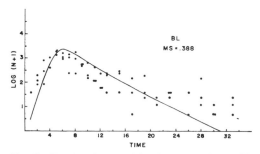

FIG. 8. Results of competition between PC and BL.

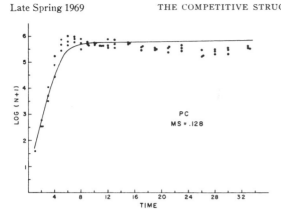

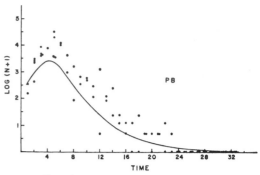

FIG. 12. PB in four-way competition.

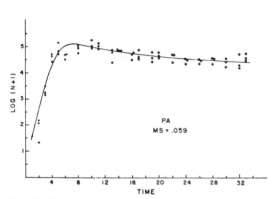

FIG. 10. Results of competition between PC and PA.

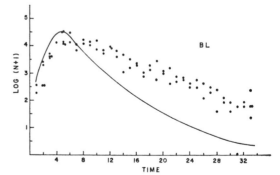

FIG. 13. PC in four-way competition.

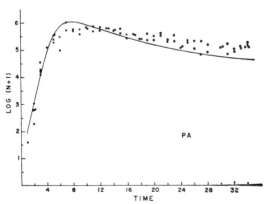

FIG. 11. PA in four-way competition.

FIG. 14. BL in four-way competition.

matrix of the last section and the r's and K's from the single species population growth experiments. Also shown are the observed data from the four-way competition experiments.

Under the first criterion of validation—only the final outcome is desired—the data are described perfectly by the model. The model predicts that PB will be extinct; BL will be at a very low popu-

lation density; PA and PC will have higher densities. As can be easily seen from the figures, the data correspond exactly to these predictions.

Likewise under the second criterion the model is validated. The general trends are predicted in every case, and the rank order of abundance is as predicted, at virtually all points in time. The general trends of growth, peaking, and decay are predicted fairly accurately. The early peak for PC reflected in the experiments is essentially the

370   JOHN H. VANDERMEER   Ecology, Vol. 50, No. 3

same as that species exhibits in two-way competition (see Fig. 6, 8, and 10).

Finally, consider the third level of validation. Somewhere in between the second and third level the validation becomes questionable. The third level requires that precise predictions be made. The three species PC, PA and PB conform quite well, but BL seems to deviate excessively from the expected. However, as discussed previously, it is not sufficient to examine the data and expectation alone, but they must be considered in the light of the pair-wise experiments, since the correspondence between expected and observed can only be as good as that of the basic model. It is for this reason that the mean square deviations were computed earlier (Fig. 5–10). In Table 3 are listed the pooled mean squares for each species pooled over all two-way competitions in which that species was a part—and the mean

TABLE 3. Mean squares for two-way and four-way competition experiments, and ratio between the two (taking largest mean square as numerator). Numbers in italics are degrees of freedom

Statistic	Species			
	PA	PC	PB	BL
Pooled mean square for 2-way competition	0.11 *212*	0.15 *197*	0.43 *230*	0.26 *209*
Mean square for 4-way competition	0.10 *74*	0.22 *74*	0.36 *74*	1.28 *74*
Ratio	1.1	1.5	1.2	4.9

squares for the same species in four-way competition. The first three species show very similar mean squares between two-way and four-way competition; indeed PA and PB are even better predicted in the case of four-way than in the case of two-way competition. The case of BL is somewhat less encouraging. The mean square is roughly five times greater in four-way than in two-way. The fact that the experimental points are always greater than predicted, implies that whatever higher order interaction there may be, they act solely to decrease the competitive effect felt by BL. They are noncoalition type higher order interactions.

However, it might be argued that if one is interested in this model from the standpoint of the community, it is not really valid to compare predictions on a specific basis. We should instead be interested in the general performance of the model in predicting community dynamics. To this end the sum of squares pooled over all two-way competition cases was computed as 0.246 and the

same pooled over all four-way cases was 0.487, only a two-fold difference. Thus, the four-way model predicted about half as well as the two-way model, implying that, from the point of view of the community as a whole, the higher order interactions are rather unimportant.

The above discussion is similar to the problem of multiple comparisons tests in statistics. Given a group of means, we wish to test for differences. In setting an error rate we are faced with deciding on the relative importance of pair-wise or experiment-wise rates. That is, is it more important to minimize the probability of making any mistake at all, or the probability that a given pair of means may be judged different when they are truly the same. Similarly in the present context, we are faced with deciding whether we want the model to make predictions about the community as a whole, or about the components of the community. The former criterion provides a single judgment of the model. The latter provides as many judgments as there are components.

Thus, if we are concerned with judging the present model under the third criterion—absolutely precise predictions—and if we are interested in the individual components, the noninteractive model is "good" three times and "bad" once. On the other hand, it appears to be simply "good" on the level of the whole community.

CONCLUSION

The conclusions of these experiments were basically stated in the previous section. However, it might be well at this point to recall statements made earlier about the choice of the experimental system. It was the author's prejudice that higher order interactions are very important in nature. Thus, the experimental system was chosen to maximize the chance of not finding higher order interactions, hoping, of course, that even then such interactions would be significant. Then, because of the way the experimental system was put together, one could make a fairly powerful statement about the universal occurrence of higher order interactions in nature. Unfortunately, the basic models proved to be excellent predictors (i.e., higher order interactions did not seem important), and one can only cite this study as a single example of the insignificance of higher order interactions.

ACKNOWLEDGMENTS

I am deeply indebted to Nelson G. Hairston who provided guidance and encouragement throughout the course of this work. I wish to thank James T. MacFadden, Lawrence B. Slobodkin, and Donald W. Tinkle for constructive comments and criticisms. I also am thankful to Douglas J. Futuyma for much valuable advice, and to my wife Jean for assistance in tabulating data and typing the manuscript. Computer time was granted by the University of Michigan Computation Center. This research was supported by an NSF summer fellowship for graduate teaching assistants, and an NIH predoctoral fellowship.

LITERATURE CITED

Cunningham, W. J. 1954. A nonlinear differential-difference equation of growth. Proc. Nat. Acad. Sci. 40: 708–713.

Ehrlich, P. R. and L. C. Birch. 1967. The "balance of nature" and "population control." Amer. Naturalist 101: 97–107.

Fisher, R. A., A. S. Corbet, and C. B. Williams. 1943. The relation between the number of species and the number of individuals in a random sample from an animal population. J. Animal Ecol. 12: 42–58.

Garfinkel, D. 1967. A simulation study of the effect on simple ecological systems of making rate of increase of population density-dependent. J. Theoret. Biol. 14: 46–58.

Gause, G. F. 1934. The struggle for existence. Williams and Wilkins, Baltimore.

Hairston, N. G. 1959. Species abundance and community organization. Ecology 40: 404–416.

Hairston, N. G. and G. W. Byers. 1954. The soil arthropods of a field in southern Michigan. A study in community ecology. Contrib. Lab. Vert. Biol., Univ. Mich. 64: 1–37.

Hairston, N. G., F. E. Smith, and L. B. Slobodkin. 1960. Community structure, population control, and competition. Amer. Naturalist 94: 421–425.

Hairston, N. G. and S. L. Kellermann. 1965. Competition between varieties 2 and 3 of Paramecium aurelia: the influence of temperature in a food limited system. Ecology 46: 134–139.

Hairston, N. G., J. D. Allan, R. K. Colwell, D. J. Futuyma, J. Howell, J. D. Mathias, and J. H. Vandermeer. 1969. The relationship between species diversity and stability: an experimental approach with protozoa and bacteria. Ecology 49: 1091–1101.

International Business Machines, 1968; System/360 scientific subroutine package (360A-CM-03X) Version III Programmer's manual. IBM Technical Publications Department, White Plains, N. Y.

Kendall, D. G. 1948. On some modes of population growth leading to R. A. Fisher's logarithmic series distribution. Biometrika 35: 6–15.

Leslie, P. H. 1957. An analysis of the data for some experiments carried out by Gause with populations of the Protozoa, Paramecium aurelia and Paramecium caudatum. Biometrika 44: 314–327.

Levins, R. 1968. Evolution in changing environments. Some theoretical explorations. Monographs in population biology, Princeton Univ. Press.

MacArthur, R. H. 1955. Fluctuations of animal populations, and a measure of community stability. Ecology 36: 533–536.

———. 1960. On the relative abundance of species. Amer. Naturalist 94: 25–36.

MacArthur, R. H. and R. Levins. 1967. The limiting similarity, convergence, and divergence of coexisting species. Amer. Naturalist 101: 377–385.

Margalef, R. 1957. Information theory in ecology. (English trans. by Hall, W.) Gen. Systems 3: 36–71.

Murdoch, W. W. 1966. Community structure, population control, and competition—a critique. Amer. Naturalist 100: 219–226.

Pielou, E. C. 1966a. Species-diversity and pattern-diversity in the study of ecological succession. J. Theor. Biol. 10: 370–383.

———. 1966b. The measurement of diversity in different types of biological collections. J. Theor. Biol. 13: 131–144.

Preston, F. W. 1948. The commonness, and rarity, of species. Ecology 29: 254–283.

———. 1962a. The canonical distribution of commonness and rarity. Ecology 43: 185–215.

———. 1962b. The canonical distribution of commonness and rarity. Part II. Ecology 43: 410–432.

Ralston, A. and H. S. Wilf. 1960. Mathematical methods for digital computers. Wiley, New York.

Slobodkin, L. B., F. E. Smith, and N. G. Hairston. 1967. Regulation in terrestrial ecosystems, and the implied balance of nature. Amer. Naturalist 101: 109–124.

Smith, F. E. 1952. Experimental methods in population dynamics: a critique. Ecology 33: 441–450.

Sonneborn, T. M. 1950. Methods in the general biology and genetics of Paramecium aurelia. J. Exper. Zool. 113: 87–147.

Vandermeer, J. H. 1968. The structure of communities as determined by competitive interactions: a theoretical and experimental approach. Ph.D. thesis, Univ. Mich., Ann Arbor.

Wangersky, P. J. and W. J. Cunningham. 1956. On time lags in equations of growth. Proc. Nat. Acad. Sci. Wash. 42: 699–702.

Watt, K. E. F. 1964. Comments on fluctuations of animal populations and measures of community stability. Can. Entomol. 96: 1434–1442.

LIGHT CONTROL OF AQUATIC INSECT ACTIVITY AND DRIFT

JOHN E. BISHOP[1]

Department of Biology, University of Waterloo, Waterloo, Ontario, Canada

(Received November 27, 1968; accepted for publication February 4, 1969)

Abstract. Investigations on aquatic insect activity, measured as drift in an artificial stream system with rigid light and temperature conditions, demonstrated a light-controlled, labile, exogenously-mediated activity rhythm. A threshold value for light, incident at the water surface, which when decreased led to high drift rates, and when increased suppressed activity,

[1] Present address: School of Biological Sciences, University of Malaya, Kuala Lampur, Malaysia.

Vol. 115, No. 2   The American Naturalist   February 1980

COMPETITIVE EXCLUSION

ROBERT A. ARMSTRONG* AND RICHARD McGEHEE

Department of Biology, University of California, San Diego, La Jolla, California 92093; School of
Mathematics, University of Minnesota, Minneapolis, Minnesota 55455

Submitted October 24, 1977; Accepted September 21, 1978

Volterra (1928) was apparently the first to use a mathematical model to suggest that the indefinite coexistence of two or more species limited by the same resource is impossible. This theme, which has been expanded by several authors into the statements that n species cannot coexist on fewer than n resources (MacArthur and Levins 1964; Levins 1968) or in fewer than n "niches" (Rescigno and Richardson 1965) or when limited by fewer than n "limiting factors" (Levin 1970), has become known as the "competitive exclusion principle" (Hardin 1960).

The "principle" has been the center of much heated debate. Slobodkin (1961) has argued that it is not really a principle at all, but rather a tautology, and can serve only as a "rule of ecological procedure" to be followed in examining cases of species coexistence. Hutchinson (1961, p. 143) has phrased essentially the same thought in a more positive manner: "Just because the theory is analytically true and in a certain sense tautological, we can trust it in the work of trying to find out what has happened" to allow coexistence. In Hutchinson's view the principle is useful precisely because it is believed to be a tautology, a statement which is logically true and therefore not subject to empirical falsification.

It is therefore not surprising, given Hutchinson's influential view on the utility of the competitive exclusion principle, that a number of authors have attempted to extend the range of the tautology by generalizing Volterra's model to cases of more than one resource or limiting factor (MacArthur and Levins 1964; Rescigno and Richardson 1965; Levins 1968; Levin 1970; Haigh and Maynard Smith 1972; Haussman 1973; Armstrong and McGehee 1976a, 1976b; Kaplan and Yorke 1977; McGehee and Armstrong 1977). Our purpose in this paper to review these attempts, to examine the basis for recent results, and to provide a general framework for examining theoretical problems of competitive exclusion.

In Section 1 we examine Volterra's original proof of competitive exclusion, paying particular attention to the biological assumptions underlying Volterra's model. In this section we also introduce the various models which have been used in attempts to extend Volterra's model to cases of more than one resource, and discuss the results of previous authors. Section 2 contains a detailed discussion of

* Present address: Department of Ecology and Evolution, State University of New York at Stony Brook, Stony Brook, New York 11794.

Am. Nat. 1980. Vol. 115, pp. 151–170.

the work of Koch (1974*b*) and ourselves (Armstrong and McGehee 1976*a*; McGehee and Armstrong 1977), which proves that two species can indeed coexist on one resource in a time-invariant and spatially homogeneous environment.

In Section 3 we discuss the mathematical notion of an attractor and restate the general question of competitive exclusion. Section 4 summarizes current knowledge on all aspects of this general question. In Section 5 we explore the special problem of coexistence at fixed densities. We conclude that, in this special case, n species cannot coexist on fewer than n resources.

1.   THE BASIC MODELS

We begin with a close examination of Volterra's (1928) original model. Volterra first assumed that the dynamics of competing species can be described by the use of differential equations. This assumption is very important, and has been almost universally adopted by those who have studied competitive exclusion from a mathematical point of view. We shall discuss this most basic assumption more fully in Section 3.

For the moment, postulate the existence of n species $x_i, i = 1, \ldots, n$, competing for the same resource R. Let the specific (or per capita) growth rate of each species increase linearly with the amount of resource present, so that

$$\frac{1}{x_i}\frac{dx_i}{dt} = \gamma_i R - \sigma_i, \qquad i = 1, \ldots, n, \tag{1}$$

where $\sigma_i > 0$ is the rate at which the population would decline in the absence of resource and $\gamma_i > 0$ relates increased resource abundance to increased growth. Next assume that the amount of resource available to any competitor at time t is diminished by the presence of the competitors such that at any time t,

$$R = R_{\max} - F(x_1, \ldots, x_n). \tag{2}$$

Here $F(x_1, \ldots, x_n)$ is an unbounded increasing function of the population densities x_i, with $F(0, \ldots, 0) = 0$. Substituting (2) into (1) and replacing $\gamma_i R_{\max} - \sigma_i$ by ϵ_i yields Volterra's original equations:

$$\frac{dx_i}{dt} = x_i[\epsilon_i - \gamma_i F(x_1, \ldots, x_n)], \qquad i = 1, \ldots, n. \tag{3}$$

Volterra showed that, as $t \to \infty$, the species with the largest value of ϵ_i/γ_i will approach a finite nonzero density, and the remaining species will all approach extinction, provided that $\epsilon_i > 0$ and $x_i(0) \neq 0$ for the winning species.

Several simplifying assumptions are implicit in the above model. (i). The organisms under consideration are "simple" in the sense that the dynamics of the system can be adequately described by the species densities x_i. Complications arising from age structure or physiological state are assumed unimportant. (ii) The species interact only through the resource, so that their specific growth rates are functions of R alone, not of the x_i. (iii) The system under consideration is spatially homogeneous. (iv) The resource is uniform in quality. For example, if the resource consists of particles of food, these are uniform in size and nutritional value.

(v) There is no explicit time dependence to the interactions, either in terms of time-dependent interaction parameters or external forcing. There are no time lags.

Coexistence has been shown to be possible in many cases where one or more of these assumptions are violated. For example, Haigh and Maynard Smith (1972) showed that two predators could coexist on one prey species if they utilized different life stages of the prey (contra assumption [i]); and Stewart and Levin (1973) and Koch (1974a), following the suggestion of Hutchinson (1961), demonstrated that two species can coexist on a single resource in a time-varying environment (contra [v]).

The Volterra model (1)–(3) is an example of a "linear abiotic resource" model: "linear" because the specific growth rates of the competitors are linear functions of resource densities, and "abiotic resource" because the resource regenerates according to an algebraic relationship. At any given time a parcel of abiotic resource exists either in a "free" state or in a "bound" state. In the free state it is available for use by any individual, while in the bound state it is in use by some individual. For example, if the resource were a chemical nutrient, that part of the nutrient pool which is currently in use by living individuals is in a bound state; the remainder is in a free state. A second example of an abiotic resource is space, a parcel of which is in the bound state if it is occupied by some individual and in the free state if it is not occupied. A parcel of abiotic resource is regenerated from the bound state to the free state through the death of the individual by which it was bound. Such regeneration is assumed to occur instantaneously.

The Volterra model (1)–(3) is easily generalized to include k resources (Rescigno and Richardson 1965). It can be further generalized by relaxing the assumption of linearity in equation (1), yielding the class of "abiotic resource" models:

$$\frac{dx_i}{dt} = x_i u_i (R_1, \ldots , R_k), \qquad i = 1, \ldots , n, \tag{4}$$

$$R_j = R_{j\ max} - F_j(x_1, \ldots , x_n) \tag{5a}$$

$$= s_j(x_1, \ldots , x_n), \qquad j = 1, \ldots , k. \tag{5b}$$

Since the R_j are to be considered resources, it is assumed that species growth rates will increase with resource availability, and that resource densities will decrease with species densities. These conditions are specified by

$$\frac{\partial u_i}{\partial R_j} \geq 0 \qquad \text{and} \qquad \frac{\partial s_j}{\partial x_i} \leq 0, \tag{6}$$

where the equalities hold if and only if a particular species i does not use a particular resource j. A large class of chemostat models can be reduced to the abiotic resources model (Canale 1970; Waldon 1975; see also Appendix A). The "limiting factor" equations of Levin (1970) are exactly (4) and (5b) without the monotonicity restrictions (6).

Another important class of models concerns "biotic" resources, resources which regenerate according to their own differential equations, as would prey species (MacArthur and Levins 1964; Koch 1974b; Armstrong and McGehee 1976a; McGehee and Armstrong 1977). The defining equations for this class of models are

$$\frac{dx_i}{dt} = x_i u_i(R_1, \ldots, R_k), \qquad i = 1, \ldots, n,$$

$$\frac{dR_j}{dt} = R_j g_j(R_1, \ldots, R_k, x_1, \ldots, x_n), \qquad j = 1, \ldots, k. \tag{7}$$

The monotonicity conditions

$$\frac{\partial u_i}{\partial R_j} \geq 0 \qquad \text{and} \qquad \frac{\partial g_j}{\partial x_i} \leq 0, \tag{8}$$

analogous to equations (6), are expected to apply to this model.

Various authors have used these models in attempting to prove that n species cannot indefinitely coexist on $k < n$ resources or limiting factors (Volterra 1928; MacArthur and Levins 1964; Rescigno and Richardson 1965; Levins 1968; Levin 1970). All such early attempts contained the assumption that the specific growth rates u_i of the competing species are linear functions of resource or factor densities. In addition, these authors (with the exception of Levin 1970) restricted their attention to coexistence at fixed densities.

More recently, several authors (Koch 1974b; Zicarelli 1975; Armstrong and McGehee 1976a, 1976b; Kaplan and Yorke 1977; McGehee and Armstrong 1977) have shown that when these two restraints are simultaneously relaxed the coexistence of n species on $k < n$ resources becomes possible. In the following sections we detail the conditions under which this coexistence is possible and provide a coherent framework for viewing problems of competitive exclusion.

2. COEXISTENCE OF TWO SPECIES ON ONE BIOTIC RESOURCE

Koch (1974b) was the first to point out via computer simulation that two species could coexist on one biotic resource. This coexistence occurred along what appeared to be a periodic orbit, not at an equilibrium point. The coexistence of two species on one biotic resource was later confirmed analytically (McGehee and Armstrong 1977) and expanded to the case of n species coexisting on one biotic resource (Zicarelli 1975).

In this section we provide insight into the mechanism behind this coexistence. We will present an intuitive look at why the coexistence depends both on the nonlinearity of the growth functions u_i and on the lack of system equilibrium. Those who desire complete proofs should consult the papers of McGehee and Armstrong (1977) and Zicarelli (1975).

Consider a system composed of two species x_1 and x_2 competing for the same biotic resource R. Let the defining equations for this system be:

$$\frac{dx_1}{dt} = x_1\left(-m_1 + \frac{c_1 \eta_1 R}{R + \Gamma}\right), \tag{9a}$$

$$\frac{dx_2}{dt} = x_2(-m_2 + c_2 \eta_2 R), \tag{9b}$$

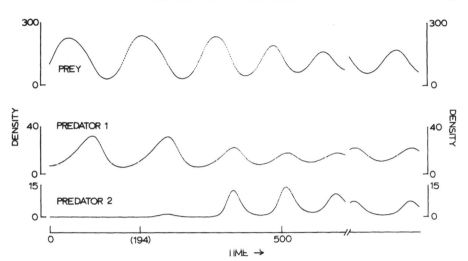

Fig. 1.—Computer simulation of two predators coexisting on a single prey (biotic resource). The model is that of eqq. (9) with $m_1 = .1$, $c_1 = .3$, $\eta_1 = .5$, $\Gamma = 50$, $r = .1$, $K = 300$, $m_2 = .11$, $c_2 = .33$, $\eta_2 = .003$. The system was started at $x_1 = 1$, $R = 400$, $x_2 = 0$, a point very near the two-species periodic orbit of x_1 and R. At $t = 194$, a small amount (.01) of predator x_2 is added; it readily invades the limit cycle. To the right of the break in the axis the apparent periodic behavior of the three-species system is shown. This limiting configuration is also reached if a small amount of x_1 is added to the two-species system (x_2, R) near its stable equilibrium.

$$\frac{dR}{dt} = R\left[r\left(1 - \frac{R}{K}\right) - \frac{\eta_1 x_1}{R + \Gamma} - \eta_2 x_2\right]. \qquad (9c)$$

In these equations m_1 and m_2 are the death rates of the competitors in the absence of resource; η_1 and η_2 are rate constants for resource consumption (per unit competitor and per unit resource); c_1 and c_2 are conversion efficiencies of resource biomass into competitor biomass; r and K are, respectively, the maximum growth rate and carrying capacity of the prey; and Γ is a half-saturation constant in the functional response of competitor 1. Note that competitor 2 is of the Lotka-Volterra type.

Computer simulation of the above system (fig. 1) suggests that the three species coexist along a periodic orbit for appropriate parameter values. How is this coexistence effected?

Consider first the species pair (x_1, R) in the absence of species 2. This pair has an equilibrium point (x_1^*, R^*) defined by

$$-m_1 + \frac{c_1 \eta_1 R^*}{R^* + \Gamma} = 0, \qquad r\frac{1 + R^*}{K} - \frac{\eta_1 x_1^*}{R^* + \Gamma} = 0.$$

The equilibrium point (x_1^*, R^*) may be stable or unstable, depending on parameter values.

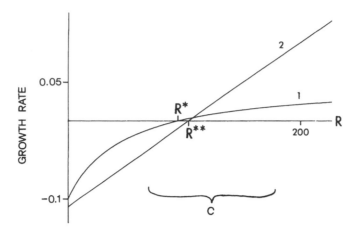

FIG. 2.—Growth rates of the predators of fig. 1 as functions of prey density R. The region C represents the approximate range of variation of R over one cycle in the three-species system.

Consider next the species pair (x_2, R) in the absence of species 1. This pair has an equilibrium point (x_2^{**}, R^{**}) defined by

$$-m_2 + c_2\eta_2 R^{**} = 0, \qquad r\frac{1 - R^{**}}{K} - \eta_2 x_2^{**} = 0.$$

The equilibrium point (x_2^{**}, R^{**}) will be globally stable for all parameter values (McGehee and Armstrong 1977).

Now think of a small amount of species 1 introduced into the system (x_2, R) near its equilibrium point. Species 1 can successfully invade if and only if $dx_1/dt > 0$ when x_1 is small and (x_2, R) is near (x_2^{**}, R^{**}). From equation (9a) we see that invasion can occur if and only if $-m_1 + c_1\eta_1 R^{**}/(R^{**} + \Gamma) > 0$. Since the specific growth rate of species 1 is an increasing function of R, this inequality holds if and only if $R^{**} > R^*$. In the simulation (fig. 1) the parameters were chosen such that $R^* = 100$ and $R^{**} = 110$ (fig. 2), insuring that species 1 can indeed invade the two-species equilibrium point (x_2^{**}, R^{**}).

Why, then, does competitor 1 not exclude competitor 2? By reasoning similar to that given above, we know that $dx_2/dt < 0$ near the two-species equilibrium point (x_1^*, R^*) if $R^{**} > R^*$. However, we also know that this equilibrium point may not be stable. In fact, for certain parameter values the two-species system (x_1, R) approaches a periodic orbit. Furthermore, around this orbit the average prey density $\bar{R}$ will be greater than R^* (Appendix B). Species 2 will be able to invade the system (x_1, R) along this periodic orbit if and only if its average rate of increase along the orbit is positive. That is, invasion is possible if and only if

$$\frac{1}{\tau}\int_0^\tau \frac{1}{x_2}\frac{dx_2}{dt}\, dt = \frac{1}{\tau}\int_0^\tau [-m_2 + \eta_2 c_2 R(t)]\, dt > 0, \tag{10}$$

where τ is the period of the two-species limit cycle of x_1 and R. The right-hand side of (10) reduces to $-m_2 + \eta_2 c_2\bar{R} > 0$, where $\bar{R} \equiv (1/\tau)\int_0^\tau R(t)\, dt$, so that invasion is possible if and only if $\bar{R} > R^{**}$. Therefore, if parameters in the model (9) are chosen such that $R^* < R^{**} < \bar{R}$, mutual invasibility is insured. Such is the case in the example of figure 1.

Although the foregoing argument shows that each species is able to invade the other near its equilibrium configuration, it does not constitute a proof of coexistence (see Appendix C). Such a proof requires consideration of the global properties of the three-species system, as in McGehee and Armstrong (1977). The above argument does expose two crucial points, however. (1) We could not have obtained mutual invasibility or stable coexistence if both competitors had obeyed Lotka-Volterra dynamics, since for Lotka-Volterra predators $R^* = \bar{R}$ (Appendix B). (2) The cycling of the two-species system (x_1, R) is necessary for coexistence for exactly the same reason: Without such cycling $R^* = \bar{R}$.

Although the properties of nonlinearity and lack of equilibrium are crucial for the coexistence of n species on $k < n$ resources, these properties are in no sense pathological. First, the assumption of nonlinear, saturating functional response curves is much more realistic than the assumption of Lotka-Volterra functional response, since the latter implies that the specific growth rate of a species increases indefinitely with increased resource density (Armstrong and McGehee 1976b). (We specified a Lotka-Volterra functional response for x_2 in system [9] for purely expository purposes; in Koch's (1974b) original computer simulation, both competitors had saturating functional response curves.)

Second, we follow Levin (1970) in feeling that coexistence which is not at fixed densities still deserves to be considered coexistence. For example, if a system composed of a predator species and a prey species persists indefinitely, even if this persistence is along a cycle, we would assert that the two species are indeed coexisting.

3.   REPHRASING THE PROBLEM

With the realization that two species could be made to coexist on a single biotic resource, we were led to rephrase the problem of competitive exclusion in more fundamental terms (McGehee and Armstrong 1977).

Our basic assumption is that the population dynamics of a community consisting of m species is adequately described by a set of ordinary differential equations of the form

$$\frac{dx_i}{dt} = x_i f_i(x_1, \ldots, x_m), \qquad i = 1, \ldots, m. \tag{11}$$

Here x_i is the density of species i and $f_i(x_1, \ldots, x_m)$ is its per capita growth rate. For the purposes of this section, the k resource equations of the biotic resource model (7) are not explicitly distinguished from the n competitor equations of the same model. Thus, for biotic resources, $m = n + k$.

The decision to use differential equations is not totally innocuous. Cole (1960) has objected to the competitive exclusion principle on the basis that all species with finite population sizes (i.e., all real species) are doomed to eventual extinction because of statistical fluctuations in population size. We must recognize, therefore, that we cannot use differential equation models to prove indefinite coexistence. Rather, we seek from differential equation models indications of strong tendencies towards coexistence.

In terms of the model (11) we can now define the term *persistence*.

Definition 1.—The system (11) is said to exhibit "persistence at fixed densities" if it possesses an asymptotically stable equilibrium point $x^* = (x_1^*, \ldots, x_m^*)$ with $x_i^* > 0$ for all $i = 1, \ldots, m$.

If the system is started near its equilibrium point x^*, then each species in a system satisfying definition 1 will tend asymptotically to its equilibrium density x_i^*; that is, the distance between the state vector x and the equilibrium position x^* will tend to zero as $t \to \infty$. Since all species are present at the final equilibrium, we say they are coexisting at fixed densities.

Definition 1 is far too restrictive to serve as a general definition of persistence. For example, a predator and prey can coexist with neither species ever approaching either extinction or constant density. Such a system should be considered persistent. To include possibilities other than coexistence at fixed densities we use a notion common in the mathematical theory of dynamical systems, namely, that of an "attractor." Roughly speaking, we define an "attractor block" to be a region in the state space $\{(x_1, \ldots, x_m)\}$ such that solutions starting on the boundary of the region pass into its interior. (A precise definition of "attractor block" can be found in a previous paper [McGehee and Armstrong 1977].)

Definition 2.—The system (4) is said to exhibit "persistence" if it has an attractor block bounded away from the m faces $\{x_i = 0\}$, $i = 1, \ldots, m$.

If the species start initially with densities in the attractor block, their densities will remain in the block for all future time. If $x_i > 0$, $i = 1, \ldots, m$, at all points within the block, then no species will ever approach extinction for any solution in the block, and the system is considered persistent (fig. 3).

Note that persistence at fixed densities (definition 1) is a special case of persistence (definition 2). In the first case, the densities are either constant or are approaching constant values. In the second case, the densities may be fluctuating in a seemingly unpredictable way.

We are interested in imposing certain structures on the system (11) and in determining whether those structures imply the impossibility of persistence.

Definition 3.—A given structure will be said to exhibit "competitive exclusion" if no system with such a structure is persistent.

The Volterra model described in the introduction is an example of such a structure. The parameters of the model are the constants n, γ_i, σ_i and $R_{\max}$, and the function F. For different parameter values we obtain different systems, but these systems all have the same structure. Volterra proved that no system with this structure can be persistent, and hence that this structure exhibits competitive exclusion.

More concretely, consider a two-species Volterra model (3) with $F(x_1, x_2) = \alpha x_1 + \beta x_2$. If $\epsilon_1/\gamma_1 \neq \epsilon_2/\gamma_2$, then there are no equilibrium points with both species present (fig. 4a). If $\epsilon_1/\gamma_1 = \epsilon_2/\gamma_2$, then there is a whole line of equilibrium points (fig. 4b). If $\epsilon_1/\gamma_1 < \epsilon_2/\gamma_2$, species 1 will go extinct (i.e., $x_1 \to 0$ as $t \to \infty$), as shown

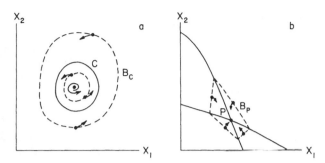

FIG. 3.—Two examples of attractor blocks bounded away from the faces $x_1 = 0, x_2 = 0$. a, A predator-prey cycle C surrounded by an attractor block B_C. b, Isocline diagram and stable equilibrium point P of a two-competitor system. The attractor block B_P has been drawn around the equilibrium point.

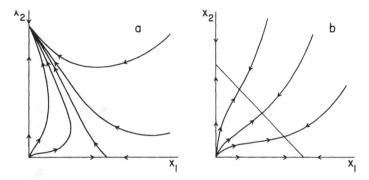

FIG. 4.—a, Exclusion of species 1 (x_1) in the Volterra model when $\epsilon_1/\gamma_1 < \epsilon_2/\gamma_2$. b, Phase portrait of the case $\epsilon_1/\gamma_1 = \epsilon_2/\gamma_2$. Note the line of critical points, none of which is asymptotically stable. In neither case can an attractor block bounded away from the faces $x_1 = 0, x_2 = 0$ be constructed.

in figure 4a. If $\epsilon_1/\gamma_1 > \epsilon_2/\gamma_2$, species 2 will go extinct. These extinctions will occur except in the trivial case where one species is initially absent.

This example illustrates the exclusion principle. Note first that, except in the case that $\epsilon_1/\gamma_1 = \epsilon_2/\gamma_2$, one or the other species exhibits a strong deterministic tendency towards extinction. In the exceptional case $\epsilon_1/\gamma_1 = \epsilon_2/\gamma_2$, the system will approach the line of equilibrium points. Since the line of equilibrium points intersects both axes, no attractor block bounded away from the axes can be constructed. Since no model of the Volterra type exhibits a deterministic tendency toward persistence, models of the Volterra type are said to exhibit competitive exclusion.

We should emphasize that our criterion for competitive exclusion is essentially a negative one: a system which does not exhibit a deterministic tendency toward coexistence is considered to exhibit exclusion. An alternative approach would have been to define competitive exclusion by the extinction of "excess" species. An example follows.

Definition 4.—A system of n species and $k < n$ resources or limiting factors is said to exhibit "strict competitive exclusion" if at least $n - k$ species become asymptotically extinct.

We feel that the weaker definition 3 is superior because "excess" species may not go extinct asymptotically. For example, consider the Volterra model described above. Except when $\epsilon_1/\gamma_1 = \epsilon_2/\gamma_2$, all models with this structure exhibit strict competitive exclusion. When $\epsilon_1/\gamma_1 = \epsilon_2/\gamma_2$, all trajectories approach a line of fixed points (fig. 4b) and hence the system does not exhibit strict competitive exclusion. However, no single fixed point is asymptotically stable, and it can be argued that small external forces may move the system from one fixed point to another. Eventually, such disturbances will move the system close to the boundary, and one or the other species may be considered extinct.

This example shows that there exist systems which exhibit no strong deterministic tendencies toward coexistence, but also in which no species becomes asymptotically extinct. Our criterion for competitive exclusion (definition 3) categorizes such systems as nonpersistent.

4.   RESULTS OF MORE RECENT INVESTIGATIONS

The attractor block formulation of the competitive exclusion problem was first applied to the coexistence of two competitors on one biotic resource. McGehee and Armstrong (1977) constructed an attractor block for this three-species system, proving that a strong deterministic tendency towards coexistence could exist in such a system (fig. 5).

Since that time, several further results have been proved.

4.1.   *Coexistence of Any Number of Species on One Biotic Resource*

Zicarelli (1975) has extended the proof of McGehee and Armstrong (1977) to show that any number of species can be made to stably coexist on one biotic resource. Therefore, the set of abiotic resource models (eqq. [7] and [8]) does not exhibit competitive exclusion.

4.2.   *Coexistence of Species on "Limiting Factors"*

Zicarelli's (1975) proof implies that any number of species can coexist on as few as two of Levin's (1970) limiting factors. To see this, note that Zicarelli's model is a special case of equations (7) and (8) with one resource; i.e., $dx_i/dt = x_i u_i(R)$, $i = 1, \ldots, n$, and $dR/dt = Rg(R, x_1, \ldots, x_n)$. The two limiting factors are the density of resource R and the function $g(R, x_1, \ldots, x_n)$. (See Levin 1970; Armstrong and McGehee 1976a.) Zicarelli's result thus implies that the set of limiting factors models does not obey the competitive exclusion principle for $k > 2$ limiting factors. Kaplan and Yorke (1977) have provided an independent proof of this fact for $k > 3$ limiting factors.

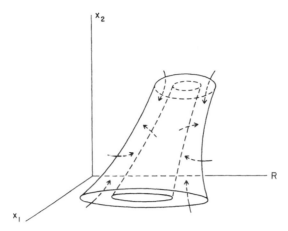

Fig. 5.—Attractor block for a two-predator, one-prey model. Once inside the solid torus the trajectory never leaves, implying a strong deterministic tendency toward indefinite coexistence.

The case of one limiting factor has also been solved. McGehee and Armstrong (1977) showed that two species cannot coexist on one limiting factor. More recently, Nitecki (1978) has shown that persistent systems with $n \geqslant 3$ species and one limiting factor can be constructed. Therefore, the only case in which a limiting factors model exhibits competitive exclusion is the case of two species on one limiting factor.

4.3. *Coexistence of Species on Abiotic Resources*

N species on four abiotic resources.—Armstrong and McGehee (1976b) have shown that it is possible to construct persistent systems of $n > 4$ species on four abiotic resources. To prove this point, we first constructed a system in which any number of species could coexist on a single resource in a time-varying environment. Next we asserted that it is possible to construct three-species, three-abiotic resource competition systems whose solutions tend to asymptotically stable periodic orbits. Smale (1976) has proved that such systems exist; Strobeck (1973, p. 652) has constructed a three-species competition system, unstable near its three-species equilibrium point, which in computer simulations appears to approach a periodic orbit.

We then combined these two subsystems, making the three-species–three-resource subsystem provide a periodic environment for the n-species–one-resource subsystem. The resulting system has $n > 4$ species coexisting on $k = 4$ conservative resources.

N species on one abiotic resource: Volterra revisited.—Volterra (1928) showed that n species could not coexist on one abiotic resource. Volterra's proof, however, is plagued by the same drawback as many of the succeeding models: the assumption that the $u_i(R)$ in equations (4) are linear. It is somewhat surprising, then, that in the one-abiotic resource case the linearity assumption is unnecessary. The monotonicity conditions on the u_i (eqq. 6) are sufficient to insure that all species except one will become extinct. When n species compete for a single limiting resource, the species which can exist at the lowest level of available resource will prevail. A proof is given in Appendix D.

An open question.—The behavior of models of n species on two or three abiotic resources remains unknown. It seems clear to us that the methods used by Zicarelli (1975) could be used to show the existence of persistent systems of n species on three abiotic resources. The case of two abiotic resources seems much more delicate. However, the work of Nitecki (1978) leads us to conjecture that there exist persistent systems of n species on two abiotic resources.

5.   COEXISTENCE AT FIXED DENSITIES

McGehee and Armstrong (1977) have shown that coexistence at fixed densities of n species on $k < n$ resources or limiting factors is impossible. Assertions to this effect have been made before. MacArthur and Levins (1964) noted that it is "infinitely unlikely" that n planes drawn in a k-dimensional space will intersect at a point. They interpreted this fact to mean that the coexistence at fixed densities of n species on $k < n$ resources is "infinitely unlikely." Kaplan and Yorke (1977) have added mathematical precision to these statements.

Whittaker and Levin (1976) have taken issue with MacArthur and Levins, stating that convergent evolutionary pressures may indeed result in point coexistence of n species on $k < n$ resources with probability greater than zero. They assert, however, that although such coexistence is deterministically possible, the fact that the community matrix at the equilibrium point is singular (has at least one eigenvalue equal to zero) means that the system will be extremely vulnerable to stochastic perturbations. Hence they too conclude that the point coexistence of n species on the $k < n$ resources or limiting factors is impossible.

We would argue, however, that a dominant eigenvalue of zero does not necessarily imply vulnerability to perturbation. Consider, for example, the model $dx/dt = x(k - x)^3$. This model has a dominant eigenvalue of zero, and yet is asymptotically stable near $x^* = k$. Furthermore, the system is rather insensitive to perturbation, in the sense that small environmental variations will not move the system far from its deterministic equilibrium. For example, consider any sort of environmental change which causes the carrying capacity k to vary in time. Let this variation be bounded, so that at any time t, $0 < k_1 < k(t) < k_2 < \infty$, where k_1 and k_2 are, respectively, the minimum and maximum values of the carrying capacity $k(t)$. It is evident that if the population density x lies in the interval (k_1, k_2) at time $t = 0$, then $x(t)$ will remain in that interval for all time. Furthermore, any trajectory which starts outside the interval (k_1, k_2) at $t = 0$ will eventually enter the interval, provided $x(0) \neq 0$. Thus the Whittaker-Levin argument does not always apply.

We have been able to show that n species cannot coexist on $k < n$ resources, in the sense of definition 1. That is, we have been able to show that attracting point equilibria cannot exist for $k < n$. The first key word in this statement is *attracting*. Point equilibria may exist, even for $k < n$. However, attracting point equilibria, which we consider the proper criterion for coexistence at fixed densities, cannot exist for $k < n$.

The second key part of this statement is *cannot exist*. Attracting point equilibria are literally impossible, not just "very unlikely," when $k < n$.

Our proof rests on some rather technical mathematical points. The basic scenario is outlined below. Those readers wishing a more precise statement of the proof should consult McGehee and Armstrong (1977).

Recall that Levin's limiting factors model is defined by

$$\frac{dx_i}{dt} = x_i u_i(R_1, \ldots, R_k), \qquad i = 1, \ldots, n,$$
$$R_j = s_j(x_1, \ldots, x_n), \qquad j = 1, \ldots, k. \tag{12}$$

Note further that the class of limiting factors models contains both the class of abiotic resource models and the class of biotic resource models (McGehee and Armstrong 1977). Therefore, if we can prove that coexistence at fixed densities of n species on $k < n$ limiting factors is impossible, we will automatically have proved the same result for both abiotic resource and biotic resource models.

We first note that the set of systems of the form (12), and with no equilibrium points, is "dense" (McGehee and Armstrong 1977). In other words, if we are given a system of n species and $k < n$ limiting factors, and if this system has an equilibrium point, we can always find a "nearby" system with no equilibrium points.

We next note that a fundamental property of an attractor block is that it shares certain topological properties with the attractor within it. In particular, the Euler characteristic is shared. The Euler characteristic is an integer associated with every geometric object. The Euler characteristic of a point is 1, that of a circle is 0, that of a solid sphere is 1, and that of a solid torus is 0. An attractor block which surrounds a point attractor will be topologically equivalent to a solid sphere (fig. 3b). Both the block and the attractor have Euler characteristic 1. An attractor block which surrounds a periodic attractor will be topologically equivalent to a solid torus (figs. 3a, 5). Both have Euler characteristic 0.

A second fundamental property of attractor blocks is that they remain attractor blocks under slight perturbation, even though the attractor itself may change. Therefore, the Euler characteristic of the attractor cannot change under small perturbations. For example, imagine a system having a point attractor. Surround this point with an attractor block constructed so that the vector field is everywhere transverse to its boundary. Now perturb the system slightly, perhaps by altering one of the parameters in the original set of equations. The attractor block for the unperturbed system is still an attractor block for the perturbed system. Therefore, even though the new attractor may no longer be a point, it must have the same Euler characteristic as a point, namely 1.

Given any system of n species and $k < n$ limiting factors, we know that we can find a system arbitrarily nearby which has no equilibrium points. Therefore all atractor blocks for systems of n species and $k < n$ resources must be topologically compatible with the fact that the attractor inside the attractor block may have no equilibrium points. Mathematically, all attractor blocks for such systems must have Euler characteristic 0. (The torus of fig. 5 is an example of such a block.) Since point attractors have Euler characteristic 1, point attractors cannot exist.

DISCUSSION

We have shown that it is possible to construct systems in which n species can coexist on $k < n$ resources or limiting factors. Why, then, have many authors (MacArthur and Levins 1964; Rescigno and Richardson 1965; Levins 1968; Levin 1970) been led to assert that such coexistence is impossible? Three observations are germane.

First, in all these earlier models, species' per capita growth rates (the functions u_i in eqq. [4] and [7]) were assumed to be linear functions of resource densities. It is indeed true that if the functions u_i are linear in resource densities or in limiting factors, persistence (Sec. 3, definition 2) is impossible (Levin 1970; McGehee and Armstrong 1977). This result may apply directly to species with Holling type I functional response curves (Holling 1965) if resource densities always remain below the levels needed to saturate the functional responses of the competitors and if the competitors' growth rates are directly proportional to resource consumption rates. However, this result will not in general apply to species with other types of functional response curves, except in the case $k = 1$ (sec. 4.3 and Appendix D).

The second point is that some authors (MacArthur and Levins 1964; Levins 1968) have considered only coexistence at fixed densities. Indeed, it can be proved that in a large class of biologically reasonable models, n species cannot coexist at fixed densities on $k < n$ resources (Sec. 5). Because this last result does not require the assumption of linearity, Armstrong and McGehee (1976b) have suggested that the competitive exclusion principle will in general apply only to cases of coexistence at fixed densities.

Third, note that if the assumption of linearity and the assumption of coexistence at fixed densities are simultaneously relaxed, it becomes possible to construct examples of n species coexisting of $k < n$ resources or limiting factors. The species coexist because of internally generated cyclic behavior.

These observations allow us to clarify the relationship of Volterra's work to that of his successors. In the case of only one resource, the system must eventually approach a point equilibrium (Appendix D). Since the requirement of equilibrium is sufficient to assure that competitive exclusion will hold (Sec. 5), Volterra's linearity assumption can be relaxed without affecting his results. When more than one resource or limiting factor is involved, however, equilibrium is not assured, and the assumption of linearity becomes critical. Therefore, the assumption that species' per capita growth rates are linear in resource densities, an assumption that occurs only as a mathematical convenience in Volterra's proof, becomes critical in cases where more than one resource is involved.

SUMMARY

Recent developments in the mathematical theory of competitive exclusion are discussed and placed in historical perspective. The models which have been used in theoretical investigations of competitive exclusion are classified into two groups: those in which the resources regenerate according to an algebraic relationship (abiotic resource models), and those in which resource regeneration is governed by differential equations (biotic resource models). We then propose a mathematical framework for considering problems of competitive exclusion, and provide examples in which n competitors can coexist on $k < n$ resources (both biotic and abiotic). These systems persist because of internally generated cyclic behavior. We conclude that the competitive exclusion principle applies in general only to coexistence at fixed densities.

ACKNOWLEDGMENTS

RAA was partially supported by USPHS Postdoctoral Training grant GM07199. RM was partially supported by NSF grant MCS 76-06003A02.

APPENDIX A

Consider the following chemostat model of n species competing for k nutrients:

$$\frac{dx_i}{dt} = x_i[u_i(R_1, \ldots, R_k) - D], \quad i = 1, \ldots, n,$$

$$\frac{dR_j}{dt} = -\sum_{i=1}^{n} c_{ji} x_i u_i(R_1, \ldots, R_k) + D(C_j - R_j), \quad j = 1, \ldots, k.$$

Here x_i is the concentration of species i, R_j is the concentration of nutrient j, and $u_i(R_1, \ldots, R_k)$ is the specific growth rate of species i (cf. eqq. (4), (5), the abiotic resource model). The constant D is the dilution rate of the growth medium, the constant C_j is the concentration of nutrient j in the incoming medium, and the constant c_{ji} relates the uptake of nutrient j to the production of species i.

Following Canale (1970) and Waldon (1975), we introduce the variables ϕ_j which measure the total concentration of nutrient j in the chemostat: $\phi_j = R_j + \sum_{i=1}^{n} c_{ji} x_i$. One easily computes that $d\phi_j/dt = D(C_j - \phi_j)$, from which it follows that ϕ_j exponentially approaches C_j, regardless of the concentrations of any of the species.

Therefore, when considering any sort of ultimate behavior of the model (such as steady-state behavior), one may assume that $\phi_j = C_j$. The model then reduces to

$$\frac{dx_i}{dt} = x_i[u_i(R_1, \ldots, R_k) - D], \quad i = 1, \ldots, n,$$

$$R_j = C_j - \sum_{i=1}^{n} c_{ji} x_i, \quad j = 1, \ldots, k,$$

which has the form of equations (4) and (5a) with $u_i - D$ replacing u_i and $F_j(x_1, \ldots, x_n) = \sum_{i=1}^{n} c_{ji} x_i$.

The results discussed in Sections 4 and 5 for abiotic resource models now can be applied directly to this class of chemostat models. For example, if $k < n$, then any attractor for the above model must have Euler characteristic zero. In particular, there can be no point attractors. Also, if $4 \leq k < n$, then there exist chemostat models of the above form with periodic attractors; therefore the chemostat structure does not imply competitive exclusion. Finally, if $k = 1$, then the model does exhibit competitive exclusion. Indeed, as shown in Appendix D, most models of this form exhibit strict competitive exclusion. This last result generalizes theorems of Hsu et al. (1976) and Hsu (1978).

APPENDIX B

We first show that a species whose growth rate increases linearly with resource density requires the same average resource density to maintain itself in an environment where the resource density varies as it does to maintain itself in an environment where the resource level is constant in time. Second, we show that if the growth response curve saturates with increasing prey density, the average prey density required for maintenance in a time-varying environment is higher than that required in a constant environment.

Consider a species x whose growth rate is determined by the density of some limiting resource R. Assume that the growth of species x is determined by the equation $dx/dt = xf(R)$, where $f(R)$ is the specific growth rate of species x as a function of resource density R. The response of the resource to utilization will be determined by a separate equation. However, the dynamics of resource regeneration are unimportant for the present argument.

We first consider a constant environment and determine the resource level R which will allow the population to maintain a constant size. Assuming that $f(0) < 0, f(\infty) > 0$, and that f is strictly increasing, we see that there is a unique value R^* such that $f(R^*) = 0$; R^* is the resource density required for x to maintain itself in a constant environment.

We next consider a time-varying environment and assume that the resource density is given by $R(t)$. Assume that, after some time τ, the population $x(t)$ returns to its initial value. (For example, $R(t)$ and $x(t)$ might both be periodic with period τ.) Now define $\bar{R} = (1/\tau) \int_0^\tau R(t)\, dt$, the average value of the resource density over the interval $0 \to \tau$. Also define $\overline{f(R)} = (1/\tau) \int_0^\tau f[R(t)]\, dt$, the average value of the specific growth rate of x. Since x returns to its initial value after time τ, we have

$$\int_0^\tau \frac{1}{x}\frac{dx}{dt}\, dt = \log x(\tau) - \log x(0) = 0.$$

Therefore,

$$\overline{f(R)} = \frac{1}{\tau}\int_0^\tau \frac{1}{x}\frac{dx}{dt}\, dt = 0.$$

We now ask: What is the relation between $\bar{R}$, the average value of resource density required by x to maintain itself in a time-varying environment, and R^*, the value of resource density required by x to maintain itself in a constant environment?

The answer is immediate if f is linear, since if $f(R) = aR + b$ then $f(\bar{R}) = \overline{f(R)} = 0$. Therefore $\bar{R} = R^*$. This result was first noted by Volterra (1928) in the special case of Lotka-Volterra population oscillations.

Consider now the case in which the growth response $f(R)$ saturates with increasing R; i.e., assume that f is concave downward. Draw the tangent line $L(R)$ to the curve $f(R)$ at $R = R^*$ (fig. 6). Since $L(R) > f(R)$ for $R \neq R^*$, we have that $\overline{L(R)} > \overline{f(R)} = 0$ in a time-varying environment. Since L is linear, $L(\bar{R}) = \overline{L(R)} > 0$. Therefore, since $L(R^*) = 0$, $\bar{R} > R^*$. Therefore, a species possessing a growth curve which is concave downward requires a higher average resource supply to persist when the resource level varies in time than it does when the resource level is maintained at $R = R^*$.

APPENDIX C

Does mutual invasibility imply coexistence? For a simple two-species Lotka-Volterra competition model, mutual invasibility does imply coexistence, since coexistence will occur if each species is able to invade the equilibrium of the other. However, in higher dimensions the situation is more complicated.

Consider a competition model of three species. Suppose that, in the absence of any one species, the remaining two come to an equilibrium. Suppose further that in each case the third species is able to invade the equilibrium of the other two. Do these conditions imply that the system is persistent?

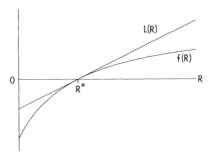

FIG. 6.—Graphs of the functions $f(R)$ and $L(R)$ discussed in Appendix B. The function $f(R)$ is concave downward; $L(R)$ is the tangent line to $f(R)$ at $R = R^*$.

The answer is not simple, as is illustrated in the example of three competing species discussed by May and Leonard (1975). For any given species pair, one species excludes the other so that each pairwise equilibrium point has one species absent. The third species can always invade this pairwise equilibrium. Therefore this example satisfies the conditions stated in the previous paragraph. However, the system oscillates wildly, with each species infinitely often coming arbitrarily close to extinction, followed by a recovery to a large population density. Thus the system is not persistent and illustrates the subtlety of the original question.

Now consider the system of two predators and one prey discussed in Section 2. One can imagine that the following behavior might occur. The system starts with x_1 and R near their periodic orbit and with x_2 small. Species 2 first invades, but then declines in such a way that the system approaches the equilibrium point (x^*, R^*). Then, with x_2 small, the system approaches the periodic orbit in the (x_1, R) plane, and the process starts anew. With each successive occurrence the minimum value of x_2 becomes smaller, so that species 2 comes arbitrarily close to extinction infinitely often. However, each near-extinction of species 2 is followed by a successful invasion. If one could construct a system whose only attractor contains such an orbit, this system would not be persistent.

Although the above argument is not precise, it does illustrate the danger of supposing that mutual invasibility always implies coexistence.

APPENDIX D

Consider equations (4) and (5a) with $k = 1$:

$$\frac{dx_i}{dt} = x_i u_i(R), \qquad i = 1, \ldots, n,$$

$$R = R_{max} - F(x_1, \ldots, x_n).$$

We wish to state conditions which imply that this system exhibits strict competitive exclusion. Nitecki (1978) showed that if no assumptions are made on the partial derivatives of the functions u_i and F, then there are persistent systems with this form. Volterra (1928) showed that if the u_i are linear functions, then the system exhibits strict competitive exclusion for most values of the parameters. The following assumptions are more general than Volterra's but restrictive enough to exclude Nitecki's example.

Assume a slightly stronger version of (6): $\partial u_i / \partial R > 0$ and $\partial F / \partial x_i > 0$. Assume that each species has a positive value of R for which it is just able to maintain its density, i.e., for each i there is an $R_i^* > 0$ such that

$$u_i(R) < 0 \qquad \text{for } R < R_i^*,$$

$$u_i(R) = 0 \quad \text{for} \quad R = R_i^*,$$
$$u_i(R) > 0 \quad \text{for} \quad R > R_i^*.$$

Assume that only one of the species has the smallest value of R_i^*. If necessary, relabel the species so that R_1^* is the smallest, i.e. $R_1^* < R_i^*$, $i = 2, \ldots, n$. Let K_1 be the carrying capacity of species 1 in the absence of all other species, i.e. K_1 is the unique density such that $F(K_1, 0, \ldots, 0) = R_{\max} - R_1^*$. Given the above assumptions, we prove the following theorem.

THEOREM D1. If $x_1(0) > 0$, then, as $t \to \infty$, $x_1(t) \to K_1$ and $x_i(t) \to 0$ for $i = 2, \ldots, n$.

The intuitive idea behind the proof is the following. If the available resource R is below the critical level R_1^*, then all the species will decline until R exceeds this critical level. At that point species 1 will start to increase while the others continue to decline. Species 1 will then asymptotically reach its carrying capacity while the others asymptotically approach extinction.

We prove the theorem by constructing two Liapunov functions. For the basic definitions and examples of Liapunov functions, see LaSalle and Lefschetz (1961).

Write $\mathbf{x} = (x_1, \ldots, x_n)$, $\mathbf{x}_e = (K_1, 0, \ldots, 0)$. Define the three sets

$$\mathcal{B} = \{\mathbf{x}: x_i \geq 0 \; \forall_i \quad \text{and} \quad F(\mathbf{x}) < R_{\max} - R_1^*\},$$

$$\mathcal{C} = \{\mathbf{x}: x_i \geq 0 \; \forall_i \quad \text{and} \quad F(\mathbf{x}) > R_{\max} - R_1^*\},$$

$$\Delta = \{\mathbf{x}: x_i \geq 0 \; \forall_i \quad \text{and} \quad F(\mathbf{x}) = R_{\max} - R_1^*\}.$$

The set $\mathcal{B}$ consists of all points in the state space for which the concentration of available resource R is greater than the critical concentration R_1^*, $\mathcal{C}$ is the set where $R < R_1^*$, and Δ is the set where $R = R_1^*$. The union of these three sets is the entire positive orthant, and Δ forms the common boundary between the two regions $\mathcal{B}$ and $\mathcal{C}$.

On Δ, we have $dx_1/dt = 0$. Therefore

$$\frac{dF}{dt} = \sum_{i=2}^{n} \frac{\partial F}{\partial x_i} \frac{dx_i}{dt} = \sum_{i=2}^{n} \frac{\partial F}{\partial x_i} x_i u_i(R_1^*).$$

Since $u_i(R_1^*) < 0$, $i = 2, \ldots, n$, we know that $dF/dt < 0$ on Δ unless $\mathbf{x} = \mathbf{x}_e$. We have therefore proved the following:

LEMMA D2. If $\mathbf{x}(t_1) \in \Delta - \{\mathbf{x}_e\}$, then $dF/dt < 0$ when $t = t_1$.

In other words, we have shown that, except for the equilibrium point $\mathbf{x}_e$, any solution which at some time hits Δ passes immediately into $\mathcal{B}$. Note that D2 also implies that $\mathcal{B}$ is positively invariant, i.e. any solution which gets into $\mathcal{B}$ stays in $\mathcal{B}$ for all future time. More precisely,

LEMMA D3. If $\mathbf{x}(t_1) \in \mathcal{B}$, then $\mathbf{x}(t) \in \mathcal{B}$ for $t \geq t_1$.

Now consider the function on $\mathcal{B}$: $V(x_1, \ldots, x_n) = -x_1$. This is a Liapunov function on $\mathcal{B}$, since

$$\frac{dV}{dt} = -\frac{dx_1}{dt} = -x_1 u_1(R) < 0, \quad \text{if} \quad x_1 \neq 0.$$

Therefore D2 and D3 imply

LEMMA D4. If $x(t_1) \in \mathcal{B}$ and if $x_1(t_1) \neq 0$, then $\mathbf{x}(t) \to \mathbf{x}_e$ as $t \to \infty$.

That is, any solution which gets into $\mathcal{B}$ must approach the equilibrium point $\mathbf{x}_e$. We have therefore proved the conclusion of theorem D1 for any solution which starts in $\mathcal{B}$ or Δ. We have only left to prove the result for a solution starting in $\mathcal{C}$.

Consider the function F on $\mathcal{C}$, which is a Liapunov function since

$$\frac{dF}{dt} = \sum_{i=1}^{n} \frac{\partial F}{\partial x_i} x_i u_i(R) < 0 \quad \text{in} \; \mathcal{C}.$$

Therefore the following is true:

COMPETITIVE EXCLUSION   169

Lemma D5. If $\mathbf{x}(t_1) \in \mathscr{C}$, then either (a) there exists a $t_2 > t_1$ such that $\mathbf{x}(t_2) \in \Delta - \{\mathbf{x}_e\}$, or (b) $\mathbf{x}(t) \to \mathbf{x}_e$ as $t \to \infty$.

In other words, any solution which starts in $\mathscr{C}$ either approaches the equilibrium point $\mathbf{x}_e$ or hits Δ. If it approaches $\mathbf{x}_e$, then the conclusion holds; if it hits Δ, then we have shown above that it passes into $\mathscr{B}$ and then approaches $\mathbf{x}_e$. In either case the conclusion of theorem D1 holds, and the proof is complete.

LITERATURE CITED

Armstrong, R. A., and R. McGehee. 1976a. Coexistence of two competitors on one resource. J. Theor. Biol. 56:499–502.

———. 1976b. Coexistence of species competing for shared resources. Theor. Popul. Biol. 9:317–328.

Canale, Raymond P. 1970. An analysis of models describing predator-prey interactions. Biotechnol. Bioeng. 12:353–378.

Cole, L. C. 1960. Competitive exclusion. Science 132:348–349.

Haigh, J., and J. Maynard Smith. 1972. Can there be more predators than prey? Theor. Popul. Biol. 3:290–299.

Hardin, G. 1960. The competitive exclusion principle. Science 131:1292–1298.

Haussman, U. 1973. On the principle of competitive exclusion. Theor. Popul. Biol. 4:31–41.

Holling, C. S. 1965. The functional response of predators to prey density and its role in mimicry and population regulation. Mem. Entomol. Soc. Can., no. 45.

Hsu, S. B. 1978. Limiting behavior for competing species. Soc. Ind. Appl. Math. J. Appl. Math. 34:760–763.

Hsu, S. B., S. P. Hubbell, and P. Waltman. 1976. A mathematical theory for single-nutrient competition in continuous cultures of micro-organisms. Soc. Ind. Appl. Math. J. Appl. Math. 32:366–383.

Hutchinson, G. E. 1961. The paradox of the plankton. Am. Nat. 95:137–145.

Kaplan, J. L., and J. A. Yorke. 1977. Competitive exclusion and nonequilibrium coexistence. Am. Nat. 111:1032–1036.

Koch, A. L. 1974a. Coexistence resulting from an alternation of density dependent and density independent growth. J. Theor. Biol. 44:373–386.

———. 1974b. Competitive coexistence of two predators utilizing the same prey under constant environmental conditions. J. Theor. Biol. 44:387–395.

LaSalle, J., and S. Lefschetz. 1961. Stability by Liapunov's direct method with applications. Academic Press, New York.

Levin, S. A. 1970. Community equilibria and stability, and an extension of the competitive exclusion principle. Am. Nat. 104:413–423.

Levins, R. 1968. Evolution in changing environments. Princeton University Press, Princeton, N.J.

MacArthur, R., and R. Levins. 1964. Competition, habitat selection, and character displacement in a patchy environment. Proc. Natl. Acad. Sci. USA 51:1207–1210.

May, R. M., and W. J. Leonard. 1975. Nonlinear aspects of competition between species. Soc. Ind. Appl. Math. J. Appl. Math. 29:243–253.

McGehee, R., and R. A. Armstrong. 1977. Some mathematical problems concerning the ecological principle of competitive exclusion. J. Differ. Equations 23:30–52.

Nitecki, Z. 1978. A periodic attractor determined by one function. J. Differ. Equations 29:214–234.

Rescigno, A., and I. W. Richardson. 1965. On the competitive exclusion principle. Bull. Math. Biophys. Suppl 27:85–89.

Slobodkin, L. B. 1961. Growth and regulation of animal populations. Holt, Rinehart, & Winston, New York.

Smale, S. 1976. On the differential equations of species in competition. J. Math. Biol. 3:5–7.

Stewart, F. M., and B. R. Levin. 1973. Partitioning of resources and the outcome of interspecific competition: a model and some general considerations. Am. Nat. 107:171–198.

Strobeck, C. 1973. N species competition. Ecology 54:650–654.

170   THE AMERICAN NATURALIST

Volterra, V. 1928. Variations and fluctuations of the number of individuals in animal species living together. J. Cons. Cons. Int. Explor. Mer. 3:3–51.

Waldon, M. G. 1975. Competition models. Am. Nat. 109:487–489.

Whittaker, R. H., and S. A. Levin, eds. 1976. Niche: theory and application. Dowden, Hutchinson & Ross, Stroudsburg, Pa.

Zicarelli, J. 1975. Mathematical analysis of a population model with several predators on a single prey. Ph.D. thesis, University of Minnesota.

Ecology (1977) **58**: pp. 338–348

RESOURCE COMPETITION BETWEEN PLANKTONIC ALGAE: AN EXPERIMENTAL AND THEORETICAL APPROACH[1]

DAVID TILMAN[2]

Division of Biological Sciences, Department of Ecology and Evolutionary Biology, University of Michigan, Ann Arbor, Michigan 48109 USA

Abstract. The results of 76 long-term competition experiments between two species of freshwater algae (*Asterionella formosa* and *Cyclotella meneghiniana*) grown along a resource gradient agree with the predictions of two different models of resource competition. Both models are based on the functional resource-utilization response of each species to limiting resources. The Monod model and the Variable Internal Stores model of competition made similar predictions. *Asterionella* was observed to be competitively dominant when both species were phosphate limited; *Cyclotella* was dominant when both species were silicate limited; and both species stably coexisted when each species was growth-rate limited by a different resource. Almost 75% of the variance in the relative abundances of these two species along a natural silicate-phosphate gradient in Lake Michigan is explained by the Monod model.

Key words: Competition models; diatoms; functional response; Michaelis-Menten; Monod; phosphate; resource competition; resource gradient; resource utilization; silicate.

INTRODUCTION

Many recent theoretical studies of interspecific competition have dealt with the dependence of competition on the resource utilization abilities of each species. The models used in these studies may be classified as being of two types. One uses the classical Lotka-Volterra competition equations, estimating the coefficient of competition, alpha, from some measure of the resource utilization overlap between species (MacArthur 1969, 1970; Orians and Horn 1969; May 1975). The other approach employs models which explicitly include the availability (concentration) of the resource and the functional dependence of growth of each species on the availability of the resource (Stewart and Levin 1973; Greeney et al. 1973; MacArthur 1972; O'Brien 1974; Petersen 1975; Taylor and Williams 1975; Titman 1976). The work reported here is an experimental test of the latter type of mechanistic model. The results of 76 long-term competition experiments between two species of freshwater algae grown under controlled-culture laboratory conditions along a two-resource gradient are compared with the predictions of resource-utilization models of interspecific competition. This is a test of the utility of species-specific resource acquisition and utilization information in predicting the steady-state outcome of competition between two species potentially limited by two resources.

Asterionella formosa Hass. and *Cyclotella meneghiniana* Kutz., freshwater diatoms which are seasonally abundant in midlatitude, mesotrophic lakes (Kopczynska 1973; Stoermer and Kopczynska 1967), were the two species studied. The two potentially limiting resources for this study were phosphate

and silicate, the nutrients which most often limit algal growth in such lakes (Lund et al. 1963; Powers et al. 1972; Schelske and Stoermer 1971; Kilham 1971).

The nutrient kinetics of the clones of *Asterionella* and *Cyclotella* used in these competition experiments are reported elsewhere (Tilman and Kilham 1976). With this information on the ability of each species to acquire and utilize resources, two different models of resource competition are used to predict the steady-state outcome of interspecific competition for potentially limiting silicate and phosphate. The first model (Model I) is based on the Monod equations (Monod 1950; Herbert et al. 1956; Taylor and Williams 1975). The second model (Model II) is based on a variable internal stores model of growth (Droop 1974). Both competition models are founded on models of the functional acquisition and growth response of a single species to a single resource.

The experiments were designed to test how well such single species physiological information can predict the outcome of competition. To avoid variability in the outcome of competition that could be attributed to genetic changes in the populations (cf., Park et al. 1964), the same clone of each species was used throughout all the experiments. Both clones were bacteria-free isolates. Sterile technique was used throughout the experiments to eliminate the possible complication of bacterial competition for limiting resources. Competition experiments were allowed to proceed for between 30 to 40 days, to assure that the results observed were the steady-state outcome of interspecific competition. Two single-species culture controls were performed for all conditions at which competition experiments were performed, thus verifying that each species could exist by itself under all conditions tested. It was hoped that these precautions would minimize the variance in the outcome of the competition experiments, allowing a better test of the relationship between the resource utilization abilities of a species and its competitive abilities.

[1] Manuscript received 19 April 1976; accepted 21 September 1976.

[2] Present address: Department of Ecology and Behavioral Biology, 108 Zoology Building, University of Minnesota, Minneapolis, Minnesota 55455 USA.

The mathematical treatment of both models is a steady-state analysis, limited to those aspects that are relevant to the steady-state results of the competition experiments performed. The competition experiments were designed to determine the long-term outcome of interspecific competition under conditions in which each species, by itself, would be able to maintain a stable population. They were not designed for short-term dynamic analysis of competitive displacement. Thus the dynamic aspects of competitive displacement are not included in the analysis of the two models. The steady-state experimental results and theoretical predictions are compared with each other and with some observations on the relative abundances of *Asterionella* and *Cyclotella* along a natural gradient (in space and time) of phosphate and silicate in Lake Michigan.

MODEL I: MONOD MODEL OF RESOURCE COMPETITION

The Monod model of growth of a single species limited by a single resource was proposed by Monod (1950), and extensively developed and tested on single species cultures of bacteria (Herbert et al. 1956). The Michaelis-Menten model of enzyme kinetics has the same formulation as the Monod model. Dugdale (1967) proposed that the Michaelis-Menten model be used to describe nutrient use by marine phytoplankton, and O'Brien (1974) and Petersen (1975) proposed that it be used to describe nutrient competition between algae. Numerous other workers (Eppley and Thomas 1969; Guillard et al. 1973; Kilham 1975) have reported close agreement between experimental observations on a single species and the Monod or Michaelis-Menten equations. Although the Monod equation can be derived from enzyme kinetic theory, I consider it to be a simple equation that provides a reasonable approximation to the functional relation between growth rate and resource availability. For a continuous flow system, the equations for the ith of n total species and for the jth of m different resources are as follows:

$$dN_i/N_i dt = \min_{1 \le j \le m} [r_i S_j/(K_{ij} + S_j) - D] \qquad (1)$$

$$dS_j/dt = D({}_0 S_j - S_j) - \sum_{i=1}^{n} N_i r_i S_j/[(K_{ij} + S_j)Y_{ij}], \qquad (2)$$

where

r_i = maximal growth rate of species i

K_{ij} = half saturation constant for species i limited by resource j (that is, the nutrient concentration at which it has half its maximal growth rate)

Y_{ij} = yield of species i limited by resource j (number of cells produced per unit of resource j)

N_i = number of cells of species i per unit volume

S_j = concentration of resource j external to the cells

${}_0 S_j$ = influent concentration of resource j

D = steady-state growth rate (true dilution rate)

n = number of species present

m = number of potentially limiting resources.

Equation 1 states that the growth rate of a species will be completely determined by that one nutrient that is most limiting, of all those that are potentially limiting. This has been shown experimentally by Droop (1974) for a marine alga potentially limited by phosphate and vitamin B-12. The switching that occurs as a species changes from being growth-rate limited by one resource to being limited by another resource is not easily dealt with analytically. In this analysis, I employ a method that is suitable for steady-state conditions. Because the only cases of experimental interest are those in which each species can exist by itself, boundary condition pecularities (such as washout of a species) are ignored.

For species i limited by resource j, Eqs. 1 and 2 provide the following steady-state relations (mathematical steady-state occurs when time derivatives are equal to zero)

$$N^*_{i(j)} = Y_{ij}({}_0 S_j - S^*_j) \qquad (3)$$
$$S^*_j = D K_{ij}/(r_i - D), \qquad (4)$$

where

$N^*_{i(j)}$ = steady-state population size of species i when it *alone* is limited by resource j (This is comparable to the carrying capacity of environment j for species i)

S^*_j = steady-state external concentration of resource j when only species i is present.

The boundary between a species being growth-rate limited by resource 1 or by resource 2 is calculated by setting $N^*_{i(1)} = N^*_{i(2)}$. This provides an expression of the influent concentrations of resources 1 and 2 at which species i is equally limited by both resources at steady state (Eq. 5):

$${}_0 S_1 = S^*_1 + ({}_0 S_2 - S^*_2)(Y_{i2}/Y_{i1}). \qquad (5)$$

The physiological constants in Table 1 were used to determine the boundary between phosphate and silicate limitation of *Asterionella*, for the concentrations of phosphate and silicate used in the experiments. This boundary (from Eq. 5) is shown in Fig. 1. For any silicate to phosphate ratios greater than 90 (to the left of the boundary), *Asterionella* should be phosphate limited. For $[S_i/P] < 90$ (to the right of the boundary), *Asterionella* should be silicate limited. The boundary between *Cyclotella* being phosphate or silicate limited was also determined from the kinetic parameters of Table 1. For $[S_i/P] > 6$ (to the left of the boundary), *Cyclotella* is phosphate limited. To the right of the boundary, for $[S_i/P] < 6$, *Cyclotella* is silicate limited (Fig. 1). The slight curvature in the boundaries comes from the dependence of S^*_1 and S^*_2 on dilution rate. When several species are grown together, these

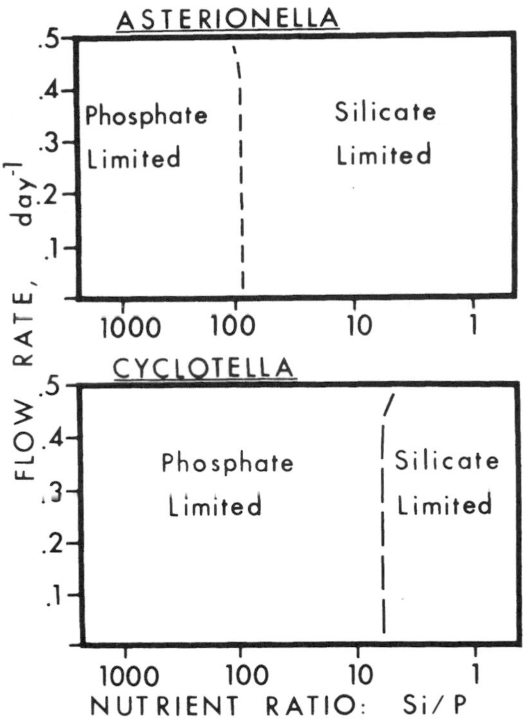

340   DAVID TILMAN   Ecology, Vol. 58, No. 2

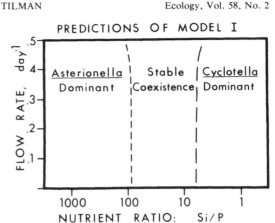

FIG. 1. The upper drawing shows the silicate to phosphate ratios for which *Asterionella* should be silicate or phosphate limited, as calculated using the physiological constants of Table 1 and Eq. 5. The lower drawing shows the same relationship for *Cyclotella*. The position of the curves reflects the resource gradient used.

FIG. 2. The predictions of the Monod model of resource competition, made using the physiological constants of Table 1, are shown.

competitive dominant. Maximal growth rates of *Asterionella* and *Cyclotella* (Table 1) are not significantly different ($P \geq 0.95$). The lower half saturation constant of *Asterionella* (0.02 μM PO$_4$) compared to *Cyclotella* (0.25 μM PO$_4$) means that *Asterionella* should be competitively dominant over *Cyclotella* when both species are phosphate limited. When both species are silicate limited, the significantly lower ($P \geq 0.95$) half saturation constant for growth of *Cyclotella* (1.44 μM SiO$_2$) compared to *Asterionella* (3.94 μM SiO$_2$) indicates that *Cyclotella* should be the superior competitor. These predictions are shown in Fig. 2.

For the case in which species 1 is limited by resource 1 and species 2 is limited by resource 2, the steady-state solution for the system of equations represented by Eqs. 1 and 2 gives the following:

$$S_1^* = DK_{11}/(r_1 - D), \qquad (7)$$
$$S_2^* = DK_{22}/(r_2 - D), \qquad (8)$$
$$N_{1(1)}^* = N_1 + N_2 (Y_{11}/Y_{21}) \qquad (9)$$
$$N_{2(2)}^* = N_2 + N_1 (Y_{22}/Y_{12}). \qquad (10)$$

Equations 9 and 10 are directly analogous to the steady-state form of the Lotka-Volterra equations. Note that $N_{i(j)}^*$ is the "carrying capacity" of species i when it alone is limited by resource j (Eq. 3). Y_{11}/Y_{21} is directly analogous to alpha of the Lotka-Volterra equations at steady state; Y_{22}/Y_{12} is beta of the Lotka-Volterra equations. An analysis comparable to that used for the Lotka-Volterra equations at steady state shows that both species should coexist stably under the conditions defined by Eqs. 7, 8, 9 and 10 and by the parameters of Table 1. Coexistence of two species, when each species is limited by a different resource, has been demonstrated theoretically by Stewart and Levin (1973), Petersen (1975) and Taylor and Williams (1975).

Figure 2 summarizes the outcomes of competition predicted by Model I. In the region in which both

boundaries should define, at steady state, the regions in which each species is limited by either silicate or phosphate. The boundaries for *Asterionella* and *Cyclotella* divide the nutrient ratio/flow rate plane into three regions (Fig. 2). For [S$_i$/P] > 90, both species should be phosphate limited. In the region to the right of the *Cyclotella* boundary ([S$_i$/P] < 6), both species should be silicate limited. In the region between the two boundaries, for 6 < [S$_i$/P] < 90, both species should be limited by different resources: *Asterionella* by silicate and *Cyclotella* by phosphate.

Three cases of competition need be considered. For two species limited by the same nutrient, the species which is able to lower the external nutrient concentration the most will competitively displace all other species at steady state (Taylor and Williams 1975). From Eq. 4, species 1 will be competitively superior to species 2 when

$$K_{11}/(r_1 - D) < K_{21}/(r_2 - D). \qquad (6)$$

Both species will be able to coexist stably only if $K_{11}/(r_1 - D) = K_{21}/(r_2 - D)$. For two species with the same maximal growth rates, the species with the lower half saturation constant for growth (K_{ij}) should be the species are phosphate limited, it is predicted that *Asterionella* should be dominant. In the region in which both species are limited by different resources, coexistence is predicted. In the region in which both species are silicate limited, it is predicted that *Cyclotella* should be dominant.

MODEL II: VARIABLE INTERNAL STORES MODEL

The variable internal stores physiological model proposed by Droop (1974) to describe the response of a single species of algae to nutrient limitation was used in a modified form by Lehman et al. (1975) to model interspecific competition for resources. A simplified form of their equations is presented here. The model assumes that internal nutrient concentration (Q) determines growth rate and that internal nutrient concentrations are determined by the joint processes of nutrient uptake, assumed to follow a Michaelis-Menten process, and growth;

$$dN_i/N_i dt = \underset{1 \le j \le m}{\text{MIN}} [r_i(1 - g_{ij}/Q_{ij}) - D], \quad (11)$$

$$dQ_{ij}/dt = V_{ij}(S_j/(S_j + k_{ij})) - r_i(Q_{ij} - g_{ij}), \quad (12)$$

$$dS_j/dt = D(_0S_j - S_j) - \underset{i=1, n}{\Sigma} [N_i V_{ij}(S_j/(S_j + k_{ij}))], \quad (13)$$

where

r_i = maximal growth rate of species i

g_{ij} = minimal internal stores of resource j by species i; i.e., internal nutrient concentration at which growth ceases. This is identical to k_Q of Droop (1974)

V_{ij} = maximal rate of uptake of nutrient j by species i

k_{ij} = half saturation constant for uptake of nutrient j by species i

Q_{ij} = internal stores or "cell quotient"; amount of nutrient j internal to each cell of species i

N_i = number of cells of species i per unit volume

S_j = external concentration of resource j

$_0S_j$ = influent concentration of resource j

n = number of species

m = number of resources.

Equation 11 includes the assumption that growth rate is determined solely by the internal nutrient concentration (Q_{ij}) which is lowest relative to the minimal internal stores for that nutrient (g_{ij}). This has been experimentally shown by Droop (1974).

The following relations hold at steady state:

$$N^*_{i(j)} = (_0S_j - S_j)/Q^*_{ij} \quad (14)$$

$$Q^*_{ij} = g_{ij}r_i/(r_i - D) \quad (15)$$

$$S^*_j = r_i g_{ij} k_{ij} D/(V_{ij}(r_i - D) - r_i g_{ij} D). \quad (16)$$

The boundary between a species being growth-rate limited at steady state by nutrient 1 or 2 is derived by setting $N^*_{i(1)} = N^*_{i(2)}$. This gives Eq. 17:

$$_0S_1 = S^*_1 + (_0S_2 - S^*_2)(g_{11}/g_{12}). \quad (17)$$

The predicted boundary between *Asterionella* being

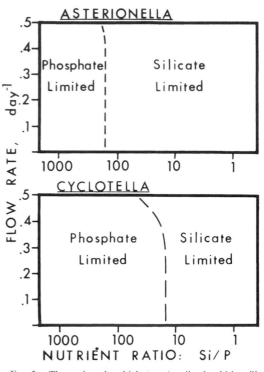

FIG. 3. The regions in which *Asterionella* should be silicate or phosphate limited, as predicted by Model II (Eq. 17) using the physiological constants of Table 2, are shown in the upper drawing. The lower drawing shows the same relationship for *Cyclotella*.

phosphate limited or silicate limited is shown in Fig. 3. This was calculated using Eq. 17 and the physiological constants of Table 2. The boundary is at [Si/P] of ≈ 170. For *Cyclotella*, the predicted boundary, calculated similarly, is at [Si/P] of ≈ 15 (Fig. 3).

The variable internal stores model predicts that the nutrient ratio/flow rate plane will be divided into three regions: a region in which both species are phosphate limited, one in which each species is limited by a different nutrient, and one in which both species are silicate limited. According to Model II, two species limited by the same nutrient can only coexist if the steady-state nutrient concentrations caused by each species alone are equal. The species which causes the lower steady-state nutrient concentration will competitively displace all other species at steady state. Under limitation by resource j, species 1 will be dominant over species 2 if Eq. 18 holds:

$$r_1 g_{1j} k_{1j} D/[V_{1j}(r_1 - D) - r_1 g_{1j} D] < \\ r_2 g_{2j} k_{2j} D/[V_{2j}(r_2 - D) - r_2 g_{2j} D]. \quad (18)$$

This can be approximated by Eq. 19 when D is much less than the washout rate:

$$r_1 g_{1j} k_{1j} D/[V_{1j}(r_1 - D)] < r_2 g_{2j} k_{2j} D/[V_{2j}(r_2 - D)]. \quad (19)$$

342   DAVID TILMAN   Ecology, Vol. 58, No. 2

TABLE 1. Physiological constants needed for Model I (from Tilman and Kilham 1976). Maximal growth rates (r) of *Asterionella* and *Cyclotella* were not significantly different ($P \geq 0.95$). Yield (Y) is for cultures with $0.5r$, to avoid bias from dependence of growth on internal stores. K is the half saturation constant

Species	Nutrient	K (μM)	r (doublings/day)	Y (cells/μmole)
Asterionella formosa	PO_4	0.02	0.9	2.18×10^8
Asterionella formosa	SiO_2	3.94	1.1	2.51×10^6
Cyclotella meneghiniana	PO_4	0.25	0.8	2.59×10^7
Cyclotella meneghiniana	SiO_2	1.44	1.3	4.20×10^6

The physiological constants of Table 2 indicate that, for all flow rates and nutrient concentrations at which neither species would be washed out of a single species culture, *Asterionella* should be the superior competitor under phosphate limitation. Because there are not variance estimates for all the parameters of Table 2, a test for significant differences between the two sides of Eq. 18 is not possible. However, the two sides differ by a factor of four, which seems to be great enough to predict that *Asterionella* should be the superior competitor under phosphate limitation. When both species are silicate limited, Eq. 18 predicts that *Asterionella* should have a slight advantage over *Cyclotella*. However, the two sides of Eq. 18 differ by < 15%. Because of the errors in estimating the parameters of Table 2 used in Eq. 18, I believe that this must be considered a prediction of no significant difference between the two species. It is unlikely that both species are identical, which is the condition needed for stable coexistence of both species when they are limited by the same resource. The "prediction" shown in Fig. 4 is that either *Asterionella* or *Cyclotella* should be the superior competitor when both species are silicate limited.

For species 1 limited by resource 1, and species 2 limited by resource 2, with no luxury consumption of a nonlimiting resource, the following equations hold at steady state:

$$S_1^* = r_1 g_{11} k_{11} D / [V_{11}(r_1 - D) - r_1 g_{11} D];$$
$$S_2^* = r_2 g_{22} k_{22} D / [V_{22}(r_2 - D) - r_2 g_{22} D],$$
$$Q_{11}^* = g_{11} r_1 / (r_1 - D) \qquad Q_{12}^* = Q_{11}^*(g_{12}/g_{11})$$
$$Q_{21}^* = Q_{22}^*(g_{21}/g_{22}) \qquad Q_{22}^* = g_{22} r_2 / (r_2 - D)$$
$$N_{1(1)}^* = N_1 + (Q_{21}^*/Q_{11}^*)N_2$$
$$N_{2(2)}^* = N_2 + (Q_{12}^*/Q_{22}^*)N_1 \qquad (20)$$

TABLE 2. Physiological constants needed for Model II (from Tilman and Kilham 1976). Symbols are: g = minimal internal stores of nutrient; r = maximal growth rate; K = half saturation constant; V = maximal rate of nutrient uptake (see text)

Species	Nutrient	g (μM/cell)	r (ln/day)	K (μM)	V (μM $\cdot$ cell$^{-1} \cdot$ h^{-1})
Asterionella formosa	PO_4	1.75×10^{-9}	0.7	2.8	9.85×10^{-9}
Asterionella formosa	SiO_2	2.96×10^{-7}	1.2	7.7	3.58×10^{-8}
Cyclotella meneghiniana	PO_4	1.07×10^{-8}	0.7	0.8	5.51×10^{-9}
Cyclotella meneghiniana	SiO_2	1.57×10^{-7}	1.2	7.5	1.51×10^{-8}

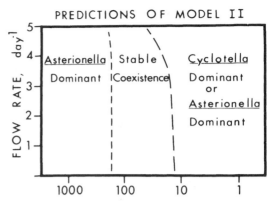

FIG. 4. The steady-state predictions of the variable internal stores model of competition, made using the physiological constants of Table 2, are shown.

These equations predict stable coexistence of two species if each is growth-rate limited by a different resource and each is a superior competitor for one of the two resources. These predictions are shown in Fig. 4.

The ability of these two models to predict the steady-state outcome of competition will be judged by three criteria. (1) The agreement between the predicted outcomes (dominance, coexistence) and those experimentally observed in the three general regions. (2) The agreement between the predicted placement of the boundaries and the observed placement. (3) The agreement between the predicted shape of the boundaries and the observed shape of the boundaries.

LONG-TERM COMPETITION EXPERIMENTS: MATERIALS AND METHODS

Asterionella formosa Hass. (clone FraAF) was isolated from Frains Lake, Michigan into axenic (bacteria-free) culture by S. S. Kilham. *Cyclotella meneghiniana* Kutz. (clone CyOh) was isolated from Lake Ohrid, Yugoslavia, by S. S. Kilham and obtained in axenic condition by V. McAlister. The general methods used for the competition experiments are described elsewhere (Titman 1976; Tilman and Kilham 1976). I will give the basic details.

A freshwater medium of double-distilled H_2O and inorganic salts and vitamins was used for all cultures ("WC" of Guillard and Lorenzen 1972). The concentrations of silicate and phosphate were varied so that silicate to phosphate ratios would range from $\approx 1{,}000$ to 1 (micromoles per micromoles). Concentrations of phosphate and silicate were never so low that steady-state populations of each species, grown singly, were not maintained at all nutrient ratios and flow rates tested. Influent phosphate concentrations ranged from ≈ 0.10 μM to ≈ 15 μM. Influent silicate concentrations ranged from ≈ 100 μM to ≈ 9 μM. Concentrations chosen were low enough that either silicate or

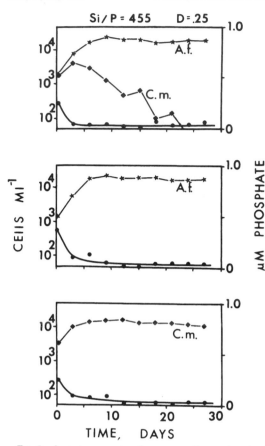

FIG. 5. Long-term semicontinuous growth experiments at [Si/P] of 455 (μM/μM) and flow rate of 0.25/day. Upper graph: stars–*Asterionella* (A.f.), diamonds–*Cyclotella* (C.m.), and closed circles–phosphate concentration, throughout 27-day competition experiments. Middle graph: Number of *Asterionella* (A.f.) (stars) and phosphate concentrations (closed circles) throughout a 27-day experiment with only *Asterionella* present (control). Lower graph: Number of *Cyclotella* (C.m.) (diamonds) and phosphate concentration (closed circles) throughout a *Cyclotella* growth experiment (control).

phosphate should have been the growth-rate-limiting nutrient through all experiments.

Cultures were grown in a culture box at 20°C with 100 μein m^{-2} S^{-1} illumination provided by "cool-white" fluorescent lights. All cell counts were done with a microscope using a Sedgwick-Rafter counting chamber, with samples preserved in Lugol's acetate solution (Guillard 1973). Cell counts and measurements of reactive extracellular phosphate and silicate were performed periodically. Phosphate and silicate were determined with the methods of Strickland and Parsons (1972), with absorbance read on a spectrophotometer with either 50- or 10-mm quartz cells. All samples were filtered through Millipore® filters which had been presoaked in double-distilled H$_2$O.

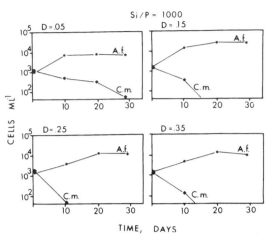

FIG. 6. Competition (two species) experiments for a silicate to phosphate ratio in the influent medium of 1,000, at four flow rates (D) (0.05, 0.15, 0.25, and 0.35/day). Symbols as in Fig. 5.

Long-term competition experiments were performed in flow-through (semicontinuous) culture. Cultures were started with each species in approximately equal abundance, generally at ≈ 1,000 cells/ml. Innocula were grown in media low in phosphate and silicate for a week or more before being used. Cultures were diluted manually daily. The flow rate (f) is reported as the ratio of the volume removed per day to the total culture volume. This may be converted to the true steady-state growth rate by the conversion D = ln(1/(1-f)). The flow rates generally used were 0.05, 0.15, 0.25, 0.35, and 0.50/day.

For each competition experiment at a particular silicate to phosphate ratio and flow rate, two single-species control cultures were also run. This was done both to gain physiological information (as in Table 2)

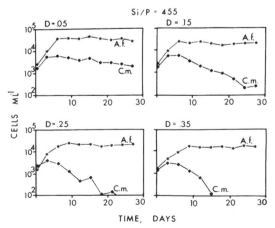

FIG. 7. Competition experiments for [Si/P] = 455 at flow rates of 0.05, 0.15, 0.25 and 0.35/day. Symbols as in Fig. 5.

DAVID TILMAN   Ecology, Vol. 58, No. 2

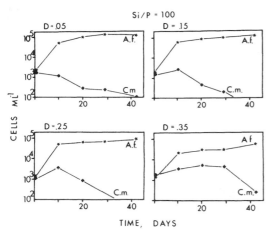

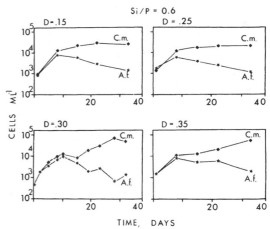

FIG. 8. Competition experiments for [Si/P] = 100 at four flow rates. Symbols as in Fig. 5.

FIG. 10. Competition experiments for [Si/P] = 0.6 for flow rates of 0.15, 0.25, 0.30, 0.35/day. Symbols as in Fig. 5.

and to assure that each species was able to grow to a stable, steady-state population under all conditions tested. Any displacement of one species from a two species competition culture must be caused by interactions between the two species. The controls assure that the results observed are due to competition.

RESULTS

For the purposes of these experiments, a species was considered to be competitively dominant when it comprised 95% or more of the total number of cells in the culture. If neither species had reached dominance by the predetermined end of an experiment, the results were termed coexistence. Because of the arbitrary, but necessary, nature of this definition, the actual time course of competition is shown for numerous cases throughout the range of conditions tested.

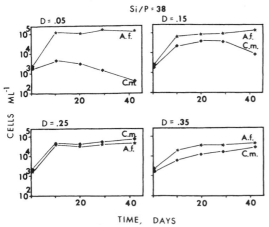

FIG. 9. Competition experiments for [Si/P] = 38 at four flow rates. Symbols as in Fig. 5.

Figure 5 shows the results of one competition experiment at [Si/P] = 455 and f = 0.25/day, along with the two control (single-species) cultures. Phosphate concentrations are shown. Silicate concentrations were also periodically measured, but were never low enough to be limiting. A silicate to phosphate ratio of 455 is within the region for which both species are predicted by Model I and Model II to be phosphate limited. Both models predict that Asterionella should be the superior competitor. This was the case (Fig. 5). By day 24, no Cyclotella were observed in the 1 ml sample counted, but > 10^4 Asterionella were counted.

For [Si/P] ≥ 100, Asterionella was dominant at all flow rates tested. Time series of competition for [Si/P] of 1,000, 455, and 100 are shown in Figs. 6, 7 and 8. At [Si/P] of 38, Asterionella was dominant at f = 0.05/day in triplicate competition experiments and at f = 0.5/day in two of three triplicates. Both species coexisted in the other triplicate experiments at intermediate flow rates for [Si/P] = 38. The time course of competition is shown in Fig. 9. At [Si/P] = 0.6, Cyclotella was dominant at all flow rates tested (Fig. 10).

The times to 99% dominance by Asterionella are shown in Fig. 11 for all cultures in which Asterionella became dominant. In general, it takes more time for Asterionella to become dominant the closer the silicate to phosphate ratio is to the observed boundary between dominance by Asterionella and coexistence of both species. Competitive displacement occurs more rapidly at higher flow rates, indicating the potential influence of mortality on the competitive process. There is a discontinuity in the trend of slower competitive displacement nearer the boundary of coexistence. This can be seen for the silicate to phosphate ratios <100, for which displacement is slightly faster than for ratios of 100. This trend was even more pronounced in the time to 95% and to 85% dominance. It should be

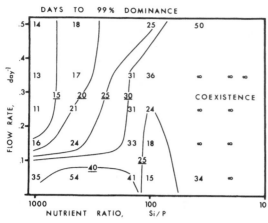

FIG. 11. The days to 99% dominance of *Asterionella* (by number of cells) over *Cyclotella* in cultures in which *Asterionella* became dominant. The times shown are the averages, rounded to integral days, computed by linear regression through log-transformed cell counts. The curves shown (with their labels underlined) are hand drawn.

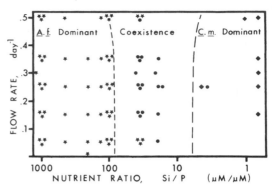

FIG. 12. The steady state results of all 76 long-term competition experiments are compared with the predictions of the Monod model (Model I). Stars represent cultures in which *Asterionella* (*A.f.*) was competitively dominant. Diamonds represent cultures in which *Cyclotella* (*C.m.*) was dominant. Closed circles represent stable coexistence of both species.

noted that competitive exclusion to the level of 99% dominance is a fairly slow process, often requiring 25 to > 40 days, even at higher flow (turnover) rates.

The steady-state results of all 76 competition experiments between *Asterionella* and *Cyclotella* are shown in comparison with the predicted results (Figs. 12 and 13). Stars symbolize dominance by *Asterionella*; diamonds, dominance by *Cyclotella*; dots, coexistence of both species. The major trends observed in the competition experiments are: (a) A region at [Si/P] generally greater than $\approx$ 80, but with boundaries curving toward lower [Si/P] at both high and low flow rates, in which *Asterionella* was dominant. (b) A region for 6 < [Si/P] < 80 in which both species coexisted. (c) At region for [Si/P] generally less than about 6 in which *Cyclotella* was dominant.

Tilman et al. (1976) have observed that *Asterionella* colonies exhibit morphometric changes in the number of cells per colony when they are limited by silicate and phosphate. The average number of cells per colony in *Asterionella* grown to steady state in flow-through cultures under silicate limitation decreased from > 20 cells/colony at low flow rates to 8.0 cells/colony at flow rates near the washout rate. Under phosphate limitation, a completely different trend was observed. The number of cells per colony in *Asterionella* increased from < 2 at low flow rates to 8 cells/colony as flow rate approached the maximal growth rate. At low to moderate flow rates, *Asterionella* never averaged > 6 cells/colony under phosphate limitation, nor < 8 cells/colony under silicate limitation.

For the competition experiments reported here, in the region in which *Asterionella* was observed to be dominant over *Cyclotella* (Fig. 12 or 13), no cultures

were observed at flow rates of < 0.35/day to have > 8 cells/colony. Throughout this region, *Asterionella* averaged 5.7 cells/colony at flow rates from 0.15 to 0.40/day. This indicates that these *Asterionella* were phosphate limited. In the other competition cultures at these same flow rates (in which both species coexisted or *Cyclotella* was dominant), *Asterionella* averaged 8.9 cells/colony. These *Asterionella* were probably silicate limited. This provides some indirect evidence that *Asterionella* cells were phosphate limited in the region in which they were dominant over *Cyclotella* and silicate limited in the regions in which both species coexisted or *Cyclotella* was dominant.

DISCUSSION

The general agreement between the predictions of Models I and II and the observed experimental results indicates the potential power of resource-based theories of competition to predict the outcome of in-

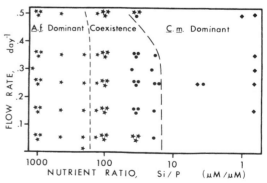

FIG. 13. The steady-state results of all 76 long-term competition experiments are compared with the predictions of Model II. Symbols are the same as for Fig. 12.

346   DAVID TILMAN   Ecology, Vol. 58, No. 2

terspecific competition. The results also support a contention of mathematical models of competition: two species can coexist at steady state when each is limited by a different resource (Levins 1968; Stewart and Levin 1973; Petersen 1975; Taylor and Williams 1975).

Three criteria are used to judge the similarity of the experimental results and the predictions of Models I and II. The first is the agreement between the predicted outcomes of competition (dominance, coexistence) and the observed outcomes, independent of their quantitative placement. The Monod model (Model I) correctly predicted the steady-state outcomes in all three regions. The variable internal stores model (Model II) correctly predicted the outcome in two of the three regions (Fig. 13), with one region being "too close to call". The second criterion is the agreement between the quantitative placement of the predicted and observed boundaries. The predicted boundaries were determined using a complex algebraic function. Because variance estimates are not available on all the parameters used, it is difficult to determine if the boundaries predicted by each model differ significantly either from each other or from those observed. It is only possible to make qualitative comparisons. From Fig. 12, Model I seems to predict accurately the boundary between dominance by *Asterionella* and coexistence of both species. (Agreement of shape will be discussed later.) The other boundary, between coexistence and dominance by *Cyclotella*, is harder to judge, for lack of sufficient data points, but the predicted and observed outcomes seem in agreement. From Fig. 13, Model II seems to be a less accurate predictor of the boundary between dominance by *Asterionella* and coexistence of both species. For the number of data points there are to base a judgement on, it seems to be just as good a predictor as Model I of the other boundary. The remarkable agreement between the predictions of Model I and the observed results must be viewed with caution. The placement of the boundary depends on the ratio of yields (Y_{11}/Y_{12}). The yields used (Table 1) were obtained from steady-state single-species cultures at flow rates of half the maximal flow (growth) rate. This minimized the error caused by the increasing efficiency with which algae can utilize nutrients at low concentrations (viz. the variable internal stores model). If the classical estimate of yield had been used (Herbert et al. 1956), the boundary between *Asterionella* being limited by phosphate or silicate would have been at ≈ 150. The third criterion is the agreement between the shape of the predicted and observed boundaries. As Figs. 12 and 13 show, neither model correctly predicted the observed curvature of the boundary between dominance by *Asterionella* and coexistence. The other boundary was not defined well enough for any comparison.

Of the models used, the Monod is the conceptually simpler. It requires fewer parameters and those needed are more easily obtained. The steady-state comparisons of the two models indicate that the Monod model is apparently superior for one or two of the three criteria of judgement. This may be the case merely because of the increased experimental error necessitated by the greater number of parameters in Model II. Though more complex, a possible advantage of the variable internal stores model is its ability to easily include luxury consumption of resources (hoarding). Luxury consumption may not only affect the steady-state population densities of the two species and the dynamics of population growth, but also the outcome of competition. For instance, at low steady state growth rates under silicate limitation, *Asterionella* can store phosphate 80× in excess of that needed at that growth rate. This luxury storage decreases with increasing growth rate (Tilman and Kilham 1976). Thus, at low flow rates in the region in which it is predicted that each species should coexist because *Asterionella* is silicate limited and *Cyclotella* is phosphate limited (Figs. 2 and 4), *Asterionella*, by its luxury consumption of phosphate, may lower the concentration of phosphate such that *Cyclotella* would be competitively displaced. Luxury consumption may thus be a factor causing curving of the boundaries describing the regions of Figs. 2 and 4. Further competition experiments will be needed to test the relative merits of these two models, and their variations, in describing both the steady-state and dynamic aspects of interspecific competition for resources.

Another possible criterion might have been to determine if each species were growth-rate limited by the predicted nutrient in the three observed regions of competitive dominance and coexistence. The morphometric changes in the number of cells per colony of *Asterionella* indicate that *Asterionella* was phosphate limited in the region in which it was dominant over *Cyclotella* and that it was silicate limited in the regions in which both species coexisted or *Cyclotella* was dominant. This agrees with the predictions of both Models I and II.

There are some observational data for Lake Michigan which indicate that competition for phosphate and silicate may be important in determining the distribution and relative abundance of *Asterionella* and *Cyclotella*. These are data from an intensive inshore sampling (Kopczynska 1973) and from transects across the lake (E. F. Stoermer, *personal communication*). In order to interpret these data in terms of the laboratory experiments performed, it is necessary to consider the difference between the silicate to phosphate ratios used up to this point, and those which are available from the Lake Michigan data. The silicate to phosphate ratios of Figs. 1 to 13 are ratios of the supply rates of these nutrients, but the data available for Lake Michigan are measurements of the concentrations of phosphate and silicate actually in the lake samples. Assuming that the measured concentrations are at or near some steady-state values, the Monod model may be used to predict the outcome of competition. In theory, a species should be equally limited by external, steady-state concentrations of two resources when Eq. 21 holds:

$$[S_1]/[S_2] = K_1/K_2, \qquad (21)$$

where S_1 and S_2 are the steady-state concentrations of resources 1 and 2 and K_1 and K_2 are the half-saturation constants for growth of this species limited by resources 1 and 2, respectively (Titman 1976). Using the half-saturation constants of Table 1, *Asterionella* should be dominant for ratios of external silicate to phosphate concentrations >200; *Cyclotella* should be dominant for ratios <5.7; and both species should coexist for ratios between 5.7 and 200. This prediction is shown across the top of Fig. 14. The relative abundance of *Cyclotella* along the observed silicate-phosphate gradient is also shown. As Fig. 14 shows, *Cyclotella* is very abundant relative to *Asterionella* at stations which have [Si/P] less than about 10. These are generally near-shore stations. *Asterionella* is very abundant relative to *Cyclotella* at stations which have [Si/P] greater than about 100. These are open lake stations. Coexistence (by the definition used previously) occurs, generally, for silicate to phosphate ratios between about 10 and 100. An analysis of variance on the combined data of Kopczynska (1973) and E. F. Stoermer (*personal communication*) revealed that 74.3% of the variance in the relative abundance of these two species can be explained by the Monod model of resource competition. Considering the non-steady-state nature of these samples, any agreement with a trend predicted from a laboratory study on two clones which were not isolated from Lake Michigan is especially encouraging. Many factors other than resource competition must influence the natural distribution of phytoplankton. Trophic interactions, spatial and temporal heterogeneity, and various physical factors must all play a part. However, the ability of the Monod model to explain over 70% of the observed variance in the natural distribution of these two species supports the view that resource competition is a major factor determining the distribution and abundance of species in nature.

Analysis of the two models of resource competition indicates that, at steady state, they are directly analogous to the steady-state form of the classical Lotka-Volterra equations of competition. This relationship is encouraging in that it demonstrates that resource-based models simplify to the classical equations at steady state. However, there is a danger in using resource-based models to estimate the parameters of the Lotka-Volterra equations. Resource-based models of competition are mechanistic, with the formulation of the model stating explicitly the mechanisms of competition in terms of the resource acquisition and utilization abilities of each species.

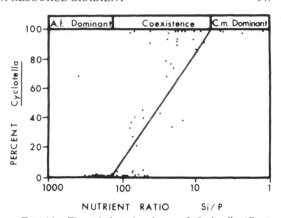

FIG. 14. The relative abundance of *Cyclotella* (C.m.) (compared to the total of *Cyclotella* and *Asterionella* [A.f.]) in samples from Lake Michigan plotted against ambient silicate to phosphate ratios in the same samples ($n = 78$). Over 70% of the observed variance in the relative abundance of these two species is explained by the steady state form of the Monod model. Notation at the top of the figure shows predicted outcomes of competition; the broken line shows the expected proportion of *Cyclotella* in the region of coexistence.

The Lotka-Volterra equations are purely descriptive and their parameters (alpha, beta, carrying capacities) are assumed to be constants. Resource-based models indicate that those parameters are only constant at steady state. For instance, the analog of alpha in the resource-based models varies with growth rate and past history, only being a constant at steady state. The steady-state estimate of alpha which can be obtained from resource-based models may only be used in the Lotka-Volterra equations at steady state. This use of an estimated alpha eliminates both the explicit statement of the mechanisms of resource competition and the ability to analyze competition under non-steady-state conditions. Moreover, use of resource models emphasizes the role of resources in determining competitive interactions and the need for information on the abilities of each species to acquire and use the potentially limiting resources. Although descriptive models of competition may be useful tools for understanding some aspects of community structure, mechanistic models may also be used for that purpose and are superior in that they may validly be used to analyze non-steady-state situations.

In conclusion, two currently used models of algal nutrient physiology seem able to make significant predictions about the nature of interspecific competition for resources. Both models are mechanistic. Both use the functional resource acquisition and/or utilization response of each species to potentially limiting resources to predict the outcome of interspecific competition for resources. Both models correctly predicted three regions of competition. The two regions in which each species should have been limited by the

348   DAVID TILMAN   Ecology, Vol. 58, No. 2

same resource were dominated by a single species (Figs. 12 and 13). Both species coexisted, as predicted, in the region in which each species should have been limited by a different resource. The relative abundance of *Asterionella* and *Cyclotella* along a natural gradient of silicate and phosphate in Lake Michigan agrees with that predicted by these models. Over 70% of the observed variance in the distribution of these two species in Lake Michigan can be explained by the Monod model (Fig. 14). Although indirect, this supports the view that interspecific competition for resources may be important in structuring natural communities. These results demonstrate the power that resource-based theories of interspecific interaction may have in explaining the distribution and abundances of species in nature.

ACKNOWLEDGMENTS

This work was supported by a University of Michigan Rackham Predoctoral Fellowship and a Rackham Dissertation Grant to D. Titman, and by National Science Foundation Grant GB-41315 to P. Kilham. This is part of a dissertation submitted in partial fulfillment of the Ph. D. degree at the University of Michigan. I thank Drs. Peter Kilham, Susan Kilham, Julian Adams, John Vandermeer, Stephen Hubbell, Nelson Hairston and Francis Evans for their advice, comments and guidance, and D. Wethey, D. Brambilla and others for their comments on this manuscript. I thank Gene Stoermer for graciously providing unpublished data on *Asterionella* and *Cyclotella* in Lake Michigan. I thank C. Kott and D. Morast for assistance as this work progressed. Please note my name was formerly David Titman.

LITERATURE CITED

Droop, M. R. 1974. The nutrient status of algal cells in continuous culture. J. Mar. Biol. Assoc. United Kingdom **54**:825–855.

Dugdale, R. C. 1967. Nutrient limitation in the sea: Dynamics, identification and significance. Limnol. Oceanogr. **12**:685–695.

Eppley, R. W., and W. H. Thomas. 1969. Comparison of the half-saturation constants for growth and nitrate uptake of marine phytoplankton. J. Phycol. **5**:375–379.

Greeney, W. J., D. A. Bella, and H. C. Curl. 1973. A theoretical approach to interspecific competition in phytoplankton communities. Am. Nat. **107**:405–425.

Guillard, R. R. L. 1973. Division rates, p. 289–311. *In* J. R. Stein [ed.] Phycological Methods. Cambridge Press, New York.

———, P. Kilham, and T. A. Jackson. 1973. Kinetics of silicon-limited growth in the marine diatom *Thalassiosira pseudonana* Hasle and Heimdal (=*Cyclotella nana* Hustedt). J. Phycol. **9**:233–237.

Guillard, R. R. L., and C. J. Lorenzen. 1972. Yellow-green algae with chlorophyllide *c*. J. Phycol. **8**:10–14.

Herbert, D., R. Elsworth, and R. C. Telling. 1956. The continuous culture of bacteria; A theoretical and experimental study. J. Gen. Microbiol. **14**:601–622.

Kilham, P. 1971. A hypothesis concerning silica and the freshwater planktonic diatoms. Limnol. Oceanogr. **16**:10–18.

Kilham, S. S. 1975. Kinetics of silicon-limited growth in the freshwater diatom *Asterionella formosa*. J. Phycol. **11**:396–399.

Kopczynska, E. E. 1973. Spatial and temporal variations in the phytoplankton and associated environmental factors in the Grand River outlet and adjacent waters of Lake Michigan. Ph. D. thesis. Univ. Michigan, Ann Arbor. 487 p.

Lehman, J. T., D. B. Botkin, and G. E. Likens. 1975. The assumptions and rationales of a computer model of phytoplankton population dynamics. Limnol. Oceanogr. **20**. 343–364.

Levins, R. 1968. Evolution in changing environments. Princeton University Press, Princeton. 120 p.

Lund, J. W. G., F. J. H. Mackereth, and C. H. Mortimer. 1963. Changes in the depth and time of certain chemical and physical conditions and of the standing crop of *Asterionella formosa* Hass. in the north basin of Windermere in 1947. Philos. Trans. R. Soc. London Ser. B Biol. Sci. **246**:255–290.

MacArthur, R. H. 1969. Species packing, or what competition minimizes. Proc. Natl. Acad. Sci. **64**:1369–1375.

———. 1970. Species packing and competitive equilibrium for many species. Theor. Pop. Biol. **1**:1–11.

———. 1972. Geographical Ecology. Harper and Row Publishers, New York. 269 p.

May, R. M. 1975. Some notes on estimating the competition matrix, α. Ecology **56**:737–741.

Monod, J. 1950. La technique de culture continue; theorie et applications. Ann. Inst. Pasteur Lille **79**:390–410.

O'Brien, W. J. 1974. The dynamics of nutrient limitation of phytoplankton algae: A model reconsidered. Ecology **55**:135–141.

Orians, G. H., and H. S. Horn. 1969. Overlap in food of four species of blackbirds in the potholes of Washington. Ecology **50**:930–938.

Park, T., P. H. Leslie, and D. B. Mertz. 1964. Genetic strains and competition in populations of *Tribolium*. Physiol. Zool. **37**:97–162.

Petersen, R. 1975. The paradox of the plankton: An equilibrium hypothesis. Am. Nat. **109**:35–49.

Powers, C. F., D. W. Schults, K. W. Malueg, R. M. Brice, and M. D. Schuldt. 1972. Algal responses to nutrient additions in natural waters. II. Field experiments, p. 141–156. *In* G. E. Likens [ed.] Nutrients and Eutrophication. Am. Soc. Limnol. Oceanogr. Lawrence, Kansas.

Schelske, C. L., and E. F. Stoermer. 1971. Eutrophication, silica depletion and predicted changes in algal quality in Lake Michigan. Science **173**:423–424.

Stewart, F. M., and B. R. Levin. 1973. Partitioning of resources and the outcome of interspecific competition: A model and some general considerations. Am. Nat. **107**:171–198.

Stoermer, E. F., and E. E. Kopczynska. 1967. Phytoplankton populations in the extreme southern basin of Lake Michigan, 1962–1963, p. 88–106. *In* F. Elder [ed.] Proc. 10th Conf. Great Lakes Res. Int. Assoc. Great Lakes Res. Univ. Michigan Press, Ann Arbor, Michigan.

Strickland, J. D. H., and J. R. Parsons. 1972. A practical manual of seawater analysis. Fish. Res. Board Canada Bull. No. 167. 311 p.

Taylor, P. A., and J. L. Williams. 1975. Theoretical studies on the coexistence of competing species under continuous-flow conditions. Canadian J. Microbiol. **21**: 90–98.

Tilman, D., and S. S. Kilham. 1976. Phosphate and silicate growth and uptake kinetics of the diatoms *Asterionella formosa* and *Cyclotella meneghiniana* in batch and semi-continuous culture. J. Phycol. **12**:375–483.

Tilman, D., S. S. Kilham, and P. Kilham. 1976. Morphometric changes in *Asterionella formosa* colonies under phosphate and silicate limitation. Limnol. Oceanogr. **21**: 883–886.

Titman, D. 1976. Ecological competition between algae: Experimental confirmation of resource-based competition theory. Science **192**:463–465.

TAXONOMIC DIVERSITY OF ISLAND BIOTAS

Daniel S. Simberloff

Department of Biological Science, Florida State University, Tallahassee, Florida 32306

Received March 5, 1969

Students of biogeography since Darwin have focused disproportionately on oceanic islands. The prime bases for this interest have been the distinct forms which have evolved in the genetic isolation provided by islands and the ecological situation pertaining because the species successfully colonizing any island are but a small subset of the mainland species pool.

One aspect of the latter effect which has received attention is that, within any higher taxon, the mean number of species per genus (S/G) on an island is usually lower than S/G for its presumed source area (MacArthur and Wilson, 1967). If it is assumed that congeneric species tend to resemble one another more in any measurable biological characteristic than do less closely related species, then the lower S/G on an island implies a more "diverse" biota on the island than on the source area. Although Williams (1964) pointed out that a random subset of any species pool has an expected S/G lower than that of the entire pool, Moreau (1966) and Grant (1966) attach significance to the lower S/G *per se* on islands, without regard for whether this S/G is lower or higher than expected, and attribute the lower insular value to ecological and/or evolutionary phenomena.

In this paper I will first treat qualitatively the general distribution of the S/G ratio for random subsets of any species pool, then analyze the data for a series of well-studied island groups, and finally reassess the ecological and evolutionary ideas formulated on this subject in the light of the statistical treatment.

THEORETICAL CONSIDERATIONS

For any species pool the exact expected value of S/G[1] for a subset of given size is readily derived from the hypergeometric distribution, but unless the subset is very small the computations are not feasible. Williams (1964) determined $E(S/G)_N$ empirically by placing chips (for species) with numbers (for genera) into a box, shaking the box, and withdrawing N chips. He did this three times and took the mean value for mean species per genus as $E(S/G)_N$. A more precise version of this method involves generating uniform random numbers on a digital computer to draw the species. With a large number of simulated drawings, it is easy not only to place confidence limits on $E(S/G)_N$, but also to determine empirically the distribution of S/G_N and so to compare a set of N species found on a real island with the distribution of sets of N species drawn randomly from the source biota.

To demonstrate general features of the curve $E(S/G)_N$ vs. N the complete curve for $1 \leqq N \leqq \sigma$, where σ = the total number of species, was plotted by 200 random drawings at each integral N for the birds of Morocco (Etchécopar and Hüe, 1964). A similar but less complete curve was derived for the vascular plants of Great Britain (Perring and Walters, 1962). Results are shown in Figures 1 and 2.

It should be noted first that for $N = 1$, $E(S/G)_N = 1$, and for $N = \sigma$, $E(S/G)_N = \sigma/\gamma$, where γ = the number of genera. At both these points there is no variance. Between these points the discrete curve is apparently monotonic non-decreasing with

---

[1] Hereafter in the text, S/G refers to the expected mean, $\overline{S/G}$.

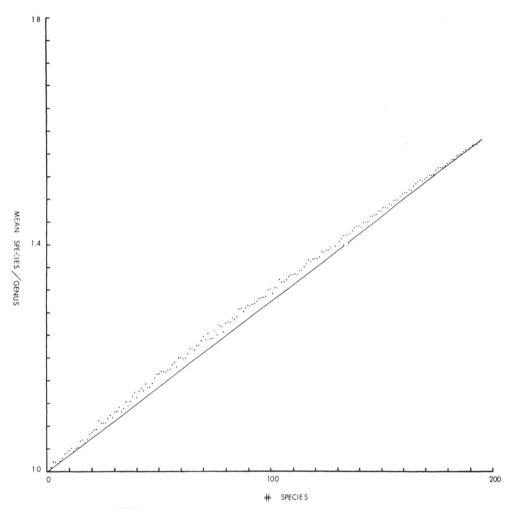

FIG. 1. Plot of $E(\overline{S/G})_N$ vs. N (number of species) for subsets of land birds of Morocco drawn "randomly" by computer. Line is straight-line approximation described in text.

the exact shape determined by the species-genus distribution of the source biota. Williams (1964) observes that this distribution in nature frequently approximates the logarithmic series. This was not borne out well in the examples cited below, but the results of the computer simulation are independent of this distribution.

Another point of interest is that if a straight line is plotted on this curve between $(1, 1)$ and $(\sigma, \sigma/\gamma)$, a fast approximation can be given for N species by locating the y-coordinate of this line for $x = N$. It is:

$$E(S/G)_N = \frac{(N - 1) \, ((\sigma/\gamma) - 1)}{\sigma - 1} + 1.$$

An investigator who knows only the number of species and genera on an island and wishes to compare S/G with $E(S/G)_N$ for the appropriate species pool can approximate the latter measure from the equation if he knows only the numbers of species and genera in the species pool.

Experience, including the data analyzed below, shows that the approximation is remarkably good for most species pools up to a few hundred species, regardless of

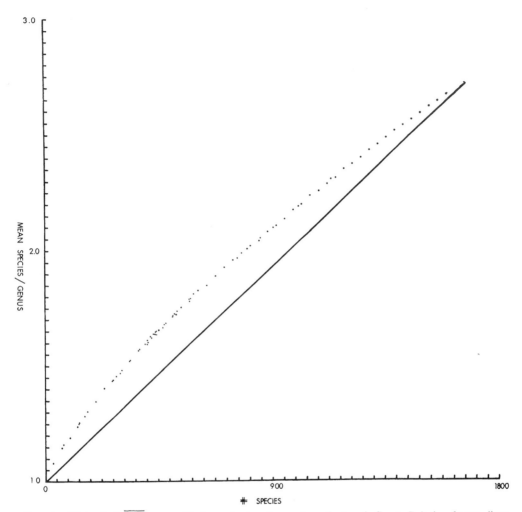

Fɪɢ. 2. Plot of $E(\overline{S/G})_N$ vs. N for subsets of vascular plants of Great Britain drawn "randomly" by computer, with straight-line approximation.

their distribution into genera (see Fig. 1). For the few tests made with larger species pools the approximation usually underestimated the true value by 10–20%. Further generalizations about the approximation are unwarranted since a situation where a determination of S/G relative to $E(S/G)_N$ is critical would almost certainly involve sufficient data so that the simulation could be used.

Sᴏᴜʀᴄᴇꜱ ᴏꜰ Eʀʀᴏʀ

Lack of adequate biological knowledge in three areas limits the number of ex-amples which can be studied: taxonomy, completeness of species lists, and definition of the source area for a given island. Somewhat inaccurate taxonomy, particularly with regard to questions of lumping or splitting closely related populations, does not lead to greatly altered results so long as a single taxonomy is consistently applied to both the source area and the island. If a source area contains σ species in γ_1 genera, and an island $N < \sigma$ species in λ_1 genera, according to a split taxonomy, one can then compare N/λ_1 to $E(S/G)_N$. If a different, relatively lumped taxonomy is used, the number of genera in the source and island biotas, respectively, will be $\gamma_2 \leq \gamma_1$ and $\lambda_2 \leq \lambda_1$. Although the mean number of species per genus on the island, N/λ_2, is now greater than its previous value (N/λ_1), it generally bears the same relation to the (new) $E(S/G)_N$ as did the former value. There are limits to the distortion of the taxonomy which can be permitted without changing the results materially, and the degree of change depends not only on the particular taxonomic revision but also on the characteristics of the species-genus distributions. The general conclusion, however, has been validated by deliberate distortions of hypothetical cases as well as by analyzing several of the sets of ornithological data to be described later using other classification systems than the most generally accepted ones. Although, as will be seen, most of the S/G for real islands vary but slightly from $E(S/G)_N$, the direction of variation was preserved in all tests run with published taxonomies different from the ones originally used and all but the most extreme distortions of hypothetical classifications.

Imprecision in defining the proper source area for a given island is more troublesome. Operationally this problem is often trivial; e.g.: for the numerous islands around Britain whose land birds and vascular plants are analyzed in the next section, both geographic considerations and the fact that virtually all the species on any island are found in adjacent Britain imply that Britain is the functional source area. A similar situation must hold for Cuba and the Isle of Pines or Hispaniola and Gonave.

The choice of a source area may not be so obvious, however. The published data make it unlikely that the chosen source area is too small. First, the great majority of island species must be found in a tentative source or else it must surely be enlarged. Second, many published mainland lists are not finely subdivided and can only be used whole. Can this in some sense invalidate an $E(S/G)_N$ by causing the N species to be drawn from too large a pool? Let us consider a hypothetical situation where most of the propagules landing on an island in fact originate in a small section of the adjacent mainland, and we wish to determine the error incurred in using a much larger section of the mainland, including the small "true" source, as the source area in a simulation. The species pool of the large area contains that of the sub-area as a proper subset, and also contains many other species. In picking N of the σ species in the pool to colonize the island, we are picking $\sigma - N$ species in the pool *not* to colonize the island. Suppose there are σ' $(\sigma > \sigma' > N)$ species in the sub-area which is the actual source. Then if we use only these σ' species as the species pool we have chosen $\sigma - \sigma'$ species not to colonize the island before the simulation and during the simulation we will choose the remaining $\sigma' - N$ which will not colonize the island. In short, knowing the sub-area and its pool of σ' species only establishes a given order of choosing species not to colonize the island—it does not in itself change the numerical result. Put another way, if we have σ species and wish to choose N randomly and then to compare N/λ with some actual island value, it does not matter whether we first choose σ' of the σ species and then choose our final N species from the σ', *so long as* the σ' species are themselves chosen randomly from the σ species.

Returning to a real sub-area of a larger area, the crux of the matter is whether the biota of a sub-area is ever a truly random subset of the larger area. The protagonists in the controversy over whether S/G for islands is greater or less than expected have implied that the trend is a general one, and have often likened sub-areas of a large area to islands. If this were really true then use of a larger source than the actual one should still allow the trend to be manifested, since it should make no difference whether the selection of species is done in one step or two—whether the biota of a small sub-area or island is compared directly to that of the large source or to that of a large sub-area of the source. It would not even matter that the σ' species of the large sub-area were not a random set of the σ species of the whole source so long as the trend were completely general.

A study currently in progress indicates that there is little geographic variation in S/G values for land birds, particularly when localities of similar latitude are compared. This fact plus the claimed analogy of continental sub-areas and oceanic islands and generality of the trend might lead one to feel that the proper course is to compile a world or at least continental list for the group of interest and to draw the N species for comparison to those on a real island directly from this all-inclusive pool.

However, the uniformly small deviations of actual from expected S/G described in the next section show that this would be a dangerous approach. As more and more area is added to the true source, the species of the true source eventually cease to be a random subset of the increased source— the shape of the species-genus distribution changes. For as one increases the area, and consequently the environmental heterogeneity, beyond the immediate source, he is liable to add entirely new habitats not found on the island or in the immediate source area. These new habitats will have biotas at least partially different from those of the island and its immediate source; so long as the species-genus distribution that results from the addition of these new biotas is not qualitatively different from that of the original biota there is no distortion incurred in drawing species randomly from the enlarged biota. If the distribution is changed, however (say, if the new biotas include in the aggregate a disproportionate number

of monotypic genera), then the new $E(S/G)_N$ may differ from the original— the size and direction of the change depends on the species-genus distribution. Several comparisons were made between actual and expected S/G for land birds of large sub-areas of North America. Unlike the oceanic island results to be described, the actual value tended, if anything, to be lower than expected. A probable reason is that by the time areas the size of a large continental section are reached, various parts of the biota have evolved to some extent in geographic isolation. The added species are therefore inordinately disjoint taxonomically from the original ones, which tends to lower S/G (Appendix).

In short, neither of the two possible preconditions for use of a vastly expanded source area hold—the biota of a sub-area is usually not a random subset, and the relationship between actual and expected S/G is not uniform throughout the entire size scale of sources and sub-areas or islands. But $E(S/G)_N$ is usually not greatly distorted by, say, doubling the area of the source through the process described above unless the species added to the pool are distributed into genera in a highly unusual fashion.

A second consideration which favors use of a limited source area when such can safely be delimited concerns the dispersal powers of the species in question. Obviously a species which preferentially or exclusively inhabits the coast is preadapted to colonization of nearby offshore islands (Wilson, 1959, 1961), but in some sense all species inhabiting a restricted source area (not just the littoral ones) can be considered as having ample, if not equal, opportunity to colonize the island. This is not to say, of course, that all species disperse equally well—indeed, it will be postulated later that differences among genera in dispersal ability may help to explain observed differences in actual from expected S/G—but if a restricted source area is used we may assume that large numbers of species are not precluded from colonizing the island by the sheer distance of their ranges from the island. This does not eliminate the utility of a somewhat enlarged source area in the simulation; if the species which would be hindered by distance were distributed into genera in such a way that their removal would not greatly affect the species-genus distribution their distance from the island should have little numerical effect.

Incomplete species lists are similarly difficult. In the first instance, assume that only the island list is incomplete. Two difficulties immediately arise:

i) The actual S/G, which is almost certainly an increasing function of the number of species, cannot be given.
ii) $E(S/G)_N$, which is a non-decreasing function of N, must be based on an estimated N.

If we knew the shape of the species-genus distribution for the island, as well as the total number of species, then (ii) would be solved immediately and we could estimate actual S/G. But the shape of the species-genus distribution cannot be assumed with sufficient confidence, Williams notwithstanding. Furthermore, the accuracy of estimating total number of species from area and habitat diversity measures is not great enough to be useful here. The relatively small deviations of actual from expected S/G on the islands for which species lists are quite complete indicate that the approximations necessary to extrapolate incomplete lists would be too crude to warrant faith in their results.

It is not worthwhile to skip the extrapolation of the incomplete list and simply to compare the actual S/G for the species listed with $E(S/G)$ for that many species drawn randomly from the species pool. If we were certain that the species-genus distribution of the subset of species which is known truly represents the distribution for all island species then such a comparison

might be justified. But this assumption seems risky, at least if a large fraction of the island forms are unrecorded. In certain groups, for example, one might expect that monotypic genera would be the focus of a disproportionate amount of study and that the species not listed would include an overly large number of additional forms falling in genera which are already well represented. As with the other inaccuracies discussed, the amount of distortion inherent in the use of incomplete island lists depends on the species-genus distribution involved, and it is impossible to give a general estimate of this distortion. The relatively small differences between actual and expected S/G encountered in the well-studied examples and the fact that these have been construed as proof of ecological and evolutionary effects indicate that it would be foolhardy to use lists which were not comprehensive. If less than, say, 80% of the island species are believed recorded the results should be incorporated into a hypothesis only with extreme caution. For a result to be considered a certainly valid indication of the true state of affairs on an island 90% or more of the insular biota must be known. Most of the examples discussed in the next section are in the latter category.

Incomplete source lists are similarly troublesome. From a mathematical standpoint the problem is identical to that involved in choosing either a large source or a restricted source within the large one—that is, if the known species are a random subset of the entire source species pool, then drawing N species from the known subset to approximate $E(S/G)_N$ should not yield a different value than if the drawing were from the entire pool. But biologically the assumption that the known subset is a random sample (with regard to the species-genus distribution) is dubious. As with the island biotas, one might expect certain genera to be studied disproportionately. Or if a discrete geographic region of the source is understudied and contains all of a major habitat within the source, then addition of the unknown to the known species might distort the species-genus distribution greatly, particularly if the two sets of species were largely in disjoint genera. The amount of distortion varies and a general rule cannot be postulated, but for the examples discussed below the source lists are believed to be at least 90% complete.

SIMULATION RESULTS

The simulation was performed on a series of well-studied taxa from several sources and their islands. For comparison to the N species on a given island, N species were drawn randomly 100 times from the appropriate species pool. The animal data were processed using the RANNO random number generator on the Harvard Computing Center IBM 7094; the RN1 subroutine on the Florida State University Computing Center CDC 6400 was used for the plant data. Significance of the deviation of S/G on a given island from the empirical $E(S/G)$ was tested by a two-tailed Student's t-test, and a 95% confidence belt for true $E(S/G)$ was calculated using t with the appropriate number of degrees of freedom. The distribution of S/G is not normal, but it is believed that the robustness of the t-test justifies faith in the results of the above tests. The explicit distribution will be discussed later.

Because insular taxonomic diversity has been expressed in terms of percent congeneric species in addition to S/G (Grant, 1966), all land bird data have simultaneously been processed in terms of "% congeners." Mean number of species per genus is preferable to percentage congeneric species as an indicator of taxonomic diversity for two main reasons. First, its variance is lower. Secondly, percentage congeneric species usually tells less about a large biota since it indicates only the number of species whose genera are monotypic in the biota in question. This num-

ber as a fraction of the total number of species decreases rapidly with increasing species number. The direction of deviation of actual from expected value was identical for the two measures in 105 of the 116 examples studied.

Results are given in Table 1. Several other sets of data were analyzed but not included because the source and/or island lists were suspected to be quite incomplete. This was especially true for insects and spiders.

The Ardeidae were used in the analysis (as indicated) where the available lists included them. In many instances they are ecologically as restricted to islands as are land birds—in marked contrast to the Laridae, say. Several tests run without the Ardeidae produced no qualitative change in the results.

The immediately striking picture presented by the data in Table 1 is that for most islands in most groups the actual and expected S/G are very close. Of the 180 tabulated islands only 38 show significant differences at the 1% level and an additional 16 are added at the 5% level. And the majority of these cases concern plants of the British Isles (sections 15–23).

Despite the paucity of individual islands with S/G significantly different from that expected on the basis of chance, there is a distinct tendency toward higher than expected values. At the crudest level 70.6% of the 180 islands have higher than expected S/G ($x_1^2 = 30.4$; $P < .001$), and the fraction rises to 79.5% ($x_1^2 = 54.3$; $P < .001$) if the Shetland birds are discounted, for reasons discussed below. Furthermore, a number of the contrary islands have very small biotas and should perhaps be omitted on statistical grounds described later. The trend is not restricted to a few large island groups. Eighteen of the 23 parts of Table 1 conform to it, while only two contradict it—land birds of the Shetlands (section 11) and the Isle of Pines (section 3).

30 D. S. SIMBERLOFF

TABLE 1. *Actual mean number of species per genus, and that expected on a hypothesis of random choice, for a series of island biotas.*

Island or source	Group	No. spp.	Actual $\overline{S/G}$	$E(\overline{S/G})$	Actual % congeners	Expected % congeners	Authority
1) *Yucatan Peninsula*	Land	264	1.275				Paynter (1955)
Cozumel	birds[b]	87	1.130	1.104	23.0	17.1	
Holbox		48	1.067	1.057	12.5	10.3	
Mujeres		43	1.103	1.055	16.3	9.8	
Chinchorro		17	1.000	1.023	0.0	4.0	
Contoy[a]		16	1.000	1.020	0.0	3.7	
2) *Mexico*[1]	Land	475	1.542				Eisenmann (1955)
All Tres Marias	birds	37	1.057[†]	1.056	10.8	8.7	Grant and Cowan
Maria Madre		36	1.059[†]	1.051	11.1	9.1	(1964)
Maria Magdalena		35	1.061	1.046	11.4	9.2	
Maria Cleofas		32	1.067[†]	1.050	12.5	10.1	
San Juanito		14	1.000	1.012	0.0	2.3	
3) *Cuba*	Land	93	1.257				Bond (1950)
Isle of Pines	birds	69	1.150	1.177	24.6	28.7	
4) *Hispaniola*	Land	87	1.261				Bond (1950)
Gonave	birds	49	1.195	1.141	30.6	23.1	
Tortue[a]		30	1.200*	1.081	30.0	14.0	
5) *Panama*	Land	656	1.657				Eisenmann (1955)
Barro Colorado	birds[b]	248	1.292[†]	1.285	39.1	37.6	Eisenmann (1952)
Coiba		86	1.062	1.106	11.6	17.7	Wetmore (1957)
6) *W. Equatorial Africa*[2]	Land	708	2.088				Bannerman (1953)
Fernando Poo	birds[b]	135	1.378**	1.246	45.9*	34.4	Amadon (1953)
São Tome		48	1.067	1.083	12.5	14.3	
Principe		29	1.074	1.062	13.8	10.7	
Annobon		10	1.000	1.018	0.0	3.2	
7) *Morocco*	Land	195	1.585				Etchécopar and
All Canary Islands	birds[b]	47	1.270	1.162	36.2	25.2	Hüe (1964)
Tenerife		40	1.212	1.133	32.5	21.3	Bannerman (1963)
Grand Canary		38	1.188	1.123	29.0	20.2	Volsøe (1951)
Gomera		34	1.214	1.119	29.4	19.0	Volsøe (1955)
Palma		33	1.222	1.113	30.3	18.3	
Fuerteventura		30	1.200	1.098	30.0	16.2	
Hierro		29	1.160	1.100	24.1	17.0	
Lanzarote		28	1.167	1.098	25.0	16.1	
Madeira		25	1.250*	1.092	36.0*	16.6	
8) *Mediterranean*[3]	Land	215	1.853				Voous (1960)
Sardinia	birds	119	1.526	1.484	52.1[†]	52.6	
Corsica		109	1.434	1.448	48.6	50.8	
9) *Britain*	Land	115	1.643				Schoener (unpubl.)
Ireland	birds	74	1.480	1.432	48.6	48.8	
Bute		64	1.422	1.392	45.3[†]	46.3	

* $P < .05$
** $P < .01$
[†] Within 95% confidence band for $E(\overline{S/G})$ or E (% congeners)
[a] List quite possibly less than 80% complete
[b] Ardeidae included
[1] Within 300 miles of Tres Marias, but excluding Baja California
[2] Within 1000 miles of Annobon
[3] Within 475 miles of Sardinia

Table 1. (Continued)

Island or source	Group	No. spp.	Actual $\overline{S/G}$	$E(\overline{S/G})$	Actual % congeners	Expected % congeners	Authority
9) *Britain* (Continued)							
Islay		62	1.409	1.380	45.2†	45.5	
Man		62	1.476	1.387	48.4	45.8	
Orkney Mainland		36	1.385	1.239	44.4	33.5	
Coll		36	1.333	1.234	44.4	33.4	
Rhum		34	1.360	1.216	44.1	31.6	
Canna		31	1.240	1.199	38.7	29.8	
Tiree		31	1.292	1.208	41.9	30.6	
Barra		28	1.217	1.172	35.7	26.5	
N. Uist		27	1.286	1.186	40.7	28.0	
Eigg		27	1.227	1.165	29.6	25.4	
Scillies		25	1.136	1.171	20.0	26.3	
Shetland Mainland		24	1.600**	1.170	66.7**	26.4	
Lundy		24	1.143†	1.158	25.0†	24.0	
Bardsey		23	1.150†	1.158	21.7	24.5	
St. Kilda		10	1.111	1.065	20.0	11.1	
Mingulay		10	1.111	1.054	20.0	9.5	
Ailsa Craig		7	1.000	1.028	0.0	4.4	
Skokholm		6	1.200*	1.036	33.3*	6.0	
May		6	1.000	1.047	0.0	7.7	
Steepholm		5	1.000	1.030	0.0	4.8	
Boreray		5	1.000	1.030	0.0	4.8	
Eileon Mohr		3	1.000†	1.015	0.0†	2.0	
Soay		3	1.000†	1.020	0.0	2.7	
Dun		3	1.000	1.040	0.0	5.3	
N. Rona		2	1.000†	1.000	0.0†	0.0	
10) *Orkney Mainland*	Land	36	1.385				Schoener (unpubl.)
Hoy	birds	29	1.261	1.297	34.5	39.7	
Rousay		27	1.286†	1.273	40.7	38.0	
Shapinsay		25	1.250†	1.256	36.0†	35.9	
Eday		20	1.429*	1.212	55.0*	31.4	
South Ronaldsay		19	1.462*	1.215	52.6	32.0	
Sanday		18	1.286	1.180	44.4	27.9	
Westray		17	1.417**	1.188	52.9*	29.1	
Stronsay		16	1.231	1.169	37.5	25.8	
S. Walls		16	1.231	1.152	37.5	24.1	
Flotta		13	1.083	1.158	15.4	24.6	
Burray		13	1.182	1.146	30.8	22.9	
Papa Westray		13	1.182	1.130	30.8	20.4	
North Ronaldsay		13	1.300*	1.114	46.2*	18.9	
Fara		10	1.111†	1.117	20.0†	18.0	
Cava		9	1.000	1.081	0.0	12.4	
Swona		8	1.000	1.090	0.0	14.6	
Eynhallow		8	1.143	1.088	25.0	13.9	
Calf of Eday		7	1.000	1.082	0.0	13.1	
11) *Shetland Mainland*	Land	24	1.600				Venables and
Unst	birds	18	1.286	1.396	44.4	53.0	Venables (1955)
Bressay		17	1.308	1.365	47.1	50.4	
Yell		17	1.308	1.353	47.1	49.2	
Fetlar		16	1.231	1.325	37.5	46.1	
Foula		14	1.167	1.272	28.6	40.0	
Fair Isle		13	1.083	1.264	15.4	39.1	
Whalsay		13	1.083	1.276	15.4	40.7	
Uyea Unst		13	1.182	1.272	30.8	39.8	

32   D. S. SIMBERLOFF

TABLE 1.   (Continued)

Island or source	Group	No. spp.	Actual S/G	$E(\overline{S/G})$	Actual % con-geners	Expected % con-geners	Authority
11) *Shetland Mainland* (Continued)							
Burra, West		12	1.091	1.237	16.7	36.2	
Papa Stour		12	1.091	1.246	16.7	36.3	
Vaila		12	1.200	1.259	33.3	38.3	
Burra, East		11	1.000	1.209	0.0	32.0	
Noss		11	1.000	1.294	0.0	35.5	
Vementry		11	1.222[†]	1.221	36.4[†]	33.9	
St. Ninian's Isle		10	1.111	1.214	20.0	32.5	
Mousa		9	1.000	1.210	0.0	31.6	
Out Skerries		9	1.125	1.178	22.2	27.7	
Hascosay		8	1.000	1.135	0.0	21.5	
Havra, South		8	1.000	1.129	0.0	20.8	
Trondra		8	1.000	1.128	0.0	20.0	
Gluss Isle		8	1.143[†]	1.140	25.0[†]	22.4	
Balta		8	1.143[†]	1.158	25.0[†]	24.2	
Colsay		8	1.143[†]	1.145	25.0[†]	22.6	
Uyea Isle		7	1.167	1.139	28.6	21.9	
12) *Aegean*[4]	Passer-	130	2.097				Watson (1964)
Mytilene	ine	58	1.526[†]	1.531	55.2[†]	55.8	Peterson, Mount-
Crete	birds	51	1.457[†]	1.463	47.1	52.0	fort, and Hollom
Samothrace		44	1.517	1.415	54.6	48.2	(1954)
Rhodes		41	1.414[†]	1.404	43.9	47.8	
Icaria[a]		30	1.364	1.287	43.3	38.4	
Kos		29	1.318	1.261	41.4	36.3	
Kythera		26	1.368	1.267	38.5[†]	36.1	
Skyros		25	1.471*	1.226	56.0	32.0	
Thasos[a]		24	1.263	1.236	41.7	33.4	
Kea		23	1.211[†]	1.229	30.4	33.2	
Limnos		23	1.211[†]	1.219	30.4[†]	31.7	
Andros		22	1.294	1.201	40.9	30.1	
Scopelos		22	1.294	1.217	40.9	31.6	
Samos		20	1.176	1.202	30.0[†]	29.3	
Chios[a]		19	1.188[†]	1.185	31.9	28.0	
Imbros[a]		15	1.154[†]	1.147	26.7	22.7	
13) *Rhodes*	Passer-	41	1.414				Watson (1964)
Karpathos	ine birds	26	1.444	1.275	53.9*	36.1	Peterson, Mount-
							fort, and Hollom
							(1954)
14) *Upolu* (Samoa)	Ants	63	1.909				Wilson and
Sava'i		40	1.818	1.624			Taylor (1967)
Tutuila		35	1.750	1.550			
Manu'a		14	1.273	1.246			
15) *Britain*	Vascular	1666	2.727				Perring and
Ireland	plants	1138	2.318[†]	2.312			Walters (1962)
Wight		1008	2.230	2.194			
Anglesey		855	2.065[†]	2.056			
Man		765	2.024	1.969			
Skye		594	1.947**	1.814			
Islay		581	1.990**	1.794			
Arran		577	1.917**	1.789			
Lewis		527	1.916**	1.738			

[4] Within 680 miles of Aegean Sea

TABLE 1. (Continued)

Island or source	Group	No. spp.	Actual $\overline{S/G}$	$E(\overline{S/G})$	Actual % con-geners	Expected % con-geners	Authority
15) *Britain* (Continued)							
Scillies		523	1.855**	1.723			
Mull		517	1.853**	1.726			
Colonsay		482	1.792**	1.688			
S. Uist		470	1.843**	1.667			
Eigg		453	1.783**	1.657			
Jura		444	1.889**	1.652			
Coll		443	1.786**	1.636			
Orkney Mainland		440	1.947**	1.638			
N. Uist		433	1.827**	1.640			
Rhum		425	1.801**	1.623			
Shetland Mainland		421	1.888**	1.629			
Barra		409	1.786**	1.612			
Gigha		401	1.743**	1.596			
Tiree		378	1.758**	1.573			
Lundy		338	1.609	1.523			
Canna		300	1.587**	1.470			
Muck		284	1.614**	1.458			
Mingulay		269	1.630**	1.437			
Shiant Islands[a]		159	1.590**	1.285			
May		137	1.457**	1.256			
St. Kilda[a]		131	1.489**	1.240			
Ailsa Craig		75	1.293*	1.161			
16) *Ireland*	Vascular	1138	2.318				Perring and
Inishmore[a]	plants	430	1.648†	1.642			Walters (1962)
Clare[a]		379	1.755**	1.577			
Inishbofin[a]		361	1.736**	1.560			
Inishturk[a]		326	1.707**	1.511			
Rathlin[a]		312	1.625**	1.506			
Inishmaan[a]		239	1.475	1.410			
Aran[a]		235	1.546**	1.402			
Lambay[a]		180	1.500**	1.318			
Saltee Islands[a]		174	1.526**	1.311			
Inishkea Islands[a]		166	1.419*	1.296			
Tory[a]		165	1.435*	1.310			
Great Blasket[a]		144	1.500**	1.272			
Inishtooskert[a]		41	1.171	1.088			
17) *Orkney Mainland*	Vascular	440	1.947				Perring and
Hoy	plants	354	1.735	1.791			Walters (1962)
S. Ronaldsay		207	1.522	1.507			
Fair Isle		174	1.526	1.441			
Sanday		162	1.434	1.416			
N. Ronaldsay		131	1.379	1.361			
Westray		65	1.383*	1.198			
Stronsay		62	1.192†	1.187			
18) *Shetland Mainland*	Vascular	421	1.888				Perring and
Unst	plants	246	1.618	1.581			Walters (1962)
Fetlar		189	1.575*	1.457			
Bressay		177	1.595*	1.446			
Yell		161	1.519*	1.409			
Whalsay		158	1.491*	1.391			
Foula		149	1.355	1.377			

34   D. S. SIMBERLOFF

TABLE 1.   (Continued)

Island or source	Group	No. spp.	Actual $\overline{S/G}$	$E(\overline{S/G})$	Actual % congeners	Expected % congeners	Authority
19) *Mull*	Vascular	517	1.853				Perring and
Iona	plants	388	1.694†	1.688			Walters (1962)
Treshnish Islands[a]		135	1.406	1.300			
20) *N. Uist*	Vascular	433	1.827				Perring and
Monach Islands[a]	plants	219	1.622**	1.491			Walters (1962)
21) *Lewis*	Vascular	527	1.916				Perring and
Flannen Islands[a]	plants	22	1.294**	1.056			Walters (1962)
22) *Anglesey*	Vascular	855	2.065				Perring and
Holy Isle	plants	377	1.778**	1.580			Walters (1962)
23) *Guernsey*	Vascular	804	2.072				Perring and
Herm	plants	411	1.764**	1.644			Walters (1962)

The tendency for S/G on islands to be greater than an expected value derived from a number of random drawings from the species pool would not conflict with the hypothesis that island biotas are random subsets if the distribution of S/G from a series of random drawings were negatively skewed. The median would then be greater than the mean and the probability that a random subset would have a higher S/G than expected would be greater than 0.5.

The distributions of S/G for the 100 random subsets for many of the islands listed in Table 1 were given explicitly, and manifested the following trend. For no N in the table is there negative skewness. Very low values of N yield positive skewness (frequently only 25/100 values greater than the mean), and as N increases the distribution becomes more symmetrical. For most of the islands listed in Table 1 the distribution does not deviate widely from symmetry.

In addition to supporting the validity of the observed pattern of higher than expected S/G, this characteristic increasingly positive skewness with decreasing N precludes the use of islands with very small biotas. For several of the groups in Table 1 lists are available for a few additional islands with biotas of less than 10 species,

but the results for one or a few islands are not illuminating. For if only 1 in 4 random subsets have higher than expected S/G it would take many islands with identical (small) N to manifest the small yet real tendency which seems to exist for actual island biotas to have more species per genus than expected.

The major exception to this tendency is found in the Shetland Islands bird data (Table 1.11); 21 of the 24 islands have a *lower* S/G than expected, though no differences are significant at the 5% level. A plausible explanation is that the main avian source area for all the Shetland Islands is Britain, not Shetland Mainland. Earlier I noted that any general trend would probably not be masked by using a somewhat larger source including the true source. But use of a smaller (or different) source than the true one could easily obscure or even reverse a result. For, given that island S/G generally tends to be greater than expected, if the putative source has N' species, its actual $S/G_{N'}$ will be greater than the $E(S/G)_{N'}$ for N' species drawn from the common true source of the island and its putative source. But if the island contains $N < N'$ species, and most propagules landing on the island originate outside the putative source, then even though the island will have a higher S/G_N than would be expected if the N species were drawn randomly from the total pool, it is not certain that this actual S/G_N will be greater than expected for N species drawn randomly from the N' species in the putative source. In fact, the greater the positive difference between the actual and expected $S/G_{N'}$ of the putative source, the more likely is the actual S/G_N of the island to be lower than expected for N species randomly drawn from the N' species of the putative source, independent of the relationship between the actual S/G_N and the expected value for N species from the entire pool.

Shetland Mainland has not only by far the highest S/G of all islands of Table 1.9, but the deviation from the expected value (significant at the .001 level) is even more extraordinary. If Shetland Mainland is not the actual origin of most avian propagules landing on the other Shetland Islands then the values in Table 1.11 are hardly surprising. The experiment was repeated using Britain as the avian source for the Shetland Islands (Table 2), whereupon 17 of the 24 islands had higher than expected S/G.

If the avian source area for the other Shetland Islands is Britain and not Shetland Mainland, then on geographic grounds one would expect Britain rather than Orkney Mainland to be the avian source for the Orkneys. That the analysis of Orkney bird data did not yield as anomalous a result can be attributed to Orkney Mainland's less drastic excess of S/G over $E(S/G)$: .146 (Table 1.9). A reanalysis of the data using Britain as source would undoubtedly increase the 11 : 7 ratio of islands with greater than expected S/G to islands with lower than expected S/G.

The implications of the Shetland bird data for the Shetland and Orkney vascular plant analyses (Table 1.17–18) are ambiguous. As MacArthur and Wilson (1967) note, there is no reason why the shape of the dispersal curve need be the same for

TABLE 2. *Mean number of species per genus for Shetland land birds with Britain as source.*

Island or source	No. spp.	Actual $\overline{S/G}$	Expected $\overline{S/G}$
Britain	115	1.643	
Unst	18	1.286*	1.115
Bressay	17	1.308*	1.114
Yell	17	1.308*	1.105
Fetlar	16	1.231	1.104
Foula	14	1.167	1.100
Fair Isle	13	1.083†	1.094
Whalsay	13	1.083†	1.076
Uyea Unst	13	1.182	1.093
Burra, West	12	1.091	1.068
Papa Stour	12	1.091	1.059
Vaila	12	1.200	1.068
Burra, East	11	1.000	1.066
Noss	11	1.000	1.065
Vementry	11	1.222	1.084
St. Ninian's Isle	10	1.111	1.052
Mousa	9	1.000	1.054
Out Skerries	9	1.125	1.050
Hascosay	8	1.000	1.046
Havra, South	8	1.000	1.040
Trondra	8	1.000	1.064
Gluss Isle	8	1.143	1.042
Balta	8	1.143	1.054
Colsay	8	1.143	1.046
Uyea Isle	7	1.167	1.046

* $P < .05$

† Within 95% confidence band for $E(\overline{S/G})$

two taxa. The extreme possible distributions they give, the normal and the exponential, might well apply to birds and seeds, respectively, at least if most seeds are dispersed aerially in this geographic context. Consequently one cannot predict the major floristic source of a lesser Shetland island even with accurate knowledge of the avian source. In any event, the data for the vascular plants of these islands follow the trend of greater than expected S/G, and this tendency would only be magnified if Britain were used as the source area.

CONTRADICTORY THEORIES

Two recent authors, Grant (1966) and Moreau (1966), note simply that S/G (or % congeneric species) is lower on islands than on the adjacent mainland and partly

36   D. S. SIMBERLOFF

TABLE 3. *Mean number of species per genus for Madagascan birds and Campbell Island beetles.*

Island or source	Group	No. spp.	Actual $\overline{S/G}$	Expected $\overline{S/G}$
1) *Africa*	Passerine	1309	3.839	
Madagascar	birds	47[1]	1.567**	1.171
2) *Africa*	Group D	389	3.602	
Madagascar	birds	18[1]	2.250**	1.148
3) *New Zealand*	Beetles	2580	4.300	
Campbell Island		44	1.294**	1.114

[1] Endemics only   ** $P < .01$

base their conclusions on this fact. As we have seen, the lower S/G on islands is expected because of the smaller number of species, and in fact S/G on islands is generally greater, not less, than expected if colonists were randomly drawn from the mainland pool. Since the premise of Grant and Moreau is misused their conclusions are not warranted.

Moreau (1966) shows that the genera of passerines and group D birds (non-passerine land birds, not including raptors and game birds) endemic to Madagascar contain on the average 1.6 and 2.2 species, respectively, while the same groups on the African continent average 3.5 and 3.3 species per genus. He feels that these figures can be explained either by a limited degree of intrageneric radiation on the island or by widespread extinction on Madagascar relative to that in Africa. A detailed consideration of the entire Madagascan avifauna leads him to support the latter hypothesis.

The passerines and group D birds of Africa were enumerated using the classification of Sclater (1928–30), felt by Moreau to be comparable to the classification he used for Madagascar. The appropriate numbers of species were drawn randomly 100 times; results are given in Table 3.1–2. In both instances the expected S/G is less than the actual value for Madagascar, and it seems reasonable that the lower figures per se on the island are due to the lower total numbers of species in the two groups. The higher than expected actual values do not invalidate

Moreau's hypothesis of large-scale extinction, but they obviously cannot support it.

It is worth noting that although data on S/G can, as Moreau says, be connected with an abnormal degree of intrageneric, as opposed to higher taxon, radiation, such data do not bear directly on extinction. Random extinction would have no effect on S/G other than the lowering associated with a smaller biota. If extinction were differentially greater in certain genera, however, S/G might be an indicator of the extinction.

Grant (1966), following Elton (1946), observes that a random pair of congeners are likely to be more similar ecologically than a random pair of more distantly related species. He then proposes (1966, 1968) that because of the decreased area, increased degree of isolation, and especially decreased habitat diversity of an island relative to the mainland there will be intensified competition generally, and especially among congeners since these will, on the average, have more similar ecological requirements than will other birds. The main responses, he believes, are:

1) fewer congeneric pairs (or triplets, etc.) on islands than on mainland.

2) morphological changes (e.g.—in bill size); especially, larger differences in a given character between two sympatric island congeners than between the same two species where they are sympatric on the mainland.

His evidence for the latter effect seems sound, but the former assertion deserves further consideration. MacArthur and Wilson (1967) observe that if increased taxonomic diversity (lower S/G) is shown to be general on islands, it may be analogous to the "filtering" provided by staging habitats for dispersal—they increase the relative diversity of colonists through their decreased ecological complexity. The same authors note, however, that the lower S/G of beetles which they calculate for Campbell Island vs. a possible source (New Zealand) is probably due mainly to the difference in numbers of species. This speculation is borne out by a simulation identical to that described. Results are given in Table 3.3, and follow the trend already noted.

In order to approach Grant's assertion properly, we must first formulate disjoint alternatives. In essence, Grant says that:

1a) The species of land birds on islands are a non-random subset of possible colonists, and they deviate from randomness in the direction of fewer congeners than would be expected (because of competition).

to which the mutually exclusive alternatives are:

1b) The island species are a non-random subset, but deviate in the direction of more congeners than would be expected.

1c) The island species are a random subset.

He supports alternative 1a, and claims that the ultimate reasons for nonrandomness (not necessarily exclusive or exhaustive) are:

2a) Smaller size of island than mainland.

2b) Less habitat diversity or ecological complexity on island.

2c) Spatial isolation of island.

The data of Table 1 strongly support alternative 1b. In particular, Table 1.2 (the land birds of the Tres Marias Islands), the primary basis for Grant's assertions, indicates both by percent congeners and S/G that alternative 1a is incorrect. That is, even if taxonomic diversity is greater in absolute terms on islands than on the mainland, it is certainly not greater than expected. And this state of affairs persists in spite of the assumed increased competition on islands, not in response to it. If competition is increased on islands, for reasons 2a, b, c or any others, it would tend to drive taxonomic diversity in the opposite direction from that in which it has moved, so that some other factor(s) must oppose this tendency and must, in the aggregate, be more important than competition.

Greenwood (1968) has recently doubted that the lower percentage of congeners among Tres Marias birds than among those of a nearby section of mainland is indicative of anything other than lower species numbers. His evidence is that the species-genus distribution for both areas is approximately log series, and that a log series distribution predicts a lower fraction of congeners the smaller the biota. But S/G (and percentage congeners) for the Tres Marias is in fact slightly higher than expected if nearby Mexico were the source, and this is part of a tendency common to most islands and sources, many of which do not have species-genus distributions close to the log series.

That alternative 1b and not 1a is the generally correct one does not invalidate the possible didactic value of alternatives 2a–c, at least as a starting point in the search for the reasons for this pattern.

Both area and habitat diversity are smaller for islands than for mainland, but the separate effects of these two factors on S/G are difficult to elucidate. Since the initial effect of any factor on S/G is through its action on species number, we are first faced with the well-known problem of the relationship of area and number of species. This problem may be simply stated as: does increased area itself allow for more species, or is the increase in species number due solely to new habitats which are added with increasing area? This is an unsolved problem, but Johnson et al. (1968) have shown that, though increased ecological complexity is a major cause of increased species number, there is still some effect of increased area per se not yet explained by ecological complexity (or at least ecological complexity as expressed by maximum elevation) which also contributes to raising the number of species.

This controversy relates to the matter discussed here as follows: to the extent that increased habitat diversity explains the increase in species number with added area, the increase in percent congeners (or S/G) with species number is explained largely by alternative 2b, and alternative 2a can be eliminated. Grant's (1966) plot of % congeners vs. log area for 18 islands is not illuminating because it does not address this problem. And further, once we decide on the relative significance of alternatives 2a and 2b on increased species number, we must determine if the increased species number alone accounts for the difference in percent congeners between island and mainland. As it seems that it almost, but not completely, accounts for this difference the problem becomes: do either area or habitat diversity or both have any independent, direct effect on taxonomic diversity? We know already that any such effects will be very small; an approach to determining their existence will be outlined.

Similarly, the data claimed as support for alternative 2c—spatial isolation as an independent determinant of percent congeners—are inappropriate to the problem. Grant's comparison of the land birds of the Tres Marias Islands to those of a section of the mainland with similar area, habitat, and range of elevation, showing

percent congeners to be higher on the mainland, is not relevant. The difference is almost certainly accounted for simply by the greater number of species in the mainland section (Greenwood, 1968), and this greater number of species may in turn be partly explained by the presence in the mainland area of many populations that are essentially transient in the area proper and are maintained in adjacent, ecologically different areas (MacArthur and Wilson, 1967).

Grant proposes two alternatives for the demonstrable decrease in percent congeners with increasing distance, neither of which includes the relationship between distance and species number. His first suggestion is that habitat diversity decreases with distance, which first of all implies that distance itself has no effect and secondly reduces us to the problem stated earlier of finding whether habitat diversity has some small direct effect on percent congeners or whether its only effect is through regulation of the numbers of species. The second proposal, involving more recent connection of nearer islands to mainland, is problematic. For this to drive S/G lower than expected, as Grant claims, would imply either a lower rate of speciation on the island than on the mainland or else the differential extinction on the island of all but one member of congeneric groups—a modification of Moreau's proposals. The contrary (and observed) effect on S/G would require these two factors to act in the opposite manner to that just stated. In either case it seems likely that if such an effect is general it could be attributed to either area or habitat diversity rather than to distance, once the island was isolated.

To determine whether any of these factors—area, habitat diversity, and distance from source—affect S/G independently of species number, a multiple regression analysis was performed with S/G the dependent variable, species number the first independent variable, and

TABLE 4. *Multiple linear[1] regression of $\overline{S/G}$ on species number, maximum elevation, area, and distance from source.*

Islands	Group	# islands	r^2 for species number	max R^2
Gulf of Guinea	Land birds	4	.9728*	.9995*
Canary Islands	Land birds	7	.3138	.4725
British Isles	Land birds	10	.8632**	.8978*
Channel Islands	Vascular plants	4	.9613[2]	.9993*
British Isles	Vascular plants	13	.9443**	.9580**

* $P < .05$ ** $P < .01$

[1] A curvilinear multiple regression with the variables log-transformed did not alter the results qualitatively. A few formerly insignificant variables were raised to significance and vice versa, but with no consistent pattern.

[2] Increased to .9993 (significant at 5% level) upon addition of area to regression. For no other group did the addition of area, distance, or max. elevation, singly or multiply, add significantly to r^2.

island area, distance from source, and maximum elevation independent variables added individually, and then stepwise where possible. The analysis was done on land birds of the British Isles, islands in the Gulf of Guinea, and Canary Islands; and on vascular plants of the British Isles and Channel Islands. Results, given in Table 4, show that in only one instance, plants of the Channel Islands, do any of the other three variables add significantly to the predictability given by species number alone, and then the sample was only four islands. For all five groups the maximum numerical increase in R^2 over the r^2 given by species number alone was small.

EXPLANATION OF LOW TAXONOMIC DIVERSITY ON ISLANDS

Granted that S/G on islands is greater than expected on the basis of chance, there are two published theories to explain this fact. Williams (1964), dealing with "islands" in the broadest sense (usually sub-areas of large, heterogeneous regions, but also including oceanic islands), specifically mentions the likelihood of increased intra-generic competition acting to lower S/G on islands. But he feels that any such effect is minimal compared to the tendency of congeneric species to have similar physical requirements (weather, elevation, soil,

etc.). This similar physical ecology of the congeners tends to increase S/G above that expected for a random distribution of species.

Hairston (1964) similarly accepts that within subregions (and presumably islands) the probability of finding a congeneric pair is greater than would be expected on a chance basis, but proposes a second explanation. Following Darwin (1859), he claims that the individual species within large genera are more variable, and by inference less ecologically and geographically limited, than those of small genera. He asserts that a consideration of mono-typic genera will indicate the existence of the negative correlation between size of genus and restriction of its species; such a relationship would clearly explain the tendency of island biotas to exhibit higher than expected S/G.

A third explanation, applicable especially to oceanic islands, might be a positive correlation of dispersal ability and taxonomic affinity. That is, if one species of a genus is much more adept at long-distance dispersal than the average species of a large group, a congeneric species is likely to be similarly better adapted than average for such dispersal. Biologically this seems almost to be a truism. At issue is its relative importance. The analogy to Williams' hypothesis may be noted—he proposes that ecological requirements are distributed taxonomically, while it is suggested here that dispersal capability may be distributed taxonomically.

These three hypotheses are not mutually exclusive. For instance, it is possible that both ecological adaptability and dispersal capability are strongly correlated with taxonomy. Both could conceivably contribute equally to the greater than expected S/G on islands, or one could be insignificant compared to the other. One might expect that animals adapted to living in marginal habitats (which probably requires extraordinary ecological adaptability) would tend to have high dispersal capabilities, and so for a given set of insular colonists it could be impossible to distinguish between the effects of these two factors on taxonomic diversity.

Neither does Hairston's hypothesis necessarily exclude either of the others. If members of larger genera tended to prefer certain stringent habitats that happened to predominate on islands, and tended to succeed in a broader range of habitats (including these stringent ones) than did monotypic genera, and if these habitats were sufficiently important in determining the success or failure of potential long-term colonists on an island, then both Hairston's and Williams' hypotheses could be true simultaneously.

Let us first consider Hairston's hypothesis that ecological diversity is correlated with genus size, which can be partitioned into two propositions:

1) Is it true?
2) If true, is it sufficient to explain the higher than expected S/G on islands?

The proper way to determine if genus size and ecological diversity of an individual species are correlated is to examine all the individual species in genera of several sizes within a higher taxon. Such a study is under way but is outside the scope of this work. At the simplest level,

however, the data of Table 1, sections 1, 3–5, 7, 9, 10, and 15–23 were rederived identically except that all monotypic genera were excluded from the analysis (Table 5). Monotypic plants were determined from Willis (1966), while the authorities for birds were Peters (1931–62) for those families covered and a collection of regional avifaunas for other families. Several crude numerical facts indicate that even if genus size and ecological adaptability prove to be statistically correlated, it is an oversimplification to regard this as the sole reason for the higher than expected S/G on oceanic islands.

For example, if the relationship between sizes and distributions of genera were all-important, then it should be generally true that, for a given island, the ratio (number of monotypic genera on island/number of monotypic genera on source) is less than (number of species of polytypic genera on island/number of species of polytypic genera on source). This says, roughly, that species of monotypic genera are differentially eliminated on islands. Table 6 lists this relationship for the island groups analyzed in Table 5. The expected trend is present, but is so far from universal. It is not worth pursuing this line statistically because of the crudeness of the assumptions involved.

A second approach, not entirely independent of the first, is to consider for any given island the deviations of actual from expected S/G with monotypics included and excluded, respectively. For any island both actual and expected S/G can only increase when monotypic genera are excluded, but the change in the difference between the two depends on the species-genus distributions. If it were true that monotypic genera are generally less widely distributed than species of polytypic genera then a positive difference between actual and expected S/G should probably decrease, and a negative difference increase, upon exclusion of the monotypics. The differences before and after the ex-

TABLE 5. *Expected mean number of species per genus with monotypic genera excluded.*

Island or source	Group	No. spp.	Actual $\overline{S/G}$	$E(\overline{S/G})$	Change in deviation from expected
1) *Yucatan Peninsula*	Land	224	1.341		
Cozumel	birds	74	1.156	1.127	+.003
Holbox		39	1.114	1.064	+.040
Mujeres		39	1.083	1.067	−.032
Chinchorro		13	1.000	1.018	−.005
Contoy		11	1.000	1.022	+.002
3) *Cuba*	Land	76	1.333		
Isle of Pines	birds	60	1.176*	1.265	−.062
4) *Hispaniola*	Land	74	1.321		
Gonave	birds	41	1.242	1.170	+.018
Tortue		28	1.217	1.122	−.024
5) *Panama*	Land	567	1.847		
Barro Colorado	birds	210	1.364	1.348	+.009
Coiba		78	1.069	1.136	−.033
7) *Morocco*	Land	179	1.673		
Canaries	birds	43	1.303*	1.154	+.041
Tenerife		37	1.233	1.145	+.008
Grand Canary		35	1.207	1.121	+.021
Gomera		31	1.240	1.123	+.022
Palma		31	1.240	1.112	+.019
Fuerteventura		26	1.238*	1.101	+.035
Hierro		27	1.174	1.106	+.008.
Lanzarote		24	1.200	1.086	+.045
Madeira		24	1.263**	1.090	+.015
9) *Britain*	Land	110	1.692		
Ireland	birds	72	1.500	1.481	−.029
Bute		62	1.442†	1.428	−.016
Islay		60	1.429	1.407	−.007
Man		60	1.500	1.413	+.008
Orkney Mainland		34	1.417*	1.243	+.028
Coll		34	1.360	1.258	+.003
Rhum		33	1.375	1.250	−.019
Canna		29	1.261	1.199	+.021
Tiree		29	1.318	1.221	−.041
Barra		27	1.227	1.196	−.014
N. Uist		27	1.286	1.186	0
Eigg		26	1.238	1.181	−.005
Scillies		23	1.150	1.175	+.010
Lundy		23	1.150†	1.147	+.018[a]
Bardsey		22	1.158†	1.170	−.004
St. Kilda		10	1.111	1.074	−.009
Mingulay		10	1.111	1.071	−.017
Ailsa Craig		7	1.000	1.028	0
Skokholm		6	1.200	1.040	+.004
May		6	1.000	1.042	+.005
Steepholm		5	1.000	1.047	−.017
Boreray		5	1.000	1.022	+.008
Soay		3	1.000	1.025	−.005
Dun		3	1.000	1.035	+.005
N. Rona		2	1.000†	1.040	−.040

* $P < .05$
** $P < .01$
† Within 95% confidence band for $E(\overline{S/G})$
[a] Change in direction

42   D. S. SIMBERLOFF

TABLE 5.   (Continued)

Island or source	Group	No. spp.	Actual $\overline{S/G}$	$E(\overline{S/G})$	Change in deviation from expected
10) *Orkney Mainland*	Land	34	1.417		
Hoy	birds	27	1.286	1.333	−.011
Rousay		25	1.316†	1.327	−.024[a]
Shapinsay		23	1.278†	1.285	−.001
Eday		19	1.462*	1.236	+.009
S. Ronaldsay		18	1.500*	1.224	+.029
Sanday		17	1.308	1.216	−.014
Westray		16	1.455**	1.190	+.036
Stronsay		15	1.250	1.193	−.005
S. Walls		15	1.250	1.186	−.015
Flotta		12	1.091	1.123	+.043
Burray		12	1.200	1.159	+.005
Papa Westray		12	1.200	1.143	+.005
N. Ronaldsay		13	1.300	1.119	−.035
Faia		9	1.125†	1.113	+.018[a]
Cava		8	1.000	1.100	−.019
Swona		7	1.000	1.074	+.016
Eynhallow		8	1.143	1.094	−.006
Calf of Eday		7	1.000	1.094	+.012
15) *Britain*	Vascular	1628	2.841		
Ireland	plants	1112	2.391	2.377	+.008
Wight		988	2.287	2.262	−.011
Anglesey		834	2.122	2.100	+.013
Man		751	2.063	2.020	−.012
Skye		583	1.983**	1.840	+.010
Islay		571	2.025**	1.828	+.001
Arran		567	1.949**	1.830	−.009
Lewis		518	1.947**	1.773	−.004
Scillies		512	1.889**	1.760	−.003
Mull		510	1.875*	1.759	−.011
Colonsay		471	1.826*	1.715	+.007
S. Uist		462	1.870**	1.702	−.008
Eigg		445	1.809**	1.683	0
Jura		436	1.921**	1.676	+.008
Coll		433	1.819**	1.667	+.002
Orkney Mainland		432	1.982**	1.669	+.004
N. Uist		423	1.863**	1.661	+.015
Rhum		418	1.825**	1.647	0
Shetland Mainland		412	1.925**	1.647	+.019
Barra		401	1.815**	1.624	+.017
Gigha		394	1.767**	1.622	−.002
Tiree		370	1.787**	1.591	+.011
Lundy		333	1.624	1.543	−.005
Canna		293	1.610*	1.500	−.007
Muck		278	1.635**	1.471	+.008
Mingulay		263	1.654**	1.454	+.007
Shiant Islands		155	1.615**	1.310	0
May		134	1.472**	1.268	+.003
St. Kilda		128	1.506**	1.253	+.004
Ailsa Craig		74	1.298*	1.164	+.002
16) *Ireland*	Vascular	1112	2.391		
Inishmore	plants	421	1.671†	1.671	−.006
Clare		370	1.787**	1.607	+.002
Inishbofin		352	1.769**	1.575	+.018

TABLE 5. (Continued)

Island or source	Group	No. spp.	Actual S/G	$E(\overline{S/G})$	Change in deviation from expected
16) *Ireland* (Continued)					
Inishturk		316	1.746**	1.535	+.015
Rathlin		303	1.656**	1.515	+.022
Inishmaan		235	1.487	1.424	−.002
Aran		230	1.565**	1.413	+.008
Lambay		177	1.513**	1.332	−.001
Saltee Islands		169	1.550**	1.330	+.005
Inishkea Islands		159	1.446**	1.306	+.017
Tory		159	1.459**	1.313	+.021
Great Blasket		137	1.539**	1.275	+.036
Inishtooskert		40	1.176	1.091	+.002
17) *Orkney Mainland*	Vascular	432	1.982		
Hoy	plants	347	1.761	1.824	−.007
S. Ronaldsay		201	1.546	1.516	+.015
Fair Isle		169	1.550	1.455	+.010
Sanday		156	1.458	1.419	+.020
N. Ronaldsay		127	1.396	1.359	+.019
Westray		64	1.391**	1.206	0
Stronsay		59	1.204	1.188	+.011
18) *Shetland Mainland*	Vascular	412	1.925		
Unst	plants	241	1.639	1.594	+.008
Fetlar		186	1.590	1.479	−.007
Bressay		173	1.617**	1.453	+.015
Yell		158	1.534	1.426	−.002
Whalsay		153	1.515	1.413	+.002
Foula		143	1.375	1.393	+.004
19) *Mull*	Vascular	510	1.875		
Iona	plants	381	1.716	1.703	+.007
Treshnish Islands		131	1.424*	1.295	+.023
20) *N. Uist*	Vascular	423	1.863		
Monach Islands	plants	214	1.643*	1.518	−.006
21) *Lewis*	Vascular	518	1.947		
Flannen Islands	plants	21	1.313**	1.063	+.012
22) *Anglesey*	Vascular	834	2.122		
Holy Isle	plants	368	1.813**	1.601	+.014
23) *Guernsey*	Vascular	789	2.115		
Herm	plants	407	1.777**	1.662	−.005

clusion of the monotypics are not strictly comparable numerically because the sizes of species pool and island biota both change, but with the large species pools and relatively small numbers of monotypic genera in the examples treated in Table 5 it is unlikely that an uncorrected numerical comparison of magnitudes of the differences would lead to a serious error of interpretation. In Table 6 are enumerated the directions of change in difference be-

tween actual and expected S/G for all islands listed in Table 5. The two alternatives are a negative change, predicted by Hairston's hypothesis, and no change or a positive change. If any tendency is manifested it is the opposite to that expected if Hairston's hypothesis were true —there are more positive than negative changes when monotypics are excluded. If the number of islands in a given group with positive and negative change, respec-

TABLE 6. *Differential elimination of species in monotypic and polytypic genera, respectively.*

Source # (from Table 1)	Group	# islands	# islands, monotypics differentially eliminated	# islands, others differentially eliminated	Direction of change in deviation from expected 0 or +	−	Median ($\frac{\Delta \text{ deviation}}{\|\overline{S/G} - E(\overline{S/G})\|}$)
1	L. birds	5	2	3	3	2	.217
3	"	1	1	0	0	1	2.296
4	"	2	1	1	1	1	.267
5	"	2	1	1	1	1	1.018
7	"	9	7	2	9	0	.232
9	"	25[1]	19	5	12	13	.250
10	"	18[1]	6	11	9	9	.162
	Subtotals	62	37	23	35	27	
15	V. plants	30	23	7	20	10	.048
16	"	13	4	9	10	3	.077
17	"	7	1	6	6	1	1.000
18	"	6	3	3	4	2	.080
19	"	2	0	2	2	0	.692
20	"	1	1	0	0	1	.046
21	"	1	0	1	1	0	.050
22	"	1	1	0	1	0	.071
23	"	1	1	0	0	1	.042
	Subtotals	62	34	28	44	18	
	Totals	124	71	51	79	45	

[1] Island with equal elimination rates

tively, are regarded as a matched pair, then although this tendency is not significant by the sign test, it is significant at the .05 level by the Wilcoxon signed-rank test ($N = 13$, $T = 16$).

Too much should not be made of the above two sets of data. The statistical tests cited are not entirely appropriate and further work on the entire question is required. All that can be said here is that for monotypic versus polytypic genera the hypothesis is not trivially true, and its validity as regards polytypic genera of different sizes—say, six-membered genera vs. seven-membered genera—is even more questionable.

On a biological plane, a detailed examination of the bird data shows that monotypic genera fall predominantly into two classes. First are the geographically and ecologically restricted species of the sort that Hairston (and Darwin before him) was primarily considering—birds restricted to single mountain ranges, the vicinity of one lake, etc. These species are differentially excluded from oceanic islands. By contrast, several monotypic genera of large, usually strong-flying birds, especially in the Accipitridae (and Ardeidae, where these were counted) are found on virtually all islands—that is, they are differentially included in the faunas of oceanic islands. Although the former class may be larger numerically, the broad and inclusive geographical ranges of many of the latter class cause many islands, especially those with few bird species, to have a disproportionately large number of monotypic genera. As size of island avifauna increases, these omnipresent monotypic genera become less important numerically.

Whether or not Hairston's proposition is valid, and whatever the biological factors involved, the data in Table 5 indicate that the general excess of actual over expected S/G on islands cannot be completely explained by differential geographic restric-

TABLE 7. *Partition of islands of equal area (explained in text).*

		Distance from Source	
		Low	High
	Low	A	B
Habitat Diversity			
	High	C	D

tions of different-sized genera. I have already shown that the direction of change of this excess upon elimination of monotypic genera from the analysis seems more generally the opposite of that predicted by the differential restriction theory. But more importantly, the sizes of these changes compared to the original differences between actual and expected S/G were, as a rule, small (Table 6). Most of the large values result from small original difference in the denominator rather than numerically large change upon elimination of the monotypics.

The distinction between the other two hypotheses (taxonomic distribution of ecological requirements and dispersal abilities, respectively) is more difficult. One might expect the previously mentioned multiple regression analysis to be enlightening. If the dispersal aspect overweighed the ecological one, then distance should increase the coefficient of determination more than either maximum elevation or area, and vice versa. The increases in R^2 given by all three of these were so low as to render them useless; furthermore, there was no trend.

A qualitative approach rests on the use of pairs of similarly sized islands. If we construct Table 7 and use maximum elevation as an indicator of ecological diversity, then each member of an appropriate pair of islands will fall in one of the four boxes. Depending on which of the two possible hypotheses is correct, the difference between actual and expected S/G would either increase, decrease, or be unaffected in a move from one box to another. For example, if one island fell in box A and the other in box B, then in going from A to B the distance hypothesis would predict an increase in the difference between actual and expected S/G, while the ecological diversity hypothesis would predict no change. Since S/G will be very near expected the result from such an island pair would carry little weight.

But the competing hypotheses predict opposite changes in one of the six possible pairs (AD). A transition from box A to box D yields an increase on the distance hypothesis and a decrease on the ecological diversity hypothesis. Eleven island pairs from several sections of Table 1 with area approximately equal were tested. Six instances indicated the ecological diversity hypothesis and five the distance hypothesis. Numerically many of the differences in deviation were very small ($<$ .01 in three cases) and the area equalities were only approximate so that a firm conclusion cannot be drawn. As was stated earlier, the two hypotheses are not mutually exclusive and the indication here is that both factors are effective, probably to varying degrees in different situations, but with neither being generally much more important than the other. The unequal restriction of different-sized genera may also contribute in some instances to the difference between actual and expected S/G, but its general effect is questionable and its numerical contribution small.

SUMMARY

1) For any taxon higher than genus, the expected mean number of species per genus ($E(S/G)$) for a randomly drawn subset is a non-decreasing function (approximately linear for many taxa) of the size of the subset, and is therefore lower than the mean number of species per genus for the entire taxon.

2) The relationship of actual S/G for an island biota to S/G expected for a random subset of identical size drawn from the species pool of the presumed

source area is relatively insensitive to modifications of taxonomy and to use of a source area slightly larger than the real one. This relationship is subject to more drastic change upon use of incomplete species lists for island or source or a presumed source area smaller than or different from the real source.

3) Pseudo-random drawings of biotas of identical size to those of a number of islands show that, in general, the mean number of species per genus, though lower than in the source biota, is higher than would be expected on a hypothesis of random colonization.

4) The deviation of actual from expected S/G is not strongly correlated with island area, maximum elevation, and distance from source.

5) The claimed positive correlation between species range and genus size, even if it should be shown to exist, is probably insufficient to account for the magnitude of these deviations.

6) The main causes of the excess of insular S/G's over those predicted by chance are probably two simultaneous tendencies:

a) similarity of congeneric species in ecological requirements.

b) similarity of congeneric species in dispersal capabilities.

Acknowledgments

William H. Bossert, Michael P. Johnson and Edward O. Wilson provided thoughtful comment and various kinds of biological and statistical assistance on many aspects of this work. Thomas W. Schoener made available unpublished lists of the avifauna of the British Isles. Computing costs were partially borne by NSF grant GJ 367 to the Florida State University Computing Center.

Literature Cited

Amadon, D. 1953. Avian systematics and evolution in the Gulf of Guinea. Bull. Amer. Mus. Nat. Hist. 100, No. 3.

Bannerman, D. A. 1953. The birds of west and equatorial Africa. Oliver and Boyd.

——. 1963. Birds of the Atlantic islands. Vol. 1. Oliver and Boyd.

Bond, J. 1950. Check-list of birds of the West Indies. 3d ed. Acad. Nat. Sci. Philadelphia.

Darwin, C. 1859. The origin of species. John Murray.

Eisenmann, E. 1952. Annotated list of birds of Barro Colorado Island, Panama Canal Zone. Smithsonian Inst. Misc. Coll. 117, No. 5.

——. 1955. The species of Middle American birds. Trans. Linnaean Soc. N. Y. 7:1–128.

Elton, C. S. 1946. Competition and the structure of animal communities. J. Anim. Ecol. 15:54–68.

Etchécopar, R. D., and F. Hüe. 1964. Les oiseaux du Nord de l'Afrique de la Mer Rouge aux Canaries. Boubée.

Grant, P. R. 1966. Ecological compatibility of bird species on islands. Amer. Natur. 100: 451–462.

——. 1968. Bill size, body size, and the ecological adaptations of bird species to competitive situations on islands. Syst. Zool. 17: 319–333.

Grant, P. R., and I. McT. Cowan. 1964. A review of the avifauna of the Tres Marias Islands, Nayarit, Mexico. Condor 66:221–228.

Greenwood, J. J. D. 1968. Coexistence of avian congeners on islands. Amer. Natur. 102:591–592.

Hairston, N. G. 1964. Studies on the organization of animal communities. J. Anim. Ecol. 33(Suppl.):227–239.

Johnson, M. P., L. G. Mason, and P. H. Raven. 1968. Ecological parameters and plant species diversity. Amer. Natur. 102:297–306.

MacArthur, R. H., and E. O. Wilson. 1967. The theory of island biogeography. Princeton Univ. Press.

Moreau, R. E. 1966. The bird faunas of Africa and its islands. Academic Press.

Paynter, R. A. 1955. The ornithogeography of the Yucatan Peninsula. Bull. Yale Peabody Mus. Natur. Hist. 9:1–347.

Perring, F. H., and S. M. Walters. 1962. Atlas of the British flora. Nelson.

Peters, J. L. 1931–62. Check-list of birds of the world. Vols. 2–7, 9–10, 15. Harvard Univ. Press.

Peterson, R. T., G. Mountfort, and P. A. D. Hollom. 1954. A field guide to the birds of Britain and Europe. Houghton Mifflin.

Sclater, W. L. 1928–30. Systema avium aethiopicarum. 2 vols. Taylor and Francis.

Venables, L. S. V., and U. M. Venables. 1955. Birds and Mammals of Shetland. Oliver and Boyd.

Volsøe, H. 1951. The breeding birds of the

Canary Islands. Vidensk. Meddr. Dansk. Naturh. Foren. 113:1–153.

——. 1955. The breeding birds of the Canary Islands. Vidensk. Meddr. Dansk. Naturh. Foren. 117:117–177.

Voous, K. H. 1960. Atlas of European birds. Nelson.

Watson, G. 1964. Ecology and evolution of passerine birds on the islands of the Aegean Sea. Ph.D. Thesis, Dept. Biol., Yale Univ.

Wetmore, A. 1957. The birds of Isla Coiba, Panama. Smithsonian Inst. Misc. Coll. 134, No. 9.

Williams, C. B. 1964. Patterns in the balance of nature and related problems in quantitative ecology. Academic Press.

Willis, J. C. 1966. A dictionary of the flowering plants and ferns. 7th ed. Cambridge Univ. Press.

Wilson, E. O. 1959. Adaptive shift and dispersal in a tropical ant fauna. Evolution 13: 122–144.

——. 1961. The nature of the taxon cycle in the Melanesian ant fauna. Amer. Natur. 95: 169–193.

Wilson, E. O., and R. W. Taylor. 1967. The ants of Polynesia. Pacific Insects Monogr. 14:1–109.

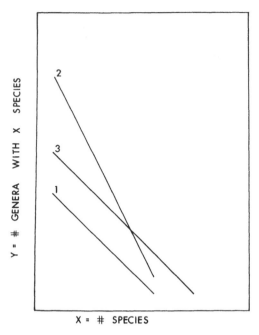

Fig. 3. Illustration of notion of two subsets drawn randomly from same set of species. 1 = species-genus distribution of one subset. 3 = species-genus distribution resulting from addition of two subsets drawn randomly from same set. 2 = species-genus distribution resulting from addition of two subsets drawn randomly from disjoint sets with identical species-genus distribution.

Appendix

The concept of a species-genus distribution which is "unchanged" upon merging of several biotas to form a single biota, or the removal of a set of species from a biota, deserves elucidation. Two truly random subsets drawn from a set with species-genus distribution of given shape will:

a) Have species-genus distributions similar in shape to one another and to the original distribution.

b) Not be inordinately overlapped or disjoint.

Several notions in the text can be related to the above requirements. For instance, in the addition of new species with increasing size of source area, it is not sufficient that the new species be distributed into genera similarly to the original ones. These genera must partially overlap the original ones so as to preserve the shape of the species-genus distribution. A graphic example is given in Fig. 3 assuming a linear species-genus distribution; it could as well be done for the log series and other distributions. If the original species-genus distribution is depicted as line 1 and an equal number of species are added, distributed identically but into disjoint genera, line 2 results (decreased slope, x-intercept unchanged). If two sets are random subsets of some larger set their addition should result in a distribution represented by line 3 (slope unchanged, x-intercept increased).

An identical interpretation can be given to the problem of random extinction. Here one set is that of extinguished species and the other set comprises those which remain. In the question of incomplete lists, the first set consists of recorded species and the second of unrecorded ones.

THE
AMERICAN NATURALIST

Vol. 104, No. 940   The American Naturalist   November–December 1970

HERBIVORES AND THE NUMBER OF TREE SPECIES IN TROPICAL FORESTS

DANIEL H. JANZEN

Department of Biology, University of Chicago

Wet lowland tropical forests characteristically have many tree species and low density of adults of each species compared with temperate-zone forests in habitats of similar areal extent, topographic diversity, and edaphic complexity (Black, Dobzhansky, and Pavan 1950; Richards 1952; Poore 1968; Ashton 1969). Despite reports that adults of some species of lowland tropical trees show clumped distributions (Poore 1968; Ashton 1969), I believe that a third generalization is possible about tropical tree species as contrasted with temperate ones: for most species of lowland tropical trees, adults do not produce new adults in their immediate vicinity (where most seeds fall). Because of this, most adults of a given tree species appear to be more regularly distributed than if the probability of a new adult appearing at a point in the forest were proportional to the number of seeds arriving at that point. This generalization is based on my observations in Central and South American mainland forests, on discussions with foresters familiar with these forests, on discussions with J. H. Connell about Australian rain forests, and on data given in the papers cited above.

I believe that these three traits—many tree species, low density of each species, and more regular distribution of adults than expected—are largely the result of two processes common to most forests: (1) the number of seeds of a given species arriving at a point in the forest usually declines with distance from the parent tree(s) and varies as the size of the viable seed crop(s) at the time of dispersal, and (2) the adult tree and its seeds and seedlings are the food source for many host-specific plant parasites and predators. The negative effect of these animals on population recruitment by the adult tree declines with increasing distance of the juvenile trees from their parent and from other adult trees. A simple model summarizes these two processes (fig. 1). It will lead us to examine the effects of different kinds of plant predators on juveniles, ecological distance between parents, dispersal agents, environmental predictability and severity—among other factors—on the number of tree species in a habitat, their den-

502   THE AMERICAN NATURALIST

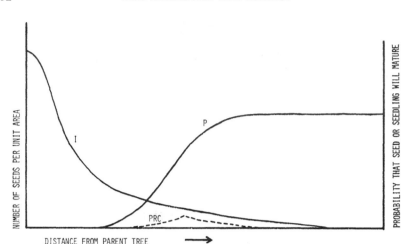

F<small>IG</small>. 1.—A model showing the probability of maturation of a seed or seedling at a point as a function of (1) seed-crop size, (2) type of dispersal agents, (3) distance from parent tree, and (4) the activity of seed and seedling predators. With increasing distance from the parent, the number of seeds per unit area (I) declines rapidly, but the probability (P) that a dispersed seed or seedling will be missed by the host-specific seed and seedling predators, before maturing, increases. The product of the I and P curves yields a population recruitment curve (PRC) with a peak at the distance from the parent where a new adult is most likely to appear; the area under this curve represents the likelihood that the adult will reproduce at all, when summed over all seed crops in the life of the adult tree. In most habitats, P will never approach 1, due to nonspecific predation and competition by other plants independent of distance from the parent. The curves in this and the following figures are not precise quantifications of empirical observations or theoretical considerations, but are intended to illustrate general relationships only.

sities, and their spatial juxtaposition. Almost none of the many hypotheses generated by this examination can be tested with data currently available in the literature. While I am at present testing some of these in Central American forests, they are all offered here in the hope of stimulating others to examine them as well.

It is my intention in studies of tropical species diversity to shift the emphasis away from the utilization and manipulation of diverse resources to generate a diverse consumer community (Are niches narrower in the tropics?), and toward an examination of the ability of the consumer community to generate and maintain a diverse resource base. Thus I am not so much concerned with Where did all the tropical tree species come from? (as Haffer [1969] has asked for birds), as I am in raising the question How do you pack so many into a forest? In short, this study is an extension to the plant community of Paine's (1966) suggestion that "local animal species diversity is related to the number of predators in the system and their efficiency in preventing single species from monopolizing some important, limiting, requisite" (see Spight 1967; Murdoch 1969, for elaborations of this statement). The same concept was applied by Barbehenn (1969) to the interactions of tropical mammals with their predators and parasites, and by Lowe-McConnell (1969) to tropical fishes. MacArthur (1969) has generalized the system for vertebrates that prey on other animals in tropical communities.

This paper presents a major problem in terminology. Words, such as "carnivore," "graminivore," "herbivore," "frugivore," and "sanguivore," designate clearly the type of host or prey of an animal. The terms, "predator" and "parasite," describe the effect of the animal on its host or prey. Unfortunately, both "predator" and "parasite" have conventionally been used as synonyms for "carnivore" and "sanguivore," thereby excluding those animals that feed on plants. However, "plant parasite" is appearing in the literature to denote animals or plants that feed on a plant but do not kill it. Similarly, I wish to use "seed predator" or "seedling predator" to cover those animals that eat entire plants, or at least eat enough so that the plant dies immediately. The act of a fox seeking out and eating mice differs in no significant way from a lygaeid bug seeking out and eating seeds, or a paca seeking out and eating seedlings. Words, such as "herbivore," "frugivore," or "graminivore," are inadequate substitutes for "seed predator" or "seedling predator" since they do not tell the fate of the juvenile plants.

HOST SPECIFICITY

The degree of host-specificity displayed by the seed and seedling predators strongly influences the model in figure 1. Without host-specificity, the P curve in figure 1 would be horizontal, offspring would more likely mature close to their parents, and regulation of tree density by seed predators would depend on the distance between seed-bearing trees of any species serving as foci for these predators. All tree species would be affected by physical environmental conditions favoring certain plant predators, and it is unlikely that these would make any particular tree species very rare or extinct. That the vast majority of insects that prey on seeds (of various ages) are host-specific in tropical communities must be inferred from three sources (the literature is sterile on the subject):

1. From 1963 to 1970, I have reared insects from the seeds of better than 300 lowland Central American plant species. Almost without exception, a given insect species reproduces on only a small subset (one to three species) of the hundreds of potential host species available in the habitat. This is not negated by the observation that some of these insects have different hosts at different times of the year or in different habitats, and may feed on a wide variety of water and food sources (e.g., *Dysdercus fasciatus* bugs feeding on dead insects [Janzen 1970c]) that do not result in egg production (e.g., Sweet 1964).

2. Detailed studies of the life histories of insects that prey on fruits and seeds of forest and orchard trees in the United States suggest strong host-

504   THE AMERICAN NATURALIST

specificity, both in terms of field censuses and the specific behavior of the insects themselves (e.g., Bush 1969; Schaefer 1962, 1963; and a voluminous forestry literature). However (and this is where temperate forests appear to differ from tropical ones), many of these are complexes of species, such as acorn weevils (*Curculio* spp.) on oaks (*Quercus* spp.), that feed on all members of a genus in a habitat and therefore will not necessarily result in extreme rarity of any of the prey species, unless all are made rare. Such complexes are also present in tropical forests, but do not appear to constitute as large a population of the total herbivore complex as in temperate forests. However, even if host specificity were no greater in the tropics than in temperate zones, it is clearly high enough in both areas to allow for a model such as that in figure 1.

3. While there are many cases of strong host specificity by predators on seeds and fruits (e.g., Janzen 1970*a*, 1970*b*, 1970*c*, 1971), examples of the opposite case are rare among insects in nature. This statement is not negated by the long host list that may be compiled for some insects by a study such as Prevett's (1967) or with a catalogue such as the new edition of Costa Lima's (1967–68) catalogue of insects that live on Brazilian plants, where host records are summed across many habitats, seasons, and geographic areas. In the latter reference, no distinction is made between insects reproducing on a plant or those merely feeding on it. For the purposes of this paper, a species of insect will be considered to be host-specific if most of its population feeds on one (or very few) species of seed or seedling in a habitat undisturbed by man.

Seed-eating vertebrates may show comparatively little host-specificity but, as will be discussed later, may be facultatively host-specific and thus can be included in the model in some cases.

SEED DISPERSAL

Seed dispersal to sites near parents is affected by two different groups of predators on dispersed juveniles, the distance-responsive and density-responsive predators. The probability that a juvenile plant will be eaten by a *distance-responsive* predator is primarily a function of the ecological distance between that juvenile and adult trees of the same species. Distance-responsive predators are commonly parasites on the adults, but predators on seedlings. This is because seedlings cannot withstand the loss of leaves and shoot tips to the degree that adult trees can. The probability that a juvenile plant will be eaten by a *density-responsive* predator is primarily a function of the ecological distance between that juvenile and other juveniles. Density-responsive predators rely primarily on the presence of one juvenile to survive long enough to find another or to be stimulated to search hard to find another. Any given species of predator can belong to both categories, but in general the activities of herbivores can be profitably viewed with this dichotomy in mind.

Seed Immigration and Distance-responsive Predators

Intensities and patterns of seed shadows cast by parent trees are functions of seed crop size, seed predation before dispersal, and characteristics of the dispersal agents. From figure 2, it is obvious that increasing the predation on seeds before dispersal (i.e., lowering I, the number of seeds dispersed per unit area) may (1) reduce immigration proportionately more, far from the parent than close to it, and therefore reduce the distance of new adults from their parents if juvenile trees are not subject to predation; (2) reduce the number of seeds that escape the distance-responsive herbivore because they fall sufficiently far from their parent; hence the probability that the adult will reproduce at all during its lifetime is lowered, as is the population density of adults in the habitat; and (3) reduce the likelihood that the tree species in question will competitively displace other tree species or reduce their population densities.

The similar immigration curves in figure 2 are only one of several possible sets that could result from lowering the size of the viable seed crop. If the seeds are killed after they are nearly mature, and therefore imbedded in an intact fruit, they may not be distinguished from viable

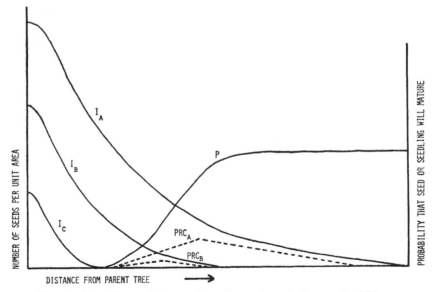

FIG. 2.—The effects of increased predispersal predation on the PRC curve when the predators are distance-responsive. The seed crop of I_B is about one-half of that of I_A, and the seed crop of I_C is about one-ninth that of I_A. This figure should be contrasted with figure 3, where a reduction in seed crop affects the PRC curve quite differently when the predators act in a density-responsive manner.

ones by the dispersal agent. Thus immigration curves produced by pre-dispersal seed mortabilty (e.g., fig. 2, curve I_C) are of similar form but lower than those of an intact crop (e.g., fig. 2, curve I_A). But if a potential dispersal agent selects and discards viable seeds at the parent tree (e.g., capuchin monkeys feeding on *Apeiba* fruits [Oppenheimer 1968]), the immigration curves may become steeper as seed destruction increases. On the other hand, if seed predation is early in the development of the fruit, and only fruits with viable seeds mature, then the immigration curves flatten out and accentuate the distance between new adult trees. This is because the reduced amount of fruit will result in a lower proportion of the total crop being ignored by satiated dispersal agents, and therefore falling to the ground beneath the parent tree. However, very small seed crops are often ignored by dispersal agents, leading to extreme truncation of the immigration curve at the right tail.

The most important predators on immature and mature seeds before dispersal are insects such as bruchid (Hinckley 1961; Wickens 1969; Prevett 1967; Parnell 1966; Janzen 1969a, 1970a), curculionid (e.g., De-Leon 1941; Janzen 1970b; Barger and Davidson 1967), and scolytid beetles (e.g., Shaefer 1962, 1963), lygaeid and pyrrhocorid bugs (e.g., Myers 1927; Yonke and Medler 1968; Eyles 1964; Janzen 1970c), Lepidoptera larvae (e.g., Breedlove and Ehrlich 1968; Hardwick 1966; Janzen 1970a; Coyne 1968; Dumbleton 1963), aphids (e.g., Phillips 1926b), and fly larvae (e.g., Pipkin, Rodriquez, and Leon 1966; Gillett 1962; Brncic 1966; Knab and Yothers 1914), birds such as parrots, and mammals, such as squirrels and monkeys (e.g., Smythe 1970; Smith 1968; Struhsaker 1967; Oppenheimer and Lang 1969). Insects are generally obligatorily host-specific, while vertebrates may be facultatively host-specific (for a given short time, they concentrate their foraging on or under the seed-breeding tree, but are not restricted to this species). A conservative estimate, based on large seed crop collections by Gordon Frankie and myself in Central America, is that at least 80% of the woody plants in lowland forest have mild to severe predispersal predation on reproductive parts by obligatorily or faculta-tively host-specific animals.

Any factor that increases the ability of these seed predators to move between seed crops in time or space, and to eat seeds more rapidly once there, will generally increase the number of tree species that can coexist in a given habitat (following the argument presented in an earlier section). Such a process would not result when the number of new adult trees produced by a parent is independent of the number of viable seeds dis-persed. Such systems are not easy to imagine when the absolute number of surviving seeds is small. While the distance between new adults and their parents will be reduced by predispersal seed mortality (e.g., figs. 2, 3), two other processes would increase this distance. First, seed and seed-ling predators acting after dispersal are likely to prey more heavily on juveniles near the parent or near other juveniles (see later section). Second, any pair of exceptionally close adult trees will mutually contribute

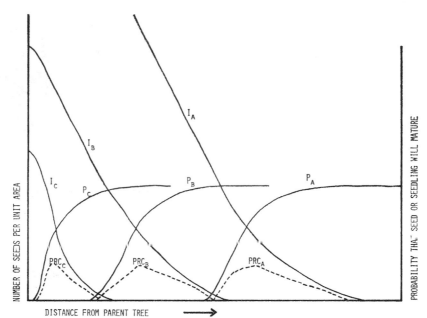

FIG. 3.—The effects of increased predispersal seed predation (progression from I_A to I_C) on the PRC curve when the predators are density-responsive. The PRC_C curve is slightly less peaked than would be the case if the density-responsive predators were identical for all three I curves; I have assumed that some density-responsive predators would be completely absent for the small I_C curve because there are not enough seeds or seedlings to attract them in the first place.

seed-crop predators, each greatly lowering the other's chances of reproducing at all, and lowering the probability of a third adult appearing at the same site.

There are two important characteristics of the dispersal agents. First, the faster they remove the seeds, the greater the survival of the seed crop that was subject to predispersal mortality. For example, in many legumes dispersed by birds and mammals, the second generation of bruchids in the seed crop kills virtually all seeds not yet dispersed (e.g., Janzen 1969a, 1970a).

Second, a large, but less intense, seed shadow can increase the distance between new adults and the parent. As used here, the intensity of a seed shadow is measured by the number of seeds falling per unit area and the size of a seed shadow is measured by the area over which the seeds are dispersed. My observations of dispersal around tropical forest trees, and the numerous anecdotes in Ridley's (1930) compendium of tropical seed dispersal systems, indicate that considerable variation in shape of immigration curves is possible (fig. 4). For example, the negative exponential (I_A) in figure 4 may be produced by wind dispersal (rare in tropical forest habitats but more common in dry areas [Smythe 1970; Croat 1970; Ridley

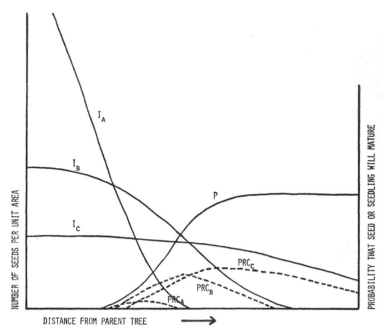

Fɪɢ. 4—The effects of different dispersal agents on the *PRC* curve when the predators are distance-responsive (see fig. 5 for the same effects associated with density-responsive predators). Each of the three *I* curves is generated by a viable seed crop of approximately the same size. For further explanation, see text.

1930]) or secondary dispersal by large mammals (e.g., *Persea*, Lauraceae; *Carapa*, Meliaceae) after the seeds have fallen. When water, steep topography, or seasonal winds are involved, the seed shadow is probably greatly skewed in one direction. Curve I_B may be associated with birds and rodents with short seed retention time (e.g., regurgitation of lauraceous fruits by trogons, toucans, and cotingas, and scatter-hording [Smythe 1970; Morris 1962] by rodents such as agoutis and agouchis). Curves of type I_C may be produced by vertebrates such as birds, bats and terrestrial mammals with long seed retention in the intestine, and by burs that stick to feathers and fur.

Thus dispersal agents may be responsible for the survival of a given tree species in a habitat that greatly favors the seed predators. In the progression from I_A to I_C there is (1) an increase in distance between new adults but at a decreasing rate, (2) an increase in seed survival (since for the present the P curve is held constant) which may lead to more adults surviving, (3) an increase in skewness of the PRC curve, leading to greater variation in adult tree dispersion and density, and higher rates of invasion of unoccupied habitats.

It is important to notice, however, that a major change in dispersal, as between the I_B and I_C curves, has relatively little influence on the height or location of the peak of the *PRC* curve.

If we add the complication that some dispersal agents are also major seed predators (e.g., agoutis [Smythe 1970]), it becomes apparent that the tremendous variety of fruit shapes, sizes, flavors, hardnesses, toxicities, and other traits are all probably highly adapted to taking advantage of those aspects of the dispersal agents that will yield an optimal seed shadow, counteracting the predispersal seed predators and the postdispersal predators discussed below.

A major constraint on the ways the adult may enhance seed escape is that dispersal must also get the seed to areas in the habitat where competitive and nutrient conditions are optimal for seedlings, a so-called safe site (Harper et al. 1961). While seed size is a major factor in determining the percentage of predation on a seed crop, it also influences strongly the suitability of a particular site for seedling survival. As seed size (or seed protection) decreases and seed number therefore increases (possibly leading to an increase in absolute numbers of surviving seeds through predator satiation [Janzen 1969a]), the number of safe sites in the habitat automatically decreases at an undetermined rate. Safe sites may disappear faster than new seeds are produced through reduction in seed size, and therefore this means of predator escape may yield no real increase in adult plant density. This complication does not, however, modify the interaction between the seed predators and the dispersal agents for any given regime of safe sites, edaphic heterogeneities, successional stages, and so forth.

Seed Dispersal and Density-responsive Predators

When the size of the seed crop is increased over evolutionary time, in the face of density-responsive predators on seeds and seedlings (e.g., fig. 3), the peak of the PRC curve moves rapidly outward. This and associated modifications of the PRC curve result in (1) only a slight increase in rate of new adult tree production for a large increase in viable seed crop size, (2) increased distance between newly appearing adults, and (3) a new equilibrium density lower than, or the same as, before the increase in seed crop size.

As conditions become more favorable for seed predation before dispersal, a given set of density-responsive predators will be less able to bring about wide spacing of new adults than distance-responsive seed and seedling predators (fig. 2). Also, density-responsive predators will not reduce the tree's total chance of reproducing to the degree that distance responsive predators can when predispersal predation is increased.

When seed shadows resulting from different sets of seed dispersal agents are compared against a background of density-responsive predators on juveniles (fig. 5), the PRC curves are dramatically different. As with the distance-responsive predator complex, the PRC is shifted outward with the progression of I_A to I_C, but unlike figure 4, it remains high. In general, it appears that the density-responsive predators will allow higher densities of adult trees in response to a change in the dispersal agents than is the

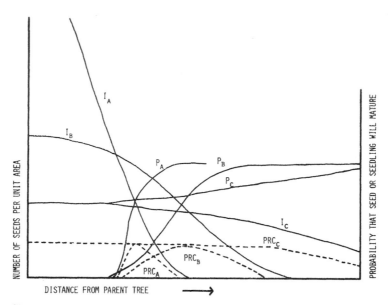

Fɪɢ. 5.—The effects of different dispersal agents on the *PRC* curve when the predators are density-responsive. A shift in the patterns of dispersal agents has a very marked effect on the shape of the *PRC* curves (in contrast with the effects of the distance-responsive seed predators depicted in fig. 4).

case with the distance-responsive predators. Curve I_C also shows that some dispersal agents can minimize the distance between new adults, if they lower sufficiently the seed density near the adult.

As mentioned previously, many predators on seeds and seedlings may act in either a distance- or density-responsive manner, depending (among other factors) on season, availability of alternate foods, and relative density of juvenile plants. When the immigration curve is viewed against the total array of predators, the actual outcome will depend on the relative proportions of these two types of predation activity. Enough data are not yet available to predict these proportions for various habitats.

PREDATION ON DISPERSED JUVENILES

Distance-responsive Predators

Once the juvenile trees have been dispersed, any factor increasing the effective distance to which the distance-responsive seed and seedling predators search will augment the distance between new adults and hence lower the density of new adults (fig. 6). An increase in the distance of effective searching may have a variety of causes, such as: (1) change in the search behavior of the predators, (2) increase in the size of the population of parasites (which act as predators to juveniles) feeding on the parent tree, (3) increased proximity of parent trees which may be due to increased

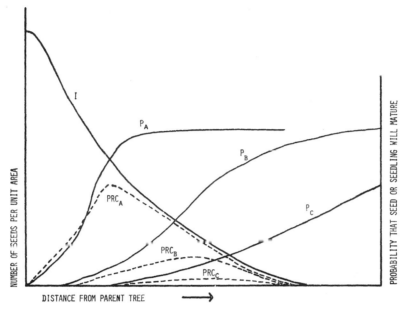

Fig. 6.—The impact of increasing the effective distance at which distance-responsive predators can act on a seed crop of fixed size. Were curve P_A simply to be shifted to the right, rather than changing in slope as well (as shown in curves P_B and P_C), the PRC_A curve would become lower and shift to the right, but would retain its sharp peak.

competitive ability or other factors, and (4) decreased synchrony of parent and offspring vegetative growth cycles, etc.

As implied earlier, the distance-responsive predators are primarily insects that fed on the canopy of the parent. For example, the crown on a mature woody vine, *Dioclea megacarpa*, in lowland deciduous forests in Costa Rica harbors a large population of apparently host-specific erebine noctuid larvae that feed on shoot tips. They harvest as much as 50% of the new branch ends. There is a steady but slow rain of these caterpillars on the forest floor; most return to the crown or wander off to pupate. However, they will feed on any intact shoot tip of a seedling of *D. megacarpa* they encounter. The young plant has sufficient reserves to produce about three main axes; further decapitation kills the seedling (the previous decapitations slowed its development, which probably would also be fatal over a longer period). For this reason, there is no survival of seedlings directly under the parent. But seedlings more than about 5 m from the edge of the parent's crown show only slight damage from these caterpillars (Janzen 1971). The well-known ability of the larvae of the shoot-tip borer *Hypsipyla grandella* to prevent plantation plantings of *Cedrela mexicana* (Meliaceae) on Caribbean islands (e.g., Beard 1942; Holdridge 1943; Cater 1945) is a similar case, and this moth is likely responsible for the wide spacing of adults of this lumber tree in natural forests in Central America.

Rodents may use the parent tree as "flags" indicating the presence of juveniles, and therefore function as distant-responsive predators. For example, in evergreen primary forest on the Osa Peninsula in southwestern Costa Rica, the large and winged seeds of *Huberodendron allenii* (Bombacaceae) are heavily preyed upon by numerous rodents on the forest floor. Any seed placed near the base of the parent, sterile or fertile, is invariably eaten within two nights. Seeds placed more than 50 m from adult *H. allenii* are found much more slowly, some lasting at least 7 days.

Host-specific fungi with resistant spores may also serve as distance-responsive predators (lethal parasites) since they do not wander off in search of more food as the seedling population is decimated. Even fungi without resistant spores may act in this fashion, if remnants of the seedling crop persist from year to year.

Distance-responsive predators may be very effective at producing wide spacing and low density of new adults near old parents, but they should be ineffective at far distances, compared with density-responsive predators. The danger of a "flag" or reservoir of predators in the crown of the parent tree declines rapidly with distance from the parent since the distance-responsive predators do not leave it to search for seeds or seedlings. Second, local patches of juveniles, resulting from overlap or concentration of the seed shadow far from a parent, are less likely to be located by distance-responsive predators than by density-responsive ones. This should be especially important for tree species in early succession. Some environments favor continual parasitism on the adult plant (e.g., tropical wet forest); habitats in these environments should have many widely spaced tree species. While this effect may be magnified for some tree species in seasonal habitats where deciduousness of adults can result in host-specific insects searching for more succulent juveniles, in general, seasonality probably allows more escape closer to the parent.

The three different dispersal distributions in figure 4 are differentially influenced by an increase in predation range by distance-responsive predators. This is shown in figure 7, where changing the survival probability curves from curve P_A to P_B (1) makes I_C the dispersal curve that yields the highest peak rather than I_A as before, (2) increases the spread between peaks in the PRC curves, and (3) lowers the peaks of the PRC curves.

For a given predation range, figure 7 illustrates that changes in the dispersal agents can dramatically alter the location and size of the PRC curve.

Density-responsive Predators

The density-responsive predators should be much superior to distance-responsive ones at causing new adults to appear far from their parent (e.g., figure 3), since the distance-responsive predators do not search past a distance that is representative of some yearly average seed density. No matter how large the seed crop in a given year, or how far the seed from a

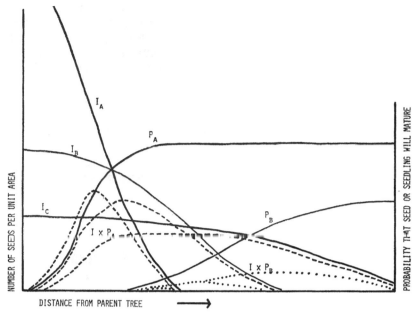

Fig. 7.—The effect of increasing the range of operation of the distance-responsive predators on the different I curves produced by different types of dispersal agents. As more of the seed is dispersed further from the parent (I_A to I_C) the PRC curve generated by the P_A curve is lowered slowly and broadened rapidly. For distance-responsive predators that act at far distances (P_B), however, the same change of I_A to I_C results in an increase in the height of the PRC curve and no gross change in its general shape.

parent, density-responsive predators will pursue seeds and seedlings until their density is so low that search is no longer profitable. This group comprises numerous insects that are host-specific on dispersed seeds and seedlings, and facultatively host-specific mammals and insects. As Phillips (1924) said when describing the reproductive biology of African *Ocotea*, "Aggregation is responsible for very few seedlings [surviving]. This is probably on account of the prevalence of pathogenic fungi and destructive insects in the neighborhood of a large accumulation of fruits." Numerous forest-floor mammals may subsist almost entirely on fallen seeds and fruits at certain times (e.g., peccaries, agoutis, pacas, coatis, deer, rats, etc.) and they tend to concentrate on the fruit under a particular tree (e.g., Kaufman 1962). That they will wander off when full, but return later, creates nearly the same effect as though they were obligatorily host-specific to that tree species. Incidentally, this heavy predation (e.g., African wild pigs killing most, but not all, of the capsules of *Platylophus* they ingest [Phillips 1925]) should not be allowed to obscure the extremely important roles these same mammals play in dispersing undigested or unopened seeds away from the parent (e.g., Phillips 1926a; Smythe 1970). This seed predation is one of the costs the plant pays for dispersal. Fungal, bacterial, and viral diseases can also be density-responsive predators, whether airborne or in-sect-borne. As Fournier and Salas (1967) found, the disease may enter at the point of insect damage, though it is not clear to what degree such diseases are host-specific. Even if not host-specific in the strict sense, they may yield a local epidemic because many hosts are available in the con-centration of seedlings around a parent.

It is obvious that the change in dispersal agents from I_A to I_C in figure 5 can yield a major increase in the distance of new adults from parents, in the face of a given set of density-responsive predators. However, the shallower the dispersal curve, the more seeds will fall past a distance representing critical density for continued predation. As more seeds escape in this manner, the new adults should build up near the parent to the point where seed shadow overlap around parents makes the seed and seed-ling density as high with a I_C dispersal curve as with I_A or I_B curves (fig. 5). However, the better the conditions for searching by predators on seeds and seedlings, and the more time they have available to search, the less effective a flattening of the dispersal curve will be as a means of escape from them.

THE SYSTEM AS A WHOLE

Population Recruitment Surface

The adults of any tree species produce a total seed shadow that may be represented as a gently undulating surface with tall peaks of various shapes centered on the reproductive adults and occasional low rises where seed shadows overlap or dispersal agents concentrate owing to habitat heterogeneity (fig. 8). The general height of the entire surface, and the height of the peaks around the parents, will be a function of the efficiency of the dispersal agents, of the predispersal seed predators, and of the parents' productivity.

The distance- and density-responsive predators should produce an un-dulating probability surface for survival of the seeds. Depressions in this surface, ranging from large pits to shallow basins, should be centered on fertile adults of the tree species in question. The diameters and depths of basins will vary with the proximity of other fertile adults and the suita-bility of the habitat to survival of the predators while they are on the parent, or moving between juveniles. There may also be shallow basins representing increased survival wherever there are low rises in the total seed shadow.

Multiplying these two surfaces together yields a population recruitment surface (its cross-section is the *PRC* of figs. 1–7) which will generally be very low and flat, but also has low "crater rims" ringing the parents and slight rises in areas of multiple seed shadow overlap far from the parents (fig. 9). In determining the impact of any specific seed or seedling preda-tor, dispersal agent, or rise in parental productivity on the population recruitment surface, we must consider that a specific change will often

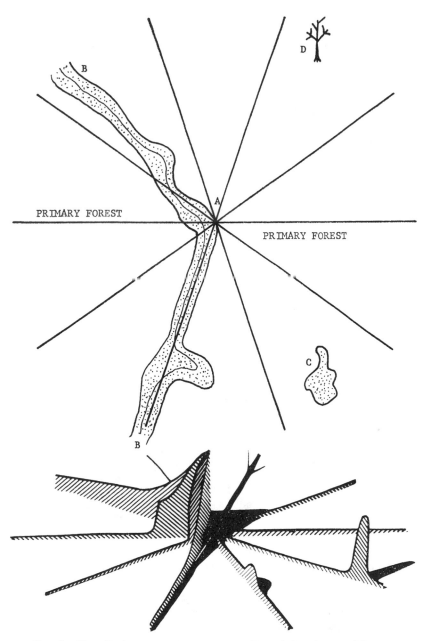

Fig. 8.—Hypothetical complex seed shadow (below) based on seed frequency plotted against distance from the parent for ten radial sectors around the parent at *A* (map shown above). It is assumed that there is only one parent of this species in the general area, and, for example, that the large seed crop is dispersed by birds living in early stages of primary plant succession. Environmental heterogeneity is represented by the river and accompanying narrow strip of primary succession (*B*: stippled), the small patch of primary succession where a large tree was windthrown (*C*), and the large dead tree emergent over the primary forest canopy where birds moving between vegetation in early stages of succession might rest (*D*).

244   Part Two

516   THE AMERICAN NATURALIST

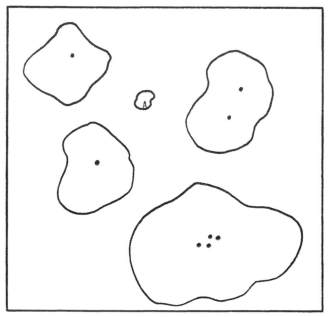

Fig. 9.—Vertical view of the population recruitment surface for a tree species with reproductive adults represented by solid dots. Heavy rings surrounding the adults represent the areas in the habitat where new trees are most likely to appear. The area inside a ring is very unlikely to produce another adult and is thus available to other species of trees, irrespective of the competitive ability of adults or seedlings of the first tree species. The area outside the rings is variably prone to production of new adults, depending on the dispersal agents. Area A represents a local rise in the population recruitment surface due to seed shadow overlap of the four uppermost trees.

yield a compensatory change in another aspect of the surface. For example, an evolutionary change that doubles seed number and thereby enhances predispersal predator satiation (Janzen 1969a), will result in smaller seeds and perhaps more effective dispersal. On the other hand, as mentioned earlier, it may also reduce the number of safe sites in the habitat, lower the number of surviving seedlings, and finally bring about no change in total adult density. Likewise, the evolution of a costly chemical defense of seeds may result in fewer seeds (provided that there is no concomitant increase in parental productivity), which may produce no net change in total juvenile survival to adulthood per parent (although the part of the habitat where these juveniles survive may be changed). The only safe prediction appears to be that conditions favoring the seed and seedling predators are likely to lower the density of reproducing adults of a given species and increase the distance of newly produced adults from their parents.

General Predictions

The general model described in figures 1–7, and the resultant population recruitment surface described above, generate several hypotheses that can be tested by field experimentation or observations. If these hypotheses are found to be generally false for any particular habitat, we must depend on competition, interference, and edaphic interactions to explain the low density of most tree species, high number of tree species, and wide spacing of adult trees in that habitat, be it tropical or temperate.

1. If seeds are placed or planted at various distances from a parent tree at low density (to avoid density-responsive predators), their survival to the well-developed sapling stage should increase with distance from the parent. The mortality agents should be predators on seeds and seedlings. Such observations should be possible on naturally dispersed seeds as well.

2. The percentage of seed mortality on a parent tree should be inversely correlated with its distance to other fertile adults of the same species (in this and previous years). In considering tests of this hypothesis, the distance between seed crops may also be measured in units of time (see example of *Hymenaea courbaril* below). This hypothesis is most relevant for predators that move directly from one seed crop to the next.

3. Where historical accident has produced various densities of reproducing adults of a given species, the average seed mortality on these parents should be an inverse function of the density of reproducing adults. This hypothesis is most relevant to predators that first spread out over the general habitat from an old seed crop, and subsequently find a seed crop.

4. If seeds or seedlings are placed in small patches of various densities in the usual seedling habitat, the survival of any one juvenile should be an inverse function of the number of juveniles in its group.

5. If we categorize the adults of the tree species in an area as either regularly spaced, distributed at random, or clumped, the regularly spaced species should show the best agreement with the first and fourth of these hypotheses.

Problems in Testing the Hypotheses

There are sampling problems as well as problems brought on by alternative population recruitment strategies available to the tree. Some problems are discussed below, primarily to emphasize the complexity of the interaction system being analyzed.

Sampling problems are the biggest impediment at present. The predators must be identified, but are extremely difficult to observe in action. In the case of *Dioclea megacarpa* mentioned previously, an entire crop of 58 seedlings can be killed by 16 "caterpillar-hours" spread over a 4-month period; a caterpillar is on a seedling only 0.6% of the time. A rodent pauses only a few seconds to pick up a seed from the forest floor. Since postdispersal predators exist by harvesting from a small population of small plants with little ability for self-repair, they must have low population densities at most places at most times. If seed or seedling densities are artificially increased to speed up an experiment, there may be a concentration of facultatively host-specific predators that normally feed on other, more abundant, species. Many parasites may have a synergistic effect with predators; for example, Homoptera feeding on phloem of shoot tips may have no direct effect, yet may weaken the plant to later or simultaneous microbial or insect attack. Also, just as Connell (1961) found barnacles weakened by competition to be more susceptible to natural catastrophes, a seedling weakened by mild parasitism by an insect is likely to be an inferior competitor when compared with undamaged sibs (Bullock 1967).

In contemporary tropical communities, usually lightly to heavily disturbed by European types of agriculture, the absence of many dispersal agents and superabundance of others may yield highly unadaptive seed shadows and dispersal timing (e.g., the heavy mortality of undispersed *Cassia grandis* seeds by bruchid beetles where the natural vertebrate dispersal agents have been hunted out [Janzen 1970*a*]). The seed and seedling predators are likely to be affected in the same manner, in addition to the direct destruction of seeds and seedling by humans. Introduced plants often serve as alternate or superior hosts. This results in very different patterns of survival of the predators than is the case in natural forests, at times when their native host plant is nonproductive or seedlings are missing. For example, the introduction of a cotton field into a habitat grossly changes the population structure of the wild cotton-stainer bugs (*Dysdercus* spp.) which are dependent on Malvales for reproduction (Bebbington and Allen 1936; Janzen 1970*c*). Selective logging, extremely common in the tropics, directly changes the density and distribution of adult trees, rendering examination of the population dynamics of adult trees nearly impossible.

Alternative population recruitment strategies, habitat heterogeneity, and differential competitive ability of the seedlings must also be considered when testing these hypotheses.

Allelopathic systems may have two very different effects. Webb, Tracey, and Haydock (1967) have shown that the roots of adult *Grevillea robusta* trees release a compound that kills their own seedlings in Australian rain forests, leading to wide spacing of adults. While this behavior cannot lead to spacing of new adults much past the root territory of the parent tree, it certainly will affect the postdispersal predators. Incidentally, this allelopathy may also serve as an effective escape mechanism from a very effective predispersal seed predator that has great difficulty moving between adults. On the other hand, the "allelopathic" activity of ants on ant-plants in killing other plants around the parent tree often aids in producing local pure stands of *Cecropia,* swollen-thorn acacias, *Tachigalia,* etc. (Janzen 1969*b*).

While geometric distance may be adequate for first approximations to the ecological distance between parent trees, such units are clearly inadequate for specific cases. Two trees 300 m apart along a dry ridge top are clearly not the same ecological distance from each other as they are from a tree 300 m away in the adjacent riparian bottom lands. An individual tree that regularly has a small seed crop may have much less influence on local seed predation than one much farther away that always has a large crop. Alternate host trees may provide enough food to maintain the metabolism of the predator during its dispersal, but not enough for its reproduction. Such trees may greatly shorten the ecological distance between two hosts. Edaphic conditions between two reproductive adults may prevent any maturation of seedlings, but seedlings surviving on seed reserves and some environmental resources may provide a continuous area between adult trees of habitat suitable for the predators.

Parents may compete more with their own seedlings than with those of other trees. Challinor (1968) has shown that different species of temperate forest trees produce differential shortages of inorganic nutrients under their canopies, and it is well known that nutrient requirements vary with the species of tree. This may also be the case in tropical forests to the degree that these trees depend on inorganic nutrients not tied up in the tree-leaf-mycorrhiza-root-tree cycle. However, just as with allelopathy, the negative effect of this competition on seedlings would not extend past the root territory of the individual adult, and might well be counterbalanced by the advantage to the seedling of having its own specific mycorrhizae (Went and Stark 1968) present at the time of germination. Further, the ions required by an insolated reproducing adult may be quite different from those required by a shaded seedling. Finally, I know of no evidence that the shade cast by a parent tree is likely to be more inimical to the growth of its own seedlings than to those of other trees.

Perhaps, the most difficult problem in examining this system is the escape through time by juveniles. Tropical forest trees have a reputation of fruiting at intervals of two or more years (e.g., Ashton 1969; Richards 1952; Janzen 1670b), as do many temperate trees (e.g. Salisbury 1942; Smith 1968; Sharp 1958; Sharp and Sprague 1967). This behavior may yield larger seed crops at greater ecological separation in time, and less predictable time intervals, than is the case with annual fruiting. However, the freedom from predators that may result from this behavior is bought at the cost of the adult not placing seeds in the habitat during many years of its life. Infrequent seed setting is therefore most likely where suitable habitat for seedling survival (in the physiological sense) is not in short supply or erratically available (unless, of course, a dormant seed or stunted seedling can survive until a site becomes available). In short, missing seed crops imply that the geometric distance between parents is not the main relevant variable determining the impact of predators on the presence of a tree species in the habitat. For example, individuals of *Hymenaea courbaril* (Leguminosae) fruit every 3–5 years in Costa Rica, and thus achieve moderate escape from the seed predator *Rhinochenus stigma* (Curculionidae). This behavior means that to the weevil, trees of *H. courbaril* are three to five times as sparse as would be indicated by the total density of adult trees. Where this curculionid weevil is absent (El Salvador north through southern Mexico, and in Puerto Rico), the tree usually fruits every year (Janzen 1970b). It is of interest in this connection that the dipterocarp forests of southeast Asia are well known for both clumped distributions of many species and long intervals between fruiting periods which are sometimes synchronized (Richards 1952; Federov 1968; Ashton 1969). They suffer very heavy seed predation (Ashton 1969), but by seed predators that are most likely generalists feeding on one or more genera of trees.

Since male trees do not bear seed crops, dioecious trees (more common in tropical than in temperate forests [Ashton 1969]) must likewise be censused with caution. For example, dioecious palms in the genus *Scheelea* suffer more than 90% seed predation by the bruchid, *Caryobruchus buscki*, in some Costa Rican forests (Janzen, unpublished). To the bruchid, the density of adult palms is only half that recorded by an observer who does not recognize that only female trees bear seeds. It is tempting to hypothesize that if the palm were hermaphroditic or monoecious, its equilibrium density of adult palms would be considerably lower than at present.

The discussion so far has focused on relatively host-specific predators, but some animals show relatively little response to changes in the density of juveniles of a given species. These animals have the same effect on population structure as a slight lowering of the total seed crop, except where they happen to be locally abundant (for reasons other than the seed crop). Predators of intermediate host-specificity may have an extremely confusing influence on field experiments. If one species of major predator is host-specific on two tree species that fruit slightly out of phase with each other, the first to fruit will have a more negative effect on seed survival of the second than vice versa. When censusing the adult and juvenile trees to evaluate natural experiments, all the prey species of the predator must be recorded.

DISCUSSION

Had biologists generally followed the brief introductory comments in Ridley's (1930) compendium of tropical seed dispersal, the body of the present paper might have been written many years ago.

In almost every plant the greatest number of its seeds fall too near the mother plant to be successful, and soon perish. Only the seeds which are removed to a distance are those that reproduce the species. Where too many plants of one species are grown together, they are apt to be attacked by some pest, insect, or fungus. It is largely due to this . . . that one-plant associations are prevented and nullified by better means for dispersal of the seeds. When plants are too close together, disease can spread from one to the other, and can become fatal to all. Where plants of one kind are separated by those of other kinds, the pest, even if present, cannot spread, and itself will die out, or at least become negligible.

Gillett (1962) and Bullock (1967) reemphasized this. Their brief papers emphasized that (1) previously unoccupied habitats are very important in the production of new populations because an invading species often leaves its predator and pest parasite population behind, allowing it to enter a new habitat more readily than a resident species can expand its population to use the same resources; and (2) that insect attack should strongly affect a seedling's competitive abilities. Van de Pijl (1969) followed with the statement that "dense stands of any species are thinned out by pests, the vacant places becoming occupied by other species," but did not develop the idea further.

These papers, and the plant-herbivore interactions hypothesized here, all point in one direction: as conditions become more favorable for the seed and seedling predators in a habitat (for example, in moving from moist temperate to moist tropical forests), that habitat will support more species of trees because no one species can become common enough to competively oust most of the others. The obvious corollary is that as the number of species of trees in a habitat increases, interspecific competitive ability of seedlings and saplings declines in proportional importance in determining the apportionment of the total biomass among those present. This is not to say that interspecific competition is unimportant in tropical forests; a tree may persist in the face of very heavy predation if the occasional surviving seedling is a very superior competitor, and a tree with very light predation may be a very poor competitor yet survive by repeated trials at establishment. Where the tree is free from predation on juveniles (for some or all of its reproductive life) and a superior competitor as well, we can expect to find conditions closer to those of the few-species stands characteristic of temperate forests. The single-species stands in tropical mangrove swamps are excellent examples of this. Numerous seed samples of the large and very abundant seeds or undispersed seedlings of red mangrove (*Rhizophora mangle*), black mangrove (*Avicennia nitida*), mangle pinuela (*Pelliciera rhizophora*), and white mangrove (*Laguncularia racemosa*) in Costa Rica have failed to reveal host-specific seed predators. The seeds are probably high in tannin content as are the vegetative parts of the plants from these four widely separated families (e.g., Allen 1956). To exist in such pure stands and bear seeds almost continually, the plants must have extremely good chemical defenses against insects. Despite the successional nature of the mangrove community, mangrove seeds or seedlings cannot even escape in space since an earlier sere than any given stage of mangrove succession is usually present within a few meters as the mangrove forest advances into the estuary.

A word of caution is in order regarding the spacing influence of seed and seedling predators on new adult trees, as supplementary to the effect of predators on the density of the adult tree population. Ashton (1969) and Poore (1968) have recently stressed that the trees of many species of southeast Asian rain forests are distributed with various amounts of contagion, but their figures do not distinguish between reproductive and sterile adult-sized individuals, or between males and females. In some cases, subadults are not distinguished from adults. Secondly, if the location of all members (seeds, saplings, adults) of the population of a given species of tree are recorded over a short period of time including a period of reproduction, I predict that their data will show that adults are not nearly so clumped as the total population (over all age classes). Nevertheless, it must be recognized that even a forest with all species having clumped distributions would have its equilibrium number of tree species greatly increased by a large complex of host-specific predators on juvenile plants.

The comments in this paper apply directly to other organisms where the survival of a juvenile to maturity requires the death of a long-lived adult at the same point in space, and where the juveniles are very susceptible to predation. Sessile marine invertebrates (Paine 1966)and ant colonies fall in this category. As Connell (1961, 1963) stresses, regular spacing of the adults of these types of organisms is generally attributed to strong intraspecific interference through a third agent, the predator.

Relative freedom from predation, and therefore a low number of tree species in the forest, may come about in at least three major ways.

1. Both Southwood (1961) and Gillet (1962) have emphasized that invading plant species may leave their predators and parasites behind. If my observations of Puerto Rican forests are representative (the situation on Hawaii is apparently similar [W. H. Hatheway, personal communication]), tree population structures on tropical islands differ strikingly from those on adjacent mainlands. Here, trees such as *Trophis* (Moraceae), have extremely dense stands of seedlings and saplings under the canopy of the parent; these forests lack the native rodents and large terrestrial birds (e.g., *Tinamus, Crypturellus, Crax*) that thoroughly remove *Trophis* seeds from under the parent in Costa Rican lowland forests. Puerto Rican forests have many fewer species of adult trees per unit area than do mainland Central American areas of similar weather regime. My cursory plot censuses in Puerto Rico indicated a structure much more similar to hardwood forests in the southeastern United States than to Central America, in that seedlings and saplings of the canopy member trees are very common in the vicinity of putative parents.

In the second growth vegetation in Puerto Rico, a similar case is presented by *Leucaena glauca*, a mimosaceous shrub that loses upward of 90% of its seed crop to bruchid beetle predation in central America, and occurs there as a scattered adult with rare seedlings. In Puerto Rico, no bruchids attack this species, virtually every seed is viable, the adults are surrounded by dense stands of seedlings and intermediate-aged juveniles, and are extremely common.

2. In predation-rich habitats, newness to the habitat may be an adequate defense mechanism for some species for a period of time. The critical point, however, is not so much that the previous predators were left behind, but that the new ones cannot deal with the chemical defenses of the seeds and seedlings, or lack the behavioral traits to attempt such attacks. This is unlikely to be a permanent condition, but when it lasts for a long time, it may

result in a very common tree species. One example is provided by ferns; they are notorious for being extremely free from insect attack both as juveniles and adults (which can only be for chemical reasons), and for occurring in large pure stands in tropical forests. Another is the tree *Pentaclethra macroloba*. A mimosaceous legume, it is extremely abundant in the wet lowland forests of northeastern Costa Rica. Mimosaceous legumes with finely divided leaves are generally very rare in tropical lowland ever-green forests and *P. macroloba* has likely left behind a major set of preda-tors and parasites, if it came originally from the deciduous forests where mimosaceous legumes are very abundant. The large seeds, produced vir-tually throughout the year by the population, are preyed upon only very slightly by squirrels before dispersal, and by terrestrial rodents after dis-persal. Its seeds and seedlings have only very slight predation by insects, when compared to other forest trees, and the insects involved appear to be general foragers. When a hole opens in the forest canopy, the chance of there being a seedling of *P. macroloba* below it is much greater than for any other single species in the forest. I predict that the introduction of an insect host-specific on *P. macroloba* seeds would both reduce the population of adults dramatically and allow either invasion of other species, or ex-pansion of the adult populations of resident species. This latter event is unlikely, however, since pressure from predators, rather than competition with *P. macroloba,* is probably holding their populations down.

The *Pentaclethra* example stresses the importance of the invaded habitat having different predators from the habitat of origin. It should be much harder for a legume to invade a habitat rich in species attacked by bruchids than one in which bruchids are rare. A corollary of this is that a resident species may prevent the invasion of a closely related species, by serving as the source of a predator that finds seeds of the invader to be suitable prey (Janzen 1970a). To become established, the invading tree may have to exist at an even lower density than in its native habitat. Federov (1966) has recently stressed the sympatric existence of congeners as characteristic of tropical forests (although sympatry of congeners is certainly characteristic of temperate forests, too). In the light of the activity of the predators, this is easily understood. First, the density of congeners is held low enough so that they have no chance of directly com-petitively excluding each other. Second, they must be species that either do not share major seed and seedling predators, or can survive at the lower densities that will be produced by a predator that treats both as one species.

The minimal density at which a tree population can exist is of great importance for understanding how many tree species can ultimately be packed into a forest habitat. The major deterrent to low density of adults appears to be reduction in outcrossing. Several authors have concluded that self-pollination is probably the rule in tropical rain forest trees (Baker 1959; Corner 1954; Richards 1952, 1969). There are numerous pollinators in tropical forests with the ability to provide outcrossing at long interplant distances (Ashton 1969; Janzen 1968, unpublished). Second, contrary to popular belief, there is a major reason why outcrossing is of utmost importance in the relatively uniform climate of tropical forests. In more stringent and unpredictable climates, the physical climate is the major challenge (aside from intertree competition, a problem in all forests), and in great part can be met through vegetative plasticity as well as through genetic change. Most important, a genotype optimal for weather conditions now is most likely to be optimal or nearly optimal for a considerable number of generations (at least until the weather changes). However, the challenge of a seed or seedling predator can be met only through behavioral or chemical changes, the success of which cannot be monitored directly by the parent. Such change can be brought about by genetic change alone. The new challenge of a predator capable of breaching the current chemical defenses of the adult or its seeds may occur abruptly at any time, and will greatly lower the fitness of the current tree biochemical phenotype. In other words, the more favorable the physical environment to the predators, the more frequently in evolutionary time the chemical defenses of the plant will have to be modified through genetic change if the plant is to persist in the community, and therefore the more important will be outcrossing. Van Steenis's (1969) recent suggestion that extinction of trees ''is a common feature in tropical rain forest'' and Ashton's (1969) comment that interspecific tree hybrids do extremely poorly in rain forest thus take on new meaning.

3. Temporal heterogeneity and unpredictability of the physical environment may both lead to freedom from predation on juvenile trees at certain times. This is best reflected in the low number of tree species in temperate forests. Weather changes of a regular type are indirectly responsible for regular large fluctuations in insects that prey upon seeds and seedlings. To the degree that an adult tree can produce juvenile plants when the population of its predators is low, the juveniles will have only intertree competition and edaphic conditions to deal with (a major challenge irrespective of intraspecific seed and seedling proximity). While a tropical tree may put a new crop of seeds into the habitat once a year (similar to temperate trees), the predators may have as many as 12 months in the year to search for food, in contrast with the considerably shorter period in temperate forests or strongly seasonal tropical ones. The occasional unpredictably hard seasons for predators (e.g., Barrett 1931; Parnell 1966) may result in a wave of juveniles of a tree species passing through the habitat, especially if it is coupled with a very large crop of seeds (Smith 1968). This again leads to conditions in which adult tree community composition is primarily a function of the competitive ability of the seedlings and saplings, allowing a few competitively superior tree species to dominate the community.

SUMMARY

A high number of tree species, low density of adults of each species, and long distances between conspecific adults are characteristic of many lowland tropical forest habitats. I propose that these three traits, in large part, are the result of the action of predators on seeds and seedlings. A model is presented that allows detailed examination of the effect of different predators, dispersal agents, seed-crop sizes, etc. on these three traits. In short, any event that increases the efficiency of the predators at eating seeds and seedlings of a given tree species may lead to a reduction in population density of the adults of that species and/or to increased distance between new adults and their parents. Either event will lead to more space in the habitat for other species of trees, and therefore higher total number of tree species, provided seed sources are available over evolutionary time. As one moves from the wet lowland tropics to the dry tropics or temperate zones, the seed and seedling predators in a habitat are hypothesized to be progressively less efficient at keeping one or a few tree species from monopolizing the habitat through competitive superiority. This lowered efficiency of the predators is brought about by the increased severity and unpredictability of the physical environment, which in turn leads to regular or erratic escape of large seed or seedling cohorts from the predators.

ACKNOWLEDGMENTS

This study was supported by NSF grants GB-5206, GB-7819, and GB-7805, and the teaching program of the Organization for Tropical Studies. The manuscript has profited greatly through discussions with the staff and students in that organization. Special thanks are due to H. G. Baker, A. Bradshaw, R. K. Colwell, J. H. Connell, W. H. Hatheway, E. Leigh, R. Levins, R. C. Lewontin, M. Lloyd, C. D. Michener, G. H. Orians, N. J. Scott, N. Smythe, R. R. Sokal, and J. H. Vandermeer.

LITERATURE CITED

Allen, P. H. 1956. The rain forests of Golfo Dulce. Univ. Florida Press, Gainesville. 417 pp.

Ashton, P. S. 1969. Speciation among tropical forest trees: some deductions in the light of recent evidence. Biol. J. Linnean Soc. London 1:155–196.

Baker, H. G. 1959. Reproductive methods as factors in speciation in flowering plants. Cold Spring Harbor Symp. Quant. Biol. 24:177–191.

Barbehenn, K. R. 1969. Host-parasite relationships and species diversity in mammals: a hypothesis. Biotropica 1:29–35.

Barger, J. H., and R. H. Davidson. 1967. A life history study of the ash seed weevils, *Thysanocnemis bischoffi* Blatchley and *T. helvole* Leconte. Ohio J. Sci. 67:123–127.

Barrett, L. I. 1931. Influence of forest litter on the germination and early survival of chestnut oak, *Quercus montana* Willd. Ecology 12:476–484.

Beard, J. S. 1942. Summary of silvicultural experience with *Cedrela mexicana* Roem., in Trinidad. Caribbean Forest. 3:91–102.

Bebbington, A. G., and W. Allen. 1936. The food-cycle of *Dysdercus fasciatus* in acacia savannah in Northern Rhodesia. Bull. Entomol. Res. 27:237–249.

Black, G. A., T. Dobzhansky, and C. Pavan. 1950. Some attempts to estimate species diversity and population density of trees in Amazonian forests. Bot. Gaz. 111:413–425.

526   THE AMERICAN NATURALIST

Breedlove, D. E., and P. R. Ehrlich. 1968. Plant-herbivore coevolution in lupines and lycaenids. Science 162:671–672.

Brncic, D. 1966. Ecological and cytogenetic studies of *Drosophila flavopilosa*, neotropical species living in *Cestrum* flowers. Evolution 20:16–29.

Bullock, J. A. 1967. The insect factor in plant ecology. J. Indian Bot. Soc. 46:323–330.

Bush, G. L. 1969. Mating behavior, host specificity, and the ecological significance of sibling species in frugivorous flies of the genus *Rhagoletis* (Diptera—Tephritidae). Amer. Natur. 103:669–672.

Cater, J. C. 1945. The silviculture of *Cedrela mexicana*. Caribbean Forest. 6:89–113.

Challinor, D. 1968. Alteration of surface soil characteristics by four tree species. Ecology 49:286–290.

Connell, J. H. 1961. Effects of competition, predation by *Thais lapillus*, and other factors on natural populations of the barnacle *Balanus balanoides*. Ecol. Monogr. 31:61–104.

———. 1963. Territorial behavior and dispersion in some marine invertebrates. Jap. Soc. Population Ecol. 5:87–101.

Corner, E. J. H. 1954. The evolution of tropical forests. *In* J. Huxley, A. C. Hardy, and E. B. Ford [ed.], Evolution as a process. 2d ed. Humanities, New York. 367 pp.

Costa Lima, A. M. 1967–1968. Quarto catalogo dos insectos que vivem nas plantas do Brasil, seus parasitos e predadores. Pts. 1 and 2. D'Araujo e Silva et al. [ed.] Dept. Def. Insp. Agro., Minist. Agric., Rio de Janeiro. 2215 p. total.

Coyne, J. F. 1968. *Laspeyresia ingens*, a seed worm infesting cones of longleaf pine. Ann. Entomol. Soc. Amer. 61:1116–1122.

Croat, T. 1970. Seasonal flowering behavior in Central Panama. Ann. Mississippi Bot. Gardens (in press).

DeLeon, D. 1941. Some observations on forest entomology in Puerto Rico. Caribbean Forest. 2:160–163.

Dumbleton, L. J. 1963. The biology and control of *Coleophora* spp. (Lepidoptera—Coleophoridae) on white clover. New Zealand J. Agr. Res. 6:277–292.

Eyles, A. C. 1964. Feeding habits of some Rhyparochrominae (Heteroptera: Lygaeidae) with particular reference to the value of natural foods. *Roy. Entomol. Soc. London, Trans.* 116:89–114.

Federov, A. A. 1966. The structure of the tropical rain forest and speciation in the humid tropics. J. Ecol. 54:1–11.

Fournier, L. A., and S. Salas. 1967. Tabla de vida el primer año de la población de *Dipterodendron costaricense* Radlk. Turrialba 17:348–350.

Gillett, J. B. 1962. Pest pressure, an underestimated factor in evolution. Systematics Association Pub. No. 4. Pp. 37–46.

Haffer, J. 1969. Speciation in Amazonian forest birds. Science 165:131–137.

Hardwick, D. F. 1966. The life history of *Schinia niveicosta* (Noctiudae). J. Lepidoptera 20:29–33.

Harper, J. L., J. N. Clatworthy, I. H. McNaughton, and G. R. Sagar. 1961. The evolution and ecology of closely related species living in the same area. Evolution 15:209–227.

Hinckley, A. D. 1961. Comparative ecology of two beetles established in Hawaii: an anthribid, *Araecerus levipennis*, and a bruchid, *Mimosestes sallaei*. Ecology 42:526–532.

Holdridge, L. R. 1943. Comments on the silviculture of *Cedrela*. Carribbean Forest. 4:77–80.

Janzen, D. H. 1968. Reproductive behavior in the Passifloraceae and some of its pollinators in Central America. Behavior 32:33–48.

———. 1969a. Seed-eaters versus seed size, number, toxicity and dispersal. Evolution 23:1–27.

———. 1969b. Allelopathy by myrmecophytes: the ant *Azteca* as an allelopathic agent of *Cecropia*. Ecology 50:147–153.

————. 1970a. *Cassia grandis* L. beans and their escape from predators: a study in tropical predator satiation. Ecology (in press).

————. 1970b. Escape in time by *Hymenaea courbaril* (Leguminosae from *Rhinochenus stigma* (Curculionidae). Ecology (submitted for publication).

————. 1970c. Escape of *Sterculia apetala* seeds from *Dysdercus fasciatus* bugs in tropical deciduous forest. Ecology (in press).

————. 1971. Predator escape in time and space by juveniles of the vine, *Dioclea megacarpa*, in tropical forests. Amer. Natur. (in press).

Kaufman, J. H. 1962. Ecological and social behavior of the coati *Nasau nasau* on Barro Colorado Island, Panama. Univ. of Calif. Pub. Zool. 60:95–222.

Knab, F., and W. W. Yothers. 1914. Papaya fruit fly. J. Agr. Res. 2:447–453.

Lowe-McConnell, R. H. 1969. Speciation in tropical freshwater fishes. Biol. J. Linnean Soc. London 1:50–75.

MacArthur, R. H. 1969. Patterns of communities in the tropics. Biol. J. Linnean Soc. London 1:19–30.

Morris, D. 1962. The behavior of green acouchi (*Myoprocta pratti*) with special reference to scatter-hoarding. Zool. Soc. London, Proc. 139:701–732.

Murdoch, W. W. 1969. Switching in general predators: experiments on predator specificity and stability of prey populations. Ecol. Monogr. 39:335–354.

Myers, J. G. 1927. Ethological observations on some Pyrrocoridae of Cuba. Ann. Entomol. Soc. Amer. 20:279–300.

Oppenheimer, J. R. 1968. Behavior and ecology of the white-faced monkey, *Cebus capucinus*, on Barro Colorado Island, C.Z. Ph.D. thesis. Univ. Michigan, Ann Arbor.

Oppenheimer, J. R., and G. E. Lang. 1969. *Cebus* monkey: effect on branching of *Gustavia* trees. Science 165:187–188.

Paine, R. T. 1966. Food web complex and species diversity. Amer. Natur. 100:65–75.

Parnell, J. R. 1966. Observations on the population fluctuations and life histories of the beetles *Bruchidius ater* (Bruchidae) and *Apion fuscirostre* (Curculionidae) on broom (*Sarothamnus scoparius*). J. Amer. Ecol. 35:157–188.

Phillips, J. F. V. 1924. The biology, ecology, and sylviculture of "stinkwood", *Ocotea bullata* E. Mey: introductory studies. South Afr. J. Sci. 21:275–292.

————. 1925. *Platylophus trifoliatus* D. Don: a contribution to its ecology. South Afr. J. Sci. 22:144–160.

————. 1926a. General biology of the flowers, fruits, and young regeneration of the more important species of the Kenyan forests. South Afr. J. Sci. 23:366–417.

————. 1926b. Biology of the flowers, fruits, and young regeneration of *Olinia cymosa* Thub. Ecology 7:338–350.

Pipkin, S. B., R. L. Rodriquez, and J. Leon. 1966. Plant host specificity among flower-feeding neotropical *Drosophila* (Diptera: Drosophilidae). Amer. Natur. 100:135–156.

Poore, M. E. D. 1968. Studies in Malaysian rainforest. I. The forest on Triassic sediments in Jengka Forest Reserve. J. Ecol. 56:143–196.

Prevett, P. F. 1967. Notes on the biology, food plants and distribution of Nigerian Bruchidae (Coleoptera), with particular reference to the northern region. Bull. Entomol. Soc. Nigeria 1:3–6.

Richards, P. W. 1952. The tropical rainforest. Cambridge Univ. Press, New York. 450 p.

————. 1969. Speciation in the tropical rainforest and the concept of the niche. Biol. J. Linnean Soc. London 1:149–153.

Ridley, H. N. 1930. The dispersal of plants throughout the world. L. Reeve & Co., Ashfort, England. 744 p.

Salisbury, E. J. 1942. The reproductive capacity of plants. Bell, London. 244 p.

Schaefer, C. H. 1962. Life history of *Conophthorus radiatae* (Coleoptera: Scolytidae) and its principal parasite, *Cephalonomia utahensis* (Hymenoptera: Bethylidae). Ann. Entomol. Soc. Amer. 55:569–577.

528   THE AMERICAN NATURALIST

Schaefer, C. H. 1963. Factors affecting the distribution of the Monterey pine cone beetle (*Conophthorus radiatae* Hopkins) in central California. Hilgardia 34:79–103.

Sharp, W. M. 1958. Evaluating mast yields in the oaks. Pennsylvania State Univ. Agr. Exp. Sta. Bull. 635:1–22.

Sharp, W. M., and V. G. Sprague. 1967. Flowering and fruiting in the white oaks. Pistillate flowering, acorn development, weather, and yields. Ecology 48:243–251.

Smith, C. C. 1968. The adaptive nature of social organization in the genus of tree squirrels, *Tamiasciurus*. Ecol. Monogr. 38:31–63.

Smythe, N. 1970. Relationships between fruiting seasons and seed dispersal methods in a neotropical forest. Amer. Natur. 104:25–35.

Southwood, T. R. E. 1961. The number of species of insects associated with various trees. J. Anim. Ecol. 30:1–8.

Spight, T. M. 1967. Species diversity: a comment on the role of the predator. Amer. Natur. 101:467–474.

Struhsaker, T. T. 1967. Ecology of vervet monkeys (*Cercopithecus aethiops*) in the Masai-Amboseli Game Reserve, Kenya. Ecology 48:891–904.

Sweet, M. 1964. The biology and ecology of the Rhyparochrominae of New England (Heteroptera: Lygaeidae). Entomol. Amer. 43:1–124; 44:1–201.

Van der Pijl, L. 1969. Evolutionary action of tropical animals on the reproduction of plants. Biol. J. Linnean Soc. London 1:85, 96.

Van Steenis, C. G. G. J. 1969. Plant speciation in Malasia, with special reference to the theory of non-adaptive saltatory evolution. Biol. J. Linnean Soc. London 1:97–133.

Webb, L. J., J. G. Tracey, and K. P. Haydock. 1967. A factor toxic to seedlings of the same species associated with living roots of the non-gregarious subtropical rain forest tree *Grevillea robusta*. J. Appl. Ecol. 4:13–25.

Went, F. W., and N. Stark. 1968. Mycorrhiza. Bioscience 18:1035–1038.

Wickens, G. E. 1969. A study of *Acacia albida* Del. (Mimosoideae). Kew Bull. 23:181–202.

Yonke, T. R., and J. T. Medler. 1968. Biologies of three species of *Alydus* in Wisconsin. Ann. Entomol. Soc. Amer. 61:526–531.

Vol. 111, No. 982   The American Naturalist   November–December 1977

MECHANISMS OF SUCCESSION IN NATURAL COMMUNITIES AND THEIR ROLE IN COMMUNITY STABILITY AND ORGANIZATION

Joseph H. Connell and Ralph O. Slatyer

Department of Biological Sciences, University of California, Santa Barbara,
California 93106; and Department of Environmental Biology, Research School of
Biological Sciences, Australian National University, Canberra, Australia

It is in changing that things find repose. [Heraclitus]

Succession refers to the changes observed in an ecological community following a perturbation that opens up a relatively large space. The earliest studies described the sequence of species that successively invade a site (Cowles 1899; Cooper 1913; Clements 1916); more recent studies have described changes in other characteristics such as biomass, productivity, diversity, niche breadth, and others (see review in Odum 1969). In this paper we will discuss only changes in species composition.

Clements (1916) proposed a theory of the causes of succession so satisfying to most ecologists that it has dominated the field ever since (see Odum 1969). Although doubts were raised earlier (Gleason 1917; Egler 1954), queries and objections have recently increased in number (McCormick 1968; Connell 1972; Drury and Nisbet 1973; Colinvaux 1973; Niering and Goodwin 1974, etc.). This paper will review the theory and the evidence and propose alternative testable models. We consider first the mechanisms which determine the changes during succession and second the relationships between succession and community stability and organization.

The mechanisms producing the sequence of species have not been elucidated for several reasons. First, direct evidence is available only for the earliest stages when many species are short lived and amenable to experimentation (Keever 1950). The sequence later in succession has not been directly observed for the obvious reason that these later-appearing species persist for much longer than the usual ecological study or even than the investigator. Therefore, the later sequence has had to be reconstructed from indirect evidence of various sorts, such as by tabulating the vegetation found on sites abandoned after cultivation at various past times (Oosting 1942) or by dating the living and dead trees on one site (Cooper 1913; Henry and Swan 1974).

Second, some possible mechanisms have been ignored, particularly the effects of grazing animals. The study of succession has in the past been carried out mainly by persons working solely with plants. This can be justified, in the sense

Amer. Natur. 1977. Vol. 111, pp. 1119–1144.

that plants not only are the primary producers but also usually constitute both the greatest amount of biomass and the structural form of a community (sessile animals also play this role in many aquatic communities). However, it has meant that the mechanisms conceived have usually been restricted to the interactions of plants with the physical environment or with other plants (Langford and Buell 1969). The interactions with organisms that consume plants have always been included as one of the many factors influencing succession, but again most of the attention has been given to the consumers involved in the cycling of mineral nutrients, particularly the decomposers such as microorganisms and fungi, rather than to animal herbivores.

The result has been to focus attention on the resources of plants so that the biological interaction regarded as being of overriding importance is competition. This has coincided with the development of a theory of community structure based almost entirely on competition (Hutchinson 1958; MacArthur 1972 and previous work; Levins 1968; Vandermeer 1972; for a contrasting view, see Connell 1975). As a result the most recent critical reviews of ecological succession have designated physical stresses to plants and competition for resources between plants as the main mechanisms determining the course of succession (Drury and Nisbet 1973; Colinvaux 1973; Horn 1974). In this paper we suggest that in addition to the competitive interactions between plants or sessile animals, interactions with herbivores, predators, and pathogens are of critical importance to the course of succession.

Third, the mechanisms that determine succession have not been defined clearly or stated in the form of hypotheses testable by controlled field experiments. In this paper we have tried to do this as well as to suggest certain experiments as tests.

We will direct our attention here to the succession of species that occupy the surface and modify the local physical conditions, e.g., plants and sessile aquatic animals. Other organisms, such as herbivores, predators, pathogens, etc., will be included only when they affect the distribution and abundance of the main occupiers of space. Species that depend upon the shelter of the larger occupants (e.g., understory species of plants, various animals such as those that live beneath mussel beds, etc.) will not be dealt with. We define a community as the set of organisms that occur together and that significantly affect each other's distribution and abundance. It is the interactions that make a community a unit worthy of study. Lastly, we will consider only those changes in species composition that would occur in the absence of significant trends in the physical regime, or in Tansley's (1935) terminology, "autogenic" succession.

MECHANISMS DETERMINING THE SEQUENCE OF SPECIES

Three Alternative Models

Figure 1 describes three different models of mechanisms that would bring about a successional change after a perturbation, assuming no further significant changes in the abiotic environment. Between the first two steps in the

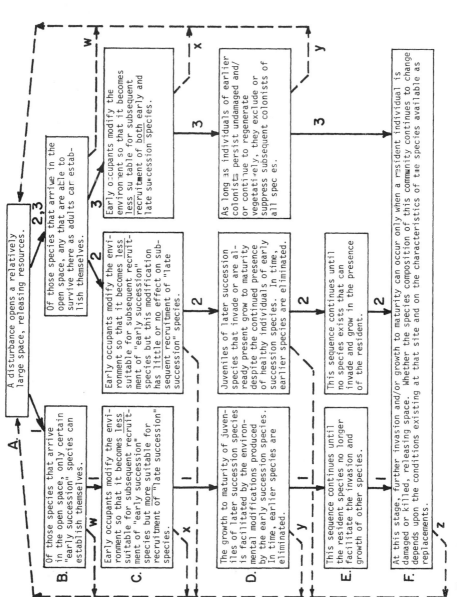

Fig. 1.—Three models of the mechanisms producing the sequence of species in succession. The dashed lines represent interruptions of the process, in decreasing frequency in the order w, x, y, and z.

diagram (A to B) there is a major dichotomy between alternative models of succession. Model 1 assumes that only certain "early successional" species are able to colonize the site in the conditions that occur immediately following the perturbation. Models 2 and 3 assume that any arriving species, including those which usually appear later, may be able to colonize. Egler (1954) was the first to distinguish this latter process, which he calls "initial floristic composition," from the "relay floristics" of model 1. This dichotomy emphasizes the fundamental difference between the original conception of succession proposed by Clements (1916) and the alternative ones described here. We will refer to model 1 as the "facilitation" model.

Up to this point all models agree that certain species will usually appear first because they have evolved "colonizing" characteristics such as the ability to produce large numbers of propagules which have good dispersal powers, to survive in a dormant state for a long time once they arrive (Marks 1974), to germinate and become established in unoccupied places, and to grow quickly to maturity. They are not well adapted to germinating, growing, and surviving in occupied sites, where there is heavy shade, deep litter, etc., so that offspring seldom survive in the presence of their parents or other adults. Thus in all models, early occupants modify the environment so that it is unsuitable for further recruitment of these early-succession species.

Where the models differ is in the mechanisms that determine how new species appear later in the sequence. In the facilitation model the early-succession species modify the environment so that it is more suitable for later-succession species to invade and grow to maturity (steps C and D in fig. 1). In describing how the exposed surface of a landslide may be recolonized, Whittaker (1975, p. 171) outlined the steps of the facilitation model: "One dominant species modified the soil and microclimate in ways that made possible the entry of a second species, which became dominant and modified environment in ways that suppressed the first and made possible the entry of a third dominant, which in turn altered its environment." This sequence continues until the resident species no longer modifies the site in ways that facilitate the invasion and growth of a different species (step E).

In model 2, the modifications wrought on the environment by the earlier colonists neither increase nor reduce the rates of recruitment and growth to maturity of later colonists (steps C and D). Species that appear later are simply those that arrived either at the very beginning or later and then grew slowly. The sequence of species is determined solely by their life-history characteristics. In contrast to the early species, the propagules of the later ones are dispersed more slowly and their juveniles grow more slowly to maturity. They are able to survive and grow despite the presence of early-succession species that are healthy and undamaged. As stated by MacArthur and Connell (1966, p. 168), "In the case of forest succession, each species is able to stand deeper shade than the previous one, and as the forest grows the canopy becomes thicker and casts an even deeper shade. In this new, deeper shade other species are more successful . . . . [T]olerant species are those that are successful in shade. As expected the climax forests are composed of the most tolerant species." The end point is reached when the most shade-tolerant species available occupies the site and casts shade so deeply (or removes other resources to such a low level) that its own offspring cannot survive. Although we have used shade tolerance as an example here, tolerance to other environmental factors, such as moisture, nutrients, allelochemicals, grazing, etc., may be equally or more important in other circumstances. We will refer to this as the "tolerance" model. It serves as an intermediate case between the first and third models.

In contrast to the first model, the third holds that once earlier colonists secure the space and/or other resources, they inhibit the invasion of subsequent colonists or suppress the growth of those already present. The latter invade or grow only when the dominating residents are damaged or killed, thus releasing resources (steps C and D). We will refer to model 3 as the "inhibition" model.

At this point (step D) in model 3, the possibility exists that the very first colonists, by interfering with further invasion, may have prevented any further succession. In contrast to the other two, in model 3 the species of individual that replaces a dying resident need not have life-history characteristics different from the original resident. It need not be a different species adapted to conditions modified in a particular way by former residents (model 1) or one that is more tolerant of reduced levels of resources (model 2). This being the case, it is possible that a resident may be replaced by another of the same species or of a different species also having "early succession" characteristics. Then the traditional successional sequence won't occur. If, on the other hand, the replacement happens to be a species having "late succession" characteristics, then the traditional successional sequence will be observed. Since the early-succession species are shorter lived, they will be replaced more often than would the longer-lived late-succession species. If propagules of these later species are available for invasion, then after several years of transitions the latter species will tend to accumulate, with the result that the early species will gradually decrease in relative abundance. In model 3, the great tolerance of late-succession species is of importance, not in allowing net growth beneath earlier species (as suggested in model 2), but in allowing the late species to survive through long periods of suppression. In effect, tolerance compensates for lower vagility of propagules, increasing the chances that a seedling of a late species will be available on the site to replace a dying earlier individual. In this way the operation of the inhibition model 3 will produce a succession of species leading from short-lived to long-lived species, as is commonly observed.

In summary, the mechanisms producing the sequence of species observed are as follows. In all models the earlier species cannot invade and grow once the site is fully occupied by their own or later species. However, the models differ in the way later species become established after their propagules arrive. In the "facilitation" model 1, the later ones can become established and grow only after earlier ones have suitably modified the conditions. In the "tolerance" model 2, later species are successful whether earlier species have preceded them or not; they can become established and grow to maturity in the presence of other species because they can grow at lower levels of resources than can earlier ones. In the "inhibition" model 3, later species cannot grow to maturity in the presence of earlier ones; they appear later because they live longer and so gradually accumulate as they replace earlier ones. Another distinction between the models is in the cause of death of the early colonists. In models 1 and 2, they are killed in competition with the later species. The latter grow up and shade or otherwise deprive the former of resources. In model 3, however, this cannot happen; the early species are killed by local disturbances caused by physical extremes or natural enemies such as herbivores, parasites, or pathogens. We will now consider the evidence for each model.

Evidence

The mechanisms of the facilitation model probably apply to most heterotrophic successions of consumers feeding on carcasses, logs, dung, litter, etc. Savely (1939) pointed out that certain insect species that bore into logs must precede others that attack the inner tissues. Similarly some species of insects appear in dung and carcasses only after these have been decomposed to a certain degree by earlier colonists (Payne 1965). No experimental investigation has been carried out to demonstrate the details of the process, but the evidence seems to support the application of this model. In the absence of primary producers such localized successions finally exhaust the energy source.

Evidence in support of model 1 for autotrophs comes from primary successions on newly exposed surfaces. For example, Crocker and Major (1955) and Lawrence et al. (1967) have suggested that the characteristics of soils newly exposed by a retreating Alaskan glacier probably make the establishment of plants extremely difficult. However, those "pioneer" species that are able to colonize will ameliorate these conditions, reducing pH, increasing nitrogen content, adding a layer of organic soil over the hardpan, reducing desiccating winds, etc. Seedlings of spruce trees then appear in these new conditions, seldom if ever in the original exposed sites (Reiners et al. 1971). Therefore it is reasonable to conclude that the spruce could not have invaded until the pioneers had ameliorated the original conditions. A second example of the operation of model 1 in primary succession is the colonization of sand dunes on lake shores (Cowles 1899; Olson 1958). The pioneer plants stabilize the moving sands which otherwise would not be suitable for colonization by later-appearing species. More conclusive evidence would require a set of field experiments, manipulating separately the various factors to determine which contributed most to the establishment of the later successional species. However, even without such experiments these cases seem to support model 1.

Field experimental tests of the facilitation model are few. The only terrestrial example we have found involves the giant saguaro cactus. Experimental broadcasting of seeds, transplanting of seedlings, and observations of survival of natural seedlings showed that they survive only in the shade of other species of "nurse plants," or, in a few instances, in the shade of rocks (Niering et al. 1963; Steenbergh and Lowe 1969; Turner et al. 1969). As in the other instances described, the mechanisms of model 1 apply in the early stages of colonization of very rigorous extreme environments. Whether this model applies to replacements at later stages of terrestrial succession remains to be seen; we are not aware of any such evidence at present.

In a review of marine benthic successions, Connell (1972) searched for evidence from field experiments supporting model 1. The only evidence he found was that of Scheer (1945), whose experimental evidence indicates that sessile marine animals (hydroids) probably attached more readily to glass plates immersed in the sea if these had previously been coated by bacteria in the laboratory. Another possible example of this model is provided by the mussel *Mytilus* which seldom appears very early in recolonization of rocky shores. Bayne (1965) and others have noted that larval mussels often attach preferentially to filaments provided by previously settled algae, hydroids, etc. However, Seed (1969) has found that they do not require such organisms and will attach to rough surfaces or crevices in bare rock. Harger and Tustin (1973) suggest that the large alga *Eklonia* may colonize only after filamentous organisms have become established. In none of the many other marine examples reviewed (Connell 1972) was there evidence that earlier species facilitated the establishment of later ones.

The evidence in support of the first step (B in fig. 1) in models 2 and 3 is that late successional species of land plants are often able to become established without any preparation of the site by earlier species (Egler 1954; Drury and Nisbet 1973). The later steps (C to E) of model 2 require that later species be able to invade and grow at lower levels of resources than earlier species. This is usually expressed in terms of greater tolerance by later species to shade or to reduction in other resources. In effect, this model specifies that later species are superior to earlier ones in exploiting resources. Even if the earlier ones reduce resources enough to depress the rate of growth of the later species, the latter will still be able to grow to maturity in the presence of the former. Necessary and sufficient evidence in support of model 2 would consist of observations or experiments showing that invasion and growth to maturity of later species neither require conditions produced by earlier species (model 1) nor are inhibited by them (model 3). Although this is theoretically possible, we have found no convincing examples. In the invasion and growth to complete dominance by mussels on marine rocky shores, no experiments have been performed testing the effects of the previous occupants on this process. The observation of Bayne (1965) cited above suggests that they may fit model 1; experiments would be welcome. Likewise in terrestrial successions, the effects of previous residents have not been elucidated. In fact, if the more shade-tolerant species are intolerant to full sun, as with saguaro cactus, they may be examples of model 1.

Evidence supporting model 3 consists of observations that early species suppress the establishment of later ones, inhibit their growth, and reduce their survival. Keever (1950) and Parenti and Rice (1969) have shown experimentally that early-colonizing land plants reduce the rates of germination and growth of other species arriving later. Niering and Egler (1955) and Niering and Goodwin (1974) found that a closed canopy of shrubs prevented the invasion of trees for periods up to 45 yr. Webb et al. (1972) found that 12 yr after an experimental clearing in montane rain forest, the sprawling shrub *Lantana* had occupied a large area, excluding and suppressing tree species. Besides these data from land plants, there is also evidence from marine organisms living on hard substrates, that the first colonists prevent later ones from attaching. O'Neill and Wilcox (1971) got opposite results from those of Scheer (1945) in marine species; on glass plates, a thick coating of bacteria apparently inhibited attachment of diatoms. Likewise, Sutherland (1974) found that once sedentary marine invertebrates had covered the undersurface of tiles suspended from a wharf, other species invaded only after the occupants had died and sloughed off.

Field experimental demonstrations showing that early species exclude or suppress later ones come from several sources. For the earliest stages, McCormick (1968) found that removing the pioneering annual plants resulted in faster growth and earlier flowering of perennials. As yet no data from this unique study have been published. In an experimental study in the marine rocky intertidal zone, W. P. Sousa (unpublished manuscript) has found that removing early succession algae resulted in a much greater abundance of later succession algae.

For intermediate stages the best evidence comes from some of the first controlled field experiments ever done, trenching in forests. During the earlier stages of succession in forests with trees less than 50 yr old, more light penetrates to the ground than in old climax forests. However, several series of trenching experiments in these early succession forests showed that young trees grew only when the root competition with older trees was removed by trenching (Fricke 1904; Toumey and Kienholz 1931; Korstian and Coile 1938). Thus even with the greater light levels of early succession forests, the late succession seedlings are suppressed by root competition.

These observations and experiments indicate that in many instances the high tolerance of later succession species to low levels of resources still does not allow them to grow to maturity if they are dominated by a stand of early species. Studies by Vaartaja (1962), Grime and Jeffrey (1965), and others have shown that late species maintain themselves in the presence of dominating earlier species by having a lower metabolic rate, by repairing damages, and by fending off attacks of herbivores, soil pathogens, etc. The later species simply survive in a state of "suspended animation" until more resources are made available by the damage or death of an adjacent dominating individual.

Even though earlier species may continue to exclude or suppress later ones for long periods, the former eventually are damaged or killed and are then replaced. For example, in succession on prairies, annual weeds and grasses are gradually replaced by perennial ones. In Oklahoma, an annual species of grass lasted up to 15 yr before a perennial species of bunchgrass replaced it, to survive and dominate for another 50 yr as others slowly invaded (Booth 1941). The perennial never grew more rapidly than the annual at any level of mineral nutrients (Rice et al. 1960), so it could not displace it by exploitation competition as required by model 2. Rather it presumably simply filled in the space opened up by the death of the annual and held it thereafter. The seedlings of sugar maple, one of the dominant late succession species in North American deciduous forests, become established mainly in the light gaps opened up when trees die (Bray 1956; Westman 1968).

This evidence suggests that in many situations, early and mid-succession life forms (e.g., perennial grasses and shrubs, green algae, etc.) may quickly secure the space opened up after a disturbance and then hold it, excluding typical late-succession species. This is especially true when the former can propagate vegetatively as well as sexually. The opportunities for a new seedling of any species to become established in a dense perennial grass sward or shrub thicket are virtually zero. By vegetative reproduction the dominant species can persist for a very long time.

PREDICTIONS AND TESTS ON THE MODELS

We predict that the facilitation model 1 will commonly apply to situations in which the substrate has not been influenced by organisms beforehand. It should apply to many primary successions, since soils newly exposed by receding glaciers, shorelines, etc., may have extreme properties of nutrients, structure, pH, etc., that render them difficult for most species to invade. In contrast, in secondary succession the soils have already supported plants and so present fewer difficulties to colonists. Therefore, we predict that models 2 and 3 apply to most secondary successions. If the previous occupation has not influenced the substrate (e.g., on marine rock surfaces), however, model 1 may apply. The facilitation model should also hold in heterotrophic successions in logs, corpses, etc., where there are barriers to initial penetration through bark or skin, so that specialist scavengers must bore through these barriers before other species can enter.

Rather than the purely observational evidence that is usually adduced, much better tests of the models could be made with controlled field experiments. For example, the best test of the hypothesis given in step B in figure 1 would consist of excluding early species from sites to see whether late species could colonize. The only published account of such an experiment is that of McCormick (1968), but no data were included.

Experimental tests of later stages are more difficult, because of the longer life spans of later species. However, the processes at step D, figure 1, could be investigated in the following way. Seeds and/or seedlings of later species could be transplanted and grown with and without earlier species. If later species grew better when early species were absent, models 1 and 2 would be rejected; if much worse, models 2 and 3 would be rejected; if there were little or no difference, models 1 and 3 would be rejected. The trenching experiments described earlier indicate that the first alternative (models 1 and 2 rejected) seems to apply to many forests in the intermediate stages of succession.

Model 3 could be tested by observing whether later succession species could invade a stand of early species that was either left intact (protected from fire, grazing, etc.) or in which gaps were created by removing some early individuals.

If later species invaded and grew only in the gaps, model 3 would be supported, models 1 and 2 rejected.

Careful experimental study of the early stages of primary successions would be particularly welcome. Present evidence suggests that here is where the traditional facilitation model 1 may be expected to apply most closely. Broadcasting and planting of seeds and transplanting seedlings of later-succession species are crucial field experiments that are feasible on such places as recently exposed moraines of receding glaciers, lava and ash beds from recent volcanic eruptions, newly exposed sand bars and dunes, recent landslides, and newly uncovered rocky shores. Such introductions with and without associated early-succession species would test the different models. If the later species become established without the early ones being present, the facilitation model will be rejected. If not, further field experiments could be done to determine what sort of modifications of the environment are necessary to ensure their establishment.

These field experiments were suggested by the statements in figure 1, which in most instances were stated as testable hypotheses. Some guidelines to the proper design and limitations of controlled field experiments have recently been described by Connell (1974).

SUCCESSION AND COMMUNITY STABILITY

In many communities, major disturbances occur frequently enough that succession will usually be cut short and started all over again, as indicated by pathways x and y in figure 1. Under what circumstances would we expect this to happen? Disturbance by man dates back to preneolithic cultures. In Britain, prehistoric man set fires to drive out game and cut vegetation to clear land for agriculture (Smith 1970; Turner 1970). Other disturbances not associated with man are natural fires, landslides, severe storms, and various biological causes such as intense grazing (e.g., the bison on North American plains) or predation on sessile marine organisms. For example, within the past several thousand years much of the forest of North America has been badly damaged or destroyed by fire at least once every few hundred years, within the life span of the dominant conifers (Heinselman and Wright 1973). These major sources of perturbation are so widespread as to suggest that even before man's interference became common, in relatively few natural communities did succession ever stop.

After a severe disturbance or during a short respite from normally heavy and continuous grazing or predation, there is usually a burst of regeneration that, once established, suppresses later regeneration. Thus a single age-class emerges that may dominate the scene for long periods. Henry and Swan (1974) found that the white pine trees that got established after catastrophes in the late seventeenth century dominated the forest for 200 to 250 yr thereafter, suppressing almost all later tree invasion. Similar waves of regeneration of a single age-class have been demonstrated in forests after large grazers were reduced

(Peterkin and Tubbs 1965) and after spruce budworm epidemics (Morris 1963). The existence of dominant, widely spaced age-classes resulting from such episodic regeneration after perturbations is an indication that succession has not yet stopped in an equilibrium assemblage.

If no such catastrophes have intervened we will have arrived at an assemblage of long-lived individuals that would usually be regarded as late-successional, or "climax," species, step F in figure 1. We will now consider the second question posed at the beginning, "Under what conditions will the species composition remain in a steady-state equilibrium?"

Theory

Stated simply, a system is stable if it persists despite perturbations. It is impossible to discover whether a system is stable if it is not tested with a disturbance. In real communities this is not a problem because challenges are being continually offered to the system's stability in the form of variations in physical conditions, invasions of competing species, natural enemies, etc.

Margalef (1969) pointed out that systems persist either by giving way to the perturbation and subsequently recovering to the original state or by not giving way at all. He suggested that these could be called, respectively, "adjustment or lability," vs. "conservatism, endurance, or persistence." Since we have equated stability with persistence, we will refer to the two sorts of mechanisms as adjustment and resistance, respectively. In his discussion of the application of the theory of stability to ecological systems, Lewontin (1969) discussed the nature of the field of transformations in which the system moves. If there exists a point at which the transformation vector is zero, so that the system does not change, it is called a stationary point. Whether it is also a stable point can be decided only by observing that, in the region nearby, all the transformation vectors point toward it. If the system returns to a stable point from any other point in the vector field, i.e., after any degree or extent of perturbation, it is globally stable. If it returns to it only after small perturbations and to another stable point after a large perturbation, the system exhibits only neighborhood stability. Each stable point has its own basin of attraction, the neighborhood in which the system returns to the original point.

In Margalef's (1969) terminology, the process of succession represents "adjustment" stability. If all successions on a site led to a similar species composition at equilibrium, as postulated by Clements (1916), this would be global stability. If quite different species compositions were reached, the system would have multiple stable points. Only by observing the process of adjustment after perturbation can such judgments be made.

If a community resists perturbation, there will be no succession since there is no change. Therefore, we need not consider this mechanism in detail, except to point out that individuals resist perturbations by defenses against stresses from physical factors, attacks by natural enemies, and invasions by competitors.

Succession as Adjustment Stability: The Importance of Scale

In recovery from a perturbation, it is the maintenance of species composition that we are considering. However, before the stability of any real community can be discussed, three scales must be specified: the time, space, and intensity of perturbations. In other words, to judge stability we need to decide how long and over what space the present species composition must persist in the face of a given intensity of perturbation. The reason for this proviso can be illustrated by the following example: Horn (1974, p. 28) states, "Early successional patches are by definition ephemeral, while tracts of the climax remain relatively unchanged for several generations." But this statement holds only under certain scales of time and space. In a small area of forest, early-succession stages are individually ephemeral because the species are short lived and seldom persist for many generations. But if the early-succession species are not to go extinct, somewhere either disturbances must occur close enough in time and space to provide open sites or else such sites must exist continually (e.g., ridges with much drier conditions than the surrounding forest) in which they can produce successive generations and so perpetuate their species. Therefore a large enough tract of forest must be included in the system under consideration so that the kind of habitat recognized as "early succession patches" remains in existence, somewhere, for generations.

In contrast, the assertion that tracts of the climax remain unchanged for several generations is supported only by general impressions. The long-enduring climax tree lends an air of permanence, but as Frank (1968) points out, this implies nothing about self-perpetuation. So on the scale of generation times and over a large enough tract, if both early- and late-succession stages persist despite perturbations, both are stable.

Thus, to be able to judge the degree of stability of the species composition of a community, the following site characteristics must be met: (1) an area large enough to ensure either sufficient site diversity or that disturbances opening up new sites occur at intervals no longer than an early succession species persists (including the period that the seeds lie dormant [Marks 1974]). This ensures that the early-succession species are able to persist somewhere in the system. (2) An observation period at least as long as the longest generation time of any of the species and also long enough so that the whole range of kinds and intensities of perturbations will have had a chance to occur. This would allow enough time to see how much the species composition varied over at least one complete turnover of generations.

These requirements may be so stringent as to make it virtually impossible to determine the stability of communities composed of long-lived species. But unless these scales of time, space, and intensity of disturbance are defined in relation to the organisms comprising the community, any pronouncements about stability are of limited value.

We have already incorporated variations in the frequency of disturbance in figure 1; pathways w, x, y, and z represent a sequence of decreasing frequency of major disturbances. Let us now consider the general effects of varying the

MECHANISMS OF SUCCESSION   1131

TABLE 1

The Effect of the Size of Area Disturbed and Intensity of Disturbance
on the Course of Succession

Intensity of Disturbance	
Extreme	Slight

Large area:
I. *A long succession*: Assuming no survivors, all colonists must come from outside, so will consist mainly of early succession species with high vagility of propagules. Since late-succession species have low vagility, they will spread in slowly from the borders. Meanwhile the early species may go through several generations.

Small area:
III. *Some succession*: Surrounding adults colonize, with propagules or vegetative growth. Propagules of distant early-succession species also colonize, but since resources are reduced by the neighboring adults, the early species may not grow as quickly as usual.

Large area:
II. *A moderate amount of succession*: The area will be refilled: (*a*) by individuals growing from propagules that arrived from distant early-succession species or those that were present in the soil before the disturbance; (*b*) by growth of surviving juveniles of late-succession species that before had been living suppressed in the shade of the adults; and (*c*) by resprouting of damaged adults.

Small area:
IV. *No succession*: The small gap is refilled by growth of surrounding adults and/or of previously suppressed offspring of the late-succession species.

scales of intensity of disturbance and size of area disturbed (see table 1). First, if the disturbance is both intense and also extends over a large area, such as extensive cultivation, or a large fire severe enough to kill all of the plants in the forest, all recruitment must come from outside. The pioneer species with highest vagility of propagules will then secure and hold the ground for a long time, with the climax species only slowly spreading in from the edges. Obviously, return to the original forest will take a long time. Similarly, observations of colonization on very extensive new surfaces exposed in the sea, such as on new sea walls, show an initial colonization within a few weeks of diatoms and green algae, whereas the larger long-lived algae may not appear for 2 or 3 yr (Moore 1939; Rees 1940).

Second, if the disturbance is less severe but affects a very large area, such as extensive damage from a hurricane which often kills large trees but not the undergrowth, regrowth of survivors as well as recruitment from seeds will occur. Opportunists, whose seeds have either been present in the soil or newly arrived from surrounding areas, may germinate and rapidly grow up, suppressing the seedlings of climax species that have survived either as seeds or seedlings from the original forest. Alternatively, surviving shrubby undergrowth may suppress these seedlings. However, some members of the climax species may have survived as taller saplings or as portions of adults that send up sprouts. These may be too tall to be suppressed by the pioneers (Webb et al. 1972). Thus the return to the original state will not be delayed as much as in the first case by the dominance of early-succession species.

Third, if the disturbance is severe over a small area, such as a lightning strike that kills all individuals in a small space, recruitment must come from outside, either by seeds or vegetative growth of neighbors. Because the area is small, seeds of both low vagility from nearby climax trees and greater vagility from more distant early-succession species will colonize the gap. In small gaps resources of light and soil nutrients are reduced by the neighboring trees so that the early-succession pioneers may not grow quickly enough to suppress the growth of the climax offspring. Climax seedlings may even grow faster than those of earlier stages in small gaps; data in Horn (1971, p. 33) suggest this. A similar case has been demonstrated on marine rocky shores (Pyefinch 1943). A quite small area of surface was cleared in the midst of a bed of large long-lived algae; offspring of these large species soon became established and filled in the gap within the first year, in marked contrast to their slow invasion on extensive new surfaces as described above.

In the fourth case, the disturbance is slight over a small area, such as when a single adult dies. Light and soil water and nutrients are only slightly increased over a small area, and few individuals are killed by the disturbance. The gap is filled either by vegetative growth of the surrounding adults or by replacement of the dead adult by growth of offspring of late-successional species that are already present as suppressed individuals. Few early-succession species invade successfully because the area is small and the level of resources is low. In this case the whole process takes place within step F, figure 1.

Patterns of Stability Following Recovery from Major Disturbances

Succession, as represented by steps A through F in figure 1, is the process by which a community recovers from a perturbation. Two questions are relevant here: (1) What determines the rate of recovery after major perturbation, and (2) how closely will the species composition return to the original state?

Regarding the first question, the three models produce different rates of recovery. In the facilitation model 1, early-succession species enhance the invasion and growth of late-succession species, so the former increase the rate of recovery. In the tolerance model 2 the early species reduce the rate of recovery since they suppress the rates of invasion and growth of late species. In the inhibition model 3, the early species prevent recovery completely until they die or are damaged. Thus the rate of recovery, i.e., degree of stability, drops in the order of models 1, 2, and 3.

In terms of the management of either natural or disturbed sites, the correct plan to encourage a quicker recovery from perturbation depends on the type of community it is desired to develop and upon the likely model pathway that succession would follow. Assuming that a situation like the original community is desired, and if model 1 tends to operate, early succession species should be encouraged. If model 2, they should probably be discouraged, and if model 3, they must be discouraged. In the latter two, the best plan would be to replant the species that were there originally and remove any early ones that invade. In fact, it may be necessary, in order to preserve some communities, to tolerate some events that ordinarily would be regarded as unmitigated catastrophes. The long-term maintenance of alluvial redwood groves may depend upon the Heraclitean forces of fire and flood to remove the trees that suppress young redwoods (Stone and Vasey 1968). Person and Hallin (1942) pointed out that natural regeneration of redwood requires removal of competing species. The second question is, how closely will the species composition tend to return to the original state? In the four cases just described (see table 1) we suggest that the probability of a close return increases in the order I, II, III, IV. Considering a large tract of land, the more extensive and/or intense the disturbance, the longer the succession and the less probable that the final composition will resemble the original.

Does the Species Composition Ever Reach a Steady-State Equilibrium?

Let us now consider communities that are subjected only to slight disturbances over small areas (step F, fig. 1). Here the future course of events will consist of a series of very small-scale changes as individuals die and are replaced. We now ask the question, does the species composition remain constant over several generations? We will answer this on two different spatial scales. The smallest scale is the individual organism, so we will first discuss how species may vary during a plant by plant replacement process. Second, we will consider whole tracts of land containing a number of species.

The pattern of small-scale changes will depend upon whether individuals are more likely to be replaced by a member of their own or another species. The species of replacing individual will depend upon how the conditions at the spot had become modified during the previous occupation. In relation to the success of their own offspring, three types of conditions could be produced in the immediate vicinity of the individual being replaced.

In the first, the conditions are such that offspring of the same species will be favored over those of other species. That is, offspring of the same species may be concentrated near the adult so that when it dies there is a very high probability that it will be replaced by one of them. Such precise self-replacement would mean that not only the species composition but also their relative abundance and spatial pattern would remain constant. This would represent the highest possible degree of stability.

The most likely instance of this occurring would be one in which the late-succession species reproduce vegetatively from root or stump sprouts. The situation described by Horn (1975) in which, at the calculated equilibrium state, American beech was strongly dominant may be an example, since every beech offspring was a root sprout. The same pattern occurs in other species such as *Tilia americana* and in Antarctic beech forests in Australia, the trees sprouting in circles around old stumps.

A second and opposite alternative is a species in which conditions in the vicinity of a late-successional individual become modified in such a way that its offspring can no longer survive there. Then when the adult dies, it would be replaced by another species. For example, seedlings of the cedars of Lebanon "thrive under hardwood trees and shrubs but not under cedar trees, so that some disturbance is apparently necessary for cedar-forest regeneration" (Beals 1965, p. 694). Another example of this was indicated by Florence (1965), who suggested that in old-growth redwood forests conditions in the soil may change gradually for the worse as resistant portions of the litter (e.g., lignins) accumulate and, perhaps as a consequence, pathogenic microorganisms increase while saprophytes decrease. Thus redwoods, the epitome of the long-enduring climax species, may not replace themselves unless the soil environment is changed. Florence (1965) found that redwood seedlings grew poorly in soil from old redwood groves unless the microorganisms were killed by irradiation. He suggests that seedlings will persist and grow only if the inimical soil environment is changed, either by new soil being brought in by stream deposition (Zinke 1961) or by a set of hardwood species intervening between redwood generations. Such "soil fatigue" has been observed in other forests (see review in Florence 1965). Other examples in which soil microorganisms have been demonstrated or implicated in the death of seedlings in the vicinity of adults of the same species are from *Eucalyptus* forests (Florence and Crocker 1962; Evans et al. 1967) and, in a rain-forest tree, *Grevillea* (Webb et al. 1967).

If only a few species are available, a "cyclic succession" may occur, each species alternating with one or two others. The first examples of such cyclic successions were pointed out by Watt (1947); others have since been studied in Alaskan flood-plain vegetation (Drury 1956), old-growth redwoods (Florence 1965), etc. Aubreville's (1938) "mosaic theory of regeneration" in tropical rain forests seems to fit this model also. Unlike the first situation, which is stable in areas the size of an adult individual, stability in the second situation can only occur on a larger scale, minimally accommodating individuals of several species.

In the third alternative the site where the adult stood remains neither more nor less favorable for offspring of the same species. The species of replacing individual will depend upon the relative abundance of propagules arriving there or of suppressed individuals already present. Since its own offspring are not more disadvantaged than those of other species, it is highly likely that they will be the commonest young in the immediate vicinity of the adult. Only if the species produces very many highly dispersed offspring, as in the planktonic larvae of marine sedentary organisms, would this likelihood be reduced. Therefore this more closely resembles the first than the second alternative. Thus the first acts as an "absorbing sink," for, when conditions of either the first or the third type are associated with the replacing individual, the probability is very high that that species will continue to occupy that site for many generations.

In that case, why do any instances of the second type exist? The answer is that this situation is produced not by the species itself, nor by competing species, but by natural enemies that attack that species in preference to another. At least two possible mechanisms could produce this result. Either the predators could be generalists that switch their attention to whichever species is commoner or else they are specialists attacking that particular species. The first mechanism might apply in instances where local patches of a single species are produced following the operation of replacement processes of the first and third type described above. This behavior has been predicted and verified in invertebrate predators by Murdoch (1969) and Murdoch and Oaten (1975). Similar studies of herbivores attacking plants would be welcome to see whether the same principles apply as in predator-prey interactions.

The second mechanism, specialist natural enemies, has been proposed for tropical forests by Janzen (1970) and Connell (1971), who have suggested that fallen seeds and young seedlings will be attacked more heavily near the parent tree than further away. Field experimental tests have rejected this hypothesis for seeds in four instances (Connell 1971, two species; Janzen 1972a; Wilson and Janzen 1972) and supported it in a fifth, involving an introduced insect seed predator in a disturbed habitat (Janzen 1972b). The hypothesis has been supported for seedlings by a field experiment (Connell 1971) and field observations (Janzen 1971). Thus in some tropical forests, the pattern of turnover of trees in climax stands may be caused by this mechanism.

Concerning changes in species composition on a larger scale, tracts bearing communities of several species, Horn (1974) states, "If stability is defined as the absence, or inverse, of species turnovers and of population fluctuations, then stability increases tautologically with succession. There is nothing magic about this invariable increase in stability because succession is defined as occurring when the specific composition of the community is changing, and it is defined as having stopped when the composition of the community is not changing."

Obviously, if in the plant-by-plant process the first alternative described earlier holds and every individual is replaced by another of the same species, the climax stage will possess both local and large-scale stability and, by Horn's definition, succession will have stopped. But if either of the other alternatives holds, so that individuals may be replaced by others of different species, then stability of species composition will not necessarily follow and so, by Horn's definition, succession may or may not ever stop. The only test of this has been performed by Horn (1975). His steady-state Markov extrapolation resulted in a species composition that resembled fairly closely an old-growth natural forest nearby. However the species that almost completely dominated both the simulated and real forest was American beech, which in Horn's example was reproducing entirely by root sprouts. If, as seems likely, these root sprouts grow to replace the main tree when it dies, this forest is composed essentially of immortal individual beeches, an extreme example of the first alternative described above.

We have found no example of a community of sexually reproducing individuals in which it has been demonstrated that the average species composition has reached a steady-state equilibrium. Until this is demonstrated, we conclude that, in general, succession never stops.

1136   THE AMERICAN NATURALIST

SUCCESSION AND COMMUNITY ORGANIZATION

The three models of succession described earlier are based upon three quite different views of the way ecological communities are organized.

The Facilitation Model 1

The idea that the presence of later-succession species is dependent upon early ones preparing a favorable environment for them implies a high degree of organization in ecological communities. Although few modern ecologists would subscribe to Clements's (1916) analogy with an individual organism, the idea is widely held that the community is a highly integrated, well-adjusted set of species. A succinct summary of this view is given by Odum (1969):

Ecological succession may be defined in terms of the following three parameters: (i) It is an orderly process of community development that is reasonably directional and, therefore, predictable. (ii) It results from modification of the physical environment by the community; that is succession is community-controlled even though the physical environment determines the pattern, the rate of change, and often sets limits as to how far development can go. (iii) It culminates in a stabilized ecosystem in which maximum biomass (or high information content) and symbiotic function between organisms are maintained per unit of available energy flow. In a word, the "strategy" of succession as a short-term process is basically the same as the "strategy" of long-term evolutionary development of the biosphere—namely, increased control of, or homeostasis with, the physical environment in the sense of achieving maximum protection from its perturbations. [P. 262]

The idea that succession is a process of "community development" led to the characterization of "immature" and "mature" stages of an ecosystem (Margalef 1963). Odum (1969) proposed a tabular model of the contrasting trends in energetics, structure, life history, nutrient cycling, selection pressure, and overall homeostasis to be expected in the developmental and mature stages of a community.

This model has been severely criticized in recent reviews by McCormick (1968), Drury and Nisbet (1973), and Colinvaux (1973). They point out that most of the proposed characteristics of "mature" communities are simply the consequence of the passage of time rather than of internal control. For example, in old-field succession biomass increases since trees take time to grow. More nutrients are tied up in the bodies of trees than in herbs. Starting from nothing, species diversity and biochemical diversity increase as colonizers arrive. As a consequence of these obvious trends others of necessity follow: changes in production/biomass and other ratios, increases in structural complexity, increase in importance of detritus in nutrient regeneration, etc.

Other predictions of the model derive from the apparently firmly held view that the mature community, like the adult organism, is a highly organized, stabilized system, with maximum homeostasis achieving maximum protection from perturbations from the environment. This view is based solely on the analogy, not, in our opinion, on evidence. Odum (1969, table 1) has proposed a series of trends. Mature communities (as contrasted to developmental stages) are presumed to have more weblike food chains, more well-organized stratification and spatial heterogeneity, narrower niche specialization, longer and more complex life cycles, selection pressures "for feedback control (K-selection)" rather than "for rapid growth (r-selection)," low entropy, and high information content.

All or most of these predicted characteristics are not findings but deductions from the concept that the mature community is in fact in a steady state that is maintained by internal feedback control mechanisms. All have been questioned by one or other of the recent reviews (Drury and Nisbet 1973; Colinvaux 1973; Horn 1974). The bases of the doubts, which we endorse, are that since the embryology analogy is unsupported, and since there is no evidence that so-called mature communities are internally controlled in a steady state, such characteristics cannot be deduced from them.

Obviously the mechanisms determining the sequence of species of model 1 may apply, for example, to heterotrophic successions and certain primary successions, even if the high degree of positive integration described above does not hold in those same communities.

The Tolerance Model 2

This view holds that succession leads to a community composed of those species most efficient in exploiting resources, presumably each specialized on different kinds or proportions of resources. Connell (1975) pointed out that this model may hold in two sorts of circumstances: (a) for certain groups of animals that have evolved a high degree of independence from the rigors of both the physical and biotic environments (warm-blooded vertebrates, large predators, social insects, etc.); (b) where natural enemies are reduced but the physical environment is not so severe as to remove most organisms directly; then the populations may be limited by resources.

Several examples of the latter situation are described in Connell (1975); the winning competitors were species that were more effective in interference rather than more efficient in exploiting resources. Another example is from those areas on coral reefs protected from hurricane damage but where predators of corals are not common. In such an area at Heron Island, Queensland, the surface is almost completely occupied by those competitors that are most effective in interfering with their neighbors, colonies of "staghorn" corals that have grown up over all neighbors and now hold the space against invaders (Connell 1976). Elsewhere the reef is damaged by frequent hurricanes and the succession is kept in an earlier stage, exemplified in figure 1 by pathways x or y.

In dense, light-limited forests, this tolerance model predicts that the set of species most tolerant (i.e., able to grow on the lowest level of resources) will eventually dominate the community at equilibrium. Predictions of the outcome of successional trends have been made for several forests. For example, Stephens and Waggoner (1970) extrapolated from transition probabilities directly measured over several decades in a forest undergoing succession. By assuming a stationary Markov process, they concluded that at equilibrium the moderately shade-tolerant species will be in the majority, rather than the forest progressing inexorably toward the very shade-tolerant species as the model predicts. This is not a consequence of disturbance setting back succession, since the transition probabilities were estimated only from undisturbed plots.

Horn (1975) estimated transition probabilities of successive generations indirectly by recording the saplings underneath each species of mature trees. Assuming that each sapling has an equal probability of replacing that mature tree in the next generation of the canopy, he used the proportions of saplings of each species as transition probabilities. In the subsequent extrapolation, the forest at equilibrium was dominated by the very shade-tolerant beech, a different result from that of Stephens and Waggoner. This result apparently supports the model of increasing competitive ability. It is probably a consequence of the fact that, since every young beech recorded was a root sprout of an adult and since the adult probably contributes energy to the root sprout, these "offspring" have a great competitive advantage over independent saplings in the shade.

Thus the prediction of the competition model has so far not been verified in forests replacing themselves by independent offspring (Stephens and Waggoner 1970), but only in one that apparently is doing so mainly by vegetative reproduction.

The Inhibition Model 3

In this model no species necessarily has competitive superiority over another. Whichever colonizes the site first holds it against all comers. After all the empty space is filled, invasion is possible only if the new colonist brings along its own resources, such as a large seed with enough stored energy to sustain the seedling for awhile in an undisturbed stand of vegetation where no spare energy is available.

Since replacement occurs only when resources are released by the damage or death of the previous occupant, the species composition shifts gradually and inexorably (given no further major disturbances) toward species that live longer. This is not because these species are more likely to colonize; quite the opposite. It is because once a long-lived species becomes established, it persists by definition. This, as Frank (1968) has pointed out, is succession by tautology! No directional mechanism (as in models 1 and 2) need be invoked for model 3. Simply by these life-history characteristics, long-lived species eventually dominate the ecological scene.

The ability to survive a long time is a function of having defenses against all the inevitable hazards. Examples abound of the defensive adaptations that enable late-succession species to survive better than early species. Billings (1938) showed that, as compared to early-succession species, the juveniles of late-succession species develop deeper and more extensive root systems, allowing them to persist through drought periods better than early-succession species. Stone and Vasey (1968) point out that several species of trees that invade gaps and suppress young redwoods in alluvial groves are killed by fires or by the alluvium deposited by floods, whereas the redwoods are not harmed. The allocation of energy and matter into harder, denser wood must cause a tree to grow more slowly. But harder wood is a better defense against storm damage and wood borers. Likewise, some species of corals produce a dense, massive skeleton at the expense of slower growth and occupation of space than other corals that produce a less dense, branched skeleton and quickly secure space. Connell (1973) found that two species of fast-growing corals had many more mollusks and sponges boring into and weakening their skeletons than did a slower-growing massive species.

Adaptations against natural enemies include various morphological (hard wood, spines, fibers, etc.) and chemical defenses (secondary substances such as alkaloids, tannins, etc.). Cates and Orians (1975) found that generalist herbivorous slugs ate early-succession species in preference to species that occurred in late-successional stages in the coniferous forests of the northwestern United States. In apparent conflict with these results, Otte (1975) found that generalist herbivorous grasshoppers preferred late-succession bushes, vines, and trees to early-succession herbs in Texas. This conflict may perhaps be resolved with the following argument. The grasshoppers studied by Otte were three species of *Schistocerca* that can disperse widely. In contrast to the forest habitat of slugs, such mobile insects are more characteristic of open savannah or xeric grassland. Thus the species of plants that persist in the grasshopper habitats may be herbs rather than shrubs and trees. The grasshoppers, along with other grazing insects and vertebrates, plus seasonal fires, may be eliminating shrubs and trees and preserving the herbaceous vegetation as the "climax" stage. In their own habitat they may be behaving in the same way that slugs do in their wet forest, attacking and eliminating certain species and not eating the climax, i.e., persistent, species.

Lest this reconstruction seem farfetched, we would like to emphasize that controlled field experiments have demonstrated in several instances that natural enemies have eliminated species which were superior competitors capable of holding space against invasion. Sea urchins often clear algal mats (Paine and Vadas 1969), and predatory starfish and snails eliminate mussels (Paine 1966, 1974; Dayton 1971). These natural enemies are important components of the community and often determine the species composition of the climax.

Model 3 emphasizes that "possession is eleven points in the law" (Cibber 1777, p. 121); once an individual secures the space it resists the invasion of competitors. Eventually it may be damaged or killed and invaders may replace it. In this model, early-succession species may be just as resistant to invasion by competitors as late species, so the "climax" species are those most resistant to being damaged or eliminated by fires, storms, natural enemies, etc.

SUMMARY

The sequence of species observed after a relatively large space is opened up is a consequence of the following mechanisms. "Opportunist" species with broad dispersal powers and rapid growth to maturity usually arrive first and occupy empty space. These species cannot invade and grow in the presence of adults of their own or other species. Several alternative mechanisms may then determine which species replace these early occupants. Three models of such mechanisms have been proposed.

The first "facilitation" model suggests that the entry and growth of the later species is dependent upon the earlier species "preparing the ground"; only after this can later species colonize. Evidence in support of this model applies mainly to certain primary successions and in heterotrophic succession.

A second "tolerance" model suggests that a predictable sequence is produced by the existence of species that have evolved different strategies for exploiting resources. Later species will be those able to tolerate lower levels of resources than earlier ones. Thus they can invade and grow to maturity in the presence of those that preceded them. At present there exists little evidence in support of this model.

A third "inhibition" model suggests that all species resist invasions of competitors. The first occupants preempt the space and will continue to exclude or inhibit later colonists until the former die or are damaged, thus releasing resources. Only then can later colonists reach maturity. A considerable body of evidence exists in support of this model.

In the majority of natural communities succession is frequently interrupted by major disturbances, such as fires, storms, insect plagues, etc., starting the process all over again. However, if not interrupted, it eventually reaches a stage in which further change is on a small scale as individuals die and are replaced. The pattern of these changes will depend upon whether individuals are more likely to be replaced by a member of their own or another species. If the former, stability will be assured. However, in terrestrial communities, conditions in the soil in the immediate vicinity of long-lived plants may become modified in such a way that offspring of the same species are much less favored than those of other species. A likely cause is the buildup of host-specific pathogenic soil organisms near a long-lived plant. In this case, the species at each local site keep changing, producing local instability. Whether the average species composition of the whole tract does not change, exhibiting global stability, or whether it keeps changing has not yet been decided for any natural community.

ACKNOWLEDGMENTS

We are grateful to the following persons for comments on various versions of this paper: P. Abrams, S. Avery, M. Connell, J. Cubit, J. Dixon, E. Ebsworth, M. Fawcett, L. Fox, R. Howmiller, A. Kuris, D. Landenberger, M. Macduff, P. McNulty, W. Murdoch, W. Niering, A. Oaten, C. Peterson, D. Potts, S. Schroeter, W. Sousa, and R. Warner.

LITERATURE CITED

Aubreville, A. 1938. La forêt coloniale: les forêts de l'Afrique occidentale française. Ann. Acad. Sci. Coloniales 9:1–245.
Bayne, B. L. 1965. Growth and delay of metamorphosis of the larvae of *Mytilus edulis*. Ophelia 2(1):1–47.
Beals, E. W. 1965. The remnant cedar forests of Lebanon. J. Ecol. 53:679–694.

Billings, W. D. 1938. The structure and development of old field shortleaf pine stands and certain associated physical properties of the soil. Ecol. Monogr. 8:437–499.

Booth, W. E. 1941. Revegetation of abandoned fields in Kansas and Oklahoma. Amer. J. Bot. 28:415–422.

Bray, J. R. 1956. Gap phase replacement in a maple-basswood forest. Ecology 37:598–600.

Cates, R. G., and G. H. Orians. 1975. Successional status and the palatability of plants to generalized herbivores. Ecology 56:410–418.

Cibber, C. 1777. Woman's wit (or, the lady in fashion). Pages 99–200 in The dramatic works of Colley Cibber Esq. Vol. 1. J. Rivington, London; reprinted, AMS, New York, 1966.

Clements, F. E. 1916. Plant succession. Carnegie Inst. Washington Pub. 242. 512 pp.

Colinvaux, P. A. 1973. Introduction to ecology. Wiley, New York. 621 pp.

Connell, J. H. 1971. On the role of natural enemies in preventing competitive exclusion in some marine animals and in rain forest trees. Pages 298–312 in P. J. den Boer and G. R. Gradwell, eds. Dynamics of populations. Centre for Agricultural Publication and Documentation, Wageningen, Netherlands.

———. 1972. Community interactions on marine rocky intertidal shores. Annu. Rev. Ecol. Syst. 3:169–192.

———. 1973. Population ecology of reef-building corals. Pages 205–245 in R. E. Endean and O. A. Jones, eds. Biology and geology of coral reefs. Vol. 2, Biol. 1. Academic Press, New York.

———. 1974. Field experiments in marine ecology. Pages 21–54 in R. Mariscal, ed. Experimental marine biology. Academic Press, New York.

———. 1975. Some mechanisms producing structure in natural communities: a model and evidence from field experiments. Pages 460–490 in M. Cody and J. Diamond, eds. Ecology and evolution of communities. Harvard University Press, Cambridge, Mass.

———. 1976. Competitive interactions and the species diversity of corals. Pages 51–58 in G. O. Mackie, ed. Coelenterate ecology and behavior. Plenum, New York.

Cooper, W. S. 1913. The climax forest of Isle Royale, Lake Superior, and its development. Bot. Gaz. 55:1–44, 115–140, 189–235.

Cowles, H. C. 1899. The ecological relations of the vegetation on the sand dunes of Lake Michigan. Bot. Gaz. 27:97–117, 167–202, 281–308, 361–391.

Crocker, R. L., and J. Major. 1955. Soil development in relation to vegetation and surface age at Glacier Bay, Alaska. J. Ecol. 43:427–448.

Dayton, P. K. 1971. Competition, disturbance, and community organization: the provision and subsequent utilization of space in a rocky intertidal community. Ecol. Monogr. 41:351–389.

Drury, W. H. 1956. Bog flats and physiographic processes in the upper Kuskokwim river region, Alaska. Contrib. Gray Herb. Harvard Univ. no. 178. Pages 1–13.

Drury, W. H., and I. C. T. Nisbet. 1973. Succession. J. Arnold Arboretum 54:331–368.

Egler, F. E. 1954. Vegetation science concepts. 1. Initial floristic composition—a factor in old-field vegetation development. Vegetatio 4:412–417.

Evans, G., J. B. Cartwright, and N. H. White. 1967. The production of a phytotoxin, nectrolide, by some root-surface isolates of Cylindrocarpon radicicola, Wr. Plant and Soil 26:253–260.

Florence, R. G. 1965. Decline of old-growth redwood forests in relation to some soil microbiological processes. Ecology 46:52–64.

Florence, R. G., and R. L. Crocker. 1962. Analysis of blackbutt (Eucalyptus pilularis Sm.) seedling growth in a blackbutt forest soil. Ecology 43:670–679.

Frank, P. W. 1968. Life histories and community stability. Ecology 49:355–357.

Fricke, K. 1904. "Licht und schattenholzarten" ein wissenschaftlich nicht begrundetes dogma. Cent. f.d. gasamte Fortwesen 3:315–325.

Gleason, H. A. 1917. The structure and development of the plant association. Bull. Torrey Bot. Club 43:463–481.

1142   THE AMERICAN NATURALIST

Grime, J. P., and D. W. Jeffrey. 1965. Seedling establishment in vertical gradients of sunlight. J. Ecol. 53:621–642.
Harger, J. R. E., and K. Tustin. 1973. Succession and stability in biological communities. I. Diversity. Int. J. Environmental Stud. 5:117–130.
Heinselman, M. L., and H. E. Wright, eds. 1973. The ecological role of fire in natural conifer forests of western and northern North America. Quarternary Res. 3:317–513.
Henry, J. D., and J. M. A. Swan. 1974. Reconstructing forest history from live and dead plant material—an approach to the study of forest succession in southwest New Hampshire. Ecology 55:772–783.
Horn, H. S. 1971. The adaptive geometry of trees. Princeton University Press, Princeton, N.J. 144 pp.
———. 1974. The ecology of secondary succession. Annu. Rev. Ecol. Syst. 5:25–37.
———. 1975. Markovian properties of forest succession. Pages 196–211 in M. Cody and J. Diamond, eds. Ecology and evolution of communities. Harvard University Press, Cambridge, Mass.
Hutchinson, G. E. 1958. Concluding remarks. Cold Spring Harbor Symp. Quant. Biol. 22: 415–427.
Janzen, D. H. 1970. Herbivores and the number of tree species in tropical forests. Amer. Natur. 14:501–528.
———. 1971. Escape of juvenile Dioclea megacarpa (Leguminosae) vines from predators in a deciduous tropical forest. Amer. Natur. 15:97–112.
———. 1972a. Association of a rainforest palm and seed-eating beetles in Puerto Rico. Ecology 53:258–261.
———. 1972b. Escape in space by Sterculia apetala seeds from the bug Dysdercus fasciatus in a Costa Rican deciduous forest. Ecology 53:350–361.
Keever, C. 1950. Causes of succession on old fields of the Piedmont, North Carolina. Ecol. Monogr. 20:229–250.
Korstian, C. F., and T. S. Coile. 1938. Plant competition in forest stands. Duke Univ. School Forest. Bull. 3:1–125.
Langford, A. N., and M. F. Buell. 1969. Integration, identity and stability in the plant association. Adv. Ecol. Res. 6:84–136.
Lawrence, D. B., R. E. Schoenike, A. Quispel, and G. Bond. 1967. The role of Dryas drummondi in vegetation development following ice recession at Glacier Bay, Alaska, with special reference to its nitrogen fixation by root nodules. J. Ecol. 55: 793–813.
Levins, R. 1968. Evolution in changing environments. Princeton University Press, Princeton, N.J. 120 pp.
Lewontin, R. C. 1969. The meaning of stability. Pages 13–24 in Diversity and stability in ecological systems. Brookhaven Symp. Biol. no. 22.
MacArthur, R. H. 1972. Geographical Ecology. Harper & Row, New York. 269 pp.
MacArthur, R. H., and J. H. Connell. 1966. The biology of populations. Wiley, New York. 200 pp.
McCormick, J. 1968. Succession. Pages 22–35, 131, 132 in VIA 1. Student publication, Graduate School of Fine Arts, University of Pennsylvania, Philadelphia.
Margalef, R. 1963. On certain unifying principles in ecology. Amer. Natur. 97:357–374.
———. 1969. Diversity and stability: a practical proposal and a model of interdependence. Pages 25–37 in Diversity and stability in ecological systems. Brookhaven Symp. Biol. no. 22.
Marks, P. L. 1974. The role of pin cherry, (Prunus pensylvanica L.) in the maintenance of stability in northern hardwood ecosystems. Ecol. Monogr. 44:73–88.
Moore, H. B. 1939. The colonization of a new rocky shore at Plymouth. J. Anim. Ecol. 8: 29–38.
Morris, R. F., ed. 1963. The dynamics of epidemic spruce budworm populations. Mem. Entomol. Soc. Can. no. 31. 332 pp.

Murdoch, W. W. 1969. Switching in general predators: experiments on predator specificity and stability of prey populations. Ecol. Monogr. 39:335–354.

Murdoch, W. W., and A. Oaten. 1975. Predation and population stability. Adv. Ecol. Res. 9:1–131.

Niering, W. A., and F. E. Egler. 1955. A shrub community of *Viburnum lentago*, stable for twenty-five years. Ecology 36:356–360.

Niering, W. A., and B. H. Goodwin. 1974. Creation of relatively stable shrublands with herbicides: arresting "succession" on rights-of-way and pastureland. Ecology 55: 784–795.

Niering, W. A., R. H. Whittaker, and C. H. Lowe. 1963. The saguaro: a population in relation to environment. Science 142:15–23.

Odum, E. P. 1969. The strategy of ecosystem development. Science 164:262–270.

Olson, J. S. 1958. Rates of succession and soil changes on southern Lake Michigan sand dunes. Bot. Gaz. 119:125–170.

O'Neill, T. B., and G. L. Wilcox. 1971. The formation of "primary film" on materials submerged in the sea at Port Hueneme, California. Pacific Sci. 25:1–12.

Oosting, H. J. 1942. An ecological analysis of the plant communities of Piedmont, North Carolina. Amer. Midland Natur. 28:1–126.

Otte, D. 1975. Plant preference and plant succession. A consideration of evolution of plant preference in *Schistocerca*. Oecologia 18:129–144.

Paine, R. T. 1966. Food web complexity and species diversity. Amer. Natur. 100:65–75.

———. 1974. Intertidal community structure. Oecologia 15:93–120.

Paine, R. T., and R. I. Vadas. 1969. The effects of grazing by sea urchins, *Strongylocentrotus* spp. on benthic algal populations. Limnol. Oceanogr. 14:710–719.

Parenti, R. I., and E. L. Rice. 1969. Inhibitional effects of *Digitaria sanguinalis* and possible role in old-field succession. Bull. Torrey Bot. Club 96:70–78.

Payne, J. 1965. A summer carrion study of the baby pig *Sus scrofa Linnaeus*. Ecology 46: 592–602.

Person, H. L., and W. Hallin. 1942. Natural restocking of redwood cutover lands. J. Forest. 4:683–688.

Peterkin, G. F., and C. R. Tubbs. 1965. Woodland regeneration in the New Forest, Hampshire, since 1650. J. Appl. Ecol. 2:159–170.

Pyefinch, K. A. 1943. The intertidal ecology of Bardsey Island, North Wales, with special reference to the recolonization of rock surfaces, and the rock-pool environment. J. Anim. Ecol. 12:82–108.

Rees, T. 1940. Algal colonization at Mumbles Head. J. Ecol. 28:403–437.

Reiners, W. A., I. A. Worley, and D. B. Lawrence. 1971. Plant diversity in a chronosequence at Glacier Bay, Alaska. Ecology 52:55–69.

Rice, E. L., W. T. Penfound, and L. M. Rohrbaugh. 1960. Seed dispersal and mineral nutrition in succession in abandoned fields in central Oklahoma. Ecology 41: 224–228.

Savely, H. E. 1939. Ecological relations of certain animals in dead pine and oak logs. Ecol. Monogr. 9:321–385.

Scheer, B. T. 1945. The development of marine fouling communities. Biol. Bull. 89:13–21.

Seed, R. 1969. The ecology of *Mytilus edulis* L. (Lamellibranchiata) on exposed rocky shores. I. Breeding and settlement. Oecologia 3:277–316.

Smith, A. G. 1970. The influence of mesolithic and neolithic man on British vegetation: a discussion. Pages 81–96 *in* D. Walker and R. G. West, eds. Studies in the vegetational history of the British Isles. Cambridge University Press, Cambridge.

Steenbergh, W. F., and C. H. Lowe. 1969. Critical factors during the first years of life of the saguaro (*Cereus giganteus*) at Saguaro National Monument, Arizona. Ecology 5: 825–834.

Stephens, G. R., and P. E. Waggoner. 1970. The forests anticipated from 40 years of natural transitions in mixed hardwoods. Bull. Connecticut Agr. Exp. Sta. no. 77. Pages 1–58.

1144   THE AMERICAN NATURALIST

Stone, E. C., and R. B. Vasey. 1968. Preservation of coast redwood on alluvial flats. Science
        159:157–161.
Sutherland, J. P. 1974. Multiple stable points in natural communities. Amer. Natur. 108:
        859–873.
Tansley, A. G. 1935. The use and abuse of vegetational concepts and terms. Ecology 16:
        284–307.
Toumey, J. W., and R. Kienholz. 1931. Trenched plots under forest canopies. Yale Univ.
        School Forest. Bull. no. 3:5–31.
Turner, J. 1970. Post-neolithic disturbance of British vegetation. Pages 96–116 *in* D.
        Walker and R. G. West, eds. Studies in the vegetational history of the British
        Isles. Cambridge University Press, Cambridge.
Turner, R. M., S. M. Alcorn, and G. Olin. 1969. Mortality of transplanted saguaro seedlings.
        Ecology 5:835–844.
Vaartaja, O. 1962. The relationship of fungi to survival of shaded tree seedlings. Ecology 43:
        547–549.
Vandermeer, J. H. 1972. Niche theory. Annu. Rev. Ecol. Syst. 3:107–132.
Watt, A. S. 1947. Pattern and process in the plant community. J. Ecol. 35:1–22.
Webb, L. J., J. G. Tracey, and K. P. Haydock. 1967. A factor toxic to seedlings of the same
        species associated with living roots of the nongregarious subtropical rain forest
        tree *Grevillea robusta*. J. Appl. Ecol. 4:13–25.
Webb, L. J., J. G. Tracey, and W. T. Williams. 1972. Regeneration and pattern in the sub-
        tropical rain forest. J. Ecol. 6:675–695.
Westman, W. E. 1968. Invasion of fir forest by sugar maple in Itasca Park, Minnesota. Bull.
        Torrey Bot. Club 95:172–186.
Whittaker, R. H. 1975. Communities and ecosystems. 2d. ed. Macmillan, New York. 385 pp.
Wilson, D. E., and D. H. Janzen. 1972. Predation on *Scheelea* palm seeds by bruchid
        beetles: seed density and distance from the parent palm. Ecology 53:955–959.
Zinke, P. J. 1961. Chronology of the Bull Creek sediments and the associated redwood
        forests. Pages 22–25 *in* Annual report, Redwood Ecology Project. Wildland
        Research Center, University of California, Berkeley.

Vol. 111, No. 978   The American Naturalist   March–April 1977

MAINTENANCE OF HIGH DIVERSITY IN CORAL REEF
FISH COMMUNITIES

Peter F. Sale

School of Biological Sciences, University of Sydney, Sydney, N.S.W. 2006, Australia

The high diversity typical of tropical communities has been of interest for some time (Fischer 1960; Connell and Orias 1964; MacArthur 1965, 1969; Paine 1966; Pianka 1966; Whittaker 1969). Colwell (1973), remarking on the causes of this high diversity, pointed to "rather general agreement on theoretical grounds that tropical species should be more specialized ecologically than species at higher latitudes." He may have overstated the degree of consensus, since a number of alternative hypotheses have been proposed and not all of them predict greater specialization in the tropics (see summaries in Krebs 1972; MacArthur 1972; McNaughton and Wolf 1973; Pianka 1974). Pianka (1974) provides a table of 10 hypothetical mechanisms for increasing diversity. Of the 10, five act by reducing niche breadth, and two others act by allowing greater resource overlap and competition among coexisting species. A fair statement of the current consensus is that the tropical community is a mature equilibrium community of numerous species whose coexistence is satisfactorily explained in the theory based on Lotka-Volterra competition and predation equations. This "equilibrium view" takes no account of environmental change or patch structure and emphasizes ways of efficiently partitioning resources.

This view of the tropical community has developed through theoretical considerations of resource partitioning (e.g., MacArthur and Levins 1967; Roughgarden 1974; Roughgarden and Feldman 1975) and through field investigations, primarily of terrestrial biota, particularly birds (Diamond 1973; MacArthur 1969), but including insects (Janzen 1973) and forest trees (Janzen 1970; Whittaker 1969). This emphasis on terrestrial groups might be expected to make it a view which most accurately describes the tropical rain forest. However, not all scientists working in rain forest communities accept it (e.g., Connell 1971).

While we know little about rain forests, we know still less about coral reef communities. Until recently, Kohn's (1959, 1968, 1971) data represented practically the only thorough study of resource partitioning in an assemblage of coral reef species. They showed a high degree of specialization and little overlap in food requirements of coexisting species of *Conus*, an important genus of reef gastropod. These results are consistent with the view that tropical communities are comprised of numerous specialist species effectively dividing

Amer. Natur. 1977. Vol. 111, pp. 337–359.

the resources available, and in the absence of information to the contrary it has seemed best to assume that coral reef communities are organized in this way.

In fact, the limited data now available on reef fishes appear not to support the equilibrium view. Inasmuch as reef fishes may be representative of other reef fauna, these data must force us to reconsider the maintenance of high within-habitat diversity in the coral reef community. Current data indicate that many reef fishes do not finely partition resources of food or living space, that they are often, if not usually, limited by the supply of suitable living space, and that they live in an environment which, with respect to the supply of living space, is unpredictable in time and space. This paper is an attempt to bring together and evaluate these data and to reexamine in their light the maintenance of high within-habitat diversity in reef fish communities. The conclusions reached here can only be tentative, but my hope is that they will stimulate a search for more data and experimental tests of the proposed hypotheses.

REEF FISHES—HIGH DIVERSITY AND HIGH NICHE OVERLAP

That reef fish communities are diverse is well known (Bakus 1966; Smith and Tyler 1972; Talbot and Goldman 1972). For example, the numbers of species on the Great Barrier Reef range from 1500 in the north to 900 at One Tree Reef close to the southern limit of reef development.

Part of this high diversity is of a between-habitat type, since reef fishes tend to be sedentary and, to a limited extent, habitat specialists. Nevertheless there remains a high diversity within habitats. For example, Smith and Tyler (1972) reported 53 resident species and a total of 75 species ($H'_{[base\ E]} = 3.3$) on a single patch reef approximately 3 m in diameter and 1.6 m high. Smith (1973) collected up to 67 species in single small rotenone samples in the Bahamas. Goldman and Talbot (1976) cite collections from single rotenone stations which contained 150 (One Tree Reef) and 200 (Palau) species of fish. Coexistence of such large numbers of species of fish on a small area of reef demands explanation. An attempt at explanation might begin with an examination of the degree of specialization and pattern of resource partitioning among coexisting species of reef fish.

If reef fishes are predominantly specialists which finely partition the resources available to them, it follows that an examination of their use of the major resources of food and habitat should disclose two things. Most species will be relatively narrow niched (i.e., specialized) with respect to one or both of these resource categories. More important, they will be specialized upon different resources, exhibiting low overlap in their use of food or habitat space. Some species may be more generalized, but these would be expected to be in the minority. As will be shown below, however, the degree to which reef fishes are narrow niched and nonoverlapping in requirements is not great. It appears insufficient to account completely for their high diversity within habitats.

General feeding habits of reef fishes have been examined by Hiatt and Strasburgh (1960) and Randall (1967). Bakus (1969) has reviewed feeding processes in shallow marine waters. Herbivorous fishes, uncommon outside coral reef communities, comprise roughly 22% by weight and a similar percentage of species of reef fishes (Bakus 1966). They are notably generalists in their feeding habits. Choat (1969) could find no differences in foods taken by Scaridae on Heron Reef, Great Barrier Reef; Jones (1968), although emphasizing those differences which did exist, found few differences in foods taken by Acanthuridae in Hawaii. In both of these major families of herbivores, individuals feeding in the same place take the same foods, and multispecies feeding schools are commonly seen.

Predatory and omnivorous species are somewhat more specialized in diet. Roughgarden (1974) used Randall's (1967) data to demonstrate partitioning by food particle size in the Serranidae, and Vivien and Peyrot-Clausade (1974) have demonstrated a partitioning of prey by taxonomic category among three species of Holocentridae. Nevertheless it remains true that differences between the foods of coexisting, related species are usually minor (see Hiatt and Strasburg 1960, Vivien 1973), and the degree of partitioning of food resources demonstrated in *Conus* (Kohn 1959) has not been approached in any study of an assemblage of reef fishes.

Bakus (1969) compared several species of Labridae and claimed that tropical species were more specialized in diet than closely related temperate ones. His measure of niche breadth yields low values (high specialization) when small proportions of the sample of fishes examined contain each type of food, even if many different types of food are taken. Considering this bias, the small number of species examined, and the relatively slight difference in niche breadth demonstrated between tropical and temperate forms, his data can only be taken as suggestive. They urgently require substantiation.

There are a few pronounced food specialists among reef fishes, just as there are fish which feed upon items not available to fish in the temperate zone. Their existence should not be permitted to detract from the fact that the majority of reef fishes are generalist feeders. Even among the food specialists it is common for two or three sympatric species to show nearly identical specialization, thus overlapping greatly in the foods they consume (table 1). This would not be expected if the specializations had evolved as a means of partitioning the food resources available and suggests instead that those food sources specialized upon are in abundant supply.

A fine partitioning of living space appears, at first, a more likely characteristic of reef fishes. They are predominantly sedentary animals. Bardach (1958), Reese (1973), and Springer and McErlean (1962) have provided data on the range of movements of many species. Smith and Tyler (1972) provided estimates of the area used (none greater than 7 m radius) by individuals of 63 species, although individuals of 12 other species ranged more widely. Russell et al. (1974) have listed 58 of 85 species which colonized their artificial reefs as residents. Resident fish restricted their movements to the immediate vicinity of a single reef, 1.6 × 0.6 × 0.6 m in size. Ogden and Buckman (1973) demonstrated that the apparently wide-ranging feeding schools of *Scarus croicensis* restricted their movements to about $\frac{1}{2}$ hectare of a reef and that their membership included

TABLE 1

SOME REPRESENTATIVE EXAMPLES OF GROUPS OF SYMPATRIC REEF FISHES WHICH ARE FOOD SPECIALISTS WITH
SIMILAR DIETARY SPECIALIZATIONS

Species in Group	Comments	Source
Holacanthus ciliaris, H. tricolor, Pomacanthus arcuatus, P. paru, Cantherines marcrocerus	Sponges comprise at least 70% of diet of all species; all consume a wide range of species of sponge.	Randall and Hartman 1968
Chaetodon citrinellus, C. lunula, C. triangulum, C. trifasciatus, Megaprotodon strigangulum	All feed principally upon polyps of *Acropora* spp. *M. strigangulum* and *C. triangulum* defend feeding territories where they coexist with *C. trifasciatus* at Heron Reef	Reese 1973
	M. strigangulum, C. lunula and *C. citrinellus* occur together at Eniwetok Atoll	Hiatt and Strasburg 1960
Epibulus insidiator, Gomphosus varius.............	Both adapted (though in different ways) to feed upon crustacea found within ramose coral heads	Hiatt and Strasburg 1960
Labroides bicolor, L. dimidiatus, L. rubrolabiatus	All are exclusively "cleaner" species which coexist in the Society and Tuamotu Islands, where they were seen occupying the same cleaning station	Randall 1958
	L. bicolor and *L. dimidiatus* were observed with contiguous cleaning areas at Heron Reef	Robertson 1974

individuals that successively joined and then departed from the school, remaining members only while the school traversed their own home sites of up to 50 m².

Such sedentary habits are conducive to the development of narrow habitat requirements by reef fishes, especially when one considers the diverse array of habitats provided on a coral reef. As habitat specialists, reef fishes could finely partition the living space available to them as long as each species used a slightly different type of space.

Habitat partitioning does occur, to the extent that few species of reef fish occur over all regions of a reef. Thus it is profitable, as has long been recognized, to list the species likely to occur in broadly defined habitats such as "surge channel" and "reef flat" (see Choat 1969; Fishelson et al. 1974; Hiatt and Strasburg 1960; Jones 1968; Talbot and Goldman 1972). The resulting lists are lengthy, and little further separation of species is attained by a finer resolution of habitats. Clarke has recently quantified habitat overlap for pomacentrid and chaetodontid fishes near Bimini. Despite clear partitioning of space by fish, five of the eight habitats examined each contained 17 of the 22 species recorded, leading him to conclude that coexistence must largely be explained by factors which operate within habitats (Clarke, in press).

When we examine the space requirements of fish which coexist within these broadly defined habitats, we do not find, as a general rule, a high degree of specialization on microhabitats and a resulting fine partitioning of living space. A small minority of primarily inquiline forms are highly specialized and are discussed below, but the species listed in table 2 are representative of the majority of species for which data exist. Consider, for example, *Dascyllus aruanus* with a home range of only 1 m radius. It requires living branched coral adjacent to a patch of sand, yet on Heron Reef it uses 12 species from four genera of coral and sand patches from at least a 15-m depth on the reef slope, to the reef flat, and lagoonal patch reefs (Sale 1970, 1971a). It is easy to visualize many ways in which this sedentary species might have developed more specialized requirements. Yet it has not and coexists with several other species with similar habitat needs.

An alternative approach to the question of how specialized the space requirements of reef fishes are is to compare them with fishes of temperate rocky shores. Adopting this approach entails an obvious danger in trying to compare the degree of specialization of species from different places (the koala-opossum paradox of Colwell and Futuyma 1971), and furthermore, quantitative data are sparse. The data in table 3 are representative of what is available for temperate species. I suggest that the reef fishes in table 2 are not notably more specialized in requirements for space than the temperate shore fishes in table 3, but a definitive statement must await a comparative study of reef and temperate communities.

As noted above, there are a minority of species on reefs with highly specialized habitat needs, especially among fish with inquiline habits. Yet, just as feeding habits of sympatric food specialists may coincide, the habitat requirements of coexisting habitat specialists often overlap greatly (table 4). The partitioning

TABLE 2

Some Recent Attempts to Characterize the Habitat Requirements of Particular Reef Fishes (Compare with Table 3)

Species	Known Requirements	Source
Acanthuridae (20 spp)...............	Assigned to midwater, surge zone, sand patch, and subsurge reef habitats	Jones 1968
Acanthurus triostegus	Juvenile requires shallow water, cover within 1 m, and algal food on the substratum	Sale 1969
Dascyllus aruanus	Requires living, branched coral (12 spp. at Heron Reef) adjacent to sand patch, and 0–15 m depth of water	Sale 1970, 1971a
D. trimaculatus	Shelters among large coral boulders of many species, in shallow water; juveniles also use anemone, Gyrostoma, and sea urchin, Diadema	Fricke 1973
Pomacentrus flavicauda..............	Require dead coral rubble interspersed with sand; occurrence is correlated with amount of rubble-sand interface	Low 1971
Opistognathus aurifrons	Build burrows in substratum which must contain rubble mixed with sand; within 1 m of rock outcrops, 2–50 m depth	Colin 1973

TABLE 3

EXAMPLES OF HABITAT REQUIREMENTS HELD BY TEMPERATE MARINE FISHES

Species	Known Requirements	Source
Chromis dispilus	Aggregate above rocky outcrops or cliffs in 10–30 m depth; nest sites on smooth even rocky surfaces	Russell 1971
Hypsypops rubicunda	Favor high-relief rocky substrata in 0–20 m depth, use large holes or crevices for shelter	Clarke 1970
Crenilabrus melanocercus	Occurred in 15–18 m depth at foot of cliff where *Posidonia* covered slope commenced	Potts 1968
Hypsoblennius gilberti, H. jenkinsi, H. gentilis	*Gilberti* occurs from intertidal to 5 m depth over any hard substratum; *jenkinsi* occurs subtidally to 10 m, using holes or small crevices, esp. pholadid burrows, *Serpulorbis* masses and fouling masses; *gentilis* at intermediate depths, usually subtidal, most common in *Mytilus* beds	Stephens et al. 1970
Paralabrax clathratus	Requires some bottom relief whether provided by rock, kelp, or debris; ranges 0–30 m depth	Quast 1968
Chasmodes bosquianus	Shelter enclosures must have a somewhat restricted opening and firm walls, dead oyster shells the most common shelter	Phillips 1971

TABLE 4

EXAMPLES AMONG REEF FISHES OF COEXISTING HABITAT SPECIALISTS

Species	Comments	Source
Evermannichthys metzelaari, *Pariah scotius*	Obligate sponge-dwelling gobies collected from same 5 specimens of *Sphecio-spongia vesparia*; *E. metzelaari* occurred only in that species	Tyler and Bohlke 1972
Gobiosoma chancei, G. horsti, *Risor ruber, Phaeoptyx xenus*	All are obligate sponge dwellers; *G. horsti* and *P. xenus* were collected from the same chimneys of specimens of 2 species of sponge; *G. chancei* occurred in the same chimneys of specimens of a third species of sponge as *P. xenus* and as *R. ruber* but not in the same chimneys	Tyler and Bohlke 1972
Amphiprion chrysopterus, *A. melanopus, A. tricinctus*	Obligate commensals of anemones, all occur at Eniwetok Atoll where *A. tricinctus* occupies both species used by *A. melanopus,* and 2 of 3 used by *A. chrysopterus*	Allen 1972
Gobiodon ceramensis, G. erythrospilus, *G. histrio, G. quinquestrigatus*	Obligate commensals of living branched corals, all were collected from *Acropora corymbosa* more often than from other species, and overlapped extensively in the other species in which they occurred	Tyler 1971

of living space is not achieved by the coexistence of specialists that are specialized on the same types of space.

It is surprising that reef fishes have failed to partition their habitats more finely than their foods, because there is considerable evidence that reef fishes are far more likely to compete for suitable space than for food (Smith and Tyler 1972). For example, sea grasses away from the shelter of reef outcrops in Puerto Rico are not grazed by reef fishes (Randall 1965), and fish are rapidly recruited to artificial reefs established in reef lagoons (Russell et al. 1974). The refilling of space when territorial (Low 1971; Sale 1974, 1975), or nonterritorial (Sale and Dybdahl 1975) fish are experimentally removed, and the existence, in at least one species, of behavioral mechanisms which achieve an efficient dispersion of individuals over the available habitat (Sale 1972a, 1972b) all suggest that suitable space is important and frequently limiting. Species vary in the degree to which the individual maintains exclusive occupancy of a living site. Many fish require only a suitable shelter site within which they sleep, while some carry on all activities within a territory from which their own or many species may be excluded.

Because of the high overlap in habitat and food requirements among many reef fishes, the concept of a guild (Root 1967) is useful. A guild is a group of species which use the same environmental resources in similar ways. The members of a guild of reef fishes may be spatially separated to some extent, but the high diversity of fish within reef habitats derives from the simultaneous presence of two or three species of each of a number of guilds sharing their common space and food resources. Examples of guilds where habitat overlap is known to occur are listed in table 5.

The guild of territorial Pomacentridae present on Heron Reef (Sale 1974) can serve as an example. Some spatial separation exists among its eight members, but at least three species are potentially resident on any site suitable for this guild, and six species can come into contact on the reef crest. Most information is available for the three species which occur on the upper reef slope. These are *Eupomacentrus* (= *Pomacentrus*) *apicalis*, *Pomacentrus wardi*, and *Plectroglyphidodon* (= *Abudefduf*) *lacrymatus*. Within rubble patches on the upper slope, all available space is occupied by a series of contiguous and usually nonoverlapping territories held by individuals of these three species. All individuals hold territories throughout juvenile and adult life. When fishes die the vacated space is rapidly refilled by residents and by new colonists. A site within a rubble patch can be used by any of these three species, and I have found no tendency for space initially held by one species to be taken up, following mortality, by the same species. There is no evidence of a successional sequence of ownership of a site by the three species in turn (Sale 1975), nor have I yet detected any differences between *E. apicalis* and *P. lacrymatus* in abilities to capture space and to hold onto space once gained. *Pomacentrus wardi* shows higher rates of recruitment and mortality in rubble patches and uses a refuge in nonpreferred space off rubble patches to maintain its occupancy of space inside rubble patches in the face of competition with the other two species (Sale 1974, 1975). Even the differences in competitive ability between *P. wardi* and

TABLE 5

EXAMPLES OF GUILDS OF REEF FISHES WITHIN WHICH SOME DEGREE OF HABITAT OVERLAP IS KNOWN TO OCCUR

Species in Group	Comments	Source
Acanthurus (12 spp), *Ctenochaetus* (2 spp), *Naso* (4 spp), *Zebrasoma* (2 spp)	20 species of Acanthuridae from Hawaii and Johnston Islands fall into 3 guilds based on foraging methods; there is extensive habitat overlap in all guilds; heterotypic feeding schools common	Jones 1968 (see also Barlow 1974)
Scarus (15 spp)	All species forage on the reef slope at Heron Reef; at least 5 species forage in each of the other habitats recognized; heterotypic schools are common and may include Acanthuridae and Siganidae	Choat 1969
Glyphidodontops (*Abudefduf*) *biocellatus*, *Plectroglyphidodon* (*A.*) *lacrymatus*, *Eupomacentrus* (*Pomacentrus*) *apicalis*, *E.* (*P.*) *fasciolatus* (*jenkinsi*), *E. gascoynei*, *Pomacentrus bankanensis* (*dorsalis*), *P. flavicauda*, *P. wardi*	All are territorial herbivores on coral rubble substrata at Heron Reef; territories are defended interspecifically; *P. wardi* occurs wherever any member of the guild might occur; the other 7 species are partially spatially separated but at least 3 species may occur at any rubble patch	Sale 1974
Chromis atripectoralis, *C. caeruleus*, *C. dimidiatus*, *C. lepidolepis*, *C. leucurus*, *C. ternatensis*	All are midwater planktivores which hover 1–2 m above the substratum while feeding; heterotypic feeding schools of at least 3 species occur at Eniwetok Atoll; *C. atripectoralis* and *C. caeruleus* are nearly identical in ecology, as are *C. dimidiatus* and *C. leucurus*	Swerdloff 1970
Chaetodon triangulum, *C. trifasciatus*, *Megaprotodon strigangulum*	All feed on coral polyps and maintain feeding territories or home ranges on the Heron Reef slope	Reese 1973
Holocentrus diadema, *H. lacteoguttatus*, *H. scythrops*, *H. spinifer*, *H. xantherythrus*, *H. tiere*, *Holotrachys lima*, *Myripristis argyromus*, *M. berndti*, *M. multiradiatus*	All are nocturnal predators feeding upon a wide range of invertebrates; some species are morphologically very similar; all were taken in one or both collections made in 8–12 and 10–25 m water in Hawaii; heterotypic schools occur	Gosline 1965

NOTE.—That these guilds each contain species from a single family may indicate the approach of the investigator rather than the true extent of the guild. Roughgarden (1974) has discussed a scarid-acanthurid guild.

the other two species are minimal, and I have observed newly settled juveniles of all three species maturing while surrounded by adults of the other species (Sale 1974). Coexistence of the three species on rubble patches is thus maintained despite virtually identical space requirements and, for two of them, nearly identical ability to acquire and hold space.

To summarize, all of the data presented above suggest that reef fishes are habitat specialists to the extent that broadly defined habitats on a reef contain different assemblages of species. Within habitats, however, reef fishes, with few exceptions, do not show extreme specialization with respect either to food or to space requirements. Furthermore, those few species with highly specialized requirements often coexist with other species showing the same specializations. Characteristically, several members of each guild of fishes coexist in the same habitat, often foraging in the same places at the same time. There is evidence that living space, in particular, may be limiting to reef fish populations, yet competition for space has not resulted in a fine degree of partitioning of space among the species present.

It is not a necessary requirement of the equilibrium view that reef species finely partition the resources available. Theoretical studies indicate that more overlap in resource requirements may be tolerated in tropical communities if coexisting species have leptokurtic resource utilization curves (Roughgarden 1974) or if predation is more intense (Roughgarden and Feldman 1975). No field data exist to indicate that either of these possibilities applies in reef communities, and other theoretical work rules out the possibility that overlap can be greater in the tropics simply because environmental conditions are less variable there (May and MacArthur 1972). The data considered in the remainder of this paper provide strong support for an alternative view.

THE REEF ENVIRONMENT—BENIGN BUT UNPREDICTABLE

Conditions in an environment vary. Environments can be severe, showing large variations in conditions, or they can be benign, showing only slight variations. The changes in conditions will be patterned in a predictable environment and without pattern in an unpredictable one. While severity and predictability may be correlated (Slobodkin and Sanders 1969), it is best to consider them as separate attributes.

Variations in environmental conditions may occur temporally or spatially and on a larger or a smaller scale. Reef fishes, because they are sedentary and have life spans of 1 to several years, will be most influenced by small-scale spatial (over meters rather than kilometers) and temporal (over months or years rather than decades) changes. There is evidence that, on this scale, the reef environment is both spatially and temporally unpredictable in topography, despite being a generally benign environment.

A considerable small-scale spatial patchiness is evident on reefs primarily because of the differing growth forms of various species of coral. In addition, living coral is interspersed with a variety of other substrata. Superimposed upon this small-scale patchiness is a larger-scale pattern: the topographic zonation to which considerable attention has been given in the past (see Stoddart 1969a). This zoned pattern is quite coarse. For example Loya and Slobodkin (1971) recognized only six zones to a depth of 30 m at Eilat, Red Sea. To date a zoned pattern has not been demonstrated on the small scale where topographic variation is due principally to the presence of corals of differing form. It is this unzoned small-scale variation which influences reef fishes.

A variety of temporal changes occur on reefs which have a direct effect on the availability of particular types of living space for fishes. Many of these changes are unpredictable in time of occurrence and affect only a small area of reef. Growth processes of the corals themselves change the relative availability of particular kinds of living space. These changes are predictable, but they are frequently and unpredictably interrupted by events causing destruction of coral. Small-scale changes due directly or indirectly to wave action include the breaking off or turning over of colonies or parts of colonies, silting, and the shifting of sand or unconsolidated rubble. In addition, competition with other corals (Lang 1970), and the action of borers and other coral predators have an important effect. Connell (1973) has documented such small-scale changes for corals at Heron Reef.

Changes on a larger spatial scale, which remain unpredictable temporally, are produced by cyclonic storms. The varied effects of storms have been documented by Stoddart (1969b), Glynn et al. (1964), and Connell (1973). Physical abrasion of coral cover, rolling and tossing of coral boulders, and extensive movements of sand and rubble all cause pronounced changes to the small-scale topography over large portions of an affected reef.

The changes considered so far affect the availability of living space by directly creating or destroying habitat. Those brought about by storms and wave action must have their most pronounced effects on the shallower parts of a reef. Those due to biotic factors will be most important in the deeper, less physically disturbed regions. However, another type of change may be more important in all habitats than any of these. Predators of reef fishes affect the availability of living space by releasing space for reoccupancy. Predators may hunt in particular habitats and at particular times of day, but as far as the prey fishes are concerned the time and place of capture of each food item is unlikely to be predictable. And the removal of one prey fish creates a vacant living space of a particular type.

ADAPTATIONS TO AN UNPREDICTABLE SUPPLY OF SPACE

Observations on the life cycles of reef fishes provide evidence compatible with the claim that the unpredictability of the supply of living space is important to them. In a situation in which the supply of living space is limiting and vacant living space is generated unpredictably in space and time, the production of numerous offspring scattered widely in space and time appears to be the only satisfactory way of getting some offspring successfully to living sites. Similarly, successful offspring should stay put. (Selection for offspring with the ability to oust prior residents from suitable space might be an alternative possibility, but since this has not occurred among reef fishes, discussion of this point is deferred until later).

In fact, with very few exceptions, reef fishes are sedentary animals which breed often, producing numerous clutches of dispersive eggs or larvae. Breeding seasons are long, if not year round, and individuals breed frequently during the season. Munro et al. (1973) have documented the times of occurrence of reproductive ripe fish for 83 species off Jamaica. In many of these there is a pronounced peak of activity in February and March, but extended seasons are usual, and in some cases (e.g., Pomadasyidae, Lutjanidae, *Abudefduf saxatilis*) ripe fish are present throughout the year. Their data do not permit determining the frequency of spawning by individuals, but I would expect multiple spawning to occur in most cases. *Labroides dimidiatus*, whose adults breed every day just after high tide for 7 months at Heron Reef (Robertson 1974) represents an extreme, but data on a number of other species are listed in table 6. The extended seasons and the frequent clutches by each individual are the strongest evidence that these life cycles are in response to the unpredictability of the supply of living space in the reef environment. There could be other reasons for producing dispersive larvae.

The larvae of reef fishes are presumably dispersed by means of currents. To this extent, dispersal is not unpredictable. It remains, however, beyond the control of the adult population. Larval life may last as long as $2\frac{1}{2}$ months (*Acanthurus triostegus*, Randall 1961) and usually for at least a week, so that offspring can be dispersed widely.

As noted earlier, reef fishes do not often have highly specialized habitat requirements. Perhaps the necessity of finding living space provided so unpredictably precludes narrowing of the choice by development of stringent space requirements. Thus, despite a shortage of space, reef fishes are seldom extreme space specialists, and the result is that guilds exist with similar requirements for space.

MAINTENANCE OF HIGH DIVERSITY WITHIN HABITATS

The data summarized in the preceding pages demonstrate that most reef fishes are closely tied to the substratum and require living sites on a reef and that the supply of living space is often short and is in any event unpredictable. The fish succeed in obtaining this unpredictably supplied resource by (1) dispersing numerous pelagic larvae widely, both in space and time, so as to maximize chances of getting some offspring to suitable space; (2) having requirements for space that remain general enough for there to be some chance of finding a suitable site; and (3) remaining in a site, once found, for extended periods or throughout adult life. But as a result, guilds of reef fish exist with similar requirements for space. Competition within guilds for space can be expected to occur and is well documented in one case (Sale 1974, 1975). It is not immediately clear how diversity within reef habitats is maintained under these circumstances.

TABLE 6

PATTERNS OF REPRODUCTION IN SOME REPRESENTATIVE REEF FISHES

SPECIES	DURATION OF SEASON (Months)*	FREQUENCY OF SPAWNING BY INDIVIDUAL	PELAGIC LIFE†		SOURCE
			Eggs	Larvae	
Pomacentridae:					
Amphiprion chrysopterus	9[a]	Monthly	No	Yes	Allen 1972
Acanthochromis polyacanthus	4[b]	?	No	No	Robertson 1973
Chromis caeruleus	8[b]	Up to twice per week	No	Yes	Sale 1971b
Glyphidodontops biocellatus	2[b]	Each 2 weeks	No	Yes	Keenleyside 1972
A. saxatilis	5[c]	Multiple	No	Yes	Fishelson 1970
Eupomacentrus partitus	12[d]	Multiple	No	Yes	Myrberg 1972a
Labridae:					
Crenilabrus (8 spp)	?[e]	Weekly	No	Yes	Fiedler 1964
Thalassoma bifasciatum	12[d]	Many times during year	Yes	Yes	Feddern 1965
24 species in family	(most spp)[b]	Multiple	Yes	Yes	Choat 1969
Labroides dimidiatus	7[b] 6	Daily	Yes	Yes	Robertson 1974
Scaridae:					
Scarus (15 spp)	6[b]	Multiple	Yes	Yes	Choat 1969
Sparisoma rubripinne	12[f]	?	Yes	Yes	Randall and Randall 1963
Acanthurus triostegus	8[g], 12[a]	Multiple	Yes	Yes	Randall 1961
Gobiidae:					
Gobiosoma oceanops	5[d]	2–3 times per month	No	Yes	Valenti 1972

NOTE.—Species selected are from five of the most important families in numbers of species on reefs. Except for the Acanthuridae, there is considerably more information available on species of these families than is listed. However, I know of no species other than *Acanthochromis polyacanthus* without pelagic larvae. Pomacentridae have been particularly well studied (see Fishelson et al. 1974; Reese 1964; Russell 1971), but they are one of a few families (others are Apogonidae, Gobiidae, Blenniidae) with demersal eggs. Breder and Rosen (1966) indicate that extended seasons and pelagic eggs are usual among reef fishes, but little further information exists for most families. Observation sites are shown as follows: a = Eniwetok, b = Heron Reef, c = Eilat, d = Florida, e = Naples, f = Virgin Islands, g = Hawaii.

* Duration of season varies with latitude.

† Duration of larval life has been determined as 2½ months in *Acanthurus triostegus* (Randall 1961). Morphology of newly hatched larvae in many forms indicates a minimum of ½ to 1 month larval life.

I believe the maintenance of the high diversity of reef fish communities is a direct result of the unpredictability of the supply of living space. This unpredictability has forced fish to adopt the strategy of producing numerous, highly dispersive larvae, and therefore the recruitment of young to any place on a reef is largely independent of the composition of the population of adult fishes already there. Because of the requirement that they be numerous and dispersive, the young recruited are small, frail organisms unlikely to be very good at ousting other residents from spaces they hold. It is a measure of the unpredictability of the reef environment that no fish has evolved the production of a clutch of one or two nondispersive, but competitively superior, space-grabbing larvae. When colonists of two species of a guild compete for space it is probably a competition in which no particular species is favored. Instead, each interaction that occurs is likely to be decided in favor of the prior resident—the individual which is already at home has a psychological advantage (Braddock 1949; Frey and Miller 1972; Greenberg 1947; Myrberg 1972b; Phillips 1971), and frequently a size advantage, in any struggle. Thus, the species of a guild are competing in a lottery for living space in which larvae are tickets and the first arrival at a vacant site wins that site. The lottery operates within habitats and at the level of the individual fish. Single vacant sites are unpredictably generated and are filled by individual colonists, to be held until these fish also disappear and are replaced by new colonists of their own or of other species. In this lottery, there is no reason to expect that the number of chances at winning—that is, the number of larvae seeking to settle—will be closely correlated with the size of the population of adults of a species present in an area. Fisheries biologists find either no correlation or a very weak one between numbers of recruits to a fishery and the size of the adult population (Beverton and Holt 1957; Cushing 1973). Thus, chance successes by one species need not lead to that species progressively usurping all living spaces within habitats of that type from other species in the lottery.

That chance processes are important in allocating space among species coexisting within habitats does not imply that there is no pattern to the distribution of fishes over the reef. As discussed above, reef fishes are habitat specialists on a broad scale, and we can expect that newly settling larvae of each species will show appropriate habitat preferences. Each species competes in only some of the many lotteries operating for different kinds of living sites in different habitats on a reef.

A different sort of competition may develop between residents if, as they grow in size, they require larger living spaces. Fish which maintain exclusive use only of a shelter site may have to move to larger quarters, and territorial forms might have to increase the size of the area defended. In both circumstances, certain species may consistently win over others of their guild, but several factors act to prevent this competition from becoming very lopsided. For example, all residents will have the psychological advantage of being at home when neighbors attempt to encroach upon them, and smaller individuals may use the topography of their home sites by seeking shelter in places inaccessible to larger neighbors rather than fleeing from the area. In any event, this competition is for space for an individual—space that will become vacant again as soon as that individual dies.

In those guilds whose member species are morphologically quite similar, it is likely that the advantage of being at home will be sufficient to ensure that residents of any species usually outcompete invading individuals. However, even in guilds in which considerable morphological or behavioral differences exist between member species, the competitively inferior species may be able to survive by adopting a fugitive strategy. *Pomacentrus wardi* makes use of a fugitive strategy to coexist with the competitively superior members of its guild (Sale 1974, 1975). So long as all species of a guild win some of the time and in some places, they will continue to put larvae into the plankton and hence into the lottery for new sites.

As intimated above, data concerning the coexistence of *P. wardi* and other members of its guild (Sale 1974, 1975) are fully compatible with this lottery hypothesis. In addition, recent study of another guild provides further support for the concept. A wide range of species of fish on Heron Reef makes use of the interstices in living or dead coral as shelter. Isolated coral colonies are occupied by small groups of resident fishes belonging to several species. A study of the patterns by which species of fish were distributed among colonies of coral (Sale and Dybdahl 1975) indicated that chance was the principle determinant of species composition within single colonies of coral. These results indicate that the lottery hypothesis has validity for species that are not strictly territorial.

There are many similarities between the lottery hypothesis presented here, and the theory of island biogeography (MacArthur and Wilson 1967). In both systems the species composition of a single patch of habitat is a function largely of chance colonization and chance extinction, while the diversity within patches depends, in addition, on the species richness of the region that serves as a source of colonists—the mainland source area in island theory, and the totality of patches of similar type in the lottery system. However, one major difference exists between these systems. In the lottery system there is no provision for population growth within patches of habitat other than through colonization of the patch by additional individuals of the same species. All reproduction by residents is exported to other patches, and the analogues of immigration and extinction of species in island theory are the colonization and loss of individuals in the lottery hypothesis. In addition, the patches considered are smaller and more ephemeral than those considered in island theory.

This system of lotteries for space has many points in common with those mechanisms believed to explain the maintenance of diversity in other potentially space-limited communities. Considerable information now exists concerning the sessile fauna of the rocky intertidal zone (Connell 1961a, b; Dayton 1971). Much importance had been placed on predation by starfishes in maintaining the diversity of this community (Paine 1966, 1971), but this predation is only one of several ways in which vacant living space is generated here. Dayton (1971) has pointed to the importance of drifting logs in creating living space on some coasts, and it appears that the important characteristic is that, as on the coral reef, the production of new living space is unpredictable. Fager (1971) has emphasized the importance of chance patterns of settlement in determining the diversity and structure of communities on fouling plates, and Porter (1972) has suggested that predation by *Acanthaster* and other unpredictably timed events maintain the diversity of corals on reefs on the Pacific coast of Panama by generating vacant space.

Levins and Culver (1971), Horn and MacArthur (1972), Levin (1974), and Slatkin (1974) have all recently examined the theory of interactions of competing species in a patchy environment. While there are important differences in the assumptions of their various models, all are agreed that in some cases in which it would not be possible in a homogeneous environment, the coexistence of quite similar species is permitted in patchy environments because of migration between patches. Levins and Culver (1971) and Horn and Mac-Arthur (1972) considered the survival of a competitively inferior, fugitive species in the presence of a superior competitor. For the fugitive to survive, its rate of successful migration between patches must be greater than the rate of extinction of its populations within patches. Rate of extinction is higher in those patches occupied by the superior species. Horn and MacArthur (1972) point to the interesting case in which two different sorts of patch existed. Here, coexistence is sometimes possible even when the fugitive is inferior to its competitor in both types of patch. Horn and MacArthur (1972) state that it is possible to build up a community of fugitives so long as each species has a rate of migration between patches that is greater than that of any species competitively superior to it. Their assumption of random migration between patches seems likely to be applicable to the reef environment. For communities of fugitive species, the limit on the number of species is reached when migration rates become great enough that the environment ceases to be treated as patchy. The limit is higher in environments in which rates of extinction of local populations are greater. Slatkin (1974) modified the model of Levins and Culver (1971) to remove the assumption that species were independently distributed among habitat patches. His model predicts that similar species can never exclude each other from a region made up of patches between which migration occurs. If the species differ sufficiently for one to exclude the other from the region, the system is determinate in the sense that there is no possibility of a priority effect whereby the first species to invade the region can always exclude the other.

Levin (1974) considered the survival of species similar enough that, once established in a patch, either could prevent the colonization of that patch by the other species. Again, migration between patches did not prevent coexistence, although his model tolerates less migration than that of Slatkin (1974). Levin points out that it is the patchy quality of the environment, rather than the number of kinds of patch, that permits coexistence.

The results of Horn and MacArthur (1972) and Slatkin (1974) in particular seem applicable to the coral reef environment and may provide a basis for a numerical treatment of the lottery hypothesis. The life histories of reef fishes are such that migration between patches occurs at each generation, and the variety of small-scale disturbances on reefs makes extinction rates of local populations (i.e., within patches) high. These important facets of the ecology of reef fishes are ignored in the more classical theories supporting the equilibrium view.

CONCLUSIONS

If the system I advocate is correct, we can make two predictions about reef fish communities. The first is that at the level of the species, rather than the guild, reef fish communities have an unstable structure (that is, the species composition of the fish at a site will not tend to recover following disturbance caused by the addition or removal of fish). The relative abundance of the species of a guild at any site is largely a result of the chance recruitment of young to that site and will change from time to time. Neighboring sites may show different patterns of change. The selective experimental removal of individuals of one species of a guild should not be followed rapidly by recovery of the original species composition of that site.

Changes in abundance of species of reef fishes have not been noted, partly because appropriate data have not been collected and partly because the reef environment is benign. It is unlikely that naturally occurring changes in abundance will be pronounced or occur over an extensive area of reef. But this is constancy (Lewontin 1969), not stability.

The second prediction is that the diversity of reef fish communities is directly correlated with the rate of small-scale, unpredictable disturbances (including predation) to the supply of living space. This can be tested by measuring the diversity of fishes in similar habitats in sheltered and in exposed reef areas. Diversity should be higher in the exposed sites, since rates of small-scale disturbances will be higher there. It can also be tested experimentally by manipulating the rate of disturbance, perhaps by increasing the rate of predation through a program of spearing randomly selected resident fishes.

SUMMARY

Data have been drawn together to demonstrate that reef fishes by and large are food and habitat generalists with a large amount of overlap in requirements among coexisting species. Suitable living space is the resource most likely to be in short supply for them, and their environment, although benign, is one in which the supply of living space is both spatially and temporally unpredictable. The argument is developed that reef fishes are adapted to this unpredictable supply of space in ways which make interspecific competition for space a lottery in which no species can consistently win. Thus, the high diversity of reef fish communities may be maintained because the unpredictable environment prevents development of an equilibrium community.

ACKNOWLEDGMENTS

This research was supported through a grant from the Australian Research Grants Committee. I thank R. Dybdahl for his assistance and Drs. A. J.

Underwood, G. J. Caughley, and A. C. Hodson for discussion and comments on the manuscript. Dr. J. Roughgarden and several anonymous referees also provided constructive criticism.

LITERATURE CITED

Allen, G. R. 1972. The Anemonefishes, their classification and biology. T.F.H. Publications, Neptune City, N.J.

Bakus, G. J. 1966. Some relationships of fishes to benthic organisms on coral reefs. Nature 210:280–284.

————. 1969. Energetics and feeding in shallow marine waters. Int. Rev. Gen. Exp. Zool. 4: 275–369.

Bardach, J. E. 1958. On the movements of certain Bermuda reef fishes. Ecology 39:139–146.

Barlow, G. W. 1974. Contrasts in social behavior between Central American cichlid fishes and coral-reef surgeon fishes. Amer. Zool. 14:9–34.

Beverton, R. J. H., and S. J. Holt. 1957. On the dynamics of exploited fish populations. Fishery Invest. Min. Agr. Fisheries Food (Gr. Brit.) Ser. II, 19:1–533.

Braddock, J. C. 1949. The effect of prior residence upon dominance in the fish *Platypoecilus maculatus*. Physiol. Zool. 22:161–169.

Breder, C. M., Jr., and D. E. Rosen. 1966. Modes of reproduction in fishes. American Museum Natural History, New York.

Choat, J. H. 1969. Studies on labroid fishes. Ph.D. diss. University of Queensland, Australia.

Clarke, R. D. In press. Habitat distribution and species diversity of pomacentrid and chaetodontid fishes near Bimini, Bahamas. Marine Biol.

Clarke, T. A. 1970. Territorial behavior and population dynamics of a pomacentrid fish, the garibaldi, *Hypsypops rubicunda*. Ecol. Monogr. 40:189–212.

Colin, P. L. 1973. Burrowing behavior of the yellowhead jawfish *Opistognathus aurifrons*. Copeia 1973:84–90.

Colwell, R. K. 1973. Competition and coexistence in a simple tropical community. Amer. Natur. 107:737–760.

Colwell, R. K., and D. J. Futuyma. 1971. On the measurement of niche breadth and overlap. Ecology 52:567–576.

Connell, J. H. 1961a. The influence of interspecific competition and other factors on the distribution of the barnacle *Chthamalus stellatus*. Ecology 42:710–723.

————. 1961b. Effect of competition, predation by *Thais lapillus*, and other factors on natural populations of the barnacle *Balanus balanoides*. Ecol. Monogr. 31:61–104.

————. 1971. On the role of natural enemies in preventing competitive exclusion in some marine animals and in rain forest trees. Pages 298–312 *in* P. J. den Boer and G. R. Gradwell, eds. Proceedings of the Advanced Study Institute on Dynamics of Numbers in Populations. Center for Agricultural Publishing and Documentations, Wageningen, Netherlands.

————. 1973. Population ecology of reef-building corals. Pages 205–245 *in* O. A. Jones and R. Endean, eds. Biology and Geology of coral reefs. Vol. 2, Biology 1. Academic Press, New York.

Connell, J. H., and E. Orias. 1964. The ecological regulation of species diversity. Amer. Natur. 98:399–414.

Cushing, D. H. 1973. Dependence of recruitment on parent stock. J. Fisheries Res. Board Can. 30:1965–1976.

Dayton, P. K. 1971. Competition, disturbance and community organization: the provision and subsequent utilization of space in a rocky intertidal community. Ecol. Monogr. 41:351–389.

Diamond, J. M. 1973. Distributional ecology of New Guinea birds. Science 179:759–769.

Fager, E. W. 1971. Pattern in the development of a marine community. Limnol. Oceanogr. 16:241–253.

356   THE AMERICAN NATURALIST

Feddern, H. A. 1965. The spawning, growth, and general behavior of the bluehead wrasse, *Thalassoma bifasciatum* (Pisces: Labridae). Bull. Marine Sci. 15:896–941.

Fiedler, K. 1964. Verhaltensstudien an Lippfischen der Gattung *Crenilabris*. (Labridae, Perciformes). Z. Tierpsychol. 21:521–591.

Fishelson, L. 1970. Behaviour and ecology of a population of *Abudefduf saxatilis* (Pomacentridae, Teleostei) at Eilat (Red Sea). Anim. Behav. 18:225–237.

Fishelson, L., D. Popper, and A. Avidor, 1974. Biosociology and ecology of pomacentrid fishes around the Sinai Peninsula, northern Red Sea. J. Fish Biol. 6:119–133.

Fischer, A. G. 1960. Latitudinal variations in organic diversity. Evolution 14:64–81.

Frey, D. F., and R. J. Miller. 1972. The establishment of dominance relationships in the blue gourami, *Trichogaster trichopterus* (Pallas). Behaviour 42:8–62.

Fricke, H. W. 1973. Ecology and social behaviour of the coral reef fish *Dascyllus trimaculatus* (Pisces, Pomacentridae). Z. Tierpsychol. 32:225–256.

Glynn, P. W., L. R. Almodovar, and J. G. Gonzales. 1964. Effects of Hurricane Edith on marine life in La Parguera, Puerto Rico. Caribbean J. Sci. 4:335–345.

Goldman, B., and F. H. Talbot. 1976. Aspects of the ecology of coral reef fishes. Pages 125–154 *in* O. A. Jones and R. Endean, eds. Biology and geology of coral reefs. Vol. 3, Biology 2. Academic Press, New York.

Gosline, W. A. 1965. Vertical zonation of inshore fishes in the upper water layers of the Hawaiian Islands. Ecology 46:823–831.

Greenberg, B. 1947. Some relations between territory, social hierarchy, and leadership in the green sunfish (*Lepomus cyanellus*). Physiol. Zool. 20:267–299.

Hiatt, R. W., and D. W. Strasburg. 1960. Ecological relationships of the fish fauna on coral reefs of the Marshall Islands. Ecol. Monogr. 30:65–127.

Horn, H. S., and R. H. MacArthur. 1972. Competition among fugitive species in a harlequin environment. Ecology 53:749–752.

Janzen, D. H. 1970. Herbivores and the number of tree species in tropical forests. Amer. Natur. 104:501–528.

———. 1973. Sweep samples of tropical foliage insects: effects of seasons, vegetation types, elevation, time of day, and insularity. Ecology 54:687–708.

Jones, R. S. 1968. Ecological relationships in Hawaiian and Johnston Island Acanthuridae (Surgeonfishes). Micronesica 4:309–361.

Keenleyside, M. H. A. 1972. The behaviour of *Abudefduf zonatus* (Pisces: Pomacentridae) at Heron Island, Great Barrier Reef. Anim. Behav. 20:763–775.

Kohn, A. J. 1959. The ecology of *Conus* in Hawaii. Ecol. Monogr. 29:47–90.

———. 1968. Microhabitats, abundance, and food of *Conus* on atoll reefs in the Maldive and Chagos Islands. Ecology 49:1046–1061.

———. 1971. Diversity, utilization of resources, and adaptive radiation in shallow-water marine invertebrates of tropical oceanic islands. Limnol. Oceanogr. 16:332–348.

Krebs, C. J. 1972. Ecology. The experimental analysis of distribution and abundance. Harper & Row, New York.

Lang, J. C. 1970. Inter-specific aggression within the scleractinian reef corals. Ph.D. diss. Yale University.

Levin, S. A. 1974. Dispersion and population interactions. Amer. Natur. 108:207–228.

Levins, R., and D. Culver. 1971. Regional coexistence of species and competition between rare species. Proc. Nat. Acad. Sci. (U.S.A.) 68:1246–1248.

Lewontin, R. C. 1969. The meaning of stability. Brookhaven Symp. Biol. 22:13–23.

Low, R. M. 1971. Interspecific territoriality in a pomacentrid reef fish. *Pomacentrus flavicauda* Whitley. Ecology 52:648–654.

Loya, Y., and L. B. Slobodkin. 1971. The coral reefs of Eilat (Gulf of Eilat, Red Sea). Symp. Zool. Soc. London 28:117–139.

MacArthur, R. H. 1965. Patterns of species diversity. Biol. Rev. 40:410–533.

———. 1969. Patterns of communities in the tropics. Biol. J. Linnaean Soc. (London) 1: 19–30.

———. 1972. Geographical ecology: patterns in the distribution of species. Harper & Row, New York.

MacArthur, R. H., and R. Levins. 1967. The limiting similarity, convergence and divergence of coexisting species. Amer. Natur. 101:377–385.

MacArthur, R. H., and E. O. Wilson. 1967. The theory of island biogeography. Princeton University Press, Princeton, N.J.

McNaughton, S. J., and L. L. Wolf. 1973. General ecology. Holt, Rinehart & Winston, New York.

May, R. M., and R. H. MacArthur. 1972. Niche overlap as a function of environmental variability. Proc. Nat. Acad. Sci. (U.S.A.) 69:1109–1113.

Munro, J. L., V. C. Gaut, R. Thompson, and P. H. Reeson. 1973. The spawning seasons of Caribbean reef fishes. J. Fish Biol. 5:69–84.

Myrberg, A. A., Jr. 1972a. Ethology of the bicolor damselfish, Eupomacentrus partitus (Pisces: Pomacentridae): a comparative analysis of laboratory and field behaviour. Anim. Behav. Monogr. 5:199–283.

———. 1972b. Social dominance and territoriality in the bicolor damselfish, Eupomacentrus partitus (Poey) (Pisces: Pomacentridae). Behaviour 41:207–231.

Ogden, J. C., and N. S. Buckman. 1973. Movements, foraging groups, and diurnal migrations of the striped parrotfish Scarus croicensis Bloch (Scaridae). Ecology 54:589–596.

Paine, R. T. 1966. Food web complexity and species diversity. Amer. Natur. 100:65–75.

———. 1971. A short-term experimental investigation of resource partitioning in a New Zealand rocky intertidal habitat. Ecology 52:1096–1106.

Phillips, R. R. 1971. The relationship between social behavior and the use of space in the benthic fish Chasmodes bosquianus Lacépède (Teleostei, Blenniidae). II. The effect of prior residency on social and enclosure behavior. Z. Tierpsychol. 29:389–408.

Pianka, E. R. 1966. Latitudinal gradients in species diversity: a review of concepts. Amer. Natur. 100:33–46.

———. 1974. Evolutionary ecology. Harper & Row, New York.

Porter, J. W. 1972. Predation by Acanthaster and its effect on coral species diversity. Amer. Natur. 106:487–492.

Potts, G. W. 1968. The ethology of Crenilabrus melanocercus, with notes on cleaning symbiosis. J. Marine Biol. Assoc. U.K. 48:279–293.

Quast, J. C. 1968. Observations on the food and biology of the kelp bass, Paralabrax clathratus, with notes on its sportfishery at San Diego, California. Pages 81–108 in W. J. North and C. L. Hubbs, eds. Utilization of kelp-bed resources in Southern California. Fish Bulletin No. 139, Calif. Dep. Fish Game.

Randall, J. E. 1958. A review of the labrid fish genus Labroides, with descriptions of two new species and notes on ecology. Pacific Sci. 12:327–347.

———. 1961. A contribution to the biology of the convict surgeonfish of the Hawaiian Islands, Acanthurus triostegus sandvicensis. Pacific Sci. 15:215–272.

———. 1965. Grazing effect on sea grasses by herbivorous reef fishes in the West Indies. Ecology 46:255–260.

———. 1967. Food habits of reef fishes of the West Indies. Stud. Trop. Oceanogr. 5:665–847.

Randall, J. E., and W. D. Hartman. 1968. Sponge-feeding fishes of the West Indies. Marine Biol. 1:216–225.

Randall, J. E., and H. A. Randall. 1963. The spawning and early development of the Atlantic parrotfish, Sparisoma rubripinne, with notes on other scarid and labrid fishes. Zoologica 48:49–60.

Reese, E. S. 1964. Ethology and marine zoology. Oceanogr. Marine Biol. Annu. Rev. 2:455–488.

———. 1973. Duration of residence by coral reef fishes on "home" reefs. Copeia 1973:145–149.

Robertson, D. R. 1973. Field observations on the reproductive behaviour of a pomacentrid fish, *Acanthochromis polyacanthus*. Z. Tierpsychol. 32:319–324.

———. 1974. The ethology and reproductive biology of *Labroides dimidiatus*. Ph.D. diss. University of Queensland, Australia.

Root, R. B. 1967. The niche exploitation pattern of the blue-grey gnatcatcher. Ecol. Monogr. 37:317–350.

Roughgarden, J. 1974. Species packing and the competition function with illustrations from coral reef fish. Theoret. Pop. Biol. 5:163–186.

Roughgarden, J., and M. Feldman. 1975. Species packing and predation pressure. Ecology 56:489–492.

Russell, B. C. 1971. Underwater observations on the reproductive activity of the demoiselle *Chromis dispilus* (Pisces: Pomacentridae). Marine Biol. 10:22–29.

Russell, B. C., F. H. Talbot, and S. Domm. 1974. Patterns of colonisation of artificial reefs by coral reef fishes. Pages 207–215 *in* A. Cameron et al., eds. Second International Symposium on Coral Reefs, Proceedings. Vol. 1. Great Barrier Reef Committee, Brisbane.

Sale, P. F. 1969. Pertinent stimuli for habitat selection by the juvenile manini, *Acanthurus triostegus sandvicensis*. Ecology 50:616–623.

———. 1970. Behaviour of the humbug fish. Australian Natur. Hist. 16:362–366.

———. 1971a. Extremely limited home range in a coral reef fish, *Dascyllus aruanus* (Pisces: Pomacentridae). Copeia 1971:324–327.

———. 1971b. The reproductive behaviour of the pomacentrid fish, *Chromis caeruleus*. Z. Tierpsychol. 29:156–164.

———. 1972a. Influence of corals in the dispersion of the pomacentrid fish, *Dascyllus aruanus*. Ecology 53:741–744.

———. 1972b. Effect of cover on agonistic behavior of a reef fish: a possible spacing mechanism. Ecology 53:753–758.

———. 1974. Mechanisms of co-existence in a guild of territorial fishes at Heron Island. Pages 195–206 *in* A. Cameron et al., eds. Second International Symposium on Coral Reefs, Proceedings. Vol. 1. Great Barrier Reef Committee, Brisbane.

———. 1975. Patterns of use of space in a guild of territorial reef fishes. Marine Biol. 29: 89–97.

Sale, P. F., and R. Dybdahl. 1975. Determinants of community structure for coral reef fishes in an experimental habitat. Ecology 56:1343–1355.

Slatkin, M. 1974. Competition and regional coexistence. Ecology 55:128–134.

Slobodkin, L. B., and H. L. Sanders. 1969. On the contribution of environmental predictability to species diversity. Brookhaven Symp. Biol. 22:82–93.

Smith, C. L. 1973. Small rotenone stations: a tool for studying coral reef fish communities. Amer. Mus. Novitates 2512:1–21.

Smith, C. L., and J. C. Tyler. 1972. Space resource sharing in a coral reef fish community. Bull. Natur. Hist. Mus. Los Angeles County. 14:125–170.

Springer, V. C., and A. J. McErlean. 1962. A study of the behavior of some tagged South Florida coral reef fish. Amer. Midland Natur. 67:386–397.

Stephens, J. S., Jr., R. K. Johnson, G. S. Key, and J. E. McCosker. 1970. The comparative ecology of three sympatric species of California blennies of the genus *Hypsoblennius* Gill (Teleostomi, Blenniidae). Ecol. Monogr. 40:213–233.

Stoddart, D. R. 1969a. Ecology and morphology of recent coral reefs. Biol. Rev. 44:433–498.

———. 1969b. Post-hurricane changes in the British Honduras reefs and cays. Atoll Res. Bull. 131:1–25.

Swerdloff, S. N. 1970. Behavioral observations on Eniwetok damselfishes (Pomacentridae: *Chromis*) with special reference to the spawning of *Chromis caeruleus*. Copeia 1970:371–374.

Talbot, F. H., and B. Goldman. 1972. A preliminary report on the diversity and feeding relationships of the reef fishes of One Tree Island, Great Barrier Reef System.

Pages 425–442 *in* Proceedings of the Symposium on Corals and Coral Reefs, 1969. Marine Biol. Ass. India.

Tyler, J. C. 1971. Habitat preferences of the fishes that dwell in shrub corals on the Great Barrier Reef. Proc. Acad. Natur. Sci. Philadelphia. 123:1–26.

Tyler, J. C., and J. E. Bohlke. 1972. Records of sponge-dwelling fishes primarily of the Caribbean. Bull. Marine Sci. 22:601–642.

Valenti, R. J. 1972. The embryology of the neon goby, *Gobiosoma oceanops*. Copeia 1972: 477–482.

Vivien, M. L. 1973. Régimes et comportements alimentaires de quelques poissons des récifs coralliens de Tuléar (Madagascar). Terre et la vie, revue d'écologie applique. 27: 551 577.

Vivien, M. L., and M. Peyrot-Clausade. 1974. A comparative study of the feeding behaviour of three coral reef fishes (Holocentridae), with special reference to the polychaetes of the reef cryptofauna as prey. Pages 179–192 *in* A. Cameron et al., eds. Second International Symposium on Coral Reefs, Proceedings. Vol. 1. Great Barrier Reef Committee, Brisbane.

Whittaker, R. H. 1969. Evolution of diversity in plant communities. Brookhaven Symp. Biol. 22:178–195

30 March 1979, Volume 203, Number 4387   **SCIENCE**

Tree Dispersion, Abundance, and Diversity in a Tropical Dry Forest

That tropical trees are clumped, not spaced,
alters conceptions of the organization and dynamics.

Stephen P. Hubbell

A widely held generalization about tropical tree species is that most occur at very low adult densities and are of relatively uniform dispersion, such that adult individuals of the tree species are thinly and evenly distributed in space. If true, this generalization has potentially profound consequences for the reproductive biology, population structure, and evolution of tropical tree species (*1*). In this article the adequacy of this generalization is judged with respect to a particular tropical forest, a large tract of which has been mapped in detail (*2*).

The origins of this generalization can be traced back at least to Wallace (*3*), who stated the following concerning his impressions of species densities in Malaysian forests:

If the traveller notices a particular species and wishes to find more like it, he may often turn his eyes in vain in every direction. Trees of varied forms, dimensions, and colours are around him, but he rarely sees any one of them repeated. Time after time he goes toward a tree which looks like the one he seeks, but a closer examination proves it to be distinct. He may at length, perhaps, meet with a second specimen half a mile off, or may fail altogether, til on another occasion he stumbles on one by accident.

Dobzhansky and co-workers (*4*) enumerated the species in several 1-hectare stands of Amazonian rain forest, and concluded that "the population density of a half or more of the tree species in Amazonian forests is likely to be less

The author is an associate professor in the Department of Zoology, and in the Program in Ecology and Evolutionary Biology, University of Iowa, Iowa City 52242.

than one individual per hectare." One or both parts of this generalization (low density, uniform dispersion) now appear in most ecology texts (*5*), and theories have been proposed to explain both the causes and consequences of low density and uniform dispersion of adult tropical trees. Janzen (*6*) and Connell (*7*) independently proposed theories to explain low density and spacing between adults. Janzen focused attention on the effects of host-specific herbivores that attack

seeds. He noted that a high proportion of seeds falling under the parent tree are killed by such seed "predators," so called because the death of the seed is virtually assured; and he suggested that only those viable seeds transported some distance away from the parent would escape discovery and manage to germinate. The predicted result: "most adults of a given tree species appear to be more regularly distributed than if the probability of a new adult appearing at a point in the forest were proportional to the num-

ber of seeds arriving at that point" (*6*, p. 501).

Connell, in his earlier rain-forest studies, focused more attention on the dispersion and survival of young tree seedlings. In experimental studies of the fate of seedlings in an Australian rain forest, Connell, Tracy, and Webb (*8*) showed that survival was better in seedlings planted under adults of different species than under adults of the same species. They did not identify the causes of the differential mortality, but did suggest that herbivores attracted by adjacent adults would more often tend to defoliate and kill nearby seedlings rather than distant seedlings. Janzen and Connell both argued that such host-specific attack by herbivores would reduce the local density of any given species, open up habitat to invasion by additional species, and thereby maintain high species diversity (*9*).

Many explanations have been offered for the high species diversity in tropical forests, and these have been classified into equilibrium and nonequilibrium hypotheses by Connell (*10*), who now believes that high diversity is only maintained because of frequent disturbance.

Summary. Patterns of tree abundance and dispersion in a tropical deciduous (dry) forest are summarized. The generalization that tropical trees have spaced adults did not hold. All species were either clumped or randomly dispersed, with rare species more clumped than common species. Breeding system was unrelated to species abundance or dispersion, but clumping was related to mode of seed dispersal. Juvenile densities decreased approximately exponentially away from adults. Rare species gave evidence of poor reproductive performance compared with their performance when common in nearby forests. Patterns of relative species abundance in the dry forest are compared with patterns in other forests, and are explained by a simple stochastic model based on random-walk immigration and extinction set in motion by periodic community disturbance.

Potential consequences of a low-density uniform dispersion of adult trees in tropical species might include lower outcrossing success, reduction in deme size, and requirements for long-distance pollination. Thus, the generalization that adults of tropical tree species are widely spaced has also spawned a number of hypotheses about unusual breeding systems in tropical trees (*11*), or special pollinator movements over long distances (*12*). It now appears that the majority of tropical tree species is facultatively or obligately outcrossed; and the frequency of dioecy in tropical trees is very high by temperate zone standards (*13, 14*). Animals rather than wind, in most cases, are the agents of cross-pollination.

Characteristics of the Dry Forest

Many of the ideas about low density and uniform tree dispersion developed from observations in the rain forest. Therefore, the major results of this study are presented with the caveat that they may apply better to deciduous tropical forests or monsoon forests than to rain forests. Nonetheless the patterns found in the dry forest appear to be fully consistent with the available information on rain forests, with the exception that the dry forest has only a third to half the number of tree species. Conclusions about forest dynamics are necessarily tentative because they are based on the circumstantial evidence provided by one census at a single point in time.

The dry forest results were obtained from a detailed map of 13.44 hectares of forest in Guanacaste Province, Costa Rica, in which all woody plants with stem diameters at breast height (dbh) ≥ 2 centimeters were located to the nearest meter. A separate map of the population of each species was drawn by computer, with each plant marked by a letter to indicate position and dbh, and whether it was juvenile or adult. Maps were then analyzed in an attempt to answer, in part or in whole, the following questions about each of the tree species: (i) Is the dispersion of the adult tree population uniform? (ii) Are adult trees less clumped than the juveniles, or than the population as a whole? (iii) How do the densities of adult and juvenile trees change with greater distance from a given adult in the population? (iv) Is there evidence that spacing from a given adult to other trees increases from the juvenile to the adult tree classes? (v) How distant is the nearest adult, and how many adults can be expected within *m* meters of a given adult? (vi) What, if any, is the relationship between rarity of a tree species and its dispersion pattern or its population (size, age) structure? (vii) Does mode of seed dispersal or breeding system relate to abundance or dispersion pattern? (viii) Is there any evidence for density dependence in the per capita reproductive performance of adult trees? Questions of interest about the dry-forest community in general include: (ix) What tentative conclusions can be drawn about the equilibrium or nonequilibrium status of the forest? (x) What are the pat-

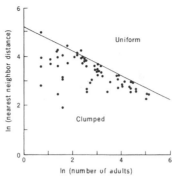

Fig. 1. Clumping in 61 tropical dry-forest tree species, as measured by mean nearest neighbor distance in meters. Diagonal line is the expected nearest neighbor distance for randomly dispersed populations [see (*37*)].

terns of relative species abundance in the dry forest, and how do they compare with the patterns in forests at other latitudes? Are the patterns of similar mechanistic origin?

Questions (i) through (iv) are of importance in evaluating the Janzen-Connell hypothesis. Question (v) is concerned with the minimum distance that a potential pollinator must fly to encounter another conspecific adult, and relates to the general question of deme size in tropical dry forest tree species. The answer to question (vi) is relevant to a discussion of the minimal density at which adult outcrossed trees of a species are capable of self-replacement. One way that a rare species can persist, in theory at least, even in the absence of long-distance pollination, would be through a higher degree of adult clumping than found in common species (*15*). Answers to questions (v) and (vi) may also suggest which of several alternative explanations of tree rarity is most probable (*16*).

The answer to question (vii) should reveal whether there are systematic differences in the adult or juvenile dispersion patterns of small-seeded species with wind or bird dispersal and large-seeded species with water or mammal dispersal. It should also reveal whether dioecious species differ systematically in density or dispersion pattern from self-compatible or hermaphroditic species. Janzen (*6*) and Bawa and Opler (*14*) hypothesized that, given similar pollination systems, the average interadult distance might be less in dioecious species because deme size would be roughly halved from that of an equally abundant hermaphroditic species. Skewed sex ratios, to the extent that they occur (*17*), should exacerbate the problem for the locally more abundant sex.

Question (viii) addresses the issue of what regulates the population size of tropical tree species. Does the number of nearby conspecific adults have a measurable influence on the reproductive performance of a given adult tree? The Janzen-Connell hypothesis would lead us to expect greater losses per adult in seeds and seedlings when adults are closer together, provided the per capita attractiveness of adult trees to herbivores is greater when the adults are in groups. Grouped adults may also compete more strongly for pollinator attention (*14, 18*) and encounter root and crown competition if they are actually adjacent to one another (*19*). Alternatively, there may be no detectable effect of adult density on per capita reproductive success (*20*).

Question (ix) relates to questions (vi) and (viii) and the reproductive success of rare trees. If the forest and its component species are in equilibrium, there should be found evidence that the rare species as well as the common species are self-replacing. In equilibrium theory, each species is presumed to be competitively superior at exploiting a particular microhabitat (*10*), with relative species abundance determined by the relative abundance of the microhabitat types. However, if the forest is in a nonequilibrium state, there should be evidence that some species are increasing in numbers while others, most likely the currently rare species, are declining.

Finally, question (x) asks whether the patterns of dominance and diversity in tropical, temperate, and boreal forests reveal any similarities that might suggest similar underlying control mechanisms. Using a simple stochastic model, I show that it is relatively easy to generate the patterns of relative species abundance in natural forest communities along a disturbance gradient. The model adds to the current theory of island biogeography a new statistical explanation for the relative abundance of species in arbitrarily defined habitat "islands" (*21*).

Study Site

The tract of mapped forest in Guanacaste Province, Costa Rica, lies approximately 8 kilometers west of Bagaces and 2 kilometers south of the Pan American Highway (*22*). The tract has been the site of studies of the pollinator community (*23*), and of the breeding systems and seasonal phenology of a number of tree species (*14, 18*). As a result of these and other studies (*24*), the pollination biology and the phenological cycles of leafing, shedding, flowering, and fruiting are known for many of the deciduous forest tree species. Most species are obligately outcrossed by insects, birds, or bats. The climate of the Dry Forest Life Zone has also been described (*25*).

A total of 135 species of woody plants with stem diameters ≥ 2 cm dbh were found in the mapped area. These include 87 overstory and understory trees, 38 shrubs, and 10 vine species (*26*). About two-thirds of the tree species are deciduous through part or all of the dry season (December to May). The canopy is somewhat broken (15 to 25 meters high), and consists primarily of trees with medium-length boles and spreading crowns. Canopy cover is approximately 87 percent; the remainder consists of light gaps made by recent tree falls, and a narrow

grassland corridor through the center of the site (7 percent of the mapped area). The understory consists of a diverse array of tree seedlings and shrubs (*27*). Vines are common, but epiphytes are infrequent. Insect flower visitors, especially bees, are abundant, as are night pollinators such as bats and hawkmoths. Several types of vertebrate seed dispersers are common (*28*), and there is evidence of heavy insect and vertebrate seed predation in a number of tree species (*29*).

The mapped forest tract was a rectangular area, 420 m by 320 m, gridded into 336 quadrats 20 m on a side. All woody plants ≥ 2 cm dbh were identified to species (*30*), measured for diameter, and mapped to within ± 1 m of their true position within each quadrat (*31*). Sepa-

rate data were taken on the dbh of trees and shrubs in flower or fruit to establish a lower bound on the diameter of reproductive individuals (adults) for each species (*32*). The data for each quadrat were transcribed to computer cards; programs sorted individuals to species and recomputed the coordinates of each plant from its local quadrat to its coordinates in the study area as a whole. Maps were then drawn by a Calcomp plotter for each species (*33*).

Juvenile and adult dispersion patterns were examined over a range of quadrat sizes from 4 m^2 to 38,416 m^2 (196 m on a side), by Morisita's index of dispersion, I_δ (*34*). Determining juvenile and adult densities at different distances from each adult tree in the population was done from the Calcomp maps. Average den-

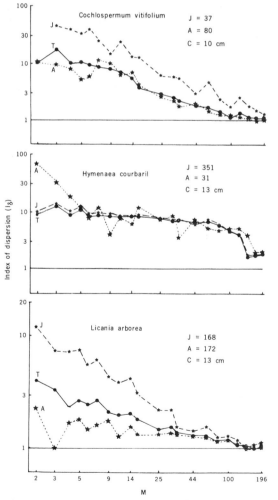

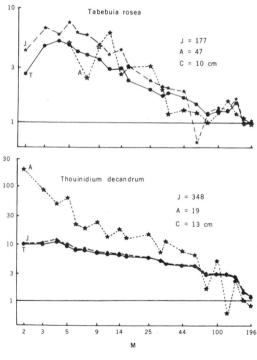

Fig. 2. Morisita's index of dispersion, I_δ, as a function of quadrat size for five sample dry-forest tree species. Numbers of the x-axis are lengths of the quadrat sides in meters. The spacing of the quadrat sizes along the x-axis is in terms of log quadrat area. Log I_δ values are plotted on the y-axis to make more visible the changes in I_δ that occur at larger quadrat sizes. The horizontal line through $I_\delta = 1$ indicates the expected value for randomly dispersed populations. J, juveniles; A, adults; T, total population; C, cutoff dbh for adults.

sites were computed for increments of 5 m out to a distance of 100 m (35). The maps were also analyzed to obtain the distribution of nearest neighbor distances between adults.

Tests for density dependence in per capita reproductive performance were made on 30 of the most abundant species. The gridded area was divided into 80 subplots of equal area, and the densities of adult and juvenile trees were noted for each subplot. The ratio of juveniles to adults was plotted as a function of adult density. Significant departures of slopes from zero were determined by regression analysis of variance. A necessary assumption of this test is that the adults counted in a subplot are the parents of the juveniles in the same subplot. Validity of this assumption varies with tree species and mode of seed dispersal (36).

Adult and Juvenile Dispersion Patterns

The adults of tropical dry-forest tree species are not uniformly dispersed in the forest, as shown in Fig. 1 for the 61 species on which information could be obtained on threshold adult diameter. The diagonal line in Fig. 1 is the expected nearest neighbor distance for randomly dispersed adults at the given mean density (37). Of these species, 44 (72 percent) exhibit significant adult clumping ($P < .05$, F test). The remaining 17 species (28 percent) have adult dispersion patterns which cannot be distinguished from random. No species has a significantly uniform adult dispersion.

When the dispersion of both adults and juveniles together is considered, the clumping is again pronounced. Of the 114 identified tree, shrub, and vine species having at least two individuals in the mapped area (38), fully 102 species exhibited significant clumping ($P < .05$, F test) in quadrats < 196 m on a side. As many as 95 species still showed significant clumping even in quadrats as small as 14 m on a side. No species showed a significantly uniform pattern of dispersion of its total population.

The pattern of change in Morisita's index of dispersion as quadrat size is increased from 2 to 196 m on a side, is shown (Fig. 2) for five species chosen at random from the 30 most common species (39). The dispersion indices for juveniles, adults, and total population are plotted separately. In all cases, including the species not illustrated, the dispersion index, I_δ, drops from its highest values in the smaller quadrat sizes irregularly downward toward unity as quadrat size

increases. Such I_δ patterns are typical of populations having "point sources" of relatively high population density, surrounded by more diffuse clouds of individuals diminishing in density away from the centers (32). Small quadrats may contain the high-density centers, thereby producing large I_δ values, whereas large

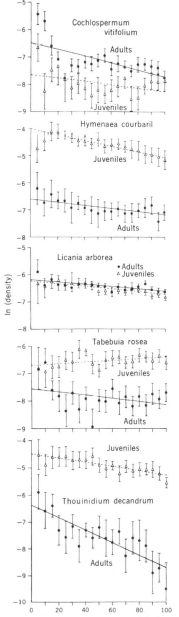

quadrats tend to have lower I_δ values because they average the density of the concentrated centers with the density of the more sparsely populated surroundings.

Question (ii) asks if the adults are less clumped than the population as whole, as would be expected from the Janzen-Connell hypothesis (6, p. 522). At a quadrat size of 14 m on a side and for the 30 most common tree species, adults are less clumped than juveniles and than the population as a whole in 16 species, equally clumped in nine species, and more clumped in five species. The results change somewhat depending on quadrat size and the scale of pattern resolution as well. Of the five species (Fig. 2), two species (Cochlospermum and Licania) show more clumping in juveniles than in adults at small quadrat sizes, two species (Hymenaea and Thouinidium) show more clumping in adults at small quadrat sizes, and one species (Tabebuia) shows approximately equal clumping in adults and juveniles.

Because of such clumping many species appear to be quite rare (less than one individual per hectare) when half or more of the study area is considered, but they turn out to be quite common when the rest of the study area is included. These patchy tree distributions help to explain data on tree species densities for relatively small plots of tropical forest (1 or 2 hectares) (40).

Demographic Neighborhood of the

Average Adult

The "demographic neighborhood" of a tree may be defined as the population of adults and juveniles of the same species occurring within a specified radius of the tree. The "average adult" may be considered for the purposes of this article as one that exhibits the expected demographic neighborhood for an adult of the given species.

Question (iii) asks how the densities of adults and juveniles change with increasing distance from the average adult in the population. If the Janzen-Connell hypothesis is correct, densities should be lower near the adult than at some intermediate distance. What constitutes an

Fig. 3. Adult and juvenile densities (natural logarithm of numbers per square meter) at various distances from the average adult in the population, showing the negative slopes away from the average adult (the slope in Tabebuia rosea for juveniles is not different from zero) in the five sample species. The bars are 95 percent confidence limits for the means.

Table 1. Relative adult (A) and juvenile (J) densities in 5-m distance intervals away from the average adult, in the five sample species. Relative densities within adult and juvenile classes have been normalized to a sum of 100 percent for ease of visual comparison. In each distance interval, χ^2 tests were performed on the numbers of juveniles and adults counted.

Species		Relative densities at distance interval in meters:									
		0 to 4.9	5 to 9.9	10 to 14.9	15 to 19.9	20 to 24.9	25 to 29.9	30 to 34.9	35 to 39.9	40 to 44.9	45 to 49.9
Cochlospermum vitifolium	J	27.43	5.72	12.34	14.62	8.89	6.36	7.71	6.12	4.82	5.99
	A*	31.11	24.59	9.92	5.93	3.03	4.71	4.57	5.36	5.06	5.72
Hymenaea courbaril	J	7.23	9.30	12.77	12.15	11.45	10.36	9.44	9.29	8.21	9.80
	A†	16.41	9.90	13.19	10.35	10.69	8.06	9.75	7.93	7.09	6.63
Licania arborea	J	10.17	9.18	10.93	11.60	9.64	10.13	10.33	10.76	9.42	7.84
	A†	15.44	9.47	11.79	9.63	9.35	8.53	10.02	9.83	9.75	6.19
Tabebuia rosea	J	7.74	7.12	8.14	8.35	10.35	9.99	15.90	15.68	9.06	7.67
	A*	21.24	26.61	9.81	7.92	4.44	8.82	4.90	7.12	2.54	6.60
Thouinidium decandrum	J	11.44	7.98	10.78	10.86	11.02	9.50	9.72	9.89	11.08	7.73
	A*	26.94	13.48	16.84	6.74	5.25	7.86	3.60	6.57	5.18	7.54

*Significant difference ($P < .05$).   †Not significant ($P > .05$).

"intermediate distance" will depend on the tree species, but clearly it should not exceed the mean nearest neighbor distance between adults. The 30 most common species were analyzed, and the results do not support the Janzen-Connell expectation. Either adult and juvenile densities decline approximately exponentially away from the average adult, or densities remained unchanged, to and beyond the intermediate distances appropriate for the species. In 67 percent of the species, there were negative slopes for the log-transformed juvenile densities as a function of distance; and the slopes in all remaining species were not distinguishable from zero (*41*). An even greater percentage (90 percent) of these species also showed negative slopes for adult densities, and there were no positive slopes. Thus, at least for the 30 most common species, the average adult is clearly found in a clump with other adults and juveniles.

The adult and juvenile density curves for the five sample species (Fig. 3) reveal that the densities after log-transformation generally exhibited equal variance at all distances. In 15 of the 30 most common species, the highest mean juvenile density occurred in the 0 to 5-m annulus, closest to the adult. In ten additional species with horizontal density curves, mean juvenile density was no lower in the annulus closest to the adult than in other annuli. However, in the five remaining species, maximal juvenile density was achieved between 5 and 15 m from the adult. *Hymenaea* is one of these species in which there was a notable scarcity of juveniles immediately under the adult canopy. This species suffers heavy losses of seeds to predators (such as specialized insects and generalized vertebrates), but in spite of these losses, the maximum recruitment of juveniles is still quite close to the adult (15 m), only a

few meters beyond the crown perimeter (*42*).

It is possible that the distance at which the density of juveniles is maximal is not the distance with the highest probability of producing an adult. If juveniles die more often when they are growing close to an adult, there should be an increase in the mean distance between an adult and successively older cohorts of neighboring trees [question (iv)]. Alternatively, if mortality is a random thinning process regardless of distance from adult trees, there should be no significant change with increasing cohort age in the proportion of cohort individuals at a given distance. Seeds, seedlings, and saplings < 2 cm dbh—stages in which most of the mortality occurs—were not mapped. However, whatever the postulated mortality patterns in seeds and seedlings, the greatest density of surviving juveniles that remain after this mortality has taken its toll is nevertheless usually in the annulus closest to the adult. Therefore, if appreciable spacing is occurring, it must occur as a result of differential mortality among censused cohorts of juveniles ≥ 2 cm dbh.

Relative densities in juvenile and adult cohorts were computed to a distance of 50 m in the five sample species (Table 1). Two species (*Hymenaea* and *Licania*) do not show significant differences in the distributions of relative adult and juvenile densities, and for these species the null hypothesis of random thinning cannot be rejected. In the three remaining species, adult relative densities are shifted significantly closer to the average adult than are juvenile relative densities. This suggests that, contrary to prediction, the more distant juveniles suffer greater losses in these species (*43*). One explanation for such a result could be that adults are already growing in the sites most favorable to the species, with

outlying areas generally of lower microhabitat suitability. This pattern is repeated in the 30 most common tree species (including the five discussed above), of which 13 species have relative adult densities shifted closer to the average adult than relative juvenile densities. In the remaining 17 species the null hypothesis of random thinning could not be rejected. No species showed the predicted shift in relative adult densities to greater distances.

Question (v) concerns tree spacing from the point of view of pollinator movement. Although adults generally occur in clumps with other adults and juveniles, it is more specifically the absolute distance from one adult to the next that is important to pollination success. For the 30 most common species, the mean nearest neighbor distances are within 20 m in 16 species, and within 40 m in all species (*44*). These 30 species average 5.6 ± 1.3 adults within 50 m of a given adult, and 12.7 ± 2.7 adults within 100 m. A fourfold increase in area produces, on average, only a 2.3-fold increase in the number of adults, a further indication of adult clumping (Fig. 4).

The demographic neighborhood of the average adult is a composite of all the adults and juveniles in the population. Therefore, the density curves away from the average adult cannot be equated to the seed shadows of single adult trees since the curves result from the superposition of several overlapping shadows of neighboring adults. Nevertheless, the approximately negative exponential character of the composite density curves away from the average adult constitutes strong circumstantial evidence that relatively simple physical and biological mechanisms govern seed dispersal in these species.

One might expect that the effects of different seed sizes and modes of dispersal would be reflected in the slopes of these density curves [question (vii)]. In log transformation the slopes of the exponential density curves conveniently become independent of the absolute abundance and seed production of the species, permitting cross-species comparisons. Comparison of the slopes for adult and juvenile log density as a function of distance, in mammal-, wind-, and bird- or bat-dispersed species, is shown in Fig. 5. Mammal-dispersed species show the steepest slopes; shallower slopes are found in wind-dispersed species; and the shallowest slopes, on average, characterize the bird- and bat-dispersed species. All three pairwise contrasts of juvenile slopes are significantly different; and two of three adult slopes (except wind compared to bird) are significantly different ($P < .05$, Mann-Whitney U test).

Early work on dispersal by Dobzhansky and Wright (45) suggested that a bivariate normal might typically describe dispersal patterns. However, dispersal is frequently strongly leptokurtic, with a pronounced peak near the point of propagule origin (46). These results for dry-forest tree species suggest that seed dispersal as well as juvenile survival are much more leptokurtic in distribution in mammal-dispersed species than in either wind- or bird-dispersed species.

Dispersion, Abundance, and Density Dependence

In the preceding discussion I have dealt primarily with the most common third of the tree species. What is the dispersion of rare species? If individuals of rare species were dispersed at random, or spaced uniformly, nearest neighbor distances should increase with decreased density as fast or faster than the inverse square root of mean density (37); but this is not the case in adults (Fig. 1). When total population size is considered, the trend is toward increased clumping with decreased abundance (Fig. 6). A few species (outliers) do not conform to the general species sequence. I believe that these nonconforming species are probably "accidentals" or last survivors of once more abundant species (see below). Greater clumping in rare species has also been reported by Hairston (47) in old-field communities of soil arthropods.

The second part of question (vii), whether differences in breeding system explain any of the variation in dispersion, independently of population size, is answered in Fig. 6 (48). Self-compatible species are not overrepresented among rare species, nor are dioecious species more clumped for their abundance than hermaphrodite species. That most rare species are highly clumped might suggest that they are at least locally successful (reproducing themselves) when their microhabitat requirements are satisfied. However, the data on size (dbh) class suggest that per capita reproduction in rare species is considerably less per unit time than in common species. The coefficient of skewness of the dbh distribution about the midpoint diameter for each species was computed

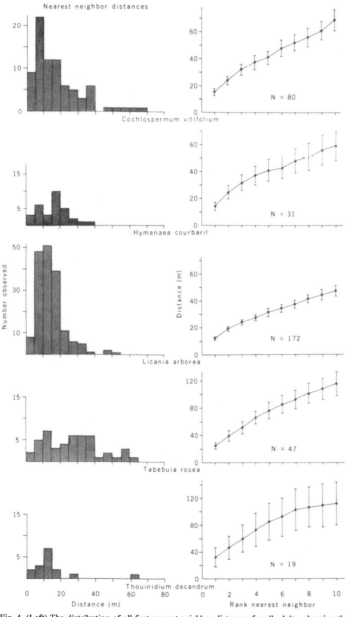

Fig. 4. (Left) The distribution of all first nearest neighbor distances for all adults, showing the closeness of the nearest adult, as well as the tendency to skew near (mode < mean). (Right) The mean distances, and their 95 percent confidence limits, to the first ten nearest neighbor adults, in the five sample species.

(49); species having an excess of large adults show positive or zero skewing, and species with an excess of juveniles show negative skewing.

Rare species exhibit positive or zero skewing about the midpoint dbh, whereas common species exhibit negative skewing (Fig. 7). No biology is needed to explain negative skewing in common species because it would be physically impossible to fit so many trees into the available space if all were large adults. However, skewing in rare species is not constrained in either direction, yet no rare species is negatively skewed. This pattern does not prove that the rare species are failing to reproduce themselves—for example, successful reproduction might be extremely episodic *(16)*—but it is clear evidence for a much slower per capita rate of juvenile establishment in rare species than in common species *(50)*.

Some of the rare species in the forest may be last survivors of species once more common in earlier successional stages, and some may be accidentals that became established through good fortune outside the habitat to which they are optimally adapted. All seven of the outlier species (Fig. 6) may be such accidentals because they are represented solely by very large, randomly scattered adults *(51)*. These and other rare species are common on sites elsewhere in Guanacaste Province *(52)*, and, where common, they exhibit strong negative dbh class skewing, as would be expected. Unless one is prepared to accept radically different life history strategies in adjacent plant populations of the same species, these results point to reproductive failure in the rare species in the tract of forest described here *(53)*. Thus, the available circumstantial evidence suggests that the forest is in a nonequilibrium state (question ix).

Rare species might persist longer in a given forest stand than otherwise if the abundance of common tree species is limited, short of complete space monopoly, by density-responsive herbivory or seed predation, or by other density-dependent processes. There is no question that seed predation by specialized herbivores is intense and results in considerable thinning. Even if seed and seedling predation is a random thinning process in relation to distance from the average adult, it is still quite possible that seed predation can lower the per capita reproductive performance of whole clumps of adults compared to more isolated adults.

Density dependence can be detected in the tree species [question (viii)]. Of

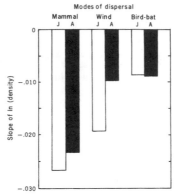

Modes of dispersal

Fig. 5. Relation between the mode of seed dispersal and the steepness of the slope with which the natural log of the density (numbers per square meter) in adult (*A*) and juveniles (*J*) decreases with distance (meters) from the average adult in the population, showing the steeper slopes found in the heavier seeded, mammal-dispersed species. Sample sizes: 9 mammal-dispersed species; 12 wind-dispersed species; 9 bird- or bat-dispersed species.

the 30 most common species, 17 species (57 percent) exhibited significantly negative slopes in the regression of juveniles per adult on adult density. All 13 remaining species also showed a negative slope, but the null hypothesis of zero slope could not be rejected ($P > .05$). Density dependence was detected more frequently in species producing a large number of small seeds (generally the wind- and bird-dispersed species) as compared to species producing a smaller number of large seeds (mammal-dispersed species). Two-thirds of the wind- and bird-dispersed species exhibited density dependence, whereas only one-third of the mammal-dispersed species did *(54)*.

What is the source of this apparent density dependence? If it is primarily density-responsive seed and seedling predation, one might expect that the species attacked most heavily would in general be those exhibiting the strongest density dependence. However, in general, the large-seeded species are more frequently attacked by host-specific seed predators (commonly bruchid weevils) than are the smaller seeded species. An alternative hypothesis is that the density dependence is occurring via reduced per capita seed output in crowded adults, perhaps because of competition for pollinator attention or root and crown competition. It is also possible that some of these intraspecific effects are more apparent than real. The apparently greater frequency of density dependence among wind- and bird-dispersed tree species may be spurious *(36)*. Moreover, if the adults of several species are positively associated in space, the shade and root competition they collectively produce would result in "diffuse" competition against the seedlings of all species growing beneath them *(55)*.

Dominance-Diversity Relationships

Although it has long been clear that tropical land plant communities are far richer in species number than their temperate or boreal counterparts, sufficiently large data sets from which to make quantitative comparisons of relative species abundance have become available only in the last 25 years. When the species of a plant community are arranged in a sequence of importance (using a measure such as standing crop, basal area, or annual net production) from most to least important, they form a smooth progression without major discontinuities from the common to the rare

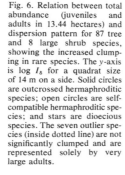

Fig. 6. Relation between total abundance (juveniles and adults in 13.44 hectares) and dispersion pattern for 87 tree and 8 large shrub species, showing the increased clumping in rare species. The *y*-axis is log I_δ for a quadrat size of 14 m on a side. Solid circles are outcrossed hermaphroditic species; open circles are self-compatible hermaphroditic species; and stars are dioecious species. The seven outlier species (inside dotted line) are not significantly clumped and are represented solely by very large adults.

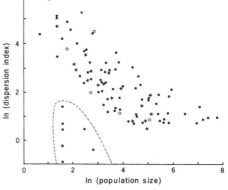

species. Different plant communities produce characteristic "dominance-diversity" curves when the importance values are log-transformed and plotted against the rank of the species in importance. Simple communities with few species generally yield almost straight ("geometric") lines on these semilog plots, whereas species-rich communities characteristically exhibit S-shaped ("lognormal") progressions (56).

Figure 8 compares the dominance-diversity curve for the tropical dry forest at 10°N (this study) with the curves for an equatorial (Amazonian) forest (57), for a rich temperate forest, and for a species-poor temperate montane forest similar to boreal forest (56). A number of factors distort the quantitative differences between the temperate and tropical dominance-diversity curves (58), but the qualitative pattern is clear: tropical forests exhibit the same general lognormal curve characteristic of rich temperate forests, but the distributions differ in location (mean) and scale (variance).

The rank-1 species in the dry forest has an importance value of 11 percent, compared with only 4.7 percent in the Amazonian forest. At 35°N, the rank-1 species in rich temperate forest has an importance value of 36 percent, which increases to about 65 percent in the montane spruce-fir forest. If the latter dominance-diversity pattern is comparable to patterns in the boreal coniferous forest at 50° to 60°N, then, as a general rule, the importance value of the dominant species in neotropical and neoarctic forests increases by approximately one percentage point for every latitudinal degree of northward movement, starting from a base of a few percent at the equator. Factors such as topographic diversity, elevation, the frequency and magnitude

of disturbance, and physical harshness will likely cause local deviations from this general pattern (59).

Relative Species Abundance in Nonequilibrium Communities

The qualitative similarity between the dominance-diversity curves for temperate and tropical forests suggests that similar processes control the relative abundance of tree species in the two regions (question x). May (60) has cautioned against reading too much significance into lognormal species abundance patterns, noting that lognormal distributions may be expected when many random variables compound multiplicatively, given the "nature of the equations of population growth" and the central limit theorem. Recently, Caswell (61) built several "neutral" models of community organization and relative species abundance, based on neutral-allele models in population genetics; but lognormal relative abundance patterns were not obtained from any of them.

Another line of reasoning, however, does generate lognormal relative abundance patterns under one set of circumstances, as well as geometric patterns under other circumstances. The model is essentially a dynamic version of MacArthur's "broken stick" hypothesis (62), and is based on a nonequilibrium interpretation of community organization. Suppose that forests are saturated with trees, each of which individually controls a unit of canopy space in the forest and resists invasion by other trees until it is damaged or killed. Let the forest be saturated when it has K individual trees, regardless of species. Now suppose that the forest is disturbed by a wind storm,

landslide, or the like, and some trees are killed. Let D trees be killed, and assume that this mortality is randomly distributed across species, with the expectation that the losses of each species are strictly proportional to its current relative abundance (63). Next let D new trees grow up, exactly replacing the D "vacancies" in the canopy created by the disturbance, so that the community is restored to its predisturbance saturation until the next disturbance comes along (64). Let the expected proportion of the replacement trees contributed by each species be given by the proportional abundance of the species in the community after the disturbance (65). Finally, repeat this cycle of disturbance and resaturation over and over again.

In the absence of immigration of new species into the community, or of the recolonization of species formerly present but lost through local extinction, this simple stochastic model leads in the long run to complete dominance by one species. In the short run, however, the model leads to lognormal relative abundance patterns, and to geometric patterns in the intermediate run. The magnitude of the disturbance mortality, D, relative to community size, K, controls the rate at which the species diversity is reduced by local extinction: the larger D is relative to K, the shorter the time until extinction of any given species, and the faster the relative abundance patterns assume an approximately geometric distribution (66).

The random differentiation of relative species abundance in a 40-species community closed to immigration is illustrated in Fig. 9 (67). Given equal abundances at the start, after 25 disturbances the species form the set of approximately lognormal dominance-diversity curves;

Fig. 7 (left). Coefficient of skewness of dbh distribution about midpoint dbh, as a function of total species abundance. Rare species are positively skewed toward large dbh, suggesting infrequent or highly episodic reproduction. Fig. 8 (right). Comparison of the dominance-diversity curves for two tropical forests and two temperate forests. Importance values for the temperate forests are based on annual net production; importance values for the dry forest in Costa Rica are from basal area (cross-sectional area of all stems of a given species); importance values in the Amazonian forest are based on above-ground biomass [see (58) for additional comments].

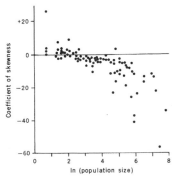

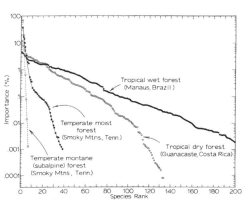

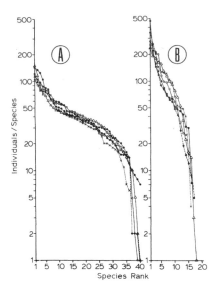

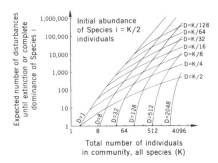

Fig. 9 (left). Randomly generated patterns of relative species abundance in a 40-species community closed to immigration; $K = 1600$, $D = 400$. (A) Dominance-diversity curves after 25 disturbances, showing an approximately lognormal pattern. (B) Dominance-diversity curves after 250 disturbances, showing an approximately geometric pattern. Five independent simulations are plotted to show the high predictability of the patterns for given values of K and D. Fig. 10 (right). Expected number of disturbances of size D in a community of K individuals until a species starting with $K/2$ individuals either goes extinct or becomes completely dominant, in a community closed to immigration. The actual expectations are points for discrete (integral) values of D and K. These points have been connected for fixed D, or D as a fixed fraction of K, for graphical clarity.

whereas after 250 disturbances with no immigration, the species have formed the set of approximately geometric curves. These transient distributions are not perfectly lognormal or geometric, but exhibit a "tailoff" phenomenon in the abundances of rare species. Rare species are somewhat less common than would be expected from the symmetric lognormal distribution, resulting from the continual attrition of rare species through local extinction. The large data set for the dry forest reveals this same rare-species tailoff in the dominance-diversity curve (Fig. 8).

If total community size is expanded (K larger), the number of disturbances to extinction of any given species gets rapidly larger for any fixed disturbance size, D (Fig. 10) (68). If D is small and K is moderate-sized to large, time to extinction or complete dominance can be very long. For example, if eight trees are killed with each disturbance out of a community total of 512, it will on average take about 90,000 such disturbances to remove or "fix" a species which begins with 256 individuals (69).

When immigration of new or old species is allowed, however, eventual complete dominance by one species is precluded; and a stochastic equilibrium is established between species immigration and local species extinction. The resulting relative abundance patterns can be either more nearly lognormal or more nearly geometric, depending on the relative importance of immigration in replenishing species diversity reduced by disturbance. If local extinctions outpace immigrations, local relative abundance patterns become increasingly geometric with time. As the rarer species drop out, the remaining species fill the gaps and grow more abundant, and thereby become less prone to extinction per unit time for a given level of disturbance. This causes the local extinction rate to drop into balance with the immigration rate. In contrast, if species immigrations outpace local extinctions, species must accumulate in the community, and local relative abundance patterns become more lognormal with time. Since more and more species are being packed into a finite space of size K, however, some species inevitably random walk to such rarity that they have a high risk of local extinction per time as compared with common species (70). A point is reached when sufficient numbers of species are rare and locally going extinct per unit time that the immigration rate is balanced, and the number of species in the community stops increasing.

Observed latitudinal changes in dominance-diversity patterns of local forest communities are predicted by the model provided that there are (i) fewer species in the source pool of potential immigrants, and (ii) more frequent or severe disturbances (or both) in boreal and temperate forests than in tropical forests. Whether these conditions are met is not yet clear. Persistently lower species richness in boreal and temperate forests may be due to elevated rates of extinction during the Pleistocene (71). In historical times boreal and temperate forests have been disturbed frequently by fires, plagues, and strong winds (72). On somewhat longer time scales, paleoecological evidence reveals that interglacial periods have been too brief for equilibria to be achieved in temperate deciduous forests before glacial periods set in again (73).

Disturbance in equatorial forests tends to be a localized, rather than a large-scale phenomenon, with blowdowns generally involving only a few trees at a time (74). Indeed, the observed clumped dispersion pattern of dry-forest tree species is just what one would expect if the forest is essentially a palimpsest of small, regenerating light gaps of different ages. However, stability of tropical climates over Pleistocene times is not certain, in view of mounting evidence of long dry periods during full-glacial and mid-Holocene times in equatorial regions (75).

Obviously this model is an oversimplified representation of the dynamics of natural communities (76), but it does provide a number of important lessons. First, we may expect to observe substantial differentiation of the relative abundance of species in natural communities as a result of purely random-walk processes—a kind of "community drift" phenomenon. Second, we cannot necessarily conclude that, just because a species is of rank-1 importance in a community, its current success is due to competitive dominance or "niche pre-emption" (56), stemming from some superior adaptation to the local environment. Because such a simple model generates the basic patterns of relative species abundance in natural plant communities, it would perhaps be preferable to use departures from the lognormal or geometric distributions as evidence for competitive dominance. Finally, whether or not forest communities are at equilibrium, our understanding of community organization would profit from more study of processes of disturbance, immigration, and local extinction, in conjunction with the more traditional studies of the biotic interactions of species (such as competition, niche differentiation, and seed predation).

References and Notes

1. P. S. Ashton, *Biol. J. Linn. Soc.* **1**, 155 (1969).
2. A companion paper (S. P. Hubbell, J. E. Klahn, G. Stevens, R. Ferguson, in preparation) describes the forest site in detail, and discusses the patterns for each tree species.
3. A. R. Wallace, *Tropical Nature and Other Essays* (Macmillan, London, 1878), p. 65.
4. G. A. Black, Th. Dobzhansky, C. Pavan, *Bot. Gaz.* **111**, 413 (1950); J. M. Pires, Th. Dobzhansky, G. A. Black, *ibid.* **114**, 467 (1953).
5. P. A. Colinvaux, *Introduction to Ecology* (Wiley, New York, 1973), p. 477ff.; J. M. Emlen, *Ecology: An Evolutionary Approach* (Addison-Wesley, New York, 1973), p. 400; C. J. Krebs, *Ecology: The Experimental Analysis of Distribution and Abundance* (Harper & Row, New York, 1972), p. 520; R. H. MacArthur and J. H. Connell, *The Biology of Populations* (Wiley, New York, 1967), p. 37; R. H. MacArthur, *Geographical Ecology* (Harper & Row, New York, 1972), p. 191; E. R. Pianka, *Evolutionary Ecology* (Harper & Row, New York, ed. 2, 1978), p. 296; P. Price, *Insect Ecology* (Wiley Interscience, New York, 1975), p. 49; J. L. Richardson, *Dimensions of Ecology* (Williams & Wilkins, Baltimore, 1977), p. 219; R. E. Ricklefs, *Ecology* (Chiron, Portland, 1973), p. 721. Most texts present several of the competing theories to explain the low density of tropical tree species.
6. D. H. Janzen, *Am. Nat.* **104**, 501 (1970).
7. J. H. Connell, in *Dynamics of Populations*, P. J. den Boer and G. R. Gradwell, Eds. (PUDOC, Wageningen, Netherlands, 1970), pp. 298–312.
8. J. H. Connell, J. G. Tracy, O. O. Webb, unpublished result cited in (*7*).
9. D. H. Janzen, in *Taxonomy and Ecology*, V. H. Heywood, Ed. (Systematics Association, special volume No. 5), (Academic Press, New York, 1973), chap. 10; *Annu. Rev. Ecol. Syst.* **2**, 465 (1971).
10. J. H. Connell, *Science* **199**, 1302 (1978).
11. H. G. Baker, *Cold Spring Harbor Symp. Quant. Biol.* **24**, 177 (1959); A. A. Fedorov, *J. Ecol.* **54**, 1 (1966); A. Kaur, C. O. Ha, K. Jong, V. E. Sands, H. T. Chan, E. Soepadmo, P. S. Ashton, *Nature (London)* **271**, 440 (1978).
12. D. H. Janzen, *Science* **171**, 203 (1971); G. W. Frankie, in *Tropical Trees: Variation, Breeding, and Conservation*, J. Burley and B. T. Styles, Eds. (Linnean Society Symposium Series No. 2, 1976), pp. 151–159.
13. K. S. Bawa, *Evolution* **28**, 85 (1974).
14. ——— and P. A. Opler, *ibid.* **29**, 167 (1975).
15. I do not mean to suggest that clumping is any sort of evolved adaptation to rarity.
16. For example, if a rare species occurs only as widely scattered, very large adults, we may suspect that the population is a relict of an earlier successional episode when the species was once more abundant. Alternatively, if a rare species is locally abundant with a high proportion of juveniles, it may be self-replacing and may be rare only because its microhabitat is local and rare. This sort of evidence is only circumstantial, at best, since it is always possible to construct a life table for any observed age distribution consistent with population growth, constancy, or decline. However, the reasonableness of some life tables necessary to balance some age distributions may be questionable. For example, to balance a population consisting only of extremely old individuals requires (i) highly episodic reproduction so that there is very little overlap of generations, (ii) extremely delayed age at maturity, (iii) high survival from seed to adult, or else extremely high adult fecundities, and (iv) synchronous fruiting, at long intervals, corresponding to (i). [Also see (*53*).]
17. K. S. Bawa and P. A. Opler (*14*), and M. N. Melampy and H. F. Howe [*Evolution* **32**, 867 (1978)] have reported that, although sex ratios may be skewed from 1:1, the staminate and pistillate plants are distributed at random with respect to one another in the species they studied. Determining what "sex ratio" means in plants can be difficult at times, particularly if there is temporal variation in the number of staminate and pistillate flowers offered by individual plants. See also R. W. Cruden and S. W. Hermann-Parker [*ibid.* **31**, 863 (1977)] and K. S. Bawa (*ibid.*, p. 52).
18. G. W. Frankie, H. G. Baker, P. A. Opler, *J. Ecol.* **62**, 881 (1974).
19. Information on the effects of root and crown competition and thinning on the growth of plantation stands is available in references on tropical trees [T. C. Whitmore, *Tropical Rainforests of the Far East* (Clarendon, Oxford, 1975)].
20. Potentially positive effects of density on reproductive success are also possible, within certain density ranges. For example, if pollinators are limiting seed set, a group of adult trees might exhibit greater per capita seed set than isolated adults if per capita rates of pollinator visitation were greater in the group.
21. The mathematical model is similar in aspects to verbal models developed by J. H. Connell (*10*) and J. Terborgh [*Am. Nat.* **107**, 956 (1973)].
22. The site is a nearly level plain at 100 m elevation, 10°32′N, 85°18′W. Soils are uniform, pale orange-brown silty clays derived from low bluffs of rhyolitic tuff, and exposed basaltic flows 200 to 600 m north of the site. Soils are 1.4 to 2.2 m deep, underlain by basaltic flows. Because the forest is seasonally dry for 6 months, a shallow layer of leaf litter (A-0) and humus (A-1) persists all year. A small seasonal creek flows through the site in a channel cut to the basaltic bedrock.
23. E. R. Heithaus, *Ann. Mo. Bot. Gard.* **61**, 675 (1974); L. K. Johnson and S. P. Hubbell, *Ecology* **56**, 1398 (1975); S. P. Hubbell and L. K. Johnson, *ibid.* **58**, 949 (1977); *ibid*, in press.
24. D. H. Janzen, *Evolution* **21**, 620 (1967); R. Daubenmire, *J. Ecol.* **60**, 147 (1972).
25. L. R. Holdridge, *Life Zone Ecology* (Tropical Science Center, San José, Costa Rica, 1967); J. O. Sawyer and A. A. Lindsey, *Vegetation of the Life Zones of Costa Rica* (Indiana Academy of Sciences, Indianapolis, 1971); C. E. Schnell, Ed., *O. T. S. Handbook* (Organization for Tropical Studies, San José, Costa Rica, 1971).
26. This total represents approximately a third of all woody species reaching a size of 2 cm dbh or larger in the Dry Forest Life Zone of Costa Rica.
27. Cattle enter the forest on occasion and cause some damage to the shrub understory by trampling and browsing. This disturbance is discussed in relation to the dispersion patterns found (*2*); cattle disturbance cannot have produced the clumped dispersion patterns of adult trees, nor the concentrations of juveniles around adults found in nearly all species (*2*).
28. Potential mammalian seed dispersers common in the forest include deer, howler monkeys, variegated squirrels, and agoutis. Visits by white-faced monkeys, spider monkeys, tapirs, peccaries, and pacas have also been recorded. Other mammals include armadillos, coatimundis, and kinkajous. Frugivorous birds are also common.
29. It is difficult to find intact seeds of species such as *Hymenaea courbaril*, *Enterolobium cyclocarpum*, *Cassia* spp., and many others that have not been attacked by weevils.
30. Many of the unknown plants were identified in the field by P. A. Opler and confirmed by voucher specimens sent to W. Burger (Field Museum, Chicago) or R. Leisner (Missouri Botanical Gardens, St. Louis). A few specimens could not be identified to species because they lacked reproductive structures.
31. Individual maps for each quadrat were drawn on graph paper (scale 2 m/cm) in the field. Plants within 5 m of the quadrat perimeter were mapped first; plants inside the inner square (10 by 10 m) were added afterward, the perimeter plants being the reference points. Independent checks of the accuracy of mapping showed that two people could locate most plants to within 1 m of each other, and of the plant's true position. Mapping took two people 3 months. Gross forest structure, by dbh class, is as follows: 2.0 to 4.9 cm, 9788 stems; 5.0 to 9.9 cm, 2606 stems; 10.0 to 14.9 cm, 1359 stems; 15.0 to 19.9 cm, 650 stems; 20.0 to 29.9 cm, 768 stems; 30.0 to 39.9 cm, 411 stems; 40.0 to 49.9 cm, 117 stems; 50.0 to 99.9 cm, 163 stems; > 100 cm, 27 stems.
32. It was not possible to check the reproductive capacity of every tree or species. Information was obtained on 61 tree species. The somewhat arbitrary rule of setting the lower limit for adult size at the smallest-sized individual of the species in flower or fruit was used. In general, this rule means that we probably have virtually all of the adults in the "adult" class. By the same token, however, some nonreproductive subadults are probably included with the adults. For purposes of this study, it was better to risk overestimating the number of adults since, in general, there were many more juveniles than adults.
33. Maps were drawn to a scale of 1:450. Locations of individual plants were marked by a letter. The letter A represented plants < 2.5 cm dbh; B represented plants between 2.5 and 5.0 cm dbh; and thereafter, letters represented 5 cm dbh increments.
34. M. Morisita, *Mem. Fac. Sci. Kyushi Univ. Ser. E* **2**, 215 (1959). Morisita's dispersion index is a ratio of the observed probability of drawing two individuals randomly without replacement from the same quadrat (over q quadrats), to the expected probability of the same event for individuals randomly dispersed over the quadrats. The index is unity when individuals are randomly dispersed, regardless of quadrat size or the mean density of individuals per quadrat. Values greater than unity indicate clumping, and values between 0 and 1 indicate uniformity. An F statistic can be computed to test for significant departure of the index (symbolized by I_δ) from unity (randomness).
35. Circular, clear plastic overlays with concentric rings drawn to scale every 5 m were made by Thermofax copier. These overlays were centered in turn over each adult tree on the map, and the numbers of adults and juveniles were counted in each successive 5-m annulus, out to 100 m. Density was computed by dividing the number of counts by the area of the annulus in question.
36. Small-seeded species with wind, bird, or bat dispersal should exhibit greater dispersal distance than large-seeded species dispersed by ground mammals. The greater dispersal distance expected in small-seeded species should tend (i) to obscure the local difference in seed production by both crowded and scattered adults and (ii) to increase the apparent per capita reproduction of scattered versus crowded adults. Consider, for example, the limiting case in which regional averaging of juvenile production is so complete that all quadrats have an equal density of juveniles. Then quadrats that have a small number of adults will show a high ratio of juveniles to adults, whereas quadrats that have many adults will show a low ratio, regardless of any density dependence.
37. The expected nearest neighbor distance, d, is given by $d = 1/(2\sqrt{\rho})$, where ρ is the mean density in number of adults per square meter [P. Clark and F. C. Evans, *Ecology* **35**, 445 (1954)].
38. Of the 21 remaining species of trees, shrubs, or vines represented by one individual, eight species could not be identified from our vegetative samples.
39. *Cochlospermum vitifolium* Spreng. (Cochlospermaceae) is a deciduous species found in large light gaps. It grows rapidly and has wood of low density. Its large dry-season flowers (see cover) are pollinated by big anthophorid bees, and its seeds are wind-dispersed. *Hymenaea courbaril* L. (Leguminosae) is a slow-growing, evergreen species commonly found in riparian areas, and it has dense wood. It has fairly large flowers which are bat-pollinated. Its seeds are large and encased in a thick, woody pod, are attacked on the tree before dispersal by bruchid weevils, and are food for a number of mammals [D. H. Janzen, *Science* **189**, 145 (1975)]. *Licania arborea* Seem. (Rosaceae) is a slow-growing, evergreen, high-canopy species of the mature dry forest, with very dense wood. It has small, bee-pollinated flowers, and its seeds are dispersed by bats, birds, and monkeys. *Tabebuia rosea* DC (Bignoniaceae) is a tree of intermediate stature characteristic of forest edge and large light gaps. It is relatively fast growing, with wood of intermediate density. It is deciduous and produces a showy display of large flowers in the dry season. Pollination is by large anthophorid bees. Its pods are straplike, containing numerous small, flat, winged, wind-dispersed seeds. *Thouinidium decandrum* Radlk. (Sapindaceae) is a medium-sized, deciduous tree of mature forest, with a moderate growth rate and medium to dense wood; its seeds are wind-dispersed.

40. P. S. Ashton (*1*) and M. E. D. Poore [*J. Ecol.* **56**, 143 (1968)] also report clumping in rain-forest tree species in Malaysian dipterocarp forests. Their large rain forest maps cannot be used very effectively to test the Janzen-Connell hypothesis since only trees with a circumference ≥ 30 cm were mapped, and they did not distinguish between adult and juvenile trees.

41. Because the demographic neighborhoods of adjacent adults are not completely independent (some juveniles are counted in the neighborhoods of more than one adult), compensation for this partial nonindependence should be made by a downward adjustment of the degrees of freedom of the regression analysis of variance. Accordingly, the average redundancy of juvenile counts was determined by the ratio of apparent number of juveniles in the neighborhoods of all adults to the actual number of juveniles in the union of all adult neighborhoods. The number of degrees of freedom was reduced by dividing the number of adults by the mean juvenile redundancy. This procedure results in a conservative estimate of the true degrees of freedom (R. Lenth, Department of Statistics, Univ. of Iowa, personal communication), and corresponds to the case of completely dependent demographic neighborhoods. For example, suppose that there are only three coincident adults in the population, which consequently have completely identical demographic neighborhoods in which every juvenile is counted three times. Therefore, mean juvenile redundancy is 3, and the number of independent data sets is found by dividing the number of adults by 3, giving one degree of freedom. A similar procedure was followed in testing the slopes of the adult density curves, with mean adult redundancy being used to adjust the degrees of freedom. However, even without the downward adjustment of degrees of freedom, the significance tests are conservative since individual trees counting in the neighborhoods of more than one adult should always bias the slopes in the positive, not in the negative, direction. The slopes found were all negative or zero.

42. Fifteen meters away from the parent tree might be enough to permit the seeds to escape from discovery by seed predators. However, D. E. Wilson and D. H. Janzen [*Ecology* **53**, 954 (1972)] found that in *Scheelea* palm there was no reduction in the percentage of seeds attacked under the palm and at a distance of 8 m, when seed density was held constant.

43. This analysis, of course, is analogous to constructing a vertical life table, which assumes that the population size has been approximately stationary for some time. While the conclusions are not necessarily invalid if the population is increasing, their validity cannot be confirmed.

44. In the 30 most common tree species, immediately adjacent adults with touching crowns are not infrequent.

45. Th. Dobzhansky and S. Wright, *Genetics* **28**, 304 (1943).

46. J. A. Endler, *Geographic Variation, Speciation, and Clines* (Princeton Univ. Press, Princeton, N.J., 1977).

47. N. G. Hairston, *Ecology* **40**, 404 (1959).

48. The self-compatible species may do little if any selfing (R. Cruden, personal communication). K. S. Bawa (personal communication) has shown that most of the hermaphroditic species in the dry forest he has studied are self-incompatible. Therefore, in Fig. 6, hermaphroditic species for which self-compatibility data are lacking are pooled with the obligately outcrossed hermaphrodites.

49. I chose the midpoint of the dbh range as the pivotal size in order to be conservative in my estimate of the number of species showing positive skewness. This means that, when positive skewing is detected, it is actually very pronounced.

50. Similar patterns of skewing have been reported in rare species by Connell (*10*) and by D. H.

Knight [in *Tropical Ecological Systems* (Ecology Series No. 11), F. B. Golley and E. Medina, Eds. (Springer-Verlag, New York, 1975), pp. 53–59]. Knight studied late second-growth stands of semi-evergreen forest on Barro Colorado Island, Panama.

51. In 7 years of work at the study site, I have seen no juveniles produced by these species in spite of repeated flowering (seed set was not observed).

52. Of the 59 species in the forest studied, which occur with an average density (ignoring dispersion) of greater than one individual per hectare, I know of at least 42 species that occur locally in much higher densities elsewhere in Guanacaste; and I expect the same is true for most of the remaining species.

53. The most parsimonious explanation for the data on rare species in the dry-forest stand described here is general reproductive failure. However, undoubtedly there are some rare species in the forest that are replacing themselves, and there are probably some species that are everywhere rare. G. S. Hartshorn [in *Tropical Trees as Living Systems*, P. B. Tomlinson and M. H. Zimmerman, Eds. (Cambridge Univ. Press, New York, 1978)] has suggested that "non-regenerating" rare species may simply require particular types of light gaps in order to regenerate, and that rarity is due to the infrequency of creation of such gaps.

54. Of 21 wind- or bird-dispersed species, 14 showed density dependence, but of nine mammal-dispersed species, only three showed density dependence.

55. Diffuse competition is a term coined by R. H. MacArthur [*Geographical Ecology* (Harper & Row, New York, 1972)]. It refers to the sum of competition from all interspecific competitors in the community acting on a given species. I have not yet analyzed the association of species to check this possibility.

56. R. H. Whittaker, *Science* **147**, 250 (1965).

57. Data of H. Klinge, as was reported by E. F. Brunig [*Amazoniana* **4**, 293 (1973)].

58. The curves for the Smoky Mountains are derived from analysis of single, 0.1-hectare stands in the cove and spruce-fir forests, respectively. Also, they represent all vascular plants, not just woody plants; and the importance values are based on annual net primary production. The respective dry-forest and rain-forest curves, however, are based on larger quadrats (13.44 and 1.0 hectares), but only woody plants; and the respective importance values are based on basal area and above-ground biomass. The quantitative effects of these differing methods are difficult to assess, but fortunately the effects are partially canceling (larger plots mean more species, but eliminating nonwoody species means fewer species). If there is a greater percentage of nonwoody plants in temperate forests, the difference between the species richness of temperate and tropical forests may be somewhat underestimated in Fig. 8.

59. M. P. Johnson and P. H. Raven, *Evol. Biol.* **4**, 127 (1970); S. J. McNaughton and L. L. Wolf, *Science* **167**, 131 (1970).

60. R. M. May, in *Theoretical Ecology: Principles and Applications*, R. M. May, Ed. (Saunders, Philadelphia, 1976).

61. H. Caswell, *Ecol. Monogr.* **46**, 327 (1977).

62. R. MacArthur, *Am. Nat.* **94**, 25 (1960); G. Sugihara (unpublished result) has also developed a "sequential breakage" model that generates lognormal patterns.

63. Losses are governed by the hypergeometric distribution. Thus, for the *i*th species the probability of losing *j* individuals, $j \le D \le N_{it}$, where N_{it} is the population of species *i* at the current time *t*, is given by

$$\binom{N_{it}}{j}\binom{K - N_{it}}{D - j} \bigg/ \binom{K}{D}.$$

64. A disturbance can be as small as the death of a single tree.

65. Recruitment to fill the *D* disturbance vacancies is governed by the binomial distribution. Let N'_{it} be the number of the *i*th species at time *t* after disturbance. For the *i*th species, the probability of contributing *m* replacement individuals, $m \le D$, is:

$$\binom{D}{m}\left(\frac{N'_{it}}{K - D}\right)^m \left(\frac{K - N'_{it} - D}{K - D}\right)^{D - m}$$

66. It is easy to prove that rare species are more likely to go extinct per unit time than are common species: If $N_{it} > D$, there is no chance that species *i* will go extinct in the next disturbance; but if $N_{it} \le D$, there is a nonzero chance of extinction in the next disturbance. Therefore, the mean time to extinction of a species *j* with $N_{jt} > D$ must exceed that of a species *i* with $N_{it} \le D$ by at least the mean time it takes for N_j to decrease to *D*. If a "rare" species is one for which $N \le D$, then increasing *D* will increase the number of rare species going extinct per disturbance over successive disturbances.

67. In this example, I chose a large *D* simply to speed up the process of random-walk extinction. These transient distributions obey the "canonical hypothesis" of F. W. Preston [*Ecology* **43**, 185 and 410 (1962)]. It has been shown (*21*) that the canonical lognormal is the result of imposing a fixed ceiling, *K*, on the total number of individuals of all species in the community. This result has also been discovered independently by G. Sugihara (personal communication). The Monte Carlo simulations were performed at the University of Iowa Computer Center on an IBM 360/70.

68. Because the species can random walk up or down in abundance, either outcome (extinction or complete dominance) is possible. I chose to illustrate a species with *K/2* individuals because it represents the abundance at which a species is equally likely to go to either outcome. It is also the abundance with the longest mean transient time, both outcomes considered.

69. The model is a Markovian random walk of the abundance of the *i*th species between 0 and *K*. Mean transient times from a starting abundance of *K/2* individuals were found from the fundamental matrix determined for particular values of *D* and *K*.

70. There is an added risk of extinction for rare species if they are more clumped than common species, such that a single disturbance might kill all individuals in a given local area.

71. The extinction of many temperate tree species is well documented in Europe (*73*).

72. For example, see M. L. Heinselman, *Quat. Res.* **3**, 329 (1973); J. D. Henry and J. M. A. Swan, *Ecology* **55**, 772 (1974).

73. M. B. Davis, *Geosci. Man.* **13**, 13 (1976).

74. R. Foster, personal communication; S. P. Hubbell, personal observations.

75. B. S. Vuilleumier, *Science* **173**, 771 (1971); J. E. Damuth and R. W. Fairbridge, *Bull. Geol. Soc. Am.* **81**, 189 (1970); J. Haffer, *Science* **165**, 131 (1969); B. S. Simpson and J. Haffer, *Annu. Rev. Ecol. Syst.* **9**, 497 (1978).

76. The model in its simplest form as presented here corresponds to the "equal chance hypothesis" discussed by Connell (*10*), provided that per capita chances of reproduction or death are made the same for all species. For greater realism, species differences in dispersal, fecundity, competitive ability, and resistance to environmental stresses need to be treated.

77. Help from the following people was vital to the completion of this study; Jeffrey Klahn, George Stevens, Paul Opler, Richard Ferguson, Ronald Leisner, William Burger, Daniel Janzen, Joseph Connell, Leslie Johnson, Robin Foster, and Douglas Futuyma. Others have also made contributions which are appreciated. The study was supported by the National Science Foundation.

3

Productivity and Resources

Robert W. Sterner and David M. Post

Introduction

Plants and other autotrophs respond almost universally to nutrient fertilization by growing faster and achieving higher biomass. This is one of the most repeatable ecological responses known, showing unequivocally that nutrients are master variables governing ecological processes. A great many questions and complexities underlie this single cause-effect reaction. The papers in this section have made significant inroads into defining this response, outlining its characteristics, and putting it into context with other ecological processes. One central issue of both practical and theoretical importance has been to define which nutrients matter where. Another has been to integrate the response of plants to nutrients with other factors such as herbivory and overall food web structure. A third is to understand how biological factors including succession and species identity affect the relationship between productivity and resources.

What Resource Is Limiting Where?

Life bumps up against many different boundaries—physical, chemical, and biological—in its struggle for existence. A handful of differ-

ent substances have widespread importance as nutrients. These include the elements nitrogen, phosphorus, potassium, silicon, and iron. The question of which resources limit primary production in any given location is fundamental to ecology and critical for managing a wide range of ecosystems.

Humans have long relied on aquatic ecosystems to carry away their wastes, but the combination of increased human population densities and intensified use of chemicals in industry and agriculture accelerated releases of nutrients into waterways during the 1940s–1960s. By the middle of the twentieth century, the resulting high levels of primary production were causing widespread massive algal blooms and low oxygen levels in lakes and rivers around the world. To restore these degraded ecosystems and protect others, the nutrient at fault had to be identified and then controlled. So the hunt began.

Controversy quickly erupted. Long-term studies in lakes were indicating that anthropogenic inputs of nitrogen (N) and most especially phosphorus (P) were to blame (Edmondson 1970). Comparative studies also were highlighting P as the most likely limiting nutrient in lakes (Vollenweider 1975), with some role for N also acknowledged (Likens 1972). These studies were suggestive, but a higher level of proof was needed to effect social investment and motivate management action. It was then that science met industry pushback. Phosphorus was an important ingredient in most detergents, and the detergent industry cited evidence at the time, primarily from experiments conducted in small closed bottles, that other elements, such as N, carbon (C), and trace elements such as molybdenum and iron, were at least as important as P in limiting primary production in lakes.

What was missing was critical experimental proof, which was provided by one of the most famous and impactful sets of experiments in ecology. Resolving this controversy was a principal reason why the Canadian government set aside a field station where experiments at the whole-lake scale were possible. Starting in 1969, David Schindler and colleagues conducted a series of experiments to resolve the controversies surrounding elemental limitation in lakes (reviewed in Schindler 1990). Of these, Schindler's 1977 paper "Evolution of phosphorus limitation in lakes" is foundational because it summarized many of these experiments in a way that solidified the argument for why C and N can be supplied by the atmosphere, making P the ultimate limiting nutrient in lakes. Schindler (1977) addressed a pressing applied problem, but also summarized the physical, chemical, and biological mechanisms through which nutrient limitation in lakes is regulated. Similar arguments had also been made for marine production in the global ocean (Redfield 1958) and for soils (Walker and Syers 1976). This controversy about P versus other elements in limiting production has not entirely died, however, and some still wonder whether there is an important and somewhat neglected role for N or other elements in contributing to nutrient limitation in lakes (Paerl et al. 2016).

The whole-lake studies of Schindler and colleagues proved to be convincing, and in many regions of the world P is now banned or limited in detergents. The resulting positive effect on water quality has been profound. The focus of nutrient management in developed regions has shifted from point sources of P, such as detergents and human waste, to nonpoint sources of P, such as agriculture and urban landscapes (Carpenter et al. 1998). Algal blooms caused by cultural eutrophication continue to be an issue, for example in Lake Erie, where nonpoint source pollution still is a major problem (Scavia et al. 2014).

The question of elemental limitation in the ocean is fundamentally the same as for lakes, but the issues enveloping the work are entirely different. At the time John Martin and Steve Fitzwater published their 1988 paper "Iron deficiency limits phytoplankton growth in the north-east Pacific subarctic" (not included in this volume), oceanographers knew of large regions of the world's oceans that contained high nitrogen and phosphorus but low chlorophyll, so N and P weren't limiting there. Such high nutrient low chlorophyll (HNLC) regions were mysterious. One line of thinking was that grazing by herbivorous zooplankton continually kept algae in check. Another hypothesis was that resources besides N and P were responsible. This was a version of the classic predators vs. resources debate, and the implications were especially important because the HNLC regions were so large that they were significant in global C cycling and climate regulation. Trace metals, especially iron (Fe), could conceivably have been responsible, but the technical challenges of measuring minuscule concentrations of metal dissolved in the ocean from the deck of ships made of metal were considerable, and many technical refinements had to be made to make those measurements without contamination. Martin and Fitzwater (1988) hypothesized that Fe was the key limiting element in HNLC portions of the ocean because it has limited modes of movement and atmospheric Fe movement is expected only downwind of continental regions of high Fe supply. Given the tiny quantities of Fe needed to make living organisms, a small amount goes a long way. Martin (1990) suggested that there was a coupling between Fe and glacial-interglacial cycles because during glacial periods, more Fe-bearing dust came off of continents, fueling algal growth and depleting atmospheric CO_2. If this hypothesized linkage was correct, as Martin once famously proclaimed, "Give me half a tanker of iron and I'll give you an ice age" (see Strong, Cullen, and Chisholm 2009, 236).

Though the Schindler studies discussed above were conducted at the whole-ecosystem scale, which had much to do with their impact, it is interesting to recognize that the foundational work of Martin and Fitzwater (1988) was as small as Schindler's was large. They were notable technically only for the exacting precautions that were taken to minimize Fe contamination from the surroundings. Ocean water from the northeast Pacific was brought onboard ship and dispensed into two-liter plastic bottles, and several concentrations of Fe were added to different bottles. Where Fe was added, chlorophyll increased several-fold in comparison to the controls. These small experiments did not fully settle the issue, but they were the inspiration for other, larger studies later that confirmed iron limitation in large portions of the oceans.

What was needed to more firmly test the iron hypothesis were experiments at larger scale, but these were far from routine in the open ocean. The prototype such experiment was called IRONEX (Coale et al. 1996). During this work, a "Lagrangian" approach was utilized, where an inert but highly fluorescent dye was injected into the open ocean so that a ship equipped to measure the dye concentration could remain with the same parcel of water even as the parcel moved in space. Fe was added at the same time as the dye, and algal abundance, among other variables, was measured. Multiple similar Fe addition experiments were eventually performed and through these, a greatly advanced knowledge of the biogeochemistry of oceans was developed (Boyd et al. 2007), and Martin and Fitzwater's (1988) Fe hypothesis was considered confirmed. The possibility of using this finding as a basis for geoengineering Earth's climate remains highly controversial. Studies showed that most of the C taken up by Fe-stimulated phytoplankton was consumed and turned back into CO_2 before it sank into the deep ocean. These findings made the Fe-climate linkage inefficient as a climate control mechanism, and the ethical problems of purposely polluting vast regions of the Earth are considerable. Martin and Fitzwater (1988) greatly broadened interest in trace metals in general and led to widespread method development and improvement at analysis.

Integration of Resources and Food Webs

While questions of "bottom up" regulation of population growth were being sorted out, somewhat separate lines of inquiry were examining the "top down" effects of consumers on the organisms they preyed upon. It is beyond dispute that both resources and consumers play a role in ecological dynamics and the papers in this section integrated the role of resources and consumption.

In the middle of the twentieth century, it seemed that the environment was becoming greener and more productive nearly everywhere. Crop yields were increasing as a result of breeding, fertilizer application, and irrigation. As we discussed above, surface waters were undergoing cultural eutrophication and algal blooms were a common, unwanted sight. Michael Rosenzweig's 1971 paper "Paradox of enrichment: destabilization of exploitation ecosystems in ecological time" asked whether systems enriched with nutrients were simply more productive versions of unenriched ones, or instead if enrichment fundamentally changed the character of the system.

Rosenzweig (1971) approached this question by evaluating several different models of predators and prey, utilizing the mathematical technique of isocline analysis and finding the stability of the predator-prey equilibrium under a range of enrichments. These techniques were well established in theoretical ecology at the time. They had been used to help understand the oscillatory dynamics sometimes characteristic of predator-prey systems (Volterra 1926 [reprinted in Real and Brown 1991]). Rosenzweig's analysis told him that increased enrichment can be a destabilizing force, so that instead of benefiting from high prey growth rates, the predator may instead become extinct, meaning greener worlds were not just more of the same, but different things entirely. He referred to this outcome as the "paradox of enrichment."

Rosenzweig's paper carried a simple but powerful message that fit in with the preservation and conservation ethic that was emerging as a strong political force at the time. Part of the allure of this paper probably stemmed from its clear reasoning from a simple set of models. The approach was based on the belief that if one captures the essence of an ecological phenomenon in a simple model, one can draw potentially broad and general conclusions without incorporating the blizzard of detail that all real, living ecological systems obviously include but which also make every ecological situation unique (a lesson also learned from May 1972, 197 [discussed in part 4 of this volume]).

Rosenzweig studied several models that all gave the same result, which suggested to him that the paradox of enrichment was quite general. However, work since his has shown that the destabilization he discovered depends critically on "prey dependence." This means that the growth of the predator depends only on the numbers of prey in the system at the time. Geometrically, in the predator vs. prey graphs used in isocline analysis, prey dependence means that the predator isocline is perpendicular to the prey axis (see Berryman 1992 for a more detailed explanation).

Others (see review in Roy and Chattopadhyay 2007) have since followed up on Rosenzweig's study, exploring a wide range of conditions where predator growth depends on such factors as spatial heterogeneity (Scheffer and De Boer 1995), unpalatable prey (Genkai-Kato and Yamamura 1999), inducible defenses (Vos et al. 2004), predator interference (Rall et al. 2008), or trait evolution (Mougi and Nishimura 2008). Many of these conditions, it is now known, strongly affect whether enrichment causes destabilization, and we recognize today that the paradox of enrichment is not a universal outcome of greater resource supply.

The paradox of enrichment has also inspired advances in disease ecology (Sharp and Pastor 2011) and stoichiometry (Loladze et al. 2000). The foundational nature of Rosenzweig's paper lies not so much in whether it was right or whether it was wrong, but in the tremendous advances in knowledge it inspired, a critical contribution from a scholar with many, many prodigious accomplishments.

Most basic models of population change consider factors that affect growth and factors that affect loss, but they do not explicitly consider the interaction between the two, that changing one may change the other. At the time Samuel McNaughton wrote his 1979 paper "Grazing as an optimization process: grass-ungulate relationships in the Serengeti," the field of plant-animal interactions was almost entirely limited to the consideration of the deleterious effects of herbivory. Plant responses to herbivory had been noted but generally only in evolutionary time. In this foundational paper, McNaughton (1979) cited voluminous examples of how plants could adjust growth upward to compensate for loss by herbivory. Experimental ecology was also still relatively new at the time, and deploying herbivore exclosures on the African Serengeti grassland to test basic ecological theories using "charismatic megafauna" was a bold step.

McNaughton (1979) described both field and lab experiments designed to quantify how grass biomass gain (g m^{-2}) was affected by the browsing from the very abundant Serengeti ungulates. The central result in this paper is a single plot of data with 16 data points and a modestly successful statistical fit of a humped curve ($r^2 = 0.687$) demonstrating the nonintuitive result that increased herbivory could lead to increased plant productivity. Sometimes new ideas arise from scant evidence, and in spite of the rather thin empirical footing, the concepts in this paper captivated many ecologists and inspired them to think about the role of plant adjustments to herbivory. McNaughton later wrote a more comprehensive review of plant responses to herbivory (McNaughton 1983). It is important to recognize that the stimulation of growth resulting from herbivory written about here encapsulates many different mechanisms at the physiological and community scales. Even simple density-dependent response to thinning can result in the kind of optimization curve McNaughton pointed to, and that would not be a real departure in thinking compared to classical approaches. This mechanism, for example, was part of the trophic cascade model (Carpenter et al. 1985) discussed below. But as written by McNaughton and other authors since, there are many other potential mechanisms whereby plants can compensate for grazer-induced losses, such as benefiting from nutrients recycled by herbivores.

This paper led to considerable debate, much of which seemed to invoke arguments where different parties were failing to address exactly the same issues. Discussion at times moved from "Do plants compensate for being eaten?" to "Do plants benefit from being eaten?" (Owen and Wiegert 1976; Stenseth 1978). This thinking in turn led some to wonder whether it was more appropriate to think of plants and herbivores as coevolved mutualists. Discussion of the evolutionary implications of optimization curves of plant growth as a function of herbivory continues, but probably the single most long-lasting impact of McNaughton's paper is that, in the mind of the practicing ecologist, the plant has become a more active player in its interactions with herbivores than it once was.

The idea that whole trophic levels can be treated as single homogeneous units with dynamics governed by interactions with trophic levels above and below is a rather startling thought, but it has been entertained by ecologists for years, and indeed was foreshadowed by Darwin's discussion of trophic interactions among species in the metaphorical entangled bank. In 1960, Hairston et al. (1960 [reprinted in Real and Brown 1991]) put forward one of the most sweeping generalizations ever about trophic level control and concluded that "(1) Populations of producers, carnivores, and decomposers are limited by their respective resources in the classical density-dependent fashion. (2) Interspecific competition must nec-

essarily exist among the members of each of these three trophic levels. (3) Herbivores are seldom food-limited, appear most often to be predator-limited, and therefore are not likely to compete for common resources." Their hypothesis was that the major source of control of trophic levels alternates between resources and predation from the top of food chains (where there are no higher predators to exert control) downward.

Fretwell (1977) took the next logical step and asked how trophic level control varied as the number of links in the food chain changed. Like HSS, he proceeded by verbal, qualitative argumentation. Lauri Oksanen and coauthors formalized these arguments in their 1981 paper "Exploitation ecosystems in gradients of primary productivity," in which they extended the Rosenzweig isocline model discussed earlier into >2 dimensions. The Oksanen et al. (1981) model simultaneously addressed how food chain length would get longer as the productivity of the ecosystem increased (more energy and resources to support higher predators) and how control of trophic levels should vary as new trophic levels were added. The model in Oksanen et al. (1981) clarified this set of relationships, drawing much attention to the concept of trophic level control and illuminating how shifts in top down vs. bottom up control are central to food webs. It touched on some very basic questions in ecology, namely what sets the length of food chains and what is the balance of competition and predation in governing populations.

Like Hairston et al. (1960) before them, Oksanen et al. (1981) contributed to ecology not so much by making successful specific predictions (indeed, these general ideas are very hard to test in specific systems) but by providing a single relatively simple organizational scheme for thinking about multi–trophic level communities. The inherent "webbiness" of trophic interactions (see Polis 1991 [reprinted in part 4 of this volume]) would seem to render any discussion of homogeneous trophic levels immediately moot. However, as work on trophic cascades discussed below illuminates, it is sometimes the case that assemblages of species do act as homogeneous food chain links. This is a tremendous insight that ecologists are still incorporating into their thinking. Subsequent work has challenged the assumption that food chain length and productivity are closely related (Post et al. 2000) and suggests that food chain length may be regulated by ecosystem size, disturbance, and productivity (Post 2002; Takimoto and Post 2013).

For much of the twentieth century, ecologists worked under the assumption that primary production was limited by abiotic conditions such as light, water, and nutrients. By the early 1980s, it was clear that these factors could explain much but not all of the variation observed in primary production. For example, nutrient concentration could explain 50–60% of the variation in primary production in lakes (Vollenweider 1975; Smith 1982), but what accounted for the remaining variation?

At the same time there were emerging hints that food web structure could influence primary production. As discussed above, Hairston et al. (1960) had proposed that terrestrial ecosystems, with three trophic levels, were green (high plant biomass) because top predators limited the density of herbivores, releasing plants from herbivory. This idea of reciprocal trophic control was controversial at the time (see, e.g., Murdoch 1966; Ehrlich and Birch 1967), but it was complemented by growing evidence that predators could regulate community structure (Brooks and Dodson 1965; Paine 1966). Oksanen et al. (1981) further extended the thinking of Hairston et al. (1960). Paine (1980) also suggested that the strong effects of predators could cascade through food webs. Missing were papers that clearly addressed the remaining variation in primary production and provided a powerful empirical test of the theory. Those missing papers would come at first not from the terrestrial ecosystems discussed in Hairston et al. (1960) but from work on lakes.

In 1985 and 1987, Stephen Carpenter and colleagues published two papers that addressed the role of food web structure in regulating primary production. Carpenter et al. (1985) was a theoretical and synthetic paper that predicted that changes in the biomass of piscivorous fish (fourth trophic level) would "cascade" through the food web to influence primary productivity, in essence arguing that changing the top predators in a lake would alter the greenness of the water. They suggested that food web structure could explain much of the variation in primary production that remained unexplained by nutrients, and they argued that the active management of piscivorous fish could be used to manage water quality in lakes. Carpenter et al. (1985) noted that there was already some experimental evidence from lakes for cascading trophic interactions (Hrbáčke et al. 1961; Shapiro and Wright 1984) and that the use of food web manipulations as a management tool had already been termed "biomanipulation" (Shapiro and Wright 1984). It was, however, the foundational experiment reported by Stephen Carpenter and colleagues in their 1987 paper "Regulation of lake primary productivity by food web structure" that provided the most persuasive evidence. This study (Carpenter et al. 1987), conducted in small lakes at the University of Notre Dame Environmental Research Center in northern Wisconsin and Michigan, experimentally manipulated the presence of piscivorous largemouth bass and planktivorous minnows, creating lakes with three trophic levels (planktivorous fish at top of food chain) or four trophic levels (piscivorous fish at top). As predicted by theory, they found that the lake with four trophic levels had lower algal biomass and production than the lake with three trophic levels. Interestingly, the treatment in the planktivore (third trophic level) treatment broke down towards the end of the summer because the few largemouth bass that remained in the lake (the fourth trophic level) caused a behavioral shift in the planktivorous minnows. This result hinted at the importance of nonconsumptive or behaviorally mediated effects, topics that would influence much of the work on trophic cascades in later years (Schmitz et al. 1997; Werner and Peacor 2003).

The results of Carpenter et al. (1987) helped spark decades of work on trophic cascades. The strongest early support for trophic cascades came from aquatic ecosystems (Power 1990; Estes et al. 1998), leading Strong (1992) to ask "are trophic cascades all wet?" in a paper that echoed some of the earlier criticisms of Hairston et al. (1960). The original ideas about trophic cascades produced debates and expansions that revolved around the role of nutrient regeneration (Vanni and Layne 1997), resource productivity (McQueen et al. 1989), resource edibility (Leibold 1989), behavior (Schmitz et al. 1997; Werner and Peacor 2003), food web complexity (Polis and Strong 1996), and stoichiometry (Sterner et al. 1992). It also stimulated renewed interest in understanding how and why food chain length (the number of trophic levels) varied among ecosystems (Post et al. 2000). The strength of trophic cascades varies among ecosystems (Shurin et al. 2002) and with life history variation within species (Post et al. 2008), but trophic cascades are now recognized as widespread even in terrestrial ecosystems (Pace et al. 1999). One spectacular and richly documented example of a trophic cascade is in Yellowstone National Park (Ripple et al. 2001; Ripple and Beschta 2012), where the reintroduction of gray wolves after a 70-year absence created changes that rippled across the food web. Elk populations have declined and the fear of wolves has changed the distribution of elk on the landscape, allowing the recovery of heavily browsed vegetation. Multiple interlocking species have increased or declined, and the recovery of wolves has influenced both terrestrial and aquatic ecosystems across the greater Yellowstone ecosystem. Though comparatively rare, top predators often have a critical structuring role in food webs and ecosystems (Estes et al. 2011).

Ecology is built upon knowledge of the abundance, dynamics, and functional connections among species. Top predators are "hidden" forces because they are rare but other organism are hidden because they are tiny. Organisms that are too small to observe directly are very abundant and important to ecological processes but they present special challenges to those who are trying to resolve the functional connections among species. Up until the 1960s, we lacked the ability to study the details of microbial activities in situ (Hobbie 1994). The reigning paradigm about aquatic systems was that essentially all primary production not respired as CO_2 stays within the particle phase and is eaten by herbivores, such that the fate of carbon was dominated by the "grazing food chain" (algae to zooplankton to fish). It was thought that little dissolved organic matter spilled over for bacteria to use. According to this perspective, it was safe to ignore bacteria and their protozoan grazers in studying aquatic production.

A great revolution in microbial ecology overturned all those ideas. It was built on methodological advances first and on new concepts second. Through new direct-count techniques, the biomass of free-living bacteria was revised dramatically upwards. Other techniques improved the reliability of measurements of microbial activity (Fuhrman and Azam 1980). It was the 1983 paper by Farooq Azam and colleagues "The ecological role of water-column microbes in the sea" that put forward a fundamentally new concept and formed the foundation of decades of research. Azam et al. (1983) introduced the term "microbial loop" to describe the role of microbes; the introduction of this term was key to the influence of this paper (Fenchel 2008). They suggested that a microbial food chain parallel to and connected with the traditional grazing chain existed. The paper described how more carbon moves through bacterioplankton than originally thought. This work completely changed the "wiring diagram" of the aquatic ecosystem. It also elevated the importance of bacterial predators, i.e., flagellates and ciliates, which Azam et al. (1983) suggested held bacteria abundance within narrow bounds (note this is consistent with the prey dependence formulation that was part of Rosenzweig 1971). Thus, energy released as dissolved organic matter by phytoplankton is rather inefficiently returned to the main food chain via a microbial loop of bacteria-flagellates-microzooplankton. Through studies that were inspired by Azam et al. (1983) we now know that we live in a more microbially dominated world than once imagined. These ideas have continued to evolve. It is recognized today that there are blurrier boundaries between microbes and traditional plankton than described by Azam et al. (1983). Ecological connections are webbier, with C and nutrients passed back and forth between the members of the classical grazing chain as well as the bacteria and their predators (Sherr et al. 1988). Microbes have a more important ecological role than conceived pre–Azam et al. They are part of, and active participants in, the carbon and nutrient processing at the base of the aquatic food web.

Succession and Species

The last set of foundational papers on the topic of productivity and resources explores the role of succession and species identity more generally. System characteristics change over time via ecological succession, and this changes how efficiently resources are retained. Sometimes a single species exerts such strong control that it alone dramatically affects the resource-productivity relationships in a given habitat. Finally, biodiversity is related to productivity in ways that ecologists are still sorting out.

Eugene Odum (1969 [reprinted in Real and Brown 1991]) used the idea of succession to set up broad hypotheses for how ecosystem dynamics change over time. Odum's predictions about changes in production, food web structure, species diversity, selection, homeostasis, and nutrient cycling were simultaneously highly controversial and profoundly influential. Of particular influence on ecosystem ecology was his prediction that nutrient conservation (the fraction of nutrients entering an ecosystem that are retained rather than lost) would increase as ecosystems mature. Odum noted that "an important trend in successional development is the closing or 'tightening' of the biogeochemical cycles of major nutrients, such as nitrogen, phosphorus, and calcium." (Odum 1969, 265).

Peter Vitousek and William Reiners's foundational 1975 paper "Ecosystem succession

and nutrient retention: a hypothesis" greatly expanded upon Odum's prediction that nutrient retention increases with ecosystem development. First, they refined the prediction about retention, stating that nutrient retention would first increase and then decrease as plant biomass accrual increased and then decreased through succession. In other words, plant biomass increase, and therefore nutrient retention, would be greatest in intermediate aged successional stages. They also predicted that the patterns of nutrient retention would vary among essential and limiting nutrients. Retention was predicted to be greatest for limiting elements, intermediate for essential elements, and lowest for nonessential elements. Then, using data from north temperate forest watersheds, they showed that the intermediate-aged ecosystems (logged ~100 years before the study) had higher nutrient retention than old-aged ecosystems (unlogged watersheds).

In testing and refining Odum's original prediction, Vitousek and Reiners (1975) sparked decades of work on nutrient limitation and retention in terrestrial ecosystems including work on patterns of nitrogen saturation and loss from old-growth forest (Hedin et al. 1995; Aber et al. 2003) and disturbed ecosystems (Vitousek et al. 1979); and broad-scale patterns of nitrogen export into coastal ecosystems (Howarth et al. 1996).

The role of diversity and complexity in maintaining ecological function is a long-standing topic in ecology (MacArthur 1955; Elton 1958; May 1972 [discussed in part 4]). As we've seen in this section, by the early 1990s, ecology had addressed to a great extent the role of elemental limitation of primary production and had made progress towards understanding how food web structure could influence primary production, but questions remained about the role of biological diversity in regulating primary production (Ehrlich and Mooney 1983; McNaughton 1994; Vitousek and Hooper 1994). A widely restated hypothesis held that production should increase with increasing species richness, but there was only limited data to support this hypothesis.

In the final foundational paper in this section, David Tilman, David Wedin, and Johannes Knops experimentally tested the influence of species richness on productivity and nutrient utilization in Minnesota grasslands in their 1996 paper "Productivity and sustainability influenced by biodiversity in grassland ecosystems" (not included in this volume). This experiment built upon a previous study (Tilman and Downing 1994) that was influential but criticized for hidden treatments in its experimental design (Givnish 1994; Huston 1997). Tilman et al. (1996) addressed many of the previous concerns by the way they experimentally manipulated species to create treatments of species richness ranging from single-species monocultures to 25-species polyculture. They found that above-ground primary productivity and nitrogen use and retention all increased with increasing species richness. Tilman et al. (1996) speculated that the increase in production and nutrient use resulted from compensatory competitive interactions created by differences in morphology and physiology among species, also termed niche complementarity. This work was influential in part because of the conservation message it supported, namely that biodiverse ecosystems are more stable and retain nutrients better than ecosystems with reduced species diversity.

Tilman et al. (1996) and other early experiments (e.g., Naeem et al. 1994), initiated a wave of research on the effects of biodiversity on ecosystem function (BEF). In the two decades after these initial experiments, many groups improved and expanded the BEF research into new ecosystem types (Duffy 2002; Worm et al. 2006), studied temporal trends (Reich et al.

2012), expanded the mechanistic understanding of diversity effects (Doak et al. 1998; Loreau and Hector 2001), and addressed phylogenetic constraints (Cadotte et al. 2008) and multifunctionality (Hector and Bagchi 2007). Major meta-analyses show that biodiversity effects are widespread (Cardinale et al. 2006) and can be as large as resource limitation and herbivory (Hooper et al. 2012; Tilman et al. 2012). Meta-analyses also showed that species sampling effects, rather than niche complementarity, was the most strongly supported mechanism (Cardinale et al. 2006). Much of the BEF research is motivated by concerns over the global extinction crisis, but a number of authors have pointed out small-scale BEF experiments provide little value because there is limited evidence for declines in local biodiversity (Vellend et al. 2013), largely due to widespread biological homogenization (Rahel 2002; Olden et al. 2004).

In summary, during the 1970s, '80s, and part of the '90s, ecologists worked to build realistic, practical, but also theoretically powerful concepts of how multi–trophic level, species-diverse ecosystems are constrained by limiting resources. Certain resources are more likely than others to provide unsurmountable constraints to system functioning. The structure of the community (the number of trophic levels, species richness and species identity, and the stage of success) is also an important determinant of the relation between resources and the organisms that rely on them. These advances in many respects form the core of knowledge upon which current ecological thought is built.

Literature Cited

Aber, J. D., C. L. Goodale, S. V. Ollinger, M.-L. Smith, A. H. Magill, M. E. Martin, R. A. Hallett, et al. 2003. Is nitrogen deposition altering the nitrogen status of northeastern forests? BioScience 53: 375.

Azam, F., T. Fenchel, J. G. Field, J. S. Gray, L. A. Meyer-

Reil, and F. Thingstad. 1983. The ecological role of water-column microbes in the sea. Marine Ecology Progress Series 10: 257–263.

Berryman, A. A. 1992. The origins and evolution of predator-prey theory. Ecology 73: 1530–1535.

Boyd, P. W., T. Jickells, C. S. Law, S. Blain, E. A. Boyle, K. O. Buesseler, K. H. Coale, et al. 2007. Mesoscale iron enrichment experiments 1993–2005: synthesis and future directions. Science 315: 612–617.

Brooks, J. L., and S. I. Dodson. 1965. Predation, body size, and composition of plankton. Science 150: 28–35.

Cadotte, M. W., B. J. Cardinale, and T. H. Oakley. 2008. Evolutionary history and the effect of biodiversity on plant productivity. Proceedings of the National Academy of Sciences 105: 17012–17017.

Cardinale, B. J., D. S. Srivastava, J. E. Duffy, J. P. Wright, A. L. Downing, M. Sankaran, and C. Jouseau. 2006. Effects of biodiversity on the functioning of trophic groups and ecosystems. Nature 443: 989–992.

Carpenter, S. R., N. F. Caraco, D. L. Correll, R. W. Howarth, A. N. Sharpley, and V. H. Smith. 1998. Nonpoint pollution of surface waters with phosphorus and nitrogen. Ecological Applications 8: 559–568.

Carpenter, S. R., J. F. Kitchell, and J. R. Hodgson. 1985. Cascading trophic interactions and lake productivity. BioScience 35: 634–639.

Carpenter, S. R., J. F. Kitchell, J. R. Hodgson, P. A. Cochran, J. J. Elser, M. M. Elser, D. M. Lodge, et al. 1987. Regulation of lake primary productivity by food web structure. Ecology 68: 1863–1876.

Coale, K. H., K. S. Johnson, S. E. Fitzwater, R. M. Gordon, S. Tanner, F. P. Chavez, L. Ferioli, et al. 1996. A massive phytoplankton bloom induced by an ecosystem-scale iron fertilization experiment in the equatorial Pacific Ocean. Nature 383: 495–501.

Doak, D. F., D. Bigger, E. K. Harding, M. A. Marvier, R. E. O'Malley, and D. Thomson. 1998. The statistical inevitability of stability-diversity relationships in community ecology. American Naturalist 151: 264–276.

Duffy, J. E. 2002. Biodiversity and ecosystem function: the consumer connection. Oikos 99: 201–219.

Edmondson, W. T. 1970. Phosphorus, nitrogen, and algae in Lake Washington after diversion of sewage. Science 169: 690–691.

Ehrlich, P. R., and L. C. Birch. 1967. The "balance of nature" and "population control." American Naturalist 101: 97–107.

Ehrlich, P. R., and H. A. Mooney. 1983. Extinction, substitution, and ecosystem services. BioScience 33: 248–254.

Elton, C. S. 1958. The Ecology of Invasions by Animals and Plants. Methuen.

Estes, J. A., J. Terborgh, J. S. Brashares, M. E. Power, J. Berger, W. J. Bond, S. R. Carpenter, et al. 2011. Trophic downgrading of planet Earth. Science 333: 301–306.

Estes, J. A., M. T. Tinker, T. M. Williams, and D. F. Doak. 1998. Killer whale predation on sea otters linking oceanic and nearshore ecosystems. Science 282: 473–476.

Fenchel, T. 2008. The microbial loop: 25 years later. Journal of Experimental Marine Biology and Ecology 366: 99–103.

Fretwell, S. D. 1977. Regulation of plant communities by food-chains exploiting them. Perspectives in Biology and Medicine 20: 169–185.

Fuhrman, J. A., and F. Azam. 1980. Bacterioplankton secondary production estimates for coastal waters of British Columbia, Antarctica, and California. Applied and Environmental Microbiology 39: 1085.

Genkai-Kato, M., and N. Yamamura. 1999. Unpalatable prey resolves the paradox of enrichment. Proceedings of the Royal Society of London. Series B: Biological Sciences 266: 1215–1219.

Givnish, T. J. 1994. Does diversity beget stability? Nature 371: 113–114.

Hairston, N. G., F. E. Smith, and L. B. Slobodkin. 1960. Community structure, population control, and competition. American Naturalist 94: 421–425.

Hector, A., and R. Bagchi. 2007. Biodiversity and ecosystem multifunctionality. Nature 448: 188–190.

Hedin, L. O., J. J. Armesto, and A. H. Johnson. 1995. Patterns of nutrient loss from unpolluted, old-growth temperate forests: evaluation of biogeochemical theory. Ecology 76: 493–509.

Hobbie, J. E. 1994. The state of the microbes: a summary of a symposium honoring Lawrence Pomeroy. Microbial Ecology 28: 113–116.

Hooper, D. U., E. C. Adair, B. J. Cardinale, J. E. K. Byrnes, B. A. Hungate, K. L. Matulich, A. Gonzalez, et al. 2012. A global synthesis reveals biodiversity loss as a major driver of ecosystem change. Nature 486: 105–108.

Howarth, R. W., G. Billen, D. Swaney, A. Townsend, N. Jaworski, K. Lajtha, J. A. Downing, et al. 1996. Regional nitrogen budgets and riverine N & P fluxes for the drainages to the North Atlantic Ocean: natural and human influences. Biogeochemistry 35: 75–139.

Hrbáčke, J., M. Dvořakova, V. Kořínek, and L. Procházková. 1961. Demonstration of the effect of the fish stock on the species composition of zooplankton and the intensity of metabolism of the whole plankton association. SIL Proceedings, 1922–2010 14: 192–195.

Huston, M. A. 1997. Hidden treatments in ecological experiments: re-evaluating the ecosystem function of biodiversity. Oecologia 110: 449–460.

Leibold, M. A. 1989. Resource edibility and the effects of predators and productivity on the outcome of trophic interactions. American Naturalist 134: 922–949.

Likens, G. E., ed. 1972. Nutrients and eutrophication: the limiting-nutrient controversy. Proceedings of the Symposium on Nutrients and Eutrophication: The Limiting-Nutrient Controversy. Special symposia. American Society of Limnology and Oceanography, Publ. Office.

Loladze, I., Y. Kuang, and J. J. Elser. 2000. Stoichiometry in producer-grazer systems: linking energy flow with element cycling. Bulletin of Mathematical Biology 62: 1137–1162.

Loreau, M., and A. Hector. 2001. Partitioning selection and complementarity in biodiversity experiments. Nature 412: 72–76.

MacArthur, R. H. 1955. Fluctuations of animal populations and a measure of community stability. Ecology 36: 533.

Martin, J. H. 1990. Glacial-interglacial CO_2 change: the iron hypothesis. Paleoceanography 5: 1–13.

Martin, J. H., and S. E. Fitzwater. 1988. Iron deficiency limits phytoplankton growth in the north-east Pacific subarctic. Nature 331: 341–343.

May, R. M. 1972. Will a large complex system be stable? Nature 238: 413–414.

McNaughton, S. J. 1979. Grazing as an optimization process: grass-ungulate relationships in the Serengeti. American Naturalist 113: 691–703.

McNaughton, S. J. 1983. Physiological and ecological implications of herbivory. Pages 657–677 in O. L. Lange, P. S. Nobel, C. B. Osmond, and H. Ziegler, eds. Physiological Plant Ecology III. Springer.

McNaughton, S. J. 1994. Biodiversity and function of grazing ecosystems. Pages 361–383 in E.-D. Schulze and H. A. Mooney, eds. Biodiversity and Ecosystem Function. Springer.

McQueen, D. J., M. R. S. Johannes, J. R. Post, T. J. Stewart, and D. R. S. Lean. 1989. Bottom-up and top-down impacts on freshwater pelagic community structure. Ecological Monographs 59: 289–309.

Mougi, A., and K. Nishimura. 2008. The paradox of enrichment in an adaptive world. Proceedings of the Royal Society B: Biological Sciences 275: 2563–2568.

Murdoch, W. W. 1966. "Community structure, population control, and competition": a critique. American Naturalist 100: 219–226.

Naeem, S., L. J. Thompson, S. P. Lawler, J. H. Lawton, and R. M. Woodfin. 1994. Declining biodiversity can alter the performance of ecosystems. Nature 368: 734–737.

Odum, E. P. 1969. The strategy of ecosystem development. Science 164: 262–270.

Oksanen, L., S. D. Fretwell, J. Arruda, and P. Niemelä. 1981. Exploitation ecosystems in gradients of primary productivity. American Naturalist 118: 240–261.

Olden, J. D., N. L. Poff, M. R. Douglas, M. E. Douglas, and K. D. Fausch. 2004. Ecological and evolutionary consequences of biotic homogenization. Trends in Ecology & Evolution 19: 18–24.

Owen, D. F., and R. G. Wiegert. 1976. Do consumers maximize plant fitness? Oikos 27: 488.

Pace, M. L., J. J. Cole, S. R. Carpenter, and J. F. Kitchell. 1999. Trophic cascades revealed in diverse ecosystems. Trends in Ecology & Evolution 14: 483–488.

Paerl, H. W., J. T. Scott, M. J. McCarthy, S. E. Newell, W. S. Gardner, K. E. Havens, D. K. Hoffman, et al. 2016. It takes two to tango: when and where dual nutrient (N & P) reductions are needed to protect lakes and downstream ecosystems. Environmental Science & Technology 50: 10805–10813.

Paine, R. T. 1966. Food web complexity and species diversity. American Naturalist 100: 65–75.

Paine, R. T. 1980. Food webs: linkage, interaction strength and community infrastructure. Journal of Animal Ecology 49: 666.

Polis, G. A. 1991. Complex trophic interactions in deserts: an empirical critique of food-web theory. American Naturalist 138: 123–155.

Polis, G. A., and D. R. Strong. 1996. Food web complexity and community dynamics. American Naturalist 147: 813–846.

Post, D. M. 2002. Using stable isotopes to estimate trophic position: models, methods, and assumptions. Ecology 83: 703–718.

Post, D. M., M. L. Pace, and N. G. Hairston. 2000. Ecosystem size determines food-chain length in lakes. Nature 405: 1047–1049.

Post, D. M., E. P. Palkovacs, E. G. Schielke, and S. I. Dodson. 2008. Intraspecific variation in a predator affects community structure and cascading trophic interactions. Ecology 89: 2019–2032.

Power, M. E. 1990. Effects of fish in river food webs. Science 250: 811–814.

Rahel, F. J. 2002. Homogenization of freshwater faunas. Annual Review of Ecology and Systematics 33: 291–315.

Rall, B. C., C. Guill, and U. Brose. 2008. Food-web connectance and predator interference dampen the paradox of enrichment. Oikos 117: 202–213.

Real, L. A., and J. H. Brown, eds. 1991. Foundations of Ecology: Classic Papers with Commentaries. University of Chicago Press.

Redfield, A. C. 1958. The biological control of chemical factors in the environment. American Scientist 46: 230A, 205–221.

Reich, P. B., D. Tilman, F. Isbell, K. Mueller, S. E. Hobbie, D. F. B. Flynn, and N. Eisenhauer. 2012. Impacts of biodiversity loss escalate through time as redundancy fades. Science 336: 589–592.

Ripple, W. J., and R. L. Beschta. 2012. Trophic cascades in Yellowstone: the first 15 years after wolf reintroduction. Biological Conservation 145: 205–213.

Ripple, W. J., E. J. Larsen, R. A. Renkin, and D. W. Smith. 2001. Trophic cascades among wolves, elk and aspen on Yellowstone National Park's northern range. Biological Conservation 102: 227–234.

Rosenzweig, M. L. 1971. Paradox of enrichment: destabilization of exploitation ecosystems in ecological time. Science 171: 385–387.

Roy, S., and J. Chattopadhyay. 2007. The stability of ecosystems: A brief overview of the paradox of enrichment. Journal of Biosciences 32: 421–428.

Scavia, D., J. D. Allan, K. K. Arend, S. Bartell, D. Beletsky, N. S. Bosch, S. B. Brandt, et al. 2014. Assessing and addressing the re-eutrophication of Lake Erie: central basin hypoxia. Journal of Great Lakes Research 40: 226–246.

Scheffer, M., and R. J. De Boer. 1995. Implications of spatial heterogeneity for the paradox of enrichment. Ecology 76: 2270–2277.

Schindler, D. W. 1977. Evolution of phosphorus limitation in lakes. Science 195: 260–262.

Schindler, D. W. 1990. Experimental perturbations of whole lakes as tests of hypotheses concerning ecosystem structure and function. Oikos 57: 25.

Schmitz, O. J., A. P. Beckerman, and K. M. O'Brien. 1997. Behaviorally mediated trophic cascades: effects of predation risk on food web interactions. Ecology 78: 1388–1399.

Shapiro, J., and D. I. Wright. 1984. Lake restoration by biomanipulation: Round Lake, Minnesota, the first two years. Freshwater Biology 14: 371–383.

Sharp, A., and J. Pastor. 2011. Stable limit cycles and the paradox of enrichment in a model of chronic wasting disease. Ecological Applications 21: 1024–1030.

Sherr, B. F., E. B. Sherr, and C. S. Hopkinson. 1988. Trophic interactions within pelagic microbial communities: indications of feedback regulation of carbon flow. Hydrobiologia 159: 19–26.

Shurin, J. B., E. T. Borer, E. W. Seabloom, K. Anderson, C. A. Blanchette, B. Broitman, S. D. Cooper, et al. 2002. A cross-ecosystem comparison of the strength of trophic cascades: strength of cascades. Ecology Letters 5: 785–791.

Smith, V. H. 1982. The nitrogen and phosphorus dependence of algal biomass in lakes: an empirical and theoretical analysis. 1: Chlorophyll model. Limnology and Oceanography 27: 1101–1111.

Stenseth, N. C. 1978. Do grazers maximize individual plant fitness? Oikos 31: 299.

Sterner, R. W., J. J. Elser, and D. O. Hessen. 1992. Stoichiometric relationships among producers, consumers and nutrient cycling in pelagic ecosystems. Biogeochemistry 17.

Strong, A. L., J. J. Cullen, and S. W. Chisholm. 2009. Ocean fertilization: science, policy, and commerce. Oceanography 22: 236–261.

Strong, D. R. 1992. Are trophic cascades all wet? Differentiation and donor-control in speciose ecosystems. Ecology 73: 747–754.

Takimoto, G., and D. M. Post. 2013. Environmental determinants of food-chain length: a meta-analysis. Ecological Research 28: 675–681.

Tilman, D., and J. A. Downing. 1994. Biodiversity and stability in grasslands. Nature 367: 363–365.

Tilman, D., P. B. Reich, and F. Isbell. 2012. Biodiversity impacts ecosystem productivity as much as resources, disturbance, or herbivory. Proceedings of the National Academy of Sciences 109: 10394–10397.

Tilman, D., D. Wedin, and J. Knops. 1996. Productivity and sustainability influenced by biodiversity in grassland ecosystems. Nature 379: 718–720.

Vanni, M. J., and C. D. Layne. 1997. Nutrient recycling and herbivory as mechanisms in the "top-down" effect of fish on algae in lakes. Ecology 78: 21.

Vellend, M., L. Baeten, I. H. Myers-Smith, S. C. Elmendorf, R. Beausejour, C. D. Brown, P. De Frenne, et al. 2013. Global meta-analysis reveals no net change in local-scale plant biodiversity over time. Proceedings of the National Academy of Sciences 110: 19456–19459.

Vitousek, P. M., J. R. Gosz, C. C. Grier, J. M. Melillo, W. A. Reiners, and R. L. Todd. 1979. Nitrate losses from disturbed ecosystems. Science 204: 469–474.

Vitousek, P. M., and D. U. Hooper. 1994. Biological diversity and terrestrial ecosystem biogeochemistry. Pages 3–14 in E.-D. Schulze and H. A. Mooney, eds. Biodiversity and Ecosystem Function. Springer.

Vitousek, P. M., and W. A. Reiners. 1975. Ecosystem succession and nutrient retention: a hypothesis. BioScience 25: 376–381.

Vollenweider, R. A. 1975. Input-output models with special reference to phosphorus loading concept in limnology. Swiss Journal of Hydrology 37: 53–84.

Volterra, V. 1926. Fluctuations in the abundance of a species considered mathematically. Nature 118: 558–560.

Vos, M., A. M. Verschoor, B. W. Kooi, F. L. Wäckers, D. L. DeAngelis, and W. M. Mooij. 2004. Inducible defenses and trophic structure. Ecology 85: 2783–2794.

Walker, T. W., and J. K. Syers. 1976. The fate of phosphorus during pedogenesis. Geoderma 15: 1–19.

Werner, E. E., and S. D. Peacor. 2003. A review of trait-mediated indirect interactions in ecological communities. Ecology 84: 1083–1100.

Worm, B., E. B. Barbier, N. Beaumont, J. E. Duffy, C. Folke, B. S. Halpern, J. B. C. Jackson, et al. 2006. Impacts of biodiversity loss on ocean ecosystem services. Science 314: 787–790.

Evolution of Phosphorus Limitation in Lakes

Natural mechanisms compensate for deficiencies of nitrogen and carbon in eutrophied lakes.

D. W. Schindler

Algal physiologists have conclusively demonstrated that freshwater phytoplankton require several elements for growth and reproduction (1). Most limnologists have considered that all of these elements might act as potential limiting nutrients in freshwater lakes. While there is now general agreement that phytoplankton will be phosphorus-limited in the majority of lakes (2), eutrophication management based entirely on phosphorus control is still viewed by some as too simplistic (3, 4). Yet in past years evidence has mounted for an astoundingly precise relationship between the concentration of total phosphorus and the standing crop of phytoplankton in a wide variety of lakes (5), including many in which low nitrogen-to-phosphorus ratios should favor limitation by nitrogen. In this article, evidence from whole-lake experiments is used to reexamine the nutrient-control question, with particular attention to factors that might be missing in experiments on a smaller scale. Evidence for the two elements, carbon and nitrogen, that have received widespread attention as alternatives to phosphorus as limiting nutrients (6) will be considered.

Control of Phytoplankton Populations in Lakes by Carbon

Five years ago, many scientists believed that carbon might control eutrophication in some lakes (7). The scientific evidence for this view came largely from bioassay experiments done in small bottles (8), where phytoplankton were stim-

D. W. Schindler is scientific leader of the Experimental Limnology Project at the Freshwater Institute, Environment Canada, 501 University Crescent, Winnipeg, Canada R3T 2N6.

ulated by addition of carbon, but not by phosphorus or nitrogen. This was interpreted by many as indicating that eutrophication was caused by overfertilization with carbon. Whole-lake experiments have now shown that these experiments provided misleading evidence by excluding processes occurring in natural bodies of water (9, 10). In experimentally fertilized lakes, invasion of atmospheric carbon dioxide supplied enough carbon to support and maintain phytoplankton populations that were proportional to phosphorus concentrations over a wide range of values (Fig. 1). Yet in bottle bioassay experiments, phytoplankton showed evidence of extreme carbon limitation throughout the period when algal populations were responding in proportion to total phosphorus concentrations. But more important, there was a strong tendency in every case for lakes to correct carbon deficiencies, maintaining concentrations of both chlorophyll and carbon that were proportional to the phosphorus concentration (Figs. 1 and 2).

Bottle bioassay experiments to test the carbon-limitation hypothesis were inadequate in two respects. First, experiments were done in small, closed or semiclosed containers, where turbulence of the water and interaction with the overlying atmosphere were restricted. Second, the proportion of alkalinity supplied by hydroxyl ions has been found to affect the rate at which carbon invades lake ecosystems (11), and no attempt was made in the bottle experiments to simulate such conditions (12).

These observations have led me to conclude that carbon control no longer deserves consideration as a method for managing eutrophication of natural waters (13).

Control of Phytoplankton Populations in Lakes by Nitrogen

Because of the problems our group encountered in extrapolating from small-scale experiments to whole-lake management of carbon, we designed several whole-lake experiments to investigate the nitrogen management question. Because of the important role played by atmospheric carbon, we paid particular attention to fixation of atmospheric nitrogen by algae as a supplementary source of this element.

In lake 227, the site of our first carbon experiments, the fertilizers added to the lake had contained adequate or excess nitrogen. The ratio by weight of nitrogen to phosphorus was 14. Fertilizer with this ratio was applied for 6 years, 1969 through 1974. During the entire period, phytoplankton were dominated by the green alga Scenedesmus and other algae incapable of fixing gaseous nitrogen. I hypothesized that by reducing the ratio of nitrogen to phosphorus in fertilizer, blue-green algae, which are capable of fixing atmospheric nitrogen, might be favored. Therefore, in a second experiment, the northeast basin of lake 226 was fertilized with a nitrogen-deficient fertilizer (4). The ratio of nitrogen to phosphorus by weight was only 5. In every year that this fertilizer was applied (1973 to 1975), nitrogen-fixing blue-green algae of the genus Anabaena dominated the lake. Fixation of nitrogen accounted for a substantial proportion of the total income of nitrogen to the lake (14), and the ratio of total nitrogen to total phosphorus in the lake remained similar to that in other lakes (Fig. 3).

Another whole-lake experiment showed that this difference between lakes 226 and 227 was no accident. In 1975, the nitrogen-to-phosphorus ratio of fertilizer to lake 227 was reduced to 5, duplicating the ratio used in lake 226. For the first time in the 8 years we have studied the lake, a nitrogen-fixing blue-green alga, this time Aphanizomenon gracile, became dominant in midsummer (Fig. 4). Also for the first time, substantial nitrogen fixation was detected. At this time I cannot explain why different genera of nitrogen fixers should be dominant in the two lakes. This difference may be due to differences in other micronutrients or growth inhibitors and requires further investigation (15).

A final experiment has provided a somewhat different result. Only phosphorus fertilizer was applied to lake 261 for 3 years. No nitrogen fixers appeared in the phytoplankton, but a luxuriant growth of attached algae (periphyton) which are capable of high fixation of nitrogen was observed (*16*). In spite of this, the phytoplankton chlorophyll increased somewhat, and the chlorophyll-to-phosphorus ratio was only slightly lower than in other lakes (Fig. 1).

Discussion

The experiments described above clearly demonstrate the existence in lakes of biological mechanisms which are capable of eventually correcting algal deficiencies of carbon and, at least in some cases, of nitrogen. It is, however, noteworthy that the ratios of carbon to phosphorus and nitrogen to phosphorus in our lakes are maintained at 174 and 31, respectively. These are much higher than the values in inputs to any of our lakes, which suggests that internal factors, as well as atmospheric compensation, may also favor high C/P and N/P ratios (*17*). No external mechanisms exist for phosphorus, which has no gaseous atmospheric cycle. Experiments in lakes 304 and 226 southwest, both phosphorus-deficient, show that there is no appreciable internal compensation in our Precambrian Shield lakes (*18*). While a sudden increase in the phosphorus input, as is common during cultural eutrophication, may cause algae to exhibit symptoms of limitation by either nitrogen or carbon or both, there are long-term processes at work in the environment which may cause the deficiencies to be corrected eventually, once again leaving phytoplankton growth proportional to the concentration of phosphorus. If our results are typical, they explain why phosphorus limitation or phosphorus proportionality is commonly observed in lakes, even where the nutrient ratios of geochemical and cultural sources might be expected to favor limitation by nitrogen or carbon. I hypothesize that only lakes which have experienced very recent increases in phosphorus input, without corresponding increases in nitrogen and carbon, or those receiving enormous influxes of phosphorus will not show the correlation between total phosphorus and standing crop.

This "evolution" of appropriate nutrient ratios in freshwaters involves a complex series of interrelated biological, geological, and physical processes, including photosynthesis, the selection for species of algae that can utilize atmospheric nitrogen, alkalinity, nutrient supplies and concentrations, rates of water renewal, and turbulence. It is impossible to

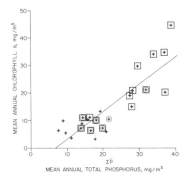

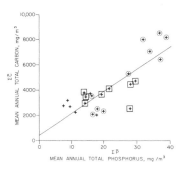

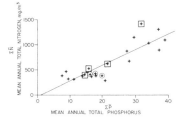

Fig. 1 (top). Relationship between mean annual concentrations of total phosphorus and chlorophyll a in lakes of the Experimental Lakes Area. All averages are weighted for morphometry of the lakes. Both fertilized and unfertilized lakes are included. ($\circ$) Data for fertilized lakes where nutrients supplied by fertilizer, precipitation, and runoff were deficient in nitrogen (N/P < 8 by weight). ($\square$) Data for lakes where nutrients from the same three sources were deficient in carbon (C/P < 50 by weight). Points (+) without either circles or squares represent lakes where C/P and N/P ratios in inputs are either natural or higher than natural. The linear regression equation chlorophyll $a = 0.987\Sigma\bar{P} - 6.520$ fits the points with the correlation coefficient $r = .86$.   Fig. 2 (middle). Mean annual total carbon and total phosphorus concentrations in fertilized lakes of the Experimental Lakes Area, illustrating how the carbon content of these lakes has increased because of addition of phosphorus. Total carbon includes seston, dissolved inorganic carbon, and methane, but excludes dissolved organic carbon (DOC). Concentrations of DOC have not changed as a consequence of eutrophication. ($\circ$) Data for lakes that received no carbon with fertilizer. ($\square$) Data for lakes that received low fertilizations with sucrose (C/P $\sim$ 6 by weight in lake 226, $\sim$ 14 in lake 304, and $\sim$ 85 in lake 226 southwest). Points without circles or squares represent lakes with either natural or higher than natural C/P inputs. The regression equation $\Sigma\bar{C} = 173.6\Sigma\bar{P} + 420.3$ fits the points with $r = .83$. Fig. 3 (bottom). Mean annual concentrations of total nitrogen and total phosphorus in fertilized lakes, illustrating that the nitrogen content of a lake increases when phosphorus input is increased, even when little or no nitrogen is added with fertilizer. ($\circ$) Data for lake 261, which received only natural nitrogen. ($\square$) Data for lakes 226 northeast and 227. The N/P in input to lake 226 northeast in 1973 to 1975 and lake 227 in 1975 were less than 8. Points without circles or squares represent lakes with N/P equal to or greater than the natural ratio. The regression equation $\Sigma\bar{N} = 30.9\Sigma\bar{P} - 29.3$ fits the points with $r = .86$.

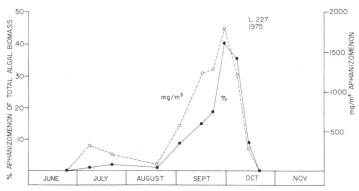

Fig. 4. Summer population of *Aphanizomenon* in lake 227 in 1975, after the N/P ratio in fertilizer had been cut back to 5 by weight. The genus had not been observed in quantifiable numbers in any of the six previous years, when an N/P ratio of 15 had been used in fertilizer. High fixation of atmospheric nitrogen accompanied the bloom of this species.

visualize a laboratory bioassay experiment that could realistically represent all of these parameters.

On the basis of data from several studies of the carbon, nitrogen, and phosphorus cycle, I hypothesize that schemes for controlling nitrogen input to lakes may actually affect water quality adversely by causing low N/P ratios, which favor the vacuolate, nitrogen-fixing blue-green algae that are most objectionable from a water quality standpoint. Conversely, when phosphorus control causes an increase in the N/P ratio, the resulting shift from "water bloom" blue-green algae to forms that are less objectionable may be as important as quantitative decreases in algal standing crop. Several authors have observed such species shifts with changing N/P ratios (19).

It is clear that management decisions on nutrient control measures must be based on controlled field tests as well as simple laboratory bioassays.

References and Notes

1. For a summary of these, see J. Vallentyne, *Can. Fish. Mar. Serv. Misc. Spec. Publ. No. 22* (1974), p. 162.
2. The general assemblies of both the International Limnological Congress and the International Ecology Congress unanimously passed resolutions recommending widespread phosphorus control as a solution to eutrophication. Almost all of the freshwater scientists in the world were represented.
3. For example, see J. W. G. Lund [*Nature (London)* 249, 797 (1974)] for a critique of phosphorus control, including my report of the same year (4).
4. D. W. Schindler, *Science* 184, 897 (1974).
5. P. Dillon and F. Rigler, *J. Fish. Res. Board Can.* 32, 1519 (1975); R. A. Vollenweider, *Schweiz. Z. Hydrol.* 37, 53 (1975); D. W. Schindler, *Limnol. Oceanogr.*, in press.
6. See papers in G. E. Likens, Ed., *Am. Soc. Limnol. Oceanogr. Spec. Symp. No. 1* (1972).
7. For example, see articles in *Can. Res. Dev.* 3, 19 (1970).
8. For example, W. Lange, *Nature (London)* 215, 1277 (1967); *J. Phycol.* 6, 230 (1970); M. Sakamoto, *J. Fish. Res. Board Can.* 28, 203 (1971); A. Christie, *Ont. Water Resour. Comm. Res. Publ. No. 32* (1968).
9. D. W. Schindler, G. Brunskill, S. Emerson, W. Broecker, T-H. Peng, *Science* 177, 1192 (1972); D. W. Schindler (10); S. Emerson, W. Broecker, D. W. Schindler, *J. Fish. Res. Board Can.* 30, 1475 (1973).
10. D. W. Schindler, *Int. Ver. Theor. Angew. Limnol. Verh.* 19, 3221 (1975).
11. _____ and E. Fee, *J. Fish. Res. Board Can.* 30, 1501 (1973).
12. S. Emerson [*Limnol. Oceanogr.* 20, 743 (1975); *ibid.*, p. 754] showed that gas exchange is roughly proportional to the square of the wind velocity at the lake surface. He also reported that chemical enhancement of gas exchange in softwater lakes may yield values five to ten times higher than unenhanced values, once nutrient additions have caused depletion of inorganic carbon, so that alkalinity is dominated by hydroxyl ions.
13. These values are summarized by D. W. Schindler (10).
14. In lake 226, nitrogen fixation contributed 38 percent of the total nitrogen income in 1974 and 19 percent in 1975 (R. Flett, University of Manitoba, thesis, 1976).
15. For example, T. P. Murphy, D. R. S. Lean, and C. Nalewajko [*Science* 192, 900 (1976)] showed that *Anabaena* requires iron for fixation of atmospheric nitrogen and that this genus can suppress the growth of other species of algae by excretion of a growth-inhibiting substance.
16. M. Turner and R. Flett, unpublished data. As yet no quantitative estimates of nitrogen fixation for an entire season are available. G. Persson, S. K. Holmgren, M. Jansson, A. Lundgren, and C. Anell [in *Proceedings of the NRC-CNC (SCOPE) Circumpolar Conference on Northern Ecology (Ottawa, 15 to 18 September 1975)*] reported similar results for a lake in Sweden that was fertilized with phosphorus.
17. Possible additional mechanisms are outlined by D. W. Schindler [in *Environmental Biogeochemistry*, J. O. Nriagu, Ed. (Univ. of Michigan Press, Ann Arbor, 1976), pp. 647–664]. In particular, nitrogen appears to be more efficiently recycled from sediments than phosphorus.
18. Whole-lake experiments with phosphorus-deficient fertilizations in lakes 226 southwest and 304 have confirmed the lack of either biological or geochemical mechanisms for enhancing inputs of phosphorus.
19. For example, see P. Sze, *Phycologia* 14, 197 (1975); M. Michalski and K. Nicholls, *Phosphorus Removal and Water Quality Improvements in Gravenhurst Bay, Ontario* (Ontario Ministry of Environment, Rexdale, Ontario, 1975); M. Michalski and N. Conroy, *Proc. 16th Conf. Great Lakes Res.* (1973), p. 934; W. T. Edmondson, *Verh. Int. Ver. Limnol.* 18, 284 (1972). Other members of our staff have recently been able to cause shifts in dominance from blue-green to green algae in hypereutrophic lakes by adding nitrogen (J. Barica and H. Kling, personal communication).
20. My thanks to T. Ruszczynski, who performed the calculations for Figs. 1, 2, and 4, to D. Findlay, whose plankton identifications and counts allowed these interpretations, and to J. Prokopowich for chemical analyses. The critical comments of K. Patalas, R. Flett, and E. Fee are greatly appreciated.

The Biosphere Reserve Program in the United States

A program has been developed to select key sites for environmental research and monitoring.

Jerry F. Franklin

Biosphere reserves are major elements in Unesco's "Man and the Biosphere" (MAB) program and in the U.S.–U.S.S.R. Environmental Agreement. They are part of an international system of reserves with the primary objectives of conservation of genetic diversity, environmental research and monitoring, and education.

The scientific community must be aware of the existence and potential of the biosphere reserves if they are to fulfill their intended functions. I will outline the conceptual development of the Unesco effort, the philosophy guiding its implementation in the United States, and the utilization and expansion of U.S. biosphere reserves expected in the future. The views presented are those of the U.S. National Committee for Man and the Biosphere.

Development of the Biosphere Reserve Concept

The concept of biosphere reserves was developed as a major element of Project 8, *Conservation of Natural Areas and of the Genetic Material They Contain*, in the Unesco-sponsored Program on Man and the Biosphere (1). This project, which emerged as an important component early in the MAB planning, was initially considered in detail by an expert panel, which met in Morges, Switzerland, in September 1973. Establishment of a worldwide network of biosphere reserves was this panel's first recommendation. A task force with the responsibility of defining "criteria and guidelines for the selection and establishment of biosphere reserve" (2, p. 9) met in Paris in May 1974. The task force report is the source of the following information on the international program.

Biosphere reserves have three basic purposes or objectives: (i) conservation or preservation—"to conserve for present and future use the diversity and integrity of biotic communities of plants and animals within natural ecosystems, and to safeguard the genetic diversity of species on which their continuing evolution depends" (2, p. 6); (ii) research and

The author is chief plant ecologist, Pacific Northwest Forest and Range Experiment Station, Forest Service, U.S. Department of Agriculture, Forestry Sciences Laboratory, Corvallis, Oregon 97330. He was chairman of the U.S. Man and the Biosphere Committee on Project 8 (Biosphere Reserves) and is U.S. chairman of Project V-4.1 (Biosphere Reserves) under the U.S.–U.S.S.R. Environmental Agreement.

N.D.) for ten locations at China Lake and for two locations at Boron are a direct result of interference due to VO_3^- and other anionic metal complexes. It also is significant that the data on As content from single-sweep polarography are in good agreement with the SDDC data only for a single well location, namely, location 29 of Table 1. In view of the projected importance of water pollution analysis, several conclusions have been derived from the work reported:

1) The underground waters of the Boron and China Lake basins are associated with heterogeneous rock formations, many of which contain large amounts of water-soluble minerals. Thus, the subsurface water more closely resembles dilute brines or even brackish seawater (1) than it does the usual surface waters for which the colorimetric SDDC method was developed. In the SDDC test procedure (4) the analyst is cautioned (paragraph 1.2) about interference with AsH_3 evolution caused by certain metals; all the metals listed in paragraph 1.2 were found in our mineralized water samples. Stratton and Whitehead (9) and Ballinger et al. (10) developed and evaluated the SDDC method for relatively pure river drinking water, and neither of these groups of authors found it necessary to use a second method of known accuracy and specificity for As to verify their results.

2) It is recommended that the SDDC method be used for the routine analysis of desert water only after verification of the As content by means of x-ray or single-sweep polarographic measurements. Angino et al. (11) cited the presence of the same interfering metals in detergents as those reported here but they evidently ignored the cau-

tion in paragraph 1.2 of the SDDC test procedure in their statement that a high degree of precision in the determination of As was attained. No mention was made of the degree of accuracy obtained.

3) Accurate and highly specific analytical methods should be employed by concerned agencies to correctly detect the presence of and to define the extent of water pollution. The accurate analysis of trace toxicants in water in the parts-per-billion range is difficult, at best, and research efforts in this area are vitally needed, as advocated by Smith (12).

G. C. WHITNACK
Michelson Laboratory, Naval Weapons Center, China Lake, California 93555
H. H. MARTENS
Boron Community Services District, Boron, California 93516

References

1. State of California, 6th Regional Water Quality Control Board, Bishop, Calif., "Report on Arsenic Occurrence in the North Muroc Hydrologic Basin" (Feb. 1969).
2. Naval Weapons Center, China Lake, Calif., *Tech. Mem.*, "The Determination of Arsenic in Drinking Water" (March 1969).
3. G. C. Whitnack and R. G. Brophy, *Anal. Chim. Acta* **48**, 123 (1969).
4. *Standard Methods for the Examination of Water and Wastewater* (American Public Health Association, New York, ed. 12, 1965), p. 56.
5. *Federal Register*, Title 42, Public Health Service, Part 72 (6 March 1966).
6. F. Marcie, *Environ. Sci. Technol.* **1**, 165 (1967).
7. E. B. Sandell, *Colorimetric Determination of Traces of Metals* (Interscience, New York, ed. 3, 1959); O. Menis and T. C. Rains, *Anal. Chem.* **41**, 952 (1969); K. E. Burke and M. M. Yanak, *ibid.*, p. 963; A. Arnold et al., *Analyst* **94**, 664 (1969).
8. D. E. Robertson, *Anal. Chem.* **40**, 1067 (1968).
9. G. Stratton and H. C. Whitehead, *J. Amer. Water Works Ass.* **54**, 861 (1962).
10. D. G. Ballinger, R. J. Lishka, M. E. Gales, Jr., *ibid.*, p. 1424.
11. E. E. Angino, L. M. Magnuson, T. C. Waugh, O. K. Galle, J. Bredfeldt, *Science* **168**, 389 (1970).
12. R. G. Smith, *Anal. Chem.* **40**, No. 7, 24A (1968).

7 August 1970; revised 21 October 1970 ∎

man (1) reported destabilization of a stable exploitation ecosystem which resulted in the extinction of both the exploiter (an acarophagous mite) and its victim (an herbivorous mite). They produced this result by trebling the herbivore's food density. By using a variety of realistic models, I predict that instability should often be the result of nutritional enrichment in two-species interactions.

Rosenzweig and MacArthur (2) showed that exploitation (or predator-prey) ecosystems do not necessarily exhibit any oscillations. Furthermore, even if there are oscillations, they do not last under ordinary circumstances. If the exploiter is quite proficient at reproducing in the presence of few of its victims, then the ecosystem does not persist. If, however, the victims are relatively proficient at escape or their exploiters have a relatively poor reproductive efficiency or digestive efficiency, then the system will persist in ecological time (3).

The dividing line between persistent and explosive systems is definable from a general graph of exploitation (2). The victim's density V is plotted against P, the exploiter's density. The collection of graph points at which $dV/dt = 0$ is called the victim's isocline. The collection of points at which $dP/dt = 0$ is called the exploiter's isocline. Any point of intersection between the two isoclines is an ecosystem equilibrium, but not all such equilibria will result in a steady state. The usual form of the prey isocline is a hump (4). If the equilibrium is at a point on the left side of the hump, the predator is too proficient and the system will ordinarily not persist. If equilibrium is at a point on the right-hand (downslope) side of the hump, the system will persist. Thus, the hump's peak is over a critical value of V, V^*. If the equilibrium value of V is larger than V^*, the system is safe. If not, it is in danger of extinction.

If the exploiters do not actually interfere with each other directly—if they never battle over the same individual victim or engage in cannibalism or territorial defense—then the P isocline is a simple vertical line $(V = J)$. The position of this line is fully determined by the phenotypes of the exploiter and its victim. It does not change with nutrient flow or energy supply.

To discover the effect of enriching a system, one needs to find how V^* changes as enrichment proceeds. If enrichment increases V^*, then it is jeopardizing the system, because eventu-

Paradox of Enrichment: Destabilization of Exploitation Ecosystems in Ecological Time

Abstract. *Six reasonable models of trophic exploitation in a two-species ecosystem whose exploiters compete only by depleting each other's resource supply are presented. In each case, increasing the supply of limiting nutrients or energy tends to destroy the steady state. Thus man must be very careful in attempting to enrich an ecosystem in order to increase its food yield. There is a real chance that such activity may result in decimation of the food species that are wanted in greater abundance.*

Schemes for increasing primary productivity by enriching an ecosystem's energy or nutrient flow are much in evidence today and are probably a re-

flection of the increasing demands of the world's population. Such schemes may end in catastrophe.

In 1963, Huffaker, Shea, and Her-

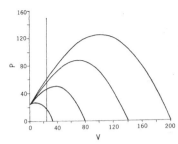

Fig. 1. Isoclines at four levels of productivity ($K = 34$, 80, 140, and 200) when model 4 is used. Symbol V is victim's density; P is exploiter's. The curved lines are V isoclines, which peak over higher values of V at higher K. The vertical line, $V = 25$, is the P isocline. From the slopes of the V isoclines at the points where they intersect the P isocline, one expects only the $K = 34$ system to have a steady state.

ally V^* will be made greater than J.

Briefly the method is this. Set $dV/dt = 0$ and solve for P. This is the algebraic equation for the V isocline. Take $\partial P/\partial V$. The value of V that satisfies $\partial P/\partial V = 0$ is V^*. If K represents the standing crop of V where $P = 0$, then K must be directly proportional to the flow rates of limiting nutrients. Thus, enrichment implies K increase. The final step, then, is to obtain $\partial V^*/\partial K$. Since this is always positive, enrichment leads toward system instability.

Each analytical model used is the difference between inherent rate of increase of V and the number of V that die (or are not born) owing to the activities of P.

Assume that each V must receive a quota of nutrients at a rate Q in order just to replace itself. Nutrients flow to the lone individual V at a rate R. Thus, this individual reproduces at a net rate $r\,(R - Q)$. As V increases, however, each individual V is subjected to intraspecific competition. One may assume that an individual V's effective "feeding" rate diminishes with increasing V^a, where a is the victim's competition constant, $1 \geqslant a > 0$ (5). Thus the per capita reproductive rate is $r\,(RV^{-a} - Q)$ and the inherent rate is $rV\,(RV^{-a} - Q)$. From the parentheses one obtains $K = (R/Q)^{1/a}$. Hence, to increase K, R must be increased (a and Q are constant), and $\partial V^*/\partial K$ has the same sign as $\partial V^*/\partial R$ (6).

In addition to the above model, I have used the traditional Pearl-Verhulst logistic $rV\,(1 - V/K)$ and the Gompertz $rV\,(\ln K - \ln V)$.

Kill rate models. Lotka and Volterra's approach was to treat the two populations like molecules: kVP [see (7) and references therein]. This has been shown to be inadequate. Gause (7) obtained a reasonable fit to a kill rate curve by taking the square root of V. We can generalize this procedure by taking V to the gth power $0 < g \leqslant 1$. Thus, a kill rate model is kPV^g.

Another model is based on observations (5, 8) that one lone exploiter will attack $k(1 - e^{-cV})$ victims in a fixed amount of time. Since the exploiters compete only by reducing each other's food supply, P exploiters will kill $kP(1 - e^{-cV})$.

Thus models for dV/dt include

$$dV/dt = rV(RV^{-a} - Q) - kP(1 - e^{-cV}) \quad (1a)$$

$$dV/dt = rV(1 - V/K) - kPV^g \quad (2a)$$

$$dV/dt = rV(RV^{-a} - Q) - kPV^g \quad (3a)$$

$$dV/dt = rV(1 - V/K) - kP(1 - e^{-cV}) \quad (4a)$$

$$dV/dt = rV(\ln K - \ln V) - kPV^g \quad (5a)$$

$$dV/dt = rV(\ln K - \ln V) - kP(1 - e^{-cV}) \quad (6a)$$

In view of the lack of convincing tests of any of the models as a general case for all systems, I have analyzed all six.

The first step in each analysis is omitted here: solution of each equation for P when $dV/dt = 0$. For example, Eq. 4a becomes

$$P = \frac{rV(1 - V/K)}{k(1 - e^{-cV})}$$

This set of equations is the set of V isoclines.

Next we obtain $\partial P/\partial V$ and determine the conditions under which this

will be zero. These are the V^* equations:

$$R = \frac{Q(V^*)^a\,(e^{cV^*} - 1 - cV^*)}{(e^{cV^*} - 1)\,(1 - a) - cV} \quad (1b)$$

$$K = \frac{(2 - g)}{(1 - g)}\,V^* \quad (2b)$$

$$R = \frac{Q(V^*)^a\,(1 - g)}{(1 - a - g)} \quad (3b)$$

$$K = V^*\frac{(2e^{cV^*} - cV^* - 2)}{(e^{cV^*} - cV^* - 1)} \quad (4b)$$

$$\ln K = \ln V^* + 1/(1 - g) \quad (5b)$$

$$\ln K = \ln V^* + 1 + \frac{cV^*}{e^{cV^*} - 1 - cV^*} \quad (6b)$$

The final step requires a small explanation. We need the sign of $\partial V^*/\partial K$ or $\partial V^*/\partial R$. Often the equation systems are easily solved for K or $\ln K$, but not V^*. However, $\partial V^*/\partial K$ is positive if and only if $\partial K/\partial V^*$ is. And $\partial K/\partial V^*$ is positive if and only if $\partial \ln K/\partial V^*$ is. Hence, we can readily proceed with these latter two partial derivatives. Three are positive for any set of values of the constants:

$$\frac{\partial K}{\partial V^*} = (2 - g)/(1 - g) \quad (2c)$$

$$\frac{\partial K}{\partial V^*} = \frac{2(e^{cV} - 1)(e^{cV} - 1 - cV - c^2V^2/2)}{(e^{cV} - 1 - cV)^2} \quad (4c)$$

$$\frac{\partial \ln K}{\partial V^*} = \frac{1}{V^*} \quad (5c)$$

Equation 4c is always positive because the MacLaurin series for e^{cV} is $1 + cV + c^2V^2/2 + c^3V^3/6 + \cdots$ (see Figs. 1 and 2).

The other three cases are not quite so readily handled. Equation 3b does not always have a positive solution for V^*. In fact V^* is negative if and only if $(1 - a - g)$ is also negative. The V isocline of Eq. 3a is humpless for such values of $(a + g)$. Values this great im-

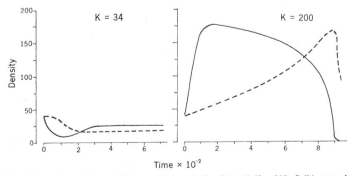

Fig. 2. Iteration of the model 4 exploitation at $K = 34$ and $K = 200$. Solid curve is V; dashed curved is P. Enrichment causes the simulated extinction of both species. The exploiter equation used was $dP/dt = AkP\,(e^{-cJ} - e^{-cV})$. Time units are in calculator cycles (10).

ply intense intraspecific producer competition and also a relatively low tendency for exploiters to become hungry or satiated and to modify their behavior accordingly. In such a system there is no tendency for extinction regardless of productivity.

However, if $(a + g)$ is less than 1, there is a positive $V*$ and

$$\frac{\partial R}{\partial V*} = \frac{aQ(1 - g) \ (V*)^{a-1}}{(1 - a - g)} \qquad (3c)$$

Clearly Eq. 3c is always positive if $V*$ is biologically real. Hence, in model 3, if there is any threat of system extinction, it is increased by enrichment.

Models 1 and 6 are similar and most complex. It turns out that Eq. 6b is satisfied by two values of V. One is $V*$. Another is a very small value of V that occurs over a trough in the V isocline. Thus, there is ambiguity in the following:

$$\frac{\partial \ln K}{\partial V*} =$$
$$\frac{(e^{eV*} - 1) \ (e^{eV*} - 1 - cV* - c^2V*^2)}{(e^{eV*} - cV* - 1)^2} \qquad (6c)$$

This equation, set to zero, holds for both $V*$ and the V under the trough. The unstable equilibrium values of V are those between V (trough) and $V*$. Model 6c is positive for $V*$ and negative for V (trough) (9). Hence, as enrichment proceeds, the range of unstable V is increasing at both ends. Therefore again, enrichment unambiguously tends to weaken the steady state. Model 1 has the same characteristics (9).

Until we are confident that the conclusions based on these systems do not apply to natural ecosystems, we must remain aware of the danger in setting enrichment as a human goal.

MICHAEL L. ROSENZWEIG

*Department of Biological Sciences,
State University of New York,
Albany 12203*

References and Notes

1. C. B. Huffaker, K. P. Shea, S. G. Herman, *Hilgardia* 34 (9), 305 (1963).
2. M. E. Rosenzweig and R. H. MacArthur, *Amer. Natur.* 97, 209 (1963).
3. Throughout, phenotypes of both species are held constant, ignoring evolution. Current hopes for a high magnitude increase in productivity would probably deny natural selection the time to act.
4. E. J. Maly, *Ecology* 50, 59 (1969); M. L. Rosenzweig, *Amer. Natur.* 103, 81 (1969). The isocline equations of this report are further evidence for the hump, though one is "humpless" in some cases and two others have, in addition, an (artifactual?) extreme left-hand rise to $+\infty$.
5. K. E. F. Watt, *Can. Entomol.* 91, 129 (1959).
6. This model is sigmoidal. Its point of inflection is at $V = (1 - a)^{1/a} K$. Thus it depends on the intensity of intraspecific competion. By L'Hospital's rule, as a approaches 0, $(1 - a)^{1/a}$ or exp ln $(1 - a)/a$ approaches $1/e$. Thus, if intraspecific competition is slight, this equation closely resembles the Gompertz curve. In

fact, I doubt if any $a < 0.5$ is readily distinguishable from the Gompertz. However, the Pearl-Verhulst yields an inflection at $K/2$, which is always greater than that predicted from this model.
7. G. F. Gause, *The Struggle for Existence* (Williams & Wilkins, Baltimore, 1934).
8. V. S. Ivlev, *Experimental Ecology of the Feeding of Fishes* (Yale Univ. Press, New Haven, 1961).
9. A proof is on file and available to those who desire it.
10. Iteration was performed with the use of difference equations on a Wang programmable calculator. I thank Dr. Fred Walz for its use.
11. I thank Drs. E. Leigh, III, S. Levin, D. McNaught, L. Segel, and M. Slatkin for valuable comments; and J. Riebesell, whose careful work was helpful in making mathematical errors scarce, if not entirely absent. Supported by the National Science Foundation and the Research Foundation of the State University of New York.

26 October 1970

Induction of Liver Acetaldehyde Dehydrogenase: Possible Role in Ethanol Tolerance after Exposure to Barbiturates

Abstract. Mice were injected twice a day for 4 days with saline or phenobarbital or ethanol. Treatment with phenobarbital, but not ethanol, increased the amount of liver acetaldehyde dehydrogenase activity. More rapid removal of acetaldehyde, which is a toxic metabolic intermediate of ethanol, may contribute to the alcohol tolerance exhibited by persons who use barbiturates regularly.

Acetaldehyde, a pharmacologically active metabolite of ethanol, appears to contribute to the actions of ingested ethanol (1). Acetaldehyde can induce nausea, vomiting, and sweating; it causes release of catecholamines and depression of oxidative phosphorylation in isolated brain tissue. Recently it was shown that acetaldehyde condenses with catecholamines (2) or potentiates a similar condensation by certain endogenous aldehydes (3) to form isoquinoline alkaloids that may possess biologic activity. Because the pharmacologic effects of acetaldehyde are generally subjectively unpleasant, it has even been suggested that "while ethanol actions may be the reason that people drink alcohol, the actions of acetaldehyde may be more related to why they stop" (1).

The purpose of our study was to see if phenobarbital, an inducer of many liver enzymes (4), would increase the levels of acetaldehyde dehydrogenase (AcDH), which is believed to be the enzyme primarily responsible for detoxication of acetaldehyde. We were interested in the tolerance to the effects of ingested alcoholic beverages exhibited by persons who use barbiturates regularly. If AcDH activity were elevated after exposure to barbiturates, the consequent more rapid removal of acetaldehyde could contribute to ethanol tolerance. Our results with mice support this hypothesis.

Male and female Paris R-III or C-57BL mice were injected intraperitoneally twice a day (at approximately 9:30 a.m. and 4:30 p.m.) for 4 days with either isotonic saline (buffered at pH 7.4 with $0.01M$ sodium phosphate) or with sodium phenobarbital (75 mg/

kg) or with ethanol (2.4 g/kg, administered as a 30 percent by volume solution), each dissolved in the isotonic buffer. On day 5, the animals were killed by cervical dislocation, and the liver was homogenized (ten strokes in a Teflon and glass homogenizer) in 20 volumes of ice-cold water. The homogenate was kept at about 0°C for the duration of the experiment. Cell debris was removed by centrifugation at $700g$ for 5 minutes. Enzyme assays were completed within 2 hours after death.

Acetaldehyde dehydrogenase was assayed at room temperature by a modification of the method of Maxwell and Topper (5). Assay tubes contained 6 ml of buffer ($0.1M$ glycine at pH 9.5, containing 4.0 mM mercaptoethanol and 2.0 mM ethylenediaminetetraacetic acid), 0.4 ml of nicotinamide adenine dinucleotide (NAD) solution (Calbiochem, 10 mg/ml), and 0.1 ml of acetaldehyde solution (Eastman, 1 percent by volume in water). A blank containing all of the above except acetaldehyde was used. The reaction was started by the addition of 0.2 ml of liver extract. Reduced NAD (NADH) concentration was measured by the optical density at 340 nm in a Gilford model 300 spectrophotometer after 3.5 and at 7.5 minutes; enzyme activity was given by the increase in optical density during the 4-minute interval, with the blank substracted.

In three separate experiments, increases in liver AcDH activity were observed after 4 days of treatment with phenobarbital, but AcDH activity was not increased after treatment with ethanol. Pooled results of the three experiments (Fig. 1) indicate that

Vol. 113, No. 5   The American Naturalist   May 1979

GRAZING AS AN OPTIMIZATION PROCESS: GRASS-UNGULATE RELATIONSHIPS IN THE SERENGETI

S. J. McNaughton

Biological Research Laboratories, 130 College Place, Syracuse, New York 13210; and
Serengeti Research Institute, P.O. Seronera via Arusha, Tanzania

Until recently (Vickery 1972; Harris 1974; Dyer 1975; Mattson and Addy 1975; Dyer and Bokhari 1976; McNaughton 1976), ecologists have tended to view plants as relatively passive participants in short-term interactions at the plant-herbivore interface, suffering tissue reduction from herbivory, and responding in evolutionary time through the evolution of novel antiherbivore chemicals and structures (Fraenkel 1959; Ehrlich and Raven 1964; Levin 1971, 1973; Caswell et al. 1973; Freeland and Janzen 1974). Agronomists, botanists, foresters, and range managers, however, have frequently emphasized the direct plant responses to herbivory of compensatory growth and assimilate reallocation (Parker and Sampson 1931; Robertson 1933; Canfield 1939; Labyak and Schumacher 1954; Stein 1955; Brougham 1961; Leonard 1962; Jameson 1963; Pearson 1965; Neales and Incoll 1968; Hutchinson 1971; Ryle and Powell 1975). This literature in applied ecology documents the frequent occurance of compensatory plant growth following tissue reduction by herbivores or management practices.

However, no straightforward generalizations are possible regarding the immediate effects of herbivores on plant growth and resource allocation. Consequences of tissue damage are under the complex control of plant genetics (Parker and Sampson 1931; Weaver and Houghen 1939; Newell and Keim 1947; Branson 1953; Robocker and Miller 1955; Baker and Hunt 1961; Hart et al. 1971), intensity and frequency of herbivore effects (Graber 1931; Canfield 1939; Jacques and Edmond 1952; Lodge 1960; Marshall and Sager 1965; Forde 1966; Gifford and Marshall 1973), plant developmental stage at the time of herbivore impact (Blaisdell and Pechanec 1949; Leopold 1949; Cook and Stoddart 1953; Jameson and Huss 1959; Hussey and Parr 1963; Tanisky 1969), plant tissues that are affected (Murneek 1925, 1926; Eaton 1931; Humphries 1958; Dunn and Engel 1971), and the modifying effects of such other environmental factors as light, nutrients, temperature, and water (Wilson and McGuire 1961; Madison 1962; Bean 1964; Griffith and Teel 1965; McKee et al. 1967; Wardlaw 1968; Sosebee and Wiebe 1971). Of course, growth may be inhibited by excessive defoliation (Milthorpe and Davidson 1965; Davidson and Milthorpe 1966a, 1966b), and an optimum defoliation level is anticipated (Noy-Meir 1975; Caughley 1976). Thus, simple statements about the effect of herbivores or other tissue damaging agents on plant growth, development, and resource allocation are fraught with error. Rather, the plant responds to a whole complex of environmental factors,

Am. Nat. 1979. Vol. 113, pp. 691–703.
© 1979 by The University of Chicago. 0003-0147/79/1305-0005$01.37

of which herbivore impact is only one. Further, the nature of the response will be conditioned by the genetic and developmental background of the plant.

Over a hundred years ago, Boussingault (1868) proposed that photosynthesis often may be limited by the accumulation of photosynthetic products. The rate at which assimilates accumulate in an actively photosynthesizing leaf depends upon the balance among photosynthesis, respiration, and outward translocation to assimilate consumption centers, commonly called sinks, elsewhere in the plant (Mason and Maskell 1928; Canny and Askham 1967; Neales and Incoll 1968). The plant is characterized as a source-sink system in which active photosynthetic tissue produces compounds which either are utilized for its own maintenance and growth, stored in situ, or translocated outward to other sites of utilization and storage (Burt 1964; Maggs 1964; Hartt 1963; Neales and Incoll 1968; King et al. 1967; Chatterton et al. 1972; Thorne and Koller 1974). Increasing the drain on leaf substrates by promoting translocation to sinks commonly results in increased photosynthetic rates in source leaves. This photosynthesis enhancement is accompanied by increased protein content, decreased mesophyll resistance, reduced starch levels, and increased sucrose translocation. Partial defoliation, by increasing assimilate demand in meristems of remaining shoots, usually stimulates photosynthetic rate per unit of remaining leaf area (Wareing et al. 1968; Gifford and Marshall 1973). In addition to the direct stimulatory effect of defoliation upon photosynthetic capacity of the remaining canopy, defoliation may allow more efficient light use by reducing mutual leaf shading (Donald and Black 1958; Jameson 1963; Hughes 1969; Heslehurst and Wilson 1971; Robson 1973). In particular, since large ungulates graze upon grasses from above, and growth is primarily from basal intercalary meristems, the older and less efficient tissues (Ludlow and Wilson 1971) are preferentially removed (Langer 1972), leading to greater light intensity on younger, previously shaded tissues (Jameson 1963).

Defoliation has a substantial effect upon assimilate allocation within the plant. There is pronounced diversion of carbohydrates from roots following defoliation (Kinsinger and Shaulis 1961; Gifford and Marshall 1973; Ryle and Powell 1975), and defoliation frequently reduces root growth (Crider 1955; Oswalt et al. 1959). Sosebee and Wiebe (1971) found that defoliation and soil moisture interact to determine patterns of assimilate translocation. Reduced water supply increased translocation to roots and crowns, while partial defoliation increased translocation to younger leaves. Therefore, the balance between root and crown storage and utilization of substrates by leaf meristems will be influenced by the balance between defoliation intensity and soil water potential. Although more severe defoliation is accompanied by greater diversion of assimilates from roots and a greater reduction in root growth, Dunn and Engel (1971) observed that more severe foliage clipping could result in a greater stimulation of subsequent root growth after a lag period of 3–4 wk.

The balance between vegetative and reproductive tissues also is influenced by herbivory. Dyer (1975) has shown that seed filling in maize may be enhanced by moderate herbivore damage to ears during certain developmental stages, and fruit and seed yield compensation following herbivory has been reported for many other plants (Taylor and Bardner 1968; Harris 1974). The converse of these effects is the inhibition of flowering and fruit set that commonly accompanies defoliation (Archbold 1942; Sprague 1954; Laude et al. 1957; Roberts 1958; Jameson 1963; Stoy 1965). The balance between reproductive and vegetative structures, like the other effects of herbivory, is regulated by plant phenological stage, which tissue is damaged, environmental parameters, and plant genetics.

Herbivory modifies hormonal balance within the plant substantially (Avery and Briggs 1968; Avery and Lacey 1968), particularly the parity between growth promoting and retarding hormones produced in the root and translocated to the shoot (Weiss and Vaadia 1965; Meidner 1967; Pallas and Box 1970; Torrey 1976). A greater flow of growth promoting hormones to residual meristems following defoliation promotes cell division and enlargement and activity in quiescent meristems, additional mechanisms of compensatory growth accompanying herbivory. A longer-term effect of modifications of hormonal equilibria following defoliation is a reduction in the rate of photosynthesis decline with leaf aging, thus maintaining assimilatory capacity of residual leaf tissue at higher levels over a longer time period (Richmond and Lang 1957; Woolhouse 1967; Neales et al. 1971; Gifford and Marshall 1973).

Two additional longer-term effects of defoliation may account for productivity increases caused by leaf-eating herbivores. First, soil water may be conserved due to reduction of the transpiration surface (Daubenmire and Colwell 1942; Baker and Hunt 1961) and because photosynthetic rate increases are associated primarily with reductions in mesophyll resistance rather than stomatal resistance (Gifford and Marshall 1973; Thorne and Koller 1974). The latter effect suggests that water-use efficiency may be increased by partial defoliation. Second, plant growth may be stimulated by nutrients recycled from dung and urine. This is so well known as to warrant little comment (Peterson et al. 1956; Lotero et al. 1966; Weeda 1967).

Finally, one potential direct stimulatory effect of grazing ruminants upon grass productivity may rise out of plant growth promoting agents in ruminant saliva (Vittoria and Rendina 1960; Reardon et al. 1972, 1974). Direct growth stimulations up to 50% above control levels have been recorded following addition of ungulate saliva to surfaces of manually clipped leaves.

In summary, productivity of herbivore affected plant tissues may be compensated or stimulated by:

1. Increased photosynthetic rates in residual tissue;

2. Reallocation of substrates from elsewhere in the plant;

3. Mechanical removal of older tissues functioning at less than a maximum photosynthetic level;

4. Consequent increased light intensities upon potentially more active underlying tissues;

5. Reduction of the rate of leaf senescence, thus prolonging the active photosynthetic period of residual tissue;

6. Hormonal redistributions promoting cell division and elongation and activation of remaining meristems, thus resulting in more rapid leaf growth and promotion of tillering;

7. Enhanced conservation of soil moisture by reduction of the transpiration surface and reduction of mesophyll resistance relative to stomatal resistance;

8. Nutrient recycling from dung and urine;

9. Direct effects from growth promoting substrates in ruminant saliva.

It is clear from reviewing the literature of herbivore damage, whether real or simulated, that an optimum tissue reduction level should occur, beyond which plant growth will be reduced (Vickery 1972; Dyer 1975; Noy-Meir 1975; Caughley 1976). This paper reports field and laboratory experiments designed to quantify the optimization curve in an ecosystem where herbivore load is intense (Stewart and Talbot 1962; Talbot and Stewart 1964; Watson and Kerfoot 1964; Hendrichs 1970; McNaughton 1976, 1979), the fauna is one of substantial antiquity (Leakey 1965; Cooke 1968; Gentry 1968), and I therefore assume that coevolution of plants and their herbivores has had an important influence on present properties of the ecosystem.

STUDY AREA AND METHODS

As Talbot and Stewart (1964, p. 815) observed, "The last known great concentrations of mixed species of plains wildlife in Africa, or in the world, are found in the Serengeti-Mara region of (Tanzania) and Kenya." The Serengeti ecosystem is defined operationally by the annual movements of large herds of migratory wildebeest (*Connochaetes taurinus Albojubatus* Thomas) between wet season occupance areas on the open Serengeti Plains and dry season occupance areas on savanna grasslands to the west near Lake Victoria and to the north in the Mara River drainage basin (Talbot and Talbot 1963; Pennycuick 1975). Movements of zebra (*Equus burchelli* Gray) are similar in scope. Those of the other major migratory species, Thomson's gazelle (*Gazella thomsonii* Gunther), are less extensive. Together, these three species constitute over 60% of the grazing mammal biomass (Stewart and Talbot 1962; Talbot and Stewart 1964; Hendrichs 1970). The experiment reported here was done in a grassland dominated by *Andropogon greenwayi* Napper (Anderson and Talbot 1965; Schmidt 1975) from January to May, 1975, while major concentrations of the three principal ungulates were present. Although the wet season is nominally from November through May (Norton-Griffiths et al. 1975), the first substantial rain on the Serengeti Plains during the 1974–1975 wet season was on January 3, 1975. As in most equatorial regions (Livingstone 1975), rainfall in the Serengeti is largely from convective storms arising from local synoptic processes superimposed on the intertropical convergence zone and is characteristically erratic (Pennycuick and Norton-Griffiths 1976). Data are presented here from 102 grazing days, defined by presence of one of the three major animal species, during the 146-day period between January 3 and May 29 when all three species had left the study site for their dry season ranges.

A permanent exclosure was built at the beginning of the period, and adjacent temporary exclosures were used to measure short-term grass regrowth subsequent to protection from grazing. Plant biomass (g/m^2) was measured as described previously (Tucker et al. 1973; Pearson et al. 1976; McNaughton 1976, 1979). Above-ground net productivity ($g/m^2 \cdot day$) was calculated from positive biomass increments. Grazing intensity (G) was calculated as $1 - g/ng$, where g was biomass in grazed areas unprotected by fencing and ng was biomass in the permanent exclosure. This index will be zero when grazing does not reduce plant biomass below control levels, and will approach one as grazing increases. Temporary exclosures were moved according to the grazing rotation patterns of the three ungulate species: Whenever an influx of grazers reduced plant biomass below control levels, they were moved. Thus, short-term protection was employed to measure actual grass growth under a grazing regimen defined by the grazers movements. The temporary exclosure approach was designed because I felt traditional approaches, such as simulating grazing by clipping, were unlikely to reproduce the manifold effects of grazing animals documented in the introduction. Actual productivity was the sum of the positive biomass increments inside temporary exclosures, control productivity was the sum of positive biomass increments inside the permanent exclosure, and grazing stimulation was actual minus control productivity. Soil water potentials (S) in -bars were back-calculated from control green biomasses after I had established that the two parameters were closely related ($r^2 = .839$ for $P < .001$ with df $= 15$). The study site was routinely assayed at 6 day intervals and as frequently as daily when warranted by high rates of grass growth or consumption.

To determine whether grazing effects on net above-ground productivity could be partially simulated in the laboratory, clones of *Kyllina nervosa* Steud. were collected and returned to Syracuse University's Biological Research Laboratories. This species is a dryland sedge dominant in that part of the Serengeti ecosystem where annual rainfall is less than about 500 mm (Anderson and Talbot 1965; McNaughton, in prep.). It is abundant in one of the most intensely grazed regions of the Serengeti: The migratory animals go there whenever there is enough rain to promote plant growth (Talbot and Talbot 1963; Watson and Kerfoot 1964; Pennycuick 1975; Kreulen 1975). Although I used temporary exclosures because I felt clipping would not reproduce the effects of grazing animals, clipping seems the only feasible approach to simulating grazing under laboratory conditions. Cloned individuals were grown in Sherer CEL 37-14 growth chambers and were clipped at 2, 4, and 6 cm heights, at frequencies of $\frac{1}{2}$, 1, 3, 5, 6, and 19 days. Growth increments were oven dried and weighed, and biomass increments were converted to grams per square meters times days.

RESULTS AND DISCUSSION

Stimulation of net above-ground primary productivity by grazing was a complex function of grazing intensity (fig. 1), as regulated by soil water potential. There was a sharp stimulation peak when defoliation was moderate, and a long tail when grazing reduced plant biomass to less than half of control levels. Productivity stimulation was described ($R^2 = .777$, $P < .001$ with df $= 13$) by

$$\Delta P = e \exp \{.76 - 3.5(\ln G) - 1.24(\ln G)^2 - .64(\ln S)\}$$

where ΔP was experimental minus control productivity and G and S were as defined previously. Grazing explained 68.7% of the variance in productivity stimulation (X, $F_{1,13} = 10.96$, $P < .01$; X^2, $F_{1,13} = 14.14$, $P < .01$) and soil water potential contributed 9% ($F_{1,13} = 5.24$, $P < .05$). Mean control productivity was 15.1 ± 4.9 (.95 interval) g/m$^2 \cdot$ day, so productivity was doubled under moderate grazing if soil moisture tension was low.

Wildebeest was the only species for which sufficient data were obtained to provide

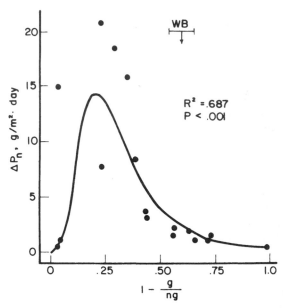

FIG. 1.—Relationship between stimulation of above-ground grassland productivity (ΔP_n) and grazing intensity $(1 - g/ng)$; the line plotted was fitted without incorporating soil moisture, which reduced the unexplained variance in stimulation by 9%. The WB indicates the mean and .95 confidence range of wildebeest grazing intensity in these grasslands throughout the wet season.

a reasonable confidence limit on G for a grazing species. On average this species overcropped the grassland significantly compared to the level that would maximize forage yield (fig. 1). Biomass yield of forage to the wildebeest population would be higher at a grazing level about half of the average they maintained.

Apparent overcropping, however, may maximize total nutrient yield to the wildebeest. Forage quality during the dry season falls below maintenance requirements of the animals with a resulting depletion of body reserves; these reserves are replenished during the wet season (Sinclair and Duncan 1972; Sinclair 1974a, 1974b). It is well known that forage quality diminishes with plant tissue age (Glover et al. 1960; Armstrong et al. 1964; Miller et al. 1965; Braun 1973) and that higher quality forage not only gives a higher nutritional yield, but can be consumed in larger quantities by ruminants (Hungate 1975). Thus, apparent overcropping by wildebeest probably maintains higher nutritional yield than would less intense grazing, allowing more rapid replenishment of body reserves than would consumption of larger quantities of lower quality forage (Baile and Forbes 1974; Hungate 1975). Most efficient exploitation of these grasslands by wildebeest, in terms of total yield, would be accomplished at lower consumption levels. Like butterfly larval grazing (Slansky and Feeny 1977), wildebeest grazing may be an optimum power system (Odum and Pinkerton 1955). The wildebeest-grass interface, however, is more interactive than the caterpillar-crucifer one (Slansky and Feeny 1977), since yield to the wildebeest is a direct consequence of their grazing and the resultant compensatory physiological responses of the dominant grasses.

GRASS-UNGULATE RELATIONSHIPS   697

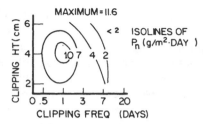

FIG. 2.—Isolines of net above-ground productivity (g/m² · day) of *Kyllinga nervosa* clipped at various heights and frequencies under controlled environment conditions. Maximum productivity of 11.6 g/m² · day occurred when plants were clipped daily at a height of 4 cm.

Two points about the form of the relationship between stimulation of above-ground productivity and grazing intensity warrant comment. First, optimization of grass growth by grazing was a complex process, reaching a peak at moderate levels of defoliation. Second, the long tail on the curve at high grazing intensities indicates that these grasslands are remarkably resistant to overgrazing; productivity was maintained at control levels even when defoliation was severe.

Although, as documented in the introduction, growth of above-ground tissues may be at the expense of substrates stored in roots and crowns, most of the species common in these grasslands cannot suffer fatal reserve depletion under the prevailing grazing regime. Many, in fact, are obligate grazophils whose occurrence depends upon the grazers. When reexamining exclosures in 1975 that were built in the early 1960's (Watson 1966), I found that most of the presently abundant grasses had substantially reduced abundances in the protected areas (McNaughton 1979). *Andropogon greenwayi*, for instance, made up 56% of the plant biomass outside, but was completely absent inside; so, in addition to having a remarkable ability to withstand heavy grazing, *A. greenwayi* is maintained in these grasslands by sustained grazing. It is one of the few sod-forming species in the flora, most species being bunchgrasses, and its prostrate growth form must allow it to sustain sufficient leaf area under most grazing intensities to prevent substantial root reserve depletion.

Net above-ground primary productivity of *Kyllina nervosa* was a complex function of clipping frequency and height (fig. 2), as revealed by a significant interaction term in the analysis of variance ($F_{8,135} = 13.9$, $P < .001$). Maximum net above-ground productivity was 11.6 g/m² · day, recorded at a 4 cm height when clipping was at daily intervals. I know of no other plant which has been reported to be able to sustain shoot growth when subjected to such intense clipping, much less have an optimum at this intensity. The productivity values reported here, however, are substantially below field data in grasslands where *K. nervosa* is abundant (McNaughton 1975). This experiment demonstrated a remarkable ability by *K. nervosa* to sustain growth under intense levels of foliage removal, undoubtedly a consequence of natural selection for compensatory growth responses during a long period of intense herbivore load.

Grasses and large grazing ungulates have a long and intimate coevolutionary history, but the Serengeti ecosystem is one of the few areas where the products of that history remain relatively undisturbed. Although the concept of optimum stocking density is well established in range management (Klipple and Costello 1960; Bement

1969; Hutchinson 1971) I would not expect the relationships between domestic forages and ungulates to necessarily resemble those documented here for a natural grazing system. *Andropogon greenwayi* and *K. nervosa* dominate grasslands subjected to intensive herbivory for a considerable period, given the antiquity of the region's fauna. A major problem in development of optimum stocking criteria for managed rangelands is the need to preserve forage for use during periods when grass growth is limited by low moisture or, in temperate grasslands, low temperature. In this system of native migratory herbivores, in contrast, a pasture rotation pattern has evolved naturally in relation to seasonal rainfall patterns.

The disappearance of *A. greenwayi* and most of the other dominant grasses from plots protected from grazing for several years indicates that they have comparatively higher fitnesses under intense grazing than other species, which became dominant when grazing was curtailed. This does not imply that the absolute fitnesses of the present dominants is enhanced by grazing. For instance, there was little seed production in grazed areas, while flowering culm density commonly exceeded $200/m^2$ in exclosures. To the extent that sexual reproduction contributes to fitness of these grasses, grazing reduced fitness of many species tangibly. However, the ability of these grasses to maintain high levels of productivity under very intense grazing is a clear adaptation to high herbivore load. Traditional standards of overgrazing, clearly applicable to domestic ungulates and forages, may have slight application to an ungulate fauna and its forages which are products of a long coevolutionary history.

Although these experiments indicate that compensatory growth responses of the plants are a significant factor in this ecosystem's energy flow, this does not imply that the relationship between the plants and herbivores is symbiotic by any conventional definition of that term (Mattson and Addy 1975). The grasses clearly pay a cost in reproductive potential as a consequence of the intense defoliation regime, and even those species whose abundance is tied to the grazers probably periodically suffer significant levels of reserve depletion. Thus, while competitive fitness is enhanced by grazing, absolute fitness may be impaired in comparison to the ungrazed condition.

SUMMARY

A substantial literature is reviewed which indicates that compensatory growth upon tissue damage by herbivory is a major component of plant adaptation to herbivores. Experiments in Tanzania's Serengeti National Park showed that net above-ground primary productivity of grasslands was strongly regulated by grazing intensity in wet-season concentration areas of the large ungulate fauna. Moderate grazing stimulated productivity up to twice the levels in ungrazed control plots, depending upon soil moisture availability. Productivity was maintained at control values even under very intense grazing, suggesting that conventional definitions of overgrazing may be inapplicable to these native plant-herbivore systems. A laboratory clipping experiment with a sedge abundant in one of the most intensely utilized regions resulted in a maximum net above-ground productivity of $11.6 \text{ g/m}^2 \cdot \text{day}$ when clipped daily at a height of 4 cm. Few plant species have been reported with the ability to maintain a significant level of productivity under such intense clipping. This suggests that the high grazing load of the Serengeti ecosystem has constituted strong selection on the plants for compensatory growth upon defoliation.

GRASS-UNGULATE RELATIONSHIPS 699

ACKNOWLEDGMENTS

I thank the trustees and director of Tanzania National Parks for permission to reside and work in the Serengeti National Park, T. Mcharo, former director of the Serengeti Research Institute for inviting me to work there, and Margaret, Sean, and Erin McNaughton for valuable assistance. This work was supported by NSF ecosystem studies program grants BMS 74-02043, DEB 74-02043, (A02) and DEB 77-20360 to Syracuse University.

LITERATURE CITED

Anderson, G. D., and L. M. Talbot. 1965. Soil factors affecting the distribution of the grassland types and their utilization by wild animals on the Serengeti Plains, Tanganyika. J. Ecol. 53:33–56.

Archbold, H. K. 1942. Physiological studies in plant nutrition. XIII. Experiments with barley on defoliation and shading of the ear in relation to sugar metabolism. Ann. Bot. (Lond.) 6:487–531.

Armstrong, D. G., K. L. Blaxter, and R. Waite. 1964. The evaluation of artificially dried grass as a source of energy for sheep. III. The prediction of nutritive value from chemical and biochemical measurements. J. Agric. Soc. (Camb.) 62:417.

Avery, D. J., and J. B. Briggs. 1968. The etiology and development of damage in young trees infested with fruit tree red spider mite, *Panonychus ulmi* (Koch). Ann. Appl. Biol. 61:277–288.

Avery, D. J., and H. J. Lacey. 1968. Changes in the growth-regulator content of plum infested with fruit tree red spider mite, *Panonychus ulmi* (Koch). J. Exp. Bot. 19:760–769.

Baile, C. A., and J. M. Forbes. 1974. Control of feed intake and regulation of energy balance in ruminants. Physiol. Rev. 54:160–214.

Baker, J. N., and O. J. Hunt. 1961. Effects of clipping treatments and clonal differences on water requirements of grasses. J. Range Manage. 14:216–219.

Bean, E. W. 1964. The influence of light intensity upon the growth of a S.37 cocksfoot (*Dactylis glomerata*) sward. Ann. Bot. (Lond.) 28:427–443.

Bement, R. E. 1969. A stocking rate guide for beef production on blue grama range. J. Range. Manage. 22:83–86.

Blaisdell, J. P., and J. F. Pechanec. 1949. Effects of herbage removal at various dates on vigor of bluebunch wheatgrass and arrowleaf balsamroot. Ecology 30:298–305.

Boussingault, J. B. 1868. Cited in T. F. Neales and L. D. Incoll. 1968. The control of leaf photosynthesis rate by the level of assimilate concentration in the leaf: a review of the hypothesis. Bot. Rev. 34:107–125.

Branson, F. A. 1953. Two new factors affecting resistance of grasses to grazing. J. Range Manage. 6:165–171.

Braun, J. M. N. 1973. Primary production in the Serengeti: purpose, methods and some results of research. Ann. Univ. Abidjan, Ser. E, Ecol. 4:171–188.

Brougham, R. W. 1961. Factors limiting pasture production. Proc. N.Z. Soc. Anim. Prod. 21:33–46.

Burt, R. L. 1964. Carbohydrate utilization as a factor in plant growth. Aust. J. Biol. Sci. 17:867–877.

Canfield, R. H. 1939. The effect of intensity and frequency of clipping on density and yield of black grama and tobosa grass. U.S. Dep. Agric. Tech. Bull. 681.

Canny, M. J., and M. J. Askham. 1967. Physiological inferences from the evidence of translocated tracer: a caution. Ann. Bot. (Lond.) 31:409–416.

Caswell, H., F. Reed, S. N. Stephenson, and P. A. Warner. 1973. Photosynthetic pathways and selective herbivory: a hypothesis. Am. Nat. 107:465–480.

Caughley, G. 1976. Plant-herbivore systems. Pages 94–113 *in* R. M. May, ed. Theoretical ecology: principles and applications. Saunders, Philadelphia.

Chatterton, N. J., G. E. Carlson, W. E. Hungerford, and D. R. Lee. 1972. Effect of tillering and cool nights on photosynthesis and chloroplast starch in *pangola*. Crop. Sci. 12:206–208.

Cook, C. W., and L. A. Stoddart. 1953. Some growth responses of crested wheatgrass following herbage removal. J. Range Manage. 6:267–270.

Cooke, H. B. S. 1968. Evolution of mammals on southern continents, II. The fossil mammal fauna of Africa. Q. Rev. Biol. 43:234–264.

Crider, F. J. 1955. Root-growth stoppage resulting from defoliation of grass. U.S. Dep. Agric. Tech. Bull. 1102.

Daubenmire, R. F., and W. E. Colwell. 1942. Some edaphic changes due to overgrazing in the *Agropyron-Poa* prairie of southeastern Washington. Ecology 23:32–40.

Davidson, J. L., and F. L. Milthorpe. 1966a. Leaf growth in *Dactylis glomerata* following defoliation. Ann. Bot. (Lond.) 30:173–184.

———. 1966b. The effect of defoliation on the carbon balance in *Dactylis glomerata*. Ann. Bot. (Lond.) 30:185–198.

Donald, C. M., and J. N. Black. 1958. The significance of leaf area in pasture growth. Herbage Abstr. 28:1–6.

Dunn, J. H., and R. E. Engel. 1971. Effect of defoliation and root-pruning on early root growth from Merion Kentucky bluegrass sods and seedlings. Agron. J. 63:659–663.

Dyer, M. I. 1975. The effects of red-winged blackbirds (*Agelaius phoeniceus* L.) on biomass production of corn grains (*Zea mays* L.). J. Appl. Ecol. 12:719–726.

Dyer, M. I., and U. G. Bokhari. 1976. Plant-animal interactions: studies of the effects of grasshopper grazing on blue grama grass. Ecology 57:762–772.

Eaton, F. M. 1931. Early defloration as a method of increasing cotton yields, and the relation of fruitfulness to fiber and boll characters. J. Agric. Res. 42:447–462.

Ehrlich, P. R., and P. H. Raven. 1964. Butterflies and plants: a study in coevolution. Evolution 18:586–608.

Forde, B. J. 1966. Translocation in grasses. 2. Perennial ryegrass and couch grass. N.Z. J. Bot. 4:496–514.

Fraenkel, G. S. 1959. The raison d'etre of secondary plant substances. Science 129:1466–1470.

Freeland, W. J., and D. H. Janzen. 1974. Strategies in herbivory by mammals: the role of plant secondary compounds. Am. Nat. 108:269–289.

Gentry, A. W. 1968. Historical zoogeography of antelopes. Nature 217:874–875.

Gifford, R. M., and C. Marshall. 1973. Photosynthesis and assimilate distribution in *Lolium multiflorum* Lam following differential tiller defoliation. Aust. J. Biol. Sci. 26:517–526.

Glover, J., D. W. Duthie, and H. W. Dougall. 1960. The total digestible nutrients and gross digestible energy of ruminant feeds. J. Agric. Sci. 55:403–408.

Graber, L. F. 1931. Food reserves in relation to other factors limiting the growth of grasses. Plant Physiol. 6:43–71.

Griffith, W. K., and M. R. Teel. 1965. Effect of nitrogen and potassium fertilization, stubble height, and clipping frequency on yield and persistence of orchardgrass. Agron. J. 57:147–149.

Harris, P. 1974. A possible explanation of plant yield increases following insect damage. Agro-ecosyst. 1:219–225.

Hart, R. H., G. E. Carlson, and D. E. McCloud. 1971. Cumulative effects of cutting management on forage yields and tiller densities of tall fescue and orchardgrass. Agron. J. 63:895–898.

Hartt, C. E. 1963. Translocation as a factor in photosynthesis. Naturwissenschaften 21:666–667.

Hendrichs, H. 1970. Schatzungen der Huftierbiomasse in der Dornbusch-Savanne nordlich und westlich der Serengeti in Ostafrika nach einem neuen Verfahren und Bemerkungen zur Biomasse der anderen pflanzenfressenden Tierarten. Saeugetierkd. Mitt. 18:237–255.

Heslehurst, M. R., and G. L. Wilson. 1971. Studies on the productivity of tropical pasture plants. III. Stand structure, light penetration, and photosynthesis in field swards of *Setaria* and green leaf *Desmodium*. Aust. J. Agric. Res. 22:865–878.

Hughes, A. P. 1969. Mutual shading in quantitative studies. Ann. Bot. (Lond.) 33:381–388.

Humphries, E. C. 1958. Effect of removal of part of the root system on the subsequent growth of the root and shoot. Ann. Bot. (Lond.) 22:251–257.

Hungate, R. E. 1975. The rumen microbial ecosystem. Annu. Rev. Ecol. Syst. 6:39–66.

Hussey, N. W., and W. J. Parr. 1963. The effect of glasshouse red-spider mite (*Tetranychus urticae* Koch.) on the yield of cucumbers. J. Hortic. Sci. 38:255–263.

Hutchinson, K. J. 1971. Productivity and energy flow in grazing/fodder conservation systems. Herbage Abstr. 41:1–10.

Jacques, W. A., and D. B. Edmond. 1952. Root development in some common New Zealand pasture plants. V. The effect of defoliation and root pruning on cocksfoot (*Dactylis glomerata*) and perennial ryegrass (*Lolium perenne*). N.Z. J. Sci. 34:231–248.

Productivity and Resources   345

GRASS-UNGULATE RELATIONSHIPS   701

Jameson, D. A. 1963. Responses of individual plants to harvesting. Bot. Rev. 29:532–594.

Jameson, D. A., and D. L. Huss. 1959. The effect of clipping leaves and stems on number of tillers, herbage weights, root weights, and food reserves of little bluestem. J. Range Manage. 12:122–126.

King, R. W., I. F. Wardlaw, and L. T. Evans. 1967. Effects of assimilate utilization on photosynthetic rates of wheat. Planta 77:262–276.

Kinsinger, F. E., and N. Shaulis. 1961. Carbohydrate content of underground parts of grasses as affected by clipping. J. Range Manage. 14:9–12.

Klipple, G. E., and D. F. Costello. 1969. Vegetation and cattle responses to different intensities of grazing on short-grass ranges on the Central Great Plains. U.S. Dep. Agric. Tech. Bull. 1216.

Kreulen, D. 1975. Wildebeest habitat selection on the Serengeti Plains, Tanzania, in relation to calcium and lactation: a preliminary report. East Afr. Wildl. J. 13:297–304.

Labyak, L. F., and F. Schumacher. 1954. The contribution of its branches to the main-stem growth of loblolly pine. J. For. 52:333–337.

Langer, R. H. M. 1972. How grasses grow. Arnold, London.

Laude, H. M., A. Kadish, and R. M. Love. 1957. Differential effect of herbage removal on range species. J. Range Manage. 10:116–120.

Leakey, L. S. B. 1965. Olduvai Gorge 1951–61. Vol. 1. Cambridge University Press, London.

Leonard, E. R. 1962. Inter-relations of vegetative and reproductive growth, with special reference to indeterminate plants. Bot. Rev. 28:353–410.

Leopold, A. C. 1949. The control of tillering in grasses by auxin. Am. J. Bot. 36:437–440.

Levin, D. A. 1971. Plant phenolics: an ecological perspective. Am. Nat. 195:157–181.

———. 1973. The role of trichomes in plant defense. Q. Rev. Biol. 48:3–15.

Livingstone, D. A. 1975. Late Quaternary climatic change in Africa. Annu. Rev. Ecol. Syst. 6:249–280.

Lodge, R. W. 1960. Effects of burning, cultivating, and mowing on the yield and consumption of crested wheatgrass. J. Range Manage. 13:318–321.

Lotero, J., W. W. Woodhouse, and R. G. Petersen. 1966. Local effect on fertility of urine voided by grazing cattle. Agron. J. 58:262–265.

Ludlow, M. M., and G. L. Wilson. 1971. Photosynthesis of tropical pasture plants. III. Leaf age. Aust. J. Biol. Sci. 24:1077–1087.

McKee, W. H., R. H. Brown, and R. E. Blaser. 1967. Effect of clipping and nitrogen fertilization on yield in stands of tall fescue. Crop Sci. 7:567–570.

McNaughton, S. J. 1975. Structure and function of Serengeti grasslands. Annu. Rep. Serengeti Res. Inst. 1974–1975:51–61.

———. 1976. Serengeti migratory wildebeest: facilitation of energy flow by grazing. Science 191:92–94.

———. 1979. Grassland-herbivore dynamics. In A. R. E. Sinclair and M. Norton-Griffiths, eds. Serengeti: studies of ecosystem dynamics in a tropical savanna. University Chicago Press, Chicago (in press).

Madison, J. H. 1962. Turfgrass ecology. Effects of mowing, irrigation, and nitrogen treatments of *Agrostis palustris* Huds., "Seaside" and *Agorstis tenuis* Sibth., "Highland" on population, yield, rooting and cover. Agron. J. 54:407–412.

Maggs, D. H. 1964. Growth rates in relation to assimilate supply and demand. I. Leaves and roots as limiting regions. J. Exp. Bot. 15:574–583.

Marshall, C., and G. R. Sagar. 1965. The influence of defoliation on the distribution of assimilates in *Lolium multiflorum* Lam. Ann. Bot. (Lond.) 29:365–370.

Mason, T. G., and E. J. Maskell. 1928. Studies on the transport of carbohydrates. II. The factors determining the rate and the direction of movement of sugars. Ann. Bot. (Lond.) 42:571–636.

Mattson, W. J., and N. D. Addy. 1975. Phytophagous insects as regulators of forest primary production. Science 190:515–522.

Meidner, H. 1967. The effect of kinetin on stomatal opening and the rate of intake of carbon dioxide in mature primary leaves of barley. J. Exp. Bot. 18:556–561.

Miller, W. J., C. M. Clifton, O. L. Brooks, and E. R. Beatty. 1965. Influence of harvesting age on digestibility and chemical composition of pelleted coastal bermudagrass. J. Dairy Sci. 48:209–212.

Milthorpe, F. L., and J. L. Davidson. 1965. Physiological aspects of regrowth in grasses. Pages 241–255 in F. L. Milthorpe and J. D. Ivins, eds. The growth of cereals and grasses. Butterworths, London.

Murneek, A. E. 1925. The effects of fruit on vegetative growth in plants. Proc. Am. Soc. Hortic. Sci. 21:274–276.

————. 1932. Growth and development as influenced by fruit and seed formation. Plant Physiol. 7:79–90.

Neales, T. F., and L. D. Incoll. 1968. The control of leaf photosynthesis rate by the level of assimilate concentration in the leaf: a review of the hypothesis. Bot. Rev. 34:107–125.

Neales, T. F., K. J. Treharne, and P. F. Wareing. 1971. A relationship between net photosynthesis, diffusive resistance, and carboxylating enzyme activity in bean leaves. Pages 89–96 in M. D. Hatch, C. B. Osmond, and R. O. Slatyer, eds. Photosynthesis and photorespiration. Wiley, New York.

Newell, L. C., and F. D. Keim. 1947. Effects of mowing frequency on the yield and protein content of several grasses grown in pure stands. Nebr. Agric. Exp. Stn. Res. Bull. 150.

Norton-Griffiths, M., D. Herlocker, and L. Pennycuick. 1975. The patterns of rainfall in the Serengeti ecosystem, Tanzania. East Afr. Wildl. J. 13:347–374.

Noy-Meir, I. 1975. Stability of grazing systems: an application of predator-prey graphs. J. Ecol. 63:459–481.

Odum, H. T., and R. C. Pinkerton. 1955. Time's speed regulator: the optimum efficiency for maximum power output in physical and biological systems. Am. Sci. 43:331–343.

Oswalt, D. L., A. R. Bertrand, and M. R. Teel. 1959. Influence of nitrogen fertilization and clipping on grass roots. Proc. Soil Sci. Soc. Am. 23:228–230.

Pallas, J. E., and J. E. Box. 1970. Explanation of the stomatal response of excised leaves to kinetin. Nature 227:87–88.

Parker, K. W., and A. W. Sampson. 1931. Growth and yield of certain Gramineae as influenced by reduction of photosynthetic tissue. Hilgardia 5:361–381.

Pearson, L. C. 1965. Primary production in grazed and ungrazed desert communities of eastern Idaho. Ecology 46:278–285.

Pearson, R. L., L. D. Miller, and C. J. Tucker. 1976. Hand-held spectral radiometer to estimate graminaceous biomass. Appl. Optics 15:416–418.

Pennycuick, L. 1975. Movements of the migratory wildebeest population in the Serengeti area between 1960 and 1973. East Afr. Wildl. J. 13:65–87.

Pennycuick, L., and M. Norton-Griffiths. 1976. Fluctuations in the rainfall of the Serengeti ecosystem, Tanzania. J. Biogeogr. 3:1–13.

Petersen, R. G., W. W. Woodhouse, and H. L. Lucas. 1956. The distribution of excreta by freely grazing cattle and its effect on pasture fertility. 2. Effect of returned excreta on the residual concentrations of some fertility elements. Agron. J. 48:444–449.

Reardon, P. O., C. L. Leinweber, and L. B. Merrill. 1972. The effect of bovine saliva on grasses. J. Anim. Sci. 34:897–898.

————. 1974. Response of sideoats grama to animal saliva and thiamine. J. Range Manage. 27:400–401.

Richmond, A. E., and A. Lang. 1957. Effect of kinetin on protein content and survival of detached Xanthium leaves. Science 125:650–651.

Roberts, H. M. 1958. The effect of defoliation on the seed producing capacity of bred strains of grasses. I. Timothy and perennial ryegrass. J. Br. Grassl. Soc. 13:225–261.

Robertson, J. H. 1933. Effect of frequent clipping on the development of certain grass seedlings. Plant Physiol. 8:425–447.

Robocker, W. C., and B. J. Miller. 1955. Effects of clipping, burning and competition on establishment and survival of some native grasses in Wisconsin. J. Range Manage. 8:117–120.

Robson, M. J. 1973. The growth and development of simulated swards of perennial ryegrass. I. Leaf growth and dry weight changes as related to the ceiling yield of a seedling sward. Ann. Bot. (Lond.) 37:487–500.

Ryle, G. J. A., and C. E. Powell. 1975. Defoliation and regrowth in the Graminaceous plant: the role of current assimilate. Ann. Bot. (Lond.) 39:297–310.

Schmidt, W. 1975. Plant communities on permanent plots of the Serengeti Plains. Vegetatio 30:133–145.

Sinclair, A. R. E. 1974a. The natural regulation of buffalo populations in East Africa. IV. The food supply as a regulating factor, and competition. East Afr. Wildl. J. 12:291–311.

————. 1974b. The resource limitation of trophic levels in tropical grassland ecosystems. J. Anim. Ecol. 44:497–520.

Sinclair, A. R. E., and P. Duncan. 1972. Indices of condition in tropical ruminants. East Afr. Wildl. J. 10:143–149.

Slansky, F., and P. Feeny. 1977. Stabilization of the rate of nitrogen accumulation by larvae of the cabbage butterfly on wild and cultivated food plants. Ecol. Monogr. 47:209–228.

Sosebee, R. E., and H. H. Wiebe. 1971. Effect of water stress and clipping on photosynthate translocation in two grasses. Agron. J. 63:14–19.

Sprague, M. A. 1954. The effect of grazing management on forage and grain production from rye, wheat and oats. Agron. J. 46:29–33.

Stein, W. I. 1955. Pruning to different heights in young Douglas fir. J. For. 53:352–355.

Stewart, D. R. M., and L. M. Talbot. 1962. Census of wildlife on the Serengeti and Loita Plains. East Afr. Agric. For. J. 28:58–60.

Stoy, V. 1965. Photosynthesis, respiration and carbohydrate accumulation in relation to yield. Physiol. Plantarum Suppl. 4:1–125.

Talbot, L. M., and D. R. M. Stewart. 1964. First wildlife census of the entire Serengeti-Mara region, East Africa. J. Wildl. Manage. 28:815–827.

Talbot, L. M., and M. H. Talbot. 1963. The wildebeest in western Masailand. Wildl. Monogr. no. 12.

Tanisky, V. I. 1969. The harmfulness of the cotton bollworm, *Heliothis obsoleta* F. (Lepidoptera, Noctuidae) in southern Tadzhikistan. Entomol. Rev. (Engl. Transl. Entomol Obozr.) 48:23 29.

Taylor, W. E., and R. Bardner. 1968. Effects of feeding by larvae of *Phaedon cochleariae* (F.) and *Pultella maculipennis* (Curt.) on the yield of radish and turnip plants. J. Appl. Biol. 62:249–254.

Thorne, J. H., and H. R. Koller. 1974. Influence of assimilate demand on photosynthesis, diffusive resistances, translocation and carbohydrate levels of soybean leaves. Plant Physiol. 54:201–207.

Torrey, J. G. 1976. Root hormones and plant growth. Annu. Rev. Plant Physiol. 27:435–459.

Tucker, C. J., L. D. Miller, and R. L. Pearson. 1973. Measurement of the combined effect of green biomass, chlorophyll, and leaf water on canopy spectroreflectance of the shortgrass prairie. Pages 601–627 *in* F. Shahrokhi, ed. Proceedings of the second annual remote sensing of Earth resources Conference, University of Tennessee, Tullahoma.

Vickery, P. J. 1972. Grazing and net primary production of a temperate grassland. J. Appl. Ecol. 9:307–314.

Vittoria, A., and N. Rendina. 1960. Fattori condizionanti la funzionalita tiaminica in piante superiori e cenni sugli effetti dell bocca dei runinanti sull erbe pascolative. Acta Med. Vet. (Naples) 6:379–405.

Wardlaw, I. F. 1968. The control and pattern of movement of carbohydrates in plants. Bot. Rev. 34:79–105.

Wareing, P. F., M. M. Khalifa, and K. J. Treharne. 1968. Rate limiting processes in photosynthesis at saturating intensities. Nature 220:453–457.

Watson, R. M. 1966. Game utilization in the Serengeti: preliminary investigations, part II. Wildebeest. Br. Vet. J. 122:18–27.

Watson, R. M., and O. Kerfoot. 1964. A short note on the intensity of grazing of the Serengeti plains by plains-game. Z. Saeugetierkd. 29:317–320.

Weaver, J. E., and V. H. Houghen. 1939. Effect of frequent clipping on plant production in prairie and pasture. Am. Midl. Nat. 21:396–414.

Weeda, W. C. 1967. The effect of cattle dung patches on pasture growth, botanical composition and pasture utilization. N.Z. J. Agric. Res. 10:150–159.

Weiss, C., and Y. Vaadia. 1965. Kinetin-like activity in root apices of sunflower plants. Life Sci. 4:1323–1326.

Wilson, D. B., and W. S. McGuire. 1961. Effects of clipping and nitrogen on competition between three pasture species. Can. J. Plant Sci. 41:631–642.

Woolhouse, H. W. 1967. The nature of senescence in plants. Symp. Soc. Exp. Bot. 21:179–213.

Vol. 118, No. 2	The American Naturalist	August 1981

EXPLOITATION ECOSYSTEMS IN GRADIENTS
OF PRIMARY PRODUCTIVITY

Lauri Oksanen,* Stephen D. Fretwell,† Joseph Arruda,† and
Pekka Niemelä*

*Kevo Subarctic Research Institute, University of Turku, SF-20500 Turku 50, Finland; †Division of
Biology, Kansas State University, Manhattan, Kansas 66506

Submitted October 17, 1978; Revised November 5, 1979; Final Revision December 9, 1980; Accepted
January 12, 1981

Formal studies on trophic exploitation can be traced back to the Lotka-Volterra predation models summarized by Gause (1934). Since Rosenzweig and MacArthur (1963) included the resource-determined carrying capacity of prey in predation models, the development of these models has been rapid. Holling's (1965) study showed how the saturation of predators can be included and Rosenzweig (1969, 1971) related the shape of the prey isocline to the productivity and other characteristics of the prey population. May (1972), Gilpin (1975), and Tanner (1975) have examined the behavior of predator-prey systems without a locally stable equilibrium point. A three-dimensional model able to deal simultaneously with exploitative herbivore-plant and carnivore-herbivore interactions outlined by Rosenzweig and MacArthur (1963) was actually developed by Rosenzweig (1973) and elaborated by Wollkind (1976).

However, the existence of definite predator and prey isoclines is a hypothesis, not a fact. The random predator implicit in exploitation models differs radically from the prudent predator of Slobodkin (1968). It is possible to relax the assumption of completely random exploitation (Rosenzweig 1977). In order to make the isoclines unambiguous the additional assumption is then needed that no significant changes occur in the proportion of the prey population that belongs to the most predation-susceptible age group. Further, it is essential to require that exploitation of victims with a positive reproductive potential occurs. If the exploiter is dependent on some product of a population and unable to affect its production rate, the prey isocline is transformed to a line perpendicular to the prey axis. Such a product could be aged or injured animals (Mech 1966), the social detritus of a population (Errington 1963), or plant organs of sufficiently high quality (Kalela 1962; Tast and Kalela 1971). Other production-regulated trophic interactions may be found in nature. The American mink studied by Errington (1963) does not seem an exceptionally inefficient carnivore. Some species that look like ordinary herbivores have turned out to be critically dependent on the quality of their resources

Am. Nat. 1981. Vol. 118, pp. 240–261.
© 1981 by The University of Chicago. 0003-0147/81/1802-0016$02.00

(Baltensweiler et al. 1977; Haukioja et al. 1978). Drawing on such observations White (1978) has presented a general theory of ecosystem structure, arguing that almost all populations are resource limited and that the significance of trophic exploitation is negligible.

Yet, there is empirical evidence that points toward the opposite direction. The *Opuntia* plague of Australia was eliminated by introducing a herbivore coevolved with cacti (see e.g., Krebs 1972, pp. 367–369). Reindeer have dramatic impact on vegetation when their densities are allowed to expand (Höglund and Eriksson 1973), and problems of overgrazing (i.e., strong impact of utilizer upon the utilized populations) are common enough to invite the application of exploitation models to range management (Noy-Meir 1975). For a proponent of White's hypothesis, the fact that depletion of prey is a common outcome of laboratory studies on trophic relationships (see references in Rosenzweig 1969, 1977) must also appear puzzling.

When stating that the quality of most plant material is too low to support herbivores, White (1978) neglected the possibility that a feeding strategy evolves which is energetically wasteful but ensures a sufficient rate of intake of limiting nutrients. Aphids are the clearest example of this, but low ecoenergetic efficiency accompanied by an ability to subsist on low-quality forage is found in diverse group of animals, e.g., sawflies (Haukioja and Niemelä 1974), lemmings (Batzli 1975) and zebras (Bell 1971).

The basis of the "prudent predator" hypothesis is not quite solid, either. Slobodkin (1974) maintained that the diversity of prey and the diversity of their antipredator strategies exceeds the diversity of predators. Thus, carnivores are exposed to conflicting selective pressures giving the prey an edge in the race to capture and to avoid being captured. Cohen's (1977) review does not support this statement. Perhaps herbivores are more troubled by conflicting selective pressures. Adaptations that make a herbivore an elusive prey (fleetness) or a submergent one (restricted and nocturnal activity period; see Maiorana [1975] for further discussion) conflict with ones required in efficient utilization of forage (large digestive tract, long activity period).

This discussion suffices to show that the importance of trophic exploitation is an open question. Arguments and counterarguments can be found, but such debate hardly leads anywhere. It appears more useful to state some set of assumptions (admitting that they may turn out to be unrealistic), to analyze their logical consequences, and to compare these with data. If the data and the predictions match, the plausibility of the assumptions is improved (Tricker 1965). If they conflict, this helps us replace them with better ones. It is indeed possible that correspondence between observations and predictions is just good luck, and the construction of meaningful null hypotheses may be impossible. This may be regarded as a reason to use the inductionistic approach, where the probability that chance alone is responsible for the results can be assessed by means of standard statistics. The merits of inductionism may be appreciable in young branches of science where new sets of data have a good chance to make all hypotheses obsolete. It appears, however, that if a branch of science is to develop beyond the initial stage it has to start using the hypothetico-deductive method. Predictions and tests with all their problems seem to be the only way to differentiate useful general statements from useless ones (see Lakatos 1972).

TOWARD A TESTABLE HYPOTHESIS OF EXPLOITATION ECOSYSTEMS

Few predictions can be deduced solely from the assumption that there are exploitative relationships between populations. In order to obtain a theory with some predictive power, it is imperative to link the isocline shapes with environmental variables. This approach has been pioneered in Rosenzweig's (1971) Paradox of Enrichment. Rosenzweig assumes that an increase in primary productivity expands the isoclines of prey species. He thus suggests that there is a simple connection between an ecosystem-level phenomenon and properties of individual populations. However, the analysis was performed before Rosenzweig (1973) developed his three-dimensional exploitation model which allows an explicit consideration of the dual role of herbivores—as predators of vegetation and as prey of carnivores. Hence the robustness of the result (that enrichment leads to destabilization) can be questioned.

A simple connection between ecosystem-level units (trophic levels) and individual populations was also assumed to exist in the paper of Hairston et al. (1960). In a debate with Murdoch (1966) and Ehrlich and Birch (1967), Slobodkin et al. (1967) appeared able to demonstrate that the proposition qualifies as a scientific hypothesis. Yet, relatively little has been built on that basis, possibly because of another question discussed by the above cited critics: Is it meaningful to use trophic levels as ecological units?

Indeed, nature is not divided into perfectly distinct trophic levels, but it is not divided into perfectly distinct local populations, either. The appropriateness of an abstraction depends on the problem to be studied. If one wants to deduce empirical consequences from the assumption that trophic interactions are exploitative, the auxiliary assumption that trophic levels can be treated as homogenous units is inviting. Exploitation models deal with populations that make linear food chains, whereas overlapping resource utilization and branched food chains appear common in nature (see e.g., Wiegert and Odum 1969). The possibility that other populations would start to utilize the resources of a strongly predator-limited prey creates profound evolutionary problems (Van Valen 1973). If we treat trophic levels as units, then, by definition, food chains are always linear and alternate utilizers cannot exist. Such an assumption also appears realistic enough to be potentially useful. Even if all organisms must be assumed to maximize their expansive energy by using whatever resources are available (see Van Valen 1975), photosynthesis, utilization of vegetative plant organs, and carnivory require adaptations too different to allow an individual organism to be efficient in more than one of these ways of energy intake. This discontinuity also implies relative homogeneity within each trophic level. Competition within resource-limited trophic levels should make the sum of utilization curves match the distribution of resources (MacArthur 1972, pp. 59–69), so that a rather homogenous exploitation pressure should be exerted upon the populations on the level below.

In a more abstract form the core of Hairston et al. (1960) can be phrased as follows. Trophic level B could severely decrease the density of populations on trophic level A, but the impact of trophic level C upon B prevents this from happening. However, is there any reason to interpret A, B, and C, respectively, as plants, herbivores, and carnivores? Fretwell (1977) asked this question and answered it negatively. He noted that the essential characteristic of C is that it is not exploited, but trophic level C's distance above the base of food chains cannot be fixed. Instead, Fretwell proposed that the distance depends on the rate of primary production and on the average ecological efficiency among consumers. Fretwell thus assumed that, given constant ecological efficiencies, increased primary productivity is both a necessary and a sufficient condition for the lengthening of food chains. While the first point appears clear and logical, the second one does not necessarily conform with implications of the exploitation approach. If we accept the idea that populations and trophic levels may be predation limited we must also be prepared to see situations where increased primary productivity does not pass smoothly to the top of the food chains. The more an ecological unit is limited by predation the less its standing crop can respond to an increased productivity of resources. Even the increase in its productivity is bounded by physiological constraints on turnover rate.

In order to remedy this problem, we now apply graphical predation models in the conceptual framework provided by Hairston et al. (1960) and Fretwell (1977). The densities (live mass per unit area) of plants (P), herbivores (H), and carnivores (C) are dimensions of our model (as in Rosenzweig [1973]). A fourth dimension, assumed to be the ultimate regulator of the system, is the maximum gross primary productivity allowed by the environment (G) which will be called "potential productivity" below. This formalization can be equally well regarded as a reanalysis of Rosenzweig's (1971) Paradox of Enrichment looking at a wider range of productivities, using the more powerful tool of three-dimensional exploitation models and considering the possibility that carnivores may be preyed upon by secondary carnivores.

PLANT ISOCLINE AND POTENTIAL PRODUCTIVITY

In the three-dimensional predation model of Rosenzweig (1973), plants are assumed to compete directly with each other for some resources which set an upper limit to their density. Because a similar growth curve is obtained if plants are limited by the amount of heterotrophic plant tissue that can be maintained with the maximum rate of photosynthesis allowed by the environment (which may be a more realistic assumption for terrestrial vegetation), we accept Rosenzweig's assumption as a technical shorthand. Rosenzweig (1973) further assumed that plants are not directly harmed by low density (no true Allee effect) and that herbivores are saturated at high plant densities (with a consequent technical Allee effect for plants). We include these assumptions and further assume that herbivore saturation occurs in accordance to Holling's (1965) type II functional response curve. (As we deal with trophic levels, alternate prey cannot exist; thus, type III functional response is not plausible.) Our final assumption about plants is that, in the absence of significant herbivory, their expansion conforms to the logistic growth curve. Given this, the rate of change in plant density can be described by Tanner's (1975) equation for the growth rate of prey:

$$dP/dt = rP(1 - P/K) - wPH/(D + P) \tag{1}$$

where P = plant density, r = intrinsic growth rate of plants, K = maximum density of plants in the absence of herbivory, H = herbivore density, w is the maximum foraging rate of an individual herbivore, and w/D is its rate of searching when unsaturated (see Holling 1965; Tanner 1975). From equation (1) the (zero) isocline of plants can be obtained by setting $dP/dt = 0$ and solving for H as

$$H(P) = \frac{-r}{wK} P^2 + \frac{r(K - D)}{wK} P + \frac{rD}{w}. \tag{2}$$

Equations (1) and (2) are biologically meaningful only when P is positive. If there are no plants ($P = 0$), their rate of change would be determined by immigration and survival of propagules which is not included in our model. Note also that $H(P)$ can be zero only once on the positive side of the P-axis (when $P = K$). When P approaches zero, $H(P)$ approaches rD/w which is positive.

Let us examine the conditions under which the plant isocline is humped (has a local maximum when $P > 0$). By differentiating equation (2) we obtain

$$\frac{dH}{dP} = -\frac{2r}{wK} P + \frac{r(K - D)}{wK} \tag{3}$$

which is zero when $P = (K - D)/2$. Thus the necessary and sufficient condition for the hump is that $K > D$. What does an ecosystem look like which is at the limit of meeting this condition (i.e., $K = D$)? When the last term in equation (1) is divided by H, the consumption rate of an individual herbivore (R, which can be also interpreted as the rate of energy flow per unit of herbivore biomass) is obtained as

$$R(P) = \frac{wP}{D + P} \tag{4}$$

Assume further that $P = K$ (i.e., there are so few herbivores that plants are allowed to reach a density very close to resource-determined carrying capacity). Then $P = K = D$ and $R = w/2$. Since w is the maximum foraging (capturing) rate of herbivores (see above), an ecosystem where the plant isocline is at the verge of being humped should be so sparsely vegetated, because of the adversity of the physical environment, that even if herbivores entered a previously ungrazed area, they would have to spend as much time in searching for food as in feeding activities.

In more productive environments, the plant isocline is humped. The height of the hump is obtained by substituting $P = (K - D)/2$ into equation (2) and solving for H as

$$H_{\max} = \frac{r(K - D)^2}{4wK} + \frac{rD}{w}. \tag{5}$$

Now, we need to express K and r as functions of potential productivity (G). For K

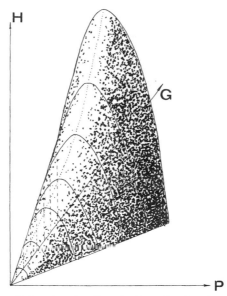

FIG. 1.—The shape of the plant isocline (horizontal axis, plant density; vertical axis, herbivore density) along a gradient of potential primary productivity (G). The ridge, formed by the humps of individual plant isoclines, is presented as a line of small dots.

this is easy: The amount of living protoplasm that can be maintained is directly proportional to the rate of photosynthesis. In plants, r is just another expression for the net productivity/biomass-ratio in early stages of secondary succession (when the plant cover is not yet closed). Hence, it appears reasonable to assume that r is directly proportional to G, too. This discussion and equation (5) yield equations (6) and (7):

$$H_{\max} = aG^2 + bG + c \qquad (6)$$

where a, b, and c are positive constants and

$$K = kG \qquad (7)$$

where k is a positive constant. Figure 1 presents this relation between the shape of the plant isocline and potential productivity. Note that the height of the isocline increases quadratically with increasing potential productivity, whereas the maximum density of phytomass, K, increases linearly.

In predation models, it has been customary to use the expressions biomass, amount of living protoplasm, or amount of edible material as if they were synonymous. Indeed, they are not, nor does the logic of the model require it. The model does require that density be expressed in units that are proportional both to the amount of living protoplasm and to the amount of edible material. If predation models are used in some cases of particular herbivore-plant interactions (as in Noy-Meir 1975; Caughley 1976), the ordinary interpretation of phytomass may pass this criterion. But in a general model we must define phytomass as the amount of herbage (grasses and herbs, leaves, shoot tips and phloem of woody

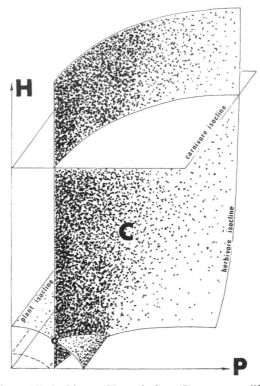

FIG. 2*a*.—Carnivore (*C*), herbivore (*H*), and plant (*P*) system at different points of the productivity (*G*) axis. Unstable equilibrium is a solid dot; stable equilibrium is a dot with a star. An unproductive system where the plant isocline (the rightward sloping surface) remains well below the carnivore isocline. The herbivore isocline (the surface bending backwards from the plane of the page) crosses the plant isocline on the right side of its hump and in the *C* = 0 plane, creating a stable herbivore-plant equilibrium.

plants) per unit of area. Wood is to be considered accumulated organic material, equivalent to peat in bogs, and not a constituent of phytomass at all.

CONSUMER ISOCLINES AND THE PREDICTED CHARACTERISTIC OF COMMUNITIES

Our model includes climate only indirectly, as one of the factors explaining the variation in potential productivity. The shape of consumer isoclines is made constant, because we assume that consumers are adapted to the climate in which they live and that costs of this adaptation are just a minor factor in the energy budget of consumers. We also make the implicit assumptions of Hairston et al. that territoriality does not limit the total density of any trophic level, that any effect of interference is insignificant within the trophic level as a whole and that no true Allee effect exists (see Caughley 1976). Finally, the consumption rate of herbivores and carnivores is assumed to conform to Holling's (1965) type II functional response curve and assimilation rate is assumed directly proportional to consumption rate.

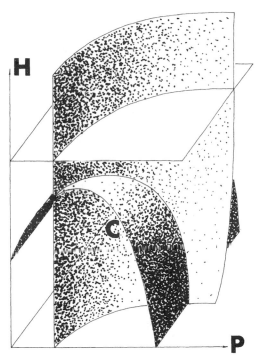

FIG. 2b.—A system where the potential primary productivity has twice the value it had in fig. 2a. The herbivore-plant equilibrium is unstable, so that sustained cyclic changes in plant and herbivore density are to be expected. Symbols as in figure 2a.

The consumer isoclines thus become similar to those in the simplest version of three-dimensional predation models proposed by Rosenzweig and MacArthur (1963) except that the herbivore isocline meets the plane of zero herbivore density along a line where the C-coordinate is an asymptotically increasing function of P (fig. 2). (Bending the isocline to the $C = 0$, $H = 0$ line, as done by Rosenzweig and MacArthur, implies a true Allee effect.) The carnivore isocline is a plane perpendicular to the H-axis. The herbivore isocline is a surface bending from the $C = 0$ plane in the direction of the C-axis, its C-coordinate being an asymptotically increasing function of P and a monotonically increasing function of H.

Imagine that figure 1 is sliced with a vast number of planes perpendicular to the G-axis, the plant isoclines thus obtained are combined with the consumer isoclines and the resulting set of three-dimensional pictures is presented as a movie. We then see a continuous gradient of ecosystems, but no amorphous continuum. Instead there are distinct break points where one trend is suddenly replaced by another one. Some scenes of this movie are presented in figure 2, and the predicted pattern of phytomass is summarized in figure 3.

In the least productive ecosystems (fig. 2a, dotted line, and interval $G < g_0$ in fig. 3), the plant isocline does not meet the herbivore isocline, and there is no grazing chain in the ecosystem. The P/G ratio is constant. With increasing potential productivity, the plant isocline reaches the herbivore isocline, and a stable

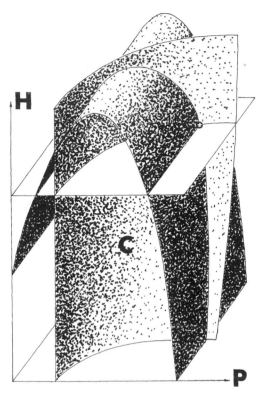

Fɪɢ. 2c.—The potential productivity is 1.25 times that of the system in fig. 2b. A locally stable carnivore-herbivore-plant equilibrium has been established. Symbols as in figure 2a.

herbivore-plant equilibrium is established. We now enter an interval (fig. 2a, interval $g_0 < G < g_1$ in fig. 3) where increased potential productivity only increases the equilibrium density of herbivores and their impact on vegetation. The phytomass remains constant. In communities with a high average ecological efficiency, the intersection of plant and herbivore isoclines can pass above the carnivore isocline before it reaches the hump of the plant isocline or even before such a hump is formed. It is also possible that the hump is reached first. Then, we would have an interval where no stable equilibrium point exists (fig 2b, interval $g_1 < G < g_2$ in fig. 3). In the less productive part of this interval, herbivores cycle with plants but may cross the carnivore isocline during phases of high density. Thus, the system may support nomad carnivores such as the snowy owl (*Nyctea scandica*) and jaegers (*Stercorarius spp.*) which move around searching for areas with high herbivore densities or survive in other kinds of ecosystems during phases of low herbivore density. As potential productivity increases further, the carnivore isocline is passed, a still unstable carnivore-herbivore-plant equilibrium is established and the role of carnivores as proximate regulators of herbivore density becomes successively more important.

Our herbivores are assumed to exhibit weak "technical mutualism" (because of

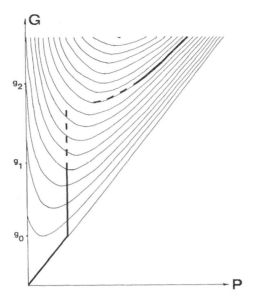

FIG. 3.—A contour map presentation of the herbivore-plant-productivity system of figure 1. The contours represent herbivore densities on the plant isocline. The path of the equilibrium point is a solid line when the equilibrium is locally stable and a dashed line when it is unstable.

carnivore saturation and lack of significant interference) at all densities. Hence the steepness of the plant isocline at the carnivore-herbivore-plant equilibrium must exceed some minimum before the triple equilibrium (fig. 2c) can be stable (Rosenzweig 1973, pp. 285–288). This occurs at $G = g_2$ (fig. 3). In more productive environments, we see stable herbivore populations, subjected to heavy carnivory but without a shortage of resources, which seems to correspond to the ideas of Hairston et al. The relation between equilibrium phytomass and potential productivity is nearly linear again and, as the potential productivity further increases, an ever decreasing fraction of primary productivity will be used by grazers (fig. 2c, interval $G > g_2$ in fig. 3). However, this scenario is conditional. If the technical mutualism among herbivores is strong enough, the three-link equilibrium point will not be stabilized and population cycles persist. Their characteristics just change from cycles in which both carnivory and depletion of vegetation are important (see Keith 1974) to relatively pure carnivore-herbivore cycles.

So far, our model has rather faithfully reiterated most conclusions of Fretwell (1977), demonstrating that the ambiguities of his approach are rather unimportant in the lower end of a productivity gradient. The main difference is that our model fails to generate the fluctuating patterns of phytomass, predicted by Fretwell. In the interval $g_0 < G < g_2$ the increasing trend in equilibrium phytomass is stopped but not reversed.

Fretwell (1977) predicted that increased primary productivity will eventually allow the formation of a fourth trophic level, with consequent relaxation of primary carnivory and intensification of herbivory. In our model this is conditional, too. The intersection of the herbivore isocline with any plane of constant herbivore density is a line where the C-coordinate is an asymptotically increasing function of plant density (see fig. 2). The isocline of primary carnivores belongs to this set of planes, and the C-coordinate of the equilibrium point thus increases asymptotically with increased potential productivity. As a consequence, the equilibrium density of primary carnivores may eventually reach the predator isocline of secondary carnivores but may also remain below this isocline forever. The stability analysis of Rosenzweig (1973) shows that as the plant isocline becomes very steep at the three-link equilibrium, exploitation ecosystems with herbivores exhibiting technical mutualism become destabilized again. Thus noninterference exploitation ecosystems where the constellation of consumer isoclines prevents the formation of equilibria with four trophic levels seem to be subjected to the destabilization-by-enrichment effect discussed by Rosenzweig (1971).

What types of isocline constellations could occur? This depends on the ecological efficiency (production/consumption ratio) of saturated herbivores, which limits the degree to which herbivores can respond to increased abundance of resources. We expect that terrestrial food chains may belong to the latter category (three links forever), whereas aquatic ones represent the former one (lengthening to four links).

If we assume that the equilibrium density of primary carnivores in the ecosystem represented by figure 2c is at the verge of reaching the predator isocline of secondary carnivores, then how would an infinitely small increase in potential productivity influence its structure? In our model, the impact of the top trophic level is to prevent the density at the lower level from increasing. Thus, the C-coordinate of the new equilibrium point with four trophic levels must remain the same as that of the three-link equilibrium point. Further, the new equilibrium point must belong to plant and herbivore isoclines. We now insert a plane perpendicular to the C-axis through the three-link equilibrium point in figure 2c and look for a point where the lines of intersection between this plane and plant and herbivore isoclines (fig. 4) meet each other. The alternative equilibrium point appears at considerably greater herbivore density and lower plant density as compared to the three-link equilibrium point. With the slightest increase in potential productivity, the three-link equilibrium point passes behind this plane and is no longer acceptable. The four-link equilibrium point remains in this plane and, with further increase in G, climbs along the herbivore isocline, the equilibrium herbivore density keeps increasing and the plant density keeps decreasing.

Instead of secondary carnivores there may be facultative secondary carnivory. This seems probable in all food chains where herbivores and carnivores have similar sizes. Secondary carnivory can be treated as an extreme form of interference within the carnivore trophic level. Following Wollkind (1976) it seems that the rise in equilibrium density of herbivores would then be more gradual than that caused by a distinct fourth trophic level. The stability of the equilibrium would be enhanced.

THE MODEL AND REAL ECOSYSTEMS

In the previous section we saw that the model allows the existence of several alternative ecosystem states. Without some numerical assumptions we cannot

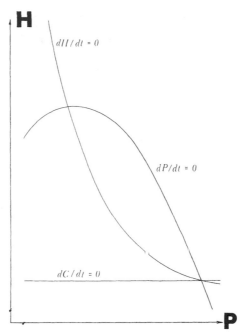

Fig. 4.—The intersections of plant, herbivore, and primary carnivore isoclines with the predator isocline of secondary carnivores (which can be imagined as a plane perpendicular to the C-axis, inserted in figure 2c through the equilibrium point of the system).

predict which one should be observed. Fortunately, also some unambiguous predictions are created. Elimination of the top trophic level or its absence because of barriers of dispersal should have a radical impact on the lower ones. In any continuous productivity gradient extending from very unproductive environments (deserts or arctic-alpine boulder fields) to relatively productive ones (e.g., broad-leaved forests), a zone characterized by intense natural grazing pressure must be found. The relationship between (equilibrium) phytomass and primary productivity along such a gradient is predicted by the model (fig. 3). We now compare these predictions to existing data.

It is a common observation of aquatic ecologists (e.g., Hrabaček 1962; Brooks and Dodson 1965; Svärdson 1976) that the presence of efficient planktivores has a dramatic impact upon the constitution of the zooplankton, as our hypothesis requires. Svärdson also notes that, besides barriers of dispersal, the composition of the planktivore community (the third trophic level) is influenced by temperature and nutrient content, the main factors determining the potential productivity of Fennoscandian lakes. He further suggests that the impact of planktivores does not stop at the second trophic level; that the decreased density of zooplankters causes increased density of phytoplankton and, consequently, increased actual primary productivity. As tentative evidence for this, he presents statistics for commercial catches showing that, at a given nutrient content, the lakes inhabited by the most efficient planktivores tend to be the most productive ones. He further notes that:

. . . the ultimate evidence, of course, is that derived from the same body of water in a normal state and later, when its fish fauna has been eradicated. Such an experiment was recently performed in the small lake, Stocklidsvattnet, in south-western Sweden. . . . After the [eradication] treatment, alkalinity fell, visibility was doubled and the primary productivity per hour sank from 15–20 mg C/m³ in 1973 and 8–12 mg C/m³ in 1974 to only 1–3 mg C/m³ in 1975. . . . In Lake Stocklidsvattnet the fish fauna was dominated by roach, which was then ultimately responsible for a primary organic production several times higher than when the same body of water was not inhabited by any fish. (Svärdson 1976, pp. 165–166)

Lake Stocklidsvattnet may have its terrestrial counterpart on the Kaibab plateau. There, the third trophic level was also strongly decimated, the apparent result being an expansion of herbivores, decrease of phytomass, and the establishment of a new equilibrium at lower plant and herbivore densities (Rasmussen 1941). Unfortunately, the original data are rather vague (see Caughley 1970). While admitting this, we cannot see why alternate hypotheses, derived from equally vague data, should be considered more credible.

On Isle Royale, the moose population irrupted before the immigration of wolves. The irruption was followed by resource depletion and decline of the moose population. After the establishment of a large wolf pack in the fifties the moose density has been relatively high, but no real irruption has occurred (Mech 1966).

The littoral communities of the Aleutians provide the most solid case to look at. In the absence of sea otters, populations of sea urchins are dense and kelp beds are almost eliminated. Recently, the sea otter has been re-established in littoral communities of some islands with consequent decrease in the density of sea urchins and dramatic increase in kelp density. The existence of treatments and controls makes the significance of predation by sea otters almost certain (Simenstad et al. 1978).

We now turn our attention to observed patterns of biomass in gradients of primary productivity. Fretwell (1977) emphasized that in relatively unproductive parts of the moisture-controlled productivity gradients of the central U.S. (from Sonoran desert to shortgrass prairies), phytomass does not appear to expand with increasing primary productivity, whereas the abundance of graminids appears to increase, suggesting intensification of grazing pressure.

The data provided by Whittaker and Niering from Sonoran deserts (1975), USA/IBP Grassland Biome data (Sims and Coupland 1979) and Hulbert's (personal communication) data from Konza Prairie confirm this impression. The aboveground phytomasses of desert communities range from 0.39 to 1.31 kg/m². On desert grassland Whittaker and Niering reported that aboveground phytomass was 0.26 kg/m² whereas the average peak aboveground biomass of the corresponding IBP site only was 0.13 kg/m². On IBP sites representing shortgrass and northern mixed prairies corresponding values ranged from 0.10 to 0.18 g/m². The average peak above ground biomasses of tallgrass prairie were again higher (0.29 g/m² on the IBP Osage site, from 0.33 to 0.50 on Konza, all values refer to sites not grazed by cattle during the study period). The range of primary productivities is about from 0.1 to 1 kgm⁻² year⁻¹.

This pattern in peak aboveground phytomass looks almost too encouraging to be true. (The decrease in aboveground phytomass from deserts to shortgrass plains is readily understandable, because the dominance of woody plants decreases, too.) However, a look at the grazing chain reveals problems. What is left from the natural grazing chain of grasslands mainly consists of invertebrates and among them, there is no clear relationship between trophic structure and primary productivity (Dyer 1979). Carnivores are abundant even on the desert grassland site, and this appears to be the case in true deserts too (S. J. Chaplin, personal communication). If natural grazers ever were important, they must have been the big ones that are now extinct or at very low densities in most remaining grassland areas, most of their niche being taken over by cattle. As cattle had been excluded from the IBP sites discussed above, the proposed impact of grazing must have been largely evolutionary. (Plants allocating most of the production below ground had a selective advantage.) Since this interpretation depends on uncertain assumptions about the past, the case of grasslands must be regarded as inconclusive.

The IBP data are also available from the temperature-controlled arctic productivity gradient (Bliss and Wielgolaski 1973; Rosswall and Heal 1975). The information on consumer biomass is fragmentary, but predictions referring to patterns of phytomass can be tested.

In the arctic-subarctic communities represented in the material (fig. 5), potential grazers of vascular plants are present. The vascular plant data (solid figures) conform with the relationship between equilibrium phytomass and productivity predicted in figure 3. The slight discrepancies (too much phytomass in the two polar semideserts of Devon Island and in the unproductive boreal bog; the steep rise of biomass in the most productive communities) are understandable because IBP data include wood in phytomass whereas our model does not. The pattern of total phytomass (open figures) is remarkably different, the Devon Island data points showing a steep and apparently linear rise of phytomass in response to increased primary productivity. However, most of this phytomass consists of moss. A geographical barrier (a barren, partially glaciated highland) isolates the study area from the range of brown lemmings (*Lemmus*) adapted to graze moss. Hence, this discrepancy actually conforms with the predictions.

The extremely isolated antarctic islands provide excellent cases to study how the absence of parts of the grazing chain influences patterns of phytomass. As a point of reference, the expected phytomass-versus-productivity curve of fig. 5 is used. We predict that data points from relatively unproductive communities without grazers are far above the line whereas those from more productive communities with grazers but without significant carnivores should be below it. The IBP data (fig. 6) behave in accordance with this prediction. Signy Island communities without herbivorous vertebrates have much more phytomass than correspondingly unproductive arctic communities (with herbivores), whereas the productive Macquairie herbfield is far below the line. On Macquairie there are introduced rabbits, and native carnivores (skuas) are probably rather inefficient rabbit predators. The introduced feral cats appear inefficient, too (unable to enter rabbit burrows). The South Georgia data points fit the line rather well, and the situation (introduced reindeer hunted by the crew of the local whaling station) allows the establishment of a semblance of a three-link ecosystem. However, the whaling station has been closed since the early 1960's and, according to the theory

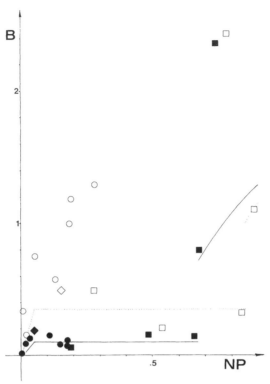

FIG. 5.—Plant communities of some arctic-subarctic IBP Tundra Project study areas, chosen on the basis of availability of data. Horizontal axis: net primary productivity (NP, kg dry wt m^{-2} year^{-1}), vertical axis: peak aboveground biomass (B, kg dry wt m^{-2}). Dots = Devon Island data points (a polar desert; two polar semidesert communities [cushion plant lichen, cushion plant moss]; three moist-to-wet meadows). Squares = Hardangervidda data points (lichen heath; dry meadow; wet meadow; willow thicket; birch forest). Diamond = Abisko data point (dwarf shrub; cottongrass bog). The expected pattern, derived from figure 3, is shown as a line (solid for vascular plants, dotted for the whole vegetation). Further explanation in the text.

the departure of the carnivore should lead to a strong increase in herbivore density and to consequent changes in the vegetation of the most productive biotopes. The report of the IBP team (in Rosswall and Heal 1975) conforms with this: Reindeer density is said to be increasing; local replacement of the giant tussock grass, *Poa flabellata*, and the shrub, *Acaena magellanica*, by light swards of *Poa annua* has occurred. Also these data suggest that mammalian grazers are the ones that make a difference.

    In continuous polar productivity gradients, biomass patterns appear to conform to figure 3 and the isolated areas deviate from this pattern in the predicted way. Indeed, the data points are so few that the observed fit can be coincidence. Thus, we suggest two further observations that one should be able to make if the fit is a real one.

    1. Since most mammals have difficulty crossing water barriers, we should find

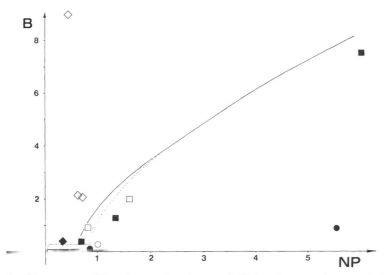

Fig. 6.—Plant communities of antarctic-subantarctic IBP study areas chosen and plotted as in figure 5. Diamonds = Signy Island data points (three moss banks [without vascular plants]; *Deschampsia antarctica*, meadow [vascular plant data only]). Dots = South Georgia data points (*Festuca contracta,* meadow; *Acaena magellanica,* shrub; stand of *Poa flabellata*). Squares = Macquairie data points (herbfield; *Poa foliosa,* grassland). In the two most productive communities, the share of cryptogams is negligible, so that total and vascular plant data could not be plotted separately.

situations resembling those of antarctic and subantarctic islands from arctic and boreal islands isolated by smaller stretches of sea. Because carnivores have much lower population densities than herbivores the risk of local extinction can also be expected to be much higher for carnivores (see MacArthur 1972). Thus, moderately isolated islands can be expected to be largely devoid of carnivores, but have the herbivore trophic level intact. Strongly isolated islands should not have mammalian grazers, either. Cases of the former type are predicted to resemble Macquairie (even relatively productive communities being subjected to heavy herbivory); those of latter type are predicted to resemble Signy (massive plant cover even on relatively unproductive sites). The outer islands of the Baltic Sea are moderately isolated. Corresponding to predictions, extremely high hare densities and depletion of vegetation has been observed (Häkkinen and Jokinen [1974] and references therein). The islands of Spitsbergen are high arctic and strongly isolated. Lemmings are absent whereas wild reindeer are present. Consequently, we predict that the islands should have massive moss banks whereas the lichen grounds of the main island should be strongly depleted and that the reindeer has, at least to some extent, started to utilize the more abundant moss resources.

2. A comparison between figure 9 and figure 6 suggests further observations referring to population cycles. In areas where the annual net aboveground productivity ranges from 50 to 150 g/m², all grazers, even rodents, should be noncyclic. The strongest population cycles should occur when the grazed community (or

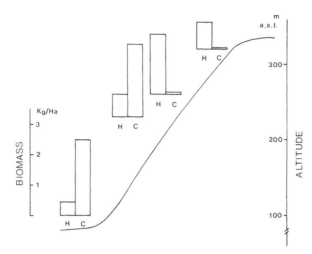

FIG. 7.—The trophic structure of invertebrates (aphids excluded because of their mutualism with ants) living in the foliage of mountain birch (*Betula pubescens* ssp. *tortuosa*) at the southern slope of Mount Jesnalvaara, Finnish Lapland (*H* = herbivores, *C* = carnivores).

stratum) produces annually 300–600 g/m^2. If strong population cycles are found in more productive communities, our model predicts that carnivory is the main proximate cause of decline in herbivore densities, because resource shortage has progressively less importance with increased primary productivity. The few studies of microtine populations in extremely unproductive environments (Fuller et al. 1975; Birney et al, 1976) appear to confirm prediction 2. Birney et al.'s study is especially interesting, because they showed that increased productivity in the environment (reflected in increased rodent populations) leads to population fluctuations. Further, the phytomass (interpreted as cover by the authors) also fluctuated in all but the lushest study area. In this area, a particularly dense stoat population was reported to exist.

Studies performed at Kevo Research Institute, northeastern Lapland, give some information about the trophic structure among invertebrates in a subarctic timberline area. In these timberline forests, the only abundant tree species (mountain birch, *Betula pubescens* ssp. *tortuosa*) is repeatedly devastated by outbreaks of the geometrid moth *Oporinia autumnata* (Tenow 1972). Normally severe outbreaks occur in highland areas which may subsequently change into tundras, whereas lower slopes and river valleys tend to remain undamaged (Kallio and Lehtonen 1973). There is also a clear gradient in the trophic structure of invertebrates associated with birch foliage (fig. 7). In valley forests, both herbivores (mainly aphids and beetles) and carnivores (mainly ants) are present. (Ants apparently get most of their energy from sugars excreted by aphids.) Along the slope, there is a transition from a three-level trophic structure into a two-level one, and on a hilltop with just scattered, shrublike birches the density of carnivores is negligibly small (more detailed data will be published by P. Niemelä and K. Laine).

Given the data presented above that seemed to stress the importance of mammals, the situation appears enigmatic. We suggest that both mammals and invertebrates are important but invertebrates, often able to complete their life cycle during the most favorable season, are much less sensitive to variations in annual primary productivity. On the other hand, ectothermal carnivores are likely to be immobilized by low temperatures. If this is the case, then the terrestrial grazing chain consists of two fundamentally different branches: one homeothermic and vertebrate; the other ectothermic and arthropod. The present version of our hypothesis is then applicable to the vertebrate branch. When dealing with arthropods it may be more useful to keep the plant isocline unchanged along a cline of declining annual productivity, as long as productivity during the most favorable month is reasonably constant. The carnivore isocline would vary in response to temperature, yielding patterns similar to those found for homeothermic food chains.

The transition from three-link to four link trophic structure, which we expect to find at least in aquatic ecosystems, offers perhaps the most interesting and counterintuitive prediction: An increase in primary productivity should lead to decreased phytomass, increased herbivore density, and unchanged density of primary carnivores. The predation pressure of primary carnivores upon herbivores should decrease, with consequent changes in life history characteristics of herbivores (from r- to K-strategy).

Svärdson (1976) reported that in the eutrophic lake, Hjälmaren, the bulk of commercial catches consists of the sander, a secondary carnivore, whereas planktivores (primary carnivores) prevail in the more oligotrophic big Swedish lakes. Increased productivity thus appears to lead to a transition from a three-link to a four-link ecosystem, as predicted by our theory. Unfortunately, Svärdson did not report whether this difference in trophic structure is reflected in the constitution of lower trophic levels.

A similar transition from three-link to four-link ecosystems in some Kansas farm ponds was studied by Arruda (1979). The trophic structure primarily associated with the macrophyte-periphyte complex, consisting of producers, herbivorous benthic invertebrates, benthic carnivores, and *Centrarchidae* fishes (top carnivores, TC) is plotted against total primary productivity in fig. 8. The Centrarchids are quite opportunistic and have no dislike of herbivorous prey, so their biomass has been added to the biomass of benthic carnivores to yield the sum of all carnivores biomass (C).

The small number of ponds and the necessity to use a part (the macrophyte-periphyte complex) as representative of the whole (the pond) may limit the generality of the results. Nevertheless, the predictions are tentatively corroborated and the data support the counterintuitive prediction that increased primary productivity results in decreased phytomass. Variation in life-history characteristics of benthic invertebrates corresponds to the idea that more productive ponds have less intense primary carnivory (Arruda 1979).

The correspondence between observations and predictions appears encouraging. Even in the case of Isle Royale which served as a counterexample against the

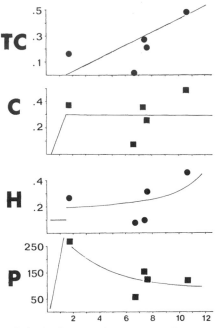

FIG. 8.—Structural trends in the farm pond ecosystems. Data source: Arruda 1979. Horizontal axis: actual gross primary productivity (g oxygen m^{-2} day^{-1}), vertical axis: biomass (g dry mass m^{-2}, P = plants, H = herbivores, C = all carnivores, TC = top carnivores alone). The expected pattern is shown as a solid line. The expected relationships are derived from figure 4 (by letting the plant isocline expand) and from accompanying discussion.

random predator approach, exploitation appears to be a major controlling force. Yet the fragmentary nature of our test material is a severe weakness. The situation calls for a detailed analysis of biomass patterns, population dynamics and life-history traits along some gradients of primary productivity and also for further theoretical work.

We are presently investigating the dichotomy between mammalian and arthropod branches of terrestrial grazing chains which require changes in the model as applied to terrestrial ecosystems. The fact that our exploitation models also ignore seasonality weakens the plausibility of our statements about stable and cyclic populations, inviting further theoretical work in this area as well.

SUMMARY

Based on the assumption that each trophic level acts as a single exploitative population, a model relating the trophic structure of ecosystems to their potential primary productivity is developed. According to the model, herbivory pressure should be most severe in relatively unproductive environments. With increased potential productivity, the role of predation in herbivore regulation should become more important and the impact of herbivory upon plant communities should

EXPLOITATION ECOSYSTEMS 259

decrease. In very productive environments, increase in herbivory pressure is again probable, at least in aquatic ecosystems. The predicted pattern of phytomass and predicted results of manipulations are compared with available data. A reasonable fit between predictions and observations is found, although the sparsity of data and methodological uncertainties weaken the corroboration in several cases. In terrestrial ecosystems, the present version of the model seems best applicable to the vertebrate branch of the grazing chain, whereas the arthropod branch may be more sensitive to temperature than to average annual productivity.

ACKNOWLEDGMENTS

An early version of the paper was thoroughly and constructively criticized by Olli Järvinen, University of Helsinki, and a later one by Erkki Haukioja, University of Turku. The data base of the paper has benefited from the notes of Ossi V. Lindquist, University of Kuopio; Heikki Henttonen, University of Helsinki; Lloyd Hulbert, Kansas State University; and Stephen Chaplin, University of Missouri. Useful comments about wording were provided by Michael Rosenzweig, Robert Colwell, and Tarja Oksanen. The work was supported by a grant from Suomen Akatemia (The Academy of Finland).

LITERATURE CITED

Arruda, J. A. 1979. A consideration of trophic dynamics in some tallgrass prairie farm ponds. Am. Midl. Nat. 102:259–264.

Baltensweiler, W., G. Benz, P. Bovey, and V. Delucchi. 1977. Dynamics of larch bud moth populations. Annu. Rev. Entomol. 22:79–100.

Batzli, G. 1975. The role of small mammals in arctic ecosystems. Pages 234–268 in F. B. Golley, K. Petrusewicz, and L. Ryskowski, eds. Small mammals, their productivity and population dynamics. IBP Synthesis 5. Cambridge University Press, Cambridge.

Bell, R. H. V. 1971. A grazing ecosystem in Serengeti. Sci. Am. 255(1):86–93.

Birney, E. C., W. E. Grant, and D. D. Baird. 1976. Importance of vegetative cover to cycles of *Microtus* populations. Ecology 57:1043–1051.

Bliss, L. C., and F. E. Wielgolaski, eds. 1973. Proc. Conf. Primary Production and Production Process. Tundra biome. Dublin, Ireland.

Brooks, J. L., and S. I. Dodson. 1965. Predation, body size and composition of plankton. Science 150:28–35.

Caughley, C. 1970. Eruption of ungulate populations, with emphasis on Himalayan thar in New Zealand. Ecology 51:53–72.

———. 1976. Plant-herbivore systems. Pages 94–113 in R. May, ed. Theoretical ecology: principles and applications. Blackwell, London.

Cohen, J. E. 1977. Ratio of prey to predators in community food webs. Nature 270:165–166.

Dyer, M. I. 1979. Consumers. Pages 78–86 in R. T. Coupland, ed. Grassland ecosystems of the world. IBP Synthesis 18. Cambridge University Press, Cambridge.

Ehrlich, P. R., and L. C. Birch. 1967. The "balance of nature" and "population control." Am. Nat. 101:97–107.

Errington, P. L. 1946. Predation and vertebrate populations. Q. Rev. Biol. 21:144–177, 221–245.

———. 1963. Muskrat populations. Iowa State University Press. Ames, Iowa.

Fretwell, S. D. 1977. The regulation of plant communities by food chains exploiting them. Perspect. Biol. Med. 20:169–185.

260   THE AMERICAN NATURALIST

Fuller, W. A., A. M. Martell, R. F. C. Smith, and S. W. Speller. 1975. High arctic lemmings, *Dicrostonyx groenlandicus*. II. Demography. Can. J. Zool. 53:867–878.

Gause, G. F. 1934. The struggle for existence. William & Wilkins, Baltimore.

Gilpin, M. E. 1975. Group selection in predator-prey communities. Princeton University Press, Princeton, N.J.

Hairston, N. G., F. E. Smith, and L. B. Slobodkin. 1960. Community structure, population control and competition. Am. Nat. 94:421–425.

Häkkinen, I., and M. Jokinen. 1974. On the winter ecology of the snow hare (*Lepus timidus*) in the outer archipelago [in Finnish, English summary]. Suom. Riista 25:5–14.

Haukioja, E., and P. Niemelä. 1974. Growth and energy requirements of larvae of *Dineura virididorsata* (Retz.) (Hym., Tenthredinidae) and *Oporinia autumnata* (Bkh.) (Lep., Geometridae) feeding on birch. Ann. Zool. Fenn. 11:207–211.

Haukioja, E., P. Niemelä, L. Iso-Iivari, H. Ojala, and E.-M. Aro. 1978. Birch leaves as a resource for herbivores. I. Variation in the suitability of leaves. Rep. Kevo Subarct. Res. Stn. 14:5–12.

Höglund, N., and B. Eriksson. 1973. Förvildare tamrenarnas inverkan på vegetationen inom Lövhögsområdet. Statens Naturvårdsverket, Forskningssekretariatet 7–61/72.

Holling, C. S. 1965. The functional response of predators to prey density and its role in mimicry and population regulation. Mem. Entomol. Soc. Can. 45:3–60.

Hrabaček, J. 1962. Species composition and the amount of zooplankton in relation to fish stock. Rozpr. Cesk. Akad. Ved. Rada Mat. Prir. Ved. 72(10):1–16.

Kalela, O. 1962. On the fluctuation in the numbers of arctic and boreal small mammals as a problem of production biology. Ann. Acad. Sci. Fenn., Ser. A IV. Biol. 66:1–38.

Kallio, P., and J. Lehtonen. 1973. Birch forest damage caused by *Oporinia autumnata* (Bkh.) in 1965–66 in Utsjoki, N. Finland. Rep. Kevo Subarct. Res. Stn. 10:55–69.

Keith, L. B. 1974. Some features of population dynamics in mammals. Pages 17–58 *in* Proc. XI Int. Congr. Game Biol., Stockholm, Sweden.

Krebs, C. J. 1972. Ecology: the experimental analysis of distribution and abundance. Harper & Row, New York.

Lakatos, I. 1972. Falsification and the methodology of scientific research programs. Pages 91–196 *in* I. Lakatos, and A. Musgrave, eds. Criticism and the growth of knowledge. Cambridge University Press, Cambridge.

MacArthur, R. 1972. Geographical ecology. Harper & Row, New York.

Maiorana, V. C. 1975. Predation, submergent behavior and tropical diversity. Evol. Theory 1:157–177.

May, R. 1972. Limit cycles in predator-prey communities. Science 177:900–902.

Mech, L. D. 1966. The wolves of Isle Royale. Fauna National Parks U.S. Government Printing Office, Washington, D.C.

Murdoch, W. W. 1966. Community structure, population control and competition—a critique. Am. Nat. 100:219–226.

Noy-Meir, I. 1975. Stability of grazing systems: an application of predator-prey models. J. Ecol. 63:459–481.

Rasmussen, D. I. 1941. Biotic communities of Kaibab Plateau, Arizona. Ecol. Monogr. 3:229–275.

Rosenzweig, M. L. 1969. Why does the prey isocline have a hump? Am. Nat. 103:81–87.

———. 1971. Paradox of enrichment: destabilization of exploitation ecosystems in ecological time. Science 171:385–387.

———. 1973. Exploitation in three trophic levels. Am. Nat. 107:275–294.

———. 1977. Aspects of biological exploitation. Q. Rev. Biol. 52:371–380.

Rosenzweig, M. L., and R. MacArthur. 1963. Graphical representation and stability conditions of predator-prey interactions. Am. Nat. 97:209–223.

Rosswall, T., and W. O. Heal, eds. 1975. Structure and function of tundra ecosystems. SNV, Forskningsråd, Bull. 20, Stockholm, Sweden.

Simenstad, C. A., J. A. Estes, and K. W. Kenyon. 1978. Aleuts, sea otters and alternate stable-state communities. Science 200:403–411.

Sims, P. L., and R. T. Coupland. 1979. Producers. Pages 49–72 *in* R. T. Coupland, ed. Grassland ecosystems of the world. IBP Synthesis 18. Cambridge University Press, Cambridge.

Slobodkin, L. B. 1968. How to be a predator? Am. Zool. 8:43–51.

―――. 1974. Prudent predator does not require group selection. Am. Nat. 108:665–678.

Slobodkin, L. B., F. E. Smith, and N. C. Hairston. 1967. Regulation in terrestrial ecosystems and the implied balance of nature. Am. Nat. 101:109–124.

Svärdson, G. 1976. Interspecific population dominance in fish communities of Scandinavian lakes. Rep. Inst. Freshw. Res. Drottningholm 55:144–171.

Tanner, J. T. 1975. The stability and the intrinsic growth rates of prey and predator populations. Ecology 56:855–867.

Tast, J., and O. Kalela. 1971. Comparisons between rodent cycles and plant production in Finnish Lapland. Ann. Acad. Sci. Fenn., Ser. A IV. Biol. 186:1–14.

Tenow, O. 1972. The outbreaks of *Oporinia autumnata* (Bhk.) and Operophthera spp. (Lep., Geometridae) in the Scandinavian mountain chain and northern Finland 1862–1968. Zool. Bidr. Upps. (Suppl.) 2:1–107.

Tricker, R. A. R. 1965. The assessment of scientific speculation. Elsevier, New York.

Van Valen, L. 1973. Pattern and the balance of nature. Evol. Theory 1:31–49.

―――. 1975. Energy and evolution. Evol. Theory 1:179–229.

White, T. C. R. 1978. The importance of a relative shortage of food in animal ecology. Oecologia 33:71–86.

Whittaker, R. H., and W. A. Niering. 1975. Vegetation of Santa Catalina mountains, Arizona. V. Biomass, production and diversity along the elevation gradient. Ecology 56:771–790.

Wiegert, R. G., and E. P. Odum. 1969. Radionuclide tracer measurements of food web diversity in nature. Pages 709–710 *in* Proc. 2d Natl. Symp. Radioecol. Clearinghouse Fed. Sci. Info., Springfield, Va.

Wollkind, D. J. 1976. Exploitation in three trophic levels: an extension allowing intraspecies carnivore interaction. Am. Nat. 110:431–447.

Ecology, 68(6), 1987, pp. 1863–1876
© 1987 by the Ecological Society of America

REGULATION OF LAKE PRIMARY PRODUCTIVITY
BY FOOD WEB STRUCTURE[1]

S. R. Carpenter
Department of Biological Sciences, University of Notre Dame, Notre Dame, Indiana 46556 USA

J. F. Kitchell
Center for Limnology, University of Wisconsin, Madison, Wisconsin 53706 USA

J. R. Hodgson and P. A. Cochran
Division of Natural Sciences, St. Norbert College, De Pere, Wisconsin 54115 USA

J. J. Elser and M. M. Elser
Department of Biological Sciences, University of Notre Dame, Notre Dame, Indiana 46556 USA

D. M. Lodge,[2] D. Kretchmer, and X. He
Center for Limnology, University of Wisconsin, Madison, Wisconsin 53706 USA

AND

C. N. von Ende
Department of Biological Sciences, Northern Illinois University, DeKalb, Illinois 60115 USA

Abstract. We performed whole-lake manipulations of fish populations to test the hypothesis that higher trophic levels regulate zooplankton and phytoplankton community structure, biomass, and primary productivity. The study involved three lakes and spanned 2 yr. Results demonstrated hierarchical control of primary production by abiotic factors and a trophic cascade involving fish predation.

In Paul Lake, the reference lake, productivity varied from year to year, illustrating the effects of climatic factors and the natural dynamics of unmanipulated food web interactions. In Tuesday Lake, piscivore addition and planktivore reduction caused an increase in zooplankton biomass, a compositional shift from a copepod/rotifer assemblage to a cladoceran assemblage, a reduction in algal biomass, and a continuous reduction in primary productivity. In Peter Lake, piscivore reduction and planktivore addition decreased zooplanktivory, because potential planktivores remained in littoral refugia to escape from remaining piscivores. Both zooplankton biomass and the dominance of large cladocerans increased. Algal biomass and primary production increased because of increased concentrations of gelatinous colonial green algae.

Food web effects and abiotic factors were equally potent regulators of primary production in these experiments. Some of the unexplained variance in primary productivity of the world's lakes may be attributed to variability in fish populations and its effects on lower trophic levels.

Key words: fish; food web; herbivory; lakes; largemouth bass; piscivory; planktivory; primary production; zooplankton.

Introduction

The relative importance of biotic and abiotic factors in the regulation of ecological systems has been debated throughout this century (McIntosh 1985). The assumptions that density-dependent factors regulate populations and that communities are near equilibrium have dominated recent ecological thought, but are presently challenged by evidence that density-independent factors are important, nonequilibrium popula-

tions are common, and long-term community stability is rare (Caswell 1983, Connell and Sousa 1983). Long-standing tradition in limnology attributes control of ecosystem processes to abiotic factors (Wetzel 1983), while the effects of biotic factors have been recognized principally at the community level (Sih et al. 1985, Kerfoot and Sih 1987). However, many studies have noted the influence of predator-mediated changes in plankton community structure on ecosystem processes (Hrbacek et al. 1961, Brooks and Dodson 1965, Shapiro 1980, DiBernardi 1981, Carpenter and Kitchell 1984, McQueen et al. 1986).

Lake ecosystem productivity depends on the supply of nutrients, especially phosphorus (Wetzel 1983).

[1] Manuscript received 2 October 1986; revised 14 January 1987; accepted 18 January 1987.
[2] Present address: Department of Biological Sciences, University of Notre Dame, Notre Dame, Indiana 46556 USA.

1864 Ecology, Vol. 68, No. 6

However, even in phosphorus-limited lakes, <50% of the variance in productivity can be explained by phosphorus loading (Schindler 1978). The cascading trophic interactions hypothesis explains differences in productivity among lakes with similar nutrient supplies but contrasting food webs (Carpenter et al. 1985). In simplified form, the cascade hypothesis states that a rise in piscivore biomass brings decreased planktivore biomass, increased herbivore biomass, and decreased phytoplankton biomass (Hrbacek et al. 1961, Shapiro 1980). Specific growth rates at each successive trophic level show opposite responses to biomass. Productivity of each trophic level is maximized at an intermediate biomass of its predators.

According to the cascade hypothesis, lake ecosystem productivity is regulated hierarchically through both biotic and abiotic mechanisms (Carpenter et al. 1985). Abiotic factors such as mixing and nutrient supply establish the potential productivity. Actual productivity is determined by food web structure, which depends largely on the strength of interspecific interactions (Paine 1980).

We tested the cascade hypothesis by performing whole-lake manipulations of fish populations and examining the effects of these manipulations on plankton community structure and primary production. Prior to manipulation, the food web of Tuesday Lake had essentially three trophic levels (primary producer, herbivore, planktivore), while the food webs of Paul and Peter lakes had a fourth trophic level, piscivores. Following a year of baseline studies, a reciprocal exchange of fish was carried out between Peter and Tuesday lakes, while Paul Lake remained unmanipulated as a reference ecosystem. Thus, we could determine effects on the plankton of interannual variation in climate and food web structure (Paul Lake) and effects of substantial changes in fish community structure (Peter and Tuesday lakes). We expected that exchange of fishes between Peter and Tuesday lakes would make the food webs more similar and thereby increase the functional similarity of these ecosystems. In Tuesday Lake, piscivore addition and planktivore removal caused major changes in the plankton and primary production that conformed with our hypothesis. In Peter Lake, however, piscivore removal and planktivore addition had surprising consequences that resulted from unexpected behavioral responses of the fish and enhancement by grazers of the growth of certain gelatinous algae.

Study Lakes

Paul, Peter, and Tuesday lakes have long histories of use in experimental limnology (Hasler 1964), and early research on them initiated the experimental approach to ecosystem studies (Likens 1985a). These private lakes with unexploited fish populations are morphometrically similar and lie within 1 km of one another in watersheds owned by the University of Notre Dame.

Recent limnological descriptions of the lakes are given by Carpenter et al. (1986), J. Elser et al. (1986), and M. Elser et al. (1986).

Prior to initial manipulation in 1951, bass were the predominant fish in Paul and Peter lakes (Johnson 1954). Between 1951 and 1976, the lakes were subjected to a series of limings (Peter Lake only) and fish manipulations (both lakes), including rotenone treatment and stocking of trout (Hasler 1964, M. Elser et al. 1986). Trout populations dwindled by 1970 and a cyprinid–*Umbra* assemblage (Tonn and Magnuson 1982; hereafter, simply "minnow assemblage") became dominant. Largemouth bass (*Micropterus salmoides*) were re-introduced to Peter Lake in 1975 and Paul Lake in 1978. By 1980 the minnow assemblage had collapsed and bass dominated the food webs of both lakes (Kitchell and Kitchell 1980). When we began our study in 1984, largemouth bass was the only fish species in the lakes, and the plankton closely resembled that found in the lakes prior to initial manipulation in 1951 (M. Elser et al. 1986).

Tuesday Lake becomes anoxic each winter, but supports an abundant minnow assemblage of redbelly dace (*Phoxinus eos*), finescale dace (*P. neogaeus*), and central mudminnows (*Umbra limi*). These planktivores flourish in lakes that winterkill and lack piscivores (Tonn and Magnuson 1982). Tuesday Lake was artificially circulated using compressed air and then stocked with trout in the summer of 1956 (Schmitz 1958). Winterkill subsequently removed the trout and restored the original minnow assemblage.

Methods

Experimental design

Fish manipulations in 1985 followed baseline studies in 1984. In 1985, reciprocal fish exchanges were carried out between Peter and Tuesday lakes, while Paul Lake remained undisturbed as a reference ecosystem. Approximately 90% of the adult bass biomass in Peter Lake was transplanted to Tuesday Lake, and ≈90% of the minnow biomass in Tuesday Lake was moved to Peter Lake. Specifically, 375 bass, weighing 45.7 kg, were moved from Peter Lake to Tuesday Lake during 23–31 May 1985, followed by 91 bass (10.1 kg) on 27 July 1985. Adult bass were spawning during the manipulation in May, and continued to do so after the manipulation in all three lakes. By 27 July 1985, bass remaining in Peter Lake had eliminated the minnow population and were feeding on young-of-the-year (YOY) bass (see Results and Discussion). The additional 91 adult and juvenile bass were moved on 27 July to reduce predation on the large cohort of YOY bass recruited in Peter Lake in 1985. During 23–31 May 1985, 44 901 minnows (39 654 *P. eos*, 2692 *P. neogaeus*, and 2655 *U. limi*), weighing 56.4 kg, were moved from Tuesday Lake to Peter Lake. Fishes were held in live cages or net enclosures for up to 7 d after capture and before release into Peter and Tuesday lakes. Mortality during holding was <3% for all fish species. Periodic searches following release detected few dead fish.

During summer stratification in 1984 and 1985, fish populations and diets, plankton community structure, and primary production were monitored. We focused our study on summer stratification because metabolic rates are greatest and lakes are least subject to variable mixing regimes during that period.

Fish populations and diets

Bass populations were estimated by mark-recapture using both hook-and-line and electrofishing methods (Bagenal 1978). All adult bass (>195 mm) captured were marked with intramuscular tags. All adult bass in Tuesday Lake, and 50–60% of those in Paul and Peter lakes, were tagged. Double-tag studies showed that tag loss was negligible. Juvenile bass (<195 mm) were marked by fin clipping. Here we report population estimates made during 14–22 August 1984 and 20–26 August 1985. Since most mortality occurred over winters, and young-of-the-year (YOY) bass were difficult to catch in June and July, August studies provided the most satisfactory annual index of bass populations, and intensive sampling efforts were made during these periods. Between 10.8 and 54.3% of marked fish were recaptured in each of these studies.

Methods for analyzing fish diets were described in detail by Hodgson and Kitchell (1987) and will be recounted briefly here. Diets of juvenile and adult bass were determined by flushing the stomachs of at least 20 bass per lake approximately every other week (Seaburg 1957). Data presented here are based on a total of 1128 bass stomach samples. Occasional dissections of sacrificed animals indicated that flushing completely removed stomach contents. The index of relative importance (IRI), which averages percentages of numbers, wet mass, and frequency of occurrence contributed by each food type, was used to summarize diets (George and Hadley 1979). The minimum IRI is zero, for items that are not eaten. The maximum IRI is 100%, for a food type that is the sole component of the diet.

Populations of minnows and YOY bass were estimated every 2 wk by overnight sets of 36–58 minnow traps per lake, 24 deployed around the perimeter and 12–34 in a midlake transect. YOY bass populations in all three lakes were estimated by mark-recapture using a series of electrofishing samples during 24–30 August 1985. In Tuesday Lake, minnow populations were analyzed by the depletion method (DeLury estimate: Bagenal 1978) during 23–31 May 1985. When it was possible to catch YOY bass and minnows, their diets were determined by microscopic examination of stomach contents dissected from 5–30 animals per lake on each sampling date.

Limnological methods

Limnological methods used in these lakes have been described in detail previously (Carpenter et al. 1986, J. Elser et al. 1986, M. Elser et al. 1986), and will be summarized briefly here. All limnological variables were measured weekly at a fixed central station in each lake, except where otherwise noted. Profiles were taken of temperature, dissolved oxygen, and light penetration. Water samples were taken with an opaque Van Dorn bottle at depths of 100, 50, 25, 10, and 1% of surface irradiance in 1984, and at those depths plus 5% of surface irradiance in 1985. Phytoplankton samples were pooled from the upper three samples, all of which were taken from the mixed layer. Zooplankton were collected by vertical hauls of a 75-μm mesh Nitex net (30 cm diameter). Single hauls were made in 1984 and duplicate hauls were pooled in 1985. Throughout this paper, "zooplankton" refers to crustaceans and rotifers, and does not include *Chaoborus*. Separate samples for *Chaoborus* were collected at night by vertical hauls of a 202-μm mesh net. Three to six replicate *Chaoborus* samples were pooled every 2 wk or at monthly intervals from each lake. The filtering efficiencies of the nets for *Chaoborus* and for each zooplankton taxon in each lake were calculated from duplicate profiles taken at seven equally spaced depths with a Schindler-Patalas trap.

Chlorophyll *a*, corrected for pheopigments, was determined fluorometrically (Parsons et al. 1984), and the data were employed in the productivity calculations (see below). Samples were collected on Whatman GF/F filters, frozen, sonicated and homogenized in methanol (Marker et al. 1980), centrifuged, and analyzed.

Carbon fixation was determined by the ^{14}C method at the depths listed above. Details of our procedure and the method for calculating daily depth-integrated productivity were given by Carpenter et al. (1986).

Phytoplankton preserved with Lugol's iodine were settled and then identified, measured, and enumerated under an inverted microscope following procedures described by J. Elser et al. (1986). Both algal unit volume and biovolume (i.e., biologically active volume, containing protoplasm and exclusive of gelatinous sheaths, loricae, etc.) were determined (J. Elser et al. 1986). Phytoplankton data were presented in two size classes: <30-μm greatest axial linear dimension (GALD) and >30-μm GALD. Bag experiments in these lakes showed contrasting responses of these size classes to changes in zooplankton biomass and community composition (Bergquist 1985, Bergquist et al. 1985, Bergquist and Carpenter 1986).

Zooplankton were identified, counted, and measured under a dissecting microscope. *Conochilus* colonies, which could not be preserved, were counted and measured in fresh samples within an hour of sampling. Dry biomasses were computed as described by Carpenter et al. (1986). Biomass was divided into contributions

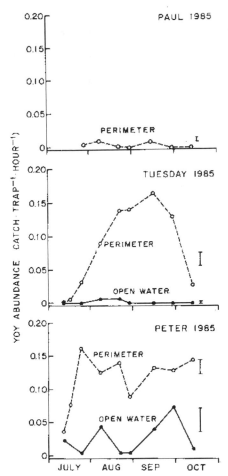

FIG. 1. Catch per unit effort of young-of-the-year (YOY) bass vs. date in 1985 for perimeter and open-water traps in three study lakes. Vertical bar indicates pooled standard error.

by large (>1 mm) and small (<1 mm) animals. Grazers larger than ≈1 mm differ substantially from smaller animals in their effects on phytoplankton (Bergquist et al. 1985, McQueen et al. 1986).

Phytoplankton and zooplankton assemblages were compared using Sorenson's similarity index (Janson

and Vegelius 1981). This index ranges from zero (for samples with no species in common) to one (for identical samples). Mean summer biomasses of each taxon in each lake each year were expressed as percentages of total mean summer biomass for similarity index calculations.

Chaoborus were counted under a dissecting microscope by instar and, except for instar I larvae, by species. Dry biomass was calculated from individual masses based on groups of 30–160 animals of each species and instar, with the following exceptions. Mass of instar II *C. flavicans* was estimated from the mass of similar-sized *C. punctipennis*. Masses of *C. trivittatus* (instars II and IV) were taken from Fedorenko and Swift (1972). Pupal masses were assumed equal to instar IV masses. No instar I larvae were included in the biomass estimates. Instar I larvae never made up >0.3% of total *Chaoborus* biomass. Instar I larvae were captured effectively by the 202-μm mesh net, but because of their small size and rapid growth to larger instars, they were a negligible component of *Chaoborus* biomass.

RESULTS AND DISCUSSION

Reference ecosystem (Paul Lake)

Regional weather patterns differed between the two study years. Responses of Paul Lake, which was not manipulated, demonstrated the limnological consequences of these interannual differences in weather, as well as the dynamics of an undisturbed food web. For example, the epilimnion of Paul Lake was cooler in 1985 than in 1984. Mean mixed-layer temperature was 0.7°C cooler in June–July 1985, and 2.7° cooler in August 1985, than in the corresponding periods in 1984. Similar interannual differences occurred in the lakes monitored by the Long-Term Ecological Research Program, located ≈20 km from Paul Lake (T. Frost and T. Kratz, *personal communication*).

Adult bass populations were lower in 1985 than in 1984, while juvenile bass were more abundant in 1985 (Table 1). Since juvenile bass preyed more heavily on zooplankton than did adult bass (Table 2), zooplanktivory by fishes was probably higher in 1985 than in 1984. YOY (young-of-the-year) bass were not common in Paul Lake, because they were cannibalized by juvenile and adult bass (Tables 1 and 2). YOY bass were

TABLE 1. Fish densities in the study lakes in August 1984 and August 1985. Standard errors are given in parentheses.

	Paul		Peter		Tuesday	
	1984	1985	1984	1985	1984	1985
Minnows (catch·trap⁻¹·h⁻¹)	0	0	0	0	0.94 (0.22)	0
Bass (fish/ha)						
YOY	—*	242 (225)	—*	5960 (1120)	0	4200 (2020)
Juveniles (<195 mm)	80 (10)	187 (48)	67 (5)	19 (10)	0	190 (80)
Adults (>195 mm)	201 (87)	162 (54)	154 (61)	16 (3)	0	293 (91)

* In 1984, young-of-the-year (YOY) bass were present in Paul and Peter lakes, but catches were too low to estimate densities.

TABLE 2.   Index of relative importance (IRI, %) for selected diet items of juvenile and adult bass in the study lakes in 1984 and 1985.*

Lake	Food item	1984		1985	
		Juveniles	Adults	Juveniles	Adults
Paul	zooplankton	23	7	27	13
	Chaoborus	7	19	11	13
	fishes	4	4	4	6
Peter	zooplankton	21	8	30	27
	Chaoborus	7	11	10	11
	fishes	4	12	23	17
Tuesday	zooplankton	−†	−	5	7
	Chaoborus	−	−	33	26
	fishes	−	−	31	22

* In all cases the standard error is <1% (n > 100 in each lake each year).
† Bass were not present in Tuesday Lake in 1984.

concentrated in littoral refugia, and were never caught in open-water traps (Fig. 1).

Chaoborus biomass (Fig. 2) was dominated by *C. flavicans. C. punctipennis* was present, and *C. trivittatus* was rare.

Zooplankton biomass was lower in 1985 than in 1984 (Fig. 3). Mean summer biomass was 2.67 g/m² in 1984 (SE = 0.36, n = 13), and 1.16 g/m² in 1985 (SE = 0.10, n = 13). This reduction was due to decreases in the densities and average sizes of the dominant cladocerans *Daphnia pulex, D. rosea,* and *Holopedium gibberum.* Biomasses of both rotifers and copepods differed little between the 2 yr. In both years, *Orthocyclops modestus* and *Mesocyclops edax* were the dominant copepods, and colonial *Conochilus* and a large *Asplanchna* species were the dominant rotifers.

No substantial change in zooplankton composition occurred between the two years (Table 3, Appendix I). The similarity of the 1984 and 1985 zooplankton assemblages in Paul Lake was greater than the interannual similarity coefficients for the two manipulated lakes (Table 3). No substantial difference in the relative abundance of large zooplankters occurred between the two years (Fig. 4).

Algal biovolume density was lower in 1985 than in 1984 (Fig. 5). Mean summer biovolume density was 0.319 × 10⁶ μm³/mL (SE = 0.041, n = 13) in 1984, and 0.183 × 10⁶ μm³/mL (SE = 0.015, n = 13) in 1985. Total biovolume in late June 1984 showed the effects of a *Synura* bloom that was not repeated in 1985. *Chrysosphaerella longispina,* another large colonial chrysophyte, was also much less abundant in 1985 than in 1984 (Appendix II). Otherwise, algal species composition was similar in 1984 and 1985, dominated by flagellated nannoplankton (*Cryptomonas, Mallomonas,* several small chlorophytes), *Gloeocystis, Oocystis,* and *Anabaena circinalis.* The phytoplankton assemblages of Paul Lake in 1984 and 1985 had greater similarity coefficients than the 1984 and 1985 phytoplankton assemblages of the manipulated lakes (Table 3).

Primary production in 1985 was consistently lower than primary production in 1984 (Fig. 6). Mean daily primary production (measured as carbon) was 70.2 mg·m⁻³·d⁻¹ (SE = 3.8, n = 78) in 1984, and 27.8 mg·m⁻³·d⁻¹ (SE = 1.14, n = 94) in 1985. In both years, substantial variability resulted from fluctuations in irradiance and phytoplankton composition and density, but there were no obvious trends in production through either summer. Both the intrasummer and interannual variability exhibited by Paul Lake are similar to the variance in primary production known from many temperate lake ecosystems (Carpenter and Kitchell 1987).

Bass addition/minnow removal (Tuesday Lake)

In Tuesday Lake, increased piscivory and decreased planktivory caused major changes in the zooplankton and phytoplankton that conformed to the predictions of our hypothesis. Minnows were abundant in Tuesday Lake in 1984, but persisted only briefly after the manipulation in 1985 (Fig. 7). Minnows were not caught in Tuesday Lake after 1 June 1985, although minnows were observed in beaver channels leading away from the lake's margin. Dietary analyses indicated that minnows were omnivorous: overall, their diets consisted of 57% periphyton and 43% zooplankton by volume.

Adult and juvenile bass populations in Tuesday Lake in 1985 exceeded those of the other lakes (Table 1). YOY bass in Tuesday Lake in 1985 were confined to the littoral zone (Fig. 1). The increased catch in perimeter traps in July and August 1985 resulted from recruitment to trappable size, while the declining catch

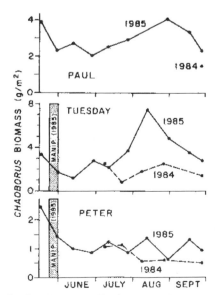

FIG. 2.   Dry biomass of *Chaoborus* (g/m²) vs. date in 1984 (– – –) and 1985 (——) in the three study lakes. Vertical bar indicates pooled standard error.

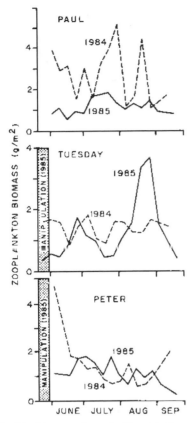

FIG. 3. Dry biomass of zooplankton (g/m²) vs. date in 1984 (- - -) and 1985 (——) in the three study lakes.

in September and October resulted from cannibalism of YOY bass by larger bass. Fishes were a relatively large diet component for Tuesday Lake bass (Table 2). Most of the predation on YOY bass occurred after 8 August 1985. Bass in Tuesday Lake ate less zooplankton than bass in Paul Lake (Table 2). Planktivory was

reduced in Tuesday Lake in 1985 because densities of planktivorous minnows were reduced, and potentially planktivorous YOY bass remained in the littoral zone because of intense predation by adult bass.

Biomass of *Chaoborus* was substantially higher in August 1985 than in 1984 in Tuesday Lake (Fig. 2). The dominant chaoborid in Tuesday Lake was *C. punctipennis. C. flavicans* was present, and *C. trivittatus* was rare.

The fish manipulations substantially changed the zooplankton community structure and size structure in Tuesday Lake. Zooplankton assemblages in 1984 and 1985 had a similarity coefficient of only 0.50, compared with the interannual similarity coefficient of 0.84 in the reference lake (Table 3). In 1985, the zooplankton of Tuesday Lake became more similar to those of Paul Lake. The abundance of zooplankton >1 mm in length increased abruptly in August 1985 (Fig. 4), and led to a 70% increase in total zooplankton biomass in August (Fig. 3). Mean zooplankton biomass during August was 1.45 g/m² (SE = 0.11, n = 4) in 1984, and 2.48 g/m² (SE = 0.66, n = 4) in 1985.

TABLE 3. Coefficients of similarity for zooplankton (above diagonal) and phytoplankton (below diagonal) in Paul, Peter, and Tuesday lakes in 1984 and 1985.

Lake and year	Paul		Peter		Tuesday	
	1984	1985	1984	1985	1984	1985
Paul						
1984	1	0.84	0.62	0.69	0.18	0.47
1985	0.59	1	0.76	0.77	0.30	0.52
Peter						
1984	0.56	0.32	1	0.71	0.36	0.47
1985	0.17	0.16	0.22	1	0.24	0.38
Tuesday						
1984	0.21	0.17	0.21	0.06	1	0.50
1985	0.19	0.21	0.11	0.40	0.32	1

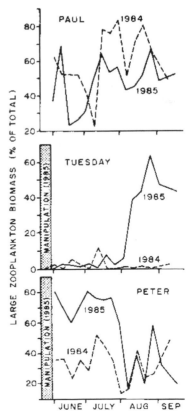

FIG. 4. Percentage of zooplankton biomass consisting of animals >1 mm in length in the three study lakes in 1984 (- - -) and 1985 (——).

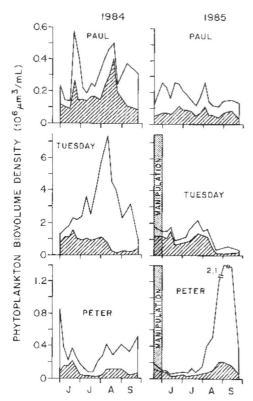

FIG. 5. Phytoplankton biovolume density (10^6 μm³/mL [parts per million]) vs. date in 1984 and 1985 in the three study lakes. Biovolume is divided into small (greatest axial linear dimension [GALD] < 30 μm; hatched area) and large (GALD > 30 μm; open area) size classes.

The increases in zooplankton size and biomass in August 1985 resulted from the replacement of a rotifer–copepod assemblage by large cladocerans. Copepods constituted 83% of the zooplankton biomass in August 1984, but only 21% in August 1985. In contrast, the percentage of zooplankton biomass accounted for by cladocerans rose from 11% in August 1984 to 78% in August 1985. During 1984, the copepods *Tropocyclops prasinus* and *Orthocyclops modestus* dominated the zooplankton (Appendix I). *Bosmina longirostris* was the dominant cladoceran, with *Daphnia pulex, Diaphanosoma leuchtenbergianum*, and *Holopedium gibberum* also present. The dominant rotifers were *Polyarthra vulgaris* and *Conochiloides dossuarius*. In 1985, copepod biomass (predominantly *T. prasinus*) peaked in late June and then declined. Biomass of all copepods was lower in 1985 than in 1984, and two species common in 1984 (*Cyclops varicans rubellus* and *Leptodiaptomus siciloides*) were not detected in 1985 (Appendix I). After late July 1985, rotifer biomass was consistently lower than in 1984. In contrast, cladoceran biomass rose during August 1985, and by the end of

the month attained values higher than those found in the other lakes. Cladocerans accounting for this rise were, in order of appearance, *Diaphanosoma leuchtenbergianum, Holopedium gibberum*, and *Daphnia pulex*. The biomass peak in August 1985 (Fig. 3) was due almost entirely to *H. gibberum* and *D. pulex*.

The manipulation altered algal size structure, biovolume density, and productivity. Mean biovolume density of phytoplankton in the mixed layer decreased from 3.45 × 10^6 μm³/mL (SE = 0.47, n = 13) in 1984 to 1.14 × 10^6 μm³/mL (SE = 0.19, n = 13) in 1985. Loss of phytoplankters > 30-μm GALD accounted for this decline (Fig. 5). Primary production declined steadily through the summer of 1985, and by August was about one-fifth as large as productivity in 1984 (Fig. 6). Mean daily primary production during August (measured as carbon) was 146 mg·m⁻³·d⁻¹ (SE = 15.8, n = 24) in 1984, and 29 mg·m⁻³·d⁻¹ in 1985.

The most conspicuous change in phytoplankton

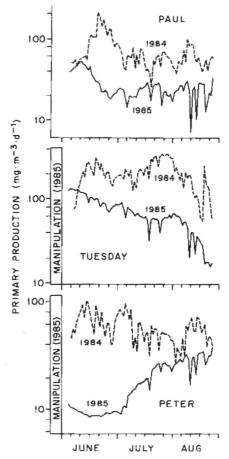

FIG. 6. Daily primary production (as carbon) vs. date in 1984 (– – –) and 1985 (——) in the three study lakes. Note semilogarithmic axes.

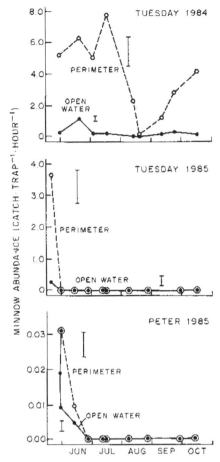

FIG. 7.   Catch per unit effort of *Phoxinus eos* (minnows) vs. date in perimeter and open-water traps in Tuesday Lake during 1984 and 1985, and Peter Lake during 1985. Catches of *P. neogaeus* and *Umbra limi* combined were <15% of the total and showed the same seasonal trends and inshore/offshore pattern as *P. eos*. No minnows were present in Peter Lake in 1984, or in Paul Lake in either year. Vertical bar indicates pooled standard error.

composition was the loss of relatively large, long-lived, slow-growing taxa (Appendix II). In 1984 and 1985, the small algal size class was dominated by *Peridinium pulsillum, Chrysochromulina, Cryptomonas, Trachelomonas,* and small flagellated chlorophytes (Appendix II). Biovolume density of small phytoplankters (Fig. 5) was similar in 1984 (mean = 0.85×10^6 μm^3/mL, SE = 0.12, $n = 13$) and 1985 (mean = 0.77×10^6 μm^3/mL, SE = 0.12, $n = 13$). The large size class was dominated by dinoflagellates in 1984: *Glenodinium quadridens,* several *Peridinium* species of 30–50 μm GALD, and the large (74–83 μm GALD) *Peridinium limbatum,* which accounted for the biovolume peak in August 1984. Two colonial taxa, *Chrysosphaerella longispina* and *Microcystis aeruginosa,* were also prominent in

the large size class during 1984. Biovolume density of the large algal class was much lower in 1985 (mean = 0.33×10^6 μm^3/mL, SE = 0.07, $n = 13$) than in 1984 (mean = 2.58×10^6 μm^3/mL, SE = 0.52, $n = 13$). Reduced densities of dinoflagellates >30-μm GALD and colonial taxa accounted for this interannual difference. Spiny, grazing-resistant algae such as *Golenkinia, Diceras,* and a large *Mallomonas* species dominated the large phytoplankton in August 1985, but their growth did not compensate for the loss of the large dinoflagellates.

Certain of these phytoplankton responses could be forecast from the results of bag experiments. Nannoplankters, which respond negatively to large increases in zooplankton biomass, are relatively insensitive to moderate changes in zooplankton biomass because growth stimulation by excreted nutrients and losses to grazing are compensatory (Carpenter and Kitchell 1984, Bergquist 1985, Bergquist et al. 1985, Bergquist and Carpenter 1986). Twofold changes in zooplankton biomass, such as those observed in Tuesday Lake (Fig. 3), have little effect on net growth rates of nannoplankton in bag experiments (Bergquist and Carpenter 1986 and J. Elser et al., *personal observation*). Shapiro and Wright (1984) also found that whole-lake manipulations had little effect on the nannoplankton.

Among larger phytoplankters, two different responses to increasing grazer biomass are evident. Dinoflagellates are negatively affected by relatively modest increases in zooplankton biomass (Bergquist 1985, Bergquist et al. 1985). In contrast, large chlorophytes and chrysophytes, especially spiny, gelatinous, or colonial forms, are stimulated by increased biomass of large cladocerans (Porter 1976, Bergquist et al. 1985, Lehman and Sandgren 1985, Bergquist and Carpenter 1986). Only a few of these grazing-enhanced taxa were present in Tuesday Lake in 1985.

Bass removal/minnow addition (Peter Lake)

The response of Peter Lake was surprising and more complex than that of Tuesday Lake. We had expected that the removal of bass and addition of minnows would increase zooplanktivory, decrease zooplankton biomass and mean size, and increase phytoplankton biomass and production. Phytoplankton biomass and production did indeed increase, but by unexpected mechanisms.

The manipulation reduced the densities of juvenile and adult bass (Table 1). Densities of YOY bass, however, were much greater in 1985 because piscivory was greatly reduced. As in the other lakes, YOY bass were more numerous in the littoral zone (Fig. 1). Following further removals of adult bass from Peter Lake in July 1985 (see Methods: Experimental Design), piscivore densities were very low. Consequently YOY bass catches in open-water traps in Peter Lake were higher than those in the other lakes after 1 August 1985.

Minnows moved offshore immediately after release

in late May 1985 (Fig. 7). For up to 4 wk following release, minnow guts contained zooplankton (especially the colonial rotifer *Conochilus*), and bass guts contained minnows. By 2 wk after the release, minnows had moved inshore, presumably to escape predation, which had greatly reduced minnow populations by mid-June. After 27 June 1985, no minnows were caught in Peter Lake, although a few minnows were seen in water <20 cm deep.

Zooplanktivory by juvenile and adult bass was less in 1985 than in 1984. Even though individual bass ate more zooplankton in 1985 (Table 2), juvenile and adult bass populations were greatly reduced (Table 1). For juvenile bass, the amount of zooplankton per gut increased ≈1.5 times, while the population declined by a factor of 3.5. For adult bass, the amount of zooplankton per gut increased ≈3.4 times, while the population declined by a factor of nearly 10. Overall, therefore, zooplanktivory by juvenile and adult bass in 1985 was ≈40% of zooplanktivory in 1984.

Zooplanktivory by YOY bass was greater in 1985 than in 1984. YOY bass were too rare to enumerate in 1984, but the population was certainly not larger than the population of YOY bass in Paul Lake in 1985. Therefore, the YOY bass population in August 1985 was at least 25 times larger than that of August 1984. YOY bass diets could not be determined in 1984, and were 48% zooplankton in 1985. The most intense feeding on pelagic zooplankton occurred after July 1985 when the YOY bass became more numerous offshore (Fig. 1).

Chaoborus biomass was slightly lower in 1984 than in 1985 (Fig. 2). The *Chaoborus* species composition of Peter Lake was dominated by *C. flavicans*. Both *C. punctipennis* and *C. trivittatus* were present.

Zooplankton biomass (Fig. 3) during July and August was greater in 1985 (mean = 1.25 g/m², SE = 0.13, $n = 9$) than in 1984 (mean = 0.96 g/m², SE = 0.12, $n = 9$). Zooplankton community composition and size structure also differed between the two years (Table 3, Fig. 4). Cladocerans, predominantly *Daphnia pulex*, had greater biomass in 1985 than 1984, while copepods had lower biomass in 1985 than in 1984 (Appendix I). In June 1984, the Peter Lake zooplankton was dominated by *Diaptomus oregonensis*, which declined steadily in biomass through June. Biomass of *D. oregonensis* and the codominant copepods *Orthocyclops modestus* and *Cyclops varicans rubellus* remained low in 1985. Rotifer biomass and species composition were similar in 1984 and 1985.

Large zooplankton (>1 mm) were more dominant in June and July 1985 than in June and July 1984 (Fig. 4). This size increase was associated with greater abundance and size of *Daphnia pulex*. During June and July, *D. pulex* were larger in 1985 (mean length with 95% confidence interval = 1.52 ± 0.11 mm) than in 1984 (mean length = 1.26 ± 0.09 mm). About 1 August 1985, the dominance of large zooplankton abruptly

declined (Fig. 4). This decline coincided with a decrease in mean size of *D. pulex* to 1.1 ± 0.08 mm. *Diaphanosoma leuchtenbergianum*, a cladoceran not recorded in Peter Lake in 1984, was common in August 1985. These changes in the zooplankton followed the onset of zooplanktivory by YOY bass in the pelagic zone of Peter Lake in July 1985. However, cladoceran biomass in August 1985 exceeded that in August 1984 despite the decline in *D. pulex* size in 1985. Mean cladoceran biomass in August was 0.338 g/m² in 1984 (SE = 0.046, $n = 4$) and 0.566 g/m² in 1985 (SE = 0.123, $n = 4$).

Algal biovolume density was lower in 1985 than in 1984 through the middle of July (Fig. 5). As in Paul Lake, the *Synura* bloom of June 1984 was not repeated in 1985, and biovolume density of *Chrysosphaerella longispina* was much less in 1985 than in 1984 (Appendix II). However, the overall similarity of the phytoplankton assemblages of Paul and Peter lakes was much lower in 1985 than in 1984 (Table 3). After mid-July, algal biovolume density in Peter Lake was greater in 1985 than in 1984. High biovolume density in August and September 1985 was due to *Sphaerocystis schroeteri*, a gelatinous colonial chlorophyte in the large size class. During this period, the net growth rate of *S. schroeteri* was positively correlated with zooplankton biomass (J. Elser et al., *personal observation*). Two colonial blue-green algae in the large size class, *Anabaena circinalis* and *Gomphosphaeria lacustris*, also had greater biovolume in 1985 than in 1984. The small size class, dominated by *Cryptomonas* and other flagellated nannoplankton, had similar biovolumes in both years. Very similar changes in algal density and composition have followed increases in cladoceran biomass in bag experiments in Peter Lake (Bergquist 1985, Bergquist and Carpenter 1986, and J. Elser et al., *personal observation*).

Primary production was low in June 1985 in Peter Lake, but increased steadily during the summer (Fig. 6). Peak primary production in 1985 coincided with the bloom of gelatinous green algae in August. In contrast, no intraseasonal trend in primary production was evident during 1984.

General Discussion

These results demonstrate the joint effects of food web structure and abiotic factors on ecosystem productivity. In Paul Lake, interannual productivity differences of about a factor of two occurred. These differences represent the background variability of an undisturbed lake. In Peter and Tuesday lakes, the effects of the manipulation were superimposed on the background variability exhibited by Paul Lake. We accounted for the interannual variability exhibited by Paul Lake by calculating differences between manipulated lake data and Paul Lake data for samples paired in time. Interlake differences in zooplankton biomass and the logarithm of primary production were then averaged by month (Fig. 8). Normal probability plots

1872 S. R. CARPENTER ET AL. Ecology, Vol. 68, No. 6

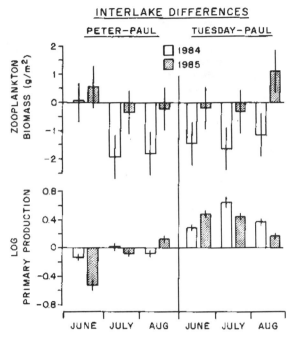

Fig. 8. Differences between manipulated and reference ecosystems in zooplankton biomass and primary production in 1984 and 1985. The vertical line through each bar is the 95% confidence interval for the mean difference between lakes.

(Box et al. 1978) showed that these variates were approximately normally distributed. A similar procedure was suggested by Stewart-Oaten et al. (1986) for comparisons of time series from unreplicated experiments.

Tuesday Lake's differences from the reference ecosystem in zooplankton biomass varied little from month to month in 1984 (Fig. 8). In 1985, however, the zooplankton biomass difference was highest in August, and it was always higher than in 1984. This increase in zooplankton biomass resulted from reduced zooplanktivory. Primary production differences between Tuesday and Paul lakes were highest in July 1984. In 1985, however, primary production differences decreased through the summer. In June, primary production differences were higher in 1985, but in July and August primary production differences were lower in 1985. This decrease in productivity relative to the reference ecosystem was caused by enhanced grazer biomass in 1985. Responses of Tuesday Lake were consistent with several other whole-lake planktivore removal experiments (Henrikson et al. 1980, Benndorf et al. 1984, Shapiro and Wright 1984).

Peter Lake's differences from the reference ecosystem in zooplankton biomass were similar in June 1984 and June 1985 (Fig. 8). In 1985, zooplankton biomass differences were more positive for 1984 during July and August. The net result of the fish manipulations in Peter Lake was therefore an enhancement of herbivore biomass relative to the reference lake. Primary production differences between Peter and Paul lakes in 1984 were highest in July. In 1985, primary production differences increased steadily from month to month, and by August exceeded those of 1984. The increase in productivity relative to the reference ecosystem was caused by enhanced zooplankton biomass in 1985. This response of Peter Lake was the opposite of that observed in Tuesday Lake.

A unimodal relationship between primary production and zooplankton biomass has been supported by theory (Carpenter et al. 1985), modeling studies (Carpenter and Kitchell 1984, 1987), in situ bag experiments (Bergquist and Carpenter 1986, and J. Elser et al., personal observation), and laboratory studies (Sterner 1986). The whole-lake data reported here provide further evidence that zooplankton biomass and primary production are positively related under certain conditions and negatively related under others. The opposite responses of Peter and Tuesday lakes to increased zooplankton biomass arose because of the contrasting phytoplankton of the lakes before manipulation. Tuesday Lake supported abundant but slow-growing dinoflagellates that flourished with the small zooplankters present before the manipulation. After the manipulation, the rising biomass of large cladocerans eliminated the dinoflagellates. A few grazing-resistant taxa (Golenkinia, a spiny Mallomonas species) increased in density in 1985, but could not compensate for the loss of the large dinoflagellates, and primary production declined. Peter Lake, in contrast, had many grazer-adapted phytoplankters that coexisted with large cladocerans before the manipulation (M. Elser et al. 1986). In bag experiments performed in 1984, 9 of 16 algal taxa responded positively or unimodally to increased zooplankton biomass because of nutrient regeneration by the herbivores (Bergquist and Carpenter 1986). The increased zooplankton biomass in Peter Lake in 1985 stimulated the growth of certain grazer-adapted algae that were already established in the lake, causing productivity to increase.

These studies demonstrate the importance of large-scale, long-term manipulations in the analysis of cascading trophic interactions. Strong predator avoidance behaviors have clear adaptive value (Werner and Gilliam 1984) and can have dramatic consequences for community structure. For example, Power and Matthews (1983) showed that periphyton densities in stream pools were determined by presence or absence of piscivorous fishes and the avoidance behaviors evoked in herbivorous minnows. The behavioral responses of fishes that led to increased zooplankton biomass in Peter Lake could not have been predicted from bag experiments, because they occurred at a lakewide scale. The plankton responses of both lakes took a full season to develop. Future changes, such as increased concentrations of grazing-resistant algae in Tuesday Lake, cannot be ruled out. The stability of ecosystems altered by fish manipulations remains an open question, but paleolimnological data (Kitchell and Kitchell 1980, Kitchell and Carpenter 1987), the few long-term records available (Mills et al. 1987, Scavia et al. 1986), models (Carpenter and Kitchell 1987), and time lags induced by ontogenic changes in fishes' diets (Werner and Gilliam 1984, Carpenter et al. 1985) all suggest that fluctuations spanning many years are likely.

Fish population fluctuations are commonly caused by variable recruitment, catastrophic mortality, and/or exploitation (Pitcher and Hart 1982, Peterson and Wroblewski 1984, Steele and Henderson 1984, Mills et al. 1987). These fluctuations may occur on a much longer time scale than the short-term effects of meteorologic variability on phytoplankton (Harris 1980, Carpenter and Kitchell 1987). Therefore, the effects of food web structure and abiotic factors on phytoplankton may be essentially independent, and together account for most of the observed variation in lake ecosystem productivity. Although fisheries biology and limnology have evolved as largely separate disciplines, they must coalesce as we establish a holistic view of factors and interactions that regulate lake ecosystem functioning.

Acknowledgments

We extend special thanks to Professor Arthur D. Hasler, whose foresight established both the ideas and the institutional commitments that make long-term experimental studies of these lakes possible. Our work was supported by NSF grants BSR-83-08918 to J. F. Kitchell, S. R. Carpenter, and J. R. Hodgson; BSR-86-04996 to J. F. Kitchell; BSR-86-06271 to S. R. Carpenter; and BSR-83-73662 to C. N. von Ende. This paper is a contribution from the University of Notre Dame Environmental Research Center. We thank C. Hughes for the drafting.

Literature Cited

Bagenal, T. B. 1978. Methods for assessment of fish production in fresh waters. Blackwell, Oxford, England.

Benndorf, J., H. Kneschke, K. Kossatz, and E. Penz. 1984. Manipulation of the pelagic food web by stocking with predacious fishes. Internationale Revue der Gesamten Hydrobiologie 69:407–428.

Bergquist, A. M. 1985. Effects of herbivory on phytoplankton community composition, size structure, and primary production. Dissertation. University of Notre Dame, Notre Dame, Indiana, USA.

Bergquist, A. M., and S. R. Carpenter. 1986. Limnetic herbivory: effects on phytoplankton populations and primary production. Ecology 67:1351–1360.

Bergquist, A. M., S. R. Carpenter, and J. C. Latino. 1985. Shifts in phytoplankton size structure and community composition during grazing by contrasting zooplankton assemblages. Limnology and Oceanography 30:1037–1045.

Box, G. E. P., W. G. Hunter, and J. S. Hunter. 1978. Statistics for experimenters. John Wiley and Sons, New York, New York, USA.

Brooks, J. L., and S. I. Dodson. 1965. Predation, body size, and composition of plankton. Science 150:28–35.

Carpenter, S. R., M. M. Elser, and J. J. Elser. 1986. Chlorophyll production, degradation, and sedimentation: implications for paleolimnology. Limnology and Oceanography 31:112–124.

Carpenter, S. R., and J. F. Kitchell. 1984. Plankton community structure and limnetic primary production. American Naturalist 124:159–172.

Carpenter, S. R., and J. F. Kitchell. 1987. The temporal scale of limnetic primary production. American Naturalist 129:417–433.

Carpenter, S. R., J. F. Kitchell, and J. R. Hodgson. 1985. Cascading trophic interactions and lake productivity. BioScience 35:634–639.

Caswell, H. 1983. Life history theory and the equilibrium status of populations. American Naturalist 120:317–339.

Connell, J. H., and W. P. Sousa. 1983. On the evidence needed to judge ecological stability or persistence. American Naturalist 121:789–824.

DiBernardi, R. 1981. Biotic interactions in freshwater and effects on community structure. Bollettino di Zoologia 48:353–371.

Elser, J. J., M. M. Elser, and S. R. Carpenter. 1986. Size fractionation of algal chlorophyll, carbon fixation, and phosphatase activity: relationship with species-specific size distributions and zooplankton community structure. Journal of Plankton Research 8:365–383.

Elser, M. M., J. J. Elser, and S. R. Carpenter. 1986. Paul and Peter lakes: a liming experiment revisited. American Midland Naturalist 116:282–295.

Fedorenko, A. Y., and M. C. Swift. 1972. Comparative biology of *Chaoborus americanus* and *Chaoborus trivittatus* in Eunice Lake, British Columbia. Limnology and Oceanography 17:721–730.

George, E. L., and W. F. Hadley. 1979. Food and habitat partitioning between rock bass (*Ambloplites rupestris*) and smallmouth bass (*Micropterus dolomieui*) young of the year. Transactions of the American Fisheries Society 108:253–261.

Harris, G. P. 1980. Temporal and spatial scales in phytoplankton ecology: mechanisms, methods, models, and

1874   S. R. CARPENTER ET AL.   Ecology, Vol. 68, No. 6

management. Canadian Journal of Fisheries and Aquatic Sciences 37:877–900.

Hasler, A. D. 1964. Experimental limnology. BioScience 14: 36–38.

Henrikson, L., H. Nyman, H. Oscarson, and J. Stenson. 1980. Trophic changes, without changes in the external nutrient loading. Hydrobiologia 68:257–263.

Hodgson, J. R., and J. F. Kitchell. 1987. Opportunistic foraging in largemouth bass (Micropterus salmoides). American Midland Naturalist, in press.

Hrbacek, J., M. Dvorakova, V. Korinek, and L. Prochazkova. 1961. Demonstration of the effect of the fish stock on the species composition of zooplankton and the intensity of metabolism of the whole plankton assemblage. Internationale Vereinigung für theoretische und angewandte Limnologie, Verhandlungen 18:162–170.

Janson, S., and J. Vegelius. 1981. Measures of ecological association. Oecologia (Berlin) 49:371–376.

Johnson, W. E. 1954. Dynamics of fish production and carrying capacity of some northern softwater lakes. Dissertation. University of Wisconsin-Madison, Madison, Wisconsin, USA.

Kerfoot, W. C., and A. Sih, editors. 1987. Predation: direct and indirect impacts on aquatic communities. University Press of New England, Hanover, New Hampshire, USA.

Kitchell, J. A., and J. F. Kitchell. 1980. Size-selective predation, light transmission, and oxygen stratification: evidence from the recent sediments of manipulated lakes. Limnology and Oceanography 25:389–402.

Kitchell, J. F., and S. R. Carpenter. 1987. Piscivores, planktivores, fossils, and phorbins. Pages 132–146 in W. C. Kerfoot and A. Sih, editors. Predation: direct and indirect impacts on aquatic communities. University Press of New England, Hanover, New Hampshire, USA.

Lehman, J. T., and C. D. Sandgren. 1985. Species-specific rates of growth and grazing loss among freshwater algae. Limnology and Oceanography 30:34–46.

Likens, G. E. 1985a. An experimental approach for the study of ecosystems. Journal of Ecology 73:381–396.

Marker, A. F., C. A. Crowther, and R. J. M. Gunn. 1980. Methanol and acetone as solvents for estimating chlorophyll a and phaeopigments by spectrophotometry. Ergebnisse der Limnologie 14:52–69.

McIntosh, R. P. 1985. The background of ecology. Cambridge University Press, New York, New York, USA.

McQueen, D. J., J. R. Post, and E. L. Mills. 1986. Trophic relationships in freshwater pelagic ecosystems. Canadian Journal of Fisheries and Aquatic Sciences 43:1571–1581.

Mills, E., J. Forney, and K. Wagner. 1987. Fish predation and its cascading effect on the Oneida Lake food chain. Pages 118–131 in W. C. Kerfoot and A. Sih, editors. Predation: direct and indirect impacts on aquatic communities. University Press of New England, Hanover, New Hampshire, USA.

Paine, R. T. 1980. Food webs, linkage interaction strength,

and community infrastructure. Journal of Animal Ecology 49:667–685.

Parsons, T. R., Y. Maita, and C. M. Lalli. 1984. A manual of chemical and biological methods for seawater analysis. Pergamon, New York, New York, USA.

Peterson, I., and S. Wroblewski. 1984. Mortality rate of fishes in the pelagic ecosystem. Canadian Journal of Fisheries and Aquatic Sciences 41:1117–1120.

Pitcher, T. J., and P. J. B. Hart. 1982. Fisheries ecology. Groom Helm, Beckenham, Kent, England.

Porter, K. G. 1976. Enhancement of algal growth and productivity by grazing zooplankton. Science 192:1332–1334.

Power, M. E., and W. J. Matthews. 1983. Algae-grazing minnows (Campostoma anomalum), piscivorous bass (Micropterus spp.), and the distribution of attached algae in a small prairie-margin stream. Oecologia (Berlin) 60:328–332.

Scavia, D., G. L. Fahnenstiehl, M. S. Evans, D. Jude, and J. T. Lehman. 1986. Influence of salmonid predation and weather on long-term water quality trends in Lake Michigan. Canadian Journal of Fisheries and Aquatic Sciences 43:435–443.

Schindler, D. W. 1978. Factors regulating phytoplankton production and standing crop in the world's lakes. Limnology and Oceanography 23:478–486.

Schmitz, W. R. 1958. Artificially induced circulation in thermally stratified lakes. Dissertation. University of Wisconsin-Madison, Madison, Wisconsin, USA.

Seaburg, K. G. 1957. A stomach sampler for live fish. Progressive Fish Culturist 19:137–139.

Shapiro, J. 1980. The importance of trophic-level interactions to the abundance and species composition of algae in lakes. Pages 105–115 in J. Barica and L. Mur, editors. Hypertrophic systems. Dr. W. Junk, The Hague, The Netherlands.

Shapiro, J., and D. I. Wright. 1984. Lake restoration by biomanipulation. Freshwater Biology 14:371–383.

Sih, A., P. Crowley, M. McPeek, J. Petranka, and K. Strohmeier. 1985. Predation, competition, and prey communities: a review of field experiments. Annual Review of Ecology and Systematics 16:269–312.

Steele, J. H., and E. W. Henderson. 1984. Modeling long-term fluctuations in fish stocks. Science 224:985–987.

Sterner, R. W. 1986. Herbivores' direct and indirect effects on algal populations. Science 231:605–607.

Stewart-Oaten, A., W. W. Murdoch, and K. R. Parker. 1986. Environmental impact assessment: "pseudoreplication" in time? Ecology 67:929–940.

Tonn, W. M., and J. J. Magnuson. 1982. Patterns in the species composition and richness of fish assemblages in northern Wisconsin lakes. Ecology 63:1149–1166.

Werner, E. E., and J. Gilliam. 1984. The ontogenetic niche and species interactions in size-structured populations. Annual Review of Ecology and Systematics 15:393–425.

Wetzel, R. G. 1983. Limnology. Saunders, Philadelphia, Pennsylvania, USA.

APPENDIX I

Zooplankton composition and mean summer dry biomass (mg/m^2) in the three study lakes during 1984 and 1985. Standard errors ($n = 13$) appear in parentheses where they exceed 1 and indicate week-to-week variability.

Taxon	Paul 1984	Paul 1985	Peter 1984	Peter 1985	Tuesday 1984	Tuesday 1985
Cladocerans						
Bosmina longirostris	P*	P	N*	N	96 (23)	1
Daphnia pulex	993 (189)	448 (84)	437 (133)	611 (102)	22 (12)	43 (19)
Daphnia rosea	267 (51)	72 (22)	19 (6)	9 (1)	P	12 (6)
Diaphanosoma leuchtenbergianum	N	N	N	2	11 (5)	2
Holopedium gibberum	709 (187)	230 (153)	195 (51)	133 (47)	3	599 (324)
Copepods						
Cyclops varicans rubellus	3	N	88 (31)	5	9 (7)	N
Diaptomus oregonensis	6	29 (2)	209 (43)	62 (8)	N	P
Epischura lacustris	20 (7)	1	4	2	N	N
Leptodiaptomus siciloides	N	N	1	P	2	N
Mesocyclops edax	74 (23)	35 (7)	6	11 (3)	N	N
Orthocyclops modestus	86 (14)	26 (5)	82 (15)	8 (1)	82 (27)	21 (12)
Skistodiaptomus pallidus	N	N	4	P	N	N
Tropocyclops prasinus	P	1	N	P	126 (20)	107 (13)
Copepodites and nauplii	308 (53)	287 (40)	401 (98)	207 (32)	873 (112)	425 (84)
Rotifers						
Ascomorpha eucadis	P	P	P	P	P	1
Asplanchna	17 (7)	7	7	2	N	N
Conochiloides dossuarius	5	2	P	P	12 (3)	3
Conochilus (colonial)	34 (8)	11 (3)	5 (1)	12 (4)	N	N
Conochilus (solitary)	N	N	N	N	P	3
Filinia terminalis	2	P	5	P	P	N
Gastropus stylifer	P	P	2	P	2	P
Kellicottia spp.	P	P	P	P	2	P
Keratella cochlearis	P	2	P	P	2	2
Keratella testudo	P	P	P	P	P	P
Ploesoma	N	N	N	P	2	1
Polyarthra vulgaris	2	5	5	8	67 (9)	35 (16)
Synchaeta	N	P	2	P	4	P
Trichocerca cylindrica	2	1	N	N	12 (4)	1
Trichocerca multicrinis	N	N	N	N	5	1

* For minor taxa, P denotes present with biomass <1 mg/m^2 and N denotes not recorded.

1876   S. R. CARPENTER ET AL.   Ecology, Vol. 68, No. 6

APPENDIX II

Phytoplankton composition and mean summer biovolume density (10^3 μm^3/mL) in the study lakes during 1984 and 1985. Standard errors ($n = 13$) appear in parentheses where they exceed 1 and indicate week-to-week variability.

Taxon	Paul 1984	Paul 1985	Peter 1984	Peter 1985	Tuesday 1984	Tuesday 1985
Cyanophyta						
Anabaena circinalis	10 (2)	11 (2)	7 (2)	13 (1)	P*	N*
Anabaena planctonica	N	N	15 (7)	N	N	N
Chroococcus dispersus	1	P	P	P	2	P
Chroococcus limneticus	P	N	6 (2)	N	N	N
Dactylococcopsis smithii	P	1	P	P	N	N
Dactylococcopsis fascicularis	N	P	N	P	2	4
Gomphosphaeria lacustris	2	N	8 (2)	17 (7)	N	N
Merismopedia tenuissima	7 (3)	7 (2)	N	1	N	N
Microcystis aeruginosa	1	7 (2)	2	N	184 (33)	1
Chlorophyta						
Ankistrodesmus	P	N	N	N	N	N
Ankyra judayi	P	P	P	P	N	P
Arthrodesmus	P	N	10 (5)	N	35 (15)	4
Cerasterias staurastroides	N	N	P	N	N	N
Closteriopsis longissima	2	P	P	P	18 (5)	8 (1)
Cosmarium	N	N	19 (5)	N	N	1
Crucigenia spp.	P	4	N	4	N	N
Diceras	N	P	N	P	1	P
Dictyosphaerium pulchellum	N	N	N	P	1	P
Gloeocystis	26 (5)	19 (6)	5	6 (3)	N	P
Golenkinia	N	N	N	N	N	2
Unicellular flagellates 3–8 μm GALD†	22 (3)	13 (2)	2	7 (2)	126 (49)	43 (7)
Oocystis	59 (16)	11 (3)	33 (11)	8 (2)	N	1
Quadrigula ssp.	P	7 (1)	P	1	N	P
Scenedesmus	N	P	N	P	N	N
Selenastrum minutum	1	P	1	4	59 (12)	37 (8)
Sphaerocystis schroeteri	12 (4)	3	14 (2)	341 (108)	N	N
Spinocosmarium	N	N	N	N	1	N
Staurastrum	P	P	P	N	5	N
Chrysophyta						
Asterionella	N	N	1	7	1	1
Chrysochromulina	N	1	N	3	199 (61)	411 (96)
Chrysosphaerella longispina	31 (10)	P	37 (9)	3	229 (54)	1
Dinobryon bavaricum	2	8 (2)	P	3	3	9 (2)
Dinobryon cylindricum	3	5	16 (8)	13 (3)	P	P
Dinobryon sertularia	N	P	9 (3)	2	3	N
Dinobryon sociale var. *americanum*	P	P	P	P	2	2
Mallomonas spp.	20 (3)	29 (7)	3	5	38 (6)	146 (38)
Rhizosolenia	N	P	N	P	14	3
Synedra	N	N	P	N	N	N
Synura	33 (18)	P	18 (12)	1	14 (6)	33 (19)
Cryptophyta						
Cryptomonas spp.	48 (9)	41 (6)	17 (8)	22 (5)	41 (9)	29 (2)
Rhodomonas minutum	10 (3)	2	4	2	5	3
Euglenophyta						
Phacus	2	N	N	N	N	N
Trachelomonas	P	2	P	P	42 (8)	30 (9)
Pyrrhophyta						
Glenodinium quadridens	P	N	1	2	372 (109)	92 (28)
Peridinium cinctum	N	N	N	N	116 (22)	1
Peridinium limbatum	N	N	11	N	1181 (299)	12
Peridinium pulsillum	7	1	P	4	199 (43)	113 (31)
Peridinium wisconsinense	N	N	2	N	304 (101)	5

* For minor taxa, P denotes present with biovolume density <10 μm³/mL and N denotes not recorded.
† Greatest axial linear dimension.

| Vol. 10: 257–263, 1983 | MARINE ECOLOGY – PROGRESS SERIES
Mar. Ecol. Prog. Ser. | Published January 20 |

The Ecological Role of Water-Column Microbes in the Sea*

F. Azam[1], T. Fenchel[2], J. G. Field[3], J. S. Gray[4], L. A. Meyer-Reil[5] and F. Thingstad[6]

[1] Institute of Marine Resources, Scripps Institution of Oceanography, La Jolla, California 92093, USA
[2] Institute of Ecology and Genetics, University of Aarhus, DK-8000 Aarhus-C, Denmark
[3] Zoology Department, University of Cape Town, Rondebosch 7700, South Africa
[4] Institutt for Marinbiologi og Limnologi, Universitetet i Oslo, Postboks 1064, Blindern, Oslo 3, Norway
[5] Institut für Meereskunde, Universität Kiel, Düsternbrooker Weg 20, D-2300 Kiel 1, Federal Republic of Germany
[6] Institute for Microbiology, University of Bergen, Bergen, Norway

ABSTRACT: Recently developed techniques for estimating bacterial biomass and productivity indicate that bacterial biomass in the sea is related to phytoplankton concentration and that bacteria utilise 10 to 50 % of carbon fixed by photosynthesis. Evidence is presented to suggest that numbers of free bacteria are controlled by nanoplanktonic heterotrophic flagellates which are ubiquitous in the marine water column. The flagellates in turn are preyed upon by microzooplankton. Heterotrophic flagellates and microzooplankton cover the same size range as the phytoplankton, thus providing the means for returning some energy from the 'microbial loop' to the conventional planktonic food chain.

INTRODUCTION

Bacteria and other micro-organisms have long been known to play a part in marine ecosystems (Sorokin, 1981), but it has been difficult to study them quantitatively. Traditional methods of enumerating marine bacteria were based on plate counts, serial dilutions or phase-contrast microscopy which gave estimates of, at best, 10 % of actual numbers and have generally been discarded for estimating bacterial biomass. Recently several chemical techniques have been used to estimate biomass, these include ATP, (see review by Karl and Holm-Hansen, 1980; see also Sorokin and Lyutsarev, 1978), muramic acid (Fazio et al., 1977; Moriarty, 1977) and lipo-polysaccharides (LPS) (Watson et al., 1977). However, for estimating the biomass of natural assemblages in different physiological states these suffer from the drawback of having varying conversion factors between the cell component measured and bacterial biomass. The technique of epifluorescence microscopy has recently come into use for directly counting bacteria. The fluorescent dyes used

include Acridine Orange (Francisco et al., 1973, Hobbie et al., 1977), DAPI (Porter and Feig, 1980) and Hoechst 33825 (Paul, 1982). The above combined with scanning or transmission electron microscopy for estimating bacterial volumes, give the best present estimates of bacterial biomass (Krambeck et al., 1981).

There have also been recent developments in estimating bacterial production rates. The most promising of these are the frequency of dividing cells (FDC) (Hagström et al., 1979) though it is tedious and difficult to calibrate satisfactorily, and Tritiated Thymidine Incorporation (TTI) (Fuhrman and Azam, 1980, 1982); it yields maximum estimates because of the conservative assumptions involved. To relate production to the active fraction of bacterial assemblages, TTI autoradiography may be used in combination with epifluorescence microscopy (Fuhrman and Azam, 1982).

Thus it is now possible to obtain realistic estimates of the biomass and productivity of bacteria in the sea, and to re-examine their role critically. This paper is confined to data based on acridine orange direct counts and electron micropic estimates of biomass; it considers the new data on bacterial production.

Traditionally, bacteria have been regarded as remineralisers, responsible for converting organic mat-

* All authors are members of the working group on bacteria and bacterivory, NATO Advanced Research Institute, Bombannes, France, May, 1982

258   Mar. Ecol. Prog. Ser. 10: 257–263, 1983

ter to inorganic and recycling nutrients to primary producers. We have considered the questions: Is this true? If so, how does it occur?

This short paper covers only the role of free bacteria in the water column; similar principles probably apply to bacteria on detrital particles and in sediments.

BACTERIAL BIOMASS AND PRODUCTION

Table 1 shows some recent estimates of bacterial numbers and biomasses in different marine environments obtained using epifluorescence microscopy for numbers combined with electron microscopy for biomass estimates. There is a general trend for increasing bacterial numbers and biomass with increasing primary productivity, a correlation confirmed by measurements in a range of natural and experimental situations (E. A. S. Linley and R. C. Newell, unpubl.[*]). The estimates for a given environment vary within 10^4 to

Table 1. Biomass estimates of marine bacteria, summarized from Meyer-Reil (1982) and Es and Meyer-Reil (1982), assuming a conversion factor of 10 % from live mass to carbon equivalent

Environment	Numbers ($\times 10^8 \, l^{-1}$)	Biomass ($\mu g \, C \, l^{-1}$)
Estuaries	50	?
Coastal waters	10–50	5–200
Offshore waters	0.5–10	1–5
Deep waters	0.1	?

5×10^6 cells ml^{-1}, a relatively small range, suggesting that some homeostatic mechanism may operate. Of the bacterial biomass, up to 10 to 20 % may be attached to particles, the majority are free bacterioplankton (Hobbie et al., 1972; Azam and Hodson, 1977). Even under unusual coastal conditions, such as in estuaries or kelp beds when much mucilage is released, only 11 to 14 % of the bacterial biomass is attached to aggregates (Linley and Field, 1982).

There are only a few reliable estimates of bacterioplankton production rates because the methods have been developed so recently. These suggest that production may be 2 to 250 $\mu g \, C \, l^{-1} \, d^{-1}$ in coastal waters, with bacterioplankton production being 5 to 30 % of primary production for both coastal and offshore waters (Es and Meyer-Reil, 1982). Fuhrman and Azam (1982) estimated that bacteria consume 10 to 50 % of total fixed carbon, assuming a carbon conversion efficiency of 50 %. However, a wide range of carbon con-

version efficiencies have been reported. Pure culture experiments reviewed by Calow in Townsend and Calow (1981) – see also Koch (1971), Payne and Wiebe (1978) and Williams (1981) – give values in the range 40 to 80 % of carbon substrate being converted to bacterial carbon. Newell and co-workers (Linley and Newell, 1981; Newell et al., 1981; Stuart et al., 1982) obtained values in the range 6 to 33 % in microcosm experiments using natural seawater on a variety of types of plant debris. High carbon conversion efficiencies appear to occur under nutrient-rich (especially nitrogen-rich) conditions (Newell, pers. comm.). If this proves to be true, it suggests that bacteria consume carbon as a source of energy while scavenging nitrogen for protein synthesis. The results of Koop et al. (1982) in an in situ microcosm experiment support this concept, since they found carbon in kelp debris was incorporated in bacteria at 28 % efficiency, whilst nitrogen was incorporated at 94 % efficiency. The rate of respiration of bacteria appears to be linearly related to their growth rate (Fenchel and Blackburn, 1979), the exact relationship depending upon the carbon conversion efficiency. The proportion of carbon respired is (1 - proportion of carbon assimilated), thus present estimates range from 40 to 90 % of absorbed carbon being respired by bacteria. This calculation assumes that a negligible amount of carbon is excreted by bacteria, an assumption that is probably generally valid although there is some qualitative evidence suggesting that molecules of small size may be excreted (Itturriaga and Zsolnay, 1981; Novitsky and Kepkay, 1981).

CONDITIONS FAVOURING BACTERIAL GROWTH

Fuhrman et al. (1980) have shown that off California there is a good correlation between the distribution of chlorophyll and bacteria on a scale of tens of kilometres. Bacteria also show seasonal patterns of abundance, presumably in response to dissolved organic matter (DOM) released by phytoplankton (Meyer-Reil, 1977; Larsson and Hagström 1979). It appears that DOM is released by some species of phytoplankton as nutrients are depleted at the end of a bloom. The fraction of labile photosynthate released is probably in inverse proportion to the concentration of the limiting nutrient, as shown for nitrogen in Fig. 1 (Joiris et al., 1982; Sakshaug, pers. comm.). DOM release (in the form of glucan) is also evident on a diel scale, glucan being synthesized and excess released during daylight and being utilised for protein synthesis that continues through the dark hours (Barlow, 1982; Sakshaug, pers. comm.). These cycles of DOM release obviously have implications for bacterial

[*] Institute for Marine Environmental Research, Plymouth, England

growth. Further evidence of inter-related diel rhythms is that the specific growth rate of heterotrophic water column bacteria increases significantly during daylight and decreases at night (Sieburth et al., 1977; Azam, 1982; Fuhrman et al., unpubl.). This is supported by free amino-acid and DOC measurements showing a similar diel pattern in the water column (Meyer-Reil et al., 1979).

Natural selection appears to favour motility amongst water column bacteria (Koop, 1982), as opposed to those in marine sediments, although there is little quantitative information on this. There is evidence that bacteria show kinesis in a field 10 to 100 μm from algal cells, close enough to take advantage of DOM (Azam, in press). Under laboratory conditions in natural seawater, bacteria were observed to remain at distances of the order of 10 μm from algal cells, possibly being repelled by antibiotics produced by healthy algae. They attach mainly to moribund algae.

FACTORS LIMITING BACTERIA IN THE SEA

Since the respiration of aerobic marine bacteria is linearly related to their growth rate (Fenchel and Blackburn, 1979), and the slope of the line appears to depend on other nutrients, principally nitrogen for protein synthesis (see p. 261), it follows that the supply of either carbon for energy, or other nutrients may limit bacterial growth. However, in phytoplankton blooms DOM is often produced, and bacteria with their large surface: volume ratio are adapted to scavenging nutrients from the water at very low concentrations. The difficulties experienced in quantifying the release of DOM by algae are almost certainly due to its rapid uptake by bacteria. The same argument applies to nutrient cycling. If bacteria are so well adapted to

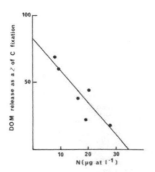

Fig. 1. Release of DOM by phytoplankton, expressed as percentage of carbon fixed in photosynthesis, and plotted against nitrogen concentration. (After Joiris et al., 1982). Similar relationships have been found for several phytoplankton species, with varying slopes

scavenging DOM and nutrients at low concentrations, what limits their population size to the biomass levels given in Table 1? Fenchel (1982a, b, c, d, in press) has shown that heterotrophic microflagellates in the size range 3 to 10 μm are effective bacteriovores in the sea, capable of filtering 12 to 67 % of the water column per day (see also Sorokin, 1979; Sieburth, 1982). These are principally choanoflagellates and colourless chrysomonads which occur ubiquitously in seawater reaching densities of more than 10^3 cells ml^{-1} (Sieburth, 1979; Fenchel, 1982a–d). Field observations have shown predator/prey oscillations between bacteria and

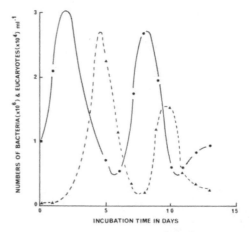

Fig. 2. Oscillations in the density of bacteria and small (3 to 10 μm) eucaryotic organisms after addition of crude oil to a 10 l sample of natural seawater. Population of eucaryotic organisms was totally dominated by small flagellates. Incubation at 15 °C in darkness. (Thingstad unpubl.)

the total flagellate fauna with a lag of 4d between bacterial and flagellate peaks (Fenchel, 1982d). In the laboratory pure cultures of flagellates show balanced growth with generation times of about 24 h at bacterial concentrations of around 10^6 cells ml^{-1} (Fenchel, 1982b).

Fig. 2 shows the predator/prey oscillations in laboratory experiments where bacterial growth in natural seawater has been stimulated by the addition of crude oil (Thingstad et al., unpubl.). In this instance oil provided an artificial and increased carbon source as nutrition for the bacteria, but in natural systems the carbon is normally produced by the release of DOM from living or moribund cells. This and mesocosm experiments conducted in 20 m high columns, 1 m in diameter, in the Lindåspolls in Norway show that as the phytoplankton bloom declined, bacterial populations built up and these in turn declined when flagellates became abundant (Thingstad, unpubl.) Newell (in press) summarizes the results of microcosm experi-

260   Mar. Ecol. Prog. Ser. 10: 257–263, 1983

ments on the degradation of DOM, kelp debris, animal faeces, phytoplankton and *Spartina* debris which all show the same successional pattern in natural seawater. Thus all the evidence to date suggests a remarkably similar pattern, with heterotrophic microflagellates controlling bacterial numbers with a lag of some 3 to 4 d between bacterial and flagellate peaks.

Physical and physiological constraints favour small organisms as bacteriovores because of their large surface-to-volume ratio which increases the probability of contact with bacteria (Fenchel, in press). A notable exception to this is provided by *Oikopleura* which have giant filters with mesh sizes of bacterial dimensions (Flood, 1978; King et al., 1980), analogous to baleen whales (Fenchel, in press). Free-living bacteria in the water column can be utilised to some extent by some larger animals such as sponges (Reiswig, 1974, 1975), and bivalves (Jørgensen, 1966; Stuart et al., 1982; Wright et al., 1982). However, bacteria are at the lower limit of efficient utilisation on an extensive scale by macrofauna – many orders of magnitude larger than the bacteria themselves. Thus while bacteria may provide nutritive supplements of some of these animals, it is doubtful whether macrofauna play a significant role in controlling bacterial populations on a large scale.

HYPOTHESIS OF THE "MICROBIAL LOOP"

Water column bacteria utilise DOM as an energy source, and this is mainly of phytoplankton origin since it has been estimated that 5 to 50 % of carbon fixed is released as DOM (see review by Larsson and Hagström, 1982). There is also a close correlation between bacterial and primary production (Es and Meyer-Reil, 1982).

Fig. 3 presents a modification of the Sheldon (1972) particle-size model to illustrate the role of bacteria and other microbes in the water column. The main feature of Sheldon's model is that organisms tend to utilise particles one order of magnitude smaller than themselves. DOM released by phytoplankton and, to a much smaller extent by animals, is utilised by bacteria. When sufficient DOM is supplied for their growth, bacteria (0.3 to 1 µm) are kept below a density of 5 to 10×10^6 cells ml^{-1}, primarily by heterotrophic flagellates which reach densities of up to 3×10^3 cells ml^{-1}. Flagellates (3 to 10 µm) probably also feed on autotrophic cyanobacteria in the same size range (0.3 to 1 µm). Cyanobacteria have rarely been counted separately but are included in most direct counts of bacteria. Flagellates eat them because they are in the particle sizes ingested. Flagellates, both autotrophic and heterotrophic, are in turn preyed upon by microzooplankton in the same size range as the larger

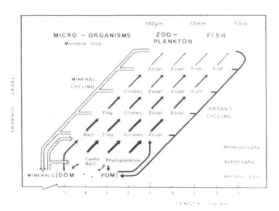

Fig. 3. Semi-quantitative model of planktonic food chains. Solid arrows represent flow of energy and materials; open arrows, flow of materials alone. It is assumed that 25 % of the net primary production is channelled through DOM and the "microbial loop", bacteria (Bact.), flagellates (Flag.) and other micro-zooplankton (e.g. ciliates). It is further assumed that the most efficient predator prey size ratio is 10:1, hence the slope of the lines relating trophic status to log body length is 1:1. The food chain base represents a size range 3 orders of magnitude (smallest bacteria 0.2 µm, largest diatoms 200 µm; therefore, any trophic level will have a size-range factor of 10^3 and conversely each size class of organisms (100 µm) will represent at least 3 trophic levels. The thickness of open arrows (left) represents the approximate relative magnitude of minerals released in excretion at each trophic level; corresponding organic losses (faeces, mucus, etc.) are shown on the right hand side

phytoplankton (10 to 80 µm). Thus energy released as DOM by phytoplankton is rather inefficiently returned to the main food chain via a microbial loop of bacteria-flagellates-microzooplankton.

DISCUSSION

An important consequence of the "microbial loop" described above, stems from the ability of bacteria to absorb mineral nutrients from the sea. Their small size and large surface-to-volume ratio allow absorption of nutrients at very low concentrations, giving them a competitive advantage over phytoplankton as documented in CEPEX bag and laboratory experiments (Vaccaro et al., 1977; Albright et al., 1980). Flagellates, by their strong predatory control of bacterial density, may play an important role in influencing the result of competition for nutrients between phytoplankton and bacteria (Thingstad, unpubl.). Bacteria sequester mineral nutrients efficiently and are consumed by flagellates with a carbon assimilation efficiency of some 60 % (Fenchel, 1982b), the remaining 40 % of carbon being egested as faeces. Furthermore, heterotrophic flagellates and other microzooplankton, excrete nitrogen and since their C/N ratios are similar to that of their food (3.5 to 4) (Fenchel and Blackburn, 1979) an amount of nitrogen corresponding to some 25 % respired carbon must also be excreted. While bacteria also excrete minerals and respire carbon, they compete efficiently to regain mineral nutrients; it therefore remains for the heterotrophic flagellates and microzooplankton to play the major role in remineralization in the sea. This is contrary to the classical view of bacteria as remineralizers, but is a view shared by Mann (1982).

The role of the 'microbial loop', including microzooplankton, is likely to be particularly important in rapidly recycling nutrients above the thermocline. This is because bacteria attached to particles and larger phytoplankton cells are likely to sink out of the mixed layer faster than the free-living (motile) bacteria (Wangersky, 1977; Kirchman et al., 1982). Evidence is provided by Fuhrman and Azam (unpubl.) who found that free and attached bacteria above the thermocline were adapted to warm temperatures by showing optimal activity at 17 °C. However, bacteria obtained from 200 m depth varied in their thermal optima, free bacteria showing optimal activity at ambient temperature implying long residence at depth while bacteria attached to particles showed optimal activity at higher temperature, suggesting recent origin from above the thermocline.

The dynamic behaviour of the 'microbial loop' is a result of several interacting ecological relationships: commensalism, competition and predation. Commensalism occurs in the production of DOM by phytoplankton and utilization by bacteria. DOM production is influenced by the availability of mineral nutrients (Fig. 1). Competition for mineral nutrients is found between phytoplankton and bacteria. This is influenced by the growth conditions for algae as well as the availability of organic substrates for bacterial growth. Predation by microflagellates on bacteria influences the outcome of the competition mentioned above. The regeneration of mineral nutrients resulting from predation will provide a feedback of some of the material flows within the "microbial loop".

From this it should be evident that carbon and mineral nutrient flows in the 'microbial loop' are tightly coupled. Information on the nature of this coupling is thus essential for an understanding of the dynamical behaviour of the 'microbial loop'. Since many marine pollution situations include the addition of mineral nutrients and/or organic carbon to the ecosystem, an understanding of the regulating mechanisms within the "microbial loop" is important.

Acknowledgements. We thank P. J. le B. Williams, C. Joiris, E. Sakshaug and R. C. Newell for stimulating discussions during the NATO Advanced Research Institute held at Bombannes, France, in May 1982. We thank Dr. M. I. Lucas, IMER, Plymouth, who generously helped design and draw Fig. 3.

LITERATURE CITED

Albright, L. J., Chocair, J., Masuda, K., Valdes, D. M. (1980). In situ degradation of the kelp *Macrocystis integrifolia* and *Nereocystis laetkeana* in British Columbia waters. Nat. Can. 107: 3–10

Azam, F. (1982). Measurement of growth of bacteria in the sea and the regulation of growth by environmental conditions. In: Hobbie, J., Williams, P. J. Le B. (eds.) Heterotrophic activity in the sea. Plenum Press, in press

Azam, F., Hodson, R. E. (1977). Size distribution and activity of marine microheterotrophs. Limnol. Oceanogr. 22: 492–501

Barlow, R. G. (1982). Phytoplankton ecology in the Southern Benguelan Current: 2. Carbon assimilation patterns. J. exp. mar. Biol. Ecol. 63: in press

Es, F. B. van, Meyer-Reil, L. A. (1982). Biomass and metabolic activity of heterotrophic bacteria. Adv. microb. Ecol., in press

Fazio, S. D., Mayberry, W. R., White, D. C. (1978). Muramic acid assay in sediments. Mar. Biol. 48: 185–197

Fenchel, T. (1982a). Ecology of heterotrophic microflagellates. 1. Some important forms and their functional morphology. Mar. Ecol. Prog. Ser. 8: 225–231

Fenchel, T. (1982b). Ecology of heterotrophic microflagellates. II. Bioenergetics and growth. Mar. Ecol. Prog. Ser. 8: 225–231

Fenchel, T. (1982c). Ecology of heterotrophic microflagellates. III. Adaptations to heterogeneous environments. Mar. Ecol. Prog. Ser. 9: 25–33

Fenchel, T. (1982d). Ecology of heterotrophic microflagellates. IV. Quantitative occurrence and importance as consumers of bacteria. Mar. Ecol. Prog. Ser. 9: 35–42

Fenchel, T. (1983). Suspended marine bacteria as food source In: Fasham, M. J. (ed.) NATO Advanced Research Institute: Flow of energy and materials in marine ecosystems. Plenum Press, in press

Fenchel, T., Blackburn, T. H. (1979). Bacteria and mineral cycling, Academic Press, London

Flood, P. R. (1978). Filter characteristics of appendicularian food catching nets. Experientia 34: 173

Francisco, D. E., Mah, R. A., Rabin, A. C. (1973). Acridine orange epifluorescence technique for counting bacteria. Trans. Am. Micros. Soc. 92: 416–421

Fuhrman, J. A. (1981). Influence of method on the apparent size distribution of bacterioplankton cells: epifluorescence microscopy compared to scanning electron microscopy. Mar. Ecol. Prog. Ser. 5: 103–106

Fuhrman, J. A., Ammerman, J. W., Azam, F. (1980). Bacterioplankton in the coastal euphotic zone: distribution, activity and possible relationships with phytoplankton. Mar. Biol. 60: 201–207

Fuhrman, J. A., Azam, F. (1980). Bacterioplankton secondary production estimates for coastal waters of British Columbia, Antarctica and California. Appl. environ. Microbiol. 39: 1085–1095

Fuhrman, J. A., Azam, F. (1982). Thymidine incorporation as a measure of heterotrophic bacterioplankton production in marine surface waters: evaluation and field results. Mar. Biol. 66: 109–120

Hagström, A., Larsson, U., Hörstedt, P., Normark, S. (1979). Frequency of dividing cells, a new approach to the deter-

262   Mar. Ecol. Prog. Ser. 10: 257–263, 1983

mination of bacterial growth rates in equatic environments. Appl. environ. Microbiol. 37: 805–812

Hobbie, J. E., Daley, R. J., Jasper, S. (1977). Use of Nucleopore filters for counting bacteria by fluorescence microscopy. Appl. environ. Microbiol. 33: 1225–1228

Hobbie, J. E., Holm-Hansen, O., Packard, T. T., Pomeroy, L. R., Sheldon, R. W., Thomas, J. P., Wiebe, W. J. (1972). A study of the distribution and activity of microorganisms in ocean water. Limnol. Oceanogr. 17: 544–555

Itturiaga, R., Zsolnay (1981). Transformation of some dissolved organic components by a natural heterotrophic population. Mar. Biol. 62: 125–129

Jørgensen, C. B. (1966). Biology of suspension feeding, Pergamon Press, Oxford

Joiris, C., Billen, G., Lancelot, C., Daro, M. H., Mommaerts, J. P., Hecq, J. H., Bertels, A. Bossicarta, M., Nijs, J. (1982) A budget of carbon cycling in the Belgian coastal zone. Relative roles of zooplankton, bacterioplankton and benthos in the utilization of primary production. Neth. J. Sea. Res., in press

Karl, D. M., Holm-Hansen, O. (1975). Methodlogy and measurement of adenylate energy charge ratios in environmental samples. Mar. Biol. 48: 185–197

King, K. R., Hollibaugh, J. T., Azam, F. (1980). Predator-prey interactions between the larvacean Oikopleura dioica and bacterioplankton in enclosed water columns. Mar. Biol. 56: 49–57

Kirchman, D., Mitchell, R. (1982). Contribution of particle bound bacteria to total microheterotrophic activity in five ponds and two marshes. Appl. environ. Microbiol. 43: 200–209

Koch, A. L. (1971). The adaptive response of Escherichia coli to a feast and famine existence. Adv. Microb. Physiol. 6: 147–217

Koop, K. (1982). Fluxes of material associated with the decomposition of kelp on exposed sandy beaches and adjacent habitats. Ph. D. thesis, University of Cape Town

Koop, K., Newell, R. C., Lucas, M. I. (1982). Microbial regeneration of nutrients from the decomposition of macrophyte debris on the shore. Mar. Ecol. Prog. Ser. 9: 91–96

Krambeck, C., Krambeck, H.-J., Overbeck, J. (1981). Microcomputerassisted biomass determination of plankton bacteria on scanning electron micrographs. Appl. environ. Microbiol. 42: 142–149

Larsson, U., Hagström, Å. (1979). Phytoplankton exudate release as an energy source for the growth of pelagic bacteria. Mar. Biol. 52: 199

Larsson, U., Hagström, Å (1982). Fractionated phytoplankton primary production, exudate release, and bacterial production in a Baltic eutrophication gradient. Mar Biol. 67: 57–70

Linley, E. A. S., Newell, R. C. (1981). Microheterotrophic communities associated with the degradation of kelp debris. Kieler Meeresforsch. 5: 345–355

Linley, E. A. S., Field, J. G. (1982). The nature and ecological significance of bacterial aggregation in a near-shore upwelling ecosystem. Estuar. coast. Shelf. Sci.

Mann, K. H. (1982). The ecology of coastal waters: a systems approach, Blackwell, Oxford

Meyer-Reil, L.-A. (1979). Bacterial growth rates and biomass production. In : Rheinheimer, G. (ed.) Microbial ecology of a brackish water environment. Springer-Verlag, Berlin, pp. 223–236

Meyer-Reil, L.-A. (1982). Bacterial biomass and heterotrophic activity in sediments and overlying waters. In: Hobbie, J. Leb, E., Williams, P. J. (eds.) Heterotrophic activity in the sea. Plenum Press, New York, in press

Meyer-Reil, L.-A., Bolter, M., Liebezeit, G., Schramm, W., (1979). Short-term variations in microbiological and chemical parameters. Mar. Ecol. Prog. Ser. 1: 1–6

Meyer-Reil, L.-A., Dawson, R., Liebezeit, G., Tiedge, H. (1978). Fluctuations and interactions of bacterial activity in sandy beach sediments and overlying waters. Mar. Biol. 48: 161–171

Moriarty, D. J. W. (1977). Improved method using muramic acid to estimates biomass of bacteria in sediments. Oecologia 26: 317–323

Newell, R. C. (1983). The biological role of detritus in the marine environment. NATO Advanced Research Institute In: Fasham, M. J. (ed.) Flow of energy and materials in marine ecosystems. Plenum Press, New York, in press

Newell, R. C., Lucas, M. I., Linley, E. A. S. (1981). Rate of degradation and efficiency of conversion of phytoplankton debris by marine microorganisms. Mar. Ecol. Prog. Ser. 6: 123–136

Novitsky, J. A., Karkay, P. E. (1981). Patterns of microbial heterotrophy through changing environments in a marine sediment. Mar. Ecol. Prog. Ser. 4: 1–7

Paul, J. H. (1982). Use of Hoechst dyes 33258 and 33342 for enumeration of attached and planktonic bacteria. Appl. environ. Microbiol. 43: 939–944

Payne, W. T., Wiebe, W. J. (1978). Growth yield and efficiency in chemosynthetic microorganisms. Ann. Rev. Microbiol. 32: 115–183

Porter, K. G., Feig, Y. S. (1981). The use of DAPI for identifying and counting aquatic microflora. Limnol. Oceanogr. 25: 943–948

Reiswig, H. M. (1974). Water transport, respiration and energetics of three tropical marine sponges. J. exp. mar. Biol. 14: 231–249

Reiswig, H. M. (1975). The aquiferous systems of three marine demospongiae. J. Morphol. 145: 493–502

Sheldon, R. W., Prakash, A., Sutcliffe, W. H. (1972). The size distribution of particles in the ocean. Limnol. Oceanogr. 17: 327–340

Sieburth, J. Mc. N. (1979). Sea microbes, Oxford University Press, New York

Sieburth, J. McN. (1982). Grazing of bacteria by protozooplankton in pelagic marine waters. In: Hobbie, J. E., Williams, A. J. Le B. (eds.) Heterotrophic activity in the sea. Plenum Press, New York, in press

Sieburth, J. McN., Johnson, K. M., Burney, C. H., Lavoie, D. M. (1977). Estimation of in situ rates of heterotrophy using diurnal changes in dissolved organic matter and growth rates of picoplankton in diffusion culture. Helgolander wiss. Meeresunters. 30: 565–570

Sorokin, Y. I. (1979). Zooflagellates as a component of eutrophic and oligotrophic communities of the Pacific Ocean. Okeanologiya 3: 476–480

Sorokin, Y. I. (1981). Microheterotrophic organisms in marine ecosystems. In: Longhurst, A. R. (ed.) Analysis of marine ecosystems. Academic Press, London, pp. 293–342

Sorokin, Y. I., Lyursarev, S. V. (1978). A comparative evaluation of two methods for determining the biomass of planktonic microflagellates. Oceanol. Acad. Sci. U.S.S.R. 18: 232–236

Stuart, V., Field, J. G., Newell, R. C. (1982). Evidence for absorption of kelp detritus by the ribbed mussel Aulacomya ater using a new ^{51}Cr-labelled microsphere technique. Mar. Ecol. Prog. Ser. 9: 263–271

Townsend, C. R., Calow, P. (eds.) (1981). Physiological ecology: an evolutionary approach to resource use, Blackwell, Oxford

Vaccaro, R. F., Azam, F., Hodson, R. E. (1977). Response of

natural marine bacterial populations to copper: controlled ecosystem pollution experiment. Bull. mar. Sci. 27: 17–22

Wangersky, P. J. (1977). The role of particulate matter in the productivity of surface waters. Helgoländer wiss. Meeresunters. 30: 546–564

Watson, S. W., Novitsky, J. A., Quinby, H. L., Valois, F. W. (1977). Determination of bacterial number and biomass in the marine environment. Appl. environ. Microbiol. 33: 940:947

Wiebe, W. J., Pomeroy, L. R. (1972). Microorganisms and their association with aggregates and detritus in the sea: a

microscopic study. In: Melchiom-Santolini, U., Hopton, J. W. (eds.) Detritus and its role in aquatic ecosystems. Mem. Ist. Ital. Idrobiol. 24 (Suppl.): 325–352

Williams, P. J. Le B. (1981). Incorporation of microheterotrophic processes into the classical paradigm of the planktonic food web. Kieler Meeresforsch. Sondh. 5: 1–28

Wright, R. T., Coffin, R. B., Ersing, C. P., Pearson, D. (1982). Field and laboratory measurements of bivalve filtration of natural marine bacterioplankton. Limnol. Oceanogr. 27: 91–98

This paper was submitted to the editor; it was accepted for printing on October 5, 1982

Ecosystem Succession and

Nutrient Retention: A Hypothesis

Peter M. Vitousek and William A. Reiners

Twenty-four broad hypotheses for trends associated with successional development of ecosystems were presented in Odum's very important paper, "The Strategy of Ecosystem Development" (1969). This series of hypotheses has helped to organize thinking on ecosystem dynamics and to focus further research. In this paper, we wish to examine one of these hypotheses, which suggests that as ecosystems mature their ability to conserve nutrients increases. Our objective is to show through logical examination of characteristics of succession, and through presentation of critical field data, that this hypothesis cannot be stated quite so simply. We also intend to show that uncritical utilization of this hypothesis, especially when combined with an unclear understanding of steady state conditions in the field, can lead to erroneous interpretations of biogeochemical behavior of ecosystems. Although our argument is developed in a forest ecosystem context, it should apply, with various modifications, to all terrestrial ecosystems.

Ecosystems are open systems in which biogeochemical functions consist of inputs from various sources, outputs to various sinks, and a variable degree of internal recycling. At the onset of succession in such a system, as in an unoccupied sand dune, for example, elemental inputs through precipitation will be more or less equaled by elemental loss through hydrologic outputs. Such a system does not have any means of conserving nutrients; any element supplied to the sand from precipitation

will be leached out and lost. The development of biomass through succession provides a mechanism for elemental uptake and, more importantly, establishes compartments, or pools, for elemental storage in biomass (Fig. 1A). As plants become established on the sand dune, elements essential to plants will accumulate in the organic matter of this simple ecosystem. As long as biomass continues to increase, uptake and storage will occur, and elemental inputs will exceed outputs (Fig. 1B). In this phase of succession, the ecosystem will demonstrate a greater capability of retaining inputs than did the unoccupied sand dune. Furthermore, since at first the growth in storage capacity of biomass and detritus pools accelerates with increasing maturity, the difference between inputs and outputs will grow, fulfilling the prediction of Odum's hypothesis.

Growth in biomass (including detritus, in this paper) cannot continue indefinitely, however. In fact, other of Odum's hypotheses describe the pattern of deceleration and cessation of growth in ecosystems (Fig. 1A). Odum suggests that ecosystems eventually reach a point where production is equal to respiration—thus net ecosystem production is equal to zero (see also Woodwell and Sparrow 1965). If this is the true pattern of biomass change and stabilization, then as storage capacity approaches a steady state, elemental outputs must again equal elemental inputs. In other words, according to this idealized view, an ecosystem will show an excess of inputs of a particular element over outputs roughly in proportion to the rate at which that particular element is bound into net ecosystem production, or the net increment of biomass (*sensu* Rodin and Bazilevich 1967). After peak net increment has been passed, this difference will decline toward zero as net growth approaches zero. Thus, a

steady state in terms of mass must show the same balance of inputs to outputs as the original unoccupied site. The presence or kind of organic matter may affect rock-weathering rates, but will not alter the essential relationships between inputs and outputs. The ability of the most mature stage of succession to conserve nutrients can be no greater than that of the original unoccupied site. Furthermore, the differences in nutrient retention between any two stages of unequal maturity will depend on their relative positions on the net increment curve as graphed in Fig. 1A. The net increment curve is skewed with a peak toward early stages of succession to be consistent with what appears to be the general pattern of net growth rate for forests (Rodin and Bazilevich 1967).

The complement of input retention—nutrient loss—is given as a family of curves in Fig. 1B. For any given element, outputs will be controlled by the amount of the element supplied through precipitation, rock weathering, and other processes, and storage changes within biomass. Both the amount of new biomass and its elemental composition would be expected to vary through succession. In forest ecosystems, for example, the elemental composition of the net biomass increment will change as the ratio of leaves to wood decreases during succession. Consequently, for forests in general the biomass composition probably becomes relatively dilute in some elements such as nitrogen.

We have used primary succession as an example in describing this model for the relationship of nutrient retention to ecosystem maturity. In primary succession, soil profile development in general represents an increase in nutrient storage capacity, contributing to input retention as ecosystems mature (Crocker and Major 1955, Olson 1958). For a whole ecosystem to be in a steady state, this process of nutrient accumulation in

Peter M. Vitousek is with the Department of Zoology, Indiana University, Bloomington, IN 47041. William A. Reiners is with the Department of Biological Sciences, Dartmouth College, Hanover, NH 03755.
This research was supported by Dartmouth College and a Sigma Xi grant-in-aid to the senior author, who himself was supported by National Science Foundation predoctoral fellowship.

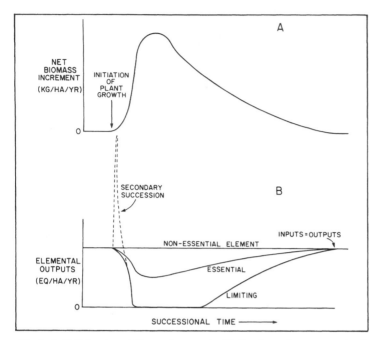

Fig. 1. A. Variation in net biomass increment with time in a primary successional sequence. Biomass accumulates rapidly at first, then the rate of biomass accretion declines gradually to zero. (From Rodin and Bazilevich 1967 and Odum 1969.) B. Variation in nutrient losses with time in a primary successional sequence. Nutrients are retained within an ecosystem as storage pools of nutrients in biomass and soil grow, but when storage pools reach steady state, nutrient outputs must equal inputs. Elements limiting to plant growth are retained most strongly; losses of these elements may decline to near zero. Elements essential but not limiting to plant growth will be significantly retained within ecosystems, but outputs will not approach zero. Outputs of elements which are nonessential to plants and which are not accumulated within ecosystems will vary little over the course of succession. The dotted line represents very high rates of nutrient loss immediately after disturbance in a secondary successional sequence. Total biomass in the ecosystem would decrease during this period of high losses.

soil must eventually approach zero so that the effect of profile development is additive to the general pattern we have hypothesized. Furthermore, in many soils, nutrient retention is to a large degree related to soil organic matter accumulation, which is included in our broad definition of biomass. The pattern of soil influences will be different for secondary succession, where increase in soil nutrient storage is far less important since the soil has already been modified by previous ecosystem processes (e.g., Mellinger[1]). Of far more im-

portance in secondary succession is the potential for temporary, rapid losses of nutrients stored in detritus and soils, particularly in organic matter which may decompose rapidly under the conditions of devegetation. Rather sensational rates of nutrient loss can be found for instances in which clear-cutting or other disturbances can permit rapid nutrient losses from detritus and soils before the rate of biomass uptake increases to where nutrients are utilized at the rate supplied by mineralization of detritus plus normal atmospheric inputs (Likens et al. 1970).

Our hypothesis for the relationship between the degree of maturity of an ecosystem and its nutrient retention properties is summarized in Figs. 1A

and 1B. We suggest that the difference between net input and output of a particular element will be proportional to the rate at which that element is incorporated into the net biomass increment for the system. Net biomass increment may be negative at the onset of secondary succession, and thus outputs of elements may be higher than inputs. In general, nutrient incorporation into biomass will increase rapidly in early succession, will reach a maximum, and will slowly decline to zero as a steady state is approached. Conversely, nutrient loss rate will be the complement of the rate of incorporation into biomass for that element.

This hypothesis can also be explained in terms of the balance between biological uptake and recycling for a given element. Recycling, the sum of all the processes that return an element from organisms to the soil, will be less than biological uptake of an element as long as that element is incorporated into net biomass increment. The difference between uptake and recycling will be greatest when net biomass increment is largest (Fig. 1A), and will decline gradually thereafter. In steady state ecosystems, recycling will equal biological uptake.

The first portion of the nutrient loss curve (Fig. 1B) describing a decrease in nutrient losses as succession proceeds has been extensively documented in a number of ecosystem studies (Johnson and Swank 1973, Pierce et al. 1972, Woodwell 1974). The inhibition of nitrification has been proposed as a mechanism controlling this decline in nutrient losses (Likens et al. 1969, Rice and Pancholy 1972, Woodwell 1974). Although there may be differences in nitrification in some of the successional sequences examined, we believe that variation in the net biomass increment is sufficient to explain most of the results reported to date.

The remainder of this paper is devoted to a test of this hypothesis through examination of our data and reexamination of other published data.

METHODS

We tested this hypothesis in watershed ecosystems in the White Mountains of New Hampshire. The plant communities in the area range from northern hardwoods forests at lower elevations through spruce-fir and fir forests at intermediate elevations to alpine tundra on the higher mountain peaks (Bliss

[1]Mellinger, M. V. 1972. Dynamics of plant succession on abandoned hay fields in central New York State. Ph.D. dissertation, Syracuse University, New York.

1963, Bormann et al. 1970). Almost all of the northern hardwoods forests and many of the spruce-fir forests have been extensively logged within the last 100 years; the high elevation fir forests and the alpine tundra, on the other hand, have been little affected by human disturbance.

We examined watersheds along the entire elevational gradient from northern hardwoods forest to alpine tundra. In this paper, we will describe results from the watersheds in the spruce-fir zone (750-1,200 meters). The spruce-fir zone is the only area where many but not all watersheds have never been subject to logging. It thus provides an area suitable for testing our hypothesis.

All of the watersheds reported on in this study were located on Mt. Moosilauke, a 1,462 m mountain only 13 km from the Hubbard Brook Experimental Forest. Watersheds on two major west-facing ridges with similar bedrock were sampled. One of the ridges has no record of logging above 750 meters, and if this ridge was ever logged, cutting must have occurred over 100 years ago (conversation, September 1974, J. W. Brown, Society for the Proctection of New Hampshire Forests). The other ridge was partially logged in 1946-1948 following severe damage in a 1938 hurricane (Brown 1958). Seven watersheds were studied on each ridge between 750 and 1,200 meters. Two of the watersheds on the recently logged ridge escaped logging, whereas the other five were in intermediate stages of succession following logging. None of the watersheds was subjected to any kind of human use at any time during this study.

Streams in all of the watersheds were sampled every two weeks during the growing season and every three to four weeks during the winter. Samples were collected in polypropylene bottles. Dissolved organic carbon concentrations were comparable with low levels found by Hobbie and Likens (1973), indicating that dissolved organic nitrogen was also low. Ammonium and nitrate were measured with an Orion ammonia electrode (Anonymous 1972). Metallic cations were analyzed by atomic absorption spectroscopy, and chloride was analyzed potentiometrically using a silver ion specific electrode.

RESULTS

Ideally, hydrologic nutrient losses from watersheds are compared in terms

of loss rate per unit area (i.e., equivalents/hectare/year). This can be done by multiplying streamwater concentrations by flow volume over time and dividing by area. If several assumptions are satisfied, however, streamwater concentrations alone can suffice for comparisons of outputs between watersheds. These assumptions are that the watersheds being compared have similar hydrologic regimes and precipitation chemistry and similar amounts of deep seepage drainage, preferably none. Hydrologic regimes include precipitation quantity and distribution in time, evapotranspiration rate, soil water retention, and storm flow characteristics. Precipitation chemistry includes the distribution of elemental concentrations by storm volume and through time. Differences in deep seepage are important if seepage alters discharge characteristics or if the chemical composition of seepage is different from that of surface runoff.

In this case, all the watersheds are located within 4 km of one another on one mountain. They are also at the same elevation, on west-facing slopes, and on the same bedrock. Their close proximity in similar physical settings minimizes the possibility of significant differences in precipitation regimes or chemistry, or in evapotranspiration. The bedrock common to all the watersheds is the massive, metamorphic Littleton Formation which is considered to be relatively free of deep seepage (Likens et al. 1967). We believe that the assumptions listed above are satisfied and that it is

therefore reasonable to use elemental concentrations as accurate indexes of outputs in comparing these watersheds.

We will emphasize nitrogen losses from these watersheds in this report because (a) nitrogen is an important, frequently limiting plant nutrient; (b) nitrification may, at times, control cation losses from some ecosystems (Likens et al. 1969); and (c) nitrogen losses changed dramatically in this region with disturbance and subsequent recovery (Likens et al. 1970, Pierce et al. 1972). Fixed nitrogen could enter these ecosystems either in precipitation or by nitrogen fixation; it could leave through denitrification or streamflow. Nitrogen fixation has been estimated by acetylene reduction in soil, litter, logs, and moss mats and found to be minimal in these highly acid coniferous forest sites (conversation, September 1974, M. Cepuran and R. L. Lambert, Department of Biological Sciences, Dartmouth College). No nitrogen-fixing lichens are present in the tree crowns. Denitrification should also be minimal in these steep, well-drained soils. Thus precipitation represents the only major input of fixed nitrogen to these ecosystems, and streamflow the only major output.

As demonstrated in Table 1, there is a substantial difference between ionic concentrations of discharge waters from the two sets of watersheds. Nitrate concentrations in particular are considerably higher in streams draining the more mature ecosystems. Although ammonium was detectable in precipitation

TABLE 1. Mean growing season (1 June-30 September 1973, 1974) streamwater concentrations from five intermediate-aged successional logged vs. nine old-aged ecosystems. Watersheds were located on west-facing slopes on Mt. Moosilauke, New Hampshire. Values are in microequivalents/liter, with standard errors of the means in parentheses.

Elements	Streamwater concentrations (μeq/1)		Ratio of concentrations	Precipitation concentrations*
	Unlogged watersheds	Logged watersheds		
NO_3^-	53(5)	8(1.3)†	6.62	40‡
K^+	13(1)	7(0.5)†	1.81	
Mg^{++}	40(4.9)	24(1.6)†	1.66	
Ca^{++}	56(4.5)	36(2.5)†	1.56	
Cl^-	15(0.3)	13(0.3)†	1.16	10
Na^+	29(2.6)	28(0.9)	1.03	

*Precipitation concentrations are derived from Likens (1970) and our own data. Concentrations are presented only when this represents the only significant input for the element.
†Difference between watershed types significant at $P < 0.01$ level.
‡Calculated assuming all fixed N in precipitation is converted to nitrate.

throughout this study, no ammonium was detectable at any time in any of these streams. Based on these concentrations, nitrate outputs from old-age ecosystems are approximately equal to precipitation inputs of fixed nitrogen when a correction for the concentrating effect of evapotranspiration is applied. This suggests that the ammonium in precipitation is quantitatively converted to nitrate within these ecosystems.

Potassium, magnesium, and calcium—all essential plant nutrients—are also retained more effectively in the successional ecosystems than in the more mature ecosystems. Concentrations of chloride, a micronutrient, are slightly higher in the streams draining the more mature ecosystems, though we doubt if this has any ecological significance. Finally, concentrations of sodium, an element not accumulated above trace levels in plant biomass in humid regions, are virtually identical in the streams draining these two ecosystem types. The similarity of concentrations of biologically less significant ions (chloride and sodium) supports our assumption that comparison of stream-water concentrations as an index of output is meaningful.

These data support the hypothesis illustrated in Fig. 1B. The variation in nitrate loss is similar to that proposed for a "limiting" nutrient. Potassium, magnesium, and calcium follow our expectations for essential but nonlimiting nutrients, and chloride and sodium (particularly sodium) follow the pattern we expected for nonessential elements.

Another line of evidence supporting the importance of biotic factors in controlling nutrient losses is seasonal variation in nutrient outputs. We would expect that elements that are retained in ecosystems by being bound up in growing biomass would have pronounced seasonal cycles with lowest losses during the growing season. Fig. 2 shows seasonal variation in nitrate concentrations of drainage waters from the watersheds described previously. The undisturbed watersheds have high, relatively constant nitrate outputs, whereas the successional ecosystems have a pronounced seasonal cycle with a growing season minimum. This seasonal cycle is consistent with our hypothesis that plant uptake regulates nitrogen losses in these ecosystems, a mechanism that would, of course, be effective only during the season of root uptake. Root uptake is particularly seasonal at the elevations studied because the soil is frozen from

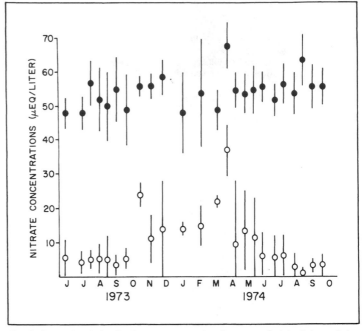

Fig. 2. Seasonal variation in nitrate concentrations in streams draining intermediate-aged successional (hollow dots) and old-aged (solid dots) forest ecosystems. Each point represents the mean of five streams for the successional forests and nine streams for the old-aged forests; the bars represent plus or minus one standard error of the mean. Nitrate concentrations are relatively constant and high in streams draining old-aged watershed ecosystems; they are low and distinctly seasonal in streams draining successional watershed ecosystems.

November to May (conversation, September 1974, T. Fahey and G. E. Lang, Department of Biological Sciences, Dartmouth College). In milder environments, the cycle may be less marked or even absent, although such a cycle is obvious at lower elevations in our region (Likens et al. 1970).

DISCUSSION

According to our hypothesis, the differences in nutrient loss rates between ecosystems of differing maturity will depend on their relative positions on the net increment curve in Fig. 1A. In the case of the data just described, the old-aged watersheds are represented somewhere near zero on the right limb of the curve, whereas the successional watersheds probably fall between the maximum on the left and the zero level on the right limb.

A similar situation is suggested by a comparison between one of the few uncut northern hardwood watersheds in New Hampshire and the younger Hubbard Brook forests (Leak 1974). Again, nitrate levels were higher in the growing

season for the old-aged forest than for the younger forests of Hubbard Brook.

A third case has been reported by Johnson and Swank (1973) for southern Appalachian forests at Coweeta Hydrologic Laboratory. They found that a 14-year-old pine watershed lost less calcium and magnesium than did a 50- to 60-year-old hardwoods watershed. They ascribed this difference to higher net productivity in younger pines. We agree, and would restate this difference as probably being caused by a higher net biomass increment in the younger vegetation, perhaps near the maximum in the curve of Fig. 1A, compared with the older hardwood forest, which is probably past the maximum for net increment. Species-specific requirements of pine relative to hardwoods may also be involved, and in this case we are probably not dealing with a single successional sequence because the white pines were planted and hardwood sprouts eliminated.

A reverse relationship between loss rates and maturity, which is consistent with this hypothesis, is expressed by data from the Hubbard Brook watershed study. In this case, loss rates from an initial, extended disturbance condition are compared with loss rates from a 60-year-old forest (Likens et al. 1970). The devegetated watershed has a negative net biomass increment, whereas the older forest is represented to the right of the maximum rate of accretion.

We would expect that the nutrients most strongly accumulated in developing ecosystems might well be the ones most rapidly lost when net biomass increment is negative. This pattern was in fact observed at Hubbard Brook: nitrate concentrations increased 57-fold after clearcutting; potassium, calcium, and magnesium concentrations increased 15.6-, 4.2-, and 4.1-fold, respectively; and sodium and chloride concentrations increased only 1.8- and 1.3-fold. The high losses of limiting and essential elements represent the inverse of the accumulation of these elements in biomass in developing ecosystems.

Based on productivity data for early stages from Marks (1974) and for the 60-year-old forest from Whittaker et al. (1974), we would predict that these very rapid loss rates will decline drastically to a minimum within 10 years after revegetation commences.

Woodwell (1974) has presented data for loss rates from two fields of abandoned agricultural land in comparison with older pine and oak forests. Nitrate losses were much greater from the two young stages than from the forest stands. The higher losses are probably caused by the imbalance of plant uptake with meteoric inputs and mineralization of stored nitrogen. Leaching of residual fertilizer may also be involved as has been demonstrated for abandoned fields in So. Carolina (Odum 1960). The abandoned fields would be represented to the left of the maximum of the net increment curve of Fig. 1A, and the forest stands to the right of the maximum point but having a higher rate of accrual than the young stages. The data presented do not permit further distinction between the two abandoned fields or between the two forest stands.

A recent noncomparative study of nitrogen losses from a 450-year-old Douglas fir-dominated watershed would seem to be a severe test of our hypothesis. Fredriksen (1972) reported that losses were low for what might seem to be a steady state system. There is no evidence, however, that total biomass has ceased to accumulate even at this old age. Old-age vegetation of this type

may still be accumulating biomass, albeit at a much lower rate than at a younger age. Indeed, this might well be the general case in the absence of natural disturbance factors. We will return to this point later.

There is another line of evidence supporting an increase in nutrient losses from steady state ecosystems as compared with intermediate-age systems. Olson (1958) examined soil and vegetation development through primary succession on Lake Michigan sand dunes. Soil nitrogen initially increased rapidly, then asymptotically approached a steady state value. Because living biomass, as indicated by basal areas, was constant over the late successional stages, it can reasonably be assumed that total ecosystem nitrogen content approached a steady state. As Olson pointed out, if the nitrogen content of the dune ecosystem is not changing, then nitrogen inputs must equal outputs. A similar, but less complete example was presented by Crocker and Major (1955) for primary succession following deglaciation in southeastern Alaska.

Evaluation of our general hypothesis requires a consideration of the nature of steady state in terrestrial, especially forest, ecosystems. One view of forest dynamics consistent with concepts of steady states in terms of mass is that forests maintain their species composition, basal area, leaf area index, and biomass in small-scale processes of individual tree death and replacement.

We suggest that the bulk of the evidence supports a contrary view: that most forests turn over in patches of various sizes controlled by intrinsic characteristics of the ecosystem or by extrinsic factors of the environment. Watt (1947) has reviewed a number of cyclic, upgrade-downgrade processes generated largely by intrinsic properties of specific plant communities. The literature is replete with evidence for sudden terminations of forest growth by natural catastrophes. Only a few may be mentioned here: Cooper (1913) for periodic blowdowns of boreal forests; Raup (1967) and Stephens[2] for hurricanes; Henry and Swan (1974) for local windstorms; Heinselman (1973), Rowe and Scotter (1974), Viereck (1973), and Kilgore (1973) for effects of fires; Morris (1963) for insect epidemics; and

Flaccus (1959) for periodic debris avalanches.

Admittedly, all of these references describe temperate forests; however, little literature is available for tropical forests to permit more widespread analysis. These examples suggest that the history of forest vegetation is one of long periods of biomass accumulation punctuated by short periods of destruction. The areas, or patches, having common origins from the most recent destructive event may vary in size from small areas resulting from the turnover of single trees to vast areas resulting from very large fires. With the present state of understanding of forest maintenance, we believe that processes generating a small patch size characteristic of tree-for-tree replacement may prevail for a few forest types all of the time, and may occur for all forest types for small periods of time, but that the prevailing processes for most forests, most of the time, generate larger scale cyclic patterns of growth and destruction.

The implications of this view of forest dynamics are obvious for the "climax" concept of community ecology (Sprugel[3], Wright 1974). Its immediate relevance to this paper, however, is in identifying situations in which watersheds may be in steady states. If the patch size of vegetation turnover is small relative to the size of a watershed, then the watershed can integrate all of the patches within it into a steady state whole. This was also the case in the old-aged spruce-fir communities that we studied, where wind-throws, a few wave-type disturbances (Sprugel[4]), and other incompletely understood natural disturbance patterns combine to generate a mosaic of small patches.

If the patch size of turnover is larger than a single watershed ecosystem, however, then it will be impossible to measure steady state conditions in a single watershed. The vegetation of such a watershed will either be growing, or it will be being rapidly destroyed. This may explain why even very old watershed ecosystems in fire-dominated regions such as the Pacific Northwest are losing less nitrogen through streamflow than they gain in precipitation (Fredriksen 1972).

[2]Stephens, E. P. 1955. The historical-developmental method of determining forest trends. Ph.D. dissertation, Harvard University, Massachusetts.

[3]Sprugel, D. G. 1974. Natural disturbance and ecosystem responses in wave-generated *Abies balsamea* forests. Ph.D. dissertation, Yale School of Forestry, Connecticut.

[4]See footnote 3.

It may be convenient to designate control watersheds in experimental studies as "mature" or "climax" in terms of a vegetational cycle, but there should be no confusion between mature or climax in the phytosociological sense and steady state in the ecosystem sense. Misuse of this terminology can lead to conceptual difficulties. For example, statements to the effect that mature systems are "tight" whereas disturbance can lead to "leakage" of nutrients are misleading. Such statements can only apply to very recently disturbed systems and should not be used to characterize immature ecosystems relative to mature ones. Furthermore, such statements only apply to certain kinds of disturbances such as those involving destruction of biomass. Steady states by definition must include elemental losses distributed in some pattern over sufficient space or time. We maintain that intermediate-aged successional ecosystems will have lower nutrient losses than either very young or very old (mature) ecosystems. We wish to emphasize that proper analysis of ecosystems, whether for elemental losses or other functions, requires a consideration of the developmental position of the ecosystem and a clear understanding of changes in net increment with time.

ACKNOWLEDGMENTS

We thank Helen Vitousek for her careful, critical reading of this paper.

REFERENCES

Anonymous. 1972. Instruction Manual, Orion Ammonia Electrode Model 95-10. Orion Research Incorporated, Cambridge, Mass. 24 p.

Bliss, L. C. 1963. Alpine plant communities of the Presidential Range, New Hampshire. *Ecology* 44:678-697.

Bormann, F. H., T. G. Siccama, G. E. Likens, and R. H. Whittaker. 1970. The Hubbard Brook ecosystem study: composition and dynamics of the tree stratum. *Ecol. Monogr.* 40:373-388.

Brown, J. W. 1958. Forest history of Mt. Moosilauke. *Appalachia* 24:23-32, 221-233.

Cooper, W. W. 1913. The climax forest of Isle Royale, Lake Superior, and its development. *Bot. Gaz.* 55:1-44, 115-140, 189-235.

Crocker, R. L., and J. Major. 1955. Soil development in relation to vegetation and surface age at Glacier Bay, Alaska. *J. Ecol.* 43:427-448.

Flaccus, E. 1959. Revegetation of landslides in the White Mountains of New Hampshire. *Ecology* 40:692-703.

Fredriksen, R. L. 1972. Nutrient budget of a Douglas-fir forest on an experimental watershed in western Oregon. Pages 115-131 *in* Proceedings–"Research on Coniferous Forest Ecosystems–A Symposium." Bellingham, Washington.

→ Heinselman, M. L. 1973. Fire in the virgin forests of the Boundary Waters Canoe Area, Minnesota. *Quat. Res. (N.Y.)* 3:329-382.

Henry, J. D., and J. M. A. Swan. 1974. Reconstructing forest history from live and dead plant material–an approach to the study of forest succession in southwest New Hampshire. *Ecology* 55:772-783.

→ Hobbie, J. E., and G. E. Likens. 1973. Output of phosphorus, dissolved organic carbon, and fine particulate carbon from Hubbard Brook watersheds. *Limnol. Oceanogr.* 18:734-742.

Johnson, P. L., and W. T. Swank. 1973. Studies of cation budgets in the southern Appalachians on four experimental watersheds with contrasting vegetation. *Ecology* 54:70-80.

Juang, F. H. F., and N. M. Johnson. 1967. Cycling of chlorine through a forested watershed in New England. *J. Geophys. Res.* 72:5641-5647.

→ Kilgore, B. M. 1973. The ecological role of fire in Sierran conifer forests. *Quat. Res. (N.Y.)* 3:496-513.

Leak, W. B. 1974. Some effects of forest preservation. *U.S. Dep. Agric. For. Serv. Res. Note* NE-186:1-4.

Likens, G. E., F. H. Bormann, N. M. Johnson, and R. S. Pierce. 1967. The calcium, magnesium, potassium, and sodium budgets for a small forested ecosystem. *Ecology* 48:772-785.

Likens, G. E., F. H. Bormann, and N. M. Johnson. 1969. Nitrification: importance to nutrient losses from a cutover forested ecosystem. *Science* 163:1205-1206.

Likens, G. E., F. H. Bormann, N. M. Johnson, D. W. Fisher, and R. S. Pierce. 1970.

Effects of forest cutting and herbicide treatment on nutrient budgets in the Hubbard Brook watershed-ecosystem. *Ecol. Monogr.* 40:23-47.

Marks, P. L. 1974. The role of pin cherry (*Prunus pensylvanica* L.) in the maintenance of stability in northern hardwood ecosystems. *Ecol. Monogr.* 44:73-88.

Morris, R. F. (ed.) 1963. The dynamics of epidemic spruce budworm populations. *Mem. Entomol. Soc. Can.* 31:1-332.

Odum, E. P. 1960. Organic production and turnover in old field succession. *Ecology* 41:34-49.

Odum, E. P. 1969. The strategy of ecosystem development. *Science* 164:262-270.

Olson, J. S. Rates of succession and soil changes on southern Lake Michigan sand dunes. *Bot. Gaz.* 119:125-170.

Pierce, R. S., C. W. Martin, C. C. Reeves, G. E. Likens, and F. H. Bormann. 1972. Nutrient losses from clear-cutting in New Hampshire. Pages 285-295 *in* Proceedings of Symposium "Watersheds in Transition." Ft. Collins, Colo., 19-22 June 1972.

Raup, H. M. 1967. American forest biology. *J. For.* 65:800-803.

Rice, E. L., and S. K. Pancholy. 1972. Inhibition of nitrification by climax ecosystems. *Am. J. Bot.* 59:1033-1040.

Rodin, L. E., and N. J. Bazilevich. 1967. Production and mineral cycling in terrestrial vegetation. Oliver and Boyd, Edinburgh. 288 p.

Rowe, J. S., and G. W. Scotter. 1973. Fire in the boreal forest. *Quat. Res. (N.Y.)* 3:444-464.

→ Viereck, L. A. 1973. Wildfire in the taiga of Alaska. *Quat. Res. (N.Y.)* 3:465-495.

Watt, A. S. 1947. Pattern and process in the plant community. *J. Ecol.* 35:1-22.

Whittaker, R. H., F. H. Bormann, G. E. Likens, and T. G. Siccama. 1974. The Hubbard Brook ecosystem study: forest biomass and production. *Ecol. Monogr.* 44:233-254.

Woodwell, G. M. 1974. Success, succession, and Adam Smith. *BioScience* 24:81-87.

Woodwell, G. M., and A. H. Sparrow. 1965. Effects of ionizing radiation on ecological systems. Pages 20-38 *in* G. M. Woodwell (ed.), Ecological Effects of Nuclear War. Brookhaven National Laboratory 917(C-43), Upton, New York.

Wright, H. E. 1974. Landscape development, forest fires, and wilderness management. *Science* 186:487-495.

4

Incorporating Trophic and Spatial Structure

Thomas E. Miller and Jordi Bascompte

Introduction

A paradigm shift occurred in ecology starting in the 1970s that greatly affected the subsequent direction of the field. The twenty or more years prior to this time was an era of competition- and niche-driven research. Many of the papers in *Ecology and Evolution of Communities* (1975) edited by Cody and Diamond are often cited as a capstone to this period, when ecologists attempted to quantify the niche and to predict patterns of species occurrence and diversity based largely on observational studies with single trophic levels ("guilds" or suites of competitors). Further, these studies often assumed communities to be at or near equilibrium, with no significant migration. For example, Pianka's (1975) work with lizards and Brown's (1975) work with rodents both provided detailed and compelling studies of resource and habitat use, quantifying the potential niche separation of species in desert guilds. Diamond (1975) proposed that species of birds adjust their niches due to diffuse competition during community assembly, predicting different species combinations at different sites. These continue to be cited as important examples of the role of competition and the niche in determining community structure.

This section of the book addresses research in two fairly different areas, trophic structure and spatial scale, each of which allowed ecology to move beyond key simplifying assumptions of earlier work. While each of these areas will be dealt with separately, what they share is allowing ecologists to scale from local units to ensembles of interacting units, either species coupled through trophic interactions or habitat patches coupled through dispersal. This, in turn led to new questions and understanding that built a more systemic view of ecological systems.

Incorporating Food Web Structure

Ecology up until 1970 largely held that more diverse communities were also more stable. Strong proponents of this view included Elton (1958) and MacArthur (1955) who, in part, argued that simple communities were likely to be more affected by disturbance or invasion. But in 1972, Robert May, building on previous numerical results by Gardner and Ashby (1970), published a very influential paper titled "Will a large complex system be stable?" [not reprinted in this volume]. He demonstrated that in randomly built model ecosystems, there is a threshold in complexity such that below that threshold communities are almost always stable, while more complex communities are almost certainly unstable. May's analytical approach allowed him to explicitly write the condition for stability as a function of both the number of species and the number and strength of their interactions. The paper stimulated intense responses, with some theoreticians finding fault in the generality of the approach (e.g., Cohen and Newman 1984) or the lack of constraints in the structure of model communities (e.g., Roberts 1974). Despite these caveats, this paper is truly classic as it shaped the way ecologists were thinking about the relationship between stability and complexity and has continued to influence the development of ecology. It is important to bear in mind that May's (1972) paper should not be read as concluding that real complex communities should be unstable, but rather that this is the baseline expectation: real food webs, therefore, must find ways to possess both complexity and stability. This is the direction taken soon by ecologists using more realistic model food webs (De Angelis 1975; Lawlor 1980) and that has continued in many current studies. May himself pointed out one path for future explorations; in the last line of the paper he noted that for a given connectance and interaction strength, a system will be more persistent if interactions are arranged in blocks or compartments and that this is "a feature observed in many natural systems" (May 1972, 414).

May's paper stimulated important work along at least two paths. One path tried to confirm or dispute the relationship between complexity and stability, while the second path built on May's suggestion about compartments and community structure. Stuart Pimm and John Lawton pursued this second path in 1980 with a study aptly titled "Are food webs divided into compartments?" They compiled a number of the empirical food webs available at the time and looked for evidence of compartments. Their approach found little support for the existence of compartments besides those associated with habitat boundaries such as those between a forest and a meadow or a lake and a forest. This

approach was clever as there was no prior clear definition of compartmentalization (or modularity as it is referred to in other fields). The problem is not only to define a measure of compartmentalization, but to assess how much would it be expected out of chance. That is, one needs to compare the empirical value with a baseline determined by random processes. To some extent, this approach is quite similar to current computational measures of modularity developed by physicists and used across many systems in complex networks (e.g., Girvan and Newman 2002; Guimerà and Amaral 2005). In this regards, Pimm and Lawton's (1980) paper was ahead of its time.

Pimm and Lawton were careful to put their results in context. Although they found little support for compartmentalization, they noted that this does not mean that food webs are not compartmentalized. They surmised that had they been able to use data on the strength of the interactions, rather than simply who interacts with whom, compartments might have been detected. They particularly cited work by Robert Paine, and this naturally links this paper with Paine's own work (Paine 1992; see discussion below) looking at the pattern of interaction strengths within real ecological communities.

There is another important aspect of Pimm and Lawton's paper: they distinguished between the possibility that compartments increase the stability of model food webs and whether those compartments are generated because they increase stability. This is an important distinction often overlooked by researchers who are quick to invoke adaptive explanations for patterns in nature (Gould and Vrba 1982). Pimm and Lawton coauthored a number of other papers on food web structure, as well individually contributing to important research in conservation biology.

Several of the caveats about food webs in Pimm and Lawton (1980) presaged later work by Gary Polis. Some of Polis's best-known work was on connections or "subsidies" between marine and terrestrial food webs (Polis et al. 1997), with fieldwork on islands in the Sea of Cortez (e.g., Polis and Hurd 1995, 1996). He was also interested in desert food webs and in how energy moved up from resources to consumers such as lizards and scorpions. His experiences led him to believe "that most cataloged webs are oversimplified caricatures of actual communities" (Polis 1991, 123). He thought that previous studies had ignored or "lumped" many species and that the diets were often assumed rather than observed, which can be especially problematic with species that experience ontogenetic diet shifts.

To make this point, Polis provided a detailed description of the interactions among the organisms found in the Coachella Valley desert in California (Polis 1991). In his paper "Complex trophic interactions in deserts: an empirical critique of food-web theory," Polis argued that this community should be fairly simple and well-documented because of its proximity to a research station. Most of his paper is a detailed description of the 175 species of plants, some 100 vertebrates, and "thousands of arthropods, parasites, and soil organisms" divided into various subwebs that make up this community. Throughout, Polis makes the point that he is necessarily simplifying his description by lumping groups or not including some whole classes of interactions. He overwhelms the reader with 6 figures of smaller portions of the community that demonstrate that this community does not fit the traditional view of simple food webs with easily identifiable trophic levels and linear food chains. Polis then compared this desert community with other published food webs. He had two orders of magnitude more species, 2–3 times longer food chains, and much higher omnivory and linkage density. In fact, it is clear that his depictions of food webs are totally different in almost all ways from those catalogued by previous researchers such as Briand and Cohen (1984).

From this study of a fairly depauperate desert community, Polis argued that previous views of food webs were "oversimplified and invalid" and that many "existing hypotheses and generalizations are simply wrong" (Polis 1991, 149). His ideas were in sharp contrast to the way most communities were viewed

and how community ecology was (and still is) taught. He continued this argument in later papers (Polis and Strong 1996; Polis et al. 2000) until his unfortunate drowning while conducting island research in the Sea of Cortez in 2000. While ecology has not completely disposed of the trophic level concept, Polis's work has stimulated ecologists to realize that real systems are much more complex than the idealized food webs of earlier work. Polis also helped to promote the recognition of omnivory as a common property of real food webs (Polis and Holt 1992; Holt and Polis 1997) and the importance of ontogenetic shifts for trophic status (Polis 1984). Work since this time, especially with stable isotope analysis, has also helped to demonstrate that trophic level might best be thought of as a continuous rather than discrete quantity (e.g., Post 2002).

Polis's extensive food webs also revealed the myriad ways in which species interactions may occur though intermediate species, which have variously been termed indirect effects and higher-order interactions (Vandermeer 1969 [reprinted in part 2 of this volume]). These "chains of interactions" are known to be important in their own right, but their presence can also make it difficult to identify and quantify simple direct interactions. This problem was both explained and partially resolved by an influential paper by Edward Bender, Ted Case, and Michael Gilpin in 1984. Their paper "Perturbation experiments in community ecology: theory and practice" continues to influence many experimental studies in community ecology.

Bender et al. (1984) began their paper by noting a common sentiment of the time: that there were significant shortcomings to observational studies of niche overlap and species interactions and that ecologists were necessarily turning to manipulative experiments. Such experiments involved manipulating the density of one species, while quantifying changes in the density or behavior of other species. The really important contribution of Bender et al. was recognizing that such experiments were being conducted in two very different ways and that these measured very different types of interactions. Press experiments removed one species and then allowed the rest of the community to achieve a new equilibrium before quantifying the effects of the removal on a remaining focal species. In doing so, press experiments measure both direct and indirect effects of one species on another through changes in the equilibrium abundance of a focal species. Pulse experiments, on the other hand, measure the short-term response in the growth rate of a focal species to the manipulation of another species' density. As such, they minimize the indirect effects of the manipulated species on the focal species that may occur via other species.

Bender et al. (1984) went on to develop mathematical models that contrasted these differences and discussed the practical applications of these concepts to experiments. They also extend these ideas to other problems with designing and interpreting experiments with multispecies systems. For example, they discuss the important problem of which species need to be considered to adequately predict community patterns ("ecological abstractions" previously discussed by Schaffer 1981) and nonlinear interactions (e.g., Abrams 1980), both of which inherently involve understanding the balance between direct and indirect effects.

While Bender et al. (1984) provided a framework for developing better future experiments, their paper also shed new light on a number of earlier experiments and theory. For example, Vandermeer's study of competition in protozoa (Vandermeer 1969) did not actually do single-species removals, but was a press experiment in that it compared different species combinations at equilibrium. Neill's (1974) experiments with aquatic communities did use press experiments, recognizing that this would necessarily include direct and indirect effects in quantifying species interactions. On the other hand, many influential competition experiments done at this time, such as Wilbur's (1972) work on higher-order interactions in salamanders, were often mixtures of press and pulse experiments because of the difficulties involved in letting species come to a new equilibrium.

The approach taken by Bender and colleagues assumed that very short-term intervals capture primarily direct effects and that longer intervals will start to include more and more indirect or higher-order effects. This illustrates both the importance of the original distinction between pulse and press experiments, but also the difficulties involved in choosing appropriate time scales for quantifying species interactions in communities. This problem was later addressed theoretically by Abrams and others (e.g., Abrams 1987, 1995; Yodzis 1988), with some guidance for practical use provided by Wootton (1994) and Laska and Wootton (1998). Also, a testimony to the vision of Bender et al.'s paper was a surge of interest in quantifying the role of indirect effects in explaining species coexistence (e.g., Saavedra et al. 2017), coevolution (e.g., Guimarães et al. 2017), and gene flow in spatially structured landscapes (e.g., Dyer and Nason 2004).

Another important paper that changed how we do experiments to quantify species interactions was Robert Paine's "Food-web analysis through field measurement of per capita interaction strength" (Paine 1992 [not reprinted in this volume]). While Bender et al. (1984) focused almost entirely on competition, Paine spent a career thinking about predation and diversity. Paine was a major figure in ecology (e.g., Paine 1966 [reprinted in Real and Brown 1991]) whose work over 40 years explicitly addressed species interactions in the context of entire food webs. Paine was primarily a field ecologist, but he strove to understand the implications of his empirical findings in the context of broad theoretical frameworks, particularly May's work on the relationship between stability and complexity. Prior to Paine (1992), experimental ecologists generally used single-species removals to estimate species interaction strengths in communities, while theoreticians had largely relied on dynamic multispecies models, often based on the approximate properties of the system very near equilibrium (captured in the Jacobian matrix [e.g., May 1973; Case 2000]). Paine wanted to integrate these approaches by proposing an experimental approach that could directly be tied to theoretical predictions, but without the assumptions inherent in what Paine called the "Jacobian approach."

Paine's (1992) paper was also a further development of his ideas about the role of predators in determining the strengths of competitive interactions and, indirectly, diversity at lower trophic levels. He originally described the disproportional effects of top predator in 1966 and proposed the term "keystone species" in 1969 (Paine 1966, 1969). He later discussed how removal experiments can quantify the strengths of interactions in an address to the British Ecological Society (Paine 1980). Like Bender et al. (1984), Paine recognized that previous theoretical and experimental work had used a number of different methods for quantifying species interactions, sometimes without recognizing the underlying mechanisms and constraints (Paine 1980). Paine wanted an experimentally practical method to quantify species interactions that recognized both biological meaning and statistical power.

The 1992 paper lays out a simple but direct manipulative approach for quantifying species interactions that is now often referred to as Paine's index (e.g., Laska and Wootton 1998; Berlow et al. 1999). To allow the effects of different species to be compared, he proposed to measure only the response of single focal basal species, such as a primary producer that outcompetes other producers for space. Single-species removals of each of the other species, especially consumers, would allow quantifying both the percent response of the focal species and the response standardized by the abundance of the removed species. He applied this method to his long-term field site in the Tatoosh Islands, outside the coast of Washington State, where sea urchins, chitins, and limpets grazed on a dominant brown alga. Perhaps the most important result was that most consumers had very little effect on the brown algae and only two species, purple urchins and a chitin, had very strong effects down the food chain to brown algae. Paine noted that the distribution of effect sizes within a community has "profound implications for community dynamics"

(Paine 1992, 73) and argued that such patterns had not been identified from previous observational studies and would require experiments such as he proposed.

While Paine's approach used a single focal species, his measure has since been extended by others to quantify many interactions within the same community (e.g., Fagan and Hurd 1994; Wootton 1997) by producing a removal matrix (Laska and Wootton 1998). Both Bender et al. (1984) and Paine (1992) encouraged ecologists to understand the differences between different ways of quantifying interactions and how they became translated in both models and practical application. This goal was continued on in later work, especially important papers by Laska and Wootton (1998) and Berlow et al. (1999), both of which continue to explain the use of Paine's index and other measures.

Dealing with Spatial Complexity

Some of the earliest experimental work recognized that coexistence may depend on spatially structured habitats (Gause 1932; Huffaker 1958 [reprinted in Real and Brown 1991]). However, most early population models made the simplifying assumptions that populations were closed to immigration and that individuals within these populations were distributed homogeneously. Richard Levins (1969) recognized that the homogeneity assumption was particularly a problem for understanding crop pests over a larger area in which there was local variation in the environmental conditions. He proposed a regional model based on the logistic equation, which predicted the equilibrium number of local sites or patches occupied based on the colonization and extinction rates of a species inhabiting such a heterogeneous landscape; in a later publication (Levins 1970), he termed this a "metapopulation" model. It represented a significant shift in how we view populations because the focus was now on local population incidence within regions, rather than on specific local abundances. This approach was also used in other publications around the same time (e.g., Cohen 1970; Levins and Culver 1971;

Horn and MacArthur 1972; Brown and Kodric-Brown 1977). Importantly, all these studies predicted that a species could exist regionally by occupying a dynamic but stable proportion of the local patches within a region. However, it appears that these early patch dynamics models were viewed as ecologically special cases and the metapopulation concept did not become widely used over the next decade. Two papers are included in this volume that expanded on the potential of the early patch dynamics models, while broadening their application and importance. In doing so, they helped to bring the metapopulation perspective to the forefront in ecology.

Ilkka Hanski is the scientist best associated with the development of the metapopulation concept. He initially published on the beetle communities that subsisted on cow dung (e.g., Hanski and Koskela 1977) and remained interested in similar fragmented landscapes throughout the rest of his life (e.g., Hanski 1994; Hanski et al. 1995). Ultimately, much of his career was spent on both theory and experimental work on metapopulations, including two books that helped to define the field. He was also very involved in conservation biology and strongly promoted the view that ecologists should participate in both the development and application of their ideas.

We could have included several different papers by Hanski in this volume. But arguably his most influential theoretical paper was "Dynamics of regional distribution: the core and satellite species hypothesis" (Hanski 1982). The earlier patch dynamics studies were interested in population persistence; how populations can achieve some equilibrium regional abundance with local patches sometimes going extinct and then being subsequently colonized. Hanski used Levins's (1970) general approach of colonization minus extinction models but allowed the probability of extinction to decrease as the proportion of patches occupied (p) increases. This seemingly small alteration in the model distinctly changes its predictions; instead of predicting an equilibrium p from between 0 and 1, the model predicts that the

equilibrium value of p will generally be either 0 or 1. Hanski described these outcomes as either being rare or satellite species (p near 0) or common core species (p near 1.0). Hanski supported this outcome with data from real communities showing that indeed species tend to either be rare or common.

Hanski (1982) used natural patterns of abundance across local environments to motivate new, testable theory. In doing so, it also linked population patterns with community patterns. Hanksi's clear logic in this and other theory papers (Hanski 1981, 1983, 1985, 1989, 1991) made him a leader in metapopulation dynamics. He also chose to test some of his own ideas: apparently, a conversation with Paul Ehrlich on butterflies stimulated Hanski to begin a very well-known and influential study of Glanville fritillary butterflies on the Åland Islands (Laine 2016). This allowed Hanksi to test metapopulation theory and has become perhaps the most cited example of a metapopulation (e.g., Hanski et al. 1995; Moilanen and Hanski 1998; Saccheri et al. 1998).

Later work by Roughgarden and Iwasa (1986), Schoener and Spiller (1987), Harrison (1991), Tilman (1994), and many others provides evidence that the metapopulation view was largely embraced by ecology, and Hanski himself has been referred to as the father of metapopulation theory. The application of the metapopulation concept became particularly important as anthropogenic land use resulted in more fragmented habitats. Hanski's later development of the incidence function model for metapopulations provided conservation biologists with a tool for predicting metapopulation persistence. Hanski was also influential in the development of the metacommunity concept (Gilpin and Hanski 1991; van Nouhuys and Hanski 2002). In fact, the patch dynamic paradigm of metacommunities (Leibold et al. 2004) builds directly on the earlier models of Levins and Hanski.

Hanski's and others' contributions inspired both ecologists and evolutionary biologists to embrace the metapopulation concept and expand the spatial scale of their studies (e.g., Dugatkin and Wilson 1991; Thompson and Burdon 1992). For example, James Brown's work extended the relationship between species abundance and distribution to explain entire species ranges (Brown 1984), Peter Karieva's experimental work with ladybugs and aphids across many host plants showed how metapopulation ideas could be applied at appropriate scales (e.g., Kareiva 1987), and Robert Holt described how the evolution of habitat selection could affect metapopulations (Holt 1985).

Ronald Pulliam (1988) modified the simple patch model to be more realistic, while also clearly expanding the importance of metapopulations for ecology and evolutionary biology. Pulliam went to graduate school when niche- and competition-based ideas were widely embraced. At the height of the influence of the Hutchinson/MacArthur niche concept, Pulliam published a wonderful study attempting to predict the coexistence of sparrow species based on bill size and predicted prey overlap (Pulliam 1975).

Pulliam's work on metapopulations, "Sources, sinks, and population regulation" (Pulliam 1988), started with the basic model used by Holt and others, which assumed generally passive dispersal among patches. In these models, individuals that disperse from a viable patch to a nonviable patch necessarily suffer a reduction in fitness. Pulliam's model instead assumed that viable patches were limited in the number of individuals they could support. Once the local habitat was past a certain density, new individuals should leave, as their expectation of individual success would be higher elsewhere, even in local patches that could be sustained only with migration.

Perhaps the most influential part of Pulliam (1988) was the clear discussion of the implications of these results. For example, this relatively simple change in the basic model meant that evolution could favor migration even in a source-sink metapopulation. Pulliam's results also meant that species could occur outside of their Grinellean niche and that the realized niche of species may be greater than their fundamental niche. Simply put, species may have persistent local populations in habitats where the resources or conditions are insufficient for their survival. In fact, the proportion of the regional population in such population sinks can even be greater than the proportion found in local population sources, which may be critical to recognize in conservation biology. This is a striking and perhaps underappreciated conclusion.

The metapopulation concept allowed ecologists to incorporate both local and regional processes into our understanding of population dynamics. A somewhat similar perspective was applied to understanding coevolutionary patterns in John Thompson's geographic mosaic theory of coevolution (Thompson 1994). In this perspective, interacting species often differ in their geographic ranges. The local populations can experience different selection, which in turn causes the evolution of traits of the interacting species to vary geographically. Extinction in local patches, along with recolonization, mutation, and genetic drift then allow trait remixing, creating patterns of selection underlying coevolving traits at the larger geographic scale.

Over the same time period, another group of ecologists was addressing similar questions but with a different perspective. Instead of viewing a landscape as having two scales (local and regional), they proposed to view scale as a continuum and proposed methods to ask what scale or scales provide the most ecological information. This was a developing area of research in the 1980s, which required a major shift in questions, field methods, and analyses (e.g., O'Neill 1986; Morris 1987), as well as the development of a new subdiscipline, landscape ecology (Naveh and Lieberman 1990).

As with several other subjects in this book, we could have chosen several foundational papers related to this view of a continuously varying spatial scale. We have included "Effects of changing spatial scale on the analysis of landscape patterns" (Turner et al. 1989) which was influential because of its clear explanation and use of the concepts grain and extent. Monica Turner and her colleagues Robert O'Neill,

Robert Gardner, and Bruce Milne analyzed the proportion of land cover types observed at a number of different landscapes across the United States, while varying both grain size (area represented by each data unit) and extent (size of the entire study area). Their primary goal was to identify general rules for comparing measures that were obtained at different spatial scales. They identify a number of problems with information loss with large grain or small extent, as well as how this loss is affected by other factors such as the spatial pattern of land types. Importantly, they propose how this loss can be predicted and then how these predictions may allow the comparison of data obtained at different spatial scales.

These concepts are further explored in three important review papers that came out soon after Turner et al. (1989). Monica Turner's review also in 1989 (Turner 1989) is both an introduction to landscape ecology and describes how to characterize the relationship between landscape patterns and ecological processes. John Wiens's (1989) and Simon Levin's (1992) papers both are convincing arguments for all ecologists to understand the relevant spatial scales of their systems. Despite these critical papers, the importance of spatial phenomena in systems and methods is often still overlooked by many ecologists.

Conclusions

The papers in this section were chosen to represent a time period when ecologists were maturing and pushing back boundaries set by previous simplifying assumptions such as isolated species and spatially closed, homogeneous systems. Collectively, these papers were somewhat ahead of their time and thus were foundational in providing guidance for later research in the natural and anthropogenic complex ecosystems around us. Many concepts from these papers, such as stability and diversity, compartmentalization, pulse experiments, source populations, and core species, are now firmly ingrained in mainstream ecology.

These papers should not be viewed as just historical documents, explaining how we got to where we are today. Instead, virtually all of these papers are as relevant today as they were when they were first published. In part, that is because some of the concepts are in areas still being developed in ecology. For example, the metapopulation ideas of Hanski are at the core of metacommunity ideas of today (Leibold and Chase 2018). In other cases, new methods of measurement and analysis are providing better tools or frameworks to quantify important parameters associated with some of these concepts. For example, the ideas inherent in Hanski's and Pulliam's papers require quantifying the rate of movement among local populations, which is benefiting from the use of new molecular methods.

Finally, there are parts of these papers that still seem undiscovered or unrecognized. One of the joys of putting together this volume has come from carefully rereading these papers and finding more than we expected. As examples, justifying the spatial scale of experiments is often neglected and many of those working with envelope or niche models for species would be well served to read Pulliam's (1988) conclusions about the relative size of fundamental versus realized niches in metapopulations.

Literature Cited

Abrams, P. A. 1980. Are competition coefficients constant? Inductive versus deductive approaches. American Naturalist 116: 730–735.

Abrams, P. A. 1987. On classifying interactions between populations. Oecologia 73: 272–281.

Abrams, P. A. 1995. Implications of dynamically variable

traits for identifying, classifying, and measuring direct and indirect effects in ecological communities. American Naturalist 146: 112–134.

Bender, E. A., T. J. Case, and M. E. Gilpin. 1984. Perturbation experiments in community ecology: theory and practice. Ecology 65: 1–13.

Berlow, E. L., S. A. Navarrete, C. J. Briggs, M. E. Power, and B. A. Menge. 1999. Quantifying variation in the strengths of species interactions. Ecology 80: 2206–2224.

Briand, F., and J. E. Cohen. 1984. Community food webs have scale-invariant structure. Nature 307: 264–267.

Brown, J. H. 1975. Geographical ecology of desert rodents. Pages 315–341 in M. L. Cody and J. M. Diamond, eds. Ecology and Evolution of Communities. Belknap.

Brown, J. H. 1984. On the relationship between abundance and distribution of species. American Naturalist 124: 255–279.

Brown, J. H., and A. Kodric-Brown. 1977. Turnover rates in insular biogeography: effect of immigration on extinction. Ecology 58: 445–449.

Case, T. J. 2000. An Illustrated Guide to Theoretical Ecology. Oxford University Press.

Cody, M. L., and J. M. Diamond, eds. 1975. Ecology and evolution of communities. Belknap.

Cohen, J. E. 1970. A Markov contingency-table model for replicated Lotka-Volterra systems near equilibrium. American Naturalist 104: 547–560.

Cohen, J. E., and C. M. Newman. 1984. The stability of large random matrices and their products. Annals of Probability 12: 283–310.

De Angelis, D. L. 1975. Stability and connectance in food web models. Ecology 56: 238–243.

Diamond, J. M. 1975. Assembly of species communities. Pages 342–444 in M. L. Cody and J. M. Diamond, eds. Ecology and Evolution of Communities. Belknap.

Dugatkin, L. A., and D. S. Wilson. 1991. Rover: a strategy for exploiting cooperators in a patchy environment. American Naturalist 138: 687–701.

Dyer, R. J., and J. D. Nason. 2004. Population graphs: the graph theoretic shape of genetic structure. Molecular Ecology 13: 1713–1727.

Elton, C. S. 1958. The Ecology of Invasions by Animals and Plants. Methuen.

Fagan, W. F., and L. E. Hurd. 1994. Hatch density variation of a generalist arthropod predator: population consequences and community impact. Ecology 75: 2022–2032.

Gardner, M. R., and W. R. Ashby. 1970. Connectance of large dynamic (cybernetic) systems: critical values for stability. Nature 228: 784–784.

Gause, G. F. 1932. Experimental studies on the struggle for existence. Journal of Experimental Biology 9: 389.

Gilpin, M., and I. Hanski, eds. 1991. Metapopulation dynamics: empirical and theoretical investigations. Academic Press.

Girvan, M., and M. E. J. Newman. 2002. Community

structure in social and biological networks. Proceedings of the National Academy of Sciences 99: 7821–7826.

Gould, S. J., and E. S. Vrba. 1982. Exaptation: a missing term in the science of form. Paleobiology 8: 4–15.

Guimarães, P. R., M. M. Pires, P. Jordano, J. Bascompte, and J. N. Thompson. 2017. Indirect effects drive coevolution in mutualistic networks. Nature 550: 511–514.

Guimerà, R., and L. A. N. Amaral. 2005. Functional cartography of complex metabolic networks. Nature 433: 895–900.

Hanski, I. 1981. Coexistence of competitors in patchy environment with and without predation. Oikos 37: 306.

Hanski, I. 1982. Dynamics of regional distribution: the core and satellite species hypothesis. Oikos 38: 210–221.

Hanski, I. 1983. Coexistence of competitors in patchy environment. Ecology 64: 493–500.

Hanski, I. 1985. Single-species spatial dynamics may contribute to long-term rarity and commonness. Ecology 66: 335–343.

Hanski, I. 1989. Metapopulation dynamics: does it help to have more of the same? Trends in Ecology & Evolution 4: 113–114.

Hanski, I. 1991. Single-species metapopulation dynamics: concepts, models and observations. Biological Journal of the Linnean Society 42: 17–38.

Hanski, I. 1994. Patch-occupancy dynamics in fragmented landscapes. Trends in Ecology & Evolution 9: 131–135.

Hanski, I., and H. Koskela. 1977. Niche relations among dung-inhabiting beetles. Oecologia 28: 203–231.

Hanski, I., T. Pakkala, M. Kuussaari, and G. Lei. 1995. Metapopulation persistence of an endangered butterfly in a fragmented landscape. Oikos 72: 21.

Harrison, S. 1991. Local extinction in a metapopulation context: an empirical evaluation. Biological Journal of the Linnean Society 42: 73–88.

Holt, R. D. 1985. Population dynamics in two-patch environments: some anomalous consequences of an optimal habitat distribution. Theoretical Population Biology 28: 181–208.

Holt, R. D., and G. A. Polis. 1997. A theoretical framework for intraguild predation. American Naturalist 149: 745–764.

Horn, H. S., and R. H. MacArthur. 1972. Competition among fugitive species in a harlequin environment. Ecology 53: 749–752.

Huffaker, C. B. 1958. Experimental studies on predation: dispersion factors and predator-prey oscillations. Hilgardia 27: 343–383.

Kareiva, P. 1987. Habitat fragmentation and the stability of predator–prey interactions. Nature 326: 388–390.

Laine, A.-L. 2016. Ilkka Hanski (1953–2016). Nature 534: 180–180.

Laska, M. S., and J. T. Wootton. 1998. Theoretical concepts

and empirical approaches to measuring interaction strength. Ecology 79: 461–476.

Lawlor, L. R. 1980. Structure and stability in natural and randomly constructed competitive communities. American Naturalist 116: 394–408.

Leibold, M. A., and J. M. Chase. 2018. Metacommunity Ecology. Princeton University Press.

Leibold, M. A., M. Holyoak, N. Mouquet, P. Amarasekare, J. M. Chase, M. F. Hoopes, R. D. Holt, et al. 2004. The metacommunity concept: a framework for multi-scale community ecology. Ecology Letters 7: 601–613.

Levin, S. A. 1992. The problem of pattern and scale in ecology: the Robert H. MacArthur award lecture. Ecology 73: 1943–1967.

Levins, R. 1969. Some demographic and genetic consequences of environmental heterogeneity for biological control. Bulletin of the Entomological Society of America 15: 237–240.

Levins, R. 1970. Extinctions. Pages 77–107 in Some Mathematical Questions in Biology, Lectures on Mathematics in the Life Sciences (Vol. 2). American Mathematical Society.

Levins, R., and D. Culver. 1971. Regional coexistence of species and competition between rare species. Proceedings of the National Academy of Sciences 68: 1246–1248.

MacArthur, R. H. 1955. Fluctuations of animal populations and a measure of community stability. Ecology 36: 533.

May, R. M. 1972. Will a large complex system be stable? Nature 238: 413–414.

May, R. M. 1973. Qualitative stability in model ecosystems. Ecology 54: 638–641.

Moilanen, A., and I. Hanski. 1998. Metapopulation dynamics: effects of habitat quality and landscape structure. Ecology 79: 2503–2515.

Morris, D. W. 1987. Ecological scale and habitat use. Ecology 68: 362–369.

Naveh, Z., and A. S. Lieberman. 1990. Landscape Ecology: Theory and Application. Springer.

Neill, W. E. 1974. The community matrix and interdependence of the competition coefficients. American Naturalist 108: 399–408.

O'Neill, R. V., ed. 1986. A Hierarchical Concept of Ecosystems. Princeton University Press

Paine, R. T. 1966. Food web complexity and species diversity. American Naturalist 100: 65–75.

Paine, R. T. 1969. A note on trophic complexity and community stability. American Naturalist 103: 91–93.

Paine, R. T. 1980. Food webs: linkage, interaction strength and community infrastructure. Journal of Animal Ecology 49: 666.

Paine, R. T. 1992. Food-web analysis through field measurement of per capita interaction strength. Nature 355: 73–75.

Pianka, E. R. 1975. Niche relations of desert lizards. Pages 292–314 in M. L. Cody and J. M. Diamond,

eds. Ecology and Evolution of Communities. Harvard University Press,

Pimm, S. L., and J. H. Lawton. 1980. Are food webs divided into compartments? Journal of Animal Ecology 49: 879–898.

Polis, G. A. 1984. Age structure component of nichewidth and intraspecific resource partitioning: Can age groups function as ecological species? American Naturalist 123: 541–564.

Polis, G. A. 1991. Complex trophic interactions in deserts: an empirical critique of food-web theory. American Naturalist 138: 123–155.

Polis, G. A., W. B. Anderson, and R. D. Holt. 1997. Toward an integration of landscape and food web ecology: the dynamics of spatially subsidized food webs. Annual Review of Ecology and Systematics 28: 289–316.

Polis, G. A., and R. D. Holt. 1992. Intraguild predation: the dynamics of complex trophic interactions. Trends in Ecology & Evolution 7: 151–154.

Polis, G. A., and S. D. Hurd. 1995. Extraordinarily high spider densities on islands: flow of energy from the marine to terrestrial food webs and the absence of predation. Proceedings of the National Academy of Sciences 92: 4382–4386.

Polis, G. A., A. L. W. Sears, G. R. Huxel, D. R. Strong, and J. Maron. 2000. When is a trophic cascade a trophic cascade? Trends in Ecology & Evolution 15: 473–475.

Polis, G. A., and D. R. Strong. 1996. Food web complexity and community dynamics. American Naturalist 147: 813–846.

Polis, G., and S. Hurd. 1996. Linking marine and terrestrial food webs: allochthonous input from the oceans supports high secondary productivity on small islands and coastal land communities. American Naturalist 147: 396–423.

Post, D. M. 2002. Using stable isotopes to estimate trophic position: models, methods, and assumptions. Ecology 83: 703–718.

Pulliam, H. R. 1975. Coexistence of sparrows: a test of community theory. Science 189: 474–476.

Pulliam, H. R. 1988. Sources, sinks, and population regulation. American Naturalist 132: 652–661.

Real, L. A., and J. H. Brown, eds. 1991. Foundations of Ecology: Classic Papers with Commentaries. University of Chicago Press.

Roberts, A. 1974. The stability of a feasible random ecosystem. Nature 251: 607–608.

Roughgarden, J., and Y. Iwasa. 1986. Dynamics of a metapopulation with space-limited subpopulations. Theoretical Population Biology 29: 235–261.

Saavedra, S., R. P. Rohr, J. Bascompte, O. Godoy, N. J. B. Kraft, and J. M. Levine. 2017. A structural approach for understanding multispecies coexistence. Ecological Monographs 87: 470–486.

Saccheri, I., M. Kuussaari, M. Kankare, P. Vikman, W. Fortelius, and I. Hanski. 1998. Inbreeding and extinction in a butterfly metapopulation. Nature 392: 491–494.

Schaffer, W. M. 1981. Ecological abstraction: the consequences of reduced dimensionality in ecological models. Ecological Monographs 51: 383–401.

Schoener, T. W., and D. A. Spiller. 1987. High population persistence in a system with high turnover. Nature 330: 474–477.

Thompson, J. N. 1994. The Coevolutionary Process. University of Chicago Press.

Thompson, J. N., and J. J. Burdon. 1992. Gene-for-gene coevolution between plants and parasites. Nature 360: 121–125.

Tilman, D. 1994. Competition and biodiversity in spatially structured habitats. Ecology 75: 2–16.

Turner, M. G. 1989. Landscape ecology: the effect of pattern on process. Annual Review of Ecology and Systematics 20: 171–197.

Turner, M. G., R. V. O'Neill, R. H. Gardner, and B. T. Milne. 1989. Effects of changing spatial scale on the analysis of landscape pattern. Landscape Ecology 3: 153–162.

van Nouhuys, S., and I. Hanski. 2002. Colonization rates and distances of a host butterfly and two specific parasitoids in a fragmented landscape. Journal of Animal Ecology 71: 639–650.

Vandermeer, J. H. 1969. The competitive structure of communities: an experimental approach with protozoa. Ecology 50: 362–371.

Wiens, J. A. 1989. Spatial scaling in ecology. Functional Ecology 3: 385.

Wilbur, H. M. 1972. Competition, predation, and the structure of the *Ambystoma–Rana sylvatica* community. Ecology 53: 3–21.

Wootton, J. T. 1994. The nature and consequences of indirect effects in ecological communities. Annual Review of Ecology and Systematics 25: 443–466.

Wootton, J. T. 1997. Estimates and tests of per capita interaction strength: diet, abundance, and impact of intertidally foraging birds. Ecological Monographs 67: 45–64.

Yodzis, P. 1988. The indeterminacy of ecological interactions as perceived through perturbation experiments. Ecology 69: 508–515.

Journal of Animal Ecology (1980), **49**, 879–898

ARE FOOD WEBS DIVIDED INTO COMPARTMENTS?*

By STUART L. PIMM† AND JOHN H. LAWTON

Environmental Sciences Division, Oak Ridge National Laboratory, Oak Ridge, Tennessee 37830, U.S.A., and *Department of Biology, University of York, Heslington, York YO1 5DD England*

SUMMARY

(1) In general, randomly constructed model food webs are less likely to be stable the more species they contain, the more interactions there are between species, and the greater the intensity of these interactions.

(2) Intriguingly, it has been argued (May 1972, 1973) that for a given interaction strength and web connectance, model food webs have a higher probability of being stable if the interactions are arranged into 'blocks' or 'compartments'; this has been coupled with the prediction that complex food webs in the real world may be similarly compartmented.

(3) Alternative food web models are briefly described. These incorporate biologically more realistic assumptions, and do not neccessarily predict that food webs are more likely to be stable if they are divided into blocks.

(4) Compartments exist in food webs if the interactions within the web are grouped into subsystems: that is, if species interact strongly only with species in their own subsystems, and interact little, if at all, with species outside it.

(5) Drawing on a number of alternative approaches, we test the null hypothesis that real food webs are not significantly more compartmented than chance alone dictates.

(6) Analyses of published food webs show that subsystems can only be detected where the webs span major habitat divisions, for example a forest and a prairie, or adjacent freshwater and terrestrial habitats. These compartments are imposed by the natural histories of the component species. There are no grounds for believing that dynamical constraints, i.e. a requirement for persistent natural food webs to be stable, play any part in imposing compartments.

(7) On a finer scale, we find no evidence for compartments in any of the food webs examined. Polyphagy in higher trophic levels may lead to a merging of detritus and grazing food chains, immediately above the level of the primary consumers. Polyphagy similarly generates non-compartmented food webs in assemblages of phytophagous insects. Several well documented food webs from other habitats are not noticeably compartmented.

(8) The implications, and limitations, of these results are discussed in the light of the general notion that loosely coupled subsystems promote ecosystem stability. On present

* Research supported by the United States National Science Foundation's Ecosystem Studies Program under Interagency Agreement No. DEB 77-25781 with the U.S. Department of Energy under contract W-7405-eng-26 with Union Carbide Corporation. Publication No. 1566, Environmental Sciences Division, ORNL.

† Current address: Department of Biological Sciences, Texas Tech. University, Lubbock, Texas 79409, U.S.A.

evidence, we conclude there are neither adequate theoretical nor convincing empirical grounds for believing that food webs are divided into compartments.

(9) These conclusions require more detailed testing using food web data which specify not only the presence, but also the strength and temporal variation of the interactions.

INTRODUCTION

Arguably the most important insight to emerge from May's (1972, 1973) analysis of model food webs is that complexity begets instability, not stability. In general, randomly constructed model food webs are less likely to be stable the more species they contain, the more interactions there are between species, and the greater the intensity of those interactions. Ecological communities persist in the real world despite, not because of, their complexity.

Real food webs are not random assemblages of species. Hence we can ask: what are the special features of real, as opposed to random food webs which tend to make the former stable and the latter unstable? The answer seems to be several things. For example, the stability of real food webs is probably enhanced by low levels of connectance between species (Rejmánek & Starý 1979); the absence of biologically absurd linkages of the type species *A* feeds on *B*, which feeds on *C*, which feeds on *A* (Pimm 1979a, b); special constraints on biomass transfer (DeAngelis 1975); a limit on the number of trophic levels (Pimm & Lawton 1977; Lawton & Pimm 1978; Pimm 1979a); a low frequency, and special patterns of omnivory (Pimm & Lawton 1978; Lawton & Pimm 1979; Pimm 1979a); and a non-random (patchy) distribution of prey and predators (Hassell 1978). One other important possibility, suggested by May (1972, 1973), has not yet been explored in detail; namely that when food webs are divided into blocks, or compartments, the probability of the resulting food webs being stable is greatly enhanced. Following May, both McNaughton (1978) and Rejmánek & Starý (1979) have recently proposed that division into discrete subsystems promotes the persistence and stability of two very different assemblages of species, plants in an African grassland (but see Lawton & Rallison 1979), and populations of aphids and their associated natural enemies. A contrary point of view has been put by Murdoch (1979) who argues that most well studied natural communities do not appear to be divided into loosely coupled subsystems.

In this paper we examine the structure of real food webs, to see if they are, or are not, divided into compartments. The webs are defined by binary data: a feeding link either exists, or it does not. Our analysis is therefore modest and preliminary, taking no account of the strength or the seasonal variation of the feeding links. If compartments can be shown to exist using binary data they will remain in more sophisticated descriptions of food webs. If compartments can not be identified using binary data, the case for the existence of subsystems within food webs is weakened, but not destroyed. We return to this problem briefly in the Discussion.

Two extreme hypotheses encompass possible web structures in the real world.

(1) The 'reticulate hypothesis': species interactions are uniformly distributed (homogeneous) throughout the system, subject to the minimal biological constraints outlined below.

(2) The 'loosely coupled subsystem hypothesis': only species within a particular subsystem interact. Between subsystems (called 'blocks' by May 1973, and 'compartments' by Pimm 1979a) there is little interaction. (In this paper, when we use the term sub-

system we shall be referring to this idea and not to any other subset of species within a system.)

The two hypotheses are illustrated in Fig. 1. Of course, food webs in the real world will neither be completely compartmented, nor totally reticulate: but if the notion that blocking promotes stability has any substance, real food webs should be more compartmented than chance alone dictates.

Deciding what chance alone dictates is not completely straightforward, because asking a computer to draw an unconstrained random food web, for example, leads to all sorts of biological absurdities: predators with nothing to feed on; autotrophes (plants) in the middle of food chains; excessively large numbers of trophic levels; thermodynamically impossible loops of the type A feeds on B, feeds on C, feeds on A, and so on. Cohen (1978) and Pimm (1979a, 1979b) discuss these problems in detail; Pimm (1979b) presents an algorithm for generating constrained random webs without biological absurdities. The key null hypothesis in this paper is that within the limits imposed by minimal biological constraints, real food webs are not significantly more blocked (compartmented) than chance alone dictates.

A closer look at the prediction that blocking enhances the stability of model food webs

May (1972, 1973) argued from considerations of connectance (the proportion of possible species interactions that are non-zero) and interaction strength (the magnitude of the non-zero interactions) that 'for a given interaction strength and web connectance (models) will do better if the interactions tend to be arranged in blocks'. By 'better' May meant more likely to be locally stable, and hence more likely to persist. His comparisons are based on the fraction of models that are asymptotically stable, the same criterion we have used elsewhere (Pimm & Lawton 1977, 1978) to make predictions about the design of food webs. However, May's comparisons are between blocked models and completely randomly organized models. When we exclude many of the biologically unreasonable phenomena found in completely random model food webs (see above) his intriguing result no longer holds. Indeed, completely compartmented models with a given level of connectance have the *least* likelihood of being stable (Pimm 1979a). The distinction between May's result and Pimm's result is important. The former specifically predicts that compartments should be present in real food webs, the latter that they should not.

Note both Pimm and May (and authors cited therein) agree that low connectance enhances a model's chances of being stable. As connectance is lowered, models will tend to become more compartmented. However, connectance and blocking are independent; when the former is fixed the latter can vary extensively (Fig. 1). The destabilizing influence of high connectance should not be attributed to a lack of compartmentalization.

Whether May's or Pimm's result is a more accurate description of the real world is probably best settled by an analysis of real webs rather than a pedantic discussion of their models' assumptions. Finding out what happens in real webs is one of the principle objectives of this paper.

May's result is a specific example of a general class of models predicting structure in food webs for dynamical reasons. It is important to realize that the same structures, or patterns (e.g. blocking) might just as easily be generated by biological, or natural history constraints. Examples in the real world of food web structures predicted by stable models would then merely be consistent with the requirements for stability; they would not be a consequence of those requirements. Compartments in food webs could presumably be

882   *Are food webs compartmented?*

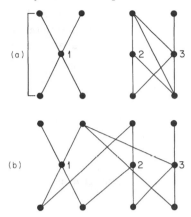

F IG. 1. Two webs with identical numbers of species, connectance, species on which no other species feed (top-predators), species which feed on nothing else within the web (basal species) and species feeding on more than one trophic level (omnivores). System (a) is a compartmented system, (b) is not.

generated, for example, because species within a habitat are more likely to interact with other species within that habitat, rather than those outside it. Each habitat requires a set of adaptations from its component species, with specialization precluding extensive interactions between habitats. For similar reasons, feeding on detritus might preclude feeding on live plant material and vice versa with the corollary that species in the grazing and detritus food chains should be separate (Odum 1962, 1963).

Obviously, subsystems whose boundaries correspond to different habitats might arise for either dynamical or biological reasons or both. For an unequivocal demonstration of dynamical constraints forcing food webs into blocks we must look at whether sets of species are compartmented within habitats. But what is a habitat? We cannot answer this objectively, but feel we can come sufficiently close for present purposes. Defining a habitat depends entirely on the size and activity of the organism(s) in question (Elton 1966). Southwood (1978) neatly illustrates the problem by correlating the size of an animal with its range. Thus, the individual grasses and shrubs in a field may represent separate habitats for herbivorous insects; but a lark will view the same grassy field as a different habitat from the nearby wood, and ignore changes in the abundance of the field's constituent grasses and shrubs; finally, a predatory hawk may include both the field and the nearby wood in its territory. What constitutes a habitat will depend very much on the taxonomic group of the organism and its trophic position. Conceivably a system may be compartmented at one trophic level but reticulate at the next.

A SEARCH FOR COMPARTMENTS COINCIDENT WITH HABITAT BOUNDARIES

In this section we will consider whether species interactions are more frequent within than between habitats. First, we will examine insects and whether their interactions form subsystems whose boundaries correspond to differing resources: plants in the first example, gall forming insects in the second. Despite widespread interest in ecosystem stability, the data available to answer our questions are few, a point to which we will

S. L. PIMM AND J. H. LAWTON   883

return below. Second, we will examine whether major habitat divisions (forest versus prairie, for example) impose a subsystem structure upon larger organisms. Finally, we will examine interactions within and between grazing and detritus food chains.

Plants and their insects as subsystems

Direct evidence

Data on the herbivorous insects and their predators and parasitoids for two or more species of plant in the same local area are few. We have two examples:

(a) The first provides quantitative as well as qualitative information about the degree of compartmentalization. Shure (1973) studied two plants, *Raphanus raphanistrum* (Linn.) and *Ambrosia artemisiifolia* (Linn.), dominant in the initial stages of an old field succession. In different experiments the two plant species were labelled with ^{32}P and the amount of label in the herbivorous and predatory insects measured over subsequent weeks.

Did the two plants and their insects behave as two separate subsystems? Figure 2 is our interpretation of the pathways taken by the ^{32}P label, based on one experiment with *Raphanus* and three with *Ambrosia*. Thus, 54% of the label placed initially on *Ambrosia* and recovered in herbivores was in species that fed on both species of plant. The corresponding figure for *Raphanus* was 66%. Shure (1973) did not indicate all the feeding preferences of the predators so we cannot be certain from which of each group of herbivores (common to both plants; unique to one plant) each of the two predatory groups (common to both plants; unique to one plant) received the label. However, some of the relationships are described in his text and these lead us to believe that predators unique to each plant species fed almost entirely on herbivores unique to each plant species. The herbivorous prey of predators common to both plants are less certain. Hence, in Fig. 2 we have indicated that predators common to both plants received the label both from herbivores common to both plants and from herbivores unique to each plant. Whichever interpretation is correct, 72% of the label placed in *Ambrosia* and subsequently recovered in predators was in species common to both plants; the corresponding figure for *Raphanus* is 28% (there is no significance in the complementarity of these percentages).

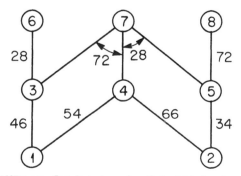

FIG. 2. Patterns of ^{32}P tracer flow in two species of plant (*Ambrosia* and *Raphanus*) in an old field system. For further discussion of how numbers were derived see text. Key: 1, *Ambrosia*; 2, *Raphanus*; 3, herbivores restricted to 1; 4 herbivores common to 1 and 2; 5, herbivores restricted to 2; 6, predators restricted to 3; 7, predators feeding on 4 and/or predators which feed on both 3 and 5; 8, predators restricted to 5. After Shure (1973).

884   *Are food webs compartmented?*

TABLE 1

Predators

	20	21	22	23	24	25	26	27	28	29	30	31	32	33	34	35	36	37
1	2	2							1		1							
2			−1	−1	2			2		1	1	1	1					
3	2		2			−1			1					1				
4			2						−1									
5						−1								1				
6																		
7			2															
8	2		2											1				
9									−1									
10	2		2															
11			−1															
12		2	2						1				1	1				1
13		2							−1							1		1
14		2							−1		1							1
15		2												1				1
16											1							
17			−1			−1			1			1	1	1		1	1	
18	2						−1		−1			1	1	1		1	1	
19	2		2				−1		1			1	1					
20									1			1	1	1		1	1	
21									1			1		1		1		1
22									1					1			1	1
23									1			1	1		1	1		
24									1			1	1		1	1		
26									1								1	1
28																	1	1
30												1						
31													1				1	1
32														1			1	1
33																		
35																1		
36																	1	
37									1					1				
39									1					1				
42																		
43														1				
45									1									
47																		1

Row and column numbers correspond to species, as described below. Table entries are: 1, if species in column is the predator or parasite of species in row; 2, if species in column kills, but does not eat species in row; −1, if species in column eats gall in row, but does not kill it (from Askew, 1961: see text for details).

Species number	Species	Species number	Species
		Cynipidae	
1	*Andricus ostreus* (Htg.)	14	*N. albipes* (Schenck)
2	*A. kollari* (Htg.)	15	*N. tricolor* (Htg.)
3	*A. curvator* (Htg.)	16	*N. aprilinus* (Giraud)
4	*A. callidoma* (Htg.)	17	*Cynips longiventris* Htg.
5	*A. inflator* (Htg.)	18	*C. divisa* Htg.
6	*A. quercus-ramuli* (L.)	19	*C. quercus-folii* L.
7	*A. albopunctatus* (Schlech.)	20	*Synergus nervosus* Htg.
8	*A. quadrilineatus* (Htg.)	21	*S. albipes* (Htg.)
9	*A. solitarius* (Fonsc.)	22	*S. gallae-pomiformis* (Boyer de Fonsc.)
10	*A. seminationis* (Giraud)	23	*S. umbraculus* (Oliv.)
11	*Biorhiza pallida* (Oliv.)	24	*S. reinhardi* (Mayr)
12	*Neuroterus quercus-baccarum* (L.)	25	*S. evanescens* (Mayr)
13	*N. numismalis* (Fourc.)	26	*S. pallicornis* Htg.
		27	*Ceroptres arator* (Htg.)

S. L. Pimm and J. H. Lawton   885

38	39	40	41	42	43	44	45	46	47	48	49	50	51	52	53	54	55	56
	1			1														
	1																	
		1																
			1															
	1	1				1												
	1												1					
	1													1	1	1		
					1													
					1													
					−1													1
																	1	
								1	1									
1																		
												1						
												1						
												1						
										1	1							

Species number	Species		Species number	Species
		Chalcidoidea		
28	*Eurytoma brunniventris* (Ratz.)		42	*O. skianeuros* (Ratz.)
29	*Megastigmus stigmatizans* (F.)		43	*Syntomaspis notata* (Walk.)
30	*Mesopolobus fuscipes* (Walk.)		44	*S. apicalis* (Walk.)
31	*Torymus cingulatus* Nees		45	*S. cyanea* (Boh.)
32	*T. nigricornis* Boh.		46	*Eupelmus uroznus* Dalman
33	*T. auratus* (Fourc.)		47	*Caenacis divisa* (Walk.)
34	*Megastigmus dorsalis* (F.)		48	*Cecidostiba leucopeza* (Ratz.)
35	*Mesopolobus fasciiventris* Westw.		49	*C. semifascia* (Walk.)
36	*M. jucundus* (Walk.)		50	*Hobbya stenonota* (Ratz.)
37	*M. tibialis* (Westw.)		51	*Pediobius lysis* (Walk.)
38	*Hobbya kollari* Askew		52	*Ormocerus latus* Walk.
39	*Olynx arsames* (Walk.)		53	*O. vernalis* Walk.
40	*O. gallaram* (L.)		54	*Pediobius clita* (Walk.)
41	*O. euedoreschus* (Walk.)		55	*Tetrastichus aethiops* (Zett.)
			56	*Eudecatoma biguttata* (Swed.)

Interpreting these figures is not easy because there is no obvious null hypothesis, i.e. we have no way of knowing what percentages of the label should be expected in each of the groups in Fig. 2 if either the reticulate or subsystem hypothesis were the more appropriate description of reality. One could easily argue that the data indicate compartmentalization, or the converse. However, since an average of 50% or more of the label goes to species shared by the two plants at each consumer level, our interpretation is that the data do not support the hypothesis of two obvious compartments.

(b) A second example is provided by the data on leaf galls and their natural enemies (mainly parasitoids and hyper-parasitoids) described by Askew (1961). All are gall formers on the oak (*Quercus robur*, Linn.) from which we have extracted the information for one locality (Wytham Woods, Oxford, England). They form one of the most detailed and carefully gathered sets of data with which we can test for the existence of compartments in food webs. Arguably an oak tree forms one habitat, in which case the following analysis more properly belongs in the next major section. Instead, we have chosen to regard each kind of gall as a potentially isolated subsystem (habitat) for its specific producers, lodgers and enemies. It makes no difference to our thesis which approach is adopted.

Although the different gall-forming species could in theory give rise to different subsystems, examination of these data (Table 1) suggest that most species share many enemies (which for convenience we will label 'predators'). However, they also show the existence of two groups with no shared interactions, and even better separation might be achieved by considering congeners as subsystems; indeed we *could* establish a number of subsystems within the data by inspection and *a posteriori* reasoning, and hence find support for the subsystem hypothesis. However, we consider this a poor procedure. The potential number of ways of organizing the interactions between these species is very large and we expect chance alone to generate some compartments. Hence, the *a posteriori* recognition of subsystems is at best a statistically risky procedure. What we need to know is whether a particular web is more compartmented than chance alone dictates. We require an objective statistical test, with a distribution we know, or at least one which we can approximate. To do this consider some features of Fig. 1.

In the system with two subsystems (a), notice that species 1 shares neither predator nor prey species with the species of the other subsystem (2 and 3). However species 2 and 3 share both predators and prey. The numbers of predators and prey shared are shown in Table 2a. In the reticulate model (b), species 1 shares predators with species 3, and shares prey with species 2. Species 2 and 3 share a prey species but they do not share predators (Table 2b). In general, there is a tendency to share prey if predators are shared and vice versa only if the system is compartmented. If the system is reticulate, then the numbers of predators shared between two species is independent, or negatively correlated with the number of prey shared.

There is one modification that must be made to this analysis before it can be applied to Askew's data. In the data for the oak-gall food webs there is a positive correlation between the number of predators a species suffers and the number of species of prey which it utilizes. A positive correlation between the number of prey types attacked by a species, and the number of predators which in turn attack it is not usual in food webs as a whole (S. L. Pimm, in prep.) but, for whatever reason, it emerges very clearly in Askew's data. The effect of species having different numbers of prey and predators can be factored out by calculating the expected number of prey (or predators) the two species should share. This requires only a knowledge of the number of species of prey (or predators)

S. L. Pimm and J. H. Lawton   887

Table 2. Observed and expected numbers of prey and predators shared by
three species in Fig. 1. Predators shared are above diagonal, prey shared
below diagonal

Observed (a)   A system with two subsystems

		Species		
		1	2	3
	1	–	0	0
Species	2	0	–	1
	3	0	2	–

Observed (b)   A system without distinct subsystems

		Species		
		1	2	3
	1	–	0	1
Species	2	1	–	0
	3	0	1	0

Expected (for both webs)

		Species		
		1	2	3
	1	–	$\frac{1}{2}$	1
Species	2	1	–	$\frac{1}{2}$
	3	1	1	–

exploited by each species and the total number of prey (or predators) within the system. Expected values have been calculated in this way for the systems in Fig. 1, and are presented in Table 2. The difference between the observed and expected values are both negative or both positive for the prey and predators of the two species being compared when and only when there are subsystems within the larger web. The differences are independent if the web is not divided into blocks. The number of cases where 'more prey are shared than expected and more predators are shared than expected' and 'fewer prey are shared than expected and fewer predators are shared than expected' can then be analysed using a standard Chi-square test.

Unlike the example shown in Fig. 1 Askew's data are further complicated by the web's complexity, which leads to the same species appearing simultaneously in the list of intermediate species, victims, and exploiters. However, ignoring this complication, which probably makes very little difference to the analysis, shows that the difference between the observed and expected number of predators shared is independent ($P > 0.05$) of the difference between the observed and expected number of prey shared. We conclude that the system is not more compartmented than we would expect by chance.

Indirect evidence

That these two examples exhaust our evidence for and against the existence of subsystems within plants and their insect faunas might seem surprising. It is therefore worth considering the sorts of information we can and cannot use to refute the hypothesis that food webs are not compartmented more than we would expect by chance.

Tests of the type used on Askew's data require detailed information on more than two trophic levels because we attempt to identify subsystems by using the interactions between two trophic levels, and then test them on a third trophic level. Adequate data are extremely scarce because few workers have the necessary time, skill and patience to collect information on large numbers of herbivores, their predators, parasitoids and hyper-parasitoids at one locality. There are numerous lists of food plants, herbivores and their natural enemies for larger geographical regions (e.g., Lawton & Schröder 1977; Lawton & Price 1979), but for obvious reasons these are not adequate to test hypotheses about food webs at particular localities within that region.

Similarly, there are many studies which list a taxonomically restricted set of herbivores and the plants they utilize. Cohen (1978) calls these 'sink webs' and presents examples. Unfortunately, sink webs cannot be used to make inferences about subsystems either. Even if the species show complete monophagy the importance of polyphagous species of different taxa is, by definition unstudied and therefore unknown. Clearly one requires a list of plant species and *all* their herbivores. These are 'source webs' and Cohen (1978) has but one example, centred upon a single species of plant.

A necessary (but not sufficient) condition for food webs to be compartmented is that data across two trophic levels indicate a subsystem structure. However, as we pointed out above, data on at least the third trophic level are needed to establish the existence of compartments. Attracted by the large number of monophagous or oligophagous phytophagous insects in certain taxa (e.g., Lawton & McNeill 1979) we earlier, tentatively suggested that food webs based on green plants might well be compartmented (Lawton & Pimm 1978). In so doing, we overlooked the fact that each of the polyphagous species of insects associated with a particular species of plant would feed on an idiosyncratic selection of alternative hosts in the habitat, a point which emerges clearly from the work of Futuyma & Gould (1979) on insect-plant associations in a deciduous forest. Hence even without the complications imposed by a third trophic level, it now appears that oligophagous and particularly polyphagous species blur any incipient compartmentalization generated by the monophagous species. This inference is supported by several studies (e.g. Gibson (1976), Hansen & Ueckert (1970); Joern (1979) and Shelden & Rogers (1978), all of which show considerable overlap in the plant species used. The data certainly do not give a clear impression of compartmentalization across two trophic levels; therefore the webs as a whole cannot be compartmented.

Vertebrates and major habitat divisions as subsystems

We have located four studies (Table 3) which describe species interactions within and between major habitat divisions. These divisions were established *a priori* by the original authors. If food webs are not divided into blocks centred on each habitat there should be as many interactions between subsystems as within subsystems, relative to the number of species in each subsystem. Figure 3 is based on the work of Summerhayes & Elton (1923) who described the major interactions within and between three habitats (terrestrial, freshwater and marine) on an arctic island. In this study the species interactions do appear to be grouped into three subsystems, though evidently some species feed across the boundaries.

In the studies we shall analyse, each species was placed in only one habitat by the original authors, even though some, or all, of its prey might come from another habitat. This suggests the following analysis:

Consider two habitats A and B, with S_A and S_B species in each. Further, suppose there are P_A and P_B interactions within each habitat and Q interactions between habitats. The average number of interactions per species within subsystems (habitats) is:

$$(P_A + P_B)/(S_A + S_B). \tag{1}$$

S. L. Pimm and J. H. Lawton

Table 3. Observed (Q) and expected ($\hat{Q}$) numbers of interactions between two habitats (A, B) with P interactions among S species in each

Habitat A	Habitat B	P_A	P_B	S_A	S_B	Q	$\hat{Q}$	χ^2	Source
Marine	Terrestrial	9	29	8	19	1	15·85	13·91**[a]	Niering (1963)
Marine	Terrestrial	21	36	26	39	11	27·36	9·78**	Koepcke & Koepcke (1952)
Aspen forest	Prairie	17	14	15	12	3	15·31	9·90**	Bird (1930)
Aspen forest	Willow forest	17	14	15	11	1	15·13	13·20**	Bird (1930)
Willow forest	Prairie	14	14	11	12	1	13·97	12·05**	Bird (1930)
Marine	Terrestrial	2	26	3	17	3	7·14	2·40NS	Summerhayes & Elton (1923)
Marine	Freshwater	2	10	3	9	0	4·50	4·00*	Summerhayes & Elton (1923)
Terrestrial	Freshwater	26	10	17	9	5	16·30	7·83**	Summerhayes & Elton (1923)

[a] NS not significant, * significant at $P < 0.05$, ** significant at $P < 0.01$.

890   *Are food webs compartmented?*

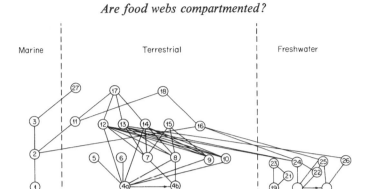

FIG. 3. Patterns of interaction described by Summerhayes & Elton (1923) within and between three arctic subsystems. Key: 1. plankton, 2. marine animals, 3. seals, 4. (a) plants (b) dead plants, 5. worms, 6. geese, 7. Colembola, 8. Diptera, 9. mites, 10. Hymenoptera, 11. sea-birds, 12. snow-bunting, 13. purple sandpiper, 14. ptarmigan, 15. spiders, 16. ducks and divers, 17. arctic fox, 18. skua and glaucous gull, 19. planktonic algae, 20. (a) benthic algae (b) decaying matter, 21. protozoa, 22. protozoa, 23. invertebrates, 24. diptera, 25. other invertebrates, 26. Lepidurus, 27. polar bear.

The number of interactions between habitats should be proportional to the relative sizes of the numbers of species in each habitat, i.e. to

$$2\left(\frac{S_A}{S_A + S_B}\right)\left(\frac{S_B}{S_A + S_B}\right). \tag{2}$$

The factor two indicates that the interactions can go from A to B and from B to A. Since the expression in (1) provides a measure of the proportion of interactions to species (an underestimate, because it ignores interactions between habitats, Q), (1) times (2) provides a conservative estimate of the expected number of interactions per species between the two habitats. Thus (1) times (2) times the number of species in the system is the expected number of interactions between habitats, $(\hat{Q})$:

$$Q = \frac{2(P_A + P_B)S_A S_B}{(S_A + S_B)^2}. \tag{3}$$

The observed value, $\hat{Q}$, and the expected value, Q can be compared using a Chi-square test; the results are shown in Table 3. For all but one comparison there are fewer interactions between the subsystems than we would expect by chance. We conclude that these large foodwebs are compartmented, with compartment boundaries matching habitat boundaries.

Grazing versus detritus food chains as subsystems

Several studies have compared the flows of energy or radioactive tracers in grazing and detritus food-chains (Odum 1962, 1963). For small invertebrates, detritus and green plants form two markedly different subsystems and so provide a basis for the compartmentalization of food chains implied by Odum's 'Y-shaped' energy flow diagram. In a typical example, Marples (1966) labelled separately both *Spartina alterniflora* (Loiseleur-Deslongchamps) and the surrounding mud with ^{32}P in a salt marsh in Georgia, U.S.A. There was virtually no overlap in the primary consumers (herbivores and detritivores) but the predators in the system, spiders, fed on both the food chains. This pattern of distinct pathways at the primary consumer level merging at the secondary consumer level (and higher) is found in a range of terrestrial and aquatic systems:

(1) the arctic habitats of Summerhays & Elton (1923);
(2) the land-sea interface of Koepcke & Koepcke (1952);
(3) the temperate forest of Varley (1970);
(4) the oligotrophic lake of Morgan (1972) and Blindloss *et al.* (1972);
(5) the eutrophic lake of Burgis *et al.* (1973) and Moriarty *et al.* (1973);
(6) the river of Mann (1964, 1965) and Mann *et al.* (1972).

In short, the grazing and detritus food chains are separate subsystems only at the primary producer/detritus and primary consumer level. Above that, the subsystems are linked by common predators, although the intensity of this linkage might repay further study. We would be surprised if there were typically as many predatory links between the detritus and green plant based food chains as within them, particularly where the detritus food chain is well separated from the plants, for example between the floor and canopy of a forest. Unfortunately, the data are not adequate for us to conduct an analysis comparable to that in the previous section. At best, detritus and green plants appear to form fuzzy compartments linked by common predators.

A SEARCH FOR SUBSYSTEMS WITHIN HABITATS

In the previous section we found some evidence for the compartmentalization of food-webs based on physically distinct habitats, but the compartments are often blurred and difficult to demonstrate. Sometimes we cannot find them at all.

Where 'habitat imposed' compartments do exist within ecosystems, they may conceivably play a part in maintaining the stability of such systems. The really important question is whether dynamical constraints force food webs into compartments independently of the habitat imposed subsystems already discussed. In other words, to test May's important and intriguing assertion (May 1972, 1973), we must discover whether subsystems exist within habitats.

Some of the problems already encountered with Askew's (1961) data also apply here; we do not know where the boundaries between the possible subsystems should be located. We know neither the number of subsystems nor the number of species per subsystem. Consequently, for any system there might be a large range of possible subsystems. Rearranging the actual interactions between the species but preserving the trophic structure produces many possible webs. Among these one might frequently recognize apparent subsystems. Our experience suggests this to be true. To circumvent this problem we must first define a statistic which measures the degree to which a system is organized into subsystems. Then, we must find how this statistic is distributed in model systems under a null hypothesis of no compartments. Only then can we test whether real webs are more compartmented than chance alone dictates. Theoretically, we could have used this approach to analyse Askew's (1961) data. We did not because for systems as large as Askew's the technique we shall now outline is impossible to compute in a reasonable amount of time with current computer facilities.

Deciding whether subsystems exist has analogies to problems in plant ecology and systematics. In plant ecology one may have 'n' locations containing one or more of 'm' plant species. The problem is to group the locations into divisions representing community types based on the presence or absence of species. In systematics one has 'm' characteristics and 'n' individuals; the latter need to be grouped so they can be given

892   *Are food webs compartmented?*

specific or subspecific labels. There is a plethora of ordination techniques for elucidating possible boundaries between groups, but we are unable to find a technique which decides whether, given the structure of the data, one should or should not recognize distinct groups. The question of whether there are actually any groups there to be recognized seems to have been appreciated (Goodall 1966) but not solved.

The compartmentalization statistic

We call our statistic describing the degree to which systems are organized into sub-systems $\bar{C}_1 : C$ for compartmentalization, '1' because it is only one of a large number of possibilities, and the bar because it is an average. The derivation of $\bar{C}_1$ is entirely heuristic. Its utility stems from it correctly distinguishing the degree to which systems are grouped into subsystems in a large number of test cases. It works well *only* when systems identical in the number of species, connectance and a variety of other properties, are compared. This proves no disadvantage here, but the statistic should not be used when systems differing in these properties are compared. An analytical understanding of the distribution of $\bar{C}_1$ is probably intractable under interesting ecological conditions, but it is amenable to numerical computation.

Consider a binary matrix, **A**, of size n by n, where n is the number of species in the matrix; the entries a_{ij} are:

$a_{ii} = 1$ all i

$a_{ij} = a_{ji} = 1$ if 'i' feeds on j and where j is fed upon by i. (We assume all feeding links are reciprocal; that is, if i feeds on j, i influences j and j influences i.

$a_{ij} = a_{ji} = 0$ otherwise.

Two examples are illustrated in Fig. 4; note both webs have identical numbers of species, interactions, trophic levels, top predators and species at the base of the web. Model a is compartmented, model b is not. The **A** matrix (Table 4) is analogous to the presence (1) or absence (0) of a species (columns) in different locations (rows) in that it shows whether a species (rows or columns) interacts with another species (columns or rows). Derived from **A** is a matrix, **C** (Table 4) which indicates the degree to which the species share predators and prey. **C** is analogous to a correlation matrix but uses an index of similarity more suitable for binary data than the Pearson product moment correlation coefficient.

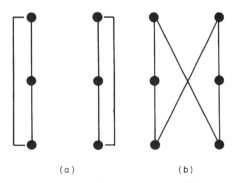

(a)   (b)

FIG. 4. Two webs: (a) is compartmented, (b) is not, for which $\bar{C}_1$ statistics are derived in Table 4.

The entries of $\mathbf{C}$, c_{ij} are obtained from:

$$c_{ij} = \frac{\text{The number of species with which both } i \text{ and } j \text{ interact}}{\text{The number of species with which either } i \text{ or } j \text{ interact}}.$$

Thus, in Fig. 4(a), species 2 interacts with species 1; so does species 3. Over and above this they only interact with each other. Hence $C_{23} = 1\cdot0$. Next, we derive $\bar{C}_1$ as the mean of the off-diagonal elements (the diagonal elements are necessarily unity); thus:

$$\bar{C}_1 = \frac{1}{n(n-1)} \sum_{i=1}^{n} \sum_{\substack{j=1 \\ j \neq i}}^{n} c_{ij}. \tag{4}$$

This statistic is on the interval 0, 1. As can be seen from Table 4 the value of $\bar{C}_1$ is greatest for the system that is most compartmented. We have verified this result by checking hundreds of matrices produced as described below and found that the values of $\bar{C}_1$ are always ranked identically with our subjective rankings based on examination of the actual food web structure.

That $\bar{C}_1$ increases with compartmentalization might seem a little surprising. In the model with subsystems each species interacts or shares predators and prey with only two species compared to four in the reticulate model. There are more zero entries in $\mathbf{C}$ in the model with subsystems (Table 4). However, the magnitude of the entries in the reticulate model are considerably reduced by two factors: (1) the numerator (in the expression for c_{ij}) is reduced because 'i' and 'j' share fewer species of predator and prey, and (2) the denominator is increased because the total number of prey and predators with which both species interact is increased.

Next we must consider the distribution of the $\bar{C}_1$ statistic. The entries in the matrix $\mathbf{A}$ must be highly non-random in any system that is ecologically reasonable. For example, we should not have predators without prey species or loops of the kind 'species A eats species B eats species C eats species A'. The problem is to generate values for $\bar{C}_1$ for a large set of webs that are reasonable but random within that reasonable set. We aim to

TABLE 4. Matrices derived from the food webs shown in Fig. 4 to illustrate derivation of compartmentalization statistic, $\bar{C}_1$

(a)

A		1	2	3	4	5	6
	1	1	1	1	0	0	0
	2	1	1	1	0	0	0
	3	1	1	1	0	0	0
	4	0	0	0	1	1	1
	5	0	0	0	1	1	1
	6	0	0	0	1	1	1

C		1	2	3	4	5	6
	1	–	1·0	1·0	0·0	0·0	0·0
	2	1·0	–	1·0	0·0	0·0	0·0
	3	1·0	1·0*	–	0·0	0·0	0·0
	4	0·0	0·0	0·0	–	1·0	1·0
	5	0·0	0·0	0·0	1·0	–	1·0
	6	0·0	0·0	0·0	1·0	1·0	–

$$\bar{C}_1 = 0\cdot4 \left(= \frac{12}{6 \times 5} \right)$$

(b)

A		1	2	3	4	5	6
	1	1	1	0	0	0	1
	2	1	1	1	0	0	0
	3	0	1	1	1	0	0
	4	0	0	1	1	1	0
	5	0	0	0	1	1	1
	6	1	0	0	0	1	1

C		1	2	3	4	5	6
	1	–	0·5	0·2	0·0	0·2	0·5
	2	0·5	–	0·5	0·2	0·0	0·2
	3	0·2	0·5	–	0·5	0·2	0·0
	4	0·0	0·2	0·5	–	0·5	0·2
	5	0·2	0·0	0·2	0·5	–	0·5
	6	0·5	0·2	0·0	0·2	0·5	–

$$\bar{C}_1 = 0\cdot28 \left(= \frac{8\cdot4}{6 \times 5} \right)$$

* The derivation of this element is discussed as an example in the text.

894   *Are food webs compartmented?*

compare actual webs with a set of random webs, constrained to be as similar as possible to the real web under consideration. Indeed, the more properties shared by our random, but constrained, webs and the real web under consideration, the less likely that differences in values of $\bar{C}_1$ will be due to factors other than the degree of grouping into subsystems. Our methods are similar (and much of the computer programme identical) to those described by Pimm (1979b) for testing other attributes of real food webs. They will be presented below only in outline.

For each real 'within-habitat' food web at our disposal we generated a random family of webs, identical in the following properties:

(1) the same number of interactions;

(2) the same number of top-predators (species on which nothing else feeds);

(3) the same number of basal species (species which feed on nothing else). These are usually plants, but sometimes are detritivores;

(4) the same number of intermediate species (species which are neither top-predators nor basal species); and

(5) the same number of modal trophic levels (the modal number of connections between each top predator and the basal species, each pathway counted separately, plus one).

The webs analysed and their parameters are shown in Table 5. The reasons for selecting these studies (Pimm 1979b) largely correspond to independent decisions made by Cohen (1978). Some additional studies have been included where these met the criteria outlined and some large webs have been included which were previously difficult to analyse on available computer facilities. For each random web we calculate $\bar{C}_1$, and for the entire set, P, the proportion of random webs with $\bar{C}_1$ less than the observed value. If the null hypothesis is correct, the mean value of P for all webs should be 0·5, with P distributed uniformly on the interval 0, 1; that is, chance alone dictates that half of the random webs should be more compartmented than the real web, and half less. Some of the individual webs in Table 1 obviously depart strongly from their random relatives, but these departures are not in any consistent direction. Some webs are more compartmented than expected, others more uniform (reticulate). We set out in this paper to discover whether food webs in general were divided into subsystems. If the mean value of P averaged over all webs in Table 5 is sufficiently large, it indicates that real webs are, on average, more compartmented than one would expect by chance. They are not; the mean of P is 0·467 and the probability that this result does not differ from 0·5 is considerably greater than 5%. We conclude that on average within habitats there is no evidence for species interactions being grouped into subsystems.

DISCUSSION AND CONCLUSIONS

Our concluding remarks are four-fold and can be brief. First we are able to present some evidence for species interactions being grouped into subsystems. However, compartmentalization does not appear to be a common phenomenon. When we observe compartments they appear for biological reasons; there is no evidence for them being generated by dynamical constraints. Encouragingly, and as we pointed out in the Introduction, the prediction that webs should be compartmented for dynamical reasons may in any case be a product of using biologically unreasonable random models. Compartmented models are no more likely to be stable than randomly organized models, if care is taken to exclude biologically unreasonable phenomena from the analyses (Pimm 1979a). Hence, a lack of compartmentalization is in accord with other predictions about food

Table 5. Degree of grouping into subsystems

N_a	M	N_{int}	N_b	N_p	$TL_{mod} - 1$				$\bar{C}_1$	P_1	System	Source
10	14	25	1	5	2	3	2	4	0.1696	0.556	Prairie	Bird (1930)
7	8	14	3	4	3	3	1		0.1508	0.400	Willow Forest	Bird (1930)
20	6	49	16	2	2	2			0.2701	1.000	Starfish	Cohen (1978) after Menge & Mauzey †
23	5	38	19	1	2				0.1440	0.727	Gastropods	Paine (1963)
12	7	26	6	1	2				0.2638	1.00	Starfish	Paine (1966)
8	11	36	2	5	3	3	3	3	0.3369	0.402	Freshwater stream	Minshall (1967)
7	11	17	1	5	3	2	2	2	0.1754	0.015	Mudflat	Milne & Dunnett (1972)
9	5	12	1	5	2	2	2	2	0.1374	0.115	Mussel Bed	Milne & Dunnett (1972)
8	8	23	4	4	2	2	1		0.2525	0.732	Freshwater stream	Jones (1949)
8	11	19	3	4	2	3			0.2239	0.33	Spring	Tilly (1968)
10	10	15	3	4	3	1	1		0.1402	0.8	Lake Fish	Zaret & Paine (1974)
6	9	12	3	6	1	1	1	1	0.1350	0.333	Cladophora	Jansson (1967)

N_a, the number of prey species, M the number of predators, N_{int} the number of interactions, N_p the number of predators, N_b the number of species at the base of the web, $TL_{mod} - 1$ the modal number of connections from each top predatory to the base of the web (one less than the modal trophic levels), $\bar{C}_1$ the degree of grouping into subsystems (see text), P_1 the proportion of model webs that have $\bar{C}_1$ that are less than the observed values.

† We have been unable to check the original paper describing this web.

896   *Are food webs compartmented?*

web structure derived from simple Lotka-Volterra equations; none of these predictions are refuted by the evidence at our disposal (Lawton & Pimm 1979; Pimm & Lawton 1978; Pimm 1979a, b).

Second, the caveat 'on the evidence at our disposal' is an important one. For example, within habitats, it is always possible for food webs to be divided into blocks, with the boundaries corresponding to the limits of most published work. That is, observers of real food webs stop recording where nature provides a convenient natural compartment. If we had seen only one or a small number of Askew's (1961) gall webs it would have been easy to argue that these represented a neatly defined, natural compartment. As it is we have access to numerous oak-gall webs and can refute the challenge. We cannot refute the argument at the next step, which is that the entire oak-gall system, or the oak tree *in toto* is the compartment, although Varley's woodland (1970) web, and our earlier comments about phytophagous insects in general make us sceptical. Those who wish to believe that webs are compartmented can always retreat to the limits of the data, but if published food webs themselves really correspond to natural compartments we see no clear way of establishing that this is so. In other words we need more and better data on food webs, specifically designed to test the compartmentalization hypothesis. On present evidence, we see no evidence for the existence of clearly defined subsystems except where they are imposed by habitat boundaries: even then they may be difficult or impossible to detect.

By using binary data (a feeding link exists, or it does not) we are, of course, unable to detect more subtle kinds of blocking. If links vary through time, for example, compartments may come and go although the whole web appears to be reticulate. Alternatively, it is possible for compartments to exist, defined by very strong interactions, and bordered by feeble ones. Paine's (1980) meticulous studies of marine intertidal food webs reveal a high variance in interaction strengths within webs, and he has further speculated: 'predictable strong interactions encourage the development of modules or subsystems embedded within the community'. If this is generally so, then food webs may still be made up of relatively autonomous subunits, despite the impression to the contrary created by binary data. We see no way of establishing the generality of Paine's proposition, or the existence of compartments which vary through time, without massive research efforts. In other words, our conclusions refer to an absence of compartments in binary food web data. They do not eliminate other, more subtle kinds of blocking.

Finally, there are some, as yet, rather hazy implications for ecosystem management in our results. If species interactions are, indeed, grouped into subsystems there will be a different effect on 'species deletion stability' (Pimm 1979c), than will be the case if the interactions are random, or are reticulate. Species deletion stability is defined as the probability that no species will become extinct following the removal of a species from the system. In model food webs species deletion stability is generally low (approximately 0·3) and details of food web design appear to have little effect on its magnitude (Pimm 1979c). Experimental removal of a top predator from an intertidal system (Paine 1966, 1980) or of a lake fish through the introduction of a superior competitor (Zaret & Paine 1974) both resulted in the loss of many species from each system. How many species 'ought' to have become extinct under the reticulate and the subsystem hypotheses respectively, we cannot say, but other things being equal, ecosystem disturbances should propagate more under the reticulate hypothesis than under the subsystem hypothesis. Hence, there may be both theoretical and practical reasons for examining species deletion stability in more detail.

S. L. PIMM AND J. H. LAWTON   897

ACKNOWLEDGMENTS

Professors R. T. Paine, L. R. Taylor and M. H. Williamson made helpful comments on the manuscript.

REFERENCES

Askew, R. R. (1961). On the biology of the inhabitants of Oak gall of Cynipidae (Hymenoptera) in Britain. *Transactions of the Society for British Entomology*, **14**, 237–269.

Bird, R. D. (1930). Biotic communities of the aspen parkland of central Canada. *Ecology*, **11**, 356–442.

Bindloss, M. E., Holden, A. V., Bailey-Watts, A. E., & Smith, I. R. (1972). Phytoplankton production, chemical and physical conditions in Loch Leven. *Productivity Problems of Freshwaters* (Ed. by Z. Kajak & A. Hillbricht-Ilkowska), pp. 639–659. Polish Scientific Publishers, (Panstwowe Wydawnictwo Naukowe).

Burgis, M. J., Darlington, J. P. E. C., Dunn, I. G., Ganf, G. G., Gwahaba, J. J., & McGowan, L. M. (1973). The biomass and distribution of organisms in Lake George, Uganda. *Proceedings Royal Society* B, 184, 271–298.

Cohen, J. E. (1978). *Food Webs and Niche Space*. Princeton University Press, Princeton, New Jersey.

DeAngelis, D. L. (1975). Stability and connectance in food web models. *Ecology*, **56**, 238–243.

Elton, C. S. (1966). *The Pattern of Animal Communities*. Methuen, London.

Futuyma, D. J. & Gould, F. (1979). Associations of plants and insects in a deciduous forest. *Ecological Monographs*, **49**, 33–50.

Gibson, C. W. D. (1976). The importance of foodplants for the distribution and abundance of some Stenodemini (Heteroptera: Miridae) of limestone grassland. *Oecologia*, **25**, 55–76.

Goodall, D. W. (1966). Hypothesis-testing in classification. *Nature* 211, 329–330.

Hansen, R. M. & Ueckert, D. N. (1970). Dietary similarity of some primary consumers. *Ecology*, **51**, 640–648.

Hassell, M. P. (1978). *The Dynamics of Arthropod Predator Prey Systems*. Princeton University Press, Princeton, New Jersey.

Jansson, A. M. (1967). The food-web of the *Cladophora*-belt fauna. *Helgalaender wissenschaftliche Meeresuntersuchungen*, **15**, 574–588.

Joern, A. (1979). Feeding patterns in grasshoppers (Orthoptera: Acrididae): factors affecting specialization. *Oecologia*, **38**, 325–348.

Jones, J. R. E. (1949). A further ecological study of calcareous streams in the 'Black Mountain' district of South Wales. *Journal of Animal Ecology*, **18**, 142–159.

Koepcke, H. W. & Koepcke M. (1952). Sobre el proceso de transformacion de la materia organica en las playas arenosas marinas del Peru. *Publicaciones Universidad Nacional Mayor San Marcos, Zoologie, Serie A*, 8.

Lawton, J. H. & McNeill, S. (1979). Between the devil and the deep blue sea: on the problem of being a herbivore. *Population Dynamics* (Ed. by R. M. Anderson, B. D. Turner & L. R. Taylor), pp. 223–244. British Ecological Society Symposium. Blackwell Scientific Publications, Oxford.

Lawton, J. H. & Pimm, S. L. (1978). Population dynamics and the length of food chains. *Nature*, **279**, 190.

Lawton, J. H. & Pimm, S. L. (1979). Are real communities unstable? *Nature*, **279**, 822.

Lawton, J. H. & Price, P. W. (1979). Species richness of parasites on hosts: Agromyzid flies on the British Umbelliferae. *Journal of Animal Ecology*, **48**, 619–638.

Lawton, J. H. & Rallison, S. P. (1979). Stability and diversity in grassland communities. *Nature*, **279**, 351.

Lawton, J. H. & Schröder, D. (1977). Effects of plant type, size of geographical range and taxanomic isolation on the number of insect species associated with British plants. *Nature*, **265**, 137–140.

Mann, K. H. (1964). The case history: the river Thames. *River Ecology and Man* (Ed. by R. T. Oglesby, C. A. Carlson, & J. A. McCann), pp. 215–231. Academic Press, New York.

Mann, K. H. (1965). Energy transformations by a population of fish in the River Thames. *Journal of Animal Ecology*, **34**, 253–275.

Mann, K. H., Britton, R. H., Kowalczewski, A., Lack, T. J., Mathews, C. P., & McDonald, I. (1972). Productivity and energy flow at all trophic levels in the River Thames, England. *Productivity Problems in Freshwaters* (Ed. by Z. Kajak & A. Hillbricht-Ilkowska), pp. 579–596. Polish Scientific Publishers, (Państwowe Wydawnictwo Naukowe).

Marples, T. G. (1966). A radionuclide tracer study of arthropod food chains in a *Spartina* salt marsh ecosystem. *Ecology*, **47**, 270–277.

May, R. M. (1972). Will a large complex system be stable? *Nature*, 238, 413–414.

898 *Are food webs compartmented?*

May, R. M. (1973). *Stability and Complexity in Model Ecosystems.* Princeton University Press, Princeton, New Jersey.

McNaughton, S. J. (1978). Stability and diversity of ecological communities. *Nature*, 274, 251–253.

Milne H., & Dunnett, G. M. (1972). Standing crop, productivity and trophic relationships of the fauna of the Ythan estuary. *The Estuarine Environment* (Ed. by R. S. K. Barnes & J. Green), pp. 86–103. Applied Science Publishers, London.

Minshall, G. W. (1967). Role of allochthonous detritus in the trophic structure of a woodland spring-brook community. *Ecology*, 48, 139–149.

Morgan, N. C. (1972). Productivity studies at Loch Leven (a shallow nutrient-rich lowland lake). *Productivity Problems of Freshwaters* (Ed. by Z. Kajak & A. Hillbricht-Ilkowska), pp. 183–205. Polish Scientfic Publishers, (Państwowe Wydawnictwo Naukowe).

Moriarty, D. J. W., Darlington, J. E. P. C., Dunn, I. G., Moriarty, C. M., & Tevlin, M. P. (1973). Feeding and grazing in Lake George, Uganda. *Proceedings of the Royal Society B*, 184, 299–319.

Murdoch, W. W. (1979). Predation and the dynamics of prey populations. *Fortschritte der Zoologie*, 25, 295–310.

Niering, W. A. (1963). Terrestrial ecology of Kapingamarangi Atoll, Caroline Islands. *Ecological Monographs*, 33, 131–160.

Odum, E. P. (1962). Relationship between structure and function in the ecosystem. *Japanese Journal of Ecology*, 12, 108–118.

Odum, E. P. (1963). *Ecology.* Holt, Rinehart, & Winston, New York.

Paine, R. T. (1963). Trophic relationships of 8 sympatic predatory gastropods, *Ecology*, 44, 63–73.

Paine, R. T. (1966). Food web complexity and species diversity. *The American Naturalist*, 100, 65–75.

Paine, R. T. (1980). Food webs: linkage, interaction strength and community infrastructure. The third Tansley Lecture. *Journal of Animal Ecology*, 49, 667–685.

Pimm, S. L. (1979a). The structure of food webs. *Theoretical Population Biology*, 16, 144–158.

Pimm, S. L. (1979b). The properties of food webs. *Ecology*, (in press).

Pimm, S. L. (1979c). Complexity and stability; another look at MacArthur's original hypothesis. *Oikos*, 33, 351–357.

Pimm, S. L., & Lawton, J. H. (1977). The number of trophic levels in ecological communities. *Nature*, 268, 329–331.

Pimm, S. L., & Lawton, J. H. (1978). On feeding on more than one trophic level. *Nature*, 275, 542–544.

Rejmánek, M. & Starý, P. (1979). Connectance in real biotic communities and critical values for stability of model ecosystems. *Nature*, 280, 311–313.

Sheldon, J. K., & Rogers, L. E. (1978). Grasshopper food habits in a shrub-steppe community. *Oecologia*, 32, 85–92.

Shure, D. J. (1973). Radionuclide tracer analysis of trophic relationships in an old-field ecosystem. *Ecological Monographs*, 43, 1–19.

Southwood, T. R. E. (1978). On the effects of size in determining the diversity of insect faunas. *Diversity of Insect Faunas* (Ed. by L. A. Mound & N. Waloff). Royal Entomological Society Symposium, 9, pp. 19–40. Blackwell Scientific Publications, Oxford.

Summerhayes, V. S. & Elton, C. S. (1923). Contributions to the ecology of Spitsbergen and Bear Island. *Journal of Ecology*, 11, 214–286.

Tilly, L. J. (1968). The structure and dynamics of Cone Spring. *Ecological Monographs*, 37, 169–197.

Varley, G. C. (1970). The concept of energy flow applied to a woodland community. *Animal Populations in Relation to Their Food Resources* (Ed. by A Watson), pp. 389–405. Blackwell Scientific Publications, Oxford.

Zaret, T. M. & Paine, R. T. (1953). Species introduction into a tropical lake. *Science*, 182, 449–455.

(*Received 5 November* 1979)

Vol. 138, No. 1   The American Naturalist   July 1991

COMPLEX TROPHIC INTERACTIONS IN DESERTS: AN EMPIRICAL CRITIQUE OF FOOD-WEB THEORY

Gary A. Polis

Department of Biology, Vanderbilt University, Box 93 Station B, Nashville, Tennessee 37235

Submitted December 28, 1988; Revised January 26, 1990;
Accepted March 23, 1990

Abstract.—Food webs in the real world are much more complex than food-web literature would have us believe. This is illustrated by the web of the sand community in the Coachella Valley desert. The biota include 174 species of vascular plants, 138 species of vertebrates, more than 55 species of arachnids, and an unknown (but great) number of microorganisms, insects (2,000–3,000 estimated species), acari, and nematodes. Trophic relations are presented in a series of nested subwebs and delineations of the community. Complexity arises from the large number of interactive species, the frequency of omnivory, age structure, looping, the lack of compartmentalization, and the complexity of the arthropod and soil faunas. Web features found in the Coachella also characterize other communities and should produce equivalently complex webs. If anything, diversity and complexity in most nondesert habitats are greater than those in deserts. Patterns from the Coachella web are compared with theoretical predictions and "empirical generalizations" derived from catalogs of published webs. The Coachella web differs greatly: chains are longer, omnivory and loops are not rare, connectivity is greater (species interact with many more predators and prey), top predators are rare or nonexistent, and prey-to-predator ratios are greater than 1.0. The evidence argues that actual community food webs are extraordinarily more complex than those webs cataloged by theorists. I argue that most cataloged webs are oversimplified caricatures of actual communities. That cataloged webs depict so few species, absurdly low ratios of predators on prey and prey eaten by predators, so few links, so little omnivory, a veritable absence of looping, and such a high proportion of top predators argues strongly that they poorly represent real biological communities. Consequently, the practice of abstracting empirical regularities from such catalogs yields an inaccurate and artifactual view of trophic interactions within communities. Contrary to strong assertions by many theorists, patterns from food webs of real communities generally do not support predictions arising from dynamic and graphic models of food-web structure.

Feeding relationships in communities are delineated in three ways. The first is the classic food web, a schematic description of trophic connections. The second quantifies energy or mass flow. Finally, interaction or functional webs experimentally identify strong links (Paine 1980; Menge and Sutherland 1987). Superficially, little work is needed to construct food webs; consequently, they most frequently represent communities. A rough, qualitative knowledge of "who eats whom" is all that is necessary to produce a simple food web, whereas experimental manipulations or quantitative measurements are necessary to construct webs of interaction or energy flow.

Several approaches analyze food webs (DeAngelis et al. 1983; May 1986; Lawton 1989; Schoener 1989). One uses models based on stability analysis. The re-

Am. Nat. 1991. Vol. 138, pp. 123–155.

sults are complex and beyond the scope of this article. However, they basically show that model systems decrease in stability with more species, more links (connectance), or greater linkage strength. The dynamic constraints needed to maintain stability are hypothesized as important in shaping the properties of webs. Stable webs are relatively simple, short (with few trophic levels), and compartmentalized and exhibit little omnivory or looping (Pimm 1982).

A second approach analyzes real food webs to determine regularities in their properties. Analyzed webs were compiled by Cohen (1978), Cohen et al. (1986, 1990), Briand (1983), and Schoenly et al. (1991). Cohen et al. (1986) published a catalog of 113 webs, and Schoenly et al. compiled 95 insect-oriented webs. Theorists (Pimm and Cohen) argue that empirically derived patterns are consistent with and validate predictions of the dynamic models above (dynamic models: Pimm 1982, Pimm and Rice 1987; cascade model: Cohen et al. 1990). "Indeed, there is a close tie between the theoretical and observational studies: real food webs have a statistical predominance of those features that, in models, increase the chance that those models will be stable. The first is that trophic interactions, though highly complex, are reasonably patterned—they demonstrate a large catalogue of assembly rules" (Pimm and Rice 1987, p. 304). Some empirical patterns and assembly rules are presented in Appendix A.

In this article, the food web of a desert community is analyzed explicitly to evaluate the patterns in Appendix A. Observed patterns are quite different from those assembled from published webs. I argue that most cataloged webs are overly simplified and poorly represent actual communities. Consequently, the practice of abstracting empirical regularities yields an inaccurate and artifactual view of trophic interactions within communities.

GENERAL PROBLEMS IN THE ANALYSIS OF EMPIRICAL FOOD WEBS

Four substantial problems beset the catalogs of webs and make them totally inadequate for the types of analyses that have been conducted (also see Glasser 1983; May 1983a; Taylor 1984; Paine 1988; Sprules and Bowerman 1988; Lawton 1989; Winemiller 1990).

1. *Inadequate representation of species diversity.*—The *major* problem is that the numbers of species in cataloged communities are far less than those in real communities. Most authors of these webs simply ignored unfamiliar species, concentrated on taxa in their expertise, and/or aggregated or "lumped" unfamiliar species into higher categories. Lumping is a severe problem. Cohen (1978) labeled lumped categories "kinds of organisms." "'Kinds' are equivalent classes with respect to trophic relations" (Cohen 1978, p. 7). Briand (1983, p. 253) clarifies and expands: "A 'kind of organism' (interchangeable henceforth with the term 'species') may be an individual species, or a stage in the life cycle of a size class within a single species, or it may be a collection of functionally or taxonomically related species." "Kinds" are also called "trophic species" (Briand and Cohen 1984) and "species" (Cohen and Newman 1985). Kinds include "basic food," "benthos," "other carnivores" (matrix 1 in Briand 1983); "algae," "plankton," "birds" (matrix 9); "zooplankton," "ice invertebrates," "fish" (matrix 21); and

"trees and bushes," "insects," "spiders," "soil insects and mites," and "parasites" (matrix 27). Only 28.7% of the total kinds in all Briand's webs are real species; nine matrices have no real species. The "kinds" simplification was criticized by Glasser (1983), May (1983*a*), Taylor (1984), Paine (1988), Lawton (1989), Lockwood et al. (1990), Winemiller (1990), and Cohen (1978) himself in a self-critique (but see Sugihara et al. 1989).

Lumping is not uniform: plants, arthropods, parasites, and organisms that live in the soil or benthos are most frequently grouped. Invertebrates are analyzed in much less detail than vertebrates (Pimm 1982), thus obscuring food-web complexity (Taylor 1984; Paine 1988). The incomplete presentation of these taxa is a serious flaw. In particular, arthropods are central to the structure of terrestrial communities. The ~800,000 identified species of insects represent ~89% of all animal species (5–50 million insect species are estimated to exist; May 1988). Soil organisms are usually ignored or lumped in spite of their importance as major pathways of energy flow in terrestrial communities (Cousins 1980; Odum and Biever 1984; Rich 1984). The tactics of ignoring and lumping species produce the depauperate webs compiled by Cohen and Briand. This is obvious from an inspection of Cohen et al.'s (1990) 113-web catalog. The number of "kinds" ranged from 3 to 48 with the average web "community" having 16.9 kinds. Real communities have more species.

This is illustrated by enumerating the species from the sandy deserts of the Coachella Valley (hereafter CV; Riverside County, Calif.): 174 species of vascular plants, 138 species of vertebrates, more than 55 species of arachnids, and a large but unknown number of lower plants, nematodes, acari, and insects (Polis 1991*a*). Insects are estimated at 2,000 to more than 3,000 species; I have identified 123 families. A still-incomplete survey in the adjacent Deep Canyon Desert Preserve identified 24 orders, 308 families, and more than 2,540 species (Frommer 1986).

2. *Inadequate dietary information.*—Published analyses of diets or lists of enemies (predators, parasites, and/or parasitoids) suggest that most species eat and are eaten by from 10^1 to 10^3 other species (see below). The inadequate incorporation of these trophic links is another major weakness of cataloged webs. The number of prey items recorded is usually a function of the amount of time and effort devoted to observation. A "yield/effort" curve (Cohen 1978) is illustrated by analyzing the diet of the scorpion *Paruroctonus mesaensis* (fig. 1). The number of prey species continues to increase with observation time. The 100th prey species was recorded on the 181st survey night; an asymptote was never reached in 5 yr and more than 2,000 person hours of field time. This suggests that the amount of effort and time needed to determine the complete diet of just the numerically dominant species is astronomical. It is unlikely that such an effort was made for most species in the cataloged webs. Thus a food web containing all species still would be an inadequate description of community trophic relations unless diets were known with more confidence.

Such inadequacy is manifested in cataloged webs. For example, they show a high proportion (28.5%, Briand and Cohen 1984; 46.5%, Schoenly et al. 1991) of top predators (consumers without predators). It is unlikely that even 1%, let alone almost one-half, of all animals do not suffer predators sometime during their lives

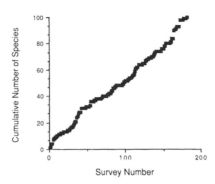

FIG. 1.—Yield effort curve for first 100 species of prey captured by the scorpion *Paruroctonus mesaensis* in the Coachella Valley.

(see below). These high figures partially result from grossly incomplete data: for example, in Cohen et al.'s (1986) catalog, 57 chains were of length one, that is, 57 herbivores were with no recorded predators! Such top predators include spiders, mites, midges, mosquitos, bees, weevils, fish larvae, blackbirds, shrews, and moles.

3. *Age structure.*—Age-related changes in food and predators are not well incorporated into web analysis. Populations are composed of age/size classes, each exhibiting significant differences in resource use, predators, and competitors (Polis 1984, 1988a; Werner and Gilliam 1984; Polis and McCormick 1986a). Age classes often eat different foods, thereby expanding diet ("life-history omnivory"; Pimm and Rice 1987). "Ontogenetic diet shifts" characterize species that undergo metamorphosis. The juveniles of at least 27 families of CV holometabolic insects eat radically different foods (live arthropods) from adults (plants). For species that grow slowly through a "wide size range," diet changes more gradually as prey size increases with predator size (e.g., arachnids, reptiles; Polis 1988a). In fact, differences in body sizes and resource use among age classes are often equivalent to or greater than differences among most "biological species" (Polis 1984). This magnitude of change is typical of wide-size-range predators (most invertebrates, hemimetabolic insects, larva of holometabolic insects, arachnids, fish, and reptiles; Polis 1984).

Predators also change during growth. Juveniles are eaten by species too small to capture adults. Such developmental "escapes" are common to all communities. For example, snakes eat eggs and newborn (but not adults) of carnivorous birds and desert tortoises. Alternately, adults are eaten by predators that do not eat small juveniles. Thus some predators (e.g., owls, kit foxes) in the CV eat only adult scorpions (Polis et al. 1981).

In summary, age/size differences in predators, prey, and competitors are the norm in terrestrial as well as aquatic habitats (contrary to Pimm and Rice's [1987] assertions) and may be major determinants of population dynamics and community structure. Unfortunately, the richness that age structure contributes has been largely ignored (but see Pimm and Rice 1987). Usually only adults are considered or the diets of all age classes are combined. Age is recognized in only 22 of 875 kinds in Cohen's (1978) catalog and in 3 of 422 in Briand's (1983). Age structure is often difficult to incorporate into studies; nevertheless it is paramount to community dynamics.

4. *Looping*.—Looping is a feeding interaction whereby A eats B and B eats C but either B (in mutual predation) or C (in a three-species loop) eats A. Cannibalism is a "self-loop" (A eats A; Gallopin 1972). Food-web theorists dismiss loops as "unreasonable structures" (Pimm 1982, p. 70; see also Gallopin 1972; Cohen 1978; May 1983*a;* Cohen et al. 1990). Pimm summarizes from the catalog of webs: "I know of no cases, in the real world, with loops" (p. 67). He has modified this view to include loops in aquatic age-structured species (Pimm and Rice 1987) but still maintains that loops are rare in terrestrial systems.

Looping is widespread. Cannibalism is reported in more than 1,300 species and is a key factor in the dynamics of many populations (Polis 1981; see also Elgar and Crespi, in press). Cannibalistic loops are frequent in the CV. Ontogenetic reversal of predation among age-structured species is the most common form of mutual predation: A juveniles are eaten by B but A adults eat B (and/or B juveniles) (Polis et al. 1989). This occurs among CV insects, spiders, scorpions, and solpugids and among predaceous lizards, snakes, and birds. For example, gopher snakes (*Pituophis*) eat eggs and young of burrowing owls whereas adult burrowing owls eat young gopher snakes.

Mutual predation can occur independently of age structure (Polis et al. 1989; Schoener 1989; Winemiller 1990). Two examples from the CV are black widow spiders and CV ants. Black widow spiders (*Latrodectus hesperus*) catch three species of scorpions by using web silk to pull them up off the ground; black widows traveling on the ground are captured by these same scorpions (Polis and McCormick 1986*b*). Second, CV ants (*Messor pergandei, Pogonomyrmex californicus, Myrmecocystus flaviceps*) regularly eat each other (Ryti and Case 1988). Killing and predation of winged reproductives (after swarming) and workers (during territorial battles) are a regular interaction among many social insects (see Polis et al. 1989).

THE COACHELLA VALLEY

The Coachella Valley is located in Riverside County, California (166°37'W, 33°54'N; area = ~780 km^2). Winters are mild; summers, dry and hot. It is a low-elevation rain shadow desert with average annual rainfall at Deep Canyon of 116 mm. The sand dune/intergrading sand flat habitat was chosen for analysis. This community is well studied because of the presence of the University of California Deep Canyon Desert Preserve. Beginning in 1969, Mayhew (1983) surveyed the vertebrates for the Deep Canyon Transect Study. This work established a list of 138 reptiles, birds, and mammals on sandy soils. Some species are not included in the web; for example, mountain lions and badgers are now absent. Further, only the 56 birds (out of 97 residents) that actually nest in the area are considered. Additional data on CV vertebrates were obtained from Weathers (1983) and Ryan (1968). Invertebrates are less well known. Much of the knowledge of them results from my long-term research (since 1973) over 17 yr and more than 4,000 h of fieldwork including more than 4,300 trap days for arthropods. Taxonomic lists are obtained from this fieldwork and from catalogs of CV arachnids (Polis and McCormick 1986b) and insects (Frommer 1986). Plants were surveyed by Zabrinskie (1979).

Food and predators were determined from the literature and my work. Since 1979, I have read ~820 papers to assemble data for the CV web. Data from the CV were used with priority; however, I was forced to supplement this information with that from other regions, some quite near (e.g., the Palen Dunes) and some much farther away (e.g., the Chihuahuan Desert). I could not find the diets (from desert areas) of 18 birds and one rodent. Finally, interviews with scientists conducting research in the CV provided taxonomic and dietary data. Because of space limitations, I include only the most important references. Additional references and the identities and diets of vertebrate species and arthropod families are given in Polis (1991a).

A series of representative subwebs depicts trophic relations in the CV sand community. Subwebs proceed from plants and detritus to various secondary consumers. Subwebs are connected so that organisms in one web consume (or are consumed by) organisms in the next. Webs are incomplete because there are far too many species to include and adequate diet data are unavailable for many species. I thus concentrate on well-studied, focal species, the trophic interactions of which are relatively well known. Webs include only species that live in the CV and only interactions for which evidence exists. Similar complexity is expected for other, less known species. Thus, these webs understate actual complexity.

Consumers are classified in terms of resource specialization (i.e., number of species eaten within one group, e.g., plants) or trophic specialization (i.e., number of different types eaten, e.g., plants, detritus, arthropods) (Levine 1980). Species vary from resource specialists that eat a few species of the same resource type to trophic generalists (omnivores) that feed on several food types. Closed-loop omnivory is a special case of omnivory in which A eats both B and C but B also eats C (Sprules and Bowerman 1988).

RESULTS

Plant-Herbivore Trophic Relations

Herbivory describes feeding interactions involving several plant products: leaves, seeds, fruit, wood, sap, nectar, roots, and tubers. Desert herbivores include microbes, nematodes, arthropods, and vertebrates. I cannot detail the herbivores of the hundreds of CV plants. Rather, I discuss broad groups with the hope of conveying the complexity in the plant-herbivore link. A wide variety of arthropods eat desert plants (Orians et al. 1977; Powell and Hogue 1979; Crawford 1981; Wisdom 1991). At least 74 families of CV arthropods are herbivorous sometime in their lives. Plants that grow in the CV are attacked by many species of insect: more than 60 eat creosote (*Larrea tridentata*; Schultz et al. 1977), more than 200 on mesquite (*Prosopis glandulosa*), and more than 89 on ragweed (*Am-*

brosia dumosa) (Wisdom 1991). Both resource specialists and generalists live in the CV. Specialists include the grasshopper *Bootettix punctatus* (on creosote; Mispagel 1978). Insects on cactus usually specialize, and generalists do not attack cactus (Mann 1969). (Damage caused by feeding on cactus facilitates several specialist and generalist fungi.) Resource generalists are more common in deserts than are specialists (Orians et al. 1977; Crawford 1981, 1986). Generalists in the CV include the harvester ant *Messor pergandei* (97 species of CV seed; Gordon 1978) and the camel cricket *Macrobaenetes valgum* (16 plant species; G. A. Polis, unpublished data).

Most herbivorous arthropods are trophic specialists on plants all their lives: for example, hemimetabolic (Orthoptera, Hemiptera, Homoptera, Thysanoptera) and some holometabolic insects (e.g., curculionid, chrysomelid, scarabid, and buprestid beetles). However, many holometabolic insects have larvae that are parasitic or predaceous on arthropods but whose adults feed on plants (Ferguson 1962; Andrews et al. 1979; Powell and Hogue 1979; Wasbauer and Kimsey 1985). Trophically flexible generalists include CV harvester ants (more than 40 categories of foods including seeds, flowers, stems, spiders, and insects from at least six orders including four ant species; Ryti and Case 1988) and camel crickets (15% plant detritus, 41% animal detritus, and <1% conspecifics).

Most CV mammals (16 of 18 species) eat plant tissue (the two bats did not). Plants (fruit) formed 0.2%–4.1% of the diet of the largest mammal, the coyote (Johnson and Hansen 1979). This is the smallest plant component for any of the 16 mammals. Over 50% of the scats of the desert kit fox contained plant material. The two rabbits and the gopher are the only trophic specialists; however, they are resource generalists eating many species. Omnivorous antelope ground squirrels fed on a seasonally changing diet (10%–60% foliage, 20%–50% seeds, 62%–95% total plants by volume; W. Bradley 1968). Rodents (*Dipodomys, Peromyscus, Perognathus*) fed on seeds and plant parts (and arthropod prey). In total, 15 mammals eat seeds (only bats and the gopher do not). Nine regularly consumed more than 50% seeds in their diet. No CV mammals specialize on particular plants. For example, pocket mice *Perognathus formosus* feed on 27 plant species; antelope ground squirrels, 24 species. With the exception of the gopher and the two rabbits, all plant-eating mammals are trophic generalists that include arthropods in their diet, for example, 1%–17% for *Dipodomys merriami* and 2%–35% for antelope ground squirrels (this species also eats vertebrates; see below).

Many (34 of 56 species) birds in the sand community feed on plant parts (seeds, nectar, flowers, and fruit). Frugivorous birds are common (e.g., cactus wren, phainopepla, verdin, and doves). Some birds eat fruit as a minor component of an omnivorous diet (e.g., roadrunner, Scott's oriole, western tanager, western bluebird, warbling vireo, Bewick's wren). Granivory is also common: 22 of 56 CV birds were reported to eat seeds (13 are primarily granivorous). Many insectivorous desert birds eat significant quantities of seed when insects are scarce (Brown et al. 1979; Brown 1986; Wiens 1991). No herbivorous birds are resource specialists. In fact, trophic specialists are rare; of 34 plant-eating birds, only five are not recorded to eat arthropods.

Two species of CV reptiles are primarily herbivorous: desert tortoise and desert iguana. Both are resource generalists eating a wide variety of plants (17–40 species for the tortoise) and plant parts. Only the tortoise is a trophic specialist. The diet of the desert iguana contains 1%–5% arthropods. Five of the nine other lizards (none of 10 snakes) consume a minor portion of plants.

Detritus, Soil Biota, and Belowground Herbivory

Detritus is a broad term applied to nonliving organic matter from living organisms. It is a universal component of food webs simply because all organisms die, plant parts senesce, and animals defecate. It may originate from plants (e.g., wood, leaves, seeds, flowers, and roots [rhizodeposition]) or animals (feces, urine, secretions, molted skin or fur, and dead animals).

Most primary productivity flows directly or indirectly through the detrital component of food webs. Herbivores process 1%–50% of net primary productivity; the rest enters the detrital system (Macfayden 1963; Odum and Biever 1984). This is particularly so in deserts, where the main energy flow often proceeds directly from autotrophs to detritivores (Seely and Louw 1980; Wallwork 1982; Crawford 1991; Freckman and Zak 1991). The plant-herbivore-carnivore link forms 12%–33% of the fate of plant production in deserts; the remainder goes through the soil/detritus chain. Nevertheless, Cousins (1980) is one of the few to incorporate detritus explicitly into food-web analysis (see also Odum and Biever 1984). He disputes placing autotrophs alone at the basal position of webs; rather, herbivory and detritivory should be considered equally important links in a "trophic continuum." Energy, produced by autotrophs and consumed during secondary production, is recycled and made available to other consumers by detritivores.

A diverse biota lives within desert soils (Crawford 1981, 1991; Wallwork 1982; Freckman and Zak 1991). Microbes (fungi, yeast, bacteria, protozoa), nematodes, mites, termites, some ants, Collembola, Thysanura, cockroaches, tenebrionid larvae, millipedes, and isopods are some of the more common of the many detritivores that live within CV soils. Although species in these taxa degrade organic material, many include facultative or obligate herbivores on belowground plant parts (Crawford 1981, 1986). Over 50% of net primary production is commonly allocated to belowground plant parts (Andersen 1987). For example, in Russian deserts, 65% of the plant biomass is belowground (Rodin and Bazilevich 1964). Species from seven orders of insects, mites, nematodes, and some rodents have adopted belowground herbivory as their primary feeding mode (Andersen 1987).

Soil organisms are quite abundant in deserts. Nematode biomass of 1–20 g/m² normally occurs (Freckman and Zak 1991). Detritivores form 37%–93% (mean = 73%) of all individual macroarthropods in four deserts analyzed by Crawford (1991). Termites are particularly abundant; their biomass is often an order of magnitude higher than that of any other desert animal (MacKay 1991; Polis and Yamashita 1991). Wallwork (1982) emphasized their importance in desert webs: termites fix nitrogen, eat large quantities of detritus, recycle nutrients within their colonies via trophallaxis and cannibalism, and ultimately release nutrients to predators.

A rich web based on detritus and underground plant parts exists within desert soils (see, e.g., Whitford 1986). Nematodes occupy several trophic roles (Freckman and Zak 1991): herbivores, plant parasites, microbial feeders, fungivores, omnivores, omnivore predators, parasites. In the Mojave, there are four trophic groups of Prostigmata soil mites: phytophages (1 family), fungivores and detritivores (3), parasites (1), and predators (8) (Franco et al. 1979; see also Santos et al. 1981). Predatory mites in litter are as common as nonpredators in nearby Joshua Tree Monument (Wallwork 1982). These mites eat nematodes, Collembola, and other mites. The large number and diversity of predatory mites led Edney et al. (1974) to conclude that two or more predator trophic levels exist in the decomposer web. Wallwork (1982) and Santos et al. (1981) suggested that decomposer pathways in desert soils were regulated by mites that prey on nematodes that feed on microorganisms.

Soil interactions become even more complex with the inclusion of macroarthropods. Most detritivorous arthropods not only eat detritus, but also feed on microorganisms (bacteria, fungus, protozoa) feeding within the detritus (Janzen 1977). For example, in the CV, the burrowing cockroach *Arenievaga* feeds below ground on living and decaying roots and ensheathed mycorrhizae. Further, most detritivores (e.g., cockroaches, tenebrionids, millipedes) host cellulolytic microbes that degrade plant detritus (Crawford 1991). Although such symbionts form a separate energetic "trophic level," they usually are not included in web analysis. Finally, several desert insects consume detritus directly or are predaceous on microarthropods and nematodes (e.g., in the CV, larvae of asilid, bombyliid, and theriviid flies, and staphylinid and clerid beetles; Edney et al. 1974; Powell and Hogue 1979).

Soil interactions are not separate from the rest of the community. Surface and subsurface herbivores are involved in competition and facilitation (Seastedt et al. 1988). Most important, energy flows in both directions via the trophic continuum. Many surface dwellers in deserts either spend part of their lives in the soil (as larvae, e.g., tenebrionids) or feed on arthropods that live permanently or temporarily below ground (see Ghilarov 1964). For example, 46% of *Paruroctonus mesaensis* prey live in soils as larvae. *Arenivaga* form 23% by weight of this scorpion's diet. Termites (10 CV species) and tenebrionids (16 species) are important conduits of energy flow from below ground when they are eaten by a diverse group of arthropod and vertebrate predators (Wallwork 1982; 35 species of known surface predators in the CV). Such predation by surface dwellers exports much of the energy recycled by detritivores and links the soil subweb to that above ground. Thus, even if herbivores and detritivores operate in distinct microhabitats, energy flowing further into the community merges into the bodies of predators common to both consumers (Odum and Biever 1984).

No study has analyzed trophic interactions within detritus and soil in the CV. The studies above were conducted in deserts (Mojave, Chihuahua) geographically adjacent to the CV. I combined information from these studies (esp. Whitford 1986; Freckman and Zak 1991) with my data to construct a soil/detritus subweb (fig. 2) using CV taxa. Note the complexity (even with extensive lumping), loops, frequent omnivory (often at nonadjacent levels), closed-loop omnivory, chains of 4–5 links, and links between belowground consumers and aboveground predators.

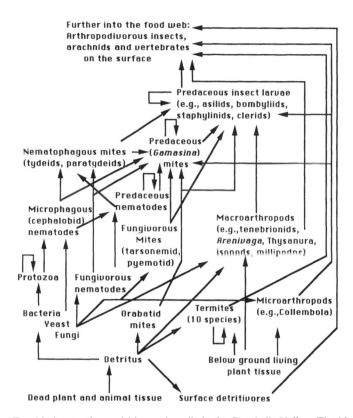

FIG. 2.—Trophic interactions within sandy soils in the Coachella Valley. The identities of a few of the important detritivorous species are as follows: Isopoda (*Venizillo arizonicus*), Collembola (Entomobryididae), Thysanura (*Leucolepisa arenaria, Mirolepisma deserticola*), Gryllacrididae (*Macrobaenetes valgum*), Blattidae (*Arenivaga investigata*), Isoptera (five *Amitermes* spp.; five other species), Tenebrionidae (16 species). Note that not all trophic links are represented (e.g., for tenebrionids and termites). An arrow returning to a taxon indicates cannibalism.

Carrion Feeders

The decomposition of carrion results from the cumulative action of microorganisms, necrophagous insects, and some vertebrates. A rich carrion fauna occurs in deserts (McKinnerney 1978; Schoenly and Reid 1983; Crawford 1991). Complex interactions involve from 28 to more than 500 species. McKinnerney's analysis of carrion from two rabbits that occur in the CV identifies 63 arthropod and four vertebrate consumers. Some specialize on particular tissue; others do not. Trophic specialists and generalists occur. Species composition and diversity change through time. Necrophagous insects, generalist and specialist predators, and omnivores are common. Vertebrates not only eat carrion but eggs and larvae of insects. Carrion feeders also consume microorganisms within the carrion (Janzen 1977). Further, many carrion species are well-known cannibals (Polis 1981). Interactions within carrion do not constitute a distinct compartment: much of the energy from carrion is exported to the rest of the community. Most organisms

DESERT FOOD WEBS   133

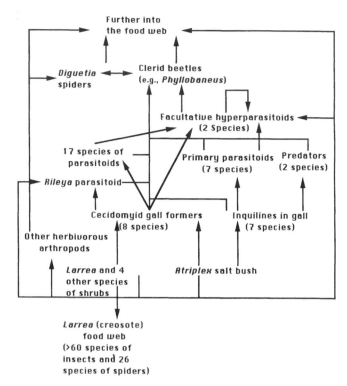

Fɪɢ. 3.—Trophic interactions within galls on saltbush *Atriplex canescens* within the Coachella Valley. In total, 67 species interact within galls formed by Cecidomyidae (midge) larvae. Many of these species are also involved in the subweb centering on creosote (*Larrea divaricata;* see Schultz et al. 1977). Note that the interactions of some species are not fully represented (e.g., *Diguetia, Phyllobaneus*). Modified from Hawkins and Goeden (1984).

associated with carrion are opportunistic: spiders, solpugids, Opiliones, ants, asilids, staphylinid beetles, reduviids, and the vertebrates not only eat insects associated with carrion (or carrion itself) but also prey on other species (McKinnerney 1978). In turn, all these species are eaten by other predators.

Arthropod Parasitoids

Parasitoids from several families of flies, wasps, and beetles are a diverse component of webs representing more than 10% of all animal species (Askew 1971). They lay eggs in or on arthropod hosts; larvae feed on and cause the death of the host (in contrast to parasites in general). Adults almost always feed on other foods (usually of plant origin). Parasitoids' trophic relations are generally quite complex (Askew 1971; Price 1975; Pimm 1982; Hawkins and Goeden 1984; Polis and Yamashita 1991). Hawkins and Lawton (1987) estimate that each species of insect herbivore is host to 5–10 species of parasitoids.

A few studies detail parasitoid-host relationships in the CV. Hawkins and Goeden (1984) studied insects associated with saltbush (*Atriplex*) galls. The system is complex with 67 species (40 common ones), at least five trophic links, and

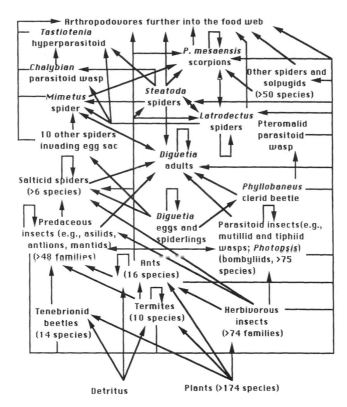

Fig. 4.—Trophic interactions above the soil surface involving a few of the predaceous arthropods living within the Coachella Valley. This subweb is focused around the spiders *Diguetia mohavea* and *Latrodectus hesperus*. Note that no vertebrates are represented.

extensive omnivory. Gall-forming midges (3 species), parasitoid Hymenoptera (26), predators (4), and inquilines (7) interact within galls (fig. 3). Midge larvae are either resource specialists on *Atriplex* or generalists on other plants. Most parasitoids are primary, attacking only midges or inquilines; seven also feed on gall tissue. Two facultative hyperparasitoids feed on gall tissue, midges, and primary and hyperparasitoids. A clerid beetle (*Phyllobaenus*) is in 10% of galls and feeds on at least 17 species from all trophic groups (and spiders; fig. 4).

The trophic relations of *Photopsis* (an abundant mutillid wasp in CV sands) are diagrammed in figure 5 (from Ferguson 1962 and my data). Females oviposit into larval cells, and *Photopsis* larvae consume the entire host. They parasitize several species of hymenopteran larva and are hyperparasitoid on parasitoids of these larva (i.e., other Hymenoptera; stylopid, meloiid, and rhipiphorid beetles; and bombyliid flies). Some hyperparasitized wasps (e.g., sphecids) also may parasitize spiders (this is likely but not established). Up to 37% are destroyed when *Photopsis* larvae themselves are parasitized by some of the same parasitoids (e.g., sphecids) that fall host to *Photopsis*. This is an example of looping via mutual parasitism and tertiary parasitism. Some *Photopsis* larvae are also host to bombyliid (e.g., *Bembix*) and stylopid parasitoids. Further, *Photopsis* larvae also are

DESERT FOOD WEBS   135

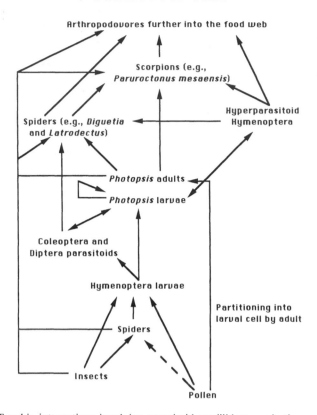

Fig. 5.—Trophic interactions involving parasitoid mutillid wasps in the genus *Photopsis* within the Coachella Valley. Note that the interactions of some species are not fully represented (e.g., scorpions, spiders). A double-headed arrow indicates looping via mutual predation.

cannibalized (and self-regulated? [Ferguson 1962]). Adults eat not only nectar, pollen, and flowers but also ground-nesting Hymenoptera. Adults are frequent prey to many arthropods (e.g., scorpions and spiders; see figs. 4 and 6).

These parasitoid subwebs are characterized by long chains and frequent omnivory, closed-loop omnivory, and looping (via cannibalism, mutual predation, and three-species loops). Further, key species export energy from this subweb when they are predators or prey in the rest of the community.

Overall, more than 20 families of insects are parasitoids of CV insects. Many prefer host species; others, however, are more generalized (see figs. 3 and 5). They sometimes cause high mortality (Mispagel 1978). Several dipteran (tachinid, bombyliid, sarcophagid) and hymenopteran (tiphiid, ichneumonid, mutillid, sphecid, and chalcidoid wasps) parasitoids are common in the CV. These develop on eggs and immature stages of Orthoptera, Neuroptera, Coleoptera, Lepidoptera, Hymenoptera, and Diptera. Velvet mite larvae (*Dinothrombium pandorae*) are ectoparasites of CV grasshoppers (Tevis and Newell 1962).

Spiders host many parasitoids (pompilid, sphecid, and ichneumonid wasps; many Diptera). Adult wasps partition nests with captured spiders; developing

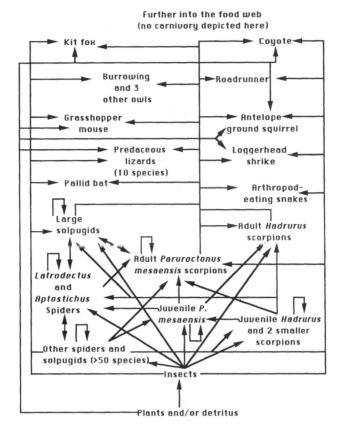

Fig. 6.—Trophic interactions centered around the prey and predators of the scorpion *Paruroctonus mesaensis*. Note that the interactions of some species are not fully represented (e.g., "spiders" and "insects") and carnivory on vertebrates is not depicted (see figs. 5 and 7).

larvae eat the moribund spiders. Adults usually eat nectar. Wasps vary from resource specialists on particular families to generalists on several families. Many pompilids (more than 11 species) occur in the CV (Wasbauer and Kimsey 1985). Abundant *Aporus hirsutus* and the less common *Psorthaspis planata* feed trap-door spiders (*Aptostichus*, Ctenizidae) to their larvae; adults drink sugar secretions from aphids and nectar from more than 10 plant species. Some CV pompilids are hyperparasitoids (e.g., *Evagetes mohave*). Spiders are also beset by a diversity of egg parasitoids/predators (Polis and Yamashita 1991). Further, kleptoparasitic insects (e.g., Drosophiloidea flies) and spiders parasitize spiders by robbing captured prey (G. A. Polis, personal observation).

Parasites

Parasites are a diverse and species-rich group that feed on almost all taxa. They can influence greatly population dynamics and community structure (May 1983*b*) and form another step in the flow of energy (some ectoparasites form yet another "gratis" level when they themselves host parasites). Many feed on several hosts during their complex life cycle. However, parasites are neither well represented in food webs nor well studied in natural communities. Of animals living in the CV, only the parasites of the coyote, lizards, and scorpions were examined in any detail. Otherwise, few data exist on the hundreds of CV endo- and ectoparasites.

Telford (1970) identified parasites in 10 CV lizards. Lizards were infected by several protozoans (mean = 7.8 species; range = 3–10) including flagellates, ciliates, amoebas, sporozoans, and gregarines. Helminth parasites (mean = 1.4; range = 0–5) included nematodes, cestodes, and Acanthocephala. Each lizard also was infested with an unknown number of mite species.

Coyote parasites include mange mites, ticks (seasonal), lice (rare), and *Pulex* fleas (on all individuals) (Gier et al. 1978). Adult fleas feed on blood; flea larvae feed on organic debris. Endoparasites, specifically the tapeworm *Taenia pistiformes*, occur in 60%–95% of all coyotes. The prime intermediate host of *Taenia* is the cottontail rabbit.

It is likely that most (all?) of the free-living animals in the CV harbor parasites. For example, 23.4% of 1,525 birds (112 species) from deserts and other areas in the southwestern United States had blood parasites (Woods and Herman 1943; Welty [1962] lists the diverse parasite fauna of birds). Inspection of CV spiders and insects usually reveals mite infestation. Many genera of CV spiders were reported with nematode parasites. Coachella Valley scorpions support nematodes and eight (pterygosomid) mite species (G. A. Polis, unpublished data).

Arthropod Predators

Arthropods are one of the most important conduits of energy flow in desert webs. Most consumed primary productivity in deserts is utilized by arthropods rather than by vertebrates (Seely and Louw 1980). These arthropods, in turn, are eaten by a host of predators, the vast majority of which are other arthropods (Crawford 1981). Many predaceous arthropods are dense and may play important community roles (Polis and Yamashita 1991); in the CV these include species of scorpions, solpugids, spiders, mantids, ant lions, robberflies, small carabid beetles, and facultatively predaceous ants.

A great variety of arthropods are predaceous sometime during their lives. In the CV, mites (8 families), arachnids (more than 23), and insects (more than 21) are predators as juveniles and adults. The complex life cycle of holometabolic insects often results in different feeding habits between stages. At least 27 families of Diptera (e.g., Tachinidae), Hymenoptera (e.g., Tiphiidae), and Coleoptera (e.g., Cleridae, Meloidae) are trophic generalists: they are predaceous as larvae and herbivorous as adults. For example, *Pherocera* (Therevidae) fly larvae are predators on beetle, fly, and moth larvae (and are cannibals) in sandy CV soils; adults drink nectar. Some parasitic Hymenoptera (e.g., sphecids, pompilids) function as predators: adults eat pollen and nectar but capture prey to feed larvae. Finally, some taxa (e.g., ants, camel crickets) are occasionally but regularly predaceous. For example, after heavy rains in the CV, the diet of *M. pergandei* included 80% *Amitermes wheeleri* termites; normally, arthropods form 2%–10% of this ant's diet (Gordon 1978).

TABLE 1

DIET CLASSIFICATION OF SOME REPRESENTATIVE PREDATORS ON ARTHROPODS

	PERCENTAGE OF ARTHROPOD TAXA IN DIET		
TAXON	Predaceous and Parasitoid	Herbivorous and Detritivorous	n
Arachnids:*			
Hadrurus arizonensis	53	47	15
Paruroctonus luteolus	60	40	10
Paruroctonus mesaensis	47	53	126
Vaejovis confusus	50	50	12
Diguetia mohavea	45	55	71
Latrodectus hesperus	54	46	35
Arachnid average ± SD	51.5 ± 5.4	48.5 ± 5.4	47.3
Lizards:			
Callisaurus draconoides	45	55	22
Cnemidophorus tigris	46	54	15
Gambelia wislizenii	33	67	15
Phrynosoma platyrhinos	28	72	18
Uta stansburiana	39	61	28
Birds:			
Blue-gray gnatcatcher	36	64	69
Burrowing owl	36	64	14
Loggerhead shrike	35	65	17
Roadrunner	35	65	23
Mammals:			
Ammospermophilus leucurus	71	29	7
Antrozus pallidus	31	69	16
Onychomys torridus	40	60	15
Pipistrellus hesperis	32	68	22
Vulpes macrotus	67	33	6
Vertebrate average ± SD	41 ± 12.9	59 ± 12.9	20.5

NOTE.—A taxon is the designated unit in which the diet was classified by the author. It varies from species to families and orders.

* The first four arachnids are scorpions; the last two are spiders.

Most are resource generalists (Polis and Yamashita 1991); for example, *P. mesaensis* is recorded to eat more than 125 prey species, *Digueta mohavea* more than 70 species, and *Latrodectus hesperus,* 35 species. In fact, some scorpions and spiders are neither true trophic specialists nor obligate predators: both scavenge dead arthropods and some spiderlings are aerial plankton feeders, eating pollen and fungal spores trapped by their web (Polis and Yamashita 1991). Facultative predators are trophic generalists eating plants, detritus, dead arthropods, and live prey. A few specialize. Adult velvet mites, *D. pandorae,* feed almost exclusively on termites (Tevis and Newell 1962). *Mimetus* spiders prey primarily on other spiders.

Trophic interactions are complex. Generalized diets are established by size relationships: predators catch what they can subdue. Consequently, smaller and/ or younger arthropods are potential prey and predators eat from all trophic levels. For example, the diet of six CV arthropods averages 51.5% other predaceous arthropods (table 1). Predator-predator feedings are particularly common in deserts because predators form a high proportion of all arthropods (Crawford 1991; Polis and Yamashita 1991). Clearly, a web representing all CV predaceous arthropods would be difficult to depict. Thus, I present subwebs (figs. 4 and 6) centered on three common species that I have studied extensively: the scorpion *P. mesaensis* (see earlier references) and the spiders *D. mohavea* and *L. hesperus* (Nuessly and Goeden 1984; Polis and McCormick 1986a, 1986b; G. A. Polis and K. H. Sculteure, unpublished manuscript).

Spiderling *D. mohavea* develop in sacs protected by the mother until her death. Eggs and spiderlings are then eaten by invaders (spiders [9 families, more than 14 species], solpugids, mites, mantispids, chrysopids, and the clerid beetle *Phyllobaenus* from fig. 3). These stages of most spiders are attacked by similar invaders (Polis and Yamashita 1991). Sibling cannibalism is also frequent (Polis 1981) and occurs among *D. mohavea* (and *L. hesperus*) spiderlings. At least three trophic levels occur in the *D. mohavea* retreat: the clerid likely eats other egg predators and is itself host to a pteromalid wasp parasitoid. Adult *Diguetia* eat more than 70 species, including 14 families of predatory insects (e.g., *Photopsis* and *Phyllobaenus*), and eight spider species, including the same species of invading *Habronatus* salticids and the araneophagous *Mimetus*. One-third of all *D. mohavea* webs include salticid prey. Adult *D. mohavea* are fed upon by *Mimetus*, *P. mesaensis*, birds, and a parasitoid pompilid. *Diguetia* is involved in at least four cases of mutual predation.

Predators form 54% of *L. hesperus*'s diet (table 1). *Latrodectus hesperus* is prey of at least eight predators including other *L. hesperus* and three predators that it eats (mutual predators = *Steatoda grossa*, *Steatoda fulva*, *P. mesaensis*). The sphecid (*Chalybian californicum*) specializes on *Latrodectus* and other theridiid spiders (e.g., *Steatoda*) (Wasbauer and Kimsey 1985). *Tastiotenia festiva* (Pompilidae) is a hyperparasitoid eating both cached theridiid spiders and developing wasps.

The biomass (g/ha) of *P. mesaensis* is the greatest of any CV predator (including vertebrates). Diet shifts during growth partially explain trophic interactions with more than 125 prey species, including 47% other predators (table 1), and mutual predation with at least 10 species (three scorpions, five solpugids, two spiders—young *P. mesaensis* are eaten by the same species eaten by adults).

Note the complexity of these webs: looping via mutual predation and cannibalism is frequent; (closed-loop) omnivory is the norm; omnivorous predators feed on herbivores, detritivores, predators, and predators of predators. Consequently, chain lengths are long even excluding parasitoids and loops (e.g., detritus–termite–*Messor* ants–ant lion–*Latrodectus*–*Steatoda*–*Mimetus*–*P. mesaensis*–Eremobatid solpugids–*Hadrurus* scorpion–[vertebrate subweb]). I strongly suspect that the depicted interactions are representative of the hundreds of other arthropod predators in the CV. Omnivory (due to age structure, opportunism, and generally catholic diets) combined with a high diversity of insect and arachnid predators necessarily creates trophic complexity. Complexity increases even further when we consider vertebrate predators of these arthropods.

Arthropodivorous Vertebrates

Arthropodivory is the consumption of arthropods. A less familiar word than insectivory, it conveys that predators eat all types of arthropods (insects, arach-

TABLE 2

FEEDING CATEGORIES OF THE VERTEBRATES RESIDENT IN THE COACHELLA VALLEY

FEEDING CATEGORY	VERTEBRATE CLASS			
	Reptiles (n = 21)	Birds (n = 56)	Mammals (n = 18)	All Vertebrates (n = 95)
Granivory:				
Primary	0 (0)	14 (25)	8 (44)	22 (23)
Secondary	0 (0)	0 (0)	0 (0)	0 (0)
Total	0 (0)	14 (25)	8 (44)	22 (23)
Herbivory:				
Primary	2 (10)	5 (9)	5 (28)	11 (12)
Secondary	5 (24)	15 (27)	1 (6)	21 (22)
Total	7 (33)	20 (34)	6 (33)	32 (34)
Arthropodivory:				
Primary	11 (52)	34 (61)	9 (50)	55 (58)
Secondary	4 (19)	15 (27)	5 (28)	24 (25)
Total	15 (71)	49 (88)	14 (78)	79 (83)
Carnivory:				
Primary	9 (43)	12 (21)	3 (17)	24 (25)
Secondary	2 (10)	2 (4)	1 (6)	5 (5)
Total	11 (52)	14 (25)	4 (22)	29 (31)

NOTE.—Species are classified as belonging primarily (food is a major component of the diet) or secondarily (food < 10% of total diet) to a feeding category. Some omnivorous species (2 reptiles, 9 birds, 7 mammals) belong to two (each > 33% of diet) or three (each >20% of diet) primary categories. The numbers in each column indicate the number and percentage (in parentheses) of species in the primary (e.g., 11 herbivorous species) or secondary (e.g., 21) feeding category and the total in this category (e.g., 32).

nids, myriapods, and terrestrial Crustacea). Most vertebrates (83% of 95 species) in the CV eat arthropods (table 2). Over half (58%) primarily eat arthropods, including 52% of the reptiles, 61% of the birds, and 50% of the mammals. Most (80%) of the 25 primary carnivores also feed on arthropods. Two-thirds (24 of 36) of the primarily herbivorous/granivorous vertebrates eat arthropods, at times in large quantities (e.g., 88%–97% of the seasonal diet of the sage sparrow in the Great Basin Desert). In total, 71% of all reptile species, 88% of birds, and 78% of mammals primarily or secondarily eat insects or arachnids (only 17 species are not reported to eat arthropods).

These vertebrates are resource generalists eating all trophic categories of arthropods (herbivores, detritivores, predators, parasitoids). For example, of 36 arthropodivorous birds whose diet is detailed sufficiently, 58% eat spiders in addition to insects. Seven of 10 lizards eat spiders and four eat scorpions. Spiders are eaten by three of 14 arthropod-eating mammals; scorpions, by five. Predaceous arthropods average 41% of the diet of the vertebrates in table 1. Further, these vertebrates tend to be trophic generalists. Most (28 of 55 species = 51%) primary arthropodivores eat plants (59% of 79 vertebrates that eat arthropods also eat plants). Of these 79, 32% are also carnivorous. A few arthropodivorous vertebrates tend to specialize. Ants form 89% by frequency of the prey of the horned lizard *Phrynosoma platyrhinos*. However, these lizards eat 17 other categories of prey (including spiders and solpugids) and 20%–50% of the diet of some

DESERT FOOD WEBS   141

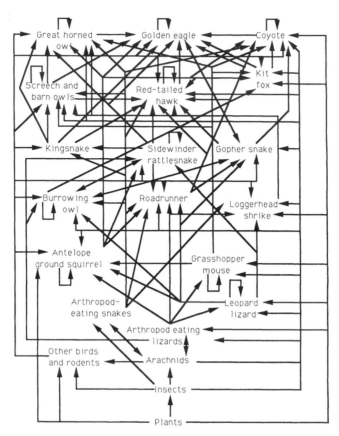

FIG. 7.—Trophic interactions involving a few of 96 vertebrates resident in the Coachella Valley. Note that the bottom of this subweb is simplified from the proceding five subwebs (figs. 2–6).

individuals is beetles. The worm snake (*Leptotyphlops humilis*) mainly eats termites and ants. No other CV vertebrates specialize on certain arthropods.

The omnivory exhibited by arthropodivorous vertebrates is illustrated in the web focused on predators of *P. mesaensis* (fig. 6). All 16 reptiles, birds, and mammals recorded to eat *P. mesaensis* eat (predaceous, detritivorous, herbivorous) insects; 44% eat plant material. Many (81%) are also carnivores (see fig. 7; three of 11 arthropods eating *P. mesaensis* also eat vertebrates).

Carnivorous Vertebrates

Carnivores kill and eat vertebrates. Of the 95 vertebrates, 24 are primarily carnivorous and five others eat vertebrates occasionally (table 2). Reptiles are the most carnivorous (9 primary, 2 secondary of 21 species = 52%; 8 of 10 snakes are primarily carnivores). The proportion of carnivores is about the same for birds (12, 2 of 56 species = 25%) and mammals (3, 1 of 18 species = 22%).

All carnivores are resource generalists preying on many vertebrates. Most (79% = 19 of 24 primary carnivores) are trophic generalists that eat arthropods (71%)

and/or plants (33%). For example, coyotes eat mammals (12 of 19 species in the CV: rabbits, rodents, gophers, antelope ground squirrels, and even kit foxes and other coyotes), birds (including eggs and nestlings; e.g., roadrunners, doves, quails), snakes (e.g., gopher and kingsnakes), lizards (e.g., horned lizards), and young tortoises as well as scorpions, insects, and fruit (see fig. 7; Johnson and Hansen 1979; G. A. Polis, unpublished data). Coyotes actively forage on nonvertebrate prey; for example, they excavate entire ant nests in the CV (Ryti and Case 1986). Great horned owls consistently eat arthropods (14.7% of diet; e.g., spiders, centipedes, Orthoptera) in addition to rodents, squirrels, lizards, snakes, and other horned owls (Ohlendorf 1971; Johnegard 1988). Many carnivores (33%) feed on carrion in addition to live prey: sidewinder rattlesnake, raven, golden eagle, horned owl, red-tailed hawk, coyote, kit fox, and antelope ground squirrel. Such scavenging means that these primary carnivores also include in their diets both microorganisms (Janzen 1977) and the rich arthropod fauna that use decaying carcasses (e.g., the horned owl).

Figure 7 is a subweb focusing on some carnivores in the CV. Note the frequency of omnivory, closed-loop omnivory, cannibalism (10 of 16 species in fig. 7), and mutual predation. In each of the three cases of mutual predation A adults eat B, but the eggs or nestling-stage individuals of A are eaten by B. For example, many snakes are nest predators (e.g., in the CV, whip snake, sidewinder, rosy boa, gopher snake). These snakes eat eggs and nestlings of species (e.g., burrowing owls) whose adults are predators upon the same snakes (e.g., gopher snakes, sidewinders). Finally, note that only the golden eagle approaches the status of top predator (i.e., a species without predators). However, even the golden eagle may not be a true top predator: at other locations, gopher snakes eat golden eagle eggs, and parasitic thrichomoniasis and sibling fratricide/cannibalism cause nestling mortality (30% and 8%, respectively; Olendoroff 1976; Palmer 1988). Thus, here is a community and food web with potentially no top predators.

DISCUSSION

Sandy areas in the Coachella Valley are the habitat for about 175 species of vascular plants, 100 vertebrates, and thousands of arthropods, parasites, and soil organisms. These species form a community connected trophically into a single food web. Although only a few components of the CV web were presented, it is possible to summarize several general trends.

1. Each subweb is complex because of the large number of interactive species, age structure, and high omnivory. A web representing all species would increase complexity even more.

2. Age structure is central to complexity. Growth necessitates and/or allows changes in diet (life-history omnivory), either gradually or radically (e.g., predaceous larvae, herbivorous adults). Predators also change with size.

3. Different microhabitats and times are connected trophically. At first inspection, distinct compartments appear (e.g., diurnal vs. nocturnal, surface vs. subsurface). However, extensive crossover exists. Foraging times change from day to night as a function of temperature; nocturnal species eat diurnal prey (e.g.,

scorpions on robberflies). Plants and detritus are eaten during all periods by species that live above and below the surface. Arthropods feeding below ground export energy when they become surface-dwelling adults (e.g., tenebrionids). "Different-channel omnivores" link compartments by feeding from different sub-webs or "energy channels" (Moore et al. 1988). Consumers eat food types rather than specialize on trophic levels (Polis et al. 1989). Such connections decompartmentalize webs and increase complexity.

The CV is not unique in its diversity or properties of its food web. Other deserts are also characterized by thousands of species (Polis 1991b). The same suite of plants, herbivores, detritivores, arthropodivores, parasitoids, parasites, and carnivores is present in all deserts. In particular, the universal existence of diverse assemblages of predaceous arthropods (Polis and Yamashita 1991) and soil organisms (Freckman and Zak 1991) must contribute to a trophic complexity similar to that of the CV. Further, omnivory is normal among desert consumers (Noy-Meir 1974; Orians et al. 1977; Seely and Louw 1980; Brown 1986).

R. Bradley (1983) also illustrates the complexity in deserts with a source web focused on predators of camel crickets in sandy areas of the Chihuahuan desert. These predators (scorpions, solpugids, burrowing owls, grasshopper mice, and pallid bats) are quite omnivorous (only 2 of 27 species pairs are noninteractive, i.e., not linked as predator or prey). Closed-loop omnivory occurs for every predator. Looping is common: six cases of mutual predation occur; six of eight species are cannibalistic. Finally, at least 18 three-species loops exist.

Comparison with Published "Empirical Generalizations"

Theorists have produced a series of generalizations derived from catalogs of published webs (see App. A). These generalizations are entering accepted ecological literature (see May 1986, 1988; Lawton and Warren 1988; Lawton 1989; Schoener 1989). The food web of the CV offers little support for these patterns.

Web patterns from the CV are now compared with those from published catalogs of webs. An argument could be made that CV web statistics should not be calculated. Even with its complexity, this web is vastly incomplete and literally scores of other subwebs could be presented. Statistics would represent only the level of complexity I arbitrarily present rather than true web parameters. However, to facilitate comparisons, I present a highly simplified web/matrix of the CV community (App. B). The thousands of species are heavily lumped into 30 "kinds of organisms" forming 22,220 chains. Table 3 compares 19 web statistics from the CV with those from published catalogs of webs (Pimm 1982; Briand 1983; Cohen et al. 1986; Schoener 1989; Schoenly et al. 1991). In the discussion below, I use the statistics from the simplified web and those from the more complete web depiction presented throughout the text.

1. The number of interactive species in the CV web is two orders of magnitude greater than the average number (17.8, Briand 1983; 16.7, Cohen et al. 1986; 24.3, Schoenly et al. 1991) from webs analyzed by theorists. In fact, Briand and Cohen's most species-rich web contains only 48 species, less diverse than any of the following CV taxa: plants, nematodes, mites, arachnids, bees, beetles, bombyliid flies, and birds.

TABLE 3

Comparison of Statistics from the Coachella Valley Food Web
with Means of the Statistics from Cataloged Webs

	Cohen et al. and Briand	Schoenly et al.	Coachella*
Total number of "kinds" or species (S)	16.7	24.3	30
Total number of links per web (L)	31	43.1	289
Number of links per species (L/S)	1.99	2.2	9.6
Number of prey per predator	2.5	2.35	10.7
Number of predators per prey	3.2	2.88	9.6
Total prey taxa/total predator taxa	.88	.64	1.11
Minimum chain length	2.22	1	3
Maximum chain length	5.19	7	12†(18)
Mean chain length	2.71, 2.86	2.89	7.34†
Connectance ($C = L/S[S - 1]/2$)		.25	.49†
Species × connectance (SC)	2–6	4.3	14.7†
Basal species (%)	19	16	10
Intermediate species (%)	52.5	38	90
Top predators (%)	28.5	46.5	0
Primarily or secondarily herbivorous (%)		14.6	60
Primarily or secondarily saprovorous (%)	21	35.5	37
Omnivorous (%)	27	22	78
Consumers with "self-loop" (%)	<1.0	<1.0	74
Consumers with mutual predation loop (%)	≪1.0	≪1.0	53

Sources.—Briand 1983; Cohen et al. 1986. See also Schoener 1989; Schoenly et al. 1991.
* The Coachella statistics are derived from the extremely simplified and highly lumped food-web matrix presented in App. B.
† Mutual links (loops) were not used to calculate the statistic.

    2. Coachella Valley chain lengths average more links than 2.86 (Briand 1983), 2.71 (Cohen et al. 1986), or 2.89 (Schoenly et al. 1991). Even excluding looping and parasites, lengths of 6 to more than 11 are common. Both J. E. Cohen (personal communication) and I interpret short chain length as an artifact of totally inadequate descriptions of real communities. Short average lengths are derived from catalogs biased toward less complex, vertebrate-centered webs. Chains are lengthened in the CV primarily during trophic interactions among the soil biota, the arthropods, and intermediate level predators. Webs including invertebrates are typically more complex than those centering on vertebrates (see 10 below). Shorter chains exist (e.g., plant–rabbit–eagle), but these are much less frequent simply because vertebrates form a relatively small proportion of all species when arthropods and soil biota are not ignored. For example, chains containing plants, insect herbivores, and insect parasitoids are estimated to contain over half of all known metazoans (Hawkins and Lawton 1987). The average length of all chains in the simplified CV web is 7.3; its maximum chain length is 18 (12 with no loops).
    3. Omnivory is frequent in the CV web but statistically "rare" in cataloged webs (Pimm 1982; Yodzis 1984). In cataloged webs, 22% (Schoenly's catalog) to 27% (Briand and Cohen's catalog) of all "kinds" are omnivorous; 78% in the simplified CV web are omnivorous. Adequate diet data, not lumping arthropods, and the inclusion of age structure partially explain the ubiquity of omnivory in the CV. It is possible that long chains in the CV exist because energy to top

consumers comes from many (lower) trophic levels in addition to adjacent upper levels (see also Sprules and Bowerman 1988).

4. Loops are "unreasonable" to modelers and purported to be "very rare in terrestrial" webs (Pimm 1982; Pimm and Rice 1987; Cohen et al. 1990). Looping also violates the assumption that body-size relations arrange species along a cascade or hierarchy, such that a species preys on only those species below it and is preyed on only by those above it ("upper triangular web structure"; Warren and Lawton 1987; Lawton 1989; Cohen et al. 1990). However, looping is neither rare nor abnormal in the CV. Most frequently, loops are a product of age structure: cannibalism and mutual predation usually result when large individuals eat smaller or younger individuals. Other factors also produce mutual predation (e.g., group predation by ants). In the simplified community matrix, 74% of the consumers show self-loops; 44% are involved in loops with another "kind."

5. Coachella Valley species interact with many more species than those in cataloged webs. Cohen (1978) and Schoenly et al. (1991) calculated the number of predators on each prey (mean = 3.2 and 2.88, respectively) and the number of prey per predator (2.5 and 2.35). Overall, cataloged species interact directly with 3.2–4.6 other species (Cohen et al. 1985). Parameters from the CV are one to two orders of magnitude greater. Higher values exist first because most CV consumers eat many species (e.g., see table 1; diets range from 15 to more than 125 items; a few arthropod herbivores and some parasites are exceptions). Second, most individual species are eaten by tens to hundreds (e.g., mesquite) of species. Such discrepancies occur because cataloged webs lump several species into "kinds" and diet data are grossly inadequate. However, even the highly lumped CV web shows that each "kind" is eaten by about 10 predators and each consumer eats about 10 prey (table 3).

6. Top predators are rare or nonexistent in the CV. Coyotes, kit foxes, horned owls, and golden eagles (the largest predators in the CV) suffer the fewest predators, but each is the reported prey of other species. This finding stands in marked contrast with cataloged webs: 28.5% (Briand and Cohen 1984; Cohen et al. 1990) to 46.5% (Schoenly et al. 1991) of kinds were top predators. This great discrepancy is undoubtedly due to the inadequacy of diet information in cataloged webs and/or to the fact that these webs only focus on a limited subset of a trophically linked community.

7. Coachella Valley data pose great difficulty to the observation that prey/predator ratios are greater than 1.0 (0.88, Briand and Cohen 1984; 0.64, Schoenly et al. 1991); that is, the number of organisms heading rows (prey) in web matrices is less than the number heading columns (predators). It is easy to show that the ratio in the CV and other real communities should be greater than 1.0. Because all heterotrophs must obtain food, every animal should head a column. Rows (prey) include plants, detritus, and all animals except those with no predators (i.e., top predators). Let x be the number of animal species that are intermediate predators (i.e., both predator and prey). Then the total number of prey is x + the number of plant species (174 in the CV) + the number of categories of detritus and carrion; the number of predators is x + the number of top predators (0 or 1 in the CV). If more species of plants exist than top predators, then the prey/

predator ratio will always be greater than 1.0. Few real (no?) communities have more top predators than autotrophs. The appearance that top predators are more frequent is an artifact (see 6 above). It appears that lumping obliterates the actual ratio primarily because more species of plants are lumped than (easily recognized) animals that are top predators. The CV community web exhibits a prey/predator ratio of 1.11. Glasser (1983), Paine (1988), and Lockwood et al. (1990) also criticize prey/predator ratios.

8. Factors 1–7 make the CV web much more complex than cataloged webs. For example, the number of trophic links varies from 31 (Cohen et al. 1986) to 43 (Schoenly et al. 1991); only 2 of Cohen et al.'s (1986) 113 webs had more than 100 links. The average CV *subweb* (figs. 2–7) has 54.7 links; the carnivore subweb alone, 107; and the simplified community web, 289.

9. The CV web questions the utility of the concept of "trophic level." A trophic level is defined as a set of organisms with a common number of chain links between them and primary producers. The nearly universal presence of omnivory and age structure makes this concept nonoperational. What trophic level are consumers that ontogenetically, seasonally, or opportunistically eat all trophic levels of arthropods in addition to plant material and (for carnivores) vertebrates? Looping further blurs the concept. If A eats B but B eats A, is B on the first, third, or (after another loop) fifth trophic level (ad infinitum)? I am not alone in criticism of this concept (see Gallopin 1972; Cousins 1980, 1987; Levine 1980; Lawton 1989).

10. Patterns 4 and 5 in Appendix A are confirmed by the CV web. First, separate compartments did not exist within one habitat. Second, arthropod-dominated systems are more complex than those dominated by vertebrates. However, few communities are not dominated (in number of individuals or species) by arthropods (Hawkins and Lawton 1987; May 1988). So not lumping arthropods, the most species-rich taxon on this planet, should increase the complexity of any web, including those cataloged by theorists.

Overall, a general lack of agreement exists between patterns from the CV and those from catalogs of published webs. Is the CV web unique or are cataloged webs so simplified that they have lost realism? That cataloged webs depict so few species, such low ratios of predators on prey and prey eaten by predators, so few links, so little omnivory, a veritable absence of looping, and such a high proportion of top predators argues strongly that they do not adequately describe real biological communities. Taylor (1984), Paine (1988), Lawton (1989), and Winemiller (1990) reach a similar conclusion.

This conclusion carries important implications. First, controversies over such issues as the causes of short chain length and omnivory, complexity versus stability of ecosystems, and the role of dynamics versus energetics in shaping web patterns (see DeAngelis et al. 1983; Yodzis 1984; Lawton 1989) are based on patterns abstracted from cataloged webs. If catalogs are an inadequate data base, these controversies may be nonissues and theorists are trying to explain phenomena that do not exist. This is a real possibility (see Lawton and Warren 1988; Paine 1988; Lawton 1989; Winemiller 1990). Second, characteristics of the CV web (complexity, omnivory, long chain lengths, looping, absence of compartments) considerably reduce stability in dynamic models of food webs (see Pimm 1982; Pimm and Rice 1987). These simplified models apparently tell us little about the structure of food webs in nature.

Self-Critique and Prospectus

Several issues need to be addressed.

1. Should all trophic links be included, or are some too weak or unusual to list (May 1983a, 1986; Paine 1988; Lawton 1989; Schoener 1989)? I included all links in the CV web. This decision was based on four factors. First, it is arbitrary and impossible to evaluate which links are and which are not important. Most CV consumers eat 20 to more than 50 items. Which should be included, which excluded? Some consumers (esp. in deserts) likely exist because they are sufficiently flexible to include a number of infrequent links that sum into an important source of energy, at least during some periods. For example, the seasonal switch by granivorous birds to arthropods provides protein and water for nestlings and may allow permanent residence for some desert birds (Welty 1962; Brown 1986).

Second, dynamics and trophic linkage are not necessarily correlated (Paine 1988; Lawton 1989). A 1% frequency of a rare species in the diet of a common species may produce high mortality; conversely, a rare species eating only one common species may scarcely affect dynamics. Further, short but intensive predation events may contribute little to the diet of a predator but may be central to prey dynamics (e.g., Wilbur et al. 1983; Polis and McCormick 1987).

Third, each link enriches the description of the community (the original purpose of food webs). We must not blur the distinction between food webs and interaction/functional webs. Food webs (such as the CV web) are based on observations of trophic relationships while interaction webs summarize the subset of all interactions that are "strong." This insight argues that all trophic links should be included because this is a food web, not an interaction web. No justification exists for excluding links from a food web. Fourth, the exclusion of certain links produces systematic bias against complexity.

2. Are desert food webs (a)typical of other webs? Deserts may differ in two main ways from other systems. First, deserts are relatively simple ecosystems characterized by low productivity and low species richness (Noy-Meir 1974; Seely and Louw 1980; Louw and Seely 1982; Whitford 1986; Polis 1991b). Should such depauperate communities translate into relatively simple webs? If so, the complexity of the CV web is much less than that of more species-rich systems.

Second, is omnivory more common in deserts than in other habitats? It is impossible to answer this question with rigor. I can only indicate that features that promote omnivory (and web complexity) in the CV are present in all systems: age structure, life-history omnivory, predators that ignore the feeding history of prey (different-channel omnivory), opportunistic feeding, and consumers that eat food in which other consumers live (e.g., scavengers, frugivores, granivores, detritivores; Polis et al. 1989). High omnivory is not restricted to deserts. Price (1975) and Cousins (1987) argue that it is normal throughout the animal kingdom. Omnivory characterizes feeding in aquatic and marine habitats (Menge and Sutherland 1987). These authors and Walter (1987) maintain that omnivory is also common in terrestrial communities. Walter and Moore et al. (1988) provide strong evidence that omnivory is one of the most frequent and dynamically important trophic links among soil and detrital webs. Sprules and Bowerman (1988) analyzed zooplankton in 515 lakes and concluded that omnivory is common. (They also note long chains, looping, cannibalism, and mutual predation.)

Regardless of whether deserts are unique, desert food webs are still of general importance. Deserts occupy at least one-quarter of the earth's land surface (Crawford 1981; Polis 1991*b*), and patterns observed in the CV thus should describe a good fraction of the terrestrial communities on this planet.

3. Not all diet data for Coachella Valley species came from the Coachella. How this affects the web is uncertain. However, the overall conclusion of great complexity should not be influenced unduly by errors arising from use of these studies.

4. The sections of this article that critique "food-web theory" tend to be nihilistic. A more satisfying article would further suggest how we can replace or revise these theories. Because of space limitations, I cannot focus on future directions other than to offer the general caveat that we cannot overlook complexity and reality in our attempts to comprehend nature. However, this article suggests that a host of important issues need to be resolved. Foremost, we need criteria to recognize which links are "sufficiently important" to include in descriptions and analyses of communities. We must establish more uniformly the spatial limits and species membership of a community and not analyze webs as different as those found in catalogs (Arctic seas to toad carrion; Briand 1983; Sugihara et al. 1989). We need to assess what easily studied subwebs (e.g., phytotelmata, plant galls, carrion) tell us about the structure of community webs and how isolated or connected these subwebs are with respect to the rest of the community. We must determine how omnivory affects the recognition of important links and how the ubiquity of omnivory affects web structure. Finally, we must incorporate age structure, changes over time and space, and recycling via the detrital loop into web analyses.

CONCLUSIONS

It appears that much "food-web theory" is not very descriptive or predictive of nature. The catalogs of webs used to abstract empirical generalizations were derived from grossly incomplete representations of communities in terms of both diversity and trophic connections. Consequently, theorists have constructed an oversimplified and invalid view of community structure. The inherent complexity of natural communities makes web construction by empiricists and analysis by theorists difficult.

So what good are food-web representations and analyses? At a minimum, they are descriptors of communities and trophic interactions. As a future goal, they may tell us how communities function and are assembled. In an ideal world, hypotheses and generalizations made by theorists could be viewed as a stage in the evolution of understanding food webs; they are definitely not finished products (J. E. Cohen, and S. L. Pimm, personal communication). Theory is designed to provoke concept-focused fieldwork that hopefully promotes the next iteration of descriptive models and generalizations that will, in turn, encourage more empirical work, all the while approaching more accurate predictions and descriptions of reality. However, a strong implication of my analysis is that many existing hypotheses and generalizations are simply wrong. For example, much theory (see Pimm 1982; DeAngelis et al. 1983; May 1983*a;* Lawton and Warren 1988) would hold that the Coachella Valley food web should be completely unstable. To advance our understanding, an adequate data base of community food webs must be assembled, experiments must be conducted, and new questions asked. This will take time. Only then will a useful, heuristic theoretical framework be constructed. Meanwhile, we should reevaluate where we now stand: it appears that much food-web theory is in critical need of revision and new direction.

ACKNOWLEDGMENTS

Many people contributed ideas, data, and energy during the 10-yr gestation of this paper. S. McCormick Carter was central to its development. S. Frommer, W. Icenogle, W. Mayhew, J. Pinto, K. Sculteure, R. Ryti, A. Muth, and F. Andrews provided data on the Coachella. The manuscript benefited greatly from suggestions by J. Cohen, C. Crawford, R. Holt, J. Lawton, B. Menge, J. Moore, C. Myers, S. Pimm, K. Schoenly, T. Schoener, and the anonymous reviewers. A special thanks to K. Schoenly for calculating many of the web statistics in table 4. Fieldwork was financed partially by the National Science Foundation and the Natural Science Committee and Research Council of Vanderbilt University.

APPENDIX A

A PARTIAL SUMMARY OF "FEATURES OBSERVED IN REAL FOOD WEBS"

These "empirical generalizations" were derived by food-web theorists from the catalog of published webs assembled by Cohen (1978), Briand (1983), and Schoenly et al. (1991; see text). All quoted phrases are from chapter 10 in Pimm 1982 (see esp. table 10.1). Roman numerals in parentheses refer to pattern number in Pimm's table 10.1.
1. Chain lengths are limited to "typically three or four" (2.86–3.71) trophic levels (iii)
    (Pimm 1982; Briand 1983; Cohen et al. 1986; Schoener 1989; Schoenly et al. 1991).
2. Omnivores are statistically "rare" (iv)
    (Pimm 1982; Yodzis 1984).
3. Omnivores feed on adjacent trophic levels (v)
    (Pimm 1982).
4. Insects and their parasitoids are exceptions to patterns 2 and 3 (vi)
    (Pimm 1982; Hawkins and Lawton 1987).
5. Webs are usually compartmentalized between but not within habitats (viii)
    (Pimm 1982).
6. Loops are rare or nonexistent and do not conform to "biological reality" (i)
    (Gallopin 1972; Cohen 1978; Pimm 1982; Pimm and Rice 1987; Lawton and Warren 1988; Cohen et al. 1990).
7. The ratio of prey species to predator species is greater than 1.0 (0.64–0.88) (x)
    (Cohen 1978; Briand and Cohen 1984; Schoenly et al. 1991).
8. The proportion of species of top predators to all species in a community averages 0.29–0.46
    (Briand and Cohen 1984; Schoenly et al. 1991).
9. Species interact directly (as predator or prey) with only 2–6 other species
    (Cohen 1978; Cohen et al. 1985; Schoenly et al. 1991).

APPENDIX B

TABLE B1

Summary Food-Web Matrix for the Coachella Valley Sand Community

	4	5	6	7	8	9	10	11	12	13	14	15	16	17	18	19	20	21	22	23	24	25	26	27	28	29	30
1	/	/		/		/	/				/	/	/	/	/	/	/	/	/	/	/		/	/		/	
2	/	/		/	/	/					/																
3	/	/		/							/									/				/	/	/	/
4	/	/	/	/	/	/	/				/									/				/	/	/	/
5			/		/																						
6			/		/																						
7				/	/			/	/	/	/	/		/		/		/	/	/	/	/	/		/	/	/
8						/		/	/	/	/	/		/		/		/	/	/	/	/	/		/	/	/
9								/	/	/	/	/		/		/		/	/	/	/	/	/		/	/	/
10								/	/	/	/	/		/		/		/	/	/	/	/	/		/	/	/
11								/	/	/	/	/	/	/		/		/	/	/	/	/	/		/	/	
12								/	/	/	/	/	/	/		/		/	/	/	/	/	/		/	/	
13								/	/	/			/	/		/		/	/	/	/	/	/		/	/	/
14								/	/	/	/	/		/		/		/	/	/	/	/	/		/	/	/
15								/	/	/	/	/		/		/		/	/	/	/	/	/		/	/	/
16								/	/	/	/	/	/		/	/		/	/	/	/	/	/		/	/	/
17								/	/	/	/	/	/		/	/											
18								/	/	/	/	/	/		/	/											
19								/	/	/	/	/	/		/	/											
20																				/				/	/	/	/
21																				/				/	/	/	/
22																			/					/	/	/	/
23																				/				/	/	/	
24												/								/	/			/	/	/	/
25											/								/	/		/	/	/	/	/	/
26																			/	/		/	/	/	/	/	/

DESER FOOD WEBS   151

TABLE B1 (*Continued*)

	4	5	6	7	8	9	10	11	12	13	14	15	16	17	18	19	20	21	22	23	24	25	26	27	28	29	30
27																		/	/				/	/	/	/	
28																							/	/		/	
29																									/	/	
30																								/			/

NOTE.—The thousands of species were lumped extensively to 30 "kinds of organisms" in order to facilitate comparison with webs contained in the various catalogs. Entries along the diagonal indicate a "self-loop" within that kind of organism. Entries below the diagonal are loops caused by mutual predation. Some statistics from this matrix are summarized in table 3. The kinds of organisms are as follows: 1, plants/plant products; 2, detritus; 3, carrion; 4, soil microbes; 5, soil microarthropods and nematodes; 6, soil micropredators; 7, soil macroarthropods; 8, soil macroarthropod predators; 9, surface arthropod detritivores; 10, surface arthropod herbivores; 11, small arthropod predators; 12, medium arthropod predators; 13, large arthropod predators; 14, facultative arthropod predators; 15, life-history arthropod omnivore; 16, spider parasitoids; 17, primary parasitoids; 18, hyperparasitoids; 19, facultative hyperparasitoids; 20, herbivorous mammals and reptiles; 21, primarily herbivorous mammals and birds; 22, small omnivorous mammals and birds; 23, predaceous mammals and birds; 24, arthropodivorous snakes; 25, primarily arthropodivorous lizards; 26, primarily carnivorous lizards; 27, primarily carnivorous snakes; 28, large, primarily predaceous birds; 29, large, primarily predaceous mammals; and 30, golden eagle.

LITERATURE CITED

Andersen, D. 1987. Below ground herbivory in natural communities: a review emphasizing fossorial animals. Quarterly Review of Biology 62:261–286.

Andrews, F., A. Hardy, and D. Giuliani. 1979. The coleopterous fauna of selected California sand dunes. California Department of Food and Agriculture Report, Sacramento, Calif.

Askew, R. 1971. Parasitic insects. Elsevier, New York.

Bradley, R. 1983. Complex food webs and manipulative experiments in ecology. Oikos 41:150–152.

Bradley, W. 1968. Food habits of the antelope ground squirrel in southern Nevada. Journal of Mammalogy 49:14–21.

Briand, F. 1983. Environmental control of food web structure. Ecology 64:253–263.

Briand, F., and J. Cohen. 1984. Community food webs have invariant-scale structure. Nature (London) 5948:264–267.

Brown, J. 1986. The role of vertebrates in desert ecosystems. Pages 51–71 in W. Whitford, ed. Pattern and process in desert ecosystems. University of New Mexico Press, Albuquerque.

Brown, J., O. J. Reichman, and D. Davidson. 1979. Granivory in desert ecosystems. Annual Review of Ecology and Systematics 10:201–227.

Cohen, J. E. 1978. Food webs and niche space. Monographs in Population Biology 11. Princeton University Press, Princeton, N.J.

Cohen, J. E., and C. Newman. 1985. A stochastic theory of community food webs. I. Models and aggregated data. Proceedings of the Royal Society of London B, Biological Sciences 224:421–448.

Cohen, J. E., C. Newman, and F. Briand. 1985. A stochastic theory of community food webs. II. Individual webs. Proceedings of the Royal Society of London B, Biological Sciences 224:449–461.

Cohen, J. E., F. Briand, and C. Newman. 1986. A stochastic theory of community food webs. III. Predicted and observed lengths of food chains. Proceedings of the Royal Society of London B, Biological Sciences 228:317–353.

———. 1990. Community food webs: data and theory. Springer, Berlin.

Cousins, S. 1980. A trophic continuum derived from plant structure, animal size and a detritus cascade. Journal of Theoretical Biology 82:607–618.

————. 1987. The decline of the trophic level concept. Trends in Ecology & Evolution 2:312–316.

Crawford, C. S. 1981. Biology of desert invertebrates. Springer, New York.

————. 1986. The role of invertebrates in desert ecosystems. Pages 73–92 in W. Whitford, ed. Pattern and process in desert ecosystems. University of New Mexico Press, Albuquerque.

————. 1991. Macroarthropod detritivores. Pages 89–112 in G. A. Polis, ed. Ecology of desert communities. University of Arizona Press, Tucson.

DeAngelis, D., W. M. Post, and G. Sugihara. 1983. Current trends in food web theory: report on a food web workshop. Technical Memo 5983. Oak Ridge National Laboratory, Oak Ridge, Tenn.

Edney, E., J. McBrayer, P. Franco, and A. Phillips. 1974. Distribution of soil arthropods in Rock Valley, Nevada. International Biological Program, Desert Biome Research Memorandum 74 32:53–58.

Elgar, M. A., and B. J. Crespi. In press. The ecology and evolution of cannibalism. Oxford University Press, Oxford.

Ferguson, W. 1962. Biological characteristics of the mutellid subgenus *Photopsis* Blake and their systematic value. University of California Publications in Entomology 27:1–82.

Franco, P., E. Edney, and J. McBrayer. 1979. The distribution and abundance of soil arthropods in the northern Mojave Desert. Journal of Arid Environments 2:137–149.

Freckman, D., and J. Zak. 1991. Desert soil communities. Pages 55–88 in G. A. Polis, ed. Ecology of desert communities. University of Arizona Press, Tucson.

Frommer, S. I. 1986. A hierarchic listing of the arthropods known to occur within the Deep Canyon Desert Transect. Deep Canyon, Riverside, Calif.

Gallopin, G. 1972. Structural properties of food webs. Pages 241–282 in B. Patton, ed. Systems analysis and simulations in ecology. Vol. 2. Academic Press, New York.

Ghilarov, M. 1964. Connection of insects with soil in different climatic zones. Pedobiologia 4:310–315.

Gier, H., S. Kruckenberg, and R. Marler. 1978. Parasites and diseases of coyotes. Pages 37–69 in M. Bekoff, ed. Coyotes: biology, behavior and management. Academic Press, New York.

Glasser, J. 1983. Variation in niche breadth and trophic position: the disparity between expected and observed species packing. American Naturalist 122:542–548.

Gordon, S. 1978. Food and foraging ecology of a desert harvester ant, *Veromessor pergandei* (Mayr). Ph.D. diss. University of California, Berkeley.

Hawkins, B., and R. Goeden. 1984. Organization of a parasitoid community associated with a complex of galls on *Atriplex* spp. in southern California. Ecological Entomology 9:271–292.

Hawkins, B., and J. Lawton. 1987. Species richness of parasitoids of British phytophagous insects. Nature (London) 326:788–790.

Janzen, D. 1977. Why fruits rot, seeds mold, and meat spoils. American Naturalist 111:691–713.

Johnegard, P. 1988. North American owls: biology, and natural history. Smithsonian Institute Press, Washington, D.C.

Johnson, M., and R. Hansen. 1979. Coyote food habits on the Idaho National Engineering Laboratory. Journal of Wildlife Management 43:951–956.

Lawton, J. 1989. Food webs. Pages 43–78 in J. Cherrett, ed. Ecological concepts. Blackwell Scientific, Oxford.

Lawton, J., and P. Warren. 1988. Static and dynamic explanation of patterns in food webs. Trends in Ecology & Evolution 3:242–245.

Levine, S. 1980. Several measures of trophic structure applicable to complex food webs. Journal of Theoretical Biology 83:195–207.

Lockwood, J., T. Christiansen, and D. Legg. 1990. Arthropod prey-predator ratios in a sagebrush habitat: methodological and ecological implications. Ecology 71:996–1005.

Louw, G., and M. Seely. 1982. Ecology of desert organisms. Longmans, New York.

Macfayden, A. 1963. The contribution of the fauna to the total soil metabolism. Pages 3–17 in J. Doeksen and J. van der Drift, eds. Soil organisms. North-Holland, Amsterdam.

MacKay, W. 1991. Ants and termites. Pages 113–150 in G. A. Polis, ed. Ecology of desert communities. University of Arizona Press, Tucson.

Mann, J. 1969. Cactus feeding insects and mites. Smithsonian Institution of the United States, National Museum Bulletin 256:1–158.

May, R. 1983*a*. The structure of food webs. Nature (London) 301:566–568.

———. 1983*b*. Parasitic infections as regulators of animal populations. American Scientist 71:36–45.

———. 1986. The search for patterns in the balance of nature: advances and retreats. Ecology 67:1115–1126.

———. 1988. How many species are there on earth? Science (Washington, D.C.) 241:1441–1448.

Mayhew, W. W. 1983. Vertebrates and their habitats on the Deep Canyon Transect. Deep Canyon, Riverside, Calif.

McKinnerney, M. 1978. Carrion communities in the northern Chihuahuan desert. Southwestern Naturalist 23:563–576.

Menge, B., and J. Sutherland. 1987. Community regulation: variation in disturbance, competition, and predation in relation to environmental stress and recruitment. American Naturalist 130:730–757.

Mispagel, M. 1978. The ecology and bioenergetics of the acridid grasshopper, *Bootettix punctatus*, on creosotebush, *Larrea tridentata*, in the northern Mojave Desert. Ecology 59:779–788.

Moore, J., D. Walter, and H. Hunt. 1988. Arthropod regulation of microbiota and mesobiota in below ground detrital webs. Annual Review of Entomology 33:419–439.

Noy-Meir, I. 1974. Desert ecosystems: higher trophic levels. Annual Review of Ecology and Systematics 5:195–213.

Nuessly, G., and R. Goeden. 1984. Aspects of the biology and ecology of *Diguetia mojavea* Gertsch (Araneae, Diguetidae). Journal of Arachnology 12:75–85.

Odum, E., and L. Biever. 1984. Resource quality, mutualism, and energy partitioning in food chains. American Naturalist 124:360–376.

Ohlendorf, H. 1971. Arthropod diet of western horned owl. Southwestern Naturalist 16:124–125.

Olendoroff, R. 1976. The food habits of North American golden eagles. American Midland Naturalist 95:231–236.

Orians, G., R. Cates, M. Mares, A. Moldenke, J. Neff, D. Rhoades, M. Rosenzweig, B. Simpson, J. Schultz, and C. Tomoff. 1977. Resource utilization systems. Pages 164–224 *in* G. Orians and O. Solbrig, eds. Convergent evolution in warm deserts. Dowden, Hutchinson & Ross, Stroudsburg, Pa.

Paine, R. T. 1980. Food webs: linkage, interaction strength and community infrastructure. Journal of Animal Ecology 49:667–685.

———. 1988. On food webs: road maps of interaction or grist for theoretical development? Ecology 69:1648–1654.

Palmer, R. 1988. Handbook of North American birds. Vol 5. Diurnal raptors. Pt. 2. Yale University Press, New Haven, Conn.

Pimm, S. L. 1982. Food webs. Chapman & Hall, New York.

Pimm, S. L., and J. Rice. 1987. The dynamics of multispecies, multi-life-stage models of aquatic food webs. Theoretical Population Biology 32:303–325.

Polis, G. A. 1981. The evolution and dynamics of intraspecific predation. Annual Review of Ecology and Systematics 12:225–251.

———. 1984. Age structure component of niche width and intraspecific resource partitioning: can age groups function as ecological species? American Naturalist 123:541–564.

———. 1988. Exploitation competition and the evolution of interference, cannibalism and intraguild predation in age/size structured populations. Pages 185–202 *in* L. Perrson and B. Ebenmann, eds. Size structured populations: ecology and evolution. Springer, New York.

———. 1991*a*. Food webs in desert communities: complexity, diversity and omnivory. Pages 383–438 *in* G. A. Polis, ed. Ecology of desert communities. University of Arizona Press, Tucson.

———. 1991*b*. Desert communities: an overview of patterns and processes. Pages 1–26 *in* G. A. Polis, ed. Ecology of desert communities. University of Arizona Press, Tucson.

Polis, G. A., and S. McCormick. 1986*a*. Patterns of resource use and age structure among species of desert scorpion. Journal of Animal Ecology 55:59–73.

———. 1986*b*. Scorpions, spiders and solpugids: predation and competition among distantly related taxa. Oecologia (Berlin) 71:111–116.

———. 1987. Intraguild predation and competition among desert scorpions. Ecology 68:332–343.

Polis, G. A., and T. Yamashita. 1991. Arthropod predators. Pages 180–222 *in* G. A. Polis, ed. Ecology of desert communities. University of Arizona Press, Tucson.

Polis, G. A., W. D. Sissom, and S. McCormick. 1981. Predators of scorpions: field data and a review. Journal of Arid Environments 4:309–327.

Polis, G. A., C. A. Myers, and R. Holt. 1989. The evolution and ecology of intraguild predation: competitors that eat each other. Annual Review of Ecology and Systematics 20:297–330.

Powell, J., and C. Hogue. 1979. California insects. University of California Press, Berkeley and Los Angeles.

Price, P. 1975. Insect ecology. Wiley, New York.

Rich, P. 1984. Trophic-detrital interactions: vestiges of ecosystem evolution. American Naturalist 123:20–29.

Rodin, L., and N. Bazilevich. 1964. Doklady Akademii Nauk SSSR 157:215–218 (from E. J. Kormandy, 1969, Concepts of ecology, Prentice-Hall, New York).

Ryan, M. 1968. Mammals of Deep Canyon. Desert Museum, Palm Springs, Calif.

Ryti, R., and T. Case. 1986. Overdispersion of ant colonies: a test of hypotheses. Oecologia (Berlin) 69:446–453.

———. 1988. Field experiments on desert ants: testing for competition between colonies. Ecology 69:1993–2003.

Šantos, P., J. Phillips, and W. Whitford. 1981. The role of mites and nematodes in early stages of litter decomposition in the desert. Ecology 62:664–669.

Schoener, T. W. 1989. Food webs from the small to the large: probes and hypotheses. Ecology 70:1559–1589.

Schoenly, K., and W. Reid. 1983. Community structure in carrion arthropods in the Chihuahuan desert. Journal of Arid Environments 6:253–263.

Schoenly, K., R. Beaver, and T. Heumier. 1991. On the trophic relations of insects: a food web approach. American Naturalist 137:597–638.

Schultz, J., D. Otte, and F. Enders. 1977. *Larrea* as a habitat component for desert arthropods. Pages 176–208 *in* T. Mabry, J. Hunziker, and D. DiFeo, eds. Creosote bush: biology and chemistry of *Larrea* in new world deserts. US/IBP (International Biological Program) Synthesis Ser. 6. Dowden, Hutchinson & Ross, Stroudsburg, Pa.

Seastedt, T., R. Ramundo, and D. Hayes. 1988. Maximization of densities of soil animals by foliage herbivory: empirical evidence, graphical and conceptual models. Oikos 51:243–248.

Seely, M., and G. Louw. 1980. First approximation of the effects of rainfall on the ecology and energetics of a Namib Desert dune ecosystem. Journal of Arid Environments 3:25–54.

Sprules, W., and J. Bowerman. 1988. Omnivory and food chain lengths in zooplankton food webs. Ecology 69:418–426.

Sugihara, G., K. Schoenly, and A. Trombla. 1989. Scale invariance in food web properties. Science (Washington, D.C.) 245:48–52.

Taylor, J. 1984. A partial food web involving predatory gastropods on a Pacific fringing reef. Journal of Experimental Marine Biology and Ecology 74:273–290.

Telford, S. 1970. A comparative study of endoparasitism among some California lizard populations. American Midland Naturalist 83:516–554.

Tevis, L., and I. Newell. 1962. Studies on the biology and seasonal cycle of the giant red velvet mite, *Dinothrombium pandorae*. Ecology 43:497–505.

Wallwork, J. 1982. Desert soil fauna. Praeger, New York.

Walter, D. 1987. Trophic behavior of "mycophagous" microarthropods. Ecology 68:226–229.

Warren, P., and J. Lawton. 1987. Invertebrate predator-prey relationships: an explanation for upper triangular food webs and patterns in food web structure? Oecologia (Berlin) 74:231–235.

Wasbauer, M. S., and L. Kimsey. 1985. California spider wasps of the sub-family Pompilinae. Bulletin of the California Insect Survey 26:1–130.

Weathers, W. 1983. Birds of Southern California's Deep Canyon. University of California Press, Berkeley and Los Angeles.

Welty, J. 1962. The life of birds. Saunders, Philadelphia.

Werner, E., and J. Gilliam. 1984. The ontogenetic niche and species interactions in size-structured populations. Annual Review of Ecology and Systematics 15:393–426.

Whitford, W. G. 1986. Decomposition and nutrient cycling in deserts. Pages 93–118 *in* W. G. Whitford, ed. Pattern and process in desert ecosystems. University of New Mexico Press, Albuquerque.

Wiens, J. A. 1991. Birds. Pages 278–310 *in* G. A. Polis, ed. Ecology of desert communities. University of Arizona Press, Tucson.

Wilbur, H., P. Morin, and R. Harris. 1983. Salamander predation and the structure of experimental communities: anuran responses. Ecology 64:1423–1429.

Winemiller, K. O. 1990. Spatial and temporal variation in tropical fish trophic networks. Ecological Monographs 60:331–367.

Wisdom, C. 1991. Herbivorous insects. Pages 151–179 *in* G. A. Polis, ed. Ecology of desert communities. University of Arizona Press, Tucson.

Woods, A., and C. Herman. 1943. The occurrence of blood parasites in birds from southwestern United States. Journal of Parasitology 29:187–196.

Yodzis, P. 1984. How rare is omnivory? Ecology 65:321–323.

Zabriskie, J. 1979. Plants of Deep Canyon and the central Coachella Valley, California. Deep Canyon, Riverside, Calif.

Ecology, 65(1), 1984, pp. 1–13
© 1984 by the Ecological Society of America

PERTURBATION EXPERIMENTS IN COMMUNITY ECOLOGY: THEORY AND PRACTICE[1]

EDWARD A. BENDER[2]
Department of Mathematics, C-012

TED J. CASE AND MICHAEL E. GILPIN
*Department of Biology, C-016, University of California at San Diego,
La Jolla, California 92093 USA*

Abstract. We analyze perturbation experiments performed on real and idealized ecological communities. A community may be considered as a black box in the sense that the individual species grow and interact in complicated ways that are difficult to discern. Yet, by observing the response (output) of the system to natural or human-induced disturbances (inputs), information can be gained regarding the character and strengths of species interactions. We define a perturbation as selective alteration of the density of one or more members of the community, and we distinguish two quite different kinds of perturbations. A PULSE perturbation is a relatively instantaneous alteration of species numbers, after which the system is studied as it "relaxes" back to its previous equilibrium state. A PRESS perturbation is a sustained alteration of species densities (often a complete elimination of particular species); it is maintained until the unperturbed species reach a new equilibrium. The measure of interest in PRESS perturbation is the net change in densities of the unperturbed species. There is a very important difference between these two approaches: PULSE experiments yield information only on direct interactions (e.g., terms in the interaction matrix), while PRESS experiments yield information on direct interactions mixed together with the indirect effects mediated through other species in the community. We develop mathematical techniques that yield measures of ecological interaction between species from both types of experimental designs. Particular caution must be exercised in interpreting results from PRESS experiments, particularly when some species are lumped into functional categories and others are neglected altogether in the experimental design. We also suggest mathematical methods to deal with temporal and random variation during experiments. Finally, we critically review techniques that rely on natural variation in numbers to estimate species interaction coefficients. The problems with such studies are formidable.

Key words: community structure; competition; disturbance; perturbation; species interaction.

INTRODUCTION

To be effective, community ecology must have answers to the following questions. First, which species are dynamically connected to which other species? Second, which further species, though not mechanistically connected, are ultimately connected through interactions involving intermediate species? Third, what are the signs of these interactions? That is, for a pair of species A and B, how will a change in the density of one affect the other, and what is the relationship between immediate and ultimate effects? Fourth, in the face of interactive loops of arbitrary length going in many directions, how is one to delimit subsystems for study? That is, what is the effect of ignoring species and what rules can be given concerning which species to ignore? Fifth, what are the consequences of lumping species into a single dynamic entity, e.g., all ants or all granivores?

In the 1960's and 1970's, the fashionable and conventional approach to the study of interspecific competition was niche analysis (Mac Arthur 1958, Levins 1968). Here one measures the ecological overlap between species' niches, for example, the similarity of their diet, habitat use, or time of foraging. The assumption is that overlap can be equated with an interaction coefficient in a dynamic model such as that of Lotka and Volterra. This method has the virtue of being rapid and convenient, but it suffers from a number of defects (Abrams 1975, 1980*a, b*, Hurlbert 1978, Case and Casten 1979, Lawlor 1980, Thomson 1980). Two stand out. First, the resources that the consumer species share may not limit consumer population growth or density, and therefore the link between overlap and dynamical effects may not be justified. Second, even if the measured niche axes are relevant from a dynamic perspective, how does one numerically convert niche overlap to dynamic impact? A number of potential formulae exist in the literature, yet all are ad hoc, and none has received any empirical validation.

Given the shortcomings of this passive, observational approach to the study of interspecies competition, ecologists have been turning to experimental approaches, in particular to perturbation studies. Here

---

[1] Manuscript received 25 August 1981; revised 13 December 1982; accepted 17 January 1983.

[2] Authors are listed in alphabetical order.

2   EDWARD A. BENDER ET AL.   Ecology, Vol. 65, No. 1

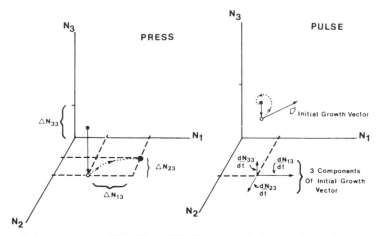

FIG. 1. A geometrical representation of PRESS and PULSE type perturbation experiments for three species communities. Each axis represents the numerical abundance of one of three species. The large open circles (○) represent the community composition before perturbation. Perturbations are represented by a solid arrow ending in an open circle. The dotted line shows the trajectory of the community following perturbation. In both experiments only species 3 has been perturbed. In the PRESS experiment, species 3 has been removed and is maintained at zero density, while species 1 and 2 adjust numerically to a new equilibrium (large filled circle [●] at the end of the dotted trajectory). The resulting values of ΔN_{i3} correspond to those used in Eq. 8. This single PRESS experiment yields estimates for the third column of the ΔN matrix. After similar perturbation experiments for species 1 and 2, the entire ΔN matrix can be calculated, then inverted to yield the terms required to calculate the interaction coefficients, i.e., the A matrix (Eq. 9).

In the PULSE experiment we require estimates for the initial growth rate of all three species immediately after the one-time removal of some species 3 individuals. These rates may be represented geometrically as the magnitude of each of the three component vectors depicted here on the $N_1 N_2$ plane. The dN_{i3}/dt correspond to the terms in Eq. 5. This single PULSE experiment directly yields estimates for the third column of the A matrix. Similar experiments involving perturbations in species 1 and 2 would yield the remaining two columns of A.

one experimentally alters the density of some species in a community and determines consequent changes in the densities or behaviors of other species. Such studies have been made by Connell (1961), Dayton (1973), Schroeder and Rosenzweig (1975), Seifert and Seifert (1976, 1979), Brown, Davidson and Reichman (1979), Holmes et al. (1979), Hairston (1981), Wise (1981), Smith (1981), Smith and Cooper (1982), and Tinkle (1982). The hope is that the resulting community dynamics will produce conclusive proof regarding the presence or absence and strength of present-day biotic interactions, such as competition or predation between species pairs. It is recognized by most workers, however, that interactions in the past may, through coevolution, become so attenuated that present-day interactions are weak, yet the community maintains structural attributes of its past history (see, e.g., Connell 1980). Detecting "the ghost of interactions past" is a subject beyond the scope of this paper.

TWO TYPES OF PERTURBATION: AN OVERVIEW

In what follows we assume that we have communities available for investigation that are fully open to undisturbed enumeration, that the logistics of actually manipulating the population sizes of one or more species is feasible, and that suitable replicates and controls pose no problems. Thus, we wish to discuss irreducible limits to knowledge about such systems rather than real-world problems of observation.

Suppose that we have a community of n species that we perturb by changing the density of some subset s (normally a single one) of the species. We wish to stress that there are very different designs that can be applied to such an experiment. In the first type (hereafter referred to as PULSE experiments) perturbation is made and then the community is allowed to respond. As it does so, we measure the initial rates of population growth of each of the n species (including the perturbed species). In the second type (hereafter referred to as PRESS experiments), the densities of perturbed species are altered and maintained at predetermined levels by adding or removing individuals as needed. In the face of this maintained alteration, the other $n - s$ unmanipulated species will readjust numerically. After this adjustment ends and the community has reached a new equilibrium, one measures the net change in density of each of the $n - s$ species. Included in PRESS experiments are experimental designs in which one or more species are totally removed from experimental plots. Fig. 1 gives a graphical interpretation of these two different techniques.

The proper interpretation of results from PULSE and PRESS experiments are often confused. To insure a proper interpretation, one must have a clear model of community interactions. We use a mathematical model to avoid the ambiguities inherent in a verbal model. The state variables of this model are numbers (or densities) of each species, and the parameters of the model are the unknowns that we wish to deduce from our experiments. The generalized Lotka-Volterra equations provide a suitable starting point in this exercise. The per capita rate of growth of species i is:

$$\frac{dN_i}{N_i dt} = c_i\left[K_i - \sum_{j=1}^{n} a_{ij}N_j\right], \quad \text{for } i = 1 \text{ to } n, \quad (1)$$

where $a_{ii} = 1$. Unless there is no self-interaction of i, this assumption on a_{ii} does not reduce the generality of Eq. 1 since a scale factor can be absorbed in c_i. Depending on the signs of K_i and a_{ij}, we could describe predator-prey, mutualistic, competitive, or amensal interactions between any two species i and j (MacArthur 1972, May 1973, Case and Casten 1979). These equations are most often used as descriptions of competitive interactions, in which case $a_{ij} > 0$ is the competition coefficient of species j on i, K_i is the carrying capacity of species i, and $c_i = r_i/K_i$, where r_i is the intrinsic rate of growth of species i.

We assume that a stable, feasible equilibrium solution $\hat{N}_i$ to Eq. 1 exists and is found by setting growth rates equal to zero, yielding

$$\sum_{j=i}^{n} a_{ij}\hat{N}_j = K_i. \quad (2)$$

When this equilibrium is substituted into Eq. 1 we obtain

$$\frac{dN_i}{N_i dt} = c_i \sum_{j=1}^{n} a_{ij}(\hat{N}_j - N_j), \quad \text{for } i = 1 \text{ to } n. \quad (3)$$

In the next two sections we derive protocols by which the parameters in Eq. 3 (i.e., the c_i's and the a_{ij}'s) may be determined by PULSE and PRESS perturbation experiments. Much of the mathematics underlying what we call PULSE and PRESS experiments has been discussed before in the ecological literature (May 1973, Richardson and Smouse 1975, Levine 1976, Holt 1977, Lawlor 1980, Schaffer 1981). However, heretofore, no one has extended this theory to develop a protocol for community perturbation experiments in ecology.

PULSE perturbation experiments

Consider a set of species whose dynamics follow Eq. 1 and whose pre-perturbation equilibrium densities $\hat{N}_i$ are known. A perturbation is made by adding or subtracting individuals of one or more of these species, and the resulting values of dN_i/dt and N_i are measured (for all i) immediately after perturbation. These values are then substituted into Eq. 3. This gives n equations in the c_i's and a_{ij}'s. If n independent perturbation experiments are done, the resulting n^2 equations can be solved for all the c_i's and a_{ij}'s. One

straightforward set of experiments results in equations that are particularly easy to solve; in experiment f only species f is perturbed from its equilibrium value, and it is perturbed to a density N_f. The immediately resulting values of dN_i/dt (for $i = 1$ to n) are denoted by dN_{if}/dt (Fig. 1). Eq. 3 becomes:

$$\frac{dN_{if}}{dt} = \hat{N}_i c_i a_{if}(\hat{N}_f - N_f), \quad \text{where } i \neq f, \quad (4a)$$

and

$$\frac{dN_{ff}}{dt} = N_f c_f(\hat{N}_f - N_f), \quad (4b)$$

where we have used the fact that $a_{ff} = 1$. It is apparent from Eq. 4a that when species compete directly (and so $a_{if} > 0$), the reduction in one species will be followed by an initial increase in the other (i.e., $dN_{if}/dt > 0$). The stronger the direct competition, the greater the response. Notice, however, that if the response of species i exceeds the response of species k (i.e., $dN_{if}/dt > dN_{kf}/dt$), then we cannot conclude that the direct competitive effect of f on i exceeds the direct competitive effect of f on k (i.e., $a_{if} > a_{kf}$). This is because the interaction terms in Eq. 4a are also multiplied by the c_i term, which includes the intrinsic growth rate.

It is important here to distinguish between direct and indirect interactions. The interpretation of the a_{ij} in Eqs. 2 and 3 depends on whether we allow the N's to represent only species on a single trophic level or on mixed trophic levels. Consumer species directly affect the resource species that they eat and vice versa, so the a_{ij}'s between consumer i and its prey j are always nonzero, but two consumers or two prey will only have nonzero interaction terms between them if they directly interfere with one another. In a single trophic level description, like the typical Lotka-Volterra competition equations, resource dynamics are not explicitly considered, and competition for resources, as well as interference competition, is parameterized into the a_{ij}'s. Thus, in this parameterization, even pure exploitation competitors without interference will have nonzero a_{ij}'s between them. Whether or not PULSE experiments will detect such exploitation competition depends critically on whether or not resource dynamics occur on a much faster time scale than consumer dynamics. We defer discussion of this problem (see Overlooked Species).

We now solve Eq. 4a, b for the unknown a_{ij}'s and the c_i's. Setting $f = j$ in Eq. 4a and $f = i$ in Eq. 4b and combining, we obtain:

$$a_{ij} = \frac{\dfrac{dN_{ij}}{dt}N_i(\hat{N}_i - N_i)}{\dfrac{dN_{ii}}{dt}\hat{N}_i(\hat{N}_j - N_j)}, \quad \text{and} \quad (5a)$$

$$c_i = \frac{\dfrac{dN_{ii}}{dt}}{N_i(\hat{N}_i - N_i)}. \quad (5b)$$

Every term on the right-hand side is experimentally measurable. A problem arises, however, if species i is not self-limiting (e.g., a prey species limited only by predators), then $a_{ii} = 0$ and $dN_{ii}/dt = 0$, and a_{ij} for $j \neq i$ is not well defined; however, $c_i a_{ij}$ is well defined. In this case we use

$$c_i a_{ij} = \frac{\dfrac{dN_{ij}}{dt}}{\hat{N}_i(\hat{N}_j - N_j)} \qquad (5c)$$

instead of Eq. 5a, b.

Notice that by performing a single experiment, we can obtain a column of $c_i a_{ij}$ terms, and thus after n such experiments, each perturbing a different species, we obtain an entire matrix containing all n^2 of the $c_i a_{ij}$ terms. At this stage we may solve for all the a_{ij} terms themselves (unweighted by c_i) because $a_{ii} = 1$. Let A denote the matrix of the a_{ij}'s. It may be of interest to determine only some of the entries in A by performing less than n experiments. Determining either a single row or column of A requires the full set of n experiments. Two experiments are required in practice even to determine a single a_{ij}. A more important case is the following: suppose we are only interested in a subset of s species. We can determine a_{ij} for $i, j = 1$ to s by doing experiments for all $f = 1$ to s and measuring dN_{ij}/dt. This result is important because we may be uncertain if other species are involved in the interactions or be unable to measure or perturb those species.

It is also possible to perform n experiments, each involving the simultaneous perturbation of q species ($q < n$); however, one must insure that all n perturbations are linearly independent. As before, at least n experiments have to be performed before one can solve for the entire A matrix.

In practice, of course, there are always unknown error terms associated with the measurement of the N_i's and dN_i/dt's. Replicate experiments will be required to assess these errors, increasing many fold the number of needed experiments. Regression methods may be adopted to find the best estimates of the a_{ij}'s. Seifert and Seifert (1976, 1979) adopt just this PULSE approach in their experimental analysis of the insect communities living in *Heliconia* plants. Their regression technique results in estimates of $c_i a_{ij}$ terms (but not c_i or a_{ij} alone) and should not be confused with attempts by others to apply regression to natural (i.e., unmanipulated) variation in numbers (e.g., Schoener 1974, Hallett and Pimm 1979). These later studies are discussed below (see Natural Variation Techniques).

To achieve the greatest accuracy in Eq. 5, large perturbations are desirable since they will make dN_{ij}/dt, dN_{ii}/dt, $\hat{N}_i - N_i$ and $\hat{N}_j - N_j$ large and thus less sensitive to errors in measurements and random fluctuations. On the other hand, N_i should not be too small since it is a factor in the ratios of Eq. 5. A reasonable time interval is needed to estimate dN_i/dt, but the N_i's

must not change so much that dN_i/dt changes appreciably over the measurement period. We cannot remove a species completely in PULSE perturbation designs because its per capita growth rate ($[dN_{ii}/dt]/N_i$) would then be impossible to determine.

PRESS perturbation experiments

Again imagine a set of n species obeying Eq. 1 and at a stable equilibrium. These equilibrium densities are given in matrix form as

$$\hat{N} = \text{A}^{-1}K, \qquad (6)$$

where A^{-1} is the inverse matrix of A; the latter contains the direct interaction terms a_{ij}. $\hat{N}$ is the column vector of equilibrium densities, and K is the column vector of K_i's. From Eq. 6 it is apparent that if K_j is varied and the K's for $i \neq j$ are fixed, then

$$\frac{\Delta \hat{N}_i}{\Delta K_j} = a_{ij}^{(-1)}. \qquad (7)$$

Thus the selective alteration of different species' carrying capacities allows for the direct solution of the $a_{ij}^{(-1)}$ terms (i.e., the elements of the inverse matrix). By performing n such experiments, one could solve for the entire A^{-1} matrix and by inversion find A. For most communities, however, an investigator has no prior knowledge of the K's, and selective manipulation of these variables is impossible. It is easier and more practical to manipulate species numbers directly, even though the derivation of the meaning of the results is more circuitous.

Suppose species f is experimentally maintained at a level N_{ff} and that the remaining $n - 1$ species are allowed to reach new equilibrium levels denoted as N_{jf}. Then at this new equilibrium $dN_i/dt = 0$ for $i \neq f$ and so Eq. 3 reduces to

$$\sum_{j=1}^{n} a_{ij}\Delta N_{jf} = 0, \qquad \text{for } i \neq f, \qquad (8)$$

where the matrix ΔN_{jf} is the observable change in density of species j following a PRESS perturbation of species f (i.e., $\Delta N_{jf} = \hat{N}_j - N_{jf}$). If this experiment is repeated for all n species, we obtain $n(n - 1)$ equations that can be solved for A (using the fact that $a_{ii} = 1$ for all i) as shown in Appendix 1. Two equivalent formulae are derived:

$$a_{ij} = \Delta N_{ij}^{(-1)}/\Delta N_{ii}^{(-1)}, \qquad \text{and} \qquad (9)$$

$$\frac{\Delta N_{if}}{\Delta N_{jf}} = a_{if}^{(-1)}/a_{jf}^{(-1)}, \qquad (10)$$

where $a_{ij}^{(-1)}$ and $\Delta N_{ij}^{(-1)}$ represent elements from the inverse of the A matrix and the ΔN matrix, respectively. To recapitulate, to measure the interaction coefficients a_{ij}, we do n experiments that yield the observable quantities ΔN_{jf}. ΔN^{-1} is the inverse of the matrix. Eq. 9 yields a_{ij}, and Eq. 10 shows that the experimental responses in density are directly related to the elements of the inverse of A, not A itself (Fig. 1).

Although we have implied that N_{ff} is arbitrary, this is not quite true. A species must never be held at such a level that another species goes extinct. For example, if N_{ff} were such that $N_{if} = 0$, then Eq. 8 would not necessarily hold. (In this case the analysis breaks down because the mathematically relevant equilibrium value for the ith species is negative.) Unlike PULSE experiments, $N_{ff} = 0$ is permissible (provided no other species goes extinct).

How can we achieve the greatest accuracy in Eq. 9? This is done by making the perturbations as large as possible with the caveat that a species must never be reduced to such a level that another species goes extinct. This caveat can present problems since we usually cannot predict nonextinction. Analysis of the effect of experimental error in Eq. 9 is much harder than in Eq. 5 because $\Delta/N_{ij}^{(-1)}$ and $\Delta N_{ii}^{(-1)}$ result from inverting the matrix ΔN of measurements. If ΔN is "ill conditioned," then even minor changes in its entries can cause major changes in the entries of ΔN^{-1}. A discussion of ill conditioning can be found in Fadeev and Fadeeva (1963), and Pomerantz (1981) illustrates the problem with ecological data. Generally speaking, a matrix with a small determinant is likely to be ill conditioned, and this might arise when two or more species interact in a similar fashion with other members of the community (i.e., are nearly equivalent members of the same guild).

Contrasts between PULSE and PRESS experiments

Fundamental differences.—A fundamental difference between PULSE and PRESS experiments is the relationship between the a_{ij}'s and the strength of a species' response. For PULSE perturbations, response strength, at least around equilibrium, is measured by dN_{ij}/dt, which is related to a_{ij} in a simple way by Eq. 4a. For PRESS perturbations response strength is measured by $\hat{N}_i - N_{if}$, which is directly coupled to the elements of the inverse matrix, the $a_{ij}^{(-1)}$'s as visible in Eq. 10. That is, the response of species i following a PRESS perturbation in species f, depends on terms connecting species i with f in the inverse of the A matrix, not the A matrix itself. Even for a pure competition community where all the a_{ij} terms are greater than or equal to zero, the off-diagonal terms in A^{-1} may take on any sign. Each of the $a_{ij}^{(-1)}$ terms is a function of all the elements in the A matrix not just the a_{ij} term. These $a_{ij}^{(-1)}$ terms may be interpreted as the total effect that species j has on the equilibrium density of species i and thus includes terms for the direct competitive effect via a_{ij}, plus terms for effects which are mediated through paths (the "loops" of Levins [1975]) involving additional species in the community (see Lawlor 1979). For example, species j may affect k, which in turn affects l, which finally

affects i; this would be one possible path of length 3. The fact that Eq. 10 involves A^{-1} rather than A has been a source of considerable confusion. We devote the remainder of this section to that subject.

Consider, for example, the classical niche-theoretic example of three species whose niches are aligned symmetrically along a single niche axis. Each species has a Gaussian-shaped niche with overlap $a < 1.0$ with its nearest neighbor. The two end species have overlap of $\approx a^4$ with one another (May 1975). For example, if $a = 0.5$, the A matrix is:

$$A = \begin{vmatrix} 1 & 0.5 & 0.0625 \\ 0.5 & 1 & 0.5 \\ 0.0625 & 0.5 & 1 \end{vmatrix},$$

and the inverse of A is:

$$A^{-1} = \begin{vmatrix} 1.422 & -0.889 & 0.356 \\ -0.889 & 1.889 & -0.889 \\ 0.356 & -0.889 & 1.422 \end{vmatrix}.$$

From the sign of $a_{13}^{(-1)}$ and $a_{31}^{(-1)}$ we see that the net effect of the two end species on the equilibrium density of one another is positive even though these species are direct competitors. The removal of either end species from the three-species community causes an initial increase in the density of the other end species at a rate given by Eq. 4a, but soon the trajectory turns around. At the new equilibrium, there is a net decrease in the density of the remaining end species (but an increase in the middle species). In fact, if $K_1 = K_2 = K_3 = 1$, then $\hat{N}_1 = \hat{N}_3 = 0.889$ and $\hat{N}_2 = 0.111$, while setting $N_{11} = 0$ gives $N_{21} = N_{31} = 0.667$.

Such indirect or gratuitous "mutualism" easily crops up in mathematical descriptions of multispecies competition communities when $n > 2$. Holt (1977) describes examples whereby alternate prey species reduce each other's equilibrium abundance, whether or not they compete, in the presence of a food-limited predator. He calls this indirect effect "apparent competition." In other examples, a predator may have a net positive effect on its prey, or the prey may have a negative effect on some of its predators, within the context of all the community interactions. Colwell and Fuentes (1975) provide other biological examples of such counterintuitive interactions.

In species-rich guilds, the results of a species removal experiment may be complex. The particular case described in Fig. 2 is highly symmetrized, but the character of the conclusion is robust for any richly connected competitive system. We set up a system of 18 competing species following the Lotka-Volterra equations (Eq. 1). All equilibrium densities are normalized to unity. The character of the competition matrix is that studied by May and MacArthur (1972). Species have Gaussian niches aligned along a single niche dimension, and nearest neighbors have $a = 0.8$; at time $t = 10$, an arbitrary member of the community

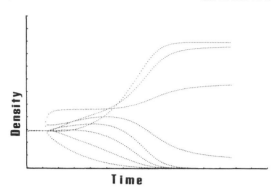

Fig. 2. Time trajectories of an 18-species competition described in the text. All densities are initially normalized to unity and are thus coincident for the first 10 time periods. At $t = 10$, a single species is removed; since the system is symmetrical, it makes no difference which one. The behavior for the next 140 time periods is shown. Ten additional species ultimately go extinct.

is removed. The time trajectories of the remaining 17 species are shown in the remainder of the figure. Because of the symmetry, all but one of the species move as pairs. As this figure shows, half the species go extinct following this perturbation experiment, negating the application of a PRESS analysis.

Just as the removal of one species may result in a population decrease in its competitor species, so might the removal of a predator result, through indirect effects, in a numerical decrease in some of its prey. Porter (1976) performed experiments enclosing freshwater algae with and without their zooplankton grazers. When zooplankton were excluded, certain species of algae increased dramatically in abundance. Other species, however, showed significant declines. Porter observed that the types of algae that increased in the presence of grazers had durable cell walls that protected them from damage during passage through the gut of the zooplankton. She attributed their increase under grazing to a direct effect from the zooplankton, namely a fertilizer effect from zooplankton excrement. Since these experiments lasted over many generations of algal growth, they may be interpreted as PRESS experiments. Thus, the possibility exists that certain alga species increased with grazing because of indirect effects. If the different algae compete for resources, then those algae which are less preferred prey will benefit indirectly from grazers because they gain a relative competitive advantage over more heavily-grazed algae.

As we have seen, except for two-species communities, the presence of a net density gain in one species (following the removal of a presumed competitor) is neither a necessary nor sufficient demonstration of competition between those two species. Similarly, the absence of a density gain is neither a necessary nor

sufficient demonstration of the absence of competition.

Procedural differences.—PRESS experiments contrast with PULSE perturbations in a variety of ways. First, one can remove a population completely in PRESS designs (i.e., set $N_{ff} = 0$). Second, the growth rate parameters c_i cannot be determined in PRESS experiments (and need not be to find A) since they affect only the rate at which equilibrium is reached and not its actual value.

On the other hand, c_i must be determined in PULSE experiments to obtain A. In fact, radically different population growth rates for the species can cause serious problems in estimating dN_{if}/dt since we need two separate measurements, such that N_i will have changed enough to estimate dN_{if}/dt but the percentage change in $K_i - \Sigma a_{ij}N_j$ will have been small.

We foresee other potential difficulties with the application of PULSE and PRESS experimental designs. PULSE experiments are more prone to experimental error since it is generally more difficult to measure the rate at which a function changes than the magnitude of that function at equilibrium. The initial rate at which species i responds to a perturbation in species j is proportional to c_i. In seasonal environments c_i may fluctuate severely, even reaching zero. Thus, it is important that each experiment have simultaneous controls. By adopting a PRESS design, we eliminate this ambiguity, but each experiment may become prohibitively long and expensive.

In PULSE experiments, the value of a_{ij} can be computed after two single species perturbation experiments; in PRESS we must have the entire ΔN matrix so that we can invert it in order to obtain a single a_{ij}, thus n experiments are required. There is, however, an important exception to the need to determine the entire ΔN matrix. Suppose one set R of species does not numerically respond following perturbations in another set P. Specifically, suppose the first p experiments have been performed, and we have $N_{jf} = \hat{N}_j$ for $f = 1$ to p and $j = p + 1$ to n. Then ΔN has the block form

$$\Delta N = \begin{vmatrix} \Delta N_{PP} & \Delta N_{PR} \\ 0 & \Delta N_{RR} \end{vmatrix},$$

which is inverted to obtain

$$\Delta N^{-1} = \begin{vmatrix} \Delta N_{PP}^{-1} & U \\ 0 & \Delta N_{RR}^{-1} \end{vmatrix},$$

where $U = -\Delta N_{PP}^{-1}\Delta N_{PR}\Delta N_{RR}^{-1}$. Note that by doing only p experiments we have found p entire columns of A, and that many of the entries are zero. (See below for more discussion.)

This argument is fundamental to Hairston's (1981) interpretation of his long-term study on a guild of seven salamander species in the southeastern United States. Hairston followed a PRESS experimental design. The experiment consisted of continual removals of the most numerically dominant species (*Plethodon jordani*) from one set of replicate plots and the second most dominant species (*P. glutinosus*) from another set. Plots were unfenced, and the experimental removals maintained the focal species at densities close to zero. Compared to control plots, Hairston observed that the removal of either of these two abundant species increased the numerical abundance of the other. However, the remaining five salamander species showed no significant increase. Hairston concluded that the two abundant species compete strongly with one another but not with the other salamander species. This conclusion is correct as long as ΔN_{RP} is zero. ΔN_{RP} is not exactly zero in Hairston's results; rather, the remaining five species' numerical responses were simply not statistically significant. The conclusions Hairston reached will still be valid unless A_{RR} (i.e., the subblock of A containing the interactions among the five unperturbed salamander species) contain entries much larger than one in absolute value. If this condition is violated then, theoretically, diffuse interactions might yield a relatively small change in unperturbed species following perturbation of others despite strong direct interactions between the species involved.

OVERLOOKED SPECIES

A perennial problem in field experiments is identifying the complete set or guild of species so as to delimit the experimental design. In PULSE experiments this problem is not serious since, as noted earlier, if there are n total species, one may focus on a subset of s species and, by doing s experiments, find all the a_{ij}'s (for $i, j = 1$ to s). In a PRESS design this neglect can be catastrophic because the elements of ΔN are shaped by the direct and indirect "links" between all the various species. There is no reason why these links should be limited to species within the same taxon or of similar body size. The magnitude and even the sign of many calculated a_{ij}'s can be affected by neglecting a single species.

Although it may be extremely difficult to identify a complete guild of interacting species, we may be able to at least identify a subset of species that strongly interact. Suppose F is this set of s species on which we focus, and R is the remaining $n - s$ species. The interaction matrix A may then be partitioned into four blocks, such that the interactions among the focal species are in the $s \times s$ block A_{FF}, the interactions among the neglected species are in block A_{RR}, etc. In Appendix 2 we derive the following result: the effects of the ignored $n - s$ species are negligible if either the interaction coefficients of ignored species on focal species (A_{FR}) are all very small or the interaction coefficients of the focal species on ignored species (A_{RF}) are all very small.

For example, suppose we are interested in understanding the interactions between two species of ants. Suppose we know elephants also have a big effect on

these ants because they trample on their colonies, yet we are fairly confident that the direct effects of ants on elephants are low. If we conduct PRESS experiments on the ants only (ignoring elephants), how much error will result in our estimates of the a_{ij}'s between ant species? A hypothetical interaction matrix with these features is the following:

$$A = \begin{array}{c} \\ \\ \\ \end{array} \begin{array}{ccc} Ant\ 1 & Ant\ 2 & Elephants \\ \left| \begin{array}{ccc} 1 & 0.7 & 1.2 \\ 0.5 & 1 & 1.3 \\ -0.01 & 0.02 & 1 \end{array} \right| \end{array}$$

Here both ant species have very small effects on elephants relative to their effects on each other. The effect of the elephants on the ants is very large (1.2 and 1.3); the exact values used are not critical as long as either A_{FR} or A_{RF} is very small. The inverse of A is

$$A^{-1} = \left| \begin{array}{ccc} \mathbf{1.524} & \mathbf{-1.058} & -0.454 \\ \mathbf{-0.803} & \mathbf{1.584} & -1.096 \\ 0.031 & -0.04 & 1.017 \end{array} \right|$$

We take this for our ΔN matrix. If we ignored elephants and adopted a PRESS perturbation design to estimate the ant-ant interactions, our interaction estimates will be given by Eq. 9, where the $\Delta N_{ij}^{(-1)}$'s are entries in the inverse of the 2×2 matrix formed by the top left corner of ΔN (i.e., the boldface numbers above). This is

$$\Delta N_{FF} = \left| \begin{array}{cc} 1.524 & -1.058 \\ -0.803 & 1.584 \end{array} \right|.$$

Thus, taking the inverse and normalizing the diagonal elements to 1 yields our estimate for A_{FF}

$$A_{FF} = \left| \begin{array}{cc} 1 & 0.668 \\ 0.527 & 1 \end{array} \right|,$$

which agrees very well with the corresponding ant-ant corner of A itself.

Overlooked species and ecological abstraction

In many cases, we purposely wish to consider the a_{ij} terms as potentially containing terms arising from indirect interactions through loops involving neglected species. The clearest example of this occurs in studies aimed at uncovering competitive interactions in a guild of consumer species. Rarely will we wish explicitly to consider all resource species in our experimental design. Imagine a community of n consumer species and m resources. By performing $n + m$ perturbations we could deduce the $n + m$ dimensional A matrix (via Eqs. 5a or 9) explicitly containing terms for all consumer-resource, resource-consumer, consumer-consumer, and resource-resource interactions. The off-diagonal terms in the $n \times n$ subblock of A dealing only with the direct consumer-consumer interactions will be zero or small unless the consumers exhibit strong interspecific interference interactions. Interspecific consumer competition will show up as the similarity between different consumer's effects on different resources, not in the consumer-consumer interactions themselves. If we ignore the resource species in our experimental design, can we somehow deduce the competitive indirect effects between consumer species that are mediated through shared resources? That is, by purposely neglecting some species, can we experimentally parameterize their effects into the a_{ij} terms of the focal species?

We consider three cases, depending on whether the relaxation times (dynamics) of the neglected species are (1) much smaller than, (2) much larger than, or (3) the same order of magnitude as the relaxation times of the species on which we focus.

Neglected species have small relaxation times.— This case was discussed by Schaffer (1981), and we have little new to add. Schaffer (1981) refers to the problem of overlooked species as "ecological abstraction." He asks what set of parameters will reasonably predict the dynamics of an *s* species Lotka-Volterra approximation if the interactions are actually described by an *n* species Lotka-Volterra model ($s < n$). He concludes that when the neglected species have much smaller relaxation times than the focal species, a Lotka-Volterra model can be used, and PRESS experiments will give correct estimates (namely the ΔN_{FF}^{-1} of Eq. A2.1). Notice that Schaffer's condition is not at all related to our condition A2.2, because we have different goals in mind. Our condition derives from requiring accuracy in the estimate of interaction coefficients. Schaffer is concerned with finding a parameterization for the reduced system that yields dynamic accuracy of that subsystem. We emphasize, therefore, that estimates for A_{FF} from PRESS experiments may be inaccurate even if the *s* species model accurately reflects community dynamics. In other words, if we estimate A_{FF}, disturb the entire community and find that its dynamic behavior is accurately predicted by our estimate, then we still cannot conclude that our estimate of A_{FF} is accurate!

What happens with PULSE experiments? As noted earlier, these are hard to perform correctly when species relaxation rates differ radically. However, if we allow a time between the two measurements for estimating dN_{ij}/dt at least comparable to the relaxation times of the neglected species, reasonably accurate estimates will be obtained for the reduced Lotka-Volterra model; that is, the estimated a_{ij}'s will "absorb" reasonably well the indirect effects between focal species that are mediated through neglected species.

Neglected species have longer relaxation times.— If the time differential is great enough, we can treat these species as constant and use a Lotka-Volterra model that does not contain any effects of the neglected species. Either PULSE or PRESS experiments can be used. We must then ask how much error is made in applying the c_i and a_{ij} estimates at a much later time when the levels of neglected populations have changed? Surprisingly, the error is small since the neglected species only affect the "carrying capacities" in the reduced model. This can be seen by regrouping the bracketed term in Eq. 1:

$$K_i - \sum_{j=1}^{n} a_{ij} N_j = \left(K_i - \sum_{j=s+1}^{n} a_{ij} N_j \right) - \sum_{j=1}^{s} a_{ij} N_j. \quad (11)$$

The parenthetic term is the "carrying capacity" for the model in which the last $n - s$ species are neglected. We can view this as an example of temporal variation in K_i (see Temporal and Random Variation).

As an example of slow relaxation, consider two species of rabbits feeding on lichens. Suppose we are interested in the amount of competition among these rabbits. We perform PRESS experiments on rabbits only; in separate experiments, we depress and maintain the numbers of each rabbit at some new level. After a while a quasi-equilibrium will be reached whereby rabbit numbers are relatively stable. Lichens, however, are still slowly adjusting. If we make our PRESS measurements of N_{ij} at this quasi-equilibrium, our conclusions regarding rabbit-rabbit interactions will not be seriously in error.

Suppose we adopt a PULSE design for this same system by perturbing the numbers of one rabbit species and immediately afterward measuring the change in the numbers of the two rabbits. This growth rate should be measured over a brief time interval, approximately one rabbit generation. If the rabbit species do not exhibit direct interference between each other, our PULSE experiments will correctly indicate no direct rabbit-rabbit interactions, but we cannot conclude from this that rabbits do not compete. The competition through lichens could be uncovered by measurements of both lichen density and single-species rabbit density. To do this we could parameterize the effects of lichens into variable "carrying capacities" for the rabbits. If the rabbits are species 1 and 2, and we lump all lichens as species 3, then

$$K_i(t') - K_i(t) = a_{i3} [N_3(t) - N_3(t')],$$
$$\text{for } i = 1,2. \quad (12)$$

Thus, we can estimate the effects of lichens on rabbits and thus rabbit exploitation competition if we can determine K_i's and N_3 at two separate times. The time interval required will be longer than that normally used for a typical PULSE experiment.

Neglected and focal species have comparable relaxation rates.—In this case it is not possible to find a Lotka-Volterra model for the focal species alone. PULSE and PRESS experiments can be expected to give radically different estimates for A, with PULSE giving correct estimates for the direct interactions only in the A_{FF} block of the full *n* species matrix and PRESS giving estimates for the combined direct and indirect effects through focal and neglected species (i.e., the $\Delta N_{FF}{}^{-1}$ of Eq. A2.1).

LUMPED SPECIES

Suppose we are only interested in the interactions between broad categories of species on the same trophic level. If we lump species together into categories and perform PRESS experiments involving categories instead of species, would the qualitative interpretation of the interactions between species necessarily be correct?

This was the tack taken experimentally by Brown and Davidson (1978) and Brown et al. (1979) in their studies on the competitive effects of granivorous desert ants and rodents. For purposes of their removal experiments, all ant species were considered as one category and all such rodents as another. By lumping species into only two broad categories and by conducting experiments involving the total removal of one category or the other, it is possible that the net qualitative interaction between species categories can be confused. We illustrate how this might happen with the following hypothetical example. Suppose we have a single rodent species and two ant species whose A matrix and $\mathbf{k}$ vector are

$$
A = \begin{vmatrix} Rodent & Ant\ 1 & Ant\ 2 \\ 1 & \frac{1}{2} & 0 \\ 0 & 1 & 2 \\ \frac{1}{8} & 0 & 1 \end{vmatrix} \quad K = \begin{vmatrix} 36 \\ 36 \\ 9 \end{vmatrix}.
$$

We have contrived ant 2 to have a large depressive effect on ant 1, but ant 2 has no effect on the rodents. Ant 1, on the other hand, has a competition coefficient of $\frac{1}{2}$ on the rodents. This system has a stable equilibrium at

$$
\hat{N} = \begin{vmatrix} 24 \\ 24 \\ 6 \end{vmatrix}.
$$

If we lump both ant species into a single category "ants" and then remove them from experimental plots, rodents will increase to their carrying capacity of 36. On the other hand, if we remove rodents, the two ant species will reach a new stable equilibrium with densities of 18 and 9, giving a summed density of 27 for ants. That is, the removal of rodents causes ant numbers to decrease also. Using Eq. 9 to calculate the a_{ij}'s for the lumped system yields:

$$
A = \begin{vmatrix} Rodent & Ants \\ 1 & \frac{2}{5} \\ -\frac{1}{8} & 1 \end{vmatrix}.
$$

Thus, our interpretation of these PRESS experiments would be one of a predator (rodents) and their prey (ants) rather than the strictly competitive system that we have modeled it to be. As a rule of thumb, lumping appears to be inappropriate when the species

in any one category have grossly different effects on one another and on species in the other category. For the case of Brown et al.'s rodent-ant system, we doubt such inequalities exist and believe that their original conclusion of ant-rodent competition is probably correct.

NONLINEAR MODELS

Many have argued that the dynamics of real communities is probably not adequately described by equations as simple as the Lotka-Volterra equations (Eq. 1), since the a_{ij} terms are probably not constant but themselves functions of population density and species frequencies (see Abrams [1980b] and Thomson [1980] for recent reviews). Even so, we still expect Eq. 1 to be valid for perturbations in a small neighborhood of an equilibrium. In this case, one would design PULSE and PRESS experiments with small perturbations. Unfortunately, as already noted, this makes it more difficult to obtain accurate results. How might one check the validity of Eq. 1 far from equilibrium? An investigator performing a PRESS experiment involving large perturbations could collect data on both the initial rate of change of the $n - 1$ unperturbed species, as well as their net density change once the community comes to rest. The values of a_{ij} could be estimated from Eq. 9 and then used in Eq. 4a to estimate c_i:

$$
c_i = \frac{\dfrac{dN_{if}}{dt}}{\hat{N}_i a_{if}(\hat{N}_f - N_{ff})}, \quad f \neq i. \tag{13}
$$

This provides $n - 1$ estimates of each of the c_i's. The degree of similarity between the various estimates for each c_i is indicative of the adequacy of the generalized Lotka-Volterra equations as a description of the observed community dynamics. However, if the central purpose of the experiments is to test for the presence of "higher-order interactions," more definitive experimental designs are available (Case and Bender 1981).

PRESS experiments do not allow rate constants to be determined because only equilibrium values are studied. The underpinnings of PULSE experiments are closely tied to the adequacy of differential equations as descriptions of the population dynamics. PRESS experiments, however, do not make this assumption. Since PRESS experiments rely only on the equilibrium condition $K = A\hat{N}$, our analysis is correct as stated for a variety of models such as Lotka-Volterra difference equations, discrete Ricker equations, and time-delay logistic differential equations.

TEMPORAL AND RANDOM VARIATION

One might argue that since all the foregoing assumes that communities are at or approaching some ideal "equilibrium," the results have little bearing on the

10　　　　　　　　　　　EDWARD A. BENDER ET AL.　　　　　　　Ecology, Vol. 65, No. 1

real world where populations are in a greater state of flux.

Suppose the Lotka-Volterra equations (Eq. 1) still hold, but c_i and/or K_i are functions of time. PULSE experiments will be virtually unaffected. We simply compare experimental populations with controls rather than with the preperturbed state. Imagine a control population with population levels N_i^c and per capita growth rates P_i^c at some time t. At this time we experimentally perturb species f in experiment f to a population level N_f. We then measure the resulting new per capita growth rate P_{if}. Subtract the control population value of Eq. 1 from that for the perturbation of species f to obtain:

$$P_{if} - P_i^c = c_i(t)a_{if}(N_f^c - N_f) \qquad (14)$$

in place of Eq. 4. Proceeding as in the derivation of Eq. 5 we obtain:

$$a_{ij} = \frac{(P_{ij} - P_i^c)(N_i^c - N_i)}{(P_{ii} - P_i^c)(N_j^c - N_j)}; \qquad (15a)$$

$$c_i(t) = \frac{P_{ii} - P_i^c}{N_i^c - N_i}. \qquad (15b)$$

In the event that the $K_i(t)$'s are constants, we may take the control population to be in equilibrium, and so $N_i^c = \hat{N}_i$ and $P_i^c = 0$. Since $P_{ij} = dN_{ij}/\hat{N}_i dt$, if $i \neq j$ and $P_{ii} = dN_{ii}/\hat{N}_i dt$, Eq. 15 reduces to Eq. 5.

PRESS experiments appear to be useless if K_i is highly variable since we always assume that the population is in equilibrium. On the other hand, if the c_i's are variable, then our analysis of PRESS experiments is unaltered since these rate parameters will not affect equilibria.

NATURAL VARIATION TECHNIQUES

Techniques based on natural variation are a third category of methods for estimating interaction coefficients. They are not experimental in the sense that the investigator manipulates populations; rather, they rely on nature doing the perturbations. Yet, how should this stochasticity be interpreted? Should we visualize the natural variation as affecting only the N's while the basic parameters describing species interactions remain constant? Or should we visualize the perturbations as affecting the parameters themselves? In the latter case, which parameters are affected? The mathematical analysis needed to proceed depends on our answers to these questions.

Can anything be done if natural variation affects only the N's? Hallett and Pimm (1979) considered the following scenario. Imagine two competitor species obeying discrete forms of Eq. 1, at least in the neighborhood of the equilibrium $\hat{N}_1, \hat{N}_2$. Using primes to denote the value of N_i at the next generation,

$$N_i' = N_i + r_i N_i (K_i - a_{i1}N_1 - a_{i2}N_2)/K_i,$$
$$i = 1,2. \qquad (16)$$

The system begins in equilibrium. Hallett and Pimm then randomly altered N_1 and N_2 by amounts x and y, respectively, where x and y are random variables with zero means, equal variances, and zero covariance. Following these random perturbations N_1 and N_2 grow by amounts Δx and Δy, respectively, in a single generation where these are given by Eq. 16. It is assumed that only $N_1 = (\hat{N}_1 + x + \Delta x)$ and $N_2 = (\hat{N}_2 + y + \Delta y)$ are observed by the investigator. After observing many such pairs of N_1 and N_2, the regression coefficients K_1^* and a_{12}^* in

$$N_1 = K_1^* - a_{12}^* N_2$$

may be estimated. A similar regression provides K_2^* and a_{21}^*.

Hallett and Pimm (1979) explored this stochastic process, using Monte Carlo simulations for different values of a_{12}, a_{21}, K_1, and K_2, with $r_1 = r_2 = 1.0$. They suggest that census data yielding the number of individuals of two species across different sites or at the same site but over different times could be used to calculate these regression coefficients, claiming that a_{ij}^* and K_i^* are estimates for a_{ij} and K_i and that the assumption of only one time step is not critical.

Actually, there is no need to perform Monte Carlo simulations since the expected regression coefficients for this single-step process can be calculated analytically; they do not equal a_{12} or a_{21}. Using a linearized form of Eq. 16 near equilibrium:

$$N_1 = \hat{N}_1 + x - \frac{r_1 \hat{N}_1}{K_1}(x + a_{12}y);$$

$$N_2 = \hat{N}_2 + y - \frac{r_2 \hat{N}_2}{K_2}(y + a_{21}x).$$

(The use of the exact form of Eq. 16 complicates the mathematics but does not alter the conclusions.) The regression coefficient a_{12}^* is $\text{cov}(N_1, N_2)/\text{var } N_2$, which is

$$a_{12}^* = \frac{K_2}{K_1} \frac{(K_1 - r_1 \hat{N}_1)r_2 \hat{N}_2 a_{21} + (K_2 - r_2 \hat{N}_2)r_1 \hat{N}_1 a_{12}}{(K_2 - r_2 \hat{N}_2)^2 + (r_2 \hat{N}_2 a_{21})^2}. \qquad (17)$$

Of course, a_{21}^* is obtained from Eq. 17 by interchanging the indices 1 and 2 wherever they occur. It is apparent from Eq. 17 that $a_{12} \neq a_{12}^*$. (This holds even if $r_1 = r_2 = 1.0$, as in Hallet and Pimm's simulations.) In cases where the two species being compared have very different r's and K's, the discrepancy between a_{12} and a_{12}^* will be large. The values of a_{12} and a_{12}^* can even be of opposite sign. We might try to obtain a_{12}, a_{21}, K_1, and K_2 by estimating a_{12}^*, a_{21}^*, K_1^*, and K_2^* and then solving equations such as Eq. 17. Unfortunately, this requires a knowledge of r_1 and r_2 and a guarantee that only a single time step was taken. A further complication arises when more than two species are involved, since, as in PRESS experiments, the calculated interaction coefficients between any two species are functions of the interactions of these species with all other species in the guild. We can think of no safe method for utilizing natural variation in the N's to estimate the parameters in Eqs. 1 or 16.

Another way of visualizing observed spatial (or temporal) variation in species numbers is to assume that density differences between sites arise because different sites have somewhat different sets of K's, and each community is in equilibrium. Schoener (1974) mentioned the situation in which there are two species in equilibrium at two sites where K_1 and a_{12} are constant and K_2 differs. If $\hat{N}_{ij}$ denotes the equilibrium population density of species i at site j, then

$$\hat{N}_{1j} = K_1 - a_{12}\hat{N}_{2j} \qquad (18)$$

for $j = 1, 2$. These equations can be solved for K_1 and a_{12}. If more than two sites are available, regression can be used to find the best estimates of K_1 and a_{12}. This readily extends to n species and n sites with K_1 and a_{1j} constant. A major defect of this method is locating sites where certain parameters differ and others do not. Such studies can probably only be conducted under carefully controlled laboratory conditions. Application to field situations is generally impossible.

Davidson (1980) assumed that all K's could vary from site to site in her study of desert ant communities. She hoped to estimate the elements of $a_{ij}{}^{(-1)}$ using Eq. 7 and the assumption that the observed spatial correlations between ant species i and j would be proportional to $a_{ij}{}^{(-1)}$. Since equilibrium at all sites is implicitly assumed, the observed covariance between $\hat{N}_i$ and $\hat{N}_j$ is independent of the r_i terms and the amplitude of the variation in the K's. (Without this assumption it is necessary to model the community dynamics with stochastic difference or differential equations (May 1975a, Turelli 1979). Yet, even with these simplifying assumptions, the relationship between the covariance in species K's and the resulting covariance in $\hat{N}_i$'s is not simple. For a linear system $K = A\hat{N}$,

$$\text{COV}(\kappa) = A \, \text{COV}(\hat{N})A^T \qquad (19a)$$

and

$$\text{COV}(\hat{N}) = A^{-1} \, \text{COV}(\kappa)A^{-1T} \qquad (19b)$$

where $\text{COV}(\hat{N})$ and $\text{COV}(\kappa)$ are the covariance matrices such that the i,j^{th} entry of $\text{COV}(\hat{N})$ gives the covariance between $\hat{N}_i$ and $\hat{N}_j$ when $i \neq j$ and the variance $(\sigma_i)^2$ when $i = j$. A simple proof of Eq. 19 is in Feller (1966: 81–82). In our case, we would like to solve for A (or A^{-1}) given the observed $\text{COV}(\hat{N})$, but clearly this is impossible unless we also know $\text{COV}(\kappa)$. Even if $\text{COV}(\kappa)$ is given, Eq. 19 has an infinite number of very different solutions, and even the sign pattern of A (or A^{-1}) is not determinable. This is true even in the simplest possible case where $\text{COV}(\kappa) = \sigma^2 I$ (I being the identity matrix).

CONCLUSIONS

As community ecology emerges from a largely descriptive science, controlled experiments are more and more used to decide issues instead of simply interpreting existing patterns in nature. With respect to perturbation experiments, there is still much confusion regarding the proper interpretation of results and how they bear on the relative strength of present-day species interactions as well as the forces shaping community structure. Some believe that such experiments can provide "necessary and sufficient" conditions to demonstrate the presence or absence of particular biotic interactions such as competition (e.g., Connell 1980, Newton 1980, Tinkle 1982) and that perturbation experiments are the only road to ecological knowledge. But we find that too much of the thinking about perturbation experiments is "two dimensional," that is, that it sees only one course of reaction following the experimenters' action. In the foregoing, we discuss the often hidden assumptions behind perturbation studies, the shortcuts that are usually taken by investigators, the pitfalls that may arise, and some of the rather counter-intuitive conclusions that emerge. For example, with PRESS designs, two species may each reach higher equilibrium densities in the absence of the other, yet the two may not compete or, for that matter, directly interact at all. The adjustment in their equilibrium densities is a result of the mediation of additional, unstudied species. Conversely, two species that do compete intensely may show no net increase in density when one or the other is removed or experimentally maintained at some low density, again due to the possible buffering effects of species intermediaries. Using PULSE experimental designs, it is possible that neither of two resource competitors will immediately increase in density following the removal or suppression of the other. This would be the case if resource dynamics are very slow and if the competitors do not exhibit overt interference.

If, through PRESS perturbation experiments, we arrive at estimates for the interaction coefficients between some set of species, and if these interaction coefficients accurately predict community dynamics in still further experiments, we still cannot conclude that our estimates even approximately represent the actual mechanistic coefficients of interaction.

In practice, no community ecologist can measure the density of every potentially interacting species in a community, yet once some species are neglected and others lumped into composite categories, there is a real danger that indirect effects can confuse and confound the interpretation of results. It is not easy to prescribe how one should look for sets of species between which interactions are strong, for the importance of one species on another may be independent of its body size, taxonomic position, or biomass; microscopic parasites may radically alter the nature of

12   EDWARD A. BENDER ET AL.   Ecology, Vol. 65, No. 1

the measured interaction between alternative host species in PRESS designs. A classic example involves the semicontrolled introduction of white-tailed deer into Nova Scotia in 1894 and the consequent extinction of caribou (Embree 1979). Interpreting the result as a PRESS experiment, involving only two focal species, deer and caribou, it is easy to reach the conclusion that these two focal species severely compete to the extinction of one. In fact, the extinction of caribou probably had nothing to do with competition but rather with differential susceptibility of these two ungulates to a meningeal parasitic nematode (Anderson 1965). A proper three-species PRESS design (e.g., one including the removal of the parasite) would have revealed this result, but in practice, we can probably never define the complete set of interacting species in an area. Unfortunately, without such knowledge, we are prone to misinterpret the interaction between any arbitrary species pair. This is a sobering thought, and one that underscores the importance of supplementing ecological experiments with the fullest amount of descriptive natural history and common sense. There is probably no one "royal road" to ecological knowledge; insights will best be gained by a pluralistic approach.

Acknowledgments

We thank Will Thomas and Eric Pianka for helpful comments. We acknowledge the support of National Science Foundation grant DEB-79-05085 to M. E. Gilpin.

Literature Cited

Abrams, P. A. 1975. Limiting similarity and the form of the competition coefficient. Theoretical Population Biology 8:356–375.

———. 1980a. Are competition coefficients constant? Inductive versus deductive approaches. American Naturalist 116:730–735.

———. 1980b. Some comments on measuring niche overlap. Ecology 61:44–49.

Anderson, R. C. 1965. *Cerebrospinae nematodiasis* in North American cervids. Transactions of the North American Wildlife Conference and Natural Resources 30:156–167.

Brown, J. H., and D. W. Davidson. 1977. Competition between seed-eating rodents and ants in desert ecosystems. Science 196:880–882.

Brown, J. H., D. W. Davidson, and O. J. Reichman. 1979. An experimental study of competition between seed-eating desert rodents and ants. American Zoologist 19:1129–1143.

Case, T. J., and E. A. Bender. 1981. Testing for higher order interactions. American Naturalist 118:920–929.

Case, T. J., and R. G. Casten. 1979. Global stability and multiple domains of attraction in ecological systems. American Naturalist 113:705–714.

Colwell, R. K., and C. Fuentes. 1975. Experimental studies of the niche. Annual Review of Ecology and Systematics 9:281–310.

Connell, J. H. 1961. The influence of interspecific competition and other factors on the distribution of the barnacle *Chthamalus stellatus*. Ecology 42:710–723.

———. 1980. Diversity and the coevolution of competitors, or the ghost of competition past. Oikos 35:131–138.

Davidson, D. W. 1980. Some consequences of diffuse competition in a desert ant community. American Naturalist 116:92–105.

Dayton, P. K. 1973. Two cases of resource partitioning in an intertidal community: making the right prediction for the wrong reason. American Naturalist 107:662–670.

Diamond, J. M. 1975. Assembly of species communities. Pages 342–444 *in* M. L. Cody and J. M. Diamond, editors. Ecology and evolution of communities. Harvard University Press, Cambridge, Massachusetts, USA.

Embree, D. G. 1979. The ecology of colonizing species, with special emphasis on animal invaders. *In* D. J. Horn, G. R. Stairs, and R. D. Mitchell, editors. Analysis of ecological systems. Ohio State University Press, Columbus, Ohio, USA.

Fadeev, P. K., and V. N. Fadeeva. 1963. Computational methods of linear algebra. Freeman Press, San Francisco, California, USA.

Feller, W. 1966. An introduction to probability theory and its applications. Volume II. J. Wiley and Sons, New York, New York, USA.

Hairston, N. G. 1981. An experimental test of a guild: salamander competition. Ecology 62:65–72.

Hallett, J. G., and S. L. Pimm. 1979. Direct estimation of competition. American Naturalist 113:593–600.

Holmes, R. T., J. C. Schultz, and P. Nothnagle. 1979. Bird predation on forest insects: an enclosure experiment. Science 206:462–463.

Holt, R. D. 1977. Predation, apparent competition, and the structure of prey communities. Theoretical Population Biology 12:197–229.

Hurlbert, S. L. 1978. The measurement of niche overlap and some relatives. Ecology 59:67–77.

Lawlor, L. R. 1979. Direct and indirect effects of n-species competition. Oecologia (Berlin) 43:355–364.

———. 1980. Overlap, similarity, and competition coefficients. Ecology 61:245–251.

Levine, S. 1976. Competitive interactions in ecosystems. American Naturalist 110:903–910.

Levins, R. 1968. Evolution in changing environments. Princeton University Press, Princeton, New Jersey, USA.

———. 1975. Evolution in communities near equilibrium. *In* M. L. Cody and M. Diamond, editors. Ecology and evolution of communities. Harvard University Press, Cambridge, Massachusetts, USA.

MacArthur, R. H. 1958. Population ecology of some warblers of northern coniferous forests. Ecology 39:599–519.

May, R. M. 1973. Stability and complexity in model ecosystems. Princeton University Press, Princeton, New Jersey, USA.

———. 1975. Some notes on measurement of the competition matrix. Ecology 56:737–741.

May, R. M., and R. H. MacArthur. 1972. Niche overlap as a function of environmental variability. Proceedings of the National Academy of Science (USA) 69:1109–1113.

Newton, I. 1980. The role of food in limiting bird numbers. Ardea 68:11–30.

Pomerantz, M. J. 1980. Do "higher order interactions" in competition systems really exist? American Naturalist 117:583–591.

Pomerantz, M. J., W. R. Thomas, and M. E. Gilpin. 1980. Asymmetries in population growth regulated by intraspecific competition: empirical studies and model tests. Oecologia (Berlin) 47:311–322.

Porter, K. G. 1977. The plant-animal interface in freshwater ecosystems. American Scientist 65:159–170.

Richardson, R. H., and P. E. Smouse. 1975. Ecological specialization of Hawaiian *Drosophila*. II. The community matrix, ecological complementation, and phyletic species packing. Oecologia (Berlin) 22:1–13.

Schaffer, W. 1981. Ecological abstraction: the consequences of reduced dimensionality in ecological models. Ecological Monographs 51:383–401.

Schoener, T. W. 1974. Competition and the form of habitat shift. Theoretical Population Biology 6:265–307.

Schroeder, G. D., and M. L. Rosenzweig. 1975. Perturbation analysis of competition and overlap in habitat utilization between *Dipodomys ordii* and *Dipodomys meriani*. Oecologia (Berlin) 19:9–28.

Seifert, R. P. 1979. A *Heliconia* insect community in a Venezuelan cloud forest. Ecology 60:462–467.

Seifert, R. P., and F. H. Seifert. 1976. A community matrix analysis of *Heliconia* insect communities. American Naturalist 110:461–483.

Smith, D. C. 1981. Competitive interactions of the striped plateau lizard (*Sceloporus virgatus*) and the tree lizard (*Urosaurus ornatus*). Ecology 62:679–687.

Smith, D. W., and S. D. Cooper. 1982. Competition among *Cladocera*. Ecology 63:1004–1015.

Thomson, J. D. 1980. Implications of different sorts of evidence for competition. American Naturalist 116:719–726.

Tinkle, D. W. 1982. Results of experimental density manipulation in an Arizona lizard community. Ecology 63:57–65.

Turelli, M. 1977. Random environments and stochastic calculus. Theoretical Population Biology 12:140–178.

Wise, D. H. 1981. A removal experiment with darkling beetles: lack of evidence for interspecific competition. Ecology 62:727–738.

APPENDIX 1

Given the observable quantities $\Delta \hat{N}_{if} = \hat{N}_j - \hat{N}_{if}$ from a series of n PRESS perturbation experiments, one for each species, we may solve for the interaction terms a_{ij} as follows. Let ΔN be the matrix with all the ΔN_{ij} terms from our n experiments. Define a new diagonal matrix D where

$$d_{ff} = \sum_{j=1}^{n} a_{fj} \Delta N_{jf}. \tag{A1.1}$$

In words, d_{ff} represents the amount by which species f would change in density if all the other species were held at densities N_{jf} and species f was free to reach a new equilibrium. This quantity d_{ff} is not observable since in our experiments it is species f that is held constant, yet by defining such an abstract quantity we will be able to reach a solution for a_{ij} in terms solely of observable quantities, namely the ΔN_{jf}'s. Using these definitions, we can write Eqs. 8 and A1.1 in matrix form as

$$A \Delta N = D, \tag{A1.2a}$$

and so

$$A = D \Delta N^{-1}. \tag{A1.2b}$$

Hence, $a_{ij} = d_{ii} \, \Delta N_{ij}^{(-1)}$ where $\Delta N_{ij}^{(-1)}$ is the ij^{th} element of the inverse matrix. Since a_{ii} and $d_{ii} = 1/\Delta N_{ii}^{(-1)}$ we have

$$a_{ij} = \Delta N_{ij}^{(-1)} / \Delta N_{ii}^{(-1)}. \tag{A1.3}$$

Another way of expressing the relationship between the measurable values $\hat{N}_i - N_{if}$ following the perturbation experiments and their quantitative relationship to the interaction coefficients, a_{ij}, is as follows: From Eq. A1.2a, $\Delta N = A^{-1} D$ and so

$$\Delta N_{if} = \hat{N}_i - N_{if} = a_{ii}^{(-1)} d_{ff}.$$

Hence

$$\frac{\Delta N_{if}}{\Delta N_{jf}} = a_{if}^{(-1)} / a_{jf}^{(-1)}, \tag{A1.4}$$

which is the expression in Eq. 10.

APPENDIX 2

To explore the effect of neglected species on the analysis of PRESS experiments, we first partition the matrices ΔN and ΔN^{-1} $(= M)$ using the subscript F for the first s species on which we focus and R for the remaining $n - s$:

$$\Delta N = \begin{vmatrix} \Delta N_{FF} & \Delta N_{FR} \\ \Delta N_{RF} & \Delta N_{RR} \end{vmatrix}; \quad \Delta N^{-1} = M = \begin{vmatrix} M_{FF} & M_{FR} \\ M_{RF} & M_{RR} \end{vmatrix}.$$

Since $M \Delta N = I$, we have:

$$\begin{vmatrix} M_{FF} \Delta N_{FF} + M_{FR} \Delta N_{RF} & M_{FF} \Delta N_{FR} + M_{FF} \Delta N_{RR} \\ M_{RF} \Delta N_{FF} + M_{RR} \Delta N_{RF} & M_{RF} \Delta N_{FR} + M_{RR} \Delta N_{RR} \end{vmatrix} = \begin{vmatrix} I & O \\ O & I \end{vmatrix}.$$

From the lower left entry, $M_{RF} \Delta N_{FF} + M_{RR} \Delta N_{RF} = 0$, and so $\Delta N_{RF} = -M_{RR}^{-1} M_{RF} \Delta N_{FF}$. Substituting this into the upper left entry, multiplying on the right by ΔN_{FF}^{-1} and rearranging

$$M_{FF} - \Delta N_{FF}^{-1} = M_{FR} M_{RR}^{-1} M_{RF}. \tag{A2.1}$$

The matrix M_{FF} should be the source of the numbers on the right side of Eq. 9 in determining a_{ij}; however, if we neglected to include the last $n - s$ species in our model, we would inadvertently use ΔN_{FF}^{-1} instead. Thus we need to know by how much the corresponding entries in M_{FF} and ΔN_{FF}^{-1} differ. This is given by Eq. A2.1. Thus to minimize error we need

$$M_{FF} \gg M_{FR} M_{RR}^{-1} M_{RF},$$

where the symbol $\gg$ means "is much greater than." We can divide the ith row of M by m_{ii} to convert this to an A matrix equation:

$$A_{FF} \gg A_{FR} A_{RR}^{-1} A_{RF}. \tag{A2.2}$$

If we make the rather reasonable assumption that A_{RR} is not ill conditioned, we can say that the effects of the ignored $n - s$ species are negligible if either the interaction coefficients of ignored species on focal species (A_{FR}) are all very small, or the interaction coefficients of the focal species on ignored species (A_{RF}) are all very small. Ill conditioning in A_{RR} may arise if the neglected species have very similar intra- and interspecific competitive effects on one another. For example, if

$$A = \begin{vmatrix} 1 & 1.1 \\ 0.9 & 1 \end{vmatrix},$$

$$A_{RR}^{-1} = \begin{vmatrix} 100 & -110 \\ -90 & 100 \end{vmatrix}.$$

Thus, the right-hand side statement of A2.2 may become excessively large even if neglected species do not interact very strongly with focal species.

OIKOS 38: 210–221. Copenhagen 1982

Dynamics of regional distribution: the core and satellite species hypothesis

Ilkka Hanski

Hanski, I. 1982. Dynamics of regional distribution: the core and satellite species hypothesis. – Oikos 38: 210–221.

A new concept is introduced to analyse species' regional distributions and to relate the pattern of distributions to niche relations. Several sets of data indicate that average local abundance is positively correlated with regional distribution, i.e. the fraction of patchily distributed population sites occupied by the species. This observation is not consistent with the assumptions of a model of regional distribution introduced by Levins. A corrected model is now presented, in which the probability of local extinction is a decreasing function of distribution, and a stochastic version of the new model is analysed. If stochastic variation in the rates of local extinction and/or colonization is sufficiently large, species tend to fall into two distinct types, termed the "core" and the "satellite" species. The former are regionally common and locally abundant, and relatively well spaced-out in niche space, while opposite attributes characterize satellite species. This dichotomy, if it exists, provides null hypotheses to test theories about community structure, and it may help to construct better structured theories. Testing the core-satellite hypothesis and its connection to the r-K theory and to Raunkiaer's "law of frequency" are discussed.

I. Hanski, Dept of Zoology, Univ. of Helsinki, P. Rautatiekatu 13, SF-00100 Helsinki 10, Finland.

Предлагается новая концепция для анализа регионального распределения видов и для сравнения характера распределения с соотношением ниш. Несколько серий данных показали, что величина средней локальной численности положительно коррелирует с региональным распространением, то есть с относительным количеством мозаично расположенных видовых стаций, занятых данным видом. Это наблюдение не соответствует модели регионального распределения, предложенной Левинсом. Здесь предлагается исправленная модель, в которой обсуждается вероятность локального исчезновения вида, как уменьшающаяся функция распространения и стохастическая версия новой модели. Если стохастические колебания скоростей локального исчезновения и/или колонизации достаточного велики, проявляется тенденция разделения видов на два четких типа, называемых "основным" и "сателлитным". Первые обычны в своем регионе, локально многочисленны, а сателлитные виды характеризуются противоположными признаками. Эта дихотомия если она существует, позволяет использовать нуль-гипотезу для проверки теории структуры сообщества и она может помочь в создании более совершенной теории. Обсуждаются результаты проверки гипотезы "основных-сателлитных" видов и ее связи с r-K теорией и биологическими спектрами Раункиера.

Accepted 26 March 1981

1. Introduction

It has been popular to define ecology as the study of abundance and distribution of organisms (e.g. Andrewartha and Birch 1954, Krebs 1972), to the extent that MacArthur and Wilson (1967) state there is no real distinction between ecology and biogeography. For the purposes of this paper, I define *abundance* as the number of individuals at a local population site (for other definitions see Hengeveld 1979). There are good reasons to express abundance as a fraction of the possible maximum numbers sustainable at the site (Andrewartha and Birch 1954), but this may be difficult particularly when dealing with multispecies communities. *Distribution* refers to the number of population sites occupied by the species; this again may be given as the fraction out of the suitable ones within an arbitrary or natural region. Population sites can be discrete units, like true islands; or, like habitat islands, they may have been delimited more arbitrarily from the rest of the environment; or they may be contagious, in which case distribution is simply the proportion of total area occupied. This definition of distribution does not specify the type of spatial patterns, i.e. the locations of the (occupied) sites in space, which is a related but different question.

Theoretical ecology has largely modelled local abundance (e.g. May 1976), while distribution has been left, until recently, to biogeographers, with the exception of the largely descriptive statistical work on animal and plant distributions (Patil et al. 1971, Bartlett 1975, Taylor et al. 1978, Ord et al. 1980). An exception to this rule is Levins's (1970, see also 1969a) model on extinction, which has been followed by a number of studies on interspecific competition (Cohen 1970, Levins and Culver 1971, Horn and MacArthur 1972, Levin 1974, Slatkin 1974, Hanski 1981a; see also Skellam 1951) and predation (Vandermeer 1973, Zeigler 1977) in patchy environments, all of which apply Levins's approach to regional population dynamics and underline the difference between local and regional interactions.

The spatial aspect of population interactions has recently received increasing attention (reviewed by Levin 1976; see also Smith 1974, Levin 1977, 1978, Gurney and Nisbet 1978a, b, Taylor and Taylor 1977, Crowley 1979, Comins and Hassell 1979, Hanski 1981b), and it has become clear that understanding of both spatial processes (distribution of the species in physical space) and resource partitioning (distribution of the species in niche space) are essential components to a satisfactory explanation of the perennial questions: Why are there so many species? Why are there so many rare species? (Wiens 1976, Yodzis 1978, Hanski 1979a). Indeed, some ecologists (e.g. Simberloff 1978) have gone so far as to maintain that, in many or most cases, spatial dynamics in independently developing populations explain most of the "community patterns" (see also Caswell 1976). This contrasts with the approach initiated by MacArthur (summarized in his 1972 book).

Whatever view one holds on the importance of competitive and other biotic interactions in structuring communities, it is an indisputable fact that communities consist of different kinds of species: some are widely distributed while others occur patchily; there exist locally abundant and locally rare species; and in some communities species are, at least apparently, well spaced-out in niche space, while in other communities guilds of similar species coexist. One is tempted to pose the question: Is it possible to find unifying factors to simplify this diversity?

I suggest some narrowing down of this question. It will first be shown that dynamics in local abundance and regional distribution are interdependent. Incorporating this observation into the type of models of regional distribution suggested and first analysed by Levins (1969a, 1970, Levins and Culver 1971) leads to an important structural change in the basic model. The key question in the analysis of the revised model is whether there is an internal equilibrium point on the distribution scale, which most species are approaching, or whether the species are just heading towards either maximal distribution and superabundance, or regional extinction.

2. Local abundance and regional distribution are interdependent

Four examples from different invertebrate taxa are put forward to answer the question, are local abundance and regional distribution independent of each other? The answer is no.

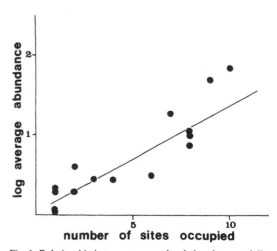

Fig. 1. Relationship between average local abundance and distribution in Anasiewicz's (1971) data on bumblebees from Lublin, Poland. While calculating average abundance only those sites were included from which the species was collected (note logarithmic y-axis). Distribution is the number of sites, maximally 10, occupied by the species. Each dot in this figure represents one species (the line has been drawn by eye).

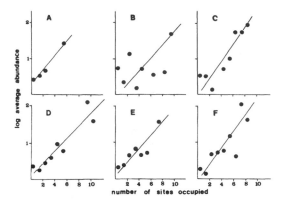

Fig. 2. Relationship between average local abundance and distribution, like in Fig. 1, in Kontkanen's (1950) data on leafhoppers from meadows in East Finland. Each dot in this figure is the average for several species, which had the same distribution. Figures A to F refer to different "communities", representing wet to dry meadows (from left to right) at early (upper row) and late summer (lower row).

Anasiewicz (1971) studied bumblebees in the parks, squares, lawns, etc. of Lublin in Poland – all good examples of discrete habitat islands in the man-made environment. Average local abundance increased with the number of sites from which the species was recorded (Fig. 1; only sites in which the species was present are included in the calculation of average local abundance).

Kontkanen (1950; see also 1937, 1957) sampled leafhoppers from meadows in East Finland. He delimited six "communities" of coexisting species, and, in

each community, a positive correlation exists between average local abundance and distribution (Fig. 2).

The third example is from Karppinen's (1958) study on soil mites (Oribatei) in two forest types in North Finland. In both habitats, a positive correlation between abundance and distribution is apparent (Fig. 3).

The final example is from my studies on dung and carrion beetles in lowland rain forest in Sarawak (Hanski unpubl.). Trapping was carried out with 10 traps for 4 nights at 12 sites, situated at least 0.5 km from each other in homogeneous virgin forest. I have restricted the analysis to the species-rich genus *Onthophagus* (Scarabaeidae). Once again, a positive correlation exists between the number of trapping sites from which the species was caught and the average catch from one site (Fig. 4). I conclude from these examples that a correlation between abundance and distribution is the rule in nature.

It is beyond the scope of this paper to go into details about the causes of this relationship, but it may be pointed out that the level of between-site movements is clearly of crucial importance. It appears to be common in nature that emigration takes place much before local carrying capacity has been reached, perhaps because of reasons discussed by Lidicker (1962) and Grant (1978).

Datum points in Figs 1 to 4 result from sampling, but because both abundance and distribution are underestimated, an increase in sample size should not change the picture qualitatively. There are, of course, truly rare yet widely distributed species, like the crane *Grus grus* in Finnish marshlands (Järvinen and Sammalisto 1976), but if communities consisting of reasonably similar species are studied, true distribution is expected to be correlated with true average abundance. The contrary

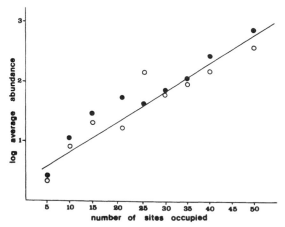

Fig. 3. Relationship between average local abundance and distribution, like in Fig. 1, in Karppinen's (1958) data on soil mites (Oribatei) from two forest types in North Finland. Each dot in this figure is the average for several species, which have been grouped into 10 distribution classes (total number of sites was 50 in both forest types). The two kinds of symbols refer to the two forest types.

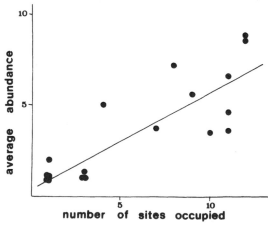

Fig. 4. Relationship between average local abundance and distribution, like in Fig. 1, in *Onthophagus* (Scarabaeidae) in the alluvial forest in Sarawak (Hanski unpubl.). A total of 12 sites was studied in homogeneous virgin forest. Each dot in this figure represents one species (note that y-axis is not logarithmic).

OIKOS 38:2 (1982)

would require spatial variance to decrease by a factor greater than approximately the squared proportional decrease in average abundance, which most certainly is not the case at least in moths and aphids (Taylor et al. 1980, see also 1978). Accepting that spatial variance is proportionally as large in rare as in abundant species, local extinctions are bound to occur (MacArthur and Wilson 1967), and distribution is, at least in rare species, less than the possible maximum.

3. Models of distribution

The paradigm for the dynamics in local abundance is still the logistic equation (Lotka 1925)

$$\frac{dN}{dt} = rN(1-N/K), \tag{1}$$

while perhaps the only similarly general model proposed for distribution is Levins's (1969a)

$$\frac{dp}{dt} = ip(1-p)-ep, \tag{2}$$

where p, a measure of distribution, is the fraction of population sites occupied by the species ($0 \le p \le 1$), and i and e are constants for a given species in a given environment. The first term in this equation is the rate of colonization of empty sites, and the second term is the rate of local extinctions. When all suitable sites in the region are occupied, p equals 1. The single internal equilibrium of Eq. (2) is stable, $\hat{p} = 1-e/i$, and regional extinction follows if $e \ge i$.

Levins (1970) subsequently analysed the stochastic version of Eq. (2): the extinction parameter, e, was assumed to be a random variable, with mean $\bar{e}$ and variance σ_e^2. Assuming no autocorrelation ("white noise"), the diffusion equation method (Kimura 1974) may be used to analyse the distribution of p, and gives (Levins 1970),

$$\Phi(p) = Cp^{2(i-\bar{e})/\sigma_e^2-2} \exp(-2ip/\sigma_e^2), \tag{3}$$

as the limiting ($t \to \infty$) distribution of $\Phi(p,t)$. This does not depend on the initial value, $p(0)$. Constant C is necessary to guarantee that $\int_0^1 \Phi(p)dp = 1$. Critical points of Eq. (3) may be found from the equation,

$$2M_{\delta p} - d/dp \, V_{\delta p} = 0, \tag{4}$$

where $M_{\delta p}$ and $V_{\delta p}$ are the mean and variance of the rate of change in the stochastic version of Eq. (2). This gives the condition,

$$i > \bar{e} + \sigma_e^2, \tag{5}$$

for a unimodal distribution $\Phi(p)$ with a peak at $p =$

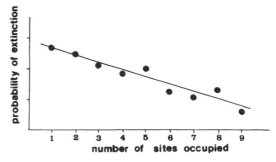

Fig. 5. Relationship between the probability of local extinction (per year) and distribution in Simberloff's (1976) data on mangrove island insects. Maximum number of sites (small mangrove islands) was nine. Each dot represents the average for several species (the line has been drawn by eye).

$1-(\bar{e}+\sigma_e^2)/i$. If (5) does not hold, $\Phi(p)$ is a decreasing function between 0 and 1. The deterministic equilibrium, obtained when $\sigma_e^2 = 0$, is always equal to or greater than the stochastic mode. It should be noted that there are two interpretations for $\Phi(p)$ (Kimura 1964). $\Phi(p)$ gives the distribution of p both for a single species during a long period of time, and for a community of similar species at a given moment.

The assumption that all the local populations are the same, implicit in model (2), is very unrealistic. To model local dynamics at each population site explicitly is out of question (though see DeAngelis et al. 1979), but the relationship found in Sect. 2 provides us with an approximative, yet qualitatively correct, non-constant relationship between p and the "average state" (abundance) in a local population: average local abundance increases with increasing p.

Probability of local extinction increases with decreasing population size (e.g. MacArthur and Wilson 1967, Christiansen and Fenchel 1977). One would expect, therefore (see Sect. 2), that e in Eq. (2) is not constant, but decreases with p. I have found 3 sets of data to test this prediction.

A re-analysis of Simberloff's (1976) results on extinction of local (island) populations of mangrove island insects shows that e in Eq. (2), a parameter related to the probability of local extinction, decreases with increasing p (Fig. 5). The same result was obtained from a similar analysis of Kontkanen's (1950) data on leafhoppers in meadows in East Finland (Fig. 6).

The third example is from Boycott's (1930, see also 1919 and 1936) study on fresh-water molluscs in small ponds in the parish of Aldenham in England. Almost a hundred ponds were surveyed for molluscs and plants in 1915 and 1925. This example is particularly important because the small size of the ponds enabled Boycott (1930: 2-3) to make accurate censuses. The extinctions observed are thus real. (Simberloff (1976) also tried to document all the populations of each island, while

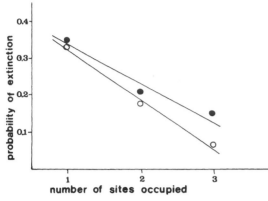

Fig. 6. Relationship between the probability of local extinction and distribution in Kontkanen's (1950) data on leafhoppers in East Finland. Maximum number of sites (meadows) was three. Black dots represent the average for several species and the probability of extinction in five years; open circles give the corresponding annual extinction probabilities (these results indicate that "re-colonization" was frequent, which is partly an artefact of the relatively small sample size, leaving some small populations unnoticed; cf. Kontkanen 1950).

Kontkanen (1950) probably missed many small populations.) My re-analysis (Fig. 7) of Boycott's (1930) data closely agrees with the above results: e is not constant but decreases with p. This result is not quite accurate, because more than one extinction-colonization event may have taken place in 10 years (cf. Diamond and May 1977), but the trend is very clear.

At present we may accept the simplest hypothesis about the rate of extinction: $e'(1-p)p$ (note that $e' \approx e$ when p is small). On this assumption Eq. (2) is replaced by

$$\frac{dp}{dt} = ip(1-p)-e'p(1-p). \qquad (6)$$

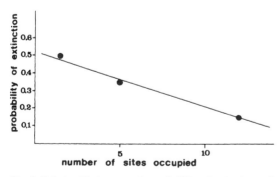

Fig. 7. Relationship between the probability of extinction and distribution in Boycott's (1930) data on fresh-water molluscs in small ponds in England. This figure gives the "net" extinction rate in 10 years. While constructing the figure, I have included the 34 ponds in Boycott's (1930) classes A and B which did not dry up during the study period. Species have been divided into three distribution classes, namely 1 or 2, 3 to 7, and 10 to 14 ponds occupied, and the dots in the figure give the average for each group (altogether there were 16 species).

I shall next explore two simple assumptions about the colonization process.

(1) If i in Eq. (6) does not depend on p, we obtain the logistic equation (1). This has been analysed, in the ecological context, by Levins (1969b; see also May 1973 and Leigh 1975). In the deterministic case, there is one stable equilibrium point, either at $\hat{p} = 1$, if $i > e'$, or at $\hat{p} = 0$, if $i < e'$. For the stochastic version, assume that $s \equiv i - e'$ is a random variable, and that there is no autocorrelation. The diffusion equation method gives,

$$\Phi(p) = Cp^{2(\bar{s}/\sigma_s^2-1)}(1-p)^{-2(\bar{s}/\sigma_s^2+1)}. \qquad (7)$$

$\Phi(p)$ is bimodal if $\sigma_s^2 > \bar{s}$. If, on the other hand, $\bar{s} > \sigma_s^2$, all populations approach maximal distribution, $p = 1$.

(2) Let us assume that the rate of colonization is $(s'p+e')p(1-p)$; the model then becomes,

$$\frac{dp}{dt} = s'p^2(1-p), \quad s' > 0. \qquad (8)$$

Evidently, there is only one stable equilibrium point, $\hat{p} = 1$. If s' is a random variable, and there is no autocorrelation, we can again use the diffusion equation method, which gives

$$\Phi(p) = C\exp(-2\bar{s}'/\sigma_{s'}^2 p)p^{2(\bar{s}'/\sigma_{s'}^2-2)}(1-p)^{-2(\bar{s}'/\sigma_{s'}^2+1)}. \qquad (9)$$

This distribution is bimodal if $\bar{s}' < \sigma_{s'}^2/3$. If the mean is greater than a third of the variance, all populations become maximally distributed.

A biological justification for assumption (2) about the rate of colonization is the probably increasing number of emigrants with increasing local abundance (e.g. Dempster 1968, Johnson 1969); presumably, more emigrants means more colonizations.

Addendum

During the preparation of this paper it escaped my notice that there may be certain mathematical problems in the use of the diffusion equation technique in the analysis of the models in Sect. 3 (Levins's 1970 analysis is erroneus; see Boorman, S. A. and Levitt, P. R. 1973. Theor. Pop. Biol. 4: 85–128; and see Roughgarden, J. 1979, pp. 384–391. Theory of population genetics and evolutionary ecology: an introduction. MacMillan). A supplementary numerical analysis of a discrete time version of Eq. (6) indicates, nonetheless, that the result presented here is qualitatively correct (Hanski 1982).

4. Ecological appraisal

The present modification of Levins's model led to a radically different conclusion from the one originally drawn by Levins: assuming stochastic variation in the rate of local extinction and/or colonization, populations

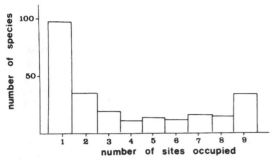

Fig. 8. Frequency distribution of species' distributions in Simberloff's (1976) data on mangrove island insects. Maximum number of sites was nine.

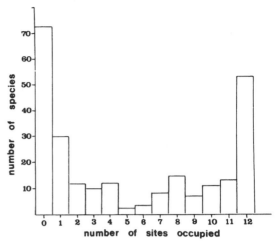

Fig. 10. Frequency distribution of species' distributions in Linkola's (1916) data on anthropochorous plants in Russian Karelia, U.S.S.R. The sites in question are isolated old houses and small villages, numbering 12 (sites 8 to 19 in Linkola 1916).

tend towards either one or the other of the (deterministic) boundary equilibria, $\hat{p} = 0$ and $\hat{p} = 1$, while in Levins's model p would hover around a stable internal equilibrium, $0 \leq \hat{p} \leq 1$. Which result more correctly reflects reality?

We recall that $\Phi(p)$ may be interpreted either as the distribution of p values in one species in a long period of time, or as the distribution of p values in many similar species at one moment of time (cf. Kimura 1964 for analogous interpretations in population genetics). To test the latter qualitatively, we require that all the species may establish local populations at the same sites, and that interspecific influences on model parameters are density- and frequency-independent. My model then predicts bimodality of p's, peaks close to unity and zero, while Levins's model predicts unimodality with the peak not very close to unity or zero.

Simberloff's (1976) data (cf. Fig. 5) support the present model; the distribution of the number of mangrove islands occupied by different species of insects appears bimodal (Fig. 8). To test this formally, we observe that

12 species were found from 6 islands, and 32 species from all 9 islands. The null hypothesis that the number of species is equal in these two classes is rejected: $\chi^2 = 9.09$, $P < 0.01$.

Onthophagus species in the lowland rain forest in Sarawak (cf. Fig. 4) also show a clear dichotomy into two sets of species (Fig. 9). I shall use the term "core" species for the locally abundant and regionally common species, and the term "satellite" species for locally and regionally rare species. In the case of mangrove island insects (Fig. 8), the same terms can be used, although "intermediate" species are now frequent.

Another data set to test this prediction is due to a study by Linkola (1916) on the occurrence of anthropochorous vascular plants near houses and villages, isolated by natural forest, in Russian Karelia (then Finland, study area ca. 10 000 km²). There was a clear size effect, large villages having more species than small ones (Linkola 1916), and colonization of isolated houses and small villages was perhaps not random, because some species were lacking systematically from them (though some would do so by chance only). For these reasons, I have restricted the analysis to 12 similar sites in Linkola's (1916) material. The frequency distribution of occurrences at the 12 sites is clearly bimodal (Fig. 10), which strongly supports the present model (see Hanski 1982 for a full analysis of Linkola's material).

It suffices to mention here that Kontkanen's (1950) results on leafhoppers and Anasiewicz's (1971) results on bumblebees also support the present model. A full analysis of these two studies will be presented elsewhere (Hanski unpubl.).

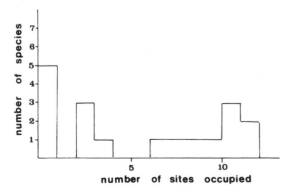

Fig. 9. Frequency distribution of species' distributions in *Onthophagus* in tropical lowland forest in Sarawak (Hanski unpubl.). Maximum number of sites was 12.

5. More about testing the core-satellite hypothesis

The core-satellite species hypothesis is a simple null hypothesis to explain regional rarity. There are two important premises in the model: (1) regional population dynamics *is* important, and (2) stochasticity in regional population dynamics *is* important. Stochastic variation in the parameters of regional dynamics (extinction, colonization) may be due to either demographic or environmental stochasticity.

As the examples given in the previous section showed, testing the main prediction of the model is simple: Is the distribution of species' regional distributions bimodal? If the distribution is clearly bimodal, there are grounds for a dichotomy, and for the use of the concept in a strong sense. Otherwise, one is left with the option of labelling the opposite ends of a continuum, like the r-K species distinction (MacArthur and Wilson 1967) has been replaced by the r-K species continuum (Pianka 1970, 1972, 1974, Southwood 1977).

Other models than the present one may predict a bimodal distribution of regional distributions, and to more rigorously test this model necessitates intensive studies on the rates of local extinction and colonization. If data are available on the rate of change in distribution, validity of Eq. (6) may be tested directly. Alternatively, one may try to document changes in species' status from the core to the satellite class, or vice versa, between regions or in time. Such changes are predicted to occur even if the pattern of environmental stochasticity remains stationary. An alternative model, which we may call an "adaptation" hypothesis, states that core species are better adapted to the environment than are satellite species, and does not predict changes from core to satellite class, or vice versa. Note that also the present model allows for interspecific differences in $\bar{s}$ and σ_s^2 (Sections 3 and 8).

L. R. Taylor's work (1974, 1978, Taylor and Taylor 1977, Taylor et al. 1978, 1980, Taylor and Woiwod 1980) on insect abundance and distribution has demonstrated the ever-changing patterns of regional distributions, anticipated by Andrewartha and Birch (1954), and Taylor's work has shown the interdependence between abundance and distribution. His results imply that local extinctions and colonizations are frequent phenomena (see also Den Boer 1977, Ehrlich 1965, Ehrlich et al. 1980), and that spatial population dynamics is important in all species.

The gravest difficulty in testing the core-satellite species hypothesis in a multispecies context is habitat selection. How does one identify the "population sites" suitable for the species? The reason for insisting on "similar species" in testing the multispecies prediction is just this; if the species are so similar that they have similar habitat selection, there is no problem. Then the number of potentially inhabitable sites is the same and the denominator in calculating the p's is the same for all species. Identical species are, of course, an unattainable

abstraction, but in many groups of closely related species habitat selection may be sufficiently similar. An extension of the core-satellite hypothesis into analyses of niche relations (in Sect. 8) requires, in any case, a distinction between habitat and niche (Whittaker et al. 1973).

There is the danger that counter-evidence (unimodal distribution of p's) is dismissed on the basis that the species were not, after all, similar enough. This would be wrong; the question about habitat selection must be resolved before the test is performed. In any case it is safest to include only similar population sites (habitat patches) in the analysis, which, to some extent, removes this problem. These restrictions do not mean that this theory could not work on any species. Careful selection of the species and of the sampling sites is required only because of problems in testing the hypothesis.

To repeat, the core-satellite hypothesis should be tested only with sets of species which may establish populations at the same sites; or, if such data are available, with records for single species in the long course of time. I believe that the former requirement was fulfilled in the above examples. A counter-example is the distributional ecology of water-striders *(Gerris)*. Vepsäläinen has shown, in a series of papers (see especially 1973, 1974a, b, 1978, Järvinen and Vepsäläinen 1976), that *Gerris* species, nine of which occur in Finland, show significant ecological and morphological differences in their adaptations to living in different kinds of lakes, ponds, streams, etc., including wing dimorphism in many species. In this case habitat selection is clearly different in different species, and the core-satellite hypothesis should *not* be used for the whole set of species, though it could be used for each species separately. In the latter context, one could talk about core and satellite populations, and comparisons should be made between regions or times.

Assuming the reality of core and satellite species, one may rephrase Hutchinson's (1959) question and ask: Why, in a given community, are there n core and m satellite species? Constraints on core and satellite species diversity are entirely different, which warrants two questions (n and m) instead of one ($n+m$). Nevertheless, satellite species may become, besides regionally extinct, also core species, and a core species can move to regional extinction only through a stage as a satellite species. Therefore the numbers of species in the two kinds are not independent of each other.

A specific prediction may be derived for the most universal trend in species diversity, namely that the number of species tends to increase with area, whether the region in question is an island or part of the mainland. Because satellite species survive as a set of small populations, their regional existence should hinge on the size of the region (e.g. Hanski 1981a). Therefore, as regions – e.g. islands – become smaller, the proportion of satellite species should decline. If Diamond's (1975, see also 1971, 1973, and others) series of species from

D-tramps to High-S species is matched with the continuum from core to satellite species, the proportion of satellite species indeed decreases with decreasing island size. Diamond (1975: 358), however, considers his D-tramps to be "r-selected", and High-S species to be "K-selected", which is in contrast with the present conceptualization – core species are certainly not "r-selected".

6. The core-satellite hypothesis and r-K theory

There exists a common basis for the core-satellite hypothesis and the by now well established r-K species theory (MacArthur and Wilson 1967, Gadgil and Bossert 1970, Pianka 1970, 1972; but see Wilbur et al. 1974, Southwood 1977, Christiansen and Fenchel 1977, Schaffer 1979). Both hypotheses stem from the same model – the logistic equation – which has been applied at the level of local abundance in the r-K model, and at the level of regional distribution in the core satellite model.

Nonetheless, the r-K species concept is used in a deterministic fashion to predict properties of species from the properties of their environment (e.g. Pianka 1970, 1974, Southwood 1977, Vepsäläinen 1978), while the core-satellite distinction is caused in the model by stochastic variation in spatial population dynamics. Unlike the satellite species, r-species are thought to be frequently locally abundant in comparison with K-species, but this does not follow from the mathematical model (logistic equation).

Although the two concepts are fundamentally different, core species are related to K-species, and, less obviously, satellite species are related to r-species.

7. A historical perspective

It is common in ecology that authors – or their readers – find "new" ideas preceded by earlier workers (Hutchinson 1978, McIntosh 1980), and nowadays preferable by Darwin. This may be viewed as a mark of soundness in the argument – or is McIntosh (1962) correct in claiming that "certain ideas seem to be invulnerable to attack and persist although subjected to multiple executions"? The core-satellite hypothesis is not an exception to the rule. The irony here is that the idea McIntosh was executing in 1962 was nothing else but bimodality of the distribution of spatial occurrences – the very prediction from the models in Sect. 3.

G. F. Gause (1936a: 323, see also 1936b) wrote: "The most important structural property of biocoenosis is the existence of definite quantitative relations between the abundant species and the rarer ones." One such relation, which Gause (1936a) discussed at length, is Raunkiaer's "law of frequency" (Raunkiaer 1913, 1918, 1934; a pioneering work by Jaccard in 1902),

which has been much used especially in plant ecology until the 1960's (e.g. Oosting 1956, Hanson and Churchill 1961, Mueller-Dombois and Ellenberg 1974), and which is of special relevance here. To see this, divide p into 5 segments of equal length (-0.2, $0.21-0.4$, etc.), and denote by A to E the numbers of species falling into the 5 classes. Raunkiaer's "law of frequency" states that $A > B > C \gtrless F < E$. Quite unexpectedly, the simple theory suggested in Sect. 3 predicted Raunkiaer's "law of frequency".

Nonetheless, with papers by Gleason (1920, 1929) and Romell (1930), criticism of the "law of frequency" started to accumulate (Gause 1936a, Preston 1948, Williams 1950), culminating in the above-mentioned "execution" by McIntosh (1962). It had been shown that the frequency distribution of species' frequencies depends, in Williams's (1950) words, on "the number of quadrats, the size of the quadrats, and on the Index of Diversity of the population." This criticism is justified. In view of the connection to the core-satellite hypothesis, one significant difference between the "laws" should be pointed out (see also Hanski 1982).

Frequency is the fraction of (usually small) samples, typically quadrats, in which the species occurs, all samples having been taken from the same homogeneous community. *Distribution,* as it was defined in the introduction and used in the models (Sect. 3), is a measure of occurrence on the between-site scale. Although the "true" population level may be difficult to specify (for an extreme example see Brussard and Ehrlich 1970a, b), the distinction between distribution and frequency is an important one whenever regional population dynamics are important, i.e. whenever many local populations are studied. It has been pointed out that the highest (E) of Raunkiaer's frequency classes is more inclusive than the lower ones, because the frequency classes include unequal density classes (Gleason 1929, Ashby 1935, McIntosh 1962). But unlike between density (abundance) and frequency classes in homogeneous communities, there is no simple statistical relationship between distribution and local abundance, the correlations in Figs 1 to 4 (Sect. 2) being due to ecological processes (notwithstanding problems of sampling; Sect. 2). In fact, the purpose of using the "law of frequency" was to determine the homogeneity of a stand of vegetation (or a community of animals; see e.g. Kontkanen 1950); bimodality (D < E) was namely expected only in homogeneous stands, which is an interesting convergence to my independently thought requirement of similar habitat selection in the species to be analysed (Sect. 5).

8. Concluding remarks: visiting Hutchinson's niche space

After these observations and theorizing, the reader may ask: What is gained by calling regionally common and locally abundant species core species, and rare species satellite species?

My answer is twofold. If the frequency distribution of species' regional distributions is indeed bimodal, this is interesting for its own sake, because it appears not to be the null hypothesis for many ecologists, who rather expect the kind of unimodality predicted by Levins's model. Secondly, and more importantly, if such a dichotomy exists in many natural communities, this should help us to provide a functional explanation for patterns of abundance and distribution. To take an example, if the core-satellite hypothesis is upheld, one may proceed by restricting the application of the equilibrium theory (MacArthur 1972, May 1973, 1976) to the core species, and employing appropriate non-equilibrium models for the satellite species. Caswell (1978) presumably had a similar idea in mind when he, after discussing the virtues of equilibrium and non-equilibrium models in ecology, conjectured: "Perhaps a community consists of a core of dominant species, which interact strongly enough among themselves to arrive at equilibrium, surrounded by a larger set of non-equilibrium species playing their roles against the background of the equilibrium species."

In the introduction I referred to a third structural property of communities besides abundance relations (Engen 1978) and spatial distributions (Simberloff 1978): distribution of species, or strictly speaking their "niches", in Hutchinson's (1957) niche space. The perennial question is how well spaced-out niches are in niche space. Intuition says and theory (e.g. MacArthur 1972, Lawlor and Maynard Smith 1976) predicts that interspecific competition causes better spacing-out, and ultimately and ideally leads to a uniform distribution of niches in niche space. In view of the controversy about the importance of competition in structuring communities (Paine 1966, Harper 1969, Janzen 1970, Dayton 1971, Connell 1975, 1978, Caswell 1976, Glasser 1979), this is an important question. The problem is that, in practice, other factors besides competition come into the play, making any "test" difficult. It is not surprising, therefore, that this kind of argument has led to widely varying conclusions (MacArthur 1972, Schoener 1974, Sale 1974, Inger and Colwell 1977, Southwood 1978, Strong et al. 1979, Pianka et al. 1979, Hanski 1979a, Lawlor 1980). The difficulty is in the formulation of a proper null hypothesis (see especially Lawlor 1980; the null hypothesis is *not* necessarily a random distribution of niches in niche space, Hanski 1979a, Grant and Abbott 1980), and in the multitude of factors potentially – and in practice – causing changes in niche position.

This is where the core-satellite hypothesis may prove useful. The following heuristic argument shows that there exists, after all, at least one unequivocal null hypothesis: if interspecific competition is important in structuring communities, core species should be better spaced-out in niche space than satellite species.

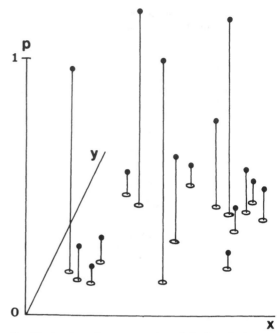

Fig. 11. A schematic representation of the hypothesis, explained in Sect. 8, that core species are better spaced-out in niche space than a random sub-set of the same size of all species, and better spaced-out than satellite species. x and y are two niche dimensions, and *p* denotes distribution, as in the rest of the paper, which varies from 0 to 1. Open circles represent niche positions, and the black dots give the position of species in the 3-dimensional space. It should be recalled that the argument is stochastic (see text), and the situation depicted in this figure is a static picture of a dynamic process.

Recall that the model is:

$$\frac{dp}{dt} = s_i p_i (1-p^i), \quad i = 1, \ldots, n \qquad \text{(Eq. 6)}$$

where n is the number of species, and s_i is a random variable with mean $\bar{s}_i$ and variance $\sigma_{s_i}^2$. The probability that species i is a core species at a given time is an increasing function of $\bar{s}_i / \sigma_{s_i}^2$. How does interspecific competition influence this ratio? Competition should increase the rate of extinction, it should decrease the rate of colonization – hence competition will decrease $\bar{s}_i$ – and it will probably increase $\sigma_{s_i}^2$. Consequently, $\bar{s}_i / \sigma_{s_i}^2$ will decrease, and the probability of species i staying/becoming a satellite species increases. The closer the competitor(s), the stronger the effect. Naturally, if there are two close competitors, both of which are core species, this model only predicts that one of them is likely to become a satellite species. The stochastic nature of the single species model is preserved in the multispecies context.

We may visualize species in a space constructed of Hutchinson's (1957) niche space and of one extra axis,

218

giving the extent of spatial distribution, p, which correlates, as we saw in Sect. 2, with abundance (Fig. 11). The greater the density of other species in the neighbourhood of any species in niche space, the smaller the probability that this species is, at a given time, a core species. Hence, core species are not expected to be a random sub-set of all the species with respect to niche position; we expect core species to comprise such a subset within which species are better spaced-out from each other than species are within a truly random sub-set. It follows that core species are also better spaced-out than satellite species.

A within-community analysis, such as suggested here, provides a concrete point of reference to test null hypotheses about community structure. Still, the present model cannot be but one step towards better understanding of community structure and its evolution. For instance, one could argue that even if core species are better spaced-out than satellite species in a community, this is perhaps not due to competition but to predation. Other studies are necessary to establish which explanation is correct. Theoretical work is also needed to clarify the expectations in multispecies communities. The value of the present approach can only be judged after several ecologists have tried it independently on their own data. Tests on dung beetles (Hanski 1980a) and bumblebees (Hanski unpubl.) have given encouraging results (see also Hanski 1979b, 1980b).

Acknowledgements – I wish to thank several colleagues for their useful comments on the different versions of the manuscript: R. Alatalo, J. Cartes, E. Connor, B. Don, Y. Haila, O. Järvinen (twice), Y. Mäkinen, E. Ranta, U. Safriel, D. Simberloff, and K. Vepsäläinen. O. Prŷs-Jones drew my attention to papers by Gause.

References

Anasiewicz, A. 1971. Observations on the bumble-bees in Lublin. – Ekol. Polska 19: 401–417.

Andrewartha, H. G. and Birch, L. C. 1954. The distribution and abundance of animals. – Chicago Univ. Press, Chicago.

Ashby, E. 1935. The quantitative analysis of vegetation. – Ann. Botany 49: 779–802.

Bartlett, M. S. 1975. The statistical analysis of spatial patterns. – Monogr. Appl. Prob. Statistics, Chapman and Hall, London.

Boycott, A. E. 1919. The freshwater mollusca of the parish of Aldenham: an introduction to the study of their oecological relationships. – Trans. Hertf. Nat. Hist. Soc. 17: 153–200.

– 1930. A re-survey of the fresh-water mollusca of the parish of Aldenham after ten years with special reference to the effect of drought. – Trans. Hertf. Nat. Hist. Soc. 19: 1–25.

– 1936. The habitats of fresh-water Mollusca in Britain. – J. Anim. Ecol. 5: 116–180.

Brussard, P. F. and Ehrlich, P. R. 1970a. Adult behaviour and population structure in *Erebia epipsodea* (Lepidoptera: Satyridae). – Ecology 51: 880–885.

– and Ehrlich, P. R. 1970b. The population structure of *Erebia epipsodea* (Lepidoptera: Satyridae). – Ecology 51: 119–129.

Caswell, H. 1976. Community structure: a neutral model analysis. – Ecol. Monogr. 46: 327–354.

– 1978. Predator-mediated coexistence: a non-equilibrium model. – Am. Nat. 112: 127–154.

Christiansen, F. B. and Fenchel, T. M. 1977. Theories of populations in biological communities. – Springer, Berlin.

Cohen, J. E. 1970. A Markov contingency-table model for replicated Lotka-Volterra systems near equilibrium. – Am. Nat. 104: 547–560.

Comins, H. N. and Hassell, M. P. 1979. The dynamics of optimally foraging predators and parasitoids. – J. Anim. Ecol. 48: 335–353.

Connell, J. H. 1975. Some mechanisms producing structure in natural communities. – In: Cody, M. L. and Diamond, J. M. (eds.), Ecology and evolution of communities, The Belknap Press of Harvard Univ. Press, Cambridge.

– 1978. Diversity in tropical rain forests and coral reefs. – Science 199: 1302–1310.

Crowley, P. H. 1979. Predator-mediated coexistence: an equilibrium interpretation. – J. Theor. Biol. 80: 129–144.

Dayton, P. K. 1971. Competition, disturbance, and community organization: the provision and subsequent utilization of space in a rocky intertidal community. – Ecol. Monogr. 41: 351–389.

DeAngelis, D. L., Travis, C. C. and Post, W. M. 1979. Persistance and stability of seed-dispersed species in a patchy environment. – Theor. Pop. Biol. 16: 107–125.

Dempster, J. P. 1968. Intra-specific competition and dispersal: as exemplified by a psyllid and its anthocorid predator. – In: Southwood, T. R. E. (ed.), Insect abundance. Symp. R. Ent. Soc. London, No. 4, Blackwell Scient. Publ., Oxford.

Den Boer, P. J. 1977. Dispersal power and survival. Carabids in a cultivated countryside. – Misc. papers 14, Land. Wageningen, The Netherlands.

Diamond, J. M. 1971. Comparison of faunal equilibrium turnover rates on a tropical and a temperate islands. – Proc. Natl Acad. Sci. USA 68: 2742–2745.

– 1973. Distributional ecology of New Guinea birds. – Science 179: 759–769.

– 1975. Assembly of species communities. – In: Cody, M. L. and Diamond, J. M. (eds.), Ecology and evolution of communities, The Belknap Press of Harvard Univ. Press, Cambridge.

– and May, R. M. 1977. Species turnover rates on islands: dependence on census interval. – Science 197: 266–276.

Engen, S. 1978. Stochastic abundance models. – Chapman and Hall, London.

Ehrlich, P. R. 1965. The population biology of the butterfly *Euphydryas editha* II. The structure of the Jasper Ridge colony. – Evolution 19: 327–336.

– Murphy, D. D., Singer, M. C., Sherwood, C. B., White, R. R. and Brown, I. L. 1980. Extinction, reduction, stability and increase: the responses of checkerspot butterfly *(Euphydryas)* populations to the California drought. – Oecologia (Berl.) 46: 101–105.

Gadgil, M. and Bossert, W. 1970. Life history consequences of natural selection. – Am. Nat. 104: 1–24.

Gause, G. F. 1936a. The principles of biocoenology. – Quart. Rev. Biol. 11: 320–336.

– 1936b. Some basic problems in biocoenology. (In Russian, summary in English) – Zool. Zh. 15: 363–381.

Glasser, J. W. 1979. The role of predation in shaping and maintaining the structure of communities. – Am. Nat. 113: 631–641.

Gleason, H. A. 1920. Some applications of the quadrat method. – Bull. Torrey Bot. Club 47: 21–33.

– 1929. The significance of Raunkiaer's law of frequency. – Ecology 10: 406–408.

Grant, P. R. 1978. Dispersal in relation to carrying capacity. – Proc. Natl Acad. Sci. USA 75: 2854–2858.

– and Abbott, I. 1980. Interspecific competition, island biogeography and null hypotheses. – Evolution 34: 332–341.

Gurney, W. S. C. and Nisbet, R. M. 1978a. Predator-prey fluctuations in patchy environments. – J. Anim. Ecol. 47: 85–102.
– and Nisbet, R. M. 1978b. Single-species population fluctuations in patchy environments. – Am. Nat. 112: 1075–1090.
Hanski, I. 1979a. The community of coprophagous beetles. – D. Phil. thesis, Univ. of Oxford, Oxford (unpubl.).
– 1979b. Structure of communities. (In Finnish, summary in English.) – Luonnon Tutkija 83: 132–137.
– 1980a. The community of coprophagous beetles (Coleoptera, Scarabaeidae and Hydrophilidae) in northern Europe. – Ann. Ent. Fenn. 46: 57–74.
– 1980b. Structure of insect communities: the core and the satellite species. – In: Abstracts of the XVI Int. Congress of Entomology, Kyoto.
– 1981a. Exploitative competition in transient habitat patches. – In: Chapman, D. G., Gallucci, V. and Williams, F. M. (eds.), Quantitative population dynamics. Statistical ecology, Vol. 13, Intern. Co-op. Publ. House, Fairland, Maryland.
→ – 1981b. Coexistence of competitors in patchy environment with and without predation. – Oikos 37: 306–312.
– 1982. Distributional ecology of anthropochorous plants in villages surrounded by forest. – Ann. Bot. Fenn. 19 (in press).
Hanson, H. C. and Churchill, E. D. 1961. The plant community. – Reinhold, New York.
Harper, J. L. 1969. The role of predation in vegetation diversity. – Brookhaven Symp. Biol. 22: 48–62.
Hengeveld, R. 1979. On the use of abundance and species-abundance curves. – In: Ord, J. K., Patil, G. P. and Taillie, C. (eds.), Statistical distributions in ecological work. Statistical ecology, Vol. 4, Intern. Co-op. Publ. House, Fairland, Maryland.
Horn, H. S. and MacArthur, R. H. 1972. Competition among fugitive species in a harlequin environment. – Ecology 53: 749–752.
Hutchinson, G. E. 1957. Concluding remarks. – Cold Spring Harbour Symp. Quant. Biol. 22: 415–427.
– 1959. Homage to Santa Rosalia or Why are there so many kinds of animals? – Am. Nat. 93: 145–159.
– 1978. An introduction to population ecology. – Yale Univ. Press., New Haven, Conn.
Inger, R. F. and Colwell, R. K. 1977. Organization of contiguous communities of amphibians and reptiles in Thailand. – Ecol. Monogr. 47: 229–253.
Jaccard, P. 1902. Lois de distribution florale dans la zone alpine. – Bull. Soc. Vand. Sci. Nat. (Lausanne), 38.
Janzen, D. H. 1970. Herbivores and the number of tree species in tropical forests. – Am. Nat. 104: 501–528.
Järvinen, O. and Sammalisto, L. 1976. Regional trends in the avifauna of Finnish peatland bogs. – Ann. Zool. Fenn. 13: 31–43.
– and Vepsäläinen, K. 1976. Wing dimorphism as an adaptive strategy in water-striders *(Gerris)*. – Hereditas 84: 61–68.
Johnson, C. G. 1969. Migration and dispersal of insects by flight. – Methuen, London.
Karppinen, E. 1958. Untersuchungen über die Oribatiden (Acari) der Waldböden von Hylocomium-Myrtillus-Typ in Nord-Finnland. – Ann. Ent. Fenn. 24: 149–168.
Kimura, M. 1964. Diffusion models in population genetics. – J. Appl. Prob. 1: 177–232.
Kontkanen, P. 1937. Quantitative Untersuchungen über die Insektenfauna der Feldschicht auf einigen Wiesen in Nord-Karelien. – Ann. Zool. Soc. 3, No. 4.
– 1950. Quantitative and seasonal studies on the leafhopper fauna of the field stratum on open areas in North Karelia. – Ann. Zool. Soc. 13, No. 8.
– 1957. On the delimitation of communities in research on animal biocoenotics. – Cold Spring Harbour Symp. Quan. Biol. 22: 373–378.
Krebs, C. J. 1972. Ecology: the experimental analysis of distribution and abundance. – Harper & Row, New York.
Lawlor, L. R. 1980. Structure and stability in natural and randomly constructed competitive communities. – Am. Nat. 116: 394–408.
– and Maynard Smith, J. 1976. The coevolution and stability of competing species. – Am. Nat. 111: 79–99.
Leigh, E. G. Jr. 1975. Population fluctuations, community stability and environmental variability. – In: Cody, M. L. and Diamond, J. M. (eds.), Ecology and evolution of communities, The Belknap Press of Harvard Univ. Press, Cambridge.
Levin, S. A. 1974. Dispersion and population interactions. – Am. Nat. 108: 207–228.
– 1976. Population dynamic models in heterogeneous environment. – Ann. Rev. Ecol. Syst. 7: 287–310.
– 1977. Spatial patterning and the structure of ecological communities. – Some mathematical questions in biology, VII, Vol. 8. Am. Math. Soc., Providence, R. I.
– 1978. Pattern formation in ecological communities. – In: Steele, J. H. (ed.), Spatial pattern in plankton communities. Plenum, New York.
Levins, R. 1969a. Some demographic and genetic consequences of environmental heterogeneity for biological control. – Bull. Ent. Soc. Am. 15: 237–240.
→ – 1969b. The effect of random variation of different types of population growth. – Proc. Natl Acad. Sci. U.S. 62: 1061–1065.
– 1970. Extinction. – In: Gerstenhaber, M. (ed.), Some mathematical problems in biology. Lectures on mathematics in the life sciences 2. Am. Math. Soc., Providence, R. I.
– and Culver, D. 1971. Regional coexistence of species and competition between rare species. – Proc. Natl Acad. Sci. U.S. 68: 1246–1248.
Lidicker, W. Z. Jr. 1962. Emigration as a possible mechanism permitting the regulation of population density below carrying capacity. – Am. Nat. 96: 29–33.
Linkola, K. 1916. Studien über den Einfluss der Kultur auf die Flora in den Gegenden nördlich vom Ladogasee. I. – Acta Soc. Fauna et Flora Fenn. 45.
Lotka, A. J. 1925. Elements of physical biology. – Williams and Wilkins, Baltimore.
MacArthur, R. H. 1972. Geographical ecology. – Harper & Row, New York.
– and Wilson, E. O. 1967. The theory of island biogeography. – Princeton Univ. Press, Princeton.
May, R. M. 1973. Stability and complexity in model ecosystems. – Princeton Univ. Press, Princeton.
– (ed.) 1976. Theoretical ecology. Principles and applications. – Blackwell Sci. Publ., Oxford.
McIntosh, R. P. 1962. Raunkiaer's "law of frequency". – Ecology 43: 533–535.
→ – 1980. The background and some current problems in theoretical ecology. – Synthese 43: 195–255.
Mueller-Dombois, D. and Ellenberg, H. 1974. Aims and methods of vegetation ecology. – John Wiley, New York.
Oosting, H. J. 1956. The study of plant communities. – Freeman, San Francisco.
Ord, J. K., Patil, G. P. and Taillie, C. (eds.) 1980. Statistical distributions in ecological work. – Statistical ecology, Vol. 4. Intern. Co-op. Publ. House, Fairland, Maryland.
Paine, R. T. 1966. Food web complexity and species diversity. – Am. Nat. 100: 65–75.
Patil, G. P., Pielou, E. C. and Waters, W. E. (eds.) 1971. Spatial patterns and statistical distributions. – Statistical ecology, Vol. 1. Penn. State Univ. Press, Pennsylvania.
Pianka, E. R. 1970. On r- and K-selection. – Am. Nat. 104: 592–597.

- 1972. r and K selection or b and d selection? – Am. Nat. 106: 581–588.
- 1974. Evolutionary ecology. – Harper & Row, New York.
- Huey, R. B. and Lawlor, L. R. 1979. Niche segregation in desert lizards. – In: Horn, D. J., Mitchell, R. D. and Stairs, G. R. (eds), Analysis of ecological systems. Ohio State Univ. Press, Columbus.
Preston, F. W. 1948. The commonness and rarity of species. – Ecology 29: 254–283.
Raunkiaer, C. 1913. Formationsstätistiske Undersøgelser paa Skagens Odde. – Bot. Tidskr. Kobenhavn 33: 197–228.
- 1918: Recherches statistiques sur les formations vegetales. – Biol. Medd. 1: 1–80.
- 1934. The life forms of plants and statistical plant geography. – Clarendon Press, Oxford.
Romell, L. G. 1930. Comments on Raunkiaer's and similar methods of vegetation analysis and the "law of frequency". – Ecology 11: 589–596.
Sale, P. F. 1974. Overlap in resource use, and interspecific competition. – Oecologia (Berl.) 17: 245–256.
Schaeffer, W. M. 1979. The theory of life-history evolution and its application to Atlantic salmon. – Symp. Zool. Soc. London 44: 307–326.
Schoener, T. W. 1974. Resource partitioning in ecological communities. – Science 185: 27–39.
Simberloff, D. 1976. Experimental zoogeography of islands: effects of island size. – Ecology 57: 629–648.
- 1978. Colonization of islands by insects: immigration, extinction, and diversity. – In: Mound, L. A. and Waloff, N. (eds.), Diversity of insect faunas. Symp. R. Ent. Soc. London, No. 9, Blackwell Sci. Publ., Oxford.
Skellam, J. G. 1951. Random dispersal in theoretical populations. – Biometrika 38: 196–218.
Slatkin, M. 1974. Competition and regional coexistence. – Ecology 55: 128–134.
· Smith, A. T. 1974. The distribution and dispersal of pika: consequences of insular population structure. – Ecology 55: 1112–1119.
Southwood, T. R. E. 1977. Habitat, the templet for ecological strategies. – J. Anim. Ecol. 46: 337–365.
- 1978. The components of diversity. – In: Mound, L. A. and Waloff, N. (eds.), Diversity of insect faunas. Symp. R. Ent. Soc. London, No 9, Blackwell Sci. Publ., Oxford.
Strong, D. R., Szyska, L. A. and Simberloff, D. S. 1979. Tests of community-wide character displacement against null hypotheses. – Evolution 33: 897–913.
Taylor, L. R. 1974. Monitoring change in the distribution and abundance of insects. – Report Rothamsted Exp. Station for 1973, Part 2.
- 1978. Bates, Williams, Hutchinson – a variety of diversities. In: Mound, L. A. and Waloff, N. (eds.), Diversity of insect faunas. Symp. R. Ent. Soc. London, No. 9, Blackwell Sci. Publ., Oxford.
- and Taylor, R. A. J. 1977. Aggregation, migration and population mechanics. – Nature, Lond. 265: 415–421.
- and Woiwod, I. P. 1980. Temporal stability as a density-dependent species characteristics. – J. Anim. Ecol. 49: 209–224.
- Woiwod, I. P. and Perry, J. N. 1978. The density-dependence of spatial behaviour and the rarity of randomness. – J. Anim. Ecol. 47: 383–406.
- Woiwod, I. P. and Perry, J. N. 1980. Variance and the large scale spatial stability of aphids, moths and birds. – J. Anim. Ecol. 49: 831–854.
Vandermeer, J. H. 1973. On the regional stabilization of locally unstable predator-prey relationship. – J. Theor. Biol. 41: 161–170.
Vepsäläinen, K. 1973. The distribution and habitats of Gerris Fabr. species (Heteroptera, Gerridae) in Finland. – Ann. Zool. Fenn. 10: 419–444.
- 1974a. The wing lengths, reproductive strategies and habitats of Hungarian Gerris Fabr. species (Heteroptera, Gerridae). – Ann. Acad. Sci. Fenn. A 202.
- 1974b. The life cycles and wing lengths of Finnish Gerris Fabr. species (Heteroptera, Gerridae). – Acta Zool. Fenn. 141.
- 1978. Wing dimorphism and diapause in Gerris – determination and adaptive significance. – In: Dingle, H. (ed.), The evolution of migration and diapause in insects. Springer, New York.
Whittaker, R. H., Levin, S. A. and Root, R. B. 1973. Niche, habitat, and ecotope. – Am. Nat. 107: 321–338.
Wiens, J. A. 1976. Population responses to patchy environments. – Ann. Rev. Ecol. Syst. 7: 81–120.
Wilbur, H. M., Tinkle, D. W. and Collins, J. P. 1974. Environmental certainty, trophic level, and resource availability in life history evolution. – Am. Nat. 108: 805–817.
Williams, C. B. 1950. The application of the logarithmic series to the frequency of occurrence of plant species in quadrats. – J. Ecol. 38: 107–138.
Yodzis, P. 1978. Competition for space and the structure of ecological communities. – Lecture notes in Biomath., Springer, Berlin.
Zeigler, B. P. 1977. Persistence and patchiness of predator-prey systems induced by discrete event population exchange mechanisms. – J. theor. Biol. 67: 677–686.

Vol. 132, No. 5   The American Naturalist   November 1988

SOURCES, SINKS, AND POPULATION REGULATION

H. RONALD PULLIAM

Institute of Ecology and Department of Zoology, University of Georgia, Athens, Georgia 30602

Submitted December 9, 1986; Revised August 3, 1987; Accepted February 11, 1988

Many animal and plant species can regularly be found in a variety of habitats within a local geographical region. Even so, ecologists often study population growth and regulation with little or no attention paid to the differences in birth and death rates that occur in different habitats. This paper is concerned with the impact of habitat-specific demographic rates on population growth and regulation. I argue that, for many populations, a large fraction of the individuals may regularly occur in "sink" habitats, where within-habitat reproduction is insufficient to balance local mortality; nevertheless, populations may persist in such habitats, being locally maintained by continued immigration from more-productive "source" areas nearby. If this is commonly the case for natural populations, I maintain that some basic ecological notions concerning niche size, population regulation, and community structure must be reconsidered.

Several authors (Lidicker 1975; Van Horne 1983) have discussed the need to distinguish between source and sink habitats in field studies of population regulation; however, most theoretical treatments (Gadgil 1971; Levin 1976; McMurtie 1978; Vance 1984) of the dynamics of single-species populations in spatially subdivided habitats have not explicitly addressed the maintenance of populations in habitats where reproduction fails to keep pace with local mortality. Holt (1985) considered the dynamics of a food-limited predator that occupied both a source habitat containing prey and a sink habitat with no prey. He demonstrated that passive dispersal from the source can maintain a population in the sink and that the joint sink and source populations can exceed what could be maintained in the source alone. Furthermore, he showed that "time-lagged" dispersal back into the source from the sink can stabilize an otherwise unstable predator-prey interaction. Holt argued, however, that passive dispersal between source and sink habitats in a temporally constant environment is usually selectively disadvantageous, implying that sink populations will be transient in evolutionary time.

In this paper, I consider the consequences of active dispersal (i.e., habitat selection based on differences in habitat quality) on the dynamics of single-species populations in spatially heterogeneous environments. I argue that active dispersal from source habitats can maintain large sink populations and that such dispersal may be evolutionarily stable.

Am. Nat. 1988. Vol. 132, pp. 652–661.

BIDE MODELS

One approach to modeling spatially heterogeneous populations is to employ *BIDE* models (Cohen 1969, 1971), which simultaneously consider birth (B), immigration (I), death (D), and emigration (E). Normally, in *BIDE* models, the parameters are considered random variables but not spatially heterogeneous. In this paper, I make the opposite assumptions, namely, that rates of birth, death, immigration, and emigration are deterministic but may differ between habitats.

First, consider a spatially distributed population with m subpopulations, each occupying a discrete habitat or "compartment." If b_j and d_j are, respectively, the number of births and the number of deaths occurring over the course of a year in compartment j, then the total number of births and deaths during that year in all compartments is given, respectively, by

$$B = \sum_{j=1}^{m} b_j \quad \text{and} \quad D = \sum_{j=1}^{m} d_j, \tag{1}$$

since every birth and every death takes place in some compartment.

Now, let i_{jk} be the number of individuals immigrating from compartment k *into* compartment j. Each immigrant into j must come from one of the other $m - 1$ compartments or come into j from outside the m compartments that constitute the ensemble of interest. That is, immigration into compartment j is given by

$$i_j = \sum_{k=1}^{m} i_{jk} + i_{j0} = \sum_{k=0}^{m} i_{jk},$$

where i_{j0} represents immigration from outside the ensemble into compartment j and i_{jj} is zero.

Similarly, if e_{jk} represents the number of emigrants from j into k, then

$$e_j = \sum_{k=1}^{m} e_{jk} + e_{j0} = \sum_{k=0}^{m} e_{jk}.$$

Note that $e_{kj} = i_{jk}$ for all j, $k \neq 0$. Finally, to complete the definitions of the *BIDE* parameters, let

$$I = \sum_{j=1}^{m} i_{j0} \quad \text{and} \quad E = \sum_{k=1}^{m} e_{k0}.$$

The ensemble of all compartments is said to be in dynamic equilibrium in ecological time when the number of individuals (n_j) in each and every compartment does not change from year to year. This occurs only if the number of births plus the number of immigrants exactly equals the number of deaths plus the number of emigrants for every compartment. That is,

$$b_j + i_j - d_j - e_j = (bide)_j = 0, \tag{2}$$

for every j, and $BIDE = 0$. Source and sink compartments can now be defined in terms of the *BIDE* parameters. A source compartment (or habitat) is one for which

$$b_j > d_j \quad \text{and} \quad e_j > i_j \tag{3}$$

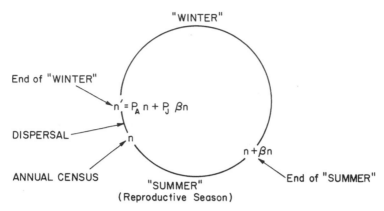

Fig. 1.—An annual census is taken in each habitat or "compartment" in the spring at the initiation of the breeding season (summer). Each individual breeding in the habitat produces β juveniles that are alive at the end of the breeding season. There is no adult mortality during the breeding season; adults survive the nonbreeding (winter) season with probability P_A and juveniles survive with probability P_J.

when $(bide)_j = 0$. A sink compartment (or habitat) is one for which

$$b_j < d_j \quad \text{and} \quad e_j < i_j \tag{4}$$

when $(bide)_j = 0$.

The above definitions apply strictly for equilibrium populations only. A more general definition of a source is a compartment that, over a large period of time (e.g., several generations), shows no net change in population size but, nonetheless, is a net exporter of individuals. Similarly, a sink is a net importer of individuals.

HABITAT-SPECIFIC DEMOGRAPHICS

To see how the *BIDE* parameters relate to habitat-specific survival probabilities and per capita birthrates, consider a simple annual cycle for a population in a seasonal environment (see fig. 1). An annual census is taken in the spring at the initiation of the breeding season. Each individual breeding in habitat 1 produces (on the average) β_1 juveniles that are alive at the end of the breeding season. There is no adult mortality during the breeding season; adults survive the nonbreeding (winter) season with probability P_A and juveniles survive with probability P_J. Thus, the expected number of individuals alive at the end of the winter and just before spring dispersal is given by

$$n_1(t + 1) = P_A n_1(t) + P_J \beta_1 n_1(t) = \lambda_1 n_1.$$

If there were only one compartment (habitat), λ_1 for a small population would be the finite rate of increase for the population. In a multi-compartment model, the λ_j's indicate which compartments are sources and which are sinks. For a simple example, consider two habitats that do not differ in either adult or juvenile survival probabilities but that do differ in per capita reproductive success. If habitat 1 is a source and habitat 2 a sink, then by definition,

$$\lambda_1 = P_A + P_J\beta_1 > 1 \tag{5a}$$

and

$$\lambda_2 = P_A + P_J\beta_2 < 1. \tag{5b}$$

The finite rate of increase for a multi-compartment model depends on the fraction of the population in each habitat. This, in turn, depends on how individuals distribute themselves among available habitats at the time of dispersal, before the onset of breeding (fig. 1).

Before considering how habitat dispersal between source and sink habitats influences population regulation, I briefly discuss population regulation in a source habitat in the absence of a sink habitat. To do this, I must specify the nature of density dependence in the source habitat. For the model discussed below, the critical feature of density dependence is that some individuals in the source habitat do predictably better than others in terms of fitness. Simple assumptions reflecting this feature are that the number of breeding sites is limited and that some individuals obtain breeding sites and others do not. A more general, and more realistic, model of the distribution of habitat quality is discussed briefly in the next section.

I assume that there are only $\hat{n}$ breeding sites available in the source habitat. If the total population size is N and no other breeding sites are available, $N - \hat{n}$ individuals either stay in the source habitat as nonbreeding "floaters" or migrate to nearby sink habitats. In either case, they fail to reproduce but survive with the same probability (P_A) as do breeding individuals. (The qualitative features of the model are unchanged by the assumption that nonbreeding individuals survive with higher or lower probability than breeders.) Since the average reproductive success of an individual securing a breeding site is β_1, the average reproductive success for the entire population is given by

$$\beta(N) = \begin{cases} \beta_1 & \text{if } N \le \hat{n}, \\ (\hat{n}/N)\beta_1 & \text{if } N > \hat{n}. \end{cases} \tag{6}$$

Thus, according to the definition of a source habitat (eq. 5a), the population increases when rare and continues to grow at the rate $\lambda_1 = P_A + P_J\beta_1$ until all breeding sites are occupied. The population will be regulated when

$$\lambda(N) = P_A + (\hat{n}/N)P_J\beta_1 = 1$$

or

$$N^* = \hat{n}P_J\beta_1/(1 - P_A). \tag{7}$$

Again, from the definition of a source, $P_J\beta_1/(1 - P_A)$ is greater than one; thus, the equilibrium population density (N^*) exceeds the number of breeding sites $(\hat{n})$, implying the existence of a nonbreeding surplus.

Assume that, adjacent to the source habitat, is a large sink habitat, where breeding sites are abundant but of poor quality. According to the definition of a sink (eq. 5b), $\beta_2 < (1 - P_A)/P_J$; thus, the sink population declines and eventually disappears altogether in the absence of immigration from the source. Individuals unable to find a breeding site in the source emigrate to the sink because a poor-quality breeding site is better than none at all. If the source is saturated and there are sufficient breeding sites in the sink, the entire nonbreeding surplus from the source emigrates, yielding an increase in the growth rate of the total population:

$$
\begin{aligned}
\lambda(N) &= (n_1/N)\lambda_1 + (n_2/N)\lambda_2 \\
&= (\hat{n}/N)\lambda_1 + \lambda_2(N - \hat{n})/N \qquad (8) \\
&= \lambda_2 + (\hat{n}/N)P_J(\beta_1 - \beta_2).
\end{aligned}
$$

The total population equilibrates when $\lambda(N) = 1$; and, according to equation (8),

$$
N^* = P_J\hat{n}(\beta_1 - \beta_2)/(1 - P_A - P_J\beta_2). \qquad (9)
$$

A relatively simple way to determine the equilibrium populations that will inhabit the source and sink habitats under this model is to note that, since the annual census is taken after the emigration of the reproductive surplus, $\hat{n}$ individuals remain in the source and $\hat{n}(\lambda_1 - 1)$ immigrate. Therefore, in terms of the BIDE model, $i_{21} = \hat{n}(P_A + P_J\beta_1 - 1)$. The local reproduction and survival in the sink is supplemented by this immigration, so that

$$
n_2(t + 1) = (P_A + P_J\beta_2)n_2(t) + i_{21} = \lambda_2 n_2(t) + \hat{n}(\lambda_1 - 1).
$$

At equilibrium, $n_2^* = i_{21}/(1 - \lambda_2)$, or

$$
n_2^* = \hat{n}(\lambda_1 - 1)/(1 - \lambda_2). \qquad (10)
$$

Notice that $\lambda_1 - 1$ is the per capita reproductive surplus in the source and $1 - \lambda_2$ is the per capita reproductive deficit in the sink.

If there are many habitats, the total population reaches an equilibrium when the total surplus in all source habitats equals the total deficit in all sink habitats. That is,

$$
\sum_{j=1}^{m_1} e_j = \sum_{j=1}^{m_1} n_j^*(\lambda_j - 1) = \sum_{k=1}^{m_2} n_k^*(1 - \lambda_k) = \sum_{k=1}^{m_2} i_k,
$$

where there are m_1 source habitats and m_2 sink habitats.

ECOLOGICAL AND EVOLUTIONARY STABILITY

In the preceding analysis, I calculated the equilibrium population sizes in source and sink habitats without addressing the stability of this equilibrium. A local-stability analysis involves finding the slope (b) of $\lambda(N)$ evaluated at the equilibrium population size N^*. If $-bN$ is less than one, the equilibrium is locally stable and approached monotonically (Maynard Smith 1968). The rate of increase for the combined source-sink population is given by equation (8). Differentiating, one obtains

$$
d\lambda(N)/dN = -\hat{n}P_J(\beta_1 - \beta_2)/N^2.
$$

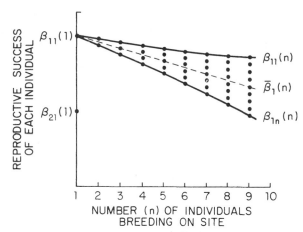

FIG. 2.—The reproductive success of each individual breeding in a particular habitat depends on the number of other individuals in that habitat. $\beta_{1j}(n)$, the expected reproductive success of the individual occupying the jth-best breeding site in habitat 1 when there are a total of n individuals breeding there; $\bar{\beta}_1(n)$, the average reproductive success in the habitat.

Noting the equilibrium population size given in equation (9), the value of $-bN$ is easily seen to be $1 - P_A - P_J\beta_2$ or $1 - \lambda_2$. Since habitat 2 is a sink, λ_2 is less than 1, and, therefore, $-bN$ is also less than one. As for the case of a source habitat with no sink, β_2 equals zero and $-bN$ equals $1 - P_A$. Thus, with or without a sink habitat, the equilibrium population size is locally stable. For the simple cases analyzed above, there is only one nonzero equilibrium, and this equilibrium is approached monotonically from any positive initial population size.

A different question of stability concerns the evolutionary stability of the dispersal rule that determines the proportion of individuals in each habitat. Holt (1985) argued that passive dispersal between a source and a sink is evolutionarily unstable. Two essential differences between the current model and that of Holt are passive versus active dispersal and unequal versus equal fitnesses within a habitat. In my model, individuals choose to leave the source whenever their expected reproductive success is higher in the sink. This never happens in Holt's model because all individuals in the source have equal fitness and the mean fitness in the source never drops to less than one. Since the mean fitness in the sink is always less than one, it always pays for individuals not to immigrate and the evolutionarily stable strategy is no dispersal.

In my model, when the local population in the source exceeds the number of breeding sites available, it pays all surplus individuals to emigrate because they can achieve a higher fitness by doing so. The habitat-selection rule built into the model is a special case of the more general rule "never occupy a poorer breeding site when a better one is still available." Assuming no habitat-specific differences in survival probability, this is the evolutionarily stable habitat-selection rule because no individual can do better by changing habitats (see Pulliam and Caraco 1984; Pulliam 1989).

A more general application of an evolutionarily stable habitat rule that also results in stable occupancy of sink habitats is illustrated in figure 2. In this figure,

$\beta_{11}(n)$ is the expected breeding success of the individual using the best breeding site in habitat 1, and $\beta_{1n}(n)$ is the expected success of the individual using the poorest site occupied when there are n individuals in habitat 1 (the source). Assuming that individuals never occupy a poorer site when a better one is still available, habitat 2 (the sink) will not be occupied as long as $\beta_{1N}(N) > \beta_{21}(1)$. That is to say, the sink habitat will not be occupied as long as all N members of the population can enjoy greater reproductive success in the source. However, if $\beta_{21}(1)$ exceeds $\beta_{1N}(N)$ before the average reproductive success in the source reaches one, the sink will be occupied and the habitat distribution will be evolutionarily stable. Of course, the relative numbers of individuals in the source and sink habitats depend on details of how reproductive success changes with crowding in each habitat. If good breeding sites in the source are rare and poor sites in the sink are relatively common, a large population may occur in the sink.

IMPLICATIONS

Sink habitats may support very large populations despite the obvious fact that the sink population would eventually disappear without continued immigration. Consider the simple situation in which each year i individuals are released into a habitat where local reproduction is incapable of keeping up with local mortality. The equilibrium population maintained in this sink habitat would be $i/(1 - \lambda)$. Thus, if no individuals survived the winter ($\lambda = 0$), only the i recently released individuals would be censused each year. If adults survived winter with probability ½, $2i$ individuals would be censused each year. If, in addition, each adult produced an average of 0.4 juveniles that survived to the following spring, the equilibrium population would be 10 times i, even though the population could not be maintained without an annual subsidy.

In some circumstances, only a small fraction of the population may be breeding in a source habitat. Figure 3 shows the fraction of the equilibrium population in source habitat based on the assumptions of the model developed above and calculated according to equation (10). Clearly, if the reproductive surplus of the source is large and the reproductive deficit of the sink is small, a great majority of the population may occur in the sink habitat. For example, with a per capita source surplus of 1.0 and a sink deficit of 0.1, less than 10% of the population occurs in habitats where reproductive success is sufficient to balance annual mortality.

The concept of niche.—Joseph Grinnell is often credited with introducing the niche concept into ecology. James et al. defined the Grinnellian niche as "the range of values of environmental factors that are necessary and sufficient to allow a species to carry out its life history"; under normal conditions, "the species is expected to occupy a geographic region that is directly congruent with the distribution of its niche" (1984, p. 18). Though James et al. suggested that a species with limited dispersal may not occur in some areas where its niche is found, they clearly implied that the species will not occur where its niche is absent. A sink habitat is by definition an area where factors are not sufficient for a species to carry out its life history, but as discussed above, some species may be more

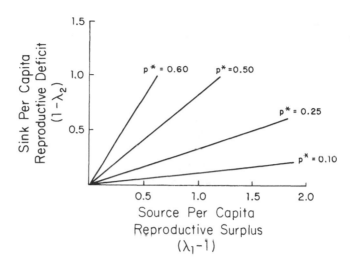

FIG. 3.—The equilibrium proportion (p^*) of the population in the source habitat depends on both the per capita surplus in the source and the per capita deficit in the sink. A large proportion of the population may occur in the sink habitat if the source surplus is large and the sink deficit is small.

common in sink habitats than in the source habitats on which sink populations depend.

Hutchinson's (1958) particularly influential formulation of the niche concept differentiated between the fundamental niche and the realized niche. Hutchinson argued that the realized niche of most species would be smaller than the fundamental niche as a result of interspecific competition. I have argued in this paper that reproductive surpluses from productive sources may immigrate into and maintain populations in population sinks. If this is commonly true in nature, many species occur where conditions are not sufficient to maintain a population without continued immigration. Thus, in such cases, it can be said that the fundamental niche is smaller than the realized niche.

Species conservation.—Given that a species may commonly occur and successfully breed in sink habitats, an investigator could easily be misled about the habitat requirements of a species. Furthermore, autecological studies of populations in sink habitats may yield little information on the factors regulating population size if population size in the sink is determined largely by the size and proximity of sources.

Population-management decisions based on studies in sink habitats could lead to undesirable results. For example, 90% of a population might occur in one habitat. On the basis of the relative abundance and breeding status of individuals in this habitat, one might conclude that destruction of a nearby alternative habitat would have relatively little impact on the population. However, if the former habitat were a sink and the alternative a source, destruction of a relatively small habitat could lead to local population extinction.

Community structure.—What is a sink habitat for one species may be a source for other species. Thus, a "community" may be a mixture of populations, some of which are self-maintaining and some of which are not. Attempts to understand phenomena such as the local coexistence of species should, therefore, begin with a determination of the extent to which the persistence of populations depends on continued immigration.

Many attempts to understand community structure have focused on resource partitioning and the local diversity of food types. The diversity and relative abundance of the organisms in any particular habitat may depend as much on the regional diversity of habitats as on the diversity of resources locally available. In extreme cases, the local assemblage of species may be an artifact of the type and proximity of neighboring habitats and have little to do with the resources and conditions at a study site. This is not to imply that local studies of the mechanisms of population regulation and species coexistence are unnecessary, but rather that they need to be done in concert with "landscape" studies of the availability of habitat types on a regional basis.

My goal is to draw attention to some of the implications of habitat-specific demographic rates. In many ways, they may be ecologically more important than the age-specific demographic rates that have received so much attention in the ecological and evolutionary literature.

SUMMARY

Animal and plant populations often occupy a variety of local areas and may experience different local birth and death rates in different areas. When this occurs, reproductive surpluses from productive source habitats may maintain populations in sink habitats, where local reproductive success fails to keep pace with local mortality. For animals with active habitat selection, an equilibrium with both source and sink habitats occupied can be both ecologically and evolutionarily stable. If the surplus population of the source is large and the per capita deficit in the sink is small, only a small fraction of the total population will occur in areas where local reproduction is sufficient to compensate for local mortality. In this sense, the realized niche may be larger than the fundamental niche. Consequently, the particular species assemblage occupying any local study site may consist of a mixture of source and sink populations and may be as much or more influenced by the type and proximity of other habitats as by the resources and other conditions at the site.

ACKNOWLEDGMENTS

I wish to acknowledge the assistance of G. Reynolds and J. Nelms in the preparation of the manuscript and the financial support of the National Science Foundation (BSR-8415770).

LITERATURE CITED

Cohen, J. 1969. Natural primate troops and a stochastic population model. Am. Nat. 103:455–477.
———. 1971. Casual groups of monkeys and men: stochastic models of elemental social systems. Oxford University Press, London.

SOURCES, SINKS, AND POPULATION REGULATION   661

Gadgil, M. 1971. Dispersal: population consequences and evolution. Ecology 52:253–261.

Holt, R. D. 1985. Population dynamics in two-patch environments: some anomalous consequences of an optimal habitat distribution. Theor. Popul. Biol. 28:181–208.

Hutchinson, G. E. 1958. Concluding remarks. Cold Spring Harbor Symp. Quant. Biol. 22:415–427.

James, F. C., R. F. Johnston, G. J. Niemi, and W. J. Boecklen. 1984. The Grinnellian niche of the wood thrush. Am. Nat. 124:17–47.

Levin, S. A. 1976. Population dynamic models in heterogeneous environments. Annu. Rev. Ecol. Syst. 7:287–310.

Lidicker, W. Z., Jr. 1975. The role of dispersal in the demography of small mammals. Pages 103–128 in F. B. Golley, K. Petrusewicz, and L. Ryszkowski, eds. Small mammals: their productivity and population dynamics. Cambridge University Press, New York.

Maynard Smith, J. 1968. Mathematical ideas in biology. Cambridge University Press, Cambridge.

McMurtie, R. 1978. Persistence and stability of single-species and predator-prey systems in spatially heterogeneous environments. Math. Biosci. 39:11–51.

Pulliam, H. R. 1989. Individual behavior and the procurement of essential resources. Pages 25–38 in J. Roughgarden, R. M. May, and S. Levin, eds. Perspectives in ecological theory. Princeton University Press, Princeton, N.J.

Pulliam, H. R., and T. Caraco. 1984. Living in groups: is there an optimal group size? Pages 122–147 in J. R. Krebs and N. B. Davies, eds. Behavioural ecology: an evolutionary approach, 2d ed. Sinauer, Sunderland, Mass.

Vance, R. R. 1984. The effect of dispersal on population stability in one-species, discrete-space population growth models. Am. Nat. 123:230–254.

van Horne, B. 1983. Density as a misleading indicator of habitat quality. J. Wildl. Manage. 47:893–901.

Landscape Ecology vol. 3 nos. 3/4 pp 153-162 (1989)
SPB Academic Publishing bv, The Hague

Effects of changing spatial scale on the analysis of landscape pattern

Monica G. Turner[1], Robert V. O'Neill[1], Robert H. Gardner[1] and Bruce T. Milne[2]
[1]*Environmental Sciences Division, Oak Ridge National Laboratory, Oak Ridge, TN 37831, USA;*
[2]*Department of Biology, University of New Mexico, Albuquerque, NM 7131, USA*

Keywords: spatial scale, grain, extent, resolution, landscape ecology, diversity, dominance, contagion,
          spatial pattern

Abstract

The purpose of this study was to observe the effects of changing the grain (the first level of spatial resolution possible with a given data set) and extent (the total area of the study) of landscape data on observed spatial patterns and to identify some general rules for comparing measures obtained at different scales. Simple random maps, maps with contagion (*i.e.*, clusters of the same land cover type), and actual landscape data from USGS land use (LUDA) data maps were used in the analyses. Landscape patterns were compared using indices measuring diversity (H), dominance (D) and contagion (C). Rare land cover types were lost as grain became coarser. This loss could be predicted analytically for random maps with two land cover types, and it was observed in actual landscapes as grain was increased experimentally. However, the rate of loss was influenced by the spatial pattern. Land cover types that were clumped disappeared slowly or were retained with increasing grain, whereas cover types that were dispersed were lost rapidly. The diversity index decreased linearly with increasing grain size, but dominance and contagion did not show a linear relationship. The indices D and C increased with increasing extent, but H exhibited a variable response. The indices were sensitive to the number (m) of cover types observed in the data set and the fraction of the landscape occupied by each cover type (P_k); both m and P_k varied with grain and extent. Qualitative and quantitative changes in measurements across spatial scales will differ depending on how scale is defined. Characterizing the relationships between ecological measurements and the grain or extent of the data may make it possible to predict or correct for the loss of information with changes in spatial scale.

Introduction

The range of spatial and temporal scales at which ecological problems are posed has expanded dramatically in recent years, and the need to consider scale in ecological analyses has often been noted (*e.g.*, Allen and Starr 1982; Delcourt *et al.* 1983; O'Neill *et al.* 1986; Addicott *et al.* 1987; Getis and Franklin 1987; Meentemeyer and Box 1987; Morris 1987). Parameters and processes important at one scale are frequently not important or predictive at

another scale, and information is often lost as spatial data are considered at coarser scales of resolution (Henderson-Sellers *et al.* 1985; Meentemeyer and Box 1987). Ecological problems often require the extrapolation of fine-scale measurements for the analysis of broad-scale phenomena. Therefore, the development of methods that will preserve information across scales or quantify the loss of information with changing scales has become a critical task. Such methods are necessary before ecological insights can be extrapolated between

154

(a) Increasing grain size

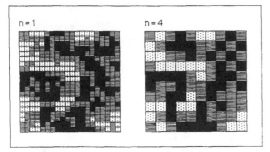

(b) Increasing extent

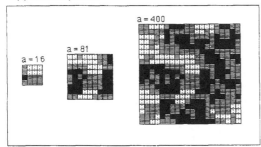

Fig. 1. Schematic illustration of (a) increasing grain size and (b) increasing extent in a landscape data set. The number of cells aggregated to form a new data unit are indicated by n; total area is indicated by a; see Methods for complete explanation.

spatial and temporal scales.

The spatial scale of ecological data encompasses both grain and extent. Grain refers to the resolution of the data, *i.e.*, the area represented by each data unit. For example, a fine-grain map might organize information into 1-ha units, whereas a map with an order of magnitude coarser resolution would have information organized into 10-ha units. Extent refers to the overall size of the study area. For example, maps of $100 km^2$ and $100,000 km^2$ differ in extent by a factor of 1000. In studies of landscape structure or function (*e.g.*, Naveh and Lieberman 1984; Forman and Godron 1986; Turner 1987a), information may be available at a variety of levels of resolution, data must often be compared across large geographic regions, and it may be necessary to extrapolate information from local to regional scales. Applications of geographic information sys-

tems (GIS) also frequently require the integration of data obtained at different spatial scales. The effects of grain and extent are thus of particular concern, and the response of ecological attributes measured on the landscape to changes in spatial scale is not known. It has been demonstrated (Gardner *et al.* 1987) that the number, size and shape of patches in randomly simulated landscapes varies with extent, as indicated by the linear dimension of the landscape. It has also been suggested (Allen *et al.* 1987) that information can be transferred across scales if both grain and extent are specified.

The purpose of this study was to observe the effects of changing the grain and extent of landscape data on observed spatial patterns and to identify some general rules for comparing measures obtained at different scales. The effects of changing grain size on the proportion of a landscape occupied by a particular land cover type were first predicted analytically then tested against randomly generated landscapes. The grain and extent of real and simulated landscapes were also experimentally varied and indices of diversity, dominance and contagion were used to compare resulting landscape patterns.

Methods

Landscape data

The USGS digital land use and land cover data base (Fegas *et al.* 1983), which provides landscape maps interpreted from NASA U2/R8−57 high altitude aerial photo coverage obtained in 1973, was used for the analyses. The original aerial photographs were hand digitized into 37 land cover categories. Because the original USGS data set divides 1:250,000 quadrangles into 24 sections, a special computer program was written to remove section boundaries and convert the polygon data to grid format. This created a single quadrangle for each landscape with linear dimensions of 650 by 950 pixels (each pixel or site has an area of 4.0 ha). The largest square (measuring 618 by 618 pixels) in the map that was free of edge effects was used for the analysis.

Table 1. Proportion of land cover type k that will be observed as a function of resolution in a random two-phase landscape. The coarse scales of resolution assume that an aggregate of n sites will be assigned to k if the majority of the sites are of type k.

Grain size (n)	$P_k(n)$				
1	0.10	0.30	0.50	0.70	0.90
3	0.028	0.216	0.500	0.784	0.972
7	0.003	0.126	0.500	0.874	0.997
11	0.0	0.078	0.500	0.922	1.0
15	0.0	0.050	0.500	0.950	1.0
19	0.0	0.033	0.500	0.967	1.0

Changing scale: Grain and extent

The grain of the landscape data was changed by aggregating groups of n adjacent pixels into a single data unit. The land cover type of the majority of pixels in an aggregate was assigned to the new data unit (shown schematically in Fig. 1a). If two land covers in an aggregate occurred in equal proportions (e.g., two pixels each of agriculture and forest), the assignment of a landscape type was done at random. Grain was varied through 19 separate aggregations such that an exact number of aggregates fit into the central portion of the map and no artifacts resulted from omitting a few rows or columns. The finest aggregation was 2×2 ($n = 4$ original pixels forming each aggregate) and the coarsest aggregation was 50×50 ($n = 2500$ original pixels per aggregate data unit). This method of aggregation simulates cases in which only coarse resolution data about land cover are available.

The extent (a) of the landscape map was altered without affecting the grain size using a nested quadrat design (shown schematically in Fig. 1b). Pattern analysis began using a group of pixels ($a = 40$) located in the center of the map. The area was then gradually increased until the entire map was covered ($a = 381,924$).

Analysis of landscape pattern

The fraction of the landscape occupied by each cover type was calculated from the LUDA data,

and several indices (O'Neill *et al.* 1988) based on information theory (Shannon and Weaver 1962) were used to describe the pattern of each landscape map. The first index, H, is a measure of diversity:

$$H = - \sum_{k=1}^{m} (P_k) \log(P_k), \qquad (1)$$

where P_k is the proportion of the landscape in cover type k, and m is the number of land cover types observed. The larger the value of H, the more diverse the landscape.

The second index, D, is a measure of dominance, calculated as the deviation from the maximum possible diversity:

$$D = H_{max} + \sum_{k=1}^{m} (P_k) \log(P_k), \qquad (2)$$

where m = number of land use types observed on the map, P_k is the proportion of the landscape in land use k, and $H_{max} = \log(m)$, the maximum diversity when all land uses are present in equal proportions. Inclusion of H_{max} in Eq. 2 normalizes the index for differences in numbers of land cover types between different landscapes. Large values of D indicate a landscape that is dominated by one or a few land uses, and low values indicate a landscape that has many land uses represented in approximately equal proportions. However, the index is not useful in a completely homogeneous landscape (i.e., $m = 1$) because D then equals zero.

The third index, C, measures contagion, or the adjacency of land cover types. The index is calculated from an adjacency matrix, Q, in which $Q_{i,j}$ is the proportion of cells of type i that are adjacent (diagonals are excluded) to cells of type j, such that:

$$C = K_{max} + \sum_{i=1}^{m} \sum_{j=1}^{m} (Q_{i,j}) \log(Q_{i,j}), \qquad (3)$$

where $K_{max} = 2\, m \log(m)$ and is the absolute value of the summation of $(Q_{i,j})\log(Q_{i,j})$ when all possible adjacensies between land cover types occur with equal probabilities. K_{max} normalizes landscapes with differing values of m and causes C to be

156

Table 2. Number of land cover types retained at each level of aggregation for seven landscapes.

	Landscape scene							
Grain size (n)	Knoxville	Athens	Natchez	Macon	Greenville	Goodland	Waycross	Mean
1	7	6	7	7	6	7	6	6.6
4	7	6	7	7	6	6	6	6.4
16	7	6	7	7	6	5	6	6.3
25	7	6	7	6	5	4	5	5.7
50	6	6	6	6	5	4	4	5.3
80	5	6	6	6	5	3	4	5.0
100	5	6	6	6	5	3	4	5.0
160	4	6	6	5	5	3	4	4.7
200	4	6	6	4	5	3	4	4.6
320	4	5	6	4	4	3	4	4.3
400	3	5	6	4	5	2	4	4.1
500	3	4	5	4	4	2	4	3.7
625	3	4	5	4	4	2	4	3.7
800	3	4	5	4	4	2	4	3.7
1000	3	3	3	4	4	2	4	3.3
1250	3	3	3	4	4	2	4	3.3
2000	3	3	3	4	4	2	3	3.1
2500	3	2	3	4	4	2	3	3.0

Table 3. Proportion of land cover type k observed as a function of grain size in random multiphase landscapes. The coarse scales of resolution assume that an aggregate of n sites will be assigned to k if the majority of the sites are of type k.

A. Initial proportions taken from Goodland, KS

Grain size (n)	$P_k(n)$			
1	0.001	0.004	0.29	0.70
4	0.0	0.001	0.205	0.795
16	0.0	0.0	0.044	0.956
50	0.0	0.0	0.015	0.985
80	0.0	0.0	0.001	0.999
100	0.0	0.0	0.0	1.0

B. Initial proportions taken from West Palm Beach, FL

Grain size (n)	$P_k(n)$						
1	0.051	0.053	0.062	0.082	0.182	0.259	0.310
4	0.031	0.033	0.040	0.058	0.174	0.288	0.376
16	0.002	0.003	0.005	0.011	0.125	0.333	0.520
25	0.001	0.001	0.001	0.003	0.095	0.328	0.572
50	0.0	0.0	0.0	0.0	0.052	0.296	0.652
80	0.0	0.0	0.0	0.0	0.022	0.271	0.707
100	0.0	0.0	0.0	0.0	0.018	0.252	0.730

Table 4. Contagion values $Q_{j,k}$ for which there will be no change in P_k with aggregation.

P_k	$Q_{k,j}$				
	0.10	0.30	0.50	0.70	0.90
0.1	0.011	0.033	0.055	0.077	0.100
0.2	0.025	0.075	0.125	0.175	0.225
0.3	0.043	0.128	0.214	0.300	0.386
0.4	0.066	0.200	0.333	0.460	0.600
0.5	0.100	0.300	0.500	0.700	0.900
0.6	0.150	0.450	0.750	*	*
0.7	0.230	0.700	*	*	*
0.8	0.400	*	*	*	*
0.9	0.900	*	*	*	*

* There is no feasible value (*i.e.*, $Q_{j,k} < 1.0$)

zero when $m = 1$ or all possible adjacencies occur with equal probability. When $m \geq 2$, large values of C will indicate a landscape with a clumped pattern of land cover types.

Results

Effects of changing grain

1. *Analytical solution for* P_k *at different scales in random landscapes*
Consider a landscape with two randomly distributed land cover types ($m = 2$), each of which occupies a proportion P_k of the landscape. The probability of having r sites of cover type k in an aggregate of n cells is given by the binomial distribution:

$$b\ (r;\ n,\ P_k) = (n!/r!\ (n-r)!)\ P_k^r\ (1-P_k)\ (n-r)\ (4)$$

An aggregate is assigned to type k using the majority rule if at least 50% of the cells are of type k. The proportion of aggregates that will contain 50% or more sites of type k can be determined by using a table of the cumulative binomial distribution for $P = b = 0.50$.

The proportion $P_k(n)$ of data units formed by aggregating n cells that will be assigned to cover type k in a two phase landscape (cover type k and all non-k; $m = 2$) is shown in Table 1. If $P_k(1)$ on the random landscape is exactly 0.5, this proportion will be maintained at all levels of aggregation (Table 1). If $P_k(1)$ is even slightly less than 0.5,

land type k will be lost at coarser resolutions; the smaller the initial P_k, the faster the disappearance of land type k. If $P_k(1)$ is greater than 0.5, cover type k will dominate the landscape at coarser resolutions until it covers the entire landscape. These results (Table 1) apply only if k is independently and randomly distributed on the landscape. Although this condition will seldom be satisfied exactly, the general pattern holds for a variety of actual landscapes (Table 2). At increasingly coarser scales of resolution, cover types with a small $P_k(1)$ disappear.

Random landscapes with the same proportions of cover types observed in the LUDA maps for Goodland, Kansas (dominated by agriculture and rangeland) and West Palm Beach, Florida (having a more even distribution of cover types) were generated to further test the relationship between P_k and grain (Table 3). Although the rare land cover types do indeed disappear, when the fractions of the landscape occupied by two cover types are similar in size and there is no extreme dominance by either (e.g., P_k's of 0.259 and 0.310 in West Palm Beach, Table 3) both values increase while rare cover types are lost. Thus, P_k values do not necessarily change unidirectionally in landscapes occupied by numerous cover types. However, with continued aggregation, the land cover type with the greatest P_k will eventually dominate the landscape.

2. *Analytical solution for* P_k *at different scales in landscapes with contagion*
The results for a simple random landscape can be extended by analyzing a landscape on which cover type k is distributed with contagion. Contagion simulates the tendency of land cover types to form clusters or patches rather than to occur at random. The probability that two adjacent sites are of type k, $Q_{k,k}$ can be calculated by moving through the landscape along transects and observing the frequency at which land cover types are adjacent to each other. The proportion of the landscape occupied by cover type k when two cells are aggregated, $P_k(2)$, is then given by:

$$P_k(2) = P_k(1)\ Q_{k,k} + 0.5\ [P_k(1)\ Q_{k,j} + P_j(1)\ Q_{j,k}]\ (5)$$

158

Table 5. Regression of three landscape indices with grain (log aggregate size) for seven landscape scenes. Values are the slopes and (r^2).

Scene	Landscape parameter		
	Diversity (H)	Dominance (D)	Contagion (C)
Goodland, KS	−0.047 (0.98)	−0.144 (0.80)	−1.737 (0.77)
Natchez, MS	−0.055 (0.96)	−0.090 (0.48)	−1.666 (0.46)
Knoxville, TN	−0.068 (0.98)	−0.112 (0.75)	−2.775 (0.76)
Greenville, SC	−0.036 (0.94)	−0.037 (0.50)	−0.808 (0.43)
Waycross, GA	−0.034 (0.92)	−0.053 (0.50)	−0.915 (0.37)
Macon, GA	−0.038 (0.94)	−0.076 (0.66)	−2.038 (0.68)
Athens, GA	−0.090 (0.94)	−0.066 (0.32)	−1.517 (0.46)
Random Goodland, KS	−0.052 (0.53)	−0.116 (0.62)	−0.576 (0.68)
Random West Palm, FL	−0.235 (0.99)	−0.67 (0.23)	−1.57 (0.45)

where $P_j(1) = 1.0 - P_k(1)$, representing the landscape sites occupied by something other than k. Cover type k will be assigned to an aggregate if both pixels are of type k. In addition, half of the k pixels adjacent to j and half of the j pixels adjacent to k will also be assigned to type k. Thus, of the sites that were type k at $n = 1$, $P_k(1)Q_{k,k}$ are retained at $n = 2$, 0.5 $P_k(1)Q_{k,j}$ are reclassified as j, and 0.5 $P_j(1)Q_{j,k}$ are reclassified from j to k. The balance between sites reclassified from j to k and *vice versa* determines whether $P_k(2)$ is greater or less than $P_k(1)$. The net change, Δ, in P_k is thus:

$$\Delta = 0.5 [P_j(1)Q_{j,k} - P_k(1)Q_{k,j}]. \tag{6}$$

Consider now the special case in which j is not contagiously distributed, *i.e.*, $P_j(1) = Q_{j,j}$. Recall that $P_j(1) = 1 - P_k(1)$ and $Q_{j,k} = 1 - Q_{j,j} = 1 - P_j(1) = 1 - (1 - P_k(1)) = P_k(1)$. Equation 3 then becomes:

$$\Delta = 0.5 P_k(1) [Q_{k,k} - P_k(1)] \tag{7}$$

Cover type k will occupy an increasing proportion of sites at coarser resolution if $P_k(1)$ is contagiously distributed, *i.e.*, $Q_{k,k} > P_k(1)$ and $P_k(2) > P_k(1)$. Type k will occupy less of the landscape at coarser resolution if $P_k(1)$ is dispersed, *i.e.*, $Q_{k,k} < P_k(1)$ and $P_k(2) < P_k(1)$. The proportion of k will not change with aggregation if the landscape is symmetric, *i.e.*, if $P_k(1) = Q_{k,k}$ and $P_k(2) = P_k(1)$. This result can be generalized because $P_k(2)$ will

tend to be larger than expected for a random distribution if $P_k(1)$ is contagiously distributed for any value of $Q_{j,j}$. Thus, the rate at which rare land cover types decreases with decreasing resolution depends on their spatial arrangement. If a rare land cover type is highly aggregated, it tends not to disappear or to diminish slowly; if it is well-dispersed in small patches, it disappears rapidly.

If the landscape is symmetric, the probability of adjacency of j to k is equal to the probability of adjacency of k to j, *i.e.*,

$$P_k(1)Q_{k,j} = P_j(1)Q_{j,k}. \tag{8}$$

This symmetry occurs if the elements of the Q matrix are calculated irrespective of a specific direction, *i.e.*, $Q_{i,j}$ represents the probability of i being adjacent to j in any direction, horizontally, vertically, or diagonally. No change will occur in P_k as grain increases under the conditions shown in Table 4. Cover type k will increase with aggregation if the value of $Q_{j,k}$ is larger than the entry in the table, and P_k will decrease if the value of $Q_{j,k}$ is less than the table entry. For the situations indicated by an asterisk, $P_k(2)$ will decrease for any feasible value of $Q_{j,k}$, *i.e.*, any value less than or equal to 1.0.

The value of $P_k(n)$ for coarser scales of resolution can be determined analytically by considering the set of adjacency probabilities that would result in an aggregate being classified as k. For example:

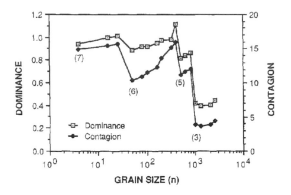

Fig. 2. Landscape indices as a function of increasing grain size as indicated by log(*n*) using the Natchez, Mississippi, quadrangle. Note the sensitivity of the dominance and contagion indices to the number of cover types present, *m*, indicated in parentheses. Other landscape scenes also showed this stair-step pattern.

$$P_k(3) = P_k(1) \, [Q_{k,k}Q_{k,k} + Q_{k,k}Q_{k,j} + Q_{k,j}Q_{j,k}] + P_j(1) \, Q_{j,k}Q_{k,k} \quad (9)$$

However, the algebraic manipulations become quite laborious as *n* increases.

3. *Effects on landscape indices*

Regressions of the landscape indices against log (*n*) (Table 5) were calculated to examine the relationship of the indices to different grain sizes. Milne *et al.* (*unpublished ms.*) demonstrated that a simple linear relationship existed between diversity, *H*, and log (scale) with consistent negative slopes and r^2 values that exceed 0.9. In contrast, the r^2 values for the regressions of *D* and *C* range from 0.32 to 0.80. The range of values for the slopes is also larger for *D* and *C*, indicating less consistency among landscapes. The linear relationship between diversity and grain does not assure a similarly simple relationship between grain and other landscape indices.

The dominance and contagion indices both depend on a maximum value, which is determined by *m*, the number of land cover types present. The number of land cover types observed decreases as resolution becomes increasingly more coarse and rare cover types disappear (Table 2). Indices *D* and *C* are sensitive to loss of cover types, and break

points occur when the indices are plotted against grain (Fig. 2). *C* and *D* increase with grain size until another cover type is lost, at which point the indices decline sharply. The net change in the indices over all aggregations is negative (Table 5). The sensitivity to *m* suggests that knowledge of the number of cover types at each scale may permit extrapolation of dominance and contagion from one level of resolution to another. Direct, linear extrapolation appears possible if *m* does not change with particular grain sizes; as *m* changes, the slope and intercept of the extrapolation function vary.

Effects of changing extent

The number of land cover types present (*m*) increases with increasing extent (Table 6), as expected. The observed P_k values of the land cover types vary with changes in extent (Table 7), asymptotically approaching the final values observed on the entire map. We further expected that the landscape indices would remain constant as extent increased until a natural boundary (*e.g.*, mountains, rivers, etc.) was crossed. The variance in the indices would increase rapidly in cases where these boundaries occur because the spatial patterning across the region would no longer be similar. This expectation held for diversity; a linear relationship was not observed between diversity and extent, as measured by regression of *H* against log (*a*) (Table 8). The slopes are small and often change sign. The r^2 values indicate that the linear trend through the data explains little of the total variance. The regressions *D* with extent show small positive slopes, but the r^2 values range from 0.019 to 0.926, indicating that a linear relationship exists for some landscapes but is not consistent across landscape scenes. Regressions of *C* show greater consistency, with relatively steep slopes and larger r^2 values. The slopes, however, still range from 0.0669 to 3.253 and the r^2 values are all below 0.90.

The indices are also sensitive to the number of land uses present (Fig. 3). Dominance and contagion appear to decrease with increasing extent until another cover type is included, at which point there is a large increase. The overall change in the

160

Table 6. Number of land cover types observed with increasing extent for seven landscapes.

| | Landscape scene | | | | | | | |
Extent (*a*)	Knoxville	Athens	Natchez	Macon	Greenville	Goodland	Waycross	Mean
40	2	1	2	4	2	2	2	2.1
126	4	2	2	4	3	2	2	2.7
360	4	3	3	4	5	2	2	3.3
1000	6	3	3	4	5	2	2	3.6
2146	6	4	3	5	5	2	2	3.9
2924	6	4	4	5	5	2	4	4.3
4452	6	4	5	5	5	3	5	4.7
6300	6	4	5	5	6	3	5	4.9
9000	6	5	6	6	6	4	6	5.6
14,440	6	5	6	6	6	4	6	5.6
20,340	6	5	6	6	6	4	6	5.6
49,000	6	5	6	6	6	5	6	5.7
100,902	6	5	6	6	6	7	6	6.0
150,430	6	6	6	6	6	7	6	6.1
225,000	7	6	6	7	6	7	6	6.4
305,026	7	6	7	7	6	7	6	6.6
361,000	7	6	7	7	6	7	6	6.6

Table 7. Proportion of land cover type k (P_k) observed as a function of extent in two sample landscapes.

A. Goodland, KS

Extent (*a*)	P_1	P_2	P_3	P_4	P_5	P_6	P_7
entire map	0.005	0.718	0.276	0.001	0.0001	0.0001	0.0003
40	0.0	0.450	0.550	0.0	0.0	0.0	0.0
126	0.0	0.722	0.277	0.0	0.0	0.0	0.0
360	0.0	0.744	0.255	0.0	0.0	0.0	0.0
1000	0.0	0.715	0.285	0.0	0.0	0.0	0.0
2146	0.0	0.741	0.259	0.0	0.0	0.0	0.0
2924	0.0	0.759	0.241	0.0	0.0	0.0	0.0
4452	0.0	0.794	0.206	0.0	0.0	0.0	0.0
6300	0.002	0.814	0.186	0.0	0.0	0.0	0.0
9000	0.002	0.800	0.199	0.0	0.0	0.0	0.0
14,440	0.002	0.800	0.196	0.0	0.0	0.0	0.001
20,340	0.003	0.805	0.190	0.0	0.0	0.0	0.001
49,000	0.008	0.779	0.212	0.0	0.0001	0.0	0.001
100,902	0.005	0.773	0.221	0.0001	0.0001	0.0001	0.001

B. Greenville, SC

Extent (*a*)	P_1	P_2	P_3	P_4	P_5	P_6	P_7
entire map	0.066	0.308	0.0	0.599	0.021	0.001	0.005
40	0.0	0.175	0.0	0.825	0.0	0.0	0.0
126	0.040	0.175	0.0	0.786	0.0	0.0	0.0
360	0.072	0.236	0.0	0.675	0.006	0.0	0.011
1000	0.057	0.202	0.0	0.583	0.140	0.0	0.018

Table 7. Continued.

2146	0.057	0.266	0.0	0.434	0.178	0.0	0.064
2924	0.051	0.262	0.0	0.432	0.186	0.0	0.069
4452	0.044	0.303	0.0	0.417	0.165	0.0	0.071
6300	0.037	0.332	0.0	0.431	0.137	0.001	0.062
9000	0.037	0.369	0.0	0.424	0.114	0.001	0.056
14,440	0.031	0.390	0.0	0.424	0.114	0.001	0.040
20,340	0.033	0.376	0.0	0.425	0.136	0.001	0.029
49,000	0.050	0.413	0.0	0.429	0.093	0.001	0.013
100,902	0.068	0.389	0.0	0.482	0.053	0.001	0.009

dominance and contagion indices is generally positive.

Discussion

The effect of changing grain can be compared with the effect of changing extent by comparing Tables 5 and 8. In general, it appears that indices of dominance and contagion decrease as grain size increases, but these indices increase as extent increases. Changing the meaning of 'scale' from grain to extent can have important qualitative and quantitative effects on how measurements change across scales. We hypothesized that dominance and

Table 8. Regression of three landscape indices with extent (log area). Values are the slopes and (r^2).

Scene	Landscape parameter		
	Diversity (H)	Dominance (D)	Contagion (C)
Goodland, KS	0.008 (0.19)	0.197 (0.88)	3.253 (0.78)
Natchez, MS	0.127 (0.78)	−0.015 (0.02)	1.535 (0.72)
Knoxville, TN	−0.058 (0.38)	0.111 (0.87)	0.938 (0.47)
Greenville, SC	0.023 (0.10)	0.067 (0.70)	0.067 (0.67)
Waycross, GA	0.044 (0.84)	0.081 (0.50)	1.519 (0.57)
Macon, GA	−0.006 (0.06)	0.077 (0.93)	1.373 (0.66)
Athens, GA	0.041 (0.42)	0.081 (0.57)	1.569 (0.83)

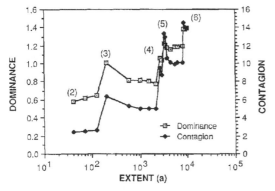

Fig. 3. Landscape indices as a function of increasing extent as indicated by log (area) in the Natchez, Mississippi, quadrangle. Note the sensitivity of the dominance and contagion indices to the number of cover types present, *m*, indicated in parentheses. Other landscape scenes also showed this stair-step pattern.

contagion would increase as resolution was decreased. This effect was observed as long as *m* (the number of land cover types) did not change, but the overall effect of losing cover types dominated. Similarly, the number of land uses tends to increase as extent increases, and dominance appears to increase as *m* increases.

Meentemeyer and Box (1987) proposed that increasing the extent of a study area would tend to increase the range of values for a landscape variable. Our results support this hypothesis, as the P_k values vary considerably as extent increases. Meentemeyer and Box further suggested that apparent detail is lost with increased area (subject to the density of patchiness), and decreased area involves newly apparent detail. Our results also support this state-

ment, as the apparent proportion of the landscape in different cover types changes with grain. The grain size at which a land cover type *k* disappears depends on P_k of the original data, *i.e.*, $P_k(1)$. The smaller the initial probability, the sooner type *k* is lost as aggregation proceeds. However, the changes in each land cover type are not necessarily unidirectional through all aggregations. Furthermore, the spatial arrangement of the land cover also influences the rate at which it disappears. Rare cover types with patchy arrangements disappear more rapidly with decreasing resolution than contagious ones. Patch density and the appropriate scale of analysis may thus be inversely related (Meentemeyer and Box 1987).

The quantification of spatial pattern is necessary to link the effects of landscape heterogeneity with ecological function and to use remotely sensed data to measure change in large spatial units. Patch size and distribution (*e.g.*, Gardner *et al.* 1987), fractal dimension (*e.g.*, Krummel *et al.* 1987), edges (*e.g.*, Turner 1987b), diversity (*e.g.*, Romme 1982; Romme and Knight 1982), and indices of dominance or contagion can all be used to quantify landscape pattern. Our results demonstrate that the spatial scale at which these patterns are quantified influences the result, and measurements made at different scales may not be comparable. Furthermore, qualitative and quantitative changes in measurements across spatial scales will differ according to how scale is defined. Thus, the definition and methods of changing scale must always be explicitly stated. It is important to define the scale of ecological data in terms of both grain, S_g, and extent, S_e. The identification of properties that do not change or change predictably across scales would simplify the extrapolation of measurements from fine scales to broad scales. However, although it may be possible to identify simple relationships between landscape parameters measured at different scales (*e.g.*, diversity, Table 4), the exact relationship varies across landscapes and does not permit extrapolation from one region to another.

Our results verify that information is lost in coarser grained spatial data. In general, information about the less frequent land cover types is most easily lost, but the rate of loss depends upon their

162

spatial arrangement. It is possible to predict the loss of information if the contagion of cover types is known. These results may have important implications for comparing information obtained at different grain sizes. If cover type k is randomly distributed, knowledge of P_k may permit extrapolation of P_k to other scales of resolution. This may allow direct comparison between landscape data of different grain sizes. The ability to extrapolate will break down when P_k reaches 0.00 or 1.00, however, because these values can be reached through aggregation from many starting points. It is not then possible to extrapolate to finer scales. Characterizing the relationships between ecological measurements and the grain and extent of the data may make it possible to predict or correct for the loss of information with changes in spatial scale. More importantly, the ability to predict how ecological variables change with scale may open the door to extrapolating information to larger scales and to comparing data measured in different regions.

Acknowledgements

We are grateful to B.H. Jackson for computer programming and to D.L. DeAngelis and A.W. King for critically reviewing the manuscript. This research was supported by the Ecological Research Division, Office of Health and Environmental Research, U.S. Department of Energy, under Contract No. DE−AC05−84OR21400 with Martin Marietta Energy Systems, Inc. and through an Alexander P. Hollaender Distinguished Postdoctoral Fellowship, administered by Oak Ridge Associated Universities, to M.G. Turner. Support for B.T. Milne was provided, in part, by the Ecosystem Studies Program, National Science Foundation, under grant No. BSR8614981. This is publication no. 3379 of the Environmental Sciences Division, Oak Ridge National Laboratory.

References

Addicott, J.F., Aho, J.M., Antolin, M.F., Padilla, D.K., Richardson, J.S. and Soluk, D.A. 1987. Ecological neighborhoods: scaling environmental patterns. Oikos 49: 340−346.

Allen, T.F.H. and Starr, T.B. 1982. Hierarchy: Perspectives for ecological complexity. University of Chicago Press, Chicago.

Allen, T.F.H., O'Neill, R.V. and Hoekstra, T.W. 1987. Inter-level relations in ecological research and management: some working principles from hierarchy theory. J. Appl. Syst. Anal. 14: 63−79.

Delcourt, H.R., Delcourt, P.A. and Webb III, T. 1983. Dynamic plant ecology: the spectrum of vegetation change in space and time. Quat. Sci. Rev. 1: 153−175.

Fegas, R.G., Claire, R.W., Guptill, S.C., Anderson, K.E. and Hallam, C.A. 1983. Land use and land cover digital data. Geological Survey Circular 895−E. U.S. Geological Survey, Cartographic Information Center, Reston, Virginia.

Forman, R.T.T. and Godron, M. 1986. Landscape ecology. John Wiley & Sons, New York.

Gardner, R.H., Milne, B.T., Turner, M.G. and O'Neill, R.V. 1987. Neutral models for the analysis of broad-scale landscape pattern. Landsc. Ecol. 1: 19−28.

Getis, A. and Franklin, J. 1987. Second order analysis of mapped point patterns. Ecology 68: 473−477.

Henderson-Sellers, A., Wilson, M.F. and Thomas, G. 1985. The effect of spatial resolution on archives of land cover type. Climatic Change 7: 391−402.

Krummel, J.R., Gardner, R.H., Sugihara, G., O'Neill, R.V. and Coleman, P.R. 1987. Landscape patterns in a disturbed environment. Oikos 48: 321−324.

Meentemeyer, V. and Box, E.O. 1987. Scale effects in landscape studies. In Landscape heterogeneity and disturbance. pp. 15−34. Edited by M.G. Turner. Springer-Verlag, New York.

Morris, D.W. 1987. Ecological scale and habitat use. Ecology 68: 362−369.

Naveh, Z. and Lieberman, A.S. 1984. Landscape ecology. Springer-Verlag, New York.

O'Neill, R.V., DeAngelis, D.L., Waide, J.B. and Allen, T.F.H. 1986. A hierarchical concept of ecosystems. Princeton University Press, Princeton, New Jersey.

O'Neill, R.V., Krummel, J.R., Gardner, R.H., Sugihara, G., Jackson, B., DeAngelis, D.L., Milne, B.T., Turner, M.G., Zygmunt, B., Christensen, S.W., Dale, V.H. and Graham, R.L. 1988. Indices of landscape pattern. Landsc. Ecol. 1: 153−162.

Romme, W.H. 1982. Fire and landscape diversity in subalpine forests of Yellowstone National Park. Ecol. Monogr. 52: 199−221.

Romme, W.H. and Knight, D.H. 1982. Landscape diversity: the concept applied to Yellowstone Park. BioScience 32: 664−670.

Shannon, C.E. and Weaver, W. 1962. The mathematical theory of communication. University of Illinois Press, Urbana.

Turner, M.G. (ed.) 1987a. Landscape heterogeneity and disturbance. Springer-Verlag, New York.

Turner, M.G. 1987b. Spatial simulation of landscape changes in Georgia: a comparison of 3 transition models. Landsc. Ecol. 1: 29−36.

5 Studies of Distribution and Abundance and the Rise of Conservation Ecology

Ben Bolker and Mary Ruckelshaus

The Rise of Conservation Ecology

The papers in this section ushered in modern conservation biology, demonstrating the importance of ecological theory in understanding and addressing dramatic species- and community-level changes in ecosystems. These papers brought ecological and statistical rigor to analyses of population growth and extinction for species of conservation concern. They acknowledged the difficulty of parameterizing complex models in a messy, data-poor world, offering alternative methods based on simpler models (Ludwig et al. 1978; Shaffer 1981; Crouse et al. 1987). In contrast to less applied subfields, they also retreated from the complexities of biotic interactions, generally attempting to characterize the dynamics of single populations rather than disentangling community interactions. Despite this simplicity, they also contributed to general ecological synthesis; for example, methods for estimating the strength of density dependence significantly clarified the long-running debate between proponents of density-independent and density-dependent factors.

These papers enhanced basic ecology by widening the spatial scale of analysis. In contrast to the small-scale, carefully designed field experiments that empiricists promoted in the 1970s and early 1980s (Paine 1974, 1980; Connell 1975; Werner and Mittelbach 1981), these studies tackled the difficulties of real-world observational data on entire populations and species.

On the theoretical side, many of these papers emphasize quantitative estimation of parameters and prediction of stochastic outcomes rather than considering only the qualitative dynamics of deterministic systems. Although these early papers using ecological theory to provide insights into the causes and consequences of species imperilment did not immediately take hold in practice (Caughley 1994; Doak and Mills 1994; Brook et al. 2000), they made lasting contributions to both basic and applied ecology.

Nonlinear Dynamics of Ecological Systems

The papers by Anderson and May (1978), "Regulation and stability of host-parasite population interactions: I. Regulatory processes," and Ludwig et al. (1978), "Qualitative analysis of insect outbreak systems: the spruce budworm and forest," are the earliest in this part of the volume; in contrast to most of the other papers, they (and Turchin and Taylor 1992, "Complex dynamics in ecological time series") focus on qualitative factors that tip ecological systems from persistence to extinction, from stable equilibria to cycles, or from cycles to chaos. In contrast to the other papers discussed in this part of the volume, these three papers emphasize the nonlinear and unstable nature of deterministic ecological systems.

Anderson and May (1978) explored simple, general deterministic models, without detailed application to specific biological communities. They presented some empirical data in order to "build the models on biological assumptions that are rooted in empirical evidence" (Anderson and May 1978, 220); however, they emphasized the *qualitative* behavior of systems, rather than trying to quantify the dynamics of any particular system.

Anderson and May's models include *only* parasite-related host regulation (density-dependence). They "assume that if the parasite fails to control host population growth, exponential increase of the host population occurs until resource limitation results in the gradual approach to a carrying capacity" (Anderson and May 1978, 223). This simplification suited their purposes of qualitative exploration; if they could show that the host population reaches a stable equilibrium (i.e., is regulated), then the regulation must be due to parasite pressure. This assumption would give incorrect *quantitative* predictions for any real system unless host resource limitation is "density vague," i.e., with unchanging growth rates below a density threshold (Strong 1986). This starkly simple model was inspired by skepticism from empiricists that eukaryotic parasites, which generally have mild effects on host fitness, could actually regulate host populations (Tompkins and Begon 1999).

The historical importance of Anderson and May (1978) stems less from its particular conclusions about stability and regulation than from its emphasis on parasites as first-class citizens of ecological communities. Surprisingly to modern ecologists—who learn in introductory courses to categorize natural enemies into parasites, parasitoids, and predators—Anderson and May spent a full page introducing this distinction. Models of the dynamics of parasites had long been connected with the broader class of predator-prey models. However, parasites and the broader spectrum of natural enemies were slower to enter the mainstream of ecology. The full-scale incorporation of nontraditional natural enemies into ecology did not begin until the early 1990s (Crawley 1992). In contrast to a sudden jump in disease- and parasite-focused ecology publications around 1990, Anderson and May's paper has seen a steady growth in citations over the years since its publication.

Ludwig et al. (1978) showed ecologists how to understand the fundamental nonlinear dynamics of a complex ecological system. They explicitly aimed to reconcile theoretical and applied perspectives, and to explore the benefits of reducing complex simulations to simpler models. In doing so, they gained basic insights (and computational efficiency) without losing too much predictive capability or important natural historical detail for the economically important and well-studied spruce budworm/forest interactions in eastern North America. This paper established the important conceptual connections (and the divide) between exploitation-oriented mathematical ecology problems and preservation/conservation framings of species and habitat persistence or risk (e.g., Ludwig et al. 1993).

By modern standards this paper is unusually pedagogical, and mathematical, for a mainstream journal. Ludwig et al. also differs from modern ecological studies in its approach to parameter estimation. They aim for qualitative insight rather than precise quantitative results; the section "Level II: general quantitative information" discusses order-of-magnitude approximations of the parameter values based on natural history knowledge of the system, while "Level III: empirical quantitative information" adapts more precise parameter values from previous field work and from a carefully parameterized simulation model of the system. While this provides useful insight into the qualitative dynamics of a system, and is an important first step in any analysis, more recent ecological models (especially those designed for management) attempt to estimate parameters directly from the observed dynamics of the system (e.g., Kendall et al. 1999).

Two decades later, Turchin and Taylor (1992) explored nonlinear dynamics from a more statistical direction. While Robert May's (1974) paper (reprinted in Real and Brown 1991) had introduced the idea of deterministic chaos to ecologists, it took nearly two decades to develop useful tools for quantifying chaos in ecological systems. Early attempts to categorize ecological chaos using methods taken from physics were initially promising (Schaffer and Kot 1985), but turned out to be too data-hungry for most short (dozens to hundreds of observations) ecological time series, and it handled the ubiquitous and complicated noise of ecological systems poorly (Schaffer et al. 1986; Wilson and Rand 1993).

Turchin and Taylor fitted models of logged population growth rates as a function of the population size at multiple times in the past (lags). They inferred that most of the populations they surveyed showed damped or sustained oscillations, while only one appeared chaotic. This conclusion contradicted earlier work by Hassell et al. (1976), who had used simpler single-lag models to conclude that most natural populations were stable. While more recent work has argued that Turchin and Taylor's "deterministic skeleton" approach (i.e., inferring dynamics from deterministic models fitted to the data) may not accurately reflect the true dynamics of a population (Coulson et al. 2004; Ellner and Turchin 2005), their broader conclusions have held up well. Turchin (2003) has further advanced the analysis of oscillatory dynamics in single populations.

From Qualitative to Quantitative; Quantitative Becomes Qualitative

While sophisticated mathematics has long been a part of ecology (Skellam1951 [reprinted in Real and Brown 1991]) on plant dispersal contains partial differential equations that would frighten most modern ecologists), the late 1980s and 1990s saw an increase in *statistical* sophistication, driven partly by applied conservation problems. Fisher's development of analysis of variance was an early example of biological application that expanded the frontiers of statistics. In the 1970s, field ecologists enthusiastically adopted Fisher's methods to analyze the results of controlled, randomized field experiments (Underwood 1997). In the 1980s, conservation biologists turned to quantitative methods to help them understand observational data on population dynamics. Conservation biologists needed to understand the behavior of unreplicated, uncontrolled time series of population densities; were populations increasing or decreasing over the long run? How likely were they to go extinct? Did increasing population density limit their growth?

Dennis and Taper (1994) ("Density dependence in time series observations of natural populations: estimation and testing") helped resolve these practical questions by introducing a statistically rigorous technique for deciding among the hypotheses that populations were (1) undergoing a random walk; (2) experiencing exponential growth or decay; or (3) experiencing density-dependent growth limitation. At the same time, their improved quantitative methodology helped resolve a fierce qualitative debate, perhaps the longest-running and most contentious in ecology, about the importance of density-independent processes (limitation) vs. density-dependent processes (regulation).

Dennis and Taper showed that their three hypotheses (random walks, exponential growth/ decline, or density-dependent growth) correspond to special cases of a simple model for the logarithm of the ratio of year-to-year geometric growth rates (i.e., $\log(N_{t+1}/N_t)$), which can be tested statistically. In itself, this was not new (e.g., Berryman 1991), but Dennis and Taper provided a rigorous method for estimating confidence intervals on the estimates, or equivalently for computing reliable p-values, which was not previously available.

Although Dennis and Taper's immediate

goal was to improve estimation procedures for applied conservation, their improved quantitative methodology and results also helped settle ecology's long-running debate over density dependence and density independence. This struggle was embodied through the perspective of field ecologists such as Andrewartha, Birch, and Davidson who noted the huge fluctuations in demographic rates of natural populations, and experimentalists such as Nicholson who noted the logical necessity of limitations on sufficiently large populations (Nicholson 1933; Davidson and Andrewartha 1948; Andrewartha and Birch 1954). Dennis and Taper identified three threads in the debate: one focusing on the need for accurate quantification of density dependence from time series, one arguing for the logical necessity of limitation, and one arguing about the precise definitions of density dependence.

The history of the density dependence debate and the ultimate success of Dennis and Taper's emphasis on quantification supports Peters's (1991) contention that progress in ecology sometimes comes through a narrow focus on predicting observed patterns. Viewed from a distance, the debate over density-dependent "versus" density-independent processes seems arcane. The important question is not whether the strength of density dependence is exactly zero, or whether it can be statistically distinguished from noise; rather, we want to know what proportion of the variation in density from place to place or time to time is driven by variation in density-independent (e.g., supply rate of larvae to coral reefs or climate-driven mortality in thrips) vs. density-dependent (e.g., resource competition) processes. Osenberg and coworkers (Schmitt et al. 1999; Osenberg et al. 2002) exemplify this more fruitful line of inquiry by focusing on the relative magnitudes of the effects of density-independent and -dependent factors, rather than simply testing whether the null hypothesis of "no density dependence" can be rejected.

Dennis and Taper's work did not settle the debate—despite Turchin's (1995) comment that

"after five decades and several generations of participants in this debate, many ecologists are tired of it" (Turchin 1995, 21), arguments over the semantics of density dependence continued through the late 1990s into the twenty-first century (e.g., Cooper 2001)—but Dennis and Taper may have heralded the beginning of the end.

Just as Dennis and Taper (1994) and Turchin and Taylor (1992) were establishing methods for rigorously quantifying nonlinear dynamics (density dependence and oscillatory patterns) in unstructured populations, Werner and Caswell's (1977) paper, "Population growth rates and age *versus* stage-distribution models for teasel (*Dipsacus sylvestris* Huds.)," illustrated how compartmental (age- or stage-structured) linear models, without any nonlinearity at all, could provide practical insights into the demography of an invasive species across eight populations. Although demographic matrix models date back to Leslie (1945 [reprinted in Real and Brown 1991]), Werner and Caswell were among the first researchers to demonstrate their practical utility and to clearly discuss the important modeling choices that ecologists must make when investigating the life history of a species or the demography of a population.

Werner and Caswell (1977) directly compared stage- and age-based characterizations of populations. While they concluded that stage-based models were more appropriate for their system, they also detected patterns that suggested that age needed to be taken into account (Slobodkin 1953); exploration of this topic continues up to the present, especially in the analysis of evolution of life histories, although with increasingly complex models (Childs et al. 2003; Caswell and Salguero-Gomez 2013).

Crouse et al. (1987) ("A stage-based population model for loggerhead sea turtles and implications for conservation") raised the profile of matrix population models in conservation biology, using a stage-based model and the barest of life table information to identify critical life stages for the management of imperiled loggerhead sea turtles. This paper opened the door for the use of stage-based matrices in conservation and management—much like island biogeography theory did for reserve design (MacArthur and Wilson 1967; Higgs 1981; Kingsland 2002). It also made a real-world impact on the listing and recovery planning of loggerhead sea turtles under the US Endangered Species Act (Heppell et al. 2000). Its finding that conservation efforts focused on protecting eggs and nestlings may have been misguided shifted management focus to using turtle excluder devices to improve survival of juveniles and subadults in nearshore shrimp fisheries (Conant et al. 2009).

Crouse et al.'s paper transformed species management by showing how population dynamic models could identify appropriate conservation strategies even for endangered, data-poor species. It demonstrated practically how elasticity and sensitivity analyses can highlight stages where management interventions are most likely to boost populations, and where data gaps are most limiting to specific conservation guidance (Caswell 2001).

The innovative use of demographic modeling to estimate population trends in species with scant data led to rapid growth of such models in conservation applications and targeted attention on collecting critical life history and dispersal data upon which model results hinged (Doak and Mills 1994). Ecologists expanded the use of matrix models to conservation applications such as identifying most vulnerable life stages for amphibians (Pechmann and Wilbur 1994) and plants (Schemske et al. 1994), controlling nonindigenous species (Byers et al. 2002), and designing marine protected areas (Roberts et al. 2003).

Enthusiasm for demographic models in conservation was tempered by critical assessments of their use and interpretation (Beissinger and Westphal 1998; Caswell 2001; Fieberg and Ellner 2001). Cautionary notes on using demographic models to estimate population trends hinged primarily on misleading results that can arise when data are scant, compromising estimates of variance in vital rates and dispersal.

Quantifying Population Persistence

The next set of papers broke new ground by illustrating how simple models could be used to estimate population persistence, leading to an explosion of population viability modeling that continues today, along with a rich literature on both the promise and perils of using models to estimate extinction risk (Boyce 1992; Brook et al. 2000; Fieberg and Ellner 2001; Beissinger and McCullough 2002). Conservation scientists enthusiastically applied the minimum viable population (MVP) concept for imperiled species; in so doing the field recapitulated the classic debate in ecology about the value of simple vs. complex models (Brown 1991). The arguments in favor of simpler models for basic ecology—transparency of assumptions, feasibility in data-poor cases, ease of validation—are just as applicable in conservation biology.

Shaffer (1981) ("Minimum population sizes for species conservation") introduced the MVP, a concept that has transformed conservation science and practice ever since. The MVP definition lays out an explicit set of risk criteria for a population based on specific intrinsic and extrinsic processes contributing to its risk. It built on previous work on the roles of different sources of variation in population dynamics (Berry 1971; May 1973; Roughgarden 1975; Soulé 1980). Shaffer identified four sources of variation: (1) *demographic stochasticity*, deriving from chance events in survival and reproduction in a finite population, (2) *environmental stochasticity*, caused by variation in habitat condition and species interactions, (3) *natural catastrophes*, such as floods or fire, and (4) *genetic stochasticity*, or changes in gene frequencies as a result of genetic drift or inbreeding. Shaffer's definition states: "A minimum viable population for any given species in any given habitat is the smallest isolated population having a 99% chance of remaining extant for 1000 years despite the foreseeable effects of demographic, environmental, and genetic stochasticity, and natural catastrophes." The paper synthesized conclusions from ecological and genetic theory in highlighting the processes driving species extinctions.

Shaffer (1981) also provided novel practical guidance for a growing cohort of conservation scientists. He pointed out that the criteria for the likelihood (99%) and timeframe (1000 years) of persistence are partly policy driven and that they require discussion between scientists and practitioners. He identified five possible approaches to estimating MVPs (experiments, analysis of biogeographic patterns, analytical models (e.g., MacArthur and Wilson 1967; Richter-Dyn and Goel 1972), simulation models, and genetic criteria (Franklin 1980's 50/500 rule)). Newer methods like simulation and analytical modeling took off around this time as useful approaches for species that are both imperiled and data-poor.

The MVP guidelines in Shaffer (1981) became nearly dogmatic rules of thumb for species conservation in the US and internationally (Mace and Lande 1991; Boersma et al. 2001; Mace et al. 2008). Unfortunately, early applications of the concept were sloppy, choosing to consider only some of the criteria for persistence, and too literally interpreted numbers intended simply as guidance. As genetic methods improved, overattention was often given to effective population size criteria, with misleading results (Lande 1993). Simulation approaches used to estimate MVP could be overly complex, as well as much less transparent than analytical models (Simberloff 1988). A focus on single species or populations, and neglect of spatial dynamics, led to limited conservation strategies in some cases (e.g., Menges 1991). Metapopulation approaches accounting for spatial variation and methods to account for source-sink dynamics began to address these shortcomings in the late 1980s (Simberloff 1988; Murphy et al. 1990).

Scientific critiques have improved application of population viability analyses, although the use of generic ranges for MVP estimates in data-poor situations remains both popular and controversial. Although estimates of absolute risk can be extremely sensitive to assumptions or to limited data, population viability analyses

are relatively robust when used to estimate the *relative* risk of extinction under different management options (Soulé 1987; Beissinger and Westphal 1998; Brook et al. 2000; Traill et al. 2007; Flather et al. 2011). Although such rankings are useful in some applications, the implementation standards for the US Endangered Species Act (Tear et al. 1995) and IUCN Red List guidelines (Mace and Lande 1991; Mace et al. 2008) demand absolute numbers, so the science of estimating MVPs is likely to remain an active debate.

Dennis et al. 1991 ("Estimation of growth and extinction parameters for endangered species") provided a much-needed approach to estimating population parameters from time series for species with poor data. Species of conservation concern are often poorly censused, and Dennis et al. (1991) provided an important advance in applying theory to gain the greatest insights possible into population status of imperiled taxa. The authors used diffusion approximations, a mathematical technique that treats population size as continuous but explicitly accounts for the greater amount of noise at small population sizes when populations are controlled by demographic stochasticity. They showed that a linear regression of transformed time series data can accurately estimate the time to a *quasi-extinction threshold*, the time when a population falls below a single member. Dennis et al.'s method describes complex population dynamics with only three parameters: the distance (on a logarithmic scale) from an initial population size to a user-specified threshold of quasi-extinction, and two parameters estimated from data, the mean (trend) and variance of the diffusion process. They illustrated the practicality of the approach by testing it with a diverse set of seven threatened species, spanning grizzly bears to California condors. In this way, the authors provided a virtual how-to guide for fitting a linear regression model to observational data and interpreting the resulting estimates of extinction risk.

Dennis et al.'s (1991) approach of combining diffusion approximations with a stochastic exponential model to estimate extinction probability and the mean and median times to extinction has become the standard practice for data-poor species. Researchers before and after Dennis et al. have developed more complex population models, but this paper permanently elevated the quality of extinction risk assessments for endangered species. This paper is an excellent example of how observation of species—in this case their decline—leads to important scientific questions and advancement of theory. Similar to the interest stimulated by Shaffer's (1981) MVP paper, Dennis et al. kicked off lively debates (e.g., Ludwig 1999; Fieberg and Ellner 2000) and constructive guidance (e.g., Morris and Doak 2002) about the conditions under which estimating extinction of a species is legitimate.

Lande (1993) ("Risks of population extinction from demographic and environmental stochasticity and random catastrophes") used diffusion equations more generally, without reference to specific examples, to synthesize and correct previous results (including Shaffer 1981) on the importance of demographic stochasticity, environmental stochasticity, and catastrophes for population viability. Lande drew general conclusions by asking how the expected persistence time of a population scales with its size under different forms of stochasticity— whether the relationship is logarithmic, linear, power-law, or exponential. Each of these forms represents successively more persistent populations. A population whose persistence time scales logarithmically with population size will always, for sufficiently large population size, out-persist a population whose persistence grows only linearly; a population with exponentially scaling persistence will eventually out-persist a power-law-scaling population. (These general conclusions are only precisely correct for *large* populations; in particular cases, small populations could still be more threatened by demographic than by environmental or catastrophic stochasticity.)

Contrary to previous assertions by Shaffer (1981) and Goodman (1987), Lande found that in populations with density-dependent growth and positive intrinsic growth rates, persistence

time scales exponentially with population size when only demographic stochasticity is present, and scales as a power-law when either environmental stochasticity or random catastrophes occur. Lande also found that random catastrophes need not be worse for a population than "ordinary" environmental stochasticity. Lande's analyses showed that theory cannot support any general statement about whether catastrophes or continuous variability have a greater impact on population viability; sufficiently strong temporal environmental variability can outweigh the effects of catastrophes.

Beyond population extinction risk, Brander (1981) ("Disappearance of common skate *Raia batis* from Irish Sea" [not reprinted in this volume]) was an early influential paper using population modeling to examine recovery potential of a harvested species. It used analytical modeling to document the first clear case of a species brought to the brink of extinction by commercial fishing. Before this work, understanding serious declines or local disappearances of marine species was unreliable because scant observations led to major lags in detection (Dulvy et al. 2003). Brander demonstrated for a fished species—and thus for fishery managers—how life history differences among skate species affect their vulnerability. The focal species (common skate; *R. batis*) is most vulnerable to fishing mortality because of its relatively low fecundity and late age at maturity; its associated larger body size also made it a prize for fishermen.

Brander's paper does more than just describe fishing effects; it also predicts how much fishing will drive the species to collapse and shows that current pressures are well beyond what the species can sustain. This paper showed that fishing can, in fact, cause the local extirpation of a species. Brander also pointed out that overfishing one species was likely to lead to declines in other species, a prediction that was later borne out in Irish Sea harvested species (Dulvy et al. 2003). Drawing attention to the interactions between harvested species opened the door to greater ecological influences on research—for example, examining how location and timing of fishing affected its impacts (Thrush and Dayton 2002), fishing's broader effects on marine ecosystems (Jennings and Kaiser 1998), and the subsequent effects of ecosystem condition on fisheries and aquaculture (Folke and Kautsky 1989; Jennings 2004).

Brander's insights were not new to fisheries scientists, who had already developed and adopted simple analytical and simulation approaches to tracking status of harvested species (beginning with Ricker 1954; and Beverton and Holt 1957). Yet this paper was an important bridge between fisheries and ecology; by highlighting a pressing conservation problem, this and subsequent literature motivated application of integrated approaches drawing from both disciplines (Folke and Kautsky 1989; Jennings 2004; Worm et al. 2009; Salomon et al. 2011).

The Dialectic of Science and Practice

The foundational papers in this part of the volume illustrate the value of an ongoing dialectic between basic ecology and the science and practice of conservation. Insights from ecological theory, experiments, and observation clearly have influenced methods, metrics, and rules of thumb used in applied conservation. In turn, testing ecological principles in practice has led to an explosion of interesting and novel questions for basic science.

Thanks to scientists willing to cross disciplinary and basic-applied divides, the dialectic continues productively today. Some of the papers represent key advances in ecological theory stimulated by carefully documented observations and dire reports from the field about species and habitat declines. Generally, simulation approaches to population viability analyses have replaced diffusion approximation approaches (Lacy 1993; Lindenmayer et al. 1995; but see Fieberg and Ellner 2000 for caveats). In turn, challenges to population viability methods led to a growth of studies using available population data in computer simulations to test extinction probabilities. The relative value of empirical observations and modeling for estimating extinction risk continues to be tested (Keith et al. 2004; Maclean and Wilson 2011).

The rich body of work that sprang from these papers has driven scientific advancements and practical guidance in how to apply population models under data-poor conditions; identification of where data are most important to reduce model uncertainty; and much-improved interpretation of model results, including the conditions under which quantitative or qualitative conclusions about species or habitat status are most appropriate (Soulé 1987; Pimm and Gilpin 1989; Burgman et al. 1993; Doak and Mills 1994). These papers highlighted the ever-present need for conservation scientists to think critically about applying theory and to understand the underlying assumptions that constrain the use of theory in practice (Werner and Caswell 1977; Doak and Mills 1994).

Ecological theorists came through with scientifically rigorous, simpler modeling approaches that are useful for providing management guidance in a complex world in which addressing environmental declines is urgent (Werner and Caswell 1977; Ludwig et al. 1978; Crouse et al. 1987; Dennis et al. 1991). Informative and reliable metrics of population status (e.g., Dennis et al. 1991) have become standard practice. The penchant of conservation practitioners for useful rules of thumb and simple modeling approaches also led to critical tests of population viability analyses for endangered species and robustness of general conclusions about minimum viable population size and extinction risk (Lande 1993; Brook et al. 2000; Reed et al. 2003; Keith et al. 2004; Traill et al. 2007). In addressing conservation needs, such as improving estimation methods, several of the papers featured here also answered long-standing debates in basic ecology.

Literature Cited

Anderson, R. M., and R. M. May. 1978. Regulation and stability of host-parasite population interactions: I. Regulatory processes. Journal of Animal Ecology 47: 219–247.

Andrewartha, H. G., and C. Birch. 1954. The Distribution and Abundance of Animals. University of Chicago Press.

Beissinger, S. R., and D. R. McCullough, eds. 2002. Population Viability Analysis. University of Chicago Press.

Beissinger, S. R., and M. I. Westphal. 1998. On the use of demographic models of population viability in endangered species management. Journal of Wildlife Management 62: 821.

Berry, R. J. 1971. Conservation aspects of the genetical constitution of populations. Pages 177–206 in E. D. Duffey and A. S. Watt, eds. The Scientific Management of Animal and Plant Communities for Conservation. Blackwell.

Berryman, A. A. 1991. Stabilization or regulation: what it all means! Oecologia 86: 140–143.

Beverton, R. J., and S. J. Holt. 1957. On the Dynamics of Exploited Fish Populations. Her Majesty's Stationery Office.

Boersma, P. D., P. Kareiva, W. F. Fagan, J. A. Clark, and J. M. Hoekstra. 2001. How good are endangered species recovery plans? BioScience 51: 643.

Boyce, M. S. 1992. Population viability analysis. Annual Review of Ecology and Systematics 23: 481–497.

Brander, K. 1981. Disappearance of common skate Raia batis from Irish Sea. Nature 290: 48–49.

Brook, B. W., J. J. O'Grady, A. P. Chapman, M. A. Burgman, H. R. Akçakaya, and R. Frankham. 2000. Predictive accuracy of population viability analysis in conservation biology. Nature 404: 385–387.

Brown, J. H. 1991. New approaches and methods in ecology. Pages 445–455 in L. A. Real and J. H. Brown, eds. Foundations of Ecology: Classical Papers with Commentary. University of Chicago Press.

Burgman, M. A., S. Ferson, and H. R. Akçakaya. 1993. Risk assessment in conservation biology. Population and community biology series (1st ed.). Chapman & Hall.

Byers, J. E., S. Reichard, J. M. Randall, I. M. Parker, C. S. Smith, W. M. Lonsdale, I. A. E. Atkinson, et al. 2002. Directing research to reduce the impacts of nonindigenous species. Conservation Biology 16: 630–640.

Caswell, H. 2001. Matrix population models: construction, analysis, and interpretation (2nd ed.). Sinauer.

Caswell, H., and R. Salguero-Gomez. 2013. Age, stage and senescence in plants. Journal of Ecology 101: 585–595.

Caughley, G. 1994. Directions in conservation biology. Journal of Animal Ecology 63: 215.

Childs, D. Z., M. Rees, K. E. Rose, P. J. Grubb, and S. P. Ellner. 2003. Evolution of complex flowering strategies: an age- and size-structured integral projection model. Proceedings of the Royal Society of London, Series B: Biological Sciences 270: 1829–1838.

Conant, T. A., P. H. Dutton, T. Eguchi, S. P. Epperly, C. C. Fahy, M. H. Godfrey, S. L. MacPherson, et al. 2009. Loggerhead Sea Turtle (Caretta caretta) 2009 Status Review under the U.S. Endangered Species Act: Report of the Loggerhead Biological Review Team to the National Marine Fisheries Service. National Marine Fisheries Service.

Connell, J. H. 1975. Some mechanisms producing structure in natural communities: a model and evidence from field experiments. Pages 460–490 in M. L. Cody and J. M. Diamond, eds. Ecology and Evolution of Communities. Harvard University Press.

Cooper, G. 2001. Must there be a balance of nature? Biology & Philosophy 16: 481–506.

Coulson, T., P. Rohani, and M. Pascual. 2004. Skeletons, noise and population growth: the end of an old debate? Trends in Ecology & Evolution 19: 359–364.

Crawley, M. J., ed. 1992. Natural Enemies: The Population Biology of Predators, Parasites, and Diseases. Blackwell.

Crouse, D. T., L. B. Crowder, and H. Caswell. 1987. A stage-based population model for loggerhead sea turtles and implications for conservation. Ecology 68: 1412–1423.

Davidson, J., and H. G. Andrewartha. 1948. Annual trends in a natural population of Thrips imaginis (Thysanoptera). Journal of Animal Ecology 17: 193.

Dennis, B., P. L. Munholland, and J. M. Scott. 1991. Estimation of growth and extinction parameters for endangered species. Ecological Monographs 61: 115–143.

Dennis, B., and M. L. Taper. 1994. Density dependence in time series observations of natural populations: estimation and testing. Ecological Monographs 64: 205–224.

Doak, D. F., and L. S. Mills. 1994. A useful role for theory in conservation. Ecology 75: 615–626.

Dulvy, N. K., Y. Sadovy, and J. D. Reynolds. 2003. Extinction vulnerability in marine populations. Fish and Fisheries 4: 25–64.

Ellner, S. P., and P. Turchin. 2005. When can noise induce chaos and why does it matter: a critique. Oikos 111: 620–631.

Fieberg, J., and S. P. Ellner. 2000. When is it meaningful to estimate an extinction probability? Ecology 81: 2040–2047.

Fieberg, J., and S. P. Ellner. 2001. Stochastic matrix models for conservation and management: a comparative review of methods. Ecology Letters 4: 244–266.

Flather, C. H., G. D. Hayward, S. R. Beissinger, and P. A. Stephens. 2011. Minimum viable populations: is there a "magic number" for conservation practitioners? Trends in Ecology & Evolution 26: 307–316.

Folke, C., and N. Kautsky. 1989. The role of ecosystems for a sustainable development of aquaculture. Ambio 18: 234–243.

Goodman, D. 1987. The demography of chance extinction. Pages 11–34 in M. E. Soulé, ed. Viable Populations for Conservation (1st ed.). Cambridge University Press.

Hassell, M. P., J. H. Lawton, and R. M. May. 1976. Patterns of dynamical behaviour in single-species populations. Journal of Animal Ecology 45: 471.

Heppell, S. S., D. T. Crouse, and L. B. Crowder. 2000. Using matrix models to focus research and management efforts in conservation. Pages 148–168 in S. Ferson and M. Burgman, eds. Quantitative Methods for Conservation Biology. Springer.

Higgs, A. J. 1981. Island biogeography theory and nature reserve design. Journal of Biogeography 8: 117.

Jennings, S. 2004. The ecosystem approach to fishery management: a significant step towards sustainable use of the marine environment? Marine Ecology Progress Series 274: 279–282.

Jennings, S., and M. J. Kaiser. 1998. The effects of fishing on marine ecosystems. Advances in Marine Biology 34: 201–352

Keith, D. A., M. A. McCarthy, H. Regan, T. Regan, C. Bowles, C. Drill, C. Craig, et al. 2004. Protocols for listing threatened species can forecast extinction: testing extinction forecasts. Ecology Letters 7: 1101–1108.

Kendall, B. E., C. J. Briggs, W. W. Murdoch, P. Turchin, S. P. Ellner, E. McCauley, R. M. Nisbet, et al. 1999. Why do populations cycle? A synthesis of statistical and mechanistic modeling approaches. Ecology 80: 1789–1805.

Kingsland, S. E. 2002. Creating a science of nature reserve design: perspectives from history. Environmental Modeling and Assessment 7: 61–69.

Lacy, R. C. 1993. VORTEX: a computer simulation model for population viability analysis. Wildlife Research 20: 45.

Lande, R. 1993. Risks of population extinction from demographic and environmental stochasticity and random catastrophes. American Naturalist 142: 911–927.

Leslie, P. H. 1945. On the use of matrices in certain population mathematics. Biometrika 33: 183–212.

Lindenmayer, D. B., M. A. Burgman, H. R. Akçakaya, R. C. Lacy, and H. P. Possingham. 1995. A review of the generic computer programs ALEX, RAMAS/space and VORTEX for modelling the viability of wildlife metapopulations. Ecological Modelling 82: 161–174.

Ludwig, D. 1999. Is it meaningful to estimate a probability of extinction? Ecology 80: 298–310.

Ludwig, D., R. Hilborn, and C. Walters. 1993. Uncertainty, resource exploitation, and conservation: lessons from history. Science 260: 17–36.

Ludwig, D., D. D. Jones, and C. S. Holling. 1978. Qualitative analysis of insect outbreak systems: the spruce budworm and forest. Journal of Animal Ecology 47: 315–332.

MacArthur, R. H., and E. O. Wilson. 1967. The Theory of Island Biogeography. Princeton University Press.

Mace, G. M., N. J. Collar, K. J. Gaston, C. Hilton-Taylor, H. R. Akçakaya, N. Leader-Williams, E. J. Milner-Gulland, et al. 2008. Quantification of extinction risk: IUCN's system for classifying threatened species. Conservation Biology 22: 1424–1442.

Mace, G. M., and R. Lande. 1991. Assessing extinction threats: toward a reevaluation of IUCN threatened species categories. Conservation Biology 5: 148–157.

Maclean, I. M. D., and R. J. Wilson. 2011. Recent ecological responses to climate change support predictions of high extinction risk. Proceedings of the National Academy of Sciences 108: 12337–12342.

May, R. M. 1973. Qualitative stability in model ecosystems. Ecology 54: 638–641.

May, R. M. 1974. Stability and Complexity in Model Ecosystems. Princeton University Press.

Menges, E. 1991. The application of minimum viable population theory to plants in D. Falk and E. Holsinger, eds. Genetics and Conservation of Rare Plants. Oxford University Press.

Morris, W. F., and D. F. Doak. 2002. Quantitative Conservation Biology: Theory and Practice of Population Viability Analysis. Sinauer.

Murphy, D. D., K. E. Freas, and S. B. Weiss. 1990. An environment-metapopulation approach to population viability analysis for a threatened invertebrate. Conservation Biology 4: 41–51.

Nicholson, A. J. 1933. Supplement: the balance of animal populations. Journal of Animal Ecology 2: 131.

Osenberg, C. W., C. M. St. Mary, R. J. Schmitt, S. J. Holbrook, P. Chesson, and B. Byrne. 2002. Rethinking ecological inference: density dependence in reef fishes: inference and density dependence. Ecology Letters 5: 715–721.

Paine, R. T. 1974. Intertidal community structure: experimental studies on the relationship between a dominant competitor and its principal predator. Oecologia 15: 93–120.

Paine, R. T. 1980. Food webs: linkage, interaction strength

and community infrastructure. Journal of Animal Ecology 49: 666.

Pechmann, J. H. K., and H. M. Wilbur. 1994. Putting declining amphibian populations in perspective: natural fluctuations and human impacts. Herpetologica 50: 65–84.

Peters, R. H. 1991. A Critique for Ecology. Cambridge University Press.

Pimm, S. L., and M. E. Gilpin. 1989. 20. Theoretical issues in conservation biology. Pages 287–305 *in* J. Roughgarden, R. M. May, and S. A. Levin, eds. Perspectives in Ecological Theory. Princeton University Press.

Real, L. A., and J. H. Brown, eds. 1991. Foundations of Ecology: Classic Papers with Commentaries. University of Chicago Press.

Reed, D. H., J. J. O'Grady, B. W. Brook, J. D. Ballou, and R. Frankham. 2003. Estimates of minimum viable population sizes for vertebrates and factors influencing those estimates. Biological Conservation 113: 23–34.

Richter-Dyn, N., and N. S. Goel. 1972. On the extinction of a colonizing species. Theoretical Population Biology 3: 406–433.

Ricker, W. E. 1954. Stock and recruitment. Journal of the Fisheries Research Board of Canada 11: 559–623.

Roberts, C. M., G. Branch, R. H. Bustamante, J. C. Castilla, J. Dugan, B. S. Halpern, K. D. Lafferty, et al. 2003. Application of ecological criteria in selecting marine reserves and developing reserve networks. Ecological Applications 13: 215–228.

Roughgarden, J. 1975. A simple model for population dynamics in stochastic environments. American Naturalist 109: 713–736.

Salomon, A. K., S. K. Gaichas, O. P. Jensen, V. N. Agostini, N. Sloan, J. Rice, T. R. McClanahan, et al. 2011. Bridging the divide between fisheries and marine conservation science. Bulletin of Marine Science 87: 251–274.

Schaffer, W. M., S. Ellner, and M. Kot. 1986. Effects of noise on some dynamical models in ecology. Journal of Mathematical Biology 24: 479–523.

Schaffer, W. M., and M. Kot. 1985. Nearly one dimensional dynamics in an epidemic. Journal of Theoretical Biology 112: 403–427.

Schemske, D. W., B. C. Husband, M. H. Ruckelshaus, C. Goodwillie, I. M. Parker, and J. G. Bishop. 1994. Evaluating approaches to the conservation of rare and endangered plants. Ecology 75: 584–606.

Schmitt, R. J., S. J. Holbrook, and C. W. Osenberg. 1999. Quantifying the effects of multiple processes on local abundance: a cohort approach for open populations. Ecology Letters 2: 294–303.

Shaffer, M. L. 1981. Minimum population sizes for species conservation. BioScience 31: 131–134.

Simberloff, D. 1988. The contribution of population and community biology to conservation science. Annual Review of Ecology and Systematics 19: 473–511.

Skellam, J. G. 1951. Random dispersal in theoretical populations. Biometrika 38: 196–218.

Slobodkin, L. B. 1953. An algebra of population growth. Ecology 34: 513–519.

Soulé, M. E. 1980. Thresholds for survival: maintaining fitness and evolutionary potential. Pages 151–170 *in* M. E. Soulé and B. A. Wilcox, eds. Conservation Biology: An Evolutionary-Ecological Perspective. Sinauer.

Soulé, M. E., ed. 1987. Viable Populations for Conservation (1st ed.). Cambridge University Press.

Strong, D. R. 1986. Density-vague population change. Trends in Ecology & Evolution 1: 39–42.

Tear, T. H., J. M. Scott, P. H. Hayward, and B. Griffith. 1995. Recovery plans and the Endangered Species Act: are criticisms supported by data? Conservation Biology 9: 182–195.

Thrush, S. F., and P. K. Dayton. 2002. Disturbance to marine benthic habitats by trawling and dredging: implications for marine biodiversity. Annual Review of Ecology and Systematics 33: 449–473.

Tompkins, D. M., and M. Begon. 1999. Parasites can regulate wildlife populations. Parasitology Today 15: 311–313.

Traill, L. W., C. J. A. Bradshaw, and B. W. Brook. 2007. Minimum viable population size: A meta-analysis of 30 years of published estimates. Biological Conservation 139: 159–166.

Turchin, P. 1995. Population regulation: old arguments and a new synthesis. Pages 19–40 *in* N. Cappuccino and P. Price, eds. Population Dynamics. Elsevier.

Turchin, P. 2003. Complex Population Dynamics: A Theoretical/Empirical Synthesis. Princeton University Press.

Turchin, P., and A. D. Taylor. 1992. Complex dynamics in ecological time series. Ecology 73: 289–305.

Underwood, A. J. 1997. Experiments in Ecology: Their Logical Design and Interpretation Using Analysis of Variance. Cambridge University Press.

Werner, E. E., and G. G. Mittelbach. 1981. Optimal foraging: field tests of diet choice and habitat switching. American Zoologist 21: 813–829.

Werner, P. A., and H. Caswell. 1977. Population growth rates and age *versus* stage-distribution models for teasel (*Dipsacus sylvestris* Huds.). Ecology 58: 1103–1111.

Wilson, H. B., and D. A. Rand. 1993. Detecting chaos in a noisy time series. Proceedings of the Royal Society of London. Series B: Biological Sciences 253: 239–244.

Worm, B., R. Hilborn, J. K. Baum, T. A. Branch, J. S. Collie, C. Costello, M. J. Fogarty, et al. 2009. Rebuilding global fisheries. Science 325: 578–585.

REGULATION AND STABILITY OF HOST-PARASITE POPULATION INTERACTIONS

I. REGULATORY PROCESSES

By ROY M. ANDERSON and ROBERT M. MAY

Zoology Department, King's College, London University, London WC2R 2LS, and Biology Department, Princeton University, Princeton, N.J. 08540, U.S.A.

SUMMARY

(1) Several models describing the dynamics of host-parasite associations are discussed.

(2) The models contain the central assumption that the parasite increases the rate of host mortalities. The parasite induced changes in this rate are formulated as functions of the parasite numbers per host and hence of the statistical distribution of the parasites within the host population.

(3) The parameters influencing the ability of the parasite to regulate the growth of its host's population, and the stability of parasite induced equilibria, are examined for each model.

(4) Three specific categories of population processes are shown to be of particular significance in stabilizing the dynamical behaviour of host-parasite interactions and enhancing the regulatory role of the parasite.

These categories are overdispersion of parasite numbers per host, nonlinear functional relationships between parasite burden per host and host death rate, and density dependent constraints on parasite population growth within individual hosts.

INTRODUCTION

Eucaryotic parasites of one kind or another play a part in the natural history of many, if not most, animals. Man is not exempt: considering helminth parasites alone, roughly 200 million people are affected by the trematode species which cause schistosomiasis and 300 million by the filarial nematode parasites. A multitude of other internal and external parasites of man produce effects ranging from minor irritations to major diseases.

The relation between populations of such parasites and their hosts can be regarded as a particular manifestation of the general predator-prey interaction. Predator-prey theory has received much attention since the work of Lotka and Volterra in the 1920s, and all contemporary ecology texts give a lot of coverage to the subject. This work, however, tends to draw its inspiration, and its empirical basis, from the world of moose and wolves (with differential equations for continuously overlapping generations) or of arthropod predators and their prey (with difference equations for discrete, non-overlapping generations); see, e.g. the reviews by May (1976, ch. 4.) and Hassell (1976). That is, the extensive predator-pray literature pertains mainly to situations where (at least in effect) the predators kill and eat their prey.

Analogous studies of the special kinds of predator-prey relations that are of parasito-

0021–8710/78/0200–0219 $02.00

220   *Regulation and stability of host-parasite interactions. I*

logical interest are comparatively few. Thus relatively little is known about the effects that parasites may have upon the population dynamics of their hosts.

In this paper we explore the dynamical properties of some simple models which aim to capture the essential biological features of host-parasite associations. In analogy with earlier work on arthropod systems (Hassell & May 1973; Hassell, Lawton & Beddington 1976; Beddington, Hassell & Lawton 1976), we have tried to build the models on biological assumptions that are rooted in empirical evidence.

The paper is organized as follows. After giving a more precise definition of what we mean by the term 'parasite', we outline the biological assumptions upon which the 'Basic model' rests, and expound the dynamical properties of this model. The Basic model assumes parasites to be distributed independently randomly among hosts; various patterns of non-random parasite distribution are then discussed, and introduced into 'model A', which is then investigated. The effects of varying the way the host's mortality depends on the density of parasites ('model B'), and of introducing density dependence into the parasites' intrinsic death rate ('model C') are explored. The following paper (May & Anderson 1978) goes on to discuss various destabilizing influences and encompasses the interaction between host and parasite reproduction rates, transmission factors, and time delays.

Throughout, the text aims to be descriptive with results displayed graphically and the emphasis on the biology of host-parasite associations. The mathematical analysis of the stability properties of the various models is relegated to appendices.

THE TERM 'PARASITISM'

Parasitism may be regarded as an ecological association between species in which one the parasite, lives on or in the body of the other, the host. The parasite may spend the majority of its life in association with one or more host species, or alternatively it may spend only short periods, adopting a free-living mode for the major part of its developmental cycle. During the parasitic phase of its life cycle, the organism depends upon its host for the synthesis of one or more nutrients essential for its own metabolism. The relationship is usually regarded as obligatory for the parasite and harmful or damaging for the host. To classify an animal species as parasitic we therefore require that three conditions be satisfied: utilization of the host as a habitat; nutritional dependence; and causing 'harm' to its host.

When one considers such interactions at the population level, the terminology now used for labelling animal associations appears rather confusing and imprecise (see Starr 1975; Askew 1971; Dogiel 1964). This is particularly apparent when one tries to formalize the nature of the harmful effect of a parasitic species on the population growth of its host. These difficulties do not always arise. For example, insect parasitoids as a developmental necessity invariably kill their host, the parasite surviving the death it induced by the adoption of a free-living mode of life in the adult phase. Askew (1971) has termed such species 'protean parasites' since they are parasitic as larval forms and free-living when adult.

Other eucaryotic parasitic organisms such as lice, fleas, ticks, mites, protozoa and helminths exhibit the nutritional and habitat requirements of a parasite but appear to do very little harm to their hosts, unless present in very large numbers. Such species do not kill their hosts as a prerequisite for successful development; indeed, in contrast to parasitoid insects, these organisms are often themselves killed if they cause the death of their host. Although parasite induced host deaths usually result when heavy infections occur, the precise meaning of the word heavy will very much depend on the size of the parasite in relation to its host, and the niche and mode of life adopted by the parasite within or on the host.

In this paper, we use the term parasite to refer to species which *do not kill their host as a prerequisite for successful development*. This distinguishes parasites from parasitoids.

Parasites exhibit a wide degree of variability between species in the degree of harm or damage they cause to their hosts. At one extreme of the spectrum, parasites merge into the parasitoid type of association with their close relationship (in population terms) with predator-prey interactions. At this extreme, death will invariably result from parasite infection, but in contrast to parasitoids such host deaths will also kill the parasites contained within. At the other end of the spectrum lie the symbiotic forms of association in which the symbiont lives on or in the host with a degree of nutritional dependence akin to a permissive gastronomic hospitality. Species at this end of the spectrum cause negligible, if any, harm to the host even when present in very large numbers.

In terms of their population dynamics, there will be differences between parasites at the two ends of this spectrum; between the parasitoid like parasites and the symbionts. Crofton (1971a, b) has stressed the importance of quantifying these notions, and has suggested that a useful first step lies in the definition of a 'lethal level', which measures the typical number of parasites required to kill a host. As Crofton notes, quantitative information about such lethal levels is hard to come by.

We shall return to these quantitative questions below: for the present we propose that an organism only be classified as a parasite if it has a detrimental effect on the intrinsic growth rate of its host population.

BASIC MODEL: BIOLOGICAL ASSUMPTIONS

We define $H(t)$ and $P(t)$ to be the magnitudes of the host and parasite populations, respectively at time t; the average number of parasites per host is then $P(t)/H(t)$. We assume that the vast majority of protozoan and helminth parasites exhibit continuous population growth, where generations overlap completely. The equations describing the way the host and parasite populations change in time are thus formulated as differential equations.

The basic model is for parasite species which do not reproduce directly within their definitive or final host, but which produce transmission stages such as eggs, spores or cysts which, as a developmental necessity, pass out of the host. This type of parasite life cycle is shown by many protozoan, helminth and arthropod species. In the following paper (May & Anderson (1978) in model E) this basic framework is modified to encompass species, such as some parasitic protozoa, which have a reproductive phase which directly contributes to the size of the parasite population within the host.

We assume that all parasitic species are capable of multiply infecting a proportion of the host population and that the birth and death rates of infected hosts are altered by the number of parasites they harbour. The precise functional relationship between the number of parasites harboured and the host's chances of surviving or reproducing varies greatly among different host-parasite associations. The rate of parasite induced host mortalities may increase linearly with parasite burden or as an exponential or power law function. Some examples of these functional relationships for protozoan, helminth and arthropod parasites of both vertebrate and invertebrate host species are shown in Fig. 1. Sometimes the relationship may be of a more complex form than suggested by these examples. The parasite, for instance, may act in an all-or-nothing manner where low burdens do not influence the hosts survival chances but at a given threshold burden the death rate rises very rapidly resulting in certain death for the hosts. In general, however, where quantitative rate estimates are available the 'harmful' effect of the parasite is usually of the more gradual forms indicated in Fig. 1.

In the majority of host-parasite associations it appears to be the death rate rather than the reproductive rate of the host which is influenced by parasitic infection. Exceptions to this general pattern are particularly noticeable in the associations between larval digenean parasites and their molluscan intermediate hosts. Many parasitic arthropods also decrease the reproductive power of their hosts, and in certain cases complete parasitic castration occurs.

Accordingly, the majority of our models assume that the parasite increases the host death rate. Attention is given to the population consequences of parasite induced reduction of host reproductive potential (in Model D) in the following paper (May & Anderson 1978).

TABLE 1. Description of the principal population parameters used in the models

Parameter	Description
a	Instantaneous host birth rate (/host/unit of time).
b	Instantaneous host death rate, where mortalities are due to 'natural causes' (/host/unit of time).
α	Instantaneous host death rate, where mortalities are due to the influence of the parasite (/host/unit of time).
λ	Instantaneous birth rate of parasite transmission stages where birth results in the production of stages, which pass out of the host, and are responsible for transmission of the parasite within the host population (i.e. eggs, cysts, spores or larvae) (/parasite/unit of time).
μ	Instantaneous death rate of parasites within the host, due to either natural or host induced (immunological) causes (/parasite/unit of time).
H_0	Transmission efficiency constant, varying inversely with the proportion of parasite transmission stages which infect members of the host population.
r	Instantaneous birth rate of parasite, whose birth results in the production of parasitic stages which remain within the host in which they were produced (/parasite/unit of time).

The two basic equations, for dH/dt and dP/dt, are constructed from several components, each of which represents specific biological assumptions. A summary of our notation is given in Table 1. These components will now be discussed, one by one.

The growth of the host population

We assume that the rate of growth of the host population is simply determined by the natural intrinsic rate of increase in the absence of parasitic infection minus the rate of parasite induced host mortalities. Both the host reproductive rate a, and the rate of 'natural' mortalities b, are represented as constants unaffected by density dependent constraints on population growth. We use the term 'natural' mortalities to encompass all deaths due to causes other than parasitic infection, e.g. predation and senescence.

Our omission of density dependent constraints on host population growth is deliberate. We recognize that in the real world host population growth will be limited by, among other factors, intraspecific competition for finite resources. Since our aim, however, is to provide qualitative insights into the mechanisms by which parasites regulate host population growth, we have excluded the concept of a carrying capacity of the hosts' environment to simplify algebraic manipulations. Such simplification clarifies predictions of biological interest.

ROY M. ANDERSON AND ROBERT M. MAY   223

We assume that if the parasite fails to control host population growth, exponential increase of the host population occurs until resource limitation results in the gradual approach to a carrying capacity.

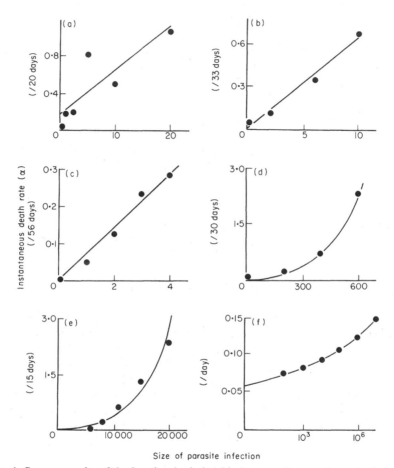

Size of parasite infection

FIG. 1. Some examples of the functional relationship between the rate of parasite induced host mortalities (α) and the parasite burden (i) per host. (The straight lines in graphs (a), (b) and (c) are the least squares best fit linear models and the curves in the graphs (d), (e) and (f) are the least squares best fit exponential models). (a) Snail host *Lymnea gedrosiana* (Annandale and Prashed) parasitized by the larval stages of the digenean *Ornithobilharzia turkestanicum* (Skrjabin) (data from Massoud 1974); (b) aquatic Hemipteran *Hydrometra myrae* (Bueno) parasitized by the mite *Hydryphantes tenuabilis* (Marshall) (data from Lanciani 1975); (c) laboratory mouse parasitized by the digenean *Fasciola hepatica* (L.) (data from Hayes, Bailer & Mitovic 1973); (d) laboratory mouse parasitized by the nematode *Heligmosomoides polygyrus* (Dujardin) (data from Forrester 1971); (e) laboratory rat parasitized by the nematode *Nippostrongylus brasiliensis* (Yokogawa) (data from Hunter & Leigh 1961); (f) laboratory mouse parasitized by the blood protozoan *Plasmodium vinckei* (Vinke and Lips) (data from Cox 1966).

Parasite induced host mortalities

In the Basic model we assume that the rate of parasite induced host mortalities is linearly proportional to the number of parasites a host harbours (see Fig. 1(a)–(c)). The number of host deaths in a small interval of time δt, among those with i parasites, may

224   *Regulation and stability of host-parasite interactions. I*

then be represented as $\alpha i \, \delta t$, where α is a constant determining the pathogenicity of the parasite to the host; the corresponding death rate among hosts with i parasites is αi. The total rate of loss of hosts in a population of size $H(t)$ is therefore

$$\alpha H(t) \sum_{i=0}^{\infty} i \cdot p(i).$$

Here $p(i)$ is the probability that a given host contains i parasites. Clearly $p(i)$ will depend on i and on various parameters characterizing the parasite distribution within the host population. The above sum is, by definition, the average number of parasites per host at time t:

$$\sum_{i=0}^{\infty} i \cdot p(i) \equiv E_t(i) = P(t)/H(t).$$

In short , under the above assumptions the net rate of parasite induced host mortality is

$$\alpha P(t). \tag{1}$$

Parasite fecundity and transmission

The rate of production of transmission stages (such as eggs, spores or cysts) per parasite is defined as λ, leading to a net rate for the toal parasite population of

$$\lambda H(t) \sum_{i=0}^{\infty} i \cdot p(i) = \lambda P(t). \tag{2}$$

In the case of direct life cycle parasites where only a single species of host is utilized, the transmission stages will pass out of the host into the external environment and will survive in this habitat as resistant stages or free-living larvae awaiting contact with or ingestion by a member of the host population. While in the external habitat, they will be subject to natural mortalities due to senescence or predation and thus only a proportion of those released will be successful in gaining entry to a new host. The magnitude of this proportion will depend on the density of the host population in relation to other 'absorbers' of the transmission stages, and the proportion may be characterized by the transmission factor (cf. MacDonald 1961)

$$H(t)/(H_0 + H(t)).$$

Here H_0 is a constant which, when varied, inversely determines the efficiency of transmission. When $H(t)$ is large and H_0 small, the efficiency approaches unity, where all the transmission stages produced gain entry to the host population. Conversely when $H(t)$ is small and H_0 large only a small proportion are successful.

The net rate at which new parasites are aquired within the host population is thus

$$\lambda P(t) . H(t)/(H_0 + H(t)). \tag{3}$$

This term contains the assumption that transmission is virtually instantaneous, no time delay occurring due to developmental processes between the birth of a transmission stage and successful contact with a new host. In some parasite life cycles, transmission stages are immediately infective to a new host, but in the majority certain developmental processes have to occur before the stage becomes fully infective. The influence of time delays in the transmission term will be examined as a modification to the basic model in the following paper (Model F in May & Anderson 1978).

The assumptions incorporated in the two equation host-parasite model are most closely linked to direct life cycle parasitic species. The population dynamics of indirect life cycle species can also be interpreted in light of the model's predictions if the population processes acting on the intermediate host or hosts and the parasitic larval stages are subsumed into the transmission term (i.e. into the factor H_0 in eqn (3)). This is obviously a major simplifying assumption, particularly in respect to the time delays which will occur during a parasite's development in its intermediate host (for a review of this subject, see May 1977). The dynamical properties of the basic model with time delays will thus be more akin to the population dynamics of indirect life cycle parasites.

Parasite mortalities

The death rate for parasites within the host population has three components.

First, there are losses due to natural host mortalities. With the intrinsic per capita host mortality rate b, such parasite losses are at the net rate

$$b . H(t) \sum_{i=0}^{\infty} i . p(i) = b . P(t). \tag{4}$$

Second, there are losses from parasite induced host deaths, where the per capita host loss rate (discussed above) is taken to be αi. The consequent net mortality rate of parasites from this cause is

$$\alpha H(t) \sum_{i=0}^{\infty} i^2 p(i) \equiv \alpha H(t) E_t(i^2). \tag{5}$$

Here $E(i^2)$ is the mean-square number of parasites per host, the precise value of which depends on the form of the probability distribution of parasite numbers per host, $p(i)$. That is, $E(i^2)$ depends on the mean parasite load, $P(t)/H(t)$, and also on the parameter(s) that specify this distribution. Appendix 1 lists the values of this 'second moment' for some commonly used discrete probability distributions, giving $E(i^2)$ in terms of the mean parasite load and measures of over- and under-dispersion.

Third, there is a component of the parasite death rate generated by natural parasite mortality within the host. Assuming a constant per capita parasite intrinsic mortality rate μ, these losses make a net contribution of

$$\mu P(t) \tag{6}$$

to the overall parasite mortality rate. These 'natural' parasite mortalities include deaths due to host immunological responses, as well as more conventional losses from parasite senescence.

BASIC MODEL: DYNAMICS

The biological ingredients discussed above can now be drawn together to give two differential equations, one describing the rate of change of the host population,

$$dH/dt = (a-b)H - \alpha P \tag{7}$$

and the other describing the parasite population dynamics,

$$dP/dt = (\lambda PH/(H_0+H)) - (b+\mu)P - \alpha H E_t(i^2). \tag{8}$$

If the parasites are distributed independently randomly among hosts, eqn (8) then becomes (see Appendix 1):

226   *Regulation and stability of host-parasite interactions. I*

$$dP/dt = P(\lambda H/(H_0 + H) - (b + \mu + \alpha) - \alpha P/H). \qquad (9)$$

Eqns (7) and (9) readily yield the equilibrium $(dH/dt = dP/dt = 0)$ host and parasite population values, H^* and P^*. From eqn (7) the equilibrium mean parasite burden is

$$P^*/H^* = (a-b)/\alpha, \qquad (10)$$

whence from eqn (9) H^* is

$$H^* = \frac{H_0(\mu + \alpha + a)}{\lambda - (\mu + \alpha + a)}. \qquad (11)$$

Provided the host population's intrinsic growth rate is positive $(a-b>0)$, eqn (11) reveals that the parasites are capable of regulating the growth of the host population only if

$$\lambda > \mu + \alpha + a. \qquad (12)$$

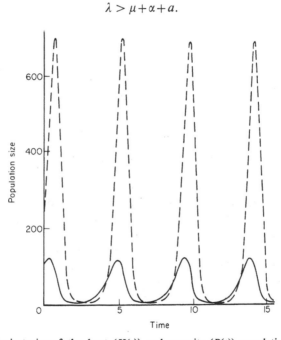

FIG. 2. The trajectories of the host $(H(t))$ and parasite $(P(t))$ populations in time, as predicted by the Basic model (eqns (7) and (9)). The model generates cyclic changes in both $H(t)$ and $P(t)$ and is neutrally stable. Solid line $H(t)$, stippled line $P(t)$. ($a = 3\cdot0$, $b = 1\cdot0$, $\mu = 0\cdot1$, $H_0 = 10\cdot0$, $\alpha = 0\cdot5$, $\lambda = 6\cdot0$).

This corresponds to the sensible requirement that the parasite 'birth rate' be in excess of the host birth rate (a) plus parasite death rates (both intrinsic, μ, and due to parasite induced host deaths, α). If this inequality (12) is not satisfied, the host population grows exponentially and will eventually be constrained by other regulatory processes such as finite resources. The parasite will also grow exponentially but at a slower rate than the host population and thus the mean number of parasites per host will tend to zero.

Equations (7) and (9) may be viewed as a general pair of predator-prey equations, and the following comments made. (i) In the 'prey' equation, (7), the 'predators' have a constant 'attack rate'; compared with the classic, but unrealistic, Lotka-Volterra model, this is a destabilizing feature, for it prevents the predators being differentially more effective at high prey densities. (ii) In the 'predator' equation, (9), we have a kind of logistic equation, with a predator carrying capacity' proportional to the prey density; this has a stabilizing effect (for a more full discussion and review along these general lines, see May (1976), ch. 4). In order to see how these countervailing tendencies are resolved, we need to make a stability analysis of the system of eqns (7) and (9).

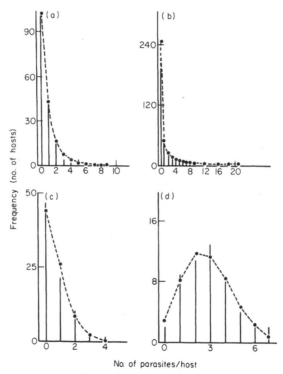

FIG. 3. Some examples of observed frequency distributions of parasite numbers per host which are empirically described by the negative binomial ((a) and (b)) or the Poisson ((c) and (d)) models. (a) The distribution of the tick *Ixodes trianguliceps* (Birula) on a population of the field mouse *Apodemus sylvaticus* (L) (data from Randolph 1975); (b) the distribution of the tapeworm *Caryophyllaeus laticeps* (Pallas) within a population of the bream *Abramis brama* (L.) (data from Anderson 1974a); (c) the distribution of larval stages of the nematode *Cammallanus oxycephalus* (Ward and Magath) in a population of young gizzard shad fish *Dorosoma cepedianum* (Day) sampled in mid summer (data from Stromberg and Crites 1974); (d) the distribution of the nematode *Ascaridia galli* in a population of chickens which had been artificially infected in the laboratory (data from Northam and Rocha 1958).

For this biologically derived pair of equations, a rigorous and fully nonlinear stability analysis may be given elegantly. This is done in Appendix 2. The outcome is surprising: the equilibrium point defined by eqns (10) and (11) is neutrally stable. That is, once perturbed from its equilibrium point the system oscillates with a period determined by the parameters of the model but with an amplitude dictated for ever after by the initial conditions of the displacement. The result is rigorously 'global', which is to say it holds

Abundance and the Rise of Conservation Ecology   527

228   *Regulation and stability of host-parasite interactions. I*

true for displacements of arbitrary magnitude (provided the initial $H_{(0)}$ and $P_{(0)}$ are positive!). Figure 2 illustrates the dynamical behaviour of such a host-parasite system.

This pathological neutral stability property means that the model defined by eqns (7) and (9) is *structurally unstable*: the slightest alteration in the form of the various underlying biological assumptions will precipitate the system either to stability (disturbances damping back to the equilibrium point) or to instability (oscillations growing

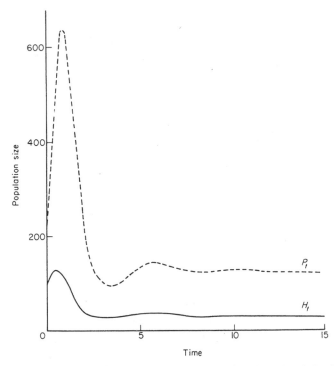

FIG. 4. Overdispersion of parasite numbers per host. The time dependent behaviour of the host ($H(t)$—solid line) and parasite ($P(t)$—stippled line) populations predicted by model A (eqns (7) and (13)). The populations exhibit damped oscillations to globally stable equilibria ($a = 3\cdot0, b = 1\cdot0, \mu = 0\cdot1, H_0 = 10\cdot0, \alpha = 0\cdot5, \lambda = 6\cdot0, k = 2\cdot0$).

until the host population begins to encounter resource limitation effects as discussed above, whereupon stable limit cycles typically ensue). Such structural instability makes the model biologically unrealistic. The Lotka-Volterra model (although having a different biological and mathematical form from eqns (7) and (9)) also has this pathological and structurally unstable property of neutral stability, for which it has been trenchantly criticized (e.g. May 1975).

The Basic model defined above is nevertheless very useful as a point of departure. We now proceed to modify it by introducing various kinds of density dependences, and of nonrandom parasite distribution among hosts. In each case, the relation between the underlying biology and the overall population dynamics is made clear by comparison with the razor's edge of the Basic model.

DISTRIBUTION OF PARASITES AMONG HOSTS

The parasitological literature contains a great deal of information concerning the distribution of parasite numbers within natural populations of hosts. Almost without exception these observed patterns are over dispersed, where a relatively few members of the host population harbour the majority of the total parasite population. It has become customary for parasitologists to fit the negative binomial model to such data, since this probability distribution has proved to be a good empirical model for a large number of observed parasite distributions. Figure 3 demonstrates the adequacy of this model in describing patterns of dispersion within host populations.

Crofton (1971a) used this observed constancy in the form of parasite distributions as a basis for a quantitative definition of parasitism at the population level. The author based this definition on the large number of parasitological processes which could in theory generate the negative binomial pattern. It is difficult, if not impossible, however, to try to reach conclusions about the biological mechanisms generating a particular distribution pattern by simply examining the resultant observed distribution of parasite numbers per host. A very large number of both biological and physical processes can generate the negative binomial model. Many of these processes have little relevance to the biologies of parasite life cycles while others are important to both parasitic and free-living organisms (Boswell & Patil 1971).

The precise mechanisms giving rise to the patterns shown in Fig. 3(a) and (b) are many and varied. Two major factors are most probably heterogeneity between members of the host population in exposure to infection (Anderson 1976a) and the influence of past experiences of infection on the immunological status of a particular host (Anderson 1976b). Such comments, however, are purely speculative since detailed experimental work is required to suggest generative processes which give rise to observed field patterns.

Consideration of the precise mechanisms which generate a particular pattern is not of overriding importance when examining the qualitative influence of overdispersion on the dynamical properties of a particular host-parasite interaction. We assume that heterogeneity in the distribution of parasite numbers per host is the rule rather than the exception and then proceed to analyse the consequences of such patterns on the population biology of the system.

From the purely phenomenological standpoint, the negative binomial has the virtue of providing a 1-parameter description of the degree of overdispersion, in terms of the parameter k: the smaller k, the greater the degree of parasite clumping. The distribution is discussed more fully in this light by Bradley & May (1977).

It is worth noting, however, that a few reports of random and undispersed distributions of parasites exist in the literature (Fig. 3(c) and (d)). These patterns are often observed either within laboratory populations (Northam & Rocha 1958; Anderson, Whitfield & Mills 1977), or within specific strata of a wild host population such as a particular age class (Anderson 1974b). In other cases random patterns may be observed at a particular point in time where the initial invasion of a host population has been captured by a sampling programme (Stromberg & Crites 1974).

MODEL A: NONRANDOM PARASITE DISTRIBUTIONS

Overdispersed distributions

If the parasites are distributed among the hosts according to a negative binomial, we can use Appendix 1 to get an explicit expression for $E(i^2)$ in eqn (8), which then reads

230   *Regulation and stability of host-parasite interactions. I*

$$dP/dt = P\{\lambda H/(H_0+H)-(\mu+b+\alpha)-\alpha(k+1)P/(kH)\} \tag{13}$$

As mentioned above, k is the parameter of the negative binomial distribution which gives an inverse measure of the degree of aggregation or contagion of the parasites within the hosts.

The equilibrium host and parasite population values then follow from eqns (7) and (13):

$$H^*/P^* = (a-b)/\alpha \tag{14}$$

$$H^* = \frac{H_0\{\mu+a+\alpha+(a-b)(k+1)/k\}}{\lambda-\{\mu+a+\alpha+(a-b)(k+1)/k\}}. \tag{15}$$

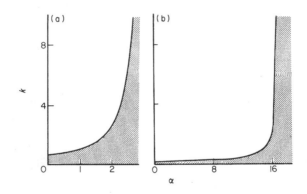

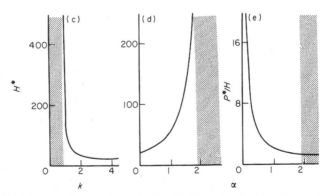

Fɪɢ. 5. Model A—overdispersed parasite distributions. Graphs (a) and (b)—the solid lines denote the boundaries in the k and α parameter space which separate regions in which parameter values give rise to globally stable parasite regulated host equilibria (unshaded areas), and unregulated host population growth (shaded areas). These boundaries are shown in terms of k and α for two values of λ. In graph (a) $\lambda = 6\cdot0$, and in (b) $\lambda = 20\cdot0$. In both (a) and (b), $a = 3\cdot0$, $b = 1\cdot0$, $\mu = 0\cdot1$, $H_0 = 10\cdot0$. In graphs (c), (d) and (e) the solid lines denote the equilibrium size of the host population, H^* (c) and (d)), and the mean parasite burden, P^*/H^* (e) for various values of k and α. The shaded regions denote areas in which the parameter values of the model lead to unregulated host population growth. The parameter values are as indicated for graphs (a) and (b) except that in (c), $\alpha = 0\cdot5$, $\lambda = 6\cdot0$, and in (d), and (e), $k = 2\cdot0$, $\lambda = 6.0$.

This equilibrium point can exist only if

$$\lambda > \mu+a+\alpha+(a-b)(k+1)/k. \tag{16}$$

The stability of the equilibrium is studied in Appendix 3, where it is shown that disturbances undergo damped oscillations back to the equilibrium population values, provided
only the aggregation parameter k is finite and positive. (In the limit $k \to \infty$, which
corresponds to the Poisson limit, the damping times becomes infinitely large, and we
recover the neutral stability of the Basic model.)

Thus if eqn (16) is satisfied, the host population is effectively regulated by the parasites.
Figure 4 illustrates the dynamical behaviour of such a system.

If the inequality (16) is not fulfilled, then (as discussed previously) the host population
grows exponentially until it encounters environmental carrying capacity limitations. The
parasite population will also grow exponentially but at a slower rate than the hosts, and
thus the mean parasite burden decreases in size.

The criterion (16) which determines the boundary between host populations that can
be regulated by their parasites, and those that cannot, is illustrated in Fig. 5.

Figure 5(a) and (b) are two slices through the k-α parameter space, for two different
values of λ (and other quantities held constant); they show that as λ, the production rate
of transmission stages, increases a larger range of k and λ values lead to a parasite
regulated host equilibrium. When the parasite's distribution is highly overdispersed
($k \to 0$), regulation is difficult to achieve for any value of the rate of parasite induced host
mortalities (α). As α increases, the degree of overdispersion must decrease, otherwise too
many parasites are lost due to parasite induced host mortalities.

An indication of the influence of α and k on the size of the equilibrium host and parasite populations, H^* and P^*, is given in Figure 5 (c), (d) and (e). Figure 5 (d) appears
odd at first glance: an increase in the rate of parasite induced host mortality
makes for an increase in H^*. The reason is that those hosts which die from parasite
infection harbour above average parasite burdens; such deaths thus have relatively
more effect on the parasite population than on the host population. As α increases,
this effect becomes increasingly pronounced, leading to a decrease in the mean
parasite burden per host (Fig. 5(e)) and an increase in H^* (Fig. 5(d)). Eventually,
for α sufficiently large, the parasites can no longer regulate the host population. Notice
also that as k increases, the parasites are spread more evenly, and hence the net rate of
loss of hosts due to parasite induced mortality rises; i.e. H^* decreases as k increases
(Fig. 5(c)).

Quantitative estimates of the parameters in eqn (16) are not easy to come by for
natural populations.

Table 2 lists reported values of k for a variety of parasite species. The numbers show
that many parasites, particularly helminths, are highly overdispersed within their host
populations, the majority of values tending to be less than 1·0. This typically high degree
of overdispersion tends to confer stability; it also suggests that net losses from host
populations due to parasite infections may be low, since only a few hosts harbour heavy
infections. This is a result of the first importance, to which we will return in the concluding discussion.

The rate of production of transmission stages by the parasite, λ, must exceed the sum
of the hosts reproductive rate, a, plus the host intrinsic growth rate multiplied by $(k+1)/k$,
plus μ and α. Particularly if k is small, this requires that the parasite must have a much
higher reproductive rate than the host, which accords with traditional beliefs. Reproductive rates of parasitic species, particularly protozoa and helminths, are invariably
high and almost without exception very much greater than the host's potential for
reproduction. For example, at the top end of the spectrum lie certain namatode species

Regulation and stability of host-parasite interactions. I

TABLE 2. Values of the negative binomal parameter k observed in natural populations of hosts and parasites

Taxonomic group of parasite	Parasite species	Host species	Range of k values	Author(s)
Platyhelminthes, Monogenea	*Diclidophora denticulata* (Olsson)	*Gadus virens* (L.)	0·7	Frankland (1954)
Digenea	*Diplostomum gasterostei* (Williams)	*Gasterosteus aculeatus* (L.)	0·1-0·7	Pennycuick (1971)
Cestoda	*Caryophyllaeus laticeps* (Pallas)	*Abramis brama* (L.)	0·1-0·5	Anderson (1974a)
	Schistocephalus solidus (Muller)	*Gasterosteus aculeatus*	0·7-2·4	Pennycuick (1971)
Nematoda	*Chandlerella quiscoli* (von Linstow)	*Culicoides crepuscularis* (Malloch)	0·5	Schmid & Robinson (1972)
	Toxocara canis (Werner)	*Vulpes vulpes* (L.)	0·5	Watkins & Harvey (1942)
	Ascaridia galli (Schrank)	*Gallus gallus* (L.)	0·7	Northam & Rocha (1958)
Acanthocephala	*Polymorphus minutus* (Groeze)	*Gammarus pulex* (L.)	0·6-3·1	Crofton (1971a)
	Echinorhynchus clavula (Dujardin)	*Gasterosteus aculeatus* (L.)	0·07-0·5	Pennycuick 1971
Arthropoda Copeopoda	*Lepeophtheirus pectoralis* (Muller)	*Pleuronectes platessa* (L.)	0·3-10·0	Boxhall (1974)
	Chondracanthopsis nodosus (Muller)	*Sebastes marinus* (L.)	0·6	Williams (1963)
	Chondracanthopsis nodosus	*Sebastes mentella* (L.)	0·2	Williams (1963)
Arachnida	*Ixodes trianguliceps* (Birula)	*Apodemus sylvaticus* (L.)	0·04-0·4	Randolph (1975)
	Liponysue bacoti (Hirst)	*Rattus rattus* (L.)	0·2	Cole (1949)
Insecta	*Pediculus humanus capitis* (L)	*Homo sapiens*	0·14	Buxton (1940)

such as *Haemonchus contortus* (Rudolphi), a parasite of sheep, which is capable of producing 10 000 eggs per day (Crofton 1966).

Estimates of α, the rate of parasite induced host mortalities, are more difficult to extract from the literature. As a rather broad generalization is appears to be widely accepted in the parasitological literature that the majority of protozoan and helminth parasites do not cause the death of many hosts in natural populations. However, where quantitative rate estimates are available from laboratory experiments (Fig. 1), it appears as though many parasite species have the capability of causing high host death rates. This is particularly apparent in parasite life cycles which utilise invertebrate hosts. Infected individuals within a host population have extremely poor survival characteristics when compared with uninfected hosts (Fig. 6). In natural populations, counter to popular opinion, parasites may play a crucial role in regulating the size of their host populations due to high parasite induced host mortality rates, coupled with small k values.

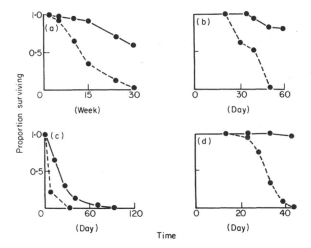

FIG. 6. Some examples of the survival characteristics of infected hosts as compared with uninfected hosts. (a) Snail host *Australorbis glabratus* (Say) infected with larval stages of the digenean *Schistosoma mansoni* (Sambon) (data from Pan 1965); (b) snail host *Lymnaea gedrosina* infected with larval stages of the digenean *Ornithobilharzia turkestanicum* (data from Massoud 1974); (c) mosquito *Aedes aegypti* (L.) infected with microfilariae of the nematode *Dirofilaria immitis* (Leidy) (data from Kershaw, Lavioipierre & Chalmers 1953); (d) snail host *Australorbis glabratus* infected with the nematode *Daubaylia potimaca* (Chitwood and Chitwood) (data from Chernin 1962) (solid lines—uninfected hosts, stippled lines—infected hosts).

A present, however, it is difficult to substantiate this conjecture, because of the difficulty in assessing α in natural populations. Our model suggests that a rough, indirect estimate of α may be obtained from eqn (14):

$$\alpha = (a-b)(H^*/P^*). \qquad (17)$$

That is, the rate α is equal to the natural intrinsic growth rate of the host population divided by the mean parasite burden P^*/H^*. This is a useful result in itself. It suggests that when overdispersed burdens of parasites per host, with small mean, are observed in the field, then α is likely to be relatively high.

Underdispersed distributions

As mentioned earlier, parasites may occasionally be under dispersed, i.e. more evenly distributed than if they were independently randomly distributed among hosts. Such underdispersed distributions may be described by a positive binomial, with a parameter k' that is inversely proportional to the degree of under-dispersion: in the limit $k' \to 0$, all hosts carry identical parasite burdens.

As shown in Appendix 1, the appropriate version of eqn (8) is obtained by simply replacing k with $-k'$ in eqn (14). It is easily shown (see Appendix 3) that such a system, which amounts to model A with a negative value of k, can have *no* stable equilibrium.

In short, an overdispersed pattern of parasite numbers per host can enable the parasites stably to regulate the host population. The exact form of the pattern of overdispersion is immaterial; the above analysis based on the negative binomial can be repeated for other qualitatively similar distributions such as the Neyman type A, with the same conclusion.

MODEL B: NONLINEAR PARASITE INDUCED MORTALITIES

So far, the parasite induced host mortality rate has been taken to be linearly proportional to the parasite burden (cf. Figs 1(a)–(c)). Often, however, the relationship between host death rate and parasite burden is of a more severe form, as indicated in Fig. 1(d)–(f).

In model B, we examine the dynamical consequences of a more steeply density dependent parasite induced death rate. For specificity, we typify such steeper density dependence by assuming that the parasite induced mortality rate varies as the square of the number of parasites in a host, i.e. as αi^2. By recapitulating the arguments that led to eqns (7) and (8), we have now

$$dH/dt = (a-b)H - \alpha HE(i^2) \tag{18}$$

$$dP/dt = \lambda PH/(H_0 + H) - (\mu + b)P - \alpha H\dot{E}(i^3). \tag{19}$$

The exact form of the average values of i^2 and i^3 depends on the parasite distribution among hosts. In Appendix 3, we indicate the equilibrium population values, and their stability character, under the various assumptions that the parasites are overdispersed (negative binomial), independently randomly distributed (Poisson), and underdispersed (positive binomial).

The nonlinearly severe parasite induced host mortality has two consequences, both of which are illustrated by Fig. 7.

On the one hand, the effect enhances the dynamic stability of an equilibrium point, provided one exists. Thus if the parameters are such that equilibrium populations H^* and P^* exist, then this point is stable for overdispersed or Poisson parasite distributions (the latter in contrast to the Basic model), and also for underdispersed distributions so long as k' is not too small (in contrast to model A, where all such underdispersed cases are unstable).

On the other hand, the domain of parameter space for which the equilibrium exists (i.e. satisfying the analogue of eqns (12) and (16)) is always smaller for model B than for the corresponding model A. This can be seen by comparing Figs 5 and 7.

All this can be explained intuitively. Compared with model A, the more steeply severe parasite induced host mortality makes it relatively harder for the parasite to check the host population's intrinsic propensity to growth. But if it can do so, this density dependence results in relatively faster damping of perturbations.

MODEL C: DENSITY DEPENDENCE IN PARASITE POPULATION GROWTH

The growth of a parasite population within a single host is often constrained by the influence of density dependent death or reproductive processes. (see Anderson 1976c). Such regulatory mechanisms may be due to either parasite induced host immunological responses or intraspecific competition for finite resources such as space or nutrients

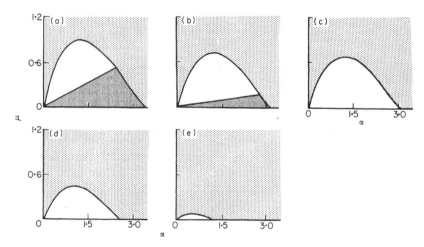

FIG. 7. Model B—non-linear parasite induced mortality rates. The influence of the distribution pattern of the numbers of parasites per host, and the rate of parasite induced host mortalities (α) on the numerical size of the host equilibrium population and its global stability properties. The solid line denotes the values of H^*, while the unshaded regions denote areas of parameter values in which the host population growth is regulated by the parasite and globally stable equilibria arise. The lightly shaded areas define regions where the parasite fails to regulate the growth of the host population; the darkly shaded areas define regions where the parasite regulates host population growth but the equilibria generated are unstable. In (a) and (b), the parasites are underdispersed following the positive binomial model. The value of k' is 6·00 in (a) and 20·0 in (b); in graphs (c) the parasites are randomly distributed (Poisson model); in (d) and (e) the parasites are overdispersed following the negative binomial model where in (d) $k = 8·0$ and in (e) $k = 2·0$ ($a = 3·0$, $b = 1·0$, $\mu = 0·1$, $H_0 = 10·0$, $\lambda = 6·0$).

within or on the host. The severity of an immunological response by the host is usually functionally related to the degree of antigenic stimulation received (number of parasites). Such responses, whether humoral or cell mediated, tend to increase the death rate of a parasite population and/or reduce its reproductive potential (Anderson & Michel 1977; Bradley 1971). Some examples of density dependent parasite death rates for helminth species in vertebrate hosts are illustrated in Fig. 8.

The influence of such processes on the dynamical properties of host-parasite interactions may be examined by appropriate modification of the basic eqns (7) and (8). For specificity, we assume the density dependence to be such that natural parasite mortality occurs at a rate μi^2, in hosts harbouring i parasites. The model is then eqn (7) for dH/dt, together with

$$dP/dt = \lambda PH/(H_0+H) - bP - (\mu+\alpha)HE(i^2). \qquad (20)$$

Apart from this modification, the model is exactly as for model A: again the detailed form of eqn (20) will depend on the pattern of parasite distribution.

The dynamical behaviour of model C is elucidated in Appendix 3. For a negative binomial distribution (with parameter k), equilibrium host and parasite populations exist only if

$$\lambda > b+\mu+\alpha+(\alpha+\mu)(a-b)(1+k)/(\alpha k). \qquad (21)$$

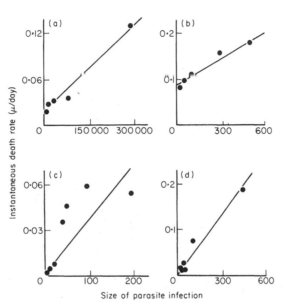

FIG. 8. Some examples of the relationship between the instantaneous parasite 'natural' death rate (μ) and parasite density (i) within individual hosts. The solid lines are the least squares best fit linear models of the form $\mu\ (i) = \hat{a}+\hat{b}i$ (a) Calves infected with the gut nematode *Ostertagia ostertagi* (Stiles); $\hat{a} = 0\cdot0171$, $\hat{b} = 0\cdot0308 \times 10^{-5}$ (rate/day) (data from Anderson & Michel 1977). (b) Chickens infected with the fowl nematode *Ascaridia lineata* (Schneid); $\hat{a} = 0\cdot0856$, $\hat{b} = 0\cdot00019$ (rate/day) (data from Ackert, Graham, Xolf & Porter 1931). (c) Rats infected with the tapeworm *Hymenolepis diminuta* (Rud); $\hat{a} = 0.00636$, $\hat{b} = 0\cdot00032$ (rate/day) (data from Hesselberg & Andreassen 1975). (d) Rats infected with the nematode *Heterakis spumosa* (Schneider); $\hat{a} = 0\cdot0265$, $\hat{b} = 0\cdot000037$ (rate/day) (data from Winfield 1932).

For independently randomly distributed parasites, the Poisson result follows as the limit $k \to \infty$ in eqn (21); for underdispersion (positive binomial with parameter k') one simply replaces k with $-k'$ in eqn (21). If the equilibrium exists, it is stable for overdispersed or Poisson distributions, and also for underdispersion provided the parameter k' is not too small (explicitly, provided $k' > (\alpha+\mu)/\mu$).

The general pattern of relationship between model C and model A is similar to that between model B and model A. As is made clear by the comparison between eqns (16) and (21), λ must be relatively larger if an equilibrium is to exist (i.e. if parasites are to be able to check host population growth) in model C; this is particularly marked if over-

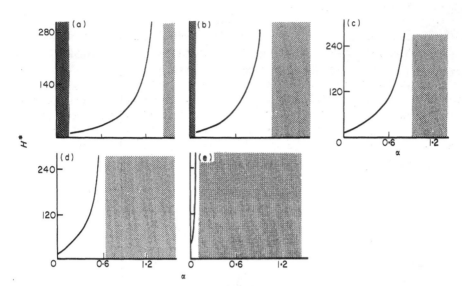

FIG. 9. Model C—density dependent constraints on parasite population growth; The solid lines enclose unshaded regions in which the parameter values lead to parasite regulated host population equilibria which are globally stable. These boundaries are shown in terms of μ and α for various patterns of dispersion of parasite numbers per host; the lightly shaded regions indicate the parameter values which give rise to unregulated host population growth. In the darkly shaded areas the parasite regulates host population growth but the equilibria produced are unstable. (a) Positive binomial, $k' = 6\cdot0$; (b) positive binomial, $k = 20\cdot0$; (c) Poisson; (d) negative binomial, $k = 4\cdot0$; (e) negative binomial, $k = 1.0$; ($a = 3\cdot0$, $b = 1\cdot0$, $H_0 = 10\cdot0$, $\lambda = 6\cdot0$).

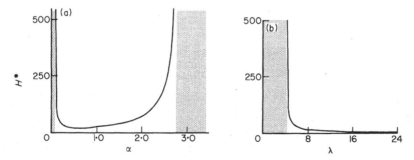

FIG. 10. Model C—density dependent constraints on parasite population growth. (a) The influence of the parameter α on the numerical size of the host population equilibrium H^* (parasites randomly distributed—Poisson model); the shaded region defines parameter values which lead to unregulated host population growth, while in the unshaded regions the parasite regulated host equilibria are globally stable. (b) The influence of the rate of parasite reproduction, λ, on H^*; the shaded and unshaded regions are as defined for (a); ($a = 3\cdot0$, $b = 1\cdot0$, $\mu = 0\cdot1$, $H_0 = 10\cdot0$, $\lambda = 6\cdot0$, $\alpha = 0\cdot5$).

Abundance and the Rise of Conservation Ecology   537

238   *Regulation and stability of host-parasite interactions. I*

dispersion is high (small k). But if the equilibrium can exist, for given k, it is more stable in model C than in model A.

These remarks are further bourne out by Figs 9 and 10. The stable domain in the μ, α parameter space increases in size when the pattern of dispersion of the parasites within the host population moves from regular to random (Fig. 9(a)–(c)), and then decreases as the pattern changes from random to aggregated (Fig. 9 (c)–(e)). When overdispersion is marked (k small), as is the case for many parasitic species (Table 2), the region of parameter space in which the parasite regulates the host population is rather small (Fig. 9(e)). The model's predictions therefore suggest that when the parasite population is tightly controlled by density dependent constraints (i.e. when μ is large) and parasites are overdispersed within the host population, other factors than parasite induced host mortalities will tend to stabilise and regulate host population growth.

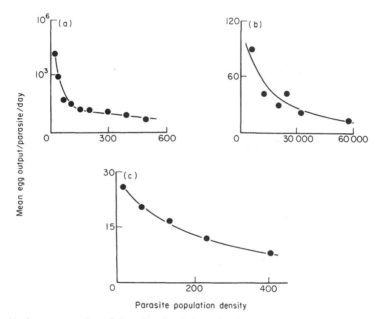

Fig. 11. Some examples of functional relationships between the rate of production of parasite transmission stages (λ) and the parasite population density (i) within individual hosts; the dots are observed points while the solid lines are fitted by eye. (a) Laboratory mice infected with the tapeworm *Hymenolepis nana* (Siebold) (data from Ghazal & Avery 1974); (b) calves infected with *Ostertagia ostertagi* (data from Michel 1969); (c) sheep infected with the fluke *Fasciola hepatica* (data from Boray 1969).

Mammalian hosts which possess well developed immunological responses to parasitic invasion, creating tight density dependent controls on parasite population growth, may well fall into this category.

When parasites are independently randomly distributed, however, the main effect of the density dependence of model C is to help stabilize the system. In the case of Poisson distributed parasites, a further insight, concerning the influence of α upon H^*, is illustrated in Fig. 10(a): too high or too low values of α lead to exponential runaway of the host population.

The above discussion has been for density dependence in the natural parasite mortality rate. For many host-parasite associations, the 'fecundity' λ is also known to be a function of the parasite population density due, as in the case of the death rate, to either immunological processes or intraspecific competition. Some quantitative examples of such responses are shown in Fig. 11 for helminth parasites of vertebrates. The inclusion of such processes in the basic model leads to qualitatively similar dynamical properties to those outlined above for density dependent parasite death rates.

CONCLUSIONS

Our population models of host-parasite interactions are all characterized by the central assumption that parasites cause host mortalities. In particular, we have assumed that the net rate of such mortalities is related to the average parasite burden of the members of a host population, and therefore to the statistical distribution of the parasites within a host population. We regard the inducement of host mortalities and/or reduction in host reproductive potential as a necessary condition for the classification of an organism as parasitic.

Species which exhibit such characteristics may in certain circumstances play an important role in regulating or controlling the growth of their host population. In such cases, the parasite will play an analogous role to a predator which suppresses the growth of its prey population. We have demonstrated that three specific categories of population processes are of particular significance in stabilizing the dynamical behaviour of a host-parasite interaction and enhancing the regulatory influence of the parasite.

The *first* of these categories of population processes concerns the functional relationships between the rate of parasite induced host deaths and the parasite burden of individual hosts. It is interesting to compare the significance of these relationships on the dynamical properties of a given host-parasite association, with the role played by functional responses in predator-prey, and host-insect parasitoid interactions (Solomon 1949; Holling 1959; Hassell & May 1973). The presence of a complex sigmoid function response between the attack rate of a predator and the density of prey (type III response in the terminology of predator-prey interactions as recently reviewed by Murdoch & Oaten 1975) may, or may not, have a stabilizing influence on the dynamics of such interactions. There will be a range of prey densities over which the death rate of the prey imposed by the predator is an increasing function of prey density and hence such a response will be stabilizing since it acts in a density dependent manner. However, in contrast, if the prey death rate is constant over a wide range of prey densities (as in type II responses), then the predator will not play a major stabilizing role in the dynamics of the prey. In such cases other constraints, such as intraspecific competition for finite resources amongst the prey population, will stabilise the interaction.

We have demonstrated, in the case of host-parasite associations, that when the parasites are randomly distributed within the host population, a linear relationship between host death rate and parasite burden gives rise to the pathological condition of neutral stability (Basic model). If however, the rate of parasite induced host mortalities is an increasing function of parasite burden, perhaps following a power law or exponential form, random distributions of parasites may lead to globally stable equilibria where the parasite is effectively regulating the growth of its host population (model B). Thus, for a given distribution pattern of parasites, certain types of functional response may stabilize the dynamics, while others lead to instabilities as in the case of predator-prey associations.

240   *Regulation and stability of host-parasite interactions. I*

The *second* category of population processes, which influence the regulatory role of parasites, concerns the statistical distribution of parasite numbers per host. When the rate of parasite induced host mortalities is linearly dependent on the number of parasites harboured, we have demonstrated that parasites may regulate host population growth provided they exhibit an overdispersed pattern of distribution within the host population (model A). The accumulating information in the literature on the overdispersed character of parasite distributions in natural populations of hosts (*vide* Table 2) lends support to the importance of this insight. The stabilizing influence of parasite contagion is somewhat analogous to the influence of aggregated distributions of insect parasitoids on the stability of host-parasitoid associations (Hassell & May 1974). Parsitoid aggregation was shown, in certain circumstances, to stabilize models which otherwise would have been quite unstable.

The *third* category of processes which enhance the regulatory role played by a parasite act within each individual parasite population (or subpopulation) harboured in each member of the host population. Model C demonstrated that density dependent constraints on parasite population growth, caused either by intraspecific competition for finite resources or immunological attack by the host, can in the absence of overdispersion or non-linear functional responses stabilise the dynamics of an interaction.

On the other hand, there are features of host-parasite interactions which tend to destabilize the system. These are discussed in the following paper (May & Anderson 1978), and a general review of pertinent field and laboratory data on host parasite associations is then presented.

ACKNOWLEDGMENTS

We thank the Natural Environment Research Council for support of R.M.A. and the National Science Foundation for support of R.M.M. (Grant No DEB77 01563).

REFERENCES

Ackert, J. E., Graham, G. L., Nolf, L. O. & Porter, D. A. (1931). Quantitative studies on the administration of variable numbers of nematode eggs (*Ascaridia lineata*) to chickens. *Transactions of the American Microscopical Society*, **50**, 206–14.

Askew, R. R. (1971). *Parasitic Insects*. Heinemann, London

Anderson, R. M. (1974a). Population dynamics of the cestode *Caryophyllaeus laticeps* (Pallas, 1781) in the bream (*Abramis brama L.*). *Journal of Animal Ecology*, **43**, 305–21.

Anderson, R. M. (1974b). An analysis of the influence of host morphometric features on the population dynamics of *Diplozoon paradoxum* (Nordmann, 1832). *Journal of Animal Ecology*, **43**, 873–87.

Anderson, R. M. (1976a). Seasonal variation in the population dynamics of *Caryophyllaeus laticeps*. *Parasitology*, **72**, 281–305.

Anderson, R. M. (1976b). Some simple models of the population dynamics of eucaryotic parasites *Mathematical Models in Medicine* (Ed. by J. Berger, W. Buhler, R. Repges and P. Tautu). *Lecture Notes in Biomathematics*. **11**, 16–57. Springer-Verlag, Berlin.

Anderson, R. M. (1976c). Dynamic aspects of parasite population Ecology. *Ecological Aspects of Parasitology* (Ed. by C. R. Kennedy) pp. 431–62 North-Holland Publishing Company, Amsterdam.

Anderson, R. M. & Michel J. F. (1977). Density dependent survival in populations of *Osteragia ostertagi*. *International Journal of Parasitology*, **7**, 321–9.

Anderson, R. M., Whitfield, P. J. & Mills, C. A. (1977). An experimental study of the population dynamics of an ectoparasitic digenean, *Transversotrema patialense* (Soparker): The cercarial and adult stages. *Journal of Animal Ecology*, **46**, 555–80.

Beddington, J. R., Hassell, M. P. & Lawton, J. H. (1976). The components of arthropod predation. II. The predator rate of increase. *Journal of Animal Ecology*, **45**, 165–86.

Boray, J. C. (1969). Experimental Fascioliasis in Australia. *Advances in Parasitology*, **7**, 95–210.

Boswell, M. T. & Patil, G. P. (1971). Chance mechanisms generating the negative binomial distribution. *Random Counts in Models and Structures* (Ed. by G. P. Patil) pp. 21–38. Pennsylvania State University Press, London.

Boxhall, G. A. (1974). The population dynamics of *Lepeophtheirus pectoralis* (Muller): dispersion pattern. *Parasitology*, **69**, 373–90.

Bradley, D. J. (1971). Inhibition of *Leishmania donovani* reproduction during chronic infections in mice. *Transactions of the Royal Society of Tropical Medicine and Hygiene*, **65**, 17–8.

Bradley, D. J. & May, R. M. (1977). Consequences of helminth aggregation for the dynamics of schistosomiasis. *Transactions of the Royal Society of Tropical Medicine and Hygiene*, (in press).

Buxton, P. A. (1940). Studies on populations of head-lice (*Pediculus humanus capitis*: Anoplura). III. Material from South India. *Parasitology*, **32**, 296–302.

Chernin, E. (1962). The unusual life-history of *Daubaylia potomaca* (Nematoda: Cephalobidae) in *Australorbis glabratus* and in certain other freshwater snails. *Parasitology*, **52**, 459–81.

Cole, L. C. (1949). The measurement of interspecific association *Ecology*, **30**, 411–24.

Cox, F. E. G. (1966). Acquired immunity to *Plasmodium vinckei* in mice. *Parasitology*, **56**, 719–32.

Crofton, H. D. (1966). *Nematodes*. Hutchinson, London

Crofton, H. D. (1971a). A quantitative approach to parasitism. *Parasitology*, **62**, 179–94.

Crofton, H. D. (1971b). A model of host-parasite relationships. *Parasitology*, **63**, 343–64.

Dogiel, V. A. (1964) *General Parasitology*. Oliver and Boyd, Edinburgh.

Forrester, D. J. (1971). *Heligmosomoides polygyrus* (= *Nematospiroides dubius*) from wild rodents of Northern California: Natural infections, host specificity, and strain characteristics. *Journal of Parasitology*, **57**, 498–503.

Frankland, H. M. T. (1954). The life history and bionomics of *Diclidophora denticulata* (Trematoda: Monogenea). *Parasitology*, **45**, 313–51.

Ghazal, A. M. & Avery, R. A. (1974). Population dynamics of *Hymenolepis nana* in mice: fecundity and the 'crowding effect'. *Parasitology*, **69**, 403–15.

Hassell, M. P. (1976). Arthropod predator-prey systems. Ch. 5 in May 1976 (*op. cit*).

Hassell, M. P., Lawton, J. H. & Beddington, J. R. (1976). The components of arthropod predation. I. The prey death rate. *Journal of Animal Ecology*, **45**, 135–64.

Hassell, M. P. & May, R. M. (1973). Stability in insect host-parasite models. *Journal of Animal Ecology*, **42**, 693–726.

Hassell, M. P. & May R. M. (1974). Aggregation of predators and insect parasites and its effect on stability. *Journal of Animal Ecology*, **43**, 567–94.

Hayes, T. J. Bailer, J. & Mitrovic, M. (1973). The pattern of mortality in mice experimentally infected with *Fasciola hepatica*. *International Journal of Parasitology*, **3**, 665–9.

Hesselberg, C. A. & Andreassen, G. (1975). Some influences of population density on *Hymenolepis diminuta* in rats. *Parasitology*, **71**, 517–23.

Holling, C. S. (1959). Some characteristics of simple types of predation and parasitism. *Canadian Entomologist*, **91**, 395–8

Hunter, G.C. & Leigh, L. C. (1961). Studies on the resistance rats to the nematode *Nippostrongylus muris* (Yokogawa, 1920). *Parasitology*, **51**, 347–51.

Kershaw, W. E., Lavoipierre, M. M. J. & Chalmers, T. A. (1953). Studies on the intake of micro filarial by their insect vectors, their survival, and their effect on the survival of their vectors. I. *Dirofilaria immitis* and *Aedes aegypti*. *Annals of Tropical Medicine and Parasitology*, **47**, 207–24.

Lancinani, C. A. (1975). Parasite induced alterations in host reproduction and survival. *Ecology*, **56**, 689–95.

Macdonald, G. (1961). Epidemiologic models in studies of vector-bourne diseases. *Public Health Reports, Wash. D.C.* **76**, 753–764 (Reprinted as Ch. 20 in *Dynamics of Tropical Disease*. (Ed. by L. J. Bruce-Chwatt and V. J. Glanville) Oxford University Press, Oxford, 1973).

Massoud, J. (1974). The effect of variation in miracidial exposure dose on laboratory infections of *Ornithobilharzia turkestanicum* in *Lymnea gedrosiana*. *Journal of Helminthology*, **48**, 139–44.

May, R. M. (ed.) (1976). *Theoretical Ecology: Principles and Applications*. Saunders, Philadelphia and Blackwell Scientific Publications, Oxford.

May, R. M. (1977). Togetherness among schistosomes: its effect on the dynamics of the infection. *Mathematical Biosciences*, (in press).

May, R. M. & Anderson, R. M. (1978). Regulation and stability of host-parasite population interactions: II. destabilizing processes. *Journal of Animal Ecology*, **47**, 249–67.

Michel, J. F. (1969). The regulation of egg output by *Ostertagia ostertagi* in calves infected only once. *Parasitology*. **59**, 767–74.

Murdoch, W. W. & Oaten, A. (1975). Predation and population stability. *Advances in Ecological Research*, **9**, 1–131.

Northam, J. I. & Rocha, U. F. (1958). On the statistical analysis of worm counts in chickens. *Experimental Parasitology*, **7**, 428–438.

Pan, C. (1965). Studies on the host-parasite relationship between *Schistosoma mansoni* and the snail *Australorbis glabratus*. *American Journal of Tropical Medicine & Hygiene*, **14**, 931–75.

Pennycuik, L. (1971). Frequency distributions of parasites in a population of three-spined sticklebacks,

Gasterosteus aculeatus L., with particular reference to the negative binomial distribution. *Parasitology*, **63**, 389–406.

Randolph, S. E. (1975). Patterns of distributions of the tick *Ixodes trianguliceps* Birula, on its host. *Journal of Animal Ecology*, **44**, 451–74.

Schmid, W. D. & Robinson, E. J. (1972). The pattern of a host-parasite distribution. *Journal of Parasitology*, **57**, 907–10.

Solomon, M. E. (1949). The natural control of animal populations. *Journal of Animal Ecology*, **18**, 1–35.

Starr, M. P. (1975). A generalised scheme for classifying organismic associations. *Symbiosis* (Ed. by P.H. Jennings and D. L. Lee), pp. 1–20. Cambridge University Press, London.

Stromberg, P. C. & Crites, J. L. (1974). Survival, activity and penetration of the first stage larvae of *Cammallanus oxycephalus*, Ward and Magath, 1916. *International Journal of Parasitology*, **4**, 417–21.

Watkins, C. V. & Harvey, L. A. (1942). On the parasites of silver foxes on some farms in the South West. *Parasitology*, **34**, 155–79.

Williams, I. C. (1963). The infestations of the redfish *Sebastes marinus* (L) and *S. mentella* Travin (Scleroparei; Scorpaenidae) by the copeopods *Peniculus clavatus* (Muller), *Sphyrion lumpi* (Kroyer) and *Chondracanthopsis nodosus* (Muller) in the eastern North Atlantic. *Parasitology*, **53**, 501–26.

Winfield, G. F. (1932). Quantitative studies on the rat nematode *Heterakis spumosa*, Schneider, 1866. *American Journal of Hygiene*, **17**, 168–228.

(*Received* 16 *May* 1977)

APPENDIX 1

This appendix lists the probability generating functions (p.g.f.'s), $\Pi(Z)$, of the probability $p(i)$ of observing (i) parasites in a host, for three discrete probability distributions. In addition, the expectations of i^2 and i^3 are outlined for each of the distributions in terms of a few parameters such as the mean number of parasites per host, m, and measures of overdispersion k and underdispersion k'.

(a) *Positive binomial*

$$\Pi(Z) = (q - pZ)^{k'} \tag{1.1}$$

where p is the probability of a successful infection occurring, and $q = 1 - p$. The parameter k' represents the maximum number of parasites a host can harbour (see Anderson 1974b).

The mean and variance of this distribution are

$$m \equiv E(i) = k'p \tag{1.2}$$

$$\mathrm{var}(i) = k'pq. \tag{1.3}$$

The expectations of i^2 and i^3 are

$$E(i^2) = (k'p)^2 + k'pq. \tag{1.4}$$

that is,

$$E(i^2) = m^2(k'-1)/k' + m. \tag{1.4a}$$

$$E(i^3) = (k'p)^3(k'-1)(k'-2)/k'^2 + 3(k'p)^2(k'-1)/k' + k'p. \tag{1.5}$$

(b) *Poisson*

$$\Pi(Z) = \exp\{m(Z-1)\} \tag{1.6}$$

where m is the sole parameter of the distribution and

$$E(i) = m \tag{1.7}$$

$$\mathrm{var}(i) = m. \tag{1.8}$$

The expectations of i^2 and i^3 are

$$E(i^2) = m^2 + m \tag{1.9}$$

$$E(i^3) = m^3 + 3m^2 + m. \tag{1.10}$$

The Poisson distribution corresponds to the limit $k' \to \infty$ (positive binomial), or $k \to \infty$ (negative binomial).

(c) *Negative binomial*

$$\Pi(Z) = (q - pZ)^{-k} \tag{1.11}$$

where p is as defined for the positive binomial, $q = 1 + p$, and the parameter k is a measure which varies inversely with the degree of aggregation of the parasites within the host population.
The mean and variance are given by

$$m \equiv E(i) = kp \tag{1.12}$$

$$\mathrm{var}(i) = kpq. \tag{1.13}$$

The expectations of i^2 and i^3 are

$$E(i^2) = (kp)^2(k+1)/k + kp. \tag{1.14}$$

That is,

$$E(i^2) = m^2(k+1)/k + m \tag{1.14a}$$

$$E(i^3) = (kp)^3(k+1)(k+2)/k^2 + 3(kp)^2(k+1)/k + kp. \tag{1.15}$$

Notice that (as follows from comparison of the defining functions (1.1) and (1.11)) the negative binomial results can all be obtained from the positive binomial ones by the simple transformation $p \to -p$ and $k' \to -k$, or equivalently, $m \to m$ and $k' \to -k$. In particular, positive and negative binomial expressions for $E(i^2)$ are interrelated by $k' \to -k$.

APPENDIX 2

This appendix gives a global account of the stability properties of the Basic model defined by eqns (7) and (9).

The expressions for the equilibrium populations H^* and P^* are given by eqns (10) and (11), and the criterion that this point be biologically sensible ($H^* > 0$) is expressed by eqn (12).

For a fully nonlinear discussion of the stability properties of this model, it is sufficient to notice that the function

$$V(H, P) \equiv e^{\alpha P/H}(H_0 + H)^\lambda H^{-(b+\mu+\alpha)} P^{(b-a)} > 0 \tag{2.1}$$

is a Lyapunov potential for this system (Minorsky 1962), and furthermore that

$$dV/dt = 0 \tag{2.2}$$

Abundance and the Rise of Conservation Ecology 543

244 *Regulation and stability of host-parasite interactions. I*

for all t. Thus V is a positive constant, determined by the initial values of H and P; the population values $H(t)$ and $P(t)$ will endlessly cycle around some closed trajectory. In other words, the Basic model corresponds to a neutrally stable or conservative dynamical system, with the consequences discussed in the main text.

Although the potential $V(t)$ of eqn (2.1) was found by analytical trickery, rather than being revealed in a vision, simply writing it down saves space while preserving rigour. Its properties may be verified by differentiating to get dV/dt in terms of dH/dt and dP/dt, whereupon use of eqns (7) and (9) lead to the central result (2.2).

APPENDIX 3

This appendix outlines the analysis of the stability properties of the various host-parasite models presented in this paper; similar analysis pertains to those in the following paper (May & Anderson 1978).

With the exception of model B (which involves $E(i^2)$ in the equation for dH/dt, and $E(i^3)$ in that for dP/dt), and model F in the following paper (which involves time delays), the models under consideration all have the mathematical form

$$dH/dt = c_1 H - c_2 P \qquad (3.1)$$

$$dP/dt = P\{\lambda H/(H_0+H) - c_3 - c_4 P/H\}. \qquad (3.2)$$

Here c_i ($i = 1,2,3,4,$) are constants which depend upon the particular biological assumptions in a given model.

Specifically, using eqn (1.14a) to express $E(i^2)$ in eqns (8) and (20) in terms of P/H and k, for a negative binomial distribution of parasites we may tabulate c_i for models A and C as follows:

	Model A	Model C	
	$c_1 = a-b$	$c_1 = a-b$	(3.3)
	$c_2 = \alpha$	$c_2 = \alpha$	(3.4)
	$c_3 = b+\mu+\alpha$	$c_3 = b+\mu+\alpha$	(3.5)
	$c_4 = \alpha\{k+1\}/k$	$c_4 = \{\alpha+\mu\}\{k+1\}/k.$	(3.6)

The corresponding expressions for a Poisson distribution of parasites are obtained by putting $k\to\infty$, and for a positive binomial by putting $k\to-k'$.

From eqns (3.1) and (3.2), the equilibrium populations H^* and P^* are

$$P^*/H^* = c_1/c_2 \qquad (3.7)$$

$$H^* = H_0 (c_3+c_1 c_4/c_2)/(\lambda-c_3-c_1 c_4/c_2). \qquad (3.8)$$

This provides the first important constraint, namely that for a biologically sensible equilibrium to be possible it is required that

$$\lambda > c_3+c_1 c_4/c_2. \qquad (3.9)$$

A linearized stability analysis of this equilibrium is carried out along standard lines (see, e.g. May 1975). Writing $H(t) = H^*+x(t)$ and $P(t) = P^*+y(t)$, and linearizing by neglecting terms of order x^2, xy and y^2, we get from eqns (3.1) and (3.2)

$$dx/dt = c_1 x - c_2 y_1 \qquad (3.10)$$

$$dy/dt = c_5 x - (c_1 c_4/c_2)y. \qquad (3.11)$$

ROY M. ANDERSON AND ROBERT M. MAY   245

Here for notational convenience, c_5 is defined as

$$c_5 = (\lambda H_0 H^*/(H_0+H^*)^2 + c_1 c_4/c_2)(c_1/c_2). \qquad (3.12)$$

The temporal behaviour of $x(t)$ and $y(t)$ then goes as exp (Λt), where the stability—determining damping rates (or eigen values) Λ are given from eqns (3.10) and 3.11) by the quadratic equation

$$\Lambda^2 + A\Lambda + B = 0. \qquad (3.13)$$

Here

$$A \equiv (c_1/c_2)(c_4-c_2) \qquad (3.14)$$

$$B \equiv c_2 c_5 - c_1^2 c_4/c_2 \equiv c_1 \lambda H_0 H^*/(H_0+H^*)^2. \qquad (3.15)$$

The requirement for neighbourhood stability is that the real part of both eigen values Λ be negative; the necessary and sufficient condition for this to be so is given by the Routh-Hurwitz criterion $A>0$ and $B>0$. From eqn (3.15) we see that B is always positive. Therefore the equilibrium point will be locally stable if $A>0$, that is if

$$c_4 - c_2 > 0. \qquad (3.16)$$

This provides the second important constraint, determining the stability character of the equilibrium point (if it exists).

Since the above models are modifications of the globally neutrally stable Basic model, it is reasonable to assume that if they are locally stable, they are globally stable; and conversely that if they are locally unstable, they are globally unstable. A rigorous proof follows from the observation that, for the general pair of eqns (3.1) and (3.2), the function

$$V = \{\exp(c_2 P/H)\}(H_0+H)^\lambda P^{-c_1} H^{\{c_1-c_3-c_1c_4/c_2\}}$$

is a global Lyapunov function. That is, $V>0$ (provided the initial host and parasite populations are positive), and

$$dV/dt = (c_2-c_4)(c_1-c_2 P/H)^2 c_2^{-1},$$

as may be verified by differentiating $V(H,P)$, and using eqns (3.1) and (3.2) to express dH/dt and dP/dt. Local stability requires $c_4 > c_2$ (see eqn (3.16)), whence $dV/dt < 0$, connoting global stability. Conversely local instability requires $c_4 < c_2$, whence $dV/dt > 0$, connoting global instability. The structurally unstable razor's edge of $c_4 = c_2$ gives both local (eqn (3.16)) and global ($dV/dt = 0$) neutral stability, as discussed more fully in Appendix 2.

In brief, local and global stability properties march together in these models.

We now proceed to indicate how the basic criteria (3.9) and (3.16) may be applied to the various models to derive the dynamical properties discussed in the main text, and illustrated in Figs 5, 6, 7, 9 and 10.

Model A

For a negative binomial distribution of parasites, the expressions (3.3) through (3.6) may be substituted in eqn (3.9) to arrive at the criterion for an equilibrium solution to be possible. This expression is given and discussed as eqn (16) in the main text. The stability criterion (3.16) here reads.

$$\alpha/k > 0 \qquad (3.17)$$

which is always satisfied; the equilibrium point is necessarily stable.

A Poisson distribution of parasites (as in the Basic model) is obtained as the limit $k \to \infty$, thus producing eqn (12) as the criterion for an equilibrium point to exist. In this case, $c_4 = c_2 = \alpha$, so that A$=0$ and the eigen value equation (3.13) becomes $\Lambda^2 + B = 0$, or

$$\Lambda = \sqrt{-B}. \tag{3.18}$$

That is, both eigen values are purely imaginary, leading to neutral stability in the linearised analysis; this, of course, is the local version of the global result established in Appendix 3.

For a positive binomial distribution, the stability criterion (3.16) becomes

$$-\alpha/k' > 0 \tag{3.19}$$

which cannot be satisfied. This model is ineluctably unstable.

Model C

For an overdispersed distribution, we substitute the appropriate expressions (3.3) through (3.6) into eqn (3.9) to arrive at eqn (21) for the criterion for an equilibrium point to be possible. The stability criterion (3.16) is here

$$\{\alpha + \mu(k+1)\}/k > 0 \tag{3.20}$$

which is always satisfied.

For a Poisson distribution, putting $k \to \infty$ in eqn (21) produces the criterion for an equilibrium to exist. Similarly, putting $k \to \infty$ in eqn (3.20) gives the stability condition $\mu > 0$, which always holds.

For an underdispersed distributions, we put $k \to -k'$ in eqn (21) to get the condition for equilibrium to exist; see Fig. 10. By likewise putting $k \to -k'$ in the stability criterion (3.20), we see that such an equilibrium will be stable if

$$\mu k' - \mu - \alpha > 0. \tag{3.21}$$

Stability is helped by high natural parasite mortality compared with parasite induced host mortality (i.e. by μ large compared with α), and hindered by high underdispersion (i.e. by small k').

Model B

In this model, the equation for dH/dt involves $E(i^2)$ and thence a term in P^2/H, and the equation for dP/dt involves $E(i^3)$ and thence a term in P^3/H^2. Consequently these host-parasite equations are not exactly of the form of eqns (3·1) and (3·2), and require separate treatment.

For a negative binomial distribution of parasites, use of the expressions (1.14) and (1.15) for $E(i^3)$ and $E(i^2)$ in eqns (18) and (19), respectively, leads to

$$dH/dt = (a-b)H - \alpha P - \alpha\{(k+1)/k\}P^2/H, \tag{3.22}$$

$$dP/dt = P\{\lambda H/(H_0 + H) - (\mu + b + \alpha) - 3\alpha\{k(+1)/k\}P/H - \alpha\{(k+1)(k+2)/k^2\}P^2/H^2\}. \tag{3.23}$$

As ever, the corresponding equations for a Poisson distribution are obtained as the limit $k \to \infty$, and for a positive binomial by the substitution $k \to -k'$.

The equilibrium mean parasite burden per host, $m^* = P^*/H^*$, follows from eqn (3.22):

$$m^* = \tfrac{1}{2}\{k/(k+1)\}[\{1+4(a-b)(k+1)/(k\alpha)\}^{\frac{1}{2}} - 1].   \tag{3.24}$$

It then follows from eqn (3.23), after some tedious algebraic manipulations, that the equilibrium host population is

$$H^* = H_0\zeta/(\lambda-\zeta),   \tag{3.25}$$

with ζ defined for convenience as

$$\zeta = \mu+a+\alpha+2(a-b)/k+\{(2k+1)/k\}\alpha m^*   \tag{3.26}$$

A biologically sensible equilibrium is possible only if

$$\lambda > \zeta.   \tag{3.27}$$

This relation among the parameters is illustrated in Fig. 7; remember that for a Poisson distribution the above equations are to be read with $k\to\infty$, and for a positive binomial with $k\to-k'$.

The stability of such an equilibrium point is determined by a linearized stability analysis of the kind described above (and, e.g. in May 1975). Linearizing eqns (3.22) and (3.23) about the equilibrium point defined by eqns (3.24) and (3.25), we eventually obtain the quadratic eigen value equation (3.13), where the coefficients A and B are

$$A = \alpha m^*\{k(2k+3)+4(k+1)m^*\}/k^2,   \tag{3.28}$$

$$B = \alpha\{1+2m^*(k+1)/k\}\{\lambda H_0 P^*/(H_0+H^*)^2\}.   \tag{3.29}$$

From eqn (3.24) it follows that $B>0$ for all positive m^*.

Clearly $A>0$ for overdispersed or Poisson ($k\to\infty$) distributions, so their equilibrium points are stable. For underdispersed distributions (positive binomial, $k\to-k'$), the stability condition $A>0$ leads to the requirement that $k'>1\cdot5$ and

$$4m^* < k'(2k'-3)/(k'-1).   \tag{3.30}$$

Thus underdispersed distributions can have stable equilibria, provided the mean parasite burden is relatively low (small m^*) and the underdispersion is not pronounced (large k'): Fig. 7 bears out these remarks.

REFERENCES TO APPENDICES

Anderson, R. M. (1974). An analysis of the influence of host morphometric features on the population dynarius of *Diplozoon paradoxom* (Nordmann, 1832). *Journal of Animal Ecology*, **43**, 873–87.

May, R. M. (1975). *Stability and Complexity in Model Ecosystems* (2nd edn). Princeton University Press, Princeton.

May, R. M. & Anderson, R. M. (1978). Regulation and stability of host-parasite interactions: II. Destabilizing processes. *Journal of Animal Ecology*, **47**, 249–67.

Minorsky, N. (1962). *Nonlinear Oscillations*. Van Nostrand, Princeton.

Journal of Animal Ecology (1978), **47**, 315–332

QUALITATIVE ANALYSIS OF INSECT OUTBREAK SYSTEMS: THE SPRUCE BUDWORM AND FOREST

By D. LUDWIG*, D. D. JONES† and C. S. HOLLING†

* Institute of Applied Mathematics and Statistics and † Institute of Resource Ecology, University of British Columbia, Vancouver, B.C., Canada V6T 1W5

SUMMARY

(1) A procedure has been described for the qualitative analysis of insect outbreak systems using spruce budworm and balsam fir as an example. This consists of separating the state variables into fast and slow categories.

(2) The dynamics of the fast variables are analysed first, holding the slow variables fixed. Then the dynamics of the slow variables are analysed, with the fast variables held at corresponding equilibrium values. If there are several such equilibria, there are several possibilities for the slow dynamics.

(3) In the case of the budworm, this analysis exhibits the possibility of 'relaxation oscillations' which are familiar from theory of the Van der Pol oscillator. In more modern terminology, the jumps of the system are governed by a cusp catastrophe.

(4) Such an analysis can be made on the basis of qualitative information only, but additional insight emerges when parameter ranges are defined by the kind of information typically available from an experienced biologist.

(5) At the least this can be a guide to assess subsequent priorities for both research and policy.

INTRODUCTION

As in all sciences, ecology has its theoretical and its empirical school. Perhaps because of the complexity and variety of ecological systems, however, both schools seem, at times, to have taken particularly extreme positions. And so the empiricists have viewed the theoretical school as designing misleading constructs and generalities with no relation to reality. The theoreticians, in their turn, have viewed the empirical school as generating mindless or mind-numbing analysis of specifics and minutiae.

This paper aims to apply some of the tools of the theoretician—specifically the qualitative theory of differential equations—to one of the most detailed and exhaustive empirical studies of an ecological system that has ever been attempted—the spruce budworm/forest interaction in eastern North America.

It has two purposes; the first is to demonstrate how far these mathematical tools can be pushed to give insight when information is available for a specific system at three different levels. The first level is purely qualitative and non-numerical. The second includes rough estimates of parameter values that are typically known by the informed biologist if he is asked the appropriate question. The third includes highly detailed quantitative data that, while rarely available, are provided in the extensive monograph of spruce budworm dynamics prepared by Morris (1963).

316   *Spruce budworm and forest*

The development of the analysis described here in fact followed precisely that sequence. The first version of the equations was prepared by one of us (Ludwig) after hearing a 1-h lecture that was totally non-numerical and qualitative. Thereafter he relied on a one-half page summary description of the system (in Holling 1973, p. 14). After an afternoon discussion some modest modifications were made (particularly Step 5, in what follows) to complete the qualitative analysis as far as it could go.

We then moved to the next level of very general and easily obtained quantitative information. Our rule was to confine ourselves to guesses of parameter values that an informed entomologist or forester might reasonably be expected to have prior to the establishment of Morris' spruce budworm project.

The final step was to use the data from that detailed study to see what additional insights were added.

The second purpose emerges from that last step. Morris' detailed study has independently provided the basis for the development and rigorous testing of a simulation model (Jones 1976). Hence, the final set of differential equations, their parameter estimates, and the topological analysis could be directly compared to the functional content and behaviour of the full simulation model. The key question was to determine if there was value in compressing the detailed explanation contained in a simulation model into an analytically tractable set of three differential equations.

The paper is organized into the three levels of information. Since the approach seems to have considerable generality, we have also identified the specific steps as a kind of 'how-to-do-it' sequence.

LEVEL 1: QUALITATIVE INFORMATION

Step 1: divide the state variables into fast and slow categories

Associated with each state variable is a characteristic time interval over which appreciable changes occur. The budworm can increase its density several hundred fold in a few years. Therefore, in a continuous representation of this process, a characteristic time interval for the budworm is of the order of months. Parasites of the budworm may be assigned a similar, or somewhat slower scale. Avian predators may alter their feeding behavior (but not their numbers) rather quickly and may be assigned a fast time scale similar to budworm. The trees cannot put forth foliage at a comparable rate, however: a characteristic time interval for trees to completely replace their foliage is on the order of 7–10 years. Moreover, the life span of the trees themselves is between 100 and 150 years, in the absence of budworm, so that their generation time is measured in decades. We first conclude, therefore, that the minimum number of variables will include budworm as a fast variable and foliage quantity (and perhaps quality) as a slow variable.

In the case of the budworm, the main limiting features are food supply, and the effects of parasites and predators. In order to describe the former, we choose a logistic form: if B represents the budworm density, then, in the absence of predation B satisfies

$$\frac{dB}{dt} = r_B B\left(1 - \frac{B}{K_B}\right). \tag{1}$$

The carrying capacity K_B is assumed to depend upon the amount of foliage available. The logistic equation is chosen here because it involves only two parameters. The later mathematical analysis is facilitated by this choice, but the results would be analogous if any other form of self-limited growth were assumed.

The effect of predation is included by subtracting a term $g(B)$ from the right-hand side of eqn (1). A feature of predators is that their effect saturates at high prey densities; i.e. there is an upper limit to the rate of budworm mortality due to predation. The consumption of prey by individual avian predators is limited by saturation, and the number of birds is limited by such factors as territorial behaviour. Similarly, parasites have a low searching capacity that prevents a rapid build-up of their numbers during an outbreak. Thus their impact does not appreciably increase with increasing budworm density.

We conclude that $g(B)$ should approach an upper limit β as $B \to \infty$. This limit β may depend upon the slow variables (i.e. the forest variables), but that possibility is deferred until Step 5, below. There is also a decrease in the effectiveness of predation at low budworm densities. This is a characteristic of a number of predators and arises in birds in part because of the effects of learning. Birds have a variety of alternative foods, and when one of them is scarce, that particular prey item is encountered only incidentally. As the item becomes more common, however, the birds begin to associate reward with that prey and they begin to search selectively for it. Thus we may assume that $g(B)$ vanishes quadratically as $B \to 0$. A convenient form for $g(B)$ which has the properties of saturation at a level β and which vanishes like B^2 is

$$g(B) = \beta \frac{B^2}{\alpha^2 + B^2}. \tag{2}$$

This represents a Type-III S-shaped functional response (Holling 1959). The parameter α in eqn (2) determines the scale of budworm densities at which saturation begins to take place. The addition of vertebrate predation to eqn (1) thus produces a total equation for the rate of change of B:

$$\frac{dB}{dt} = r_B B \left(1 - \frac{B}{K_B} \right) - \beta \frac{B^2}{\alpha^2 + B^2}. \tag{3}$$

We emphasize that this particular form was chosen to require as few parameters as possible; our final conclusions are not dependent upon the specific form of the equation, but only upon its qualitative properties.

Step 2: analyse the long-term behaviour of the fast variables when the slow variables are held fixed

In the present case, this analysis is relatively simple, since only one fast variable is considered explicitly. In more complicated situations, phase plane or other more elaborate methods might be required (Bazykin 1974). The first step in the analysis is to identify the equilibria (where $dB/dt = 0$) and determine their stability. Equilibrium values of B must satisfy

$$r_B B \left(1 - \frac{B}{K_B} \right) - \beta \frac{B^2}{\alpha^2 + B^2} = 0. \tag{4}$$

Clearly, $B = 0$ is one such value. If B is near zero, the first term (growth) dominates the second term (predation). The derivative dB/dt is positive for B slightly greater than zero, and therefore $B = 0$ is an unstable equilibrium. The remaining roots of eqn (4) satisfy

$$r_B \left(1 - \frac{B}{K_B} \right) - \beta \frac{B}{\alpha^2 + B^2} = 0. \tag{5}$$

318   *Spruce budworm and forest*

The number of roots for eqn (5) depends upon the four parameters r_B, K_B, β, and α. The next step is to combine some of these parameters where possible by scaling.

Step 2(a): Scale the equations to reduce the number of parameters

We introduce the scaled budworm density $\mu = B/\alpha$. Equation (5) takes the form

$$r_B\left(1 - \frac{\alpha\mu}{K_B}\right) - \frac{\alpha\beta\mu}{\alpha^2(1+\mu^2)} = 0. \tag{6}$$

We multiply through by α/β and (6) becomes

$$\frac{\alpha r_B}{\beta}\left(1 - \frac{\alpha\mu}{K_B}\right) - \frac{\mu}{1+\mu^2} = 0. \tag{7}$$

Now eqn (7) involves just two combinations of the original four parameters. We set

$$R = \frac{\alpha r_B}{\beta}, \quad Q = \frac{K_B}{\alpha} \tag{8}$$

and rewrite eqn (7) as

$$R\left(1 - \frac{\mu}{Q}\right) = \frac{\mu}{1+\mu^2}. \tag{9}$$

The interpretation of eqn (9) is both simple and important. The left-hand side of eqn (9) is the *per capita* growth rate of the scaled variable μ (with respect to a scaled time $t' = \beta/\alpha\, t$). The right-hand side of eqn (9) is the *per capita* death rate due to predation, also in scaled variables. The points where these curves intersect are the non-zero equilibria for μ (and equivalently, B). The two sides of eqn (9) are plotted in Fig. 1. The left-hand side is a straight line, with intercepts R and Q. The right-hand side is a curve which passes through the origin and is asymptotic to the μ axis at high densities.

Step 2(b): Examine the equilibria of the fast variables as a function of the parameter values

The equilibria for the budworm variable are defined where the straight growth curve

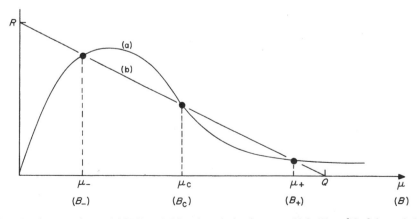

FIG. 1. The growth rate (a) $[R(1 - \mu/Q)]$ and predation loss rate (b) $[\mu/(1+\mu^2)]$ of the scaled budworm density μ. Stable equilibria occur at μ_- and μ_+; an unstable equilibrium is at μ_c.

intersects the peaked predation curve (Fig. 1). The number and location of these intersections depends on the two parameters R and Q. In this section we examine the nature
of this dependence.

Equation (9) provides a minimum of one and a maximum of three equilibria. A case
of three is shown in Fig. 1 with the lower, middle, and upper values labelled as μ_-, μ_c,
and μ_+, respectively. Although eqn (9) and Fig. 1 are, strictly speaking, in terms of
the scaled variable μ we shall henceforth substitute the original variable B, keeping in
mind that they differ only by a constant factor, $1/\alpha$.

It should be clear from Fig. 1 that the locations of the intersections depend upon the
relative positions of the growth and predation curves. In this particular choice of scaling,
both of the system parameters, R and Q, appear in the straight (growth) function. This
makes it easier to visualize how changes in R and Q will change the number and location
of the equilibria.

Although we call R and Q parameters, we assume that they may turn out to be functions of the slow variables of forest development. The original purpose in separating
slow from fast variables was to allow us temporarily to treat the slow ones as parameters.
The definition of 'fast' is synonymous with the assumption that for any (R, Q) the value
of B will converge rapidly to one of the stable equilibria, either B_- or B_+.

The dynamics of the system can be visualized in Fig. 1 by imagining that initially
$B = B_-$ and R is low. R is then slowly increased while keeping Q fixed. That is, the
straight line is rotated clockwise about its right-hand intercept. The values of B_- and B_c
will converge in an accelerating manner while B_+ increases only slightly. At the value
of R where B_- and B_c coincide, the lower equilibrium is lost and the next increase in R
sends the insect density to B_+. If we now reverse the path of R, the level of B_+ will
decrease very slowly, even beyond the time where B_- and B_c are recreated. It is only
when R assumes even lower values that B_c and B_+ coincide and the upper equilibrium
is lost. Very similar geometric arguments would illuminate the effects of changing Q.

It is clear that all the action occurs when the intermediate, unstable equilibrium, B_c,
coalesces with either the upper or the lower equilibrium. When this happens the density
may either jump from a low value to a high one or the reverse. This behaviour is similar

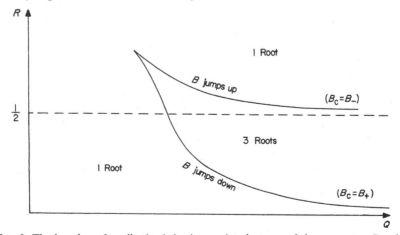

FIG. 2. The location of qualitative behaviour points in terms of the parameters R and Q.
Regions with one root have one stable equilibrium. The region with three roots has two
stable and one unstable equilibrium (as in Fig. 1). The critical curves separating these
regions locate conditions where the budworm density changes radically.

to the sudden outbreaks and collapse that characterize the spruce budworm populations. The equilibria B_- and B_+ correspond to budworm limitation by predators and food, respectively. As forest conditions improve, budworm growth exceeds the control by predators and an outbreak occurs. On the other hand, if the forest is destroyed far enough, the predators can again regain control. Note that the conditions under which upward and downward jumps occur may be quite different since the critical combinations of R and Q may be widely separated.

The critical values of R and Q are where two roots of eqn (9) coalesce and disappear. This translates into the two critical curves of Fig. 2. (The details of the calculation are shown in the Appendix.) The upper curve is when B_c and B_- join and the lower equilibrium is lost, and the lower curve is for $B_c = B_+$, which eliminates the upper equilibrium.

The point where both critical curves meet is that unique combination of R and Q where $B_- = B_c = B_+$.

The two critical curves define a critical region, inside of which there are three equilibria—two stable separated by one unstable. Above this region there is one (high) equilibrium and below it there is one (low). The critical region can be thought of as two overlapping surfaces of stable equilibria. An R-value moving upward with $B = B_-$ must pass completely through the critical region before the upward jump occurs. It must then return completely through, past the lower curve, before B collapses.

The type of phenomena we have presented readily fits into the arena of Catastrophe Theory (Thom 1975). The application of this theory to dynamical systems can be found in Zeeman (1972, 1976) and Jones (1975). The particular case of two parameters and one fast variable has been given the name of a 'cusp catastrophe'. In fact a cusp appears in Fig. 2 where the two critical curves join. The important generality provided by that body of theory is that we may use all the lessons learned from other 'cusp catastrophies' in our current case. Thom's theory says that, at the appropriate qualitative level, all such systems are the same. The equivalence, seated in deep mathematical theorems, is in harmony with our opening assertion that the exact form of our equations was not important so long as they met certain biologically necessary, qualitative criteria regarding their shape.

Step 3: Decide upon the response of the slow variables when the fast variables are held fixed

In order to characterize the state of the balsam fir forest, one ought to keep track of the age or size distribution of the trees, their foliage quantity, and their physiological condition. However, periodic budworm outbreaks synchronize the development of the trees, and the age distribution may be replaced by a single variable S, which gives the average size of the trees. S will be identified with the total surface area of the branches in a stand. Similarly, the condition of the foliage and health of the trees will be summarized in a single variable E, which may be analogously identified with an 'energy reserve'.

Since the maximum value of surface area is bounded, we shall choose a logistic form for S,

$$\frac{\mathrm{d}S}{\mathrm{d}t} = r_S S\left(1 - \frac{S}{K_S} \times \frac{K_E}{E}\right) \qquad (10)$$

which allows S to approach its upper limit, K_S. The additional factor K_E/E is inserted into eqn (10) because S does not inevitably increase under conditions of stress; surface area may decline through the death of branches or even whole trees. However, during endemic

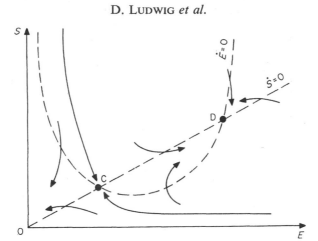

Fɪɢ. 3 The plane of the forest variables S and E. When there are few budworm the isoclines for no change in S and E intersect at the two points C and D

times E will be close to its maximum value K_E and S will grow to its maximum, K_S. We assume the energy reserve E also satisfies an equation of the logistic type:

$$\frac{dE}{dt} = r_E E \left(1 - \frac{E}{K_E} \right) - P \frac{B}{S} \qquad (11)$$

If B is small, then E will approach its maximum K_E. The second term on the right-hand side of eqn (11) describes the stress exerted on the trees by the budworm's consumption of foliage. This stress is proportional to B/S. Since B has units of number per acre and S has units of branch surface area per acre, B/S is the number of budworms per branch. This is the natural density measure for the feeding process. The factor P may be regarded as constant for our present purposes.

Step 4: Analyse the long-term behaviour of the slow variables, with the fast variables held at their corresponding equilibria

The isoclines for the systems (10), (11), are obtained by setting their left-hand sides equal to zero. Thus $dS/dt = 0$ if

$$S = \frac{K_S}{K_E} E, \quad \text{or} \quad \text{if } S = 0 \qquad (12a)$$

and $dE/dt = 0$ if

$$S = \frac{PB}{r_E E \left(1 - \dfrac{E}{K_E} \right)} = \frac{PBK_E}{r_E} \times \frac{1}{E(K_E - E)}. \qquad (12b)$$

These curves are sketched in Fig. 3, for the case when B, and therefore PBK_E/r_E, is small and there are two equilibria for S and E at C and D. The point C is a saddle point, and hence unstable. There is a single pair of trajectories which reach C and which form a separatrix (heavy arrows). If E and S start out to the right of the separatrix, then (E,S) will approach D as $t \to \infty$. If E and S start out to the left of the separatrix, they move off into the direction of $E = -\infty$. While this is not realistic, it is a consequence of the form

322   *Spruce budworm and forest*

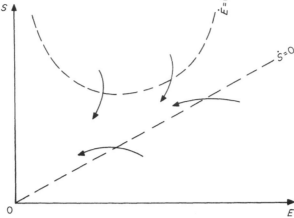

FIG. 4. The plane of the forest variables S and E. When there are many budworm the isoclines for no change in S and E separate and all trajectories head toward low E values.

of eqns (10) and (11). In full completeness we expect E to be limited by $E = 0$ and the origin
of Fig. 3 would be a stable equilibrium. In a later section we shall patch up eqns (10)
and (11) for low values of E and S. For the time being it is enough to know whether the
point heads towards point D or toward the left. In the latter case we argue on logical
grounds that it will eventually reach the origin. Even then we could conceptualize a
source term that would regenerate a new forest after the old had collapsed. This circumstance is beyond the time frame of our present model.

Now if B increases, then the U-shaped isocline will move up, and the points C and D
will approach each other, and the region to the left of the separatrix will take up more
and more of the (E,S) plane. Finally, the U-shaped curve can move entirely above the
straight S isocline, as shown in Fig. 4.

Now every trajectory converges to the left, presumably to the origin.

Step 5: Combine the preceding results to describe the behaviour of the complete system and identify the needs for additional coupling of the equations

Figures 2 and 3 imply the possibility of periods during which budworm populations
are low and stable (at B_-). This condition holds if such populations start at low values
and the system resides below the upper critical curve of Fig. 2. S and E will increase to
the equilibrium condition at point D, noted in Fig. 3. The budworm will be limited
primarily by predation.

However, if the upper branch of the critical curve of Fig. 2 is crossed, budworm
populations will begin to increase on a fast time scale towards B_+ If this happens, Fig. 3
must give way to Fig. 4. Given a complete separation of the two isoclines as shown in
Fig. 4, both E and S will decrease. If, as a consequence, the lower critical curve (Fig. 2)
is crossed, there will be a rapid decline of budworm density. The particular pattern
described above represents one complete outbreak/decline cycle of the budworm/balsam
interaction.

A number of other possibilities also exist, however. For example, if budworm populations are partially controlled, the two isoclines of Fig. 3 might never separate to the
extent shown in Fig. 4. In such a case, the complete system would reach an equilibrium,

with B, S, and E all at relatively high values. This corresponds to the phenomenon of 'perpetual outbreak' that has been observed in New Brunswick as a consequence of insecticide spraying. Conditions under which this may occur will be given below.

Moreover, the above patterns are only generated if the critical curves are crossed in a particular direction. That depends on the movement across the (R,Q) plane of Fig. 2. In the present form R and Q seem to be constants. As a first departure we expect that Q would increase as the forest grew. Movement in the R direction would seem to be possible if an external driving variable changed one of its component parameters. For example, weather might increase r_B, the instantaneous rate of increase of budworm, enough to carry R across the upper curve in Fig. 2.

However, for sake of clarity, and our step-by-step format, we have left the issue of careful coupling of these equations to this point.

Let us first determine whether the budworm eqn (3) has any terms that should be expressed as functions of the slow variables. In order to do this, we refine our interpretation of B and S to represent quantities per acre of forest. Since the amount of foliage available per acre is roughly proportional to S, K_B should be proportional to S.

$$K_B = K' S. \tag{13}$$

Thus K_B measures carrying capacity in larvae per acre, while K' measures carrying capacity in larvae per unit of branch area.

Similarly, terms in the predation rate (eqn 2) are also dependent upon the branch surface area. Predators, such as birds, search units of foliage, not acres of forest, so that the relevant density is larvae per unit of surface area. Thus the half-saturation density for B is also proportional to S:

$$\alpha = \alpha' S. \tag{14}$$

The new parameter α' is measured in larvae per unit of branch area.

If eqns (13) and (14) are substituted into eqn (8), the result is

$$R = \frac{\alpha' r_B}{\beta} S, \quad Q = \frac{K'}{\alpha'}. \tag{15}$$

Note that Q is independent of S, while R is proportional to S. When the forest is young, R will be small, but Q may be quite large. Thus R and Q will be below the critical region in Fig. 2. Budworm densities will be low, not only in larvae per acre, but in larvae per branch. The densities per branch will be low because the predators will find it easy to search the small number of branches per acre. As the forest grows, control by predators becomes more uncertain, because of satiation of their appetites. Finally, the upper critical curve in Fig. 2 will be crossed and control by predators becomes ineffective as B rapidly increases to a high level (B_+) being limited now by food.

The effect of a rapid increase in B may be to change the dynamics of the slow variables from that depicted in Fig. 3 to that in Fig. 4. If this happens, the budworm outbreak will lead to a collapse of the forest. From eqns (3) and (13), we see that B/S is close to K' ($B \simeq K_B$) during a budworm outbreak. Can the forest reach an equilibrium in that case? Equation (12b) may be rewritten as

$$\frac{E}{K_E}\left(1 - \frac{E}{K_E}\right) = \frac{P}{r_E K_E}\frac{B}{S} \sim \frac{PK'}{r_E K_E}. \tag{16}$$

324   *Spruce budworm and forest*

Since the left-hand side of eqn (16) is dimensionless, we denote the right-hand side as a dimensionless parameter M:

$$M = \frac{PK'}{r_E K_E}.$$ (17)

M is easily interpreted as the ratio of a rate of energy consumption by budworm (by eating foliage) to the rate of energy production by the trees. The maximum possible value of the left-hand side of eqn (16) is $1/4$. We conclude that no equilibrium is possible if $M > 1/4$. The preceding analysis indicates that an equilibrium is likely if $M < 1/4$.

Now, assuming that $M > 1/4$, a budworm outbreak must lead to a decline of the forest. Since R depends upon S, eventually the lower branch of the critical curve in Fig. 2 will be crossed, and the budworm population must collapse. However, it is not clear whether the budworm collapse will occur in time to save the forest from complete destruction. On the other hand, if the budworm density begins collapsing too soon, the effect on the forest may be reduced enough to establish a stable equilibrium for all three variables. In fact, the numerical values of the parameters in our equations will determine which of these behaviours will occur.

When realistic parameter values are substituted into our eqns (3), (10) and (11), budworm outbreaks lead to collapse of the forest. That is, E and S decline sharply during a budworm outbreak, and eventually E becomes negative. Negative values of E are unrealistic, and our model is not valid when extensive tree deaths occur. This situation can be remedied in two ways. The first (simplest) way is to recognize that the system eqns (3), (10) and (11) represents a cohort of trees and its resident budworm population. This cohort is only capable of going through one severe decline. If this were to happen, then the model must be started again with a small value of S and with E near K_E in order to generate the next outbreak. Regeneration from one cohort of trees to the next is a discontinuous process, and one might as well represent this fact explicitly in the model.

On the other hand, for mathematical convenience, it might be desirable to devise a system of differential equations which adjusts the behaviour of E and S for small values in order to simulate the growth of a new cohort. We have no need for such an adjustment in the present investigation, but we shall indicate how it might be carried out.

According to eqn (13), the carrying capacity for the budworm is independent of E, i.e. the trees put forth the same amount of foliage, regardless of their physiological condition. It would seem more reasonable that K' should depend upon E, and that K' should decline rather sharply if E falls below a certain threshold T_E. Therefore, we may replace eqn (13) by

$$K_B = K'S \frac{E^2}{T_E{}^2 + E^2}.$$ (18)

If T_E is small compared with K_E, then K_B will show a sharp decline near $E = T_E$. A corresponding change should also be made in eqn (11). We may set

$$P = P' \frac{E^2}{T_E{}^2 + E^2}$$ (19)

because the stress on the trees is related to the amount of foliage consumed.

The resulting equations are as follows:

$$\frac{dB}{dt} = r_B B \left(1 - \frac{B}{K'S} \cdot \frac{T_E^2 + E^2}{E^2} \right) - \beta \frac{B^2}{(\alpha'S)^2 + B^2} \tag{20}$$

$$\frac{dS}{dt} = r_S S \left(1 - \frac{S}{E} \frac{K_E}{K_S} \right) \tag{21}$$

$$\frac{dE}{dt} = r_E E \left(1 - \frac{E}{K_E} \right) - P' \frac{B}{S} \frac{E^2}{T_E^2 + E^2}. \tag{22}$$

This system appears to be much more complicated than eqns (3), (10) and (11), but its qualitative behaviour will be exactly the same, except for small values of E. If the parameter T_E is chosen properly, the system eqns (20)–(22) will exhibit a regeneration similar to that obtained by re-starting eqns (3), (10) and (11). However, we shall use eqns (3), (10) and (11) in the sequel.

This is as far as we can go if we restrict ourselves exclusively to qualitative information. Some effort is now needed to quantify parameters in order to define, more precisely, the behaviour of the system. Since we are interested in determining how far one can predict with different levels of information, it is useful to identify two levels of quantitative information to add to the first qualitative level—one very general and based on estimates by experienced field naturalists and one more detailed and specific.

LEVEL II—GENERAL QUANTITATIVE INFORMATION

In order to complete our model, we must estimate its parameters. Especially important are the combinations of parameters which form Q, R and M, eqns (15) and (17). These will determine the qualitative behaviour of the system. Most of the other parameters are rate constants which determine the speed with which certain processes occur, but do not alter the basic qualitative picture.

The parameter K' in eqn (13) measures the carrying capacity of the forest in larvae per branch. An entomologist with cursory knowledge of budworm can confidently state that from 100 to 300 larvae can be supported by an average branch of balsam foliage in good condition. The parameter α' is likely to be low. Knowing, roughly, the speed of movement and distance of perception of birds for insect prey, α' can be estimated as one to two larvae per branch. [This particular analysis has been expanded by Holling, Jones & Clark (in preparation).] These ranges then permit a calculation of the likely range for Q from eqn (15). The results range from $Q = 50$ to 300, and strongly suggest that the system resides in or below the critical region (Fig. 2) during the endemic phase.

But still, outbreaks will only occur if R increases above $1/2$, and crosses the upper critical curve. Again, rough estimates of the elements determining the value of R (eqn (15)) can be obtained as follows:

r_B The budworm is capable of a five-fold increase in density per year. Since we are using a continuous time model, we set $e^{r_B} = 5$, and conclude that $r_B = 1\cdot6$/year.

β The value of β has been estimated in the literature using the most general information on size of birds, their maximum daily consumption, the proportion of budworm in their diet, and their ranges of densities (Kendeigh 1947; George & Mitchell 1948; Mitchell 1952; Dowden, Jaynes & Carolin 1953; Morris *et al.* 1958). These estimates of the maximum consumption range from 20 000 to 36 000 larvae per acre per year.

326   *Spruce budworm and forest*

S A maximum value for S is K_S.

K_S A fully recovered forest contains about 24 000 average branches per acre.*

If the preceding estimates are combined (eqn (15)), we find that R, for a mature forest, lies between 1·1 and 3·8. Because R need only be 0·5 we have considerable leeway in the value of S below K_S that will initiate an outbreak. A forest is fully mature at age 80, while outbreaks have a period of about 40 years. Actually, numerical results show that an outbreak is not immediate when the upper branch of the cusped curve is crossed. It requires a number of years for the budworm to show an appreciable rise in density. Although predation cannot control the budworm when $R > 0·5$, the predation does appreciably slow the rate of growth. Hence, a value of 1·1 for R at $S = K_S$ is not unreasonable.

Now we turn to the estimation of M. As mentioned above, M is a ratio of energy consumed by budworm to energy produced by trees. The critical value for M is 0·25 as trees collapse if $M > 0·25$. The time required for such a collapse will depend upon M. A forest can withstand defoliation for approximately 4 years, which implies that Mr_E must be approximately 0·3, since B does not in fact reach the value $K'S$ as assumed in the derivation of M (eqn (17)).

Some rates which should be estimated are r_S and r_E. The time of regeneration of the forest after an outbreak depends on r_S. It can be adjusted to make the period between outbreaks approximately 40 years. A value of $r_S = 0·15$/year gives satisfactory results. Likewise, r_E sets the rate at which trees recover from the stress of defoliation. Since this recovery is fairly rapid, a value of 1/year is assumed for r_E. Since E is a synthetic variable we can set its maximum value as $K_E = 1$.

The only remaining parameter is P, the rate of energy consumption by budworm. From eqn (17)

$$P = \frac{Mr_E K_E}{K'} \simeq 1·5 \times 10^{-3}.$$

As an independent measure, it is known that 150 to 200 larvae per branch can consume approximately 25% of the foliage. Therefore

$$P = \frac{0·25}{150} = 1·7 \times 10^{-3}.$$

All the parameters for eqn (20) through eqn (22) are summarized in Table 1.

LEVEL III—EMPIRICAL QUANTITATIVE INFORMATION

Level I, the development of the model, utilized only qualitative information about the system's behaviour. In Level II we made a first attempt at estimating parameter values, but restricted ourselves to general quantitative information. This is the type of information that an experienced biologist might provide without specifically examining the New Brunswick budworm.

In Level III we examine the field data that have been collected over many decades and determine the best values for the parameters as we have defined them. This step is made easier in this particular instance because much of the work has already been done in

* The standard field measure for an 'average branch' is one that can be circumscribed by a polygon of 10 ft² area.

D. LUDWIG *et al.*   327

TABLE 1. Parameter values for Level II and Level III information

Symbol	Description	Units	Level II	Level III
r_B	intrinsic budworm growth rate	/year	1·6	1·52
K'	maximum budworm density	larvae/branch	100–300	355
β	maximum budworm predated	larvae/acre/year	20 000–36 000	43 200
α'	½ maximum density for predation	larvae/branch	1–2	1·11
r_S	intrinsic branch growth rate	/year	0·15	0·095
K_S	maximum branch density	branches/acre	24 000	25 440
K_E	maximum E level	–	1·0	1·0
r_E	intrinsic E growth rate	/year	1·0	0·92
P'	consumption rate of E	/larvae	0·0015	0·00195
R	$\alpha' r_B S/\beta$	–	1·07–3·84	0·994 (S/K_S)
Q	K'/α'	–	50–300	302
M	$PK'/r_E K_E$	–	0·15–0·45	0·71

connection with the construction of a detailed simulation model of this system. That simulator is a central element in a program of ecological policy design—a program to synthesize the methodologies and concepts of systems ecology and modelling, optimization, and decision theory in a case study framework. A review of that project and some of the lessons learned can be found in Holling *et al.* (1976).

The primary source of data for the simulator was Morris (1963), with considerable additional expert opinion from the personnel and files of Environment Canada's Maritimes Forest Research Centre. The simulator mimics the univoltine character of the insect as a difference model with yearly time steps. As a result, its parameters are not all appropriate for a continuous model formulation without some adjustments. These adjustments could be made in the original data, but for convenience we choose to let the simulation serve as a surrogate for the real world, and we consult it for measures that are analogous to the parameters needed for our model—eqns (3), (10) and (11). The errors and discrepancies generated by going through this 'middle-man' are on the same order as those when we assume the simple continuous form that we have.

Specific details about the budworm simulation model can be found in Jones (1976). We now briefly check off the parameter values suggested by that reference and indicate the discrepancies with Level II values. First, consider the intrinsic role of growth, r_B. We find in the simulation that the maximum growth rate between generations in a mature forest with low budworm densities is 4·56. Thus, $r_B = \ln (4·56) = 1·52$/year, in close agreement with that found above.

The hypothetical carrying capacity per branch K' is the most difficult to interpret. The simulation has a comparable equilibrium at $K' = 335$ larvae/branch. However, numerical experience shows that there can be a transient overshoot to values of 600 or more. This wide range is a consequence of the discrete nature of the insect population. We adopt K' as 335 because of the conceptual parallel of that value, but note that the continuous model will not overshoot to the high values seen in the simulation and in nature.

The parameters of bird predation are taken from the data summarized in Holling, Jones & Clark (in preparation). That paper specifically identifies three groups of insectivorous birds that represent three distinct size classes and, to some degree, three different modes of search. This more detailed, but still qualitative, analysis of field data identifies an expected value for β of 43 200 larvae/acre/year. This maximum consumption level is significantly higher than that found in the literature and reported in Level II. That literature only considered one class of birds—the arboreal feeders (e.g. warblers). Two other classes of birds were not previously recognized as important because they are normally ground nesters (e.g. juncos and grosbeaks) and because their numbers do not increase during an outbreak. However, their large size and appetite make them at least equally as important as the smaller species and β is increased accordingly. Because the density of budworm which produces half-saturation of predation is different for each bird class we take an average for each, weighted by their contribution to the total predation. Thus, $\alpha' = 1 \cdot 11$ larvae/branch.

The parameters for tree growth are taken by fitting eqn (10) (with $E = K_E$) to a typical history of branch surface area following the collapse of an outbreak. This gives a growth rate of $r_S = 0 \cdot 095$/year and an asymptotic level of $K_S = 25 \cdot 4 \times 10^3$ branches/acre.

It was recognized early in the simulation development that something analogous to an 'energy reserve' was affecting the response dynamics of trees. However, there were insufficient data to incorporate this process adequately. The solution taken was to deputize foliage for this function, and we continue that here. As E is an intensive factor we lose no generality by defining $K_E = 1$. The value of r_E is evaluated from the rate of increase in foliage. The maximum that foliage can increase in the simulation is $1 \cdot 26$-fold per iteration. Thus $\exp r_E = 1 \cdot 26$ gives $r_E = 0 \cdot 23$ yr^{-1}. This maximum occurs when foliage is about half its maximum; and so, as a consequence of the logistic form, $r_E = 4 \times 0 \cdot 23 = 0 \cdot 92$/year.

P is the maximum rate that an individual feeds on 'E', which is $P = 1 \cdot 95 \times 10^{-3}$/larvae.

Using the above values, the three aggregate parameters R, Q and M assume the following values. First

$$R = 1 \cdot 72 \, (S/K_S)$$
$$Q = 302$$

(or $Q = 540$ if $K' = 600$).

Finally,

$$M = \frac{PK'}{r_E K_E} = 0 \cdot 71$$

(or $M = 1 \cdot 27$, if $K' = 600$).

The parameter values for Level III are also summarized in Table 1. There is remarkable agreement between the values found from extensive and intensive collection and analysis of field data and those estimated from a first field estimation.

One can go a lot further than traditionally assumed with informed, but qualitative, insight into ecological process. Extensive data collection efforts need not always be carried out and completed before the system is abstracted into an analysable model.

We have been emphasizing that the central and important aspect of this analysis is a process and not a product. The actual numerical integration of our model using the estimated parameters is a final, though anticlimactic, step to be performed for completeness. Because the parameter values from Level II and Level III information are so similar we adopt only Level III and use these values in our model (eqns (3), (10) and (11)). The integrated time course is shown in Fig. 5 through one outbreak cycle (ending in year 43). Fig. 6 shows a typical outbreak cycle exhibited by the simulation model.

The qualitative behaviour is similar, as this analysis predicted. The major difference in the appearance is between the graphs of surface area S. This is an expected discrepancy resulting from our attempt to mimic a 75 age class model of tree population with a single

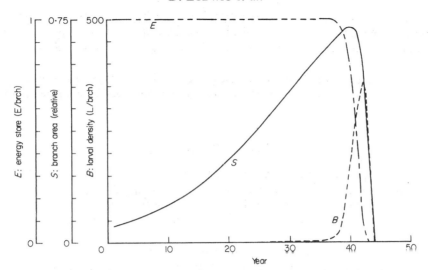

FIG. 5. A numerical solution of the differential equations using the parameter values of Level III information.

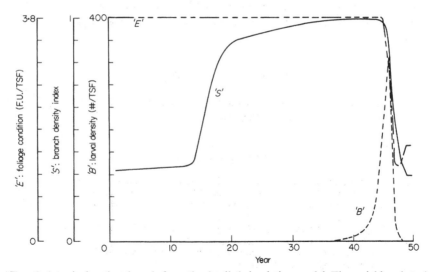

FIG. 6. A typical outbreak cycle from the detailed simulation model. The variables plotted are the ones most analogous to the differential equation model: '*E*' is the total foliage available per unit of branch surface area in arbitrary 'foliage units'/10 ft² (F.U./TSF); '*S*' is the density of surface area in TSF/acre scaled such that '*S*' = 1 is equivalent to $S = 0.94$ in Fig. 5; '*B*' is the density of third instar larvae in number 1 TSF

state variable. Other discrepancies between Figs 5 and 6 were also anticipated as they result from fundamental differences between discrete and continuous models. First, the maximum level of S is lower in Fig. 5 than in 6. Equivalently, the outbreak is triggered at a lower 'threshold' of S in Fig. 5. In the simulation model the process of competition between birds (not included in our differential equations) enhances their effectiveness in young forests (low S). Thus S must reach a higher value in the simulation model before

330 *Spruce budworm and forest*

the control by predators is overcome (Holling, Jones & Clark, in preparation). Secondly, the differential equations as conceptualized are capable of only one outbreak while the simulation model output is periodic because the regeneration of new trees is explicitly modelled.

The differential equations could be adjusted, expanded, and otherwise 'tuned' so that their time traces more closely matched the simulation model, but we elect not to follow that procedure.

DISCUSSION

An obvious requirement for the success of this procedure is a good understanding of the basic phenomena, to know which are the most important variables, and to know the main features of their interactions. Such knowledge is never completely available at the beginning of an investigation. In the case of the budworm we had behind us extensive field investigations and simulation experience. The present methods have been exceedingly useful when applied in conjunction with the simulation. One can narrow the reasonable parameter ranges using these procedures. Perhaps most important, the analytic model is likely to extend our understanding of the phenomena, since the full armory of mathematical techniques is available (Ludwig, Weinberger & Aronson, in preparation). This raises a final question about the level of mathematical training or ability required to carry out such a program. In principle, the methods which were applied to the budworm do not go beyond first year calculus; however, their effective use requires considerable mathematical confidence.

ACKNOWLEDGMENTS

The authors are indebted to Dr M. Levandowsky for much advice and encouragement Research was supported in part by NRC Grant No A9239.

REFERENCES*

Bazykin, A. D. (1974). Volterra's system and the Michaelis–Menten equation (in Russian). *Problems in Mathematical Genetics* (Ed. by V. A. Ratner), pp. 103–42. U.S.S.R. Academy of Sciences, Novosibirsk. (Available in English as *Structural and Dynamic Stability of Model Predator–Prey Systems*. 1975. Institute of Resource Ecology, Report R-3-R.)

Dowden, P. B., Jaynes, H. A. & Carolin, V. M. (1953). The role of birds in a spruce budworm outbreak in Maine. *Journal of Economic Entomology*, **46**, 307–12.

George, J. L. & Mitchell, R. T. (1948). Calculations on the extent of spruce budworm control by insectivorous birds. *Journal of Forestry*, **46**, 454–5.

Holling, C. S. (1959). The components of predation as revealed by a study of small mammal predation on the European pine sawfly. *Canadian Entomologist*, **91**, 293–320.

Holling, C. S. (1973). Resilience and stability of ecological systems. *Annual Review of Ecology and Systematics*, **4**, 1–23.

* In an effort to make the bibliography as current as possible, we have cited many works available at present only as publications of the International Institute for Applied Systems Analysis (IIASA) and our own Institute of Resource Ecology (IRE). These may be obtained from the following addresses:

Documents and Publications
International Institute for Applied Systems Analysis
Schloss Laxenburg
A-2361 Laxenburg
Austria

Publications (att'n: Ralf Yorque)
Institute of Resource Ecology
University of British Columbia
Vancouver, British Columbia
Canada V6T 1W5

Jones, D. S. (1975). The application of catastrophe theory to ecological systems. International Institute for Applied Systems Analysis Report RR-75-15. (To appear in 1977 in *Simulation in Systems Ecology* (Ed. by G. S. Innis), Simulation Councils Proceedings.)

Jones, D. D. (1976). *The Budworm Site Model.* Institute of Resource Ecology, Working Paper W-13.

Kendeigh, S. C. (1947). Bird population studies in the coniferous forest biome during a spruce budworm outbreak. *Biological Bulletin,* 1, Division of Research, Ontario Department of Lands and Forests.

Morris, R. F. (Ed.) (1963). *The Dynamics of Epidemic Spruce Budworm Populations.* Memoirs of the Entomological Society of Canada, No. 21. 332 pp.

Morris, R. F., Cheshire, W. F., Miller, C. A. & Mott, D. G. (1958). The numerical response of avian and mammalian predators during a gradation of the spruce budworm. *Ecology,* 34, 487–94.

Mitchell, R. T. (1952). Consumption of spruce budworm by birds in a Maine spruce-fir forest. *Journal of Forestry,* 50, 387–9.

Thom, R. (1970). Topological models in biology. *Towards a Theoretical Biology* (Ed. by C. H. Waddington). 3: Drafts. Edinburgh University Press.

Thom, R. (1975). *Structural Stability and Morphogenesis.* (English translation by D. H. Fowler). Benjamin, Reading, Mass.

Zeeman, E. C. (1972). Differential equations for the heartbeat and nerve impulse. *Towards a Theoretical Biology* (Ed. by C. H. Waddington), 4: Essays. Edinburgh University Press.

Zeeman, E. C. (1976). Catastrophe theory. *Scientific American,* 234, 65–83.

(*Received* 13 *June* 1977)

APPENDIX

The double roots of equation (9)

Equation (9) has a double root if the straight line given by the left-hand side of eqn (9) is tangent to the predation term of the right side. We adopt the following expressions for the two sides of eqn (9):

$$f(\mu) = R\left(1 - \frac{\mu}{Q}\right) \tag{A1}$$

$$g(u) = \frac{\mu}{1 + \mu^2}.$$

A double root occurs if

$$f(\mu) = g(\mu)$$

and

$$\frac{df}{d\mu} = \frac{dg}{d\mu}. \tag{A2}$$

We treat eqn (A2) as a pair of simultaneous equations and solve for R and Q parametrically with respect to μ. These relationships can be written as

$$\frac{R}{Q}(Q - \mu) = \frac{\mu}{1 + \mu^2}$$

and

$$\frac{R}{Q} = \frac{\mu^2 - 1}{(1 + \mu^2)^2}.$$

332   *Spruce budworm and forest*

It follows after some algebra that

$$R = \frac{2\mu^3}{(1+\mu^2)^2} \tag{A3}$$

and

$$Q = \frac{2\mu^3}{(\mu^2-1)}. \tag{A4}$$

The cusp point of Fig. 2 where both critical curves meet (or begin) is the point where the derivatives of R and Q with respect to μ both vanish. This corresponds to the inflection point of $g(\mu)$, which occurs at $\mu = \sqrt{3} = 1\cdot73$. This value in eqn (A3) and (A4) gives the cusp point for (R,Q) as

$$R_c = \frac{3^{3/2}}{8} = 0\cdot650 \tag{A5}$$

$$Q_c = 3^{3/2} = 5\cdot196.$$

Further, the axis of the cusp is oriented as

$$\frac{dR}{dQ} = -1/16$$

Equations (A3) and (A4) also generate two limiting conditions. When $\mu \to \infty$, $Q \to \infty$ and $R \to 0$. As $\mu \to 1$, $Q \to \infty$, but $R \to 1/2$. Figure 2 is based upon this information.

Ecological Monographs, 64(2), 1994, pp. 205–224
© 1994 by the Ecological Society of America

DENSITY DEPENDENCE IN TIME SERIES OBSERVATIONS OF NATURAL POPULATIONS: ESTIMATION AND TESTING[1]

Brian Dennis

*Department of Fish and Wildlife Resources[2] and Department of Mathematics and Statistics,
University of Idaho, Moscow, Idaho 83844 USA*

Mark L. Taper

Department of Biology, Montana State University, Bozeman, Montana 59717 USA

Abstract. We report on a new statistical test for detecting density dependence in univariate time series observations of population abundances. The test is a likelihood ratio test based on a discrete time stochastic logistic model. The null hypothesis is that the population is undergoing stochastic exponential growth, stochastic exponential decline, or random walk. The distribution of the test statistic under both the null and alternate hypotheses is obtained through parametric bootstrapping. We document the power of the test with extensive simulations and show how some previous tests in the literature for density dependence suffer from either excessive Type I or excessive Type II error. The new test appears robust against sampling or measurement error in the observations. In fact, under certain types of error the power of the new test is actually increased. Example analyses of elk (*Cervus elaphus*) and grizzly bear (*Ursus arctos horribilis*) data sets are provided. The model implies that density-dependent populations do not have a point equilibrium, but rather reach a stochastic equilibrium (stationary distribution of population abundance). The model and associated statistical methods have potentially important applications in conservation biology.

Key words: bootstrapping; conservation biology; density dependence; elk; equilibrium; grizzly bear; likelihood ratio; logistic model; nonlinear autoregressive model; population regulation; statistical power; stochastic difference equation; stochastic population model; time series analysis.

Introduction

Whether or not populations in nature tend to have growth rates regulated by their own densities has long been a key but frustrating problem of ecological research (The Biological Laboratory 1957, McLaren 1971, Colinvaux 1973, Kingsland 1985). It is widely acknowledged that possible answers to this question have broad theoretical implications for the structure of communities and the evolution of the species that comprise them (Pianka 1974, Cody and Diamond 1975, May 1976, Roughgarden 1979, Emlen 1984, Diamond and Case 1986). Moreover, assessment of density dependence has attained great practical importance in conservation biology because the strength and form of density dependence have a large influence on the expected survival times of natural populations (Ginzburg et al. 1990, Stacey and Taper 1992).

The availability in recent years of "long term" (≈ 20 yr) data on population abundances has steered part of the density dependence controversy into a debate about statistical concepts and methods (Eberhardt 1970, Reddingius 1971, 1990, Bulmer 1975, Royama 1977, 1981, Slade 1977, Vickery and Nudds 1984, 1991, Gaston and Lawton 1987, Pollard et al. 1987, den Boer

and Reddingius 1989, Reddingius and den Boer 1989, den Boer 1990, 1991, Solow 1990, 1991, Turchin 1990, Turchin et al. 1991, Crowley 1992, Turchin and Taylor 1992). One conceptual argument, exemplified by den Boer (1991), asserts that if density dependence is to be a cornerstone of ecological theory, a certain burden of proof needs to be satisfied. It is the role of density dependence theorists to demonstrate convincingly that time series abundance data can be distinguished statistically from the trajectories of a density-independent stochastic growth model or even of a random walk model. Another argument discounts the need for statistical detection of density dependence, claiming that population regulation is a purely logical consequence of the ecological processes by which populations grow (Royama 1977, Berryman 1991a). A third argument questions the ecological meaning of existing statistical methods for density dependence testing, because the concept of a population equilibrium is unclear when the environment fluctuates (Wolda 1989). Throughout this conceptual debate, statistical methods for detecting density dependence have been introduced and refined in a steady stream (Reddingius 1971, Bulmer 1975, Slade 1977, Vickery and Nudds 1984, Pollard et al. 1987, Reddingius and den Boer 1989, Turchin 1990, den Boer 1991). The conclusions of various data analysis studies about the prevalence of density dependence seem to vary depending on which methods are used.

[1] Manuscript received 31 August 1992; revised 1 June 1993; accepted 12 July 1993.
[2] Address reprint requests to the author at this department.

206   BRIAN DENNIS AND MARK L. TAPER   Ecological Monographs
Vol. 64, No. 2

One group of studies has used a statistical model of population growth with a density dependence term proportional to the logarithm of population abundance (Reddingius 1971, Bulmer 1975, Gaston and Lawton 1987, den Boer and Reddingius 1989, Reddingius and den Boer 1989, den Boer 1990, Vickery and Nudds 1991, Crowley 1992). While different particular methods for testing whether the density dependence term should be included in this model have different powers (Pollard et al. 1987, Vickery and Nudds 1991), these analyses frequently suggested that density dependence is not as prevalent as expected by ecological theory. The related concepts of "stabilization" (den Boer 1968, 1990) and "density vagueness" (Strong 1986a, b) have been suggested to account for such findings; the concepts essentially take population growth to be density independent (but noisy) over a wide range of densities, with density dependent regulation occurring more or less sharply at very high densities.

However, in contrast to the above studies, Woiwod and Hanski (1992) and Holyoak and Lawton (1992) detected frequent density dependence using tests based on the same logarithmic density dependence model (among other tests). Woiwod and Hanski (1992) analyzed thousands of insect data sets, many of which exceeded 20 observations in length; Holyoak and Lawton (1992) treated 32 insect data sets of 8 or 12 observations. In these studies, longer time series showed increased prevalence of density dependence. Earlier results of Hassell et al. (1989) and Solow and Steele (1990) had also highlighted the importance of sample size to the statistical power of density dependence tests.

Another set of studies employed a model with a density dependence term proportional to population abundance (Turchin 1990, Berryman 1991a, Turchin et al. 1991, Turchin and Taylor 1992). Woiwod and Hanski (1992) and Holyoak and Lawton (1992) used the model as well. These studies found widespread density dependence, sometimes in the form of delayed regulation (second order lags; see Turchin 1990). The statistical methods used to test whether the density dependence term(s) should be included in the model were based on standard results from ordinary regression analysis. Generalization of these analyses to multiple species systems has been reported (Berryman 1991b).

Still other analyses have been based on various statistical properties of random walks (Vickery and Nudds 1984, den Boer 1991, Crowley 1992). The studies regarded a random walk model as the null hypothesis to be rejected by data according to some criterion. While Pollard et al. (1987) suggested that an exponential growth model (containing the random walk model as a special case) makes a more biologically interesting null hypothesis, the possibility that real data sets often cannot be distinguished from random walk trajectories remains unsettling to density dependence proponents. Indeed, den Boer (1991) concludes that Nicholson's

(1933) hypothesis that populations "exist in a state of balance because densities fluctuate about a relatively stable norm" is not supported by random walk comparisons or other statistical tests. Though den Boer (1991) does caution that these analyses do not mean that populations obey random walk models, his results should inspire some rereadings of Birch's (1957) and Andrewartha's (1957) earlier density independence arguments.

In this paper, we introduce a new test for density dependence in time series data of population abundances. We propose that a discrete time stochastic logistic model used by Turchin (1990) and Berryman (1991a) can serve as a useful and descriptive model for such testing in a variety of ecological situations. Statistical inference methods for this model, however, have not been well understood in the past. We develop parameter estimation methods and hypothesis testing methods for the model and focus on a likelihood ratio hypothesis test of density-independent vs. density-dependent population growth. Because the distribution of the test statistic is intractable, we show how its critical values can be estimated with a parametric bootstrapping method. The power properties of this new test are documented here with extensive simulations. We illustrate the use of the test with examples. The results of past empirical studies are likely influenced by the statistical testing methods used. In particular, we show that the randomization test of Pollard et al. (1987) has low power (excessive Type II error) compared to the new test. Also, we find that the regression tests of Turchin (1990) and Berryman (1991a) suffer from inflated size (excessive Type I error). The likelihood ratio test proposed here, by contrast, is a size 0.05 test and represents the practical limits of power that can be attained for the stochastic logistic growth model. We discuss the effects of sampling variability, the ecological interpretation of density dependence testing, the concept of a stochastic equilibrium, and the potential use of the new test in population viability analysis.

MODEL DESCRIPTION

Let N_t represent population abundance (as censused, estimated, or indexed) at time t, where $t = 0, 1, 2, \ldots$. The model we present relates N_{t+1} to N_t:

$$N_{t+1} = N_t \exp(a + bN_t + \sigma Z_t). \tag{1}$$

Here a and b are constants, σ is a positive constant, and σZ_t is a random shock to the population growth rate. In this paper, we are mostly concerned with values of b such that $b \leq 0$. We assume that Z_t has a normal distribution with a mean of 0 and a variance of 1 [we write $Z_t \sim \text{normal}(0, 1)$], and that $Z_0, Z_1, Z_2, \ldots$ are uncorrelated. The model involves two essential ideas. First, the per-unit-abundance growth rate is defined in discrete time as $\ln N_{t+1} - \ln N_t$, analogous to $(1/n) \, dn/$

$dt = d \ln n/dt$ in continuous time. Second, that rate so defined is taken to be a linear function of N_t plus noise.

The constant b is the slope of the linear function. If $b = 0$, the per-unit-abundance growth rate does not depend on N_t. When $b < 0$, the per-unit-abundance growth rate decreases as N_t becomes larger. An increasing per-unit-abundance growth rate, or Allee effect, results when $b > 0$ (Dennis 1989a).

The type of variability inherent in the model (Eq. 1) is "environmental" as opposed to "demographic." Models with demographic variability become essentially deterministic as population size becomes large (see discussion by Dennis et al. 1991). A population governed by Eq. 1, however, fluctuates at large as well as small sizes. The distinctions between environmental and demographic variability have ramifications in conservation biology (Leigh 1981, Shaffer 1981, Goodman 1987, Simberloff 1988, Dennis et al. 1991, Wissel and Stöcker 1991).

The population abundances $N_0, N_1, N_2, \ldots$ are not independent under this model, even though the random shocks (Z_t) are independent. As we show later in this paper, failure to account for the dependence among the N_t values is the source of flaws in some previous statistical tests for density dependence. The stochastic process N_t defined by Eq. 1 is a Markov process: given that the population has attained some particular size n_t at time t, the future distribution of population sizes depends on n_t, but not on past sizes.

The Markov property is a fairly general assumption applicable in many ecological situations. The deterministic analogue of the Markov property is simply that population abundance can be described by a first-order difference equation. Even in populations with overlapping generations or age structure, some index of population abundance can behave as if governed by a first-order difference equation. For example, Livdahl and Sugihara (1984) and Barlow (1992) document systems in which complex, nonlinear life histories give rise to simple linear dependence of per-unit-abundance growth rate on abundance. Also, Cushing (1989) has provided a theoretical justification of how a simple nonlinear difference equation can emerge from a population projection matrix model (such as a Leslie matrix) in which there is nonlinear dependence of demographic rates on population abundance. We discuss later the evaluation of the model for a given data set by residual analysis and by testing for second-order lags (see *Hypothesis testing* and *Discussion*).

We point out that the mean population abundance at time $t + 1$ under the model is not given by Eq. 1 with $\sigma = 0$. Because $E[\exp(\sigma Z_t)] = \exp(\sigma^2/2)$ and because of the Markov property, the mean population abundance at time $t + 1$ given $N_t = n_t$ is

$$E(N_{t+1} \mid N_t = n_t) = n_t \exp(r + bn_t). \qquad (2)$$

Here $r = a + (\sigma^2/2)$.

A deterministic analogue to Eq. 2 is a type of discrete time logistic model that has been analyzed extensively in population ecology (May 1976):

$$n_{t+1} = n_t \exp(r + bn_t). \qquad (3)$$

This model is known also as a Ricker equation from its similarity to the Ricker stock-recruitment relationship in fisheries (Ricker 1954). The linear form $r + bn_t$ is a simple way of representing density dependent feedback in the per-unit-abundance growth rate (as defined by $\ln n_{t+1} - \ln n_t$). The deterministic model has a positive point equilibrium at

$$\tilde{n} = -r/b, \qquad (4)$$

provided $b < 0$.

Eq. 3 may seem an overly simplified representation of the complex processes of density dependence in natural populations. However, the linear relationship $r + bn_t$ can be regarded as a Taylor series approximation near $\tilde{n}$ of a more biologically detailed rate function (Dennis and Patil 1984, Dennis and Costantino 1988). The stability properties of $\tilde{n}$ and the dynamic behaviors of the deterministic model up to and including chaos are well known (May 1976). The growth model defined by Eq. 1 is a stochastic generalization of Eq. 3 and can be regarded as a type of stochastic, discrete time logistic model.

This stochastic logistic model becomes a first-order nonlinear autoregression model when transformed to a logarithmic scale. By letting $X_t = \ln N_t$, we obtain the following time series model from Eq. 1:

$$X_{t+1} = X_t + a + be^{X_t} + \sigma Z_t. \qquad (5)$$

Transforming the model to a logarithmic scale has three main advantages. First, theoretical statistical knowledge about such nonlinear autoregressive models has increased in recent years (Tong 1990). Use of Eq. 5 provides connections between ecological time series data, mathematical population modeling, and established results in mathematical statistics. Second, valid point estimates (but not confidence intervals or hypothesis tests) of the parameters can be obtained with ordinary linear regression packages (see *Parameter estimation*). Third, for some parameter values, one can obtain a diffusion process approximation to X_t. Such an approximation provides simple expressions for long-run statistical properties of X_t, including the stationary distribution and mean first-passage times (see *Discussion*).

The model can be altered to include second- or higher order lags. The model would take the form

$$N_{t+1} = N_t \exp(a + b_1 N_t + b_2 N_{t-1} + \ldots + b_m N_{t-m+1} + \sigma Z_t) \qquad (6)$$

for incorporating time lags up to order m. Using this model, Turchin (1990) and Turchin et al. (1991) have argued for the prevalence of second-order lags in ecological populations. We make some preliminary recommendations in this paper (see *Discussion*) about how

208   BRIAN DENNIS AND MARK L. TAPER   Ecological Monographs
                                                                                      Vol. 64, No. 2

testing for second-order lags might be accomplished. A full account of the statistical properties of Eq. 6 and of statistical inference methods for the model must be deferred to a future paper.

An alternative first-order population model was introduced by Reddingius (1971):

$$N_{t+1} = N_t \exp(a + b \ln N_t + \sigma Z_t). \quad (7)$$

Royama (1981) modified this model to incorporate higher order time lags. On a logarithmic scale, Eq. 7 becomes

$$X_{t+1} = X_t + a + bX_t + \sigma Z_t, \quad (8)$$

where, as before, $X_t = \ln N_t$. A statistical motivation for use of this model is that it can be written in the form of a linear, first-order autoregressive model (AR(1)):

$$X_{t+1} - \mu = \beta(X_t - \mu) + \sigma Z_t, \quad (9)$$

with $\mu = -a/b$ and $\beta = 1 + b$. The AR(1) model has well-known statistical properties and established, packaged inference procedures. Written in the form of Eq. 7, the AR(1) model is seen to be a type of discrete time, stochastic Gompertz model [the Gompertz growth equation is $(1/n)dn/dt = a + b \ln n$]. While the statistical convenience of this model is a desirable quality, its biological postulate is that growth rate depends, if at all, only logarithmically on population density. By contrast, the stochastic logistic model (Eq. 1) structurally allows for stronger density dependence.

Computer-generating a time series from the stochastic logistic model using Eq. 5 is a simple procedure. Given numerical values of a, b, and σ^2, and starting at a fixed value $X_0 = x_0$, one can easily calculate X_1, X_2, . . . recursively with the help of a routine for generating standard normal random variables. The simplicity of generating trajectories from the hypothesized stochastic mechanism underlying the data turns out to be a key for convenient and powerful statistical inferences (see *Parameter estimation* and *Hypothesis testing*).

Distinguishing three cases of the model (Eq. 5) is important to density dependence testing. The cases form a series of three nested hypotheses. The simplest is Model 0:

$$H_0: a = 0, b = 0. \quad (10)$$

Model 0 defines X_t as a discrete time Brownian motion process (or Wiener process) with zero drift ($a = 0$). This is the classic "random walk" model; X_t has a normal distribution centered at x_0 with a variance of $\sigma^2 t$. No feedback of population density to the growth rate takes place ($b = 0$).

Model 1 is

$$H_1: a \neq 0, b = 0. \quad (11)$$

Model 1 also defines X_t as a discrete time Brownian motion process, but this time with a positive or neg-

ative drift parameter ($a \neq 0$). Under Model 1, X_t has a normal distribution with a mean of $x_0 + at$ and a variance of $\sigma^2 t$. Model 1 in the original population abundance scale (Eq. 1) is a type of stochastic exponential growth or decay model. It is identical to the model described by Dennis et al. (1991) for estimating extinction risks for endangered species (their parameter μ is the same as a in this paper; σ^2 is the same quantity in both papers). In Model 1, no density dependent feedback occurs ($b = 0$).

Finally, it is Model 2 that contains full-fledged density dependence:

$$H_2: a \neq 0, b \neq 0. \quad (12)$$

In many cases, a one-sided variant of Model 2 is of greatest interest, and we can redefine Model 2 as

$$H_2: a \neq 0, b < 0. \quad (13)$$

Testing for density dependence can be regarded as determining whether the added parameter in Model 2 produces noticeably improved description of the data. The first step in such determination is estimating the unknown parameters from the data.

Parameter Estimation

Statistically, the problem of connecting the model (Eq. 1) with data amounts to specifying a likelihood function. Let $n_0, n_1, \ldots, n_q$ be the recorded population abundances, so that q is the number of one-step transitions and $q + 1$ is the total number of observations in the time series. Let $x_0 = \ln n_0$, $x_1 = \ln n_1$, . . . , $x_q = \ln n_q$ denote the log-transformed abundances. The likelihood function gives the probability that, under the stochastic mechanism defined by Eq. 1, the outcome of the process N_t would be the observed time series. The likelihood function is defined as the joint probability density function (pdf) for the random variables $N_0, N_1, N_2, \ldots, N_q$ evaluated at $n_0, n_1, n_2, \ldots, n_q$. It is more convenient to specify the likelihood function for the log-transformed observations, because of the autoregressive structure indicated by Eq. 5. Given that log-population size is at x_{t-1} at time $t - 1$, the distribution of X_t is, according to Eq. 5, normal with a mean of $x_{t-1} + a + b \exp(x_{t-1})$ and a variance of σ^2. Thus, the pdf for X_t, given $X_{t-1} = x_{t-1}$, is a normal curve:

$$p(x_t | x_{t-1}) = (\sigma^2 2\pi)^{-1/2} \exp[-(x_t - x_{t-1} - a \\ - be^{x_{t-1}})^2/(2\sigma^2)]. \quad (14)$$

Because of the Markov property, the joint likelihood of the data is the likelihood of a transition from x_0 to x_1, multiplied by the likelihood of a transition from x_1 to x_2, etc. This joint likelihood is just a product of normal pdf's of the form given by Eq. 14. It is a function of the data values $x_0, x_1, \ldots, x_q$, and, more importantly, of the unknown model parameters a, b, and σ^2:

TABLE 1. Maximum likelihood estimates for unknown parameter σ_0^2 in Model 0 (random walk), parameters a_1 and σ_1^2 in Model 1 (stochastic exponential growth), and parameters b_2, a_2, and σ_2^2 in Model 2 (stochastic logistic growth). Observed population abundance at time t is n_t; also $x_t = \ln n_t$, $y_t = x_t - x_{t-1}$, $\bar{y} = (y_1 + y_2 + \cdots + y_q)/q$, and $\bar{n} = (n_0 + n_1 + \cdots + n_{q-1})/q$.

Model 0	$\hat{\sigma}_0^2 = \dfrac{1}{q} \displaystyle\sum_{t=1}^{q} y_t^2$		
Model 1	$\hat{a}_1 = \bar{y}$	$\hat{\sigma}_1^2 = \dfrac{1}{q} \displaystyle\sum_{t=1}^{q} (y_t - \bar{y})^2$	
Model 2	$\hat{b}_2 = \dfrac{\displaystyle\sum_{t=1}^{q} (y_t - \bar{y})(n_{t-1} - \bar{n})}{\displaystyle\sum_{t=1}^{q} (n_{t-1} - \bar{n})^2}$	$\hat{a}_2 = \bar{y} - \hat{b}_2\bar{n}$	$\hat{\sigma}_2^2 = \dfrac{1}{q} \displaystyle\sum_{t=1}^{q} (y_t - \hat{a}_2 - \hat{b}_2 n_{t-1})^2$

$L(a, b, \sigma^2)$

$$= \prod_{t=1}^{q} p(x_t | x_{t-1})$$

$$= \sigma^2 2\pi^{-q/2} \exp\left[-\frac{1}{2\sigma^2} \sum_{t=1}^{q} (x_t - x_{t-1} - a - be^{x_{t-1}})^2 \right].$$
(15)

This likelihood function $L(a, b, \sigma^2)$ plays a central role in parameter estimation and hypothesis testing.

We must note that Eq. 15, strictly speaking, is not the joint pdf of X_0, X_1, . . . , X_q evaluated at the observations (it is not the full likelihood). Rather, it is the joint pdf of X_1, X_2, . . . , X_q, conditional on $X_0 = x_0$, and evaluated at the observations. We recommend conditioning on the initial observed population size (and using Eq. 15 as the likelihood) because in practice the probabilistic mechanism producing the observation x_0 is typically unknown.

Maximum likelihood (ML) parameter estimates have numerous desirable statistical properties (Stuart and Ord 1991). ML estimates are defined as the parameter values, denoted $\hat{a}$, $\hat{b}$, and $\hat{\sigma}^2$, that jointly maximize $L(a, b, \sigma^2)$ (or equivalently, $\ln L(a, b, \sigma^2)$). ML estimates are asymptotically efficient (they have the smallest variances in large samples), are consistent (variances approach zero as $q \to \infty$), are asymptotically unbiased (biases approach zero as $q \to \infty$), and have distributions that approach normal distributions for large samples. Standard mathematical statistics books only list these properties for independent, identically distributed observations (Stuart and Ord 1991, Rice 1988). We point out that these desirable properties of ML estimates have been demonstrated for time series models (dependent data) of this type as well (Bhat 1974, Tong 1990).

Obtaining ML estimates for the random walk (Model 0) and exponential growth (Model 1) models is easy. Let $L_0(\sigma^2)$ represent the likelihood function for Model 0, that is, Eq. 15 with $a = 0$ and $b = 0$. Let $L_1(a, \sigma^2)$ represent the likelihood function for Model 1 ($b = 0$). Also, let $y_t = x_t - x_{t-1}$, $t = 1, 2, \ldots, q$. Note that y_t

can be thought of as the per-unit-abundance growth rate observed in the population for transition t. The value of σ^2, denoted $\hat{\sigma}_0^2$, that maximizes $L_0(\sigma^2)$ appears in Table 1. It is easily found by setting $\partial \ln L_0(\sigma^2)/\partial\sigma^2$ equal to zero and solving for σ^2. The values of a and σ^2 that jointly maximize $L_1(a, \sigma^2)$, denoted $\hat{a}_1$ and $\hat{\sigma}_1^2$, also appear in Table 1. They are found by setting $\partial \ln L_1(a, \sigma^2)/\partial a$ and $\partial \ln L_1(a, \sigma^2)/\partial\sigma^2$ equal to zero simultaneously.

The estimates for Models 0 and 1 are familiar ones. Let Y_1, Y_2, . . . , Y_q denote the one-step differences in logarithmic population sizes: $Y_t = X_t - X_{t-1}$ (the value realized by Y_t in the data is y_t). Because the Y_t's are increments of Brownian motion under Models 0 or 1, much is known about their statistical properties (see Dennis et al. 1991). Under Model 0, $Y_t \sim$ normal(0, σ^2) and Y_1, Y_2, . . . , Y_q are independent. Under Model 1, $Y_t \sim$ normal(a, σ^2) and Y_1, Y_2, . . . , Y_q are independent. Both models reduce to simple cases of random sampling from normal distributions. Thus, standard confidence intervals from normal theory for σ^2 (Model 0) or a (Model 1) are valid and exact. Dennis et al. (1991) give further details.

For the full stochastic logistic model (Model 2), obtaining ML estimates is also easy. To keep notation consistent, we will write $L_2(a, b, \sigma^2)$ instead of $L(a, b, \sigma^2)$ for the likelihood function of Model 2 (Eq. 15). The values of b, a, and σ^2 that jointly maximize $L_2(a, b, \sigma^2)$ are listed in Table 1. The Model 2 estimates are also familiar ones. The estimates of a and b are least squares estimates obtained by performing a linear regression of y_t on n_{t-1}, $t = 1, 2, \ldots, q$. Thus, $\hat{a}_2$, $\hat{b}_2$, and $\hat{\sigma}_2^2$ can be calculated with any standard regression package.

Confidence intervals printed by standard regression packages, however, are not valid for Model 2. Printed hypothesis tests for parameter values are not valid either. For the printed intervals and tests to be valid, the Y_t's would have to have independent normal($a + bn_{t-1}$, σ^2) distributions (the standard linear regression model). Under Model 2, though, the Y_t's are not independent, due to the autoregressive structure of the model, nor do they have unconditional normal distributions. The overlap between the standard linear regression model and Model 2 stops at point estimation; both models happen to have identical ML estimates. The statistical distributions of the ML estimates under the two models are radically different, though, and so interval estimates and hypothesis tests for Model 2 cannot be based on the ordinary regression model.

Instead, approximate confidence intervals for parameters in Model 2 can be found by either bootstrapping or jackknifing. Bootstrapping involves estimating the distributions of $\hat{a}_2$, $\hat{b}_2$, and $\hat{\sigma}_2^2$ in some fashion using the data. The following parametric bootstrap method makes efficient use of the information present in modest-sized samples ($q \approx 20$). The procedure is simple but requires some computer programming. First, from the data calculate the ML parameter estimates for Model 2 (Table 1). Second, repeatedly (say, 2000 times) computer-generate time series of the same length as the original data from the *estimated* model:

$$X_t = X_{t-1} + \hat{a}_2 + \hat{b}_2 \exp(X_{t-1}) + \hat{\sigma} Z_t. \quad (16)$$

A given series generated from the estimated model ("bootstrapped" series) will be denoted with asterisks: x_0^*, x_1^*, . . . ,x_q^*. Each series should be started at the observed initial value: $X_0 = x_0^* = x_0$. To each such bootstrapped time series, fit Model 2 by calculating bootstrapped ML estimates, denoted $\hat{a}_2^*$, $\hat{b}_2^*$, and $\hat{\sigma}_2^{2*}$, using the expressions in Table 1. Third, for an approximate 95% confidence interval for a, take the 2.5[th] and the 97.5[th] sample percentiles of the 2000 $\hat{a}_2^*$ values. Likewise, use the $\hat{b}_2^*$ and the $\hat{\sigma}_2^{2*}$ values to construct confidence intervals for b and σ^2. An adjustment known as the bias-corrected percentile method (see Efron and Gong 1983) might make the coverage rate of the bootstrap confidence interval closer to 95%.

Jackknifing also involves using the data to estimate the distributions of $\hat{a}_2$, $\hat{b}_2$, and $\hat{\sigma}_2^2$, and it typically requires less computing time. Lele (1991) describes a jackknifing technique for dependent data. For Model 2, Lele's technique entails dropping transitions from the data one by one and refitting the model each time to the remaining transitions. A transition here is a change in the data over one time step, that is, from x_{t-1} to x_t. One performs the regressions on the pairs $(y_1, n_0), (y_2, n_1), . . . , (y_q, n_{q-1})$, each time leaving out (y_t, n_{t-1}), where $t = 1, 2, . . . , q$. Lele (1991) provides expressions for consistent estimates of the variances and covariances of $\hat{a}_2$, $\hat{b}_2$, and $\hat{\sigma}_2^2$.

We have focused our computer simulations in this paper on evaluating hypothesis testing methods and cannot make any recommendations at this time concerning which type of confidence intervals for Model 2 parameters have superior properties. A large-scale evaluation of the coverage probabilities for bootstrapped and jackknifed confidence intervals is a topic for future research.

Missing data can be handled in the ML estimates by simply incorporating in the analysis all the one-step transitions present in the data. Thus, if the jth year (or whatever time period) population size, n_j, was missing, one would perform the regression calculations using $(y_1, n_0), (y_2, n_1), . . . , (y_{j-1}, n_{j-2}), (y_{j+2}, n_{j+1}), . . . , (y_q, n_{q-1})$. One missing year means that there are $q - 2$ one-step transitions present in the data (two transitions missing). The ML formulas (Table 1) would have $q - 2$ instead of q as a divisor, and $\bar{n}$ (Table 1) would include n_0, n_1, . . . , n_{j-2}, n_{j+1}, n_{j+2}, . . . , n_{q-1} in the sum (but not n_{j-1} or n_q).

HYPOTHESIS TESTING

Statistical theory draws a careful distinction between a statistical hypothesis and a scientific hypothesis (for instance, see Stuart and Ord 1991). A statistical hypothesis is an assumption about the form of a probability model, and a statistical hypothesis test is the use of data to make a decision between two probability models. A scientific hypothesis, on the other hand, is an explanatory assertion about some aspect of nature.

For density dependence studies, a general scientific hypothesis of interest is the assertion that a population's abundance produces a negative feedback effect on its growth rate (Berryman 1991a). From this assertion, we expect that time series observations of a density-dependent population would lead us to favor Model 2 over Model 1 as a model of the population's abundance. However, investigators should be aware that other stochastic mechanisms besides ecological feedback can produce observations that pass statistical density dependence tests, including the test described here (see *Discussion*).

When deciding between two models, the likelihood ratio (LR) test originating with Neyman and Pearson provides the benchmark for test power (Neyman and Pearson 1933, Stuart and Ord 1991). If the two probability models are completely specified (no unknown parameters), the LR test has power greater than or equal to any other size α test, according to the Neyman–Pearson Lemma (Stuart and Ord 1991). Here, however, Models 0, 1, and 2 are not completely specified, that is, they contain unknown parameters. LR tests that are modified to accommodate unknown parameters are sometimes called "generalized LR tests." The statistical criteria for choosing test methods are more complex when one or more of the models is not completely specified. In some simple textbook cases (for example, a one-sided t test), the LR test is the uniformly most powerful test. In other cases, the power of the LR test tends to compare quite favorably to other tests according to various definitions of "asymptotic relative efficiency" (these criteria are reviewed by Serfling 1980 and Stuart and Ord 1991). Thus, LR tests for comparing Models 0 and 1, Models 1 and 2, or Models 0 and 2, if feasible to construct, would likely offer desirable power properties.

In general, the LR test for comparing two models, i and j say, is constructed as follows. The model with the smallest number of unknown parameters typically forms the null hypothesis, while the more complex model becomes the alternate hypothesis. Let $\hat{L}_i$ denote the likelihood function for the null hypothesis (Model i), evaluated at the ML estimates of all unknown parameters in the model. Essentially, $\hat{L}_i$ represents the estimated joint probability density of the data (or estimated likelinood that the data would have arisen) under Model i. Likewise, let $\hat{L}_j$ denote the alternate hypothesis (Model j) likelihood function, maximized over all parameter values permissible in the model. The LR test is to choose between Model i and Model j on the basis of the value of the LR test statistic:

$$\Lambda_{ij} = \hat{L}_i / \hat{L}_j. \tag{17}$$

The decision is made in favor of Model i if $\Lambda_{ij} > c$, where c is some constant cutoff point selected by the investigator and is made in favor of Model j if $\Lambda_{ij} \leq c$. The value of c is selected so that the probability of wrongly choosing Model j when the data in fact arise from Model i (that is, the probability of a Type I error) is fixed at some small number, α (the size of the test). The study of such LR tests occupies a prominent portion of any mathematical statistics text (e.g., Bain and Englehardt 1987, Rice 1988).

The essential problem in constructing an LR test is finding the value of c corresponding to the desired test size, α. In the normal-based linear models of analysis of variance and regression, the test statistic Λ_{ij} is a monotone function of the more familiar variance ratio statistic. Under the null hypothesis, the variance ratio statistic has an F distribution. The value of c then is calculated by transforming the $100(1 - \alpha)^{\text{th}}$ percentile of the appropriate F distribution. In a broad class of other models, including many nonlinear regression models, time series models, and loglinear models, the statistic $G_{ij}^2 = -2 \ln \Lambda_{ij}$ has, under the null hypothesis, a distribution that converges to a chi-square distribution as sample sizes increase. The value of c is obtained (approximately) from the $100(1 - \alpha)^{\text{th}}$ percentile of the chi-square distribution. Unfortunately, for testing among Models 0, 1, and 2, blind application of these traditional results can lead to erroneous inferences.

In the case of testing Model 0 (random walk) vs. Model 1 (exponential growth), the LR statistic does reduce to a statistic with an F distribution (with 1 and $q - 1$ degrees of freedom), or equivalently, a Student's t distribution ($q - 1$ degrees of freedom). Under Model 0, $\hat{L}_0 = L_0(\hat{\sigma}_0^2)$ is the likelihood function evaluated at the ML estimate of σ^2 (Table 1), and $\hat{L}_1 = L_1(\hat{a}_1, \hat{\sigma}_1^2)$ is the likelihood function evaluated at the ML estimates of a and σ^2 (Table 1). The LR test statistic can be written in several algebraically equivalent forms:

$$\Lambda_{01} = L_0(\hat{\sigma}_0^2)/L_1(\hat{a}_1, \hat{\sigma}_1^2) = (\hat{\sigma}_0^2/\hat{\sigma}_1^2)^{-q/2}$$
$$= \{1 + [T_{01}^2/(q - 1)]\}^{-q/2}. \tag{18}$$

Here,

$$T_{01} = \hat{a}_1/[q\hat{\sigma}_1^2/(q - 1)]^{1/2} \tag{19}$$

is the familiar t statistic for testing whether the mean is zero or not for independent normal(a, σ^2) random variables. Under Model 0, T_{01} has a Student's t distribution with $q - 1$ degrees of freedom. Alternatively, $T_{01}^2 = F_{01}$ has an F distribution with 1 and $q - 1$ degrees of freedom. Thus, the cutoff point c for Λ_{01} is a simple function of an F or a Student's t percentile. Model 1 is favored over Model 0 if $|T_{01}| \geq t_{\alpha/2}$, where $t_{\alpha/2}$ is the $100[1 - (\alpha/2)]^{\text{th}}$ percentile of the appropriate Student's t distribution.

In some circumstances, a one-sided hypothesis about a might be of interest. For instance, a test of H_0: $a = 0$ vs. H_1: $a < 0$ could be used to determine if an endangered species is in decline or not, provided the species is not abundant enough for density dependence effects to be important. The one-sided test in this case would reject H_0 if $T_{01} \leq t_{1-\alpha}$.

Unfortunately, for testing Model 1 vs. Model 2, the distribution of the LR statistic is unknown. The likelihood function maximized under Model 2 (Eq. 15) is given by $\hat{L}_2 = L_2(\hat{a}_2, \hat{b}_2, \hat{\sigma}_2^2)$. The LR test statistic can be written in the following forms:

$$\Lambda_{12} = L_1(\hat{a}_1, \hat{\sigma}_1^2)/L_2(\hat{a}_2, \hat{b}_2, \hat{\sigma}_2^2)$$
$$= (\hat{\sigma}_1^2/\hat{\sigma}_2^2)^{-q/2} = \exp(-G_{12}^2/2)$$
$$= \{1 + [T_{12}^2/(q - 2)]\}^{-q/2}. \tag{20}$$

Here $G_{12}^2 = -2 \ln \Lambda_{12}$, and

$$T_{12} = \hat{b}_2[(q - 2) \sum_{t=1}^{q} (n_{t-1} - \bar{n})^2/(q\hat{\sigma}_2^2)]^{1/2}. \tag{21}$$

The statistic T_{12} is identical to the familiar t statistic used for testing whether the slope parameter is nonzero in a linear regression. However, T_{12} (Eq. 21) does not have a Student's t distribution, not even approximately, due to the time dependence of the observations. Testing for density dependence based on an assumed Student's t distribution for T_{12} produces unacceptably inflated Type I error rates (see *Discussion*).

One should not even be lulled into using the traditional chi-square approximation for the distribution of G_{12}^2. Under the exponential growth model of the null hypothesis, the population is not ergodic (N_t does not probabilistically tend to return to any given abundance level). The value $b = 0$ is at the edge of the set of values ($b < 0$) for which the stochastic process N_t is ergodic. Without ergodicity, the theorems of mathematical statistics that give the chi-square approximation for G_{12}^2 do not apply. Simulations (not reported here) indicated that the use of the chi-square approximation produces inflated Type I error rates. As can be seen from Eq. 9, the situation is akin to testing whether $\beta = 1$ in an AR(1) process, a well-known case in which the chi-square approximation for G_{ij}^2 fails (Dickey and Fuller 1981).

Instead, the distribution of Λ_{12} (or T_{12}, or $G_{12}{}^2$) can be estimated from the data through parametric bootstrapping. The critical percentile, c, is an unknown function of the two unknown parameters, a and σ^2, in the null hypothesis (Model 1). If Model 1 did indeed give rise to the data, then the ML estimates (Table 1) are in principle quite good estimates of a and σ^2. In fact, the ML estimate $\hat{a}_1$ and the bias-corrected estimate of σ^2 given by

$$\tilde{\sigma}_2{}^2 = q\hat{\sigma}_1{}^2/(q-1) \tag{22}$$

are the uniformly minimum variance unbiased estimates. We would expect therefore that the time series model given by

$$X_t = X_{t-1} + \hat{a}_1 + \tilde{\sigma}_1 Z_t, \tag{23}$$

where $Z_1, Z_2, \ldots$ are independent normal$(0,1)$ random variables, represents a reasonably good estimate of the mechanism that produced the data under the null hypothesis. We have found in our simulations a slight but detectable advantage to using the unbiased estimate, $\tilde{\sigma}_1{}^2$, in place of $\hat{\sigma}_1{}^2$ (see *Test validation*).

The bootstrap idea is straightforward. Generate data sets repeatedly from the *estimated* null hypothesis model (Eq. 23). For each of these "bootstrap" data sets, fit Models 1 and 2 and calculate an LR statistic (Λ_{12}, T_{12}, or $G_{12}{}^2$). The resulting 2000 or so LR statistic values constitute a random sample from the *estimated* distribution of the test statistic under the null hypothesis. The appropriate sample percentile of those values becomes the estimated critical value for the test. We use the term "parametric bootstrap test" instead of "Monte Carlo test" to emphasize the fact that the model under H_1 is being estimated (Beran 1986, Efron 1986). The terminology is common in the statistics literature (e.g., Schork 1992).

The parametric bootstrap likelihood ratio (PBLR) test is quite simple to conduct using the following steps. The two-sided test of H_1: $b = 0$ vs. H_2: $b \neq 0$ is described first. (1) Obtain ML estimates for all parameters in Models 1 and 2 using the expressions in Table 1. (2) Calculate $T_{12}{}^2$ (or $G_{12}{}^2$, or Λ_{12}) as in Eq. 21. (3) Generate 2000 or more data sets in the form $x_0{}^*$, $x_1{}^*$, $\ldots$, $x_q{}^*$ from the estimated null model (Eq. 23). Each of these bootstrap data sets should start at $x_0{}^* = x_0$ and be the same length as the original set. (4) Calculate for each bootstrap data set the parameter estimates for Models 1 and 2 (Table 1), obtaining $\hat{a}_1{}^*$, $\hat{\sigma}_1{}^{2*}$, $\hat{a}_2{}^*$, $\hat{b}_2{}^*$, $\hat{\sigma}_2{}^{2*}$. (5) Obtain in this fashion 2000 (or so) bootstrap values of the LR test statistic, $T_{12}{}^{2*}$ (Eq. 21; or $\Lambda_{12}{}^*$ or $G_{12}{}^{2*}$). Each of these $T_{12}{}^{2*}$ values represents an independent observation from the *estimated* distribution of $T_{12}{}^2$ (likewise Λ_{12}, $G_{12}{}^2$). (6) Take the $100(1-\alpha)^{\text{th}}$ sample percentile, $\hat{f}_\alpha$, as the estimate of the critical percentile of the distribution of $T_{12}{}^2$. (7) Reject H_1: $b = 0$ in favor of H_2: $b \neq 0$ if the original value of $T_{12}{}^2$ is greater than $\hat{f}_\alpha$.

Alternatively, at step 6 one can estimate a P value

for the test with the proportion of $T_{12}{}^{2*}$ values that are greater than or equal to $T_{12}{}^2$. The null hypothesis would be rejected if $\hat{P} \leq \alpha$, where $\hat{P}$ is the estimated P value.

Note that in each bootstrap cycle of the calculations, only the value of $T_{12}{}^{2*}$, the original data, the original ML estimates, and the original test statistic need to be retained; the values $\hat{a}_1{}^*$, $\hat{b}_1{}^*$, $\hat{a}_2{}^*$, $\hat{b}_2{}^*$, $\hat{\sigma}_2{}^{2*}$, and $x_0{}^*$, $x_1{}^*$, $\ldots$, $x_q{}^*$ do not need to be stored. The procedure is easily programmed and is acceptably fast: our program, written in the GAUSS matrix language (Aptech Systems 1991), completes the calculations, using 8000 bootstrap samples, for one moderately sized ($q \approx 16$) data set in ≈ 11 min on an old IBM AT/286. (A short SAS program written by the authors for conducting the PBLR test takes ≈ 60 min to run on a 286 machine.)

The one-sided PBLR test of H_1: $b = 0$ vs. H_2: $b < 0$ is straightforward. In step 1 above, reject H_2 outright if $\hat{b}_2 \geq 0$; otherwise continue. In step 5 above, calculate $T_{12}{}^*$ using Eq. 21 for each of the bootstrap samples, instead of $T_{12}{}^{2*}$. In step 6, estimate the critical percentile of the distribution of T_{12} with the $100\alpha^{\text{th}}$ sample percentile, $\hat{f}_{1-\alpha}$, of the $T_{12}{}^*$ values. Alternatively, estimate a P value by the proportion of $T_{12}{}^*$ values that are less than T_{12}. In step 7, reject H_1 if $T_{12} \leq \hat{f}_{1-\alpha}$, or if $\hat{P} < \alpha$.

We point out that the distribution of T_{12} is not symmetric, nor is it centered at zero. The two-sided test conducted with the $100(\alpha/2)^{\text{th}}$ and $100[1-(\alpha/2)]^{\text{th}}$ sample percentiles of the $T_{12}{}^*$ values is different from the previously described two-sided test that uses the $100(1-\alpha)^{\text{th}}$ percentile of the $T_{12}{}^{2*}$ values. The power properties of both tests have not been compared.

A PBLR test of Model 0 against Model 2 might be of interest in some studies. Other methods for distinguishing a drift-free random walk from a density-dependent process (e.g., den Boer 1991) are likely not as powerful. The one-sided test of H_0: $a = 0$, $b = 0$ vs. H_2: $a \neq 0$, $b < 0$ would be conducted as follows. (1) Obtain ML estimates for all parameters in Models 0 and 2 using Table 1. Reject H_2 outright if $\hat{b}_2 \geq 0$; otherwise continue. (2) Calculate the LR statistic as

$$\begin{aligned}
\Lambda_{02} &= L_0(\hat{\sigma}_0{}^2)/L_2(\hat{a}_2, \hat{b}_2, \hat{\sigma}_2{}^2) \\
&= (\hat{\sigma}_0{}^2/\hat{\sigma}_2{}^2)^{-q/2}. \tag{24}
\end{aligned}$$

(3) Generate bootstrap data sets in the form $x_0{}^*$, $x_1{}^*$, $\ldots$, $x_q{}^*$ from the estimated random walk model,

$$X_t = X_{t-1} + \hat{\sigma}_0 Z_t, \tag{25}$$

starting at $x_0{}^* = x_0$. (4) For each bootstrap data set, calculate the parameter estimates for Models 0 and 2: $\hat{\sigma}_0{}^{2*}$, $\hat{a}_2{}^*$, $\hat{b}_2{}^*$, $\hat{\sigma}_2{}^{2*}$. From these parameter estimates, calculate a value of the LR statistic as

$$\Lambda_{02}{}^* = \begin{cases} 1, & \hat{b}_2{}^* \geq 0; \\ (\hat{\sigma}_0{}^{2*}/\hat{\sigma}_2{}^{2*})^{-q/2}, & \hat{b}_2{}^* < 0. \end{cases} \tag{26}$$

(5) Obtain in this fashion 2000 (or so) values of $\Lambda_{02}{}^*$.

May 1994   DENSITY DEPENDENCE   213

The provision of setting $\Lambda_{02}{}^*$ equal to 1 when $\hat{b}_2{}^* \geq 0$ will result in a one-sided test. (6) Take the $100\alpha^{th}$ percentile, $\hat{\lambda}_\alpha$, as the estimate of the critical percentile of Λ_{02}. Alternatively, estimate a P value by the proportion of $\Lambda_{02}{}^*$ values that are less than Λ_{02}. (7) Reject H_0 if $\Lambda_{02} \leq \hat{\lambda}_\alpha$, or if $\hat{P} < \alpha$.

Of course, Model 1 contains Model 0 as a special case. The previously described test of Model 1 against Model 2 thus implicitly includes Model 0 in the null hypothesis. Normally, the test of Model 1 against Model 2 should be used, unless there are specific reasons for restricting the null hypothesis to a pure random walk.

The PBLR tests can accommodate missing data. The fundamental "observation" in the likelihood function (Eq. 15) is not a (logarithmic) population size, x_t, but a *transition* from x_{t-1} to x_t. If x_j is missing from the data, it means that two transitions are missing: x_{j-1} to x_j, and x_j to x_{j+1}. The likelihood function (Eq. 15) with $p(x_j|x_{j-1})$ and $p(x_{j+1}|x_j)$ omitted from the product is the joint pdf of X_1, X_2, . . . , X_{j-1} given x_0 and X_{j+2}, X_{j+3}, . . . , X_q given x_{j+1}. This likelihood is used as the fundamental building block for parameter estimates and hypothesis tests.

When observations are missing, the ML parameter estimates for Model 2 are easiest to calculate with a least squares approach. The formulae (Table 1) otherwise must be altered in messy ways. Take as the fundamental data set the logarithmic differences y_1, y_2, . . . , y_{j-1}, y_{j+2}, y_{j+3}, . . . , y_q (y_j, y_{j+1} missing). The sample mean of the y_i^2 values provides the ML estimate of σ^2 for Model 0. The sample mean of the differences, $\bar{y}$, and the sample mean of the $(y_t - \bar{y})^2$ values provide respectively the ML estimates of a and σ^2 for Model 1. The linear regression of y_t on n_{t-1} (using all transitions present in the data) provides ML estimates of a and b for Model 2. The ML estimate for σ^2 in Model 2 would be the sum of squared errors (from the regression) divided by the number of transitions present ($q - 2$, if just one population size is missing). The bootstrap data sets are generated (Eq. 23 or 25) as a series of one-step transitions starting at x_0 (x_0, $X_1{}^*$, $X_2{}^*$, . . . , $X_{j-1}{}^*$), followed by a series starting at x_{j+1} (x_{j+1}, $X_{j+2}{}^*$, . . . , $X_q{}^*$). All parameter estimates and tests are thus conditioned on the observed starting values, x_0 and x_{j+1}, of series of one-step transitions.

An alternative approach to testing with missing data is to condition only on x_0. Parameter estimates and the test statistic are computed as described above using all one-step transitions present in the data. However, bootstrap data sets are generated starting at x_0 for all times (x_0, $X_1{}^*$, $X_2{}^*$, . . . , $X_q{}^*$), including missing times. The bootstrap values of the test statistic are then calculated after omitting from the bootstrap data sets the generated observations occurring at missing times (that is, omit $X_j{}^*$ before calculating $T_{12}{}^*$).

The two approaches to handling missing data are subtly different. The first treats the uninterrupted time series essentially as separate series, but assumes the series are governed by the same (density-independent or -dependent) model. The first approach would be preferred if, for instance, the population was restarted at size x_{j+1} after some drastic change (a harvest or catastrophe). The second approach treats the uninterrupted series as one single series with some observations (the missing ones) simply unknown. Which approach is most appropriate will be case specific, although the resulting tests are not likely to differ much unless the number of missing transitions is large. The statistical properties of the two approaches have not yet been compared.

Because the PBLR test is a parametric test, some additional model checking is advised in any application. Judicious use of model diagnostic techniques will help minimize problems associated with "Type III error" (fitting the wrong model). In particular, serious departures of the data from the Markov property or from the model could likely be detected through some form of residual analysis. For the stochastic logistic model, diagnostic techniques would center around the conditional residuals: $\hat{e}_t = x_t - x_{t-1} - \hat{a} - \hat{b} \exp(x_{t-1})$, $t = 1, 2, . . . , q$. Under the model assumptions, these residuals should be approximately normal white noise. The residuals can be subjected to the customary analysis techniques of linear time series modeling (see Tong 1990:322 for discussion). We often use the Lin-Mudholkar test for normality against asymmetric alternatives (Lin and Mudholkar 1980, Tong 1990:324) in addition to the standard normal plots and autocorrelation tests. There is a caveat, however: in the nonlinear setting, the adequacy of the normal-white noise approximation is unknown and varies from model to model. The properties of these diagnostic techniques for the stochastic logistic model would be a worthwhile topic for future study.

According to statistical principles of asymptotic relative efficiency, the PBLR tests represent approximate size α tests having power functions that cannot be exceeded by much. The principles rest on large-sample theorems of statistics (Serfling 1980). But do these asymptotic assurances apply to the data sets of moderate lengths likely to be encountered in ecological practice? A large-scale, Monte Carlo power assessment of the PBLR test of Model 1 against Model 2, along with comparative studies of other available tests, provides some answers.

TEST VALIDATION

It is desirable that statistical tests have several properties. First, their size should be close to their nominal size, that is, if the null hypothesis is true the probability of rejecting the null hypothesis should be close to what the investigator thinks it is. Second, the test should be powerful enough to detect scientifically interesting deviations from the null hypothesis. Third, the test should be robust to measurement error. This last property is

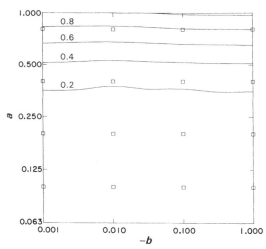

FIG. 1. A contour plot showing estimated power of the parametric bootstrap likelihood ratio test of density dependence as a function of model parameters a and b. The null hypothesis for the test is Model 1 ($b = 0$), and the alternate hypothesis is Model 2 ($b < 0$). Each □ indicates location of 1000 simulated tests; each test used 200 bootstrap samples and the values $q = 19$, $n_0 = -a/b$, and $\sigma = 0.2$. The power surface was estimated using a distance weighted least squares algorithm (McLain 1974).

particularly important when dealing with population abundance data, which commonly contain substantial uncertainty.

We have investigated the qualities of the PBLR test of Model 1 against Model 2 (both one- and two-sided) with Monte Carlo simulation. Using a known set of parameters we generate a time series of simulated population densities according to Eq. 1. This simulated time series is then subjected to the PBLR test exactly as if it were data from field observations. For each set of parameters (a, b, σ, q, and n_0: we have couched the simulations in terms of σ rather than σ^2) chosen for study, this process was repeated a large number of times, usually 1000. All tests were conducted at a nominal 0.05 level. The proportion of times the null hypothesis was rejected was recorded for each parameter set. If the parameter b was zero, that is if the null hypothesis of no density dependence was true, this proportion represents an estimate of the size of the test. If b was not zero then the proportion of rejections is an estimate of the power of the test under the set of parameters. We denote by $\phi(a, b, \sigma, q, n_0)$ the probability of rejection of the null hypothesis as a function of model parameters (power function), and by $\hat{\phi}(a, b, \sigma, q, n_0)$ its estimate from simulations.

Test size

We examined the size of the PBLR test (one- and two-sided) under a broad range of parameters. The parameters a and σ ranged from 0.05 to 1.6, while q

ranged from 4 to 64. In all, 64 different sets of parameters were tested each with two separate simulations of 1000 trials. The sample mean of these 128 values of $\hat{\phi}$ for the one-sided tests was 0.0504 with a sample variance of 0.0000470. The underlying population mean of the $\hat{\phi}$ values is thus not significantly different from 0.05 ($Z = 0.66$, $N = 128$, $P = .25$). The sample variance was close to the variance expected under binomial sampling, 0.0000475 ($= 0.05(0.95)/1000$). Further, despite the wide variety of parameters used, the distribution of $\hat{\phi}$ values observed was not significantly different from a normal distribution with a mean and variance of 0.05 and 0.0000475, according to a Kolmogorov–Smirnov test ($D = 0.0611$, $N = 128$, $P > .5$). Results for the two-sided tests (based on T_{12}^2) were similar. Thus there is no reason to believe that the true size of the PBLR test is different from its nominal size. If any deviations do exist, they are of insignificant magnitude.

The size results reported above were obtained for the PBLR test that uses the unbiased estimate, $\tilde{\sigma}_1^2$, instead of the ML estimate, $\hat{\sigma}_1^2$, in the estimated null hypothesis model (Eq. 23). We detected through simulations a slight but noticeable increase in the test size over the nominal size of 0.05 when the ML estimate is used. While the increase is small enough to be of little practical importance, it is easily corrected simply by using $\tilde{\sigma}_1^2$.

Test power

As with the size of the test, we investigated power extensively with simulations. The power of a test depends in general on the specific true values of the parameters. We have assessed the influence of b, a, σ, q, and n_0 on power. A number of our results contradict unreflective intuition.

First, the probability, $\phi(a, b, \sigma, q, n_0)$, of rejecting the null hypothesis is nearly independent of the parameter b as long as b is not zero (Fig. 1). Thus, the influence of b on power is not continuous; instead, b acts as a switch to change the qualitative behavior of the model. This discontinuity in the power function is, from the standpoint of statistical theory, unusual (e.g., Bain and Engelhardt 1987:373). One would normally expect the power function to increase smoothly, starting from a level of α, as the parameter in question becomes farther from its hypothesized null value.

However, such smooth textbook dependence of power on b would in fact be an undesirable property. The parameter b is related to the level around which N_t is fluctuating according to the density-dependent model (Eq. 1). While the concept of point equilibrium (carrying capacity) is of questionable meaning in a stochastic model (Dennis and Costantino 1988, Wolda 1989), we can see that the level $-a/b$ (Eq. 5) represents a center for the return tendencies of N_t. If $N_t > -a/b$, then $\ln N_{t+1}$ is expected to decrease (Eq. 5), while if $N_t < -a/b$, $\ln N_{t+1}$ is expected to increase. One presum-

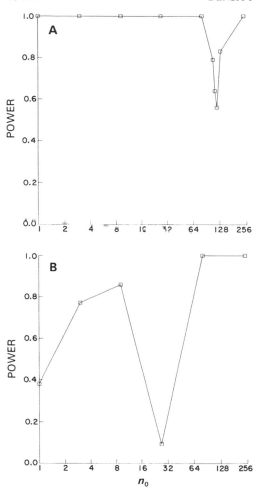

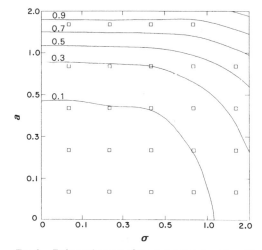

FIG. 2. Estimated power of the parametric bootstrap likelihood ratio test of density dependence as a function of initial population density, n_0. The null hypothesis for the test is Model 1 ($b = 0$), and the alternate hypothesis is Model 2 ($b < 0$). Each square represents 1000 simulated tests; each test used 200 bootstrap samples and the values $q = 9$, $b = -0.01$, $\sigma = 0.05$. (a) $a = 1.2$. (b) $a = 0.3$.

the population density to move toward a central value when displaced from it. It makes sense that a deviation in the initial population size would increase the test power. However, if the parameter a is low, then power decreases again if n_0 is too far below the value of $-a/b$ (Fig. 2b). With a low a and a low initial size, the population will tend to increase for a number of time steps, making it difficult to distinguish the time series from one that would be produced by a population undergoing exponential growth.

We turn now to the effect of environmental variation in growth rate on the power of the PBLR test. The parameter that measures environmental variation is σ, the standard deviation of $Y_t = \ln (N_t/N_{t-1})$ conditional on $N_{t-1} = n_{t-1}$. (The standard deviation of Y_t conditional on $N_0 = n_0$ is an unknown, increasing function of σ that depends on t as well.) Recall that σ is estimated by the root-mean-squared error in a linear regression of the y_t values on the n_{t-1} values (Table 1). The influence of σ on power is as intriguing as the influence of n_0. As σ increases so does the power of the PBLR test, although the effect is minimal until σ is around the magnitude of a (Fig. 3). This increase in power is counterintuitive to investigators accustomed to thinking about "error" in standard regressions. However, the above discussion of n_0 and power resolves the apparent contradiction. In a nutshell, the test works by detecting return toward an abundance level from deviations away from that level. Stochastic fluctuations provide some of these deviations.

As would be expected, increasing the length of the time series, $q + 1$, increases power. Fig. 4 shows $\hat{\phi}$ as

FIG. 3. Estimated power of the parametric bootstrap likelihood ratio test of density dependence as a function of model parameters a and σ. The null hypothesis for the test is Model 1 ($b = 0$), and the alternate hypothesis is Model 2 ($b < 0$). Each square represents 1000 simulated tests; each test used 200 bootstrap samples and the values $q = 9$, $b = -0.01$, $n_0 = -a/b$. Contours were drawn as in Fig. 1.

ably would not want the power of a density dependence test to depend on whether the population was fluctuating around 1000 or 100, but rather only on whether the population was fluctuating around (that is, was showing some tendency to return to) some unspecified level. It is thus entirely reasonable that the numerical value of b does not affect the power of the test under Model 2, because b is simply a reflection of the units in which population size is measured.

The interaction of initial population size, n_0, and power is also interesting. Power is quite low when n_0 is near $-a/b$. As n_0 deviates from this value, power increases (Fig. 2a). What distinguishes the density dependent model from the null model is the tendency for

216 BRIAN DENNIS AND MARK L. TAPER Ecological Monographs
Vol. 64, No. 2

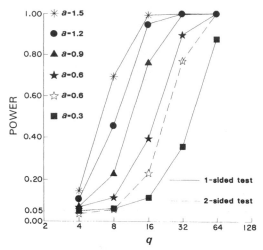

FIG. 4. Estimated power of the parametric bootstrap likelihood ratio test of density dependence as a function of sample size, q (number of one-step transitions in the time series). The null hypothesis for the test is Model 1 ($b = 0$). Each symbol represents 1000 simulated tests; each test used 200 bootstrap samples and the values $b = -0.01$, $\sigma = 0.05$, $n_0 = -a/b$.

a function of q for a spectrum of a values ranging from 0.3 to 1.2. All simulations in this figure had initial sizes fixed at $n_0 = -a/b$, and thus power is at a local minimum. Nonetheless, power becomes quite reasonable by the time q is 16. Even time series as short as eight transitions can have a nontrivial probability of rejecting the false null hypothesis if the rate parameter a is not small. In the figure, the two curves marked with stars portray power for a series of runs with a equal to 0.6. The solid line is for one-sided tests and the dashed line represents two-sided tests. The difference between the two types of tests in the probability of correctly rejecting the null hypothesis may be important, particularly when power is modest. In the figure when q is 16, the power of the one-sided test is almost twice the power of the two-sided test. Many of the abundance records for natural populations are only 10–30 yr long. Power is expected to only be moderate. Thus we strongly recommend the use of the one-sided test.

Knowledge of a test's power is extremely useful when interpreting results. If power is low then failure to reject the null hypothesis contributes only weak evidence in favor of the null model. True power can only be quantified if the real parameters are known. For the density dependence test, the power can be estimated in a statistically consistent fashion by substituting the empirically estimated parameters for the true parameters and conducting a Monte Carlo simulation such as described above. However, we suggest some caution in that such power estimates will only approximate the true power in small samples. The maximum likelihood parameter estimates under the density dependent model have a finite sample bias. While this has no influence on the

size of the PBLR test for significance testing, it will affect power estimates.

Measurement error

The sensitivity of regression analysis to measurement errors in the predictor variables is well known (Fuller 1987). We were particularly concerned about the impact of measurement error on the PBLR test because the "predictor" variable in the regression is population density. Populations are frequently estimated rather than censused, and in some cases the time series consist of relative population indices (light trap counts, redd counts, etc.). Sampling error could potentially be an important source of variability in the time series in these situations. What happens if the PBLR test is applied directly to such time series data?

We have investigated the consequences of sampling error with simulations similar to those discussed above in estimating size and power. As above, we generated a time series of simulated population densities according to Eq. 1 using a known set of parameters. This simulated time series was then contaminated with a noise variable representing measurement error and subjected to the PBLR test exactly as if it were data from field observations. Since many population density estimates or relative indices are based in some fashion on binomial or Poisson sampling, the variance of the noise contaminating N_t was itself made proportional to N_t (the observations entering the data set had zero-truncated normal(N_t, cN_t) distributions). The results of this study are summarized in Table 2. Remarkably, the size of the test is hardly influenced even by massive amounts of sampling error. Even more remarkably, power is somewhat increased by this common type of sampling error. We have noted earlier that

TABLE 2. Estimated power of the parametric bootstrap likelihood ratio test of density dependence in the presence of sampling or measurement error. Each observation, n_t, simulated from the stochastic logistic model was replaced by an observation generated from a zero-truncated normal(n_t, cn_t) distribution before testing. Power estimates are based on 1000 trials. In each trial, $q = 9$, $a = 0.5$, $\sigma = 0.1$, and $n_0 = 50$.

b	$c^{1/2}$	Rejection probability
0	0.00	0.051
0	0.05	0.055
0	0.10	0.038
0	0.20	0.052
0	0.30	0.046
0	0.40	0.040
0	0.50	0.037
−0.01	0.00	0.102
−0.01	0.05	0.098
−0.01	0.10	0.099
−0.01	0.20	0.129
−0.01	0.30	0.137
−0.01	0.40	0.147
−0.01	0.50	0.180

TABLE 3. Parametric bootstrap likelihood ratio test of density dependence for the grizzly bear population in the Yellowstone region. The null hypothesis for the test is Model 1 ($b = 0$), and the alternate hypothesis is Model 2 ($b < 0$). Number of bootstrap samples = 8000.

Population size data, N_t*	33	36	34	39	35	34	38	36	37	41	39	51
	47	57	48	59	64							

Maximum likelihood parameter estimates
under Model 1 (density independence) $\hat{a}_1 = 4.14 \times 10^{-2}$; $\hat{\sigma}_1^2 = 1.49 \times 10^{-2}$
Maximum likelihood parameter estimates
under Model 2 (density dependence) $\hat{a}_2 = 1.41 \times 10^{-1}$; $\hat{b}_2 = -2.41 \times 10^{-3}$; $\hat{\sigma}_2^2 = 1.45 \times 10^{-2}$
Likelihood ratio statistics§ $T_{12} = -0.60$; $t_{0.95} = -2.9$; $P = 0.72$

* Original data consist of the yearly counts of adult females seen with cubs, from 1973–1991 (Eberhardt et al. 1986; R. R. Knight, *personal communication*). Values listed here for N_t are calculated from the original data as a 3-yr moving sum (sum of 1973, 1974, and 1975 counts, sum of 1974, 1975, and 1976 counts, etc.).
§ Likelihood ratio test statistic (Eq. 21), estimated fifth percentile of the test statistic distribution under Model 1, and estimated P value for the test.

deviations increase the power of this test. Apparently, these deviations do not even have to be entirely real.

EXAMPLES

We present, in this section, several worked examples of the PBLR test of density dependence. Numerous examples of density dependence testing in the literature have involved insect populations. To the scarce supply of large mammal examples in the density dependence debate, we add a few more (Tables 3–5). Fowler (1984, 1987) has given additional information and insights about density dependence in large mammals. In this section we also analyze 16 insect data sets assembled by den Boer and Reddingius (1989) in order to compare results of the PBLR test to earlier published results.

The grizzly bear (*Ursus arctos horribilis*) population of the greater Yellowstone ecosystem shows no evidence of density dependence in time series abundance data (Table 3). The data (Table 3) consist of a 3-yr running sum of adult females seen with cubs. An adult female produces cubs on average every 3 yr, so the 3-yr running sum of this relatively visible component of the population represents an estimate of the minimum number of adult females in the population (see Knight and Eberhardt 1984, 1985, Eberhardt et al. 1986, and Dennis et al. 1991 for discussion). Table 3 reflects the counts from 1973 through 1991 (Eberhardt et al. 1986, R. R. Knight, *personal communication*). Dennis et al. (1991) found increased variability in the data after 1971, possibly related to the garbage dump closures in 1970–1971 or to institution of new aerial survey methods. The outcome of the density dependence test suggests that the population has not yet reached carrying capacity, that is, the time series gives no reason to favor Model 2 over Model 1.

Results of model diagnostic procedures for the grizzly data are mixed. The residuals from both models are normally distributed, according to the Lin-Mudholkar test (Model 1: LM = -0.28, $P = .78$; Model 2: LM = -0.78, $P = -.43$; see Tong 1990:324). However, the residuals from both models have some autocorrelation, according to standard tests with the first- and second-order sample autocorrelation statistics

(Model 1: $\sqrt{q}\hat{\rho}_1 = -2.73$, $P = .0064$; $\sqrt{q}\hat{\rho}_2 = 1.77$, $P = .077$; Model 2: $\sqrt{q}\hat{\rho}_1 = -2.21$, $P = .027$; $\sqrt{q}\hat{\rho}_2 = 1.71$, $P = .088$; see Tong 1990:324). While the properties of these or other white noise tests have not been investigated for the residuals of Model 2, the results suggest that the grizzly female population has a higher order autocorrelation structure not accounted for by either Model 1 or Model 2. Oscillations from year-class imbalances in the population could cause such autocorrelation. The large variability of the population, though, gives reason for concern about its long-term viability (Dennis et al. 1991).

Two elk (*Cervus elaphus*) populations in the greater Yellowstone ecosystem have noticeable density dependence (Tables 4 and 5). The data on the northern Yellowstone population (Table 4) are winter census records from Houston (1982:17); the data on the central valley population in Grand Teton National Park (Table 5) are from Boyce (1989) and represent summer mark–recapture estimates.

The northern Yellowstone population increased rapidly after artificial removals from the park were ended in 1969. The population appears to have subsequently attained a stochastic equilibrium. The power of the density dependence test was enhanced because the initial population was far from equilibrium (see *Test validation*). Residual plots and tests show no outliers, no significant first- or second-order autocorrelation, and no significant departures from normality (Model 1: LM = 0.91, $P = .36$; $\sqrt{q}\hat{\rho}_1 = 0.89$, $P = .37$; $\sqrt{q}\hat{\rho}_2 = 1.26$, $P = .26$; Model 2: LM = -1.08, $P = .28$; $\sqrt{q}\hat{\rho}_1 = -1.53$, $P = .13$; $\sqrt{q}\hat{\rho}_2 = 0.21$, $P = .83$; see Tong 1990:324).

The central valley population in Grand Teton National Park fluctuates substantially, and the estimated P value for the test is just under .05. The Grand Teton population has a missing observation in year 1983, and so the test was conditioned on n_{21} (= 1453) in addition to n_0 (= 1627) (see *Hypothesis testing*). The second transition (1527 to 824) is a possible outlier for both Model 1 and Model 2, with a standardized residual of ≈ 2.7 for Model 2. No significant first- or second-order autocorrelation is evident in the first 19 consecutive

218　　　　　　　　　　　　BRIAN DENNIS AND MARK L. TAPER　　　　　　　Ecological Monographs
Vol. 64, No. 2

TABLE 4. Parametric bootstrap likelihood ratio test of density dependence for the northern Yellowstone elk population. The null hypothesis for the test is Model 1 ($b = 0$), and the alternate hypothesis is Model 2 ($b < 0$). Number of bootstrap samples = 8000.

Population size data, N_t*	3172	4305	5543	7281	8215	9981	10 529
	12 607	10 807	10 741	11 855	10 768		

Maximum likelihood parameter estimates under Model 1 (density independence)	$\hat{a}_1 = 1.11 \times 10^{-1}$; $\hat{\sigma}_1{}^2 = 2.06 \times 10^{-2}$
Maximum likelihood parameter estimates under Model 2 (density dependence)	$\hat{a}_2 = 4.68 \times 10^{-1}$; $\hat{b}_2 = -4.14 \times 10^{-5}$; $\hat{\sigma}_2{}^2 = 4.89 \times 10^{-3}$
Likelihood ratio statistics§	$T_{12} = -5.37$; $t_{0.95} = -2.8$; $P = 0.0025$

* Data are winter census records from 1968–1979 listed by Houston (1982). The 1977 value is an adjusted value given by Houston (1982:23).

§ Likelihood ratio test statistic (Eq. 21), estimated fifth percentile of the test statistic distribution under Model 1, and estimated P value for the test.

residuals (Model 1: $\sqrt{19}\hat{\rho}_1 = -0.69$, $P = .49$; $\sqrt{19}\hat{\rho}_2 = -0.95$, $P = .34$; Model 2: $\sqrt{19}\hat{\rho}_1 = 0.48$, $P = .63$; $\sqrt{19}\hat{\rho}_2 = -0.32$, $P = .75$). With the outlier removed, the residuals are acceptably normal (Model 1: LM = 1.95, $P = .051$; Model 2: LM = 1.91, $P = .056$).

den Boer and Reddingius (1989) used the Pollard et al. (1987) randomization test to look for density dependence in 16 insect populations. Their paper provides a table with the original data. The randomization test did not flag a single population as density dependent. By contrast, the PBLR test rejects density independence for two of the populations at the .05 significance level (Table 6). If the regression test for $b < 0$ based on the Student's t distribution is used, the density dependent count jumps to eight (Table 6).

The den Boer and Reddingius (1989) data illustrate the role of test power. The regression test is obviously more powerful but is inappropriate because it is not a size 0.05 test (see *Discussion*). Both the randomization and the PBLR are close to size 0.05 tests. Because of the asymptotic relative efficiency of LR tests, the PBLR test probably represents the practical limit of power for testing $b < 0$ in the stochastic logistic model. Even though the power can exceed that of the randomization test by 50% (see *Discussion*), the basic thrust of den Boer and Reddingius' results remains intact. If there is density dependence in these 16 populations, it is difficult to detect from time series data alone.

Could the two density dependent cases have oc-

curred simply by chance? If one assumes that all 16 populations follow the null hypothesis model (stochastic exponential growth), then two or more "successes" out of 16 trials is a plausible outcome when the success probability is .05 (if $W \sim$ binomial (16, 0.05), then $P[W \geq 2] \approx .19$). However, there is more information present in the collection of P values than simply the number of them < .05. If all the populations were realizations of the null hypothesis model, the P values would represent 16 independent observations from a uniform(0, 1) distribution. Then $-2 \ln P_i$ would be an observation from a chi-square(2) distribution, and the sum of k such values would be an observation from a chi-square($2k$) distribution (Fisher's test; see Fisher 1958). From the P values in Table 6, we find that $\Sigma - 2 \ln P_i = 46.1$, a value that is just below the 95th percentile of the chi-square(32) distribution ($P' \approx .051$). Enough of the P values in this meta-analysis are "leaning" toward the alternate hypothesis end so as to cast doubt upon the assumption that all 16 populations are realizations of the density-independent model, though one would not reject that assumption at a strict .05 significance level.

DISCUSSION

The PBLR test can easily be adapted, if desired, to testing for density dependence under the Gompertz model (Eqs. 7 and 8). The modification would use x_t values in place of n_t values in the parameter estimates

TABLE 5. Parametric bootstrap likelihood ratio test of density dependence for the elk population in the central valley of Grand Teton National Park. The null hypothesis for the test is Model 1 ($b = 0$), and the alternate hypothesis for the test is Model 2 ($b < 0$). Number of bootstrap samples = 8000.

Population size data, N_t*	1627	1527	824	891	1140	1322	1431	1733	1131
	1611	1644	1991	1762	1076	1442	1800	1667	1558
	1396	1753	···	1453	1804				

Maximum likelihood parameter estimates under Model 1 (density independence)	$\hat{a}_1 = 1.45 \times 10^{-2}$; $\hat{\sigma}_1{}^2 = 6.85 \times 10^{-2}$
Maximum likelihood parameter estimates under Model 2 (density dependence)	$\hat{a}_2 = 7.31 \times 10^{-1}$; $\hat{b}_2 = -4.93 \times 10^{-4}$; $\hat{\sigma}_2{}^2 = 4.65 \times 10^{-2}$
Likelihood ratio statistics§	$T_{12} = -2.92$; $t_{0.95} = -2.88$; $P = 0.044$

* Data are summer mark–recapture population estimates from 1963–1985 (1983 missing) listed by Boyce (1989).

§ Likelihood ratio test statistic (Eq. 21), estimated fifth percentile of the test statistic distribution under Model 1, and estimated P value for the test.

(Table 1) for the alternate hypothesis model. The resulting test is essentially the modification of Vickery and Nudd's (1984) "simulation" test that was suggested by Pollard et al. (1987). The test procedure is otherwise identical to the test we have proposed. Mountford (1988) employed such a Gompertz-based PBLR test using Model 0 (random walk) as the null hypothesis.

Reddingius (1971) recognized the theoretical desirability of LR tests, and he developed a small table of critical values for the Gompertz-based LR statistic using Monte Carlo simulation. In practice, the investigator must estimate parameters from the data before referring to the table. Use of Reddingius' table thus is quite similar to conducting a PBLR test for Gompertz-type density dependence.

Many subsequent investigators have taken (implicitly or explicitly) the Gompertz model as the alternate hypothesis in density dependence tests, but have used different test statistics (Bulmer 1975, Royama 1977, Slade 1977, Vickery and Nudds 1984, 1991, Gaston and Lawton 1987, Pollard et al. 1987, den Boer and Reddingius 1989, Reddingius and den Boer 1989, den Boer 1990, Holyoak and Lawton 1992, Woiwod and Hanski 1992). Crowley (1992) modified the Gompertz to include sampling variability. The statistics include the sample correlation of the y_t's and x_{t-1}'s (Pollard et al. 1987), the slopes of principal and reduced major axes (Slade 1977), the reciprocal of von Neumann's ratio (Bulmer 1975), and the number of times that the one-step transitions have moved toward (or away) from a given abundance level (Crowley 1992).

Of the tests studied by these investigators, the randomization test of Pollard et al. (1987) based on the sample correlation coefficient appears the most powerful (Vickery and Nudds 1991, Crowley 1992). This is not surprising; the LR statistic for testing whether $b = 0$ in the Gompertz model (or in the logistic model) is a monotone function of the squared sample correlation coefficient:

$$\Lambda_{12} = [1 - R^2]^{q/2}. \qquad (27)$$

Here R is the sample correlation coefficient of the y_t's and x_{t-1}'s (or n_{t-1}'s if the test is adapted to the logistic model given by Eq. 1). The randomization procedure proposed by Pollard et al. (1987) estimates the distribution of R or R^2 by taking random permutations of the y_t's to construct new time series data sets. A new value of R^2 is obtained from each set. This is essentially a form of nonparametric bootstrapping. However, the resulting estimate of the distribution of R^2 (or R, or Λ) under the null hypothesis does not make use of the sufficient statistics $\hat{a}$ and $\hat{\sigma}^2$, and therefore does not make the most efficient use of the data ("sufficient statistics" contain all the information about model parameters that is present in the data; see Rice 1988).

Our simulations reveal that the randomization test of Pollard et al. (1987) is considerably less powerful

TABLE 6. Results of two density dependence tests performed on 16 insect data sets listed by den Boer and Reddingius (1989). Shown are the values of the likelihood ratio test statistic (T_{12}), the number of one-step transitions (q), the P values estimated by parametric bootstrapping ($\hat{P}$), and the P values resulting from a Student's t distribution with $q - 2$ degrees of freedom (P_{student}). The null hypothesis is Model 1 ($b = 0$), and the alternate hypothesis is Model 2 ($b < 0$). Order of entry corresponds to order in den Boer and Reddingius' (1989) Table 1.

T_{12}	q	$\hat{P}$	P_{student}
−1.84	18	.243	.042
−2.54	18	.089	.011
−3.14	14	.033	.004
−3.07	14	.033	.005
−1.51	13	.436	.080
−2.15	13	.154	.027
−2.52	12	.095	.015
−0.91	28	.776	.185
−1.45	18	.349	.083
−1.79	19	.457	.107
−1.61	14	.378	.067
−2.07	13	.198	.031
+2.17	11	.999	.971
−0.62	11	.759	.275
−1.86	11	.228	.048
−1.66	10	.397	.067

than the PBLR test when the data are generated from the stochastic logistic model given by Eq. 1 (Table 7). The parametric test attains as much as a 50% increase in power over a range of parameter values. Even if the randomization test is adapted to the logistic model, by using n_{t-1}'s instead of x_{t-1}'s in Eq. 27, the power levels of the parametric test are not approached (Table 7).

There might be some nonparametric benefits in terms of robustness of the randomization test, though such benefits have not been assessed. The stochastic logistic (Eq. 1) that we have assumed as the alternate hypothesis is a fairly general model that can describe many situations. Possible modifications of the model for studying robustness of tests might include use of a heavy-tailed distribution for the Z_t's instead of the normal distribution, or allowing the Z_t's to be autocorrelated. Few abundance data sets have more than 30 observations, however, and it is likely that the full benefits of nonparametric approaches (which depend heavily on large-sample consistency theorems) to density dependence testing will not be realized in practice.

Use of the Gompertz stochastic model in and of itself could involve a loss of power. Detecting growth-rate feedback that is proportional to $\ln N_t$ instead of N_t will likely require data covering wider ranges of abundance. As noted before, a PBLR test can be constructed for the Gompertz model; such a test would represent the practical limit of power that can be attained under that model. If this PBLR-Gompertz test is performed on data arising from a stochastic logistic, the size of the test remains ≈ 0.05, but a detectable loss of power ensues (Table 7). The difference in power between the PBLR-logistic and the PBLR-Gompertz quantifies the

220   BRIAN DENNIS AND MARK L. TAPER   Ecological Monographs
Vol. 64, No. 2

TABLE 7. Estimated sizes and powers of various one-sided density dependence tests, when data arise from the stochastic logistic model.* Tests are the parametric bootstrap likelihood ratio (PBLR) tests based on the stochastic logistic and the stochastic Gompertz and randomization tests based on the stochastic logistic and stochastic Gompertz. All tests were conducted at a nominal significance level of .05 and used 200 bootstrap or randomized samples where appropriate. Each estimate was obtained from 1000 trials. For all tests, $q = 9$.

b	n_0	σ	a	PBLR (logistic)	PBLR (Gompertz)	Randomization (logistic)	Randomization (Gompertz)
0	100	0.25	0.5	0.039	0.053	0.041	0.039
0	100	0.25	1.0	0.051	0.039	0.052	0.052
0	100	0.25	1.5	0.045	0.060	0.050	0.051
0	100	0.50	0.5	0.052	0.029	0.038	0.050
0	100	0.50	1.0	0.050	0.048	0.048	0.044
0	100	0.50	1.5	0.046	0.045	0.046	0.039
0	100	0.75	0.5	0.041	0.045	0.035	0.042
0	100	0.75	1.0	0.049	0.045	0.063	0.059
0	100	0.75	1.5	0.056	0.047	0.052	0.044
0	100	1.00	0.5	0.046	0.055	0.040	0.049
0	100	1.00	1.0	0.065	0.050	0.043	0.042
0	100	1.00	1.5	0.054	0.040	0.039	0.052
−0.01	−a/b	0.25	0.5	0.104	0.101	0.086	0.084
−0.01	−a/b	0.25	1.0	0.362	0.354	0.304	0.290
−0.01	−a/b	0.25	1.5	0.759	0.742	0.699	0.692
−0.01	−a/b	0.50	0.5	0.152	0.112	0.104	0.092
−0.01	−a/b	0.50	1.0	0.424	0.399	0.322	0.270
−0.01	−a/b	0.50	1.5	0.807	0.608	0.682	0.665
−0.01	−a/b	0.75	0.5	0.191	0.121	0.106	0.103
−0.01	−a/b	0.75	1.0	0.494	0.269	0.333	0.250
−0.01	−a/b	0.75	1.5	0.775	0.458	0.675	0.421
−0.01	−a/b	1.00	0.5	0.237	0.131	0.107	0.095
−0.01	−a/b	1.00	1.0	0.562	0.220	0.336	0.188
−0.01	−a/b	1.00	1.5	0.788	0.298	0.629	0.308

* The model is $Y_t = a + bN_{t-1} + \sigma Z_t$, where N_t is population abundance at time t, $Y_t = \ln(N_t/N_{t-1})$, $Z_t \sim$ normal(0,1), and $t = 1, 2, \ldots, q$.

intrinsic effect of looking for density dependence with a model of log-density dependence.

Of course, one could just as easily take the Gompertz as the "true" model in the simulations. The logistic, though, is fundamentally a nonlinear dynamic model and possesses a wider range of dynamic behaviors (such as limit cycles and chaos). The Gompertz is fundamentally a linear dynamic model (see Eq. 9) and therefore has a restricted repertoire of dynamic behaviors. The logistic would seem a more flexible choice for modeling the dynamic behavior of natural populations, and we have therefore centered our investigations around statistical properties arising from the logistic.

Some investigators using Gompertz-based methods have concluded that density dependence is frequently weak and not as widely prevalent as theoretical ecologists might expect (den Boer and Reddingius 1989, den Boer 1990). Without extensive logistic-Gompertz model evaluations and analyses using many data sets, it is not clear whether their results arise from nature or from increased Type II error rates inherent in the Gompertz-based statistical methods. The low power of such methods has been acknowledged (Vickery and Nudds 1991), though not explained. We speculate that searching for density dependence with a model of log-density dependence might be akin to trying to photograph a distant bird with a wide-angle instead of a telephoto lens.

On the other hand, investigators using the stochastic

logistic model, particularly with modifications for second-order lags (Eq. 6), have reported pervasive evidence of density dependence (Turchin 1990, Turchin et al. 1991, Berryman 1991a). Some of these results are probably influenced by the properties of the regression-based statistical testing methods they used. As we have noted, the t and F statistics for testing whether slope parameters are zero in Eq. 6 do not have t or F distributions. Our simulations indicate that the Type I error rates in one-sided tests of first- or second-order lags using Student's t distributions are markedly inflated over the nominal rate of 0.05 (Table 8). Users of such tests will find density dependence too often when it in fact is weak or absent. Extensions of the t and F tests to multispecies versions of the logistic (Berryman 1991b) must be called into question as well, until such tests receive further study. In addition, the regression-based, two-sided test for a first-order lag has inflated Type I error rates for some parameter values (Table 8). The regression-based, two-sided test for a second-order lag appears to possess reasonable Type I error rates (Table 8). It is noteworthy that the investigators have tended to use the two-sided regression test for second-order lags. Thus, their results concerning the prevalence of second-order lags might hold up when the data are analyzed with techniques that have received thorough evaluation.

The PBLR procedure can be adapted to test for second-order lags. Such a test provides a check on the

Markov assumption implicit in the stochastic logistic model. One would use Model 2 (Eq. 1) as the null hypothesis and the second-order lag model (Eq. 6 with $b_3, b_4, \ldots$ set to zero) as the alternate hypothesis. The test is conditioned on n_0 and n_1, so the ML estimates for the null hypothesis are based on the time series starting with n_1. The likelihood function for the alternate hypothesis is a product of conditional normal pdf's of the form $p(x_t | x_{t-1}, x_{t-2})$, because X_t conditioned on x_{t-1} and x_{t-2} has a normal($x_{t-1} + a + b_1 \exp(x_{t-1}) + b_2 \exp(x_{t-2})$, σ^2) distribution. It can be shown that the ML estimates of a, b_1, and b_2 are the least squares estimates obtained by performing a multiple linear regression of y_t on n_{t-1} and n_{t-2}. The bootstrap data sets would be obtained from the estimated null hypothesis model, and bootstrap values of the LR statistic would be obtained by fitting both models to each bootstrap data set. A study of the power properties of this test is in progress.

What is the ecological interpretation of rejecting Model 1 in favor of Model 2? Essentially, the outcome results when the data contain sufficient information to estimate an additional parameter. That parameter, b, imparts an ergodic behavior to the model (when b is negative): large populations tend to decline, and small populations tend to increase. Model 2 thus quantifies a *return tendency* in the data. The return point of the population is $-a/b$; this represents the population abundance at which the *average* change in $\ln N_t$, conditional on N_{t-1}, is zero (Eq. 5). Failure to reject Model 1 can occur when the return point, if one exists, is simply too large or too small to be estimated (out of the range of the data). According to Model 1, a "density-vagueness" (Strong 1986a, b) prevails over the range of the data. Note that for a growing population, it might only be a matter of time before the return point can be estimated. Similarly, a population declining toward the return point at first also resembles Model 1.

The return point, $-a/b$, is not an equilibrium. The "equilibrium" of the discrete time stochastic logistic model is not a point; rather, it is a long-term stationary probability distribution of population sizes. Wolda (1989, 1991) has questioned the meaning of density dependence tests that rely on high densities being above and low densities being below an "equilibrium," because of the impossibility of separating "fluctuating equilibrium values" from "fluctuating deviations from those equilibrium values" (Wolda 1991). Indeed, it is not likely that any statistical method will be able to distinguish these mechanisms of fluctuation from time series data alone. The model (Eq. 1) instead accommodates both of these mechanisms; the noise represented by Z_t describes in a phenomenological fashion a population's growth rate fluctuating for whatever reasons. Once noise is admitted, an ecological Rubicon of sorts is crossed: there no longer is a point equilibrium, conceptually, mathematically, or empirically. It is not correct to claim that a point equilibrium may

TABLE 8. Estimated sizes of density dependence tests in which the critical percentiles of the test statistics are taken from Student's t distributions used in regression analysis. Each estimate was obtained from 1000 trials; all tests were conducted at a nominal significance level of .05.

		First-order lag*		Second-order lag†	
σ	a	One-sided	Two-sided	One-sided	Two-sided
0.25	0.5	0.110	0.076	0.091	0.054
0.25	1.0	0.064	0.047	0.077	0.035
0.25	1.5	0.052	0.051	0.073	0.046
0.50	0.5	0.158	0.107	0.088	0.052
0.50	1.0	0.105	0.075	0.080	0.034
0.50	1.5	0.077	0.064	0.063	0.032
0.75	0.5	0.227	0.141	0.088	0.041
0.75	1.0	0.136	0.094	0.086	0.049
0.75	1.5	0.095	0.069	0.063	0.035
1.00	0.5	0.258	0.162	0.078	0.041
1.00	1.0	0.173	0.122	0.076	0.048
1.00	1.5	0.115	0.088	0.057	0.031

* The base model is $Y_t = a + bN_{t-1} + \sigma Z_t$, where N_t is population abundance at time t, $Y_t = \ln(N_t/N_{t-1})$, Z_t normal(0,1), and $t = 1, 2, \ldots, q$. Null hypothesis is H_1: $b = 0$, one-sided alternate hypothesis is H_2: $b < 0$, and two-sided alternate hypothesis is H_2: $b \neq 0$. For all simulations, $b = 0$, $n_0 = 100$, and $q = 9$. Critical percentiles were obtained from a Student's t distribution with $q - 2$ df.

† The base model is $Y_t = a + b_1N_{t-1} + b_2N_{t-2} + \sigma Z_t$, where $t = 2, 3, \ldots, q + 1$. Null hypothesis is H_0: $b_2 = 0$, one-sided alternate hypothesis is H_1: $b_2 < 0$, and two-sided alternate hypothesis is H_1: $b_2 \neq 0$. For all simulations, $b_2 = 0$, $b_1 = -0.01$, $n_0 = -a/b$, and $q = 15$. Critical percentiles were obtained from a Student's t distribution with $q - 3$ df.

emerge as a result of density dependence analyses (Berryman 1991a). Wolda (1989) has stated it well: "Equilibrium is not a point but a cloud of points." The stationary distribution of the stochastic logistic model can be approximated by a positively skewed distribution known as a gamma distribution (Dennis and Patil 1984, Dennis 1989b). Some ecological ramifications of this concept of a stochastic equilibrium have been discussed elsewhere (May 1974, Dennis and Patil 1984, Dennis and Costantino 1988, Dennis 1989b, Desharnais et al. 1990, Costantino and Desharnais 1991, Kemp and Dennis 1993).

Investigators must carefully distinguish the statistical hypothesis of density dependence (as exemplified by Model 2) from the ecological hypothesis (biological mechanism of negative feedback on growth rate). Analyses of time series data have all the pitfalls of any observational studies. Other stochastic mechanisms with stationary distributions, such as a series of independent, identically distributed population sizes drawn from some statistical distribution, are better described by Model 2 (and similar models such as the stochastic Gompertz) than Model 1 or Model 0 (Wolda and Dennis 1993). Indeed, if $b = -1$ in the stochastic Gompertz (Eq. 8), then population sizes *are* independent, identically distributed lognormal random variables. Quantities such as annual rainfall and spring snowpack levels thus qualify as "density dependent" under statistical tests (see Wolda and Dennis 1993), but it is questionable whether ecologists would consider such quantities to be density dependent in an ecological sense. We believe the PBLR test can be a useful component of a case for ecological density dependence, but should not be the sole component.

One important application of density dependence testing is in conservation biology. A critical question is how to estimate population trends and properties of the first-passage distribution from time series data (e.g., Dennis et al. 1991). The first-passage distribution is the probability distribution of the time it will take for a stochastic process (population size) to first attain some lower (or higher) value. Properties of interest include the mean time to reach a lower value and the probability of reaching it before reaching a given higher value. Preliminary evidence indicates that estimates of first-passage properties can vary substantially depending on whether or not a density dependent model is used (Ginzburg et al. 1990, Stacey and Taper 1992). The PBLR test represents a potentially valuable aid to deciding whether or not to account for density dependence in population viability analysis.

ACKNOWLEDGMENTS

We thank J. Reddingius, H. Wolda, S. Orzack, T. Shenk, R. F. Costantino, D. Kelt, P. A. Marquet, K. P. Burnham, and A. A. Berryman for their many helpful comments. The paper has also benefitted from stimulating discussions with P. den Boer, W. P. Kemp, S. R. Lele, and P. L. Munholland. Work by B. Dennis was supported in part by grants from USDA-ARS (number 58-91H2-2-237) and USDA Forest Service (number INT-92688-RJVA). Work by M. L. Taper was supported in part by grants from NSF (number BSR-8821458) and US-EPA (number CR-820086). This paper is contribution number 683 of the Forest, Wildlife, and Range Experiment Station of the University of Idaho.

LITERATURE CITED

Andrewartha, H. G. 1957. The use of conceptual models in population ecology. Cold Spring Harbor Symposia on Quantitative Biology 22:219–232.

Aptech Systems. 1991. GAUSS. Aptech Systems, Kent, Washington, USA.

Bain, L. J., and M. Engelhardt. 1987. Introduction to probability and mathematical statistics. PWS Publishers, Boston, Massachusetts, USA.

Barlow, J. 1992. Nonlinear and logistic growth in experimental populations of guppies. Ecology 73:941–950.

Beran, R. 1986. Discussion of the paper by C. F. J. Wu. Annals of Statistics 14:1295–1298.

Berryman, A. A. 1991a. Stabilization or regulation: what it all means! Oecologia 86:140–143.

———. 1991b. Can economic forces cause ecological chaos? The case of the northern California Dungeness crab fishery. Oikos 62:106–109.

Bhat, B. R. 1974. On the method of maximum-likelihood for dependent observations. Journal of the Royal Statistical Society Series B 36:48–53.

The Biological Laboratory. 1957. Population studies: animal ecology and demography. Cold Spring Harbor Symposia on Quantitative Biology. Volume 22. Long Island Biological Association, New York, New York, USA.

Birch, L. C. 1957. The role of weather in determining the distribution and abundance of animals. Pages 203–215 in The Biological Laboratory. Population studies: animal ecology and demography. Cold Spring Harbor Symposia on Quantitative Biology. Volume 22. Long Island Biological Association, New York, New York, USA.

Boyce, M. S. 1989. The Jackson elk herd: intensive management in North America. Cambridge University Press, Cambridge, England.

Bulmer, M. G. 1975. The statistical analysis of density dependence. Biometrics 31:901–911.

Cody, M. L., and J. M. Diamond, editors. 1975. Ecology and evolution of communities. Harvard University Press, Cambridge, Massachusetts, USA.

Colinvaux, P. 1973. Introduction to ecology. John Wiley & Sons, New York, New York, USA.

Costantino, R. F., and R. A. Desharnais. 1991. Population dynamics and the Tribolium model: genetics and demography. Monographs on Theoretical and Applied Genetics 13. Springer-Verlag, New York, New York, USA.

Crowley, P. H. 1992. Density dependence, boundedness, and attraction: detecting stability in stochastic systems. Oecologia 90:246–254.

Cushing, J. M. 1989. A strong ergodic theorem for some nonlinear matrix models for the dynamics of structured populations. Natural Resource Modeling 3:331–357.

den Boer, P. J. 1968. Spreading of risk and stabilization of animal numbers. Acta Biotheoretica 18:165–194.

———. 1990. On the stabilization of animal numbers. Problems of testing. 3. What do we conclude from significant test results? Oecologia 83:38–46.

———. 1991. Seeing the trees for the wood: random walks or bounded fluctuations of population size? Oecologia 86:484–491.

den Boer, P. J., and J. Reddingius. 1989. On the stabilization of animal numbers. Problems of testing. 2. Confrontation with data from the field. Oecologia 79:143–149.

Dennis, B. 1989a. Allee effects: population growth, critical density, and the chance of extinction. Natural Resource Modeling 3:481–537.

———. 1989b. Stochastic differential equations as insect population models. In L. McDonald, B. Manly, J. Lockwood, and J. Logan, editors. Estimation and analysis of insect populations. Lecture Notes in Statistics 55:219–238.

Dennis, B., and R. F. Costantino. 1988. Analysis of steady-state populations with the gamma abundance model: application to Tribolium. Ecology 69:1200–1213.

Dennis, B., P. L. Munholland, and J. M. Scott. 1991. Estimation of growth and extinction parameters for endangered species. Ecological Monographs 61:115–143.

Dennis, B., and G. P. Patil. 1984. The gamma distribution and weighted multimodal gamma distributions as models of population abundance. Mathematical Biosciences 68:187–212.

Desharnais, R. A., B. Dennis, and R. F. Costantino. 1990. Genetic analysis of a population of Tribolium. IX. Maximization of population size and the concept of a stochastic equilibrium. Genome 33:571–580.

Diamond, J., and T. J. Case, editors. 1986. Community ecology. Harper and Row, New York, New York, USA.

Dickey, D. A., and W. A. Fuller. 1981. Likelihood ratio statistics for autoregressive time series with a unit root. Econometrica 49:1057–1072.

Eberhardt, L. L. 1970. Correlation, regression, and density dependence. Ecology 51:306–310.

Eberhardt, L. L., R. R. Knight, and B. M. Blanchard. 1986. Monitoring grizzly bear population trends. Journal of Wildlife Management 50:613–618.

Efron, B. 1986. Discussion of the paper by C. F. J. Wu. Annals of Statistics 14:1301–1304.

Efron, B., and G. Gong. 1983. A leisurely look at the bootstrap, the jackknife, and cross-validation. American Statistician 37:36–48.

Emlen, J. M. 1984. Population biology. Macmillan, New York, New York, USA.

Fisher, R. A. 1958. Statistical methods for research workers. Hafner, New York, New York, USA.

Fowler, C. W. 1984. Density dependence in cetacean populations. Reports of the International Whaling Commission Special Issue 6:373–379.

———. 1987. A review of density dependence in populations of large mammals. Pages 401–441 *in* H. H. Genoways, editor. Current mammalogy. Volume 1. Plenum, New York, New York, USA.

Fuller, W. A. 1987. Measurement error models. John Wiley & Sons, New York, New York, USA.

Gaston, K. J., and J. H. Lawton. 1987. A test of statistical techniques for detecting density dependence in sequential censuses of animal populations. Oecologia 74:404–410.

Ginzburg, L. R., S. Ferson, and H. R. Akçakaya. 1990. Reconstructibility of density dependence and the conservative assessment of extinction risks. Conservation Biology 4:63–70.

Goodman, D. 1987. Consideration of stochastic demography in the design and management of biological reserves. Natural Resource Modeling 1:205–234.

Hassell, M. P., J. Latto, and R. M. May. 1989. Seeing the wood for the trees: detecting density dependence from existing life-table studies. Journal of Animal Ecology 58:883–892.

Holyoak, M., and J. H. Lawton. 1992. Detection of density dependence from annual censuses of bracken-feeding insects. Oecologia 91:425–430.

Houston, D. B. 1982. The northern Yellowstone elk: ecology and management. Macmillan, New York, New York, USA.

Kemp, W. P., and B. Dennis. 1993. Density dependence in rangeland grasshoppers (Orthoptera: Acrididae). Oecologia 96:1–8.

Kingsland, S. E. 1985. Modeling nature: episodes in the history of population ecology. University of Chicago Press, Chicago, Illinois, USA.

Knight, R. R., and L. L. Eberhardt. 1984. Projected future abundance of the Yellowstone grizzly bear. Journal of Wildlife Management 48:1434–1438.

Knight, R. R., and L. L. Eberhardt. 1985. Population dynamics of Yellowstone grizzly bears. Ecology 66:323–334.

Leigh, E. G. 1981. The average lifetime of a population in a varying environment. Journal of Theoretical Biology 90:213–239.

Lele, S. 1991. Jackknifing linear estimating equations: asymptotic theory and applications in stochastic processes. Journal of the Royal Statistical Society Series B 53:253–267.

Lin, C.-C., and G. S. Mudholkar. 1980. A simple test for normality against asymmetric alternatives. Biometrika 67:455–461.

Livdahl, J. P., and G. Sugihara. 1984. Non-linear interactions of populations and the importance of estimating per capita rates of change. Journal of Animal Ecology 53:573–580.

May, R. M. 1974. Stability and complexity in model ecosystems. Second edition. Princeton University Press, Princeton, New Jersey, USA.

———, editor. 1976. Theoretical ecology: principles and applications. W. B. Saunders, Philadelphia, Pennsylvania, USA.

McLain, D. H. 1974. Drawing contours from arbitrary data points. Computer Journal 17:318–324.

McLaren, I. A., editor. 1971. Natural regulation of animal populations. Atherton, New York, New York, USA.

Mountford, M. D. 1988. Population regulation, density dependence, and heterogeneity. Journal of Animal Ecology 57:845–858.

Neyman, J., and E. S. Pearson. 1933. On the problem of the most efficient tests of statistical hypotheses. Philosoph-

ical Transactions of the Royal Society Series A 231:289–337.

Nicholson, A. J. 1933. The balance of animal populations. Journal of Animal Ecology 2:132–178.

Pianka, E. R. 1974. Evolutionary ecology. Harper and Row, New York, New York, USA.

Pollard, E., K. H. Lakhani, and P. Rothery. 1987. The detection of density-dependence from a series of annual censuses. Ecology 68:2046–2055.

Reddingius, J. 1971. Gambling for existence. A discussion of some theoretical problems in animal population ecology. Acta Biotheoretica 20 (Supplement):1–208.

———. 1990. Models for testing: a secondary note. Oecologia 83:50–52.

Reddingius J., and P. J. den Boer. 1989. On the stabilization of animal numbers. Problems of testing. 1. Power estimates and estimation errors. Oecologia 78:1–8.

Rice, J. A. 1988. Mathematical statistics and data analysis. Wadsworth, Belmont, California, USA.

Ricker, W. E. 1954. Stock and recruitment. Journal of the Fisheries Research Board of Canada 11:559–623.

Roughgarden, J. 1979. Theory of population genetics and evolutionary ecology. an introduction, Macmillan, New York, New York, USA.

Royama, T. 1977. Population persistence and density dependence. Ecological Monographs 47:1–35.

———. 1981. Fundamental concepts and methodology for the analysis of population dynamics, with particular reference to univoltine species. Ecological Monographs 51:473–493.

Schork, N. 1992. Bootstrapping likelihood ratios in quantitative genetics. Pages 389–396 *in* R. LePage and L. Billard, editors. Exploring the limits of bootstrap. John Wiley & Sons, New York, New York, USA.

Serfling, R. J. 1980. Approximation theorems of mathematical statistics. John Wiley & Sons, New York, New York, USA.

Shaffer, M. L. 1981. Minimum population sizes for species conservation. BioScience 31:131–134.

Simberloff, D. 1988. The contribution of population and community ecology to conservation science. Annual Review of Ecology and Systematics 19:473–511.

Slade, N. A. 1977. Statistical detection of density dependence from a series of sequential censuses. Ecology 58:1094–1102.

Solow, A. R. 1990. Testing for density dependence: a cautionary note. Oecologia 83:47–49.

———. 1991. Response. Oecologia 86:146.

Solow, A. R., and J. H. Steele. 1990. On sample size, statistical power, and the detection of density dependence. Journal of Animal Ecology 59:1073–1076.

Stacey, P. B., and M. Taper. 1992. Environmental variation and the persistence of small populations. Ecological Applications 2:18–29.

Strong, D. R. 1986a. Density vagueness: abiding the variance in the demography of real populations. Pages 257–268 *in* J. Diamond and T. J. Case, editors. Community ecology. Harper and Row, New York, New York, USA.

———. 1986b. Density-vague population change. Trends in Ecology and Evolution 1:39–42.

Stuart, A., and J. K. Ord. 1991. Kendall's advanced theory of statistics. Volume 2: classical inference and relationship. Fifth edition. Oxford University Press, New York, New York, USA.

Tong, H. 1990. Non-linear time series: a dynamical system approach. Oxford University Press, Oxford, England.

Turchin, P. 1990. Rarity of density dependence or population regulation with lags? Nature 344:660–663.

Turchin, P., P. L. Lorio Jr., A. D. Taylor, and R. F. Billings. 1991. Why do populations of southern pine beetles (Coleoptera: Scolytidae) fluctuate? Environmental Entomology **20**:401–409.

Turchin, P., and A. D. Taylor. 1992. Complex dynamics in ecological time series. Ecology **73**:289–305.

Vickery, W. L., and T. D. Nudds. 1984. Detection of density-dependent effects in annual duck censuses. Ecology **65**: 96–104.

Vickery, W. L., and T. D. Nudds. 1991. Testing for density-dependent effects in sequential censuses. Oecologia **85**:419–423.

Wissel, C., and S. Stöcker. 1991. Extinction of populations by random influences. Theoretical Population Biology **39**: 315–328.

Woiwod, I. P., and I. Hanski. 1992. Patterns of density dependence in moths and aphids. Journal of Animal Ecology **61**:619–629.

Wolda, H. 1989. The equilibrium concept and density dependence tests. What does it all mean? Oecologia **81**:430–432.

———. 1991. The usefulness of the equilibrium concept in population dynamics. A reply to Berryman. Oecologia **86**: 144–145.

Wolda, H., and B. Dennis. 1993. Density dependence tests, are they? Oecologia **95**:581–591.

Ecology, 73(1), 1992, pp. 289–305
© 1992 by the Ecological Society of America

COMPLEX DYNAMICS IN ECOLOGICAL TIME SERIES[1]

PETER TURCHIN AND ANDREW D. TAYLOR
Southern Forest Experiment Station, 2500 Shreveport Highway, Pineville, Louisiana 71360 USA

Abstract. Although the possibility of complex dynamical behaviors—limit cycles, quasiperiodic oscillations, and aperiodic chaos—has been recognized theoretically, most ecologists are skeptical of their importance in nature. In this paper we develop a methodology for reconstructing endogenous (or deterministic) dynamics from ecological time series. Our method consists of fitting a response surface to the yearly population change as a function of lagged population densities. Using the version of the model that includes two lags, we fitted time-series data for 14 insect and 22 vertebrate populations. The 14 insect populations were classified as: unregulated (1 case), exponentially stable (three cases), damped oscillations (six cases), limit cycles (one case), quasiperiodic oscillations (two cases), and chaos (one case). The vertebrate examples exhibited a similar spectrum of dynamics, although there were no cases of chaos. We tested the results of the response-surface methodology by calculating autocorrelation functions for each time series. Autocorrelation patterns were in agreement with our findings of periodic behaviors (damped oscillations, limit cycles, and quasiperiodicity). On the basis of these results, we conclude that the complete spectrum of dynamical behaviors, ranging from exponential stability to chaos, is likely to be found among natural populations.

Key words: autocorrelation function; chaos; complex deterministic dynamics; delayed density dependence; dynamical behaviors of populations; insect population dynamics; limit cycles; long-term population records; nonlinear time-series modelling; quasiperiodicity; time-series analysis.

INTRODUCTION

The relative importance of density-dependent vs. density-independent factors in determining population abundances and dynamics is a central issue in ecology. Much of the debate over this question has focused on two opposing viewpoints (e.g., Nicholson 1954, Andrewartha and Birch 1954). According to the first viewpoint populations are regulated around a stable point equilibrium by density-dependent mechanisms, while the second one maintains that population change is largely driven by density-independent factors. There is, however, a third possibility. In addition to stable point equilibria, density-dependent processes can produce complex population dynamics—limit cycles, quasiperiodic oscillations, and aperiodic chaos. While the possibility of such dynamics has been recognized theoretically since the 1970s (May 1974, 1976), most ecologists have remained skeptical of their importance in nature.

One well-known attempt to determine the frequency of various kinds of dynamic behaviors in insect populations was made by Hassell et al. (1976). They concluded that most natural populations show monotonic damping (the most stable kind of equilibrium behavior), with only 1 case (out of 24) of damped oscillations, 1 case of a limit cycle, and no cases of chaos. Despite a number of caveats listed by Hassell et al. (1976), this

result was very influential in convincing ecologists that complex dynamics are rarely found in nature (e.g., Berryman and Millstein 1989, Nisbet et al. 1989). In this paper we argue that the results obtained by Hassell et al. (1976) largely resulted from their overly simple method of analysis. Most importantly, they used a single-species model that lacked delayed density dependence. Delayed density dependence, however, is expected to arise as a result of biotic interactions in multispecies communities and as a result of population structure (Royama 1981, Murdoch and Reeve 1987; L. R. Ginzburg and D. E. Taneyhill, *unpublished manuscript*), and in fact is found in many insect populations (Turchin 1990). Using a single-species model without delayed density dependence biases the results in favor of stability, since complex dynamics are more likely in higher-dimensional systems, and mistakenly analyzing such systems in fewer dimensions will tend to hide this complexity (Guckenheimer et al. 1977, Schaffer and Kot 1985).

One approach to higher-dimensional analysis of ecological time series has been advocated by Schaffer and co-workers (Schaffer 1985, Schaffer and Kot 1985, 1986, Kot et al. 1988), who used the method of "phase-space reconstruction" in which unknown densities of interacting populations are represented with lagged densities of the studied population. Schaffer and Kot (1986) examined time series of several natural populations and concluded that reconstructed dynamics of these populations resembled chaos. The major weakness of such analysis, however, is its reliance on visual (and

[1] Manuscript received 18 June 1990; revised 20 December 1990; accepted 7 February 1991; final version received 11 March 1991.

therefore inherently subjective) examination of recon-structed attractors (Berryman and Millstein 1989, Ell-ner 1989).

In this paper we build on ideas of both Hassell et al. (1976) and Schaffer and Kot (1986). Our goal is to develop an objective methodology for extracting de-terministic dynamics from short and noisy ecological time series. Unlike Hassell et al. (1976) who specified a particular equation with which to model data, we used a general and flexible methodology described by Box and Draper (1987), the response-surface meth-odology (RSM). We followed Schaffer and Kot (1985) by using lags to represent the multidimensional dy-namics of the system (e.g., unknown densities of in-teracting species or age structure). We used our meth-odology to reconstruct deterministic dynamics from long-term records of population fluctuations of 14 in-sects (with some further comparisons to 22 mammal and bird species).

Since the methodology proposed here is new, we do not know how well it succeeds at reconstructing com-plex dynamics from data. This is especially true for detecting chaos. However, methodologies for detecting periodic behaviors (e.g., limit cycles) are well under-stood. Accordingly, we begin by using one of these methodologies, which is based on estimating the au-tocorrelation function (ACF) for each data set. We use ACF patterns to characterize presence or absence of periodic behaviors in natural populations, and then compare ACF results to conclusions reached with the response-surface methodology. Our logic is that if RSM is not capable of extracting limit cycles from data, then there is little hope that we can use it to detect chaos. If, on the other hand, we can accurately reconstruct one kind of nonequilibrium behavior, limit cycles, then confidence in our ability to reconstruct another kind, chaos, is correspondingly enhanced:

METHODS

The data set

We collected and analyzed every terrestrial animal population time series we could obtain, subject to the following criteria: (1) Data were annual and continu-ous; if a time series had missing data, only the longest uninterrupted period was used. (2) Time series had to contain at least 18 yr of continuous census data, so that no less than half the total degrees of freedom would always be available for the error term in our response surface model (see *Reconstructing endogenous dynam-ics . . . ,* below). (3) The data were for a single locality (spatial scale having been determined by the original author). Where several time series were available for the same species, we selected the longest one, to avoid overrepresentation of much-studied species.

We exercised no selectivity beyond applying these criteria. Nonetheless, our data set cannot be regarded as representative of natural populations, since the orig-

inal investigators' selection of populations for study is inherently biased. In particular, forest pests exhibiting outbreaks clearly are over-represented.

Considerations of space prevent us from fully dis-cussing our results for all 36 time series (Table 1 and Table 2). As a compromise, we show the complete spectrum of results for all series in one group—insects. We selected insects for detailed discussion partly be-cause we are most familiar with this group. More im-portantly, insect data sets tend to be more reliable, since the majority of insect data were collected with the specific goal of quantifying insect population fluc-tuations, unlike the data extracted from fur returns or bag records. Nevertheless, as will be seen later, many of the patterns found in mammal and bird data sets are very similar to insect patterns.

Investigating time series with autocorrelation functions

As the first step in our analysis of the population time series, we used the qualitative diagnostic tech-niques based on estimating the autocorrelation func-tion (ACF; Box and Jenkins 1974; for discussions of ACF in ecological context see Finerty [1980], Nisbet and Gurney [1982]). Prior to the analysis the values of population density at each year, N_t, were log-trans-formed, $L_t \equiv \log N_t$. The autocorrelation function is estimated by calculating the correlation coefficient be-tween pairs of values $L_{t-\tau}$ and L_t separated by lag τ ($\tau = 1, 2, \ldots$). These correlation coefficients are then plotted as a function of lag τ.

The shape of the estimated ACF provides insights regarding two aspects of population dynamics: station-arity and periodicity. A process is stationary if its dy-namical properties do not change during the period of the study. Stationary processes fluctuate around con-stant mean levels, with constant variances. As will be seen later, our ability to reconstruct the endogenous dynamics of a system depends considerably on whether they are stationary, or not. ACFs of stationary pro-cesses are characterized by an exponential decay to zero, either monotonic or oscillatory (Box and Jenkins 1974).

Other ACF patterns indicate various forms of non-stationarity. A possible cause of nonstationary dynam-ics is density independence, perhaps arising because density regulation only occurs at extreme levels—"floors" and "ceilings"—that were not encountered by the population during the study. Such a population undergoes a "random walk," in which the population gradually wanders away from its initial density. There is, then, no true mean around which fluctuations occur. Alternatively, environmental changes occurring on a time scale comparable to the length of the observed time series could produce a gradual trend in the mean. In either of these situations, the ACF will decay slower than exponentially, and will become increasingly neg-ative at long lags (Fig. 1A).

February 1992   COMPLEX DYNAMICS IN ECOLOGICAL DATA   291

TABLE 1. Summary of insect time series.

Species	Time period	Reference
Phyllopertha horticola (garden chafer)	1947–1975	Milne 1984
Choristoneura fumiferana (spruce budworm)	1945–1972	Royama 1981
Dendrolimus pini (pine spinner moth)	1881–1940	Schwerdtfeger 1941
Hyloicus pinastri (pine hawkmoth)	1881–1930	Schwerdtfeger 1941
Dendroctonus frontalis (southern pine beetle)	1958–1987	Turchin et al. 1991
Panolis flammea (pine beauty moth)	1881–1940	Schwerdtfeger 1941
Lymantria monacha (nun moth)	1900–1941	Bejer 1988
Bupalus piniarius (pine looper)	1881–1940	Schwerdtfeger 1941
Hyphantria cunea (fall webworm)	1937–1958	Morris 1964
Vespula spp. (wasps)	1921–1946	Southwood 1967
Drepanosiphum platanoides (sycamore aphid)	1969–1987	Dixon 1990
Lymantria dispar (gypsy moth)	1954–1979	Montgomery and Wallner 1987
Zeiraphera diniana (larch budmoth)	1949–1986	Baltensweiler and Fischlin 1987
Phyllaphis fagi (beech aphid)	1969–1987	Dixon 1990

Nonstationarity can also be caused by externally driven periodic changes in the mean. The resulting dynamics have been called "phase-remembering quasi-cycles" (Nisbet and Gurney 1982), since the exogenous forcing factor maintains the regularity of the oscillation despite random perturbations in abundances. The ACF of such a system might look like the one in Fig. 1C: it does not decay to zero, but rather oscillates around zero with constant amplitude. The period of oscillation of the ACF is determined by the periodicity of the external forcing factor. In ecology the most important such periodic factor is seasonality. By using only data sets that reported population densities on a yearly basis, however, we have avoided the complications of seasonality.

In addition to externally driven nonstationary periodicity, stationary periodicity may arise from the endogenous dynamics of the system. Population fluctuations with an endogenous periodic component ("phase-forgetting quasi-cycles," Nisbet and Gurney 1982) will be produced when the deterministic dynamics are damped oscillations (around a stable point equilibrium), a limit cycle, or "weak" chaos (Poole 1977). The ACF of these systems is characterized by an oscillatory decay to zero (Fig. 1D). In contrast, a nonperiodic stationary system, resulting from exponential stability (of a point equilibrium), will have a monotonically decaying ACF (Fig. 1B).

As a diagnostic tool the estimated ACF is much more useful than "eyeballing" the observed time series. By averaging over, and thus smoothing, the noisy time series, ACF reveals the periodic pattern in the data if it is present. The average period of oscillations is readily determined by observing at which lags ACF achieves its maxima. The speed with which ACF maxima approach zero reveals the strength of the periodic component, that is, how long the process "remembers" its history. Finally, a quick, although crude, test of the hypothesis that there is a periodic component in population fluctuations can be performed by determining whether ACF at the lag equal to one period is greater than the 95% confidence limit.

Using lags to represent multidimensional dynamics of the system

Numerical changes of a population typically affect and are in turn affected by the population abundances of resources, natural enemies, and competitors. Thus, in order to understand and predict how a population changes with time, we need information about the abundances of interacting species. However, usually data are available only for a single population, and we never know abundances of all populations in the community. This difficulty can be overcome by considering the population change from the previous year $t - 1$ to the current year t as a function of not only previous year's density, N_{t-1}, but also of densities N_{t-2}, N_{t-3}, ... The mathematical justification for this methodology is provided by a theorem proved by Takens (1981), and the method has been successfully used in many physical and chemical applications (e.g., Argoul et al. 1987), where it is called "attractor reconstruction

TABLE 2. Summary of vertebrate time series.

Species	Time period	Reference
Lynx	1821–1934	Moran 1953
Foxes (two species)	1879–1930	Elton 1942
Colored fox	1834–1925	Elton 1942
Marten	1834–1925	Elton 1942
Arctic fox	1834–1925	Elton 1942
Mink	1914–1957	Keith 1963
Muskrat	1914–1957	Keith 1963
Coyote	1914–1957	Keith 1963
Snowshoe hare	1847–1903	Leigh 1968
Varying hare	1884–1908	Naumov 1972
Squirrel	1933–1955	Naumov 1972
Belyak hare	1932–1954	Labutin 1960
Lynx	1932–1954	Labutin 1960
Fox	1932–1954	Labutin 1960
Wolf	1932–1954	Labutin 1960
Parchment beaver	1752–1849	Jones 1914
Wolverine	1752–1911	Jones 1914
Rabbit	1862–1932	Middleton 1934
Ruffed Grouse	1915–1972	Keith 1963
Black-capped Chickadee	1958–1983	Loery and Nichols 1985
Heron	1934–1952	Lack 1954
Great Tit	1912–1941	Lack 1954

292   PETER TURCHIN AND ANDREW D. TAYLOR   Ecology, Vol. 73, No. 1

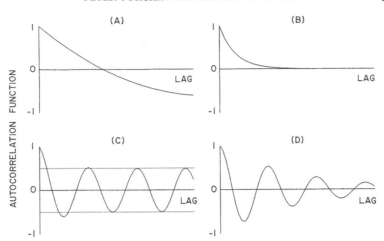

Fig. 1. Theoretical shapes of autocorrelation functions for (A) a process with nonstationary mean and no periodicity; (B) a stationary process with exponential return to equilibrium; (C) a process driven by an exogenous periodic force, or phase-remembering quasi-cycle; (D) a stationary process with endogenously generated periodicity, or phase-forgetting quasi-cycle.

in time delay coordinates" (Schaffer 1985, Ellner 1989). Representing the unknown densities of interacting species with delayed density dependence is also a venerable tradition in population ecology (Hutchinson 1948, Moran 1953, Berryman 1978, Royama 1981). Essentially, one replaces the "true" multivariate system describing deterministic population change

$$N_t^1 = G^1(N_{t-1}^1, N_{t-1}^2, \ldots, N_{t-1}^k)$$
$$N_t^2 = G^2(N_{t-1}^1, N_{t-1}^2, \ldots, N_{t-1}^k)$$
$$\vdots \qquad \vdots$$
$$N_t^k = G^k(N_{t-1}^1, N_{t-1}^2, \ldots, N_{t-1}^k)$$

(where N_t^i is the density of species i at time t, and G^i is a function describing the change in the density of species i with respect to the densities of interacting species) with a single equation for one species that involves lags

$$N_t^1 = F^1(N_{t-1}^1, N_{t-2}^1, \ldots, N_{t-p}^1). \qquad (1)$$

It is important to note that N^i can refer not only to populations of interacting species, but also to abundances of different cohorts of the same species, if the population has age, physiological, or spatial structure.

The above argument leads us to the following general model:

$$N_t = F(N_{t-1}, N_{t-2}, \ldots, N_{t-p}, \epsilon_t), \qquad (2)$$

where we have added the exogenous component ϵ_t to the equation for population change. ("Endogenous" refers to dynamical feedbacks affecting the system, including those that involve a time lag, e.g., natural enemies. "Exogenous" refers to density-independent factors that are not a part of the feedback loop.) We will model the exogenous component as a random, normally distributed variable with mean zero, and variance σ^2. The quantity p is the *order* of the process, that

is, the maximum lag time beyond which a past value of population density has no direct effect on the current population change (autocorrelations can persist much longer than p, since past values of N affect intermediate values, which in turn affect present).

Reconstructing endogenous dynamics with response-surface methodology

Our major goal in this paper is to develop a methodology that would objectively determine the type of the dynamic behavior that characterizes the endogenous component of population change. Several approaches have been suggested, all based on the method of reconstructing the attractor in time-delayed coordinates described in the preceding section. One is to estimate the dimensionality of the reconstructed attractor (for the explanation of this approach see Ellner [1989]). This approach appears to work for perfectly accurate data even with relatively short time-series (50 points), although dealing with noisy data sets remains problematic (Ellner 1989). Another approach relies on the direct estimation of Lyapunov exponents from experimental time series (Eckmann and Ruelle 1985, Wolf et al. 1985; for an explanation of Lyapunov exponents see Abraham and Shaw [1983]). This method requires enormous amounts of data: a minimum of several thousand data points is needed to characterize a low-dimensional attractor (Vastano and Kostelich 1986). The method of Sugihara and May (1990), which uses nonlinear forecasting to detect chaos in noisy time series, also requires substantial amounts of data (500–1000 points in their applications).

Making as few assumptions as possible about the nature of the process that has produced the observed time series is a powerful feature of the above methods, but it is also their weakness. Such nonparametric, mod-

February 1992   COMPLEX DYNAMICS IN ECOLOGICAL DATA   293

el-independent approaches typically require plentiful data points. In ecology, where the length of time series rarely exceeds 20–30 yr, one is forced to use a parametric approach, which is much more frugal with data points.

The approach that we followed in this paper consists of approximating the function F in Eq. 2. This function describes the behavior of trajectories on the reconstructed attractor, and thus knowing its properties gives us a complete description of the system dynamics. For example, the dynamic behavior of F could be formally characterized by calculating its dominant eigenvalue and dominant Lyapunov exponent. Alternatively, one may determine the type of dynamics simply by iterating Eq. 2 on the computer, and observing the resulting dynamics. We have followed the latter course in this paper.

A potential problem associated with using a parametric approach, however, is that one may happen to choose an inappropriate model with which to approximate F. This possibility can be minimized by using the general method of response-surface fitting described by Box and Draper (1987). Briefly, this method is similar to regular regression in that it employs polynomials for approximating the shape of F. However, both the response (dependent) variable and the predictor (independent) variables are transformed using the Box-Cox power transformation (Box and Cox 1964), with the transformation parameter (the exponent) being also estimated from the data. In the following paragraphs we describe the logic and details of the approach with which we have extracted endogenous dynamics from ecological time series.

The first step is to decide on the number of lags p to include in the model, that is, the "embedding dimension" (Schaffer 1985). Ideally, since the correct p is unknown, one should start with a low-dimensional model and then increase the dimension until the result does not depend on further increase in dimensionality. In practice, due to data limitations (primarily the length of a time series) only a few lags can be examined. In their attempt to extract deterministic dynamics from data, Hassell et al. (1976) used a model with only one lag (only direct density dependence):

$$N_t = \lambda N_{t-1}(1 + aN_{t-1})^{-\beta}. \tag{3}$$

We took the next step and used a model with two lags (in other words, we added delayed density dependence). Thus, the general model (Eq. 2) becomes

$$N_t = F(N_{t-1}, N_{t-2}, \epsilon_t). \tag{4}$$

Biological considerations indicate that F can be represented as a product of N_{t-1} and the per-capita replacement rate $f(N_{t-1}, N_{t-2}, \epsilon_t)$. In general, f will have a simpler form, and can be approximated with a polynomial of one order lower than F. For example, if f is a monotonically decreasing function of N_{t-1} and N_{t-2}, it can be approximated with a first-order polynomial

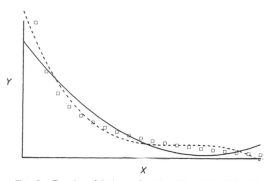

Fig. 2. Results of fitting a function $Y = aX^{-\frac{1}{2}}$ (□) with polynomials of second (——) and third order (– – –).

(together with appropriate transformations of the predictor variables), while the function Γ will have a maximum and will need to be approximated with a quadratic polynomial. These considerations lead us to the following model:

$$N_t = N_{t-1}f(N_{t-1}, N_{t-2}, \epsilon_t). \tag{5}$$

We are now in position to estimate f by fitting a response surface to the observed replacement rate N_t/N_{t-1} as a function of N_{t-1} and N_{t-2}. Highly nonlinear dependence of the replacement rate on lagged population densities in several data sets and in some theoretical models (P. Turchin and A. D. Taylor, *personal observation*), necessitated using polynomials of at least second order (see Box and Draper 1987). However, polynomials by themselves are notoriously bad at approximating both the function and its derivative, especially for log-like functions that are characterized by rapidly changing derivatives. Consider, for example, data plotted in Fig. 2. Fitting a quadratic polynomial to the nonlinear function represented by points, we find that at high values of the predictor variable, the fitted function has a positive slope, while the actual function has a negative slope. Correct estimation of the slope of f is crucial to the success of accurately reconstructing endogenous dynamics, since whether an equilibrium is stable or not, and whether the attractor is periodic or chaotic will depend on the derivatives of f. Using higher-degree polynomials does not help, even though they provide a progressively better approximation to f, since higher-degree polynomials "oscillate" around the true function (e.g., the cubic polynomial in Fig. 2). In addition, such an approach is very wasteful of degrees of freedom. A better approach, proposed by Box and Cox (1964), is to power-transform either predictor, or response variables, or both. The logarithm is naturally embedded in the power transformation family, since letting $\theta \rightarrow 0$ is equivalent to a log transformation (Box and Cox 1964; see also Sokal and Rohlf 1981: 423–426).

While transforming predictor variables affects only

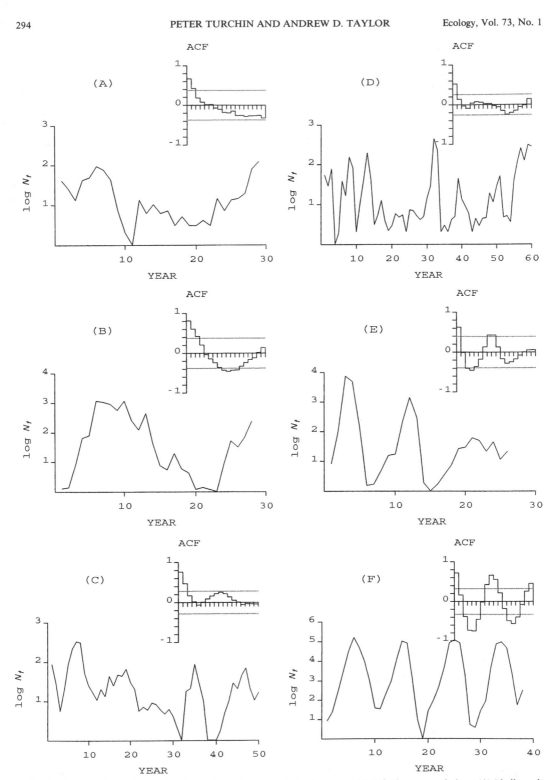

FIG. 3.   Log-transformed (base 10) time series and autocorrelation functions (ACF) for insect populations: (A) *Phyllopertha horticola*, (B) *Choristoneura fumiferana*, (C) *Hyloicus pinastri*, (D) *Panolis flammea*, (E) *Lymantria dispar*, (F) *Zeiraphera*

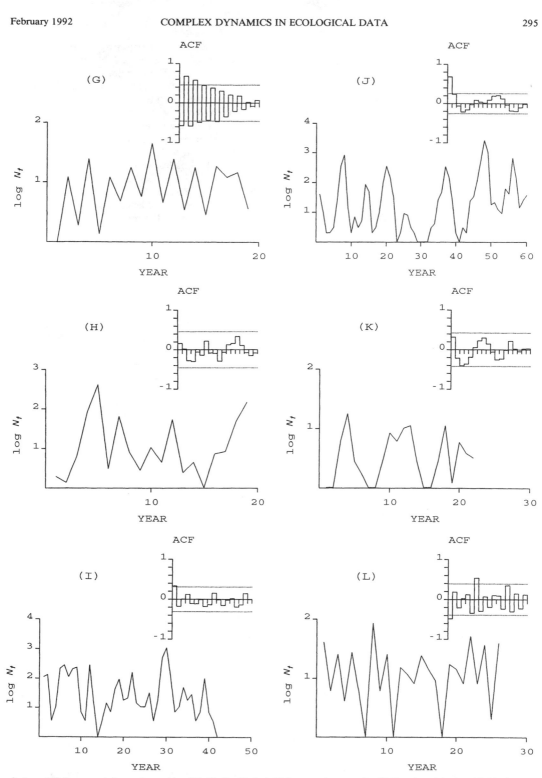

diniana, (G) *Drepanosiphum platanoides*, (H) *Phyllaphis fagi*, (I) *Lymantria monacha*, (J) *Bupalus piniarius*, (K) *Hyphantria cunea*, (L) *Vespula* spp.

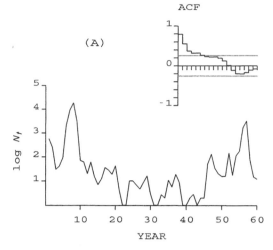

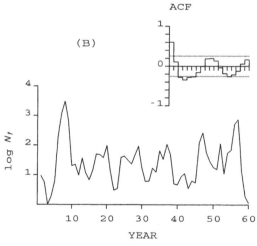

Fig. 4. The time series of *Dendrolimus pini* and its autocorrelation function (ACF). (A) actual data; (B) quadratically detrended data.

This model has a total of eight parameters (six parameters defining the quadratic surface and two transformation exponents). The best transformations (θ-values) of the predictor variables for any specific system are unknown, and need to be estimated from the data. The transformations were estimated by fitting the model (Eq. 6) by least squares for all combinations of θ_1 and θ_2 equal to $\{-1, -0.5, 0, \ldots, 2.5, 3\}$ (using log-transform when $\theta_i = 0$) and selecting the θ-values that resulted in the smallest residual sum of squares (Box and Draper 1987). The farther the estimated θ_i is from 1, the more nonlinear is the transformation.

The type of RSM-extracted dynamics was determined by iteration of the model (Eq. 6) on the computer. The initial values N_1 and N_2 were set equal to the mean population density of the observed series. This procedure decreased the likelihood of being misled by multiple attractors, if any were present. The simulated trajectory was plotted as an N_t vs. N_{t-1} phase plot. If the trajectory approached a single point, the system was classified as stable. If the trajectory settled onto several points, the dynamics were classified as a limit cycle. In many cases the trajectory would not settle onto a finite number of points, but instead all the points in the phase space would be lying on an ellipse (after discarding transients). Such dynamical behavior, called "quasiperiodic" in mathematical literature, results when the period of the oscillation is irrational, so that the solution never repeats itself exactly (Schaffer and Kot 1985). This kind of behavior is commonly found in discrete models of order >1, such as the model (Eq. 6). From the ecologist's point of view, the distinction between limit cycles and quasiperiodic dynamics is not very important, so we will treat them together as a single category. Finally, a "strange" at-

the functional shape of *f*, transformation of the response variable also affects the error structure. Population data are non-negative, often right-skewed, and more variable when the mean is large. Taking log-transforms of the response variable, a standard procedure in population ecology (Moran 1953, Finerty 1980, Royama 1981, Pollard et al. 1987), tends to alleviate all these problems at the same time (Ruppert 1989). Accordingly, we log-transformed the replacement rate N_t/N_{t-1}, obtaining the rate of population change, $r_t \equiv \log(N_t/N_{t-1})$. Defining $X \equiv N_{t-1}^{\theta_1}$ and $Y \equiv N_{t-2}^{\theta_2}$, the above argument leads to the following model for extracting deterministic dynamics from data:

$$r_t = a_0 + a_1 X + a_2 Y + a_{11} X^2$$
$$+ a_{22} Y^2 + a_{12} XY + \epsilon_t. \qquad (6)$$

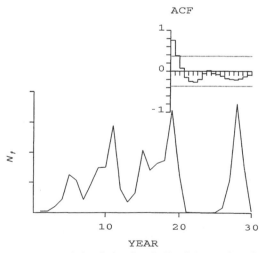

Fig. 5. Population fluctuations in *Dendroctonus frontalis* and its autocorrelation function (ACF). Data are plotted on the arithmetic (not log-transformed) scale.

TABLE 3. Summary of reconstructed dynamics.

Species	Autocorrelation function	Response-surface model result
Phyllopertha horticola	Non-stationary Non-periodic	No regulation
Choristoneura fumiferana	Non-stationary or a very long cycle	Exponentially stable
Dendrolimus pini	Non-stationary Non-periodic*	Exponentially stable†
Hyloicus pinastri	Non-stationary Non-periodic*	Damped oscillations
Dendroctonus frontalis	Non-stationary Non-periodic	Damped oscillations‡
Panolis flammea	Stationary Non-periodic	Exponentially stable
Lymantria monacha	Stationary Non-periodic	Damped oscillations
Bupalus piniarius	Stationary Suggestive of periodicity	Damped oscillations
Hyphantria cunea	Stationary Suggestive of periodicity	Damped oscillations
Vespula spp.	Stationary Suggestive of periodicity	Damped oscillations
Drepanosiphum platanoides	Stationary Periodic (2 yr)	Limit cycle (2 yr)
Lymantria dispar	Stationary Periodic (8.5 yr)	Quasiperiodicity ($\approx$ 7 yr)
Zeiraphera diniana	Stationary Periodic (9 yr)	Quasiperiodicity ($\approx$ 8 yr)
Phyllaphis fagi	Stationary Suggestive of periodicity	Chaos

* Autocorrelation function of the detrended series suggests periodicity.
† Damped oscillations extracted from the detrended series.
‡ Diverging oscillations and chaos extracted from the first and second half of the series, respectively.

'tractor, indicating chaotic dynamics, can look much like an ellipse that has been stretched and then folded. Another possibility is for a strange attractor to be separated into several discontinuous pieces (see Schaffer [1987] for the explanation of many routes to chaos, and examples of phase graphs for various kinds of attractors).

When iterating the model (Eq. 6) using an estimated response surface with noise, or a chaotic response surface without noise, the trajectory occasionally jumps

outside the range of observed N_t values. This causes a difficulty, because the shape of the response surface where it is not constrained by data points may be quite strange, e.g., the surface could blow up to infinity. In order to prevent such occurrences, the values of the function $f(N_{t-1}, N_{t-2})$, at the boundary of the box in the $N_{t-1} - N_{t-2}$ phase space defined by the maximum and the minimum of the observed series, were extrapolated for areas outside the box. In other words, when the simulated trajectory left the minimum–maximum box,

TABLE 4. Estimated response-surface parameters, as defined by Eq. 6.

Species	θ_1	θ_2	a_0	a_1	a_2	a_{11}	a_{22}	a_{12}
Phyllopertha horticola	−1.0	0.5	−2.637	0.399	6.280	−0.003	−2.887	−0.661
Choristoneura fumiferana	0.0	−1.0	0.028	−0.282	−0.007	−0.081	0.000	−0.002
Dendrolimus pinni	3.0	1.0	0.163	0.034	−0.665	0.000	0.089	−0.024
Hyloicus pinastri	1.5	2.0	0.339	−0.217	−0.326	0.029	0.005	−0.009
Dendroctonus frontalis	0.0	1.5	0.291	−1.157	0.193	−0.211	−0.125	0.386
Panolis flammea	0.5	3.0	1.306	−2.660	−0.010	0.700	−0.000	0.013
Lymantria monacha	0.0	−1.0	−1.370	−0.262	0.062	0.054	−0.002	−0.037
Bupalus piniarius	3.0	1.0	0.655	0.003	−1.522	0.000	0.185	−0.005
Hyphantria cunea	3.0	−1.0	0.408	−0.061	−0.632	0.001	0.176	−0.046
Vespula spp.	0.5	0.0	5.241	−10.251	−0.646	3.986	0.303	1.385
Drepanosiphum platanoides	1.0	−1.0	2.722	−3.665	0.292	0.646	−0.027	−0.108
Lymantria dispar	−0.5	0.5	2.894	−0.208	−8.170	0.005	1.753	0.297
Zeiraphera diniana	0.5	0.0	−4.174	4.349	−1.790	−1.280	−0.124	0.437
Phyllaphis fagi	2.5	3.0	1.130	−2.532	−0.452	0.564	0.001	−20.363

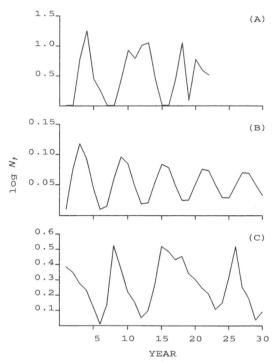

FIG. 6. *Hyphantria cunea*: observed time series (A), and trajectories predicted by response-surface model (RSM) without noise (B), and with noise (C) (ϵ_t is normally distributed with mean zero and standard deviation $\sigma = 0.2$).

the computer program evaluated the function $f(N_{t-1}, N_{t-2})$ at the point on the box boundary nearest to the point (N_{t-1}, N_{t-2}).

RESULTS

Autocorrelation function (ACF) patterns

Several insect data sets exhibit ACFs suggestive of nonstationarity. The ACF of *Phyllopertha horticola,* the garden chafer, does not decay to zero, but instead becomes progressively more negative as the lag increases (Fig. 3A). Two other populations appear to oscillate around a nonstationary mean (Figs. 4 and 3C). The level around which the *Dendrolimus pini* population is fluctuating appears to have first decreased, and then increased, while the population of *Hyloicus pinastri* exhibited a downward trend. We estimated the long-term trend for *Dendrolimus* by fitting a quadratic polynomial to N_t as a function of time. Subtracting the estimated trend from the time series ("detrending"), we obtain a series that appears to fluctuate around a constant level, and whose ACF is of the periodic phase-forgetting kind (although not quite significantly periodic at 95% level) (Fig. 4). This example shows that the ability of the ACF to detect periodic behaviors is sensitive to whether the underlying process is stationary or not (see also Box and Jenkins 1974).

The *Dendroctonus frontalis* population exhibited a different kind of nonstationarity, in which the mean stayed more or less constant, but the amplitude of the oscillation increased with time, with both the peaks becoming higher and the troughs becoming lower (Fig. 5). One possible explanation of this pattern is increased instability of the *Dendroctonus* population as a result of a several-fold enrichment of this beetle's food base over the last 30 yr (Turchin et al. 1991).

Several insect populations appeared to have periodic dynamics: significant periodicity was found in the ACF's of three populations (Fig. 3E, F, and G), and the ACF was suggestive of an oscillation in an additional four cases (Fig. 3H, J, K, and L). In each of the three periodic cases the ACF was of the phase-forgetting kind, that is, the peaks in ACF decayed at higher lags. This suggests that oscillations in these populations are driven not by an exogenous periodic force, but by endogenous dynamics.

Reconstructed endogenous dynamics

Applying response-surface methodology (RSM) to the insect time series indicated the following spectrum of endogenous dynamics: no regulation (one case); sta-

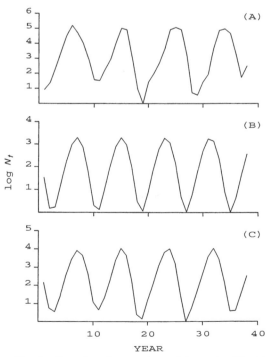

FIG. 7. *Zeiraphera diniana*: observed time series (A), and trajectories predicted by response-surface model (RSM) without noise (B), and with noise (C) (ϵ_t is normally distributed with mean zero and standard deviation $\sigma = 0.2$). Note that the deterministic trajectory in (B) does not exactly repeat itself every oscillation. This is an example of quasiperiodic behavior, in which the period of oscillation is an irrational number.

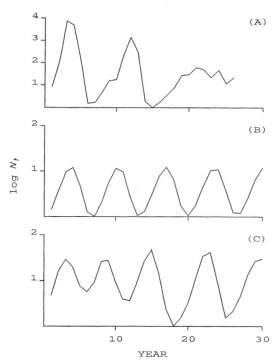

FIG. 8. *Lymantria dispar*: observed time series (A), and trajectories predicted by the response-surface model (RSM) without noise (B), and with noise (C) (ϵ_t is normally distributed with mean zero and standard deviation $\sigma = 0.2$).

ble, exponential damping (three cases); stable, oscillatory damping (six cases); limit cycle (one case); quasiperiodicity (two cases); and chaos (one case) (Table 3). The parameters defining estimated response surfaces are listed in Table 4 for each insect species. In the following subsections we will consider each of these categories in turn, paying particular attention to whether the RSM results are consistent with those from the ACFs, and to the effects of nonstationarity.

No regulation. —Iteration of the model estimated by RSM for the garden chafer, *Phyllopertha horticola,* exhibited unstable behavior: at first the population grew at a very slow rate, and then it suddenly crashed (population density decreasing by about five orders of magnitude). Regressions of r_t on lagged population densities N_{t-1} and N_{t-2} (Turchin 1990) did not indicate any density regulation in this population, suggesting that the garden chafer population may undergo a density-independent "random walk." This conclusion is supported by the nonstationary shape of the ACF (Fig. 3A), and is in agreement with the previous analysis of Milne (1984). With a very large number of data points that were generated by a density-independent population process, RSM would fit a level plane to the scatterplot of r_t as a function of N_{t-1} and N_{t-2}. Since we had to deal with a limited amount of data, it appears

that RSM fitted the vagaries of the data rather than the actual relationship, producing a meaningless result.

Equilibrium dynamics: exponential stability. —Of the three cases classified by RSM as exponentially stable, one (*Dendrolimus pini*) had an ACF that exhibited evidence of nonstationarity. When the *Dendrolimus* data were made stationary with quadratic detrending, RSM suggested that this population may be in the oscillatory damping regime, which agrees with the periodicity exhibited by the ACF of the detrended series (Fig. 4). This result demonstrates the sensitivity of RSM results to nonstationarity.

The case of *Choristoneura fumiferana* presents a puzzle. Although it was suggested that this population undergoes periodic outbreaks (Royama 1984), regressing r_t on lagged population densities did not detect any signs of density-dependent regulation. The shape of the ACF is consistent with either of the two hypotheses: that the budworm population cycles with a very long period, or that it is nonstationary (for example, the population could be tracking a long-term oscillatory trend in its food base). It is clear that data on more than a single outbreak will be needed before we are able to reach any conclusions about this insect's dynamics.

The final case for which RSM indicates exponential stability is *Panolis flammea.* This finding is consistent

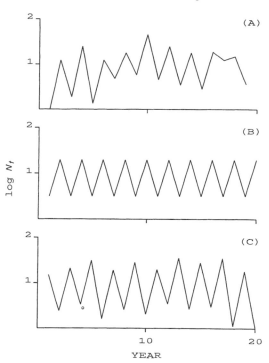

FIG. 9. *Drepanosiphum platanoides*: observed time series (A), and trajectories predicted by the response-surface model (RSM) without noise (B), and with noise (C) (ϵ_t is normally distributed with mean zero and standard deviation $\sigma = 0.2$).

300   PETER TURCHIN AND ANDREW D. TAYLOR   Ecology, Vol. 73, No. 1

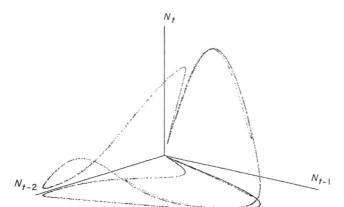

FIG. 10.   The strange attractor extracted from the *Phyllaphis* time series. This graph was produced by iterating Eq. 6 2200 times on the computer. The first 200 points were discarded, and the last 2000 points were plotted in the N_t vs. N_{t-1} vs. N_{t-2} phase space. To aid in visualizing the attractor, a projection of the attractor onto the N_{t-1} vs. N_{t-2} plane is also shown.

with the shape of the ACF, which rapidly decays to zero and does not show any signs of periodicity thereafter. Thus, our result suggests that density fluctuations of almost three orders of magnitude observed in this population were produced by density-independent factors. Nevertheless, the population is regulated around

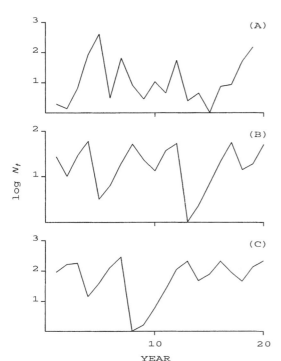

FIG. 11.   *Phyllaphis fagi*: observed time series (A), and trajectories predicted by the response-surface model (RSM) without noise (B), and with noise (C) (ϵ_t is normally distributed with mean zero and standard deviation $\sigma = 0.2$).

an equilibrium, as indicated by the RSM result of exponential stability and significant regressions of r_t on both N_{t-1} ($F_{1,55} = 7.31$, $P < .01$) and N_{t-2} ($F_{1,55} = 7.00$, $P < .05$).

Equilibrium dynamics: damped oscillations. — Two of the six cases classified as damped oscillations were nonstationary. One, the southern pine beetle (*Dendroctonus frontalis*), may have been misclassified, since we do not know how to detrend a series with the kind of nonstationarity exhibited by the southern pine beetle (constant mean but increasing amplitude of oscillations). The second nonstationary case, *Hyloicus pinastri* showed a trend in the mean. Removing the trend did not alter the RSM result, but did produce stronger evidence for periodicity in the ACF (ACF was significantly negative at the half-period, but not significantly positive at the full period).

The damped-oscillation dynamics reconstructed by RSM for the stationary cases ran the complete spectrum from rapid to slow convergence to the equilibrium. The slowest convergence to equilibrium was found in the fall webworm population (Fig. 6B), which is one of the populations with ACF suggestively, but not significantly periodic. It is known that populations characterized by oscillations slowly converging to an equilibrium will behave like noisy limit cycles in a stochastic environment (e.g., Poole 1977). Thus, adding a modest amount of stochastic variation to the deterministic dynamics extracted by RSM produces sustained pseudoperiodic oscillations (Fig. 6C).

Complex dynamics: limit cycles and quasiperiodicity. — The three insect time series that were classified by RSM as limit cycles or quasiperiodic dynamics were also the ones for which the ACF had a significantly periodic component (Fig. 3E, F, and G and Table 3). Moreover, the period of extracted oscillations was very close to the observed period: 8 vs. 9 yr for larch budmoth (Fig. 7), 7 vs. 8.5 yr for gypsy moth (Fig. 8), and

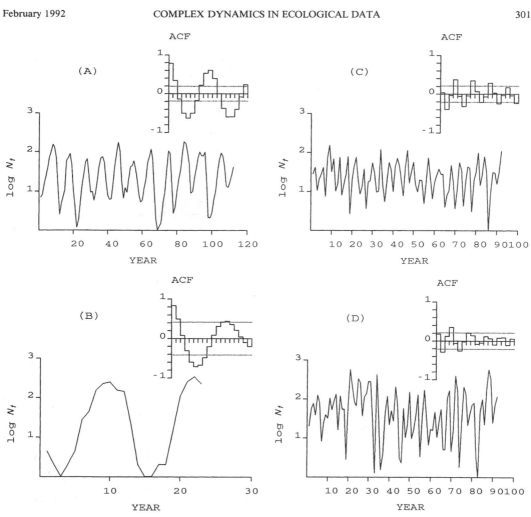

FIG. 12. Four mammal time series for which the autocorrelation function (ACF) was significantly periodic. (A) lynx, (B) belyak hare, (C) colored fox, (D) arctic fox.

2 vs. 2 yr for sycamore aphid (Fig. 9). The relative amplitude of the oscillation in the larch budmoth and the sycamore aphid was also matched by the RSM trajectories (Figs. 7 and 9), although RSM underestimated the amplitude of gypsy moth oscillations. Such a close correspondence between patterns observed in actual time series and the time series generated by response surfaces is a strong indication that RSM is at the very least capable of correctly reconstructing periodic complex dynamics.

Complex dynamics: chaos. —Finally, in one case, *Phyllaphis fagi,* RSM-extracted dynamics were of the chaotic kind. The "strange" nature of the attractor extracted from this time series is apparent when it is plotted in the $N_t - N_{t-1} - N_{t-2}$ phase space (Fig. 10). It is not clear, however, how robust this result is. Does the prediction of chaos depend on a delicate balance

of RSM-estimated parameters? We addressed this question by performing a sensitivity analysis on the data set. We excluded each data point in turn, estimated the response surface for the reduced data set, and determined its qualitative behavior. Our results indicate that the prediction of chaos in this case was not due to a freak combination of "just right" data values, since chaos was extracted in 9 out of 17 reduced data sets, with the rest divided between limit cycles (2 cases, with periods of 8 and 5), stability (3 cases), and diverging oscillations leading to extinction (3 cases).

RSM-predicted dynamics (Fig. 11B and C) were characterized by periods of exponential growth for 3–4 yr (lines of constant slope on the log scale) followed by crashes, as well as by periods of rapid oscillations. Some features of the observed trajectory were similar to RSM dynamics. Observed time series had two pe-

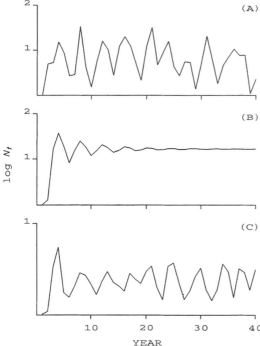

FIG. 13. Colored fox: observed time series (A) (only the middle 40 yr are shown), and trajectories predicted by the response-surface model (RSM) without noise (B), and with noise (C) (ϵ_t is normally distributed with mean zero and standard deviation $\sigma = 0.2$).

riods of almost exponential growth (3 and 4 yr), with the first period followed by a crash (what happened after the second period is unknown), and there was a period of rapid oscillations during the middle portion of the time series (Fig. 11A). On the other hand, the actual trajectory did not exhibit a rigid regulatory "ceiling" that was a characteristic feature of simulations. Another point of similarity between the observed and extracted dynamics was that ACFs of both exhibited weak periodicities, with a period of 7 yr in the data and 5 yr in the RSM model.

At this time we know little about the ability of RSM to extract deterministic chaos from data. In addition, the data are sparse. Therefore we cannot make any definite statements about whether endogenous dynamics of the *Phyllaphis* population are chaotic or not. However, the fact that RSM did extract chaotic dynamics in at least one case indicates that the region within the parameter space where the model (Eq. 6) is chaotic overlaps with the region enclosing parameter estimates for actual insect populations. In other words, one does not need to postulate biologically unrealistic values of parameters to obtain chaotic dynamics within the framework of the model (Eq. 6).

Vertebrate data sets

The time series of vertebrate populations exhibited a similar spectrum of ACF patterns. In particular, examination of ACFs suggested that there were four cyclic mammal populations (Fig. 12). Vertebrate populations also exhibited many of the same dynamic behaviors that we found among insects: 3 cases exhibited unstable oscillations leading to extinction, 6 cases were classified as exponential damping, 11 cases as damped oscillations, and 2 as quasiperiodic dynamics. There were no cases of chaos. Of the four mammal populations that had significantly periodic ACF, two were found to have quasiperiodic RSM dynamics (lynx and belyak hare). RSM-reconstructed dynamics for the colored fox and the arctic fox were damped oscillations. The damped oscillations regime is more plausible than a four-point cycle for these populations because ACF peaks were of rather small magnitude: ACF at the first peak, 4 yr (ACF[4]) was <0.4 (compare this with the lynx ACF[10] = 0.6, or the larch budmoth ACF[9] = 0.7). Such a sharp drop-off in ACF reflects a much noisier-looking time series of the two foxes, compared to either lynx or belyak hare, and therefore is more consistent with RSM-indicated oscillatory damping, than with limit cycles. The period of damped oscillations predicted by RSM was 4 yr (Fig. 13), the same as the pattern in the ACF. This result once again demonstrates the ability of RSM to accurately mimic the patterns observed in actual time series.

DISCUSSION

Our results are very different from those of Hassell et al. (1976), who concluded that all but 2 of their 24 insect populations had exponentially stable point equilibria. By contrast, our response-surface methodology (RSM) found exponential stability in only 3 of our 14 insect populations. The remaining populations were classified as unregulated (one case), damped oscillations (six cases), limit cycles (one case), quasiperiodic oscillations (two cases), and chaos (one case). The vertebrate examples exhibited a similar spectrum of dynamics, although there were no cases of chaos. We do not wish to claim that all of these classifications (especially the two most extreme ones, *Phyllaphis* and *Phyllopertha*) are correct. This fairly small number of examples does, however, include convincing cases of periodic dynamics (damped oscillations, limit cycles, and quasiperiodicity) and one case with parameter values at least approaching those producing chaos. We conclude, then, that the complete spectrum of dynamical behaviors, ranging from exponential stability to chaos, is likely to be found among natural populations.

The contrast between our findings and those of Hassell et al. (1976) resulted from three important differences in methodology: (1) fitting actual time-series data instead of the two-step method of Hassell et al., (2) using a model with a much more flexible functional

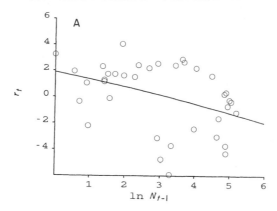

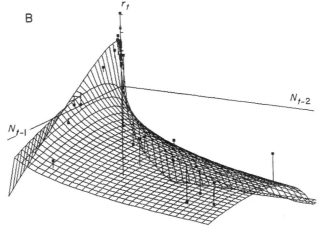

FIG. 14. Response function estimated for larch budmoth: (A) r_t is a function of N_{t-1} only, and (B) r_t is a function of both N_{t-1} and N_{t-2}.

form, and (3) accounting, albeit indirectly, for the multidimensional nature of population dynamics that could be due to interactions with other populations within the community, or to population structure. We believe that the last of these differences is the most critical. Indeed, fitting data with a first-order model (only terms involving N_{t-1} [the previous year's density]), which in all other respects was identical to model (Eq. 6), produced results very similar to those of Hassell et al.: 11 cases of exponential stability, 2 cases of damped oscillations, and one limit cycle (the sycamore aphid). It is revealing that this truncated model, as well as the analysis of Hassell et al. (1976), classified the larch budmoth population in the Engadine Valley in Switzerland (see Baltensweiler and Fischlin 1987) as exponentially stable, although this population is arguably the most convincing example of a quasiperiodic attractor in our data set. This misclassification happened because in this population there is little effect of N_{t-1} on r_t (the rate of population change), and a large effect of N_{t-2} (compare Fig. 14A to 14B). When we reduce

the dimensionality of the model by ignoring N_{t-2}, we turn a clean, strongly nonlinear response surface in three dimensions into a cloud of largely uninformative points in two dimensions. Fitting the model to these points then yields a gentle slope (Fig. 14A), indicating mild direct density dependence and thus stability.

The flexibility of our RSM model, provided by inclusion of both the Box-Cox transformation and quadratic terms, also was essential for correct classification. For example, the quadratic term (although not the second lag) was necessary for an accurate reconstruction of the sycamore aphid dynamics. Fitting a model with either one or two lags but no quadratic terms leads to a classification of damped oscillations, in contrast to the conclusion of a two-point limit cycle obtained by a quadratic RSM (with either one or both lags). Simulations of the RSM model with no quadratic terms produced an ACF that decayed to zero much faster than the ACF of either the data or the full RSM model. In addition, the full model came much closer to reproducing the perfect alternation of increases and decreases seen in the observed series. Thus, the quadratic term was essential for reaching the correct conclusion in this case.

The preceding examples suggest that leaving out important factors, such as delayed density dependence or strong nonlinearities, may lead to incorrectly classifying a population as more stable than it actually is. In other words, use of overly simple models for reconstructing endogenous dynamics from data may be biased in favor of finding stability. This may well apply to our own analysis, since regressions of r_t on lagged population densities indicate that lags of order higher than two are not infrequent (P. Turchin and A. D. Taylor, *unpublished analysis*). Analysis of cases with higher dimensional response surfaces might well result in additional findings of complex dynamics, though the feasibility of such expanded analysis will be limited by the relatively short length of a typical ecological time series.

The methodology used in this paper is by no means perfect. For instance, it cannot effectively handle systems with multiple equilibria. By applying a standard model to each case, we also risk misclassifying some instances by using an inappropriate model. As we noted above, inclusion of additional lags may be appropriate in a number of cases (subject to data constraints). However, our model may be more complex than needed for some systems; whether such overfitting in any way biases the results is unknown but under investigation.

Another limitation is that our approach currently lacks any means for determining "confidence intervals" around our dynamical predictions. Confidence limits can be obtained for each parameter estimate of the model (Eq. 6), but they tell us nothing about how variation in parameter estimates will affect our conclusion about the type of extracted dynamics. Another potential problem is the estimation bias that arises when models such as Eq. 6 are fitted to data with substantial observation errors (Walters and Ludwig 1987).

In closing we note that much controversy surrounding the issue of population regulation stems from the one-dimensional viewpoint, held by many, that attempts to place all populations within the spectrum ranging from tight control around a stable point equilibrium (regulation) to little or no dynamical feedback in population density (no regulation). Bias against complex endogenous dynamics is so strong that most discussions or criticisms of population regulation do not mention (e.g., Wolda 1989)—or even dismiss outright—the possibility that populations may undergo cyclic or chaotic fluctuations. The following quotations show that this view is shared both by experimentalists: "the rarity with which populations fluctuate cyclically in nature . . ." (Hairston 1989:6), and theoreticians: "deterministic stability is the rule rather than the exception, at least with insect populations" (Nisbet and

Gurney 1982:55). Our results suggest otherwise. We argue, therefore, that natural populations cannot be ranked within a one-dimensional spectrum going from no regulation to tight regulation around a point equilibrium. Instead, a two-dimensional scheme needs to be employed, with one axis indicating the relative strength of the exogenous (density-independent) component, and the other axis indicating the type of endogenous dynamics.

ACKNOWLEDGMENTS

We thank J. Allen, A. Berryman, L. Ginzburg, J. Hayes, P. Kareiva, and W. Morris for reading and commenting on the manuscript. A. Berryman, A. Dixon, and W. Morris were very generous in sharing data sets.

LITERATURE CITED

Abraham, R. M., and C. D. Shaw. 1983. Dynamics: the geometry of behavior. Part 2. Chaotic behavior. Aerial Press, Santa Cruz, California, USA.

Andrewartha, H. G., and L. C. Birch. 1954. The distribution and abundance of animals. University of Chicago Press, Chicago, Illinois, USA.

Argoul, F., A. Arnedo, P. Richetti, J. C. Roux, and H. L. Swinney. 1987. Chemical chaos: from hints to confirmation. Accounts of Chemical Research **20**:436–442.

Baltensweiler, W., and A. Fischlin. 1987. The larch budmoth in the Alps. Pages 331–351 *in* A. A. Berryman, editor. Dynamics of forest insect populations: patterns, causes, implications. Plenum, New York, New York, USA.

Bejer, B. 1988. The nun moth in European spruce forests. Pages 211–231 *in* A. A. Berryman, editor. Dynamics of forest insect populations: patterns, causes, implications. Plenum, New York, New York, USA.

Berryman, A. A. 1978. Population cycles of the douglas-fir tussock moth (Lepidoptera: Lymantriidae): the time-delay hypothesis. Canadian Entomologist **110**:513–518.

Berryman, A. A., and J. A. Millstein. 1989. Are ecological systems chaotic—and if not, why not? Trends in Ecology and Evolution **4**:26–28.

Box, G. E. P., and D. R. Cox. 1964. An analysis of transformations. Journal of the Royal Statistical Society **B26**: 211–252.

Box, G. E. P., and N. R. Draper. 1987. Empirical model-building and response surfaces. John Wiley & Sons, New York, New York, USA.

Box, G. E. P., and G. M. Jenkins. 1976. Time series analysis: forecasting and control. Holden Day, Oakland, California, USA.

Dixon, A. F. G. 1990. Population dynamics and abundance of deciduous tree-dwelling aphids. Pages 11–23 *in* A. D. Watt, S. R. Leather, M. D. Hunter, and N. A. C. Kidd, editors. Population dynamics of forest insects. Intercept, Andover, England.

Eckmann, J. P., and D. Ruelle. 1985. Ergodic theory of chaos and strange attractors. Review of Modern Physics **57**:617.

Ellner, S. 1989. Inferring the causes of population fluctuations. Pages 286–307 *in* C. Castillo-Chavez, S. A. Levin, and C. A. Shoemaker, editors. Mathematical approaches to problems in resource management and epidemiology. Lecture Notes in Biomathematics. Volume 81. Springer-Verlag, Berlin, Germany.

Elton, C. 1942. Voles, mice and lemmings. Problems in population dynamics. Clarendon, Oxford, England.

Finerty, J. P. 1980. The population ecology of cycles in small mammals. Yale University Press, New Haven, Connecticut, USA.

Guckenheimer, J., G. Oster, and A. Ipaktchi. 1977. The

dynamics of density-dependent population models. Journal of Mathematical Biology 4:101–147.

Hairston, N. G. 1989. Ecological experiments: purpose, design, and execution. Cambridge University Press, Cambridge, England.

Hassell, M. P., J. H. Lawton, and R. M. May. 1976. Patterns of dynamical behavior in single species populations. Journal of Animal Ecology 45:471–486.

Hutchinson, G. E. 1948. Circular causal systems in ecology. Annals of the New York Academy of Sciences 50:221–246.

Jones, J. W. 1914. Fur-farming in Canada. Commission of Conservation, Ottawa, Canada.

Keith, L. B. 1963. Wildlife's ten-year cycle. University of Wisconsin Press, Madison, Wisconsin, USA.

Kot, M., W. M. Schaffer, G. L. Truty, D. J. Grase, and L. F. Olsen. 1988. Changing criteria for order. Ecological Modelling 43:75–110.

Labutin, Yu. V. 1960. Predators as a factor of change in the number of Belyak hares. Pages 192–190 in S. P. Naumov, editor. Research into the cause and natural dynamics of the numbers of Belyak hares in Yakutia. USSR Academy of Sciences, Moscow, USSR (in Russian).

Lack, D. 1954. The natural regulation of animal numbers. Clarendon Press, Oxford, England.

Leigh, E. 1968. The ecological role of Volterra's equations. Pages 1–61 in M. Gerstenhaber, editor. Some mathematical problems in biology. American Mathematical Society, Providence, Rhode Island, USA.

Loery, G., and J. D. Nichols. 1985. Dynamics of a Black-capped Chickadee population, 1958–1983. Ecology 66:1195–1203.

May, R. M. 1974. Biological populations with nonoverlapping populations: stable points, stable cycles, and chaos. Science 186:645–647.

———. 1976. Simple mathematical models with very complicated dynamics. Nature 261:459–467.

Middleton, A. D. 1934. Periodic fluctuations in British game populations. Journal of Animal Ecology 3:231–249.

Milne, A. 1984. Fluctuation and natural control of animal population, as exemplified in the garden chafer (*Phyllopertha horticola* (L.)). Proceedings of the Royal Society of Edinburgh 82B:145–199.

Montgomery, M. E., and W. E. Wallner. 1987. The gypsy moth: a westward migrant. Pages 353–375 in A. A. Berryman, editor. Dynamics of forest insect populations: patterns, causes, implications. Plenum, New York, New York, USA.

Moran, P. A. P. 1953. The statistical analysis of the Canadian lynx cycle. Australian Journal of Zoology 1:163–173.

Morris, R. F. 1964. The value of historical data in population research, with particular reference to *Hyphantria cunea* Drury. Canadian Entomologist 96:356–368.

→ Murdoch, W. W., and J. D. Reeve. 1987. Aggregation of parasitoids and the detection of density dependence in field populations. Oikos 50:137–141.

Naumov, N. P. 1972. The ecology of animals. University of Illinois Press, Urbana, Illinois, USA.

Nicholson, A. J. 1954. An outline of the dynamics of animal populations. Australian Journal of Zoology 2:9–65.

Nisbet, R., S. Blythe, B. Gurney, H. Metz, and K. Stokes. 1989. Avoiding chaos. Trends in Ecology and Evolution 4:238–239.

Nisbet, R. M., and W. S. C. Gurney. 1982. Modelling fluctuating populations. John Wiley & Sons, Chichester, England.

Pollard, E., K. H. Lakhani, and P. Rothery. 1987. The detection of density-dependence from a series of annual censuses. Ecology 68:2046–2055.

Poole, R. W. 1977. Periodic, pseudoperiodic, and chaotic population fluctuations. Ecology 58:210–213.

Royama, T. 1981. Fundamental concepts and methodology for the analysis of animal population dynamics, with particular reference to univoltine species. Ecological Monographs 51:473–493.

———. 1984. Population dynamics of the spruce budworm *Choristoneura fumiferana*. Ecological Monographs 54:429–462.

Ruppert, D. 1989. Fitting mathematical models to data: a review of recent developments. Pages 274–284 in C. Castillo-Chavez, S. A. Levin, C. A. Shoemaker, editors. Mathematical approaches to problems in resource management and epidemiology. Lecture Notes in Biomathematics 81:274–284.

Schaffer, W. M. 1985. Order and chaos in ecological systems. Ecology 66:93–106.

———. 1987. Perceiving order in the chaos of nature. Pages 313–350 in M. Boyce, editor. Life histories: theory and patterns from mammals. Yale University Press, New Haven, Connecticut, USA.

Schaffer, W. M., and M. Kot. 1985. Are ecological systems governed by strange attractors? BioScience 35:342–350.

Schaffer, W. M., and M. Kot. 1986. Chaos in ecological systems: the coals that Newcastle forgot. Trends in Ecology and Evolution 1:58–63.

Schwerdtfeger, F. 1941. Ueber die Ursachen des Massenwechsels der Insekten. Zeitung Angeweine Entomologie 28:254–303.

Sokal, R. R., and F. J. Rohlf. 1981. Biometry. W. H. Freeman, New York, New York, USA.

Southwood, T. R. E. 1967. The interpretation of population change. Journal of Animal Ecology 36:519–529.

Sugihara, G., and R. M. May. 1990. Nonlinear forecasting as a way of distinguishing chaos from measurement error in time series. Nature 344:734–741.

Takens, F. 1981. Detecting strange attractors in turbulence. Pages 366–381 in D. A. Rand and L. S. Young, editors. Dynamical systems and turbulence. Springer-Verlag, New York, New York, USA.

Turchin, P. 1990. Rarity of density dependence or population regulation with lags? Nature 344:660–663.

Turchin, P., P. L. Lorio, A. D. Taylor, and R. F. Billings. 1991. Why do populations of southern pine beetles (Coleoptera: Scolytidae) fluctuate? Environmental Entomology 20:401–409.

Vastano, J. A., and E. J. Kostelich. 1986. Comparison of algorithms for determining Lyapunov exponents from experimental data. Pages 100–107 in G. Meyer-Kress, editor. Dimensions and entropies in chaotic systems. Springer Verlag, Berlin, Germany.

Walters, C. J., and D. Ludwig. 1987. Effects of measurement errors on the assessment of stock–recruitment relationships. Canadian Journal of Fisheries and Aquatic Science 38:704–710.

Wolda, H. 1989. The equilibrium concept and density dependence tests. What does it all mean? Oecologia (Berlin) 81:430–432.

Wolf, A., J. B. Swift, H. L. Swinney, and J. A. Vastano. 1985. Determining Lyapunov exponents from a time series. Physica 16D:285–317.

Ecology (1977) **58**: pp. 1103–1111

POPULATION GROWTH RATES AND AGE *VERSUS* STAGE-DISTRIBUTION MODELS FOR TEASEL (*DIPSACUS SYLVESTRIS* HUDS.)[1]

Patricia A. Werner

*W. K. Kellogg Biological Station and Department of Botany, Michigan State University,
Hickory Corners, Michigan 49060 USA*

AND

Hal Caswell

Biological Sciences Group, University of Connecticut, Storrs, Connecticut 06268 USA

Abstract. Mathematical models are developed to examine the population-level response of an herbaceous plant species (teasel, *Dipsacus sylvestris* Huds.) which was experimentally introduced into several habitats and monitored for 5 yr. Models based on morphological stages (size) rather than chronological age give more satisfactory results. Population growth rates (λ_m) range from 0.63 to 2.60, which are likely typical for fugitive plants. Values are interpreted as responses to both external and internal factors. Grass litter, and the presence of other dicotyledonous species, and the overall primary productivity of the rest of the community are important factors determining the success or failure of an attempted colonization by teasel. Individual plant and population-level growth rates seem to be determined independently.

Key words: Age-distribution; biennial; colonization; Dipsacus; *fugitive species; mathematical models; Michigan; morphological stages; old-fields; plant populations; population dynamics; population growth rates; population projection matrix.*

Introduction

To a large extent, ecological studies of single plant species or its populations have been concerned with elucidating evolutionary radiation within a group, the plasticity of responses to the environment, or selection and fitness for particular sites. Recently there has been a great interest in basic population dynamics, including estimates of growth rates, age-specific (or size-specific) mortality rates, birth rates, and the factors influencing these. (See reviews by Willson 1972, Harper and White 1974 and references in Werner 1976.) Studies of this type are a foundation for the development of general theory in population biology.

Populations respond to both their outer environment and their own internal state. Their response is complex and multivariate, including changes in birth, death, immigration and emigration rates, growth rates at both the individual and the population level, and ultimately genetic and evolutionary changes. In this paper we use mathematical models to examine the population-level response of a plant (teasel, *Dipsacus sylvestris* Huds.). Our effort differs from other studies of plant population dynamics (e.g., Hartshorn 1972, 1975, but see Sarukhan and Gadgil 1974) in examining several populations of a single species under different environmental conditions. Our estimates of population parameters (e.g., growth rate, transient response, reproductive value) can thus be interpreted as responses to both external and internal factors. In a related paper (Cas-

well and Werner 1977), we examine the effects of hypothetical changes in the life history of teasel, by modifying the internal structure of the models.

The Populations

Teasel (*Dipsacus sylvestris* Huds.) is a strictly semelparous plant with no vegetative propagation. Seeds germinate mainly in the spring, producing a large-leaved rosette which requires vernalization before forming a single tall (0.5–2.5 m) flowering stalk in some subsequent summer. Although commonly classified a biennial, the length of the rosette phase is variable, and may last >5 years (Werner 1975*a*).

In eastern North America, teasel usually is found in the later stages of old-field succession (before shrubs), in meadows, and in ruderal habitats where periods between disturbances are >1 yr. Populations are sparsely distributed regionally, but, where found, may dominate the vegetation in terms of biomass. The regional rarity of populations may be due partly to the dispersal pattern of seeds (achenes): >99% of them fall passively to the ground within a 1.5-m radius of the parent plant (Werner 1975*b*). Teasel is native to Europe, where its populations are also restricted to disturbed habitats, and are regionally rare but locally very dense; the few (insect) herbivores that feed on teasel are found in both N. America and Europe. It seems unlikely that the selective pressures operating on teasel now are vastly different from those under which it has been evolving for a long time.

Research on the population dynamics of teasel was

[1] Manuscript received 5 October 1976; accepted 21 March 1977.

1104   PATRICIA A. WERNER AND HAL CASWELL   Ecology, Vol. 58, No. 5

Table 1. Vegetation characteristics of the study fields. See Werner 1975c for species composition and abundances

| Field | \multicolumn{4}{c}{Relative levels of abundance of:} | | | | Mean[2] annual net primary production (g/m[2]) |
	Grass	Litter	Herbaceous dicots	Woody dicots[1]	
A	Medium	Low	Low	Negligible	343
B	Medium	Low	Low-Medium	Neglible	250
C	Medium	Low	Low	High	279
D	Low	Low	High	Negligible	233
J	Low	Negligible	High	High	287
K	High	High	Negligible	None	312
L	Medium	Low	Low	Negligible	377
M	Medium	Low	Low	Negligible	245

[1] Woody dicots create an overhead shading canopy.
[2] ANPP of total vegetation in control plots within the fields in year 2. Method for determining ANPP is described in Werner 1977.

initiated in 1969 in Kalamazoo County, Michigan, USA, within a set of contiguous, small fields (each 25 × 25 m) which were set aside for succession studies in 1964 and monitored annually. Eight fields were used; all were 2-yr fallow (Fields J, K, L, M) or 3-yr fallow (Fields A, B, C, D), although they differed greatly in vegetative structure (Table 1). Within each of the 8 fields, 26 plots 0.5 × 0.5 m in size were permanently marked and handsown with teasel seed at the rate of 150 seeds per plot in the winter of 1968–69. Subsamples of the seed lot prior to counting and sowing showed a 99% viability. Upon germination, individual seedlings were marked and their fates monitored for five growing seasons. Some plots were sacrificed each year to obtain estimates of primary production (cf. Werner 1977).

It is important to note that teasel was absent from the study area from 1964 until 1972, except where it was deliberately introduced for this study. The nearest natural population was 11.2 km from the study site. In order to follow accurately the initially introduced seed cohorts, any flowering heads that were produced prior to the third year of the study were removed before their seeds could be dispersed. By the third spring, germination of seeds from the initial cohort was negligible and thereafter seed heads of flowering plants were allowed to remain.

These teasel populations are artificial only in the sense that they were purposely sown in the fields. In a more important sense, they are natural; the introductions are simulated colonizations of habitats where teasel happened to be absent but of a type which it commonly colonizes. We note that the density of seeds used (600/m²) is typical of the density of seed-rain in a natural population. The fact that colonization consisted of a specified number of seeds, sown into marked pieces of ground in order to accurately assess the fate of the various seed cohorts, is irrelevant to the subsequent growth and behavior of the populations.

THE MODELS

Because production of seeds was counted, but not allowed to feed back into the population, the repeated

counts in each quadrat provide a cohort record of both survivorship and reproduction. With this information we can construct parameters for a linear, time-invariant, matrix model of population dynamics, a so-called Leslie matrix or population-projection matrix. The form of the model is:

$$\bar{x}(t + 1) = \bar{A}\bar{x}(t) \qquad (1)$$

where $\bar{x}(t)$ is a vector of state variables (age-classes or physiological stage classes) and $\bar{A}$ is a square matrix which determines the dynamics of the population. The time unit, in our case, is 1 yr.

The linearity of this model implies that there are no effects of density on population growth. Ultimately, of course, this cannot be true. However, teasel populations are capable of surviving for long periods of time, slowly increasing to much higher densities than those encountered during the course of this experiment. It is unlikely that density had any serious effects on the population during the period in which the parameters were being estimated. How long this state of affairs would persist, were the population to continue to grow under these dynamics, is another question (Caswell and Werner 1977).

$\bar{A}$ is a constant matrix, implying that there are no changes in the environmental conditions confronting the population. We assume that successional and other changes in the eight fields were slow relative to the year-to-year dynamics of survival and reproduction, at least over the course of this experiment. This assumption is hard to test, but is probably justified (cf. Werner 1972).

State variables

The state variables, $\bar{x}(t)$, cannot be chosen arbitrarily. They must fulfill some very distinct requirements (Caswell et al. 1972). Their role is to encapsulate the relevant history of the system so that the current environmental stimulus, together with the current state, uniquely determines the system's response. An inadequate choice of the state variable fails to satisfy this requirement, and the resulting system description will be indeterminate. There will be more than one

TABLE 2. State vectors

Age classification	Stage classification
$x(1)$ seeds	$x(1)$ seeds
$x(2)$ dead or dormant seeds, year 1	$x(2)$ dead or dormant seeds, year 1
$x(3)$ dead or dormant seeds, year 2	$x(3)$ dead or dormant seeds, year 2
$x(4)$ rosettes, year 1	$x(4)$ small rosettes (<2.5 cm)
$x(5)$ rosettes, year 2	$x(5)$ medium rosettes (2.5–18.9 cm)
$x(6)$ rosettes, year 3	$x(6)$ large rosettes (≥19.0 cm)
$x(7)$ rosettes, year 4	$x(7)$ flowering plants
$x(8)$ flowering plants	

response possible to a given stimulus, the particular response occurring determined by the past history of the system.

The simplest state variable for population dynamic models is the number of individuals. Much ecological theory is built on the qualitative properties of models using this state variable, but except in special situations, it is not adequate for a detailed description of population dynamics. Populations of the same total size can behave very differently, depending on their internal composition. The effect of age structure is the most obvious example; a population of prereproductive individuals responds differently to a given environment than a population of the same number of reproductive or postreproductive individuals, yet all have the same population size. Lotka (1924) showed that in a constant environment, the age distribution would eventually become stable at which point it would legitimately be collapsed into a scalar measure of population size. Leslie (1945) developed a discrete technique to deal with the dynamics of the age distribution itself when the conditions of the stable age distribution theorem are not met.

Growth plasticity, however, makes even the complete age distribution an inadequate state variable for many organisms, higher plants in particular. Individuals of the same age may respond in very different ways to the same environment, depending on their history. Morphological stages based on size may encapsulate the history of a plant more accurately than a chronological age classification.

Because we have information on both the age and the size of the individual plants in this study, we have a unique opportunity to compare 2 choices of state variables for the same populations. The first (Table 2) is an age classification, consisting of seeds, dead or dormant seeds (explained below) of 2 ages, rosettes aged 1, 2, 3, and 4 yr, and flowering plants. Our second choice for a state variable is a vector of morphological stages: seeds, dead or dormant seeds, small (<2.5 cm), medium (2.5–18.9 cm) and large (≥19.0 cm) rosettes, and flowering plants. Neither model subdivides the flowering plant category; all flowering plants are considered identical with regard to seed output, regardless

of size or age. In the fields, the number of seeds produced per flowering plant varied little with the age of the plant, and varied (linearly) with the size of the previous-year's rosette only across sizes within the "large rosette" category.

The age vector has 8 elements, the stage vector only 7; thus the $\overline{A}$ matrices for the 2 models are of dimension 8×8 and 7×7, respectively. By comparing the behavior of these 2 models, we hope to determine which of these state variables most accurately describes the dynamics of the populations.

Parameter estimation

In the $\overline{A}$ matrix of Eq. (1), the (i, j) entry gives the number of individuals of stage i produced in year $t + 1$, per individual of stage j in year t. Estimation of these parameters requires a knowledge of the fate, in the next year, of individual plants of each stage in the current year. For the purposes of parameter estimation, each of the 26 quadrats (each 0.25 m²) in a field were considered replicate populations. The transition probabilities in the $\overline{A}$ matrices are means taken over these replicates. Table 3 shows the resulting matrices, for the age and stage classifications, for each field.

Probabilities of seed germination were based on germination measurements made every 3 days in each field (Werner 1977). The ungerminated seeds remaining out of the 150 sown in each quadrat were assigned to the "first-year dead or dormant seed" pool. This category contains dormant seeds capable of germinating the next year as well as a number of seeds whose fate is unknown—dead or eaten. It was impossible to separate dormant from dead seeds in the field. The "second-year dead or dormant seeds" category was calculated similarly, based on seeds in the first pool that failed to germinate their second year. Since third-year germination was negligible, it was assumed in the model that all seeds not germinating after their second year of dormancy were in fact dead.

Seed germination of teasel occurs in the spring. The figures for seed germination (the first 3 columns of $\overline{A}$ in Table 3) give the proportion of seeds (or dormant seeds) that both germinate and survive until fall. For these columns, the difference between one and the sum of a column gives the proportion of plants that die between one autumn and the next, a year later.

The probabilities of transition from one age or size class to another were obtained by following marked individual plants in the plots from one year to the next (Werner 1977). Seed production per flowering plant was estimated from regressions of seed number on head size (Werner 1975c).

The averaging of transition probabilities over replicate quadrats within a field smooths out small-scale spatial variation and demographic stochasticity. In addition, some of the parameters (e.g., in the stage model) were estimated in more than 1 yr. Hence, the elements of the matrices in Table 3 are averages over

1106   PATRICIA A. WERNER AND HAL CASWELL   Ecology, Vol. 58, No. 5

TABLE 3. Matrices of transition probabilities (see Table 2 for state vectors)

Field		Age Classification									Stage Classification						
		x(1)	x(2)	x(3)	x(4)	x(5)	x(6)	x(7)	x(8)		x(1)	x(2)	x(3)	x(4)	x(5)	x(6)	x(7)
A	x(1)	...	...	...	...	...	...	...	431.	x(1)	...	...	...	...	...	...	431.
	x(2)	.748	...	...	...	...	...	...	...	x(2)	.748	...	...	...	...	...	...
	x(3)	...	.966	...	...	...	...	...	...	x(3)	...	.966	...	...	...	...	...
	x(4)	.079	.021	.010	...	...	...	...	...	x(4)	.008	.013	.010	.125	...	...	...
	x(5)	...	...	...	.137	...	...	...	...	x(5)	.070	.007	.000	.125	.238	...	...
	x(6)	...	...	...	...	.396	...	...	...	x(6)	.002	.008	.000	.038	.245	.167	...
	x(7)	...	...	...	...	...	.167	...	...	x(7)	...	...	...	.000	.023	.750	...
	x(8)	...	...	...	.003	.234	.583	.250	...								
B	x(1)	...	...	...	...	...	...	...	465.	x(1)	...	...	...	...	...	...	465.
	x(2)	.761	...	...	...	...	...	...	...	x(2)	.761	...	...	...	...	...	...
	x(3)	...	.925	...	...	...	...	...	...	x(3)	...	.925	...	...	...	...	...
	x(4)	.078	.059	.018	...	...	...	...	...	x(4)	.009	.044	.016	.000	...	...	...
	x(5)	...	...	...	.287	...	...	...	...	x(5)	.066	.016	.001	.074	.269	...	...
	x(6)	...	...	...	...	.681	...	...	...	x(6)	.004	.000	.000	.000	.462	.367	...
	x(7)	...	...	...	...	...	.443	...	...	x(7)	...	...	...	.000	.000	.592	...
	x(8)	...	...	...	.002	.136	.233	.324	...								
C	x(1)	...	...	...	...	...	...	...	1,132	x(1)	...	...	...	...	...	...	1,132
	x(2)	.756	...	...	...	...	...	...	...	x(2)	.756	...	...	...	...	...	...
	x(3)	...	.940	...	...	...	...	...	...	x(3)	...	.940	...	...	...	...	...
	x(4)	.024	.033	.003	...	...	...	...	...	x(4)	.011	.033	.003	.000	...	...	...
	x(5)	...	...	...	.172	...	...	...	...	x(5)	.014	.000	.000	.338	.268	...	...
	x(6)	...	...	...	...	.706	...	...	...	x(6)	.000	.000	.000	.045	.513	0.167	...
	x(7)	...	...	...	...	...	.000	...	...	x(7)	...	...	...	.000	.000	0.833	...
	x(8)	...	...	...	.000	.053	.800	.200	...								
D	x(1)	...	...	...	...	...	...	...	1,105	x(1)	...	...	...	...	...	...	1,105
	x(2)	.567	...	...	...	...	...	...	...	x(2)	.567	...	...	...	...	...	...
	x(3)	...	.916	...	...	...	...	...	...	x(3)	...	.916	...	...	...	...	...
	x(4)	.066	.033	.018	...	...	...	...	...	x(4)	.031	.053	.018	.029	...	...	...
	x(5)	...	...	...	.226	...	...	...	...	x(5)	.035	.018	.000	.327	.394	...	...
	x(6)	...	...	...	...	.739	...	...	...	x(6)	.000	.000	.000	.059	.376	.000	...
	x(7)	...	...	...	...	...	.514	...	...	x(7)	...	...	...	.000	.000	.667	...
	x(8)	...	...	...	.000	.016	.208	.278	...								
J	x(1)	...	...	...	...	...	...	...	476	x(1)	...	...	...	...	...	...	476
	x(2)	.423	...	...	...	...	...	...	...	x(2)	.423	...	...	...	...	...	...
	x(3)	...	.987	...	...	...	...	...	...	x(3)	...	.987	...	...	...	...	...
	x(4)	.070	.009	.006	...	...	...	...	...	x(4)	.024	.009	.006	.007	...	...	...
	x(5)	...	...	...	.038	...	...	...	...	x(5)	.044	.000	.000	.050	.158	...	...
	x(6)	...	...	...	...	.287	...	...	...	x(6)	.001	.000	.000	.002	.008	.000	...
	x(7)	...	...	...	...	...	.000	...	...	x(7)	...	...	...	.000	.000	.250	...
	x(8)	...	...	...	.001	.000	.000	1.000	...								
K	x(1)	...	...	...	...	...	...	...	.000	x(1)	...	...	...	...	...	...	.000
	x(2)	.800	...	...	...	...	...	...	...	x(2)	.800	...	...	...	...	...	...
	x(3)	...	.959	...	...	...	...	...	...	x(3)	...	.959	...	...	...	...	...
	x(4)	.008	.040	.001	...	...	...	...	...	x(4)	.001	.034	.001	.000	...	...	...
	x(5)	...	...	...	.067	...	...	...	...	x(5)	.007	.006	.000	.048	.275	...	...
	x(6)	...	...	...	...	.400	...	...	...	x(6)	.000	.000	.000	.000	.025	.000	...
	x(7)	...	...	...	...	...	.000	...	...	x(7)	...	...	...	.000	.000	.000	...
	x(8)	...	...	...	.000	.000	.000	1.00	...								
L	x(1)	...	...	...	...	...	...	...	503	x(1)	...	...	...	...	...	...	503
	x(2)	.430	...	...	...	...	...	...	...	x(2)	.430	...	...	...	...	...	...
	x(3)	...	.970	...	...	...	...	...	...	x(3)	...	.970	...	...	...	...	...
	x(4)	.046	.024	.005	...	...	...	...	...	x(4)	.010	.021	.005	.000	...	...	...
	x(5)	...	...	...	.152	...	...	...	...	x(5)	.036	.003	.000	.190	.253	...	...
	x(6)	...	...	...	...	.312	...	...	...	x(6)	.000	.000	.000	.070	.105	.150	...
	x(7)	...	...	...	...	...	.000	...	...	x(7)	...	...	...	.000	.002	.517	...
	x(8)	...	...	...	.0003	.136	.000	1.000	...								

TABLE 3. (Continued)

Field		Age Classification									Stage Classification							
M	x(1)	...	...	...	...	...	...	...	635	x(1)	...	...	...	...	...	...	...	635
	x(2)	.634	...	...	...	...	...	...	...	x(2)	.634	...	...	...	...	...	...	...
	x(3)	...	.973	...	...	...	...	...	...	x(3)	...	.974	...	...	...	...	...	...
	x(4)	.128	.024	.013	...	...	...	...	...	x(4)	.013	.017	.011	.000	...	...	...	...
	x(5)	...	...	...	.128	...	...	...	...	x(5)	.109	.004	.002	.077	.212	...	...	...
	x(6)	...	...	...	...	.396	...	...	...	x(6)	.006	.003	.000	.038	.281	.000	...	...
	x(7)	...	...	...	...	...	.627	...	...	x(7)	...	...	...	.000	.063	1.000	...	...
	x(8)	...	...	...	.029	.259	.160	.213	...									

both small-scale spatial and short-term temporal variability, and are currently the best estimates we can make for populations of teasel.

RESULTS AND DISCUSSION

Population growth rates

The solution to Eq. 1 can be written

$$\bar{x}(t) = \bar{A}^t \bar{x}(0).$$

When the matrix function $\bar{A}^t$ is expanded in terms of eigenvalues and eigenvectors (Frame 1964) the resulting expression for $\bar{x}(t)$ is a series

$$\bar{x}(t) = \sum_{i=1}^{n} c_i \bar{v}_i \lambda_i^t,$$

where the c_i are constants which express the initial conditions, and the λ_i and $\bar{v}_i$ are the eigenvalues (assumed distinct) and right eigenvectors, respectively, of $\bar{A}$. The eigenvalues and eigenvectors are determined by the equation

$$\bar{A}\bar{v}_i = \lambda_i \bar{v}_i.$$

As t grows large, the contribution to $\bar{x}(t)$ of any of the λ_i of modulus <1 will decay toward zero, while that of any λ_i of modulus >1 will increase geometrically.

If $\bar{A}$ has a maximal eigenvalue, λ_m, which is larger (in modulus) than all the other eigenvalues, the long-term behavior of the population is given by

$$\lim_{t \to \infty} \bar{x}(t) = c_m \bar{v}_m \lambda_m^t.$$

The population will ultimately grow geometrically at the rate λ_m per year, with a structure (age or stage) defined by $\bar{v}_m$.

Sykes (1969) has summarized the conditions under which the existence of such an eigenvalue can be guaranteed when $\bar{A}$ is a Leslie matrix. In the case of more general matrices, such as ours, much of this theory does not apply. The Perron-Frobenius Theorem, in its weak form applicable to reducible nonnegative matrices (Gantmacher 1959:80), guarantees the existence of a nonnegative eigenvalue of modulus greater than or equal to that of all the other eigenvalues. The only disturbing possibilities not ruled out are a maximal eigenvalue of exactly zero or of exactly the same modulus as one or more of the other eigenvalues. However, neither of these situations is structurally stable in general, and they can be neglected when the elements of the matrices are estimated from real data. Thus we are justified in applying theory based on the existence of a maximal eigenvalue, and, in fact, all our matrices possess such an eigenvalue.

The biological interest in λ_m arises partly from its identification with evolutionary fitness by Wright (1937) and Fisher (1930; Fisher actually used $r = \ln\lambda_m$). They assigned values of λ_m to the different genotypes within a population and demonstrated that those genotypes with a higher λ_m would ultimately dominate the population. Fisher's Fundamental Theorem asserts that the value of λ_m for the population (the average of the values for the different genotypes) will continue to increase under the impact of selection, at a rate proportional to the genetic variance in individual fitness. In an unlimited, competition-free, optimal environment, the per capita growth rate of a genotype is known as r or r_m, the "intrinsic rate of increase," and selection operating in such an environment will result automatically in increasing the intrinsic rate of increase of the population. This result is not incompatible with the notion of K-selection introduced by MacArthur (1962, MacArthur and Wilson 1967) to describe population growth in a limited, competitively full environment. In both environments, selection operates to increase individual fitness, and still results in an increase in population fitness, but it may do so through parameters other than r_m.

Table 4 shows the population growth rates λ_m, and the instantaneous values $r = \ln\lambda_m$, for each of the eight populations, for both the age and stage models.

TABLE 4. Eigenvalues (λ_m) and instantaneous growth rates ($r = \ln \lambda_m$) for the age and stage models

Field	Age model		Stage model	
	λ_m	r	λ_m	r
A	1.263	0.233	1.797	0.586
B	1.462	0.380	1.989	0.688
C	1.401	0.337	1.875	0.629
D	1.350	0.300	2.071	0.728
J	0.333	−1.100	0.628	−0.465
K	0.0004	−7.752	0.275	−1.291
L	0.891	−0.115	1.195	0.178
M	1.679	0.518	2.605	0.957

1108   PATRICIA A. WERNER AND HAL CASWELL   Ecology, Vol. 58, No. 5

TABLE 5. Stable age and size distribution for stage and age models

Model		Fields					
		A	B	C	D	L	M
	AGE						
$x(1)$	seeds	46.87	51.13	51.55	56.35		59.13
$x(2)$	dead or dormant seeds, year 1	27.74	26.60	27.83	23.68		22.33
$x(3)$	dead or dormant seeds, year 2	21.21	16.83	18.67	16.06		12.95
$x(4)$	rosettes, year 1	3.56	4.01	1.60	2.96		4.93
$x(5)$	rosettes, year 2	0.38	0.79	0.20	0.49		0.37
$x(6)$	rosettes, year 3	0.12	0.37	0.10	0.27		0.09
$x(7)$	rosettes, year 4	0.02	0.11	0.00	0.10		0.03
$x(8)$	flowering	0.14	0.16	0.06	0.07		0.16
	STAGE (size)						
$x(1)$	seeds	58.48	61.15	61.14	68.76	58.21	71:45
$x(2)$	dead or dormant seeds, year 1	24.36	23.38	24.67	18.84	20.96	17.39
$x(3)$	dead or dormant seeds, year 2	13.09	10.90	12.37	8.33	17.01	6.50
$x(4)$	small rosettes	0.53	0.87	0.81	1.59	0.93	0.50
$x(5)$	medium rosettes	2.79	2.59	0.69	1.96	2.45	3.30
$x(6)$	large rosettes	0.50	0.88	0.23	0.40	0.31	0.56
$x(7)$	flowering	0.24	0.26	0.10	0.13	0.14	0.29

Table 5 lists the corresponding stable distributions (age or size, respectively).

A value of $\lambda_m > 1$ (corresponding to an $r > 0$) implies that the population will grow geometrically. A value less than one results in a population decline to extinction. Using either state variable, the populations in our eight fields exhibit a spectrum of growth rates from rapid extinction (J, K) through borderline persistence (L) to rapid population growth (A, B, C, D and especially M).

Clearly, the *Dipsacus* populations cannot continue growing at rates as high as $\lambda_m = 2.6$ forever. Either their own density must ultimately limit population growth, or the environment will change in such a way as to eliminate them. Both factors are probably important; density effects are considered in another paper (Caswell and Werner 1977). The nonequilibrium values of λ_m underline teasel's ecological position as a fugitive species.

There are few other valid estimates of population growth rates for natural plant populations. Hartshorn (1972, 1975) used a size class matrix model to study *Pentaclethra macroloba* and *Stryphnodendron excelsum*, 2 rain forest tree species in Costa Rica. His values of λ_m for the 2 species are 1.002 and 1.047, respectively. These are as expected for large organisms in equilibrium with a very stable environment. Sarukhan and Gadgil (1974) developed a model similar to ours to describe population growth in 3 species of *Ranunculus*. They obtained values of λ_m from 0.743–1.801 for *Ranunculus repens*, 0.095–4.665 for *Ranunculus bulbosus*, and 0.350–2.328 for *Ranunculus acris*. The ranges reported are over 5–10 quadrats in a single Welsh meadow. The occurrence of values $\ll 1$ and $\gg 1$, and the range of values found, are similar to our results for teasel and are to be expected in species living in nonequilibrium situations.

The stable age and size distribution (Table 5) are similar among fields and between models. They have been calculated before only by Hartshorn (1972, 1975) and are presented here for their possible value in later comparative studies. The stage model consistently predicts higher proportions of seeds and flowering plants in the population, and a smaller proportion of rosette plants. This is, no doubt, related to the higher population growth rates exhibited by the stage model (see below). Stable distributions are not presented for fields J and K, or for L in the age model, because these populations decline to extinction rather than grow with a stable distribution.

The high proportion of the population in the dead or dormant seed category is somewhat artificial. As mentioned earlier, there is no way of knowing what fraction of these seeds are dormant and what fraction are dead. The very low germination rates (Table 3) and the lack of long-term dormancy mechanisms suggest that most are dead, rather than dormant. Thus these components of the stable vectors are artificially inflated by the inclusion of a portion of the population that is no longer extant.

Age vs. stage as state variables

We cannot directly compare observed and predicted population trajectories using the 2 models, because reproduction was prevented by removing the seed heads before the seeds could be dispersed. There are, however, several interesting indirect comparisons to be made between the two state variables.

First, Werner (1975a) showed that, when all 8 populations are lumped together, size is a better predictor than age of plant fate from year to year. Our model supports this on a field-by-field basis as shown in Table 6. Here we examine the accuracy with which the 2 models predict the first occurrence of flowering in each

TABLE 6. Observed and predicted year of first flowering. Numbers in parentheses give the number of flowering plants predicted for that year

Field	Predicted		Observed
	Age model	Stage model	
A	2 (0.03)	2 (0.42)	2
	8 (1.03)	3 (2.15)	
B	2 (0.02)	2 (0.33)	3
	8 (2.41)	3 (2.81)	
C	3 (0.03)	3 (0.93)	3
	9 (1.33)	7 (6.69)	
D	3 (0.03)	3 (1.52)	4
	9 (1.00)		
J	2 (0.008)[1]	2 (0.42)[1]	2
K	never	never	never
L	2 (0.003)	2 (0.12)	2
	>25	15 (1.03)	
M	2 (0.56)	2 (1.74)	2
	5 (1.53)		

[1] Declines thereafter.

of the fields. Since the model deals in real numbers, while fractional plants are impossible, we have shown both the time when the model predicts the appearance of the first flowering plants and the time when the predicted number of flowering plants first exceeds 1. The stage model is clearly more accurate. In only 2 cases was its prediction more than 1 yr off. One of these cases (field C) was so close (a prediction of 0.93 flowering plants) to 1 at the correct year that flowering plants would surely be expected at this time. The other (field L) is a marginally persistent population: λ_m for the age model is 0.891, for the stage model 1.195. Both the age and stage models predict a very long time to the first flowering, which actually occurred in the 2nd year. Over all 8 populations, the deviation between observed and predicted first flowering is 1.62 ± 1.64 ($\bar{x} \pm$ SE) years for the stage model as opposed to 6.00 ± 2.58 for the age model.

Because of the plastic nature of growth in teasel plants, it is not surprising to discover that the stage-based models are superior to the age-based models in predicting population behavior. As with most sessile organisms, selective advantage has been conferred on those individuals able to modify growth rates, to switch from vegetative to reproductive modes, or to change age of first reproduction in response to a changing environment from which they cannot move. The large body of theory and analytical techniques developed by animal demographers cannot be applied directly to studies of plants with the possible notable exceptions of very long-lived perennials where sometimes variation in growth rates, age to first reproduction, etc., can be smoothed over long time periods, or annuals where growth and reproduction is canalized into a definite time period. Even in the case of annuals, however, stage-based models may also prove more useful than

age-based models when analyses rightly include the dynamics of seeds in the soil; for most annuals, the length of the dormancy period of seeds is plastic to some high degree and dependent upon environmental conditions.

Although the stage distribution is clearly superior to the age distribution as a state variable for teasel, this is not the complete story. The values of λ_m vary from field to field in the same way in either model (Spearman rank correlation coefficient $r_s = 0.93$), and there is only 1 case (field L) where the 2 models make contradictory predictions about population persistence. However, the maximum eigenvalues for the age model are in every case smaller than those for the stage model. If the age and stage classification were simply regroupings of one another, our use of cohort survival data should have resulted in identical values for λ_m (give or take sampling error). This is so because when the population reaches its stable distribution, at which time it is growing at the rate λ_m, *any* grouping of the states (including just summing all classes to obtain population size) will grow at λ_m. Thus any matrix generated by grouping categories must have a maximum eigenvalue equal to λ_m. The fact that there is a consistent pattern in which the eigenvalues of the age model are always less than those of the stage model implies that the 2 ways of classifying the population are not independent. The age-transition dynamics are affected by the size distribution, and *vice versa*. A complete population description would require at least an age × stage classification as a state variable. Such state variables for populations have been proposed on theoretical grounds (cf. Caswell et al. 1972) and have been applied by Slobodkin (1953) and Sinko and Steiffer (1969) to invertebrate animals.

This interaction between the 2 classifications is also suggested in Table 1 of Werner (1975a) where, for example, the probability of death varies by as much as a factor of 8 (from .02 to .17) between rosettes of the same size but different ages. However, the interaction was not detectable in that paper; statistical tests cannot be used to examine age × size interaction since the data are in the form of single numbers per block.

Further evidence for an age × stage interaction comes from the variances of the elements of the $\bar{A}$ matrices. Since each quadrat within a field was treated as a replicate population, these variances measure the degree of uncertainty in the prediction of the future state from the value of the current state. Eliminating the seed, dormant seed, and flowering plant categories from consideration because they are the same for both models, the average variances ($\pm$ SE) for the remaining matrix elements are $.068 \pm .009$ for the age model and $.072 \pm .010$ for the stage model. The values are not significantly different. This means that the variation, from one replicate population to another, in the description of transition dynamics, is about the same whether the age or stage alone is used as a state variable.

One hypothesis is that this interaction involves an

1110   PATRICIA A. WERNER AND HAL CASWELL   Ecology, Vol. 58, No. 5

age-dependence of mortality for large rosettes. Large rosettes older than 5 yr seem to experience a dramatically increased mortality. This effect cannot be incorporated into the stage model, which is blind to the age of a rosette, and it might account for the higher λ_m values generated by the stage model. However, if this were entirely the answer, one would expect that fields with high survivorship in large rosettes should show the largest discrepancy between λ_m (stage) and λ_m (age). In fact, this is not the case (cf. Tables 3 and 4).

It would be of considerable interest to construct a combined age × stage distribution model for *Dipsacus*. The existence of the age × size interaction suggests that such a model would be an improvement over either the age or stage classifications alone. Unfortunately, parameterizing the resultant 16-dimensional model would require considerably more data than we have available. Moreover, Werner's (1975*a*) demonstration of the superiority of size over age as a predictor of rosette transitions, and the superiority of the stage model over the age model in predicting the occurrence of flowering suggest that the improvement over the stage model might not be great. Because of the demonstrated superiority of the stage model, and the theoretical reasons for preferring size to age as a stage variable for organisms with plastic growth, we will carry out the rest of our analyses on the stage model only.

Environmental effects on population growth

The biotic environments confronted by the *Dipsacus* populations in the eight fields differ in many respects. In Table 1 we have summarized some of these differences; we relate them here to the growth rate of the populations. The direct action of the environment is on the individual plant, which germinates, grows, lives, and dies in relation to its surroundings. The results of this action appear at successively higher levels of organization: the population (through birth rates, death rates, growth and extinction), the community (dominance by particular species or lifeforms, diversity, etc.) the ecosystem (nutrient cycling, microclimate alteration, etc.), and the biosphere (atmospheric homeostasis, global element cycles and energy budgets, etc.). The population growth rate, λ_m, is a natural parameter (although only one of several) for measuring environmental impact at the population level, since it integrates the effects of birth and death rates in a way intimately related both to population dynamics (growth, extinction) and evolution (as the mean fitness of the population).

The populations in fields J and K have growth rates considerably less than one, indicating that these fields are closed to continuance by *Dipsacus*. The reasons are different for each field. Field K had a high level of grass and grass litter (*Agropyron repens* L.). Werner (1975*d*) has shown that *Agropyron* litter inhibits the germination of *Dipsacus* seeds in both the field and laboratory. Examination of the matrices (Table 3) reveals that the

TABLE 7. Correlation coefficients (Spearman's rank correlation coefficient, r_s) between community productivity (annual net primary production in control plots; cf. Table 1), population growth rate (λ_m) and individual plant growth rate (mean diameter of rosettes in August; Werner 1977)

	λ_m	Community productivity
Individual growth rate	0.000[1]	.476
		$P \simeq .10$
λ_m	—	$-.79$
		$P < .05$

[1] A rarity in statistics: a correlation coefficient of exactly zero.

population in Field K had the lowest germination probability (.199), and also the lowest survivorship of seedlings from spring until fall (.042) of any of the populations.

Field J had high levels of herbaceous dicots, which compete strongly with *Dipsacus* (Werner 1977), and was also heavily shaded by woody dicots (mainly *Rhus typhina* L.). Germination probability was highest (0.577) in this field, and 1st-year seedling survivorship was not unusually low (0.120, rank 5th). However, the heavy shading and dicot cover depressed the growth rates of rosettes, resulting in high mortalities and very low transition probabilities from one size class to the next (the subdiagonal terms in the $\bar{A}$ matrix, Table 3). The result is the same as that of the germination depression in field K: ultimate extinction of the population.

Beyond these 2 extreme cases, there is no obvious relationship between the vegetation factors and λ_m. Teasel is able to colonize successfully a range of old-field communities, being barred from invasion by the combination of extreme shading and competition from dicot herbs (e.g., Field J) or by litter effects on seeds and seedlings (e.g., Field K). See Werner (1977) for further discussion of the effect of these factors on individual plants.

Another possible measure of the competitive stress faced by *Dipsacus* is the annual net primary production of the community being invaded. Table 1 shows this figure, which was measured in control quadrats for each field (Werner 1977). The productivity in an environment is inversely correlated with λ_m for *Dipsacus* (Spearman rank correlation $r_s = -0.79$, $P < .05$). Thus, there is strong indication that the growth rate of *Dipsacus* populations is suppressed by the "success" of the rest of the community.

Population growth vs. individual plant growth

There are some interesting comparisons between these results on population growth rate and the growth rates of individual plants (Werner 1977). It is well known, on mathematical grounds (Lewontin 1965), that increases in individual growth rate can dramatically increase population growth rate by shortening developmental time. It is not necessarily true, however, that individual and population growth rates in real populations always respond in the same way to the same factors.

Table 7 shows the rank correlation coefficients between λ_m, individual plant growth rates, and community productivity. Community productivity and λ_m are negatively correlated, but community productivity and individual growth rate are positively correlated. There is a correlation of exactly zero between the individual plant growth rates and population growth rates.

Over our sample of 8 environments, then, it appears that individual and population level growth rates are determined independently of one another. This does not rule out the possibility of a distinct relation between the 2, as seems to be the case in field J where low individual growth rates due to competition with other dicots are associated with a low value of λ_m. It means that such a relationship cannot be assumed as a general rule.

Correlation analysis is clearly not going to unravel the causal mechanism relating individual and population growth rates. We have spoken earlier of the integrative nature of such measures as λ_m. The price for such integration is the loss of detailed information on internal mechanisms. The point remains, however, that these analyses clearly demonstrate "emergent" properties at the population level that are not predicted by knowledge of the mechanisms operating at the individual level.

Acknowledgments

Assistance in the collection of field data by C. D. Tallon is greatfully acknowledged. Various aspects of this research were supported by National Science Foundation grants BMS 74-01602 to P. A. Werner, GB 43378 to H. Caswell, and GI-20 to H. E. Koenig and W. E. Cooper. W. K. Kellogg Biological Station contribution number 307.

Literature Cited

Caswell, H., and P. A. Werner. 1977. Transient response and life history analysis of teasel (*Dipsacus sylvestris* Huds.). Ecology (*In Press*).

Caswell, H., H. E. Koenig, J. Resh, and Q. Ross. 1972. An introduction to systems science for ecologists, p. 4–78. *In* B. C. Patten [ed.] Systems analysis and simulation in ecology, Vol. II. Academic Press, New York.

Fisher, R. A. 1930. The genetical theory of natural selection. Dover Publ., New York. 272 p.

Frame, J. S. 1964. Matrix functions and applications. IV. Matrix functions and constituent matrices. IEEE Spectrum 1:123–132.

Gantmacher, F. R. 1959. Applications of the theory of matrices. Interscience Publ., New York. 317 p.

Harper, J. L., and J. White. 1974. The demography of plant Annu. Rev. Ecol. Syst. 5:419–463.

Hartshorn, G. S. 1972. The ecological life history and population dynamics of *Pentaclethra macroloba*, a tropical wet forest dominant and *Stryphnodendron excelsum*, an occasional associate. Ph.D. thesis, University of Washington, Seattle. 119 p.

———. 1975. A matrix model of tree population dynamics, p. 41–51. *In* F. B. Golley and E. Medina [eds.] Tropical ecological systems: trends in terrestrial and aquatic research. Springer-Verlag, New York.

Leslie, P. H. 1945. On the use of matrices in certain population mathematics. Biometrika 33:183–212.

Lewontin, R. C. 1965. Selection for colonizing ability, p. 77–94. *In* H. G. Baker and G. H. Stebbins [eds.] The genetics of colonizing species. Academic Press, New York.

Lotka, A. J. 1924. Elements of mathematical biology. Dover Publ., New York. 465 p.

MacArthur, R. 1962. Some generalized theorems of natural selection. Proc. Natl. Acad. Sci. 48:1893–1897.

MacArthur, R., and E. O. Wilson. 1967. The theory of island biogeography. Princeton University Press, Princeton, N.J. 203 p.

Sarukhan, J., and M. Gadgil. 1974. Studies on plant demography: *Ranunculus repens* L., *R. bulbosus* L., and *R. acris* L. III. A mathematical model incorporating multiple modes of reproduction. J. Ecol. 62:921–936.

Sinko, J. W., and W. Streiffer. 1969. Applying models incorporating age-size structure to *Daphnia*. Ecology 50:608–615.

Slobodkin, L. B. 1953. An algebra of population growth. Ecology 34:513–519.

Sykes, Z. M. 1969. On discrete stable population theory. Biometrics 25:285–293.

Werner, P. A. 1972. Effect of the invasion of *Dipsacus sylvestris* on plant communities in early old-field succession. Ph.D. thesis, Michigan State University. University Microfilms, Ann Arbor, Michigan. 140 p.

———. 1975a. Predictions of fate from rosette size in teasel (*Dipsacus fullonum* L.) Oecologia 20:197–201.

———. 1975b. A seed trap for determining patterns of seed deposition in terrestrial plants. Canadian J. Bot. 53:810–813.

———. 1975c. The biology of Canadian weeds: 12. *Dipsacus sylvestris*. Huds. Canadian J. Plant. Sci. 55:783–794.

———. 1975d. The effects of plant litter on germination in teasel, *Dipsacus sylvestris* Huds. Am. Midl. Nat. 94:470–476.

———. 1976. Ecology of plant populations in successional environments. Syst. Bot. 1:246–268.

———. 1977. Colonization success of a "biennial" plant species: experimental field studies in species coexistence and replacement. Ecology (*In Press*).

Willson, M. F. 1972. Evolutionary ecology of plants: a review. II. Ecological life histories. The Biologist 54:148–162.

Wright, S. 1937. The distribution of gene differences in populations. Proc. Natl. Acad. Sci. 23:307–320.

Ecology, 68(5), 1987, pp. 1412–1423
© 1987 by the Ecological Society of America

A STAGE-BASED POPULATION MODEL FOR LOGGERHEAD SEA TURTLES AND IMPLICATIONS FOR CONSERVATION[1]

DEBORAH T. CROUSE
Department of Zoology, University of Wisconsin, Madison, Wisconsin, 53706 USA

LARRY B. CROWDER[2]
Department of Zoology, North Carolina State University, Raleigh, North Carolina 27695-7617 USA

AND

HAL CASWELL
Biology Department, Woods Hole Oceanographic Institution, Woods Hole, Massachusetts 02543 USA

Abstract. Management of many species is currently based on an inadequate understanding of their population dynamics. Lack of age-specific demographic information, particularly for long-lived iteroparous species, has impeded development of useful models. We use a Lefkovitch stage class matrix model, based on a preliminary life table developed by Frazer (1983*a*), to point to interim management measures and to identify those data most critical to refining our knowledge about the population dynamics of threatened loggerhead sea turtles (*Caretta caretta*). Population projections are used to examine the sensitivity of Frazer's life table to variations in parameter estimates as well as the likely response of the population to various management alternatives. Current management practices appear to be focused on the least responsive life stage, eggs on nesting-beaches. Alternative protection efforts for juvenile loggerheads, such as using turtle excluder devices (TEDs), may be far more effective.

Key words: Caretta caretta; *demography; endangered species; management; (marine) turtles; stage class matrix projection models; southeastern United States.*

INTRODUCTION

Increases in the human population and degradation of habitats have caused many species that were formerly common to decline to near extinction. As a result, we have become increasingly involved in attempting to preserve populations of rare or endangered species. Current management decisions may be critical in preventing their extinction. But to make effective management decisions for any species, we must estimate the population's response to various management alternatives. Unfortunately, many management decisions seem to be based more on ease of implementation or accessibility of particular life stages than a priori clear expectations of population responses to management.

Many marine turtle populations are threatened with extinction (Federal Register 1978, Groombridge 1982), and recently much attention and effort have been focused on their conservation (Bjorndal 1982, Hopkins and Richardson 1985). Nearly all of the conservation efforts have focused on a single life stage: eggs on the nesting-beach. Turtle nests are readily accessible and protectable, and losses and protection successes are

easily monitored. But, given our poor understanding of turtle population dynamics, it is not clear whether egg protection efforts will ultimately prevent marine turtle extinction.

Recently some authors have suggested that reductions in juvenile and/or adult mortality may be important to the enhancement of dwindling loggerhead populations (Richardson 1982, Richardson and Richardson 1982, Frazer 1983*a*), but this idea has not been explored systematically and no quantitative predictions have been made. In this paper, we use recent demographic data for threatened loggerhead turtles (*Caretta caretta*) from the southeastern United States to develop a stage class population model. We then test the sensitivity of the model to variations in parameter values and compare the model predictions with what is known about marine turtle population dynamics. We also explore the potential effects of several different management scenarios on the long-term survival of loggerhead turtles.

STAGE CLASS MODELS

Lefkovitch (1965) demonstrated that the Leslie matrix population projection technique (Lewis 1942, Leslie 1945) was actually a special case of the more general matrix A such that

$$A n_t = n_{t+1}$$

[1] Manuscript received 30 May 1986; revised 12 January 1987; accepted 13 January 1987.
[2] Address reprint requests to L. B. Crowder.

or

$$\begin{bmatrix} a_{11} & a_{12}\cdots a_{1s} \\ a_{21} & a_{22}\cdots a_{2s} \\ \cdot & \cdot\cdot\cdot\cdot \\ a_{s1} & a_{s2}\cdots a_{ss} \end{bmatrix} \begin{bmatrix} n_1 \\ n_2 \\ \cdot \\ n_s \end{bmatrix}_t = \begin{bmatrix} n_1 \\ n_2 \\ \cdot \\ n_s \end{bmatrix}_{t+1}$$

where n_i gives the abundance of individuals in a particular life stage at time t.

A is known as the "population projection matrix" and describes the number of offspring born to each stage class that survive a given time period as well as the proportion of individuals in each stage class that survive and remain in that stage vs. those that survive and enter another stage, otherwise known as the "transition probability." Thus the elements of the matrix A incorporate the fecundity, mortality, and growth rates of each stage class. The Leslie matrix divided the population into equal age classes. In the Lefkovitch matrix, there is no necessary relation between stage and age; the fundamental assumption is that all individuals in a given stage are subject to identical mortality, growth, and fecundity schedules. The technique of population projection, postmultiplying this matrix by the population vector, is used to forecast future population states. The dominant eigenvalue λ_m of the Lefkovitch stage class matrix is equal to e^r, where r is the intrinsic rate of increase of the population in the equation

$$N_t = N_0 e^{rt}.$$

Thus, if $\lambda_m = e^r = 1$, then $r = 0$, and the population size remains stable.

In a constant environment, the proportion of individuals in different age classes of a population tends toward a stable age distribution (Lotka 1925). Similarly, each population matrix A has a corresponding right eigenvector w_m that represents the stable stage distribution of the population such that

$$Aw_m = \lambda_m w_m.$$

For the matrices considered here, any initial population stage structure projected forward will approach the stable stage distribution w_m, where each stage class increases in size λ_m times each time period. The reproductive value of each stage is given by the elements of the left eigenvector v corresponding to λ_m, defined by

$$v'A = \lambda_m v'.$$

These reproductive values estimate the expected reproductive contribution of each stage to population growth.

Thus, the primary differences between the Leslie age class matrix and Lefkovitch's stage class matrix are that the stage classes may differ in their duration *and* that individuals may also remain in a stage from one time to the next. Vandermeer (1975, 1978) has clarified the theoretical constraints, resulting from errors due to

sample size and the distribution of individuals, on the selection of stage categories when they are not biologically apparent (as with insect instars).

Lefkovitch (1965) originally derived the stage class matrix to model an animal population, the cigarette beetle (*Lasioderma serricorne*). But its more frequent use by botanists, to investigate several plant species with widely diverse life history patterns (Hartshorn 1975, Werner and Caswell 1977, Meagher 1982, Caswell 1986), testifies to the versatility and power of the technique.

LOGGERHEAD DEMOGRAPHIC PARAMETERS

To construct a stage class population matrix for any species, data on fecundity and survival rates for the individuals in each stage are necessary. In addition, some measure of the probability of remaining in a stage vs. that of moving on to another stage is required. Unfortunately, such demographic parameters are very difficult to measure in long-lived, mobile organisms. In marine turtles only the adult nesting females, eggs and hatchlings, and stranded, dying turtles are ever seen on the beaches. Turtles often travel great distances (Carr 1967), occasionally nesting on more than one beach (Stoneburner and Ehrhart 1981), while a given female may nest only once every several years (=remigration rate: Carr and Carr 1970). Thus long-term monitoring of individual animals, often over a number of beaches, is necessary to obtain accurate estimates of fecundity and survival.

Furthermore, no method has yet been devised to obtain accurate ages of sea turtles. Rapid juvenile growth rates quickly obscure notches cut in the shell margin, so these marks are useful only for short-term studies, and the shells are generally too thin and fragile to hold a reliable tag for long. In various attempts at shell and flipper tagging in the past, the tag return rates have been abysmally low, resulting in inadequate estimates of nesting remigration (and thus fecundity) rates, survival and growth rates, and age at reproductive maturity (Pritchard 1980, Richardson 1982, Frazer 1983b).

The lack of reliable information on age-specific rates in marine turtles precludes the use of age-based population models, so stage class models must be employed. A few stages (eggs, hatchlings, and mature, nesting adults) are biologically distinct and easily recognized, and some estimates of survival rates are becoming available for these stages. But the long juvenile period between the hatchling and adult stages presents more difficulties. Most researchers have used size (as measured by carapace length) as an index of age in marine turtles (Uchida 1967, Mendonca 1981), so size-based stage classes seem appropriate.

Only a few studies on particular nesting-beaches (Hughes 1974, Carr et al. 1978, Richardson et al. 1978) have lasted long enough to generate the data necessary to model population dynamics in marine turtles. Estimates for the various components of loggerhead fe-

1414 DEBORAH T. CROUSE ET AL. Ecology, Vol. 68, No. 5

TABLE 1. Loggerhead fecundity components (ranges based on published literature). Little Cumberland Island (LCI), Georgia, estimates are given in the rightmost column. The source for each datum is given in parentheses.

Component	Low estimate	High estimate	LCI estimate	
			Probability	Year
Remigration rate*	every 5th yr (Frazer 1983)	every year (Hughes 1974, and others)	.0358	1
			.4989	2
			.3221	3
			.1119	4
			.0313	5
			(weighted ave., Frazer 1984)	
Clutch frequency†	1 per season (Richardson 1982)	7 per season (Lenarz et al. 1981)	2.99 per season (Richardson 1982)	
Mean clutch size‡ (eggs/clutch)	100, Florida (Davis and Whiting 1977)	126, South Carolina (Caldwell 1959)	120 (Richardson 1982)	
	101, Oman (Hirth 1980)	125, Florida (Gallagher et al. 1972)		

* Remigration rate is defined as the interval (yr) between nesting seasons for individual females.
† Clutch frequency is the number of clutches of eggs per female per nesting season.
‡ Mean clutch size is the average number of eggs per clutch for a given nesting-beach.

cundity vary widely (Table 1); estimates of survival are equally variable (Table 2). Recently however, Richardson's 20-yr project on Little Cumberland Island (LCI), Georgia has begun to generate defensible estimates of fecundity (Richardson 1982, Frazer 1984) and survival (Frazer 1983a, b) for loggerhead turtles in the southeastern United States (see the right-hand columns of Tables 1 and 2).

Frazer recently analyzed the LCI data (Richardson 1982, Frazer 1983a, b, 1984, 1987) as well as data from other southeastern loggerhead populations and produced a preliminary life table for a natural (=wild) loggerhead population. Frazer's age-specific life table for the LCI loggerhead population (1983a) assumes a closed population with a 1:1 sex ratio, first reproduction at 22 yr, a maximum life span of 54 yr, and a population declining at the rate of 3%/yr (Frazer 1983b). Frazer's assumptions are subject to debate among sea turtle biologists, but they are well within the published ranges for these values and seem reasonable, at least as a starting point.

Because Frazer's original data derive from survival and fecundity estimates for various size classes rather than from turtles of known ages, a stage-based life table is perhaps easier to defend than Frazer's age-specific life table. Hence we condensed Frazer's life table for the LCI loggerhead population into seven stage classes (Table 3). These classes are: (1) 1st yr (eggs and hatchlings), (2) small juveniles, (3) large juveniles, (4) subadults, (5) novice breeders, (6) 1st-yr remigrants, and (7) mature breeders. Stages 5, 6, and 7 have been considered separately, despite similar survival probabilities, because of large differences in fecundity among these three classes (Frazer 1984). Stage 7 was not subdivided further despite Frazer's carefully calculated age-specific fecundities, because the maximum deviation from 80 of ±7 eggs in any year was judged to be minor compared with other factors contributing to the population dynamics. In addition to the fact that a stage class approach is better supported by the data, one major advantage over using an age class model based on Frazer's life table is that simulations can be per-

TABLE 2. Loggerhead survivorship estimates for each life stage (from Crouse 1985). Estimates from Little Cumberland Island (LCI), Georgia, are given in the rightmost column. Sources for data are given in parentheses.

Stage	Mortality factors*	Survivorship estimates		LCI estimate (Frazer 1983a)
Eggs				
	Predation, erosion, poaching, bacteria, plant roots	6%	(Hopkins et al. 1979)	
		3 − 90%	(Talbert et al. 1980)	
		77.8%	(Hughes 1974)	1st yr = 0.6747
		$\bar{X}$ = 80%	(Hirth 1890)	
Hatchlings	Predation, ORV† ruts, pollution		unknown	
Juveniles	Predation, trawlers		unknown	small juveniles = 0.7857 large juveniles = 0.6758
Adolescents	Trawlers, predation		unknown	subadults = 0.7425
Adults	Trawlers, predation, senescence		unknown	adults = 0.8091

* Major sources of mortality for each stage are given in likely order of importance.
† ORV = off-road vehicles. The 5-cm hatchlings often become trapped in vehicle ruts, which subjects them to increased predation and desiccation.

TABLE 3. Stage-based life table for loggerhead sea turtles based on data in Frazer (1983a). These values assume a population declining at $\approx$3%/yr.

Stage number	Class	Size* (cm)	Approximate ages (yr)	Annual survivorship	Fecundity (no. eggs/yr)
1	eggs, hatchlings	<10	<1	0.6747	0
2	small juveniles	10.1–58.0	1–7	0.7857	0
3	large juveniles	58.1–80.0	8–15	0.6758	0
4	subadults	80.1–87.0	16–21	0.7425	0
5	novice breeders	>87.0	22	0.8091	127
6	1st-yr remigrants	>87.0	23	0.8091	4
7	mature breeders	>87.0	24–54	0.8091	80

* Straight carapace length.

formed with a 7 × 7 matrix, instead of an unwieldy 54 × 54 matrix, thereby minimizing error propagation caused by repeated multiplication of parameters.

THEORETICAL POPULATION PROJECTIONS

Model

The stage class matrix we have developed incorporates Frazer's fecundity, survival, and growth rates, and uses yearly iterations to make population projections for loggerhead sea turtles. Frazer (1983a) estimated or interpolated the annual survival for each of several size classes of turtles (e.g., large juveniles = 58.1–80 cm straight carapace length [SCL]; adult females = >87.1 cm SCL). He then used previously calculated (Frazer 1983a) growth curves for wild loggerheads to assign age ranges to the size classes and divided each class into the corresponding number of age classes. Finally, he assigned each age class the annual survival calculated for that entire size class, assuming that annual survival was constant for all turtles throughout that class, regardless of age. Clearly, the original data better support a stage class approach than a more traditional age-based model.

Our model divides the life cycle into the seven stages shown in Table 3. To create a stage-based projection matrix, we must estimate, for each stage, the reproductive output (F_i), the probability of surviving and growing into the next stage (G_i), and the probability of surviving and remaining in the same stage (P_i). The fecundities F_i are given in Table 3. The transition probabilities G_i and P_i can be estimated from the stage-specific survival probabilities p_i and stage duration d_i. Because we know little about the variability of survival and growth rates within a stage, we will assume that all individuals within a stage are subject to the same survival probability and stage duration. As more precise data on the growth rates and survival of turtles of various sizes become available they can be readily incorporated into the model.

Within each stage there are individuals who have been in that stage for 1, 2, . . . , d_i yr. By setting the proportion of individuals alive in the first cohort of stage class i to 1 and the probability of turtles in that cohort surviving to the next year to p_i (Frazer's annual survival probability for the entire size = stage class), the probability of those individuals surviving d years becomes p_i^d. Assuming that the population is stationary and the age distribution within stages is stable, the relative abundance of these groups of individuals then becomes 1, p_i, p_i^2, . . . , $p_i^{d_i-1}$. In the interval from t to $t + 1$, the oldest individuals in this stage will move to the next stage, if they survive. All the younger individuals will remain in the stage. Thus the proportion remaining, and surviving, is given by

$$P_i = \left(\frac{1 + p_i + p_i^2 + \ldots + p_i^{d_i-2}}{1 + p_i + p_i^2 + \ldots + p_i^{d_i-1}}\right)p_i.$$

Rewriting the geometric series $1 + p + p^2 + \ldots + p^{d-1}$ as $(1 - p^d)/(1 - p)$, we can rewrite (P_i) as

$$P_i = \left(\frac{1 - p_i^{d_i-1}}{1 - p_i^{d_i}}\right)p_i. \tag{1}$$

Thus, the number of individuals in any cohort within a stage class declines through time as a function of the stage-specific annual survival probability and the number of years spent in that stage.

That proportion of the population that grows into the next stage class and survives (G_i) is similarly given by the proportion of individuals in the oldest cohort of the stage times the annual survival for the stage, or

$$G_i = \left(\frac{p_i^{d_i-1}}{1 + p_i + p_i^2 + \ldots + p_i^{d_i-1}}\right)p_i,$$

which can be rewritten, in the same manner as before, as

$$G_i = \frac{p_i^{d_i}(1 - p_i)}{1 - p_i^{d_i}}. \tag{2}$$

Loggerhead population matrix

The resulting stage class population matrix (Table 4) takes the form

$$\begin{bmatrix} P_1 & F_2 & F_3 & F_4 & F_5 & F_6 & F_7 \\ G_1 & P_2 & 0 & 0 & 0 & 0 & 0 \\ 0 & G_2 & P_3 & 0 & 0 & 0 & 0 \\ 0 & 0 & G_3 & P_4 & 0 & 0 & 0 \\ 0 & 0 & 0 & G_4 & P_5 & 0 & 0 \\ 0 & 0 & 0 & 0 & G_5 & P_6 & 0 \\ 0 & 0 & 0 & 0 & 0 & G_6 & P_7 \end{bmatrix},$$

TABLE 4. Stage-class population matrix for loggerhead sea turtles based on the life table presented in Table 3. For the general form of the matrix and formulae for calculating the matrix elements see Theoretical Population Projections.

0	0	0	0	127	4	80
0.6747	0.7370	0	0	0	0	0
0	0.0486	0.6610	0	0	0	0
0	0	0.0147	0.6907	0	0	0
0	0	0	0.0518	0	0	0
0	0	0	0	0.8091	0	0
0	0	0	0	0	0.8091	0.8089

where F_i is the stage-specific fecundity, and P_i and G_i are the probability of surviving and remaining in the same stage vs. the probability of surviving and growing to the next stage as defined by Eqs. 1 and 2.

Population projections

For each simulation, the power method (Searle 1966, Keyfitz 1977) was used to take successively higher powers of the matrix and postmultiply by the population vectors until the resultant vectors differed from each other by only a scalar factor; this factor is λ_m (the dominant eigenvalue) and the vector is proportional to the right eigenvector w. The population vectors at this point represent the right eigenvector or stable stage distribution. The left eigenvector v can be found by applying the same procedure to the transposed matrix. For computing efficiency, an initial population vector was generated (Crouse 1985) based on a stationary population life table proposed by Frazer (1983a) and used as the initial vector for all succeeding simulations.

The eigenvalue and intrinsic rate of increase for the matrix in Table 4 are $\lambda_m = 0.9450$ and $r = -0.0565$, which are not dissimilar to Frazer's (1983a) values of $\lambda_m = 0.9719$ and $r = -0.0285$ for his 54-yr life table. This confirms that our seven-stage matrix adequately describes the population in Frazer's life table.

The stable stage distribution w and reproductive value vector v are given in Table 5. The stable stage distribution is dominated by small juveniles, eggs and hatchlings, and large juveniles; subadults and adults are very rare. The reproductive value is low for the first three stages, jumps dramatically for subadults, and is even higher for the last three stages.

Sensitivity analyses

One benefit of constructing a population matrix is that one may test how sensitive the population growth rate is to variations in fecundity, growth, or survival rates by simulating changes in these parameters and then calculating λ_m and the resultant r of the new matrix. By simulating the same proportional change for each stage successively, one can compare the relative effect on the different stages.

The ranges of population parameters for various loggerhead life history stages (Tables 1 and 2) suggest that it is not unreasonable to expect some loggerhead populations to show reductions of 50% in fecundity or

survival of specific stages relative to those in our initial population matrix. Therefore we simulated 50% reductions in these parameters for each life history stage with the remaining matrix components held constant (Fig. 1; Appendix). Changes in stages 2, 3, and 4 were effected by reducing Frazer's overall stage-specific annual survival by 50% and then calculating P_i and G_i for each stage using Eqs. 1 and 2 (see Appendix).

Although 50% reductions in fecundity and 1st-yr (eggs and hatchlings) survival reduce λ_m and cause the population to decline more swiftly (Fig. 1), a similar reduction in survival in any of the immature stages (2, 3, and 4) causes a much larger reduction in λ_m and a corresponding increase in the rate of population decline. After at least one reproductive season, such a reduction in adult survival results in more moderate reductions in λ_m and r, similar to those seen with reductions in fecundity and 1st-year survival.

What would happen if new management practices eliminated mortality in any of these stages? Of course, no management practice can promise zero mortality for any period of time, but such a simulation should help identify the life stage(s) on which management efforts would be most efficiently spent. The results of elimination of mortality for each stage class respectively are presented in Fig. 1b. Also included is a simulation of a doubling in fecundity, which is within the range of possibilities presented earlier.

Once again, the juvenile and subadult stages are most responsive to such a change. In fact, an increase in survival to 1.0 in any one of stages 2, 3, or 4 (or that of the suddenly immortal mature females) was suffi-

TABLE 5. Stable stage distribution (w_m) and reproductive values (v') for the loggerhead population matrix given in Table 4.

Stage class	Stable stage distribution (Dominant eigenvector)	Reproductive values (Left eigenvector)
1 (eggs, hatchlings)	20.65	1.00
2 (small juveniles)	66.975	1.40
3 (large juveniles)	11.46	6.00
4 (subadults)	0.66	115.845
5 (novice breeders)	0.04	568.78
6 (1st-yr remigrants)	0.03	507.37
7 (mature breeders)	0.18	587.67

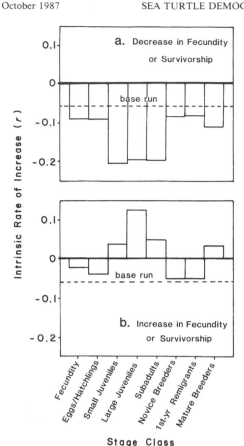

FIG. 1. Changes in rate of increase *r* resulting from simulated changes in fecundity and survival of individual life history stages in the loggerhead population matrix (remaining components held constant). The dashed line represents the *r* determined in the baseline run on the initial matrix. (a.) Simulations represent 50% decreases in fecundity or survivorship. (b.) Simulations represent a 50% increase in fecundity or an increase in survivorship to 1.0. Stages 2–4 (juveniles and subadults) show the strongest response to these simulated changes. (Specific calculations are presented in Crouse 1985.)

cient to reverse the decline of the model population. (The tremendous increase in λ_m and *r* seen in stage 3 [large juveniles] is partly artifact since survival was lowest for this stage initially [see Tables 3 and 4], making the increase to 1 in this stage proportionately larger.) More importantly, the simulation indicates that no matter how much effort was put into protecting eggs on the beach, this alone could not prevent the eventual extinction of the model population. Similarly, the turtles could not reverse their decline via increases in fecundity unless they could more than double egg production, which seems unlikely.

Since the estimates of growth and of age at first reproduction in the literature (Crouse 1985) show considerable uncertainty, another reasonable question is:

What if age at first reproduction is really only 16? or 28? These conditions, a 6-yr decrease or increase in the age of reproductive maturity, were simulated (Fig. 2) by subtracting and adding 2 yr to the calculations of P_i and G_i for each of the three immature stages. In fact, a mere 3-yr reduction in the age of first reproduction, well within the bounds of the growth estimates available, comes very close to halting the decline in this population. How flexible loggerheads might be in age at first reproduction is unknown, but clearly it would be profitable to have better estimates of age at reproductive maturity in order to forecast population changes.

One disadvantage of simulation experiments of this sort is that the results are dependent on the chosen perturbations of the original matrix. Analytical methods (reviewed by Caswell 1986) avoid this difficulty by calculating the sensitivity of λ to changes in life cycle parameters. Here we are interested in the proportional sensitivity (or "elasticity") of λ_m; that is, the proportional change in λ_m caused by proportional change in one of the life cycle parameters. These proportional sensitivities can be calculated, given the stable stage distribution *w* and reproductive value *v*. The proportional sensitivity of λ_m to a change in each matrix element a_{ij} is given by

$$\frac{\partial \ln \lambda}{\partial \ln a_{ij}} = \frac{a_{ij}}{\lambda}\frac{\partial \lambda}{\partial a_{ij}} = \frac{a_{ij}}{\lambda}\left(\frac{v_i w_j}{\langle v,\, w \rangle}\right),$$

where $\langle \rangle$ denotes the scalar product.

The elasticities of λ_m with respect to F_i, P_i, and G_i are shown in Fig. 3. Because these elasticities sum to 1 (DeKroon et al. 1986), the relative contribution of the matrix elements (F_i, P_i, and G_i) to λ_m can be compared. This supports the conclusions of our previous

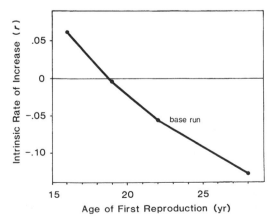

FIG. 2. Resultant *r* for model loggerhead population with different growth rates (represented by age of first reproduction). The baseline run assumed age at first reproduction of 22 yr. Increasing age at first reproduction decreases *r*. Age of first reproduction near 19 yr would lead to *r* nearly equal to 0.

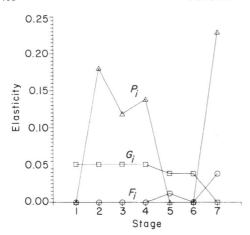

FIG. 3. The elasticity, or proportional sensitivity, of λ_m to changes in fecundity F_i (O), survival while remaining in the same stage P_i ($\triangle$), and survival with growth G_i ($\square$). Because the elasticities of these matrix elements sum to 1, they can be compared directly in terms of their contribution to the population growth rate r.

simulations: increases in fecundity have only a small effect on λ_m. Changes in the probability of survival with growth G_i are more important, while changes in the probability of survival in the same stage P_i contribute the most to λ_m. By the same token, changes in the juvenile, subadult, and mature adult stages have a greater impact on λ_m than changes in the 1st yr, novice breeders, and 1st-yr remigrants.

In this model, P_i and G_i are derived parameters; they depend on both stage-specific annual survival probability p_i and stage duration d_i. We have also calculated the elasticities of λ_m with respect to these parameters (Fig. 4). The population dynamics are very sensitive to variations in the survival probability of juveniles and subadults (Fig. 4a). The results of changes in stage duration are relatively small (Fig. 4b), and are generally negative (since increases in stage duration usually cause decreases in λ_m).

Hence, it would appear that survival, particularly in the juvenile and subadult stages, has the largest effect on population growth. Conveniently, survival is also the parameter that is most amenable to human alteration. With this in mind, we simulated several possible management scenarios.

Management scenarios

While it would be impossible to increase survival of any stage to 1.0, the National Marine Fisheries Service has recently devised a technology, the Trawl Efficiency or Turtle Excluder Device (TED), that virtually eliminates turtle mortality due to incidental capture and drowning in shrimp and fish trawls (Seidel and McVea 1982, Anonymous 1983). Such incidental capture is believed to be the major source of mortality in juvenile

and adult turtles in the southeastern United States (Anonymous 1983, Mager 1985). Since use of this technology would presumably affect all of the turtles feeding in estuaries and nearshore marine habitats, an increase in survival of all three immature stages was simulated (Table 6). Simply increasing immature survival to 0.80 would allow this population to increase.

Because the small juveniles occur less frequently in nearshore marine systems, it might be difficult to increase their survival using TEDs. Thus a simulation was performed where the survival of stage 2 was left unchanged but survival of stages 3 and 4 was increased to 0.80, and the survival of adults was increased to 0.85 (presuming that adults also would benefit from

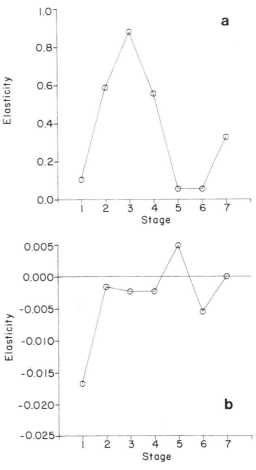

FIG. 4. (a) The elasticity, or proportional sensitivity, of λ_m to changes in annual stage-specific survival probability p_i. (b) The elasticity of λ_m to changes in stage duration d_i. Elasticity in stage duration is negative because stage duration and population growth rates r are inversely related. Overall, proportional sensitivity is much higher to survival than to stage duration.

the TED). Once again, the simulated population started to grow (Table 6).

Hatch success on specific beaches frequently falls below the 0.80 that Frazer used for his life table calculations. Therefore a simulation was performed with 1st-yr survival decreased by 50% (corresponding to a hatch success of 40%), while for stages 3 and 4 survival was increased to 0.80 and adult survival was increased to 0.85. Even with little active egg protection, these small increases in immature and adult survival caused the simulated loggerhead population to grow (Table 6).

Finally, if protection efforts were to focus on one stage over the others, perhaps because of the availability of appropriate technology or ease of access to a particular life stage, how much of an increase in the survival of any *single* stage would be necessary to produce a stable population? As noted earlier, even 100% survival in stages 1 (eggs and hatchlings), 5 (novice breeders), and 6 (1st-yr remigrants) alone would be insufficient to achieve stability. In fact, increasing stage 1 survival to 1.0 increases r to only -0.0357 (Fig. 1b), and the population continues to decline, suggesting that achieving zero mortality of eggs on nesting-beaches would likely be ineffective as a management tool if no concurrent action were taken in the juvenile stage. However, increasing stage 2, 3, 4, and 7 survival one at a time revealed that stability ($\lambda_m = 1$, $r = 0$) could be achieved by increasing stage 3 survival by just 14% (from 67.6 to 77%), whereas an increase of 16% would be necessary for stage 2, 18.5% for stage 4, or 17% for stage 7. In other words, the increase in survival necessary to achieve population stability was smallest for stage 3 (large juveniles).

DISCUSSION

Loggerhead model

Our simulations strongly suggest that if the fecundity, survival, and growth rates of loggerhead turtle populations in the southeastern United States are at all similar to those proposed by Frazer, then the key to improving the outlook for these populations lies in reducing mortality in the later stages, particularly the large juveniles.

Yet Pritchard (1980) noted that many sea turtle biologists are unclear whether to focus their conservation efforts on eggs, immatures, or breeding adults; Pritchard himself opted for saving mature females over juveniles, noting that the adults had already survived the hazards of the long juvenile stages and were ready to lay their valuable eggs. While it is true that reproductive value is highest in the adults, very few turtles actually make it to these stages to reproduce (Table 5). By increasing survival of large juveniles (who have already survived some of the worst years) a much larger number of turtles are likely to reach maturity, thereby greatly magnifying the input of the increased reproductive value of the adult stages.

TABLE 6. Three management scenarios involving changes in mortality in various life stages of loggerhead sea turtles. For each scenario, the stages are listed along with the old and the new matrix elements* (P_i, G_i) and the resulting λ_m and r.

		Change in initial matrix		Result	
Stage	Coefficient	Old	New	λ_m	r
Immature survivorship increased to 0.80					
2	P_2	0.7370	0.74695		
	G_2	0.0486	0.0531		
3	P_3	0.6610	0.7597	1.06	+0.06
	G_3	0.0147	0.0403		
4	P_4	0.6907	0.7289		
	G_4	0.0518	0.0710		
Large juveniles and subadults = 0.80; adults = 0.85					
3	P_3	0.6610	0.7597		
	G_3	0.0147	0.0403		
4	P_4	0.6907	0.7289		
	G_4	0.0518	0.0710	1.06	+0.06
5	G_5	0.8091	0.8500		
6	G_6	0.8091	0.8500		
7	P_7	0.8089	0.8500		
First-year = 0.33735; large juveniles, subadults = 0.80; adults = 0.85					
1	G_1	0.6747	0.33735		
3	P_3	0.6610	0.7597		
	G_3	0.0147	0.0403		
4	P_4	0.6907	0.7289		
	G_4	0.0518	0.0710	1.02	+0.02
5	G_5	0.8091	0.8500		
6	G_6	0.8091	0.8500		
7	P_7	0.8089	0.8500		

* P_i = the probability of survival while remaining in the same stage. G_i = the probability of surviving while growing to the next stage.

Frazer (1983a) acknowledged his parameter estimates are somewhat uncertain, particularly survival in the earlier stages. First-year survival was interpolated based on literature estimates for undisturbed natural nests. Young juvenile survival was also interpolated, based on gross survival from eggs to adults, and assuming a closed population with a stable age distribution. And Frazer noted that the LCI population clearly has been "exploited," albeit unintentionally, by trawlers in recent years and is not likely at a stable age distribution. However, although Frazer's assumption of a closed population has yet to be proven, evidence of morphological differentiation, heavy metal concentration, and genetic polymorphisms (Smith et al. 1977, Stoneburner 1980, Stoneburner et al. 1980) supports the idea of closed populations, at least on a regional basis. And Frazer's estimate of older juvenile survival seems more reliable because it was based on a catch curve analysis of data for stranded and live-caught turtles. Finally, Frazer's calculations were made on one of the longest, most complete data bases available, in which quality control has been excellent.

In fact, the results of our sensitivity analyses indicate that uncertainty in 1st-yr survival estimates, leading to a fairly large change in those estimates would have

1420   DEBORAH T. CROUSE ET AL.   Ecology, Vol. 68, No. 5

relatively little effect on λ_m, though a similar error in estimation of small juvenile survival would have a larger effect. Because the highly sensitive large juvenile stage (stage 3) is one where Frazer's estimate is stronger, the inferences we have drawn with respect to parameter variations in this stage are both more valid and more important. The most serious problem for our projections is the possibility of compounded errors, i.e., errors in the same direction in two or more stages (Wilbur 1975, Tinkle et al. 1981). While an endless variety of scenarios could be simulated using this model, we feel it is better to explore the sensitivity of the various assumptions and parameters of the model. In this way, the importance of our ignorance of the natural history or rate functions of various life stages can be evaluated.

The simulations do indicate that the model is sensitive to changes in age at reproductive maturity, suggesting that weak estimates and/or regional shifts in growth rate might be important. Balazs (1982) has reported wide variance in growth rates for green turtles feeding in different habitats. Indeed, more recently Frazer and Ehrhart (1985) have suggested that maturity in loggerheads may come as late as 30 yr, i.e., the population might actually be declining faster than 3%, and require even more attention to halting juvenile mortality. The sensitivity of the model to growth rate combined with the strong possibility of different growth rates in various estuarine habitats points to an important gap in our understanding of loggerhead natural history.

Another important lesson brought home by the population projection technique is that population strength or longevity does not come from sheer numbers alone, but rather from the integrative result of survival, fecundity, and individual growth throughout the life cycle. For example, if an initial population of 500 000 animals (including 1277 adult females) is subjected to a 50% reduction in large juvenile survival, the population will have no (< 1) adult females in just 40 yr, and is clearly headed for rapid extinction. However, the same starting population retains > 52 adult females (≈ 60 times as many), and thus some potential for recovery, after 40 yr when the same 50% reduction is induced in the egg/hatchling stage. Thus the presence of large numbers of animals can be deceptive, implying robustness, when in reality such a population might be highly susceptible to perturbations in particular life stages.

Life history strategies

Loggerheads probably evolved under conditions of high environmental uncertainty on the nesting-beaches, leading to highly variable survival in the eggs and hatchlings. Relatively low mortality in the larger stages allowed delayed maturity and iteroparity, thereby facilitating high fecundity rates to offset the egg and hatchling mortality. Unfortunately, as Wilbur (1975) noted when he published the first life table for any species of turtle, "One of the most serious gaps in the study of life histories is the analysis of long-lived, iteroparous species."

A recent attempt to develop a unified model for patterns of covariation in the life history traits of reptiles (Stearns 1984) was based entirely on snakes and lizards because of the paucity of accurate demographic information for turtles. Average adult female length was most influential on covariation in reptilian life history traits, with some additional influence attributed to phylogenetic relationships (Stearns 1984). Stearns further suggested that microevolutionary explanations are not sufficient to account for the patterns in the data.

More recently, Wilbur and Morin (1987) analyzed the life history evolution of 80 species of turtles with respect to a number of traits in addition to those that Stearns (1984) examined, concluding that female size and habitat are the most important factors determining reproductive characteristics. They also compared data from three widely scattered populations of green turtles (*Chelonia mydas*), where differing predation rates on various life stages have occurred for hundreds of years, and concluded, in contrast to Stearns, that "these different selective regimes could well have resulted in genetic differences in the life histories of isolated *Chelonia* populations."

Long-lived iteroparous animals probably adjust short-term reproductive effort to maximize lifetime reproductive success rather than short-term gains (Wilbur and Morin 1987). Thus females may become reproductive only when they have been able to store sufficient energy. The frequency of reproduction may not be the result of selection for any particular reproductive cycle, but instead the result of phenotypic plasticity, such that reproductive frequency would be expected to change in the face of environmental fluctuations (Frazer 1983a).

Thus, while more life history information for turtles has become available in recent years, it is clear that major gaps still exist and that the implications for marine turtle protection are as yet unclear. Even if adult female size is the primary determinant of reproductive strategies, differential selection histories and very recent environmental conditions may result in differential expression of reproductive traits.

Loggerhead management

These gaps in our knowledge become especially important when managers must decide where to focus protection efforts. While most marine turtle conservation projects in the southeastern United States have concentrated on reducing egg mortality on beaches, several researchers have noted recently that after 20 and even 30 yr of nest protection on some beaches we have not yet seen the increase in nesting turtles expected as a result of nest protection (Carr et al. 1978 and J. I. Richardson 1982 and *personal communication*). This may, in part, be an artifact resulting from

620   Part Five

the ever increasing estimates of age at first reproduction for several species. It may also be a consequence of less rigid "beach imprinting" than previously believed. Even so, a general increase in regional nesting populations would be expected.

Obviously, based on the uncertainties in Frazer's estimates of survival in the youngest stages and the variability possible in age at maturity, we should maintain current efforts to reduce egg mortality, particularly on beaches with consistently low egg survival. However, the low elasticity of stage 1 survival indicates that the model can tolerate considerable uncertainty in this parameter. As more data are amassed on other aspects of the sea turtle life cycle, managers need to address the uncomfortable possibility that their current conservation efforts may be focusing on the part of the turtle's life history least likely to produce noticeable, long-term results. If, as the results of this investigation imply, mortality must be reduced at other life stages, what can be done?

This analysis indicates that a 14% increase in survival of large juveniles would allow the simulated loggerhead population to grow ($\lambda_m > 1$, $r > 0$). As noted, incidental capture and drowning in shrimp trawls is believed to be the largest single source of mortality in juvenile and adult marine turtles in the southeastern United States (Anonymous 1983, Mager 1985). Data for 3 yr from North Carolina (Fig. 5; Crouse 1985) indicate that the majority of the turtles that "strand" (wash up on the beach, dead or dying) fall in the 50–80 cm SCL size range. This finding is similar to those seen in the Chesapeake Bay and Georgia (Ruckdeschel and Zug 1982, Lutcavage and Musick 1985). This range closely overlaps the large juvenile stage class, which proved most sensitive to simulated reductions in mortality.

The Trawl Efficiency Device (or Turtle Excluder Device, TED) mentioned earlier can be installed in existing trawls and virtually eliminates the capture and drowning of marine turtles (Siedel and McVea 1982 and C. Oravetz, *personal communication*). The TED has the added advantage of eliminating other large objects (bycatch) from the trawl, thereby improving the hydrodynamics of the trawl and improving fuel efficiency (Anonymous 1983). Easley (1982) found that a small but significant increase in the shrimp caught in paired tests resulted in an economic advantage to larger vessels installing the device. Smaller and lighter versions of the TED are currently being tested for performance and durability (C. Oravetz 1985 and *personal communication*). Increased use of TEDs in the trawl fishery might provide advantages to both the fishery *and* threatened loggerhead populations.

It seems clear that more information should be collected on the distribution of immature turtles in the nearshore waters, incidence of trawl-related juvenile mortality, and the potential for the TED to reduce mortality in specific size classes. Additional studies

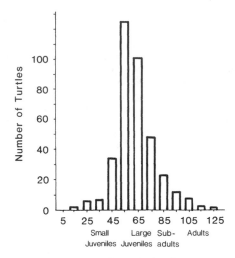

FIG. 5. Size-frequency distribution of stranded turtles in North Carolina, 1981–1983. Total number of stranded turtles reported by 10 cm (straight carapace length) size classes, June 1981–December 1983 (from Crouse 1985). The stage classes used in this analysis are named below their respective size categories.

clearly are needed to strengthen the parameter estimates in Frazer's life table, particularly in the areas of 1st-yr and small juvenile mortality, growth rates, and age at first reproduction. However the model's responses to increases in survival in the large juvenile and adult stages imply that we should not wait for these results before implementing measures, such as the TED, to reduce mortality in the larger size classes.

ACKNOWLEDGMENTS

We would like to thank Dr. Nathaniel Frazer (Mercer University) for his encouragement and support of our use of his loggerhead life table. In addition we thank Dr. Thomas Meagher (Duke University) for his stage class matrix projection computer program. Additional assistance was provided by Dr. Kenneth Pollock, Southeastern Fish and Game Statistics Project (North Carolina State University), Dan Smith, Woods Hole Oceanographic Institution, and the Zoology Department at North Carolina State University. D. T. Crouse—This paper represents one portion of a Ph.D. dissertation filed with the University of Wisconsin, Madison (1985). L. B. Crowder—This work was supported in part by the Office of Sea Grant, NOAA (NA85AA-D-SG022). This is paper #10862 in the Journal Series of the North Carolina Agricultural Research Service, Raleigh, NC 27695. H. Caswell—This work was supported in part by NSF Grants BSR82-14583 and OCE85-16177 and NOAA Grant NA83AA-D-00058. Woods Hole Oceanographic Institution Contribution 6386.

LITERATURE CITED

Anonymous. 1983. Environmental assessment of a program to reduce incidental take of sea turtles by the commercial shrimp fishery in the southeast United States. National Marine Fisheries Service, United States Department of Commerce. St. Petersburg, Florida, USA.

Balazs, G. H. 1982. Growth rates of immature green turtles

in the Hawaiian Archipelago. Pages 117–126 *in* K. A. Bjorndal, editor. Biology and conservation of sea turtles. Smithsonian Institution Press, Washington, D.C., USA.

Bjorndal, K. A., editor. 1982. Biology and conservation of sea turtles. Smithsonian Institution Press, Washington, D.C., USA.

Caldwell, D. K. 1959. The loggerhead turtles of Cape Romain, South Carolina. Bulletin of the Florida State Museum Biological Series 4:319–348.

Carr, A. 1967. So excellent a fishe. Natural History Press, Garden City, New York, USA.

Carr, A., and M. H. Carr. 1970. Modulated reproductive periodicity in *Chelonia*. Ecology 51:335–337.

Carr, A., M. H. Carr, and A. B. Meylan. 1978. The ecology and migrations of sea turtles. 7. The West Caribbean green turtle colony. Bulletin of the American Museum of Natural History 162:1–46.

Caswell, H. 1986. Life cycle models for plants. Lectures on Mathematics in the Life Sciences 18:171–233.

Crouse, D. T. 1985. The biology and conservation of sea turtles in North Carolina. Dissertation. University of Wisconsin, Madison, Wisconsin, USA.

Davis, G. E., and M. C. Whiting. 1977. Loggerhead sea turtle nesting in Everglades National Park, Florida, USA. Herpetologica 33:18–28.

deKroon, H., A. Plaisier, J. v. Groenendael, and H. Caswell. 1986. Elasticity: the relative contribution of demographic parameters to population growth rate. Ecology 67:1427–1431.

Easley, J. E. 1982. A preliminary estimate of the payoff to investing in a turtle excluder device for shrimp trawls. Final report to Monitor International and the Center for Environmental Education in cooperation with the National Marine Fisheries Service. Washington, D.C., USA.

Federal Register. 1978. Listing and protecting loggerhead sea turtles as "threatened species" and populations of green and olive Ridley sea turtles as threatened species or "endangered species." Federal Register 43(146):32800–32811.

Frazer, N. B. 1983a. Demography and life history evolution of the Atlantic loggerhead sea turtle, *Caretta caretta*. Dissertation. University of Georgia, Athens, Georgia, USA.

———. 1983b. Survivorship of adult female loggerhead sea turtles, *Caretta caretta*, nesting on Little Cumberland Island, Georgia, USA. Herpetologica 39:436–447.

———. 1984. A model for assessing mean age-specific fecundity in sea turtle populations. Herpetologica 40:281–291.

———. 1987. Survival of large juvenile loggerheads (*Caretta caretta*) in the wild. Journal of Herpetology 21:232–235.

Frazer, N. B., and L. M. Ehrhart. 1985. Preliminary growth models for green (*Chelonia mydas*) and loggerhead (*Caretta caretta*) turtles in the wild. Copeia 1985:73–79.

Gallagher, R. M., M. L. Hollinger, R. M. Inghle, and C. R. Futch. 1972. Marine turtle nesting on Hutchinson Island, Florida in 1971. Florida Department of Natural Resources Marine Research Laboratory Special Scientific Report 37:1–11.

Groombridge, B. 1982. The IUCN amphibia—reptilia red data book. Part 1. Testudines, Crocodylia, Rhynchocephalia. International Union for the Conservation of Nature and Natural Resources, Gland, Switzerland.

Hartshorn, G. S. 1975. A matrix model of tree population dynamics. Pages 41–51 *in* F. B. Golley and E. Medina, editors. Tropical ecological systems. Springer-Verlag, New York, New York, USA.

Hirth, H. F. 1980. Some aspects of the nesting behavior and reproductive biology of sea turtles. American Zoologist 20:507–524.

Hopkins, S. R., T. M. Murphy, Jr., K. B. Stansell, and P. M. Wilkinson. 1979. Biotic and abiotic factors affecting nest mortality in the Atlantic loggerhead turtle. Proceedings of the Annual Conference Southeastern Association of Fish and Wildlife Agencies 32:213–223.

Hopkins, S. R., and J. I. Richardson, editors. 1985. Final: recovery plan for marine turtles. United States Fish and Wildlife Service and National Marine Fisheries Service, Miami, Florida, USA.

Hughes, G. R. 1974. The sea turtle of South Africa. II. The biology of the Tongaland loggerhead turtle *Caretta caretta* with comments on the leatherback turtle *Dermochelys coriacea* L. and the green turtle *Chelonia mydas* L. in the study region. Investigational Report Number 36. Oceanographic Research Institute, Durban, South Africa.

Keyfitz, N. 1977. Introduction to the mathematics of populations: with revisions. Addison-Wesley, Reading, Massachussetts, USA.

Lefkovitch, L. P. 1965. The study of population growth in organisms grouped by stages. Biometrics 21:1–18.

Lenarz, M. S., N. B. Frazer, M. S. Ralston, and R. B. Mast. 1981. Seven nests recorded for loggerhead turtle (*Caretta caretta*) in one season. Herpetological Review 12:9.

Leslie, P. H. 1945. On the use of matrices in certain population mathematics. Biometrika 33:183–212.

Lewis, E. G. 1942. On the generation and growth of a population. Sankhya 6:93–96.

Lotka, A. J. 1925. Elements of mathematical biology. Dover, New York, New York, USA.

Lutcavage, M., and J. A. Musick. 1985. Aspects of the biology of sea turtles in Virginia. Copeia 1985:449–456.

Mager, A., Jr. 1985. Five-year status reviews of sea turtles under the Endangered Species Act of 1973. National Marine Fisheries Service, St. Petersburg, Florida, USA.

Meagher, T. R. 1982. The population biology of *Chamaelirium luteum*, a dioecious member of the lily family: two-sex population projections and stable population structure. Ecology 63:1701–1711.

Mendonca, M. T. 1981. Comparative growth rates of wild immature *Chelonia mydas* and *Caretta caretta* in Florida. Journal of Herpetology 15:447–451.

Pritchard, P. C. H. 1980. The conservation of sea turtles: practices and problems. American Zoologist 20:609–617.

Richardson, J. I. 1982. A population model for adult female loggerhead sea turtles (*Caretta caretta*) nesting in Georgia. Dissertation. University of Georgia, Athens, Georgia, USA.

Richardson, J. I., and T. H. Richardson. 1982. An experimental population model for the loggerhead sea turtle (*Caretta caretta*). Pages 165–176 *in* K. A. Bjorndal, editor. Biology and conservation of sea turtles. Smithsonian Institution Press, Washington, D.C., USA.

Richardson, J. I., T. H. Richardson, and M. W. Dix. 1978. Population estimates for nesting female loggerhead sea turtles (*Caretta caretta*) in the St. Andrew Sound area of southeastern Georgia, USA. Florida Marine Research Publication 33:34–38.

Ruckdeschel, C., and G. R. Zug. 1982. Mortality of sea turtles *Caretta caretta* in coastal waters of Georgia. Biological Conservation 22:5–9.

Searle, S. R. 1966. Matrix algebra of the biological sciences. John Wiley and Sons, New York, New York, USA.

Seidel, W. R., and C. McVea, Jr. 1982. Development of a sea turtle excluder shrimp trawl for the southeast U.S. penaeid shrimp fishery. Pages 497–502 *in* K. A. Bjorndal, editor. Biology and conservation of sea turtles. Smithsonian Institution Press, Washington, D.C., USA.

Smith, M. H., H. O. Hillestad, M. N. Manlove, D. O. Straney, and J. M. Dean. 1977. Management implications of genetic variability in loggerhead and green sea turtles. Pages 302–312 *in* T. J. Peterle, editor. XIIIth International Congress of Game Biologists, Atlanta, Georgia, USA, March 11–15, 1977. The Wildlife Society, Washington, D.C., USA.

Stearns, S. C. 1984. The effects of size and phylogeny on patterns of covariation in the life history traits of lizards and snakes. American Naturalist **123**:56–72.

Stoneburner, D. L. 1980. Body depth: an indicator of morphological variation among nesting groups of adult loggerhead sea turtles (*Caretta caretta*). Journal of Herpetology **14**:205–206.

Stoneburner, D. L., and L. M. Ehrhart. 1981. Observations on *Caretta c. caretta*: a record internesting migration in the Atlantic. Herpetological Review **12**(2):66.

Stoneburner, D. L., M. N. Nicora, and E. R. Blood. 1980. Heavy metals in loggerhead sea turtle eggs (*Caretta caretta*): evidence to support the hypothesis that demes exist in the western Atlantic population. Journal of Herpetology **14**: 171–175.

Talbert, O. R., Jr., S. E. Stancyk, J. M. Dean, and J. M. Will. 1980. Nesting activity of the loggerhead turtle (*Caretta caretta*) in South Carolina. I. A rookery in transition. Copeia 1980:709–718.

Tinkle, D. W., J. D. Congdon, and P. C. Rosen. 1981. Nesting frequency and success: implications for the demography of painted turtles. Ecology **62**:1426–1432.

Uchida, I. 1967. On the growth of the loggerhead turtle, *Caretta caretta*, under rearing conditions. Bulletin of the Japanese Society of Scientific Fisheries **33**:497–506.

Vandermeer, J. H. 1975. On the construction of the population projection matrix for a population grouped in unequal stages. Biometrics **31**:239–242.

———. 1978. Choosing category size in a stage projection matrix. Oecologia (Berlin) **32**:79–84.

Werner, P. A., and H. Caswell. 1977. Population growth rates and age *versus* stage-distribution models for teasel (*Dipsacus sylvestris* Huds.). Ecology **58**:1103–1111.

Wilbur, H. M. 1975. The evolutionary and mathematical demography of the turtle *Chrysemys picta*. Ecology **56**:64–77.

Wilbur, H. M., and P. J. Morin. 1987. Life history evolution in turtles. *In* C. Gans and R. Huey, editors. Biology of the Reptilia. Volume 16. Ecology B: defense and life history. A. R. Liss, New York, New York, USA.

APPENDIX

TABLE A1. Sensitivity analysis. Resultant eigenvalue (λ_m) and r after simulated 50% decreases in fecundity and survival probabilities for loggerhead sea turtles.

	Change in initial matrix			Result	
Stage	Coeffi-cient	Old	New	λ_m	r
None	Baseline run			0.945	−0.06
Fecun-dity	F_5	127	64		
	F_6	4	2	0.91	−0.09
	F_7	80	40		
1	G_1	0.6747	0.33735	0.91	−0.09
2	P_2	0.7370	0.3919	0.81	−0.21
	G_2	0.0486	0.00085		
3	P_3	0.6610	0.3378	0.82	−0.20
	G_3	0.0147	0.0001		
4	P_4	0.6907	0.3696	0.82	−0.20
	G_4	0.0518	0.0016		
5	G_5	0.8091	0.4045	0.92	−0.08
6	G_6	0.8091	0.40455	0.92	−0.08
7	P_7	0.8089	0.4045	0.90	−0.11

Minimum Population Sizes
for Species Conservation

Mark L. Shaffer

Many species cannot survive in man-dominated habitats. Reserves of essentially undisturbed habitat are necessary if such species are to survive in the wild. Aside from increased efforts to accelerate habitat acquisition for such species, the most pressing need facing conservationists is development of a predictive understanding of the relationship between a population's size and its chances of extinction.

Biologists have long known that the smaller the population, the more susceptible it is to extinction from various causes. During the current era of heightened competition for use of the world's remaining wildlands, this qualitative understanding is of limited utility to conservation and natural resource planners. The old adage that "the bigger the reserve, the better" must be replaced with more precise prescriptions for how much land is enough to achieve conservation objectives. Efforts at making such determinations have been clouded by inconsistencies in the focus on the unit to be preserved (population, species, community, ecosystem) and lack of an explicit definition of what constitutes successful preservation (persistence for 10, 100, 1000 years, etc.).

The intricate interdependencies of living things dictate that conservation efforts be focused on the community and ecosystem level. Unfortunately, the very magnitude of complexity of these systems makes such efforts difficult. Moreover, certain species are more sensitive than others to changing conditions and begin to decline prior to any noticeable degradation of the community to which they belong. Consequently, conservation efforts have been and, in many cases, will continue to be at the single-species level. Many species currently in jeopardy are large-bodied and/or specialized, two characteristics that usually

lead to low population densities. If we are successful in providing sufficient room for their survival, then other, less space-demanding members of their communities should also survive.

In this paper, I propose tentative criteria for successful preservation at the population level, discuss the various methods available for determining the population sizes and their area requirements to meet these criteria, and relate both to overall conservation strategy.

THE MINIMUM VIABLE POPULATION CONCEPT

Levins (1970) has estimated that, since the beginning of the Cambrian, species have been going extinct at the rate of about one per year (though not uniformly). Extinction thus appears to be a relatively common event. The factors leading to extinction, though varied, can be lumped into two categories: systematic pressures and stochastic perturbations. A necessary first step in the preservation of any species is to identify and, if possible, compensate for any systematic pressures threatening that species. This is not the type of problem of interest here. Rather, the focus is on those stochastic perturbations that may extinguish populations of a species even in an environment that, on average, is favorable for their growth and persistence.

In general, there are four sources of uncertainty to which a population may be subject:

• *demographic stochasticity*, which arises from chance events in the survival and reproductive success of a finite number of individuals (May 1973, Roughgarden 1975);

• *environmental stochasticity* due to temporal variation of habitat parameters and the populations of competitors, predators, parasites, and diseases (May 1973, Roughgarden 1975);

• *natural catastrophes*, such as floods, fires, droughts, etc., which may occur at random intervals through time; and

• *genetic stochasticity* resulting from changes in gene frequencies due to founder effect, random fixation, or inbreeding (Berry 1971, Soulé 1980).

Little is known about the role of any of these factors in any specific case of extinction. Because all of them increase in importance with decreasing population size, assessing the relative importance of each will always prove difficult.

The extinction of the heath hen (*Tympanuchus cupido cupido*) provides an example of the situation (Simon and Géroudet 1970). Once fairly common from New England to Virginia, the species steadily declined as European settlement progressed. By 1876 the species remained only on Martha's Vineyard, and by 1900 there were fewer than 100 survivors. In 1907 a portion of the island was set aside as a refuge for the birds, and a program of predator control was instituted. The population responded to these measures, and by 1916 had reached a size of more than 800 birds. But in that year a fire (natural catastrophe) destroyed most of the remaining nests and habitat, and during the following winter the birds suffered unusually heavy predation (environmental stochasticity) from a high concentration of goshawks (*Accipiter gentilis*). The combined effects of these events reduced the population to 100-150 individuals. In 1920, after the population had increased to about 200, disease (environmental stochasticity) took its toll, and the population was again reduced below 100. Though the species endured awhile longer, by 1932 the last survivor was gone. In the final stages of the population's decline, the birds appeared to become increasingly sterile, and the proportion of males increased (demographic, environmental, or genetic stochasticity). Which of these events, or what combination, was the critical determinant in the species disappearance is unknown.

The net effect of all these types of perturbations on a population's prospects for survival depends to a great extent on

Shaffer is a nongame biologist with the U.S. Fish and Wildlife Service, Office of Migratory Bird Management, Washington, DC 20240. ©1981 American Institute of Biological Sciences. All rights reserved.

the population's relationship to other populations of that species—what might be termed its biogeographic context. Any factor depressing the size or growth of a population may be mitigated by immigration of individuals from other populations.

Clearly then, a minimum viable population is not one that can simply maintain itself under average conditions, but one that is of sufficient size to endure the calamities of various perturbations and do so within its particular biogeographic context. Furthermore, survival (and hence, preservation) must be measured relative to some time frame and some set of conditions. Hooper (1971) has already pointed out this fact, but it does not seem to be widely recognized within the conservation field and certainly not in the minds of the general public. The problem of determining minimum viable population sizes and their area requirements is analogous to designing reservoirs to hold flood waters. A reservoir capable of holding the once-in-50-year flood may be grossly inadequate for the once-in-1000-year flood. What time frame to use and the levels of variation and catastrophe to anticipate in determining minimum viable population sizes are very much open questions, but it is critical to view the problem in this way.

Because the dedication of lands to the preservation of biotic diversity increasingly has to compete with investments of that land for the production of other goods and services, reserves should be evaluated on the basis of their utility for meeting the conservation goals set for them over some reasonable time frame. On the other hand, the uncertain nature of the factors that threaten small populations argues against too precise a set of criteria. Given this dilemma, I tentatively and arbitrarily propose the following definition of minimum viable population size: *A minimum viable population for any given species in any given habitat is the smallest isolated population having a 99% chance of remaining extant for 1000 years despite the foreseeable effects of demographic, environmental, and genetic stochasticity, and natural catastrophes.*

I must stress the tentative nature of this definition. The critical level for survival probabilities might be set at 95%, or 100%, or any other level. Similarly, the time frame of 1000 years might be lengthened to 10,000 or shortened to 100. Such criteria urgently need discussion among conservationists, planners, and natural resource managers. The impor-

tant point is that such a definition is an explicit set of performance criteria for a conservation unit under an explicit set of conditions.

DETERMINING MINIMUM POPULATION SIZES

There are five possible approaches to determining minimum viable population sizes and their area requirements: experiments, biogeographic patterns, theoretical models, simulation models, or genetic considerations. None of these is very sophisticated, and certainly none can be considered failsafe. Nevertheless, most can contribute to a better understanding of population size and survival and to more realistic estimates of the land area required to preserve populations of a particular species.

Experiments

The most straightforward approach to the problem of assessing minimum viable population sizes is simply to create isolated populations and monitor their persistence. This approach is intractable for two reasons: First, we cannot experimentally measure persistence in terms of decades and centuries; institutional abilities or willingness to support research projects are usually limited to a decade or less. Second, in most cases irreversible decisions on land use will be made in the very near term (10 to 20 years). Unless conservationists can provide useful estimates of the location, number, and size of reserves in this time period, the opportunity to do so may be permanently foreclosed. Development interests will not (in some cases, cannot) await the results of research that may take a century to complete. This should not be construed as an argument against long-term population-monitoring studies. Such studies are of great potential value in many areas of biology and ecology, but their utility for solving this particular problem is limited.

Biogeographic Patterns

Examination of the distributional patterns of species that occur in insular or patchy patterns can provide a first approximation of minimum area requirements and, provided some estimate of densities, minimum viable population sizes. This approach requires that species communities occupying such habitat patches are in equilibrium and the approximate length of their isolation is

known. If these conditions are met, such surveys can reveal both the smallest island or patch inhabited by a species and also the percent of islands or patches of a certain class supporting that species, measured either in area or species diversity.

Robbins' (1979) work on the habitat size relationships of the migrant neotropical avifauna of the eastern deciduous forest employs this approach. Although this is probably the most tractable and reliable approach to the problem, it does have its limitations. The most critical is that there is apparently no clear relationship, either theoretical or empirical, between the percent of occupied patches of a certain size and the potential longevity of the populations they support. This is a key research need.

For example, a particular species might be a breeding resident of 95% of islands or patches 50–100 km² in area. Unfortunately, knowing this fact alone reveals nothing about the frequency with which populations on such patches go extinct or recolonize. Suppose populations of this hypothetical species on patches of the given dimensions go extinct, on average, every 20 years. Relying on a single reserve of this size to maintain the species without alternate reserves to provide sources for recolonization will prove ineffective in the long term. To make this approach workable, there must be good information on both the frequency with which species occur on islands or habitat patches of different sizes and species-specific extinction/colonization rates typical of these units. Some information of this type is available for certain avian species (for an overview see Wilcox 1980 or Terborgh and Winter 1980), but much remains to be done.

An additional complication with this approach is that population characteristics (e.g., density, mortality and fecundity rates) of many species show wide variation from one part of their range to another depending on habitat quality or community structure. Two habitat patches of the same size may not support equally large or enduring populations. Such habitat differences are critical to wise conservation planning, and any research efforts employing this approach must recognize and deal with this fact.

Obviously, this approach cannot be used for those species that have contiguous distributions and do not occur either on islands or patchily distributed habitats.

132

Despite these drawbacks, analysis of biogeographic patterns, when coupled with studies of species-specific turnover rates, is one of the major means of determining minimum viable population sizes and their area requirements. The most valuable research directions to fully utilize this approach are to extend its application to nonavian species, refine analyses to reflect differences in habitat quality as well as quantity, and determine if there is any relationship between the percent occurrence of a species and its characteristic extinction/colonization rates.

Theoretical Models

There are several theoretical models for predicting the probability that a small population will go extinct and the time this will take. There are also numerous models of population growth under stochastic conditions (demographic, environmental, or both). But these models either embody unrealistic assumptions or lead to currently unresolved mathematical problems.

For example, the models of Mac-Arthur and Wilson (1967) and Richter-Dyn and Goel (1972) assume a constant carrying capacity and birth and death rates that change only in response to density. Thus, they deal only with the probability of extinction due to demographic stochasticity. The branching process theory employed by Keiding (1975), though capable of dealing with both demographic and environmental stochasticity, is restricted to exponential growth. Diffusion theory, as employed by various workers (for a review see Shaffer 1978), applies only to a totally unpredictable environment. Moreover, there has been some doubt about the appropriate method of analysis to be applied to diffusion theory equations. Turelli (1977), after a thorough review of analysis methods, emphasized that the diffusion theory models are most appropriately viewed as approximations to more realistic models that are analytically intractable.

The idiosyncratic nature of many species and the great variability inherent in nature probably preclude direct application of any single theoretical model to many real-world situations. Nevertheless, this is an important area of research and deserves increased attention. From a practical standpoint, the most fruitful approach would be the development of a small number of models to fit various scenarios of population growth and regulation and use these to determine both the relationship of population size to extinction probabilities and the sensitivity of the results to assumptions inherent in the models and key population characteristics. For example, in the classic model of logistic population growth, do survival probabilities depend primarily on mortality or fecundity rates, or carrying capacity? Is the mean or variance of these parameters more important in assessing survival? To what extent is survival affected by introduction of time lags in the density-dependent process of mortality and reproduction? Such models should also facilitate development of effective management strategies for reserves that are too small to assure persistence if left alone.

Simulation Models

Because they are not subject to the various constraints of analytical models, computer simulations employing numerical methods may provide a tractable approach to determining minimum viable population sizes and their area requirements. Aside from their greater realism, such models also provide a flexible tool for assessing the effects of changes in various parameter values (e.g., mortality and fecundity rates, etc.) and/or relationships (density-dependent versus density-independent mortality rates, etc.).

The principle drawbacks of this approach are a lack of generality (i.e., the simulations have to be altered for different species) and demand for extensive data. At a minimum, such models (for vertebrates) require knowledge of the mean and variance of age and sex-specific mortality and fecundity rates, age structure, sex ratios, dispersal, and the relationship of these various parameters to density. Such information should be gathered over a sufficiently long period to assure that it is representative of the full range of conditions the population is subject to, including cyclic behavior. (For some species this may require very long data bases.)

Based on the extensive data of Craighead et al. (1974) for the grizzly bear (*Ursus arctos* L.) in Yellowstone National Park, I (Shaffer 1978) used the simulation approach to assess minimum viable population sizes and area requirements for this species. The simulations could evaluate the effects of either or both demographic and environmental stochasticity. Natural catastrophes appear to be unimportant for the grizzly, and lack of adequate genetic information precluded testing the effects of genetic stochasticity. The results of this analysis showed that populations of less than 30–70 bears (depending on population characteristics) occupying less than 2500–7400 km² (depending on habitat quality) have less than a 95% chance of surviving for even 100 years. Survival probabilities were most affected by changes in the mean mortality rate, cub sex ratio, and age at first reproduction of females.

This type of approach should be expanded to other species when sufficient data exist. Where possible, future field population studies should be designed to gather the types of information necessary to develop simulation models. Such simulations provide the most tractable and realistic alternative to the analysis of biogeographical patterns and species-specific turnover rates.

Genetic Considerations

Several workers have based minimum population recommendations on genetic and evolutionary arguments. Franklin (1980) has suggested that, simply to maintain short-term fitness (i.e., prevent serious in-breeding and its deleterious effects), the minimum effective population size (in the genetic sense) should be around 50. He further recommended that, to maintain sufficient genetic variability for adaptation to changing environmental conditions, the minimum effective population size should be around 500. Soulé (1980) has pointed out that, above and beyond preserving short-term fitness and genetic adaptability, long-term evolutionary potential (at the species level) may well require a number of substantially larger populations.

These recommendations were based on very general applications of basic genetic principles and, consequently, are somewhat oversimplified. A more detailed approach would involve gathering information on the degree of genetic variability and the breeding structure of the species to be preserved. Given this information, it should then be possible to determine what size population would provide (at some probability level) a representative sample of the genetic diversity typical of the species and what size would be necessary to assure (at some probability level) that none of this variability would be lost due to in-breeding and genetic drift over some specified period of time. Lacking this sort of detailed work, the above recommendations should be viewed as very rough guidelines rather than specific prescriptions.

SUMMARY

Nature reserves represent the investment of land for the production of a public good—the persistence of populations of various species and the communities they form. The fact that population persistence depends, to a great extent, on population size immediately raises the issue of how persistent society wishes remnant wild populations to be.

This is not a question that can be answered solely on biological grounds. In an expanding human world, competition for use of a finite land base can only intensify. Conservationists will increasingly be pressed on the need to preserve many species and on the efficiency (in terms of land) with which such preservation can be accomplished. In this atmosphere, scientists must develop some consensus on the standards to be applied in determining what constitutes a minimum viable population for successful preservation. I have offered one tentative definition in this paper, but it is not to be taken literally. It is intended as an example for consideration, not a standard for application.

Given some consensus on the standards to be applied, several of the methods outlined here may be used to begin determining minimum viable population sizes and land area requirements for species in jeopardy. The most promising approaches are the extension and refinement of analyses of biogeographic distribution patterns and species-specific turnover rates and the use of available population data in computer simulations designed to test extinction probabilities. Theoretical mathematical models may be useful in revealing which population characteristics or processes are likely to be most important in affecting survival probabilities. Genetic determinants of minimum viable population sizes are still unclear; their resolution hinges primarily on a better knowledge of the breeding structure and genetic variability of particular species and, most importantly, the role of genetic variability in population growth and regulation.

REFERENCES CITED

Berry, R. J. 1971. Conservation aspects of the genetical constitution of populations. Pages 177–206 in E. D. Duffey and A. S. Watt, eds. *The Scientific Management of Animal and Plant Communities for Conservation*. Blackwell, Oxford.

Craighead, J. J., J. Varney, and F. C. Craighead, Jr. 1974. *A Population Analysis of the Yellowstone Grizzly Bears*. Montana Forest and Conservation Experiment Station Bulletin 40. University of Montana, Missoula.

Franklin, I. R. 1980. Evolutionary change in small populations. Pages 135–150 in M. E. Soulé and B. A. Wilcox, eds. *Conservation Biology: An Evolutionary-Ecological Perspective*. Sinauer, Sunderland, MA.

Hooper, M. D. 1971. The size and surroundings of nature reserves. Pages 555–561 in E. D. Duffey and A. S. Watt, eds. *The Scientific Management of Animal and Plant Communities for Conservation*. Blackwell, Oxford.

Keiding, N. 1975. Extinction and exponential growth in random environments. *Theoret. Pop. Biol.* 8: 49–63.

Levins, R. 1970. Extinction. Pages 77–107 in M. Gerstenhaber, ed. *Some Mathematical Questions in Biology, Vol. II*. American Mathematical Society, Providence, RI.

MacArthur, R. H., and E. O. Wilson. 1967. *The Theory of Island Biogeography*. Princeton University Press, Princeton, NJ.

May, R. M. 1973. *Stability and Complexity in Model Ecosystems*. Princeton University Press, Princeton, NJ.

Richter-Dyn, N., and N. S. Goel. 1972. On the extinction of a colonizing species. *Theoret. Pop. Biol.* 3: 406–433.

Robbins, C. S. 1979. Effect of forest fragmentation on bird populations. Pages 198–213 in USDA, Forest Service. *Management of North Central and Northeastern Forests for Nongame Birds*. Workshop Proc. U.S. Dept. Agric. For. Serv., Gen. Tech. Rep. NC-51. USDA For. Serv., North Cent. For. Exp. Stn., St. Paul, MN.

Roughgarden, J. 1975. A simple model for population dynamics in stochastic environments. *Am. Nat.* 109: 713–736.

Shaffer, M. L. 1978. Determining Minimum Viable Population Sizes: A Case Study of the Grizzly Bear (*Ursus arctos* L.). Unpublished Ph.D. dissertation, Duke University, Durham, NC.

Simon, N., and P. Géroudet. 1970. *Last Survivors*. World Publishing Co., New York.

Soulé, M. E. Thresholds for survival: maintaining fitness and evolutionary potential. Pages 151–170 in M. E. Soulé and B. A. Wilcox, eds. *Conservation Biology: An Evolutionary-Ecological Perspective*. Sinauer, Sunderland, MA.

Terborgh, J., and B. Winter. 1980. Some causes of extinction. Pages 119–134 in M. E. Soulé and B. A. Wilcox, eds. *Conservation Biology: An Evolutionary-Ecological Perspective*. Sinauer, Sunderland, MA.

Turelli, M. 1977. Random environments and stochastic calculus. *Theoret. Pop. Biol.* 12: 140–178.

Wilcox, B. A. 1980. Insular ecology and conservation, Pages 95–118 in M. E. Soulé and B. A. Wilcox, eds. *Conservation Biology: An Evolutionary-Ecological Perspective*. Sinauer, Sunderland, MA.

Ecological Monographs, 61(2), 1991, pp. 115–143
© 1991 by the Ecological Society of America

ESTIMATION OF GROWTH AND EXTINCTION PARAMETERS FOR ENDANGERED SPECIES[1]

Brian Dennis
Department of Forest Resources and Department of Mathematics and Statistics,
University of Idaho, Moscow, Idaho 83843 USA

Patricia L. Munholland
Department of Mathematical Sciences, Montana State University, Bozeman, Montana 59717-0001 USA

J. Michael Scott
United States Fish and Wildlife Service, Idaho Cooperative Fish and Wildlife Research Unit,
College of Forestry, Wildlife, and Range Sciences, University of Idaho,
Moscow, Idaho 83843 USA

Abstract. Survival or extinction of an endangered species is inherently stochastic. We develop statistical methods for estimating quantities related to growth rates and extinction probabilities from time series data on the abundance of a single population. The statistical methods are based on a stochastic model of exponential growth arising from the biological theory of age- or stage-structured populations. The model incorporates the so-called environmental type of stochastic fluctuations and yields a lognormal probability distribution of population abundance. Calculation of maximum likelihood estimates of the two unknown parameters in this model reduces to performing a simple linear regression. We describe techniques for rigorously testing and evaluating whether the model fits a given data set. Various growth- and extinction-related quantities are functions of the two parameters, including the continuous rate of increase, the finite rate of increase, the geometric finite rate of increase, the probability of reaching a lower threshold population size, the mean, median, and most likely time of attaining the threshold, and the projected population size. Maximum likelihood estimates and minimum variance unbiased estimates of these quantities are described in detail.

We provide example analyses of data on the Whooping Crane (*Grus americana*), grizzly bear (*Ursus arctos horribilis*) in Yellowstone, Kirtland's Warbler (*Dendroica kirtlandii*), California Condor (*Gymnogyps californianus*), Puerto Rican Parrot (*Amazona vittata*), Palila (*Loxioides bailleui*), and Laysan Finch (*Telespyza cantans*). The model results indicate a favorable outlook for the Whooping Crane, but long-term unfavorable prospects for the Yellowstone grizzly bear population and for Kirtland's Warbler. Results for the California Condor, in a retrospective analysis, indicate a virtual emergency existed in 1980. The analyses suggest that the Puerto Rican Parrot faces little risk of extinction from ordinary environmental fluctuations, provided intensive management efforts continue. However, the model does not account for the possibility of freak catastrophic events (hurricanes, fires, etc.), which are likely the most severe source of risk to the Puerto Rican Parrot, as shown by the recent decimation of this population by Hurricane Hugo. Model parameter estimates for the Palila and the Laysan Finch have wide uncertainty due to the extreme fluctuations in the population sizes of these species. In general, the model fits the example data sets well. We conclude that the model, and the associated statistical methods, can be useful for investigating various scientific and management questions concerning species preservation.

Key words: California Condor; conservation biology; diffusion process; endangered species; exponential growth; extinction; grizzly bear; inverse Gaussian distribution; Kirtland's Warbler; Laysan Finch; lognormal distribution; Palila; parameter estimation; Puerto Rican Parrot; stochastic differential equation; stochastic population model; Whooping Crane; Wiener process.

Introduction

The extinction of a population is a chance event. A population's growth inevitably displays stochastic fluctuations due to numerous unpredictable causes. Consequently, a species with an average negative growth rate might temporarily prosper, while a species with a positive rate might become endangered. The field of conservation biology has recognized the importance of accounting for stochastic factors in species preservation efforts (Shaffer 1981, Samson et al. 1985, Soulé 1986, 1987, Burgman et al. 1988, Lande 1988, Simberloff 1988). Such accounting in practice has proved no easy task.

[1] Manuscript received 11 December 1989; revised and accepted 26 July 1990; final version received 23 August 1990.

Many mathematical models of population growth incorporating stochastic fluctuations have been studied; Nisbet and Gurney (1982) and Goel and Richter-Dyn (1974) provide excellent surveys. The chance of extinction or waiting time to extinction is frequently the focus of theoretical or numerical analyses of stochastic growth models (Capocelli and Ricciardi 1974, Richter-Dyn and Goel 1974, Feldman and Roughgarden 1975, Keiding 1975, Leigh 1981, Tier and Hanson 1981, Ginzburg et al. 1982, Braumann 1983*a, b,* Wright and Hubble 1983, Strebel 1985, Goodman 1987, Lande 1987, Iwasa and Mochizuki 1988, Lande and Orzack 1988, Dennis 1989*a*). Ginzburg et al. (1982), in particular, described the concepts of risk analysis as a framework for such extinction studies. These studies have tended to use relatively simple stochastic models such as univariate birth–death processes or diffusion processes, since the analyses are considerably easier as compared to more complex models. Despite their apparent lack of realism, these simple stochastic models have yielded qualitative insights into general management questions, such as the determination of minimum viable population sizes, whether demographic, genetic, or environmental fluctuations are more important to a species' survival, whether single large or several small reserves afford the least extinction risk, and how Allee effects might be manifested in stochastic populations (Leigh 1981, Wright and Hubbell 1983, Wilcox 1986, Goodman 1987, Burgman et al. 1988, Simberloff 1988, Burkey 1989, Dennis 1989*a*).

The usefulness of simple stochastic models, though, seemingly diminishes for specific, real situations. First, many endangered vertebrate populations are age structured and have periodic breeding seasons, so that models containing a single state variable and/or continuous time would appear unrealistic and inappropriate for quantitative predictions concerning particular species. An alternate approach preferred by some investigators has been the construction of detailed simulation models with many variables, parameters, and stochastic components (e.g., Shaffer and Samson 1985, Mode and Jacobson 1987*a, b,* Ferson et al. 1989). Second, stochastic models are limited in practice by the quality and quantity of data on the growth of endangered species. In this regard, the improvements offered by detailed simulation models over simple analytical models can be dubious. It is a statistical fact of life that less data allows estimation of fewer parameters, and it is quite common for components (particularly stochastic components) of simulation models to be set rather arbitrarily. Finally, even in situations where good data exist, it has not been at all clear to investigators how to interface stochastic models (or deterministic ones, for that matter) with data. Appropriate statistical methods for parameter estimation, model evaluation, and hypothesis testing for stochastic growth models would enhance our understanding of biological populations.

However, mathematical studies have suggested that simple stochastic models might serve as useful approximations for various quantities pertaining to age-structured populations experiencing stochastic fluctuations. Theoretical results concerning stochastic Leslie matrices have indicated that, under very mild assumptions, the logarithm of total population size in an age-structured population can be approximated by a simple stochastic process known as the Wiener process (or Brownian motion) with drift (Tuljapurkar and Orzack 1980, Heyde and Cohen 1985, Tuljapurkar 1989). This model represents a simple stochastic exponential growth model for population size, and as such had been proposed and analyzed in the population biology literature (Capocelli and Ricciardi 1974). Most of its statistical properties, like the distribution of the waiting time until an upper or lower fixed size is attained, had been derived decades earlier by physicists and mathematicians. Population biologists have realized that these statistical properties might prove useful for extinction calculations (Capocelli and Ricciardi 1974, Ginzburg et al. 1982, Braumann 1983*b*, Levinton and Ginzburg 1984, Lande and Orzack 1988). Lande and Orzack (1988), in particular, recently emphasized the potential importance of the Wiener-drift process as a general approximation for age-structured populations, and showed with computer simulations that the associated approximations for extinction-related quantities can be quite acceptable.

Furthermore, statistical inference methods for simple stochastic processes are accumulating in the mathematical statistics literature (e.g., Basawa and Prakasa-Rao 1980), and inference for the Wiener-drift process has been developed in the specific context of population biology. Braumann (1983*b*) and Dennis (1989*b*) studied the question of how to fit this stochastic exponential growth model to data. To "fit" means to estimate the two parameters (denoted μ and σ^2 in this paper) in the model in some statistically acceptable manner, given data on some growing or declining population. Braumann (1983*b*) derived maximum likelihood (ML) estimates for time series data with observations spaced at equal intervals. Dennis (1989*b*) generalized these estimates to unequally spaced intervals and showed how the problem can be transformed to a simple linear regression, making available the whole battery of linear model diagnostics, tests, and software.

These combined developments now make possible the estimation of quantities related to growth and extinction for a variety of endangered species, using relatively straightforward statistical techniques. We describe the necessary techniques in this paper, and provide illustrative analyses of data on the Whooping Crane (*Grus americana*), grizzly bear (*Ursus arctos horribilis*), Kirtland's Warbler (*Dendroica kirtlandii*), California Condor (*Gymnogyps californianus*), Puerto Rican Parrot (*Amazona vittata*), Palila (*Loxioides bailleui*), and Laysan Finch (*Telespyza cantans*). The quantities estimated are functions of the two parameters in the stochastic exponential growth model and include the continuous rate of increase, the finite rate of increase, the probability of extinction, the mean time to extinction, and the projected population size. The model is easy to use and fits the example data well. The statistical methods discussed here for applying the model are best used with true census data, or with population estimates in situations where sampling variability is small compared to population variability. We conclude that the model, in conjunction with the statistical inference methods described here, is a potentially valuable tool for addressing scientific and management questions in conservation biology.

THE STOCHASTIC EXPONENTIAL GROWTH MODEL

Projection matrix

The Lewis-Leslie model (Lewis 1942, Leslie 1945) is a frequently used mathematical representation of density-independent growth of an age-structured population observed at discrete time intervals. The model can be written as

$$M(t + 1) = A(t)M(t), \tag{1}$$

where $M(t)$ is a column vector containing elements representing numbers of individuals (usually females) in each age class at time t ($t = 0, 1, 2, \ldots$), and $A(t)$ is a square matrix containing age-specific fecundity rates (top row), age-specific survivorship rates (subdiagonal), and zeros elsewhere (see van Groenendael et al. 1988 for a recent review). The model is easily generalized to stage-structured populations by incorporating additional positive elements into the projection matrix $A(t)$ (Lefkovitch 1965). If the elements in $A(t)$ are constant, the total population size ultimately approaches exponential growth or decline, after initial age- (or stage-) structure imbalances damp out into a stable age structure. The exponential growth is represented by

$$N(t) = n_0\lambda^t, \tag{2}$$

where $N(t)$ is the total population size [summed elements of the vector $M(t)$, $n_0 = N(0)$ is the initial population size, and λ is the dominant eigenvalue of the projection matrix (finite rate of increase).

However, the elements of a realistic projection matrix should fluctuate with time, since fecundity, survivorship, or stage transition rates are seldom constant in nature. An alternative modeling approach is to assume that the elements of $A(t)$ change with time in the form of a (multivariate) stationary time series. Tuljapurkar (1989) has given a comprehensive review of the demographic theory of populations governed by such dynamics. This modeling assumption is mathematically broad enough to include many real situations and is fundamental to the analysis methods we describe in this paper. Note that this assumption excludes populations experiencing nonstationary fluctuations in demographic rates, such as decreasing survival or reproduction rates due to diminishing habitat.

This stochastic formulation seems to have the minimum level of biological detail necessary for describing a vertebrate population. By contrast, a deterministic, single-state variable model such as Eq. 2 would not likely provide much useful information concerning survival or extinction of an endangered species.

A stochastic single-state variable model, though, can adequately approximate the statistical properties of the fluctuations in total population size resulting from the stochastic projection matrix. Results of Tuljapurkar and Orzack (1980) and Heyde and Cohen (1985), based on central limit theorems, state that the quantity $X(t) = \log N(t)$ will have, as t becomes large, an approximate normal distribution with a mean of $x_0 + \mu t$ and a variance of $\sigma^2 t$ [written $X(t) \stackrel{.}{\sim} \text{normal}(x_0 + \mu t, \sigma^2 t)$, where "$\sim$" means "is distributed as," and the dot indicates that the distribution is approximate], where $x_0 = \log n_0$. The approximation can be improved by adjusting n_0 for initial age structure imbalances (see Lande and Orzack 1988), but we confine ourselves in this paper to estimation techniques that do not require detailed knowledge of age structure. Here, μ is a real-valued constant, and σ^2 is a positive, real-valued constant. These parameters depend on properties of the underlying stochastic projection matrix. If the matrices $A(1), A(2), \ldots$, are serially uncorrelated (for example, each year or time period, elements of the projection matrix are drawn from a multivariate distribution, independent of previous years), then

$$\mu \approx \log \lambda - (\sigma^2/2) \tag{3}$$

and

$$\sigma^2 \approx \lambda^{-2}\delta'C\delta, \tag{4}$$

where λ is now the dominant eigenvalue of the average projection matrix $A^*\{= E[A(t)]\}$, C is the variance–covariance matrix of the multivariate distribution from which the elements of $A(t)$ arise, and δ is a column vector containing partial derivatives of λ with respect to each element of A^* (Tuljapurkar 1982b).

While estimating the multitudinous quantities in a projection matrix with any useful degree of precision can be exceedingly difficult, estimating μ and σ^2 is possible with just a single time series of observations on total population size. Furthermore, various quantities related to extinction are functions of μ and σ^2 and are straightforwardly estimated.

Ease of estimation arises from the fact that the approximate normal distribution of $X(t)$ is identical to the distribution of a Wiener process with drift (e.g., Goel and Richter-Dyn 1974). The Wiener-drift model is a simple type of continuous-time, continuous-state, Markov stochastic process known as a diffusion process. Taking $X(t)$ to be a Wiener-drift process, strictly speaking, imposes an additional layer of approximation on top of the results of Tuljapurkar and Orzack

118   BRIAN DENNIS ET AL.   Ecological Monographs
                                                                          Vol. 61, No. 2

(1980) and Heyde and Cohen (1985), in that the Wiener-drift process has the above-mentioned normal distribution for small values of t as well as large. Lande and Orzack (1988), though, showed with computer simulations that extinction probabilities under various hypothetical life histories were accurately predicted with the Wiener-drift approximation, provided the fluctuations in the projection matrix elements were small or moderate. We discuss diagnostic procedures later in this paper to evaluate the adequacy of the Wiener-drift model for a given data set (see *Model evaluation* section).

Diffusion approximation

We assume that the natural logarithm of total population size, $X(t) = \log N(t)$, is adequately approximated by a Wiener process with drift. This process has been extensively studied; Goel and Richter-Dyn (1974), Ricciardi (1977), and Karlin and Taylor (1981) provide lucid expositions of it and other diffusion processes. The constant μ is known as the infinitesimal mean of the process, since $\mu \Delta t$ is the (approximate) average amount of change in the process over a tiny time interval Δt. The constant σ^2 likewise is known as the infinitesimal variance. The process has a transition probability density function (pdf) corresponding to a normal($x_0 + \mu t$, $\sigma^2 t$) distribution:

$$p_X(x, t \mid x_0) = (2\pi\sigma^2 t)^{-1/2}\exp[-(x - x_0 - \mu t)^2/(2\sigma^2 t)],$$
$$-\infty < x < \infty. \quad (5)$$

The probability that $X(t)$ is between a and b at time t, given that the process starts at x_0, is the corresponding area under the pdf. This probability can be evaluated with the standard normal cumulative distribution function (cdf), $\Phi(\cdot)$:

$$\Pr[a < X(t) \le b] = \int_a^b p_X(x, t \mid x_0)\, dx$$
$$= \Phi\!\left(\frac{b - x_0 - \mu t}{\sqrt{\sigma^2 t}}\right)$$
$$- \Phi\!\left(\frac{a - x_0 - \mu t}{\sqrt{\sigma^2 t}}\right), \quad (6)$$

where

$$\Phi(z) = \int_{-\infty}^z (2\pi)^{-1/2}\exp(-y^2/2)\, dy. \quad (7)$$

The untransformed total population size, $N(t) = \exp[X(t)]$, is also a diffusion process. The transition pdf for $N(t)$ is that of a lognormal distribution:

$$p_N(n, t \mid n_0)$$
$$= n^{-1}(\sigma^2 t 2\pi)^{-1/2}\exp[-(\log n - \log n_0 - \mu t)^2/(2\sigma^2 t)],$$
$$0 < n < \infty. \quad (8)$$

The mean (or expected) population size, given that the process starts at n_0, is

$$E[N(t)] \equiv \psi(t; n_0, \mu, \sigma^2)$$
$$= n_0\exp\{[\mu + (\sigma^2/2)]t\}. \quad (9)$$

For the case of serially uncorrelated projection matrices, the mean population size is identical to the deterministic exponential growth model (Eq. 2), with λ given by Eq. 3. The mean of $X(t)$ by contrast does not depend on σ^2:

$$E[X(t)] \equiv \gamma(t; x_0, \mu) = x_0 + \mu t. \quad (10)$$

The geometric mean of $N(t)$ is defined by $\exp\{E[X(t)]\}$:

$$\exp\{E[\log N(t)]\} \equiv \beta(t; n_0, \mu) = n_0\exp(\mu t). \quad (11)$$

The geometric mean is also the median of the lognormal transition pdf (Eq. 8), which is a general property of the lognormal distribution (see Dennis and Patil 1988). Analytical (Tuljapurkar 1982a) and simulation (Slade and Levenson 1982, Nordheim et al. 1989) studies suggest that the geometric mean of $N(t)$ better characterizes the behavior of the process than does the mean (Eq. 9), due to the extreme positive skewness of the lognormal transition pdf. Other statistical properties of this process are catalogued by Dennis and Patil (1988).

Any diffusion process in general has an alternate mathematical representation as a stochastic differential equation (SDE) (e.g., Karlin and Taylor 1981). Of particular interest here is that the process $N(t)$ is the solution to an SDE version of the exponential growth model given by

$$dN(t) = rN(t)\, dt + \sigma N(t)\, dW(t), \quad (12)$$

where r is a real-valued constant and $dW(t) \sim$ normal(0, dt). The differential $dN(t)$ is defined mathematically in terms of an Ito stochastic integral (for example, Soong 1973); the constants μ and σ^2 in the transition pdf (Eq. 8) for $N(t)$ are related to r by

$$r = \mu + (\sigma^2/2). \quad (13)$$

For the case of serially uncorrelated projection matrices (see Eq. 3), $r \approx \log \lambda$.

Previous studies of extinction probabilities (Capocelli and Ricciardi 1974) and parameter estimation (Braumann 1983b) for the exponential growth SDE (Eq. 12) used a Stratonovich stochastic integral to define $dN(t)$; those results should not be used if the diffusion process $N(t)$ is intended as an approximation for an underlying age-structured, discrete-time system. The Stratonovich formulation instead more appropriately represents a system in which the state variable is fundamentally a continuous function of time (e.g., biomass). Turelli (1977), Capocelli and Ricciardi (1979), and Braumann (1983a) provide insights into the two ways of interpreting the SDE (Eq. 12) as an approximation for some underlying process, and Dennis and Patil (1988) list formulas for transforming Ito-based results to Stratonovich-based results and vice versa.

Extinction properties

Under the continuing unpredictable fluctuations of the Wiener-drift model, $X(t)$ could possibly cross any lower threshold size, x_e, starting from x_0. This event corresponds to the population size, $N(t)$, attaining a lower threshold size, $n_e = \exp(x_e)$, starting from n_0. If $n_e = 1$ (or $x_e = 0$), the event obviously represents the extinction of a closed, sexually reproducing population. Management efforts to promote survival of an endangered species might naturally hinge upon a different threshold size. Some fixed population size, $n_e > 1$, could be regarded as a policy threshold, or as a safety cushion to avoid the possibilities of Allee effects (e.g., Dennis 1989a), skewed sex ratios, or inbreeding. We use the term "extinction" in this paper to refer broadly to the attainment of some prespecified lower threshold, representing, if not the demise of the species, the demise of some management regime. The term "quasiextinction" has also been used in this context (Ginzburg et al. 1982). Let x_d represent the distance on the logarithmic scale from an initial population size to a lower threshold population size:

$$x_d = x_0 - x_e = \log(n_0/n_e). \quad (14)$$

As explained in the context of extinction by numerous authors (Capocelli and Ricciardi 1974, Ricciardi 1977, Tuljapurkar and Orzack 1980, Ginzburg et al. 1982, Lande and Orzack 1988), the probability $\pi(x_d, \mu, \sigma^2)$ that the process will ever attain the threshold is

$$\pi(x_d, \mu, \sigma^2) = \begin{cases} 1, & \mu \le 0; \\ \exp(-2\mu x_d/\sigma^2), & \mu > 0. \end{cases} \quad (15)$$

Given that the threshold is attained (i.e., conditioning on all sample paths of the process that reach the threshold), the amount of time, T, elapsing before the threshold is first reached is a positive, real-valued random variable with a continuous probability distribution. The cdf of the distribution can be written in terms of a standard normal cdf:

$$\Pr[T \le t] \equiv G(t; x_d, \mu, \sigma^2)$$

$$= \Phi\left(\frac{-x_d + |\mu|t}{\sigma\sqrt{t}}\right)$$

$$+ \exp(2x_d|\mu|/\sigma^2)\Phi\left(\frac{-x_d + |\mu|t}{\sigma\sqrt{t}}\right),$$

$$0 < t < \infty. \quad (16)$$

The pdf of the distribution is the derivative of $G(t; x_d, \mu, \sigma^2)$ with respect to t:

$$g(t; x_d, \mu, \sigma^2)$$
$$= x_d(2\pi\sigma^2 t^3)^{-1/2}\exp[-(x_d - |\mu|t)^2/(2\sigma^2 t)]. \quad (17)$$

This distribution, known as the inverse Gaussian distribution (a misnomer: it is not the distribution of the reciprocal of a normal random variable), has been extensively studied (see Folks and Chhikara 1978).

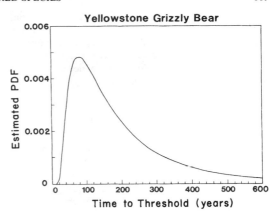

FIG. 1. Probability density function (PDF) of the inverse Gaussian distribution, plotted using maximum likelihood estimates of parameters μ and σ^2 for the Yellowstone National Park population of grizzly bears. The distribution is that of the time required for 47 female bears to decline to 10 bears.

Schrödinger (1915) originally obtained it as the first-passage time distribution for the Wiener-drift process for the case $\mu < 0$; Tweedie (1957a, b) derived many of its statistical properties. Whitmore (1978) obtained the inverse Gaussian explicitly as the conditional first-passage time distribution for the case $\mu > 0$. Whitmore and Seshadri (1987) and Lande and Orzack (1988) have given intuitive derivations of the first-passage time result.

We note that the probability distribution of the time to attain an upper threshold, given it is attained, is also the inverse Gaussian distribution (Eqs. 16 and 17), except with x_d representing $\log(n_u/n_0)$, where n_u is the upper threshold. The probability of ever attaining n_u is given by $\pi(x_d, -\mu, \sigma^2)$, that is, by (Eq. 15) evaluated at $\log(n_u/n_0)$, $-\mu$, and σ^2.

Various quantities pertaining to this distribution are of potential interest in conservation biology. The mean time until the threshold population size is reached is the expected value of T:

$$E[T] \equiv \theta(x_d, \mu) = x_d/|\mu|. \quad (18)$$

The variance of T is

$$\mathrm{Var}(T) = x_d\sigma^2/|\mu|^3. \quad (19)$$

The distribution is positively skewed and has a heavy right tail (an example shape is portrayed in Fig. 1), a fact that has implications for species preservation efforts (see *Examples* and *Discussion* sections). Percentiles and modes are quantities that help to characterize such skewed distributions. The $100 \cdot p$th percentile, $\xi_p(x_d, \mu, \sigma^2)$, is defined as the root of

$$G(\xi_p; x_d, \mu, \sigma^2) = p. \quad (20)$$

120   BRIAN DENNIS ET AL.   Ecological Monographs
Vol. 61, No. 2

For example, the 50th percentile of the distribution, or the median, represents the fixed time at which the probability of hitting the threshold before that time is 0.5. The mode of the distribution is the most likely time of hitting the threshold, and is defined as the value of t maximizing the pdf (Eq. 17):

$$t^*(x_d, \mu, \sigma^2) = \frac{x_d}{|\mu|}\left[\left(1 + \frac{9}{4\nu^2}\right)^{1/2} - \frac{3}{2\nu}\right], \quad (21)$$

where $\nu = x_d|\mu|/\sigma^2$. Note that the mode is the product of the mean and a quantity (in square brackets) between 0 and 1.

The notation used above emphasizes that these quantities are functions of the parameters x_d, μ, and σ^2. The quantity x_d, the (log-scale) distance to the threshold population, is selected by the investigator. The parameters μ and σ^2, however, are typically unknown and must be estimated from data.

Parameter Estimation

Maximum likelihood estimates

Estimates of μ and σ^2 can be obtained by observing a population at times $0, t_1, t_2, \ldots, t_q$. The recorded observations of population size will be denoted $n(0) = n_0$, $n(t_1) = n_1, \ldots, n(t_q) = n_q$, and the time intervals (not necessarily equal) between observations denoted $t_1 - 0 = \tau_1$, $t_2 - t_1 = \tau_2, \ldots, t_q - t_{q-1} = \tau_q$. For a population with a yearly breeding cycle, these observations should be spaced at least 1 yr apart and taken at the same time of year. A recommended way of fitting the stochastic exponential growth model to such data is maximum likelihood (ML) estimation.

ML estimates of the two parameters are easy to calculate and have many desirable statistical properties. The likelihood function $l(\mu, \sigma^2)$ is defined as the joint pdf for $N(t_1)$, $N(t_2)$, $\ldots$, $N(t_q)$, given $N(0) = n_0$, evaluated at the observations $n_1, n_2, \ldots, n_q$. To obtain the likelihood function, we note that $N(t)$ is a diffusion process with stationary transition probabilities. This means that the pdf of n_i given n_{i-1} depends on τ_i (the time interval since n_{i-1}) but not on t_{i-1}. Also, the diffusion process is a Markov process, meaning that the pdf of n_i given n_{i-1} does not depend on the earlier observations $n_0, \ldots, n_{i-2}$. Thus, $p_N(n_i, \tau_i|n_{i-1})$, which is the lognormal transition pdf (Eq. 8) evaluated at n_i, τ_i, and n_{i-1}, represents the likelihood of the system moving to n_i from n_{i-1} in a time interval of τ_i. The likelihood function is then the product of transition pdfs:

$$l(\mu, \sigma^2)$$
$$= p_N(n_1, \tau_1|n_0)p_N(n_2, \tau_2|n_1)\cdots p_N(n_q, \tau_q|n_{q-1}). \quad (22)$$

The ML estimates, $\hat{\mu}$ and $\hat{\sigma}^2$, are the parameter values jointly maximizing $l(\mu, \sigma^2)$ or, equivalently $\log l(\mu, \sigma^2)$, where

$$\log l(\mu, \sigma^2) = -\sum_{i=1}^{q} \log[n_i(2\tau_i\pi)^{1/2}]$$
$$- (q/2)\log \sigma^2 - [1/(2\sigma^2)]$$
$$\cdot \sum_{i=1}^{q} (1/\tau_i)[\log(n_i/n_{i-1}) - \mu\tau_i]^2. \quad (23)$$

It is straightforward to set partial derivatives of $\log l(\mu,\sigma^2)$ with respect to μ and σ^2 equal to zero and solve for the ML estimates:

$$\hat{\mu} = \left[\sum_{i=1}^{q} \log(n_i/n_{i-1})\right]\bigg/ \sum_{i=1}^{q} \tau_i$$
$$= [\log(n_q/n_0)]/t_q; \quad (24)$$

$$\hat{\sigma}^2 = (1/q) \sum_{i=1}^{q} (1/\tau_i)[\log(n_i/n_{i-1}) - \hat{\mu}\tau_i]^2. \quad (25)$$

We mention that the ML estimate (Eq. 25) of σ^2 is biased (e.g., Dennis 1989b); an unbiased estimate is

$$\tilde{\sigma}^2 = q\hat{\sigma}^2/(q - 1). \quad (26)$$

The difference between $\hat{\sigma}^2$ and $\tilde{\sigma}^2$ is negligible when q is large. Often q will not be very large in data on endangered populations, though.

Linear regression approach

These identical ML estimates of μ and σ^2 can also be calculated by a regression approach. This approach offers the following practical advantages. First, information about the statistical distributions of the ML estimates is easily obtained (e.g., confidence intervals). Second, a battery of diagnostic procedures for linear regression models becomes available for evaluating the diffusion model. Finally, the analyses can be accomplished with most of the standard computer packages for linear regression.

The approach involves transforming the observations so that a normal linear model applies. Let $W_i = \log[N(t_i)/N(t_{i-1})] = X(t_i) - X(t_{i-1})$, so that W_i represents the change in $X(t)$ (the Wiener-drift process) between times t_{i-1} and t_i. Thus, the variables $W_1, W_2, \ldots, W_q$ are increments of a Wiener-drift process, and are therefore normal, independent, and stationary (e.g., Ricciardi 1977). In fact, if $\mathbf{w} = [W_1, W_2, \ldots, W_q]'$ and $\mathbf{\tau} = [\tau_1, \tau_2, \ldots, \tau_q]'$ are defined as column vectors, the distribution of $\mathbf{w}$ becomes a multivariate normal, with mean $\mu\mathbf{\tau}$ and variance–covariance matrix $\sigma^2\mathbf{V}$:

$$\mathbf{w} \sim \text{normal}(\mu\mathbf{\tau}, \sigma^2\mathbf{V}). \quad (27)$$

Here $\mathbf{V} = \text{diag}(\mathbf{\tau})$ is a matrix with the elements of $\mathbf{\tau}$ on the main diagonal and zeros elsewhere. Let $\mathbf{G} = \text{diag}(\sqrt{\tau_1}, \ldots, \sqrt{\tau_q})$, that is, $\mathbf{V} = \mathbf{G}'\mathbf{G}$. A transformation of $\mathbf{w}$ produces an ordinary normal linear model (e.g., Graybill 1976:207):

$$\mathbf{Y} = \mathbf{G}^{-1}\mathbf{w} \sim \text{normal}(\mu\mathbf{D}, \sigma^2\mathbf{I}), \quad (28)$$

where $\mathbf{D} = [\sqrt{\tau_1}, \ldots, \sqrt{\tau_q}]'$ and $\mathbf{I}$ is the $q \times q$ identity matrix. This is a model for a simple linear regression without intercept.

In practice, the data are transformed by $y_i = [\log(n_i / n_{i-1})]/\sqrt{\tau_i}$, $i = 1, \ldots, q$. The regression approach uses $y_1, y_2, \ldots, y_q$ as values of the "dependent variable," $\sqrt{\tau_1}, \ldots, \sqrt{\tau_q}$ as values of the "independent variable," and a linear regression without intercept is performed. The formula (Eq. 24) for $\hat{\mu}$ is recognized as the slope parameter estimate, and $\tilde{\sigma}^2$ (Eq. 26) is the (unbiased) estimate of the error variance parameter. The ML estimate of σ^2 is $\hat{\sigma}^2$ (Eq. 25).

The usual linear model theory yields the distributions of $\hat{\mu}$, $\hat{\sigma}^2$, and $\tilde{\sigma}^2$ (Graybill 1976):

$$\hat{\mu} \sim \text{normal}(\mu, \sigma^2/t_q), \tag{29}$$

$$q\hat{\sigma}^2/\sigma^2 = (q - 1)\tilde{\sigma}^2/\sigma^2 \sim \text{chi-square}(q - 1). \tag{30}$$

Also, $\hat{\mu}$ is independent of $\hat{\sigma}^2$ or $\tilde{\sigma}^2$. An estimate of the standard error of $\hat{\mu}$ is $(\tilde{\sigma}^2/t_q)^{1/2}$, and a $100(1 - \alpha)\%$ confidence interval for μ is given by

$$(\hat{\mu} - t_{\alpha/2, q-1}\sqrt{\tilde{\sigma}^2/t_q}, \; \hat{\mu} + t_{\alpha/2, q-1}\sqrt{\tilde{\sigma}^2/t_q}). \tag{31}$$

Here $\Pr[|T_{q-1}| \leq t_{\alpha/2, q-1}] = 1 - \alpha$, where T_{q-1} has a Student's t distribution with $q - 1$ degrees of freedom. Confidence intervals for σ^2 can be calculated with either the ML or the unbiased estimate. For example, a $100(1 - \alpha)\%$ confidence interval for σ^2 based on the ML estimate is

$$(q\hat{\sigma}^2/\chi^2_{\alpha_1, q-1}, \; q\hat{\sigma}^2/\chi^2_{1-\alpha_2, q-1}), \tag{32}$$

where $\alpha_1 + \alpha_2 = \alpha$, $0 < \alpha_1$, α_2, and $\Pr[X_{q-1} \leq \chi^2_{\alpha, q-1}] = 1 - \alpha$, where X_{q-1} has a chi-square distribution with $q - 1$ degrees of freedom. The confidence interval using the unbiased estimate would simply substitute $(q - 1)\tilde{\sigma}^2$ for $q\hat{\sigma}^2$ in Eq. 32.

Model Evaluation

The equivalence of the Wiener-drift model for $X(t)$ and the linear regression model for the transformed increment variables Y (see Eq. 28) is extremely useful for assessing the adequacy of the diffusion approximation. Numerous diagnostic procedures are available for evaluating the adequacy of linear models; Chatterjee and Hadi (1988), Cook and Weisburg (1982), Belsley et al. (1980), and Draper and Smith (1981) are excellent references. Furthermore, changes in the Wiener-drift model, before and after a fixed time, can be detected using regression methods. The relevance of these procedures to evaluating the diffusion model is discussed.

Evaluation of model assumptions

Our assumption of a Wiener-drift model for $X(t)$ corresponds to the following regression model for the transformed increment data (from Eq. 28):

$$Y = \mu D + \epsilon, \tag{33}$$

where the $q \times 1$ vector of errors $\epsilon = (\epsilon_1, \epsilon_2, \ldots, \epsilon_q)'$;

the errors are assumed to be independent normal random variables with common mean 0 and variance σ^2. Thus, a violation of the diffusion model assumptions can be detected by evaluating the assumptions of the regression analysis.

Notice that the transition is the fundamental observation in the likelihood function (Eq. 22); the population observations $n_0, n_1, \ldots, n_q$ appear only in the context of transitions from one to the next.

Graphical and analytical methods for checking the regression model assumptions are invariably based on the residuals or transformations of the residuals, where the jth residual is the difference between the jth observation and its predicted value: $e_j = y_j - \hat{\mu}\sqrt{\tau_j}$ (e.g., Draper and Smith 1981, Cook and Weisburg 1982). Although the residuals are subject to some restrictions (e.g., correlations among the e_j may be nonzero), they can be loosely interpreted as the observed errors if the model is correct. Hence, they should exhibit behaviors that tend to confirm the model assumptions or at least that do not contradict these assumptions (Draper and Smith 1981).

Residuals can be used to check the independent increments property, an assumption imposed on $X(t)$ via the Wiener-drift approximation. This assumption is crucial to the analyses presented here. If the assumption is reasonable, then serial autocorrelations among the increment variables W_j (see Eq. 27) are negligible and hence the limiting distributional results of Tuljapurkar and Orzack (1980) and Heyde and Cohen (1985) can be applied to the transitions; that is, the W_j are approximately independent normal random variables. Correlations among the transitions suggest that the Wiener-drift process for $X(t)$ is inappropriate. In this case, while the least squares estimates of μ and σ^2 are statistically consistent (i.e., the estimates get "closer" to the true parameter values as the sample size increases), they may not be the best estimates. We refer to Heyde and Cohen (1985) for more appropriate estimators when the errors are correlated.

We note that even with serially uncorrelated environmental fluctuations, the W_j in most age-structured populations have a theoretical serial correlation. The cause of this correlation is the time lag inherent in age-structured survival and reproduction; a big pulse of reproduction, for instance, leads to another pulse some years later when the "baby boom" reaches reproductive maturity. The theoretical Wiener-drift approximation strictly applies only for time intervals encompassing many generations. Use of a Wiener-drift likelihood (Eq. 27) to model the W_j will result, in the case of serially uncorrelated environmental fluctuations, in an underestimate of the infinitesimal variance of the theoretical Wiener-drift approximation.

There is as yet no entirely satisfactory way to correct for this bias with a single time series, without additional knowledge of age-specific demographic parameters. Methods discussed by Heyde and Cohen (1985) and

122   BRIAN DENNIS ET AL.   Ecological Monographs
                                                                                        Vol. 61, No. 2

Lande and Orzack (1988) involve estimating high order autocovariances or pooling transitions into a few long-term transitions, with the price of high standard errors for the estimates of σ^2. The situation warrants further investigation. In the meantime, use of the Wiener-drift model for a population time series should rely heavily on diagnostic techniques, such as discussed here, in order to minimize potential bias.

A popular test for detecting first-order autocorrelation among the errors of a regression model is the Durbin-Watson test. The test statistic for testing the null hypothesis of uncorrelated errors is

$$d = \sum_{j=2}^{q} (e_j - e_{j-1})^2 \Big/ \sum_{j=1}^{q} e_j^2. \qquad (34)$$

Draper and Smith (1981) provide tables of upper and lower critical values of d for a number of significance levels. This test is available in many statistical computing packages and therefore can be easily incorporated in the analysis. Higher order autocorrelations can be detected by subjecting the residuals to standard time-series analyses (e.g., Pankratz 1983), provided enough data are available.

Other analytical, as well as graphical methods are available for checking the remaining model assumptions (e.g., normality, constant variance) and we refer the reader to Draper and Smith (1981) and Cook and Weisburg (1982) for a discussion of these standard procedures. It is worth noting that if the population observations $n_0, n_1, \ldots, n_q$ are taken at equally spaced time points, so that the τ_j are equal, many of these graphical methods are uninformative.

Sensitivity analysis

In recent years, numerous statistical measures have been developed for detecting unusual (e.g., outliers) or highly influential observations (Belsley et al. 1980, Chatterjee and Hadi 1988). As pointed out by Belsley et al. (1980:3), such transitions are not necessarily "bad" data points; rather, they may contain some of the most interesting sample information. However, they may also reflect an unknown recording error or they may have resulted from circumstances different from those common to the remaining data (e.g., a population reduction due to an unusual catastrophic storm or removal of population members into captivity by managers). Since these observations can have a substantial effect on the parameter estimates (that is, the estimates are extremely sensitive to them), we recommend screening for them in the analysis; many statistical software packages will calculate these measures.

An observation is an outlier if it is not successfully accommodated by the fitted regression model. Residuals or transformed residuals corresponding to outliers are large compared to those of other observations in the data set. However, a small residual does not necessarily imply that the corresponding observation is a typical one; the method of least squares avoids large residuals in fitting the model and thus may accommodate an atypical transition at the expense of the other data points. These transitions are commonly referred to as influential observations since they excessively influence the parameter estimates as compared to the remaining data. Note that a transition may be judged as an outlier, an influential observation, or both.

Outliers are identified via statistical measures based on residuals and can be detected using formal testing procedures or informal comparisons of relative magnitude. One measure, which Chatterjee and Hadi (1988: 74) call the internally studentized residual, is the jth residual divided by its estimated standard deviation:

$$I_j = \frac{e_j}{\sqrt{\tilde{\sigma}^2(1 - h_{jj})}}, \qquad (35)$$

where $h_{jj} = \tau_j/t_q$ is the jth diagonal element of the matrix $D(D'D)^{-1}D'$ (see Eq. 28). While the I_j are not independent (since the residuals are not independent), a formal test is available for detecting the presence of a single outlier. If I_{max} denotes the maximum of the values of $|I_j|$, then approximate critical values for I_{max}, at the α significance level, are given by

$$c_\alpha = \sqrt{\frac{(q - 1)f_{\alpha/q,1,q-1}}{q - 2 + f_{\alpha/q,1,q-1}}}, \qquad (36)$$

where $f_{\alpha/q,1,q-1}$ is the $100[1 - (\alpha/q)]$th percentile of an F distribution with 1 and $q - 1$ degrees of freedom. Thus, at the α level of significance, I_{max} is an outlier if $I_{max} > c_\alpha$.

The externally studentized residual (Chatterjee and Hadi 1988:74) is defined as

$$E_j = \frac{e_j}{\sqrt{\tilde{\sigma}^2_{(j)}(1 - h_{jj})}}, \qquad (37)$$

where $\tilde{\sigma}^2_{(j)}$ is the (unbiased) estimate of σ^2 when the jth transition is deleted from the analysis (see Eq. 42 below). This measure for detecting outliers has some advantages over I_j. By excluding the jth transition in estimating σ^2, $\tilde{\sigma}^2_{(j)}$ ignores gross errors in the jth observation. Also, E_j tends to reflect large deviations more dramatically than I_j. In addition, E_j has a Student's t distribution with $q - 2$ degrees of freedom, which suggests that a transition for which $|E_j| > 2$ and E_j is large in magnitude compared to those of the remaining transitions should be investigated as a possible outlier.

Measures for detecting influential observations are commonly based on the omission approach, in that they measure changes in the parameter estimates or predicted values when the jth data point is excluded from the analysis. Cook's distance, C_j (Chatterjee and Hadi 1988:117), measures the change in the slope estimate, $\hat{\mu}$, and can be expressed as a function of the internally studentized residual:

$$C_j = I_j^2 h_{jj}/(1 - h_{jj}). \qquad (38)$$

Large values of C_j indicate that the jth transition is influential. Although C_j is not a true F random variable, comparing C_j to the probability points of an F distribution with 1 and $q - 1$ degrees of freedom provides descriptive or qualitative levels of significance (Chatterjee and Hadi 1988).

Welsch and Kuh's distance, WK_j, or the $DFFITS_j$ statistic (Chatterjee and Hadi 1988:120), measures the influence of the jth observation on the predicted value $\hat{\mu}\sqrt{\tau_j}$:

$$WK_j = |E_j| \sqrt{\frac{h_{jj}}{1 - h_{jj}}}. \tag{39}$$

Again, a large value of this statistic indicates a potentially influential transition. While WK_j does not have a Student's t distribution with $q - 2$ degrees of freedom, it is a t-like statistic. This suggests that data points for which $WK_j > 2$ should be regarded as influential observations.

In some regression studies h_{jj} alone is used as an influence measure. A point is considered influential if h_{jj} exceeds some specified constant. We point out that this approach leads to some peculiarities for regression through the origin. Since $h_{jj} = \tau_j/t_q$, only transitions with longer time intervals can be influential, but never shorter time intervals. The other influence measures discussed in this subsection depend on the y_j values as well as the h_{jj} values and hence avoid this undesirable property.

Once unusual or highly influential transitions have been identified, they must be investigated to determine if they are in error or the result of some catastrophic event. If so, we recommend deleting them from the analysis. When the jth transition is deleted, the likelihood function (Eq. 22) has the corresponding transition pdf excluded from the product, and the log-likelihood (Eq. 23) has the jth term excluded from the sum. Eqs. 24 and 25 for the ML estimates then become altered to account for the missing transition:

$$\hat{\mu} = \left\{ \sum_{\substack{i=1 \\ i \neq j}}^{q} \log(n_i/n_{i-1}) \right\} \Big/ \left\{ \sum_{\substack{i=1 \\ i \neq j}}^{q} \tau_i \right\}; \tag{40}$$

$$\hat{\sigma}^2 = (1/p) \sum_{\substack{i=1 \\ i \neq j}}^{q} (1/\tau_i)[\log(n_i/n_{i-1}) - \hat{\mu}\tau_i]^2. \tag{41}$$

Here p is the number of transitions included in the estimates ($= q - 1$ when just one transition is deleted). Additional transitions can be deleted from the formulas in the same way. If transitions are deleted from the ML estimate of σ^2, then the unbiased estimate is

$$\tilde{\sigma}^2 = p\hat{\sigma}^2/(p - 1). \tag{42}$$

Note that $\tilde{\sigma}^2 = \tilde{\sigma}^2_{(j)}$ if the jth observation alone is excluded. The estimates, $\hat{\mu}$ and $\tilde{\sigma}^2$ can easily be obtained by omitting the appropriate pairs $(y_j, \sqrt{\tau_j})$ from the regression analysis. With transitions deleted, the distribution Eqs. 29–32 for the parameter estimates would

have $\Sigma\tau_i$ in place of t_q, where the sum is over all transitions included in the analysis, and p in place of q.

Change in parameters

A permanent change in the infinitesimal mean growth rate may result from a sustained change in the population's environment or from a change in conservation efforts on the part of managers. Suppose the change in conditions is known to have occurred after the jth transition. A corresponding change in μ (without a change in σ^2) is incorporated in the diffusion model by expanding the regression model (Eq. 33):

$$Y = D\mu + \epsilon, \tag{43}$$

where $\mu = (\mu_1, \mu_2)'$ and D is a $q \times 2$ matrix with first column $(\sqrt{\tau_1}, \ldots, \sqrt{\tau_j}, 0, \ldots, 0)'$ and second column $(0, \ldots, 0, \sqrt{\tau_{j+1}}, \ldots, \sqrt{\tau_q})'$. Thus, μ_1 denotes the slope parameter for the first j observations while the remaining $q - j$ transitions have slope μ_2. As before, ϵ is the $q \times 1$ vector of normal random variables, with mean 0 and common variance σ^2. The ML estimates of μ_1 and μ_2 are

$$\hat{\mu}_1 = [\log(n_j/n_0)] \Big/ \sum_{i=1}^{j} \tau_i, \tag{44}$$

$$\hat{\mu}_2 = [\log(n_q/n_j)] \Big/ \sum_{i=j+1}^{q} \tau_i, \tag{45}$$

with distributions

$$\hat{\mu}_1 \sim \text{normal}\left(\mu_1, \sigma^2 \Big/ \sum_{i=1}^{j} \tau_i\right), \tag{46}$$

and

$$\hat{\mu}_2 \sim \text{normal}\left(\mu_2, \sigma^2 \Big/ \sum_{i=j+1}^{q} \tau_i\right). \tag{47}$$

The unbiased estimate of σ^2 is

$$\tilde{\sigma}^2 = (1/(q - 2))$$
$$\cdot \left\{ \sum_{i=1}^{j} (1/\tau_i)[\log(n_i/n_{i-1}) - \hat{\mu}_1\tau_i]^2 \right.$$
$$\left. + \sum_{i=j+1}^{q} (1/\tau_i)[\log(n_i/n_{i-1}) - \hat{\mu}_2\tau_i]^2 \right\}, \tag{48}$$

while the ML estimate is

$$\hat{\sigma}^2 = (q - 2)\tilde{\sigma}^2/q. \tag{49}$$

Also,

$$q\hat{\sigma}^2/\sigma^2 = (q - 2)\tilde{\sigma}^2/\sigma^2 \sim \text{chi-square}(q - 2). \tag{50}$$

Hence, confidence intervals for the model parameters can be readily obtained (e.g., see Eqs. 31 and 32).

A formal test is available for testing whether the change in the infinitesimal growth rate is significant. The statistic

$$T_{q-2} = (\hat{\mu}_1 - \hat{\mu}_2)/\sqrt{\tilde{\sigma}^2[(1/j) + 1/(q - j)]} \tag{51}$$

124 BRIAN DENNIS ET AL. Ecological Monographs
Vol. 61, No. 2

has a Student's t distribution with $q - 2$ degrees of freedom, under the null hypothesis that $\mu_1 - \mu_2 = 0$. Note that Eq. 51 is simply the two-sample t statistic (e.g., Neter et al. 1985:13).

A sustained alteration in a population's environment or management policies may also result in a permanent change in the infinitesimal variance. The model (Eq. 43) can be used to include a change in σ^2 after the jth transition by assuming that the first j errors in ϵ are normal random variables with common mean 0 and variance σ_1^2 while the remaining $q - j$ errors are also normal with mean 0 but have common variance σ_2^2. In practice, this amounts to fitting two regression models, one to the first j transitions and another to the remaining data. Allowing a corresponding change in the mean provides the best estimates of the variance parameters, since the least squares estimate of σ^2 depends on μ (see Eqs. 52 and 53). The ML estimates of $\hat{\mu}_1$ and $\hat{\mu}_2$ are Eqs. 44 and 45. Their distributions are Eqs. 46 and 47, respectively, except with separate variance σ_1^2 and σ_2^2 substituted in place of σ^2. The unbiased estimates of the variances are

$$\tilde{\sigma}_1^2 = [1/(j - 1)] \sum_{i=1}^{j} (1/\tau_i)$$
$$\cdot [\log(n_i/n_{i-1}) - \hat{\mu}_1 \tau_i]^2, \qquad (52)$$

$$\tilde{\sigma}_2^2 = [1/(q - j - 1)] \sum_{i=j+1}^{q} (1/\tau_i)$$
$$\cdot [\log(n_i/n_{i-1}) - \hat{\mu}_2 \tau_i]^2, \qquad (53)$$

with independent distributions

$$(j - 1)\tilde{\sigma}_1^2/\sigma_1^2 \sim \text{chi-square}(j - 1), \qquad (54)$$
$$(q - j - 1)\tilde{\sigma}_2^2/\sigma_2^2 \sim \text{chi-square}(q - j - 1). \qquad (55)$$

Since the chi-square random variables Eqs. 54 and 55 are independent, the ratio of the estimated variances given by

$$F_{j-1, q-j-1} = \tilde{\sigma}_1^2/\tilde{\sigma}_2^2 \qquad (56)$$

has an F distribution with $j - 1$ and $q - j - 1$ degrees of freedom, under the null hypothesis that $\sigma_1^2 = \sigma_2^2$ (e.g., Neter et al. 1985:7).

ESTIMATING GROWTH PARAMETERS

Continuous rate of increase

We define the continuous rate of increase as the parameter r in the exponential growth SDE (Eq. 12). By adopting this definition, we abandon the notion that there is anything "intrinsic" about a deterministic formulation of exponential growth. Rather, r is simply a constant related to statistical properties of the stochastic process $N(t)$. Specifically, rn represents the infinitesimal mean of $N(t)$, that is, $rn\Delta t$ is approximately the average amount of change in $N(t)$ over a tiny time interval Δt, given that $N(t) = n$. The advantage of this definition becomes apparent when r must be estimated from time series observations.

Any parameter in a deterministic model can only be sensibly estimated from time series data by embedding the model in a statistical/stochastic framework, that is, by converting the model into a stochastic one. The parameter r in the deterministic growth equation $N(t) = n_0 e^{rt}$ is often estimated by linear or nonlinear least squares, with observations on population size serving as the "dependent variable," and observations on time as the "independent variable." The quality of such an estimate, however, depends upon: (a) an assumed statistical model of errors in the regression, and (b) whether that error model adequately represents how the data arise. The usual regression package printouts of confidence intervals for r based on least squares calculations assume independent, normal errors. This amounts to no more than fitting a stochastic growth model to the data, and a bad one at that, since the observations in a population time series are seldom independent.

Instead, the exponential growth SDE (Eq. 12) provides at the outset an explicit, realistic structure of dependence in observations of a population's size through time. Under the exponential growth SDE, log-scale increments of population size are independent, but the population's actual sizes are not. This dependence structure is directly incorporated into parameter estimates through the likelihood function (Eq. 22). If the model is an acceptable representation of the system (as evaluated by the diagnostic procedures described in the *Model evaluation* section, above), then the parameter r as defined here can be efficiently estimated from data.

The quantity r is a function, given by Eq. 13, of the parameters μ and σ^2. Substituting the ML estimates of μ and σ^2 in Eq. 13 produces the ML estimate of r:

$$\hat{r} = \hat{\mu} + (\hat{\sigma}^2/2). \qquad (57)$$

The ML estimate is biased due to the bias of $\hat{\sigma}^2$, though the bias becomes negligible when q is large. An unbiased estimate of r results from using $\tilde{\sigma}^2$ instead of $\hat{\sigma}^2$:

$$\tilde{r} = \hat{\mu} + (\tilde{\sigma}^2/2). \qquad (58)$$

Since $\hat{\mu}$ and $\tilde{\sigma}^2$ are "sufficient statistics" (all information in the data about μ and σ^2 is contained in $\hat{\mu}$ and $\tilde{\sigma}^2$), a fundamental result in statistics known as the Rao-Blackwell theorem applies (see Rice 1988:261): $\tilde{r}$ is the uniformly minimum variance unbiased (UMVU) estimate of r. In other words, no other unbiased estimate of r has a smaller variance. Curiously, $\hat{r}$ has a smaller variance than $\tilde{r}$, due to the well-known fact that $\hat{\sigma}^2$ has a smaller variance than $\tilde{\sigma}^2$. Though $\hat{r}$ will be, on the average, "closer" to r than $\tilde{r}$, it will tend to underestimate r.

The variance of $\tilde{r}$ is just the sum of the variances of $\hat{\mu}$ and $\tilde{\sigma}^2/2$, since $\hat{\mu}$ and $\tilde{\sigma}^2$ are independent:

$$\text{Var}(\tilde{r}) = \text{Var}(\hat{\mu}) + \text{Var}(\tilde{\sigma}^2/2)$$
$$= (\sigma^2/t_q) + \{\sigma^4/[2(q-1)]\}. \qquad (59)$$

The distribution of $\tilde{r}$ is complicated; it is the sum of a normal-distributed random variable, $\hat{\mu}$, and a gamma-distributed (=constant·chi-square) random variable, $\tilde{\sigma}^2/2$. However, the distribution is fairly well approximated by a normal distribution for moderately large values of q (say, 20 or so):

$$\tilde{r} \sim \text{normal}\left(r, \frac{\sigma^2}{t_q} + \frac{\sigma^4}{2(q-1)}\right). \tag{60}$$

An approximate $100(1 - \alpha)\%$ confidence interval for r is thus given by

$$\left(\tilde{r} \pm z_{\alpha/2} \sqrt{\tilde{\sigma}^2\left[\frac{1}{t_q} + \frac{\tilde{\sigma}^2}{2(q-1)}\right]}\right). \tag{61}$$

Here $z_{\alpha/2}$ is the $100[1 - (\alpha/2)]$th percentile of the standard normal distribution.

We point out that the symbol r in Braumann's (1983b) paper on estimation corresponds to the exponential growth SDE (Eq. 12) as defined by the Stratonovich stochastic integral. Braumann noted that the ML estimate of r he derived does not depend on q (the number of observations), but just on t_q (the amount of time the system has been observed). In fact, under the Stratonovich interpretation of the SDE (Eq. 12), r equals μ (the infinitesimal mean of the log-scaled process), and the infinitesimal mean of $N(t)$ is $r + (\sigma^2/2)$. In this case, the ML estimate of r is given by Eq. 24 for the ML estimate of μ.

The fundamental discrete-time nature of population growth for many vertebrate species (e.g., seasonal breeding periods), coupled with the limit theorems on stochastic projection matrices (Tuljapurkar and Orzack 1980, Heyde and Cohen 1985), suggests instead that the Ito interpretation of the SDE (Eq. 12) be used in typical endangered species contexts. Estimates of r (Eqs. 57 and 58) then become strongly dependent on q as well as t_q. Braumann (1983a) rightly observed that the differing definitions of the SDE (Eq. 12) produce semantic differences in how r is viewed as a central tendency measure. Under the Stratonovich calculus r is the rate constant in the geometric mean (Eq. 11) of $N(t)$, while under the Ito calculus r is the rate constant in the mean (Eq. 9) of $N(t)$. The mean of a lognormal random variable reflects the skewness of the distribution. The small but real possibility of large values of $N(t)$ has a strong upward influence on the mean, and so the lognormal shape parameter σ^2 appears in Eq. 9 for the mean. The estimates of r presented here (Eqs. 57 and 58) thus depend on the information available for estimating σ^2 (i.e., q) as well as μ (i.e., t_q).

Finite rate of increase

We define the finite rate of increase, denoted λ, as follows:

$$\lambda = \exp(r) = \psi(1; n_0, \mu, \sigma^2)/n_0$$
$$= \exp[\mu + (\sigma^2/2)]. \tag{62}$$

It is the mean population size after 1 yr (Eq. 9) divided by the initial size. The quantity approximates the dominant eigenvalue of the average projection matrix of the population, when the stochastic projection matrices are serially uncorrelated (see Eq. 3). More generally, λ is simply a positive constant appearing when the SDE (Eq. 12) is rewritten in discretized form as a stochastic difference equation:

$$N(t + 1) = \lambda N(t)L, \tag{63}$$

where L is a lognormal random variable with a mean of 1 and a shape parameter σ^2 (i.e., $\log L \sim \text{normal}[-(\sigma^2/2), \sigma^2]$). The distributions of $N(t)$ in Eqs. 12 and 63 coincide for integer values of t. A more descriptive term for λ might be the "discrete rate of increase," but the above term is widely used in the context of deterministic growth models.

The ML estimate of λ is obtained from the definition (Eq. 62) using the ML estimates of μ and σ^2:

$$\hat{\lambda} = \exp(\hat{r}) = \exp[\hat{\mu} + (\hat{\sigma}^2/2)]. \tag{64}$$

An alternate estimate uses the unbiased $\tilde{r}$ instead of $\hat{r}$: $\exp(\tilde{r})$. Both of these are biased estimates of λ, though the bias disappears as q and t_q become large. The bias can be eliminated with a little programming effort. Shimizu and Iwase (1981; see also Shimizu 1988) studied estimation of functions in the form $\exp(a\mu + b\sigma^2)$, where a and b are constants. With the help of their results, we find that the following expression gives the UMVU estimate of λ:

$$\tilde{\lambda} = \exp(\hat{\mu})_0F_1\left(\frac{q-1}{2}; \frac{q-1}{4}\,\hat{\sigma}^2\right). \tag{65}$$

Here $_0F_1(v; z)$ is the "zero-F-one" hypergeometric function, an easily computed infinite series:

$$_0F_1(v; z) = \sum_{j=0}^{\infty} \frac{z^j}{(v)_j(j!)}, \tag{66}$$

where $(v)_j$ denotes $v(v + 1) \ldots (v + j - 1)$, with $(v)_0 = 1$. Successive terms in the series are handily calculated with a recurrence relation: writing $Q_j = z^j/[(v)_j(j!)]$, we see that $Q_{j+1} = Q_j z/[(v + j)(j + 1)]$ and $Q_0 = 1$. The terms rapidly become small, and the sum can be truncated when adding more terms produces negligible change.

Additionally, Shimizu and Iwase's (1981) results allow us to obtain the variance of $\tilde{\lambda}$:

$$\text{Var}(\tilde{\lambda}) = \lambda^2\left[\exp\left(\frac{\sigma^2}{q}\right)_0F_1\left(\frac{q-1}{2}; \frac{(q-1)^2}{4q^2}\,\sigma^4\right) - 1\right]. \tag{67}$$

The distribution of $\tilde{\lambda}$ will converge to a normal$[\lambda, \text{Var}(\tilde{\lambda})]$ distribution as q and t_q become large. Estimates of μ and σ^2 can be substituted into the variance formula (Eq. 67) for constructing confidence intervals of the form $\{\tilde{\lambda} \pm z_{\alpha/2}[\widehat{\text{Var}}(\tilde{\lambda})]^{1/2}\}$. However, the distribution of

126   BRIAN DENNIS ET AL.   Ecological Monographs
                                                                    Vol. 61, No. 2

$\tilde{\lambda}$ is skewed, and convergence to normality might be slow. The situation should be studied further with computer simulations. In the interim, we recommend instead the use of the approximate normal distribution of $\tilde{r}$ for constructing approximate $100(1 - \alpha)\%$ confidence intervals for λ:

$$\left(\exp\left\{\tilde{r} \pm z_{\alpha/2} \sqrt{\hat{\sigma}^2\left[\frac{1}{t_q} + \frac{\hat{\sigma}^2}{2(q-1)}\right]}\right\}\right). \quad (68)$$

Mean population size

The mean population size given by Eq. 9 represents the expected value of $N(t)$ at a time t. This quantity is also a function of the unknown parameters μ and σ^2, and its ML estimate becomes

$$\hat{\psi} = \psi(t; n_0, \hat{\mu}, \hat{\sigma}^2) = n_0\exp(\hat{r}t) = n_0\hat{\lambda}^t$$
$$= n_0 \exp\{[\hat{\mu} + (\hat{\sigma}^2/2)]t\}. \quad (69)$$

As usual, the ML estimate retains a small-sample bias, as does the estimate defined using $\tilde{r}$ instead of $\hat{r}$. But the function is in the form constant$\cdot\exp(a\mu + b\sigma^2)$, and Shimizu and Iwase's (1981) results can be applied directly to obtain the UMVU estimate of the mean population size:

$$\tilde{\psi} = n_0\exp(\hat{\mu}t)_0F_1\left(\frac{q-1}{2}; \frac{t(q-t)}{4} \hat{\sigma}^2\right). \quad (70)$$

The variance of this estimate is found to be

$$\text{Var}(\tilde{\psi}) = n_0^2\exp(2\mu t + \sigma^2 t)$$
$$\cdot\left[\exp\left(\frac{t^2\sigma^2}{q}\right)_0F_1\left(\frac{q-1}{2}; \frac{t^2(q-t)^2}{4q^2} \sigma^4\right) - 1\right]. \quad (71)$$

We note that Eqs. 70 and 71 reduce to 65 and 67 when $t = 1$ (except for the constant term, n_0). Approximate interval estimates of $\psi(t; n_0, \mu, \sigma^2)$ can be constructed from the fact that the distribution of $\hat{\psi}$ converges to a normal[$\psi(t; n_0, \mu, \sigma^2)$, $\text{Var}(\hat{\psi})$] distribution. But again, the quality of the distribution approximation has not been studied. Instead, the interval given by

$$\left(n_0\exp\left\{\tilde{r}t \pm z_{\alpha/2}t \sqrt{\hat{\sigma}^2\left[\frac{1}{t_q} + \frac{\hat{\sigma}^2}{2(q-1)}\right]}\right\}\right) \quad (72)$$

should provide an acceptable approximation of a $100(1 - \alpha)\%$ confidence interval for the mean population size.

Geometric mean population size

The geometric mean of $N(t)$ (and the median of the lognormal transition pdf) given by Eq. 11 is a function of just one unknown parameter, μ. Its ML estimate is

$$\hat{\beta} = \beta(t; n_0, \hat{\mu}) = n_0\exp(\hat{\mu}t). \quad (73)$$

The ever-present small-sample bias can be removed as before with Shimizu and Iwase's (1981) results. The UMVU estimate of $\beta(t; n_0, \mu)$ is found to be

$$\tilde{\beta} = n_0\exp(\hat{\mu}t)_0F_1\left(\frac{q-1}{2}; \frac{t^2}{4} \hat{\sigma}^2\right). \quad (74)$$

The variance of $\tilde{\beta}$ becomes

$$\text{Var}(\tilde{\beta}) = n_0^2\exp(2\mu t)$$
$$\cdot\left[\exp\left(\frac{t^2\sigma^2}{q}\right)_0F_1\left(\frac{q-1}{2}; \frac{t^4}{4q^4} \sigma^4\right) - 1\right]. \quad (75)$$

While approximate confidence intervals for $\beta(t; n_0, \mu)$ could be constructed from the asymptotic normal[$\beta(t; n_0, \mu)$, $\text{Var}(\tilde{\beta})$] distribution of $\tilde{\beta}$, using the exact confidence intervals available for μ (Eq. 31) is more straightforward. A $100(1 - \alpha)\%$ confidence interval for $\beta(t; n_0, \mu)$ is

$$\{n_0\exp[(\hat{\mu} \pm t_{\alpha/2,q-1}\tilde{\sigma}/\sqrt{t_q})t]\}. \quad (76)$$

Tuljapurkar (1982a) defined a quantity, denoted α, which is the (finite or discrete) growth rate of the geometric mean population size. For the diffusion process $N(t)$, that quantity is

$$\alpha = \beta(1; n_0, \mu)/n_0 = \exp(\mu). \quad (77)$$

Just as the geometric mean $\beta(t; n_0, \mu)$ characterizes "typical" sample paths of $N(t)$ better than does the mean $\psi(t; n_0, \mu, \sigma^2)$, the quantity α gives a better portrait of the growth rate of those sample paths than does λ. Eqs. 73–76 all apply directly to estimation of α, just by setting $t = 1$ and dividing by n_0 (or n_0^2 in Eq. 75). In particular,

$$\hat{\alpha} = \beta(1; n_0, \hat{\mu})/n_0 = \exp(\hat{\mu}) \quad (78)$$

is the ML estimate, and

$$\tilde{\alpha} = \exp(\hat{\mu})_0F_1\left(\frac{q-1}{2}; \frac{\hat{\sigma}^2}{4}\right) \quad (79)$$

is the UMVU estimate.

We mention that the UMVU estimates of λ and α we have calculated so far in practice (see *Examples*) differed only slightly from the respective ML estimates. The bias in the ML estimates may turn out to be negligible upon further study.

Forecasting

The essential question in forecasting is as follows: given observations $n_0, n_1, \ldots, n_q$ of the system at times $t_0, t_1, \ldots, t_q$, what is the best prediction for the state of the system at some future time $t > t_q$? We will focus on predicting the value of $X(t) = \log N(t)$, since the mean of $X(t)$ [and the geometric mean of $N(t)$] typifies the system better than the mean of $N(t)$. The Markov property of the diffusion process implies that the expected value of $X(t)$, given that $X(t_q) = x_q$, does not depend on the earlier observations $x_0, x_1, \ldots, x_{q-1}$. Also, the stationary transitions property implies that the expected value depends on the amount of time, $s = t - t_q$, elapsed since the last observation, but not otherwise on t_q. Thus we have

$$E[X(t)\,|\,X(t_q) = x_q] = \gamma(s; x_q, \mu) = x_q + \mu s. \quad (80)$$

The ML estimate of $\gamma(s; x_q, \mu)$, found by substituting $\hat\mu$, is also the UMVU estimate:

$$\hat\gamma = \gamma(s; x_q, \hat\mu) = x_q + \hat\mu s. \quad (81)$$

This is the basic predicted value of $X(t)$.

The accuracy of this prediction can be measured with the mean squared prediction error, that is, the unconditional expectation of $[X(t) - \hat\gamma]^2$ over all possible realizations of the process. The expectation is easily calculated by first conditioning on $x_0, x_1, \ldots, x_q$ (and thus on $\hat\mu$), and subsequently averaging with respect to the distribution of $\hat\mu$. The result becomes

$$
\begin{aligned}
&E[(X(t) - \hat\gamma)^2]\\
&= E\{E[(X(t) - \hat\gamma)^2\,|\,X(0) = x_0, \ldots, X(t_q) = x_q]\}\\
&= E\{E[(X(t) - \hat\gamma + (x_q + \mu s) - (x_q + \mu s))^2\,|\,\ldots\text{etc.}]\}\\
&= E\{\sigma^2 s + (\hat\mu - \mu)^2 s^2\} = \sigma^2 s\left(1 + \frac{s}{t_q}\right).
\end{aligned}
\quad (82)
$$

Provided the model is adequate, the predictor $\hat\gamma$ is the best in the sense of minimizing the mean squared error among all linear unbiased predictors.

Prediction intervals for $X(t)$ are based on the fact that $X(t) - \hat\gamma$ has a (unconditional) normal distribution with a mean of zero and a variance given by Eq. 82. The quantity $[X(t) - \hat\gamma]/[\tilde\sigma^2 s(1 + (s/t_q))]^{1/2}$ then has a Student's t distribution with $q - 1$ df. The resulting $100(1 - \alpha)\%$ prediction interval for $X(t)$ is

$$\left[\hat\gamma \pm t_{\alpha/2, q-1} \sqrt{\hat\sigma^2 s\left(1 + \frac{s}{t_q}\right)}\right]. \quad (83)$$

A $100(1 - \alpha)\%$ prediction interval for $N(t)$ is provided by transforming the endpoints of the interval (Eq. 83) with the exponential function $[\exp(\cdot)]$.

Estimating Extinction Parameters

Probability of extinction

The probability (Eq. 15) of attaining a lower threshold is, like other extinction-related quantities, a function of the two unknown parameters, μ and σ^2. Its ML estimate is

$$
\begin{aligned}
\hat\pi &= \pi(x_d, \hat\mu, \hat\sigma^2)\\
&= \begin{cases} 1, & \hat\mu \le 0;\\ \exp(-2\hat\mu x_d/\hat\sigma^2), & \hat\mu > 0. \end{cases}
\end{aligned}
\quad (84)
$$

If the estimated probability is <1 (that is, if $\hat\mu > 0$), plotting it as a function of threshold population size is sometimes informative. The most recent population size n_q can be taken as the "initial" size by letting $x_d = \log(n_q/n_e)$ in Eq. 84 (see Eq. 14). Then

$$\hat\pi = (n_e/n_q)^{2\hat\mu/\hat\sigma^2}, \quad (85)$$

and this estimated probability of reaching n_e from n_q can be plotted as a function of n_e (or, perhaps $n_q - n_e$)

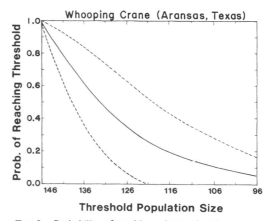

FIG. 2. Probability of reaching a lower threshold population size plotted as a function of threshold size, as estimated for the Aransas/Wood Buffalo population of Whooping Cranes. Solid line represents maximum likelihood estimate; dashed lines delimit 95% confidence interval.

to study how fast the probability decreases as n_e decreases (Fig. 2).

Expressions for variances of extinction-parameter estimates must at this time be based on approximation results. We relied on the δ method for obtaining such expressions (see Rice 1988:142, Serfling 1980:117). The method essentially involves linearizing the parameter function using a Taylor series expansion. The quality of the approximation varies from function to function, and conducting large-scale simulation studies of these results is an important task for further research. In the meantime, the δ method provides working expressions that must be regarded somewhat cautiously.

Alternative variance estimates could be calculated through bootstrapping or jackknifing (Efron 1982, Lele 1991), or through Monte Carlo simulation; we are studying these possibilities. One method of calculating bootstrap confidence intervals involves using the estimated distributions of $\hat\mu$ and $\hat\sigma^2$. From Eqs. 29 and 30, the estimated distribution of $\hat\mu$ is a normal($\hat\mu$, $\hat\sigma^2/t_q$) distribution, and the estimated distribution of $\hat\sigma^2$ is that of a chi-square($q - 1$) random variable multiplied by $\hat\sigma^2/q$. One generates a pair, $\hat\mu_B$, and $\hat\sigma^2_B$, from these distributions and then calculates the parameter of interest, say $\hat\pi_B$, with the pair only if $\hat\mu_B$ and $\hat\sigma^2_B$ both fall within their respective $100(1 - \alpha)\%$ confidence intervals (Eqs. 31 and 32). One repeats the process hundreds of times or more; the resulting set of values of $\hat\pi_B$ range across a $100(1 - \alpha)\%$ confidence interval for $\pi(x_d, \mu, \sigma^2)$. This bootstrapping method is illustrated below in connection with the distribution of extinction time. We mention that the proper observations for deletion in jackknifing are transitions, according to the theory of estimating equations (Lele 1991; discussed in the context of stochastic population models by Dennis 1989b).

Also worth mentioning is the approach of computing profile likelihoods and associated interval estimates (e.g., Kalbfleisch 1986), which bypasses the problem of variance estimation altogether.

We find, in general, the extinction parameters discussed in this section to be more poorly estimated than the growth parameters of the previous section, in that estimated variances calculated for practical examples tend to be extremely large (see *Examples* section). There are various possible explanations for this. First, time series data may intrinsically provide less information for estimating extinction parameters than for growth parameters. Second, when $\mu = 0$, the Wiener-drift process possesses so-called null recurrent behavior; although the probability of ultimately attaining a threshold is 1, the mean time to attain the threshold is infinite (see Eq. 18). Examples we have analyzed typically have estimates of μ near 0, with consequent poor performance of estimates of quantities related to the inverse Gaussian distribution. Finally, the asymptotic properties of the δ-method variance estimates may be poor, though our preliminary results with bootstrapping and jackknifing (not reported here) have yielded comparably sized variance estimates.

The approximate variance of $\hat{\pi}$, when the true value of μ is positive, is found to be

$$\text{Var}(\hat{\pi}) \approx [2x_d\pi(x_d, \mu, \sigma^2)]^2$$
$$\cdot \left\{ \frac{1}{\sigma^2 t_q} + \frac{2\mu^2(q-1)}{(\sigma^2 q)^2} \right\}. \quad (86)$$

The distribution of $\hat{\pi}$ will converge to a normal distribution with a mean of $\pi(x_d, \mu, \sigma^2)$ and a variance given by Eq. 86, provided $\mu > 0$. However, an asymptotic confidence interval for $\pi(x_d, \mu, \sigma^2)$ based on transforming a confidence interval for the quantity $2\mu x_d/\sigma^2$ in the exponent of Eq. 84 is probably better. Accordingly, an approximate $100(1 - \alpha)\%$ confidence interval for $\pi(x_d, \mu, \sigma^2)$, assuming $\mu > 0$, would take the form

$$\exp\left(-(2\hat{\mu}x_d/\hat{\sigma}^2) \pm z_{\alpha/2}\sqrt{\widehat{\text{Var}}(2\hat{\mu}x_d/\hat{\sigma}^2)}\right). \quad (87)$$

Here $\text{Var}(2\hat{\mu}x_d/\hat{\sigma}^2) = (4x_d^2/\sigma^2)[(1/t_q) + (2(q-1)\mu^2/(q^2\sigma^2))]$ is the approximate variance of $2\hat{\mu}x_d/\hat{\sigma}^2$, and $\widehat{\text{Var}}$ denotes evaluating the variance at the ML estimates of μ and σ^2. The upper bound of the confidence interval can be taken as 1 when the computed Expression (87) exceeds this value.

Distribution of extinction time

Given that the threshold will be attained, the probability of attaining it before a fixed time t is given by the cdf (Eq. 16) of the inverse Gaussian distribution. This probability is a function of μ and σ^2. Its ML estimate becomes

$$\hat{G} = G(t; x_d, \hat{\mu}, \hat{\sigma}^2). \quad (88)$$

We point out here that the Expression (16) for the inverse Gaussian cdf in terms of the standard normal cdf, while mathematically correct, can be numerically difficult to evaluate. The problem is numerical overflow of the exponential function [$\exp(\cdot)$] in the second term of the expression. We provide in the *Appendix* an easily programmed algorithm for calculating the inverse Gaussian cdf. Evaluating the cdf at $\hat{\mu}$ and $\hat{\sigma}^2$ then yields the ML estimate (Eq. 88).

Since $G(t; x_d, \mu, \sigma^2)$ is always between 0 and 1, we recommend first constructing a confidence interval for the (logit transform) quantity $H(t; x_d, \mu, \sigma^2) = \log\{G(t; x_d, \mu, \sigma^2)/[1 - G(t; x_d, \mu, \sigma^2)]\}$ and then back-transforming to obtain a confidence interval for $G(t; x_d, \mu, \sigma^2)$. According to the δ method, the large-sample variance of $H(t; x_d, \hat{\mu}, \hat{\sigma}^2)$ is

$$\text{Var}(H(t; x_d, \hat{\mu}, \hat{\sigma}^2))$$
$$\approx \text{Var}(\hat{\mu})(\partial H/\partial\mu)^2 + \text{Var}(\hat{\sigma}^2)(\partial H/\partial\sigma^2)^2$$
$$= (\sigma^2/t_q)\left(\frac{\partial G/\partial\mu}{G(1-G)}\right)^2$$
$$+ 2(q-1)(\sigma^2/q)^2\left(\frac{\partial G/\partial\sigma^2}{G(1-G)}\right)^2. \quad (89)$$

This quantity can be estimated by substituting the ML estimates $\hat{\mu}$ and $\hat{\sigma}^2$. The derivatives in Eq. 89 can be evaluated numerically, e.g., $\partial G/\partial\mu \approx [G(t; x_d, \mu + \epsilon, \sigma^2) - G(t; x_d, \mu, \sigma^2)]/\epsilon$, etc., for some small number ϵ. With $\widehat{\text{Var}}(H(t; x_d, \hat{\mu}, \hat{\sigma}^2))$ denoting the estimated variance, an approximate $100(1 - \alpha)\%$ confidence interval for $G(t; x_d, \mu, \sigma^2)$ becomes

$$\left[1 + \exp\left(-H(t; x_d, \hat{\mu}, \hat{\sigma}^2) \pm z_{\alpha/2}\sqrt{\widehat{\text{Var}}(H(t; x_d, \hat{\mu}, \hat{\sigma}^2))}\right)\right]^{-1}. \quad (90)$$

As an alternative, obtaining bootstrap confidence intervals using the method discussed earlier (*Estimating extinction parameters: Probability of extinction*) is straightforward. One can calculate the whole function $G(t; x_d, \hat{\mu}_B, \hat{\sigma}^2_B)$, that is, calculate G for a whole range of values of t, for each bootstrap pair $\hat{\mu}_B, \hat{\sigma}^2_B$. The resulting hundreds of functions of t, when plotted on one graph, will shade in a confidence region for $G(t; x_d, \mu, \sigma^2)$.

Plotting the estimated cdf (Eq. 88) and associated confidence intervals/regions as functions of t reveals how the cdf estimate becomes extremely uncertain as t becomes large, if the number of observations is small (Fig. 3). Though the confidence region as estimated for the California Condor is enormous, Fig. 3 nonetheless contains useful information. For instance, it indicates that, in 1980, the probability of the population declining to one bird within 5 yr could have been higher than 0.15.

Mean time to extinction

The mean time (Eq. 18) to reach the threshold, given the threshold is reached, is a function of just one unknown parameter, μ. Its ML estimate is

$$\theta = \theta(x_d, \hat{\mu}) = x_d/|\hat{\mu}|. \qquad (91)$$

Interestingly, when μ is positive, so that attaining the lower threshold is not certain, the mean time to reach the threshold becomes smaller as μ becomes larger (Lande and Orzack 1988). This is because any sample paths that cross the lower threshold usually do so quickly when the growth rate of the process is high. A small estimated mean time to extinction can thus occur in a rapidly growing population as well as a rapidly declining one. The difference between these two situations will be reflected in the estimated probability (Eq. 84) of reaching the threshold. While a negative estimate of μ predicts certain attainment of the threshold, a large (relative to $\hat{\sigma}^2$) positive estimate of μ predicts that attaining the threshold at all is an unlikely event.

The mean and the variance of $\hat{\theta}$ are infinite (i.e., do not exist). However, the distribution of $\hat{\theta}$ converges to a normal distribution, which possesses a mean and a variance. The mean of the asymptotic normal distribution is $\theta(x_d, \mu)$, and the variance is given by the δ method:

$$\mathrm{Var}(\hat{\theta}) \approx x_d^2 \sigma^2/(\mu^4 t_q). \qquad (92)$$

Thus, probability statements about the estimated mean time to extinction can be approximated with calculations involving a normal distribution, even though the estimate itself has a distribution without moments. In particular, an approximate $100(1 - \alpha)\%$ confidence interval for $\theta(x_d, \mu)$ would be

$$\left(\hat{\theta} \pm z_{\alpha/2}\sqrt{\widehat{\mathrm{Var}(\hat{\theta})}}\right). \qquad (93)$$

Median time to extinction

The median of the highly skewed inverse Gaussian distribution is probably a more representative measure of central tendency. The mean is inflated by the rare sample paths of $N(t)$ that take enormous amounts of time to reach the threshold. The median, denoted $\xi_{0.5}(x_d, \mu, \sigma^2)$, is defined as a root of a nonlinear equation (Eq. 20); the implicit function theorem (e.g., Rudin 1964:195) guarantees that the median is locally a differentiable function of μ and σ^2. The ML estimate of the median would be found as a root of

$$G(\hat{\xi}_{0.5}; x_d, \hat{\mu}, \hat{\sigma}^2) - 0.5 = 0. \qquad (94)$$

This equation must be solved numerically for $\hat{\xi}_{0.5}$ by using an iterative procedure such as Newton's method (see Press et al. 1986:240), coupled with a subroutine for evaluating the inverse Gaussian cdf (*Appendix*). A simple plot of $G(t; x_d, \hat{\mu}, \hat{\sigma}^2)$ over a range of values of t affords easy selection of a "close" starting value for the iterations.

The asymptotic variance of $\hat{\xi}_{0.5}$ requires some additional but straightforward computing. The δ method yields

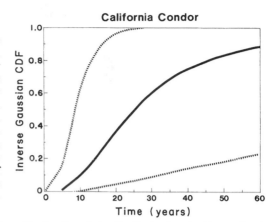

California Condor

FIG. 3. Cumulative distribution function (CDF) of the inverse Gaussian distribution, plotted using maximum likelihood estimates of μ and σ^2 for the California Condor. The distribution is that of the time required for 12 birds to decline to 1 bird. Solid line represents maximum likelihood estimate; dashed lines delimit 95% confidence region calculated with a bootstrap method.

$$\mathrm{Var}(\hat{\xi}_{0.5}) \approx (\sigma^2/t_q)\left(\frac{\partial G/\partial \mu}{g(\xi_{0.5}; x_d, \mu, \sigma^2)}\right)^2$$
$$+ 2(q - 1)(\sigma^2/q)^2\left(\frac{\partial G/\partial \sigma^2}{g(\xi_{0.5}; x_d, \mu, \sigma^2)}\right)^2. \qquad (95)$$

Here $g(\cdot; \cdot, \cdot, \cdot)$ is the inverse Gaussian pdf (Eq. 17); the derivatives can be calculated numerically. The approximate $100(1 - \alpha)\%$ confidence interval based on the asymptotic normal distribution of $\hat{\xi}_{0.5}$ is constructed in the usual manner:

$$\left[\hat{\xi}_{0.5} \pm z_{\alpha/2}\sqrt{\widehat{\mathrm{Var}(\hat{\xi}_{0.5})}}\right]. \qquad (96)$$

All quantities in $\widehat{\mathrm{Var}}(\hat{\xi}_{0.5})$ should be evaluated at $\hat{\xi}_{0.5}$, $\hat{\mu}$, and $\hat{\sigma}^2$.

ML estimates for other percentiles of the inverse Gaussian distribution are calculated as above, using the equation

$$G(\hat{\xi}_p; x_d, \hat{\mu}, \hat{\sigma}^2) - p = 0 \qquad (97)$$

to obtain the $100p$th percentile.

Most likely extinction time

Estimating the location of the mode of the inverse Gaussian distribution is also of interest, since the mode represents the most likely time that the threshold will be attained. From Eq. 21, the ML estimate of the mode is

$$\hat{t}^* = t^*(x_d, \hat{\mu}, \hat{\sigma}^2)$$
$$= \frac{x_d}{|\hat{\mu}|}\left\{\left[1 + \frac{9}{4\hat{\nu}^2}\right]^{1/2} - \frac{3}{2\hat{\nu}}\right\}, \qquad (98)$$

where $\hat{\nu} = x_d|\hat{\mu}|/\hat{\sigma}^2$. The estimate corresponds to the inflection point of the estimated inverse Gaussian cdf

130	BRIAN DENNIS ET AL.	Ecological Monographs
Vol. 61, No. 2

(Eq. 16). The variance of the asymptotic normal distribution of $\hat{t}^*$, given by the δ method, is

$$\mathrm{Var}(\hat{t}^*) \approx (\sigma^2/t_q)(\partial t^*/\partial \mu)^2$$
$$+ 2(q-1)(\sigma^2/q)^2(\partial t^*/\partial \sigma^2)^2. \quad (99)$$

When estimating this variance, the derivatives can be calculated numerically; or, an enterprising investigator might prefer to obtain analytical expressions. In either case, an approximate $100(1 - \alpha)\%$ confidence interval for $t^*(x_d, \mu, \sigma^2)$ becomes

$$\left[\hat{t}^* \pm z_{\alpha/2}\sqrt{\widehat{\mathrm{Var}(\hat{t}^*)}}\right]. \quad (100)$$

Examples

We selected seven species for detailed analyses, primarily because long-term population estimates were available for each species. Though six of the species are birds, a variety of ecological conditions and life history strategies are represented. The situations range from a species now extirpated from the wild (the California Condor) to a once extremely threatened species now undergoing a promising recovery (the Whooping Crane). In this section, we briefly review some relevant biological aspects and management efforts for each species, and we discuss estimates and predictions resulting from fitting the model.

Whooping Crane

The Whooping Crane has been the subject of protection efforts by the National Audubon Society and the Federal governments of Canada and the United States of America for more than half a century. The Whooping Crane is a long-lived bird that stands 1.5 m tall with a wingspan of 2.1 m. It becomes sexually mature on average at 5 yr of age and normally lays two eggs per clutch. Usually only one chick is raised to fledging age. One viable wild population exists, which breeds in Wood Buffalo National Park in northwestern Canada and winters at the Aransas National Wildlife Refuge on the gulf coast of Texas. As a result of major research and management programs, this population increased from 18 birds in 1938 to 146 birds in 1989. Efforts to recover the Aransas/Wood Buffalo flock include legal protection, manipulation of hunting seasons for Snow Geese and Sandhill Cranes, egg manipulation, habitat protection, and habitat improvement (USFWS 1986). Predator control programs have also been undertaken at the Grays Lake, Idaho summering area (USFWS 1986). Attempts to create a second wild flock at Grays Lake by placing Whooping Crane eggs in Sandhill Crane nests have been unsuccessful, because no breeding has occurred, and flock mortality is unusually high in comparison to the Aransas/Wood Buffalo population.

An annual census of wintering Whooping Cranes commenced in 1938 and has continued to the present (Fig. 4). The data we analyzed consist of combined annual counts of young birds and birds with adult plumage, as presented by Boyce (1987) and supplemented by recent counts (J. Lewis, *personal communication*). An overall trend of exponential growth, with fluctuations, is evident in the data.

Model parameter estimates indicate a favorable outlook for this population (Tables 1 and 2). We present extinction parameter estimates corresponding to threshold populations of $n_e = 100$ and $n_e = 10$. Thresholds for this and other populations were selected with the idea that management efforts might be planned usefully around the risks of dropping 1 and 2 orders of magnitude (e.g., from 3-digit abundances to 2, from 3 to 1, etc.). The estimates presented here were calculated with the 1940–1941 transition omitted from the data, since the decrease of 26 to 16 birds is a significant outlier ($I_j = -3.75$, $c_{0.05} = 3.2$, $E_j = -4.4$, $j = 3$) according to the outlier procedures (Eqs. 35 and 37). That transition has an inordinate influence on the parameter estimates if included in the data; the UMVU estimate of λ, for instance, decreases from 1.061 to 1.052.

Model diagnostic procedures, once the outlier transition is removed, reveal no significant additional outliers, influential observations, or first-order autocorrelation among residuals (the outlier test [Eq. 35] and the Durbin-Watson test [Eq. 34] were conducted here, and in all examples to follow, at a significance level of $\alpha = .05$). However, a spectral analysis of the residuals confirms a 10-yr cycle in the data reported by Boyce (1987), Boyce and Miller (1985), and Nedelman et al. (1987). Further refinement of the model predictions might be possible by using a time-dependent infinitesimal mean, $\mu(t)$, in the Wiener-drift process (perhaps a sine wave); we are currently developing such an approach.

We also find no evidence using the slope-change test (Eq. 51) that the value of μ changed starting with the 1957–1958 transition ($T_{48} = -0.1804$, $P = .86$). Binkley and Miller (1988) fitted a model in the form log $N(t) = \log N(0) + \mu t + \epsilon$ using ordinary least squares regression, corrected for an error structure having first-order serial correlation. They identified 1957 as a year in which the population started to recover more rapidly, because their analysis indicated a shift in μ occurred. Their model is similar to ours in that $E[\log N(t)]$ is a linear function of time in both models. The "stochastic process" aspects of our model are more explicit, though, allowing for a connection to the stochastic theory of age-structured populations and for estimation of extinction-related quantities.

The Whooping Crane recovery plan (USFWS 1986) calls for downlisting the Whooping Crane from endangered to threatened status when (among other things) 40 nesting pairs are attained in this population. Binkley and Miller (1988) estimate, using a survivorship analysis, that 40 nesting pairs would correspond to a total population of 153 birds.

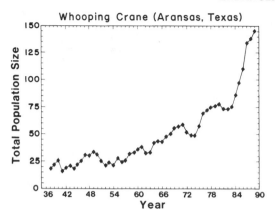

FIG. 4. Total size of the Aransas/Wood Buffalo Whooping Crane population, from 1938–1988. Data are from Boyce (1987), supplemented by more recent counts.

In order to compare Whooping Crane recovery forecasts from Binkley and Miller's (1988) model and our own, we refitted our model to the Whooping Crane data using observations only through 1986, that is, only those observations available to Binkley and Miller. The ML estimates of 4.884×10^{-2} and 1.491×10^{-2} for μ and σ^2, respectively, resulted after deleting the 1940–

1941 transition. According to the ML estimate of the mean time to reach 153 birds from 110 birds (Table 2), the population is expected under our model to attain the recovery threshold by 1993. The prediction is more optimistic than Binkley and Miller's (1988) prediction of 1997. Their estimate arises from projecting the estimated mean population size (in their model) until it hits 153. We note that in stochastic processes, the time until the mean population size hits a threshold is not generally equal to the mean time until the population hits the threshold. Also, the estimated inverse Gaussian distribution here is somewhat skewed; the ML estimates of the 5th, 50th (median), and 95th percentiles are, respectively, 1.31, 4.68, and 19.3 yr (starting from 1986). We note that the population has been progressing rapidly toward the threshold of 153 birds since 1986 (Fig. 4).

Grizzly bear

The grizzly bear was formerly found throughout most of western North America. Within the contiguous 48 states of the United States, it presently occurs in six populations (USFWS 1982, Allendorf and Servheen 1986). The numbers of bears in these populations range from fewer than 10 apiece in the North Cascade and Selway-Bitterroot populations, to 440–680 in the northern continental divide population. A 12-yr mark-

TABLE 1. Estimated growth parameters for the Whooping Crane (WC), grizzly bear (GB), Kirtland's Warbler (KW), California Condor (CC), Puerto Rican Parrot (PP), Palila (PA), and Laysan Finch (LF). Numbers in parentheses are 95% confidence limits.

Animal	$\hat{\mu}$*	$\tilde{\sigma}^2$†	$\tilde{r}$‡	$\tilde{\lambda}$§	$\tilde{\alpha}$‖	q¶	t_q (yr)#
WC**	5.157×10^{-2}	1.475×10^{-2}	5.895×10^{-2}	1.061	1.053	50	50
	(1.705×10^{-2})	(1.029×10^{-2})	(2.516×10^{-2})	(1.025)	(1.017)		
	(8.610×10^{-2})	(2.291×10^{-2})	(9.274×10^{-2})	(1.097)	(1.090)		
GB††	-7.493×10^{-3}	8.919×10^{-3}	-3.034×10^{-3}	0.9968	0.9927	27	27
	(-4.486×10^{-2})	(5.531×10^{-3})	(-3.874×10^{-2})	(0.9620)	(0.9561)		
	(2.987×10^{-2})	(1.675×10^{-2})	(3.267×10^{-2})	(1.033)	(1.030)		
KW	-1.873×10^{-2}	1.673×10^{-2}	-1.037×10^{-2}	0.9893	0.9819	20	38
	(-6.264×10^{-2})	(9.674×10^{-3})	(5.183×10^{-2})	(0.9495)	(0.9393)		
	(2.518×10^{-2})	(3.568×10^{-2})	(3.109×10^{-2})	(1.032)	(1.025)		
CC	-7.685×10^{-2}	0.1199	-1.688×10^{-2}	0.9792	0.9297	15	15
	(-0.2686)	(6.429×10^{-2})	(-0.1977)	(0.8206)	(0.7644)		
	(0.1150)	(0.2983)	(0.1639)	(1.1781)	(1.1218)		
PP‡‡	3.400×10^{-2}	1.341×10^{-2}	4.071×10^{-2}	1.041	1.035	20	20
	(-2.020×10^{-2})	(7.757×10^{-3})	(-1.023×10^{-2})	(0.9898)	(0.9800)		
	(8.820×10^{-2})	(2.861×10^{-2})	(9.165×10^{-2})	(1.096)	(1.092)		
PA	7.732×10^{-2}	0.2190	0.1868	1.190	1.094	9	13
	(-0.2220)	(9.991×10^{-2})	(-8.928×10^{-2})	(0.9146)	(0.8009)		
	(0.3766)	(0.8036)	(0.4629)	(1.589)	(1.457)		
LF	-1.058×10^{-3}	0.3662	0.1820	1.189	1.007	24	22.38
	(-0.2657)	(0.2212)	(-9.010×10^{-2})	(0.9138)	(0.7666)		
	(0.2636)	(0.7206)	(0.4542)	(1.575)	(1.302)		

* $\hat{\mu}$ = ML (maximum likelihood) estimate of diffusion process drift parameter μ (Eq. 24).
† $\tilde{\sigma}^2$ = unbiased estimate of diffusion process variance parameter σ^2 (Eq. 26).
‡ $\tilde{r}$ = UMVU (uniformly minimum variance unbiased) estimate of continuous rate of increase r (Eq. 58).
§ $\tilde{\lambda}$ = UMVU estimate of finite rate of increase λ (Eq. 65).
‖ $\tilde{\alpha}$ = UMVU estimate of geometric finite rate of increase α (Eq. 79).
¶ q = number of transitions in data set.
# t_q = length of time population has been observed.
** 1940–1941 transition deleted.
†† 1983–1984 transition deleted.
‡‡ 1972–1972 transition deleted.

132   BRIAN DENNIS ET AL.   Ecological Monographs
                                                                                          Vol. 61, No. 2

TABLE 2.   Estimated extinction parameters for the Whooping Crane (WC), grizzly bear (GB), Kirtland's Warbler (KW), California Condor (CC), Puerto Rican Parrot (PP), Palila (PA), and Laysan Finch (LF), calculated using (maximum likelihood) growth parameter estimates from Table 1. Numbers in parentheses are 95% confidence limits.

Animal	n_q*	$n_e(n_u)$†	$\hat{\pi}$‡	$\hat{\theta}$§ (yr)	$\hat{\xi}_{0.5}$‖ (yr)	$\hat{t}^*$¶ (yr)
WC	146	100	6.72×10^{-2}	7.34	5.41	2.82
			(0.00)	(2.60)	(2.76)	(1.88)
			(0.204)	(12.08)	(8.05)	(3.75)
		10	4.90×10^{-9}	52.0	49.4	44.5
			(0.00)	(18.4)	(19.0)	(20.0)
			(7.56×10^{-8})	(85.6)	(79.8)	(68.9)
	110#	(153)	1.00	6.76	4.68	2.18
			⋯	(1.93)	(2.27)	(1.41)
			⋯	(11.6)	(7.08)	(2.95)
GB	47	10	1.00	207	152	79.3
			⋯	(0.00)	(0.00)	(0.00)
			⋯	(1170)	(679)	(179)
		1	1.00	514	448	333
			⋯	(0.00)	(0.00)	(0.00)
			⋯	(2910)	(2280)	(1260)
KW	212	100	1.00	40.1	26.0	11.0
			⋯	(0.00)	(0.00)	(4.39)
			⋯	(126)	(63.1)	(17.5)
		10	1.00	163	143	109
			⋯	(0.00)	(0.00)	(0.00)
			⋯	(512)	(414)	(254)
CC	12	10	1.00	2.37	0.506	9.88×10^{-2}
			⋯	(0.00)	(0.138)	(3.07×10^{-2})
			⋯	(7.60)	(0.875)	(0.167)
		1	1.00	32.3	25.2	14.6
			⋯	(0.00)	(0.00)	(1.80)
			⋯	(104)	(68.8)	(27.4)
PP	38	10	8.04×10^{-4}	39.3	34.5	26.1
			(0.00)	(0.00)	(0.00)	(2.07)
			(9.83×10^{-3})	(96.4)	(78.8)	(50.1)
		1	3.70×10^{-9}	107	102	91.7
			(0.00)	(0.00)	(0.00)	(0.00)
			(1.17×10^{-7})	(263)	(243)	(205)
	23	10	1.17×10^{-2}	24.5	20.1	13.0
			(0.00)	(0.00)	(0.00)	(3.58)
			(9.39×10^{-2})	(60.1)	(44.2)	(22.5)
PA	4358	1000	0.311	19.0	10.5	3.58
			(0.00)	(0.00)	(0.00)	(0.00)
			(1.00)	(78.1)	(29.3)	(6.58)
		100	4.98×10^{-2}	48.8	36.9	20.2
			(0.00)	(0.00)	(0.00)	(0.00)
			(0.531)	(200)	(124)	(42.4)
LF	9349	1000	1.00	2110	30.8	4.75
			⋯	(0.00)	(0.00)	(2.12)
			⋯	(4.92×10^5)	(143)	(7.37)
		100	1.00	4290	125	19.6
			⋯	(0.000)	(0.000)	(8.73)
			⋯	(9.99×10^5)	(1020)	(30.4)

* n_q = starting population size.
† n_e = threshold population size.
‡ $\hat{\pi}$ = ML (maximum likelihood) estimate of probability of attaining threshold (Eq. 84).
§ $\hat{\theta}$ = ML estimate of (conditional) mean time to reach threshold (Eq. 91).
‖ $\hat{\xi}_{0.5}$ = ML estimate of (conditional) median time to reach threshold.
¶ $\hat{t}^*$ = ML estimate of (conditional) most likely time to reach threshold.
# No observations used after 1986.

recapture study of the Yellowstone population gave a peak population size estimate of 245 bears in 1967 (Craighead et al. 1974). The isolated Yellowstone population is the subject of our analysis.

Since the Craighead et al. (1974) study, reliable, pe-riodic estimates of total population size have not been available. However, Knight and Eberhardt (1984, 1985) and Eberhardt et al. (1986) devised an estimate of the minimal number of fully adult females in the population during the 1st yr of any 3-yr period. The estimate

Abundance and the Rise of Conservation Ecology

645

is simply a running 3-yr sum of the numbers of female grizzlies observed with cubs. The estimate is based on two points: females with cubs are probably the most observable segment of the population in aerial surveys, and the breeding interval is at least 3 yr. The resulting time series of these estimates, calculated from data published by Eberhardt et al. (1986) and from more recent unpublished data (R. R. Knight, *personal communication*), shows substantial fluctuations (Fig. 5). We mention that the Tuljapurkar and Orzack (1980) results, giving an approximate normal transition distribution to log $N(t)$, apply in general to any linear combinations of age or stage classes. Thus, segments of a population can be analyzed with the model.

Model parameter estimates for this set of data suggest that the Yellowstone grizzly population is doomed to extinction, though not in our lifetimes (Tables 1 and 2). The confidence interval for μ contains positive values (Table 1), indicating that such an interval estimate of the extinction probability would contain values <1. However, the high value of σ^2 creates a large estimated chance of extinction even if μ is slightly positive. The estimated distribution of the time required to reach a threshold of 10 bears, portrayed earlier in Fig. 1, attaches nontrivial likelihood to times ranging over hundreds of years. A level of 10 adult female bears would represent, if not extinction, a serious failure of management and preservation efforts.

The estimates given in Table 1 were calculated with the 1983–1984 transition deleted. In 1986, the observed number of females with cubs in the population jumped from 9 to 25 (R. R. Knight, *personal communication*). This large increase caused an anomalous increase in the 1983–1984 transition (the 3-yr sum jumps from 39 to 51). The residual screening procedures tag the transition as a suspect outlier; specifically, for that transition $E_j = 2.9$ (see Eq. 37), a value considerably larger than the values for remaining transitions. If the transition is included in the analysis, the estimate of μ becomes slightly positive ($\hat{\mu} = 0.002356$), the estimate of σ^2 becomes larger ($\hat{\sigma}^2 = 0.01130$), and the estimated probability of reaching 10 bears, though smaller, is still high enough to cause concern ($\hat{\pi} = 0.51$). Preservation strategies that bank upon such fortuitous increases in population size are not likely to inspire confidence.

Diagnostic indicators are otherwise acceptable, once the 1983–1984 transition is omitted. No significant outliers are present, according to the outlier test (Eq. 35), and the influence measures, Eqs. 38, 39, suggest there are no highly influential transitions. The test for first-order autocorrelation (Eq. 34) is inconclusive.

Closure of garbage dumps in Yellowstone National Park during 1970–1971 was hypothesized to have substantial negative effects on the grizzly population. Such effects have been reported for mortality and other demographic parameters (Knight and Eberhardt 1984). We find no evidence of a change in μ starting with the

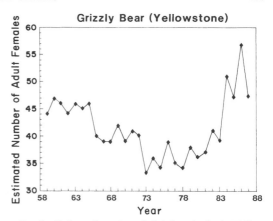

FIG. 5. Estimated number of adult females in the Yellowstone National Park grizzly bear population, 1959–1987. Data, listed by Eberhardt et al. (1986) and supplemented by recent figures, consist of a 3-yr moving sum of the yearly number of adult females seen with cubs.

1972–1973 transition ($T_{25} = 0.00838$, $P = .99$), using the slope-change test (Eq. 51). However, we do find evidence of a change in σ^2 ($F_{11,13} = 0.158$, $P < .01$), using the variance-change test (Eq. 56). In fitting the two models, the 1965–1966 transition was tagged as an outlier ($|E_j| > 2.9$) and deleted from the analysis in addition to the 1983–1984 transition. Results thus indicate greater variability in the female population size index commencing around the time of the dump closures. Two possible explanations for this variability increase are: (a) without access to the stable food supply provided by the dumps, the bears may be more influenced by fluctuations in important wild food sources such as whitebark pine nuts and ungulate carcasses; (b) the method of data collection changed from observations at dumps (Craighead et al. 1974) to aerial observations (Knight and Eberhardt 1985), which may have resulted in decreased precision. It will be interesting to determine, after more years of data accumulate, if any post-1988 fire effects can be detected with the model.

Several stochastic, age-structured simulation models of the Yellowstone grizzly population have been constructed (Shaffer 1978, Knight and Eberhardt 1984, 1985, Shaffer and Samson 1985, Suchy et al. 1985, Eberhardt et al. 1986). As best we can ascertain, these models conform to the Tuljapurkar and Orzack (1980) criteria, and the log-transformed total population size projected by these models should have the approximate statistical characteristics of a Wiener-drift process. A standing research problem is to study, using extensive simulations, how well the Wiener-drift process (and associated statistical inferences) portrays the output of these detailed projection models.

Kirtland's Warbler

The population of the Kirtland's Warbler has fluctuated around 200 singing males since 1971, after a

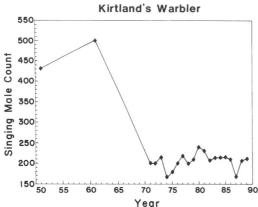

FIG. 6. Total count of Kirtland's Warbler singing males, 1951–1989. Data are from Walkinshaw (1983), supplemented by more recent counts.

precipitous population decline occurred sometime during the 1960s (Fig. 6; data from Walkinshaw 1983, and C. B. Kepler, *personal communication*). The decline triggered yearly monitoring counts of singing males in the species' breeding grounds in Michigan, which have continued to the present. Major recovery efforts on behalf of this species have included legal protection of the bird and its habitat, habitat improvement, and control of nest parasites. Limiting factors other than cowbird parasitism have not been well documented (USFWS 1985).

The model parameter estimates calculated for the singing male data give little cause for optimism (Tables 1 and 2). The results indicate that the population could persist above 10 singing males for some hundreds of years, but that the fluctuations will ultimately exterminate the population. In particular, the precipitous drop of the 1960s is well within the bounds of usual variability given by the model, since the drop occurred over a relatively extended period of 10 yr. No significant outliers, influential transitions, or first-order autocorrelation is apparent from analysis of residuals.

California Condor

The California Condor had been regarded as one of the most endangered bird species in North America (Koford 1953, Wilbur 1980, Ogden 1985). By 1987, all the condors remaining in the wild had been taken into captivity, and a captive flock of 32 birds now exists, with 11 potential breeding pairs. It is the largest North American bird, with a wingspan of 2.7 m. It lays a single egg per clutch and is not sexually mature until 5–7 yr of age. The efforts since 1980 to save the wild condor population included nest site protection, supplemental feeding, habitat protection, and a public education program to prevent loss through shooting (USFWS 1984a). In 1986, lead poisoning from ingestion of bullets in hunter-killed carcasses was identified

as a major source of mortality (Wiemeyer et al. 1988). Future release programs, in order to have a reasonable chance of success, will have to be limited to lead-free areas where supplemental feeding can be used to minimize exposure to lead-contaminated carcasses. Present recovery efforts focus on increasing the size of the captive flock prior to any release, and on development of methods for releasing birds back into the wild.

Data collected from 1965 onward indicate an inexorable population decline throughout the 1970s (Fig. 7). These data, arising from the dubious "October surveys," are problematic (Wilbur 1980). Astonishingly, no really accurate count was undertaken until the 1980s (Snyder and Johnson 1985). Nonetheless, we take a retrospective look at the October survey data, in order to discover what conclusions could have been drawn with the model in 1980 on the basis of the admittedly poor information available. We follow Snyder and Johnson (1985) in using the maximum number in each multiday October survey as reported by Wilbur (1980) for the population time series.

According to the fitted model, extinction was imminent (Tables 1 and 2). The estimated mean time of 32 yr to decline from 12 birds to 1 bird might by itself have indicated that managers had enough time to attempt to reverse the downward trend of the species in the field. However, the estimated inverse Gaussian cdf gives the probability of extinction within 20 yr to be almost 0.4 (Fig. 3), and the most likely time of extinction to be under 15 yr. Note that the growth rate estimates $\hat{\mu}$, $\tilde{r}$, $\tilde{\lambda}$, and $\tilde{\alpha}$ suggest by themselves only a slow decline, if any; the population's fate is sealed by the high value of $\tilde{\sigma}^2$. We mention that the variance estimate incorporates variability from sampling as well as population fluctuations; results about hitting times reported here apply to the population *as estimated*. This point is developed further in the *Discussion* section. No qualifications to these pessimistic conclusions

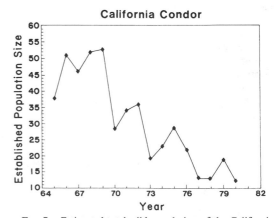

FIG. 7. Estimated total wild population of the California Condor, 1965–1980. Data are from October surveys as listed by Wilbur (1980) and Snyder and Johnson (1985).

are found in the model diagnostic procedures; no significant outliers, influential transitions, or autocorrelation can be detected. We permit ourselves a hindsight speculation that the decision to take the field population into captivity might not have been postponed so long had such estimates been considered.

Puerto Rican Parrot

The number of breeding pairs of the highly endangered Puerto Rican Parrot has not exceeded five pairs in the last 15 yr (Wiley 1985, Snyder et al. 1987, USFWS 1987). This long-lived bird exists in the wild only in the Luquillo Forest of Puerto Rico. It breeds at 2–3 yr of age and lays 2–4 eggs. The Puerto Rican Parrot is probably the most intensively managed species reviewed in this paper. Biologists protect nest sites from predators, competitors, and human disturbance. They also rebuild nest sites, excavate new sites, remove and care for young, double clutch wild birds, and repair broken bills, wings, and feathers (USFWS 1987). As a result of this intensive management, as well as the release of captive-reared animals (by replacement in nests of juveniles that have died), the population increased from 14 in 1975 to 38 in 1989 (Fig. 8). The species would almost certainly have gone extinct without the intervention.

The data we treat here consist of the largest count of adult wild birds recorded between January and April (prebreeding period) of each year, or population estimates "by reasonable inference" made by biologists on the scene when they were convinced undercounting had occurred (Fig. 8; Snyder et al. 1987, M. Wilson, *personal communication*). In 1972, 2 birds were removed from the population of 16 birds within days of the count; thus, the 1971–1972 transition is 16–16 birds, and the 1972–1973 transition is taken as 14–16 birds.

Model parameter estimates suggest the population has favorable prospects for recovery, provided the management program continues (Tables 1 and 2). We regard this prediction as unreasonably optimistic for reasons given below. The residuals from the fitted model yield no evidence of influential transitions, autocorrelation, or outliers. We tested whether any significant change in μ occurred starting with the 1982–1983 transition, since the management efforts intensified in the early 1980s. The parameter estimates differ markedly (before: $\hat{\mu}_1 = 0.01027$; after: $\hat{\mu}_2 = 0.07808$), but the difference is not statistically significant because of the large population variability and limited number of observations ($T_{18} = -1.269$, $P = .22$). Nonetheless, a substantially higher estimate of the chance of extinction results if the transitions from 1982–1983 onward are deleted [$n_e = 1$, $n_q = 22$, $\hat{\pi} = 6.59 \times 10^{-3}$, 95% CI $= (0.00, 0.21)$].

Two ecological factors not explicitly accounted for in the model give reason for pessimism about the species' chance of survival. First, the model is merely descriptive in the sense that the cause of the popula-

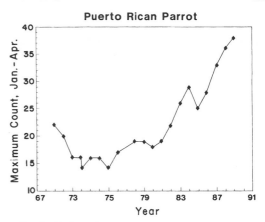

Puerto Rican Parrot

FIG. 8. Estimated total wild population of the Puerto Rican Parrot, 1969–1989. Data are from Snyder et al. (1987), supplemented by more recent counts.

tion's increase is not incorporated. This species increased because of addition of captive-reared birds, protection of juveniles, and so on; however, the breeding population has remained nearly constant. Second, environmental catastrophes are not included in the model unless such events are a regular and frequent source of variability in the time series. When this paper was being written, Hurricane Hugo had just ravaged Puerto Rico and South Carolina. A preliminary survey of the Luquillo Forest in the aftermath (M. Wilson, *personal communication*) has yielded sightings of only 8–11 birds. The birds are normally widely dispersed during the fall months, and such sightings could correspond to a population of ≈ 23 birds (M. Wilson, *personal communication*). With a starting population of 23 instead of 38, the estimated risk of declining below 10 birds increases by a factor of over 14 (Table 2). Furthermore, the values of μ and σ^2 may have changed after the storm, since extensive loss of canopy foliage cover has exposed the birds to predation from hawks. The 1989–1990 transition will no doubt be an "outlier" according to the model. The model is a supplement to, not a substitute for, basic knowledge about the population.

Palila

The Palila is a member of the Hawaiian honeycreeper family, a family known as one of the most striking examples of adaptive radiation in birds (Scott et al. 1988). This 2–2.5 cm bird weighs an average of 16 g, breeds at 1 yr of age, and lays 1–3 eggs per clutch (van Riper 1980, Berger 1981). The bird occurs only on the island of Hawaii and is restricted to portions of the upper slopes of a single mountain, Mauna Kea (Scott et al. 1984). Ongoing recovery efforts include biannual monitoring and elimination of alien species of goats and sheep, which were seriously degrading habitat on the island. Improvement of habitat quality as a result

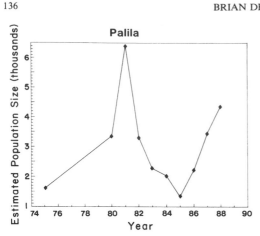

FIG. 9. Estimated total population of the Palila, 1975–1988. Data are from Scott et al. (1984), supplemented by more recent figures.

of sheep and goat removal should allow for larger population sizes more evenly distributed around the mountain. Also, plans for fire control are in place (USFWS 1978, Scott et al. 1984).

Population estimates since 1975 have varied from under 1600 to over 6000 birds (Fig. 9; Scott et al. 1984). While some of the variability is due to sampling error, fluctuations in population size account for a larger portion of the variability (Scott et al. 1984).

Model parameter estimates indicate a high propensity for population increase, but high uncertainty is present in those estimates because of the population fluctuations (Tables 1 and 2). The ML estimate of the chance of the population ever dropping below 100 birds is low, at $\approx 5\%$, but the 95% confidence interval ranges up to 50%. The chance of dropping below 1000 birds cannot be estimated with any useful degree of precision. In this situation, with a large tendency to increase but also with large stochastic fluctuations, we see the most difference between the estimates of the growth rate parameters λ and α. Recall that α discounts the small chance of the population reaching enormous sizes.

Laysan Finch

The Laysan Finch is also a Hawaiian honeycreeper. Females breed at 1 yr of age (males vary somewhat); clutch sizes range from 1 to 5 eggs and some birds produce two clutches a year (M. Morin, *personal communication*). The species is found in two populations on the Leeward islands of Hawaii in a land area of <500 ha (USFWS 1984b). Recovery efforts on behalf of this species include a translocation effort in 1967 by the United States Fish and Wildlife Service (Conant 1989), which resulted in the establishment of the Pearl and Hermes reef population. The originally small translocated population has increased to 500–700 birds. The establishment of the translocated population was an important step in reducing the chance of the Laysan

Finch becoming extinct through stochastic forces. Additional recovery efforts have included the elimination of feral rabbits from Laysan Island (in the Leeward islands). This alien species threatened to destroy all native vegetation. The islands were designated a national wildlife refuge early in this century, and formal recovery plans written more recently call for guarding against the accidental introduction of predators and nonnative organisms such as insects and weeds (USFWS 1984b). Currently, 13 Laysan Finches are in captivity.

In 22 yr of monitoring, estimates of the Leeward islands combined populations have fluctuated wildly between 5000 and 21 000 birds (Fig. 10; M. Morin, *personal communication*). As with the Palila, sampling error accounts for a part of the variability, but the fluctuations of the population itself form the largest component.

Again like the Palila, the population variability dominates the model parameter estimates (Tables 1 and 2). The extinction parameter estimates suggest that the long-term survival prospects for this species are uncertain. Though the estimated chance of dropping below 1000 birds is 1.00, great uncertainty is associated with this estimate because of the population fluctuations. More revealing is the disparity between the mean time to reach that level (over 2000 yr) and the most likely time of reaching that level (under 5 yr). Such extreme skewness of the estimated inverse Gaussian distribution indicates that explosions and catastrophes may be the rule rather than the exception in this population.

DISCUSSION

Sampling variability

When the observations in a time series are population size estimates, the data reflect sampling variability

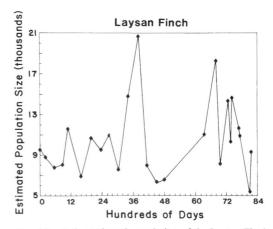

FIG. 10. Estimated total population of the Laysan Finch on Laysan Island, Hawaii, from transect samples taken at irregularly spaced time intervals, during 30 March 1969 through 16 August 1988. Data compiled by M. Morin.

June 1991   ENDANGERED SPECIES   137

as well as stochastic population fluctuations. If the sampling variation in the estimates is reasonably constant, that is, if the population sizes are estimated with relatively equal precision over time, the stochastic exponential growth model can be fitted to the series. Then the estimate of σ^2 conceptually reflects sampling variability as well as stochastic population fluctuations. The conclusions derived from fitting the model will then apply to the population (or portion of the population) as estimated, provided the model adequately describes the series. For example, in our analysis of the California Condor, $\theta(\log(12/1), \hat{\mu})$ in Table 2 is strictly described as the mean time required for the October survey population estimate to reach one bird. Such an event may or may not correspond to the population itself reaching one bird, depending on the accuracy of the population estimates.

Sampling variability can also be incorporated directly in the likelihood function for observations from a diffusion process. In general, this amounts to including an additional variance parameter. Garcia (1983), for example, included a variance component due to sampling in a diffusion model of forest growth. However, as Garcia found, one of the variance parameters becomes nearly nonidentifiable; that is, the data provide little information for its estimation. The sampling component could in principle be estimated from repeated estimates of the same population for each sampling occasion. Situations for which this is practicable are rare, though.

We mention that none of the above approaches can adequately account for sampling variation if the population sizes are estimated with widely varying precision. Thus, in analyzing the estimates of the minimal number of fully adult grizzly bear females, we are assuming that these estimates represent a reasonably constant proportion of the total population of fully adult females. This is unlikely to be the case if, for example, the aerial survey method is not standardized from year to year with respect to factors such as flying time and area covered (e.g., Seber 1982:454).

Density dependence

Perhaps the most important explicit omission in this model, and in the stochastic projection matrix formulation in general, is density dependence. Some endangered species are held in check by sheer lack of habitat. In such situations, unbridled exponential growth is hardly a likely event. The vital rates of populations near equilibrium are in balance; the fecundities and survival rates combine to produce a value of λ near 1. But to the extent that a stationary stochastic projection matrix can produce a population size appearing to fluctuate around a stable level, the diffusion model with a value of μ near 0 can describe such behavior, though not for indefinitely long periods. An example is the Laysan Finch; the population levels fluctuate but appear to have no trend up or down (Fig.

10). The Wiener-drift diffusion model adequately fits the data. The effect, however, is the introduction of large uncertainty into the parameter estimates and population predictions.

An alternate approach for modeling a population fluctuating around equilibrium is to use a diffusion process with some explicit density-dependent feedback in the infinitesimal mean. Density-dependent diffusion processes can approximate the statistical behavior of simulation models based on stochastic projection matrixes or stochastic difference equations, such as that described by Jacquez and Ginzburg (1989) and Ferson et al. (1989), or the behavior of density-dependent birth–death processes such as the model investigated by Burkey (1989). Dennis and Patil (1984) showed that an SDE version of the logistic model can serve as a general approximation for a density-dependent diffusion process having an underlying, stable equilibrium population size. Parameter estimation for this model was discussed by Dennis and Costantino (1988) and developed in more generality by Dennis (1989b); the former analyzed many examples of populations (flour beetles) fluctuating stochastically around equilibrium. Tier and Hanson (1981) derived results on the extinction (first-passage) time distribution for this logistic SDE model and others. The ramifications of using a stochastic logistic or similar model for estimating extinction-related quantities have not been investigated. Extinction parameter estimates for the Palila and the Laysan Finch might be greatly improved by such an approach.

Maximum population size

Some projection matrix models and stochastic models appearing in the conservation biology literature incorporate an upper ceiling or reflecting boundary representing a maximum population size (e.g., Suchy et al. 1985, Goodman 1987). The main effect of incorporating a ceiling is that the probability of reaching any lower boundary becomes 1; the investigations focus on properties of the extinction time (particularly the mean time to extinction). In some cases, the upper boundary is known. For instance, Suchy et al. (1985) use a value of 300 bears for the maximum grizzly population in Yellowstone, since that figure is a management target. In many cases, however, an upper boundary would have to be estimated. Such an estimate would likely take the form of an educated guess by biologists, because estimating an upper boundary from population size data in any statistically consistent fashion is extremely difficult.

We point out that extinction quantities obtained from unbounded models are often convenient approximations to quantities arising from bounded models. In the exponential growth SDE developed here, the probability of extinction (Eq. 15) is obtained as the probability of reaching x_e before reaching some upper level x_u, by taking the limit as x_u becomes large. Thus, Eq.

15 need not be regarded as implying that the population might survive eternally. Rather, Eq. 15 gives the approximate probability of reaching x_e before, say, some large upper management goal is attained. Likewise, the first-passage time, T, is approximately the waiting time required to reach x_e, given that x_e is reached before reaching the upper level.

In general, models with upper reflecting ceilings are biologically and statistically awkward. Instead, incorporation of density-dependent feedback (see above discussion, *Density dependence*) appears more promising for useful applications in conservation biology.

Types of stochastic behavior

Stochastic forces incorporated into population models have been categorized, more or less on intuitive biological grounds, as "demographic," "environmental," or "catastrophic" (May 1974, Leigh 1981, Shaffer 1981, 1987, Goodman 1987, Simberloff 1988). A fourth type of stochastic variation, "genetic" (Shaffer 1981), yields effects on population size, which would qualitatively fall into one of the above three categories (typically demographic, in the case of fluctuations in fitness from genetic drift). The fluctuations in the exponential growth SDE (Eq. 12) would appear, from our interpretation of the literature, to be of the environmental type. Unfortunately, the mathematical criteria for categorizing a given stochastic model into one of these three classes have remained vague, resulting in confusion on this point among population biologists. In particular, the type of mathematical model, such as a discrete birth–death process, continuous diffusion process, etc., does not necessarily peg the model's stochastic behavior.

We have found that the following criteria, an extension of some remarks by May (1974:33), form useful working definitions of these three types of stochastic behavior. Empirical (Pimm et al. 1988) and theoretical (Leigh 1981) studies suggest that the coefficient of variation of population size is an important component of extinction-related properties of population growth. Accordingly, let

$$\delta(t,\ n_0) = \frac{\sqrt{\operatorname{Var}(N(t)\mid N(0) = n_0)}}{\mathrm{E}[N(t)\mid N(0) = n_0]} \qquad (101)$$

be the coefficient of variation of the stochastic process $N(t)$. For fixed t, classify the stochastic fluctuations as demographic, environmental, or catastrophic based on whether $\delta(t,\ n_0)$ decreases to zero, becomes constant, or increase without bound, as a function of n_0.

This definition captures the spirit of the intuitive biological criteria. A model representing demographic forces essentially becomes a deterministic model for large population sizes. An example is the discrete, linear birth–death process explained in detail by Bailey (1964) and adopted by MacArthur and Wilson (1967). In addition, projection matrix simulations in which

members of each age class are subjected individually to survival–death Bernoulli trials would display demographic fluctuations. By contrast, environmental fluctuations in a model occur proportionately and are evident throughout all population sizes. By this definition, the exponential growth SDE (Eq. 12), and the logistic SDE used by Dennis and Patil (1984) and Dennis and Costantino (1988), represent environmental fluctuations. Stochastic projection matrices (see Eq. 1) would fall under this category. Note that the constant coefficient of variation could, in fact, be quite high. The variability of population size in some environmental-type models might thus appear to be rather catastrophic in magnitude; the key to the classification is that large populations fluctuate to the same proportional degree as smaller ones. Finally, models of truly catastrophic events (e.g., habitat destruction through fire) change a population by progressively larger proportions as the population size becomes larger.

A lingering problem with the proposed classification scheme is the terminology; the terms "demographic," "environmental," and "catastrophic" contain potentially misleading implications about the sources of the stochastic variability. However, the terms are widely used, and to attempt here to coin substitutes might confuse more than clarify. Another minor point is that $\delta(t,\ n_0)$, as a function of low values of n_0, might contain increasing, constant, and decreasing portions; realistic models might contain blends of stochastic forces, which vary in importance at different population sizes. The classification considers only limiting behavior as n_0 becomes large and is therefore somewhat coarse.

Time series models

An alternative approach to modeling population size is based on standard time series techniques (e.g., Pankratz 1983). The approach consists of transforming and/or differencing the data until obtaining a stationary series described by an autoregressive and/or moving average process. Such an approach can in fact yield useful information. Boyce (1987), for example, demonstrated the 10-yr cycle in the Whooping Crane data with such analyses.

The exponential growth SDE (Eq. 12) is a specific model for a nonstationary time series. Stationarity, according to the model, is achieved by log transforming and differencing the series. The resulting normal linear model for the increment data is the simplest particular case of the standard time series models. The diagnostic procedures aid in evaluating whether or not this particular model describes the data well.

The main advantages of the exponential growth SDE over other conventional time series models are as follows. First, the SDE model has a conceptual basis in the theory of age- or stage-structured population growth. Second, the time series available for real populations are often too short to take full advantage of the standard time series approach to model construction. Third, observations are frequently not collected at equal time intervals, a situation which is easily covered by the SDE model but difficult to handle by more usual time series methods. Finally, explicit and convenient results are available with the SDE model for estimating growth and extinction parameters.

That is not to say quantities such as the mean time to extinction could not be estimated from a conventional time series model. After developing such a model, one could conduct Monte Carlo simulations to estimate extinction parameters. If adequate data are available for identifying a particular time series model (or detailed age-structured model, for that matter), we recommend conducting the simulations and comparing the estimates to those described in this paper.

Management applications

The exponential growth SDE (Eq. 12) could prove to be useful in a variety of management contexts. Decisions about recovery programs for many endangered species must often be made with inadequate data available. Detailed information suitable for constructing a credible life table for a rare species seldom exists; management programs are fortunate if population size estimates, or even values of a relative population size index, are collected regularly. In such situations, the model might help biologists quantitatively assess what impact the usual, prevailing stochastic variation has on the long-term prospects for a species' survival. We caution that management policies should not be based solely on the results of the analyses described in this paper. Estimates of extinction-related quantities, as we have shown, are accompanied by large uncertainty, and effects of freak catastrophes are not accounted for in the model. However, even an order-of-magnitude assessment, or an assessment limited to the effects of environmental-type fluctuations, is better than none at all.

One crucial question regarding management of certain endangered species is whether, and if so when, to establish viable captive populations. The decision to capture some or all individuals of a species must be made before the wild population is reduced to a level having insufficient demographic and genetic diversity for a healthy founder population. The model conceivably could contribute information useful for this decision. The recovery programs for the Dusky Seaside Sparrow (*Ammodramus maritimus*), Guam Broadbill (*Myiagra freycineti*), and Kauaii Oo (*Moho braccatus*) are examples in which the debate over captive rearing raged interminably, only to have the wild population become extinct. Meaningful quantitative assessment of the species' recovery chances in these cases might have clarified earlier the urgent need for changes in management policies. There is little doubt that the 11th-hour decisions to capture the last remaining individuals of the black-footed ferret (*Mustela nigripes*) and the California Condor have provided opportunities to save these species. In the case of the California Condor, the model strongly suggests that emergency measures would have been justified some years sooner.

Another management question, in the United States at least, concerns the decision to list a species as officially "endangered" or "threatened" under the Endangered Species Act. Formal listing under the endangered or threatened categories empowers (and compels) government agencies to institute drastic measures to limit human activities that might impinge upon a species' survival chances. Predictably, in cases such as the Spotted Owl (*Strix occidentalis*) and Yellowstone grizzly bear population, biological grounds for such decisions are often discounted by political considerations (see, for instance, Simberloff 1987). Frequently lacking in the arguments over formal listing, though, are clear, defensible estimates of the risks of extinction. For example, recent upward trends in the Yellowstone grizzly population (as indexed by the estimated number of adult females; Fig. 5) have inspired optimistic proposals for removing the population from threatened status. Our estimates of the prospects for this population suggest that such proposals are premature.

CONCLUSIONS

The stochastic exponential growth model presented here combines minimum essential biological, stochastic, and statistical elements needed for assessing the chances of an endangered species' recovery. The model has a biological basis in the theory of age- or stage-structured populations. The stochastic fluctuations incorporated are of the environmental type, which are considered to pose greater problems for population persistence than those of the demographic type. The statistical interface of model with practice is well developed, so that parameter estimation, model evaluation, and forecasting are possible using only time series data on population sizes.

The model is no panacea. We do not intend for triage-style decisions about species preservation to be based solely on estimates arising from this model. The model does not incorporate some ecological factors, including freak catastrophic events and density dependence, which can be important in particular circumstances. Nonetheless, the examples we analyzed demonstrate that the model can provide useful supplemental information in a variety of situations.

Growth of endangered species is intrinsically stochastic. Accounting for stochastic forces in practice is a crucial problem in conservation biology. Such accounting will require, in addition to intensified efforts in population biology, further serious research attention to statistical questions of estimation and testing for mathematical models of population growth.

ACKNOWLEDGMENTS

We thank Robert Costantino, Russell Lande, Kenneth Burnham, Shripad Tuljapurkar, Lev Ginzburg, Cameron

140   BRIAN DENNIS ET AL.   Ecological Monographs
                                                                           Vol. 61, No. 2

Kepler, Greg Hayward, Marie Morin, James Lewis, Philip Dixon, and Subash Lele for their suggestions and thoughtful comments on an earlier draft of this paper. Cameron Kepler, Marie Morin, James Lewis, Jim Jacobi, Richard Knight, and Marcia Wilson generously shared unpublished information on several of the species with us; for this and other courtesies we thank them.

This research was supported by the Idaho Cooperative Fish and Wildlife Research Unit, the Forest, Wildlife, and Range Experiment Station of the University of Idaho, and the Department of Mathematical Sciences of Montana State University. The Idaho Cooperative Fish and Wildlife Research Unit is funded and supported by Idaho Department of Fish and Game, University of Idaho, USFWS, and the Wildlife Management Institute. This is contribution number 530 from the Forest, Wildlife, and Range Experiment Station of the University of Idaho.

LITERATURE CITED

Abramowitz, M., and I. A. Stegun. 1965. Handbook of mathematical functions. Dover, New York, New York, USA.

Allendorf, F. W., and C. Servheen. 1986. Genetics and the conservation of grizzly bears. Trends in Ecology and Evolution **1**:88–89.

Bailey, N. T. J. 1964. The elements of stochastic processes: with applications in the natural sciences. John Wiley & Sons, New York, New York, USA.

Basawa, I. V., and B. L. S. Prakasa-Rao. 1980. Statistical inference for stochastic processes. Academic Press, New York, New York, USA.

Belsley, D. A., E. Kuh, and R. E. Welsch. 1980. Regression diagnostics: identifying influential data and sources of collinearity. John Wiley & Sons, New York, New York, USA.

Berger, A. J. 1981. Hawaiian birdlife. Second edition. University Press of Hawaii, Honolulu, Hawaii, USA.

Binkley, C. S., and R. S. Miller. 1988. Recovery of the Whooping Crane *Grus americana*. Biological Conservation **45**:11–20.

Boyce, M. S. 1987. Time-series analysis and forecasting of the Aransas/Wood Buffalo Whooping Crane population. Pages 1–9 *in* J. C. Lewis, editor. Proceedings of the 1985 Crane Workshop. Platte River Whooping Crane Habitat Maintenance Trust, Grand Island, Nebraska, USA.

Boyce, M. S., and R. S. Miller. 1985. Ten-year periodicity in Whooping Crane census. Auk **102**:658–660.

Braumann, C. A. 1983a. Population growth in random environments. Bulletin of Mathematical Biology **45**:635–641.

———. 1983b. Population extinction probabilities and methods of estimation for population stochastic differential equation models. Pages 553–559 *in* R. S. Bucy and J. M. F. Moura, editors. Nonlinear stochastic problems. D. Reidel, Dordrecht, The Netherlands.

Burgman, M. A., H. R. Akcakaya, and S. S. Loew. 1988. The use of extinction models for species conservation. Biological Conservation **43**:9–25.

Burkey, T. V. 1989. Extinction in nature reserves: the effect of fragmentation and the importance of migration between reserve fragments. Oikos **55**:75–81.

Capocelli, R. M., and L. M. Ricciardi. 1974. A diffusion model for population growth in random environment. Theoretical Population Biology **5**:28–41.

Capocelli, R. M., and L. M. Ricciardi. 1979. A cybernetic approach to population modeling. Journal of Cybernetics **9**:297–312.

Chatterjee, S., and A. S. Hadi. 1988. Sensitivity analysis in linear regression. John Wiley & Sons, New York, New York, USA.

Cook, R. D., and S. Weisberg. 1982. Residuals and influence in regression. Chapman and Hall, London, England.

Conant, S. 1989. Saving endangered species by transloca-

tion: are we tinkering with evolution? BioScience **38**:254–256.

Craighead, J. J., J. R. Carney, and F. C. Craighead, Jr. 1974. A population analysis of the Yellowstone grizzly bears. Montana Forest and Conservation Experiment Station Bulletin 40, University of Montana, Missoula, Montana, USA.

Dennis, B. 1989a. Allee effects: population growth, critical density, and the chance of extinction. Natural Resource Modeling **3**:481–538.

———. 1989b. Stochastic differential equations as insect population models. *In* L. McDonald, B. Manly, J. Lockwood, and J. Logan, editors. Estimation and analysis of insect populations. Lecture Notes in Statistics **55**:219–238.

Dennis, B., and R. F. Costantino. 1988. Analysis of steady-state populations with the gamma abundance model: application to *Tribolium*. Ecology **69**:1200–1213.

Dennis, B., and G. P. Patil. 1984. The gamma distribution and weighted multimodal gamma distributions as models of population abundance. Mathematical Biosciences **68**:187–212.

Dennis, B., and G. P. Patil. 1988. Applications in ecology. Chapter 12 *in* E. L. Crow and K. Shimizu, editors. Lognormal distributions: theory and applications. Marcel Dekker, New York, New York, USA.

Draper, N., and H. Smith. 1981. Applied regression analysis. Second edition. John Wiley & Sons, New York, New York, USA.

Eberhardt, L. L., R. R. Knight, and B. M. Blanchard. 1986. Monitoring grizzly bear population trends. Journal of Wildlife Management **50**:613–618.

Efron, B. 1982. The jackknife, the bootstrap, and other resampling plans. Society for Industrial and Applied Mathematics, Philadelphia, Pennsylvania, USA.

Feldman, M. W., and J. Roughgarden. 1975. A population's stationary distribution and chance of extinction in a stochastic environment with remarks on the theory of species packing. Theoretical Population Biology **7**:197–207.

Ferson, S., L. Ginzburg, and A. Silvers. 1989. Extreme event risk analysis for age-structured populations. Ecological Modelling **47**:175–187.

Folks, J. L., and R. S. Chhikara. 1978. The inverse Gaussian distribution and its statistical application—a review (with discussion). Journal of the Royal Statistical Society Series B **40**:263–289.

Garcia, O. 1983. A stochastic differential equation model for the height growth of forest stands. Biometrics **39**:1059–1072.

Ginzburg, L. R., L. B. Slobodkin, K. Johnson, and A. G. Bindman. 1982. Quasiextinction probabilities as a measure of impact on population growth. Risk Analysis **2**:171–181.

Goel, N. S., and N. Richter-Dyn. 1974. Stochastic models in biology. Academic Press, New York, New York, USA.

Goodman, D. 1987. Consideration of stochastic demography in the design and management of biological reserves. Natural Resource Modeling **1**:205–234.

Graybill, F. A. 1976. Theory and application of the linear model. Wadsworth, Belmont, California, USA.

Heyde, C. C., and J. E. Cohen. 1985. Confidence intervals for demographic projections based on products of random matrices. Theoretical Population Biology **27**:120–153.

Iwasa, Y., and H. Mochizuki. 1988. Probability of population extinction accompanying a temporary decrease of population size. Researches on Population Ecology **30**:145–164.

Jacquez, G. M., and L. Ginzburg. 1989. *RAMAS*: teaching population dynamics, ecological risk assessment, and conservation biology. Academic Computing **4**:26–27, 54–56.

Kalbfleisch, J. D. 1986. Pseudo-likelihood. Pages 324–327 *in* S. Kotz, N. L. Johnson, and C. B. Read, editors. Encyclopedia of statistical sciences. Volume 7. John Wiley & Sons, New York, New York, USA.

Karlin, S., and H. M. Taylor. 1981. A second course in stochastic processes. Academic Press, New York, New York, USA.

Keiding, N. 1975. Extinction and exponential growth in random environments. Theoretical Population Biology 8: 49–63.

Knight, R. R., and L. L. Eberhardt. 1984. Projected future abundance of the Yellowstone grizzly bear. Journal of Wildlife Management 48:1434–1438.

Knight, R. R., and L. L. Eberhardt. 1985. Population dynamics of Yellowstone grizzly bears. Ecology 66:323–334.

Koford, C. B. 1953. The California Condor. National Audubon Society Research Report 4. National Audubon Society, New York, New York, USA.

Lande, R. 1987. Extinction thresholds in demographic models of territorial populations. American Naturalist 130:624–635.

———. 1988. Genetics and demography in biological conservation. Science 241:1455–1460.

Lande, R. and S. H. Orzack. 1988. Extinction dynamics of age-structured populations in a fluctuating environment. Proceedings of the National Academy of Sciences (USA) 85:7418–7421.

Lefkovitch, L. P. 1965. The study of population growth in organisms grouped by stages. Biometrics 21:1–18.

Leigh, E. G. 1981. The average lifetime of a population in a varying environment. Journal of Theoretical Biology 90: 213–239.

Lele, S. 1991. Jackknifing linear estimating equations: asymptotic theory in stochastic processes. Journal of the Royal Statistical Society Series B 53:253–267.

Leslie, P. H. 1945. On the use of matrices in certain population mathematics. Biometrika 33:182–212.

Levinton, J. S., and L. Ginzburg. 1984. Repeatability of taxon longevity in successive foraminifera radiations and a theory of random appearance and extinction. Proceedings of the National Academy of Sciences (USA) 81:5478–5481.

Lewis, E. G. 1942. On the generation and growth of a population. Sankhya 6:93–96.

MacArthur, R. H., and E. O. Wilson. 1967. The theory of island biogeography. Princeton University Press, Princeton, New Jersey, USA.

May, R. M. 1974. Stability and complexity in model ecosystems. Second edition. Princeton University Press, Princeton, New Jersey, USA.

Mode, C. J., and M. E. Jacobson. 1987a. A study of the impact of environmental stochasticity on extinction probabilities by Monte Carlo integration. Mathematical Biosciences 83:105–125.

Mode, C. J., and M. E. Jacobson. 1987b. On estimating the critical population size for an endangered species in the presence of environmental stochasticity. Mathematical Biosciences 85:185–209.

Nedelman, J., J. A. Thompson, and R. J. Taylor. 1987. The statistical demography of Whooping Cranes. Ecology 68: 1401–1411.

Neter, J., W. Wasserman, and M. H. Kutner. 1985. Applied linear statistical models. Second edition. Richard D. Irwin, Homewood, Illinois, USA.

Nisbet, R. M., and W. S. C. Gurney. 1982. Modelling fluctuating populations. John Wiley & Sons, New York, New York, USA.

Nordheim, E. V., D. B. Hogg, and S.-Y. Chen. 1989. Leslie matrix models for insect populations with overlapping generations. In L. McDonald, B. Manly, J. Lockwood, and J. Logan, editors. Estimation and analysis of insect populations. Lecture Notes in Statistics 55:289–298.

Ogden, J. 1985. The California Condor. Pages 389–399 in

R. L. Silvestro, editor. Audubon wildlife report 1985. National Audubon Society, New York, New York, USA.

Pankratz, A. 1983. Forecasting with univariate Box-Jenkins models: concepts and cases. John Wiley & Sons, New York, New York, USA.

Pimm, S. L., H. L. Jones, and J. Diamond. 1988. On the risk of extinction. American Naturalist 132:757–785.

Press, W. H., B. P. Flannery, S. A. Teukolsky, and W. T. Vetterling. 1986. Numerical recipes: the art of scientific computing. Cambridge University Press, Cambridge, England.

Ricciardi, L. M. 1977. Diffusion processes and related topics in biology. Lecture Notes in Biomathematics 14.

Rice, J. A. 1988. Mathematical statistics and data analysis. Wadsworth and Brooks/Cole, Pacific Grove, California, USA.

Richter-Dyn, N., and N. S. Goel. 1974. On the extinction of a colonizing species. Theoretical Population Biology 3: 406–433.

Rudin, W. 1964. Principles of mathematical analysis. Second edition. McGraw-Hill, New York, New York, USA.

Samson, F. B., F. Perez-Trejo, H. Salwasser, L. F. Ruggiero, and M. L. Shaffer. 1985. On determining and managing minimum population size. Wildlife Society Bulletin 13:425–433.

Schrödinger, E. 1915. Zur Theorie der Fall- und Steigversuche an Teilchen mit Brownscher Bewegung. Physikalische Zeitschrift 16:289–295.

Scott, J. M., S. Mountainspring, C. van Riper III, C. B. Kepler, J. D. Jacobi, T. A. Burr, and J. G. Giffin. 1984. Annual variation in the distribution, abundance, and habitat response of the Palila (Loxioides bailleui). Auk 101:647–664.

Scott, J. M., C. B. Kepler, C. van Riper III, and S. I. Fefer. 1988. Conservation of Hawaii's vanishing avifauna. BioScience 38:238–253.

Seber, G. A. F. 1982. The estimation of animal abundance and related parameters. Second edition. Macmillan, New York, New York, USA.

Serfling, R. J. 1980. Approximation theorems of mathematical statistics. John Wiley & Sons, New York, New York, USA.

Shaffer, M. L. 1978. Determining minimum viable population sizes: a case study of the grizzly bear (Ursus arctos L.). Dissertation. Duke University, Durham, North Carolina, USA.

———. 1981. Minimum population sizes for species conservation. BioScience 31:131–134.

———. 1987. Minimum viable populations: coping with uncertainty. Pages 69–86 in M. E. Soulé, editor. Viable populations for conservation. Cambridge University Press, Cambridge, England.

Shaffer, M. L., and F. B. Samson. 1985. Population size and extinction: a note on determining critical population sizes. American Naturalist 125:144–152.

Shimizu, K. 1988. Point estimation. Chapter 2 in E. L. Crow and K. Shimizu, editors. Lognormal distributions: theory and applications. Marcel Dekker, New York, New York, USA.

Shimizu K., and K. Iwase. 1981. Uniformly minimum variance unbiased estimation in lognormal and related distributions. Communications in Statistics, Part A—Theory and Methods 10:1127–1147.

Simberloff, D. 1987. The Spotted Owl fracas: mixing academic, applied, and political ecology. Ecology 68:766–772.

———. 1988. The contribution of population and community ecology to conservation science. Annual Review of Ecology and Systematics 19:473–511.

Slade, N. A., and H. Levenson. 1982. Estimating population growth rates from stochastic Leslie matrices. Theoretical Population Biology 22:299–308.

142   BRIAN DENNIS ET AL.   Ecological Monographs
                                                                                Vol. 61, No. 2

Snyder, N. F. R., and E. V. Johnson. 1985. Photographic censusing of the 1982–1983 California Condor population. Condor **87**:1–13.

Snyder, N. F. R., J. W. Wiley, and C. B. Kepler. 1987. The parrots of Luquillo: natural history and conservation of the Puerto Rican parrot. Western Foundation of Vertebrate Zoology, Los Angeles, California, USA.

Soong, T. T. 1973. Random differential equations in science and engineering. Academic Press, New York, New York, USA.

Soulé, M. E., editor. 1986. Conservation biology: the science of scarcity and diversity. Sinauer, Sunderland, Massachusetts, USA.

———, editor. 1987. Viable populations for conservation. Cambridge University Press, Cambridge, England.

Strebel, D. E. 1985. Environmental fluctuations and extinction—single species. Theoretical Population Biology **27**: 1–26.

Suchy, W. J., L. L. McDonald, M. D. Strickland, and S. H. Anderson. 1985. New estimates of minimum viable population sizes for grizzly bears of the Yellowstone ecosystem. Wildlife Society Bulletin **13**:223–228.

Tier, C., and F. B. Hanson. 1981. Persistence in density dependent stochastic populations. Mathematical Biosciences **53**:89–117.

Tuljapurkar, S. D. 1982a. Population dynamics in variable environments. II. Correlated environments, sensitivity analysis and dynamics. Theoretical Population Biology **21**: 114–140.

———. 1982b. Population dynamics in variable environments. III. Evolutionary dynamics of r-selection. Theoretical Population Biology **21**:141–165.

———. 1989. An uncertain life: demography in random environments. Theoretical Population Biology **35**:227–294.

Tuljapurkar, S. D., and S. H. Orzack. 1980. Population dynamics in variable environments. I. Long-run growth rates and extinction. Theoretical Population Biology **18**: 314–342.

Turelli, M. 1977. Random environments and stochastic calculus. Theoretical Population Biology **12**:140–178.

Tweedie, M. C. K. 1957a. Statistical properties of the inverse Gaussian distribution I. Annals of Mathematical Statistics **28**:362–377.

———. 1957b. Statistical properties of the inverse Gaussian distribution II. Annals of Mathematical Statistics **28**:696–705.

USFWS. 1978. Palila recovery plan. United States Fish and Wildlife Service, Portland, Oregon, USA.

———. 1982. Grizzly bear recovery plan. United States Fish and Wildlife Service, Denver, Colorado, USA.

———. 1984a. California Condor recovery plan. United States Fish and Wildlife Service, Portland, Oregon, USA.

———. 1984b. Northwestern Hawaiian Islands passerines recovery plan. United States Fish and Wildlife Service, Portland, Oregon, USA.

———. 1985. Kirtland's Warbler recovery plan. United States Fish and Wildlife Service, Minneapolis and St. Paul, Minnesota, USA.

———. 1986. Whooping Crane recovery plan. United States Fish and Wildlife Service, Rockville, Maryland, USA.

———. 1987. Recovery plan for the Puerto Rican Parrot (*Amazona vittata*). United States Fish and Wildlife Service, Atlanta, Georgia, USA.

van Groenendael, J., H. de Kroon, and H. Caswell. 1988. Projection matrices in population biology. Trends in Ecology and Evolution **3**:264–269.

van Riper, C., III. 1980. Observations on the breeding of the Palila (*Psittirostra bailleui*) of Hawaii. Ibis **122**:462–475.

Walkinshaw, L. H. 1983. Kirtland's Warbler: the natural history of an endangered species. Bulletin 58. Cranbrook Institute of Science, Bloomfield Hills, Michigan, USA.

Whitmore, G. A. 1978. Discussion of the paper by Professor Folks and Dr. Chhikara. Journal of the Royal Statistical Society Series B **40**:285–286.

Whitmore, G. A., and V. Sheshadri. 1987. A heuristic derivation of the inverse Gaussian distribution. American Statistician **41**:280–281.

Wiemeyer, S. N., J. M. Scott, M. P. Anderson, P. H. Bloom, and C. J. Stafford. 1988. Environmental contaminants in California Condors. Journal of Wildlife Management **52**: 238–247.

Wilbur, S. R. 1980. Estimating the size and trend of the California Condor population, 1965–1978. California Fish and Game **66**:40–48.

Wilcox, B. A. 1986. Extinction models and conservation. Trends in Ecology and Evolution **1**:46–48.

Wiley, J. W. 1985. The Puerto Rican Parrot and competition for its nest sites. Pages 213–223 *in* P. J. Moors, editor. Conservation of island birds. International Council for Bird Preservation, Cambridge, England.

Wright, S. J., and S. P. Hubble. 1983. Stochastic extinction and reserve size: a focal species approach. Oikos **41**:466–476.

APPENDIX

Consider the inverse Gaussian cumulative distribution function (cdf) (Eq. 16). Numerical overflow can arise in evaluating (Eq. 16) at given values $\hat{\mu}$, $\hat{\sigma}^2$, x_d, and t due to the exponential function in the second term of the cdf. Note that the second term is the product of a very large number [$\exp(\cdot)$] and a very small number [$\Phi(\cdot)$]. The overflow problem can be circumvented if the two expressions in the product can be evaluated simultaneously.

We reparameterize Eq. 16 by defining

$$a = x_d/\sqrt{\sigma^2}, \tag{A.1}$$
$$b = |\mu|/\sqrt{\sigma^2}, \tag{A.2}$$
$$y = (bt - a)/\sqrt{t}, \tag{A.3}$$
$$z = (bt + a)/\sqrt{t}, \tag{A.4}$$

so that the inverse Gaussian cdf is

$$G(t; x_d, \mu, \sigma^2) = \Phi(y) + \exp(2ab)\Phi(-z), \tag{A.5}$$

where $\Phi(\cdot)$ is the standard normal cdf (Eq. 7). Let $\phi(y)$ denote the derivative of $\Phi(y)$:

$$\phi(y) = (2\pi)^{-\frac{1}{2}}\exp(-y^2/2). \tag{A.6}$$

Several approximations are available for evaluating $\Phi(\cdot)$ as a function of $\phi(\cdot)$ (Abramowitz and Stegun 1965). We found the following formulas to be accurate and convenient to program. If $y < 4$, then from Abramowitz and Stegun (1965: 932)

$$\Phi(y) \approx 1 - \phi(y)[d_1 q_y + d_2 q_y^2 + \ldots + d_5 q_y^5], \tag{A.7}$$

where $q_y = 1/(1 + d_0 y)$ and $d_0 = 0.2316419$, $d_1 = 0.319381530$, $d_2 = -0.356563782$, $d_3 = 1.781477937$, $d_4 = -1.821255978$, and $d_5 = 1.330274429$. Alternatively, if $y \geq 4$ then we recommend approximating $\Phi(\cdot)$ by

$$\Phi(y) \approx 1 - \frac{\phi(y)}{y}$$
$$\cdot \left[1 - \frac{1}{y^2} + \ldots + \frac{(-1)^s 1 \cdot 3 \ldots (2s - 1)}{y^{2s}} \right], \tag{A.8}$$

with $s = 7$ (Abramowitz and Stegun 1965:932).

  Notice that

$$\phi(y) = \exp(2ab)\phi(z). \tag{A.9}$$

Expression A.9 is the key to calculating A.5, since it follows from A.9 that the product $\exp(2ab)\Phi(-z)$ can be approximated as a function of $\phi(y)$ and z. In particular, if $z < 4$ then

$$\exp(2ab)\Phi(-z) \approx \phi(y)[d_1 q_z + d_2 q_z^2 + \ldots + d_5 q_z^5], \tag{A.10}$$

where $q_z = 1/(1 + d_0 z)$, and $d_0, \ldots, d_5$ are given in Eq. A.7. For values of $z \geq 4$

$$\exp(2ab)\Phi(-z)$$

$$\approx \frac{\phi(y)}{z}\left[1 - \frac{1}{z^2} + \ldots + \frac{(-1)^s \cdot 3 \ldots (2s-1)}{z^{2s}}\right], \tag{A.11}$$

where we suggest $s = 7$.

Vol. 142, No. 6   The American Naturalist   December 1993

RISKS OF POPULATION EXTINCTION FROM DEMOGRAPHIC AND ENVIRONMENTAL STOCHASTICITY AND RANDOM CATASTROPHES

Russell Lande

Department of Biology, University of Oregon, Eugene, Oregon 97403

Submitted March 26, 1992; Revised October 19, 1992; Accepted December 1, 1992

Abstract.—Stochastic factors affecting the demography of a single population are analyzed to determine the relative risks of extinction from demographic stochasticity, environmental stochasticity, and random catastrophes. Relative risks are assessed by comparing asymptotic scaling relationships describing how the average time to extinction, T, increases with the carrying capacity of a population, K, under each stochastic factor alone. Stochastic factors are added to a simple model of exponential growth up to K. A critical parameter affecting the extinction dynamics is $\bar{r}$, the long-run growth rate of a population below K, including stochastic factors. If $\bar{r}$ is positive, with demographic stochasticity T increases asymptotically as a nearly exponential function of K, and with either environmental stochasticity or random catastrophes T increases asymptotically as a power of K. If $\bar{r}$ is negative, under any stochastic demographic factor, T increases asymptotically with the logarithm of K. Thus, for sufficiently large populations, the risk of extinction from demographic stochasticity is less important than that from either environmental stochasticity or random catastrophes. The relative risks of extinction from environmental stochasticity and random catastrophes depend on the mean and environmental variance of population growth rate, and the magnitude and frequency of catastrophes. Contrary to previous assertions in the literature, a population of modest size subject to environmental stochasticity or random catastrophes can persist for a long time, if $\bar{r}$ is substantially positive.

Understanding the risks of extinction affecting single populations is important in both pure and applied ecology, in the development of models of more complex, spatially distributed populations, and in the formulation of effective conservation plans for threatened and endangered species (Soulé and Simberloff 1986; Lande 1988; Karieva 1990; Gilpin and Hanski 1991).

In an important article, Shaffer (1981) suggested that stochastic demographic and genetic factors determine the minimum size of a viable population, which can be defined in terms of the probability of extinction within a specified time (e.g., a 95% probability of persistence for 100 yr, or a 99% probability of persistence for 1,000 yr). Ginzburg et al. (1982) advocated the use of stochastic demographic models as a basis for risk assessment in environmental impact statements. Concepts and methods of stochastic population modeling play an integral part in population viability analysis (Gilpin and Soulé 1986; Burgman et al. 1993).

Shaffer (1981, 1987) discussed three stochastic demographic factors that are the subject of the present article. First, *demographic stochasticity* is caused by chance realizations of individual probabilities of death and reproduction in a finite

Am. Nat. 1993. Vol. 142, pp. 911–927.

population. Because independent individual events tend to average out in large populations, demographic stochasticity is most important in small populations. Second, *environmental stochasticity* arises from a nearly continuous series of small or moderate perturbations that similarly affect the birth and death rates of all individuals (within each age or stage class) in a population (May 1974). In contrast to demographic stochasticity, environmental stochasticity is important in both large and small populations. Finally, *catastrophes* are large environmental perturbations that produce sudden major reductions in population size. Like environmental stochasticity, catastrophes are important in populations of all sizes.

Based on models of stochastic population growth and simple scaling arguments concerning the average time until extinction of a population due to each of these factors acting alone, Shaffer (1987) attempted to deduce their relative importance.

It is well-known that under demographic stochasticity alone (in a constant environment) the average time to extinction increases almost exponentially with carrying capacity (MacArthur and Wilson 1967, Richter-Dyn and Goel 1972; Leigh 1981; Gabriel and Bürger 1992). Ludwig (1976) and formulas in Leigh (1981) and Tier and Hanson (1981) show that, under density-dependent population growth in a random environment, the mean time to extinction is a power function of the carrying capacity. In contrast, Goodman (1987a, 1987b) claimed that because of environmental stochasticity alone the average time to extinction always increases less than linearly with carrying capacity. Ewens et al. (1987) developed a density-independent catastrophe model and concluded that the average time to extinction increases only logarithmically with initial population size. From the qualitative scaling relationships in Goodman (1987a, 1987b) and Ewens et al. (1987), Shaffer (1987) concluded that random catastrophes are more important than environmental stochasticity, which is more important than demographic stochasticity in determining average persistence times of populations. Later authors presented the same conclusions (Pimm and Gilpin 1989; Soulé and Kohm 1989; Hedrick and Miller 1992).

The relative importance of demographic and environmental stochasticity and random catastrophes discussed by Shaffer (1987) is intuitively appealing and eventually may be supported by empirical evidence. However, I demonstrate in this article, as shown by previous results of Ludwig (1976), Leigh (1981), and Tier and Hanson (1981), that Goodman's (1987a, 1987b) somewhat less than linear scaling of average extinction time with carrying capacity under environmental stochasticity is incorrect. I also show that the logarithmic scaling of average extinction time with initial population size under random catastrophes found by Ewens et al. (1987) does not generalize to density-dependent population growth.

I develop analytical models showing that curves of average persistence time as a function of population size may be concave or convex under the influence of either environmental stochasticity or random catastrophes, and I show that no general theoretical statement can be made concerning the relative risks of population extinction from these two stochastic factors. When the long-run growth rate of a population is negative, regardless of whether the cause is deterministic or stochastic, the average extinction time scales logarithmically with initial population size, as suggested by Ludwig (1976) and Brockwell (1985). To demonstrate these assertions, I analyze the mean time to extinction in a simple model of density-dependent growth of a population subject to different demographic risks.

DETERMINISTIC MODEL

To facilitate the analysis of stochastic factors affecting population growth, I introduce a simple deterministic model of density-dependent growth of a population without age structure. The population size, N, has a constant per capita growth rate, r, except at the carrying capacity (or ceiling), K, where growth ceases:

$$\frac{dN}{dt} = \begin{cases} rN & \text{for} \quad 1 < N < K \\ 0 & \text{for} \quad N = K. \end{cases} \tag{1}$$

For an initial population size N_0 between one and K, the population grows exponentially with time t as $N(t) = N_0 e^{rt}$. If r is positive, population growth continues until K is reached. This simple model of exponential growth up to a carrying capacity was analyzed by MacArthur and Wilson (1967), Leigh (1981), and Goodman (1987a, 1987b) in their investigations of demographic and environmental stochasticity.

If r is negative, the population declines to extinction, which is defined to occur at a population size of $N = 1$ individual. For a population initially at carrying capacity, $N_0 = K$, the time until extinction, $- (\ln K)/r$, then depends on the natural logarithm of the initial size. In the following sections we will see that a roughly logarithmic dependence of extinction time on initial population size carries over to stochastic models.

DIFFUSION THEORY FOR STOCHASTIC MODELS

Demographic and environmental stochasticity, involving a nearly continual series of small or moderate perturbations of the population numbers, can be accurately modeled as a diffusion process, provided that the mean absolute growth rate per unit time is small, $|\bar{r}| \ll 1$ (Keiding 1975; Leigh 1981). For populations with discrete generations this condition implies a low growth rate per generation. For populations with overlapping generations that reproduce at discrete times, the condition is less restrictive, requiring only a low growth rate per reproductive interval (if mortality and reproductive rates are independent of age).

Diffusion theory can then be employed to calculate the mean time to extinction of the population (Karlin and Taylor 1981, chap. 15). A diffusion process is completely described by its infinitesimal moments and by the behavior of sample paths at the boundaries. For a population of size N, the infinitesimal mean and variance, $\mu(N)$ and $\sigma^2(N)$, give, respectively, the expected change and the variance of the change in population size per unit time. Starting from a given initial size, N_0, the mean time to extinction, denoted as $T \equiv T(N_0)$, is the solution of

$$\frac{1}{2}\sigma^2(N_0)\frac{d^2T}{dN_0^2} + \mu(N_0)\frac{dT}{dN_0} = -1, \tag{2}$$

with the boundary conditions $T(1) = 0$ and a reflecting boundary at K. A reflecting boundary condition with density-independent population growth was employed by MacArthur and Wilson (1967), Leigh (1981), and Goodman (1987b). The same result can be obtained more simply by regarding equation (1) as the limit of a continuum of models with weaker forms of density dependence (see, e.g., Richter-Dyn and Goel 1972). We can then use the general solution to equation (2) in Karlin and Taylor (1981) to show that

$$T(N_0) = 2 \int_1^{N_0} e^{-G(z)} \int_z^K \frac{e^{G(y)}}{\sigma^2(y)} \, dy \, dz \,, \tag{3}$$

where

$$G(y) = 2 \int^y \frac{\mu(N)}{\sigma^2(N)} \, dN \,.$$

This formula differs slightly from that of Ludwig (1976, eq. [3.12]) and Leigh (1981, eq. [8]) because they defined extinction as occurring at a population size of $N = 0$ instead of 1, and we have incorporated the fact that N cannot exceed K in equation (1).

For populations subject to demographic or environmental stochasticity, with positive long-run growth rates and a sufficiently large initial size, Richter-Dyn and Goel (1972) and Leigh (1981) showed that the average time to extinction is nearly independent of the initial population size. A similar conclusion holds for populations subject to random catastrophes (see Appendix formula [A15a]). This occurs because such a population is most likely to grow quickly toward carrying capacity and to spend a long time fluctuating near K before stochastic factors finally cause extinction. The persistence of an established population with a positive long-run growth rate can thus be accurately described by its average time to extinction starting from carrying capacity.

We therefore follow Shaffer (1987) in considering the extinction dynamics of density-dependent populations that start at carrying capacity. Formula (3) will be used to evaluate the influence of demographic and environmental stochasticity on population persistence and to compare the results for populations with positive or negative long-run growth rates to those of earlier authors.

DEMOGRAPHIC STOCHASTICITY

In a finite population, the per capita growth rate, r, is subject to random variation due to independent chances of individual mortality and reproduction. Thus, for a population of size N, r is a random variable with mean $\bar{r}$ and variance V_1/N per unit time, with no autocorrelation. The parameter V_1 is the variance in individual fitness per unit time (Keiding 1975; Leigh 1981; Goodman 1987a, 1987b). The growth rate of a population at a particular time, r, is the mean Malthusian fitness of individuals in the population (Crow and Kimura 1970, chap. 1), and its variance follows the standard statistical formula for the sampling variance of a mean (i.e., individual variance divided by population size). The long-run growth rate of a population subject to demographic stochasticity is simply $\bar{r} = \bar{r}$.

Assuming that the continuous-time model in equation (1) represents an approximation to an actual population that reproduces at discrete time intervals, the Ito stochastic calculus is appropriate to obtain the infinitesimal mean and variance of the diffusion process (Turelli 1977; Karlin and Taylor 1981, chap. 15.14), which are

$$\mu(N) = \bar{r}N \quad \text{and} \quad \sigma^2(N) = V_1 N. \tag{4}$$

For populations initially at carrying capacity, $N_0 = K$, the average time to extinction, from formula (3), is

$$T(K) = \frac{1}{\bar{r}} \int_1^K \frac{e^{a(N-1)}}{N} \, dN - \frac{\ln K}{\bar{r}}, \tag{5a}$$

where $a = 2r/V_1$. Leigh (1981, pp. 220–221) derived a similar formula, assuming $V_1 = 1$. Figure 1 illustrates the average extinction time as a function of carrying capacity for populations with different mean growth rates.

Asymptotic scaling relationships for the average extinction time with increasing carrying capacity can be derived from formula (5a) in cases where $\bar{r}$ is positive, zero, or negative. For positive $\bar{r}$, the integral in formula (5a) can be approximated by expanding $1/N$ in a Taylor series around $1/K$ (as in Leigh 1981), and if $aK \gg 1$, we can retain only the first two terms in the series to find

$$T(K) \approx \frac{e^{a(K-1)} - 1}{\bar{r} a K} \left(1 + \frac{1}{aK} \right) \quad \text{for } aK \gg 1. \tag{5b}$$

The dominant term is proportional to e^{aK}/K. This nearly exponential scaling of average extinction time is qualitatively consistent with previous results (MacArthur and Wilson 1967; Richter-Dyn and Goel 1972; Leigh 1981; Goodman 1987b; Gabriel and Bürger 1992).

For $\bar{r} = 0$, equation (5a) reduces to a nearly linear dependence of average extinction time on carrying capacity, $T(K) = 2(K - 1 - \ln K)/V_1$, in close agreement with the linear relationship derived by Leigh (1981).

For negative $\bar{r}$, when $-aK \gg 1$ the integral in equation (5a) approaches a constant that involves the exponential integral, $E_1(-a) = \int_{-a}^{\infty} z^{-1} e^{-z} dz$ (tabulated in Abramowitz and Stegun 1972),

$$T(K) \approx \frac{-\ln K + e^{-a} E_1(-a)}{\bar{r}} \quad \text{for } -aK \gg 1 \tag{5c}$$

With a negative average (or long-run) growth in a population under demographic stochasticity, the dominant term in the asymptotic scaling of average extinction time is proportional to the logarithm of carrying capacity, as in a population undergoing a deterministic decline.

ENVIRONMENTAL STOCHASTICITY

We model the effects of a changing environment by allowing the population growth rate to fluctuate with time as a stationary time series with mean growth

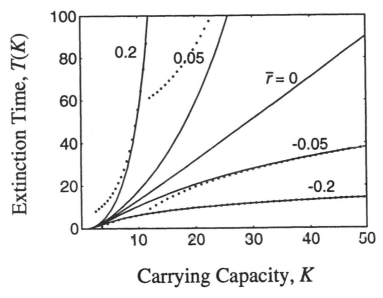

Carrying Capacity, K

FIG. 1.—Average time to extinction under demographic stochasticity for a population initially at carrying capacity. Each *curve* represents a different value of the mean population growth rate, $\bar{r}$. The sampling variance in fitness per individual per unit time due to demographic stochasticity is $V_1 = 1$. *Solid lines* show the diffusion approximation (eq. [5a]), and *dotted lines* give asymptotic approximations (eqq. [5b], [5c]).

rate $\bar{r}$, environmental variance V_e, and no autocorrelation. Again assuming that equation (1) approximates a population reproducing at discrete intervals, we can use diffusion theory to analyze the process. The infinitesimal mean and variance of the diffusion process are

$$\mu(N) = \bar{r}N \quad \text{and} \quad \sigma^2(N) = V_e N^2. \tag{6}$$

Transformation of the diffusion process to a logarithmic scale would yield the transformed infinitesimal mean and variance as $\bar{r} - V_e/2$ and V_e, respectively, in the domain $0 < \ln N < \ln K$ (Karlin and Taylor 1981, chap. 15.3). The expected value of the logarithm of population size then changes according to

$$E[\ln N(t)] = \ln N_0 + (\bar{r} - V_e/2)t \tag{7}$$

in the density-independent region. For this reason, the quantity $\tilde{r} = \bar{r} - V_e/2$, termed the long-run growth rate, can be considered as a stochastic analogue of r in the deterministic model (see Tuljapurkar 1982; Lande and Orzack 1988). Discounting the mean growth rate because of random environmental fluctuations is explained by Lewontin and Cohen (1969) in terms of the finite rate of increase, e^r, the arithmetic mean of which determines the expected population size, whereas the smaller geometric mean determines the dynamics of extinction. For the diffusion approximation in equation (6), the logarithm of the expected population size is $\ln E[N(t)] = \ln N_0 + \bar{r}t$ in the density-independent region, in contrast to equation (7).

From formula (3), the average time to extinction with environmental stochasticity is

$$T(K) = \frac{2}{V_e c}\left(\frac{K^c - 1}{c} - \ln K\right),$$
(8a)

in which $c = 2\bar{r}/V_e - 1 = (2/V_e)\hat{r}$. Figure 2 shows how the mean extinction time increases with carrying capacity for different values of c.

With a positive long-run growth rate, the average time to extinction scales asymptotically as a power function of the carrying capacity,

$$T(K) \approx 2K^c/(V_e c^2) \quad \text{for } c \ln K \gg 1.$$
(8b)

Ludwig (1976) showed for logistic population growth in a random environment that the asymptotic scaling of $T(K)$ is proportional to K^c. Formulas in Leigh (1981) and Tier and Hanson (1981) are also in agreement with this scaling of mean extinction time.

Goodman (1987b) derived a compatible result but assumed, inappropriately for environmental stochasticity, that $\bar{r} < V_e/2$, which implies a negative long-run growth rate for the population ($c < 0$) and a less than linear scaling described below. Contrary to Goodman (1987a, 1987b), in a population subject to environmental stochasticity the average time to extinction can increase faster than linearly with carrying capacity. In formula (8b), $T(K)$ is linear when $c = 1$, curves downward when $c < 1$, and curves upward when $c > 1$. The scaling of average extinction time with carrying capacity is determined by the ratio of the mean growth rate to its variance and has positive curvature when $\bar{r}/V_e > 1$. This only requires that the growth rate have a mean larger than its environmental variance.

If the long-run growth rate is zero ($\hat{r} = 0$ or $c = 0$), the average time to extinction depends on the square of the logarithm of carrying capacity, $T(K) = (\ln K)^2/V_e$. Goodman (1987b) derived this formula, but under the incorrect condition $\bar{r} = 0$ instead of $\hat{r} = 0$. When the mean growth rate is zero ($\bar{r} = 0$ or $c = -1$), the long-run growth rate is negative, $\hat{r} = -V_e/2$, and the average time to extinction is nearly proportional to the logarithm of the carrying capacity.

With a negative long-run growth rate in a population with environmental stochasticity,

$$T(K) \approx \frac{-\ln K - 1/c}{\hat{r}} \quad \text{for } -c \ln K \gg 1,$$
(8c)

the dominant term in the average time to extinction is proportional to the logarithm of carrying capacity, in agreement with Ludwig (1976).

RANDOM CATASTROPHES

Large, infrequent perturbations such as sudden catastrophes must be modeled differently than with diffusion processes such as those used above to describe demographic and environmental stochasticity. Hanson and Tuckwell (1981) introduced a population-dynamic model in which catastrophes reduce the population

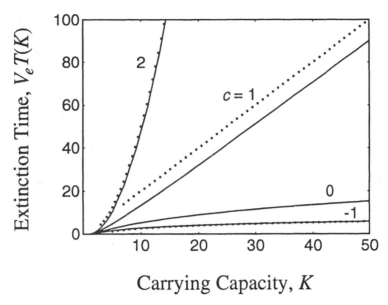

Carrying Capacity, K

Fig. 2.—Average time to extinction under environmental stochasticity for a population initially at carrying capacity. The mean and variance of the population growth rate caused by a fluctuating environment are $\bar{r}$ and V_e. Each *curve* corresponds to a different value of $c = 2\bar{r}/V_e - 1$. *Solid lines* display the diffusion approximation (eq. [8a]), and *dotted lines* are asymptotic approximations (eqq. [8b], [8c]).

by a fixed proportion at random times. In their model the population grows deterministically between catastrophes with a logistic form of density dependence. They were unable to derive a complete analytical solution but presented numerical results for some parameter values. Ewens et al. (1987) analyzed a similar model of catastrophes in a population with no density regulation. Here I show that, by modifying Hanson and Tuckwell's model to incorporate the form of density dependence in equation (1), an analytical solution can be obtained for the average time to extinction of a population subject to random catastrophes.

Each catastrophe is assumed to reduce instantaneously the population by a proportion δ, that is, a population of size N just prior to a catastrophe is reduced to size $(1 - \delta)N$ just after the catastrophe. Catastrophes occur as a Poisson process with rate parameter λ, so that the waiting time for a catastrophe has an exponential distribution, $\lambda^{-1}e^{-\lambda t}$ with mean $1/\lambda$, and the probability of i catastrophes occurring in a time interval of length t is $(\lambda t)^i e^{-\lambda t}/i!$ for $i = 0, 1, 2, \ldots$ Between catastrophes the population grows at the constant per capita rate, r, until the carrying capacity, K, is reached (eq. [1]) or a catastrophe occurs.

It is easiest to analyze this model in terms of the natural logarithm of population size, $n = \ln N$. Then from an initial value of $n_0 = \ln N_0$, n grows linearly with time at the constant rate r, $n(t) = n_0 + rt$, until it reaches $k = \ln K$ or a catastrophe strikes. Each catastrophe now reduces n by an additive amount, $\epsilon = -\ln (1 - \delta)$. Extinction occurs at $n = \ln 1 = 0$. This type of process, with additive downward jumps of constant magnitude happening at random times, was studied

EXTINCTION RISKS   919

by Hanson and Tuckwell (1978). Starting from n_0, the average time until extinction, denoted as $T(n_0)$, obeys a differential-difference equation,

$$r\frac{dT(n_0)}{dn_0} - \lambda T(n_0) + \lambda T(n_0 - \epsilon) = -1 \tag{9}$$

for $0 < n_0 \le k$, with $T(n_0) = 0$ for $n_0 \le 0$. The boundary condition differs depending on whether r is positive or negative. For $r > 0$, the appropriate boundary condition is $T(k) - T(k - \epsilon) = 1/\lambda$ because the expected time to move from the stable point k to $k - \epsilon$ is the expected waiting time until the next catastrophe, $1/\lambda$. Then $T(n_0)$ is discontinuous at the extinction boundary, $n = 0$, because starting from an infinitesimally small value, 0^+, the population will grow and persist until a (series of) catastrophe(s) causes its extinction. When r is positive, the boundary condition at $T(k)$ therefore determines $T(0^+)$ (Hanson and Tuckwell 1978). When r is negative, the appropriate boundary condition is $T(0^+) = 0$.

Exact analytical solutions of equation (9) can be obtained using Laplace transforms (Churchill 1958; Bellman and Cooke 1963), as shown in the Appendix. The exact solutions are so complex as to be uninformative, except in facilitating the construction of graphs as in figure 3. Asymptotic analysis, also given in the Appendix, produces approximate formulas that are more readily interpretable. An important parameter in both the exact and approximate solutions is the ratio of the catastrophe rate times catastrophe size to the population growth rate between catastrophes, $\gamma = \lambda \epsilon / r$. This parameter is closely related to the long-run growth rate of the population including catastrophes, $\bar{r} = r - \lambda \epsilon$. Another important parameter is the number of catastrophes from carrying capacity to extinction in the absence of population growth, k/ϵ.

Figure 3 depicts average times to extinction as a function of k/ϵ. It can be seen for $\bar{r} > 0$ (or $0 < \gamma < 1$) that $\ln(\lambda T(k))$ increases asymptotically as a linear function of k/ϵ, so that $\lambda T(k)$ increases exponentially with k/ϵ. For $\bar{r} < 0$ (or $0 > \gamma > 1$), figure 3 (bottom) shows the asymptotic linear dependence of average extinction time on k/ϵ.

For $\gamma > 0$, the asymptotic solution is

$$\lambda T(k) \approx \frac{\gamma}{1 - \gamma}\left(\frac{e^{\beta k/\epsilon} - 1}{\beta} - \frac{k}{\epsilon}\right) + O(1), \tag{10a}$$

in which β is the solution of the transcendental equation $\beta/(e^\beta - 1) = \gamma$, and $O(1)$ is a constant of order 1.

When the long-run growth rate of the population is positive, $0 < \gamma < 1$, then β is positive and the exponential term dominates in equation (10a). In this case β gives the asymptotic slope of the lines in figure 3 (top). Because $k = \ln K$, $T(k)$ is asymptotically proportional to $K^{\beta/\epsilon}$. The average time to extinction thus scales in proportion to a power of the carrying capacity and increases faster than linearly with K when $\beta > \epsilon$. For example, if each catastrophe reduces the population to $e^{-2} = 13.5\%$ of its size just prior to the catastrophe, so that the catastrophe size on the logarithmic scale is $\epsilon = -\ln(e^{-2}) = 2$, then $\beta > 2$ when $\gamma < 0.313$ or $\lambda/r < 0.157$. Thus, the average time to extinction for a population subject to

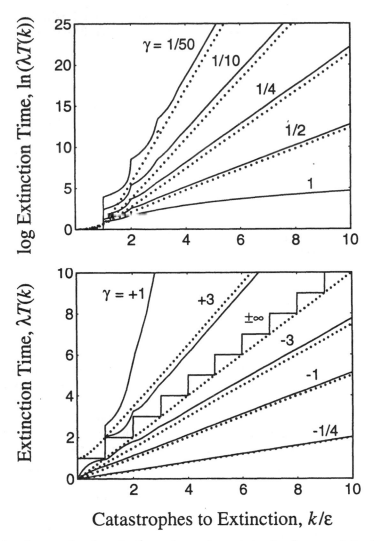

FIG. 3.—Average time to extinction under random catastrophes for a population initially at carrying capacity, $K = e^k$. Different *curves* correspond to the given values of $\gamma = \lambda\epsilon/r$ in which λ is the catastrophe rate, ϵ is the catastrophe size on the logarithmic scale, and r is the growth rate of the population between catastrophes. In both graphs the *abscissa* gives the number of catastrophes from carrying capacity to extinction for a population with $r = 0$ (or $\gamma = \pm \infty$). *Top*, double logarithmic plot for positive long-run growth rates, $r - \lambda\epsilon > 0$ (or $0 < \gamma < 1$). *Bottom*, semilogarithmic plot for negative long-run growth rates, $r - \lambda\epsilon < 0$ (or $0 > \gamma > 1$). *Solid lines* are exact solutions (Appendix), and *dotted lines* give asymptotic approximations (for $\gamma > 0$, eq. [10a] with O(1) = 1, and for $\gamma < 0$, eq. [10c]).

infrequent catastrophes can easily increase much faster than linearly with carrying capacity.

When $\gamma > 1$, the long-run growth rate is negative, and β is negative. In this case the exponential term in equation (10a) is negligible if $-\beta k/\epsilon > 2$, and the asymptotic solution becomes

$$\lambda T(k) \approx \frac{-\gamma}{1-\gamma}\left(\frac{k}{\epsilon} + \frac{1}{\beta}\right) + O(1). \qquad (10b)$$

When $\gamma < 0$, the long-run growth rate is again negative, and the asymptotic solution is simply

$$\lambda T(k) \approx \frac{-\gamma}{1-\gamma}\left(\frac{k}{\epsilon}\right). \qquad (10c)$$

Thus, when the long-run growth rate is negative in a population subject to random catastrophes, $0 > \gamma > 1$, asymptotic formulas for the average extinction time are dominated by a linear function of k, which is a logarithmic function of carrying capacity, $T(k) \approx - (\ln K)/\bar{r}$, as under demographic or environmental stochasticity (eqq. [5c], [8c]).

DISCUSSION

The average time to extinction of a population, starting from carrying capacity K, follows different scaling laws in response to demographic stochasticity, environmental stochasticity, or random catastrophes. When the long-run growth rate of the population is positive, the scaling relationships are as follows. With demographic stochasticity the per capita growth rate fluctuates because of sampling effects caused by finite population size. The average time to extinction increases almost exponentially with carrying capacity, in proportion to e^{aK}/K, where $a = 2\bar{r}/V_1$, in which $\bar{r}$ and V_1 are, respectively, the mean Malthusian fitness and its variance among individuals. The nearly exponential scaling of mean time to extinction with increasing K under demographic stochasticity alone is consistent with results of previous authors (MacArthur and Wilson 1967; Richter-Dyn and Goel 1972; Leigh 1981; Gabriel and Bürger 1992).

Under environmental stochasticity the per capita growth rate fluctuates with temporal changes in the environment. The average extinction time scales as a power of the carrying capacity, proportional to K^c, where $c = 2r/V_e - 1$, in which $\bar{r}$ and V_e are, respectively, the mean and environmental variance in r. The average extinction time therefore scales faster or slower than linearly with K, depending on whether $\bar{r}/V_e$ is greater than or less than one. Positive (upward) curvature of the scaling relationship only requires that $\bar{r} > V_e$. Formulas in Leigh (1981) and Tier and Hanson (1981) are consistent with the scaling of mean extinction time as a power function of K under environmental stochasticity.

Goodman's (1987b) formulas are also consistent with this scaling law, but he claimed that with environmental stochasticity the average extinction time always scales less than linearly with carrying capacity ($c < 1$), which implies that only extremely large populations are likely to persist for long periods in a fluctuating environment. Goodman's erroneous conclusion traces to an inappropriate analogy between environmental stochasticity and the classical birth-death model of demographic stochasticity and his assumption that $\bar{r} < V_e/2$, which implies that the long-run growth rate of the population, $\bar{r} - V_e/2$, is negative. If this assumption were generally valid, few, if any, species would exist today.

It is nevertheless true that, in sufficiently large populations, environmental stochasticity poses a greater risk of extinction than demographic stochasticity. This conclusion follows intuitively from the fact that the importance of demographic stochasticity in causing fluctuations in population growth rate, r, is inversely proportional to population size, whereas fluctuations in r caused by environmental stochasticity may be independent of population size. In the present models with a positive long-run growth rate, the average time to extinction under demographic stochasticity, proportional to e^{aK}/K, increases with K faster than that under environmental stochasticity, proportional to $K^c - 0^{c\ln K}$, regardless of the constants of proportionality or the values of a and c. If the per capita growth rate has a mean larger than its variance ($\bar{r} > V_e$ or $c > 1$), the average persistence time under environmental stochasticity may be extremely long, even for populations of modest size.

Under random catastrophes, the long-run growth rate of the population, $r - \lambda\epsilon$, is positive when the growth rate between catastrophes, r, exceeds the catastrophe rate, λ, multiplied by the catastrophe size on the natural logarithmic scale, ϵ. In this case, the average time to extinction scales in proportion to $K^{\beta/\epsilon}$, where β is the solution of an equation involving $\gamma = \lambda\epsilon/r$ (see eq. [10a]). This scaling is more than linear when the exponent is greater than one, or $\beta > \epsilon$. The average persistence time of a population subject to infrequent catastrophes can easily increase much faster than linearly with carrying capacity. Consequently, a population of modest size may persist for a long time in the presence of random catastrophes, if its long-run growth rate is positive. This analytical conclusion confirms previous numerical results obtained by Hanson and Tuckwell (1981) for a population with logistic growth subject to random catastrophes.

Because average persistence times scale as power functions of carrying capacity under both environmental stochasticity and random catastrophes, no general statement can be made about which is more important in large populations. Their relative importance depends on the values of several parameters in addition to carrying capacity, which in the present models are the mean and environmental variance of per capita growth rate and the magnitude and frequency of catastrophes. The similarity of the scaling laws for extinction risks under environmental stochasticity and random catastrophes makes intuitive sense when catastrophes are viewed as extreme manifestations of a fluctuating environment.

In the random catastrophe model analyzed by Ewens et al. (1987) the logarithmic scaling of average persistence time with population size is a consequence of their assumption that the long-run growth rate of the population is negative. This assumption was necessary in their density-independent model in order for eventual extinction to be a certain event. With density dependence, as in the present catastrophe model or that of Hanson and Tuckwell (1981), eventual extinction is certain even for a population with a positive long-run growth rate below carrying capacity.

The present analysis demonstrates that Goodman (1987a, 1987b), Ewens et al. (1987), and Shaffer (1987) seriously overestimated the risks of population extinction from environmental stochasticity and random catastrophes, as did Soulé and Kohm (1989), Pimm and Gilpin (1989), and Hedrick and Miller (1992). Prevailing views regarding the persistence of populations under environmental stochasticity or random catastrophes have been too pessimistic, at least for populations with a positive long-run growth rate. Empirical evidence may ultimately accumulate to indicate that random catastrophes pose a greater risk of population extinction than does environmental stochasticity. The only general statement supported by existing theory is that in sufficiently large populations demographic stochasticity is a less important cause of extinctions than either environmental stochasticity or random catastrophes.

Unfortunately, for many species threatened with extinction, long-run growth rates of populations are near zero or negative. For populations with a negative long-run growth rate, regardless of the stochastic factors involved, the average time to extinction scales with the natural logarithm of the carrying capacity, divided by the long-run rate of population decline, as previously noted by Ludwig (1976) and Brockwell (1985) for different forms of stochasticity and density dependence. Thus, a large initial size does little to extend the average lifetime of a population with a negative long-run growth rate.

The present results are based on a simple model of density-independent population growth below carrying capacity. They are, however, in qualitative agreement with analytical results of Ludwig (1976), Leigh (1981), and Tier and Hanson (1981) for demographic and environmental stochasticity and numerical results of Hanson and Tuckwell (1981) for random catastrophes, under the logistic model of density dependence. This concordance suggests that the qualitative scaling relationships and relative risks of extinction described here are likely to be general and robust properties of the dynamics of single-population models.

ACKNOWLEDGMENTS

I thank G. Caughley, E. G. Leigh, Jr., M. L. Shaffer, and the reviewers for helpful comments on the manuscript. This work was partially supported by National Science Foundation grant DEB-9225127.

APPENDIX

ANALYSIS OF RANDOM CATASTROPHES

Solution of the random catastrophe model is simplified by the transformations $x = n_0/\epsilon$ and $\tau(x) = \lambda T(n_0)$, so that the catastrophe size becomes one and the mean time to extinction is scaled in units of the mean waiting time between catastrophes. Equation (9) becomes

$$\frac{d\tau(x)}{dx} - \gamma\tau(x) + \gamma\tau(x-1) = -\gamma \quad \text{for } 0 < x < k/\epsilon, \tag{A1}$$

with $\tau(x) = 0$ for $x \leq 0$. The parameter $\gamma = \lambda\epsilon/r$ gives the ratio of the magnitude of the expected rate of change of n from catastrophes to the population growth rate between catastrophes. For $\gamma \geq 0$ the boundary condition is $\tau(k/\epsilon) - \tau(k/\epsilon - 1) = 1$, whereas for $\gamma < 0$ the boundary condition is $\tau(0^+) = 0$.

EXACT SOLUTION USING LAPLACE TRANSFORM

Denote the Laplace transform of $\tau(x)$ as $\bar{\tau}(s) = \int_0^\infty e^{-sx}\tau(x)dx$, where s is a complex variable. Applying the Laplace transform to equation (A1) using the tables in Churchill (1958) yields

$$\bar{\tau}(s) = \frac{-\gamma/s + \tau(0^+)}{s - \gamma(1 - e^{-s})}. \tag{A2}$$

The denominator can be expanded as a power series in $(\gamma/s)(1 - e^{-s})$, and $(1 - e^{-s})^m$ can be expanded using the binomial theorem, which gives

$$\bar{\tau}(s) = \frac{-\gamma/s + \tau(0^+)}{s} \sum_{m=0}^{\infty} \left(\frac{\gamma}{s}\right)^m \sum_{j=0}^{m} \binom{m}{j}(-1)^j e^{-js}. \tag{A3}$$

The inverse Laplace transform of $s^{-m-1}e^{-js}$ is $U(x - j)(x - j)^m/m!$, where $U(x - j)$ is the unit step function, which is zero for $x < j$ and one for $x > j$. Taking the inverse Laplace transform of equation (A3) and interchanging the order of summation, we have

$$\tau(x) = \sum_{j=0}^{\infty} \frac{(-1)^j}{j!}\{-\gamma\Sigma_A(x) + \tau(0^+)\Sigma_B(x)\}\,U(x - j), \tag{A4}$$

where

$$\Sigma_B(x) = \sum_{m=j}^{\infty} \frac{\gamma^m}{(m - j)!}(x - j)^m$$

$$= \gamma^j(x - j)^j e^{\gamma(x-j)}$$

and

$$\Sigma_A(x) = \sum_{m=j}^{\infty} \frac{\gamma^m}{(m - j)!} \frac{(x - j)^{m+1}}{m + 1}$$

$$= \int_j^x \Sigma_B(x)\,dx$$

$$= (-1)^j j!\gamma^{-1}\left\{-1 + e^{\gamma(x-j)}\sum_{i=0}^{j}\frac{(-\gamma)^i(x - j)^i}{i!}\right\}.$$

Substituting the final forms of these summations back into equation (A4), the solution is

$$\tau(x) = [x] + 1 + A(x) + \tau(0^+)B(x), \tag{A5}$$

in which $[x]$ is the integer part of x, for example, $[e] = 2$, and

$$A(x) = -\sum_{j=0}^{[x]} e^{\gamma(x-j)} \sum_{i=0}^{j} \frac{(-\gamma)^i(x - j)^i}{i!} \tag{A6}$$

$$B(x) = \sum_{j=0}^{[x]} \frac{(-\gamma)^j(x - j)^j}{j!} e^{\gamma(x-j)}. \tag{A7}$$

When γ is negative, the boundary condition at $x = 0^+$ causes the last term in equation (A5) to vanish, whereas when γ is positive, the boundary condition at $x = k/\epsilon$ determines the initial value

$$\tau(0^+) = \frac{-A(k/\epsilon) + A(k/\epsilon - 1)}{B(k/\epsilon) - B(k/\epsilon - 1)} . \tag{A8}$$

The average times to extinction for a population starting at carrying capacity, with γ positive, negative, or zero, are as follows. For $0 < \gamma < +\infty$,

$$\lambda T(k) = [k/\epsilon] + 1 + A(k/\epsilon) + \tau(0^+)B(k/\epsilon) . \tag{A9a}$$

For $-\infty < \gamma < 0$,

$$\lambda T(k) = [k/\epsilon] + 1 + A(k/\epsilon). \tag{A9b}$$

For $\gamma = \pm \infty$ (or $r = 0$), equation (A1) reduces to a difference equation with the solution

$$\lambda T(k) = [k/\epsilon] + 1 . \tag{A9c}$$

A high degree of precision is required for numerical evaluation of the exact solutions. Figure 3 was computed with 100 significant digits to evaluate the summations, using Mathematica (Wolfram 1991).

APPROXIMATE SOLUTION BY ASYMPTOTIC METHODS

Figure 3 indicates that the asymptotic form of $\tau(k/\epsilon)$ is exponential for a positive long-run growth rate, $r - \lambda\epsilon > 0$, and linear for a negative long-run growth rate, $r - \lambda\epsilon < 0$. To obtain the asymptotic behavior of $\tau(k/\epsilon)$ with increasing k/ϵ, we must first derive the asymptotic form of $\tau(x)$ with increasing x for a fixed value of k/ϵ. We try a solution to equation (A1) of the form

$$\tau(x) \approx \alpha e^{-\beta x} + ax + b , \tag{A10}$$

where α, β, a, and b are constants to be determined. Substituting formula (A10) into equation (A1) gives

$$-\alpha\{\beta - \gamma(e^\beta - 1)\}e^{-\beta x} + a(1 - \gamma) = -\gamma . \tag{A11}$$

Validity of the solution for all $0 < x < k/\epsilon$ requires that

$$\beta = \gamma(e^\beta - 1) \quad \text{and} \quad a = -\gamma/(1 - \gamma) . \tag{A12}$$

This transcendental equation for β admits a real solution only if $\gamma \geq 0$. If $\gamma < 0$, then we must set $\alpha = 0$, and equation (A10) becomes linear. The boundary condition for $\gamma \geq 0$, along with equation (A12), determines the constant

$$\alpha = \frac{-\gamma}{(1 - \gamma)\beta} e^{\beta k/\epsilon} \quad \text{for } \gamma > 0 . \tag{A13}$$

The last constant can be determined with a small error by requiring (somewhat arbitrarily) for a population with $\gamma > 0$ and $k/\epsilon = 0^+$ that $\tau(0^+) = O(1)$, which is a constant of order 1, giving

$$b = -\alpha + O(1) . \tag{A14}$$

Thus, for $\gamma > 0$ the mean time to extinction starting from x is

$$\tau(x) \approx \frac{\gamma}{1 - \gamma}\left\{e^{\beta k/\epsilon}\left(\frac{1 - e^{-\beta x}}{\beta}\right) - x\right\} + O(1) . \tag{A15a}$$

Note that $\gamma = 1$ is a singular point for the asymptotic solution, which separates two domains of behavior. For $0 < \gamma < 1$, β is positive and the exponential terms dominate for sufficiently large x. The mean extinction time is then almost independent of the initial population size (provided that $\beta x \gg 1$) and nearly equal to that starting at the carrying capacity,

$$\tau(k/\epsilon) \approx \frac{\gamma}{1 - \gamma} \left(\frac{e^{\beta k/\epsilon} - 1}{\beta} - \frac{k}{\epsilon} \right) + O(1) . \qquad \text{(A15b)}$$

For $\gamma > 1$, β is negative and the linear terms in equation (A15a) dominate when $-\beta x \gg 1$,

$$\tau(k/\epsilon) \approx \frac{-\gamma}{1 - \gamma} \left(\frac{k}{\epsilon} + \frac{1}{\beta} \right) + O(1) . \qquad \text{(A16a)}$$

For $\gamma < 0$, $\alpha = 0$ in equation (A11), and the final constant can be determined with a small error by requiring for a population with $k/\epsilon = 0^{+}$ that $\tau(0^{+}) = 0$ (since r is negative), giving $b = 0$, and

$$\tau(x) \approx \frac{-\gamma}{1 - \gamma} x . \qquad \text{(A16b)}$$

Thus, when $0 > \gamma > 1$, the asymptotic form of the mean extinction time is linear for large k/ϵ.

LITERATURE CITED

Abramowitz, M., and I. A. Stegun, eds. 1972. Handbook of mathematical functions. Dover, New York.
Bellman, R., and K. I. Cooke. 1963. Differential-difference equations. Academic Press, New York.
Brockwell, P. J. 1985. The extinction time of a birth, death and catastrophe process and of a related diffusion model. Advances in Applied Probability 17:42–52.
Burgman, M. A., S. Ferson, and H. R. Akçakaya. 1993. Risk assessment in conservation biology. Chapman & Hall, New York.
Churchill, R. V. 1958. Operational mathematics. 2d ed. McGraw-Hill, New York.
Crow, J. F., and M. Kimura. 1970. An introduction to population genetics theory. Harper & Row, New York.
Ewens, W. J., P. J. Brockwell, J. M. Gani, and S. I. Resnick. 1987. Minimum viable population sizes in the presence of catastrophes. Pages 59–68 in M. E. Soulé, ed. Viable populations for conservation. Cambridge University Press, New York.
Gabriel, W., and R. Bürger. 1992. Survival of small populations under demographic stochasticity. Theoretical Population Biology 41:44–71.
Gilpin, M., and I. Hanski, eds. 1991. Metapopulation dynamics: empirical and theoretical investigations. Academic Press, New York.
Gilpin, M., and M. Soulé. 1986. Minimum viable populations: processes of species extinction. Pages 19–34 in M. E. Soulé, ed. Conservation biology, the science of scarcity and diversity. Sinauer, Sunderland, Mass.
Ginzburg, L. R., L. B. Slobodkin, K. Johnson, and A. G. Bindman. 1982. Quasiextinction probabilities as a measure of impact on population growth. Risk Analysis 2:171–181.
Goodman, D. 1987a. The demography of chance extinction. Pages 11–43 in M. E. Soulé, ed. Viable populations for conservation. Cambridge University Press, New York.
———. 1987b. Consideration of stochastic demography in the design and management of biological reserves. Natural Resource Modelling 1:205–234.
Hanson, F. B., and H. C. Tuckwell. 1978. Persistence times of populations with large random fluctuations. Theoretical Population Biology 14:46–61.
———. 1981. Logistic growth with random density independent disasters. Theoretical Population Biology 19:1–18.
Hedrick, P. W., and P. S. Miller. 1992. Conservation genetics: techniques and fundamentals. Ecological Applications 2:30–46.

Karieva, P. 1990. Population dynamics in spatially complex environments. Philosophical Transactions of the Royal Society of London B, Biological Sciences 330:175–190.

Karlin, S., and H. M. Taylor. 1981. A second course in stochastic processes. Academic Press, New York.

Keiding. N. 1975. Extinction and exponential growth in random environments. Theoretical Population Biology 8:49–63.

Lande, R. 1988. Genetics and demography in biological conservation. Science (Washington, D.C.) 241:1455–1460.

Lande, R., and S. H. Orzack. 1988. Extinction dynamics of age-structured populations in a fluctuating environment. Proceedings of the National Academy of Sciences of the USA 85:7418–7421.

Leigh, E. G., Jr. 1981. The average lifetime of a population in a varying environment. Journal of Theoretical Biology 90:213–239.

Lewontin, R. C., and D. Cohen. 1969. On population growth in a randomly varying environment. Proceedings of the National Academy of Sciences of the USA 62:1056–1060.

Ludwig, D. 1976. A singular perturbation problem in the theory of population extinction. Society for Industrial and Applied Mathematics–American Mathematical Society Proceedings 10: 87–104.

MacArthur, R. H., and E. O. Wilson. 1967. The theory of island biogeography. Princeton University Press, Princeton, N.J.

May, R. M. 1974. Complexity and stability in model ecosystems. 2d ed. Princeton University Press, Princeton, N.J.

Pimm, S. L., and M. E. Gilpin. 1989. Theoretical issues in conservation biology. Pages 287–305 *in* J. Roughgarden, R. M. May, and S. A. Levin, eds. Perspectives in ecological theory. Princeton University Press, Princeton, N.J.

Richter-Dyn, N., and N. S. Goel. 1972. On the extinction of a colonizing species. Theoretical Population Biology 3:406–433.

Shaffer, M. L. 1981. Minimum population sizes for species conservation. BioScience 31:131–134.

———. 1987. Minimum viable populations: coping with uncertainty. Pages 69–86 *in* M. E. Soulé, ed. Viable populations for conservation. Cambridge University Press, New York.

Soulé, M. E., and K. A. Kohm. 1989. Research priorities for conservation biology. Island, Washington, D.C.

Soulé, M., and D. Simberloff. 1986. What do genetics and ecology tell us about the design of nature preserves? Biological Conservation 35:18–40.

Tier, C., and F. B. Hanson. 1981. Persistence in density dependent stochastic populations. Mathematical Biosciences 53:89–117.

Tuljapurkar, S. D. 1982. Population dynamics in variable environments. III. Evolutionary dynamics of *r*-selection. Theoretical Population Biology 21:141–165.

Turelli, M. 1977. Random environments and stochastic calculus. Theoretical Population Biology 12:140–178.

Wolfram, S. 1991. Mathematica: a system for doing mathematics by computer. 2d ed. Addison-Wesley, New York.

Associate Editor: Stephen W. Pacala

6 Evolutionary and Behavioral Ecology

Joseph Travis and F. Helen Rodd

Introduction

We often teach our students how a new discovery transforms science. From unearthing *Archaeopteryx* to describing the polymerase chain reaction, we try to convey the excitement that comes from seeing something completely new, and often wholly unexpected. We sometimes neglect to tell our students that science can be transformed by seeing something old in a new light, by developing a new way of thinking about what we have been observing.

In the period from the early 1960s through the early 1980s, substantial areas in organismal ecology were transformed not as much by discovery as by a new way of thinking. This new way of thinking, "selection thinking," was pioneered by David Lack's (1947, 1948) studies of clutch size in birds, championed by Gordon Orians (1962) in his proposal for an "evolutionary ecology," and given a mathematical framework in seminal papers by Emlen (1966) and MacArthur and Pianka (1966). Selection thinking asks *why* a certain feature might be favored in particular ecological conditions (Charnov 1982) or, to paraphrase Dobson 2013: What are the likely evolutionary advantages of particular behaviors or traits, given a particular environment?

Selection thinking brought evolutionary biology and ecology closer together (Collins 1986). The emphasis in selection thinking is on how an organism's particular trait or feature affects its fitness. For example, Lack thought about clutch sizes not just in terms of which clutch sizes would balance mortality at the population level, but in terms of how different clutch sizes affected individual fitness and how those effects promoted the optimal clutch size in a population. That we take this approach for granted is testimony to how profoundly it changed how ecologists think and how deeply it embedded evolutionary biology into ecology.

Selection thinking spawned two approaches to the study of individual traits and behaviors. In the first approach, ecologists sought to define optimal strategies for particular circumstances and recognize the causes behind associations of traits and environmental conditions. The papers by Fretwell and Lucas (1969) on territorial behavior and Charnov (1976) on foraging behavior built mathematical theories, informed by empirical observation, that generated novel, testable predictions. The papers by Grime (1977) on plant life histories, Greenwood (1980) on dispersal strategies of animals, and Coley et al. (1985) on antiherbivore defensive strategies in plants each represent a synthesis of large empirical literatures. The authors used those syntheses to find repeatable patterns with which to test competing ideas and offer provocative new hypotheses.

In the second approach, ecologists performed experimental studies in the field that directly examined the fitness consequences of different traits or strategies. The paper by Werner et al. (1983) examined the complex interplay of foraging and predation risk in governing habitat use in sunfish of different sizes. Berenbaum et al. (1986) scrutinized the genetic basis of a plant defensive compound, the fitness consequences of variation in insect herbivory and plant defense, and the constraints and trade-offs inherent in diverting resources to defense to identify the ecological standoff to which both plant and herbivore had evolved. Addicott (1986) studied the trade-offs incurred by yucca plants in their interactions with a moth that provided a benefit through pollination but also a detriment through herbivory.

A formal blending of selection thinking with organismal ecology was promoted by the methodological paper written by Lande and Arnold (1983) on the measurement of phenotypic selection. Here they developed a statistical method for integrating empirical observations on fitness or a surrogate for fitness and variation among individuals in multiple traits into the framework of theoretical quantitative genetics. Ecologists studying a wide variety of organisms used the methods in this paper to quantify the direction and intensity of selection in a variety of contexts, from surviving predators to acquiring mates. The methods also allowed ecologists to quantify certain values of individual features that could reinforce one another's effect. For example, Brodie (1992) showed that the contribution of a garter snake's escape behavior to its probability of surviving a predation attempt depended upon its dorsal pattern; some combinations had very high probabilities of survival and others, very low probabilities.

Evolutionary biology is more than simply the study of natural selection, and the con-

tribution of evolution to ecological pattern is more than just the adaptive molding of character combinations. In this period, biologists also began integrating broader evolutionary themes into ecology that put selection thinking into perspective. An outgrowth of selection thinking was a panoply of comparative studies that examined associations of characters among species within a taxonomic group. The goal of these studies was to identify patterns of correlated evolution among characters and generate hypotheses about how natural selection had molded these associations (Tinkle et al. 1970). Felsenstein (1985) and Herrera (1992) looked at comparative data differently and asked whether associations among characters might reflect, at least in part, the influence of historical evolution. Felsenstein (1985) drew attention to the role of phylogenetic relationships in creating such associations, and Herrera (1992) illustrated how biogeographic history could be responsible for them. Both papers were pioneering demonstrations that evolution influences ecology in many ways besides the action of natural selection.

The Growth of Theory in Behavioral Ecology

Stephen Fretwell and Henry Lucas, in their paper "On territorial behavior and other factors influencing habitat distribution in birds I. Theoretical development" (1969), set out to investigate a longstanding question about avian behavior: why do birds hold territories? The paper took a roundabout route to answer the question. Fretwell and Lucas began by recognizing the connection between territoriality and the distribution of individuals across a range of habitats. They asked how birds ought to distribute themselves across different habitats, given that habitats differed in fundamental suitability (defined as the expected fitness of a single individual in that habitat) and that any habitat's actual suitability would decrease as density in that habitat increased. A bird "deciding" where to settle would face a choice between a very good but crowded habitat and a less suitable but less crowded one.

Within this scenario, Fretwell and Lucas made two assumptions. First, they assumed that an individual could exercise an *ideal* choice, selecting the habitat best suited to its survival and reproduction. Second, they assumed that an individual was *free* to enter any habitat. The model they developed from these two assumptions predicted that individuals would choose the best habitat, that is, the habitat with the highest level of fundamental suitability, until the density in that habitat had lowered its suitability to the fundamental level of the next-best habitat. Individuals would then choose the second-best habitat until its suitability decreased to the fundamental level of the third-best habitat, and so on. The result would be an *ideal free* distribution of density among habitats. Fretwell and Lucas then asked how each of the postulated functions of territoriality would create deviations from the ideal free distribution in the actual distribution of birds across habitats.

This paper inspired ecologists of all stripes. One reason was that the authors brought simplicity and clarity to a subject (including terminology) that had sorely lacked them. Another was that the model synthesized fundamental concepts of ecology (density dependence), evolution (expected fitness), and behavior (territoriality). It was among the first papers to present a null model of sorts, against which empirical data could be compared to deduce a conclusion. Null models would emerge in the late 1970s and early 1980s as controversial tools for investigating ecological processes that were not amenable to ready experimentation. Finally, it was an early example of what would come to be called an eco-evo interaction; Fretwell and Lucas recognized that differences in habitat suitability would select for the habitat selection criteria used by individuals, which in turn would drive patterns of habitat occupancy as functions of density.

The ideal free distribution has been a foundation for investigating the distribution among habitats of a wide variety of organisms (Haugen et al. 2006; Moritz et al. 2014; Paterson and Blouin-Demers 2018). Some of the early applications of the ideal free distribution were subject to considerable methodological criticism (Kennedy and Gray 1993; Tregenza 1994). However, the concept has proven to be vital in ecology and new, more general models (Menezes and Kotler 2019) continue to reflect the importance of the ideas embodied in its original presentation.

This paper is also an excellent practical example for young scientists for two reasons First, a knowledge of natural history can be the foundation for important conceptual work. Fretwell's original doctoral research on earthworm brains was not going well, so he switched to thinking about the subjects of his lifelong hobby, birdwatching, and about some ideas about territorial behavior in the literature that did not sit well with his intuition about birds (Fretwell 1991). It was this contrast that motivated Fretwell to develop his own ideas.

Second, in the words of Henry Lucas: "Mathematical model building was the cure to fuzzy thinking in science" (Fretwell 1991, 8). In many cases, the implicit assumptions in a verbal hypothesis are not clear. As a result, criticisms of a hypothesis may be hard to communicate. Fretwell had grown frustrated with his inability to convey his ideas to his listeners. Once he followed Lucas's advice and started building a mathematical model, his listeners understood his points immediately. Moreover, the building of that model and the exploration of assumptions behind it revealed that there were more facets to the debates over territoriality than had been appreciated.

Not only will habitats differ in suitability, as Fretwell posited, but when food is patchily distributed, some patches will be more rewarding than others. Interest in diets and diet choices in a world of patchy environments had grown rapidly in the years since the pioneering work of Emlen (1966) and MacArthur and Pianka (1966). Eric Charnov's (1976) paper, "Optimal foraging: the marginal value theorem," took this problem to another level. Charnov recognized that a foraging animal will deplete food in a patch and he framed the question of optimal foraging differently. Given that a forager will deplete food in a patch, given that patches vary in quality, and given that there is a frequency distribution of patch quality (i.e., patches of different quality are not equally abundant), how long should an optimal forager exploit a patch before moving to a new one?

Charnov called his answer to this question *the marginal value theorem*. It produced a simple yet elegant prediction: a forager should leave a patch when the food capture rate in that patch drops to the average capture rate in the entire habitat. Charnov reinterpreted experimental data reported by Krebs et al. (1974) and showed that those data fulfilled the predictions of the marginal value theorem. A large volume of subsequent work has reinforced this theorem's validity.

There were a great many papers on foraging theory in this period (reviewed in Pyke et al. 1977); why did Charnov (1976) become so vital? For one thing, Charnov recast optimal foraging theory into a question that drew the attention of ecologists and behavioral biologists interested in animals that were choosing and lingering among patches in a heterogeneous environment. For another, the paper showed the complementarity of theory and data: data could motivate new theory and new predictions could give direction for new experiments. A third reason, articulated by Beckerman et al. (2010), is that foraging and diet are key traits linking selection with population dynamics, community structure, and the organization of food webs. While optimal foraging theory was quickly extended to incorporate the costs of travel between patches (Cowie 1977), the trade-off between foraging and vigilance against predation risk (Milinski and Heller 1978), and the effects of time constraints on foraging (Lucas 1985), the simplicity and elegance of the marginal value theorem has never lost its appeal.

The complementarity of theory and data is also illustrated in Paul Greenwood's (1980) study of gender-biased dispersal in birds and mammals, "Mating systems, philopatry and dispersal in birds and mammals" (not reprinted in this volume). When Greenwood undertook this effort, it was widely appreciated that in mammals, males typically dispersed from their natal area. Greenwood's initial contribution was to compile an extraordinary amount of banding and recapture data for birds and show that in birds, females typically dispersed from the natal area. He also reviewed data from the literature on mammals and verified that male-biased dispersal was the rule in this group.

The power of Greenwood's paper stems not just from documenting the contrast between birds and mammals, but from his elegant analysis of these patterns. He considered the fitness benefits of both dispersal and philopatry in the context of newly emerging theory on the evolution of mating systems (Bradbury and Vehrencamp 1977; Emlen and Oring 1977). In his own words:

What are the salient ecological or behavioural factors that account for the apparent differences in sex biased dispersal between birds and mammals? The problem is not one of producing alternative hypotheses for birds and mammals per se; rather it is to identify underlying similarities that may account for female biased as opposed to male biased dispersal, irrespective of the vertebrate class. (Greenwood 1980, 1149)

He considered several candidate hypotheses including constraints of phylogenetic heritage, sex chromosomes, population density, inbreeding avoidance, mate-defense polygyny, and resource-defense polygyny. Greenwood concluded that a hypothesis based on two mating systems, one (mate-defense) more characteristic of mammals and the other (resource-defense) more characteristic of birds, offered the best explanations for the fidelity and dispersal patterns seen in these two taxa as well as the notable exceptions (Doherty et al. 2002). Although Greenwood overlooked another possible factor, kin competition (Dobson 2013), his synthesis has stood as a remarkable example of the power of bringing theory to bear on large compilations of data.

For young scientists, Greenwood's paper has another appeal: it grew from his dissertation research at the University of Exeter (Dobson 2013). Like Fretwell's paper, this one illustrates how a young scientist can take a new look at what others have seen and change how everyone else thinks about the topic.

Selection thinking also inspired sweeping new ideas about the fitness advantages of having specific suites of traits; these ideas were bolstered by observations that some clusters of traits were shared by many species and associated with particular environmental conditions. Among the earliest of these efforts was Martin Cody's exploration of interspecific variation in clutch size in birds and related life-history attributes as functions of latitude (Cody 1966). This approach to trait clusters became most widely known through Eric Pianka's discussion of r- and K-selection and the suites of life-history traits expected to evolve under these contrasting ecological scenarios (Pianka 1970). While many papers of this type made important contributions to our understanding of trait clusters, four of them have been especially critical to ecology.

While some plant ecologists were also attracted to the r-K selection paradigm (Harper and Ogden 1970; Gadgil and Solbrig 1972), Phillip Grime's (1977) paper, "Evidence for the existence of three primary strategies in plants and its relevance to ecological and evolutionary theory," took a different approach. Grime described three environmental gradients: intensity of competition, intensity of stress, and frequency and intensity of disturbance. He postulated that, at certain combinations along these gradients, selection would favor certain features of plants. He postulated that when stress and disturbance were both of low intensity, selection would favor competitive ability (so-called C-selection). When the intensity of disturbance was low and intensity of stress high, selection would favor stress tolerators (S-selection); when the intensity of stress was low and disturbance high, a ruderal lifestyle would

be favored (R-selection). No plants would be viable under the high-stress, high-disturbance conditions. Grime created a graphical model in the shape of a triangle, with each side corresponding to one of his three gradients and each vertex corresponding to a combination of extremes. Grime went on to associate specific positions in this graphical model with trait combinations expected at various stages of vegetational succession and attempted to predict the circumstances under which mutualistic associations, like lichens, would evolve.

Grime's paper represented a clear departure from other papers of this type because he developed a theory specifically oriented to plants. He combined current ideas in plant ecology with selection thinking and offered an original synthesis. Grime also inspired considerable disagreement over his definition of stress and his contention that stressful environments would not be competitive ones (Tilman 1987; Grace 1991; Aerts 1999). While subsequent work suggested that these relationships are more complex than Grime's basic paradigm (Li and Watkinson 2000), Grime's paper provoked considerable new work and inspired a long history of novel insights into plant life histories.

Integrating Evolution into Organismal Ecology

In their paper, "Resource availability and plant herbivore defense," Phyllis Coley, John Bryant, and Stuart Chapin (1985) focused their attention on a vital feature of plant life histories: the allocation of energy to antiherbivore defenses and the type of defensive strategy employed. Like Grime's and other papers in this collection, they developed simple hypotheses for patterns of plant defense shared by many species that were associated with particular environmental conditions.

When Coley et al. wrote their paper, the predominant hypothesis for plant defenses was Paul Feeney's "apparency theory" (Feeney 1976; Cates and Rhoades 1977). Feeney hypothesized that long-lived plants—"apparent" because longevity increases the cumulative probability of being found and attacked—would be susceptible to generalist and specialist herbivores and would be selected to invest in high concentrations of defensive compounds. In contrast, less apparent, short-lived plants would be selected only to evade generalists with low concentrations of defensive compounds of low molecular weight like alkaloids. Data were consistent with this very simple, elegant hypothesis.

Coley et al. (1985) offered an entirely different hypothesis for the same correlations, which came to be called the resource acquisition hypothesis (RAH). They argued that these correlations emerged from the cost-benefit ratios of investing in defensive compounds. Further, they posited that those cost-benefit ratios would be different in plants found in resource-rich environments from those in plants found in resource-limited environments. As a result, the relative limitation of different resources, specifically carbon and nitrogen, would constrain the type of chemical defense employed.

The paper was an intellectual tour de force in deploying an enormous literature (there are 122 citations in the paper) to support a novel viewpoint about a well-established set of associations. The RAH remains actively debated; support has come from a meta-analysis of among-species variation (Endara and Coley 2011) but not from an analysis of intraspecific variation (Hahn and Maron 2016) or from a phylogenetically controlled analysis of defensive compounds in 56 species of oak (Moreira and Pearse 2017).

The reference above to the phylogenetically controlled analysis of Moreira and Pearse (2017) illustrates the sweeping influence of Joseph Felsenstein's (1985) paper, "Phylogenies and the comparative method" (1985) The clusters of traits that emerged in all of the early comparative analyses were presumed to be the result of natural selection. For example, Tinkle et al. wrote:

The distinct strategies that we have deduced imply that each strategy is a co-evolved complex of adaptations in morphology and ecology, as well as in reproductive physiology. (Tinkle et al., 69)

This conclusion was reinforced by observing that these trait associations usually cut across taxonomic lines, implying that they had evolved multiple times. There is a problem here, of which modern researchers are well aware. A standard statistical analysis of trait variation among species cannot distinguish a scenario in which the same cluster of traits had evolved repeatedly in many species from one in which the cluster evolved in a few ancestral species whose descendants proliferated, along with the same clusters of traits.

This problem had been acknowledged in many of the papers reporting comparative analyses, and several investigators had suggested different methods for addressing it or, sometimes, working around it. Felsenstein's paper offered a general solution based on the species' phylogenetic relationships. The key insight that Felsenstein brought to the problem was that the differences in trait values between two sister taxa are independent of the differences in trait values between another pair of sister taxa.

Felsenstein's solution, now known as the method of phylogenetically independent contrasts, has two simple premises. First, the difference in trait values between two sister taxa depends only on events occurring since they diverged from their common ancestor. Second, the magnitude of those differences will be proportional to the time elapsed since the two taxa diverged. From these premises, along with the fact that differences between sister taxa are independent, phylogenetically, from differences between other sets of sister taxa, Felsenstein derived a simple method of scaling species differences by time to estimate how strongly two variables are correlated, given the phylogeny of the species involved.

Felsenstein's paper sparked a revolution in comparative biology and inspired the development of many additional methods for taking phylogenetic heritage into account (e.g., Adams and Collyer 2019). At first reading, the paper may seem more of a contribution to evolutionary biology than ecology. However, it has had a tremendous impact in ecology. In fact, as of November 2019, it had been cited by more papers classified by Web of Science as "ecology" than by papers classified as "evolutionary biology." Ecologists have used phylogenetic comparative methods to assess the role of evolutionary history in forming present-day ecological associations. The results, not surprisingly, have been highly variable. For example, Barrow et al. (2019) found that host immunity in avian parasite-host interactions was deeply conserved phylogenetically, indicating the long reach of evolutionary history for understanding present-day host-parasite interactions. Liu et al. (2019) showed that phylogenetic affinity was a significant factor in determining patterns of coexistence of grasses on the Mongolian steppe. By contrast, Quezada et al. (2018) found that the incidence of CAM photosynthesis in Chilean bromeliads, while related to evolutionary history, was more strongly determined by climatic factors.

Our fourth paper in this category also illustrated the importance of history for creating character associations. However, in his paper "Historical effects and sorting processes as explanations for contemporary ecological patterns," Carlos Herrera (1992) looked not at phylogenetic history but at biogeographic history. This paper focused on trait clusters in the Mediterranean woody plants of southwestern Spain. Herrera showed that the appearance of "syndromes" was the result of the flora's being composed of two distinct historical groups, plants that had appeared in this region in the Pliocene or pre-Pliocene, and plants that had appeared in this region afterward. Within each group, character associations were weak (the older group) or nonexistent (the younger group). Herrera went further to use evolutionary history to demonstrate that the appearance of trait clusters was produced by differential proliferation of taxa in each period, not differential extinction.

The importance of history is widely recognized in modern ecology but, in 1992 this was not the case. Analyses that take history into account have produced some surprising results. For example, phytogeographic history, not convergent adaptation, is responsible for the similar leaf characteristics of distantly related sclerophyllous Mediterranean plants (Onstein and Linder 2016). History also helps explain paradoxical patterns. Wilman et al. (2014) showed that the puzzling age structure of ten subtropical thicket canopy species in South Africa—a rarity of seedlings and young plants—could be explained by a mismatch between current climate and the plants' germination requirements. The plants had evolved in the early Cenozoic, when the climate was much warmer and wetter, and their germination requirements for mesic soil are met only rarely in the modern climate.

The assumption underlying selection thinking about clusters of traits is that the combination of traits is adaptive. Even if a historical analysis shows that two or more traits have indeed evolved together multiple times, that analysis cannot distinguish among multiple hypotheses for why the combination has evolved. One can ask if selection works on all traits, if selection works on only one trait but the others represent correlated responses driven by genetic linkage or pleiotropy, or if there is some special property of the combination of traits that enhances the success of individuals with the best combination of trait values.

Russell Lande and Stevan Arnold (1983) presented a method for answering these questions in their paper, "The measurement of selection on correlated characters." This paper made two vital contributions (see longer discussion in Reznick and Travis 1996). First, the paper presented a statistical method for quantifying different types of phenotypic selection (directional, stabilizing, and disruptive) using multiple regression, a technique with which most ecologists were comfortable. The emphasis on phenotypic selection, rather than fitness differences among genotypes, opened the door to a wide variety of field studies on populations for which genotypic data would be difficult to obtain. Second, it placed the parameters of those regression models, which were the estimates of selection intensity, into the context of quantitative genetic models of evolution. This made it possible to bring ecological data directly into evolutionary genetic theory, and as a result, studies of selection on continuous traits increased almost exponentially.

The flood of research inspired by this paper offered insights into both evolution and ecology, particularly through reviews and meta-analyses of the accumulated results. On the evolutionary side, these analyses revealed that strong selection was often detected (Hereford et al. 2004), that temporal fluctuations in the direction of selection were common (Siepielski et al. 2009), and that selection was typically more intense on males than females (Singh and Punzalan 2018). On the ecological side, investigations focused on the agents of selection. For example, meta-analyses of selection on floral traits verified the received wisdom that the strongest agents of selection were pollinators (Caruso et al. 2019). On the other hand, meta-analyses did not validate a traditional prediction that biotic agents of selection would prove stronger than abiotic agents (Caruso et al. 2017).

Evolutionary and Ecological Trade-Offs

In some cases, correlated traits will not represent an adaptive cluster but might instead be the result of constrained evolution. Postulated trade-offs among traits have always been at the heart of selection thinking: offspring number vs. offspring size, dispersal ability vs. competitive ability, foraging vs. predator avoidance. The integration of evolutionary biology into ecology in this period included the explicit search for negative genetic correlations that would reflect inherent trade-offs that guided the direction of adaptation.

The paper by May Berenbaum, Arthur Zan-gerl, and James Nitao (1986), "Constraints on chemical coevolution: wild parsnips and the parsnip webworm," was a seminal exploration of trade-offs and selection. It examines the basic assumption beneath the plant apparency and resource acquisition hypotheses for plant secondary compounds, which is that these compounds actually deter herbivory and increase the fitness of individuals who deploy them. When Berenbaum et al. began their work, there was substantial evidence that herbivory could decrease plant survival and reproduction. However, evidence that there was genotypic variation within plant populations for resistance to herbivory was only beginning to emerge (Maddox and Root 1987), there was very little evidence that the chemical phenotypes of plants would be under measurable selection by herbivores, and little was known about the trade-offs, demographic and genetic, that might constrain the evolution of chemical defenses.

Berenbaum and colleagues had discovered a population of wild parsnip that exhibited wide phenotypic variation in the concentrations of four types of defensive chemicals, furanocoumarins, in their tissues. They planted families of plants in the field to expose them to natural levels of herbivory from parsnip webworms and found that a very high proportion of the variation in concentrations of these chemicals was genetically based and that herbivores exerted strong selection on their concentrations. The critical result was discovering the trade-offs among traits that contributed to fitness. They found negative genetic correlations between the concentrations of two furanocoumarins in seeds, bergapten and sphondin; genotypes with more bergapten had less sphondin. In addition, while increasing furanocoumarin concentrations decreased herbivory, there was a negative genetic correlation of furanocoumarin concentrations with seed production. The conclusion was that the chemical arms race between parsnip and parsnip webworms had reached a "stalemate."

The study of plant-herbivore interactions has revolved heavily around chemical defenses, as we discussed earlier in reviewing Coley et al.

(1985). Berenbaum et al.'s paper brought the direct study of evolutionary trade-offs to the scrutiny of hypotheses for variation in plant defenses. It represented one of the first complete applications of the quantitative genetic paradigm advanced by Lande and Arnold (1983), and the paper's discussion of trade-offs among defensive compounds continues to inspire new research into the incidence and nature of comparable trade-offs (Kariñho-Betancourt and Núñez-Farfán 2015).

Trade-offs of a different kind in a different plant-insect system were the critical foci of John Addicott's (1986) careful study of the interaction between yucca moths and eight species of yucca in the western United States, "Variation in the costs and benefits of mutualism: the interaction between yuccas and yucca moths." In this case, the trade-offs are the costs and benefits of the mutualism between the moth and the plant.

Most of the papers in this volume that address species interactions are focused on competition, predation, or both. Papers on interactions that provide mutual benefit between two species are much less common in the ecology literature, even though mutualisms are one of the key mechanisms in generating biodiversity (Waser and Ollerton 2006; Thompson 2013; Bascompte and Jordano 2014; Bronstein 2015a). Part of the reason for this gap is that traditionally, studies of mutualism were often descriptions of the natural history of the species, rather than tests of broader concepts. This has hampered a conceptual unification of mutualism with other kinds of species interactions (Bronstein 1994).

The resurgence of interest in mutualism (Bronstein 2015b) is an outgrowth of outstanding papers like Addicott's, which set an early standard for the quantitative study of mutualisms. Female yucca moths act as pollinators of yucca flowers but also deposit eggs into the locule of the ovary. The larvae of the moths feed on the seeds of the host plants, creating a conflict between the moth's benefit to the plant as a pollinator and its cost to the plant as a seed predator. Addicott's paper made it clear that

understanding any mutualistic interaction requires explicitly calculating the costs and benefits for the species involved. The paper is one of the first to do this, as well as providing a model study for how these costs and benefits can be studied more widely.

Although calculating costs and benefits of a mutualism seems straightforward in concept, in practice, it can be very difficult. Costs and benefits often have different currencies and can vary across space and time, making comparisons difficult. The great advantage of Addicott's paper is his choice of this particular mutualism, in which costs and benefits for the plant can be measured through the same currency, namely the number of seeds.

Addicott compared the cost-benefit ratio for eight species of yuccas that all interact with the same species of yucca moth. His approach rested on dissecting fruits and counting the total number of seeds eaten and uneaten, viable and unviable, while also counting the total number of moth larvae produced. With these simple counts, Addicott determined that the costs of this interaction range between 0 and 30% of the total benefit for the yucca plants.

This simple quantification of the ratio between cost and benefit had at least two long-lasting influences in later studies of mutualism. First, this paper goes a long way in showing that populations and communities are shaped by interactions that cannot always unequivocally be classified as of a single type or even into just competition and predation, thus adding complexity to our understanding of natural communities. This is particularly relevant in the context of complex landscapes, where the net effect of an interaction may be driven by the presence or absence of copollinators (Thompson 2005). Second, Addicott demonstrated the value of quantifying the basic economic variables in a relationship. This approach is at the core of modern approaches for understanding the ecology and evolution of mutualisms, based on concepts of trade within biological markets (see Bronstein 2019 for key references).

Finally, sometimes the failure of a theoretical model can lead to intriguing results that in turn open entirely new areas of study. This is precisely the story behind the paper "An experimental test of the effects of predation risk on habitat use in fish," by Earl Werner, James Gilliam, Donald Hall, and Gary Mittelbach (Werner et al. 1983).

The field experiment reported in the paper was motivated by the disparity between field observations about habitat use in fishes and predictions derived from a well-formulated, carefully parameterized model of optimal foraging (Hall and Werner 1977; Mittelbach 1981). Some size classes of bluegill sunfish were spending more time in less productive, shallow weed beds than expected.

Werner et al. postulated that the presence of a predator, the largemouth bass, would cause sunfish in the small, vulnerable size-classes to use a less profitable but safer habitat, and they tested this idea with an experiment that is a model of thoughtful design and execution. In their experimental pond, they examined the habitat use of bluegills of different sizes, which varied in their vulnerability to a bass, in response to the presence or absence of the predator. There were two habitat types: food-rich areas where smaller fish would be vulnerable to the predator and food-poor areas in which there was a reduced risk of predation. The result was clear: smaller fish used suboptimal habitat, which enhanced their survival in the presence of the predator, but at the cost of lower growth rates.

This paper was very influential. The trade-off between growth and the risk of mortality has proven to be a major ecological concept that plays an important role in understanding habitat use and foraging modes in animals, the timing of life-history transitions, and the coexistence of species (Gilliam and Fraser 1987; Ludwig and Rowe 1990; McPeek et al. 2001; Zhu et al. 2017). A large body of experiments on the indirect effects of predators, now usually called "nonconsumptive effects" for clarity, has

shown that those indirect effects are pervasive and can be at least as influential as their direct effects (Kenison et al. 2016; Silberbush et al. 2019). These effects are wide reaching, from effects on individuals to the structure of ecological communities (see also the papers in parts 1 and 2 in this volume).

The authorship of this paper presages the changes that were to come to the preparation of ecological papers, from one or two authors to many authors (see the general introduction to this volume). While a paper with three authors was not unusual at the time, the authors of this paper had different areas of expertise and were at different stages of their careers. In the words of all four coauthors:

Another lesson was the value of combining the talents of individuals of different abilities. The four of us (two faculty and two graduate students) each contributed different ideas and talents leading to a collaboration that blossomed into what makes a career in ecology so appealing. Through this collaboration we were able to conduct the study at a depth that would not have been possible for any of us alone. We had a great deal of fun doing the work together and it led to great friendships. (pers. comm. to F. Helen Rodd)

The papers in this section reflect a changing view of organismal ecology, from the advent of selection thinking and its application to a range of problems to the construction of sophisticated models for studying evolution and selection in ecological contexts. They also reflect how organismal ecology was influencing population and community ecology, from the nonconsumptive effects of predators to the ways that plants could influence the dynamics of their herbivores. While these papers had different immediate goals, in the long run, collectively, they paved the way for the integrated ecology we have today.

Literature Cited

Adams, D. C., and M. L. Collyer. 2019. Phylogenetic comparative methods and the evolution of multivariate phenotypes. Annual Review of Ecology, Evolution, and Systematics 50: 405–425.

Addicott, J. F. 1986. Variation in the costs and benefits of mutualism: the interaction between yuccas and yucca moths. Oecologia 70: 486–494.

Aerts, R. 1999. Interspecific competition in natural plant communities: mechanisms, trade-offs and plant-soil feedbacks. Journal of Experimental Botany 50: 29–37.

Barrow, L. N., S. M. McNew, N. Mitchell, S. C. Galen, H. L. Lutz, H. Skeen, T. Valqui, et al. 2019. Deeply conserved susceptibility in a multi-host, multi-parasite system. Ecology Letters 22: 987–998.

Bascompte, J., and P. Jordano. 2014. Mutualistic Networks. Princeton University Press.

Beckerman, A. P., O. L. Petchey, and P. J. Morin. 2010. Adaptive foragers and community ecology: linking individuals to communities and ecosystems. Functional Ecology 24: 1–6.

Berenbaum, M. R., A. R. Zangerl, and J. K. Nitao. 1986. Constraints on chemical coevolution: wild parsnips and the parsnip webworm. Evolution 40: 1215–1228.

Bradbury, J. W., and S. L. Vehrencamp. 1977. Social organization and foraging in emballonurid bats: III. Mating systems. Behavioral Ecology and Sociobiology 2: 1–17.

Brodie, E. D. 1992. Correlational selection for color pattern and antipredator behavior in the garter snake *Thamnophis ordinoides*. Evolution 46: 1284–1298.

Bronstein, J. L. 1994. Our current understanding of mutualism. Quarterly Review of Biology 69: 31–51.

Bronstein, J. L., ed. 2015a. Mutualism (1st ed.). Oxford University Press.

Bronstein, J. L. 2015b. The study of mutualism. Pages 3–19 in J. L. Bronstein, ed. Mutualism. Oxford University Press.

Bronstein, J. L. 2019. Mutualisms and Symbioses. Oxford University Press.

Caruso, C. M., K. E. Eisen, R. A. Martin, and N. Sletvold. 2019. A meta-analysis of the agents of selection on floral traits. Evolution 73: 4–14.

Caruso, C. M., R. A. Martin, N. Sletvold, M. B. Morrissey, M. J. Wade, K. E. Augustine, S. M. Carlson, et al. 2017. What are the environmental determinants of phenotypic selection? A meta-analysis of experimental studies. American Naturalist 190: 363–376.

Cates, R. G., and D. F. Rhoades. 1977. Patterns in the production of antiherbivore chemical defenses in plant communities. Biochemical Systematics and Ecology 5: 185–193.

Charnov, E. L. 1976. Optimal foraging: the marginal value theorem. Theoretical Population Biology 9: 129–136.

Charnov, E. L. 1982. The Theory of Sex Allocation. Princeton University Press.

Cody, M. L. 1966. A general theory of clutch size. Evolution 20: 174.

Coley, P. D., J. P. Bryant, and F. S. Chapin III. 1985. Resource availability and plant antiherbivore defense. Science 230: 895–899.

Collins, J. P. 1986. Evolutionary ecology and the use of natural selection in ecological theory. Journal of the History of Biology 19: 257–288.

Cowie, R. J. 1977. Optimal foraging in great tits (*Parus major*). Nature 268: 137–139.

Dobson, F. S. 2013. The enduring question of sex-biased dispersal: Paul J. Greenwood's (1980) seminal contribution. Animal Behaviour 85: 299–304.

Doherty, P. F., J. D. Nichols, J. Tautin, J. F. Voelzer, G. W. Smith, D. S. Benning, V. R. Bentley, et al. 2002. Sources of variation in breeding-ground fidelity of mallards (*Anas platyrhynchos*). Behavioral Ecology 13: 543–550.

Emlen, J. M. 1966. The role of time and energy in food preference. American Naturalist 100: 611–617.

Emlen, S., and L. Oring. 1977. Ecology, sexual selection, and the evolution of mating systems. Science 197: 215–223.

Endara, M.-J., and P. D. Coley. 2011. The resource availability hypothesis revisited: a meta-analysis: revisiting the resource availability hypothesis. Functional Ecology 25: 389–398.

Feeney, P. 1976. Plant apparency and chemical defense. Recent Advances in Phytochemistry 10: 1–40.

Felsenstein, J. 1985. Phylogenies and the comparative method. American Naturalist 125: 1–15.

Fretwell, S. 1991. Say that in algebra. Current Contents/Agricultural Biology & Environmental Sciences 8: 8.

Fretwell, S. D., and H. L. Lucas Jr. 1969. On territorial behavior and other factors influencing habitat distribution in birds: I. Theoretical development. Acta Biotheoretica 19: 16–36.

Gadgil, M., and O. T. Solbrig. 1972. The concept of *r*- and *K*-selection: evidence from wild flowers and some theoretical considerations. American Naturalist 106: 14–31.

Gilliam, J. F., and D. F. Fraser. 1987. Habitat selection under predation hazard: test of a model with foraging minnows. Ecology 68: 1856–1862.

Grace, J. B. 1991. A clarification of the debate between Grime and Tilman. Functional Ecology 5: 583.

Greenwood, P. J. 1980. Mating systems, philopatry and dispersal in birds and mammals. Animal Behaviour 28: 1140–1162.

Grime, J. P. 1977. Evidence for the existence of three primary strategies in plants and its relevance to ecological and evolutionary theory. American Naturalist 111: 1169–1194.

Hahn, P. G., and J. L. Maron. 2016. A framework for predicting intraspecific variation in plant defense. Trends in Ecology & Evolution 31: 646–656.

Hall, D. J., and E. E. Werner. 1977. Seasonal distribution and abundance of fishes in littoral zone of a Michigan lake. Transactions of the American Fisheries Society 106: 545–555.

Harper, J. L., and J. Ogden. 1970. The reproductive strategy of higher plants: I. the concept of strategy with special reference to Senecio vulgaris L. Journal of Ecology 58: 681.

Haugen, T. O., I. J. Winfield, L. A. Vøllestad, J. M. Fletcher, J. B. James, and N. C. Stenseth. 2006. The ideal free pike: 50 years of fitness-maximizing dispersal in Windermere. Proceedings of the Royal Society B: Biological Sciences 273: 2917–2924.

Hereford, J., T. F. Hansen, and D. Houle. 2004. Comparing strengths of directional selection: how strong is strong? Evolution 58: 2133.

Herrera, C. M. 1992. Historical effects and sorting processes as explanations for contemporary ecological patterns: character syndromes in Mediterranean woody plants. American Naturalist 140: 421–446.

Kariñho-Betancourt, E., and J. Núñez-Farfán. 2015. Evolution of resistance and tolerance to herbivores: testing the trade-off hypothesis. PeerJ 3:e789.

Kenison, E. K., A. R. Litt, D. S. Pilliod, and T. E. McMahon. 2016. Larval long-toed salamanders incur nonconsumptive effects in the presence of nonnative trout. Ecosphere 7.

Kennedy, M., and R. D. Gray. 1993. Can ecological theory predict the distribution of foraging animals? A critical analysis of experiments on the ideal free distribution. Oikos 68: 158.

Krebs, J. R., J. C. Ryan, and E. L. Charnov. 1974. Hunting by expectation or optimal foraging? A study of patch use by chickadees. Animal Behaviour 22: 953–964.

Lack, D. 1947. The significance of clutch-size. Ibis 89: 302–352.

Lack, D. 1948. Natural selection and family size in the starling. Evolution 2: 95.

Lande, R., and S. J. Arnold. 1983. The measurement of selection on correlated characters. Evolution 37: 1210–1226.

Li, B., and A. R. Watkinson. 2000. Competition along a nutrient gradient: a case study with Daucus carota and Chenopodium album. Ecological Research 15: 293–306.

Liu, H., C. P. Osborne, D. Yin, R. P. Freckleton, G. Jiang, and M. Liu. 2019. Phylogeny and ecological processes influence grass coexistence at different spatial scales within the steppe biome. Oecologia 191: 25–38.

Lucas, J. R. 1985. Time constraints and diet choice: different predictions from different constraints. American Naturalist 126: 680–705.

Ludwig, D., and L. Rowe. 1990. Life-history strategies for energy gain and predator avoidance under time constraints. American Naturalist 135: 686–707.

MacArthur, R. H., and E. R. Pianka. 1966. On optimal use of a patchy environment. American Naturalist 100: 603–609.

Maddox, G. D., and R. B. Root. 1987. Resistance to 16 diverse species of herbivorous insects within a population of goldenrod, Solidago altissima: genetic variation and heritability. Oecologia 72: 8–14.

McPeek, M. A., M. Grace, and J. M. L. Richardson. 2001. Physiological and behavioral responses to predators shape the growth/predation risk trade-off in damselflies. Ecology 82: 1535–1545.

Menezes, J. F. S., and B. P. Kotler. 2019. The generalized ideal free distribution model: merging current ideal free distribution models into a central framework. Ecological Modelling 397: 47–54.

Milinski, M., and R. Heller. 1978. Influence of a predator on the optimal foraging behaviour of sticklebacks (Gasterosteus aculeatus L.). Nature 275: 642–644.

Mittelbach, G. G. 1981. Foraging efficiency and body size: a study of optimal diet and habitat use by bluegills. Ecology 62: 1370–1386.

Moreira, X., and I. S. Pearse. 2017. Leaf habit does not determine the investment in both physical and chemical defences and pair-wise correlations between these defensive traits. Plant Biology 19: 354–359.

Moritz, M., I. M. Hamilton, P. Scholte, and Y.-J. Chen. 2014. Ideal free distributions of mobile pastoralists in multiple seasonal grazing areas. Rangeland Ecology & Management 67: 641–649.

Onstein, R. E., and H. P. Linder. 2016. Beyond climate: convergence in fast evolving sclerophylls in Cape and Australian Rhamnaceae predates the Mediterranean climate. Journal of Ecology 104: 665–677.

Orians, G. H. 1962. Natural selection and ecological theory. American Naturalist 96: 257–263.

Paterson, J. E., and G. Blouin-Demers. 2018. Density-dependent habitat selection predicts fitness and abundance in a small lizard. Oikos 127: 448–459.

Pianka, E. R. 1970. On r- and K-selection. American Naturalist 104: 592–597.

Pyke, G. H., H. R. Pulliam, and E. L. Charnov. 1977. Optimal foraging: a selective review of theory and tests. Quarterly Review of Biology 52: 137–154.

Quezada, I. M., E. Gianoli, and A. Saldaña. 2018. Crassulacean acid metabolism and distribution range in Chilean Bromeliaceae: influences of climate and phylogeny. Journal of Biogeography 45: 1541–1549.

Reznick, D., and J. Travis. 1996. The empirical study of adaptation in natural populations. Pages 243–289 *in* M. R. Rose and G. V. Lauder, eds. Adaptation. Academic Press.

Siepielski, A. M., J. D. DiBattista, and S. M. Carlson. 2009. It's about time: the temporal dynamics of phenotypic selection in the wild. Ecology Letters 12: 1261–1276.

Silberbush, A., N. Gertler, O. Ovadia, Z. Abramsky, and I. Tsurim. 2019. Kairomone-induced changes in mosquito life history: effects across a food gradient. Aquatic Sciences 81: 53.

Singh, A., and D. Punzalan. 2018. The strength of sex-specific selection in the wild. Evolution 72: 2818–2824.

Thompson, J. N. 2005. The Geographic Mosaic of Coevolution. University of Chicago Press.

Thompson, J. N. 2013. Relentless Evolution. University of Chicago Press.

Tilman, D. 1987. The importance of the mechanisms of interspecific competition. American Naturalist 129: 769–774.

Tinkle, D. W., H. M. Wilbur, and S. G. Tilley. 1970. Evolutionary strategies in lizard reproduction. Evolution 24: 55.

Tregenza, T. 1994. Common misconceptions in applying the ideal free distribution. Animal Behaviour 47: 485–487.

Waser, N. M., and J. Ollerton, eds. 2006. Plant-Pollinator Interactions: From Specialization to Generalization. University of Chicago Press.

Werner, E. E., J. F. Gilliam, D. J. Hall, and G. G. Mittelbach. 1983. An experimental test of the effects of predation risk on habitat use in fish. Ecology 64: 1540–1548.

Wilman, V., E. E. Campbell, A. J. Potts, and R. M. Cowling. 2014. A mismatch between germination requirements and environmental conditions: niche conservatism in xeric subtropical thicket canopy species? South African Journal of Botany 92: 1–6.

Zhu, Y., J. A. Hogan, H. Cai, Y. Xun, F. Jiang, and G. Jin. 2017. Biotic and abiotic drivers of the tree growth and mortality trade-off in an old-growth temperate forest. Forest Ecology and Management 404: 354–360.

ON TERRITORIAL BEHAVIOR AND OTHER FACTORS INFLUENCING HABITAT DISTRIBUTION IN BIRDS

I. THEORETICAL DEVELOPMENT [1]

by

STEPHEN DEWITT FRETWELL [2]

and

HENRY L. LUCAS, JR.

(Biomathematics Program, Department of Experimental Statistics, North Carolina State University, Raleigh, N.C., 27607, U.S.A.)

(Received 17.X.1968)

TABLE OF CONTENTS

I. GENERAL INTRODUCTION

HOWARD (1920) suggested several roles for territorial behavior in birds, some of which involve the dispersal of the species over available habitats. Since this classic work, there have been a large number of sometimes conflicting statements regarding the relationship of territorial behavior to the habitat distribution of a species (*e.g.*, STEWART & ALDRICH, 1951, KLUYVER & TINBERGEN, 1953, TINBERGEN, 1957, and WYNNE-EDWARDS, 1962, have all argued that territorial behavior is involved in the dispersal of birds; LACK,

1) Work done while senior author was on fellowship support from Public Health Service Grant no. GM-678 from the National Institute of General Medical Sciences.
2) Present address: Department of Biology, Princeton University, Princeton, New Jersey, 08540, U.S.A.

1954, 1964 and JOHNSTON, 1961, have rejected this view, arguing instead that the behavior only isolates individuals or pairs). LACK in particular has dealt with this problem and recently (1966, p. 136) has emphasized the absence of a clear understanding of it. The purpose of this paper is to present an interpretation of the different hypotheses considered by different authors, with the aim of clarifying and resolving the different points of view. The re-defined hypotheses will yield predictions which can be compared with field observations in order to ascertain the probable role of territorial behavior in determining the distribution of a given species.

This study will be presented in three papers. The first will provide a theoretical statement of the general problem of habitat distribution. Hypotheses of the role of territorial behavior will be defined within the context of this theory of habitat distribution, and predictions derived from each. The problem of testing these predictions will be discussed in general. The second and third papers of the series will provide sample studies which demonstrate the application of the theory.

II. INTRODUCTION TO THEORETICAL DEVELOPMENT

In offering different hypotheses about the role of territorial behavior in habitat distribution, it will be necessary to first present some ideas about the factors which lead to birds being present in one place or another, at one level of abundance or another. By considering concepts related to overcrowding and evolutionary optima, we will develop a theory to describe a particular way in which bird populations might distribute themselves over the available living places. This distribution will be called the ideal free distribution. We then will define three hypotheses for the role of territorial behavior. One hypothesis describes how territorial behavior might be part of the mechanism by which the ideal free distribution is achieved. Another describes how territorial behavior might modify the distribution from the ideal free form. A third hypothesis contends that territorial behavior has a role that is unrelated to the ideal free distribution.

III. THEORY OF HABITAT DISTRIBUTION

In order to describe these hypotheses clearly, an agreed upon statement of the problem of habitat distribution in general must be developed. Only with such agreement can the possible relationships of territorial behavior to habitat distribution be differentiated. We proceed with the following definitions.

18   S. DEWITT FRETWELL AND H. L. LUCAS

Definitions

Habitat. A habitat of a species is any portion of the surface of the earth [3]) where the species is able to colonize and live. The total area available to a species can be divided into different habitats. The area of any one habitat can be large or small, and different habitats of the same species may be of different sizes. A given habitat can consist of several subdivisions which are not contiguous. We will define habitats so that all of the area within each habitat is, at zero density of the species, essentially homogeneous with respect to the physical and biological features which we believe to be most relevant to the behavior and survival of the species. We will also frame our definitions so that different habitats are not identical with respect to those same physical and biological features.

This definition does not imply that all measurable variables within a single habitat must take constant values over all of that habitat. Some variables may be irrelevant. Others, such as temperature and humidity may compensate for one another. In the latter case, the "relevant feature" which is "homogeneous" is some function of the compensating variables. Thus evaporative heat loss rate which depends on temperature and humidity may be the relevant feature which is homogeneous, in which case temperature and humidity may still vary over a single habitat. Because of compensating variables, all of the area within a habitat does not need to *appear* homogeneous to our measuring devices. However, if an area is uniform in all measurable variables, it is in a single habitat.

Habitat distribution of a species. Suppose the total area available to a species is divided into different habitats and that the area of each habitat is known. The habitat distribution of the species is the set of numbers which state the number of individuals resident in each of the habitats. It can also be expressed as the proportions of the total population resident in the different habitats, or as the density in each habitat.

Factors affecting habitat distribution

Habitat selection. We can now consider how the habitat distribution is achieved. Habitat distribution in birds is usually based on habitat selection, at least some individual birds being exposed to a variety of habitats of which just one is chosen for residence. Therefore, the distribution may be considered as a behavioral phenomenon, involving stimuli and responses. This means that an understanding of the habitat selection responses in given en-

---

3) In a number of cases, a habitat may be restricted to a layer parallel to the surface of the earth.

vironmental circumstances will lead to an understanding of the habitat distribution. In order to understand behavioral responses, we should consider the environmental factors (excluding direct within-species individual interactions) which caused the natural selection leading to the evolution of the behavior.

Suitability. In the case of habitat choice, these factors include differences in goodness or suitability of habitat because individuals which choose relatively poor habitats are selected against. Although the stimuli directly influencing the choice of habitat may be no more than correlated with habitat goodness, it is the goodness itself which is a basic (or ultimate — see below) determinate of the behavior. To summarize, the relative suitabilities of the different habitats give rise to habitat selection which in turn determines the habitat distribution. The habitat distribution then depends on the relative goodness of the habitats.

In developing an agreed upon statement of habitat distribution, the next matter is to examine the relative suitability of the various habitats. Suppose the habitats are indexed by i, $i = 1, 2. ..., N$, where N is the total number of habitats. The goodness of each occupied habitat is related to the average potential contribution from that habitat to the gene pool of succeeding generations of the species. We are interested in some measure of that goodness, which may be called the suitability, and denoted for the ith habitat as S_i. The suitability of the habitat cannot here be precisely defined, but may be thought of as the average success rate in the context of evolution (and/or "adaptedness") of adults resident in the habitat. Stated formally, if s_{iq} is the expected success rate of the qth individual ($q = 1, 2, ..., n_i$, $n_i =$ number of birds resident in the ith habitat), then

$$S_i = \frac{1}{n_i} \sum_{q=1}^{n_i} s_{iq}. \tag{1}$$

The habitat suitability will be determined by several factors such as food supply and predators. The influence of some of these factors is density dependent so that the suitability in a habitat is affected by the density of birds there. Let us assume for the moment that the effect of density is always a decrease in suitability with an increase in density. This assumption would imply that ALLEE's principle does not operate (ALLEE, *et al.*, 1949), and it may be truly valid only when densities are not close to zero. ALLEE's principle states that survival and reproductive rates increase with population size up to some maximum. Further increase in population size leads to a decrease in survival and reproduction, as assumed here. ALLEE's principle certainly holds at very low densities. A solitary male, for example, cannot

20   S. DEWITT FRETWELL AND H. L. LUCAS

have as high a reproductive rate as a male-female pair. At moderate densities, the assumption that suitability decreases with increased density is reasonable since predators may become more active at higher densities and
competition for food more severe (LACK, 1966).

We can now define a habitat distribution which will provide a reference
for the discussion of the role of territorial behavior. This is the ideal free
distribution noted previously, and rests on assumptions about habitat suitability and the adaptive state of birds.

Ideal free distribution

Assumptions on suitability. Ignoring ALLEE's principle, if we assume that
suitability always decreases with density, then it would follow that the
maximum suitability occurs when the density approaches zero. Let us call
this maximum value the *basic suitability*, denoted for the ith habitat as B_i.
The basic suitability of the ith habitat is affected by such factors as potential
predators, food density, and cover.

These considerations lead to an equation expressing the suitability of the
ith habitat as a function of the basic suitability there, and the density (denoted d_i). We write

$$S_i = B_i - f_i(d_i), \quad i = 1, 2, \ldots, N. \tag{2}$$

The term $f_i(d_i)$ expresses the lowering effect on suitability of an increase

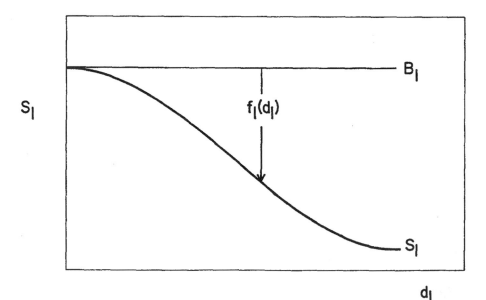

Figure 1. Suitability versus density: 1st habitat (see text).

in habitat density. Since $f_i(d_i)$ always increases with density, S_i always decreases. The Equations (2) will here be assumed to be the same through time. A possible example of Equation (2) for some value of i is plotted in Figure 1.

Before going any further, let us order the habitats in terms of their basic suitabilities, so that $B_1 > B_2 > \dots > B_n$. By definition, no two habitats have equal basic suitabilities. This is consistent with our restriction on habitat definition (stated above), that no two habitats are identical with respect to relevant features.

Assumptions on birds. A description of the suitabilities of the various habitats has been considered in order to understand the habitat selection behavior and the habitat distribution. In applying this description, we make two additional assumptions. These are (1) all individuals settle in the habitat most suitable to them, and (2) all individuals within a habitat have identical expected success rates.

The first assumption demands that the birds have habitat selection behavior which is ideal, in the sense that each bird selects the habitat best suited to its survival and reproduction. Such birds will be referred to as *ideal* individuals. It is not an unreasonable assumption since individuals which are closest to being ideal will be selected for in the evolution of the species. Therefore, if the environment has sustained the same selective pressures for a large number of generations, the behavior of actual individuals should be approximately ideal.

The second assumption demands, first of all, that the birds be *free* to enter any habitat on an equal basis with residents, socially or otherwise. For example, if a population in a habitat is limited by nest holes and if all these nest holes are occupied by residents which neither share nor are displaced, then a newly settling individual may expect to be totally unsuccessful, although the average of all residents is rather high. In this case, newly settling individuals are not free to enter the habitat on an equal basis with residents. If, when a new bird arrived, all of the occupants of the habitat came together to draw lots for the nesting holes, and those losing remained in the habitat, then the individuals would be free. This, of course, is unrealistic.

The second assumption demands also that individuals are alike, genetically and otherwise. This aspect of the assumption may heavily restrict the application of this theory, at best to local populations.

A particular difficulty which arises from this assumption concerns habitat accessibility. According to this assumption, if accessibility is relevant, every habitat must be equally accessible to all members of the species. This is absurd when we regard widely distributed birds species and only local habitats.

However, if we restrict attention to habitats as widely distributed as the bird species, or to local populations as narrowly distributed as the habitats, then this difficulty is bypassed.

Note that the second assumption, and Equation (1), imply

$$S_i = s_i.$$

Each bird expects success equal to the habitat average.

The ideal free distribution. With these assumptions we now use the suitability Equations (2) to determine the habitat selection of the individuals in the population. The ideal assumption states that each individual will go where his chance of success is highest. The assumption of homogeneity states that each individual's chance of success is highest in the habitat of highest suitability. The two assumptions together then assert that each individual will go to the habitat of highest suitability. Thus, a description of the relative habitat suitabilities to some degree determines the choices of ideal free individuals. These choices in turn determine a distribution, which may be called the *ideal free* distribution. This distribution, which is formally described below, will form a convenient basis for discussing the hypothesized relationships of territoriality to distribution.

If all individuals choose the habitat of highest suitability, then from the point of view of unsettled individuals, the suitability in all occupied habitats must be approximately equal and not less than the suitability in all occupied habitats. This is true because if some habitats had a clearly lower suitability, then some of the birds in that habitat could improve their chance of success by moving to the habitats of higher suitability. If they did not make that move, they would not have ideally adapted habitat selection behavior, contradicting the ideality assumption. The distribution is stable only when suitabilities are equal in all habitats. With Equations (2), a fixed set of habitat areas, and given population size, the condition of equal suitabilities in all occupied habitats completely determines the ideal free distribution. To prove this, let a_i be the area of the ith habitat, and M the population size. If exactly l habitats are occupied, then the equal suitabilities condition says that

$$S_1 = S_2 = \ldots = S_l. \tag{3}$$

Note that the first l habitats are the l occupied habitats. These are the l habitats with the highest basic suitabilities, B_i. In fact, if the $(l+p)^{\text{th}}$ habitat were occupied, some $(l-q)^{\text{th}}$ habitat must be unoccupied (p, q positive integers such that $0<p\leq N-l$ and $0<q<l$), since only l habitats are occupied. But, because d_{l-q} equals zero, the suitability in habitat $l-q$ is B_{l-q} and

$$S_{l-q} = B_{l-q} > B_{l+p} \geq S_{l+p},$$

or

$$S_{l-q} > S_{l+p}. \tag{4}$$

Because under our assumptions individuals settle where the habitat suitability is highest, the birds in habitat $l + p$ would move to $l—q$ where by (4) the suitability is higher. Thus, if there are l occupied habitats, they are the first l.

It is also true that M, the total population size of a given species over all its occupied habitat, is given by

$$M = a_1 d_1 + a_2 d_2 + \ldots + a_l d_l, \tag{5}$$

since

$a_i d_i$ is the number of birds in the ith habitat, and the total population is the sum of all the birds in all occupied habitats.

From Equations (2),

$$S_i = B_i — f_i(d_i), \; i = 1, \ldots, l$$

so that $S_i = S_{i+1}$ in (3) implies

$$B_i — f_i(d_i) = B_{i+1} — f_{i+1}(d_{i+1}), \; i = 1, \ldots, l—1. \tag{6}$$

There are $l—1$ of these equations in l unknowns; $d_1, d_2, \ldots, d_l$. With Equation (5), there are l equations. These l equations can be solved uniquely [4]) for the d_i ($i = 1, \ldots, l$) in terms of the constants M and the a_i. The distribution can be expressed as the proportion of birds in each habitat. Denote the proportion in the ith habitat as P_i. Then clearly

$$P_i = \frac{d_i a_i}{M},$$

and the distribution is seen to be a function of the density in each habitat. Since the densities are determined by the condition of equal suitabilities, so is the distribution.

An example of the solution of the equations is given in Figure 2 for $l = 1$, 2, and 3. The suitability curves are drawn for three habitats, 1, 2, and 3. When no birds are present, the suitability is highest in 1 and equals B_1. Therefore, if a small number of birds now settle in the habitats, they will all go to 1, because they settle where the suitability is highest. Then the density in 1 will increase from zero, and the suitability will decrease, following the

---

4) Because the $f_i(d_i)$ are always increasing.

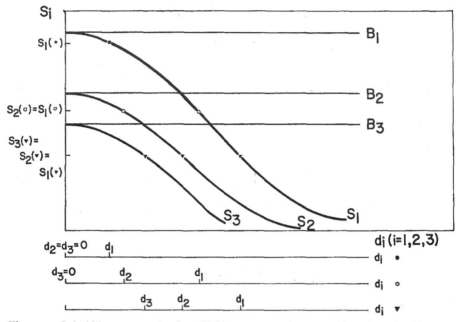

Figure 2. Suitability versus density: Habitats 1, 2, and 3. The ideal free densities are shown on the extra density coordinates at three values for the total population size, $M' < M'' < M'''$. The situation at each population size is denoted by (●) for M', (○) for M'', (▼) for M'''. At M', the lowest population size, all the population is in 1; the densities in 2 and 3 are zero. At M''', the largest population size, all three habitats are occupied.

curve labelled S_1. As the population size increases, more and more individuals will settle in 1, untill the density there is so high that the suitability is equal to the basic suitability in 2 (B_2). Now, any additional birds have a choice of habitats; 1 and 2 are equally suitable. However, these additional birds must increase the density in both habitats, and in such proportions that the suitabilities in both remain equal. Further increases in population size will raise the density in both habitats. If the population increases enough, the densities in 1 and 2 will be so high that the suitabilities in both habitats are reduced to B_3, the basic suitability in habitat 3. Any additional birds must increase the density in all three habitats in such proportions that the suitabilities in all three remain equal.

Allee-type ideal free distribution

Let us now briefly consider the effect on the distribution of ALLEE's principle which we have heretofore assumed does not apply. In this case, the S_i curves first increase with density up to a maximum then decrease.

These curves do not always have unique inverses; there are sometimes two densities corresponding to a single suitability. Therefore, the Equations (3) and (5) do not necessarily have a unique solution. Consideration of ALLEE's principle suitability curves is best done graphically as in Figure 3. At low population sizes, the birds will presumably go to habitat 1, and as the population increases, will enjoy an increasing suitability up to some maximum. Further increases in population size with all birds settling in 1, will cause a decrease in suitability in habitat 1 until at some higher population size (A), the density (d_i) in 1 is such that the suitability there equals the suitability in 2 at density 0. Now a remarkable event may occur. With a further slight increase in population size (A+), some birds will settle in 2 and perhaps some in 1. But the suitability in 2 *increases* with an increase in density, while the suitability in 1 decreases; therefore, $S_1 < S_2$, and suddenly it be-

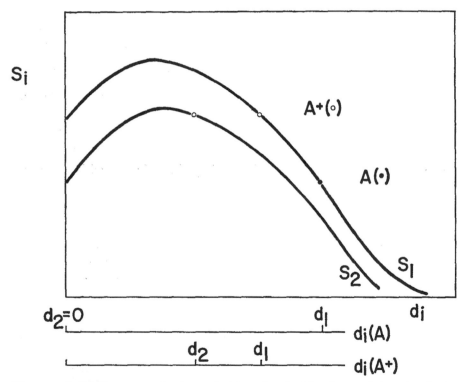

Figure 3. Suitability versus density under ALLEE's principle. At population size A the density in habitat 1 is d_1 (A). At population size A+ the density in 1 is d_1 (A+) while the density in 2 has increased to d_2 (A+). See text for explanation.

comes advantageous for birds in 1 to go to 2. Being ideal, they will so move, and may continue to move until the suitability in 2 is maximal. They would then fill the two habitats in such a way that $S_1 = S_2$, and may well become common in both (open circles in Figure 3). Thus, a very small increase in population size may result in a very large change in the distribution. By manipulating the S_1 curves, many such changes can be produced, and one can generally conclude that under the conditions of this theory, species following Allee's principle may demonstrate erratic changes in distribution with small changes in population. One can imagine curves which lead to complete shifts in population while other curves may lead to no erratic behavior at all.

Unless otherwise noted, the Allee-type ideal free distribution will not be considered in the following section.

IV. TERRITORIAL HYPOTHESES AND EFFECTS ON DISTRIBUTION

We can now discuss the possible relationships of territoriality to distribution. We will consider three distinct hypotheses and put them in the framework of the preceding discussion. In order to do this, a general definition of territorial behavior will be made which does not imply any particular role for the behavior.

Territorial behavior is defined as any site dependent display behavior that results in conspicuousness, and in the avoidance of other similar behaving individuals. Territorial behavior is specifically not restricted to defensive and/or aggressive behavior nor are they excluded. The following hypotheses to be discussed were in each case inspired by the authors cited in connection with them.

The density assessment hypothesis

This first hypothesis ascribes a role to territoriality which permits achieving the ideal free distribution (non-ALLEE). It was first described by KLUYVER & TINBERGEN (1953). Before discussing the hypothesis and its consequences in distribution, some preliminary discussion on the achieving of an ideal free distribution will be presented.

Given the total population size M, the habitat areas a_i and the equations in S_i (2), the density in each habitat is determined. The density in turn determines the proportion of the population in each habitat, or an ideal free distribution. The actual values will be expressed in terms of M and the a_i which for a given year are constants for the purpose of the present discussion. However, M, and perhaps the a_i vary with time in a somewhat irregular fashion. Therefore, the proportion of the population in a given habitat may also vary between different years. This is demonstrated in Figure 2. The suitability curves for three habitats are drawn, and the habitat densities at three population sizes shown. The relative densities in each of the three habitats are shown on the extra abscissas, and are markedly different at the three population sizes.

If the total population size does not vary, then a species could consistently achieve an ideal free distribution by being composed of individuals, a fixed proportion of which prefer each habitat at all times, or by being composed of individuals which prefer each habitat a fixed proportion of the time. However, if the variation in population size does occur and leads to considerable variation in the ideal free distribution, then such a fixed habitat selection scheme will usually not work. In this case, the habitat selection of the individuals, if it is to be ideal, must adjust to the changing conditions.

In order for individuals to be able to modify their habitat selection in accordance with changes in the ideal free distribution, there must be some cue or cues which reflect these changes. The changes come about due to variation in population size and possibly habitat areas. These variables affect the densities in the various habitats and therefore the relative suitabilities. Any cue, such as foot-print abundance, which would reflect habitat densities, or any cues which would reflect population size and perhaps habitat areas could be used by ideal individuals in achieving the ideal free distribution. The individuals would have their preference for a given habitat depend on the state of the cue.

KLUYVER & TINBERGEN (1953) suggested that the territorial behavior of resident individuals is used as a density cue by unsettled individuals so that they can avoid highly populated habitats where the chance of breeding success is presumably lower than elsewhere. These authors observed that the habitat distribution of some *Paridae* (tits) was dependent on population size. At low population levels, most of the individuals were found in a single habitat type, while at higher levels many individuals occupied another habitat type, but at a lower density (compare this with population sizes M' and M'' in Figure 2 of this paper). They emphasized that breeding success (and therefore, suitability) was not noticeably different in the two habitats. Thus, the distribution of the tits was apparently nearly ideal free, despite changing population size. No appropriate cue for density other than territoriality was observed, nor was there evidence for a cue for population size.

There is no evolutionary difficulty in supposing that territorial behavior serves as a density index. It is obviously to the residents' advantage to provide such a cue, since they suffer if their habitat is crowded to the extent that its suitability is lower than that in other habitats. It is also the advantage of the settling individuals to respond to such a cue, for by so doing they

28   S. DEWITT FRETWELL AND H. L. LUCAS

avoid habitats where high density makes their chance of success lower than elsewhere. Since the population size of tits varies considerably, the development and use of some density cue, such as territorial behavior, might be expected.

The density limiting hypothesis

Introduction. The second territorial hypothesis is based on a model from HUXLEY (1934), who described a territory as a rubber disk. The disk can be compressed, but with an increased amount of force necessary as it gets smaller. This hypothesis is relevant to the free aspect of the homogeneity assumption about birds in the ideal free model. That assumption states that any individual is free and may therefore enter any habitat on an equal basis with the birds already resident there. With the rest of the homogeneity assumption, this means that the average success of the occupants of a habitat is also the suitability that an unsettled bird will have (on the average) on settling in that habitat. The free assumption, as already mentioned, fails if the species is limited by nest holes which, once occupied by a resident, are not relinquished or shared. The second territorial hypothesis describes another possible way in which this assumption might fail. Suppose the residents of the habitat, by their territorial behavior, made it dangerous for unsettled individuals to enter the habitat. Then the average success of newly settling individuals will be lower than the habitat average, and the assumption fails. If so, ideal individuals maximizing their own success would not necessarily settle where the habitat suitability is highest, and the habitat suitabilities no longer must be equal. Since Equation (6) no longer holds, the distribution is not determined as before. There exists a new, different distribution.

The supposition that the territorial behavior of residents restricts nonresidents from settling is reasonable. In evolution such behavior effectively prevents the density in the habitat from increasing, maintaining the suitability (see Figure 4). This would give the aggressive residents a selective advantage and the behavior, as suggested by BROWN (1964), would spread throughout the population. However, if all the individuals, settled and unsettled alike, became equally aggressive, it might not then be possible for the settled birds to make a habitat less suitable to unestablished birds. This is perhaps unlikely; for example, dominance seems to depend on experience (NICE, 1936; SABINE, 1955) which for a given area should vary considerably from bird to bird. In the following discussions we will assume that all individuals are not equally aggressive, so that social dominance hierarchies are established and the free assumption does not hold.

Consequence of hypothesis on the distribution. We consider in more detail the altered ideal distribution which would arise from this hypothesis. By definition the ideal bird always goes to the habitat where his potential success is highest. In the ideal free distribution, the potential of a new individual settling in a given habitat is equal to the average of all individuals resident

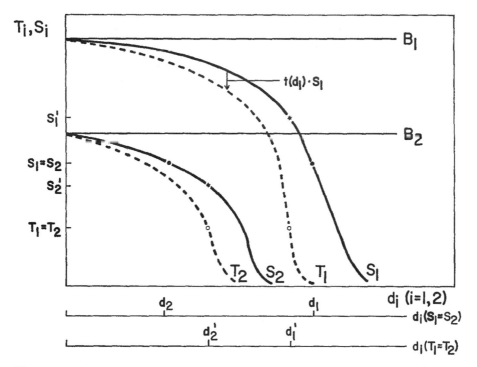

Figure 4. A comparison between ideal free and ideal dominance distributions: Habitats 1 and 2. Population size and habitat areas are constant. S_i is the suitability for established birds. T_i, the apparent suitability to unestablished birds. The filled circles represent equilibrium conditions $S_1 = S_2$ for the ideal free distribution, and densities for that distribution are given below as (d_1, d_2) on the abscissa labelled $(S_1 = S_2)$. The open circles represent equilibrium conditions $T_1 = T_2$ for the territorial distribution, and densities for that distribution are given as (d_1', d_2') on the abscissa $(T_1 = T_2)$. At $T_1 = T_2$ the actual suitabilities are given as (S_1', S_2'), marked on the S_i curves with a star. Note that in going from the ideal free to the ideal dominance distribution, at constant population size and habitat areas, the density in 1 decreases $(d_1' < d_1)$ while the density in 2 increases $(d_2' > d_2)$. Some individuals have in principle been forced from habitat 1 into habitat 2. See text for further discussion.

there, including the new one. This is the habitat suitability described by Equation (2). If the unsettled birds are restricted by the territorial behavior of residents, then their potential success is less than the average of the habitat. This suggests that we define some quantity t, $0 \leq t \leq 1$, which can be subtracted from one and multiplied into the S_i, to yield the apparent habitat suitability from the point of view of the unsettled bird. This apparent habitat suitability may be denoted by T_i, defined symbolically in Equation (7):

$$T_i = S_i \, (1-t) \tag{7}$$

Let us now consider some of the properties of t. This quantity will depend on density, and can be reasonably assumed to always increase as the density in a habitat increases. This assumption is justified because t is related to the resistance of the established birds. The resistance from each male evidently increases as the territory size diminishes (HUXLEY, 1934); that is, as the density increases. Thus, the per bird resistance always increases with density. Also, the number of birds increases with density. Therefore, the overall resistance, and t, must increase with density. We will assume that t does not vary with habitat, since it is unlikely that the territorial resistance is dependent on habitat, except as density varies. Thus, Equation (7) may be rewritten

$$T_i = S_i \, (1-t(d_i)). \tag{8}$$

The function T_i, like S_i, always decreases with density, since S_i always decreases and $t(d_i)$ always increases with density. Also, T_i is always less than or equal to S_i. The actual suitability of the habitat remains S_i; T_i is just the apparent suitability from the point of view of the unsettled bird.

The T_i in (8) are defined such that an ideally adapted individual will always go to the habitat where T is highest, assuming as before that all (unsettled) individuals are alike in their adaptation to the habitats. This leads, as with the S_i, to an equilibrium condition where the T_i are equal in all occupied habitats. The resultant equations, with (5), completely define a set of habitat densities and a new distribution. This distribution may be called the *ideal dominance* distribution. Like the ideal free distribution, the ideal dominance distribution is a useful basis for discussion, but because of the underlying assumptions, can only approximate any real life situation.

Interpretation. Some remarks about the T_i in (8) may make them easier to understand. If within an area some habitats are better than others then the territorial restrictions may just restrict newly settling individuals to the less good habitats. Then T_i represents the average success expected in the area for individuals in the less good areas. However, in a uniform habitat where no less favorable regions exist, the territories of all occupants will be, on the average, about equally suitable. In this case, $t(d_i)$ represents only an entering risk to unsettled individuals. If they can successfully settle in the habitats, they may expect to be as successful as the habitat average.

But the act of settling may involve a serious risk of physical harm. There may be a high probability of failure with accompanying mortality. Thus, the T_i may not be a measure of the success of some members of the habitat, but a measure of the average success of a hypothetical group of individuals which tried to enter a habitat until either successful or dead.

These two territorial hypotheses exemplify the distinction between ultimate and proximate determinants of behavior. These are defined as follows. Ultimate determinants of behavior are the environmental factors which produce the natural selection that leads to the genetic basis for the behavior. Proximate determinants are the stimuli that prompt the behavior. The behavior that we are considering is the habitat selection of individuals. The first hypothesis is that territorial behavior is only a proximate factor providing information about density. The second hypothesis is that territorial behavior is an ultimate factor which by directly influencing the survival of past individuals which selected certain habitats to breed in has actually caused the population-genetic basis for the habitat selection to change in some way. And since the distribution is determined by the habitat selection, the territorial behavior is an ultimate cause of the phenomenon as well. However, in the second hypothesis territorial behavior is probably also a proximate factor. It could easily provide information about itself and provide a stimulus to which unsettled individuals could respond. This is quite likely because territorial behavior generally involves display and vocal announcement as a substantial part of the "defense" of boundaries.

The spacing hypothesis

LACK (1964) and JOHNSTON (1961) have supported a third territorial hypothesis, that the territorial behavior has been evolved only to space individuals within a habitat. This means that the density in a habitat is determined independently of the behavior, but that given a certain density, the individuals separate as much as possible and have non-overlapping home ranges. For example, suppose a population of a territorial species can achieve distribution without the territorial behavior either cueing density variation or restricting habitat occupancy. Then the role of the territorial behavior may be purely behavioral, isolating mating adults to strengthen pair bonds (LACK, 1954). Or the behavior could prevent the spread of disease by a quarantine effect. Whatever its role, if the territorial behavior only spaces individuals, it has nothing to do with whether or not an individual will settle in one habitat or another. It only affects the movements of the individuals *within* the habitat. The effect of spacing is to keep the *instantaneous* distribution of individuals within the habitat fairly uniform. It does not alter the number of occupants of the habitats, nor does it affect the average number of individuals using any portion of the area within the habitat (the average being taken over some time period). Thus, the average density over any piece of land (*i.e.*, any habitat) is not influenced by territorial behavior under this hypothesis. If territorial behavior only spaces individuals, it has no effect on the habitat distribution, either ultimately or proximately.

Application of the theoretical development

Approach. The next matter is to consider the problem of identifying the actual role of the behavior relative to habitat distribution in a given species. We have defined two theoretical distributions under certain assumptions and three different territory hypotheses. We may, as a first approximation, assume that if territorial behavior has one of the hypothesized roles in a given species the actual habitat distribution will be similar to the theoretical distribution described for that hypothesis. Thus, if we can show that the actual species distribution is approximately ideal dominance, this may be taken as evidence in the support of the hypothesis that the role of the behavior is to limit density. Or if we can show that the distribution is consistently ideal free in spite of variation in population size, and can find no other evident cues for density and/or population size variation being used by the species, then this may be taken as evidence in the support of the hypothesis that the behavior is used for density assessment. If we can show that the distribution is not consistently ideal free in the face of variation in population size, nor ideal dominance, then this may be taken as evidence for a lack of any density assessment or density limiting mechanism, and spacing remains as the most tenable of the hypotheses offered. (Other hypotheses may exist, but we will not pursue them here). The approach to the problem rests on identifying the distribution as (approximately) ideal free, ideal dominance, or neither.

Density and habitat suitability. If a population has an ideal free distribution, then by definition the suitability of all habitats is equal. We used this to show that the ideal free assumptions determine a distribution. However, the densities in the different habitats are not necessarily equal (see Figure 2). Thus, if a species has a nearly ideal free distribution, habitats with different densities will show similar success rates.

If a population has an ideal dominance distribution, then the suitability of all habitats is not necessarily equal. Only the apparent suitabilities (the T_i) will be everywhere equal, assuming ideally adapted individuals. In this case, it will now be shown that, for two occupied habitats, p and q, if $d_p > d_q$ then

$S_p > S_q$: the suitability is higher in habitats of higher density. From the equilibrium condition, for occupied habitats q and p,

$$T_q = T_p. \tag{9}$$

Since, by (8)

$$T_q = S_q \, (1 - t(d_q)),$$
$$T_p = S_p \, (1 - t(d_p)),$$

then substituting in (9) we obtain

$$S_p \, (1 - t(d_p)) = S_q \, (1 - t(d_q))$$

and

$$S_p / S_q = (1 - t(d_q)) / (1 - t(d_p)). \tag{10}$$

If $d_p > d_q$, then $t(d_p) > t(d_q)$, since the t function always increases with density. Then

$$(1 - t(d_q)) > (1 - t(d_p)),$$
$$(1 - t(d_q)) / (1 - t(d_p)) > 1,$$

and therefore, by (10),

$$S_p / S_q > 1.$$

From this it follows that $S_p > S_q$ as asserted. Thus, if a species has an ideal territorial distribution, then the success rate in habitats with higher densities of residents will be higher. This assumes that the t function does not depend on habitat, and always increases with density.

It is possible using another approach to show that this relationship between density and suitability exists. Assume that territory size is inversely related to the number of birds against which it is defended (HUXLEY, 1934). The better a territory is, the more there are that are less good, and there are more birds potentially trying to overtake it. Since the territory must be defended against more birds, it must be smaller. Hence, more suitable territories are smaller, and density, which is inversely related to territory size, is seen to be higher in high suitability habitats.

Figure 4 shows examples of the S_i and T_i curves for $N = 2$, and a sample population distributed freely (filled circles) and territorially (open circles). Note that the suitability in the ideal free distribution is equal in both habitats ($S_1(.) = S_2(.)$). In the territorial distribution, the suitability and the density are higher in habitat 1 than in habitat 2 ($S_1(*) > S_2(*)$, $d_1 > d_2$).[5]

---

5) We should note the discussion of GIBB (1961) in which attention is drawn to the relationship between the role of territorial behavior and differences in habitat suitability. Gibb's remarks are not formally developed, and do not distinguish the density assessment and density limiting hypotheses. But they do foreshadow several of the ideas presented above.

These conclusions suggest that the role of territorial behavior in the habitat distribution of a territorial species can be ascertained as follows: If the high density habitats show consistently higher success rates, and if no density limiting mechanism other than territorial behavior is evident, then the role of the territorial behavior is evidently to limit density. If suitabilities in all habitats are equal, even though densities are not, if the distribution changes with changing population size, and if no alternative density or population size cue is apparent, then the role of the territorial behavior is evidently to serve as a density assessment mechanism. If neither of the above two criteria are met, then the role of territorial behavior seems to be only to space individuals. Underlying this approach is the assumption that the birds are approximately ideally adapted. This assumption may well fail, and this possibility should always be considered. There is also the possibility, ever present, of hypotheses different from those considered here.

Uncertainties in application. There are a number of uncertainties involved in these conclusions which should be given careful consideration. Some of this uncertainty is inherent in the theory, which describes the relationships of expected values or population means. Any given realization of the theory (*e.g.*, the observed habitat suitabilities in a given year) is expected to deviate from the average values, even if the assumptions of the theory are met. This deviation will alter the succeeding year's distribution somewhat (by accidental selection), and so some fluctuation in the distribution may also be expected.

Another source of uncertainty lies in inherent failure of the assumptions, particularly the ideality assumption. For example, the sensory reception of the bird cannot be perfect, and so the bird can be expected to misread whatever environmental cues it uses to assess the suitability of a habitat. Also, the correlation between these cues and suitability is probably never perfect. Finally, the birds may reasonably be expected to be always evolving towards the ideal state without ever achieving it. These failures of the ideal assumption will lead to errors in the individual judgments of habitat suitability.

Hopefully the uncertainties in our predictions will be generally independent in different years, or even in different regions in the same year. Probably most of the error-causing factors are rather local in effect. A major exception is weather, but even this factor changes from year to year. Thus, in most cases, sampling over several years, or perhaps over widely separated regions in the same year should provide a way of estimating or controlling these errors. If the errors are not independent even over years, then the species may be considered to be evolving and changes over time should be detected. Thus, results which are consistently obtained over a number of years may be reasonably considered free from these errors.

Measurement of suitability. The problem of measuring suitability remains. The suitability of a habitat is a reflection of the average genetic contribution of resident adults to the next generation, and must be closely related to the average lifetime production of reproducing offspring in the habitat. Therefore, it must depend on several components, including reproductive rate, and survival of adults and immatures. Since territoriality is normally associated with breeding behavior and habitat distribution during the breeding season, we will usually associate suitability with such things as nesting success, feeding rates, and clutch size.

V. SUMMARY OF THE THEORETICAL DEVELOPMENT

The purpose of this report is to define precisely the problem of the role of territorial behavior in habitat distribution. Models for the habitat distribution of an ideally adapted species are developed under certain hypotheses for the role of territorial behavior. These models provide well-defined statements of what each hypothesis means, and also provide under certain assumptions a way of ascertaining, for a given territorial species, what the role of territorial behavior is. The hypotheses considered were that territorial behavior (1) is part of a density assessment mechanism, (2) limits density, or (3) only spaces individuals.

VI. ACKNOWLEDGEMENTS

The theoretical developments of this paper were stimulated by correspondence with R. L. Haines and H. A. Hespenheide. T. L. Quay provided encouragement and advice relative to the field aspects of the study as reported in the next two papers; H. R. van der Vaart, C. Proctor and W. Standaert read parts or all of the manuscripts and made many valuable suggestions.

Glinda Goodwin and my wife, Armeda, prepared the manuscript.

We are extremely grateful to all of the above individuals for their interest and help.

VII. REFERENCES

ALLEE, W. C., A. E. EMERSON, O. PARK, T. PARK & K. P. SCHMIDT (1949). Principles of Animal Ecology. W. B. Saunders, Philadelphia. xii + 837 pp.

BROWN, J. L. (1964). The evolution of diversity in avian territorial systems. — Wilson Bull., 76, 160-169.

GIBB, J. (1961). Bird populations, pp. 413-446. In: A. J. MARSHALL (ed.), Biology and Comparative Physiology of Birds. II. Academic Press, New York.

HOWARD, F. (1920). Territory in Bird Life. Atheneum, London, 239 pp.

HUXLEY, J. S. (1934). A natural experiment on the territorial instinct. — Brit. Birds, 27, 270-277.

JOHNSTON, R. F. (1961). Population movements of birds. — Condor, 63, 386-388.

3*

36   DEWITT FRETWELL AND LUCAS, HABITAT DISTRIBUTION

KLUYVER, H. N. & L. TINBERGEN (1953). Territory and the regulation of density in titmice. — Arch. Neerl. Zool., 10, 265-289.

LACK, D. (1954). The natural regulation of animal numbers. Clarendon Press, Oxford, viii + 343 pp.

———. (1964). A long term study of the great tit (*Parus major*). — J. Anim. Ecol., 33 (Suppl.), 159-173.

———. (1966). Population Studies of Birds. Clarendon Press, Oxford. 341 pp.

NICE, MARGARET M. (1937). Studies in the life history of the Song Sparrow I. — Trans. Linn. Soc., N. Y., viii + 246 pp.

SABINE, WINIFRED (1955). The winter society of the Oregon Junco: the flock. — Condor, 57, 88-110.

STEWART, R. E. & J. W. ALDRICH (1951). Removal and repopulation of breeding birds in a spruce-fir forest community. — Auk, 68, 471-482.

TINBERGEN, N. (1957). The functions of territory. — Bird Study, 4, 14-27.

WYNNE EDWARDS, V. C. (1962). Animal Dispersion in Relation to Social Behavior. . Hafner Publishing Company, New York. xi + 653 pp.

THEORETICAL POPULATION BIOLOGY **9**, 129–136 (1976)

Optimal Foraging, the Marginal Value Theorem

Eric L. Charnov*

*Center for Quan. Science in Forestry, Fisheries, and Wildlife,
University of Washington, Seattle, Washington 98195; and
Institute of Animal Resource Ecology UBC, Vancouver 8, Canada*

Received December 26, 1974

There has been much recent work on foraging that derives hypotheses from the assumption that animals are in some way optimizing in their foraging activities. Useful reviews may be found in Krebs (1973) or Schoener (1971). The problems considered usually relate to breadth of diet (Schoener, 1969, 1971; Emlen 1966; MacArthur, 1972; MacArthur and Painka, 1966; Marten, 1973; Pulliam, 1974; Werner, 1974; Werner and Hall, 1974; Timmins, 1973; Pearson, 1974; Rapport, 1971; Charnov, 1973, 1976; Eggers, 1975), strategies of movement (Cody, 1971; Pyke, 1974; Smith, 1974a, b; Ware, 1975), or use of a patchy environment (Royama, 1970; MacArthur and Pianka, 1966; Pulliam, 1974; Smith and Dawkins, 1971; Tullock, 1970; Emlen, 1973; Krebs, 1973; Krebs, Ryan, and Charnov, 1974; Charnov, Orians, and Hyatt, 1976). The above list of references is provided as a beginning to this fast expanding literature and is far from exhaustive.

This paper will develop a model for the use of a "patchy habitat" by an optimal predator. The general problem may be stated as follows. Food is found in clumps or patches. The predator encounters food items within a patch but spends time in traveling between patches. This is schematically shown in Fig. 1. The predator must make decisions as to which patch types it will visit and when it will leave the patch it is presently in. This paper will focus on the second question. An important assumption of the model is that while the predator is in a patch, its food intake rate *for that patch* decreases with time spent there. The predator *depresses* (Charnov, Orians, and Hyatt, 1976) the availability of food to itself so that the amount of food gained for T time spent in a patch of type i is $h_i(T)$, where the function rises to an asymptote. A hypothetical example is shown in Fig. 2. While it is not necessary that the first derivative of $h_i(T)$ be decreasing for all T (it might be increasing at first if the predator scares up prey upon arrival in a new patch), I will limit discussion to this case since more

* (Present Address) Department of Biology, University of Utah, Salt Lake City, Utah 84112.

129

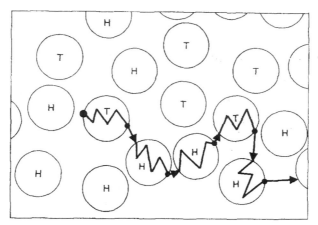

FIG. 1. A hypothetical environment of two patch types. The predator encounters prey only within a patch, but spends time in traveling between patches. Patches were labeled H or T by the flip of a coin.

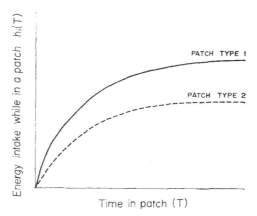

FIG. 2. The energy intake for T time spent in a patch of type i is given by $h_i(T)$. This function is assumed to rise to an asymptote.

complex functions add essentially nothing new to the major conclusions. The model is also completely deterministic, but this is not a major restriction since qualitatively similar results may be shown to follow from a corresponding stochastic model that considers the foraging as a cumulative renewal process (Charnov 1973 and in preparation). What is important is that if the environment is made up of several "patch types," the types be "mixed up" or rather distributed at random relative to one another (Fig. 1), and that many separate patches are visited in a single foraging bout with little or no revisitation. A patch type has associated with it a particular $h_i(T)$ curve. MacArthur (1972) has termed these assumptions "a repeating environment." One final assumption is necessary. The predator is assumed to make decisions so as to maximize the net rate of energy intake during a foraging bout.

The patch use model

We define as follows:

P_i = proportion of the visited patches that are of type i ($i = 1, 2,..., k$).

E_T = energy cost per unit time in traveling between patches.

E_{si} = energy cost per unit time while searching in a patch of type i.

$h_i(T)$ = assimilated energy from hunting for T time units in a patch of type i minus all energy costs except the cost of searching.

$g_i(T) = h_i(T) - E_{si} \cdot T$ = assimilated energy corrected for the cost of searching.

The time for a predator to use a single patch is the interpatch travel time (t) plus the time in the patch. Let T_u be the average time to use one patch.

$$T_u = t + \sum P_i \cdot T_i .$$

T is now written as T_i to indicate that it may be different for each patch type. The average energy from a patch is E_e.

$$E_e = \sum P_i \cdot g_i(T_i).$$

The net energy intake rate (En) is given by:

$$En = \frac{E_e - t \cdot E_T}{T_u} . \tag{1}$$

En may thus be written as

$$En = \frac{\sum P_i \cdot g_i(T_i) - t \cdot E_T}{t + \sum P_i \cdot T_i} . \tag{2}$$

It is easy to show that (2) is an energy balance equation and that En is the net rate of energy intake. With suitable interpretation of $g_i(T_i)$, (2) is identical to Schoener (1971, Eq. (2)).

The predator is assumed to control which patches it will visit and when it will leave a patch. The t is obviously a function of which patches the predator is visiting and in general should increase as more patch types are skipped over. A simple assumption would have t proportional to the distance between patches divided by the predators speed of movement. It should be noted, however, that there is no good reason to believe that t should be at all related to any of the T_i. The length of time between patches should be independent of length of time the predator hunts within any one (although the reverse statement is not true).

132   ERIC L. CHARNOV

This independence is quite important since when it holds, (2) may be written (from the standpoint of a patch type of interest j) as:

$$En = \frac{P_j \cdot g_j(T_j) + A}{P_j \cdot T_j + B},$$ (3)

where A and B are not functions of T_j. A and B are found by equating terms in (2) and (3), naming one patch type as j.

If j is being visited, the predator is assumed to control only T_j. The optimal value of T_j is given by a rather interesting theorem. For some set of patches being visited, write En as En^* when all T_i are at their optimal values. When this is true T_j satisfies the following relation.

$$\frac{\partial g_j(T_j)}{\partial T_j} = En^*, \qquad \text{for all } i = j.$$ (4)

The predator should leave the patch it is presently in when the *marginal capture rate in the patch* $(\partial g/\partial T)$ *drops to the average capture rate for the habitat.*

This rule is found by setting $\partial En/\partial T_i = 0$ for all patch types simultaneously. Since we are assuming here that the $\partial h_i(T_i)/\partial T_i$ are always decreasing, so are the $\partial g_i(T_i)/\partial T_i$ and there is a unique set of T_i that fulfills (4). This set represents a maximum as the associated Hessian matrix is negative-definite (Taha, 1971).

A graphical way of showing this result is in Fig. 3. The $g_i(T_i)$ is plotted as a

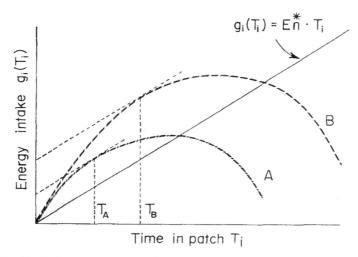

FIG. 3. Optimal use of a patchy habitat. The energy intake functions $g_i(T_i)$, are shown for a habitat with two patch types. If the ray from the origin with slope En^* is plotted, the appropriate time to spend in each patch is found by constructing the *highest* line tangent to a $g_i(T_i)$ curve and parallel to the ray. The lines and the resulting times are shown for the two patch types.

function of T_i for two patch types. If a ray from the origin with slope En^* is then plotted, the optimal T_i are easily found. To find these, simply construct lines with slope En^* and see where they become tangent to the appropriate $g_i(T_i)$ curve. In cases where the $\partial g_i(T_i)/\partial T_i$ are not strictly decreasing with T_i, more than one point of tangency may result. In these cases, the optimal T_i is that associated with the highest line of slope En^*.

DISCUSSION

Two earlier publications (Krebs, Ryan, and Charnov, 1974; Charnov, Orians, and Hyatt, 1976) derived a simplified version of the movement rule given in (4) and discussed some supporting data. Krebs, Ryan, and Charnov 1974 carried out laboratory experiments with chickadees to qualitatively test (4). They defined the time between the last capture and when an individual left a patch (several blocks of wood with mealworms in holes, suspended from an artificial tree) to go to another patch as the "giving up time" (GUT). This was taken to be a measure of the inverse of the capture rate when the bird left the patch (the marginal capture rate). The experimental design consisted of two environments. In the first, the average rate of food intake was high, in the second, it was low. Within each environment there were two or three patch types, each type having a specified number of mealworms. The predictions of the theorem, translated into the GUT measurement, were that (1) GUT should be a constant within an environment across patch types, and (2) GUT should be lower in the rich environment. Both of these predictions were supported by the data. More recent experiments using the Great Tit (*Parus major*) also support the model (R. Cowie, personal communication).

There are little other data that will allow more critical tests of the model although there are many data relating to gross predator behavior relative to clumps of prey. Smith (1974a, b), in some field experiments with thrushes, found that the tendency of a bird to remain in the area where it had already made a capture was greater the lower the overall availability of food in the habitat. The simple tendency for a predator to remain in the area where it was successful has been documented for birds (Tinbergen, Impekoven, and Frank, 1967; Krebs, MacRoberts, and Cullen, 1972); fish (Beukema, 1968); insects (Hafez, 1961; Fleschner, 1950; Laing, 1938; Mitchell, 1963; Dixon, 1959, 1970; Banks, 1957; Richerson and Borden, 1972; Hassell and May, 1973; Murdie and Hassell, 1973). Even unicellular predators exhibit increased frequency of turning following an encounter with food particles, a behavior pattern that results in a more intensive search of the vicinity of the capture (Fraenkel and Gunn, 1940; MacNab and Koshland, 1972).

The rule was used in a slightly different context by Parker and Stuart (1976),

134   ERIC L. CHARNOV

whose derivation is independent of the work here. They showed that male dung flies (*Scatophaga stercoraria*) terminate copulation at a time that maximizes the eggs fertilized/unit time for the male.

CONCLUSIONS

A rule is proposed for the movement of an optimal predator through an environment where food is found in patches, and time is expended in movement between patches. The theorem is rather general and should be useful where predators cause the prey in their immediate vicinity to become less available the longer they remain there. It receives some support from both lab and field studies, but has yet to be tested in a quantitative manner.

ACKNOWLEDGMENTS

The ideas presented here have been drawn from a thesis presented to the University of Washington. The thesis work was generously supported by the Ford Foundation (UW) and by funds supplied by the Institute of Animal Resource Ecology (University of British Columbia). I wish to thank D. G. Chapman, W. H. Hatheway, N. Pearson, C. Fowler, C. S. Holling, J. Krebs, G. Pyke, Thomas Schoener, and G. A. Parker. Gordon Orians provided his usual creative input.

REFERENCES

BANKS, C. J. 1957. The behavior of individual coccinellid larvae on plants, *J. Anim. Behav.* 5, 12–24.

BEUKEMA, J. J. 1968. Predation by the three-spined stickleback *(Gasterosteus aculeatus)*. The influence of hunger and experience, *Behavior* 31, 1–126.

CHARNOV, E. L. 1973. Optimal foraging: some theoretical explorations, Ph. D. Thesis, Univ. of Wash.

CHARNOV, E. L. 1976. Optimal foraging: attack strategy of a mantid, *Amer. Natur.* to appear.

CHARNOV, E. L., ORIANS, G. H., AND HYATT, K. 1976. Ecological implications of resource depression, Amer. Natur. to appear.

CODY, M. L. 1971. Finch flocks in the Mohave desert, *Theor. Popul. Biol.* 2, 142–158.

DIXON, A. F. G. 1959. An experimental study of the searching behavior of the predatory coccinellid beetle *Adalia decempunctata*, *J. Anim. Ecol.* 28, 259–281.

DIXON, A. F. G. 1970. Factors limiting the effectiveness of the coccinellid beetle, *Adalia bipunctata*, as a predator of the sycamore aphid, *Depanosiphum platonoides*, *J. Anim. Ecol.* 39, 739–751.

EGGERS, D. M. 1975. A synthesis of the feeding behavior and growth of juvenile sockeye salmon in the limnetic environment, Ph. D. Thesis, Univ. of Wash.

EMLEN, J. M. 1966. The role of time and energy in food preference, *Amer. Natur.* 100, 611–617.

EMLEN, J. M. 1973. "Ecology; An Evolutionary Approach," p. 493. Addison–Wesley, Reading, Mass.

FLESCHNER, C. A. 1950. Studies on search capacity of the larvae of three predators of the citrus red mite, *Hilgardia* 20, 233–265.

FRAENKEL, G. S. AND GUNN, D. L. 1940. "The Orientation of Animals," Oxford Book, New York.

HAFEZ, M. 1961, Seasonal fluctuations of population density of the cabbage aphid *Brevicoryne brassicae* (L.) in the Netherlands and the role of its parasite *Aphidius (Diaretiella) rapae* (Curtis), *Tijdschr. Plantenziekten.* 67, 445–548.

HASSELL, M. P. AND MAY, R. M. 1973. Stability in insect host-parasite models, *J. Anim. Ecol.* 42, 693–726.

KREBS, J. R. 1973. Behavioral aspects of predation, in "Perspectives in Ethology," (Bateson, P. P. G., and Klopfer, P. H., Eds.), pp. 73–111. Plenuwm, New York.

KREBS, J. R., MACROBERTS, M. H., AND CULLEN, J. M. 1972. Flocking and feeding in the Great Tit, *Parus major*, an experimental study, *Ibis* 114, 507–530.

KREBS, J. R., RYAN, J. C., AND CHARNOV, E. L. 1974. Hunting by expectation or optimal foraging? A study of patch use by chickadees, *J. Anim. Behav.* 22, 953–964.

LAING, J. 1938. Host finding by insect parasites. II. The chance of *Trichogrammae vanescens* finding its hosts, *J. Exp. Biol.* 15, 281–302.

MACARTHUR, R. H. 1972. "Geographical Ecology," p. 269. Harper and Row, New York.

MACARTHUR, R. H. AND PIANKA, E. R. 1966. On optimal use of a patchy environment, *Amer. Natur.* 100, 603–609.

MACNAB, R. M. AND KOSHLAND, D. E. 1972. The gradient-sensing mechanism in bacterial chemotaxis, *PNAS* 69, 2509–2512.

MARTEN, G. 1973. An optimization equation for predation, *Ecology* 54, 92–101.

MAYNARD SMITH, J. 1974. "Models in Ecology," p. 146. Cambridge Univ. Press, London/ New York.

MITCHELL, B. 1963. Ecology of two carabid beetles, *Bembidion lampros* and *Trechus quadristriatus*, *J. Anim. Ecol.* 32, 289–299.

MURDIE, G. AND HASSELL, M. P. 1973. Food distribution, searching success, and predator–prey models, in "The Mathematical Theory of the Dynamics of Biological Populations," (Barlett, M. S. and Hiorns, R. W. (Eds.), Academic Press, New York/ London.

PARKER, G. A. AND STUART, R. A. 1976. Animal behaviour as a strategy optimizer: evolution of resource assessment strategies and optimal emigration thresholds, *Amer. Natur.* to appear.

PEARSON, N. E. 1974. Optimal foraging theory, Quan. Science Paper No. 39, Center for Quan. Science in Forestry, Fisheries and Wildlife. Univ. of Washington, Seattle, Washington 98195.

PULLIAM, H. R. 1974. On the theory of optimal diets, *Amer. Natur.* 108, 59–74.

PYKE, G. H. 1974. Studies in the foraging efficiency of animals, Ph. D. Thesis, Univ of Chicago.

RAPPORT, D. J. 1971. An optimization model of food selection, *Amer. Natur.* 105, 575–587.

RICHERSON, J. V. AND BORDEN, J. H. 1972. Host finding of *Coeloides brunneri* (Hymenoptera: Braconidae), *Canad. Entomol.* 104, 1235–1250.

ROYAMA, T. 1970. Factors governing the hunting beavior and selection of food by the Great Tit, *Parus major*, *J. Anim. Ecol.* 39, 619–668.

SCHOENER, T. W. 1969. Models of optimal size for a solitary predator, *Amer. Natur.* 103, 277–313.

SCHOENER, T. W. 1971. Theory of feeding strategies, *Ann. Rev. Ecol. Syst.* 2, 369–404.

136   ERIC L. CHARNOV

SMITH, J. N. M. 1974. The food searching behavior of two European thrushes. I. Description and analyses of the search paths, *Behavior* **48**, 276–302; II. The adaptiveness of the search patterns, *Behavior* **49**, 1–61.

SMITH, J. N. M. AND DAWKINS, C. R. 1971. The hunting behaviour of individual Great Tits in relation to spatial variations in their food density, *Anim. Behav.* **19**, 695–706.

TAHA, H. A. 1971. "Operations Research," p. 703. Macmillan, New York.

TIMIN, M. E. 1973. A multi-species consumption model, *Math. Biosci.* **16**, 59–66.

TINBERGEN, N., IMPEKOVEN, M., AND FRANCK, D. 1967. An experiment on spacing-out as a defense against predation, *Behavior* **38**, 307–321.

TULLOCK, G. 1970. The coal tit as a careful shopper, *Amer. Natur.* **104**, 77–80.

WARE, D. M. 1975. Growth, metabolism and optimal swimming speed in a pelagic fish, *J. Fish Res. Bd. Canad.* **32**, 33–41.

WERNER, E. E. 1974. The fish size, prey size, handling time relation in several sunfishes and some implications. *J. Fish Res. Bd. Canad.* **31**, 153–1536.

WERNER, E. E. AND HALL, D. J. 1974. Optimal foraging and size selection of prey by the bluegill sunfish *(Lepomis mochrochirus)*, *Ecology* **55**, 1042–1052.

Vol. 111, No. 982   The American Naturalist   November–December 1977

EVIDENCE FOR THE EXISTENCE OF THREE PRIMARY STRATEGIES IN PLANTS AND ITS RELEVANCE TO ECOLOGICAL AND EVOLUTIONARY THEORY

J. P. Grime

Unit of Comparative Plant Ecology (NERC), Department of Botany, The University, Sheffield S10 2TN, England

The external factors limiting plant biomass in any habitat may be classified into two categories. The first, which henceforth will be described as stress, consists of conditions that restrict production, e.g., shortages of light, water, or mineral nutrients and suboptimal temperatures. The second, referred to here as disturbance, is associated with the partial or total destruction of the plant biomass and arises from the activities of herbivores, pathogens, man (trampling, mowing, and plowing), and from phenomena such as wind damage, frosts, desiccation, soil erosion, and fire.

When the four permutations of high and low stress with high and low disturbance are examined (table 1), it is apparent that only three are viable as plant habitats. This is because, in highly disturbed habitats, severe stress prevents recovery or reestablishment of the vegetation. It is suggested that each of the three remaining contingencies has been associated with the evolution of a distinct type of strategy, i.e., low stress with low disturbance (competitive plants), high stress with low disturbance (stress-tolerant plants), and low stress with high disturbance (ruderal plants). These three strategies are, of course, extremes. The genotypes of the majority of plants appear to represent compromises between the conflicting selection pressures resulting from particular combinations of competition, stress, and disturbance.

In the first part of this paper, an attempt is made to define competition, stress, and disturbance more closely and to review some of the evidence, most of which refers to vascular plants, for the existence of three primary strategies.

The second part examines the implications of the three-strategy model with regard to (1) existing theories of natural selection, especially that of the r-K continuum (MacArthur and Wilson 1967; Pianka 1970); (2) secondary strategies; (3) vegetation succession; (4) dominance; and (5) strategies in fungi and in animals.

THE COMPETITIVE STRATEGY

A Definition of Competition between Plants

Wherever plants grow in close proximity, whether they are of the same or of different species, differences in vegetative growth, seed production, and plant

Amer. Natur. 1977. Vol. 111, pp. 1169–1194.

TABLE 1

Suggested Basis for the Evolution of Three Strategies in Vascular Plants

Intensity of Disturbance	Intensity of Stress	
	Low	High
Low	Competitive strategy	Stress-tolerant strategy
High.....................	Ruderal strategy	No viable strategy

mortality are observed. It would be a mistake, however, to attribute all such differences to competition, since disparities in the performance of neighboring plants may arise from differences in capacity to exploit (or in susceptibility to) features of the physical or biotic environment. If the term competition is to be useful to the analysis of vegetation mechanisms, a definition must be found which effectively distinguishes it from other processes which influence vegetation composition and species distribution.

In a most revealing paper, Milne (1961) has traced the history of biologists' attempts to define competition. He concludes that, far from achieving the necessary distinctions, a majority of authors have not sought to make them and have often used the word as a synonym for "the struggle for existence" (Darwin 1859). Perhaps because of the resulting confusion, Harper (1961) has proposed that, for the present at least, use of the term should be abandoned. However, it may be argued that competition as a process is too important and as a term is too useful to be allowed to suffer such a fate. An alternative is to apply a strict definition and to use the term as precisely as possible. Whatever the semantic difficulties, such an exercise will be justified if it facilitates the analysis of competition mechanisms.

Here, competition is defined as "the tendency of neighboring plants to utilize the same quantum of light, ion of a mineral nutrient, molecule of water, or volume of space" (Grime 1973*b*, p. 311). This choice of words allows competition to be defined in relation to its mechanism rather than its effects, and the risk is avoided of confusion with mechanisms operating through the direct impact of the physiochemical environment or through biotic effects such as selective predation. The word "neighboring" is included to imply that a degree of physical intimacy is characteristic of competing plants.

Variation in the Competitive Ability of a Plant Species

A fundamental problem underlying any attempt to analyze the impact of competition in vegetation arises from the proposition, scarcely disputed among ecologists, that the competitive ability of a plant species varies according to the conditions in which it is growing. Variation in competitive ability may arise because environments differ in the extent to which they allow the competitive potential of a species to be realized. It is clear that competitive characteristics such as rapid growth in leaf area and root absorptive surface may be inhibited by forms of stress and damage. Second, these same attributes may be subject to genetic variation. The literature provides numerous examples of intraspecific variation in competitive attributes (e.g., Clausen et al. 1940; Bradshaw 1959; Gadgil and Solbrig 1972; Mahmoud et al. 1975).

Phenotypic and genetic variation in competitive ability is not a severe impediment to the analysis of competition since it is clear that, with care, both may be recognized and taken into account. A more serious difficulty arises from the suggestion, implicit in many ecological writings, that the nature of competition itself may vary fundamentally from one field situation to another so that, relative to other species, a particular species or genotype may be a strong competitor in one site but a weak competitor in another. Inspection of the data which have been quoted in support of this view (e.g., Newman 1973; Ellenberg and Mueller-Dombois 1974) shows that often this concept has evolved in tandem with loose definitions of competition. That is, in certain cases, shifts in the fortunes of "competitors" coincident with changes in environment may be attributed more correctly to the incursion of noncompetitive effects (e.g., differential predation) rather than to any alteration in the relative abilities of the plants to compete for resources.

When noncompetitive effects have been discounted, the remaining objection to a unified concept of competitive ability arises because competition occurs with respect to several different resources including light, water, various mineral nutrients, and space. Hence it might be supposed that the ability to compete for a given resource varies independently from the ability to compete for each of the others. Moreover, the ability to compete for a given resource might be expected to change according to its availability.

There are grounds, both theoretical and empirical, for rejecting these possibilities as major sources of variation in competitive ability. A more logical interpretation is that the abilities to compete for light, water, mineral nutrients, and space are interdependent to the extent that natural selection has caused their development to a comparable extent in any particular genotype.

In circumstances which allow rapid development to a large standing crop, the most obvious competition is that occurring above ground for space and light, and it may not be immediately obvious why success should also depend upon the effective capture of water and mineral nutrients. However, as pointed out by Mahmoud and Grime (1976), it is clear that rapid buildup of a large biomass of shoot material, a prerequisite for effective above-ground competition, is dependent upon high rates of uptake of water and mineral nutrients. Consideration of the sequence of events in competition suggests that although in productive habitats the phenomenon culminates in above-ground competition for space and light, the outcome may be strongly influenced or even predetermined by earlier competition below ground.

The importance of below-ground interactions in competitions in productive environments is well illustrated by an experiment conducted by Donald (1958) involving competition between the perennial grasses *Lolium perenne* and *Phalaris tuberosa*. In this pot experiment, a system of partitions, above and below ground, was devised so that it was possible to compare the yield of each of the two species under four treatments, i.e., competition above ground only, competition below ground only, competition above and below ground, and no competition (control). The experiment was carried out at high and low levels of nitrogen supply, but here it is only the former which is relevant to this point. The results showed that the clear advantage of *L. perenne* over *P. tuberosa* under productive conditions derived from the superior competitive ability of the species both above and below ground.

It is now necessary to turn to the vexed subject of competition in unproductive environments. The crucial problem here is to determine whether plants which occur naturally in dry, heavily shaded, or mineral deficient habitats are better able to compete for specific resources at low levels of supply than plants from productive habitats.

In searching for evidence, it is convenient first to return to Donald's (1958) experiment and in particular to consider the results obtained in the low-nitrogen treatments. These results show conclusively that the superior competitive ability exhibited by *L. perenne* under productive conditions was maintained when the two grasses competed on nitrogen-deficient soil. However, since both of the grasses in this experiment are characteristic of conditions of moderate to high fertility, it may be unwise to draw any general implication from this result.

In an experiment conducted by Mahmoud and Grime (1976), three perennial grasses of contrasted ecology were allowed to compete under productive and unproductive (low-nitrogen) conditions. One of the species, *Festuca ovina*, is restricted to unproductive sites; another, *Agrostis tenuis*, is associated with habitats of intermediate fertility; and the third, *Arrhenatherum elatius*, is a frequent dominant of productive meadows. Each species was grown in monoculture and in separate 1 : 1 mixtures with each of the other two grasses. The results obtained in the productive treatment showed that the yield of *A. elatius* in monoculture was twice that of *A. tenuis* and approximately seven times that of *F. ovina*. In the mixtures grown under productive conditions, competition in each case was conclusive in that *F. ovina* was totally eliminated by both of the other grasses, and there were no survivors of *A. tenuis* when this species was grown with *A. elatius*. In the low-nitrogen treatment all three species showed a marked reduction in yield, and this was most pronounced in *A. elatius* and *A. tenuis*. However, there was no evidence to suggest that the competitive ability of *F. ovina* increased under conditions of nitrogen stress, and no competitive impact of *F. ovina* was detected on either of the other two species.

From this result, it would appear that the ability to compete for nitrogen plays little part in the mechanism whereby *F. ovina* is adapted to survive conditions of low soil fertility. A similar conclusion may be drawn not only with respect to nitrogen but also to other essential mineral nutrients from a number of laboratory experiments (e.g., Bradshaw et al. 1964; Hackett 1965; Clarkson 1967; Rorison 1968; Higgs and James 1969) in which plants of infertile habitats have been grown on nutrient-deficient soils and solution cultures. These studies provide no convincing evidence that plants indigenous to poor soils are more efficient in the uptake of mineral nutrients when these are present at low concentrations.

An alternative explanation for the survival of plants under conditions of low soil fertility will be presented under the next heading. Essentially, this theory suggests that although competition, especially that for water and mineral nutrients, is not restricted to productive habitats, its importance in unproductive habitats is small relative to the direct impact upon the plants of those factors which cause the environment to be infertile. This argument may be extended to the full range of unproductive habitats, including those in which light and water supply are the main limiting factors. Further evidence of the decline in importance of competition in unproductive vegetation is available from a wide range of comparative studies (Ashton 1958; Pigott and Taylor 1964; Grime and Jeffrey 1965; Grime 1965, 1966; Clarkson 1967; Myerscough and Whitehead 1967; Hutchinson 1967; Loach 1970; Rorison 1968; Parsons 1968; Higgs and James 1969), from which there is accumulating evidence that, in both herbs and trees, characteristics such as rapid potential growth rate and high phenotypic plasticity with respect to the deployment of photosynthate (e.g., increase in leaf area in shade and increase in root : shoot ratio under mineral nutritional stress) become disadvantageous during extreme environmental stress.

It seems reasonable to conclude, therefore, that competition has been associated with the evolution of a distinct strategy, and that competitive ability is associated with characteristic and measurable genetic attributes (table 2) which, by maximizing the capture of resources, facilitate the exclusive occupation of fertile, relatively undisturbed environments.

As we shall see later, it is of considerable significance that no broad distinction may be drawn between herbaceous and woody species with respect to competitive ability. Perennial herbs, shrubs, and trees each appear to encompass a wide range of competitive abilities.

THE STRESS-TOLERANT STRATEGY

A Definition of Stress

Dry-matter production in vegetation is subject to a wide variety of environmetal constraints. These include not only shortages and excesses in the supply of solar energy, water, and mineral nutrients but also sub- and supraoptimal temperatures and growth-inhibiting toxins such as heavy metals from the soil or gaseous pollutants from the atmosphere. Plant species and even different genotypes may differ in susceptibility to particular forms of stress, and in consequence each stress may exercise a different effect on vegetation composition. Because many stresses may operate in the same habit over the course of a year, analyses of the impact of stress may become quite complex. Further complications arise when certain stresses either originate from or are intensified by the vegetation itself. Among the most important types of plant-induced

TABLE 2

SOME CHARACTERISTICS OF COMPETITIVE, STRESS-TOLERANT, AND RUDERAL PLANTS

	Competitive	Stress Tolerant	Ruderal
Morphology of shoot	High dense canopy of leaves; extensive lateral spread above and below ground	Extremely wide range of growth forms	Small stature, limited lateral spread
Leaf form	Robust, often mesomorphic	Often small or leathery, or needle-like	Various, often mesomorphic
Litter	Copious, often persistent	Sparse, sometimes persistent	Sparse, not usually persistent
Maximum potential relative growth rate	Rapid	Slow	Rapid
Life forms	Perennial herbs, shrubs, and trees	Lichens, perennial herbs, shrubs, and trees (often very long lived)	Annual herbs
Longevity of leaves	Relatively short	Long	Short
Phenology of leaf production	Well-defined peaks of leaf production coinciding with period(s) of maximum potential productivity	Evergreens with various patterns of leaf production	Short period of leaf production in period of high potential productivity
Phenology of flowering	Flowers produced after (or, more rarely, before) periods of maximum potential productivity	No general relationship between time of flowering and season	Flowers produced at the end of temporarily favorable period
Proportion of annual production devoted to seeds	Small	Small	Large

stress are those arising from shading and reduction of the levels of mineral nutrients in the soil following their accumulation in the plant biomass.

In order to accommodate its diverse forms, stress will be defined simply as the external constraints which limit the rate of dry-matter production of all or part of the vegetation.

Adaptation to Severe Stress

Identification of the forms of stress characteristic of particular habitats is but one aspect of the analysis which is necessary to determine the influence of stress upon vegetation. Another requirement is to estimate the extent to which stress is limiting primary production in different types of vegetation. The need for such studies arises because the role of stress changes according to its severity. The low intensity of stress, characteristic of productive habitats, functions as a modifier of competition, whereas severe stress exercises a dominant and more immediate impact upon species composition and vegetation structure through its direct effects on survival and reproduction.

The most conspicuous effect of severe stress is to eliminate or to debilitate species of high competitive ability and to cause them to be replaced by stress-tolerant species. Although the identities of stress-tolerant species vary according to the type of stress, there is abundant evidence to suggest a basic similarity in the mechanisms whereby plants are adapted to survive different forms of severe stress. Some of this evidence will now be examined by briefly considering adaptation to severe stress in four contrasted types of habitat. In two of these (arctic-alpine and arid habitats) plant production is low and stress is mainly imposed by the environment. In the third (shaded habitats) stress is plant induced, while in the fourth (nutrient-deficient habitats) stress may be due to the low fertility of the habitat and/or to accumulation of the mineral nutrients in the vegetation.

a) Arctic and alpine habitats.—Without doubt, the dominant environmental stress in arctic and alpine habitats is low temperature. In these habitats, the opportunity for growth is limited to a short summer season. For the remainder of the year growth is prevented by low temperature, and the vegetation is either covered by snow or, where it remains exposed on ridges, is subjected to extreme cold coupled with the desiccating effect of dry winds. During the growing season, production is often severely restricted not only by low temperatures but also by desiccation and mineral nutrient stress, the latter largely a result of the low microbial activity of the soil. Alpine vegetation is subject to additional stresses peculiar to high altitudes, including strong winds and intense solar radiation.

The predominant life forms in arctic and alpine vegetation are very low-growing evergreen shrubs, small perennial herbs, bryophytes, and lichens (Billings and Mooney 1968). The adaptive significance of the small stature of all of these plants appears to be related in part to the observation that during the winter when the ground is frozen and no water is available to the roots, the aerial parts of low-growing plants tend to be insulated from desiccation by a covering of snow. However, as suggested by Boysen-Jensen (1929), quite apart from the risk of winter kill of shoots projecting through the snow cover, the absence of larger plants and, in particular, erect shrubs and trees, is also related to the fact that the productivity of tundra and alpine habitats is so low that it is unlikely that there will be a sufficient surplus of photosynthate to sustain either wood production or the annual turnover of dry matter characteristic of deciduous trees and tall herbs.

A conspicuous feature of arctic and alpine floras is the preponderance of evergreen species. Annual plants are extremely rare, and the majority of bryophytes, lichens, herbs, and small shrubs remain green throughout the year. The advantage of the evergreen habit is particularly clear in long-lived prostrate shrubs. Here, Billings and Mooney (1968) suggest that the main advantage of the evergreen habit is that it obviates the necessity "to spend food resources on a wholly new photosynthetic apparatus each year" (p. 492).

It would appear, from the relatively small amount of information which is available, that both seed production and seedling establishment in arctic and alpine habitats are erratic and hazardous. However, as Billings and Mooney (1968) point out, this is compensated for first because the majority of species are long-lived perennials and second by the widespread occurrence of vegetative reproduction.

b) Arid habitats.—The comprehensive reviews by authors such as Walter (1973), Slatyer (1967), and Levitt (1975) make clear the pitfalls in attempting to generalize about the ways in which plants adapt to exploit conditions of low annual rainfall. However, if attention is confined to those species which in their natural habitats experience long periods of desiccating conditions without access to underground reservoirs of soil moisture, common adaptive features are apparent. These xerophytes are perennials in which the vegetative plant is adapted to survive for long periods during which little water is available. Several types of xerophytes may be distinguished according to the severity of moisture stress which they can survive.

In habitats which experience a short annual wet season, the most commonly occurring xerophytes are the sclerophylls. This group includes both small shrubs and trees, such as the evergreen oaks and the olive, all of which are distinguished by small hard leaves which are retained throughout the dry season. Walter (1973, p. 121) maintains that the ecological advantage of sclerophylly is related to the ability of sclerophyllous species "to conduct active gaseous exchange in the presence of an adequate water supply but to cut it down radically by shutting the stomata when water is scarce," a mechanism which enables these plants to "survive months of drought with neither alteration in plasma hydrature nor reduction in leaf area" and, when rains occur, to "immediately take up production again."

Under severely desiccating conditions, the most persistent of the xerophytes are the succulents, evergreens in which water is stored in swollen leaves, stems, or roots. During drought periods no water absorption occurs, but following rain, small, short-lived roots are produced extremely rapidly. Succulents are also distinguished by peculiarities in stomatal physiology and metabolism.

The stomata open at night and remain tightly closed during the day. Gaseous exchange occurs during the period of stomatal opening, at which time carbon dioxide is incorporated into organic acids. These are decarboxylated in the daylight to release carbon dioxide for photosynthesis. This mechanism, known as crassulacean acid metabolism, appears to represent a mechanism whereby low levels of photosynthesis may be maintained with minimal transpiration.

A well-known feature of many xerophytes and, in particular, the succulents is the rarity or erratic nature of flowing.

c) *Shaded habitats*.—The stresses associated with shade differ from those considered under the two previous headings in that they are not directly attributable to gross features of climate. However, although shade itself is not imposed by the physical environment, it becomes important only in climatic regimes which are conducive to the development of dense canopies. It may be important to bear in mind, therefore, that shade at its greatest intensities frequently coincides (1) with the high temperatures and humidities of tropical and subtropical climates or with the warm summer conditions of temperate regions and (2) with conditions of mineral nutrient depletion associated with the development of a large plant biomass. Hence, although the discussion which follows refers to adaptations to shade, there is a strong probability that some of the plant characteristics described are related to cotolerance of shade, warm temperatures, and mineral nutrient stress.

The effect of shading upon the growth rate and morphogenesis of shade-tolerant plants has been examined by growing various species, usually as seedlings, under screens of cotton, plastic, or metal (e.g., Burns 1923; Bordeau and Laverick 1958; Grime 1966; Loach 1970). The general conclusions which may be drawn from the results of these experiments is that the capacity to maximize dry-matter production in shade through modification of the phenotype (through increases in leaf area and reductions in root : shoot ratio) is most apparent in competitive species characteristic of unshaded or lightly shaded environments, while plants associated with deep shade tend to grow slowly and to show little morphogenetic response to shade treatment. The low rates of growth and the small extent of phenotypic response to shading in the shade-tolerant species suggest that adaptation to shade in these plants may be concerned more with the ability to survive for extended periods in deep shade than with the capacity to maximize light interception and dry-matter production. A similar conclusion is prompted by the observation that many of the more shade-tolerant plants, e.g., *Hedera helix* and *Deschampsia flexuosa*, have morphologies which allow considerable self-shading.

Among the shade-tolerant tree seedlings, climbing plants, and epiphytes of tropical rain forests are many which remain green throughout the year, and a similar phenology is evident in shade-tolerant herbs of temperate woodlands (Al-Mufti et al. 1977). An observation of considerable interest concerning the adaptive physiology of shade-tolerant plants is that of Woods and Turner (1971), who found in experimental studies with a range of North American trees that stomatal opening in response to increased light intensity was consistently more rapid in shade-tolerant species. The authors conclude that such rapid stomatal response may allow shaded leaves to exploit brief periods of high illumination by sun flecks.

A characteristic of many shade-tolerant species is the paucity of flowering and seed production under heavily shaded conditions. This phenomenon is particularly obvious in British woodlands, where flowering in many common plants such as *H. helix*, *Lonicera periclymenum*, and *Rubus fruticosus* agg. is usually restricted to plants exposed to sunlight at the margins of woods or beneath gaps in the tree canopy.

d) Nutrient-deficient habitats.—Under this heading attention will be confined to habitats in which mineral nutrient stress arises from the impoverished nature of the habitat. It is in these habitats that mechanisms of adaptation to mineral nutrient stress have been most closely studied. However, it is vital to the arguments developed later in this paper to recognize that severe stress may also arise under conditions in which mineral nutrients are sequestered in the living or nonliving parts of the biomass.

When the range of vegetation types characteristic of severely nutrient deficient soils is examined, certain common features are immediately apparent. Although the identity of the species involved varies in different parts of the world and according to local factors such as soil type and vegetation management, similar plant morphologies may be recognized. The herbaceous component shows a marked reduction in growth form. Among the grasses, narrow-leaved creeping or rosette species predominate. Under management conditions which allow the development of woody vegetation, nutrient-deficient soils are usually colonized by small (often coniferous) trees or sclerophyllous shrubs. In Europe and North America members of the Ericaceae such as *Calluna vulgaris*, *Vaccinium myrtillus*, and *Erica cinerea* are particularly common on highly acidic soils, while on shallow nutrient-deficient calcareous soils species such as *Helianthemum chamaecistus* and *Thymus drucei* occur. The xeromorphy evident in all these shrubs could be conceived to be an adaptation to winter or summer desiccation. However, this explanation cannot be applied in the case of the xerophytic shrubs which are known to occur on infertile soils in tropical and subtropical regions. In New Caledonia, for example, Birrel and Wright (1945) described xerophyllous shrub vegetation varying between 1–2 m in height on serpentine soil and commented upon its "unusual appearance in this region of high rainfall in which tropical forest is the ordinary plant cover" (p. 72).

Another feature which is characteristic of the vegetation of nutrient-deficient soils is the high frequency of species of inherently slow growth rate. Among the first to recognize this was Kruckeberg (1954), who noted that even when growing on fertile soils certain ecotypes of herbaceous plants adapted to survive on serpentine soils in North America grew slowly in comparison with species and ecotypes from fertile habitats. Another important early contribution was that of Beadle (1954, 1962), who recognized that slow growth rates were characteristic of species of *Eucalyptus* growing on Australian soils with low phosphorus availability. Slow rates of growth were also found to occur in *Festuca ovina* and *Nardus stricta*, two grasses of widespread occurrence on infertile soils in the British Isles (Bradshaw et al. 1964). Jowett (1964) concluded from his investigation with the grass *Agrostis tenuis* that the effect of natural selection under conditions of severe mineral nutrient deficiency had been to reduce the potential growth rate of local populations established on mine waste in Wales. Subsequent investigations (e.g., Hackett 1965; Clarkson 1967; Higgs and James 1969; Grime and Hunt 1975) have confirmed that there is a strong correlation between low potential growth rate and tolerance of mineral nutrient deficiencies.

It would appear, therefore, that adaptation for survival in infertile soils has involved in both woody and herbaceous species reductions in stature, in leaf form, and in potential growth rate. The explanation which has been put forward with varying degrees of elaboration by a number of authors (e.g., Kruckeberg 1954; Loveless 1961; Jowett 1964) to account for this phenomenon is that it is primarily an adaptation for survival under conditions of low mineral nutrient supply. This suggests that under conditions in which elements such as phosphorus and nitrogen are scarcely available, natural selection has led to the evolution of plant species and ecotypes which make low demands upon the mineral nutrient reserves of the soil and are therefore less likely to exhaust the supply in their immediate root environment. Consistent with this hypothesis are the results of a number of experiments in which species from infertile habitats have been grown under various levels of supply of major mineral nutrients (Bradshaw et al. 1964; Hackett 1965; Clarkson 1967). From these studies there is no convincing evidence that species normally restricted to infertile soils are better adapted than species from fertile habitats to maintain dry-matter production under conditions in which mineral nutrients are provided at low rates of supply. The results of an experiment by Bradshaw et al. (1964), for example, illustrate clearly that under low levels of nitrate nitrogen supply *F. ovina* and *N. stricta*, both grasses of infertile pastures, are outyielded by *Lolium perenne* and *Cynosurus cristatus*, species which are normally restricted to fertile soils.

An advantage which a low rate of dry-matter production appears to confer on a plant growing on an infertile soil is that during periods of the year when mineral nutrients are more readily available, uptake is likely to exceed the rate of utilization in growth allowing the accumulation of reserves which may be drawn upon during subsequent periods of stress (Clarkson 1967).

An additional feature of many of the slow-growing species characteristic of infertile soils is the lack of a sharply defined seasonal pattern of growth. The majority of woody and herbaceous species from infertile habitats are evergreen plants in which the leaves have a comparatively long life span. Measurements by Williamson (1976), for example, on grasses of derelict calcareous pastures in southern England have shown that in two evergreen species of nutrient-deficient soils, *Helictotrichon pratense* and *H. pubescens*, the functional life of the leaf is considerably longer than in *Arrhenatherum elatius* and *Dactylis glomerata* species which are associated with more fertile soil conditions and show a well-defined summer peak in leaf production. The most likely explanation for the greater longevity and slower replacement of leaves in plants of infertile habitats is that it represents an adaptation slowing the rate of nutrient cycling between plant and soil and hence reduces the risk of nutrient loss. It would appear, therefore, that species of infertile habitats are adapted to conserve absorbed mineral nutrients rather than to maximize the quantity captured.

General Features of Stress Tolerance

From this brief survey it is apparent that vascular plants adapted to contrasted forms of severe stress exhibit a range of features which, although varying in detailed mechanisms, represent basically similar adaptations for endurance of conditions of limited productivity. These features, listed in table 2, include inherently slow rates of growth, the evergreen habit, long-lived organs, sequestration and slow turnover of carbon, mineral nutrients and water, low phenotypic plasticity, shy flowering, and the presence of mechanisms which allow the vegetative plant an opportunistic exploitation of temporarily favorable conditions. The latter consist not only of the presence throughout the year of functional leaves (and, presumably, roots also) but, in addition, special mechanisms such as the rapid activating of stomata in sclerophylls and shade plants and the rapid sprouting of roots by succulents.

As in the case of the competitive strategy, no simple generalization can be made with regard to the stature and life form of stress-tolerant plants. Although in extremely unproductive environments stress tolerance is associated with trees, shrubs, and herbs of reduced stature, account must be taken of the fact that many of the shade-tolerant trees in temperate and tropical habitats are long lived and attain a very large size at maturity. An explanation for the wide morphological variety among stress-tolerant plants is advanced later in this paper.

An additional characteristic which may be predicted for stress-tolerant plants is that of low palatability. Because of their slow growth rates, stress-tolerant plants are likely to exhibit slow rates of recovery from defoliation, and during the protracted phase of establishment, seedlings will be particularly vulnerable to damage. It would not be surprising, therefore, to find that many stress-tolerant plants had experienced intensive natural selection for resistance to predation.

Stress Tolerance and Symbiosis

(a) *Lichens.*—Three features which under the last heading have been associated with stress tolerance in flowering plants (slow growth rate, longevity, and opportunism) are expressed in a most extreme form in lichens. This is hardly surprising in view of their ecology. Lichens are able to survive in extremely harsh environments under conditions in which vascular plants are totally excluded and in which they experience extremes of temperature and moisture supply and are subject to low availability of mineral nutrients. Although there is some diversity of opinion with regard to the longevity of lichens (see, for example, the review by Billings and Mooney [1968]), there is general agreement that many are exceedingly long lived. From experimental studies, such as those of Farrar (1976a, 1976b, 1976c), there is abundant evidence of the tendency of lichens to sequester a high proportion of the photosynthate rather than to expend it in growth. Much of the assimilate appears to be stored as sugar alcohols (polyols) in the fungal component. Lichens are able to remain alive during prolonged periods of desiccation and, on rewetting, to resume nutrient uptake and photosynthesis extremely rapidly.

(b) *Ectotrophic mycorrhizas.*—An interesting parallel may be explored between the lichens and another type of symbiosis involving fungi. In many trees and shrubs the roots develop a thick investment of fungal hyphae (ectotrophic mycorrhizas) which are connected on the inside with living cells of the root and on the outside with hyphae which are in contact with the soil and leaf litter.

A considerable amount of physiological work has been carried out to assess the ecological significance of ectotrophic mycorrhizal associations and has been reviewed extensively (Harley 1969, 1970, 1971). It has been clearly established that the uptake of mineral nutrients such as phosphorus and nitrogen may be facilitated and the yield of the host plant increased by the presence of mycorrhizal infections. Ectotrophic mycorrhizas are abundant near the soil surface, and it seems likely that they enable nutrient elements mineralized during the decay of litter to be efficiently reabsorbed into the plants.

Review of the distribution of ectotrophic mycorrhizas shows that they are strongly associated with conditions of mineral nutrient stress. These include not only vegetation types such as heathland and sclerophyllous scrub, in which the vegetation density is severely limited by the low mineral nutrient content of the habitat, but also mature temperate and tropical forests in which mineral nutrient scarcity in the soil arises from the scale of nutrient accumulation within the plant biomass.

In comparison with the roots of crop plants and species restricted to fertile soils, ectotrophic mycorrhizas appear to be long lived, and the possibility should be considered that this attribute is itself a considerable advantage to a host plant growing under conditions of severe mineral nutrient stress. In competitive herbs adapted to exploit soils with a large reservoir of available mineral nutrients, high rates of uptake are attained first through the production of a very large absorptive surface of fine roots and root hairs and second through continuous and rapid morphogenetic responses in root : shoot ratio and in the extension growth of individual parts of the root system. There is, however, continuous decay and replacement of roots, a process analogous to the rapid turnover of leaves occurring simultaneously above ground in these species. It is clear that this system of nutrient absorption, while effective in absorbing mineral nutrients, even from infertile soils, is achieved through a high expenditure of photosynthate. The main benefit of a mycorrhizal association to a host plant growing under conditions of mineral nutrient stress may be an absorptive system which, because it remains functional over a long time, allows exploitation of temporary periods of mineral nutrient availability at lower cost to the synthetic resources of the plant.

THE RUDERAL STRATEGY

A Definition of Disturbance

Reference has been made under the preceding heading to a variety of habitats in which the density of living and dead plant material is low because production is severely restricted by environmental stress. It is quite clear, however, that low vegetation densities are not confined to unproductive habitats. Some of the world's most productive terrestrial environments, including many arable fields, are characterized by a rather sparse vegetation cover. Here the low densities arise because the vegetation is subject to partial or total destruction.

Although more obvious in habitats such as arable land, the effects of damage upon vegetation are ubiquitous. The amount of vegetation and the ratio of living to dead plant material in any habitat at any point in time depend upon the balance between the processes of production and destruction.

Even within one environment there may be considerable variety in the mechanisms which bring about the destruction of living or dead vegetation components. In addition to natural catastrophies (e.g., floods and windstorms) and the more drastic forms of human impact (e.g., plowing, mowing, trampling, and burning), account must be taken of more subtle effects, i.e., seasonal, fluctuations in climate or the activities of herbivores, decomposing organisms, and pathogens.

Forms of disturbance differ with respect to their selectivity. While the effects of herbivores and decomposing organisms tend to be restricted to the living and the dead material, respectively, certain phenomena, e.g., fire, may affect both components of the vegetation. In general, there is an inverse correlation between degree of selectivity and intensity of disturbance, a relationship which may be exemplified at one extreme by molecular discrimination between litter constituents by microbial decomposing organisms and at the other by the total vegetation destruction associated with phenomena such as severe soil erosion.

Among the forms of disturbance which affect living components, a distinction may be drawn between mechanisms which involve the immediate removal of plant structures from the habitat (e.g., grazing, mowing) and those in which plant material is killed but remains in situ (e.g., frost, drought, applications of herbicides). In the latter, destruction usually proceeds in two stages. An initial rapid loss of solutes is followed by a rather longer phase in which the residue of plant structures is attacked by decomposing organisms.

To analyze the primary mechanism of vegetation, a term is required which encompasses this wide range of phenomena yet can be simply defined. The term used here is *disturbance*, which may be said to consist of the mechanisms which limit the plant biomass by causing its destruction.

Adaptation to Severe Disturbance

Just as the impact of stress upon vegetation changes according to its intensity, so also does that of disturbance. Whereas in potentially productive environments a low intensity of disturbance modifies the balance between competitive species, severe disturbance selects species with phenologies adapted to exploit temporarily favorable conditions.

Flowering plants adapted to persistent and severe disturbance (henceforth described as ruderals) have several features in common (table 2). The most consistent among these is the tendency for the life cycle to be that of the annual or short-lived perennial, a specialization clearly adapted to exploit environments intermittently favorable for rapid plant growth. A related characteristic of many ruderals is the capacity for high rates of dry-matter production (Baker 1965; Grime and Hunt 1975), a feature which appears to facilitate rapid completion of the life cycle and maximizes seed production. In ruderal plants a large proportion of the photosynthate is directed into seeds, and under conditions of stress seed production is usually maintained at the expense of a curtailment of vegetative development (Salisbury 1942; Harper 1961; Harper and Ogden 1970).

A feature of the reproductive biology of many ruderals, especially the weeds of arable fields (Barton 1961), is the ability of buried seeds to survive in the soil for long periods and, if the climate is favorable, to germinate rapidly when disturbance either exposes the seeds to light or, through removal of the insulating effect of foliage and litter, causes buried seeds of certain species to experience large diurnal fluctuations of temperature (Thompson et al. 1977).

THEORETICAL IMPLICATIONS

C-, S-, and R-Selection

The information reviewed here suggests that, during the evolution of plants, three fundamentally different forms of natural selection have occurred. The first of these (C-selection) has involved selection for highly competitive ability which depends upon plant characteristics that maximize vegetative growth in productive, relatively undisturbed conditions. The second (S-selection) has brought about reductions in both vegetative and reproductive vigor, adaptations which allow endurance of continuously unproductive conditions arising from environmental stress, severe resource depletion by the vegetation, or the combined effect of the two. The third (R-selection) is associated with a short life span and with high seed production and has evolved in severely disturbed but potentially productive environments.

It is concluded that R-selected (ruderal), S-selected (stress-tolerant), and C-selected (competitive) plants each possess a distinct family of genetic characteristics, and an attempt has been made in table 2 to list some of these.

A crucial genetic difference between competitive, stress-tolerant, and ruderal plants concerns the form and extent of phenotypic response to stress. Many of the stresses which are a persistent feature of the environments of stress-tolerant plants are experienced, although less frequently and in different contexts, by competitive and ruderal plants. From the evidence reviewed here, competitive, stress-tolerant, and ruderal plants seem to exhibit three quite distinct types of response to stress. It is concluded that such differences constitute one of the more fundamental criteria whereby the three strategies may be distinguished.

In table 3, an attempt has been made (1) to compare the circumstances in which competitive, stress-tolerant, and ruderal plants are normally exposed to major forms of stress; (2) to describe the stress response characteristic of each strategy; and (3) to predict the different consequences attending each type of response in different ecological situations.

It is concluded that the small degree of phenotypic response which is associated with endurance of protracted and severe stress in stress-tolerant plants is of low survival value in situations where stress is a prelude to either competitive exclusion or to disturbance by phenomena such as drought. Similarly, the rapid and highly plastic stress responses of competitive plants (tending to maximize vegetative growth) and of ruderals (tending to curtail vegetative growth and maximize seed production) are highly advantageous only in the specific circumstances associated, respectively, with competition and disturbance.

r- and K-Selection

It is interesting to compare the concept of C-, S-, and R-selection with those derived from previous attempts to identify strategies in plants (e.g., Harper and Ogden 1970; Gadgil and Solbrig 1972). These authors have been mainly concerned to extend to plants concepts originally applied to animals (Cody 1966; MacArthur and Wilson 1967).

The most generally accepted theory to emerge from earlier studies with both animals and plants is the concept of r- and K-selection, originally proposed by MacArthur and Wilson (1967) and expanded by Pianka (1970). This theory recognizes two types of organisms as opposite poles in the evolutionary spectrum. The first, said to be K-selected, consists of organisms in which the life expectancy of the individual is long and the proportion of the energy and other captured resources devoted to reproduction is small. The second, or r-selected type, is made up of organisms with a short life expectancy and large reproductive effort. It is now widely accepted that the majority of organisms fall between the extremes of r- and K-selection. More recent evidence (e.g., Gadgil and Solbrig 1972) suggests that genetic variation may cause populations of the same species to occupy different positions along an r-K continuum.

The diagram in figure 1 attempts to reconcile the concept of the three primary plant strategies with that of r- and K-selection. It is suggested that the ruderal and stress-tolerant strategies correspond, respectively, to the extremes of r- and K-selection, and that highly competitive species occupy an intermediate position. This relationship is consistent with the model proposed later to explain the involvement of the three strategies in the process of vegetation succession.

The most substantial way in which the three-strategy model differs from that of the r-K continuum lies in recognition of stress tolerance as a distinct strategy evolved in intrinsically unproductive habitats or under conditions of extreme resource depletion induced by the vegetation itself. Inspection of figure 1 suggests that there are two critical points along the r-K continuum. At (1), the intensity of disturbance becomes insufficient to prevent the exclusion of ruderals by competitors, while at (2) the level of supply of resources is depleted

TABLE 3

MORPHOGENETIC RESPONSES TO DESICCATION, SHADING, OR MINERAL NUTRIENT STRESS OF COMPETITIVE, STRESS-TOLERANT, AND RUDERAL PLANTS AND THEIR ECOLOGICAL CONSEQUENCES IN THREE TYPES OF HABITAT

STRATEGY	RESPONSE TO STRESS	CONSEQUENCES		
		Habitat 1*	Habitat 2†	Habitat 3‡
Competitive	Large and rapid changes in root: shoot ratio, leaf area, and root surface area	Tendency to sustain high rates of uptake of water and mineral nutrients to maintain dry-matter production under stress and to succeed in competition	Tendency to exhaust reserves of water and/or mineral nutrients both in rhizosphere and within the plant; etiolation in response to shade increases susceptibility to fungal attack	Failure rapidly to produce seeds reduces chance of rehabilitation after disturbance
Stress tolerant	Changes in morphology slow and often small in magnitude	Overgrown by competitors	Conservative utilization of water, mineral nutrients, and photosynthate allows survival over long periods in which little dry-matter production is possible	
Ruderal	Rapid curtailment of vegetative growth and diversion of resources into seed production		Chronically low seed production fails to compensate for high rate of mortality	Rapid production of seeds ensures rehabilitation after disturbance

* In the early successional stages of productive, undisturbed habitats (stresses mainly plant induced and coinciding with competition).

† In either continuously unproductive habitats (stresses more or less constant and due to unfavorable climate and/or soil) or in the late stages of succession in productive habitats.

‡ In severely disturbed, potentially productive habitats (stresses either a prelude to disturbance, e.g., moisture stress preceding drought fatalities, or plant induced, between periods of disturbance).

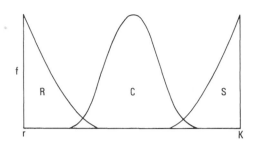

Fig. 1.—Diagram describing the frequency (f) of ruderal (R), competitive (C), and stress-tolerant (S) strategies along the r-K continuum. An explanation of the significance of critical points (1) and (2) is included in the text.

below the level required to sustain the high rates of growth characteristic of competitors, and selection begins to favor the more conservative physiologies of the stress tolerators.

The scheme represented in figure 1 is, however, both incomplete and misleading. This is because the linear arrangement of the three primary strategies does not provide for the existence of the wide range of equilibria which may exist between stress, disturbance, and competition and which has provided conditions for the evolution of the various types of "secondary" strategy described under the next heading.

Secondary Strategies

So far, it has been convenient to concentrate on plants which belong to one or another of the three primary strategies. However, it is clear that between the three extremes there are plants which are adapted to intermediate intensities of competition, stress, and disturbance.

The triangular model (Grime 1974) reproduced in figure 2 describes the various equilibria which are possible between competition, stress, and disturbance and recognizes four main types of secondary strategies which appear to have evolved in relation to particular equilibria. These consist of: (1) competitive ruderals (C-R)—adapted to circumstances in which there is a low impact of stress and competition is restricted to a moderate intensity by disturbance (e.g., fertile cattle pastures and meadows); (2) stress-tolerant competitors (C-S) —adapted to undisturbed conditions experiencing moderate intensities of stress (e.g., open forest or scrub on infertile soils); (3) stress-tolerant ruderals (S-R)—adapted to lightly disturbed unproductive habitats (e.g., droughted rock outcrops and crevices in cliffs and walls); (4) "C-S-R plants"—confined to habitats in which competition is restricted to moderate intensities by the combined effects of stress and disturbance (e.g., unfertilized pastures and meadows).

A detailed analysis of the secondary strategies is beyond the scope of this paper. However, it is possible to draw a general perspective with regard to the range of strategies encompassed by certain life forms or taxonomic groups. It

PRIMARY STRATEGIES IN PLANTS 1187

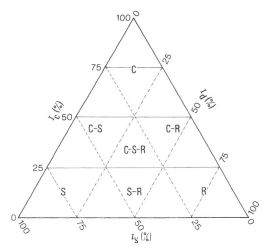

FIG. 2.—Model describing the various equilibria between competition, stress, and disturbance in vegetation and the location of primary and secondary strategies. I_c—relative importance of competition (————), I_s—relative importance of stress (— — —), I_d—relative importance of disturbance (— · — · —). A key to the symbols for the strategies is included in the text.

has been noted already that quite distinct life forms occur among the competitors and stress tolerators. In figure 3 this observation has been extended to include the secondary strategies by attempting to map the approximate distribution of selected life forms and taxa within the triangular model. This figure suggests that perennial herbs and ferns include the widest range of strategies. Annual herbs are mainly restricted to severely disturbed habitats, while biennials become prominent in the areas of the triangle corresponding to the competitive ruderals and the stress-tolerant ruderals. Trees and shrubs comprise competitors, stress-tolerant competitors, and stress tolerators. Although lichens are confined to the stress-tolerant corner of the model, bryophytes are more wide ranging with the center of the distribution in the stress-tolerant ruderals.

Vegetation Succession

Whereas the ruderal strategy comprises a fairly homogeneous group of ephemeral plants with many similarities in life history and ecology, the competitors consist of a wide range of plant forms including perennial herbs, shrubs, and trees. However, the most remarkable conjunctions in plant form and ecology occur within the stress tolerators, where we find such apparently diverse organisms as lichens and some forest trees.

In order to explain the adaptive significance of the morphological variety among the ranks of the competitors and stress tolerators, it is necessary to refer to the process of secondary (Horn 1974) vegetation succession.

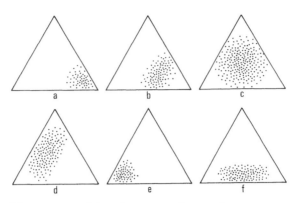

Fig. 3.—Diagrams describing the range of strategies encompassed by (*a*) annual herbs, (*b*) biennial herbs, (*c*) perennial herbs and ferns, (*d*) trees and shrubs, (*e*) lichens, and (*f*) bryophytes. For the distribution of strategies within the triangle, see figure 2.

A major factor determining the relative importance of the three primary strategies in vegetation succession is the potential productivity of the habitat. The model in figure 4 summarizes the influence of potential productivity on vegetation succession.

The model consists of an equilateral triangle in which variation in the relative importance of competition, stress, and disturbance as determinants of the vegetation is the same as that indicated by the three sets of contours in figure 2. At their respective corners of the triangle, competitors, stress tolerators, and ruderals become the exclusive constituents of the vegetation. The curves S_1, S_2, and S_3 describe, respectively, the path of succession in conditions of high, moderate, and low potential productivity, while the circles superimposed on the curves represent the relative size of the plant biomass at each stage of the succession. By comparing the course of these curves with the distributions illustrated in figure 3, it is possible to recognize the probable sequence of life forms in each succession. In the most productive habitat, the course of succession (S_1) is characterized by a middle phase of intense competition in which first competitive herbs then competitive, shade-intolerant shrubs and trees (e.g., *Rhus glabra, Ailanthus altissima*) dominate the vegetation. This is followed by a terminal phase in which stress tolerance becomes progressively more important as shading and nutrient stress coincide with the development of a large plant biomass dominated by large, long-lived forest trees. Where succession occurs in less productive habitats (S_2), the appearance of highly competitive species is prevented by the earlier onset of resource depletion; the stress-tolerant phase is associated with dominance by smaller slow-growing trees and shrubs in various vegetation types of lower biomass than those obtained at a comparable stage in S_1. In an unproductive habitat, the plant biomass remains low; the dominant life forms include lichens, bryophytes, small herbs, and dwarf shrubs; and the path of succession (S_3) moves almost

PRIMARY STRATEGIES IN PLANTS   1189

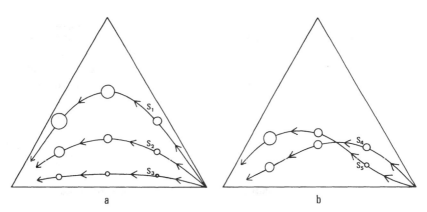

a   b

FIG. 4.—Diagrams representing the path of vegetation succession (a) under conditions of high (S_1), moderate (S_2), and low (S_3) potential productivity and (b) under conditions of increasing (S_4) and decreasing (S_5) potential productivity. The size of the plant biomass at each stage of succession is indicated by the circles. For the distribution of strategies and life forms within the triangle, see figures 2 and 3, respectively.

directly from the ruderal to the stress-tolerant phase. Under extremely unproductive conditions it seems likely that the ruderal phase may disappear altogether, and that vegetation succession will simply involve a slow progression from small to larger stress-tolerant life forms.

The curves S_1–S_3 refer to successions under conditions of fixed potential productivity, a situation which is unlikely to occur in nature. In most habitats, the process is associated with a progressive gain or loss in potential productivity. Increase in productivity, for example, may arise from nitrogen fixation or input of agricultural fertilizer. Conversely, losses of mineral nutrients may result from the harvesting of crops or from leaching of the soil. In figure 4 curves have been drawn to describe courses of succession theoretically possible under conditions of increasing (S_4) and decreasing (S_5) potential productivity.

Dominance

For both plants and animals, there is abundant evidence of the advantage of larger size in many of the interactions which occur either between species or between individuals of the same species (e.g., Black 1958; Bronson 1964; Miller 1968; Grant 1970; Grime 1973a). In natural vegetation, examples of the impact of plant size are widespread. A consistent feature of vegetation succession is the incursion of plants of greater stature, and at each stage of succession the major components of the plant biomass are usually the species with the largest life forms. It is necessary, therefore, to acknowledge the existence of another dimension—that of dominance—in the mechanism controlling the composition of plant communities and the differentiation of vegetation types.

In attempting to analyze the nature of dominance it is helpful to recognize two components: (a) the mechanism whereby the dominant plant achieves a size larger than that of its associates (this mechanism will vary according to which strategy is favored by the habitat conditions); (b) the deleterious effects which large plants may exert upon the fitness of smaller neighbors (the form of these effects does not vary substantially according to habitat or strategy, and they consist principally of the forms of stress which arise from shading, deposition of leaf litter, and depletion of mineral nutrients and water in the soil).

It is suggested, first, that dominance depends upon a positive feedback between (a) and (b), and second, that because of the variable nature of (a), different types of dominance may be recognized. In particular, a distinction can be drawn between "competitive" and "stress-tolerant" dominants. In the former, large stature depends upon a high rate of uptake of resources from the environment, while in the latter it is related to the ability to sustain slow rates of growth under limiting conditions over a very long period.

Strategies in Fungi and in Animals

The evidence reviewed here suggests that autotrophic plants conform to three basic strategies and to various compromises between them. It is therefore interesting to consider whether the same pattern obtains in heterotrophic organisms.

In fungi, there is strong evidence of the existence of strategies corresponding to those recognized in green plants. "Ruderal" life-styles are particularly characteristic of the *Mucorales*, in which most species are ephemeral colonists of organic substrates. These fungi grow exceedingly rapidly and exploit the initial abundance of sugars, but as the supply of soluble carbohydrate declines, they cease mycelial growth and sporulate profusely. "Competitive" characteristics are evident in fungi such as *Merculius lacrymans* and *Armillaria mellea*, which are responsible for long-term infections of timber and produce a consolidated mycelium which may extend rapidly through the production of rhizomorphs. Examples of "stress-tolerant" fungi appear to include the slow-growing basidiomycetes, which form the terminal stages of fungal succession on decaying matter, and the various fungi that occur in lichens and ectotrophic mycorrhizas. All of the "stress-tolerant" fungi are characterized by slow-growing, relatively persistent mycelia and low reproductive effort.

In the animal kingdom, "ruderal" life histories occur in both herbivores and carnivores. They are particularly common in insects (e.g., aphids, blowflies) where, as in ruderal plants, populations expand and contract rapidly in response to conditions of temporary abundance in food supply.

Earlier in this paper it was pointed out that the outstanding characteristic of competitive plants is the possession of mechanisms of phenotypic response which maximize the capture of resources. It was argued that these mechanisms are advantageous only in environments in which high rates of absorption of light, water, and mineral nutrients can be sustained. The same principle may apply in many animals, and a direct analogy may be drawn between the phenotypic responses of competitive plants and the activities associated with food gathering and defense of feeding and reproductive territory which are characteristic of the majority of mammals and birds in stable productive environments. Support for this hypothesis is contained in a recent review by Horn (1974), who describes as "competitively superior" the birds which characterize the second phase of avian succession and "reduce the equilibrial level of resources" during the recolonization of defaunated islands.

Consideration of the basic principles of feeding behavior (Emlen 1966; MacArthur and Pianka 1966; Schoener 1969) suggests that in extreme habitats where food is scarce, the level and frequency of food capture may be too low and the risks of environmental stress (e.g., desiccation, hypothermia) too great to permit domination of the fauna by species which indulge in very active methods of food gathering. Studies of the animals which occur in habitats such as deserts (Pianka 1966; Tinkle 1969) and coral reefs (Odum and Odum 1955; Muscatine and Cernichiari 1969; Goreau and Yonge 1971) indicate that, in the most unproductive environments, strategies of active food gathering give way to feeding mechanisms which are more conservative and appear to be analogous to the assimilatory specializations observed in stress-tolerant plants.

SUMMARY

It is suggested that evolution in plants may be associated with the emergence of three primary strategies, each of which may be identified by reference to a number of characteristics including morphological features, resource allocation, phenology, and response to stress. The competitive strategy prevails in productive, relatively undisturbed vegetation, the stress-tolerant strategy is associated with continuously unproductive conditions, and the ruderal strategy is characteristic of severely disturbed but potentially productive habitats.

A triangular model based upon the three strategies may be reconciled with the theory of r- and K-selection, provides an insight into the processes of vegetation succession and dominance, and appears to be capable of extension to fungi and to animals.

ACKNOWLEDGMENTS

I am indebted to G. J. Vermeij for his thoughtful and stimulating criticism of the manuscript. This work was supported by the Natural Environment Research Council.

LITERATURE CITED

Al-Mufti, M. M., C. L. Wall, S. B. Furness, J. P. Grime, and S. R. Band. 1977. A quantitative analysis of plant phenology and dominance in herbaceous vegetation. J. Ecol. 65 (in press).
Ashton, D. H. 1958. The ecology of *Eucalyptus regnans* F. Muell: the species and its frost resistance. Australian J. Bot. 6:154–176.
Baker, H. G. 1965. Characteristics and modes of orgin of weeds. Pages 147–168 *in* H. G.

1192   THE AMERICAN NATURALIST

Baker and G. L. Stebbins, eds. The genetics of colonizing species. Academic Press, New York. 588 pp.

Barton, L. V. 1961. Seed preservation and longevity. Hill, London. 216 pp.

Beadle, N. C. W. 1954. Soil phosphate and the delimitation of plant communities in eastern Australia. I. Ecology 35:370–375.

———. 1962. Soil phosphate and the delimitation of plant communities in eastern Australia. II. Ecology 43:281–288.

Billings, W. D., and H. A. Mooney. 1968. The ecology of arctic and alpine plants. Biol. Rev. 43:481–529.

Birrel, K. S., and A. C. S. Wright. 1945. A serpentine soil in New Caledonia. New Zeal. J. Sci. Technol. 27A:72–76.

Black, J. M. 1958. Competition between plants of different initial seed sizes in swards of subterranean clover (*Trifolium subterraneum* L.) with particular reference to leaf area and the light micro climate. Australian J. Agr. Res. 9:299–312.

Bordeau, P. F., and M. L. Laverick. 1958. Tolerance and photosynthetic adaptability to light intensity in white pine, red pine, hemlock and ailanthus seedlings. Forest Sci. 4:196–207.

Boysen-Jensen, P. 1929. Studier over Skovtracerres Forhold til Lyset. Dansk Skovforeningens Tidsskrift 14:5–31.

Bradshaw, A. D. 1959. Population differentiation in *Agrostis tenuis* Sibth. I. Morphological differentiation. New Phytol. 58:208–227.

Bradshaw, A. D., M. J. Chadwick, D. Jowett, and R. W. Snaydon. 1964. Experimental investigations into the mineral nutrition of several grass species. IV. Nitrogen level. J. Ecol. 52:665–676.

Bronson, F. W. 1964. Agonistic behavior in woodchucks. Anim. Behav. 12:270–278.

Burns, G. P. 1923. Studies in tolerance of New England forest trees. IV. Minimum light requirement referred to a definite standard. Bull. Vermont. Agr. Exp. Sta. 235.

Clarkson, D. T. 1967. Phosphorus supply and growth rate in species of *Agrostis* L. J. Ecol. 55:111–118.

Clausen, J., D. D. Keck, and W. M. Hiesey. 1940. Experimental studies on the nature of species. I. Effect of varied environments on western North American plants. Carnegie Inst. Washington Pub. 520.

Cody, M. 1966. A general theory of clutch size. Evolution 20:174–184.

Darwin, C. 1859. The origin of species by means of natural selection or the preservation of favored races in the struggle for life. Murray, London. 251 pp.

Donald, C. M. 1958. The interaction of competition for light and for nutrients. Australian J. Agr. Res. 9:421–432.

Ellenberg, J., and D. Mueller-Dombois. 1974. Aims and methods of vegetation ecology. Wiley, New York. 547 pp.

Emlen, J. M. 1966. The role of time and energy in food preference. Amer. Natur. 100:611–617.

Farrar, J. F. 1976a. Ecological physiology of the lichen *Hypogymnia physodes*. II. Effects of wetting and drying cycles and the concept of physiological buffering. New Phytol. 77:105–113.

———. 1976b. Ecological physiology of the lichen *Hypogymnia physodes*. III. The importance of the rewetting phase. New Phytol. 77:115–125.

———. 1976c. The uptake and metabolism of phosphate by the lichen *Hypogymnia physodes*. New Phytol. 77:127–134.

Gadgil, M., and O. T. Solbrig. 1972. The concept of r- and K-selection: evidence from wild flowers and some theoretical considerations. Amer. Natur. 106:14–31.

Goreau, N. I., and C. M. Yonge. 1971. Reef corals: autotrophs or heterotrophs? Biol. Bull. Marine Biol. Lab., Woods Hole 141:247–260.

Grant, P. R. 1970. Experimental studies of competitive interactions in a two-species system. II. The behavior of *Microtus*, *Peromyscus* and *Clethrionomys* species. Anim. Behav. 18:411–426.

Grime, J. P. 1965. Comparative experiments as a key to the ecology of flowering plants. Ecology 45:513–515.

———. 1965. Shade avoidance and tolerance in flowering plants. Pages 281–301 *in* R. Bainbridge, G. C. Evans, and O. Rackham, eds. Light as an ecological factor. Blackwell, Oxford.

———. 1973*a*. Competitive exclusion in herbaceous vegetation. Nature 242:344–347.

———. 1973*b*. Competition and diversity in herbaceous vegetation—a reply. Nature 244: 310–311.

———. 1974. Vegetation classification by reference to strategies. Nature 250:26–31.

Grime, J. P., and R. Hunt. 1975. Relative growth rate: its range and adaptive significance in a local flora. J. Ecol. 63:393–422.

Grime, J. P., and D. W. Jeffrey. 1965. Seedling establishment in vertical gradients of sunlight. J. Ecol. 53:621–642.

Hackett, C. 1965. Ecological aspects of the nutrition of *Deschampsia flexuosa* (L.) Trin. II. The effects of Al, Ca, Fe, K, Mn, N, P and pH on the growth of seedlings and established plants. J. Ecol. 53:315.

Harley, J. L. 1969. The biology of Mycorrhiza. 2d ed. Hill, London. 334 pp.

———. 1970. Mycorrhiza and nutrient uptake in forest trees. Pages 163–178 *in* L. Luckwill and C. Cutting, eds. Physiology of tree crops. Academic Press, New York.

———. 1971. Fungi in ecosystems. J. Ecol. 59:653–686.

Harper, J. L. 1961. Approaches to the study of plant competition. Symp. Soc. Exp. Biol. 15:1–39.

Harper, J. L., and J. Ogden. 1970. The reproductive strategy of higher plants. I. The concept of strategy with special reference to *Senecio vulgaris* L. J. Ecol. 58:681–698.

Higgs, D. E. B., and D. B. James. 1969. Comparative studies on the biology of upland grasses. I. Rate of dry matter production and its control in four grass species. J. Ecol 57:553–563.

Horn, H. S. 1974. The ecology of secondary succession. Annu. Rev. Ecol. Syst. 5:25–37.

Hutchinson, T. C. 1967. Comparative studies of the ability of species to withstand prolonged periods of darkness. J. Ecol. 55:291–299.

Jowett, D. 1964. Population studies on lead-tolerant *Agrostis tenuis*. Evolution 18:70–80.

Kruckeberg, A. R. 1954. The ecology of serpentine soils. III. Plant species in relation to serpentine soils. Ecology 35:267–274.

Levitt, J. 1975. Responses of plants to environmental stresses. Academic Press, New York. 697 pp.

Loach, K. 1970. Shade tolerance in tree seedlings. II. Growth analysis of plants raised under artificial shade. New Phytol. 69:273–286.

Loveless, A. R. 1961. A nutritional interpretation of sclerophylly based on differences in chemical composition of sclerophyllous and mesophytic leaves. Ann. Bot., N.S., 25:168–176.

MacArthur, R. H., and E. R. Pianka. 1966. On optimal use of a patchy environment. Amer. Natur. 100:603–609.

MacArthur, R. H., and E. D. Wilson. 1967. The theory of island biogeography. Princeton University Press, Princeton, N.J. 203 pp.

Mahmoud, A., and J. P. Grime. 1976. An analysis of competitive ability in three perennial grasses. New Phytol. 77:431–435.

Mahmoud, A., J. P. Grime, and S. B. Furness. 1975. Polymorphism in *Arrhenatherum elatius* (L.) Beauv. ex. J. & C. Presl. New Phytol. 75:269–276.

Miller, R. S. 1968. Conditions of competition between redwings and yellow-headed blackbirds. J. Anim. Ecol. 37:43–62.

Milne, A. 1961. Definition of competition among animals. Pages 40–61 *in* F. L. Milnthorpe, ed. Mechanisms in biological competition. Cambridge University Press, London.

Muscatine, L., and E. Cernichiari. 1969. Assimilation of photosynthetic products of zooxanthellae by a reef coral. Biol. Bull. 137:506–523.

Myerscough, P. J., and F. H. Whitehead. 1967. Comparative biology of *Tussilago farfara* L.

1194   THE AMERICAN NATURALIST

and *Epilobium adenocaulon* Hausskn. II. Growth and ecology. New Phytol. 66: 785–823.

Newman, E. I. 1973. Competition and diversity in herbaceous vegetation. Nature 244:310.

Odum, H. T., and E. P. Odum. 1955. Trophic structure and productivity of a windward coral reef community on Eniwetok Atoll. Ecol. Monogr. 25:291–320.

Parsons, R. F. 1968. The significance of growth rate comparisons for plant ecology. Amer. Natur. 102:595–597.

Pianka, E. R. 1966. Convexity, desert lizards and spatial heterogeneity. Ecology 47:1055–1059.

———. 1970. On *r*- and *K*-selection. Amer. Natur. 104:592–597.

Pigott, C. D., and K. Taylor. 1964. The distribution of some woodland herbs in relation to the supply of nitrogen and phosphorus in the soil. J. Ecol. 52:175–185.

Rorison, I. H. 1968. The response to phosphorus of some ecologically distinct plant species. I. Growth rates and phosphorus absorption. New Phytol. 67:913–923.

Salisbury, E. J. 1942. The reproductive capacity of plants. Bell, London. 244 pp.

Schoener, T. W. 1969. Models of optimal size for solitary predators. Amer. Natur. 103:277–313.

Slatyer, R. O. 1967. Plant-water relationships. Academic Press, London. 366 pp.

Thompson, K., J. P. Grime, and G. Mason. 1977. Seed germination in response to diurnal fluctuations of temperature. Nature 67:147–149.

Tinkle, D. W. 1969. The concept of reproductive effort and its relation to the evolution of life histories of lizards. Amer. Natur. 103:501–515.

Walter, H. 1973. Vegetation of the Earth in relation to climate and the ecophysiological conditions. English Universities Press, London. 237 pp.

Williamson, P. 1976. Above-ground primary production of chalk grassland allowing for leaf death. J. Ecol. 64:1059–1075.

Woods, D. B., and N. C. Turner. 1971. Stomatal response to changing light by four tree species of varying shade tolerance. New Phytol. 70:77–84.

22 November 1985, Volume 230, Number 4728

SCIENCE

Resource Availability and Plant Antiherbivore Defense

Phyllis D. Coley, John P. Bryant, F. Stuart Chapin, III

Herbivores exert a major impact on plants, both in ecological and evolutionary time scales. Insects have caused greater economic loss to American agriculture than the combined effects of damage from drought and freezing and have caused greater tree mortality than does logging. On average, more than 10 percent of the plant production in natural communities is consumed annually by herbivores (1). This loss to herbivory is

herbivory on different species can range from 0 to 100 percent during herbivore population outbreaks (4). This orders-of-magnitude range in herbivore damage among species within a single community is primarily a reflection of palatability differences among species. Although the nutritional quality of leaves and twigs can influence herbivore food choice (5), chemical and structural defenses are generally the major determinants of leaf

nature and quantity of plant defenses are determined by the resources available in the local habitat. We suggest that natural selection favors plants with slow growth rates and high levels of defense in environments with low resource availability and that plants with faster growth rates and lower defense levels are favored under conditions of high resource availability. We will first outline the proposal and present the evidence from natural systems and then discuss how these ideas compare with current theories on plant apparency and the evolution of plant defenses.

Resource Limitation and Plant Growth Characteristics

All plants are dependent on the availability of light, water, and nutrients as essential resources for growth. In nature there is a continuum of habitat types, from resource-poor habitats that support little or no plant growth, to resource-rich habitats that can potentially support rapid plant growth. This variation in habitat quality can occur over long distances, as in the change from nutrient-poor white sands forests in the northern Amazon basin to the nutrient-rich forests covering southwestern Amazonia. Habitat quality can also vary substantially over only a few meters, as, for example, when one moves from a shady forest understory, where plants are light-limited, to a sunny light gap created by a fallen tree.

The evolutionary response of plants to resource limitation has been a suite of interdependent characteristics associated with an inherently slow growth rate (Table 1) (9, 10). There are many examples of inherently slow growth rates in species from infertile sites (11), in species from shaded habitats (3, 12), and in species and even populations growing in arid areas (13). Such plants grow slowly even in the most favorable environments and have low capacities to photosynthesize and absorb nutrients (9, 10, 14). The low respiratory and photosynthetic rates in these inherently slow-growing species are associated with low levels of leaf protein (15). Slow growth resulting from a low metabolic demand may confer a greater ability to withstand chronically

Summary. The degree of herbivory and the effectiveness of defenses varies widely among plant species. Resource availability in the environment is proposed as the major determinant of both the amount and type of plant defense. When resources are limited, plants with inherently slow growth are favored over those with fast growth rates; slow rates in turn favor large investments in antiherbivore defenses. Leaf lifetime, also determined by resource availability, affects the relative advantages of defenses with different turnover rates. Relative limitation of different resources also constrains the types of defenses. The proposals are compared with other theories on the evolution of plant defenses.

greater than the average allocation to plant reproduction (2), the investment that most directly determines plant fitness. Thus herbivores exert a strong selective influence on plants by increasing plant mortality and by removing biomass that might be allocated to growth or reproduction.

Herbivory, however, is not equally distributed among all plant species. In a tropical rainforest, insects remove from 0.0003 to 0.8 percent of the leaf area per day, depending on the tree species (3). In arctic shrub tundra and boreal forests the frequency of both insect and vertebrate

Phyllis D. Coley is in the Department of Biology, University of Utah, Salt Lake City 84112. John P. Bryant and F. Stuart Chapin, III, are in the Institute of Arctic Biology, University of Alaska, Fairbanks 99701.

and twig palatability (6, 7). Plants have evolved an extraordinary array of secondary metabolites which act as antiherbivore defenses and which appear not to be waste products nor to serve any other known function in the plant (8). Clearly, the production of defenses is only favored by natural selection when the cost of production is less than the benefit of enhanced protection from herbivores.

A major goal in the study of plant-herbivore interactions is to understand why plant species differ in their commitment to defenses and hence in their susceptibility to herbivores. If plants have the potential to defend themselves effectively against herbivores, why do many species suffer high levels of herbivory? We present evidence that both the stressful environments and therefore to outcompete more rapidly growing species adapted to resource-rich environments (9, 10).

Because inherently slow-growing plants occur in environments where resources are not readily replaced, they tend to have long-lived leaves and twigs. Slow turnover of plant parts is advantageous in a low-nutrient environment because each time a plant part is shed, it carries with it approximately half of its maximum nitrogen and phosphorus pool (10). Similarly, in shady and perhaps in cold or dry environments, where the potential for energy (carbon) acquisition is low, carbon loss can be minimized by having a slow leaf turnover rate.

In contrast, resource-rich environments such as agroecosystems, old-field habitats, and many tropical regions have favored plant species that have the potential for rapid growth (9, 10). These species exhibit a characteristic set of traits (Table 1) that include a high capacity to absorb nutrients and high respiratory and light-saturated photosynthetic rates. Such species generally show a biochemical and morphological plasticity that allows them to take advantage of pulses in resource availability (9, 10, 16). Since photosynthetic rates decline with age, and older leaves are often shaded by younger ones, energy acquisition in high-resource sites is maximized by a rapid turnover of leaves (10, 14, 17, 18). The inevitable nutrient and carbon loss associated with rapid turnover of plant parts is not a strong selective influence on plants in a high-resource environment because nutrients and light are more readily available.

Growth Rates, Herbivory, and Antiherbivore Defenses

In addition to differing in general plant and leaf characteristics, inherently fast- and slow-growing plants also have consistent differences in their antiherbivore characteristics (Table 1). Fast-growers adapted to resource-rich habitats suffer higher rates of damage from herbivores and have both lower amounts and different types of defensive chemicals than slow-growing species. Observations in a variety of communities have revealed that vertebrate and invertebrate herbivores prefer feeding on fast-growing plant species of resource-rich environments (Table 2). For example, in boreal systems, herbivory by vertebrates and insects is greatest on rapidly growing trees that colonize recently disturbed

areas along rivers rather than on slowly growing species characteristic of the adjacent resource-limited sites (7, 19, 20). In a neotropical rainforest, fast-growing tree species are eaten 6 times as rapidly by insects as inherently slow-growing species in the same microhabitat (3). Leaf-eating *Colobus* monkeys from nutrient-poor forests in Africa avoid the leaves of most tree species and rely more heavily on seeds than do their congeners from forests on richer soils (21). Leaves that are eaten by *Colobus* come disproportionately from deciduous tree species as compared to evergreens (22). In feeding preference tests, fast-growing temperate plants from fertile soils were preferred by snails (23), and early successional species were preferred by slugs (24), caterpillars (25), and several species of sap- and leaf-feeders (26).

The observations that inherently slow-growing plants are less preferred by herbivores are consistent with both the amount and type of defenses (Table 1). The absolute concentrations of defenses in leaves of slow-growers from resource-limited sites tend to be at least twice as high as those in leaves of fast-growers from resource-rich sites (3, 21, 27–29). The defenses of slow-growers are primarily chemicals such as lignins or polyphenolic compounds that may have dosage-dependent effects on herbivores (28–31). In addition, lignin, or fiber content serves as structural support in the leaf (32). These types of metabolites are most often present in large concentrations (28–30) and exhibit low rates of turnover during the life of the leaf (33–34). In contrast, the chemical defenses of fast-growing species include a myriad of diverse chemicals that are present and effective in lower concentrations (28, 29, 35). These types of metabolites exhibit high turnover rates (36, 37) and thus represent a reversible commitment to defense.

Predictions for Amount of Defense

Our resource availability hypothesis suggests that the observed associations of inherent growth rates and antiherbivore defenses of plants (Table 1) is one of causality (38). We suggest that the optimal level of defense investment increases as the potential growth rate of the plant decreases (holding herbivore pressure constant) for several reasons. First, as potential growth rates become more limited by resource availability, replacement of resources lost to herbivores becomes more costly. Since this increases the relative value of limiting resources, one would expect to see higher levels of defense in resource-limited environments (39). Second, a given rate of herbivory (grams of leaf removed per day) represents a larger fraction of the net production of a slow-grower than that of a fast-grower. Therefore, because the relative impact of herbivory increases as inherent growth rate declines, we would again expect higher defenses in slower growers. And third, a percentage reduction in growth rate due to the cost of producing defenses represents a greater absolute growth reduction for fast-growing species than for slow-growing ones (40). In other words, because the relative cost of defense increases as

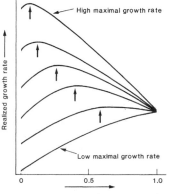

Fig. 1. Effect of defense investment on realized growth. Each curve represents a plant species with a different maximum inherent growth rate. Levels of defense that maximize realized growth are indicated by an arrow. Realized growth (dC/dt) is calculated as $dC/dt = G*C*(1 - kD^\alpha) - (H - mD^\beta)$ where G (g g^{-1} d^{-1}) is the maximum inherent growth rate permitted by the environment (without herbivores), C (g) is the plant biomass at time zero, D (g) is the defense investment, k (g d^{-1}) and α are constants that relate an investment in defense to a reduction in growth. The entire term $(1 - kD^\alpha)$ is the percentage of reduction in growth due to investment in defenses. The term H (g d^{-1}) is the potential herbivore pressure in the habitat (assuming no defense). Potential herbivory is reduced by a function of defense investment, (mD^β), where m (g d^{-1}) and β are constants that determine the shape of the defense effectiveness curve. The entire negative term $(H - mD^\beta)$ is the reduction in realized growth (g d^{-1}) due to herbivory. Since it is subtracted from growth, this assumes herbivores consume fixed amounts of leaf tissue and not fixed percentages of plant productivity. The model's results depend on the extent to which this assumption is true. To further conform to biological reality, the herbivory term $(H - mD^\beta)$ cannot be less than zero, regardless of the value of D.

growth rates increase, we would expect lower levels of defense in resource-rich environments.

Our hypothesis that the level of defense investment increases as the plant's potential growth rate decreases can be formalized mathematically (Fig. 1). We assume that in a world without herbivores, the maximum potential growth rates would be determined by the resource availability in the environment (modified slightly by allocation patterns of individual species). As noted above, evidence suggests that over evolutionary time plants have adjusted their inherent growth rates to match the degree of resource limitation in their preferred habitats. Let us now add herbivores to the model. We assume that they remove a biomass of plant material that is a function of the herbivore biomass and therefore a fixed amount, rather than a fixed percentage of the plant's productivity. Any plant that invests in defenses will reduce its losses to herbivores. The resultant plant growth rate is the balance between a growth reduction due to defense costs and a growth increase due to better protection from herbivores. The shape of this relationship between defense investment and actual growth rate is a curve with intermediate levels of defense causing maximum growth rates (Fig. 1). Below this optimal defense level (indicated by arrows), growth is reduced because of high losses to herbivores and above it, because of an excessively high cost of defense. Figure 1 shows a family of curves where only the maximum potential growth rate permitted by the environment varies. The sharp peak in the curves for fast-growing species (upper curves) suggests that deviations from the optimal defense levels have a larger negative impact on realized growth than they would for slow-growing species (lower curves). As the inherent growth rate decreases (from upper to lower curves), the optimal level of defense increases, and the level of actual herbivory decreases. These two predictions, increased defense and decreased herbivore damage in slow-growing species, have not been explained by previous models and are the major patterns observed in nature.

Predictions for Type of Defense

Inherent growth rates of plants may influence the type of defense as well as the amount. Because of the increased conservation of resources, slow-growing plants of resource-limited environments

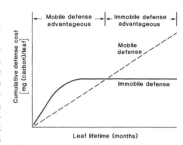

Fig. 2. The cumulative cost of defending a leaf with a large amount of an immobile defense that has negligible turnover compared to a small amount of a mobile defense that continues to turnover throughout the life of the leaf.

have longer-lived leaves than fast-growing species (Table 1) (18). We suggest that there should be a relation between the length of leaf lifetime and types of defense. Defense compounds, such as polyphenols and fiber [quantitative defenses as defined by Feeny (28)], are present in high concentrations and thus represent a high initial construction cost. They are fairly inactive metabolically, so that continued maintenance costs are small. However, because of this metabolic inactivity, these compounds are immobile, being retained in senescent leaves and lost to the plant upon leaf death (34). These types of defense, which we shall refer to as immobile defenses, would therefore be advantageous in long-lived leaves which have more time over which to spread these fixed costs (Fig. 2). Data from 41 tree species in a neotropical forest support this, showing a significant increase in polyphenol and fiber content as leaf lifetime increases (3, 41).

The other end of the defense spectrum is represented by mobile defenses such

as alkaloids, phenolic glycosides, and cyanogenic glycosides [qualitative defenses as defined by Feeny (28)], which are present in low concentrations and therefore initially represent a low total construction cost. Although the concentration of these compounds in a leaf may remain constant and small, the pool is continually turning over. For example, in several species of mint, the biological half-lives of mono- and diterpene defenses are 10 to 24 hours (42), and in several unrelated agricultural species, half-lives of various alkaloids range from 7.5 hours to 6 days (37). This high metabolic activity allows compounds to be recovered from a leaf during senescence, but also means that there is a continued metabolic cost associated with turnover. Mobile defenses are therefore not expected to be common in long-lived leaves, because the continued metabolic costs summed over leaf lifetime would likely be larger than a fixed investment in immobile defenses (Fig. 2) (43). These same arguments predict that mobile defenses would be favored in short-lived leaves. Furthermore, the metabolic turnover of mobile defenses may allow a greater plasticity in the expression of defense, as has been noted for some species (44, 45).

The types of resources available in the environment will also place constraints on the types of defenses that will be favored through evolutionary time. Clearly, in extremely nutrient-limited environments, nitrogen-based defenses would have high relative costs compared to carbon-based defenses, and should be rare (20, 46). Nitrogen-containing alkaloids are unusually common in legumes with nitrogen-fixing symbionts. Desert shrubs growing under conditions of unlimited light frequently produce such

Table 1. Characteristics of inherently fast-growing and slow-growing plant species.

Variable	Fast-growing species	Slow-growing species
Growth characteristics		
Resource availability in preferred habitat	High	Low
Maximum plant growth rates	High	Low
Maximum photosynthetic rates	High	Low
Dark respiration rates	High	Low
Leaf protein content	High	Low
Responses to pulses in resources	Flexible	Inflexible
Leaf lifetimes	Short	Long
Successional status	Often early	Often late
Antiherbivore characteristics		
Rates of herbivory	High	Low
Amount of defense metabolites	Low	High
Type of defense (sensu Feeny)	Qualitative (alkaloids)	Quantitative (tannins)
Turnover rate of defense	High	Low
Flexibility of defense expression	More flexible	Less flexible

large quantities of carbon-based terpenes that they perfume the air. Although species that grow in the forest understory, a low-carbon environment, also often have carbon-based defenses, this may reflect a compromise with other nutrient limitations and the leaf lifetime considerations discussed above. Presumably because phosphorus is limiting in almost all environments, there are no naturally occurring phosphorous-based defenses. The effectiveness of organophosphate pesticides probably arises from their novelty to herbivores.

Evolution of Plant Defenses

Another model for the evolution of plant defenses was presented by Feeny (28) and Rhoades and Cates (29). They were the first to point out many of the patterns of defense investment outlined in Table 1 and suggested that it was a plant's apparency that influenced the type of defense. They defined apparent plants as being distributed predictably in time and space, giving late successional species as an example. Because of their predictability, it was hypothesized that apparent plants were easily discovered by herbivores and should therefore show a large investment in broadly effective defenses (quantitative defenses). Unapparent plants were defined as having ephemeral or unpredictable distributions as, for example, those in early successional sites. Unapparent species were expected to rely on escaping discovery by specialist herbivores and therefore needed only to invest in less costly chemical defenses (qualitative defenses) effective against nonadapted generalist herbivores. The defense differences between apparent and unapparent plants were suggested to reflect differential effectiveness of qualitative and quantitative defenses against specialist and generalist herbivores and differential selection pressure by generalists and specialists due to plant apparency.

Because the extremes of resource availability are often associated with habitat disturbance and successional stages, considerations of resource availability or plant apparency often lead to the same predictions. Both theories suggest that successional status should be correlated with defense investment; Feeny (28) and Rhoades and Cates (29) attribute this pattern to an increase in apparency through time, whereas we suggest that it is because of a decrease in resource availability and, hence, inherent growth rates. There are, however, several studies of defense patterns of

plants that separate the effects of apparency from resource availability. In the following examples, differences in defenses (Table 1) are observed between plant species that have similar apparency in time and space but occur along a resource gradient. Grime (9) was one of the first to identify this relation, noting an increase in defenses in many British plants associated with an increase in environmental stress. In Cameroon, tree species growing in nutrient-poor soils contain twice the concentration of phenolic compounds as species in similar rainforest vegetation but growing in richer soils (21), a pattern which is probably repeated in many nutrient-poor areas (39, 47). In a neotropical forest, the mature canopy is composed of fast-growing shade-intolerant trees as well as slow-growing shade-tolerant species (48). Although both groups of species have similar apparency, the fast-growing species are eaten more by herbivores and show lower concentrations of immobile defenses than do the slow-growing species (3). In boreal communities, where species diversity is low and early

Table 2. Field studies of herbivore preferences for fast- or slow-growing plant species in natural communities. Herbivory is expressed as the relative consumption of fast-growers over slow-growers, considering only mature plants.

Herbivore	Herbi- vory	Refer- ence
Tropical forest		
Insect	6	(3)
Black *Colobus* monkey	20	(22)
Boreal forest		
Moose (winter)		
Alaska	*	(53)
Newfoundland	3	(54)
Finland	3	(55)
Moose (summer)		
Alaska	*	(56)
Snowshoe hare (winter)		
Alaska	4	(20, 57)
Michigan	10	(58)
Newfoundland	3	(54)
Snowshoe hare (summer)		
Alaska	*	(56)
Mountain hare (winter)	8	(59)
Mountain hare (summer)	28	(59)
Caribou	57	(60)
Beaver	*	(61)
Arctic tundra		
Insect	8	(19)
Microtus	*	(62)
Dicrostonyx	*	(62)
Lemmus	*	(62)
Spermophilus	4	(63)
Arctic hare	6	(64)
Musk-oxen	*	(65)
Caribou	*	(66)
Reindeer	*	(67)

*Little or no recorded use of slow-growing evergreen species.

successional riparian habitats are widespread and predictable, it is difficult to see that certain tree species would be more apparent than others. However, a gradient in resource availability and plant growth rate is well correlated with palatability to vertebrate herbivores (7, 19, 20).

We suggest that resource availability better explains the observed patterns of plant defense (Table 1) than apparency in several ways. Apparency theory argues that apparent and unapparent plants have evolved different types of defenses as a result of differential pressure from specialist and generalist herbivores. However, this is not generally supported by empirical evidence on the relative effectiveness of defense types against specialists or generalists (26, 31, 49), nor is it supported by the relative abundance of herbivore types on apparent and unapparent plants (50). Furthermore, apparency theory implies that all species should suffer similar rates of damage, with some species avoiding damage by escape and others by chemical defense. Although the mechanisms of apparency theory do not seem appropriate to explain the observed patterns of herbivory and plant defense (3, 51), the predictability of a plant in time and space may influence the degree of herbivore pressure, particularly in comparisons of species having different leaf lifetimes. In this sense, it should be included as a complementary factor when considering plant-herbivore interactions. The resource availability hypothesis, however, provides a more general and comprehensive explanation of the differences between species in herbivory and defense.

Conclusions

Other investigators have recognized the importance of resource availability in directing the evolution of a variety of plant characteristics (10, 52), and Grime (9) has made specific reference to an increase in plant defenses with an increase in habitat stress. We extend this idea and propose that resource availability in the environment is the major factor influencing the evolution of both the amount and type of plant defense. Resource limitation selects for inherently slow growth rates, which in turn favor large investments in defense. Leaf lifetime, also determined by resource availability, affects whether mobile or immobile defenses will be more advantageous. Further constraints on the types of defenses are imposed by the relative limitation of different resources.

References and Notes

1. J. R. Bray, *Ecology* 37, 598 (1956); J. J. Burdon and G. A. Chilvers, *Aust. J. Bot.* 22, 265 (1974); P. D. Coley, in *Ecology of a Tropical Forest: Seasonal Rhythms and Long-term Changes*, E. G. Leigh, A. S. Rand, D. M. Windsor, Eds. (Smithsonian Institution, Washington, D.C., 1982), pp. 123–132; L. R. Fox and B. J. Macauley, *Oecologia* 29, 145 (1977); L. R. Fox and P. A. Morrow, *Aust. J. Ecol.* 8, 139 (1983); M. D. Lowman, *Biotropica* 16, 264 (1983); S. Larsson and O. Tenow, in *Structure and Function of Northern Coniferous Forests: An Ecosystem Study*, T. Persson, Ed. (Swedish Natural Science Research Council, Stockholm, 1980), pp. 269–306; R. Misra, *Trop. Ecol.* 9, 105 (1968); B. O. Nielson, *Oikos* 31, 273 (1978); H. T. Odum and J. Ruiz-Reyes, in *A Tropical Rain Forest*, H. T. Odum, Ed. (Atomic Energy Commission, Washington, D.C., 1970), pp. 1–69; D. E. Reichle and D. A. Crossley, in *Secondary Productivity of Terrestrial Ecosystems*, K. Petrusewicz, Ed. (Panstwowe Wydawnietwo Naukowe, Warsaw, 1967), pp. 563–587; D. E. Reichle, R. A. Goldstein, R. I. Van Hook, Jr., G. J. Dodson, *Ecology* 54, 1076 (1973); B. P. Springett, *Aust. J. Ecol.* 3, 129 (1978); G. Woodwell and R. H. Whittaker, *Am. Zool.* 8, 19 (1968).
2. H. A. Mooney, *Annu. Rev. Ecol. Syst.* 3, 315 (1972).
3. P. D. Coley, *Ecol. Monogr.* 53, 209 (1983).
4. R. A. Werner, *Can. Entomol.* 11, 317 (1979); J. O. Wolff, *Ecol. Monogr.* 50, 111 (1980).
5. W. J. Mattson, *Annu. Rev. Ecol. Syst.* 11, 119 (1980); J. M. Scriber and F. Slansky, *Rev. Entomol.* 26, 183 (1981).
6. G. Fraenkel, *Science* 129, 1466 (1959); D. A. Levin, *Annu. Rev. Ecol. Syst.* 7, 121 (1976).
7. J. P. Bryant and P. J. Kuropat, *Annu. Rev. Ecol. Syst.* 11, 261 (1980).
8. J. B. Harborne, Ed., *Phytochemical Ecology* (Academic Press, New York, 1972); G. A. Rosenthal and D. H. Janzen, Eds., *Herbivores: Their Interaction with Secondary Plant Metabolites* (Academic Press, New York, 1979); R. H. Whittaker and P. P. Feeny, *Science* 171, 757 (1971).
9. J. P. Grime, *Am. Nat.* 111, 1169 (1977); J. P. Grime, *Plant Strategies and Vegetation Processes* (Wiley, Sussex, 1979).
10. F. S. Chapin, III, *Annu. Rev. Ecol. Syst.* 11, 233 (1980).
11. N. C. W. Beadle, *Ecology* 35, 370 (1954); N. C. W. Beadle, *ibid.* 43, 281 (1962); A. D. Bradshaw, M. J. Chadwick, D. Jowett, R. W. Snaydon, *J. Ecol.* 52, 665 (1964); D. Jowett, *Evolution* 18, 70 (1964); A. R. Kruckeberg, *Ecology* 35, 267 (1954); A. Mahmoud and J. P. Grime, *New Phytol.* 77, 431 (1976).
12. P. L. Marks, *Bull. Torrey Bot. Club* 102, 172 (1975).
13. R. F. Parsons, *Am. Nat.* 102, 595 (1968).
14. H. A. Mooney and S. L. Gulmon, *BioScience* 32, 198 (1982).
15. R. J. Taylor and R. W. Pearcy, *Can. J. Bot.* 54, 1094 (1976); H. A. Mooney and S. L. Gulmon, in *Topics in Plant Population Biology*, O. T. Solbrig, S. Jain, G. B. Johnson, P. H. Raven, Eds. (Columbia Univ. Press, New York, 1979), pp. 316–337.
16. H. A. Mooney, S. L. Gulmon, N. D. Johnson, in *Plant Resistance to Insects*, P. A. Hedin, Ed. (American Chemical Society, Washington, D.C., 1983), pp. 21–36.
17. C. Field and H. A. Mooney, *Oecologia* 56, 148 (1983).
18. B. F. Chabot and D. J. Hicks, *Annu. Rev. Ecol. Syst.* 13, 229 (1982).
19. S. F. MacLean and T. S. Jensen, *Oikos* 44, 211 (1985).
20. J. P. Bryant, F. S. Chapin, III, D. R. Klein, *ibid.* 40, 357 (1983).
21. D. McKey, P. G. Waterman, C. N. Mbi, J. S.

22. Gartlan, T. T. Struhsaker, *Science* 202, 61 (1978). D. B. McKey and J. S. Gartlan, *Biol. J. Linn. Soc.* 16, 115 (1981).
23. J. P. Grime, S. F. MacPherson-Stewart, R. S. Dearman, *J. Ecol.* 56, 405 (1968).
24. R. G. Cates and G. H. Orians, *Ecology* 56, 410 (1975).
25. J. M. Scriber and P. O. Feeny, *ibid.* 60, 829 (1979).
26. P. M. Reader and T. R. E. Southwood, *Oecologia* 51, 271 (1981).
27. J. S. Gartlan, D. B. McKey, P. G. Waterman, in *Recent Advances in Primatology*, D. J. Chivers and J. Herbert, Eds. (Academic Press, London, 1978), pp. 259–267.
28. P. P. Feeny, in *Recent Advances in Phytochemistry*, J. W. Wallace and R. L. Mansell, Eds. (Plenum, New York, 1976), vol. 10, pp. 1–40.
29. D. F. Rhoades and R. G. Cates, in *ibid.*, pp. 168–213.
30. P. P. Feeny, *Phytochemistry* 8, 2119 (1969); P. P. Feeny, *J. Insect Physiol.* 14, 805 (1968); D. F. Rhoades, *Biochem. Syst. Ecol.* 5, 281 (1977).
31. E. A. Bernays, *Ecol. Entomol.* 6, 353 (1981); E. A. Bernays and S. Woodhead, *Science* 216, 201 (1982).
32. T. Swain, in *Herbivores: Their Interaction with Secondary Plant Metabolites*, G. A. Rosenthal and D. H. Janzen, Eds. (Academic Press, New York, 1979), pp. 657–682.
33. J. B. Harborne, Ed. *Biochemistry of Phenolic Compounds* (Academic Press, New York, 1964); T. Swain, J. B. Harborne, C. F. Van Sumere, Eds., *Recent Advances in Phytochemistry* (Plenum, New York, 1978), vol. 12; E. Haslam, *personal communication*.
34. J. R. L. Walker, *The Biology of Plant Phenolics* (Arnold, London, 1975).
35. A. Hladik and C. M. Hladik, *Terre Vie* 31, 515 (1977); D. B. McKey, *Am. Nat.* 108, 305 (1974).
36. W. D. Loomis and R. Croteau, in *Recent Advances in Phytochemistry*, V. C. Runeckles and T. J. Mabry, Eds. (Plenum, New York, 1973), vol. 6, p. 147; D. Seigler and P. W. Price, *Am. Nat.* 110, 101 (1976); G. R. Waller and E. K. Nowacki, *Alkaloid Biology and Metabolism in Plants* (Plenum, New York, 1978).
37. T. Robinson, *Science* 184, 430 (1974).
38. Aspects of these ideas were originally developed for browsing mammals in a boreal forest (20) and for insect herbivores in a neotropical forest (3).
39. D. H. Janzen, *Biotropica* 6, 69 (1974).
40. This idea was derived independently by S. L. Gulmon and H. A. Mooney [in *On the Economy of Plant Form and Function*, T. J. Givnish, Ed. (Cambridge Univ. Press, Cambridge, in press)].
41. P. D. Coley (unpublished data) observed that leaf lifetimes for 41 species of neotropical trees in Panama are highly correlated with leaf polyphenol and fiber contents (in a multiple regression, $r = 0.74$, $P = 0.012$). Leaf lifetimes range from 4 to 36 months. Polyphenol measures include vanillin/HCl and proanthocyanidin assays, and fiber measures include neutral-detergent fiber, acid-detergent fiber, lignin, and cellulose (3). Correlation coefficients of each defense trait and leaf lifetime are between 0.41 and 0.64 ($P < 0.01$).
42. A. Breccio and R. Badiello, *Z. Naturforsch.* 22, 44 (1967); R. Croteau and M. A. Johnson, in *Biology and Chemistry of Plant Trichomes*, E. Rodriquez *et al.*, Eds. (Plenum, New York, 1984), pp. 133–185; R. Croteau and W. D. Loomis, *Phytochemistry* 11, 1055 (1972).
43. If herbivore activity is markedly seasonal, we might expect mobile defenses in long-lived leaves only during this period.
44. J. P. Bryant, *Science* 213, 889 (1981); G. Cooper-Driver, S. Finch, T. Swain, E. A. Bernays, *Biochem. Syst. Ecol.* 5, 177 (1977); D. H. Firmage, *ibid.* 9, 53 (1981); W. A. Dement and H. A. Mooney, *Oecologia* 15, 65 (1974); H. Fluck, in *Chemical Plant Taxonomy*, T. Swain, Ed. (Academic Press, New York, 1963), pp. 167–186; J. H. Lawton, *Bot. J. Linn. Soc.* 73, 187

45. (1976); N. D. Mitchell and A. J. Richards, *New Phytol.* 81, 189 (1978). We prefer the terms mobile and immobile defenses as opposed to qualitative and quantitative because the latter imply two distinct modes of action against herbivores, and these have not been well supported. The terms mobile and immobile defenses refer to physiological properties of the defenses in the plant and encompass a continuum of metabolic activity and mobility. We also think that they more accurately describe the defense characteristics of major importance to the plant.
46. J. T. Romeo, J. D. Bacon, T. J. Mabry, *Biochem. Syst. Ecol.* 5, 117 (1977).
47. E. F. Brunig, *Trop. Ecol.* 10, 45 (1969); P. D. Coley, unpublished observations.
48. N. V. L. Brokaw, in *Natural Disturbance: An Evolutionary Perspective*, S. T. A. Pickett and P. S. White, Eds. (Academic Press, New York, 1985), p. 53–69; J. S. Denslow, *Biotropica* 12 (Suppl.), 47 (1980); G. S. Hartshorn, *ibid.*, p. 23; P. W Richards, *The Tropical Rainforest* (Cambridge Univ. Press, Cambridge, 1952); T. C. Whitmore, in *Tropical Trees as Living Systems*, P. B. Tomlinson and M. H. Zimmerman, Eds. (Cambridge Univ. Press, Cambridge, 1978), pp. 639–655.
49. D. J. Futuyma, *Am. Nat.* 110, 285 (1976); D. Otte, *Oecologia* 18, 129 (1975); W. V. Zucker, *Am. Nat.* 121, 335 (1983).
50. R. G. Cates, *Oecologia* 46, 22 (1980); V. K. Brown and T. R. E. Southwood, *ibid.* 56, 220 (1983); L. R. Fox and P. A. Morrow, *Science* 211, 887 (1981); D. J. Futuyma and F. Gould, *Ecol. Monogr.* 49, 33 (1979); C. E. Holdren and P. R. Ehrlich, *Oecologia* 52, 417 (1982); V. C. Moran and T. R. E. Southwood, *J. Anim. Ecol.* 51, 289 (1982); P. A. Morrow, *Aust. J. Ecol.* 2, 89 (1977); P. Niemela, J. Tahvanainen, J. Sorjonen, T. Hokkanen, S. Neuvonen, *Oikos* 39, 164 (1982); H. F. Rowell, *Entomol. Exp. Appl.* 24, 451 (1978); D. R. Strong, J. H. Lawton, T. R. E. Southwood, *Insects on Plants* (Blackwell, Oxford, 1984).
51. L. R. Fox, *Am. Zool.* 21, 853 (1981); P. D. Coley, thesis, University of Chicago (1981).
52. T. R. E. Southwood, *J. Anim. Ecol.* 46, 337 (1977); R. H. Whittaker, in *Unifying Concepts in Ecology*, W. H. van Dobben and R. H. Lowe-McConnell, Eds. (Junk, The Hague, 1975), pp. 169–181.
53. J. O. Wolff and J. Cowling, *Can. Field Nat.* 95, 85 (1981).
54. D. G. Dodds, *J. Wildl. Manage.* 24, 53 (1960).
55. E. Pulliainen and K. Loisa, *Ann. Zool. Fenn.* 5, 220 (1968).
56. J. P. Bryant, unpublished observations.
57. D. R. Klein, *Proc. Int. Congr. Game Biol.* 13, 266 (1977).
58. I. A. Bookhout, *Mich. Dept. Conserv. Res. Develop. Rep. 38* (1965).
59. E. Pulliainen, *Ann. Zool. Fenn.* 9, 17 (1972).
60. R. D. Boertje, thesis, University of Alaska, Fairbanks (1981).
61. M. Aleksiuk, *Ecology* 51, 264 (1970).
62. G. O. Batzli and H. G. Jung, *Arct. Alp. Res.* 12, 483 (1980).
63. G. O. Batzli and S. T. Sobaski, *ibid.*, 501.
64. R. E. Pegau, G. N. Bos, K. A. Kneiland, *Caribou Food Habits* (Alaska Department of Fish and Game, Juneau, 1973), vol. 14.
65. M. A. Robus, thesis, University of Alaska, Fairbanks (1981).
66. P. J. Kuropat, thesis, University of Alaska, Fairbanks (1984).
67. G. I. Karaeuj, in *Reindeer Husbandry*, P. S. Zhigunov, Ed. (Izdatel'stvo sel'skokhozyaistvennyx Literaturnyx Zhurnalov, Plakatov, Moskva, 1961), pp. 129–176.
68. We thank M. Aide, D. Davidson, M. Geber, T. Kursar, and B. Stubblefield for discussions. Supported by NSF grant DEB-820 7170 (to F.S.C.) and by funds from the University of Utah (to P.D.C.).

Vol. 125, No. 1   The American Naturalist   January 1985

PHYLOGENIES AND THE COMPARATIVE METHOD

Joseph Felsenstein

Department of Genetics SK-50, University of Washington, Seattle, Washington 98195

Submitted November 30, 1983; Accepted May 23, 1984

Recent years have seen a growth in numerical studies using the comparative method. The method usually involves a comparison of two phenotypes across a range of species or higher taxa, or a comparison of one phenotype with an environmental variable. Objectives of such studies vary, and include assessing whether one variable is correlated with another and assessing whether the regression of one variable on another differs significantly from some expected value. Notable recent studies using statistical methods of this type include Pilbeam and Gould's (1974) regressions of tooth area on several size measurements in mammals; Sherman's (1979) test of the relation between insect chromosome numbers and social behavior; Damuth's (1981) investigation of population density and body size in mammals; Martin's (1981) regression of brain weight in mammals on body weight; Givnish's (1982) examination of traits associated with dioecy across the families of angiosperms; and Armstrong's (1983) regressions of brain weight on body weight and basal metabolism rate in mammals.

My intention is to point out a serious statistical problem with this approach, a problem that affects all of these studies. It arises from the fact that species are part of a hierarchically structured phylogeny, and thus cannot be regarded for statistical purposes as if drawn independently from the same distribution. This problem has been noticed before, and previous suggestions of ways of coping with it are briefly discussed. The nonindependence can be circumvented in principle if adequate information on the phylogeny is available. The information needed to do so and the limitations on its use will be discussed. The problem will be discussed and illustrated with reference to continuous variables, but the same statistical issues arise when one or both of the variables are discrete, in which case the statistical methods involve contingency tables rather than regressions and correlations.

THE PROBLEM

Suppose that we have examined eight species and wish to know whether their brain size (Y) is proportional to their body size (X). We may wish to test whether the slope of the regression of Y on X (or preferably of *log Y* on *log X*) is unity. Figure 1 shows a scatter diagram of hypothetical data. It is tempting to simply do

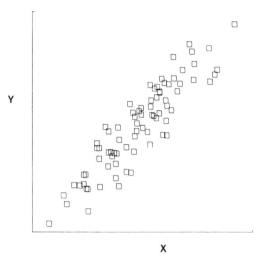

Fig. 1.—Scatter diagram of hypothetical data from 8 species, showing the relationship between Y and X.

an ordinary regression and see whether the confidence limits on the slope include unity. (Since there is random error in both X and Y this is a questionable procedure, but we leave that issue aside here.)

If we were to do such a regression, what would be the implicit statistical model on which it was based? The simplest assumption would seem to be that the points in figure 1 were drawn independently from a bivariate normal (Gaussian) distribution. What evolutionary model could result in such a distribution? The simplest is shown in figure 2: the eight species resulted from a single explosive adaptive radiation. Along each lineage there were changes in both characters. If evolution in each lineage were independent, and the changes in the two characters were drawn from a bivariate normal distribution, then our distributional assumptions would be justified. The individual species could be regarded as independent samples from a single bivariate normal distribution.

Let us accept for the moment that the changes of a set of characters in different branches of a phylogeny can be reasonably well approximated as being drawn from a multivariate normal distribution, and that changes in distinct branches are independent. Even given those assumptions, a problem arises based on the unlikelihood that the phylogeny has the form shown in figure 2.

Consider instead the phylogeny shown in figure 3. In it, the eight species consist of four pairs of close relatives. Suppose that the changes in the two characters in each branch of the tree can be regarded as drawn from a bivariate normal distribution with some degree of correlation between the characters. We might not expect the same amount of change in short branches of the tree, such as the eight terminal branches, as in longer branches such as the four that arise from the original radiation. Let us assume that the variance of the distribution of change in a branch is proportional to the length in time of the branch, much as it would be if the characters were undergoing bivariate, and correlated, Brownian motion.

COMPARATIVE METHOD   3

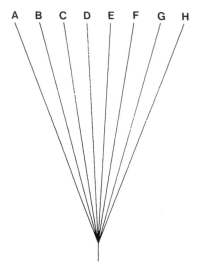

Fig. 2.—One phylogeny for the 8 species, showing a burst of adaptive radiation with each lineage evolving independently from a common starting point.

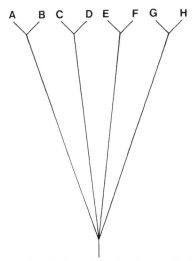

Fig. 3.—Another phylogeny for the 8 species, showing a radiation that gives rise to 4 pairs of closely related species.

This would produce a distribution like that shown in figure 4. It is apparent that instead of eight independent points we have four pairs of close relatives. If we were to carry out a statistical test based on the assumption of independence, say a test of the hypothesis that the slope of the regression of Y on X was zero, we would imagine ourselves to have 6 degrees of freedom ($8 - 2$). In fact, we very nearly have only four independent points, so that the effective number of degrees of freedom is closer to 2 ($4 - 2$). A test of the significance of the slope, or of the

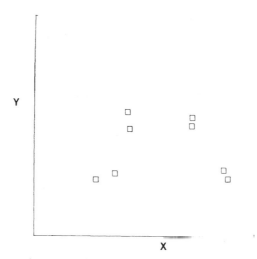

Fig. 4.—A data set simulated using the phylogeny of fig. 3, under a model of random, normally distributed, independent change in each character, where the change in each branch is drawn independently from a normal distribution with mean zero and variance proportional to the length of the branch.

extent of its difference from any preassigned value, will be excessively likely to show significance if the number of degrees of freedom is taken to be 6 rather than 2.

A worst case of sorts for the naive analysis is shown in figure 5, where the phylogeny shows that a large number of species actually consist of two groups of moderately close relatives. Suppose that the data turned out to look like that in figure 6. There appears to be a significant regression of Y on X. If the points are distinguished according to which monophyletic group they came from (fig. 7), we can see that there are two clusters. Within each of these groups there is no significant regression of one character on the other. The means of the two groups differ, but since there are only two group means they must perforce lie on a straight line, so that the between-group regression has no degrees of freedom and cannot be significant. Yet a regression assuming independence of the species finds a significant slope ($P < .05$). It can be shown that there are more nearly 3 than 40 independent points in the diagram.

One might imagine that the problem could be avoided by use of robust non-parametric statistics. In fact, nonparametric methods, unless specifically designed to cope with the problem of nonindependence, are just as vulnerable to the problem as are parametric methods. For the data of figure 6, a Spearman rank correlation finds a nearly significant correlation ($P < .065$) between the two variables, showing that it has little better ability to cope with nonindependence than do parametric methods.

One might also imagine that we could escape from the problem simply by ensuring that we sample the species at random from the species that form the tips of the phylogeny, and thus somehow escape from the nonrandomness of the pool

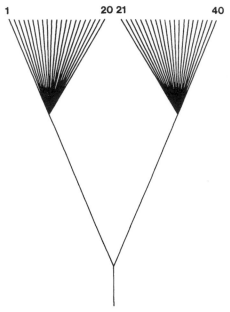

FIG. 5.—A "worst case" phylogeny for 40 species, in which there prove to be 2 groups each of 20 close relatives.

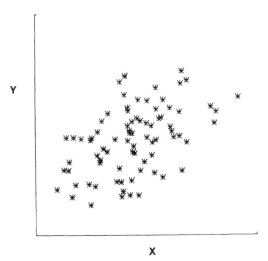

FIG. 6.—A typical data set that might be generated for the phylogeny in fig. 5 using the model of independent Brownian motion (normal increments) in each character.

6   THE AMERICAN NATURALIST

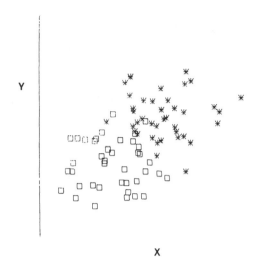

X

FIG. 7.—The same data set, with the points distinguished to show the members of the 2 monophyletic taxa. It can immediately be seen that the apparently significant relationship of fig. 6 is illusory.

of species from which we are sampling. This does not work. Imagine two species that have diverged some time ago, and thus have diverged in both brain and body weight. Clearly the correlation between those characters cannot be significant, since there are only two points. Now if each species gives rise to a group of 100 daughter species, essentially identical to it, we now have two clusters of 100 species each. Sampling species from this pool of 200 species, we are actually sampling from two species, but do not know it. Correctly analyzed, no data from this group could possibly achieve significance, but if we draw (say) 50 species at random from the 200 and analyze that data as if the points were independent we will probably conclude that there is a significant correlation between brain and body weight.

There is one case in which the problem does not arise. That is when the characters respond essentially instantaneously to natural selection in the current environment, so that phylogenetic inertia is essentially absent. In that case we could correlate a phenotype with the environment. We could also correlate two characters with each other, provided that we realized that their correlation might simply reflect response to a common environmental factor. It may be doubted how often phylogenetic inertia is effectively absent. In any case the presumption of the absence of phylogenetic inertia should be acknowledged whenever it is proposed to do comparative studies without taking account of the phylogeny.

PREVIOUS APPROACHES

The problem of correcting for nonindependent evolutionary origins has not gone unnoticed by previous workers in comparative biology. Clutton-Brock and Harvey (1977, pp. 6–8) pointed out that "if phylogenetic inertia is strong, the potential adaptations that related species may evolve will be similarly constrained, with the effect that species cannot be regarded as independent of each other." They used a nested analysis of variance to find that taxonomic level (in their case it was genera) which accounted for as much of the variation as possible, and then tried to correct for nonindependence by using genera rather than species as the units of their statistical analysis. A more complete exposition of their approach is given by Harvey and Mace (1982). Baker and Parker (1979), who were analyzing bird coloration, also pointed out the problem, and tried to correct for it by seeing whether the same relationships held within different families. Sherman (1979) and Givnish (1982) discussed the problem, though without attempting to correct for it.

Gittleman (1981) used a parsimony method to reconstruct and count the number of times parental care had evolved independently on a phylogeny derived from the classification of bony fishes. Ridley (1983) has discussed the problem of nonindependence in considerable detail, also proposing that parsimony be used to reconstruct the placements of changes on the phylogeny, to enable tests of whether the occurrence of changes in two characters are independent.

There are two problems with using a parsimony approach as suggested by Gittleman and by Ridley. The most serious is that one is usually forced to rely on a presumed monophyly of taxa in the Linnean classification system, in the absence of external evidence as to the phylogenetic relationships between the groups. The traditional classification system is, of course, only partly a monophyletic one: only about half of the classes of the Chordata are thought to be monophyletic, for example, and even within the Mammalia orders such as Insectivora and Carnivora are believed not be monophyletic. The classification system is not sufficiently detailed to show all of the structure in the phylogeny, even if all its groups were monophyletic: the relationships between some orders of mammals are undoubtedly closer than between others. The phylogenetic meaning of a given category varies from group to group: the insect genus *Drosophila* is thought to be as old as the mammalian order Primates.

A second problem with parsimony assignments is that they have only partial statistical justification: I have argued elsewhere (Felsenstein 1978) that when parsimony is used to reconstruct the phylogeny, it can have undesirable statistical properties when evolutionary rates are not small and differ sufficiently in different lineages (for a review, see Felsenstein [1983]). Even when the phylogeny is known, and parsimony is used only to reconstruct the placement of changes of character state, it is well known that this can lead to biases; for example, if two changes occur in parallel in sister lineages, the reconstruction will instead show one change occurring in their immediate ancestor. If changes in one character occurred in parallel in the sister lineages, but the change in another character occurred in the immediate common ancestor, then, although the reconstructed changes appear coincident, the actual changes in those characters were not in fact coincident.

The seriousness of the additional statistical error and statistical biases that this may cause in comparative studies has never, to my knowledge, been investigated. Nevertheless, assigning changes of characters by parsimony on a known phy-

8   THE AMERICAN NATURALIST

logeny would be immeasurably superior to simply treating the species as if independently evolved.

If we know the phylogeny and have a model of evolutionary change, it should be possible in principle to correct for the nonindependence of taxa. To see how, first let us consider the highly symmetrical phylogeny in figure 8, supposing that we know that it is the true phylogeny. Recalling that each character is being assumed to be evolving by a Brownian motion that is independent in each lineage, then taking X_i to be the phenotype X in species i it is easy to see that differences between pairs of adjacent tips, such as $X_1 - X_2$ and $X_3 - X_4$, must be independent. This is so because the difference $X_1 - X_2$ depends only on events in branches 1 and 2, while $X_3 - X_4$ depends only on events in branches 3 and 4, and these two sets of events are independent.

Brownian motion is a random process modeling the wanderings of a molecule affected by thermal noise. If we measure the position of the molecule along one axis, its successive displacements are independent. This has the effect that the displacement after time v has elapsed is the sum of a large number of small displacements, each of which is equally likely to be either positive or negative. The result is that the total displacement is drawn from a normal distribution with mean zero and a variance proportional to v. In the present model the different characters undergo Brownian motion at different rates, so that after one unit of time the change in X has variance s_X^2 and the (possibly correlated) change in Y has variance s_Y^2. After v units of time their variances are, respectively, $s_X^2 v$ and $s_Y^2 v$.

Given this model, it is straightforward to show (Felsenstein 1973, 1981b) that the contrast $X_1 - X_2$ has expectation zero and variance $2s_X^2 v_1$. Since we assume that we know the v_i, we can scale the contrast by dividing by its standard deviation, obtaining a variate that should have expectation zero and unit variance. We can similarly scale the other three contrasts $X_3 - X_4$, $X_5 - X_6$, and $X_7 - X_8$ by dividing each by the square root of $2\,s_X^2\,v_1$. Even more contrasts are available. It will be less obvious, but nevertheless true, that $(X_1 + X_2)/2 - (X_3 + X_4)/2$ is a contrast independent of the others. It will have expectation zero and variance $s_X^2(v_1 + 2v_9)$. We can continue down the tree in similar fashion, obtaining two more contrasts, $(X_5 + X_6)/2 - (X_7 + X_8)/2$ and $(X_1 + X_2 + X_3 + X_4)/4 - (X_5 + X_6 + X_7 + X_8)/4$. Their expectations are also zero, and their variances are, respectively, $s_X^2(v_1 + 2v_9)$ and $s_X^2(2v_{13} + v_9 + v_1/2)$. They too can be scaled to have unit variance.

We have now extracted from this tree seven independent contrasts on the X scale, each of which can be regarded as drawn from a normal distribution with mean zero and variance one. We can carry out the same process in the variable Y, and obtain seven independent contrasts in the same way. The X contrasts will be independent of each other but not of the Y contrasts. It can be shown that the coresponding contrasts $X_1 - X_2$ and $Y_1 - Y_2$ have covariance

$$\mathrm{Cov}[X_1 - X_2,\ Y_1 - Y_2] = 2v_1\,s_X\,s_Y\,r_{XY}, \qquad (1)$$

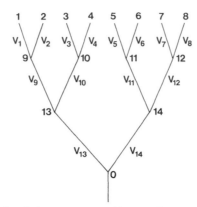

FIG. 8.—An example of a phylogeny, assumed known, from which we can define independent contrasts between taxa. This tree is highly symmetric, so that $v_1 = v_2 = v_3 = v_4 = v_5 = v_6 = v_7 = v_8$, $v_9 = v_{10} = v_{11} = v_{12}$, and $v_{13} = v_{14}$.

so that these two contrasts have the same correlation as the original variates. Since contrasts such as $X_1 - X_2$ and $X_3 - X_4$ are independent, a fortiori $X_1 - X_2$ will be independent of $Y_3 - Y_4$.

The quantities s_X and s_Y are important. It would be unreasonable to assume that characters X and Y had the same rates of evolution: s_X and s_Y are the scaling constants that convert from Brownian motion to the scales on which X and Y actually evolve. Thus X is undergoing a Brownian motion, with variance s_X^2 accumulating per unit time, and Y is undergoing a (possibly correlated) Brownian motion with variance s_Y^2 accumulating per unit time. We leave aside for the moment the problem of estimating s_X and s_Y, and assume that they are known.

By dividing each contrast by its standard deviation, we have obtained from the original eight species seven pairs of contrasts that can be regarded as drawn independently from a bivariate normal distribution with means zero, variances unity, and an unknown correlation r_{XY} between the members of a pair. Testing independence of the evolution of X and Y reduces to simply testing whether this correlation is zero. If instead we wanted to know the regression of changes in one variable on changes in another, we could use s_X and s_Y to compute

$$b_{Y.X} = s_Y \, r_{XY}/s_X \qquad (2a)$$

and

$$b_{X.Y} = s_X \, r_{XY}/s_Y. \qquad (2b)$$

These are not the usual equations for interconverting correlations and regressions, since s_X and s_Y are not observed standard deviations of X and Y, but scaling constants that are merely proportional to the standard deviations of the variables X and Y. Even though they are not standard deviations, they do allow us to correctly convert correlations into regressions. Other methods of analysis, such as principal components, can be carried out in similar fashion.

The preceding computation of contrasts depended on the phylogeny having a particular, and very unlikely, symmetric structure. Fortunately a more general procedure exists, of which the above was a special case. I have discussed its elements elsewhere (Felsenstein 1973) as part of a computational method for obtaining the likelihood of a given phylogeny. The general prescription for computing these contrasts is repeated applications of the following steps: (1) Find two tips on the phylogeny that are adjacent (say nodes i and j) and have a common ancestor, say node k. (2) Compute the contrast $X_i - X_j$. This has expectation zero and variance proportional to $v_i + v_j$. (3) Remove the two tips from the tree, leaving behind only the ancestor k, which now becomes a tip. Assign it the character value

$$X_k = \frac{(1/v_i)\,X_i + (1/v_j)\,X_j}{1/v_i + 1/v_j}. \tag{3}$$

the weighted average of X_i and X_j, the weights being proportional to the inverses of the variances v_i and v_j. (4) Lengthen the branch below node k by increasing its length from v_k to $v_k + v_iv_j/(v_i + v_j)$. This lengthening occurs because the weighted average that computes X_k in equation (3) does not compute the phenotype of the ancestor but only estimates it, and does so with an error that is statistically indistinguishable from an extra burst of evolution after node k.

After one pass through steps 1–4, we have found one contrast and reduced the number of tips on the tree by one. We continue to repeat steps 1–4 until there is only one tip left on the tree. This will extract $n - 1$ contrasts if there were originally n species. Each contrast can be divided by the square root of its variance to bring them to a common variance. Since the v_i are arbitrary, this procedure can be used on a phylogeny of any shape whatsoever, even on ones that contain multifurcations, since those can always be represented as a series of bifurcations having some branch lengths zero.

Figure 9 shows a nonsymmetric phylogeny, and table 1 the contrasts extracted from it by the above algorithm. The reader may want to try steps 1–4 on the symmetric phylogeny of figure 8, to verify the correctness of the contrasts and variances given above.

DIFFICULTIES

One might imagine that, with the ability to compute independent contrasts from any phylogeny, we have an acceptable method of correcting for the presence of the phylogeny. Unfortunately, this is not the case. A number of difficulties intervene that leave us with much work remaining to be done.

How Do We Reconstruct the Phylogeny?

In practice, we will hardly ever know the phylogeny in advance in sufficient detail to use it to obtain the contrasts. There are three sources of information likely to be used to reconstruct the phylogeny.

1. *Gene frequencies.*—A number of electrophoretic loci, blood group loci, or DNA restriction polymorphisms may be chosen and the frequencies of the alleles

COMPARATIVE METHOD   11

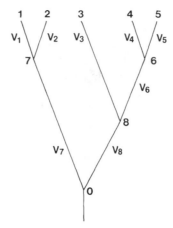

Fig. 9.—A less symmetrical phylogeny. The independent contrasts for this phylogeny are given in table 1.

TABLE 1

The Four Contrasts Extracted from the Phylogeny Shown in Figure 9, Each with Its Variance, All Computed Using Steps 1–4 in the Text

Contrast	Variance
$X_1 - X_2$	$v_1 + v_2$
$X_4 - X_5$	$v_4 + v_5$
$X_3 - X_6$	$v_3 + v_6'$
$X_7 - X_8$	$v_7' + v_8'$

where

$$X_6 = \frac{v_4 X_5 + v_5 X_4}{v_4 + v_5}$$

$$v_6' = v_6 + v_4 v_5/(v_4 + v_5)$$

$$X_7 = \frac{v_2 X_1 + v_1 X_2}{v_1 + v_2}$$

$$v_7' = v_7 + v_1 v_2/(v_1 + v_2)$$

$$X_8 = \frac{v_6' X_3 + v_3 X_6}{v_3 + v_6}$$

$$v_8' = v_7' + v_3 v_6'/(v_3 + v_6')$$

at these loci determined in a sample from each species. This permits us to make an. estimate of the phylogeny. The estimate has an error that must be taken into account when using it (see discussion below). It depends on the assumption that the evolutionary changes at the loci are predominantly due to genetic drift, and can therefore be modeled by Brownian motion of suitably transformed gene frequencies. This is more plausible the lower the taxonomic level at which we are working. A review of methods for inferring phylogenies from gene frequencies will be found in my recent paper on that subject (Felsenstein 1981*b*).

2. *Molecular sequences.*—As nucleic acid sequences become available for a wider range of organisms, it will become practical to infer the phylogeny from these, on the assumption that a "molecular clock" is valid for those sequences. I have described elsewhere (Felsenstein 1981*a*) a maximum likelihood method for inferring phylogenies from nucleic acid sequences.

3. *Quantitative characters.*—Most morphological taxonomic studies do not collect gene frequencies, but collect many morphological measurements besides those that we want to correlate. Could we use those quantitative characters to infer the phylogeny? We could do so in principle if we had a probability model for the evolution of the characters. The difficulty is that quantitative characters will evolve at different rates, and in a correlated fashion (as in the important case of size correlations). If we could find a transformation of the characters to a new set of coordinates that could be modeled as evolving independently by Brownian motion, with equal rates of accumulation of variance, we could apply the maximum likelihood methods developed for gene-frequency data. There would be no circularity involved: the new coordinates would be independent of each other and of the characters we were investigating. Unfortunately, there is no obvious way to get the information needed to untangle the skein of character correlations. Within-population samples, even when available, do not necessarily give us the required information. They allow us to estimate phenotypic covariances rather than additive genetic covariances. The latter control the covariances of evolutionary changes in characters if the characters change by natural selection or genetic drift. Even if the additive genetic covariances could be obtained by means of breeding experiments, they could not tell us whether the selection pressures for different characters were correlated. Cold weather, for example, might impose selection for large size and dark coloration while warm weather might select for small size and light coloration, leading to changes in these characters being correlated even if there were no genetic correlation between them. For the time being, transforming to remove within-population phenotypic covariance is the best that can be done, but this is necessarily an approximation, whose adequacy is unknown.

4. *The characters we are investigating.*—Sometimes the only characters available to us are the two whose relationship we are investigating. This is not only a severely limited amount of information with which to infer a phylogeny, but could result in some circularity since the phylogeny is inferred from some of the same information that is being used to reconstruct the changes of the characters along it. The matter needs a careful statistical investigation, but preliminary indications are that when we are trying to infer both the phylogeny and character correlations there is confounding between these, so that we can infer one or the other but not both.

COMPARATIVE METHOD   13

How Do We Put Confidence Intervals on Our Inferences?

If we were given a phylogeny known to be the true one, and were willing to trust the Brownian motion model of character change, we could obtain the appropriate contrasts from the phylogeny as outlined above, and use standard formulas to place confidence limits on the inferred correlations or slopes. But the phylogeny is never known without error. How are we to take the uncertainty of our knowledge of the phylogeny into account when constructing confidence intervals?

In principle this can be done by considering the phylogeny T and the set of correlations (or slopes) C to be a single multivariate quantity (T, C) being estimated by maximum likelihood. The estimate is that pair (T^*, C^*) resulting in the highest likelihood, and an approximate confidence interval is the set of all points (T, C) whose likelihood is an acceptable fraction of the maximum likelihood, as judged by the likelihood ratio test. If a single correlation or slope is being estimated, the 95% confidence interval will be all values of C for which there is a phylogeny T such that their likelihood $L(T, C) \geq 0.1465 \, L(T^*, C^*)$, since this is the ratio that would just reach significance in the likelihood ratio test with 1 degree of freedom.

It may be possible in some cases to discard the information about the phylogeny that we are least certain of, and use only those features in which we have reasonably high confidence. A student (whose name is unfortunately not known to me; see the ACKNOWLEDGMENTS section below) has pointed out to me that we could use contrasts between pairs of species that we were fairly sure had a common ancestor not shared with any member of another pair, and that these contrasts could then be safely assumed to be independent. For example, in a study of mammals we could use pairs consisting of two seals, two whales, two bats, two deer, etc. These contrasts would be independent, but would not necessarily have equal variances. We therefore could not simply compute correlations or regressions from the pairs. We could use certain nonparametric methods such as a sign test to test whether the changes in the two characters were correlated, but other nonparametric methods such as Spearman's rank correlation would not be valid because we could not assume that the pairs were drawn from a common distribution, even though they are independent. This sign test is essentially the same as that used by Baker and Parker (1979) to test whether regressions were similar in different bird families. It should be obvious that there is much statistical work remaining to be done on robust methods of using partial knowledge of phylogenies to make inferences about regressions and correlations of characters.

What If We Lack an Acceptable Statistical Model of Character Change?

All of the above has been predicated on the acceptance of the Brownian motion model as a realistic statistical model of character change. There are certainly many reasons for being skeptical of its validity. Persistence of selection pressures over time may lead to correlation of changes in successive branches of the phylogeny, and common selective regimes experienced by different populations owing to common environmental factors such as weather or predators may lead to changes in different lineages being correlated. Since the lengths of the branches of the phylogeny are given not in time units, but rather in units of expected variance of change (the v_i) the model already allows rates of change to differ in different lineages, and would allow, for example, change to be faster after speciation events than during later periods, as assumed to many punctuationists.

There is no reason to believe that the normal distribution is particularly plausible as the distribution from which changes in individual branches of the phylogeny are drawn, except insofar as the net change in a branch is the resultant of a series of bursts of change and thus might be approximately normal.

The matter of the model is an obvious point for future development and (to the extent that this is possible) empirical study. One rather serious problem that confronts comparative studies is that the relationship under study may change through time. Harvey and Mace (1982) have discussed the problem of change of the slope of the relationship between two variables with taxonomic level, which appears to be quite common. It should be possible to use the current model to study statistically whether there is any connection between the variance of a contrast and the slope of the regression of one variable on another.

What if We Do Not Take the Phylogeny into Consideration?

Some reviewers of this paper felt that the message was "rather nihilistic," and suggested that it would be much improved if I could present a simple and robust method that obviated the need to have an accurate knowledge of the phylogeny. I entirely sympathize, but do not have a method that solves the problem. The best we can do is perhaps to use pairs of close relatives as suggested above, although this discards at least half of the data. Comparative biologists may understandably feel frustrated upon being told that they need to know the phylogenies of their groups in great detail, when this is not something they had much interest in knowing. Nevertheless, efforts to cope with the effects of the phylogeny will have to be made. Phylogenies are fundamental to comparative biology; there is no doing it without taking them into account.

SUMMARY

Comparative studies of the relationship between two phenotypes, or between a phenotype and an environment, are frequently carried out by invalid statistical methods. Most regression, correlation, and contingency table methods, including nonparametric methods, assume that the points are drawn independently from a common distribution. When species are taken from a branching phylogeny, they are manifestly nonindependent. Use of a statistical method that assumes independence will cause overstatement of the significance in hypothesis tests. Some illustrative examples of these phenomena have been given, and limitations of previous proposals of ways to correct for the nonindependence have been discussed.

A method of correcting for the phylogeny has been proposed. It requires that we know both the tree topology and the branch lengths, and that we be willing to

COMPARATIVE METHOD   15

allow the characters to be modeled by Brownian motion on a linear scale. Given these conditions, the phylogeny specifies a set of contrasts among species, contrasts that are statistically independent and can be used in regression or correlation studies. The considerable barriers to making practical use of this technique have been discussed.

ACKNOWLEDGMENTS

The suggestion that pairs of closely related organisms could be used in a way that avoided the need to know the full phylogeny was made by a student during discussion following my seminar at the Department of Genetics, University College, London. Unfortunately, I have been unable to discover her name. I wish to thank Ray Huey, John Gittleman, Robert Martin, and Mart Ridley for helpful discussions and/or access to their unpublished work. I am particularly grateful to Paul Harvey for bringing this particular piece of biological real estate to my attention and for many helpful conversations about it. I also wish to thank the reviewers of this paper for many helpful suggestions and for saving me from myself on at least one point. This work was supported by Task Agreement no. DE-AT06-76EV71005 of contract number DE-AM06-76RL02225 between the U.S. Department of Energy and the University of Washington.

LITERATURE CITED

Armstrong, E. 1983. Relative brain size and metabolism in mammals. Science 220:1302–1304.
Baker, R. R., and Parker, G. A. 1979. The evolution of bird colouration. Philos. Trans. R. Soc. Ser. *B* 287:63–130.
Clutton-Brock, T. H., and P. H. Harvey. 1977. Primate ecology and social organization. J. Zool. 183:1–39.
Damuth, J. 1981. Population density and body size in mammals. Nature 290:699–700.
Felsenstein, J. 1973. Maximum-likelihood estimation of evolutionary trees from continuous characters. Am. J. Hum. Genet. 25:471–492.
———. 1978. Cases in which parsimony or compatibility methods will be positively misleading. Syst. Zool. 27:401–410.
———. 1981*a*. Evolutionary trees from DNA sequences: a maximum likelihood approach. J. Mol. Evol. 17:368–376.
———. 1981*b*. Evolutionary trees from gene frequencies and quantitative characters: finding maximum likelihood estimates. Evolution 35:1229–1242.
———. 1983. Parsimony in systematics: biological and statistical issues. Annu. Rev. Ecol. Syst. 14:313–333.
Gittleman, J. 1981. The phylogeny of parental care in fishes. Anim. Behav. 29:936–941.
Givnish, T. J. 1982. Outcrossing versus ecological constraints in the evolution of dioecy. Am. Nat. 119:849–851.
Harvey, P. H., and G. Mace. 1982. Comparisons between taxa and adaptive trends: problems of methodology. Pages 343–361 *in* Current problems in sociobiology. King's College Sociobiology Group, ed. Cambridge University Press, Cambridge.
Martin, R. D. 1981. Relative brain size and basal metabolic rate in terrestrial vertebrates. Nature 293:57–60.
Pilbeam, D., and S. J. Gould. 1974. Size and scaling in human evolution. Science 186:892–901.
Ridley, M. 1983. The explanation of organic diversity. The comparative method and adaptations of mating. Clarendon, Oxford.
Sherman, P. W. 1979. Insect chromosome numbers and eusociality. Am. Nat. 113:925–935.

| Vol. 140, No. 3 | The American Naturalist | September 1992 |

HISTORICAL EFFECTS AND SORTING PROCESSES AS EXPLANATIONS FOR CONTEMPORARY ECOLOGICAL PATTERNS: CHARACTER SYNDROMES IN MEDITERRANEAN WOODY PLANTS

Carlos M. Herrera

Estación Biológica de Doñana, Apartado 1056, E-41080 Sevilla, Spain

Submitted March 4, 1991; Revised August 28, 1991; Accepted October 9, 1991

Abstract.—Ecological patterns are not only a consequence of adaptive processes, but also influenced by phylogenetic constraints, historical effects, and sorting processes. In contrast to the attention paid to the influence of phylogeny on interspecific ecological patterns, historical effects and sorting processes have been considered less frequently. This article shows that, for the woody flora of western Andalusia, southwestern Spain, these factors may be of substantial importance for explaining covariation among life-history traits (and associated "character syndromes") in plant communities. Multivariate analysis of the covariation across genera of 10 qualitative characters (related to general habit and reproductive biology) revealed a dominant life history–reproductive gradient (called "dimension 1") defining two distinct groups of genera and associated syndromes. Syndromes may largely be explained by reference to historical effects and species sorting processes, without recourse to adaptive explanations. Lineage age (as estimated from paleontological and biogeographical data) explained a significant proportion of intergeneric variation in position along dimension 1. Many character associations contributing to the syndromes vanished after the sample was split into groups based on lineage age, and those remaining occurred exclusively within the group of "old" (pre-Mediterranean) genera. No supporting evidence was found for the contribution of differential extinction of pre-Mediterranean genera to observed syndromes. Differential diversification of lineages as a function of life-history and reproductive characteristics did contribute significantly.

Ecologists study thin temporal slices of historically dynamic systems. Evolutionary ecologists, in particular, attempt to interpret the array of ecological patterns found in these contemporary slices in terms of adaptive processes undergone by species and populations (Orians 1962; Lack 1965; Harper 1967). By using taxonomic entities (ordinarily species) as sampling units, patterns are often inferred from the distribution of ecologically relevant attributes, or their covariation, among taxa co-occurring on a particular spatial scale (e.g., local or regional). These patterns depend simultaneously on the identity of the taxa involved and on their attributes. As neither the traits of the taxa nor their involvement in a given local or regional assemblage are entirely the result of adaptive changes, ecological patterns should not be expected to be solely the consequence of adaptations. Further factors must thus be taken into consideration to explain them. For convenience, these may be grouped under the labels "phylogenetic constraints," "historical effects," and "species sorting processes," as outlined be-

Am. Nat. 1992. Vol. 140, pp. 421–446.

low. I admit, however, that the precise conceptual limits and operational relationships between these three broad categories of phenomena may often prove difficult to trace, as illustrated by the diverse usages of these and comparable terms found in the literature (see, e.g., Fowler and MacMahon 1982; Wanntorp 1983; Lechowicz 1984; Huey 1987; Gould 1989; Gould and Woodruff 1990).

Phylogenetic constraints, also sometimes referred to as "phylogenetic inertia" (Wilson 1975), "historical constraints" (Wanntorp 1983), or "historical factors" (Pearson et al. 1988), set limits on the evolution of individual species (Gould and Lewontin 1979; Felsenstein 1985; Gould 1989). They are taxon specific, and a particular taxon, regardless of where it occurs, may be forced into certain combinations of character states by virtue of these constraints. In animal species assemblages, for instance, phylogenetic constraints are responsible for a substantial proportion of observed interspecific variation in life-history attributes (see, e.g., Wiklund and Karlsson 1984; Cheverud et al. 1985; Pagel and Harvey 1988; Pearson et al. 1988; Bell 1989; Lessios 1990). In plant communities, phylogeny has been shown to be an important factor in explaining patterns of interspecific variation in seed size (Hodgson and MacKey 1986; Mazer 1989), flowering phenology (Kochmer and Handel 1986), flower duration (Stratton 1989), and features of fruit and fruiting synchrony of animal-dispersed plants (C. M. Herrera 1986, 1992; Gorchov 1990).

Historical effects become relevant explanations for ecological patterns whenever an understanding of the latter requires a knowledge of past historical states or contingencies. In contrast to phylogenetic constraints, historical effects are mainly site specific. As stated by Gould and Woodruff (1990, p. 75), "History represents a class or category of explanations, not a particular scenario," and the details of processes and mechanisms involved in historical explanations will differ from one particular study system to another. In the context of this article, historical effects are considered in relation to the dynamics of regional taxonomic assemblages and the long-term persistence of ecologically relevant characteristics of the taxa involved. As revealed by biogeography and paleontology, contemporary regional species assemblages represent transitory amalgamations of taxa that originated at different geological times and thus, presumably, in different ecological scenarios. This juxtaposition of taxa having temporally and ecologically disparate origins, and whose characteristics may have remained essentially unchanged over long time periods, may lead to contemporary patterns that are best explained as a consequence of historical effects (see Lechowicz 1984 for an example of this approach). In these instances, some present-day patterns would constitute "ecological phantoms" conceptually analogous to the "genetic phantoms" of Gould and Woodruff (1990).

Species sorting processes are also related, although for different reasons, to the long-term dynamic nature of species assemblages. They simultaneously involve taxon-specific and site-specific components. The lineages represented in a regional species pool may intrinsically differ in their rates of extinction and speciation (Stebbins 1951; Stanley 1979). In the long run, differential diversification rates (originations minus extinctions) of lineages possessing different attributes may give rise to distinct patterns in the distribution of such attributes across taxa, brought about by some process akin to what has been variously termed "selective extinction and speciation" (Fowler and MacMahon 1982), "species selection" (Stanley 1975, 1979), and lineage "sorting" (Vrba and Gould 1986). There has been a vigorous controversy in recent years with regard to the meaning of, and actual mechanisms involved in, species selection and related processes (see, e.g., Hull 1980, 1988; Hoffman 1984, 1989; Vrba 1984; Eldredge 1985; Gould 1985; Gould and Eldredge 1986; and references therein). To avoid becoming caught up in this controversy, I use the expression "sorting process" in this article to mean, generically, any process whereby lineages having different attributes acquire different proportional representations in regional species assemblages because of differences in diversification rates, regardless of the mechanism(s) involved.

In recent years, evolutionary ecologists have often examined the influence of phylogeny on interspecific ecological patterns. In contrast, the significance of historical effects and sorting processes has been considered, documented, or discussed much less frequently (but see, e.g., Fowler and MacMahon 1982; Pearson 1982; Lechowicz 1984; Brown and Maurer 1987). The objective of this article is to provide evidence demonstrating that historical effects and sorting processes may be of substantial importance in explaining patterns of covariation among life-history and reproductive features (and associated "character syndromes") observed at the regional plant community level, as illustrated by the woody flora of western Andalusia, in southwestern Spain. The flora of the Mediterranean basin is particularly well suited to an analysis of the influence of historical effects and sorting processes on contemporary patterns, as it represents a complex mixture of taxa with disparate biogeographical affinities, ages of origin, and evolutionary histories (Braun-Blanquet 1937; Raven 1971, 1974; Quézel et al. 1980; Mai 1989; Palamarev 1989). A further advantage is that our current knowledge of basic historical and ecological aspects is reasonably good for some regional floras.

It must be noted that the aim of this article is not to contribute to the literature just one more detailed analysis describing, or attempting to interpret, associations among plant reproductive and life-history traits (see references in the next section). Instead, this subject is treated as a representative example to illustrate some cautions that should be kept in mind by those attempting to interpret ecological patterns on the basis of adaptive arguments alone. Nevertheless, as shown in the next section, most of the plant characters and associations considered in this study have been previously examined for a variety of plant communities, and results obtained here are also subsidiarily relevant to some recent controversies on patterns of covariation among plant life-history and reproductive attributes.

PLANT CHARACTER SYNDROMES

The examination of plant species assemblages ranging in size from local communities to regional or world floras has often led to the identification of non-random patterns of covariation across taxa of morphological, life-history, and reproductive traits, sometimes leading to the identification of distinct character syndromes. These include associations among life form, leaf persistence, breeding system, pollinating agent, flower size, and seed dispersal mode (McComb 1966; Opler 1978; Bawa 1980; Conn et al. 1980; Freeman et al. 1980; Givnish 1980; C. M. Herrera 1982a, 1985; Sobrevila and Kalin Arroyo 1982; Flores and Schemske 1984; Bullock 1985; Fox 1985; J. Herrera 1987; Arroyo 1988; Steiner 1988; Donoghue 1989; and others). Dioecy, for instance, has been frequently found to be significantly associated with woody habit, wind pollination, small flowers, and seed dispersal by animals (see, e.g., McComb 1966; Bawa 1980; Givnish 1980; Flores and Schemske 1984). Generally, covariation between any two or more life-history and reproductive characters and the syndromes brought about by the nonrandom association of character states across taxa have been interpreted as indicative of adaptations (i.e., as being the consequence of adaptive evolutionary changes experienced by the taxa involved), and evolutionary hypotheses have frequently been advanced to explain them. In addition, some of these associations have provided the basis for formulating evolutionary explanations for the appearance of particular reproductive or life-history traits. The significant association frequently found between dioecy and seed dispersal by animals, for instance, has inspired adaptive hypotheses on the evolution of that breeding system and stirred considerable controversy (Bawa 1980, 1982a, 1982b; Givnish 1980, 1982; Thomson and Barrett 1981; C. M. Herrera 1982a, 1982b; Flores and Schemske 1984; Fox 1985; Muenchow 1987; Donoghue 1989; and references therein).

The assumption that contemporary interspecific patterns in plant life-history and reproductive traits are solely the consequence of adaptive processes affecting individual taxa (i.e., the combined importance of phylogenetic constraints, historical effects, and sorting processes is null or negligible) is common to all those adaptive interpretations of character evolution that are based on interspecific patterns of character covariation. Nevertheless, this assumption generally has been neither made explicit nor critically examined. In the few instances in which the importance of phylogenetic constraints has been examined, these have been found to play a prominent role in determining character syndromes in contemporary plant communities (Fox 1985; Muenchow 1987; Donoghue 1989), thus invalidating in part the assumption noted above. Nevertheless, the role of historical effects and sorting processes in determining plant character syndromes has not been previously examined, and this provides justification for the analyses presented here.

GENERAL METHODS

Analyses were conducted on the woody plants (angiosperms and gymnosperms) of western Andalusia, southwestern Spain, a region of 45,000 km² homogeneously characterized by a Mediterranean-type climate (Capel Molina 1981). I compiled the list of native woody plant genera according to a comprehensive regional flora (Valdés et al. 1987). From this list (83 genera) I excluded genera represented by a single species with extreme relict distributions in the region and those for which reliable information on the reproductive and life-history traits considered in this study (see below) was unavailable (for genera with geographical

HISTORY AND SORTING IN MEDITERRANEAN PLANTS   425

distributions extending beyond the Mediterranean region, I attempted to use biological information coming exclusively from studies or observations conducted in Iberian Mediterranean climatic regions; see references below). The final sample used in the analyses consisted of 66 genera belonging to 31 families. The available data on the distribution of character states among excluded genera, although fragmentary, do not suggest the possibility of any significant bias caused by omitting them from the analyses. Choosing genera, instead of species, as the units for the analyses has the advantage of minimizing the influence of a few genera with many species on overall patterns of character covariation (Fox 1985; Muenchow 1987). Similarly, restricting the survey to woody plants (shrubs, trees, and woody vines) reduces the internal heterogeneity of the sample, because life form is often correlated with reproductive traits, as noted earlier.

I scored each genus for 10 simple qualitative characters referred to general habit (deciduousness, spinescence, and sclerophylly), flowering biology (flower size and sexuality, perianth color and degree of reduction, and pollinating agent), and seed dispersal (seed size and dispersal agent). The two criteria used to select the characters were that they had been examined in some previous study of plant character syndromes and that they were readily scored on a qualitative basis. Characters and character states considered were as follows (only two possible character states were recognized for each trait): (1) spinescence: spiny (spines on stems or leaves) versus nonspiny; (2) leaf type: sclerophyllous leaves versus leaves of different characteristics; (3) habit: evergreen versus deciduous (winter or summer-facultative deciduous); (4) flower size: perianth depth times width of less than 25 mm² (5 mm × 5 mm) versus more than 25 mm²; (5) flower sexuality: hermaphroditic versus unisexual flowers; (6) perianth color: brownish or greenish versus a different color; (7) perianth reduction: perianth with at least one verticil (petals or sepals) absent or much reduced versus complete perianth; (8) pollinator type: wind pollination versus insect pollination; (9) seed size: seed length times width of less than 2.25 mm² (1.5 mm × 1.5 mm) versus more than 2.25 mm²; and (10) seed dispersal: animal dispersed (endozoochory) versus otherwise. For character 7, gymnosperms (lacking a perianth) were all scored as if they had a reduced perianth. For characters 4 and 9, cut points used to separate the "small" and "large" categories were chosen a priori, representing a subjective assessment (based on my familiarity with the regional flora) of the median values of regional distributions of flower and seed size.

A complete list of genera and character states for these qualitative variables is presented in Appendix A. Scoring of plant genera with respect to variables 1, 2, 3, 5, 6, and 7 above was mainly based on Valdés et al. (1987), with complementary information obtained from Hutchinson (1964, 1967), Ruiz de la Torre (1971), and Mabberley (1987). For characters 4, 8, 9, and 10, I relied largely on Catalán Bachiller (1978), J. Herrera (1982, 1985, 1988), C. M. Herrera (1984a, 1987, 1989), Jordano (1984), Bosch (1986), and Obeso (1986), but I also used my own unpublished observations for those genera not treated in these studies. Almost all genera were internally homogeneous with respect to all characters. In the few cases in which a genus was internally heterogeneous with regard to any variable (e.g., spinescence in *Juniperus* and flower sexuality in *Asparagus*), the rule of thumb

TABLE 1

COVARIATION OF LIFE-HISTORY AND REPRODUCTIVE CHARACTERS AMONG EXTANT GENERA
OF WESTERN ANDALUSIAN WOODY PLANTS

Character	LT	HA	FS	FSE	PC	PR	PT	SS	SD
Spinescence (SP):									
All	...	...	...	...	...	...	...	...	...
Young	...	...	...	...	...	...	...	...	...
Old	...	...	...	...	...	...	...	...	...
Leaf type (LT):									
All		*	*	*	*	*	...	...	...
Young		...	...	...	...	...	...	...	...
Old		*	*	...	...	...	...	...	...
Habit (HA):									
All			...	...	...	...	...	...	...
Young			...	...	...	...	...	...	...
Old			...	...	...	...	...	...	...
Flower size (FS):									
All				*	*	*	*	...	*
Young				...	...	...	...	...	...
Old				*	*	*	*	...	...
Flower sexuality (FSE):									
All					*	*	...	...	...
Young					...	...	...	...	...
Old					*	*	...	...	...
Perianth color (PC):									
All						*	*	...	...
Young						...	...	...	...
Old						*	*	...	...
Perianth reduction (PR):									
All							*	...	...
Young							...	...	...
Old							*	...	...
Pollinator type (PT):									
All								...	...
Young								...	...
Old								...	...
Seed size (SS):									
All									*
Young									...
Old									...
Seed dispersal (SD):									
All									
Young									
Old									

NOTE.—Pairs of characters marked with asterisks co-vary nonrandomly across genera ($P < .05$). Cells with ellipses denote nonsignificant character associations ($P \geq .05$). Analyses were conducted on the whole sample ("all," $N = 66$) and separately for genera having either contemporary intercontinental range disjunctions or Pliocene fossil records, on one hand ("old," $N = 46$), and for those lacking both features, on the other ("young," $N = 20$). For each pair of characters, a two-way contingency table was obtained, and its significance tested with a two-tailed Fisher exact probability test. In each of the three separate association tables, significance levels were adjusted by the sequential Bonferroni method (Rice 1989). Adjusted tablewide α levels for a nominal $\alpha = 0.05$ were 0.0017, 0.0012, and 0.0014 for the "all," "young," and "old" values, respectively.

used was to score it as a function of the character state prevailing among western Andalusian species.

ANALYSES AND RESULTS

Covariation of Characters

Covariation of characters across genera was first examined by constructing all possible two-way contingency tables for character pairs and testing these for significance with two-tailed Fisher exact probability tests (Zar 1984). Because many simultaneous tests were carried out, significance levels of individual tests were adjusted for increased Type I error by the sequential Bonferroni method (Rice 1989). Sixteen significant character associations were found (table 1). Spinescence was the single character for which no significant association was found, and flower size and leaf type were those exhibiting the largest number of significant associations with others. Significant associations between character pairs found among western Andalusian plant genera are analogous to those reported previously for other species assemblages. These include, among others, associations between flower sexuality and perianth coloration, flower size and seed dispersal mode, and leaf type and perianth reduction (table 1).

Analyses of the associations between character pairs are insufficient to describe overall, multivariate patterns of intergeneric variation (Fox 1985). In order to examine whether pairwise associations between characters actually reflect one or more multivariate gradients of variation, or character syndromes, a similarity matrix between pairs of genera, based on the 10 characters examined, was constructed with a simple matching coefficient. Pairs of genera similar in all characters had a similarity of one, whereas those not sharing any character state had a similarity of zero. All characters were weighted similarly in computing the similarity coefficient. Nonmetric multidimensional scaling (NMDS; Kruskal 1964; du Toit et al. 1986) was performed on the 66×66 symmetrical similarity matrix, according to procedure MDS in SYSTAT, the Kruskal algorithm, and monotonic regression (Wilkinson 1986). This geometric method determines the configuration of objects (plant genera in the present instance) in one Euclidean space of minimal dimensions that best represents the original object distances.

Models ranging from one to five dimensions yielded stress values (a measure of the correspondence between original distances and those obtained after dimensionality reduction; Kruskal 1964) of 0.15, 0.08, 0.05, 0.03, and 0.01, respectively. According to the criteria outlined by Kruskal (1964), these figures suggest that a configuration in two dimensions describes satisfactorily the relationships between plant genera on the basis of their similarities in the 10 characters considered here. The distribution of genera on the plane defined by the two dimensions is shown in figure 1 (see App. A for coordinates of individual genera). Most intergeneric variation occurs along dimension 1, and a discernible gap exists in the distribution of points along it. Genera fall into two discrete clusters, each exemplifying a characteristic association of character states, or syndrome. The group on the left side of figure 1 (Cluster I hereafter) is made up predominantly of genera with sclerophyllous, evergreen leaves, small, unisexual greenish or brownish flowers

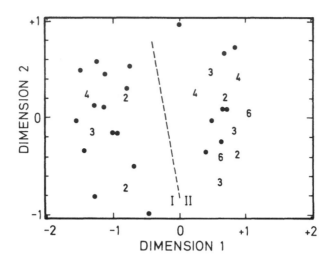

FIG. 1.—Distribution of western Andalusian woody plant genera on the plane defined by dimensions 1 and 2 obtained by nonmetric multidimensional scaling of a similarity matrix of their life-history and reproductive characteristics. *Numerals* indicate the number of tied observations at particular points. The *dashed line* is shown to illustrate the separation of genera into two distinct clusters (I and II), as described in the text.

with a reduced perianth, and large seeds dispersed by animals. Representative examples in this group are *Chamaerops* (Palmae), *Olea* (Oleaceae), *Osyris* (Santalaceae), and *Pistacia* (Anacardiaceae). In contrast, the cluster on the right side (Cluster II) is predominantly characterized by the set of complementary character states (e.g., *Cistus* [Cistaceae], *Genista* [Fabaceae], and *Rosmarinus* [Lamiaceae]).

Historical Effects

To search for evidence of historical effects in the syndromes documented above, I considered paleontological and biogeographical information.

I screened the plant paleontological literature for early Pliocene or pre-Pliocene fossil records (either pollen or macrofossils) of the extant genera of woody plants examined here. I considered only records coming from localities in the Mediterranean basin (including Portugal) and continental western Europe (France, western Germany, and Denmark). The Pliocene was chosen as a cut point for the search because it is generally agreed that current Mediterranean climatic conditions first appeared during the second half of that period (Suc 1984). In this way, extant genera with early Pliocene or pre-Pliocene fossil records are to be considered as descendants of lineages that certainly did not evolve under contemporary, Mediterranean climatic conditions. In contrast, genera failing to provide Pliocene or pre-Pliocene fossil records will more likely represent lineages appearing in the region (by origination or migration) after the beginning of current Mediterranean climatic conditions. The data used here were taken from Depape (1922), Bauzá Rullan (1961), Pons (1964, 1981), Diniz (1967), Fernández Marrón (1971), Medus

HISTORY AND SORTING IN MEDITERRANEAN PLANTS 429

TABLE 2

EFFECTS OF RANGE DISJUNCTIONS ("DISJUNCTIONS") AND
OCCURRENCE OF PLIOCENE OR PRE-PLIOCENE FOSSIL RECORDS
("FOSSILS") ON DIM1

Effect	df	Mean Square	F	P
Disjunctions	1	4.0122	6.26	.015
Fossils	1	7.7081	12.02	.001
Disjunctions × fossils	1	.0261	.04	.841
Error	62	.6433		

NOTE.—A general linear model was fitted to the data according to the GLM procedure (SAS Institute 1988) with Type III sum of squares. Disjunctions and fossils were treated as two-level (presence-absence) categorical variables. Significance of model: $F = 7.83$, df $= 3,62$, $P = .0002$, $R^2 = 0.275$.

and Pons (1980), van der Burgh (1983), Friis (1985), Vicente i Castells (1988), and Palamarev (1989; see App. A for data used).

Broadly disjunct geographical ranges of plant taxa on the world scale may be brought about through a variety of processes, but they generally attest an old origin for the lineage (Raven and Axelrod 1974; Schuster 1976; Axelrod 1983). I examined the worldwide distributional patterns of all the genera considered here and scored them for broad intercontinental disjunction of geographical range. I considered as disjunct only those geographical ranges falling into the "fragmentary North Temperate," "African-Eurasian," or "pantropical" disjunction categories as defined by Thorne (1972). Biogeographical information was taken from Hutchinson (1964, 1967), Thorne (1972), Quézel et al. (1980), and Mabberley (1987). A total of 25 genera in my sample qualified as having disjunct distributions according to the criteria used (App. A).

Paleontological and biogeographical information was examined by fitting a general linear model to the data according to procedure GLM in SAS (SAS Institute 1988). Position along the life history–reproductive gradient (DIM1), a continuous variable, was entered as the dependent (criterion) variable, and occurrence of Pliocene or pre-Pliocene fossils, and intercontinental disjunctions, as independent (predictor) variables. Results are summarized in table 2. The fitted model was statistically significant, accounting for 27.5% of observed (intergeneric) variance of DIM1. The two independent variables have significant effects on DIM1. Genera with Pliocene or pre-Pliocene fossil records or exhibiting intercontinental range disjunctions tend to score significantly lower on DIM1 than those lacking these characteristics (fig. 2). This pattern is also evidenced by the differential distribution of genera with Pliocene or pre-Pliocene fossils and exhibiting range disjunctions among the two clusters revealed by NMDS. Eighty percent of genera in Cluster I ($N = 25$) yielded Pliocene or pre-Pliocene fossil records, as compared with only 43.9% of genera in Cluster II ($N = 41$) (Fisher test, $P = .003$). Likewise, genera having disjunct distributions were unequally distributed among Clusters I and II (60% and 24.4% of genera in Clusters I and II, respectively; Fisher

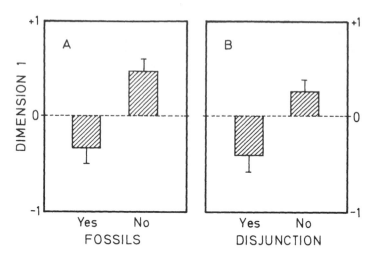

Fig. 2.—Mean location on the first axis from nonmetric scaling (score on dimension 1; fig. 1) of groups of western Andalusian plant genera defined as a function of the occurrence of (A) Pliocene or pre-Pliocene fossil records and (B) intercontinental range disjunction. Vertical bars extend over 1 SE of the mean. See table 2 for statistical analysis.

test, $P = .008$). At this point, it must be noted that the paleontological and biogeographical data used represent statistically independent lines of evidence, as there is not a significant association between the occurrence of intercontinental disjunctions and that of Pliocene or pre-Pliocene fossils in the sample of genera studied (Fisher test, $P = .21$).

Taken together, therefore, paleontological and biogeographical data show that the heterogeneity of historical origins represented in the set of genera examined has considerable explanatory power in predicting intergeneric variation in life-history and reproductive characters. To what degree are the associations between characters that determine Clusters I and II of genera only an artifact of the heterogeneity of the sample with regard to lineage age? In other words, do associations between characters still persist when the sample of genera is split as a function of presumed lineage age? Genera were split into two groups, namely "old" and "young." Those with either contemporary intercontinental range disjunction *or* Pliocene or pre-Pliocene fossil records were scored as old ($N = 46$), and those lacking both of these features were considered young ($N = 20$). The initial character-association matrix was recomputed separately for each group by the same methods used for the whole sample (table 1). Five of the 16 original significant associations between characters vanished after splitting the sample into the two age-based groups. Of the 11 original associations still remaining in the separate analyses, all occurred in the group of old genera, and none in the group of young ones.

These results suggest that historical effects are directly responsible for some of the character associations found among western Andalusian plant genera. An alternative interpretation, however, could be that modifications in the number of significant character associations are simply a consequence of the reduction in sample size (and so the power of the test) caused by splitting the sample into two smaller subsamples. I examined this possibility using a simulation scheme. The original sample of genera was randomly divided into two subsamples of a size equal to that of the old ($N = 46$) and young ($N = 20$) groups, and the significance of all pairwise character associations was tested separately in each subsample by the same methods as in table 1. This process was iterated 1,000 times, and on each occasion the number of significant character associations found in each random subsample was recorded. Critical probability values were then obtained from the cumulative frequency distributions of the number of significant-associations tests. The probability of obtaining, by chance alone, 11 or fewer significant character associations after splitting the sample of genera into the two age-based subgroups was $P = .006$, and that of failing to record even a single significant character association in the smallest group of genera (young ones) was $P = .020$. It may be concluded, therefore, that the modifications in the patterns of character associations that follow the subdivision of the sample of genera into age-based groups are real and not the consequence of a reduction in the power of statistical tests due to decreased sample size.

In conclusion, therefore, some contemporary character associations (those disappearing after sample subdivision) must be interpreted as a consequence of the historical heterogeneity of the sample of genera, whereas others (those "resistant" to subdivision) are characteristic only of a historically defined, particular subset of genera (the old ones). Examples of correlations that disappear after sample subdivision are those between flower size and seed dispersal mode and between seed size and seed dispersal mode (table 1). When the whole sample of genera is considered, animal dispersal is significantly associated with large seeds and small flowers, but none of these relations hold within each of the two age-based groups of genera. Correlations between flower sexuality, perianth color, and perianth reduction exemplify those found exclusively within the subgroup of old genera (table 1). Unisexual flowers and reduced, brownish or greenish perianths tend to be associated only in that group of genera, and these associations emerge also when the whole sample is considered.

Sorting Processes

As shown in the preceding section, the character syndromes observed among contemporary western Andalusian plant genera are partly explained by the heterogeneity of the sample with regard to historical origin. All character associations that remain after accounting for this effect apply only to the group of genera belonging to old lineages. Not a single significant character association exists within the group of more recent genera (those presumably evolved after the initiation of Mediterranean climatic conditions). Sorting processes may underlie these patterns. First, those character associations that depend exclusively on the heterogeneity of the sample with regard to historical origin may reflect differential levels of recent diversification of lineages as a function of their life-history and reproductive characteristics. Second, those character associations found exclusively within the group of old genera might be a consequence of differential post-Pliocene extinctions of pre-Mediterranean lineages occurring nonrandomly

TABLE 3

EFFECTS OF DIM1 AND GEOLOGICAL AGE OF EXTANT
WESTERN ANDALUSIAN WOODY PLANT GENERA ON THE
NUMBER OF SPECIES PER GENUS ON BOTH THE WESTERN
ANDALUSIAN AND WORLD SCALES

Effect	df	Mean Square	F	P
A. Andalusian species:*				
DIM1	1	3.5792	6.15	.016
Age	1	.6155	1.06	.31
DIM1 × age	1	1.2643	2.17	.145
Error	62	.5816		
B. World species:†				
DIM1	1	1.0998	.32	.57
Age	1	2.4698	.73	.40
DIM1 × age	1	.8224	.24	.63
Error	62	3.4038		

NOTE.—A general linear model was fitted according to the GLM procedure (SAS Institute 1988) with Type III sum of squares. Dependent variables (number of species per genus) were log-transformed for the analyses.

* Significance of model: $F = 5.84$, df $= 3,62$, $P = .0014$.

† Significance of model: $F = 0.82$, df $= 3,62$, $P = .49$.

with respect to life-history and reproductive characteristics. I will consider these two aspects in turn.

Differential diversification.—This aspect can be examined only indirectly in the data set considered in this article. The effect of potential differences in lineage diversification rate on patterns of covariation among characters could be more readily assessed if these patterns had been elucidated with species, instead of genera, as the taxonomical units for the analyses. Nevertheless, indirect evidence supporting a potential role of sorting processes in determining character syndromes could still be found in the sample of extant western Andalusian genera if, after statistically controlling for differential lineage age, (1) the number of extant species of each genus in the study region (net regional diversification, the result of originations minus extinctions) were significantly related to its score on DIM1 *and* (2) this relationship were characteristic of the region concerned, and not simply the regional expression of a pattern occurring on a broader geographical scale.

To test aspects 1 and 2 above, the influence of lineage "age" (young vs. old genera, as defined above) and life-history and reproductive characters (score on DIM1) on net diversification was examined simultaneously by fitting a general linear model to the data. The number of species comprised by each genus in western Andalusia (log transformed; based on Valdés et al. 1987) was used as the criterion variable, and age and score on DIM1 as predictor variables. One further analysis was done with the total number of species in each genus in the world (log transformed; based on Mabberley 1987) as the dependent variable. Results are summarized in table 3.

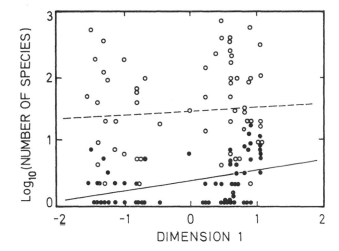

FIG. 3.—Variation in the number of extant species in western Andalusia (*filled circles, solid regression line*) and on a worldwide basis (*open circles, dashed regression line*), for each of the extant western Andalusian woody plant genera. Regression lines were fitted according to the least-squares method. See table 3 for analytical results.

In the sample of genera examined, there is a significant positive relationship between the number of species occurring in western Andalusia and DIM1 (table 3A; fig. 3). Neither the effect of age nor the interaction age × DIM1 is significant, which reveals that the regional species richness of western Andalusian plant genera is independent of lineage age and is best explained in terms of the life-history and reproductive attributes summarized by DIM1. When the total number of species of each genus on a world basis is used as the dependent variable, no significant effect is found for any of the independent variables or the inter-action term, and the model fitted is not significant (table 3B; fig. 3). These results indicate that, in western Andalusia, genera at different positions on the life history–reproductive gradient have diversified at different rates and this effect is characteristic of the region concerned and does not merely echo a general diversification pattern of the genera on the world scale. The hypothesis of a role of differential diversification in enhancing character syndromes is, therefore, supported.

Differential extinction.—I compiled the list of woody plant genera (angiosperms and gymnosperms) occurring in southwestern Europe (Iberian Peninsula and southern France) during the Pliocene and mid- to late Miocene (based on information in Depape 1922; Pons 1964; Fernández Marrón 1971; Medus and Pons 1980; Vicente i Castells 1988), exclusive of those that have since become extinct on a world basis. Genera that are not currently represented in western Andalusia but occur elsewhere in the western Mediterranean basin or in west-central Europe were likewise excluded. The sample of genera ($N = 83$) was divided into extinct ($N = 45$) and extant ($N = 38$) in the contemporary western Andalusian flora. Regionally extinct genera were scored for four of the traits examined earlier (flower sexuality, perianth reduction, pollinator type, and seed dispersal), on the

434   THE AMERICAN NATURALIST

TABLE 4

<small>Covariation of Reproductive Characters among Woody
Plant Genera That Occurred in Southwestern Europe
in the Pliocene or Mid- to Late Miocene</small>

Character	PR	PT	SD
Flower sexuality (FSE):			
Extinct	*	*	· · ·
Extant	*	· · ·	· · ·
Perianth reduction (PR):			
Extinct		*	· · ·
Extant		*	· · ·
Pollinator type (PT):			
Extinct			· · ·
Extant			· · ·
Seed dispersal (SD):			
Extinct			
Extant			

<small>Note.—Pairs of characters marked with asterisks co-vary
nonrandomly across genera ($P < .05$). Cells with ellipses denote nonsignificant ($P \geq .05$) character associations. Analyses were conducted separately for genera extant ($N = 38$)
and extinct ($N = 45$) in western Andalusia. Methods are as
in table 1. Adjusted table-wide α levels for a nominal $\alpha =
0.05$ were 0.008 for the values for extant and extinct genera.</small>

basis of characteristics currently found in extant species of these genera elsewhere (based on information provided by Hutchinson 1964, 1967; Heywood 1978;
Cronquist 1981; Mabberley 1987; and Kramer and Green 1990). These four characters were chosen because, among the traits studied, they tend to exhibit the
greatest constancy within genera. In this way, scoring for these characteristics
of genera extinct in southwestern Europe can confidently be done by reference
to the traits of their extant species elsewhere. A list of extinct genera and the
data used are shown in Appendix B.

The pattern of association of characters in the sample of extinct genera was
compared with that found in genera that occur at present in western Andalusia
and for which there are Pliocene or pre-Pliocene fossil records. With a single
exception, the pattern of association among the four characters considered is
similar in the groups of extinct and extant genera (table 4). The exception concerns the association between flower sexuality and pollinator type, which is significant among extinct genera but not among extant ones (see also table 1). These
results support the notion that, with regard to the patterns of covariation of the
characters examined, the set of extant woody genera in western Andalusia tends
to represent a random sample of the whole set of woody genera present in southwestern Europe in the Pliocene and late Miocene. The hypothesis of a role of
differential extinction in determining character syndromes thus is not supported.

DISCUSSION

This article was conceived as an exercise to illustrate that some particular
nonadaptive hypotheses deserve more consideration than they have received so

far when interspecific ecological patterns have been examined. The particular subject under consideration, plant character syndromes, was chosen only by way of illustration, and for this reason the results that refer specifically to this particular subject will not be discussed here in detail. Instead the discussion will focus mainly on the more general objective of the article, namely, documenting the importance of historical effects and sorting processes as determinants of interspecific ecological patterns.

As is found in many other plant species assemblages elsewhere, life-history and reproductive characters co-vary nonrandomly among western Andalusian plant genera. Nine out of the 10 characters examined exhibited at least one significant association with another. These pairwise character associations represent the manifestation, at the bivariate level of resolution, of a more complex multivariate gradient of variation in life-history and reproductive traits, as is found also in other studies of character syndromes (Fox 1985). In fact, some of the associations between characters found here have been previously reported from other plant assemblages. So far, therefore, this study has revealed nothing essentially different from previous investigations on plant character syndromes. In contrast to previous studies of character associations, however, the present one has shown that character associations and the multivariate syndromes they reflect may largely be explained without resorting to adaptive explanations based on the contemporary ecological conditions faced by the taxa involved.

The syndromes exhibited by contemporary western Andalusian woody plants have a strong historical component. The multivariate life-history and reproductive gradient depicted by NMDS closely reflects differential lineage age, and variables describing the latter are able to explain, despite their admitted crudeness, as much as 27% of intergeneric variation in DIM1. This historical effect is largely the consequence of (1) the internal heterogeneity of the sample with regard to lineage age (some character associations contributing to the syndromes vanish after the sample is split into groups based on presumed lineage age) and (2) the differential patterns of association of life-history and reproductive features found among lineages that originated at different geological times. Obviously, the possibility exists that the paleontological data used here are subject to some indeterminate biases (e.g., differential probability of fossilization associated with the DIM1 gradient). Because of the very nature of the data, no information is available to reject this possibility, but the observation that independent biogeographical evidence confirms the results based on paleontological data tends to suggest that biases, if any, are most likely inconsequential to the results.

All character associations that remained significant after extant genera were split into discrete old and young age groups occurred exclusively within the former category. The occurrence of these character associations in the whole set of extant genera was therefore caused exclusively by the presence within the sample of an important proportion of old, pre-Mediterranean lineages. As this group of genera represents a random subsample of those occurring in southwestern Europe during the mid- to late Miocene and Pliocene (with regard to the association of reproductive features), there is no evidence that the character associations that they exhibit at present have been brought about by post-Pliocene selective extinction processes occurring among pre-Mediterranean lineages as a function of their configuration of reproductive features. The set of character associations found at present among the regionally surviving old genera presumably were already present before the initiation of Mediterranean conditions and have persisted unchanged since the subsequent extinction of a substantial fraction of woody genera. These character associations are probably best seen as ecological phantoms, evolved in temporally and ecologically distant pre-Mediterranean, tropical-like scenarios, and evolutionarily unrelated to present ecological conditions. At this point, it is most instructive to record that those character associations that are restricted to the subset of old genera are virtually identical to those reported from present-day Neotropical forests by Bawa (1980).

In addition to historical effects, sorting processes also partly explain the patterns of covariation among characters found among extant western Andalusian plant genera. Evidence was found of differential diversification rates of genera depending on their life-history and reproductive characteristics (DIM1). After the effect of lineage age was controlled, net diversification rates of individual genera in the study region increased significantly with increasing DIM1 values. If we assume that a similar trend has also occurred at a higher level in the taxonomical hierarchy (differential origination rate of genera), this finding provides support for the idea that differential proliferation of genera along the life history–reproductive gradient has also enhanced character syndromes.

In view of the results discussed above, a parsimonious, nonadaptive explanation for observed contemporary patterns of covariation of life-history and reproductive characters among extant western Andalusian woody plant genera and associated multivariate syndromes may be synthesized as follows. Certain character associations and concomitant syndromes already existed among the set of woody plants present in southwestern Europe before the initiation of Mediterranean climatic conditions in the Pliocene, when the region was under a different, tropical-like climatic regime. Since then, and presumably influenced by the ecological changes brought about by climatic alterations, numerous extinctions and originations of genera have taken place in the area. Extinctions of old lineages occurred randomly with respect to the character associations found in them, and, consequently, the original pre-Mediterranean associations of character states across taxa are still "preserved" as ecological phantoms among those old genera that still survive. While some old lineages were becoming extinct, new ones were becoming superimposed on the surviving remains of the Pliocene species assemblage. These additions, in contrast to extinctions, did not occur randomly with respect to life-history and reproductive characters, but were more abundant on some sectors of the DIM1 gradient, hence enhancing certain associations between characters within the whole sample of extant genera. Within this group of recent genera, however, significant associations between the characters examined here no longer exist.

The contemporary woody flora of the Mediterranean basin represents a historically very heterogeneous assortment of lineages having varied origins in time and space (for details, see, e.g., Braun-Blanquet 1937; Raven 1971, 1974; Axelrod 1975; Pignatti 1978; Quézel et al. 1980; Pons 1981; Palamarev 1989). It might thus be argued that the importance of historical effects documented here is but a singular consequence of the peculiarities of this study system and cannot be readily extrapolated to other plant species assemblages. To variable degrees, however, heterogeneity in the historical background of contemporarily coexisting taxa is most likely the rule, rather than the exception, in present-day plant communities throughout the world, as is illustrated by many paleontological and biogeographical studies (e.g., Axelrod 1958, 1986; Stebbins and Major 1965; Meusel 1971; Davis 1976; Sunding 1979; Tiffney 1985). The generality of this circumstance confers on historical effects and sorting processes a potential explanatory power for a variety of interspecific ecological patterns, particularly when long-duration taxa and characters subject to appreciable phylogenetic constraints are involved. The only other study known to me in which historical effects were explicitly considered as potential determinants of ecological patterns in plant communities also provides support for this view (Lechowicz 1984).

Paying attention to the role of historical effects (or "historical explanations" of Lechowicz [1984]) as explanations of contemporary ecological patterns rests on the assumption that, for whatever reasons (e.g., phylogenetic constraints, weak selective pressures, or insufficient time), the characters under examination have remained essentially unchanged over long time periods in the face of changing ecological conditions. The natural consequence of this assumption is the expectation that contemporary patterns are "contaminated" to variable degrees by traits and trait associations that evolved in the past under different circumstances and that are not adaptations to present-day conditions. The symmetrical assumption (that present-day patterns may be explained without recourse to historical explanations), and its likewise symmetrical consequence (that observed patterns represent adaptations only to contemporary conditions), are a central tenet of the "adaptationist programme" (Gould and Lewontin 1979), and as such they have largely become quintessential features in the practice of plant evolutionary ecology (C. M. Herrera 1986). For practical reasons, it is difficult to document unambiguously the continued phenotypic stability of plant taxa over long time periods. Nevertheless, published evidence shows that the assumption of phenotypic stability that underlies historical explanations of the kind examined in this article is well supported in many instances and for a variety of traits (Stebbins and Day 1967; Goldblatt 1980; Stebbins 1982; Niklas et al. 1985; C. M. Herrera 1986; Liston et al. 1989a, 1989b; Parks and Wendel 1990; and references therein). For obvious reasons, the stability assumption cannot apply universally to all taxa, but the very fact that it sometimes does provides in itself sufficient justification for considering historical explanations as one valid alternative or at least additional hypothesis to adaptive explanations in plant evolutionary ecology research programs (see also Lechowicz 1984).

Evolutionary ecology has, by definition, many of the ingredients of a historical science, insofar as it attempts to explain contemporary patterns and processes in relation to past events. Rather paradoxically, however, it tends to exhibit in practice certain traits of an ahistorical discipline, insofar as it deliberately tends to confine itself to the examination of what is but a restricted subset of past events (adaptive evolutionary change undergone by species and populations). Apparently, the possibility that the contingencies of history play a prominent role in explaining contemporary ecological patterns has been felt by some ecologists as a threat to the survival of their scientific practice. Westoby (1988, p. 554) wrote, exemplifying what perhaps represents an extreme view, that "if history were all-important, ecologists would have to retreat to natural history. Community and ecosystem study would pass into the hands of paleontologists." These fears not only are unfounded, but also, if unchallenged, may lead to serious misinterpretations in research programs in evolutionary ecology (Gould and Lewontin 1979). Abandoning the assumption that history never has any influence on present-day ecological patterns and that everything we see can always be understood in terms of present-day environments alone does not amount to stating that history is all-important. It implies only that the potential role of historical effects should also be taken into consideration when present-day ecological patterns are interpreted, in addition to the customarily included, often ad hoc adaptive explanations. As an analogy, consider the case of the influence of phylogenetic constraints on present-day ecological patterns. Only after their potential influence on interspecific patterns was recognized by evolutionists were a variety of methods developed that allowed for a dissection of observed interspecific patterns into their potentially adaptive and phylogeny-influenced components (Huey and Bennett 1987; Pagel and Harvey 1988; Bell 1989; Burt 1989; Donoghue 1989; Grafen 1989). Likewise, only after it is recognized that past historical contingencies may often shape contemporary ecological patterns will methods be devised to quantify the relative influence of these historical effects and identify plausible adaptive hypotheses in relation to present-day environments alone. In the case of the character syndromes of western Andalusian woody plants documented here, for instance, consideration of historical effects suggests that genera falling near the right extreme of the DIM1 gradient are probably those for which adaptive explanations related to present-day environments are both most straightforward and best justified. In contrast, adaptive hypotheses and expectations would less likely apply for old genera falling near the left extreme of the DIM1 gradient. This prediction is upheld by the results of field studies on the seed dispersal ecology of species in the old, pre-Mediterranean genera *Pistacia* (Jordano 1988, 1989), *Olea* (Jordano 1987), and *Osyris* (C. M. Herrera 1984b, 1988) in southwestern Spain, which found little support for interpreting many of their reproductive features as adaptations to present-day Mediterranean climatic environments.

ACKNOWLEDGMENTS

A preliminary version of this paper was delivered at the II Jornadas de Taxonomía Vegetal, Real Jardín Botánico, Madrid, May 1990. For helpful comments, criticisms, and discussion on earlier versions of the manuscript, I am most grateful to R. Dirzo, M. J. Donoghue, P. J. Grubb, P. E. Hulme, P. Jordano, L. López-Soria, G. Stevens, M. F. Willson, R. Zamora, and an anonymous reviewer. During the preparation of this paper, I received support from Dirección General de Investigación Científica y Técnica grant PB87-0452.

APPENDIX A

TABLE A1

EXTANT WESTERN ANDALUSIAN WOODY PLANT GENERA CONSIDERED IN THIS STUDY,
AND SUMMARY OF LIFE-HISTORY, REPRODUCTIVE, PALEONTOLOGICAL,
AND BIOGEOGRAPHICAL DATA USED IN THE ANALYSES

Genus	SP	LT	HA	FS	FSE	PC	PR	PT	SS	SD	PLI	DIS	DIM1	DIM2
Adenocarpus	0	0	1	1	0	1	1	1	1	0	0	1	.82	−.14
Amelanchier	0	0	1	1	0	1	1	1	1	1	0	1	.61	−.41
Arbutus	0	1	0	1	0	1	1	1	1	1	1	1	.23	.24
Asparagus	1	0	0	0	1	0	1	1	1	1	0	1	− .68	− .50
Berberis	1	0	1	0	0	1	1	1	1	1	0	0	.61	− .67
Calicotome	1	0	0	1	0	1	1	1	1	0	0	0	.67	.09
Calluna	0	0	0	1	0	1	1	0	0	0	1	0	.84	.73
Celtis	0	0	1	0	1	0	0	1	1	1	1	1	− .80	− .73
Ceratonia	0	1	0	0	1	0	0	0	1	1	1	0	1.30	.23
Chamaerops	1	1	0	0	1	0	0	1	1	1	1	0	− 1.30	− .15
Cistus	0	0	1	1	0	1	1	1	0	0	0	0	1.06	.04
Clematis	0	0	1	1	0	0	1	1	1	0	0	1	.63	−.25
Colutea	0	0	1	1	0	1	1	1	1	0	1	0	.82	−.14
Corema	0	0	0	0	1	0	0	0	1	1	0	1	− 1.28	.12
Coriaria	0	1	0	0	0	0	1	0	1	1	1	1	− .74	.53
Coronilla	0	0	0	1	0	1	1	1	0	0	0	0	.91	.41
Cotoneaster	0	0	1	1	0	1	1	1	1	1	1	0	.61	−.41
Crataegus	1	0	1	1	0	1	1	1	1	1	1	0	.61	− .67
Cytisus	0	0	1	1	0	1	1	1	1	0	0	0	.82	− .14
Daphne	0	1	0	1	0	1	1	1	1	1	1	0	.23	.24
Dorycnium	0	0	0	1	0	1	1	1	1	0	0	0	.70	.20
Ephedra	0	1	0	0	1	0	0	0	1	1	1	1	− 1.39	.23
Erica	0	0	0	1	0	1	1	1	0	0	0	1	.91	.41
Fumana	0	0	1	1	0	1	1	1	0	0	0	0	1.06	.04
Genista	1	0	1	1	0	1	1	1	1	0	0	0	.88	− .38
Halimium	0	0	1	1	0	1	1	1	0	0	0	1	1.06	.04
Hedera	0	1	0	0	0	0	0	1	1	1	1	0	− .80	.22
Helianthemum	0	0	1	1	0	1	1	1	0	0	0	0	1.06	.04
Ilex	1	1	0	0	1	0	0	1	1	1	1	1	− 1.30	− .15
Jasminum	0	0	0	1	0	1	1	1	1	1	1	1	.48	− .03
Juniperus	1	1	0	0	1	0	0	0	1	1	1	1	− 1.55	− .03
Laurus	0	1	0	1	1	0	0	1	1	1	1	0	− .79	.30
Lavandula	0	0	1	1	0	1	1	1	0	0	0	0	1.06	.04
Lonicera	0	0	1	1	0	1	1	1	1	1	1	1	.61	− .41
Myrtus	0	1	0	1	0	1	1	1	1	1	1	1	.23	.24
Nerium	0	1	0	1	0	1	1	1	1	0	1	0	.47	.47
Olea	0	1	0	0	0	0	0	0	1	1	1	1	− 1.12	.45
Osyris	0	1	0	0	1	0	0	1	1	1	0	1	− 1.13	.11
Phillyrea	0	1	0	0	1	0	0	0	1	1	1	0	− 1.39	.23
Phlomis	0	0	0	1	0	1	1	1	1	0	0	0	.70	.20
Pinus	0	1	0	0	1	0	0	0	1	0	1	0	− 1.24	.58
Pistacia	0	1	0	0	1	0	0	0	1	1	1	1	− 1.39	.23
Prunus	0	0	1	1	0	1	1	1	1	1	1	0	.61	− .41
Pyrus	0	0	1	1	0	1	1	1	1	1	1	0	.61	− .41
Quercus	1	1	0	0	1	0	0	0	1	0	1	0	− 1.49	.49
Retama	0	1	0	1	0	1	1	1	1	0	0	0	.47	.47
Rhamnus	1	1	0	0	1	0	0	1	1	1	1	1	− 1.30	− .15
Rhododendron	0	1	0	1	0	1	1	1	1	0	1	1	.47	.47

(continued)

TABLE A1 (*Continued*)

Genus	SP	LT	HA	FS	FSE	PC	PR	PT	SS	SD	PLI	DIS	DIM1	DIM2
Rhus	0	1	1	0	1	0	0	0	1	1	1	1	−1.43	−.34
Rosa	1	0	1	1	0	1	1	1	1	1	1	0	.61	−.67
Rosmarinus	0	0	0	1	0	1	1	1	0	0	0	0	.91	.41
Rubia	0	1	0	0	0	0	0	1	1	1	0	0	−.80	.22
Rubus	1	0	0	1	0	1	1	1	1	1	1	0	.39	−.35
Ruscus	1	1	0	0	1	0	1	1	1	1	1	0	−.99	−.16
Satureja	0	1	0	1	0	1	1	1	0	0	0	0	.68	.67
Securinega	1	1	1	0	1	0	0	1	0	0	0	0	−1.27	−.81
Smilax	1	1	0	0	1	0	1	1	1	1	1	1	−.99	−.16
Sorbus	0	0	1	1	0	1	1	1	1	1	1	0	.61	−.41
Stauracanthus	1	0	0	1	0	1	1	1	1	0	0	0	.67	.09
Teucrium	0	0	0	1	0	1	1	1	0	0	1	0	.91	.41
Thymelaea	0	1	0	1	1	0	1	1	0	0	0	0	−.01	.97
Thymus	0	0	1	1	0	1	1	1	0	0	0	0	1.06	.04
Ulex	1	0	1	1	0	1	1	1	1	0	0	0	.88	−.38
Ulmus	0	0	1	0	0	0	0	0	1	0	1	1	−.46	−.99
Viburnum	0	1	0	1	0	1	1	1	1	1	1	0	.23	.24
Vitis	0	0	1	0	1	0	0	1	1	1	1	1	−.80	−.73

NOTE.—Symbols for characters, and codes for character states, are as follows: SP, spinescence (1, spiny; 0, nonspiny); LT, leaf type (1, sclerophyllous; 0, nonsclerophyllous); HA, habit (1, winter or summer-facultative deciduous; 0, evergreen); FS, flower size (1, perianth depth × width < 25 mm^2; 0, perianth length × width > 25 mm^2); FSE, flower sexuality (1, unisexual flowers; 0, hermaphroditic flowers); PC, perianth color (1, colored; 0, brownish or greenish); PR, perianth reduction (1, perianth complete; 0, at least one verticil absent or much reduced); PT, pollinator type (1, insect pollinated; 0, wind pollinated); SS, seed size (1, length × width > 2.25 mm^2; 0, length × width < 2.25 mm^2); SD, seed dispersal mode (1, endozoochorous; 0, nonendozoochorous); PLI, occurrence of Pliocene or pre-Pliocene fossil records (1, occurrence; 0, absence); DIS, intercontinental range disjunction (1, occurrence; 0, absence). DIM1 and DIM2, coordinates on the first and second dimensions generated by nonmetric multidimensional scaling of the similarity matrix of life-history and reproductive characters.

APPENDIX B

TABLE B1

Woody Plant Genera That Were Present in
Southwestern Europe in the Pliocene or Mid- to Late
Miocene and Subsequently Became Extinct There,
but Are Still Surviving Elsewhere

Genus	FSE	PR	PT	SD
Acanthopanax	1	1	1	1
Ailanthus	1	0	1	0
Aralia	1	1	1	1
Bumelia	0	1	1	1
Caesalpinia	0	1	1	0
Calycanthus	0	1	1	1
Carya	1	0	0	0
Cassia	0	1	1	0
Celastrus	0	1	1	1
Chrysophyllum	0	1	1	1
Cinnamomum	0	0	1	1
Clethra	0	1	1	0
Dalbergia	0	1	1	0
Diospyros	1	1	1	1
Empetrum	1	0	0	1
Engelhardtia	1	0	0	0
Euclea	1	1	1	1
Eucommia	1	0	0	0
Ginkgo	1	0	0	1
Glyptostrobus	1	0	0	0
Lindera	1	0	1	1
Liquidambar	1	0	0	0
Liriodendron	0	1	1	0
Magnolia	0	1	1	1
Myrica	1	0	0	1
Myrsine	1	0	1	1
Nyssa	1	0	1	1
Ocotea	1	1	1	1
Persea	0	0	1	1
Phoebe	0	1	1	1
Pisonia	1	1	1	1
Platanus	1	0	0	0
Podocarpus	1	0	0	1
Pterocarya	1	0	0	0
Robinia	0	1	1	0
Sabal	1	0	1	1
Sapindus	1	1	1	1
Sassafras	1	0	1	1
Sequoia	1	0	0	0
Symplocos	0	1	1	1
Taxodium	1	0	0	0
Torreya	1	0	0	1
Tsuga	1	0	0	0
Weinmannia	0	1	1	0
Zanthoxylum	0	1	1	1

Note.—Characters and codes for character states are as
described in App. A. See text for literature sources used.

442   THE AMERICAN NATURALIST

LITERATURE CITED

Arroyo, J. 1988. Atributos florales y fenología de la floración en matorrales del sur de España. Lagascalia 15:43–78.

Axelrod, D. I. 1958. Evolution of the Madro-Tertiary geoflora. Botanical Review 24:433–509.

———. 1975. Evolution and biogeography of Madrean-Tethyan sclerophyll vegetation. Annals of the Missouri Botanical Garden 62:280–334.

———. 1983. Biogeography of oaks in the Arcto-Tertiary province. Annals of the Missouri Botanical Garden 70:629–657.

———. 1986. Analysis of some palaeogeographical and palaeoecologic problems of palaeobotany. Palaeobotanist (Lucknow) 35:115–129.

Bauzá Rullán, J. 1961. Nueva contribución al estudio de la flora fósil de Mallorca. Boletín de la Sociedad de Historia Natural de Baleares 7:49–58.

Bawa, K. S. 1980. Evolution of dioecy in flowering plants. Annual Review of Ecology and Systematics 11:15–39.

———. 1982a. Seed dispersal and the evolution of dioecism in flowering plants—a response to Herrera. Evolution 36:1322–1325.

———. 1982b. Outcrossing and the incidence of dioecism in island floras. American Naturalist 119:866–871.

Bell, G. 1989. A comparative method. American Naturalist 133:553–571.

Bosch, J. 1986. Insectos florícolas y polinización en un matorral de romero. Tesis de licenciatura. University of Barcelona, Barcelona.

Braun-Blanquet, J. 1937. Sur l'origine des eléments de la flore méditerranéenne. Communications de la Station Internationale de Géobotanique Méditerranéenne et Alpine, Montpellier 56:8–31.

Brown, J. H., and B. A. Maurer. 1987. Evolution of species assemblages: effects of energetic constraints and species dynamics on the diversification of the North American avifauna. American Naturalist 130:1–17.

Bullock, S. H. 1985. Breeding systems in the flora of a tropical deciduous forest in Mexico. Biotropica 17:287–301.

Burt, A. 1989. Comparative methods using phylogenetically independent contrasts. Oxford Surveys in Evolutionary Biology 6:33–53.

Capel Molina, J. J. 1981. Los climas de España. Oikos-Tau, Barcelona.

Catalán Bachiller, G. 1978. Semillas de árboles y arbustos forestales. Monografías ICONA 17.

Cheverud, J. M., M. M. Dow, and W. Leutenegger. 1985. The quantitative assessment of phylogenetic constraints in comparative analyses: sexual dimorphism in body weight among primates. Evolution 39:1335–1351.

Conn, J. S., T. R. Wentworth, and U. Blum. 1980. Patterns of dioecism in the flora of the Carolinas. American Midland Naturalist 103:310–315.

Cronquist, A. 1981. An integrated system of classification of flowering plants. Columbia University Press, New York.

Davis, M. B. 1976. Pleistocene biogeography of temperate deciduous forest. Geoscience and Man 13:13–26.

Depape, G. 1922. Recherches sur la flore Pliocène de la Vallée du Rhône: flores de Saint-Marcel (Ardéche) et des environs de Théziers (Gard.). Annales des Sciences Naturelles, Botanique, Série 10 4:73–265.

Diniz, F. 1967. Une flore Tertiaire de caractère Méditerranéen au Portugal. Review of Palaeobotany and Palynology 5:263–268.

Donoghue, M. J. 1989. Phylogenies and the analysis of evolutionary sequences, with examples from seed plants. Evolution 43:1137–1156.

du Toit, S. H. C., A. G. W. Steyn, and R. H. Stumpf. 1986. Graphical exploratory data analysis. Springer, New York.

Eldredge, N. 1985. Unfinished synthesis. Oxford University Press, New York.

Felsenstein, J. 1985. Phylogenies and the comparative method. American Naturalist 125:1–15.

Fernández Marrón, M. T. 1971. Estudio paleoecológico y revisión sistemática de la flora fósil del

Oligoceno español. Publicaciones de la Facultad de Ciencias, Universidad Complutense, Madrid A 152:1–177.

Flores, S., and D. W. Schemske. 1984. Dioecy and monoecy in the flora of Puerto Rico and the Virgin Islands: ecological correlates. Biotropica 16:132–139.

Fowler, C. W., and J. A. MacMahon. 1982. Selective extinction and speciation: their influence on the structure and functioning of communities and ecosystems. American Naturalist 119:480–498.

Fox, J. F. 1985. Incidence of dioecy in relation to growth form, pollination and dispersal. Oecologia (Berlin) 67:244–249.

Freeman, D. C., K. T. Harper, and W. K. Ostler. 1980. Ecology of plant dioecy in the intermountain region of western North America and California. Oecologia (Berlin) 44:410–417.

Friis, E. M. 1985. Angiosperm fruits and seeds from the middle Miocene of Jutland (Denmark). Kongelige Danske Videnskaberne Selskab Biologiske Skrifter 24:1–165.

Givnish, T. J. 1980. Ecological constraints on the evolution of breeding systems in seed plants: dioecy and dispersal in gymnosperms. Evolution 34:959–972.

———. 1982. Outcrossing versus ecological constraints in the evolution of dioecy. American Naturalist 119:849–865.

Goldblatt, P. 1980. Systematics of *Gynandriris* (Iridaceae), a Mediterranean–southern African disjunct. Botaniska Notiser 133:239–260.

Gorchov, D. L. 1990. Pattern, adaptation, and constraint in fruiting synchrony within vertebrate-dispersed woody plants. Oikos 58:169–180.

Gould, S. J. 1985. The paradox of the first tier: an agenda for paleobiology. Paleobiology 11:2–12.

———. 1989. A developmental constraint in *Cerion*, with comments on the definition and interpretation of constraint in evolution. Evolution 43:516–539.

Gould, S. J., and N. Eldredge. 1986. Punctuated equilibrium at the third stage. Systematic Zoology 35:143–148.

Gould, S. J., and R. C. Lewontin. 1979. The spandrels of San Marco and the Panglossian paradigm: a critique of the adaptationist programme. Proceedings of the Royal Society of London B, Biological Sciences 205:581–598.

Gould, S. J., and D. S. Woodruff. 1990. History as a cause of area effects: an illustration from *Cerion* on Great Inagua, Bahamas. Biological Journal of the Linnean Society 40:67–98.

Grafen, A. 1989. The phylogenetic regression. Philosophical Transactions of the Royal Society of London B, Biological Sciences 326:119–157.

Harper, J. L. 1967. A Darwinian approach to plant ecology. Journal of Ecology 55:247–270.

Herrera, C. M. 1982a. Breeding systems and dispersal-related maternal reproductive effort of southern Spanish bird-dispersed plants. Evolution 36:1299–1314.

———. 1982b. Reply to Bawa. Evolution 36:1325–1326.

———. 1984a. A study of avian frugivores, bird-dispersed plants, and their interaction in Mediterranean scrublands. Ecological Monographs 54:1–23.

———. 1984b. The annual cycle of *Osyris quadripartita*, a hemiparasitic dioecious shrub of Mediterranean scrublands. Journal of Ecology 72:1065–1078.

———. 1985. Tipos morfológicos y funcionales en plantas del matorral mediterráneo del sur de España. Studia Oecologica 3:7–33.

———. 1986. Vertebrate-dispersed plants: why they don't behave the way they should. Pages 5–18 *in* A. Estrada and T. H. Fleming, eds. Frugivores and seed dispersal. Junk, Dordrecht.

———. 1987. Vertebrate-dispersed plants of the Iberian Peninsula: a study of fruit characteristics. Ecological Monographs 57:305–331.

———. 1988. The fruiting ecology of *Osyris quadripartita:* individual variation and evolutionary potential. Ecology 69:233–249.

———. 1989. Frugivory and seed dispersal by carnivorous mammals, and associated fruit characteristics, in undisturbed Mediterranean habitats. Oikos 55:250–262.

———. 1992. Interspecific variation in fruit shape: allometry, phylogeny, and adaptation to dispersal agents. Ecology (in press).

Herrera, J. 1982. Introducción al estudio de la biología floral del matorral andaluz. Tesis de licenciatura. University of Sevilla, Sevilla.

444   THE AMERICAN NATURALIST

———. 1985. Biología reproductiva del matorral de Doñana. Tesis doctoral. University of Sevilla, Sevilla.
———. 1987. Flower and fruit biology in southern Spanish Mediterranean shrublands. Annals of the Missouri Botanical Garden 74:69–78.
———. 1988. Pollination relationships in southern Spanish Mediterranean shrublands. Journal of Ecology 76:274–287.
Heywood, V. H., ed. 1978. Flowering plants of the world. Oxford University Press, Oxford.
Hodgson, J. G., and J. L. M. MacKey. 1986. The ecological specialization of dicotyledonous families within a local flora: some factors constraining optimization of seed size and their possible evolutionary significance. New Phytologist 104:497–515.
Hoffman, A. 1984. Species selection. Evolutionary Biology 18:1–20.
———. 1989. Arguments on evolution. Oxford University Press, New York.
Huey, R. B. 1987. Phylogeny, history and the comparative method. Pages 76–97 in M. E. Feder, A. F. Bennett, W. W. Burggren, and R. B. Huey, eds. New directions in ecological physiology. Cambridge University Press, Cambridge.
Huey, R. B., and A. F. Bennett. 1987. Phylogenetic studies of coadaptation: preferred temperatures versus optimal performance temperatures of lizards. Evolution 41:1098–1115.
Hull, D. L. 1980. Individuality and selection. Annual Review of Ecology and Systematics 11:311–332.
———. 1988. Science as a process. University of Chicago Press, Chicago.
Hutchinson, J. 1964. The genera of flowering plants: Dicotyledones. Vol. 1. Clarendon Press, Oxford.
———. 1967. The genera of flowering plants: Dicotyledones. Vol. 2. Clarendon Press, Oxford.
Jordano, P. 1984. Relaciones entre plantas y aves frugívoras en el matorral mediterráneo del área de Doñana. Tesis doctoral. University of Sevilla, Sevilla.
———. 1987. Avian fruit removal: effects of fruit variation, crop size, and insect damage. Ecology 68:1711–1723.
———. 1988. Polinización y variabilidad de la producción de semillas en Pistacia lentiscus L. (Anacardiaceae). Anales del Jardín Botánico de Madrid 45:213–231.
———. 1989. Pre-dispersal biology of Pistacia lentiscus (Anacardiaceae): cumulative effects on seed removal by birds. Oikos 55:375–386.
Kochmer, J. P., and S. N. Handel. 1986. Constraints and competition in the evolution of flowering phenology. Ecological Monographs 56:303–325.
Kramer, K. U., and P. S. Green, eds. 1990. The families and genera of vascular plants. Vol. 1. Pteridophytes and gymnosperms. Springer, Berlin.
Kruskal, J. B. 1964. Multidimensional scaling by optimizing goodness of fit to a nonmetric hypothesis. Psychometrika 29:1–27.
Lack, D. 1965. Evolutionary ecology. Journal of Ecology 53:237–245.
Lechowicz, M. J. 1984. Why do temperate deciduous trees leaf out at different times? adaptation and ecology of forest communities. American Naturalist 124:821–842.
Lessios, H. A. 1990. Adaptation and phylogeny as determinants of egg size in echinoderms from the two sides of the Isthmus of Panama. American Naturalist 135:1–13.
Liston, A., L. H. Rieseberg, and T. S. Elias. 1989a. Genetic similarity is high between intercontinental disjunct species of Senecio. American Journal of Botany 76:383–388.
———. 1989b. Morphological stasis and molecular divergence in the intercontinental disjunct genus Datisca (Datiscaceae). Aliso 12:525–542.
Mabberley, D. J. 1987. The plant-book. Cambridge University Press, Cambridge.
Mai, D. H. 1989. Development and regional differentiation of the European vegetation during the Tertiary. Plant Systematics and Evolution 162:79–91.
Mazer, S. J. 1989. Ecological, taxonomic, and life history correlates of seed mass among Indiana dune angiosperms. Ecological Monographs 59:153–175.
McComb, J. A. 1966. The sex forms of species in the flora of the south-west of western Australia. Australian Journal of Botany 14:303–316.
Medus, J., and A. Pons. 1980. Les prédécesseurs des végétaux méditerranéens actuels jusqu'au début du Miocène. Naturalia Monspelliensia, Hors Série, pp. 11–20.
Meusel, H. 1971. Mediterranean elements in the flora and vegetation of the west Himalayas. Pages

53–72 *in* P. H. Davis, P. C. Harper, and I. C. Hedge, eds. Plant life of south-west Asia. Botanical Society of Edinburgh, Edinburgh.

Muenchow, G. E. 1987. Is dioecy associated with fleshy fruit? American Journal of Botany 74: 287–293.

Niklas, K. J., D. E. Giannasi, and N. K. Baghai. 1985. Paleobiochemistry of a North American fossil *Liriodendron* sp. Biochemical Systematics and Ecology 13:1–4.

Obeso, J. R. 1986. Comunidades de passeriformes y frugivorismo en altitudes medias de la Sierra de Cazorla. Tesis doctoral. University of Oviedo, Oviedo.

Opler, P. A. 1978. Interactions of plant life history components as related to arboreal herbivory. Pages 23–31 *in* G. G. Montgomery, ed. The ecology of arboreal folivores. Smithsonian Institution Press, Washington, D.C.

Orians, G. H. 1962. Natural selection and ecological theory. American Naturalist 96:257–263.

Pagel, M. D., and P. H. Harvey. 1988. Recent developments in the analysis of comparative data. Quarterly Review of Biology 63:413–440.

Palamarev, E. 1989. Paleobotanical evidences of the Tertiary history and origin of the Mediterranean sclerophyll dendroflora. Plant Systematics and Evolution 162:93–107.

Parks, C. R., and J. F. Wendel. 1990. Molecular divergence between Asian and North American species of *Liriodendron* (Magnoliaceae) with implications for interpretation of fossil floras. American Journal of Botany 77:1243–1256.

Pearson, D. L. 1982. Historical factors and bird species richness. Pages 441–452 *in* G. T. Prance, ed. Biological diversification in the tropics. Columbia University Press, New York.

Pearson, D. L., M. S. Blum, T. H. Jones, H. M. Fales, E. Gonda, and B. R. Witte. 1988. Historical perspective and the interpretation of ecological patterns: defensive compounds of tiger beetles (Coleoptera: Cicindelidae). American Naturalist 132:404–416.

Pignatti, S. 1978. Evolutionary trends in Mediterranean flora and vegetation. Vegetatio 37:175–185.

Pons, A. 1964. Contribution palynologique à l'étude de la flore et de la végétation pliocène de la région rhodanienne. Annales des Sciences Naturelles, Botanique, Série 12 5:499–722.

———. 1981. The history of the Mediterranean shrublands. Pages 131–138 *in* F. di Castri, D. W. Goodall, and R. L. Specht, eds. Ecosystems of the world. Vol. 11. Mediterranean-type shrublands. Elsevier, Amsterdam.

Quézel, P., J. Gamisans, and M. Gruber. 1980. Biogéographie et mise en place des flores méditerranéennes. Naturalia Monspeliensia, Hors Série, pp. 41–51.

Raven, P. H. 1971. The relationships between "Mediterranean" floras. Pages 119–134 *in* P. H. Davis, P. C. Harper, and I. C. Hedge, eds. Plant life of south-west Asia. Botanical Society of Edinburgh, Edinburgh.

———. 1974. The evolution of Mediterranean floras. Pages 213–224 *in* F. di Castri and H. A. Mooney, eds. Mediterranean type ecosystems: origin and structure. Springer, Berlin.

Raven, P. H., and D. I. Axelrod. 1974. Angiosperm biogeography and past continental movements. Annals of the Missouri Botanical Garden 61:539–673.

Rice, W. R. 1989. Analyzing tables of statistical tests. Evolution 43:223–225.

Ruiz de la Torre, J. 1971. Arboles y arbustos de la España peninsular. Escuela Técnica Superior de Ingenieros de Montes, Madrid.

SAS Institute. 1988. SAS/STAT User's Guide. Release 6.03 ed. SAS Institute, Cary, N.C.

Schuster, R. M. 1976. Plate tectonics and its bearing on the geographical origin and dispersal of angiosperms. Pages 48–138 *in* C. B. Beck, ed. Origin and early evolution of angiosperms. Columbia University Press, New York.

Sobrevila, C., and M. T. Kalin Arroyo. 1982. Breeding systems in a montane tropical cloud forest in Venezuela. Plant Systematics and Evolution 140:19–37.

Stanley, S. M. 1975. A theory of evolution above the species level. Proceedings of the National Academy of Sciences of the USA 72:646–650.

———. 1979. Macroevolution, pattern and process. W. H. Freeman, San Francisco.

Stebbins, G. L. 1951. Natural selection and the differentiation of angiosperm families. Evolution 5:299–324.

———. 1982. Perspectives in evolutionary theory. Evolution 36:1109–1118.

446   THE AMERICAN NATURALIST

Stebbins, G. L., and A. Day. 1967. Cytogenetic evidence for long continued stability in the genus *Plantago*. Evolution 21:409–428.

Stebbins, G. L., and J. Major. 1965. Endemism and speciation in the California flora. Ecological Monographs 35:1–35.

Steiner, K. E. 1988. Dioecism and its correlates in the Cape flora of South Africa. American Journal of Botany 75:1742–1754.

Stratton, D. A. 1989. Longevity of individual flowers in a Costa Rican cloud forest: ecological correlates and phylogenetic constraints. Biotropica 21:308–318.

Suc, J. P. 1984. Origin and evolution of the Mediterranean vegetation and climate in Europe. Nature (London) 307:409–432.

Sunding, P. 1979. Origins of the Macaronesian flora. Pages 13–40 *in* D. Bramwell, ed. Plants and islands. Academic Press, London.

Thomson, J. D., and S. C. H. Barrett. 1981. Selection for outcrossing, sexual selection, and the evolution of dioecy in plants. American Naturalist 118:443–449.

Thorne, R. F. 1972. Major disjunctions in the geographical ranges of seed plants. Quarterly Review of Biology 47:365–411.

Tiffney, B. H. 1985. Perspectives on the origin of the floristic similarity between eastern Asia and eastern North America. Journal of the Arnold Arboretum, Harvard University 66:73–94.

Valdés, B., S. Talavera, and E. Fernández-Galiano. 1987. Flora vascular de Andalucía Occidental. Vols. 1–3. Ketres, Barcelona.

van der Burgh, J. 1983. Allochthonous seed and fruit floras from the Pliocene of the lower Rhine basin. Review of Palaeobotany and Palynology 40:33–90.

Vicente i Castells, J. 1988. Nova aportació al coneixement de la flora fòssil de Montjuïc (Barcelona). Societat d'Història Natural, Santa Coloma de Gramenet, Barcelona, Serie Monogràfica 1:5–93.

Vrba, E. S. 1984. What is species selection? Systematic Zoology 33:318–328.

Vrba, E. S., and S. J. Gould. 1986. The hierarchical expansion of sorting and selection: sorting and selection cannot be equated. Paleobiology 12:217–228.

Wanntorp, H. E. 1983. Historical constraints in adaptation theory: traits and non-traits. Oikos 41:157–160.

Westoby, M. 1988. Comparing Australian ecosystems to those elsewhere. BioScience 38:549–556.

Wiklund, C., and B. Karlsson. 1984. Egg size variation in satyrid butterflies: adaptive vs historical, 'Bauplan', and mechanistic explanations. Oikos 43:391–400.

Wilkinson, L. 1986. SYSTAT: the system for statistics. SYSTAT, Evanston, Ill.

Wilson, E. O. 1975. Sociobiology. Harvard University Press, Cambridge, Mass.

Zar, J. H. 1984. Biostatistical analysis. 2d ed. Prentice-Hall, Englewood Cliffs, N.J.

Associate Editor: F. Stuart Chapin III

Evolution, 37(6), 1983, pp. 1210–1226

THE MEASUREMENT OF SELECTION ON CORRELATED CHARACTERS

Russell Lande[1] and Stevan J. Arnold[2]

[1] *Department of Biophysics and Theoretical Biology and* [2] *Department of Biology,*
The University of Chicago, Chicago, Illinois 60637

Received June 25, 1982. Revised January 31, 1983

Natural selection acts on phenotypes, regardless of their genetic basis, and produces immediate phenotypic effects within a generation that can be measured without recourse to principles of heredity or evolution. In contrast, evolutionary response to selection, the genetic change that occurs from one generation to the next, does depend on genetic variation. Animal and plant breeders routinely distinguish phenotypic selection from evolutionary response to selection (Mayo, 1980; Falconer, 1981). Upon making this critical distinction, emphasized by Haldane (1954), precise methods can be formulated for the measurement of phenotypic natural selection.

Correlations between characters seriously complicate the measurement of phenotypic selection, because selection on a particular trait produces not only a direct effect on the distribution of that trait in a population, but also produces indirect effects on the distribution of correlated characters. The problem of character correlations has been largely ignored in current methods for measuring natural selection on quantitative traits. Selection has usually been treated as if it acted only on single characters (e.g., Haldane, 1954; Van Valen, 1965a; O'Donald, 1968, 1970; reviewed by Johnson, 1976 Ch. 7). This is obviously a tremendous oversimplification, since natural selection acts on many characters simultaneously and phenotypic correlations between traits are ubiquitous. In an important but neglected paper, Pearson (1903) showed that multivariate statistics could be used to disentangle the direct and indirect effects of selection to determine which traits in a correlated ensemble are the focus of direct selection. Here we extend and generalize Pearson's major results.

The purpose of this paper is to derive measures of directional and stabilizing (or disruptive) selection on each of a set of phenotypically correlated characters. The analysis is retrospective, based on observed changes in the multivariate distribution of characters within a generation, not on the evolutionary response to selection. Nevertheless, the measures we propose have a close connection with equations for evolutionary change. Many other commonly used measures of the intensity of selection (such as selective mortality, change in mean fitness, variance in fitness, or estimates of particular forms of fitness functions) have little predictive value in relation to evolutionary change in quantitative traits. To demonstrate the utility of our approach, we analyze selection on four morphological characters in a population of pentatomid bugs during a brief period of high mortality. We also summarize a multivariate selection analysis on nine morphological characters of house sparrows caught in a severe winter storm, using the classic data of Bumpus (1899).

Direct observations and measurements of natural selection serve to clarify one of the major factors of evolution. Critiques of the "adaptationist program" (Lewontin, 1978; Gould and Lewontin, 1979) stress that adaptation and selection are often invoked without strong supporting evidence. We suggest quantitative measurements of selection as the best alternative to the fabrication of adaptive scenarios. Our optimism that measurement can replace rhetorical claims for adaptation and selection is founded in the growing success of field workers in their efforts to measure major components of fitness in natural populations (e.g., Thornhill, 1976; Howard, 1979; Downhower and Brown, 1980; Boag and Grant, 1981; Clutton-Brock et al., 1982). The essential fact is that selection and adaptation *can* be measured.

Empirical Background

Before considering the theory, it will be useful to briefly review the kinds of data that can be analyzed with our methods. The basic data consist of a set of phenotypic measurements on individuals in a population before and after an episode of selection. In *longitudinal* studies individuals are observed (usually through a span of time) to score some component(s) of fitness, as well as a set of phenotypic characters, for each individual. *Cross-sectional* studies are performed on a population at a single point in time by comparing the phenotype distribution of cohorts of different age, to infer viability selection without actually measuring any aspect of fitness. Three features of the data are crucial to the interpretation of the results: which components of fitness are considered; whether the study is longitudinal or cross-sectional; and whether there is ontogenetic change in the characters.

Many workers have succeeded in measuring components of fitness in natural populations. Although the measurement of total lifetime fitness is a worthy goal, and one that can be achieved with some organisms such as annual plants, accurate estimates of major components of fitness also give valuable insights. The present methods for measuring selection can be applied to components of fitness as well as total lifetime fitness. Charlesworth (1980) and Lande (1982) treat the problem of measuring total fitness in populations with overlapping generations.

Longitudinal data have many advantages over cross-sectional data. Typically, cross-sectional studies are carried out by comparing the phenotype distributions of young and old individuals in a single population and assuming these to represent, respectively, samples before and after an interval of viability selection; any statistically significant change in the phenotype distribution is attributed to selection. The assumptions of cross-sectional studies are valid only if during the interval between the production of the young and old cohorts (1) there was no genetic evolution of the characters in the population, (2) the environment did not change in any way which would affect individual development of the characters, (3) differential immigration or emigration by different phenotypes did not occur, and (4) there was no ontogenetic change in the characters in the stages of the life cycle under study. Thus, cross-sectional studies are commonly applied to characters that do not change with age, such as meristic characters (Inger, 1942, 1943; Hecht, 1952; Hagen and Gilbertson, 1973; Fox, 1975; Beatson, 1976) and those with determinate growth (Kurtén, 1958; Van Valen, 1963, 1965*b*; Mason, 1964; Van Valen and Weiss, 1966). Organisms with accretionary growth can be used in cross-sectional studies to measure selection on characters of early ontogenetic stages (Weldon, 1901; di Cesnola, 1906; Sambol and Finks, 1977). Traits that change continuously during ontogeny can be analyzed with longitudinal data by using growth curve parameters of individuals as characters, instead of the original measurements (Cock, 1966; Kidwell et al., 1979; Atchley and Rutledge, 1981).

Directional Selection

Theoretical Results.—The directional aspect of selection is defined by its effect on the mean values of phenotypic characters in a population within a generation. In this section it is demonstrated that the forces of directional selection are best described by a vector of partial regression coefficients of individual relative fitness on the characters. This is a multivariate generalization of the results of Robertson (1966) and Price (1970, 1972), who showed that directional selection on a single trait can be expressed in terms of a covariance or regression of relative fitness on the character.

The change in the mean value of a phenotypic character produced by selection within a generation is known as the observed *directional selection differential* (Lush, 1945; Falconer, 1981). For a set of characters, z_1, z_2, . . . , z_n, arranged in a column vector, z, the directional selection differential is

$$s = \bar{z}^* - \bar{z} \qquad (1)$$

in which $\bar{z}^*$ denotes the vector of mean values of the characters after selection. Let the phenotype distribution in the population before selection be $p(z)$. The absolute fitness of an individual, W, can be represented by the expected fitness for its phenotype, $W(z)$, plus an individual deviation which among individuals with phenotype z has a mean of zero (so that in the population as a whole $W - W(z)$ is uncorrelated with any phenotypic measure). The mean absolute fitness in the population is

$$\bar{W} = \int W(z)p(z)dz \qquad (2a)$$

where the integration extends over all phenotypes in the population. Relative fitnesses are defined with respect to the mean fitness in the population, $w = W/\bar{W}$ so that the expected relative fitness of individuals with phenotype z is

$$w(z) = W(z)/\bar{W} \qquad (2b)$$

and $\bar{w} = 1$. After selection the phenotype distribution is $w(z)p(z)$ with mean value

$$\bar{z}^* = \int z w(z)p(z)dz. \qquad (3)$$

$\bar{z}^*$ is the average of the product of individual relative fitness with the vector of characters. Because the mean relative fitness is unity, the observed directional selection differential can be expressed from (1) and (3) as the covariance of individual relative fitness with the vector of characters,

$$s = \mathrm{Cov}[w, z]. \qquad (4)$$

Now consider the influence of selection on the genetic evolution of the mean phenotype in a population. In the absence of genotype-environment interaction and correlation, individual phenotypes before selection can be decomposed into additive genetic effects (or breeding values), x, plus independent environmental effects, e, which include nonadditive genetic effects due to dominance and epistasis,

$$z = x + e \qquad (5a)$$

with $\bar{e} = 0$ (Kempthorne, 1969; Falconer, 1981). Denoting the variance-covariance matrices of the breeding values and the environmental (plus nonadditive genetic) parts respectively as G and E, the phenotypic variance-covariance matrix is $P = G + E$, which is taken to be nonsingular, $|P| \neq 0$. The regression of breeding values on phenotypes is assumed to be linear and homoscedastic,

$$x - \bar{x} = GP^{-1}(z - \bar{z}) + \varepsilon \qquad (5b)$$

where ε is an independent residual term with zero mean. Linearity and homoscedasticity of both regressions (5a) and (5b) imply that the phenotypes and breeding values (as well as environmental effects) jointly have a multivariate normal distribution (Kendall and Stuart, 1973 Ch. 28).

An important result due to Pearson (1903) is that phenotypic selection does not change the coefficients in the regression (5b). The change in the mean breeding value produced by phenotypic selection within a generation is therefore

$$\bar{x}^* - \bar{x} = GP^{-1}(\bar{z}^* - \bar{z}). \qquad (6a)$$

If there is no change in the mean environmental effect between generations, and no change in the mean breeding value caused by recombination or mutation, then the mean breeding value of selected parents is equivalent to the mean phenotype of their offspring, and the change in the mean phenotype across one generation produced by selection is, using (1),

$$\Delta\bar{z} = GP^{-1}s \qquad (6b)$$

(Lande, 1979). Thus, with the assumption that the characters have a multivariate normal distribution, the vector $P^{-1}s$ contains all of the information concerning the forces of directional selection on the phenotypes. For quantitative (polygenic) characters, the assumption of normality is often satisfied, at least approximately, on an appropriate scale of measurement, most often logarithmic (Wright, 1968 Ch. 10; Falconer, 1981 Ch. 17).

In view of (4), the vector $\beta = P^{-1}s$ is a set of partial regression coefficients of relative fitness on the characters (Kendall and Stuart, 1973 eq. 27.42). Under quite general conditions, the method of least squares indicates that the element β_i gives the slope of the straight line that best describes the dependence of relative fitness on character z_i, after removing the residual effects of other characters on fitness (Kendall and Stuart, 1973 Ch. 27.15). There is no need to assume that the actual regressions of fitness on the characters are linear, or that the characters have a multivariate normal distribution. For this reason, the partial regression coefficients β provide a general solution to the problem of measuring the forces of directional selection acting directly on the characters.

We can now partition the observed selection differential into the parts alluded to by Pearson. Writing $s = P\beta$, the observed selection differential on the ith character contains a term due to direct selection on that character plus a set of terms due to indirect selection on all characters that are phenotypically correlated with the ith character,

$$s_i = \sum_{j=1}^{n} P_{ij}\beta_j \qquad (6c)$$
$$= P_{i1}\beta_1 + \ldots + P_{ii}\beta_i \ldots + P_{in}\beta_n$$

where P_{ii} is the phenotypic variance of the ith character and P_{ij} is the phenotypic covariance between the ith and jth characters. The term $P_{ii}\beta_i$ represents the *direct effect* of selection in changing the mean of the ith character. The remaining terms represent the *indirect effects* of selection on all other characters in changing the mean value of the ith trait.

There is an interesting interpretation of the partial regression coefficients when the phenotypic traits follow a multivariate normal distribution, although the actual regression of fitness on the characters may be nonlinear. The vector of partial regression coefficients, $\beta = P^{-1}s$, is then equivalent to the average gradient of the relative fitness surface, weighted by the phenotype distribution. To establish this, let

$$p(z) = \sqrt{(2\pi)^{-n} |P|^{-1}} e^{-\frac{1}{2}(z - \bar{z})^T P^{-1}(z - \bar{z})}$$
$$(7)$$

where the superscript T signifies matrix transposition, and define the gradient operator

$$\partial/\partial z \equiv (\partial/\partial z_1, \ldots, \partial/\partial z_n)^T. \qquad (8)$$

Using integration by parts shows that if $w(z)$ is differentiable then its average gradient, weighted by $p(z)$, is

$$\int p(z)\frac{\partial w(z)}{\partial z}\, dz = p(z)w(z)\Big|_{-\infty}^{+\infty}$$
$$- \int w(z)\frac{\partial p(z)}{\partial z}\, dz.$$

Assuming $w(z)$ is bounded, the first term on the right side vanishes, and using (1), (3) and (7), the second term is

$$\int w(z)P^{-1}(z - \bar{z})p(z)dz =$$
$$P^{-1}(\bar{z}^* - \bar{z}) = P^{-1}s. \qquad (9)$$

Thus the forces of directional selection on the characters are given by the average gradient of the surface of individual relative fitness, weighted by the phenotype distribution. We therefore refer to the vector of partial regression coefficients of individual relative fitness on the characters as the *directional selection gradient*. In the section on stabilizing selection we show how the surface of individual fitness can be approximated by estimating its average gradient and average curvature in the neighborhood of the multivariate mean phenotype.

Components of Fitness.—Selection on major components of fitness, such as lifespan, fecundity, mating success, and parental effort, can be measured by coefficients in the regression of total fitness on its components. Selection on morphological, behavioral or physiological traits can be analyzed with respect to components of fitness as well as total fitness. Major components of fitness are expected to contribute positively to total fitness, and hence to be always under directional selection to increase, given that all other components of fitness are held constant. In a regression of total fitness on its major components, the finding of negative directional selection (or stabilizing selection) on a major fitness component indicates that this component is negatively correlated with another component of fitness which is missing from the analysis.

Missing Characters.—The analysis of directional selection on a set of correlated characters within a generation will be complete only if the characters under observation are phenotypically uncorrelated with any other characters that are undergoing appreciable directional selection. An objection can always be raised that some important trait(s) has been omitted from the analysis. This hypothesis can be tested by including the new character(s) in another study. If this is not feasible, then it must be admitted that the characters under observation may be responding in part to phenotypically correlated characters that have not been measured. This criticism is not necessarily fatal to the investigation, since a partial resolution of the influence of phenotypic correlations between characters is better than none at all. It is crucial to realize that an observed directional selection gradient on a given trait can *not* be ascribed to indirect selection on a correlated trait that *has* been included in the analysis.

Highly Correlated Characters.—If a group of characters are highly correlated phenotypically, or if a large number of traits are measured, the phenotypic variance-covariance matrix may be singular, with a determinant that vanishes, $|P| = 0$, or nearly so. Then P cannot be inverted and the directional selection gradient cannot be calculated. In such cases a reduction of dimensionality should be accomplished by some method, such as by summing or averaging highly correlated characters, or by eliminating all but one trait in each highly correlated set, or by retaining only the first few principal components of phenotypic variation as characters. All of these techniques can produce a phenotypic matrix, P, that is nonsingular. Principal components have the advantage that they are uncorrelated. Consequently there can be no indirect effects of selection from one principal component

to the next and they can be analyzed separately or in any combination.

Selection on a Combination or Index of Traits.—In some situations it may be desirable to construct an index, or linear combination of characters, such that selection acting directly on that index alone would produce an observed directional selection differential. In other words, one might wish to know what combination of characters was in effect the focus of past selection. The appropriate linear combination is obtained by weighting each trait by the direct force of selection on it,

$$I = \beta^T z = s^T P^{-1} z = \beta_1 z_1 + \ldots + \beta_n z_n. \qquad (10)$$

When selection acts directly on the index I alone, the observed directional selection differential on the vector of characters z is proportional to its phenotypic covariance with I. Hence selection acting only on the index would produce an observed directional selection differential proportional to that on the original characters, since $Cov[z, I] = Cov[z, z^T]P^{-1}s = s$. This index has been employed by animal breeders as a retrospective measure of directional selection observed within a generation, and a similar index has been used to describe observed evolutionary changes between generations (Dickerson et al., 1954, 1974; Harvey and Bearden, 1962; Marcus, 1964, 1969; Magee, 1965; Yamada, 1977; Lande, 1979).

Stabilizing and Disruptive Selection

Historical Background.—Stimulated by Galton's classic work on Natural Inheritance (1889), Pearson (1896, 1903, 1911) investigated fundamental problems in the theory of selection and evolution of correlated characters (cf. Pearson, 1920). In these papers, and others by Yule (1897, 1907), the methods of partial regression and correlation were greatly advanced (Seal, 1967). Pearson failed to achieve a clear understanding of heredity and evolution because he never accepted a Mendelian basis for the inheritance of quantitative variability, which was later confirmed by the experiments of Johannsen, Nilsson-Ehle, East, Hayes and Castle, and theoretical analyses by Yule, Weinberg, Fisher and Wright (reviewed by Wright, 1968 Ch. 15; Provine, 1971). Nevertheless, Pearson derived powerful methods for analyzing the immediate impact of phenotypic selection within a generation. He was especially concerned to distinguish between directly selected characters and non-selected or indirectly selected traits which change only due to their correlations with directly selected characters.

Assuming a multivariate normal phenotype distribution for the characters in a population before selection, Pearson (1903) proved that if a trait has been modified only by indirect selection, then its partial regression coefficients on any set of other traits, provided this includes all the directly selected characters, remain unchanged by selection. In contrast, selection within a generation produces changes in the means, variances and covariances of characters that are not themselves directly selected, if they are correlated with the selected character(s). For example, consider an arbitrary form of selection on a single character, z_1, which changes the phenotypic variance from P_{11} before selection to P^*_{11} after selection, and let the proportional decrease in the variance be denoted by $\kappa = 1 - P^*_{11}/P_{11}$. Then after selection the variance of a second character, z_2, not under direct selection but having a correlation ρ_{12} with the first character, is

$$P^*_{22} = P_{22}(1 - \kappa\rho^2_{12}). \quad (11)$$

This demonstrates that selection which decreases the variance of the first character, $\kappa > 0$, also decreases the variance of the second character, whether the correlation between them is positive or negative. The covariance of the characters after selection on the first trait is $P^*_{12} = P_{12}(1 - \kappa)$. Equation (11), first derived by Pearson (1903), suggests that for a set of correlated characters under multivariate stabilizing selection, observed changes in the variance of a single character may overestimate the intensity of stabilizing selection acting directly on that trait, be-

cause of the indirect effects of stabilizing selection acting through correlated traits. This is shown more explicitly below.

Theoretical Results.—Combined with the modern view of polygenic inheritance of quantitative variation, Pearson's methods can be applied to the problem of predicting how phenotypic selection changes additive genetic variances and covariances within a generation. The formulation of this genetic change should suggest measures of stabilizing (or disruptive) selection. Consider an arbitrary form of selection acting on a multivariate normal phenotype distribution. Following Pearson (1903) and Lande (1980), equation (5b) can be used to derive the change in the variance-covariance matrix of breeding values within a generation (before mutation, segregation and recombination), as

$$G^* - G = GP^{-1}(P^* - P)P^{-1}G. \quad (12)$$

Analogy with (6a) indicates that the phenotypic part of the expression on the right, $P^{-1}(P^* - P)P^{-1}$, is a matrix measuring the forces of selection acting directly on the variances and covariances of the characters.

It is informative to express the observed change in the phenotypic variance-covariance matrix produced by selection within a generation in the form of a covariance of fitness with a function of the characters,

$$P^* - P = \int (z - \bar{z}^*)(z - \bar{z}^*)^T w(z)p(z)dz$$
$$- \int (z - \bar{z})(z - \bar{z})^T p(z)dz.$$

Substituting $\bar{z}^* = s + \bar{z}$ in the first integral, and using (4), yields

$$P^* - P = \text{Cov}[w, (z - \bar{z})$$
$$(z - \bar{z})^T] - ss^T \quad (13a)$$

which is valid for any distribution of characters under an arbitrary form of selection. The last term, $-ss^T$, gives the change in P that would be produced by a linear fitness function with the observed directional selection differential s. The covariance of relative fitness with a matrix of quadratic deviations of the characters from their mean values can therefore be taken as an observed *stabilizing selection differential*, C, which is independent of the influence of directional selection,

$$C \equiv \text{Cov}[w, (z - \bar{z})(z - \bar{z})^T]$$
$$= P^* - P + ss^T. \qquad (13b)$$

In conjunction with (12) this implies that, for characters with a multivariate normal distribution before selection, the matrix

$$\gamma \equiv P^{-1}(P^* - P + ss^T)P^{-1}$$
$$= P^{-1}CP^{-1} \qquad (14a)$$

is the *stabilizing selection gradient* which measures the forces of stabilizing (or disruptive) selection acting directly on the characters, independent of the forces of directional selection, and accounting for the phenotypic correlations between characters. The diagonal elements measure the strength of stabilizing (or disruptive) selection directly on the variance of each character, and the off-diagonal elements measure the strength of selection directly on the covariances (i.e., the intensity of selection to positively or negatively correlate pairs of characters), after allowing for the changes produced by directional selection.

Formula (14a) can also be interpreted in terms of the shape of the relative fitness surface acting on the characters. Defining the curvature operator as the matrix of second partial derivatives

$$\partial^2/\partial z^2 \equiv \begin{pmatrix} \partial^2/\partial z_1{}^2 & \cdots & \partial^2/\partial z_1 \partial z_n \\ \cdot & \cdot & \cdot \\ \cdot & & \cdot \\ \cdot & & \cdot \\ \partial^2/\partial z_n \partial z_1 & \cdots & \partial^2/\partial z_n{}^2 \end{pmatrix}$$

and using integration by parts, it can be shown from (7) that if $w(z)$ is twice differentiable, then its average curvature is

$$\int p(z) \frac{\partial^2 w(z)}{\partial z^2}\, dz = \int w(z) \frac{\partial^2 p(z)}{\partial z^2}\, dz$$
$$= \int w(z)P^{-1}(z - \bar{z})(z - \bar{z})^T P^{-1} p(z)\, dz$$
$$- P^{-1}.$$

Writing $P^{-1} = P^{-1}PP^{-1}$ and employing (13b) yields

$$\int p(z) \frac{\partial^2 w(z)}{\partial z^2}\, dz = P^{-1}CP^{-1} = \gamma. \qquad (14b)$$

Thus the stabilizing selection gradient is equivalent to the average curvature of the relative fitness surface, weighted by the multivariate normal distribution of characters.

The intensity of stabilizing or disruptive selection has often been measured using the observed effect of selection on the variance of single characters. To show how the observed changes in the phenotypic variances and covariances depend on the shape of the fitness surface acting on a multivariate normal character distribution, formula (11) can be generalized by rearranging (14a) as

$$P^* - P = P\gamma P - ss^T \qquad (15a)$$

or in terms of the ijth element

$$P^*{}_{ij} - P_{ij} = \sum_{k=1}^{n} \sum_{l=1}^{n} P_{ik}\gamma_{kl}P_{lj} - s_i s_j. \qquad (15b)$$

In the case where selection acts independently to stabilize the mean of each character toward an intermediate optimum, γ is a diagonal matrix with $\gamma_{jj} < 0$, and the observed change in the variance of the ith character is

$$P^*{}_{ii} - P_{ii} = \sum_{j=1}^{n} \gamma_{jj}P^2{}_{ij} - s^2{}_i. \qquad (15c)$$

This supports the notion that under multivariate stabilizing selection the intensity of stabilizing selection acting directly on a single character will be overestimated from observed changes in its variance, for two reasons: (i) indirect effects of stabilizing selection on every trait correlated with z_i decrease P_{ii}, regardless of the signs of the correlations, and (ii) an observed directional selection differential on z_i caused by a directional selection gradient on z_i or any correlated character also decreases P_{ii}. Thus a multivariate analysis of selection is necessary to accurately estimate the intensity of stabilizing selection acting directly on single traits, as measured by the shape of the fitness surface affecting them.

The Selective Surface Approximated by Quadratic Regression

Background.—The concept of a fitness function, a surface of selective value, or an "adaptive landscape" played a central role in the evolutionary theories of Wright (1931, 1932, 1935), Fisher (1930, 1958 Ch. 2), Dobzhansky (1937, 1951) and Simpson (1944, 1953). As early as 1903 Pearson described selection in terms of a multidimensional surface, the height of which was individual fitness as a function of metrical characters represented by the other dimensions. (In Wright's adaptive landscape the height is the mean fitness in the population as a function of gene frequencies along the other axes.) If the characters in a population follow a multivariate normal distribution both *before and after* an episode of selection within a generation, then the individual fitness surface must have a generalized Gaussian form. Pearson (1903) showed that the parameters of such a selective surface can be estimated from the observed changes in the means, variances and covariances of the characters produced by selection. An equivalent approach was applied by Karn and Penrose (1951) to describe the dependence of early survival in human neonates on birth weight and gestation time. Manly (1976, 1977) analyzed mortality selection using a multivariate fitness function with a double exponential form.

In the preceding sections we generalized Pearson's result to an arbitrary form of selection acting on a distribution of characters that is multivariate normal *before* selection. It was shown that the directional and stabilizing selection gradients are equivalent respectively to the average slope and curvature of the surface of relative individual fitness. In this section we demonstrate that for any form of selection, the coefficients in the best quadratic approximation to the selective surface measure the forces of directional and stabilizing selection on the characters, if these have a multivariate normal distribution before selection. The regression method can also be extended to the general case in which no assumption is made about the form of selection or the distribution of characters. Even in the general case, measures of selection can be obtained by linear and qua-

dratic approximations to the surface of individual relative fitness.

Theoretical Results.—To simplify subsequent notation, suppose that all characters are measured as deviations from their means before selection, so that $\bar{z} = 0$. At first it is assumed that the characters have a multivariate normal distribution in the population before selection (but not necessarily after selection), and any form of fitness function is allowed. Consider the quadratic regression of relative fitness on the characters, listing each quadratic term once,

$$w - 1 = \sum_{j=1}^{n} \beta_j z_j + \sum_{k=1}^{n} \sum_{l=k}^{n} (1 - \tfrac{1}{2}\delta_{kl})\gamma_{kl} \\ \cdot [z_k z_l - P_{kl}] + \varepsilon \qquad (16)$$

where $\delta_{kl} = 1$ if $l = k$ and zero otherwise. The method of least squares can be employed to minimize the mean squared error with respect to changes in the linear and quadratic coefficients, β_j and γ_{kl},

$$Q = \int \varepsilon^2 p(z) dz.$$

For the multivariate normal distribution all third moments vanish, hence the linear and quadratic terms are uncorrelated and $\partial Q/\partial \beta_i = 0$ implies

$$\mathrm{Cov}[w, z_i] - \sum_{j=1}^{n} P_{ij}\beta_j = 0,$$

or in matrix form

$$\beta = P^{-1}\mathrm{Cov}[w, z] \qquad (17)$$

which is equivalent·to the directional selection gradient $P^{-1}s$. The fourth moments of the multivariate normal distribution can be expressed in terms of the second moments (Kendall and Stuart, 1976 eq. 41.97),

$$\int z_i z_j z_k z_l p(z) dz = P_{ij} P_{kl} + P_{ik} P_{jl} + P_{il} P_{jk}$$

so that $\partial Q/\partial \gamma_{ij} = 0$ implies

$$\mathrm{Cov}[w, z_i z_j] - \sum_{k=1}^{n} \sum_{l=1}^{n} P_{ik} P_{jl} \gamma_{kl} = 0,$$

or in matrix form,

$$\gamma = P^{-1}\mathrm{Cov}[w, zz^T]P^{-1} \qquad (18)$$

which is equivalent to the stabilizing selection gradient (14) with $\bar{z} = 0$. The second derivatives of Q with respect to β_i and γ_{ij} are constants and demonstrate that the solutions (17) and (18) specify a unique minimum error variance.

The coefficients of the linear and quadratic terms in the regression (16) therefore measure the forces of directional and stabilizing selection acting directly on the characters, if these have a multivariate normal distribution before selection. The utility of least-squares regression is that it can be readily generalized to provide estimates of intensities of directional and stabilizing selection on correlated characters regardless of their distribution before selection.

When no assumption is made about the distribution of the characters, or the form of the fitness function, a problem exists with the use of the regression (16) alone. If the character distribution before selection displays multivariate skewness (non-zero third moments), the linear and quadratic terms are correlated and estimates of β depend on whether or not the quadratic terms are included in the regression. Stability of estimates of β can be achieved through a system of quadratic polynomials that are uncorrelated with the linear terms (see Appendix).

In practice, the simplest method of computing the measures of selection, and the quadratic approximation to the fitness surface acting on the characters, is by employing automated programs for multiple regression, which are available at most computation centers. Linear multiple regression can be used first to estimate the forces of directional selection, β, and their standard errors. Then a quadratic multiple regression (16) or (A1), can be used to estimate the forces of stabilizing selection, γ, with their standard errors. The regression (16) provides the best quadratic approximation to the selective surface (although valid estimates of β can be obtained only from a purely linear regression, or by using the orthogonal regression (A1)).

Calculation of the stabilizing selection gradient, γ, may require a rather large data set. Multiple regression can not be performed on a given sample of data unless the number of individuals in the sample, N, exceeds the number of coefficients to be estimated: n (the number of characters) for directional selection, plus $n(n + 1)/2$ for stabilizing (or disruptive) selection. If there are too few individuals in the sample, the number of coefficients to be estimated can be reduced, either by neglecting stabilizing selection, or by reducing the number of characters (e.g., by performing the selection analysis on the first few principal components of variation, instead of on the original characters). Especially when many characters are involved, limited sample size will usually make it impractical to estimate coefficients of more than second degree in the regression of relative fitness on the characters.

Cross-sectional Analysis.—Regression techniques for estimating selection coefficients are not applicable to cross-sectional data, because there are no measurements of individual fitnesses. All inferences about selection must then be drawn from a comparison of the phenotype distributions in cohorts of different age sampled from a single population at a point in time. If the basic assumptions of cross-sectional analysis (discussed above) are satisfied, the directional selection gradient which produced an observed change in the mean phenotypes between a young cohort and an old cohort can be estimated from formula (6) with $\bar{z}^*$ and $\bar{z}$ being the mean phenotype vectors in the old and young cohorts respectively, and P being the phenotypic variance-covariance matrix of the young cohort. The stabilizing selection gradient can be estimated from the observed change in the phenotypic variance-covariance matrix between cohorts, adjusted for the effect of directional selection, as in formula (14a) with P* being the phenotypic variance-covariance matrix of the old cohort.

The derivation of expressions (6) and (14), and their application in cross-sectional analysis, requires multivariate normality of the character distribution in the young cohort (before selection). For this reason, and because of the additional assumptions required, cross-sectional analysis of selection is less general and powerful than longitudinal analysis by regression techniques.

Preliminaries for Data Analysis

Transformation of Characters.—For the analysis of cross-sectional data, the most important assumption is that the distribution of characters in the young cohort (before selection) is multivariate normal. There are many types of departure from normality, but these can often be corrected with simple transformations discussed by Wright (1968 Chs. 10, 11). Normality of the separate univariate distributions for each character is a necessary, but not sufficient, condition for multivariate normality.

Major components of fitness, as measured in longitudinal studies, will hardly ever be normally distributed. However, the estimation of selection coefficients from longitudinal data does not depend on any distributional assumptions concerning fitness. No attempt should be made to transform fitness to a more normal distribution, since such transformation will lead to invalid estimates of the forces of selection. Absolute fitnesses need only be transformed to relative fitnesses by dividing each absolute measure by the mean absolute fitness, because selection differentials and gradients are relationships between relative fitness and the characters.

In practice there will almost always be some departure from multivariate normality in the distribution of characters, and it is important to realize which results and interpretations are affected. We emphasize that in longitudinal studies employing the regression methods outlined above, departure from multivariate normality of the character distribution will *not* affect the estimates of selection differentials and gradients, or their standard errors. In contrast, tests of significance for selection differentials and gradients generally assume a normal distribution of *errors from the regression* of relative fitness on the char-

acters (Kendall and Stuart, 1973 Ch. 19.10).

Standardized Selection Coefficients.— The observed selection differential on a particular character is often expressed in terms of phenotypic standard deviations, $s'_i = s_i/\sigma_i$ (Falconer, 1981). Coefficients in the regression of fitness on the characters can also be standardized by adopting units of standard deviations for every character, except fitness. The standardized measures of the intensities of directional and stabilizing selection, $\beta'_i = \beta_i\sigma_i$ and $\gamma'_{ij} = \gamma_{ij}\sigma_i\sigma_j$ respectively, are the coefficients in the linear and quadratic regressions of relative fitness on the standardized characters, $z'_i = z_i/\sigma_i$.

Selection in an Hemipteran

On the morning of April 19, 1981 adult pentatomid bugs (*Euschistus variolarius* PB) were found along the shoreline of Lake Michigan in Porter Co., Indiana. The bugs, along with several species of carabid and chrysomelid beetles, were common in the zone of waft debris, about 1 to 3 m from the water's edge. The first 94 bugs encountered in this zone were collected and of these only 39 were alive. The 55 dead bugs were in excellent condition and had apparently died recently. We suppose that the bugs were knocked down into the lake during a storm as they attempted to migrate north over Lake Michigan and were then washed ashore. There was a light rain during the morning when the collection was made, and live bugs were inactive and apparently incapable of taking flight. These circumstances suggest that we obtained an unbiased sample of bugs washed ashore during the storm.

The bugs were sorted into live and dead groups and preserved in 70% ethanol within two hours after collection. Subsequently, four linear measurements were made on each individual (head width, width of thorax [pronotum], length of scutellum, and length of forewing), using a dissecting microscope at 6-fold magnification with an ocular micrometer. All measurements were transformed to natural logarithms.

A multivariate analysis of variance gave no indication of sexual dimorphism in the mean phenotype (Wilks' lambda $= .98$; $F(4,85) = .51$, $P > 0.7$). Consequently, male and female samples were pooled for subsequent analysis. Basic statistics for the sample are shown in Table 1. The coefficients of variation of the characters range from about 3% to 6%, which is fairly typical for linear measurements.

Directional selection differentials were calculated as the covariances between relative fitness and each character; these represent the changes in character means produced by direct and indirect selection. The directional selection gradient was estimated from the partial regression coefficients of relative fitness on the characters (Table 2). The stabilizing selection gradient was estimated from a multiple regression of relative fitness on the four characters and the ten product variables (head $\times$ head, head $\times$ thorax, . . . , scutellum $\times$ wing, wing $\times$ wing). None of the quadratic coefficients were statistically significant, possibly due to the small sample size, hence the analysis of stabilizing selection is not reported here.

Interpretation of Results.—The pentatomid data provide some striking contrasts between selection differentials and gradients (Table 2). For example, there was no significant directional selection differential on the thorax, indicating that no appreciable change in mean thorax width resulted from selection. Nevertheless, in the selection gradient the coefficient for thorax is highly significant. Thus there was substantial selection directly on the thorax, but this was masked in the selection differential by indirect selection through some correlated character(s). Apparently, direct selection to decrease wing length (which is positively correlated with thorax width) produced negative indirect selection on the thorax, counteracting the direct effect of selection to increase thorax width (eq. 6b).

Selection on the scutellum is another example of the distinction between the selection differential and the selection gradient. Despite a significant decrease in

TABLE 1. *Statistics of the hemipteran population before selection, with all characters transformed to natural logarithms.*

Character	Sample size,[†] N	Mean, $\bar{z}$	Standard deviation, σ
Head	94	.880	.034
Thorax	93	2.038	.049
Scutellum	94	1.526	.057
Wing	91	2.337	.043

Correlations[‡]

Character	Head	Thorax	Scutellum	Wing
Head	1	.72	.50	.60
Thorax		1	.59	.71
Scutellum			1	.62
Wing				1

[†] Thorax and wing dimensions could not be measured on all individuals due to damage
[‡] All correlations are significant at the .001 level

mean scutellum length, there was no significant directional selection on this character. The observed selection differential for the scutellum is therefore attributable to the indirect effects of selection acting on correlated characters. Because the two traits undergoing substantial directional selection (thorax and wing) are about equally variable and have nearly the same correlation with the scutellum, it appears that the somewhat stronger selection to decrease wing length (compared with that to increase thorax width) produced the overriding indirect effect on the scutellum. These examples clearly illustrate how the observed selection differential can give a misleading picture of the selective forces acting directly on the characters.

The magnitudes of the standardized selection coefficients reveal rather strong directional selection on two of the characters. On average in the population, an increase in thorax length alone by one standard deviation would have increased relative fitness by about 58%, while an increase in wing length alone by one standard deviation would have decreased relative fitness by about 74%. The proportion of variance in relative fitness explained by the multiple regression is highly significant ($R^2 = .23$).

Two interpretations can be made of the

TABLE 2. *Directional selection differential* (s) *and gradient* (β ± *standard error) for the hemipteran population. Standardized selection coefficients* (s' *and* β') *are also given in units of phenotypic standard deviations.*

Character	s†	s'	β ± SE	β' ± SE
Head	−.004	−.11	−.7 ± 4.9	−.03 ± .17
Thorax	−.003	−.06	11.6** ± 3.9	.58** ± .19
Scutellum	−.016*	−.28*	−2.8 ± 2.7	−.17 ± .15
Wing	−.019**	−.43**	−16.6** ± 4.0	−.74** ± .18
	$\sigma_w^2 = 1.43$		$R^2 = .23**$	

* Significant at the 05 level   ** Significant at the 01 level
† Significance levels are for Spearman rank correlations of relative fitness with the characters

differing signs on the coefficients for the wing and thorax in the directional selection gradient. First, selection can be viewed as acting separately on the characters, one force favoring large thorax and another force favoring small wings. A second interpretation is that directional selection acted on a character combination or index, rather than separately on wing and thorax. The results are consistent with selection on the index 11.6 thorax–16.6 wing (eq. 10). Since the characters are measured on logarithmic scales this is nearly equivalent to saying that selection favored bugs with proportionally small wings in relation to thorax length. In turbulence and/or precipitation such bugs may have been better fliers and spent less time in the lake water during the storm. This hypothesis could be tested with aerodynamic studies.

Selection in the House Sparrow

On the morning of February 1, 1898 after a severe winter ice storm, immobilized house sparrows (*Passer domesticus*) were common on the ground in Providence, Rhode Island. Bumpus (1899) obtained a sample of 136 of these sparrows and upon bringing them into his laboratory, about half of the birds revived. Bumpus weighed all the birds and then prepared them as specimens, keeping track of the survivors and nonsurvivors. He made eight linear measurements on each specimen (total length, wing extended, skull width, and lengths of beak plus head, humerus, femur, tibiotarsus, and keel of sternum), and published the raw data in his original report.

We transformed all the linear measurements to natural logarithms. Body weight was also transformed to natural logarithms, after taking the cube root in order to make its dimensionality the same as that of the other characters (Lande, 1977). Bumpus classified males into juvenile and adult subsamples, based on plumage appearance. Multivariate Analysis of Variance showed no significant age effects on the mean vectors of the metrical traits (Wilks' lambda = .92; $F(9,77) = .74$, $P > .7$) and accordingly the juvenile and adult males were pooled for subsequent analysis. The lack of age effects probably reflects a cessation of skeletal growth soon after fledging, as has been documented for song sparrows by Smith and Zach (1979). MANOVA did reveal significant sexual dimorphism (Wilks' lambda = .52; $F(9,126) = 12.86$, $P < .0001$), and univariate analysis of variance showed significant differences between the sexes in total length, weight, wing extended, humerus, and sternum. Hence males and females were analyzed separately. The variability of the characters is consistent with the typically small coefficients of variation of linear measures in birds, between about 2% and 5%. Significant positive correlations occur for all pairs of characters, ranging from .26 to .88 in males, and from .46 to .83 in females.

Directional selection differentials and gradients were estimated for each sex. The only significant coefficients in the directional selection gradients were for weight in both sexes (male $\beta' = -.27 \pm .09$, female $\beta' = -.52 \pm .25$) and for total length in males ($\beta' = -.52 \pm .10$). The standardized selection coefficients reveal apparently strong directional selection on these traits. The proportion of total variance in relative fitness explained by the multiple regression was highly significant for males ($R^2 = .46$), while that for females was not significant ($R^2 = .20$).

The samples are not large enough to analyze stabilizing selection on all nine characters. The full quadratic regression would contain nine coefficients of directional selection and 45 coefficients of stabilizing selection. With only 49 females and 87 males there were not enough degrees of freedom to estimate all of the selection coefficients in females, or to do so precisely in males. Substantial positive correlations between all characters suggested that a reduction of dimensionality could be carried out by using the first few principal components of variation in each sex, while still retaining most of the total variation in the characters. Ordinary principal components (eigenvectors) were computed from the phenotypic variance-covariance matrices. In both cases more than 70% of the total character variance was accounted for by the first two principal components. We therefore restricted the analysis of stabilizing selection to the first two principal components of variation in each sex. All characters contributed positively to the first principal component in both sexes, which can be viewed as an index of general size. The second principal component of variation had a large positive contribution from the sternum and lesser negative contributions from humerus, femur and tibiotarsus lengths in both sexes, and can be viewed as an index of sternum size in relation to limb dimensions.

Directional and stabilizing selection gradients on the first two principal components were estimated using the method of quadratic regression described above (and in the Appendix). There was (barely) significant stabilizing selection on the first principal component of females only. The standardized selection coefficient ($\gamma'_{11} = -.45 \pm .22$) indicates that on average in the female population a random change in the first principal component alone by (plus or minus) one standard deviation would

have decreased relative fitness by about 45%.

Interpretation of Results.—The sample of sparrows obtained by Bumpus has definite limitations that should be kept in mind in interpreting our selection analysis. Because Bumpus collected birds that were immobilized by a storm, some individuals may have survived without falling to the ground and these would necessarily be missing from the sample. Thus the selection measured is conditional on the event of immobilization and may not represent the selection experienced by the entire population. In addition, the selection observed only pertains to mortality on a single night of unusually severe weather. This brief episode of selection may not reflect the forces acting throughout the year. (Similar qualifications apply to the above interpretation of selection on pentatomid bugs.)

The winter storm apparently favored small birds. This pattern of winter selection seems difficult to explain on functional grounds, since it is opposite to that which would be predicted from Bergman's rule (larger geographic races in colder climates). Johnston et al. (1972) have suggested that sparrows surviving the storm may have lost weight by metabolizing longer than those which perished, but this would not account for the highly significant decrease in male body length.

The analysis of selection on the first two principal components failed to reveal significant directional selection on either sex. Although significant directional selection was detected for some of the original characters, combination with other apparently unselected characters rendered the directional selection gradient on the first principal component nonsignificant. There was, however, strong but barely significant stabilizing selection on the first principal component of females.

DISCUSSION

We have shown that selection coefficients which appear in dynamical equations for the evolution of correlated characters can be estimated from purely phenotypic data. These same coefficients describe a surface of phenotypic selective value acting on a population. Multiple regression of individual relative fitness on the characters can be used to estimate the forces of directional selection, given by the steepest uphill slope (or gradient) in the best linear approximation to the selective surface; the forces of stabilizing or disruptive selection are given by the curvature of the best quadratic approximation to the selective surface. Even if the quadratic approximation to the selective surface has a maximum, so that an optimal phenotype exists, the population mean may differ substantially from the optimum, indicating maladaptation of the mean phenotype. Disruptive selection on one or more characters suggests that the population is in a valley or saddle between two or more adaptive peaks. Some important questions which can be approached with the techniques we have described include the following: Do selective surfaces commonly have multiple peaks? How frequent is maladaptation of the mean phenotype in a population? What is the magnitude of geographical and temporal variation in selective surfaces? Studies designed to answer such questions should be done on a number of animal and plant populations. Only then will we have adequate information on the intensity and pattern of phenotypic natural selection.

The examples we have analyzed leave much to be desired. They have, nevertheless, illustrated the application of partial regression analysis in the measurement of directional and stabilizing selection on correlated characters. The analysis of directional selection on pentatomid bugs showed that the observed change in the mean phenotype within a generation may be a very poor indicator of the actual forces of selection estimated by the selection gradient; indirect selection through correlated characters can cause the mean value of a trait to change against the force of selection on it. The analysis of Bumpus' data on sparrows demonstrated one technique for the reduction of dimensionality that may often be necessary when estimating forces of stabilizing selection on multiple

characters. Our analysis differed from most previous treatments of the Bumpus data which employed univariate selection theory and mainly compared surviving and dead samples, rather than those before and after selection (Bumpus, 1899; Harris, 1911; Calhoun, 1947; Grant, 1972; Johnston et al., 1972; O'Donald, 1973; for an exception see Manly, 1976). Grant (1972) commented that the sparrow characters were strongly correlated so that a univariate selection theory is inadequate to identify which characters were the actual target of selection.

In both the pentatomid and sparrow populations we detected apparently intense directional selection during brief periods of high mortality (see also Van Valen, 1963; Hagen and Gilbertson, 1973; Boag and Grant, 1981). Because the persistence of a population requires that the total selective mortality not exceed the reproductive rate, these results suggest that in some populations most of the directional selection within a generation may be concentrated in a few relatively short periods of high mortality.

Attempts have frequently been made to measure the intensity of stabilizing selection by its effect on the variance of single traits. We have demonstrated, however, that purely directional selection (defined by a linear fitness surface) decreases character variance (eq. 13a), and have therefore proposed to measure stabilizing selection in terms of the curvature of the best quadratic approximation to the selective surface. Furthermore, stabilizing selection on any character indirectly decreases the variance of correlated characters, whether the correlations are positive or negative (eqs. 11, 15c). For these two reasons, most previous studies (reviewed by Johnson, 1976 Ch. 7) have probably seriously overestimated the strength of stabilizing selection acting directly on single characters. This may explain why the only significant stabilizing selection we detected in the Bumpus data was on the first principal component of variation in females, although these data have been considered a classic example of stabilizing selection.

1224　　　　　　R. LANDE AND S. J. ARNOLD

SUMMARY

Multivariate statistical methods are derived for measuring selection solely from observed changes in the distribution of phenotypic characters in a population within a generation. Selective effects are readily detectable in characters that do not change with age, such as meristic traits or adult characters in species with determinate growth. Ontogenetic characters, including allometric growth rates, can be analyzed in longitudinal studies where individuals are followed through time.

Following an approach pioneered by Pearson (1903), this analysis helps to reveal the target(s) of selection, and to quantify its intensity, without identifying the selective agent(s). By accounting for indirect selection through correlated characters, separate forces of directional and stabilizing (or disruptive) selection acting directly on each character can be measured. These directional and stabilizing selection coefficients are respectively the parameters that describe the best linear and quadratic approximations to the selective surface of individual fitness as a function of the phenotypic characters.

The theory is illustrated by estimating selective forces on morphological characters influencing survival in pentatomid bugs and in house sparrows during severe weather conditions.

ACKNOWLEDGMENTS

We thank J. Antonovics, J. Bradbury, M. Bulmer, D. Burdick, T. Clutton-Brock, G. E. Dickerson, P. R. Grant, T. Nagylaki, and L. Van Valen for helpful discussions and criticisms. Jean Gladstone assisted with computations, John Steadman measured specimens, and Daniel Summers provided specific identification of the pentatomid population. This work was supported by U. S. Public Health Service grants GM27120 and 1 K04-HD-00392, and National Science Foundation grant DEB 81-11489.

LITERATURE CITED

ATCHLEY, W. R., AND J. J. RUTLEDGE. 1981. Genetic components of size and shape. I. Dynamic components of phenotypic variability and covariability during the ontogeny of the laboratory rat. Evolution 35:1161–1173.

BEATSON, R. R. 1976. Environmental and genetical correlates of disruptive coloration in the water snake, *Natrix s. sipedon*. Evolution 30:241–252.

BOAG, P. T., AND P. R. GRANT. 1981. Intense natural selection in a population of Darwin's Finches (Geospizinae) in the Galápagos. Science 214:82–85.

BUMPUS, H. C. 1899. The elimination of the unfit as illustrated by the introduced sparrow, *Passer domesticus*. Biol. Lectures, Woods Hole Marine Biol. Station 6:209–226.

CALHOUN, J. B. 1947. The role of temperature and natural selection in relation to the variations in size of the English sparrow in the United States. Amer. Natur. 81:203–228.

DI CESNOLA, A. P. 1906. A first study of natural selection in "Helix arbustorum" (Helicogena). Biometrika 5:387–399.

CHARLESWORTH, B. 1980. Evolution in Age-structured Populations. Cambridge Univ. Press, Cambridge.

CLUTTON-BROCK, T., F. E. GUINESS, AND S. D. ALBON. 1982. Red Deer: Behavior and Ecology of Two Sexes. Univ. Chicago Press, Chicago.

COCK, A. G. 1966. Genetical aspects of metrical growth and form in animals. Quart. Rev. Biol. 41:131–190.

DICKERSON, G. E., C. T. BLUNN, A. B. CHAPMAN, R. M. KOTTMAN, J. L. KRIDER, E. J. WARWICK, AND J. A. WATLEY, JR. 1954. Evaluation of selection in developing inbred lines of swine. North Central Regional Publ. No. 38, Mo. Agr. Exp. Sta. Res. Bull. 551.

DICKERSON, G. E., H. O. HETZER, E. V. KREHBIEL, AND A. E. FLOWER. 1974. Effectiveness of reciprocal selection for performance of crosses between Montana No. 1 and Yorkshire swine. III. Expected and actual response. J. Anim. Sci. 39:24–41.

DOBZHANSKY, TH. 1937. Genetics and the Origin of Species. Columbia Univ. Press, N.Y.

———. 1951. Genetics and the Origin of Species, 2nd ed. Columbia Univ. Press, N.Y.

DOWNHOWER, J. F., AND L. BROWN. 1980. Mate preferences of female mottled sculpins, *Cottus bairdi*. Anim. Behav. 28:728–734.

FALCONER, D. S. 1981. Introduction to Quantitative Genetics, 2nd ed. Longman, London.

FISHER, R. A. 1930. The Genetical Theory of Natural Selection. Clarendon, Oxford.

———. 1958. The Genetical Theory of Natural Selection, 2nd ed. Dover, N.Y.

FOX, S. F. 1975. Natural selection on morphological phenotypes of the lizard *Uta stansburiana*. Evolution 29:95–107.

GALTON, F. 1889. Natural Inheritance. MacMillan, London.

→ GOULD, S. J., AND R. C. LEWONTIN. 1979. The spandrels of San Marco and the Panglossian paradigm: a critique of the adaptationist program. Proc. Zool. Soc. London B 205:581–598.

GRANT, P. R. 1972. Centripetal selection and the house sparrow. Syst. Zool. 21:23–30.

HAGEN, D. W., AND L. G. GILBERTSON. 1973. Selective predation and the intensity of selection acting upon the lateral plates of threespine sticklebacks. Heredity 30:273–287.

HALDANE, J. B. S. 1954. The measurement of natural selection. Proc. IX Intl. Cong. Genet. 1:480–487.

HARRIS, J. A. 1911. A neglected paper on natural selection in the English sparrow. Amer. Natur. 45:314–318.

HARVEY, W. R., AND G. D. BEARDEN. 1962. Tables of expected genetic progress in each of two selected traits. U.S.D.A. Agric. Res. Serv. Publ. No. 20-12.

HECHT, M. K. 1952. Natural selection in the lizard genus *Aristelliger*. Evolution 6:112–124.

HOWARD, R. D. 1979. Estimating reproductive success in natural populations. Amer. Natur. 114:221–231.

INGER, R. F. 1942. Differential selection of variant juvenile snakes. Amer. Natur. 76:527–528.

———. 1943. Further notes on differential selection of variant juvenile snakes. Amer. Natur. 77:87–90.

JOHNSON, C. 1976. Introduction to Natural Selection. Univ. Park Press, Baltimore.

JOHNSTON, R. F., D. M. NILES, AND S. A. ROHWER. 1972. Hermon Bumpus and natural selection in the house sparrow *Passer domesticus*. Evolution 26:20–31.

KARN, M. N., AND L. S. PENROSE. 1951. Birth weight and gestation time in relation to maternal age, parity and infant survival. Ann. Eugen. 16:147–189.

KEMPTHORNE, O. 1969. An Introduction to Genetic Statistics. Iowa State Univ. Press, Ames.

KENDALL, M. G., AND A. STUART. 1973. The Advanced Theory of Statistics. Vol. 2. Inference and Relationship, 3rd ed. MacMillan, N.Y.

———. 1976. The Advanced Theory of Statistics Vol. 3. Design and Analysis, and Time-Series, 3rd ed. MacMillan, N.Y.

KIDWELL, J. F., J. G. HERBERT, AND H. B. CHASE. 1979. The inheritance of growth and form in the mouse. V. Allometric growth. Growth 43:47–57.

KURTÉN, B. 1958. Life and death of a Pleistocene cave bear, a study in paleoecology. Acta Zool. Fennica 95:1–59.

LANDE, R. 1977. On comparing coefficients of variation. Syst. Zool. 26:214–217.

———. 1979. Quantitative genetic analysis of multivariate evolution, applied to brain:body size allometry. Evolution 33:402–416.

———. 1980. The genetic covariance between characters maintained by pleiotropic mutations. Genetics 94:203–215.

———. 1982. A quantitative genetic theory of life history evolution. Ecology 63:607–615.

LEWONTIN, R. C. 1978. Adaptation. Sci Amer. 239:156–169.

LUSH, J. L. 1945. Animal Breeding Plans. Iowa State Univ. Press, Ames.

MAGEE, W. T. 1965. Estimating response to selection. J. Anim. Sci. 24:242–247.

MANLY, B. F. J. 1976. Some examples of double exponential fitness functions. Heredity 36:229–234.

———. 1977. A new index for the intensity of natural selection. Heredity 38:321–328.

MARCUS, L. F. 1964. Measurement of natural selection in natural populations. Nature 202:1033–1034.

———. 1969. Measurement of selection using distance statistics in the prehistoric orang-utan *Pongo pygmaeus palaeosumatrensis*. Evolution 23:301–307.

MASON, L. G. 1964. Stabilizing selection for mating fitness in natural populations of *Tetraopes*. Evolution 18:492–497.

MAYO, O. 1980. The Theory of Plant Breeding. Clarendon Press, Oxford.

O'DONALD, P. 1968. Measuring the intensity of natural selection. Nature 220:197–198.

———. 1970. Measuring the change of population fitness by natural selection. Nature 227:307–308.

———. 1973. A further analysis of Bumpus' data: the intensity of natural selection. Evolution 27:398–404.

PEARSON, K. 1896. Mathematical contributions to the theory of evolution. III. Regression, heredity, and panmixia. Phil. Trans. Roy. Soc. London A 187:253–318.

———. 1903. Mathematical contributions to the theory of evolution. XI. On the influence of natural selection on the variability and correlation of organs. Phil. Trans. Roy. Soc. London A 200:1–66.

———. 1911. On the general theory of the influence of selection on correlation and variation. Biometrika 8:437–443.

———. 1920. Notes on the history of correlation. Biometrika 13:25–45.

PRICE, G. R. 1970. Selection and covariance. Nature 227:520–521.

———. 1972. Extension of covariance selection mathematics. Ann. Hum. Genet. 35:485–490.

PROVINE, W. B. 1971. The Origins of Theoretical Population Genetics. Univ. Chicago Press, Chicago.

ROBERTSON, A. 1966. A mathematical model of the culling process in dairy cattle. Anim. Prod. 8:93–108.

SAMBOL, M., AND R. M. FINKS. 1977. Natural selection in a Cretaceous oyster. Paleobiology 3:1–16.

SEAL, H. L. 1967. The historical development of the Gauss linear model. Biometrika 54:1–24.

SIMPSON, G. G. 1944. Tempo and Mode in Evolution. Columbia Univ. Press, N.Y.

———. 1953. The Major Features of Evolution Columbia Univ. Press, N.Y.

SMITH, J. N. M., AND R ZACH. 1979. Heritability

of some morphological characters in a song sparrow population. Evolution 33:460–467.

THORNHILL, R. 1976. Sexual selection and nuptial feeding behavior in *Bittacus apicalis* (Insecta: Mecoptera). Amer. Natur. 110:529–548.

VAN VALEN, L. 1963. Selection in natural populations: *Merychippus primus*, a fossil horse. Nature 197:1181–1183.

———. 1965a. Selection in natural populations. III. Measurement and estimation. Evolution 19:514–528.

———. 1965b. Selection in natural populations. IV. British housemice (*Mus musculus*). Genetica 36:119–134.

VAN VALEN, L., AND R. WEISS. 1966. Selection in natural populations. V. Indian rats (*Rattus rattus*). Genet. Res. 8:261–267.

WELDON, W. F. R. 1901. A first study of natural selection in *Clausilia laminata* (Montagu). Biometrika 1:109–124.

WRIGHT, S. 1931. Evolution in Mendelian populations. Genetics 16:97–159.

———. 1932. The roles of mutation, inbreeding, crossbreeding, and selection in evolution. Proc. 6th Intl. Cong. Genet. 1:356–366.

———. 1935. Evolution in populations in approximate equilibrium. J. Genet. 30:257–266.

———. 1968. Evolution and the Genetics of Populations. Vol. 1. Genetic and Biometric Foundations. Univ. Chicago Press, Chicago.

YAMADA, Y. 1977. Evaluation of the culling variate used by breeders in actual selection. Genetics 86:885–899.

YULE, G. U. 1897. On the theory of correlation. J. Roy. Stat. Soc. 60:812–851.

———. 1907. On the theory of correlation for any number of variables, treated by a new system of notation. Proc. Roy. Soc. London A 79:182–193.

Corresponding Editor: J. Felsenstein

APPENDIX

Writing the orthogonal quadratic regression of relative fitness on the characters, setting $\bar{z} = 0$, and listing each quadratic term once,

$$w - 1 = \sum_{j=1}^{n} \beta_j z_j + \sum_{k=1}^{n} \sum_{l=k}^{n} (1 - \tfrac{1}{2}\delta_{kl})\gamma_{kl}\xi_{kl} + \varepsilon \tag{A1}$$

the quadratic polynomials are the usual quadratic deviations in (16) with an additional linear factor that corrects for skewness of the character distribution,

$$\xi_{kl} = z_k z_l - P_{kl} - \tau^T_{kl}P^{-1}z. \tag{A2}$$

$\tau^T_{kl} = (\tau_{1kl}, \ldots, \tau_{nkl})$ is a vector of all third moments of the distribution involving characters z_k and z_l,

$$\tau_{jkl} = \int z_j z_k z_l p(z)dz. \tag{A3}$$

Each of the polynomials ξ_{kl} is evidently uncorrelated with all of the z_j since

$$Cov[z, \xi_{kl}] = \tau_{kl} - Cov[z, z^T]P^{-1}\tau_{kl} = 0.$$

The coefficients in the orthogonal regression (A1) are found by minimizing the mean squared error, Q, as before. $\partial Q/\partial \beta_j = 0$ implies (17), which is what would be obtained from an ordinary linear regression (omitting the quadratic terms or polynomials in (16) or (A1)). For the coefficients of the quadratic polynomials, $\partial Q/\partial \gamma_{ij} = 0$ implies the same equations (involving fourth moments of the character distribution) that would be obtained from (16) without any distributional assumptions. The orthogonal regression (A1) is also useful for calculating separate proportions of variance in relative fitness due to directional and stabilizing selection, expressed as the squared multiple correlation coefficients, R^2, of relative fitness with the linear terms and the quadratic polynomials.

Evolution, 40(6), 1986, pp. 1215–1228

CONSTRAINTS ON CHEMICAL COEVOLUTION: WILD PARSNIPS AND THE PARSNIP WEBWORM

M. R. Berenbaum, A. R. Zangerl, and J. K. Nitao

Department of Entomology, University of Illinois, Urbana, IL 61801-3795

Abstract.—The parsnip webworm (*Depressaria pastinacella*) and the wild parsnip (*Pastinaca sativa*) together represent a potentially "coevolved" system in that throughout their ranges the plant has relatively few other herbivores and the insect has virtually no other hosts. Individual wild parsnip plants within a central Illinois population vary in their content and composition of furanocoumarins, secondary compounds with insecticidal properties. Half-sib and parent-offspring regression estimates of the heritability of furanocoumarins demonstrate that this variation is genetically based. Wild parsnip plants also vary in their resistance to damage by the parsnip webworm, which feeds on flowers and developing seeds. In an experimental garden, seed production in the primary umbel ranged from 0 to 1,664 seeds among individuals, and mean seed production of half-sib families ranged from 3.7 seeds to 446.0 seeds. Approximately 75% of the variation in resistance among half-sib families to *D. pastinacella* was attributable to four furanocoumarin characteristics—resistance is positively related to the proportion of bergapten and the amount of sphondin in seeds, and negatively related to the amount of bergapten and the proportion of sphondin in leaves. Each of the four resistance factors had significant heritability. Thus, resistance in wild parsnip to the parsnip webworm is to a large extent chemically based and genetically controlled. Genetic correlations among fitness and resistance characters, however, tend to limit coevolutionary responses between herbivore and plant. In greenhouse plants protected from herbivory, several of the resistance factors had negative genetic correlations with potential seed production. Ostensibly, highly resistant plants in the absence of herbivory would be at a competitive disadvantage in the field. The selective impact of the herbivore is also limited in this population by a negative genetic correlation among resistance factors. Selection to increase one resistance factor (e.g., the proportion of bergapten in the seed) would at the same time decrease the amount of a second resistance factor (e.g., the amount of sphondin in the seed). The wild parsnip and the parsnip webworm, then, appear to have reached an evolutionary "stalemate" in the coevolutionary arms race.

Received November 18, 1985. Accepted July 7, 1986

Though it has long been suspected that intraspecific variation in plant secondary chemistry is associated with variation in susceptibility to insect herbivory in plant populations, the nature of the association is by no means well-defined. While several studies have revealed a correlative association (Dolinger et al., 1973; Edmunds and Alstad, 1978; Chew and Rodman, 1979 [and references therein]; Sturgeon, 1979), it is impossible to determine whether observed differences in insect herbivory are the result of variation in secondary chemistry or the cause of that variation. For example, phenotypically induced alteration in secondary chemistry can occur as the result of insect attack (Green and Ryan, 1972; Rhoades, 1979; Carroll and Hoffman, 1980; Haukioja, 1980; but see Fowler and Lawton, 1985). Virtually all arguments about the evolution of resistance by defensive chemistry presume that variation in secondary chemistry, whether induced or constitutive, is at least partly genetic in origin. While natural se-

lection assuredly acts on phenotypes, change in frequencies of chemical morphs within a population over time, i.e., the type of change implicit in insect-plant coevolution (Janzen, 1980), can occur only if there is either a genetic basis for chemical variation or a genetic basis for differences in the expression or inducibility of secondary substances.

Insect-plant coevolution, as defined by Janzen (1980), also implies that herbivorous insects are capable of exerting sufficient selective pressure to change the frequency of genotypes in a plant population. There is ample evidence to support the contention that insects can reduce growth, reproductive fitness, and survivorship in plants (e.g., Morrow and LaMarche, 1978; Rausher and Feeny, 1980), and there is also evidence that susceptibility to insect attack is genetically based (Moran, 1981; Lin et al., 1984; Marquis, 1984; Service, 1984). However, there is little or no information on the selective influence of insects on the distribution of chemical phenotypes within a population.

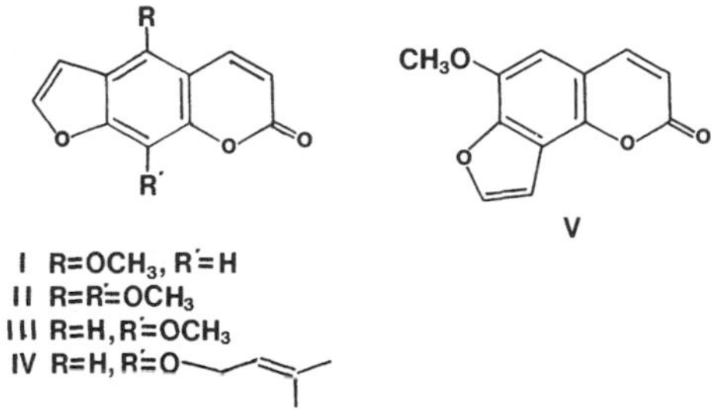

I R=OCH₃, R'=H
II R=R'=OCH₃
III R=H, R'=OCH₃
IV R=H, R'=O————

V

FIG. 1. Furanocoumarins in leaves and seeds of *Pastinaca sativa*: I) bergapten; II) isopimpinellin; III) xanthotoxin; IV) imperatorin; V) sphondin.

A correlation between plant chemistry and resistance may result not from coevolution sensu stricto but from adaptation by the plant to some other environmental variable and subsequent adjustment on the part of the insect (Jermy, 1984).

For insect-plant coevolutionary arguments to gain validity with respect to secondary plant metabolism, it is essential to demonstrate the genetic variation in both plant secondary metabolism and resistance to herbivory. To examine this question, we selected for study *Pastinaca sativa,* the wild parsnip (Umbelliferae), and *Depressaria pastinacella,* the parsnip webworm (Lepidoptera: Oecophoridae). *P. sativa,* a biennial introduced from Europe and extensively naturalized throughout eastern North America, occurs in great numbers in waste places, open fields, and along railroads (Fernald, 1950; Thompson, 1977). Like many apioid Umbelliferae, *P. sativa* produces linear and angular furanocoumarins (Berenbaum et al., 1984) (Fig. 1). A previous study of variation in furanocoumarin content in a population of *P. sativa* located on Perkins Road approximately 6 km northeast of the University of Illinois at Urbana-Champaign in Champaign County, Illinois, revealed significant differences among individual plants in the quantity and composition of furanocoumarins in the seeds of the primary umbels. The furanocoumarin content of seeds varied from 17.0 to 60.1 μg/seed, and the relative proportions of individual furanocoumarins varied from 2-fold to over 20-fold (Berenbaum et al., 1984).

Furanocoumarins have been shown to be resistance factors against insects. Linear fu-

ranocoumarins are toxic or repellent to generalized herbivores (Yajima et al., 1977; Berenbaum, 1978; Muckensturm et al., 1981), yet appear to enhance growth in *Papilio polyxenes* (Lepidoptera: Papilionidae), a specialist on Umbelliferae (Berenbaum, 1981a). Angular furanocoumarins, in contrast, markedly reduce fecundity in *P. polyxenes* (Berenbaum and Feeny, 1981). The furanocoumarins are thus likely candidates for chemical resistance factors even against a specialist such as *D. pastinacella.*

Throughout most of its range, *P. sativa* is attacked heavily by *D. pastinacella* (Hodges, 1974; Thompson, 1978; Hendrix, 1979; Berenbaum, 1981b; Gorder and Mertins, 1984). The caterpillar webs together umbels and feeds on developing flowers and seeds. After it completes larval development, it leaves the umbels to pupate in parsnip stems. Adults emerge within a month and overwinter under bark, in litter or in human habitations (Gorder and Mertins, 1984; pers. observ.). Although it has been reported to occur on *Angelica* and *Heracleum,* also members of the Umbelliferae (Hodges, 1974), *D. pastinacella* is effectively restricted to wild parsnip throughout Champaign County because it is the only host locally abundant (Thompson, 1978). In central Illinois, *D. pastinacella* is the most abundant herbivore in terms of biomass (Thompson, 1977) and can reach infestation levels of 100% of plants in a population (pers. observ.).

Pastinaca sativa and *Depressaria pastinacella,* then, together potentially represent a typical "coevolved" association (sensu Ehrlich and Raven, 1964). The plant has few other insect enemies, the majority of which are restricted to parsnip and related plants (Berenbaum, 1981b), and the insect has few other hosts, all of which are closely related and chemically similar to parsnip (Heywood, 1971). The population on Perkins Road is known to have harbored large populations of *D. pastinacella* for several consecutive years. We therefore examined this population in order to address the following questions:

1) How much of the variation in furanocoumarin composition and content in *P. sativa* is genetic in origin?

2) To what degree is resistance to attack by *D. pastinacella* genetically based and determined by the chemistry of the individual plant?

3) What is the magnitude of the selective force exerted by the herbivore among plant individuals? That is, can *D. pastinacella* exert sufficient selection pressure to effect changes in the genetic composition of the plant population?

4) If chemical resistance traits are genetically based, what maintains chemical variability? In other words, are traits conferring resistance to *D. pastinacella* associated with a fitness reduction when the herbivore is not present?

MATERIALS AND METHODS

Estimation of Genetic Control of Furanocoumarin Production

One widely accepted measure of the extent to which a trait is under genetic control is its heritability (Falconer, 1981). The heritability in the narrow sense is the proportion of the phenotypic variance in a trait that can be attributed to the additive genetic variance. The heritability is important evolutionarily because the response to selection is a function of additive genetic variance (Falconer, 1981). Covariances between maternal half-sibs and between parent and offspring were used to estimate heritabilities of both quantity and composition of furanocoumarin production (Falconer, 1981). As is the case with many umbellifers (Cruden and Hermann-Parker, 1977 and references therein), individual flowers within an umbel are protandrous; since umbels develop sequentially from primary to quaternary (Hendrix and Trapp, 1981), the plant is temporally dioecious and outbreeding in the primary umbel is ensured. Moreover, the unspecialized nature of parsnip pollinators (Bell, 1971) enhances the probability that pollen from many sources is involved in fertilization. All of these characteristics together tend to maximize the proportion of half-sibs in the primary umbel. For this reason seeds from the primary umbel were used to estimate genetic parameters. All chemical analyses were restricted to seeds of primary umbels and to leaves.

A section of the Perkins Road population was marked off, and the primary umbels of all individuals ($N = 65$) were removed and placed in individual petri dishes. The dishes were mixed, and the first twenty dishes with sufficient numbers of seed were chosen for study. This sampling technique may have eliminated from the study those plants most susceptible to insect attack, i.e., those with too few seed for analysis due to herbivory; however, accurate estimates of quantitative-genetic parameters require large sample sizes. Ten individuals from each of these twenty families were chosen at random and planted in spring 1983 in an experimental garden in Phillips Tract, a university-owned natural area approximately 4 km east of the Perkins Road population. Individuals were planted in tilled soil arranged in a completely randomized design and spaced at 30-cm intervals. The plot was weeded as necessary throughout the growing season. Seeds and leaf samples taken from plants flowering in 1984 in this garden were evaluated for furanocoumarin composition and content. Furanocoumarins were analyzed by high-pressure liquid chromatography as described in Berenbaum et al. (1984), with the exception that diethyl ether was used as the initial extraction solvent. Estimation of variance components was by standard analysis of variance of a random model (Falconer, 1981; Model II of Sokal and Rohlf, 1981); observed mean squares were set equal to their expectations.

Heritability based on half-sib families was calculated as

$$h^2 = \frac{4V_\mathrm{F}}{V_\mathrm{T}}$$

where V_F, the between-family variance (covariance between maternal half-sibs), is ¼ of the additive genetic variance and V_T is the total phenotypic variance. Since some individuals died before setting seed, the family size, k, used to estimate variance components was calculated as

$$k = \frac{N^2 - \Sigma\, n_i^2}{N(f - 1)}$$

where N is the total number of plants, f is the number of families and n_i is the number of individuals in the ith family. The possibility that the families may contain some proportion of full-sibs cannot be ruled out, and therefore the calculated heritabilites

1218   M. R. BERENBAUM ET AL.

should be considered maximum estimates. We also assume random mating and no inbreeding in the population (but see Mitchell-Olds and Rutledge, 1986). The sampling distributions of heritabilities are unknown (Tallis, 1959; Kendall and Stuart, 1963); standard errors, therefore, were estimated by the jackknife resampling procedure as shown in Efron (1982) but with the variance stabilizing transformation suggested by Arvesen and Schmitz (1970). At each iteration a family is omitted from the analysis of variance and a new heritability is calculated. Pseudovalues, θ_i, are then calculated as

$$\theta_i \cdot fh^2 - (f - 1)h^2_{-i}$$

where f is the number of families, h^2 is the heritability calculated with all 20 families, and h^2_{-i} is the heritability calculated with the ith family deleted. The average of the pseudovalues yields an estimator that is nearly unbiased and has a standard error of

$$SE_\theta = \sqrt{\frac{\Sigma(\theta - \theta_i)^2}{f(f - 1)}}$$

where θ is the average of pseudovalues.

Since seeds borne on the same primary umbel share a common maternal environment, covariances due to additive genetic maternal effects may lead to greater than expected similarities among sibs. Therefore, these effects were isolated and identified. In *Pastinaca sativa,* furanocoumarins are localized in the testa (Ladygina et al., 1970) and thus represent the maternal genotype. For each family, 10 seeds were soaked in distilled water for 24 hours, stripped of their coats, potted, and arranged randomly in a greenhouse on the UIUC campus; germination is not affected by seed coat removal (pers. observ.). The potting soil consisted of a 2:4:2:2 mixture of sand, Drummer silt loam, peat, and perlite. Seed coats were analyzed for furanocoumarins. Plants were left in the greenhouse until minimum flowering size was obtained (approximately four months [Baskin and Baskin, 1979]), then placed outdoors in a sheltered area for three months to satisfy the obligatory preflowering chilling requirement (Baskin and Baskin, 1979), and returned to the greenhouse to flower and set seed. When plants arising from the stripped seeds themselves set seed,

they were analyzed for furanocoumarin content of seed coat. This procedure provided data to estimate additive-genetic maternal effects on furanocoumarin production in the offspring by setting the covariances equal to their genetic expectations and solving for the variance components. Parent-offspring and between-family variance have expected genetic composition as follows:

$$Cov_{PO} = \frac{1}{2}\sigma_A^2 + \frac{1}{2}\sigma_{Am}^2$$

$$\sigma_F^2 = \frac{1}{4}\sigma_A^2 + \sigma_{Am}^2$$

where σ_A^2 is the additive genetic component and σ_{Am}^2 is the additive genetic maternal effect. These equations ignore variance due to nonadditive effects (Dickerson, 1969) and assume that additive-by-additive maternal effects are negligible. With two equations and two unknowns (σ_A^2, σ_{Am}^2) we can solve for the unknowns as follows:

$$\sigma_{Am}^2 = \frac{2(2\sigma_F^2 - Cov_{PO})}{3}$$

$$\sigma_A^2 = \frac{4(2\,Cov_{PO} - \sigma_F^2)}{3}.$$

The parent-offspring regression data were also used to provide a second estimate of heritability in furanocoumarin production. The formulae for calculating heritabilities from half-sib offspring on single parent regressions were those shown in Falconer (1981).

Determination of Resistance Factors and of Genetic Bases for Resistance

Half-sib families in the Phillips Tract garden were exposed to natural levels of herbivory and were censused regularly throughout 1984 for the presence of D. *pastinacella* larvae and other herbivores. Leaves of overwintering rosettes were inspected for eggs throughout May 1984, and leaf samples were collected for chemical analysis. After larvae vacated the umbels to pupate and seeds had ripened, the entire primary umbel was collected for chemical analysis and counting. Although *P. sativa* has some ability to compensate for floral loss in the primary umbel by increasing seed set in higher-order umbels (Hendrix, 1979), the compensatory ability is limited to certain size classes of plants, and the largest proportion of viable seeds is borne by the primary umbel (Hendrix, 1979, 1984; pers. observ.). Thus, the reproductive success of the primary umbel is an appropriate indicator of overall plant fitness. Estimates of heritability were obtained from plants grown in the field plot for the following characters:

1) earliest flowering date of the primary umbel;
2) final plant weight—based on oven dry weights of leaves, stems, and reproductive parts;
3) absolute amount of each of six furanocoumarins in leaves at time of oviposition (May 1984);
4) absolute amount of each of six furanocoumarins in mature seeds;
5) relative proportion of each of six furanocoumarins in leaves at time of oviposition;
6) relative proportion of each of six furanocoumarins in mature seeds;
7) total amount of furanocoumarins in leaves at time of oviposition;
8) total amount of furanocoumarins in mature seeds.

Resistance in the field to *D. pastinacella* was estimated in two ways:

1) proportion of umbellets undamaged in the primary umbel;
2) number of undamaged, filled seeds produced by each individual surviving to set seed.

The number of undamaged seeds produced is the most direct measure of fitness in that, in a biennial species, it represents a plant's reproductive contribution to the next generation; the proportion of undamaged umbellets in the primary umbel provides an index to the location and timing of insect damage. These variables were compared among individuals to identify phenotypic sources of resistance, and among families (with mean values) to identify between-family sources of resistance.

Stepwise multiple regression (Nie et al., 1975) was used to determine which concentrations or combinations of phytochemicals and other plant characters accounted for variation in seed production in the presence of herbivores; in other words, stepwise multiple regression was used to identify "resistance factors," plant characteristics (independent or explanatory variables) associated with variation in resistance estimates (the dependent or response variables). A forward stepping procedure based on F-to-enter (4.0) and F-to-remove (3.9) limits was employed. Independent variables with minimum F-to-enter value were added to the model.

Estimate of Directional Selection by *D. pastinacella on* P. sativa

The change in a character associated with selection, such as by insect herbivory, is the selection differential, the difference between the mean value of a character before and after selection. Lande and Arnold (1983) showed that the covariance between relative fitness and a character is equivalent to the selection differential for that character. Lande and Arnold's analysis was originally designed to assess the impact of a single episode of selection. In the present context, however, our experiments were designed and carried out prior to the publication of Lande and Arnold's paper and therefore are not entirely consistent with the optimum design. Thus, the measures of selection represent not one selective event but the net result of all selective events during the life of the plant. We estimated selection differentials for resistance factors in field and greenhouse populations as the covariance between relative fitness and the resistance characters. The characters used were those identified by stepwise regression as the best indicators of insect resistance. Relative fitness was calculated as the number of seeds divided by the mean number of seeds for field plants and by number of secondary rays divided by mean number of secondary rays, a correlate of potential seed production (vide infra), for greenhouse plants. Mean relative fitness in both cases is unity. Chemicals that are environmentally induced, e.g., by insect damage, are not suitable for this type of analysis; however, there is no evidence for damage-induced changes in furanocoumarins in wild parsnip, and attempts to demonstrate such changes in total furanocoumarin production have failed to yield significant effects (unpubl.).

While the selection differential measures

the effect of selection on a character, the selection gradient estimates the intensity of selection. The selection gradient is the regression of relative fitness on a character, and the steepness of the selection gradient is equivalent to the magnitude of the regression coefficient. Selection gradients were estimated by multiple regression of relative fitness on resistance factors. The partial regression coefficients thus obtained estimate the selection gradient for each character. These estimates measure the intensity of selection acting directly upon a character (Lande and Arnold, 1983).

Estimate of the Phenotypic Costs of Resistance and the Genetic Constraints on Insect Resistance in P. sativa

To estimate the cost and limits of resistance, we examined phenotypic and genetic correlations between factors associated with resistance in the field and factors providing an index to the reproductive fitness of plant individuals free of herbivory. The genetic correlation measures the degree to which two characters, in this case, furanocoumarin production and fitness, are controlled by the same gene or different linked genes (Falconer, 1981). The sign of the genetic correlation is indicative of the direction of correlated response to selection. For two traits that are positively correlated, for example, selection for an increase in one trait will result in an increase in the correlated trait. The magnitude and sign of the genetic correlation can give an indication of the effectiveness of selection by insect herbivores on plant chemical traits over time.

Genetic correlations between characters x and y were estimated as:

$$r_A = \frac{\text{Cov}_F(x, y)}{\sqrt{V_F(x) \cdot V_F(y)}}$$

where r_A, the correlation due to additive genetic effects, is the between-family covariance of characters x and y divided by the product of the between-family component of variance of x and y as determined by ANOVA (Falconer, 1981). The estimates of genetic correlation and its standard error were calculated by the jackknife procedure as described earlier but without the transformation, since no transformation has been found suitable (Arvesen and Schmitz, 1970).

In the greenhouse, where the 20 half-sib families were maintained free from insect herbivory, the following "cost" parameters, in addition to the furanocoumarin content and composition of leaves and seeds, were monitored over the course of development of the plants:

1) Photosynthetic potential, as measured by leaf conductance to water (cm/sec, measured with a Lamba leaf porometer); conductance is proportional to CO_2 flux and thus photosynthetic rate, assuming cell, chloroplast, and chemical resistances are all constant (Nobel, 1974).
2) Number of primary rays in the primary umbel—as a correlate of umbel size and potential seed production.
3) Number of secondary rays in the primary umbel—as an estimate of potential seed production. Each secondary ray bears one schizocarp, which splits at maturity into two mericarps or "seeds." Using secondary rays as a measure of seed potential obviates problems with inadequate pollination as a factor in seed production in the greenhouse, where plants were manually pollinated.
4) Final plant weight—based on oven dry weights of leaves, stems, and reproductive parts.

A high metabolic cost of production ostensibly would be reflected by a phenotypic correlation between furanocoumarin content and leaf conductance (photosynthetic potential) or by a negative correlation between furanocoumarin production and seed production or plant weight. These relationships were evaluated by examining product-moment correlations among phenotypes. In addition to cost, there may be genetic limitations on resistance. If a single gene controls the quantities of two characters that enhance resistance but increases one character and simultaneously reduces the other, overall improvement of resistance may be impeded. Genetic constraints were evaluated by examination of genetic correlations among resistance characters and fitness characters. For field plants, the only measure of metabolic costs used was plant weight, since destruction of primary umbels by *D. pastinacella* precluded enumerating potential seed or secondary ray number.

TABLE 1.   A) List of plant characters studied. B) Key for abbreviations of furanocoumarins.

A. Character	Plants measured
Flowering date	Field
Number of seeds	Field
Number of primary rays	Greenhouse
Number of secondary rays	Greenhouse
Leaf conductance	Greenhouse
Plant weight	Field and greenhouse
Furanocoumarins	Field and greenhouse

B. Furanocoumarins	Seed codes		Leaf codes	
	Amount	Proportion	Amount	Proportion
Imperatorin	IMPs	pIMPs	IMPl	pIMPl
Bergapten	BERs	pBERs	BERl	pBERl
Isopimpinellin	ISOs	pISOs	ISOl	pISOl
Xanthotoxin	XANs	pXANs	XANl	pXANl
Sphondin	SPHs	pSPHs	SPHl	pSPHl
Total	TOTs		TOTl	

RESULTS

Estimation of Genetic Control of Furanocoumarin Production

Table 1 contains a key for all abbreviations used in presenting the experimental results. Using data from both field and greenhouse populations, we were able to obtain heritability estimates on many aspects of furanocoumarin production (Table 2). Significant heritability estimates were found for nearly all of the characters, and on occasion some estimates approached unity, indicating complete genetic control over observed variation in certain traits (e.g., sphondin production in seeds) within the study environment. The heritabilities of seed furanocoumarins based on parent-offspring regression were significantly correlated with those obtained by half-sib analysis ($r = 0.657$, $P < 0.05$), suggesting that the estimates are accurate. Maternal effects on seed furanocoumarin content were negligible compared to total variation in seed content (Table 3).

Determination of Resistance Factors and of Genetic Bases for Resistance

Virtually all insect damage experienced by experimental plants was directly attributable to *D. pastinacella*; regular inspection

TABLE 2.   Heritability estimates for seed and leaf furanocoumarins in *Pastinaca sativa* from field and greenhouse half-sib families. Standard errors are in parentheses.

Seed Furanocoumarins

Source	IMPs	BERs	ISOs	XANs	SPHs	TOTs	pIMPs	pBERs	pISOs	pXANs	pSPHs
Half-sib field	0.70 (0.298)	0.54 (0.287)	0.98 (0.392)	0.61 (0.358)	0.62 (0.297)	0.67 (0.321)	0.39 (0.304)	0.94 (0.308)	0.85 (0.418)	0.39 (0.355)	0.49 (0.360)
Half-sib greenhouse	0.34 (0.258)	0.05 (0.167)	0.37 (0.222)	0.86 (0.476)	1.05 (0.421)	0.40 (0.242)	0.30 (0.228)	0.51 (0.261)	0.96 (0.356)	0.85 (0.494)	1.06 (0.467)
Offspring-parent, greenhouse	0.41 (0.166)	0.19 (0.158)	* (—)	0.65 (0.208)	1.43 (0.372)	0.28 (0.156)	1.21 (0.282)	0.45 (0.200)	1.43 (0.334)	0.53 (0.254)	1.13 (0.230)

Leaf Furanocoumarins

Source	IMPl	BERl	ISOl	XANl	SPHl	TOTl	pIMPl	pBERl	pISOl	pXANl	pSPHl
Half-sib, field	0.28 (0.310)	0.18 (0.214)	0.59 (0.419)	0.34 (0.291)	1.17 (0.473)	0.52 (0.402)	0.64 (0.254)	0.06 (0.139)	0.15 (0.195)	0.36 (0.217)	1.09 (0.342)
Half-sib, greenhouse	0.14 (0.147)	0.53 (0.225)	0.33 (0.161)	0.75 (0.283)	0.43 (0.217)	0.37 (0.222)	0.71 (0.326)	0.42 (0.197)	0.45 (0.176)	0.42 (0.252)	0.10 (0.139)

* Negative value.

TABLE 3. Estimates of maternal effects on furanocoumarin content in seeds of *Pastinaca sativa*.

Furanocoumarins	Maternal variance	Proportion of total phenotypic variance contributed by maternal effects[1]
IMPs	−3.323	0
BERs	−0.424	0
ISOs	*	*
XANs	2.291	0.060
SPHs	0.390	0.100
TOTs	−1.186	0

[1] Negative estimates of maternal variance are treated as zeros.
* Values not calculated owing to the finding of a negative covariance between offspring and parents.

revealed remarkably few other herbivores. Close examination of leaves of all plants in May 1984 revealed that approximately 99% of 187 surviving individuals in the test plot contained eggs of *D. pastinacella*. Thus, very few plants were rejected outright by ovipositing females.

In the stepwise regression analysis, two estimates of resistance were used separately as dependent variables—percentage of undamaged umbellets and the number of undamaged seeds produced. The principal independent variable accounting for a significant amount of variance in umbellet damage both among individual plants and among half-sib families was the flowering date (Table 4). Plants in the experimental garden that flowered early were able to produce a higher proportion of umbellets that escaped damage. When number of seeds was

compared among individuals irrespective of family affiliation, flowering date again accounted for the largest amount of variance (Fig. 2). However, when mean number of seeds per family was examined, four furanocoumarin variables accounted for a significant amount of variance (Table 4): the proportion of bergapten in the seeds (pBERs), the absolute amount of sphondin in the seeds (SPHs), the proportion of sphondin in the leaves (pSPHl) and the absolute amount of bergapten in the leaves (BERl). Collectively, these variables accounted for almost 75% of the variation in seed production in the presence of *D. pastinacella*. All of these resistance components have significant heritabilities in field or greenhouse environments (Table 2).

Estimate of Directional Selection by D. pastinacella on P. sativa Characters Associated with Resistance

There were pronounced family differences in seed set; mean seed production for a family in the field varied from 3.7 seeds/primary umbel to 446.0/primary umbel, and individual seed production ranged from 0 to 1,664. In the field, significant selection differentials were found for flowering date and proportion of bergapten in the seeds (Table 5), indicating that selection had shifted the distribution of both characters. The positive and negative selection differentials found for proportion of bergapten in seeds

TABLE 4. Stepwise regressions between webworm resistance estimators and plant characters.

Independent variable	% of variation explained	Cumulative variation explained	Standardized regression coefficient	Regression F (d.f.)	P
Dependent variable: proportion of undamaged umbellets					
Flowering date	14.6	14.6	0.312	11.85 (3,110)	<0.01
Plant weight	5.8	20.3	−0.235		
pBERs	4.1	24.4	−0.212		
Dependent variable: seed number					
Flowering date	24.5	24.5	−0.471	24.9 (3,110)	<0.01
Plant weight	11.7	36.2	0.349		
pISOs	4.2	40.4	0.205		
Dependent variable: mean proportion of undamaged umbellets per family					
Flowering date	31.8	31.8	−0.564	8.4 (1,18)	<0.05
Dependent variable: mean seed number per family					
pBERs	36.3	36.3	0.975	10.5 (4,15)	<0.01
SPHs	13.9	50.2	0.508		
pSPHl	10.6	60.7	−0.374		
BERl	12.9	73.6	−0.370		

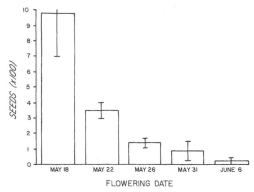

FIG. 2. Mean seed production of parsnip plants with respect to timing of first flowering in 1984 in the Phillips Tract experimental garden. Each mean is bracketed by its standard error.

and flowering date, respectively, indicate that plants with early flowering date and a high proportion of bergapten in the seeds have higher fitness. Since the partial regression coefficient for flowering time was not statistically significant, selection must have acted only indirectly on flowering date. The significant selection differential in flowering date may in part be attributed to selection acting directly on the proportion of bergapten in the seeds, one character for which a significant regression coefficient was found. The phenotypic correlation between flowering date and proportion of bergapten in seeds is negative and significant. Thus, se-

lection for increased proportion of bergapten in seeds automatically results in a negative selection differential in flowering date. Significant partial regression coefficients were found for the other three chemical resistance characters as well. In the greenhouse, a significant selection differential was found for only one of the characters, proportion of bergapten in seeds (Table 6). In this case, none of the partial regression coefficients were significant. The negative selection differential for proportion of bergapten in seeds indicates that the greenhouse plants with higher values of proportion of bergapten in seed had lower fitness.

Genetic and Phenotypic Correlations between Fitness Parameters and Resistance Factors

Phenotypic and genetic correlations between statistically significant resistance factors and fitness parameters were measured in the field and greenhouse populations (Tables 7 and 8). None of the genetic correlations in the field were significant, but seven furanocoumarin characters measured in the greenhouse—including the major resistance factors, the proportion of bergapten in seeds and the amount of bergapten in seeds—had significant negative genetic correlations with the number of secondary rays. Photosynthetic potential (leaf conductance) was not genetically correlated with any of the furanocoumarin characteristics, nor was flow-

TABLE 5. Directional selection differentials, partial regression coefficients, and phenotypic correlations among resistance characters in field parsnips. FD = flowering date; for other abbreviations, see Table 1.

Character	Selection differential[1]	Partial regression coefficient	Standardized regression coefficient
FD	−0.1891*	−1.80	−0.148
pBERs	0.0107*	21.69**	0.883**
BERl	−0.0027	−11.07*	−0.354*
pSPHl	−0.0086	−3.38*	−0.351*
SPHs	0.0106	0.32**	0.500**

	Phenotypic Correlation Matrix[2]				
	FD	pBERs	BERl	pSPHl	SPHs
FD	1.000	−0.554*	−0.031	0.079	0.263
pBERs		1.000	0.216	0.092	−0.508*
BERl			1.000	−0.082	−0.126
pSPHl				1.000	0.121
SPHs					1.000

* $P < 0.05$ ** $P < 0.01$.
[1] Significance level is for the product-moment correlation of relative fitness with the character.
[2] Significance level is for the product-moment correlation among pairs of resistance factors.

TABLE 6. Directional selection differentials, partial regression coefficients, and phenotypic correlations among resistance characters in greenhouse parsnip. FD is flowering date; for other abbreviations, see Table 1.

Character	Selection differential[1]	Partial regression coefficient	Standardized regression coefficient
pBERs	−0.00142*	−1.61	−0.463
BERl	−0.00082	−0.16	−0.044
pSPHl	−0.00091	−0.38	−0.180
SPHs	0.03104	0.01	0.158

Phenotypic Correlation Matrix[2]

	pBERs	BERl	pSPHl	SPHs
pBERs	1.000	0.321	−0.070	−0.175
BERl		1.000	0.463*	−0.165
pSPHl			1.000	−0.147
SPHs				1.000

* $P < 0.05$.
[1] Significance level is for the product-moment correlation of relative fitness with the character.
[2] Significance level is for the product-moment correlation among pairs of resistance factors.

TABLE 7. Significant phenotypic correlations between leaf or seed furanocoumarins and fitness characters, and between fitness characters in greenhouse and field plants.

	r	SE
Greenhouse ($N = 171$)		
IMPs × number of secondary rays	0.224	0.075
IMPl × number of primary rays	0.204	0.075
BERl × number of secondary rays	−0.167	0.076
BERl × number of primary rays	−0.161	0.076
Number of primary rays × number of secondary rays	0.874	0.037
Field ($N = 143$)		
ICOl × plant weight	−0.206	0.082
XANl × plant weight	0.217	0.082
TOTl × plant weight	−0.202	0.082

ering date, a significant correlate of the proportion of umbellets escaping attack by *D. pastinacella*. Potential seed production, as measured by both primary and secondary ray number, was negatively correlated phenotypically with only one furanocoumarin trait in the greenhouse plants (amount of bergapten in seeds), and it was positively correlated with imperatorin in seeds and leaves; two individual furanocoumarins in the leaves (isopimpinellin and xanthotoxin), as well as total furanocoumarin production, were negatively correlated phenotypically with plant weight in field plants exposed to herbivory (Table 7).

DISCUSSION

Stepwise multiple regression revealed that four furanocoumarin variables accounted for almost 75% of the among-family vari-

ation in seed set in the presence of parsnip webworms. Since all four variables have a significantly high heritability (Table 2), it can be concluded that variation in resistance to parsnip webworm in the wild parsnip is in part genetically based. The opposite signs of the regression coefficients for leaf and seed resistance factors suggest that the mechanism of furanocoumarin action with respect to the insect differs depending on plant part. High amounts of furanocoumarins in seed are associated with increased seed set, consistent with a defensive (allomonal) effect. High amounts of furanocoumarins in leaves, which are substantially lower in absolute terms than even low levels in seeds, are associated with reduced seed set, consistent with a host recognition (kairomonal) effect. Physiological and/or behavioral effects of furanocoumarins on *D.*

TABLE 8. Significant genetic correlations between seed or leaf furanocoumarins and fitness characters and between fitness characters in greenhouse plants.

		SE
BERl × number of secondary rays	−0.632	0.309
BERl × number of primary rays	−0.675	0.297
XANl × number of secondary rays	−0.790	0.340
XANl × number of primary rays	−0.643	0.295
SPHl × number of secondary rays	−0.688	0.267
pBERs × number of secondary rays	−1.528	0.689
pISOs × number of secondary rays	−0.888	0.408
Number of primary rays × number of secondary rays	0.809	0.183

pastinacella may thus be dose-dependent. Undoubtedly, the field environment differs from that of the greenhouse in more ways than the presence or absence of herbivores, but, given the overwhelming impact of the herbivore on seed production, it is instructive to compare field and greenhouse populations as though comparing herbivore-infested and herbivore-free treatments. That furanocoumarin chemistry is important in determining fitness in *P. sativa* in the presence of *D. pastinacella* is suggested by the fact that, while in the greenhouse several furanocoumarins are negatively correlated genetically with the number of secondary rays and hence with fitness, no furanocoumarins are significantly negatively correlated with seed set in the field. In other words, production of furanocoumarins, despite negative genetic correlations with seed production in the greenhouse, increases fitness in the presence of the herbivore in the field.

Both qualitative and quantitative aspects of furanocoumarin chemistry figure in resistance to parsnip webworm herbivory. One possible reason for the dearth of direct evidence that allelochemical variation can account for herbivore variation in other plant-insect associations (see Chew and Rodman, 1979) is that much attention to date has been focused on single-chemical or single-gene resistance factors. This approach, fostered in all probability by early studies demonstrating single gene-for-gene interactions in systems such as Hessian fly and wheat (Hatchett and Gallun, 1970), focused attention on systems in which single chemicals, often controlled at a single locus (e.g., gossypol [Wilson and Shaver, 1973]), were known to function in insect resistance. However, the applicability of the gene-for-gene concept in agricultural situations may simply be an artifact of plant breeding techniques, in which intense artificial selection removes associated protective polygenic effects from oligogenic resistance (Johnson, 1961). At present, no gene-for-gene systems have been documented in nature (Day, 1974). Gould (1983) convincingly argues that, in natural systems, resistance is likely to be quantitative and polygenic. Ostensibly, the involvement of several loci can retard the acquisition of resistance on the part of the herbivore. In the interaction between the wild parsnip and the parsnip webworm, no single furanocoumarin can account for resistance, nor can the total amount of furanocoumarins in leaf or seed. Instead, the content and composition of furanocoumarins throughout the plant are involved in determining resistance. As Feeny (1976) predicted, specialists may vary in sensitivity to representatives of a single class of chemical in their food plants; an adaptation to furanocoumarins as a class in *D. pastinacella* does not confer absolute resistance to all structural types and combinations of furanocoumarins.

While furanocoumarins are undoubtedly associated with resistance to webworms, at least one phenological factor also influences the extent of webworm damage, albeit indirectly. Flowering date is significantly correlated with the percentage of umbellets in the primary umbel escaping damage and subsequent seed set (Fig. 2). This phenological trait has, like the furanocoumarin resistance factors, a signficant heritability ($h^2 = 0.562 \pm 0.269$). Some measure of "escape in time" (Feeny, 1976) is entirely consistent with what is known of the life histories of both plant and herbivore. Parsnip webworms overwinter as adults and become active very early in spring; females were collected flying at night in March at temperatures near 10°C, and eggs were found on leaves of plants just beginning to leaf out (Nitao, unpubl.). Wild parsnip is primarily restricted to north temperate regions and is a cold-climate plant; indeed, *P. sativa* has an obligate chilling requirement prior to flowering (Baskin and Baskin, 1979). In spring, then, the opportunity exists for the wild parsnip to "outgrow" the webworm by flowering sufficiently early such that at least some umbellets of the primary umbel can flower free from insect damage.

Wild parsnip families differed greatly in their susceptibility to parsnip webworm, as shown by the wide range of seed production among families (means ranged from 4 to 446 seeds in the primary umbel). Significant selection differentials were found for several of the resistance traits (Table 5). Further analysis by multiple regression suggests that selection acts directly on all four chemical resistance factors but only indirectly on flowering time. Since flowering time has a

1226 M. R. BERENBAUM ET AL.

TABLE 9. Genetic correlations between resistance factors identified by stepwise regression. Standard errors are in parentheses. FD is flowering date; for explanation of other abbreviations, see Table 1.

	FD	pBERs	SPHs	pSPHl	BERl
FD	1.000	−0.853 (1.364)	−0.034 (0.449)	−0.086 (0.482)	0.021 (0.873)
pBERs		1.000	−0.855 (0.346)	0.677 (0.700)	0.303 (0.430)
SPHs			1.000	−0.757 (0.542)	−2.824 (2.187)
pSPHl				1.000	0.001 (0.363)
BERl					1.000

negative phenotypic correlation with proportion of bergapten in seeds, the selection differential for flowering time may be attributed to direct selection for the chemical trait with a resulting decrease in flowering time. Phenotypic correlations may also explain the absence of a selection differential even though a character is under direct selection. Selection acted directly to increase both proportion of bergapten and amount of sphondin in seeds, but a significant selection differential was found only for the proportion of bergapten in seeds. In this case, the amount of sphondin and proportion of bergapten in seeds have a negative phenotypic correlation, and stronger selection for a higher proportion of bergapten may have nullified the effect of selection for increased sphondin. In the greenhouse the only character that displayed a significant selection differential was the proportion of bergapten in seeds. In this case, however, plants with a high proportion of bergapten in seeds had reduced fitness. Since all five resistance characters have significant heritabilities, the mechanism exists to allow herbivores, with sufficient selective pressure, to cause changes in the chemical and hence genetic composition of a population, one principal requirement for demonstrating coevolution (Janzen, 1980).

Genetic correlations among resistance factors shed light on the nature of the coevolutionary relationship between *D. pastinacella* and *P. sativa* (Table 9). A favorable genetic correlation is one in which two traits that convey resistance are positively correlated. Hence, selection by herbivores for increased resistance by either trait increases both traits simultaneously. No such favorable genetic correlations were observed in *P. sativa*. Instead, the only significant genetic correlation is a negative relationship ($r_A = -0.853$) between two traits conveying resistance, proportion of bergapten and amount of sphondin in seeds. This type of correlation offers a "no-win" situation, because selection for increased resistance in the form of an increased proportion of bergapten tends to decrease resistance in the form of a decreased amount of sphondin in the seeds. The absence of a significant selection differential for the amount of sphondin in seeds in the field families is evidence of the ineffectiveness of selection in this context. Moreover, several of the resistance factors accounting for variation in herbivory have negative genetic correlations with total seed production in greenhouse plants protected from insect herbivory (Table 8). In the presence of the herbivore, this fitness reduction associated with furanocoumarin production is more than offset by increased survivorship of the offspring, so that genotypes that are high in the proportion of bergapten in the seed are favored. In a future generation when the herbivore is absent, however, such genotypes would be at a selective disadvantage because they produce fewer seeds than genotypes susceptible to the herbivore. The finding of a significant negative selection differential for the proportion of bergapten in seed in the greenhouse is consistent with this interpretation.

The nature of reduced competitive ability due to furanocoumarin production may relate to nutrient limitations on furanocoumarin biosynthesis. Light-induced biosynthesis of furanocoumarins, for example, does not take place under conditions of low nutrient availability (Berenbaum and Zangerl, unpubl.). Moreover, there are significant negative correlations between furanocoumarins that share a common biosynthetic precursor. Angular furanocoumarins are negatively genetically correlated with linear furanocoumarins, and 8-substituted furanocoumarins (xanthotoxin and imperatorin) are genetically correlated negatively with 5-substituted furanocoumarins (bergapten, isopimpinellin), suggesting some limitation on the availability of precursor molecules at branch points of a biosynthetic pathway (Berenbaum and Zangerl, unpubl.).

The interaction between the parsnip webworm and the wild parsnip, then, is at an ecological standoff of sorts. The "genetical constraints" (Berry, 1985) on the response of wild parsnip to selection—in the form of high genetic correlations among traits—tend to limit resistance in the plant population when the herbivore is present but also act to reduce resistance when the herbivore is absent. To extend the "evolutionary arms race" concept advanced by Whittaker and Feeny (1971), the interaction between wild parsnip and the parsnip webworm, at least in this central Illinois population, has reached a temporary stalemate. This situation will persist until such time that the environment changes as to favor herbivore or host or until there is a genetic change producing new resistance traits or more variation in existing resistance traits.

ACKNOWLEDGMENTS

We thank M. Rausher for valuable discussion in the planning stages of this project, E. Levine for generously allowing us access to greenhouse facilities, R. Sherman for technical assistance, and E. Heininger, M. Lynch, J. Neal, S. Passoa, S. Sandberg, and N. Wiseman for comments on the manuscript. We especially thank Michael Grossman of the Department of Animal Sciences at UIUC for continuous patient instruction and guidance in quantitative genetics. This research was supported by NSF grant BSR 82-14713 to M. R. Berenbaum.

LITERATURE CITED

ARVESEN, J. H., AND T. H. SCHMITZ. 1970. Robust procedures for variance component problems using the jackknife. Biometrics 26:677–686.

BASKIN, J. M., AND C. M. BASKIN. 1979. Studies on the autecology and population biology of the weedy monocarpic perennial, *Pastinaca sativa*. J. Ecol. 67: 601–610.

BELL, C. R. 1971. Breeding systems and floral biology of the Umbelliferae or evidence for specialization in unspecialized flowers, pp. 93–108. *In* V. Heywood (ed.), The Biology and Chemistry of the Umbelliferae. Academic Press, London, U.K.

BERENBAUM, M. 1978. Toxicity of a furanocoumarin to armyworms: A case of biosynthetic escape from insect herbivores. Science 201:532–534.

———. 1981*a*. Effects of linear furanocoumarins on an adapted specialist insect (*Papilio polyxenes*). Ecol. Entomol. 6:345–351.

———. 1981*b*. Patterns of furanocoumarin production and insect herbivory in a population of wild parsnip (*Pastinaca sativa* L.). Oecologia 49:236–244.

BERENBAUM, M., AND P. FEENY. 1981. Toxicity of angular furanocoumarins to swallowtails: Escalation in the coevolutionary arms race. Science 212: 927–929.

BERENBAUM, M. R., A. R. ZANGERL, AND J. K. NITAO. 1984. Furanocoumarins in seeds of wild and cultivated parsnip. Phytochemistry 23:1809–1810.

BERRY, R. J. 1985. The process of pattern: Genetical possiblities and constraints in coevolution. Oikos 44:222–228.

CARROLL, C. R., AND C. A. HOFFMAN. 1980. Chemical feeding deterrent mobilized in response to insect herbivory and counteradaptation by *Epilachna tredecimnotata*. Science 209:414–416.

CHEW, F. S., AND J. E. RODMAN. 1979. Plant resources for chemical defense, pp. 271–307. *In* G. A. Rosenthal and D. H. Janzen (eds.), Herbivores: Their Interaction with Secondary Plant Metabolites. Academic Press, N.Y.

CRUDEN, R. W., AND S. M. HERMANN-PARKER. 1977. Temporal dioecism: An alternative to dioecism? Evolution 31:863–866.

DAY, P. R. 1974. Genetics of Host-Parasite Interaction. Freeman, San Francisco, CA.

DICKERSON, G. E. 1969. Techniques for research in quantitative animal genetics, pp. 36–79. *In* Techniques and Procedures in Animal Science. Amer. Soc. Animal Sci., Albany, N.Y.

DOLINGER, P. M., P. R. EHRLICH, W. L. FITCH, AND D. E. BREEDLOVE. 1973. Alkaloid and predation patterns in Colorado lupine populations. Oecologia 13: 191–204.

EDMUNDS, G. F., AND D. N. ALSTAD. 1978. Coevolution in insect herbivores and conifers. Science 199:941–945.

EFRON, B. 1982. The Jackknife, the Bootstrap and Other Resampling Plans. Society for Industrial and Applied Mathematics Philadelphia, PA.

EHRLICH, P. R., AND R. H. RAVEN. 1964. Butterflies and plants: A study in coevolution. Evolution 18: 586–608.

FALCONER, D. S. 1981. Introduction to Quantitative Genetics, 2nd Ed. Longman, N.Y.

FEENY, P. 1976. Plant apparency and chemical defense. Rec. Adv. Phytochem. 10:1–40.

FERNALD, M. L. 1950. Gray's Manual of Botany, 8th Ed. American Book, N.Y.

FOWLER, S. V., AND J. H. LAWTON. 1985. Rapidly

1228   M. R. BERENBAUM ET AL.

induced defences and talking trees: The devil's advocate position. Amer. Natur. 126:181–195.

GORDER, N. K. N., AND J. W. MERTINS. 1984. Life history of the parsnip webworm *Depressaria pastinacella* (Lepidoptera: Oecophoridae), in central Iowa. Ann. Entomol. Soc. Amer. 77:568–573.

GOULD, F. 1983. Genetics of plant herbivore systems: Interactions between applied and basic study, pp. 599–653. *In* R. F. Denno and M. S. McClure (eds.), Variable Plants and Herbivores in Natural and Managed Systems. Academic Press, N.Y.

GREEN, T. R., AND C. A. RYAN. 1972. Wound-induced proteinase inhibitor in plant leaves: A possible defense mechanism against insects. Science 175:776–777.

HATCHETT, J. H., AND R. L. GALLUN. 1970. Genetics of the ability of the Hessian fly, *Mayetiola destructor*, to survive on wheats having different genes for resistance. Ann. Entomol. Soc. Amer. 63:1400–1407.

HAUKIOJA, E. 1980. On the role of plant defenses in the fluctuation of herbivore populations. Oikos 35: 202–213.

HENDRIX, S. D. 1979. Compensatory reproduction in a biennial herb following insect defloration. Oecologia 42:107–118.

———. 1984. Variation in seed weight and its effects on germination in *Pastinaca sativa* L. (Umbelliferae). Amer. J. Bot. 71:795–802.

HENDRIX, S., AND E. J. TRAPP. 1981. Plant-herbivore interactions: Insect induced changes in host plant sex expression and fecundity. Oecologia 49:119–122.

HEYWOOD, V. H., EDITOR. 1971. The Biology and Chemistry of the Umbelliferae. Academic Press, London, U.K.

HODGES, R. W. 1974. Gelechioidea: Oecophoridae. The Moths of America North of Mexico. Classey, London, U.K.

JANZEN, D. H. 1980. When is it coevolution? Evolution 34:611–612.

JERMY, T. 1984. Evolution of insect/host relationships. Amer. Natur. 124:609–630.

JOHNSON, T. 1961. Man-guided evolution in plant rusts. Science 133:357–362.

KENDALL, M. G., AND A. STUART. 1963. The Advanced Theory of Statistics, Vol. 1, 2nd Ed. Griffin, London, U.K.

LADYGINA, E. Y., V. A. MAKAROVA, AND N. S. IGNAT'EVA. 1970. Morphological and anatomical description of *Pastinaca sativa* fruit and localization of furanocoumarins in them. Farmatsyai 19:39–46. (In Russian.)

LANDE, R., AND S. J. ARNOLD. 1983. The measurement of selection on correlated characters. Evolution 37:1210–1226.

LIN, J., M. H. DICKSON, AND C. J. ECKENRODE. 1984. Resistance of *Brassica* lines to the diamond back moth (Lipidoptera: Yponomeutidae) in the field and the inheritance of resistance. J. Econ. Entomol. 77: 1293–1296.

MARQUIS, R. J. 1984. Leaf herbivores decrease fitness of a tropical plant. Science 226:537–539.

MITCHELL-OLDS, T., AND J. J. RUTLEDGE. 1986. Quantitative genetics in natural plant populations: A review of the theory. Amer. Natur. 127:379–402.

MORAN, N. 1981. Intraspecific variability in herbivore performance and host quality: A field study of *Uroleucon caligatum* (Homoptera: Aphididae) and its *Solidago* hosts (Asteraceae). Ecol. Entomol. 6:301–306.

MORROW, P. S., AND V. C. LaMARCHE. 1978. Tree ring evidence for chronic insect suppression of productivity in subalpine *Eucalyptus*. Science 201: 1244–1246.

MUCKENSTURM, B., D. DUPLAY, P. C. ROBERT, M. T. SIMONIS, AND J. C. KIENLEN. 1981. Substances antiappétantes pour insectes phytophages présentes dans *Angelica sylvestris* et *Heracleum sphondylium*. Biochem. Syst. Ecol. 9:289–292.

NIE, N. H., C. H. HULL, J. G. JENKINS, K. STEINBRENNER, AND D. H. BENT. 1975. Statistical Package for the Social Sciences. McGraw-Hill, N.Y.

NOBEL, P. S. 1974. Introduction to Biophysical Plant Physiology. Freeman, San Francisco, CA.

RAUSHER, M. D., AND P. FEENY. 1980. Herbivory, plant density and plant reproductive success: The effect of *Battus philenor* on *Aristolochia reticulata*. Ecology 61:905–917.

RHOADES, D. F. 1979. Evolution of plant chemical defense against herbivores, pp. 3–54. *In* G. A. Rosenthal and D. H. Janzen (eds.), Herbivores: Their Interaction with Secondary Plant Metabolites. Academic Press, N.Y.

SERVICE, P. 1984. Genotypic interaction in an aphid-host plant relationship, *Uroleucon rudbeckiae* and *Rudbeckia laciniata*. Oecologia 61:271–276.

SOKAL, R. R., AND F. J. ROHLF. 1981. Biometry. Freeman, San Francisco, CA.

STURGEON, K. B. 1979. Monoterpene variation in ponderosa pine xylem resin related to western pine beetle predation. Evolution 33:803–814.

TALLIS, G. M. 1959. Sampling errors of genetic correlation coefficients calculated from the analyses of variance and covariance. Aust. J. Stat. 1:35–43.

THOMPSON, J. N. 1977. Patch dynamics in the insect-*Pastinaca sativa* association: Life history tactics and population consequences. Ph.D. Diss. Univ. Illinois, Urbana-Champaign, IL.

———. 1978. Within-patch structure and dynamics in *Pastinaca sativa* and resource availability to a specialized herbivore. Ecology 59:443–448.

WHITTAKER, R. H., AND P. P. FEENY. 1971. Allelochemics: Chemical interactions between species. Science 171:757–770.

WILSON, F. D., AND T. N. SHAVER. 1973. Glands, gossypol content, and tobacco budworm development in seedlings and floral parts of cotton. Crop Sci. 13:107–110.

YAJIMA, T., N. KATO, AND K. MUNAKATA. 1977. Isolation of insect anti-feeding principles in *Orixa japonica* Thunb. Agric. Biol. Chem. 41:1263–1268.

Corresponding Editor: D. W. Schemske

Ecology, 64(6), 1983, pp. 1540–1548
© 1983 by the Ecological Society of America

AN EXPERIMENTAL TEST OF THE EFFECTS OF
PREDATION RISK ON HABITAT USE IN FISH[1]

Earl E. Werner, James F. Gilliam[2], Donald J. Hall, and
Gary G. Mittelbach[3]
*Kellogg Biological Station and Department of Zoology, Michigan State University,
Hickory Corners, Michigan 49060 USA*

Abstract. We present an experiment designed to test the hypothesis that fish respond to both relative predation risk and habitat profitability in choosing habitats in which to feed. Identical populations of three size-classes of bluegill sunfish (*Lepomis macrochirus*) were stocked on both sides of a divided pond (29 m in diameter), and eight piscivorous largemouth bass (*Micropterus salmoides*) were introduced to one side. Sizes of both species were chosen such that the small class of bluegills was very vulnerable to the bass, whereas the largest class was invulnerable to bass predation. We then compared mortality, habitat use, and growth of each size-class in the presence and absence of the bass.

Only the small size-class suffered significant mortality from the bass (each bass consumed on average about one small bluegill every 3.8 d); the two larger size-classes exhibited similar mortality rates on both sides of the pond. In the absence of the bass, we found that habitat use of all size-classes was similar and that the pattern of habitat use maximized foraging return rates (Werner et al. 1983). In the presence of the bass the two larger size-classes chose habitats to maximize return rates, but the small size-class obtained a greater fraction of its diet from the vegetation habitat, where foraging return rates were only one-third of those in the more open habitats. The small size-class further exhibited a significant depression in individual growth in the presence of the bass; the growth increment during the experiment was 27% less than that for small bluegills in the absence of the bass. Because of the reduced utilization of more open habitats by the small fish in the presence of bass, resources in these habitats were released to the larger size-classes, which showed greater growth in the presence of the bass than in its absence. We develop methods to predict the additional mortality expected on a cohort due to a reduction in growth rate (because individuals are spending a longer time in vunerable sizes), and discuss the potential for predation risk to enforce size-class segregation, which leads de facto to resource partitioning.

Key words: foraging efficiency; habitat use; Lepomis; *Michigan;* Micropterus; *Osteichthyes; predation risk; predator avoidance; size-class interactions.*

Introduction

Recently, optimality models have been usefully applied to problems in animal behavior and evolution. In general, the costs and benefits associated with particular behaviors are described and solutions derived which minimize a postulated cost/benefit function. When this approach is used to study the evolution of morphological structures or life histories, serious questions arise concerning genetic constraints and the existence of tradeoffs (e.g., Gould and Lewontin 1979). When applying the approach to the study of behavior, we are often able to measure directly the costs and benefits associated with a tradeoff and simply ask: does the individual organism over the short term have the capabilities to assess changes in its environment and have the flexibility to respond to these changes as the model predicts? A second related question, especially germane when such models are to be tested under relatively uncontrolled field situations, is: are the costs

and constraints conceived by the investigator sufficiently accurate and inclusive to account for the major selective forces that have molded the behavior(s) of interest?

We have experimentally demonstrated that fish have the capability to respond to changes in resource levels in the environment by modifying their selection of food particle size in approximate accordance with optimal foraging models (Werner and Hall 1974, Mittelbach 1981, Werner et al. 1983). We have further demonstrated that fish have the flexibility to shift habitats as relative resource levels in these habitats change and that we can predict these shifts, using the foraging models in small experimental ponds (Werner 1982, Werner et al. 1983). However, testing the predictions of such models under less controlled conditions is more difficult due to additional constraints which might be expected to modify optimal behavior, i.e., the second question above. For example, we have noted that in natural lakes small fish are restricted to weedbeds and do not conform to the predictions of models specifying optimal habitat use from the standpoint of foraging rates (Hall and Werner 1977, Mittelbach 1981). We postulate that predation risk due to piscivorous fish is responsible for this deviation from predicted behavior.

[1] Manuscript received 19 April 1982; revised 22 December 1982; accepted 29 December 1982.
[2] Present address: Department of Biological Sciences, State University of New York, Albany, New York 12222 USA.
[3] Present address: Department of Zoology, Ohio State University, Columbus, Ohio 43210 USA.

Often in natural communities a richer habitat, from the standpoint of potential foraging rate, is also one in which a forager experiences higher predation risk. Thus decisions on where to feed presumably involve some weighting of these factors according to their relative impact on fitness, i.e., there is a foraging rate/mortality risk tradeoff. It is therefore important to test whether animals assess predation risk and modify their foraging behavior accordingly, and to build this constraint into our models of optimal habitat use if they do.

A large literature documents the qualitative effects of predators on prey behavior (see Stein [1979] and Curio [1976] for reviews), but surprisingly few studies have quantified this effect, especially in the context of methods which predict the optimal behavior in the predator's absence (but see Milinski and Heller 1978, Caraço et al. 1980, Sih 1980). Clearly such studies are required if a quantitative theory of how animals adaptively balance these two conflicting demands is to be constructed.

In this paper we present an experiment designed to test the hypothesis that fish modify their habitat use under risk of predation and examine the consequences of such changes in behavior on individual growth rate, which is a major component of fitness and population dynamics in fish. Further, the magnitude of the growth depression when predators are present provides some index of the magnitude of the foraging rate/mortality risk tradeoff. In particular, we demonstrate that the presence of the largemouth bass (*Micropterus salmoides*, hereafter simply the bass) causes vulnerable sizes of the bluegill sunfish (*Lepomis macrochirus*) to utilize less profitable but safer habitats. We do this by first measuring resource levels and estimating habitat-specific foraging rates, which are used to predict optimal habitat use successfully in the absence of the bass (i.e., the control situation; see Werner et al. [1983] for details). We then contrast this case with habitat use by the bluegill in an identical environment in the presence of the bass. We further show that the changes in habitat use in the presence of the bass result in a significant decrease in growth rates of the vulnerable bluegill sizes. These results suggest that fish can balance the conflicting demands of foraging and predation risk but that this behavioral response occurs at some significant cost in terms of growth rate. We discuss the implications of these results for optimal foraging theory and the theory of species interactions.

EXPERIMENTAL DESIGN AND METHODS

The experiment was performed in a circular pond (29 m diameter, 1.8 m deep) at the Kellogg Biological Station. All macrophytes were removed from the pond except for a 3 m wide border of cattails (*Typha* spp.). This manipulation yielded three very discrete habitats: a ring of dense vegetation and an unstructured pond center of open water and bare sediments. The pond was then divided in half by a 0.6-cm mesh nylon partition, which was suspended from ropes and anchored to the pond bottom.

The experiment was initiated on 15 July 1979 and terminated by draining the pond on 28 September. Each half of the pond was stocked with identical bluegill populations: 500 small (35.5 ± 0.4 mm, average standard length measured from the tip of the snout to the posterior of the vertebral column ±1 SE), 300 medium (52.9 ± 0.4 mm), and 100 large (73.0 ± 0.8 mm) bluegills. These size-classes and relative proportions were similar to those found in bluegill populations of local lakes (Hall and Werner 1977). One-half of the pond was also stocked with eight bass (198.8 ± 2.9 mm in length). The size of the bass was carefully chosen to set up a gradient in predation risk for the different size-classes of bluegills. Using laboratory data on largemouth bass feeding (Lawrence 1957, Werner 1977) as a guide, we chose a bass size, such that the small bluegills would be extremely vulnerable and the large bluegills would be too large for the bass to catch and swallow. Bass of this size can swallow the medium size-class of bluegills in the laboratory, but it is doubtful that they could very easily capture bluegills of this size in the field. All surviving fish were recovered in September by draining the pond.

At intervals of 1 wk (or more frequently, in July), 10–20 small, 10 medium, and 10 large bluegills were seined and removed from each half of the pond for stomach analyses and determination of growth rates. We replaced the sampled fish with bluegills of identical length from a nearby holding pond. Because we wanted to make comparisons of bluegill growth in the presence and absence of the bass, we were concerned about the potentially confounding effects of reduced bluegill densities in the one pond-half due to predation by the bass. In an attempt to minimize this difference, we assumed that the bass would be eating primarily the small size-class and estimated that initially a bass would consume one small bluegill every 3 d. We further adjusted this estimate as the size of the bluegills and bass increased during the experiment. Thus on the bass side we initially added ≈20 small bluegills in addition to the replacements of the fish sampled for stomach analyses. Replacement for bass predation tapered to <10 individuals per sample date at the end of the season. Over the entire experimental period, we added an additional 144 small bluegills to the pond-half with the bass.

Habitat utilization by the fish was determined by classifying prey in the stomachs according to the habitat from which those prey originated (open water, sediments, vegetation). The vast majority of the prey in all size classes across the season could be unambiguously assigned to one of these habitat types (>90% of the diets on average across the season). Further details concerning prey sampling and the generation of predictions of optimal diet and habitat use for the

1542   EARL E. WERNER ET AL.   Ecology, Vol. 64, No. 6

fish can be found in the companion paper (Werner et al. 1983).

RESULTS

Mortality

Mortality rates were similar for the two larger size-classes of bluegills across treatments. When we recovered populations in the fall, the cumulative mortality of the medium size-class was 10% on both sides, and for the large size-class, 11% in the presence of the bass and 19% in the control half of the pond. No mortality occurred in the bass population.

As explained earlier, we added more small bluegills to the bass side to compensate for expected predation losses. Our estimates of bass feeding rates were appropriate since the numbers of small fish recovered in the fall were similar: 348 on the bass side and 359 on the control side. Thus growth data were not confounded by different bluegill densities on the two sides. From these data we estimated a mortality of 28% for small fish on the control side, which is higher than that experienced by the two larger size-classes. The mortality rate of the small bluegills in the presence of the bass, however, was 59% of the original 500 fish. (Including the 144 fish added during the experiment, 296 fish died on the bass side.) Assuming the same non-bass mortality rate (28%) on each side, we estimate that each bass consumed a small fish every 3.8 d. The small size-class, therefore, did incur significant mortality due to the presence of the bass. In the Discussion and the Appendix we further examine how this increased mortality might be apportioned between the "direct" and "indirect" effects of the presence of the bass.

Habitat use

The design of this experiment was predicated on the assumption that predation risk for the bluegill would be much reduced in the cattails compared to the unstructured water column and bare sediments. Accordingly we chose a pond with very high resource levels in the open water and bare sediments, relative to the vegetation. Only if the more profitable habitats were also more dangerous could we test whether the fish were capable of responding to a predation risk/foraging rate tradeoff.

It is generally accepted that complex habitats are safer for prey, but this has not often been quantitatively demonstrated (see, e.g., Huffaker 1958). We know of three sources of direct experimental inference that the cattail habitat should interfere with the bass' predatory efficiency. First, Glass (1971) found in laboratory pools with different densities of vertical wooden dowels that the capture rate of bass preying on guppies (*Poecilia reticulata*) decreased monotonically with an increase in dowel density (0–370 dowels/m²). Second, Savino and Stein (1982) have found that predatory success of bass feeding on bluegills declined with

increasing simulated plant density (0–1000 stems/m²). In small wading pools (2.4–3 m in diameter) containing strands of polypropylene rope, success of the bass declined most sharply between 50 and 250 stems/m². Third, in a similar set of experiments we contrasted the success of bass (100–270 mm) feeding on bluegills (20–75 mm) in open and vegetated habitats (0 and 500 stems/m² polypropylene rope). Preliminary analyses of these data indicate that the small bluegills in the present study (≈35 mm), if pursued by a 200-mm bass, would be at least twice as vulnerable in the open habitat. The density of cattails in the pond study reported here averaged 176 stems/m², but individual samples ranged as high as 400 stems/m². Thus we are confident that the small bluegills incurred much less risk in the cattail habitat of the pond.

In the companion paper (Werner et al. 1983) we examined habitat use by the fish in the absence of predation risk from the bass. We generated foraging rates for each size-class of fish in the three habitats (open water, sediments, and vegetation), using an optimal foraging model where costs and benefits were estimated from laboratory feeding experiments. Examining the return rates in the three habitats across the season, we predicted that to maximize return rates, all size-classes should begin feeding in the open water and then shift to feeding from the sediments when the profitabilities of these two habitats crossed in late July. Thus, the more open habitats were indeed the more profitable, as we had anticipated.

In the absence of the bass, the behavior of all three size-classes was in excellent accord with the model predictions (Werner et al. 1983: Fig. 5). The fish consumed >80% plankton initially and between 21 and 25 July switched dramatically to a diet of >80% sediment-dwelling prey. Utilization of sediments remained high for the remainder of the experiment except for the small size-class, which switched back to plankton in late September when profitability of the open-water habitat was again highest for this size-class (Werner et al. 1983).

While the medium and large bluegills exhibited very similar patterns of habitat use in the presence and absence of the bass, the small bluegills behaved very differently. Several lines of evidence demonstrate the effects of the bass on the foraging behavior of small bluegills. Early in the experiment *Daphnia pulex* was extremely abundant (up to 73 individuals/L) and as a consequence the profitability of the open water was 7- to 27-fold greater than that of either the vegetation or the sediments (Werner et al. 1983). All fish fed extensively on *D. pulex,* and this species very quickly disappeared. Reduced utilization of this very profitable resource by the small bluegills in the presence of the bass can be shown by comparing their use with that of the larger size-classes. (Comparisons of the small class across treatments cannot be made due to the large disparity in *D. pulex* abundance which devel-

TABLE 1. Average percent composition (± 1 SE) of the diet by habitat for the three bluegill size-classes (6 August–6 September). Row sums do not add to 100% because a small fraction ($\leq 3\%$) of prey could not be assigned to a specific habitat (see text for details).

		Vegetation	Plankton	Benthos
No predator	Small	9 ± 2	19 ± 5	69 ± 7
	Medium	14 ± 4	2 ± 0.5	81 ± 4
	Large	11 ± 3	trace	86 ± 3
Predator	Small	34 ± 10	17 ± 5	46 ± 9
	Medium	9 ± 3	16 ± 6	74 ± 7
	Large	14 ± 5	6 ± 4	78 ± 6

oped on the two sides of the pond; see below.) As *D. pulex* densities declined, its contribution to the diets declined most rapidly in the small size-class. By 30 July the small class had nearly ceased feeding on *D. pulex* (4% of diet), whereas the diet of the large and medium classes still contained 67 and 74% of this species, respectively. Over the first four dates the diet of the large size-class averaged 78% *D. pulex* and that of the small size-class only 53%. This contrasts with the control side where large, medium, and small size-classes averaged 90, 73, and 85% *D. pulex* in the diet, respectively, on those dates when they were feeding on plankton.

The reduced utilization of *D. pulex* on the side with the bass was also clearly reflected in the dynamics of the daphnids. On the control side, *D. pulex* abundances declined from 73 to 1 individual/L in 10 d, and by 25 July no *D. pulex* were found in the fish from this half of the pond. In contrast, on the side with the bass, *D. pulex* remained abundant for >20 d into the experiment and was a major part of the diet of the two larger size-classes through 6 August. Clearly the predation pressure on *D. pulex* in the open water was much reduced on this side, evidently due to the reduced utilization by the small size-class discussed above.

Direct comparisons of habitat use in the presence and absence of the bass are best made when resource levels in the three habitats are similar on the two sides. Following the demise of *D. pulex*, resource levels in each habitat were similar on the two sides for the period 6 August–6 September. Over this period then, we can compare habitat utilization of size-classes in the presence and absence of the bass, unconfounded by large differences in resource levels across treatments. Further, a large fraction of the seasonal growth occurred during this period, and the small size-classes of bluegills began to diverge in size on the two sides (see later).

Between 6 August and 6 September, a distinct pattern of habitat use emerged (Table 1). All three size-classes on the control side and the two larger size-classes on the bass side had switched to feeding from the sediments and exhibited similar diets. The small class on the bass side, however, averaged 36% vege-

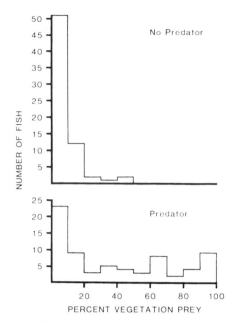

FIG. 1. Number of fish containing different fractions of prey derived from the vegetation. Data are for the small size-class in the presence and absence of largemouth bass. Numbers are for fish from 6 August and 6 September combined. The distributions differ at the $P < .001$ level (Kolmogorov-Smirnov test).

tation-dwelling prey in the diet, as compared to 9–15% for all other fish. This increased use of prey from the vegetation by the small fish was accompanied by a reduced use of prey from the sediments; all other fish were taking at least 70% benthic prey as opposed to 46% in the small class with the bass (Table 1). The increased use of vegetation-dwelling prey is clearly exhibited in Fig. 1, which indicates that use of the vegetation was not uniform among individuals in the population. In the absence of the bass, 93% of the small fish contained <20% prey from the vegetation, and no fish contained >50%. In the presence of the bass, however, 54% of the individuals contained >20% vegetation-dwelling prey. There was no relation between the fraction of vegetation-dwelling prey used and fish size, within the small size-class (regression of percent prey from vegetation against body size over range of 35–55 mm, $r^2 = 0.05$, $n = 70$, $P > .5$).

The small fish on both sides consumed more zooplankton than the large and medium classes (the high value for the medium class on the bass side is due to one anomalous date [Table 1]). The vast majority of the zooplankton eaten during this period was *Ceriodaphnia*, a form found in the vegetation as well as in the open water, and therefore it is possible that the small class in the presence of bass was also obtaining these prey from the vegetation. Thus we conclude that all fish were using habitats in very similar ways, except

1544   EARL E. WERNER ET AL.   Ecology, Vol. 64, No. 6

TABLE 2.   Mean individual dry mass (g) $\pm$ 1 SE for the three bluegill size-classes in the presence and absence of the predator. Final values are for the entire population recovered in the fall. Differences between means between predator and no predator treatments were determined by t test.

	Size-class					
	Small		Medium		Large	
	Predator	No predator	Predator	No predator	Predator	No predator
Initial mass (g)	0.28 ± 0.01		1.35 ± 0.03		3.64 ± 0.12	
	(n = 44)		(n = 27)		(n = 30)	
Final mass (g)	0.90 ± 0.02	1.13 ± 0.02**†	4.45 ± 0.05	4.35 ± 0.05	9.17 ± 0.15	8.64 ± 0.15**
	(n = 348)	(n = 359)	(n = 270)	(n = 270)	(n = 89)	(n = 81)
Increment (g)	0.62	0.85	3.10	3.00	5.53	5.00
Population increment (g)	191.0	269.5	768.8	750.0	392.6	230.0
Difference in population increment (g)	−78.5		18.8		72.6	

** $P < .01$.
**† $P < .005$.

for the small size-class in the presence of the bass, which exhibited increased use of the vegetation.

The expected foraging profitabilities provide a measure of the cost of foraging more in the vegetation (Werner et al. 1983). The predicted return rate for the small class feeding in the vegetation averaged only 32 ±6% of that for the sediments from 6 August through 6 September. We show below that, as a consequence, growth was slower in the presence of the bass.

In late August, *Daphnia ambigua* (an open-water species) appeared and increased to ≈100 individuals/ L by late September. Similarly, *Bosmina* and *Ceriodaphnia* increased in late September. These three species are much smaller than adult *D. pulex* (by at least an order of magnitude in mass), and, in general, only the small bluegills utilized these forms to any extent. If we compare the last two dates when the small size-class on both sides was taking predominantly *D. ambigua,* we find that 77% of the diet of the small fish on the control side was plankton, whereas only 52% of the diet of the small fish on the bass side consisted of plankton, although resource levels in the open water and vegetation were similar on both sides of the pond. Thus, the effect of the bass was again to reduce the small bluegills use of the open water, as was the case when the small bluegills were feeding on *D. pulex* early in the experiment.

Growth

The presence of the predator clearly caused a shift in the habitat use of the small size-class; the question now is whether this habitat shift had any effect on growth of the surviving fish. In the absence of the bass, the three size-classes each grew markedly, but at very different rates (Table 2). Larger size-classes exhibited progressively higher growth rates. However, in the presence of the bass, the small fish exhibited a significant depression in growth, whereas the medium

and large classes grew larger than in the absence of the bass (Table 2). The average growth increment of small fish was 27% less when with the bass, while those of the medium and large fish were 3 and 11% more than when no bass were present. These differences are highly significant for the small and large classes (Table 2). Examination of the size-frequency distributions of the small class on the two sides indicated no evidence of selective predation within this class. The distributions were very similar in shape, but smaller fish were recovered on the predator side, indicating a growth response. The bass increased in length from 198.8 ± 2.9 to 228 ± 3.8 mm during the experiment.

Thus, the presence of the bass significantly depressed growth rates of the small fish in accord with their increased use of the poorer habitat (vegetation). Further, because the small fish spent more time in the less profitable vegetation, this apparently released resources for the larger fish in the sediments and open water. The large class especially benefited from this release in resources (Table 2). Indeed, there was nearly equal compensation by the larger classes for the total production lost to the small class in the bass' presence. A crude estimate of total fish production (number surviving × growth increment) on the two sides only differed by 13 g or ≈1% of the production of either side (Table 2).

The resource samples also indicated the effect of the predator. We have already noted that *D. pulex* lasted nearly 2 wk longer in the predator's presence. Though not as dramatic, the effect was also apparent in the sediment habitat. *Chironomus* densities were always higher in the predator's presence after July, when the fish began feeding on this species (with the exception of two dates when they were equal). Though these differences were not large, the trend to higher midge densities in the predator's presence was very consistent, apparently due to reduced foraging pressure by the small fish. In both the open water and the sediments prey densities prior to the introduction of fish were actually slightly higher in the pond-half without the bass. Clearly, the presence of the bass had striking effects on the distribution of resources among size-classes of the bluegill.

DISCUSSION

The commonly observed fact that habitats vary temporally and spatially in foraging profitability and predation risk suggests that many animals need to balance gains and risks in their decisions on where and when to forage. Can animals assess these gains and risks, and are these factors weighted or balanced in such a way that tends to maximize fitness? This is an especially complex and critical question in the context of species that exhibit strong ontogenetic niche shifts due to the fact that relative foraging abilities and risks to predators change markedly with body size over the life history. Thus, decisions on where and when to forage must be made not only in the face of changing resource and predator dynamics, but also as these relations change with increases in body size.

We have demonstrated that the bluegill is able to assess changes in both foraging profitability and predation risk. In the companion paper (Werner et al. 1983) we showed that temporal habitat shifts by all size-classes occurred when foraging rates in another habitat became greater than those of the habitat currently used. In this paper, we further demonstrated that small, vulnerable size-classes of the bluegill showed a marked shift in foraging behavior in the presence of the bass. These data provide experimental support for the hypothesis of Hall and Werner (1977) and Mittelbach (1981) that small bluegills in natural lakes are confined to weedbeds because of a behavioral response to the greater predation risk in more open habitats. The quantitative predictions of habitat profitabilities also indicate that the small bluegills were evaluating this risk in the face of threefold greater foraging rates in the more open habitats. Whether this response also maximizes fitness remains to be tested.

The response of the small bluegills was not of an all or none nature, i.e., they did not use the unstructured or vegetated habitats exclusively on any given day. This must in part be due to the proximity of these habitats in the ponds; the small fish could feed in the open water or sediments and yet be only a matter of a metre or two from the vegetation refuge. In natural lakes the spatial separation of these habitats is usually much greater and consequently precludes this possibility. Mittelbach (1981) found that small bluegills in a natural lake did not feed on the very profitable offshore plankton prey at all except on one date when zooplankton were found within several metres of the shore. Thus the effect of the presence of the bass may be expected to be even stronger and more sharply de-

fined in natural lakes where large areas of open habitats intervene between those in which the fish feed. Of course, decreased foraging rates due to the presence of a predator may be generated not only by shifts in habitat use, but also by the necessity of greater wariness, or escape responses, which can decrease feeding rates in a particular habitat (Milinski and Heller 1978, Caraco et al. 1980). It is not known how the fish evaluate predation risk, but laboratory studies indicate that bluegills seemingly pay little attention to bass in a tank, until the bass shows subtle inclinations to begin to feed (R. Stein and J. O'Brien, *personal communication*).

The question also arises as to why the small fish exhibit such individual variation in their use of the vegetation in the presence of the predator (Fig. 1). We do not know if certain fish consistently spend more time in the vegetation than do others, i.e., if individuals tend to be risk averse or risk prone, or if this variation simply represents short-term (days, weeks) changes in habitat use by all individuals. This is an important problem, as these two hypotheses lead to very different ideas concerning individual behavior and fitness. We would expect that if risk-prone individuals existed, they would incur higher mortality rates but would also grow faster because of their use of the richer habitats. Thus, there should be a relation between size and habitat use. We noted earlier that there was no relation between body size among the small fish (ranging from ≈35 to 55 mm in length) and the fraction of their diet that came from the vegetation. The question of this individual variation in behavior deserves more detailed study.

In this study we were able to quantify the effects of predator-restricted habitat use on the growth rates of the fish. The small fish exhibited a 27% reduction in growth over part of one growing season. A growth reduction of this magnitude would certainly have far-reaching effects on the dynamics and population structure of these fish. Fish are indeterminant growers, and it is well recognized that fecundity is a direct function of size (e.g., Bagenal 1978) and that mortality rate is an inverse function of size, at least during the early part of the life history (Ricker 1979). Thus, lower growth rates protract the time spent in vulnerable stages, lower survivorship, and increase the time to reproductive maturity. Where the probability of death per day is a function of size, some of the demographic consequences of increased daily mortality rates and decreased growth rates can be assessed by examining the survivorship of fish through a size- (not age-) interval. Assume that for all sizes in the interval the presence of a predator multiplies the daily mortality rate by a factor c_μ and the growth rate by a factor c_g ($c_g = 1$ indicates no effect). It can then be shown (see Appendix) that the survivorship from size s_1 to size s_2 in the presence of a predator is given by

$$l_p(s_1, s_2) = [l_{np}(s_1, s_2)]^{c_\mu/c_g}, \qquad (1)$$

1546   EARL E. WERNER ET AL.   Ecology, Vol. 64, No. 6

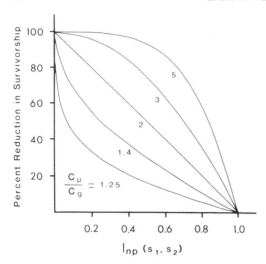

FIG. 2. Percent reduction in survivorship through a size-interval due to the presence of a predator. l_{np} (s_1, s_2) is the survivorship in the absence of the particular predator (or before a predator's density is increased). c_μ is the factor by which the predator's presence (or increase) multiplies the size-specific daily mortality rate. c_g is the factor by which the predator's presence (or increase) multiplies the forager's daily (individual) growth rate.

where l_p (s_1, s_2) = survivorship in the presence of a predator, and l_{np} (s_1, s_2) = survivorship in the absence of the predator. Thus, halving the growth rate (setting $c_g = \frac{1}{2}$, holding $c_\mu = 1$) has the same effect on survivorship through a size-interval as doubling the daily mortality rate (setting $c_\mu = 2$, holding $c_g = 1$); both result in $c_\mu/c_g = 2$. In both cases the survivorship through the size-interval is squared. The decrease in growth rate has the further consequence of increasing the time required to reach reproductive size. It appears that a predator's "indirect" effect on the prey, inducing a lowered growth rate measured by c_g, might have greater effects on population demography than the "direct" effect of raising daily mortality rates (measured by c_μ).

Fig. 2 illustrates the effect of various values of c_μ/c_g on survivorship through a size-interval. The crucial observation here is that for a given value of c_μ/c_g the impact of the presence of the predator is much stronger if the predator affects a size-interval which already exhibits low survivorship, which is the case for small fish. For example, in the case of $c_\mu/c_g = 2$, a size-interval with $l_{np}(s_1, s_2) = 0.8$ initially will have a new survivorship of $l_p(s_1, s_2) = 0.64$ by application of Eq. 1. This is a reduction of 20%. However, if $l_{np}(s_1, s_2)$ were 0.1, this would yield $l_p(s_1, s_2) = 0.01$, a reduction of 90%. Thus, unless density-dependent processes compensate in other size-intervals (which is likely), the number of fish reaching a given adult size will be

reduced by 20 or by 90%, respectively, for the two cases.

We can also use this approach to derive a crude estimate of the relative effect of the "direct" effect of adding more predators to a system (i.e., raising the probability of death/day) and the "indirect" effect of inducing lowered growth rates (which causes the prey to be in a size-interval longer and therefore accrue greater mortality through the size-interval; see Appendix). In the experiment presented here we partition the effects over the size-interval of 0.28 g (the initial mass of the small size-class of bluegills on both sides) to 0.90 g (the final mass on the predator side). We estimate that over the size-range the total reduction in survivorship in the presence of the bass was 37%. If only the "direct" effect of increased daily mortality rates (c_μ) had occurred, we estimate the reduction in survivorship would have been 23%. If only the "indirect" effect of decreased growth rates (and hence a longer time in the interval) had occurred, we estimate that the reduction would have been 10%. Thus the total reduction of 37% can be partitioned into 23% due to the c_μ alone, 10% due to c_g alone, and an additional 4% when both factors acted concurrently. Our estimates of the value of c_μ/c_g ranged from 2.34 to 3.89.

A further consequence of the predator-restricted habitat use of the small fish was the significant compensatory increase in growth of the larger size-classes. Thus, predation risk enforced a degree of intraspecific resource partitioning which exacerbated the differences in growth rates between size-classes (Table 2). This habitat segregation obviously has important consequences to population size-structure and intraspecific competition. Resources in open habitats, which often would be the preferred prey of all size-classes, may become exclusive resources for large fish in the presence of predators. The open-water plankton in particular is likely an important exclusive resource for larger bluegills in natural lakes with extensive limnetic zones. In the absence of predators, bluegills invariably develop "stunted" populations of uniformly small individuals (e.g., Swingle and Smith 1940, Wenger 1972). In such cases foraging demands evidently drive down resource levels (and mean prey size [Hall et al. 1970]), such that intense intraspecific competition prevents sustained growth. The presence of predators, which provides larger fish with exclusive resources, may be a major factor enabling them to continue to grow. Thus predation risk and its effect on habitat use may be an important mechanism mediating intraspecific competition in the field that has been largely overlooked (but see Jackson 1961). We feel that predation risk is most likely the cause of the patterns in size-class segregation noted in the bluegill, rather than an evolutionary response to intraspecific competition (e.g., Keast 1977).

We have noted that predation risk tends to concentrate the young of many species in the vegetation of natural lakes (Hall and Werner 1977, Laughlin and

Werner 1980, Mittelbach 1984). The observation that juvenile diets of several sunfishes are more similar than adult diets (Laughlin and Werner 1980, Mittelbach 1984) is likely the result of this habitat restriction of juveniles. Thus if vegetation is relatively rare and/or resources in it low, risk of predation can create significant competitive bottlenecks for these species at this point in their life histories. Identifying these bottlenecks is obviously central to considerations of niche packing in such systems, but the effects of these sorts of bottlenecks have been virtually unexplored. Our understanding of community structure in fish and other organisms with size-structured populations will be very limited until we systematically explore the consequences of these sorts of interactions.

Our experimental work suggests that competition and predation may interact in subtle but critical ways in fish communities. Shifts in competitive advantages between species due to the presence of a predator may be a great deal more subtle than simply accounting for the differential removal of each species. If bottlenecks of the sort we have hypothesized are important, the mere presence of a predator population could result in extinction of prey that are not even eaten by the predator due largely to the predator's indirect effects of increasing competition in protected habitats. These ideas connote very different mechanisms of "predator-mediated" coexistence than those ordinarily considered. Clearly the effects of a predator on the spatial distribution of its prey and the resultant changes in the strength of interactions within and between prey species may have profound effects that would not be evident from measures of prey removal rates or energy flows.

Acknowledgments

We thank Julie Bebak and Laura Riley for their help with endless samples and the data analysis, and Scott Gleeson, Carol Folt, Craig Osenberg, David Hart, Tim Ehlinger, and James Wetterer for comments on an earlier draft. Ed Turanchik and Beth Hutchison assisted in the field. John Gorentz did the computer programming. This work was generously supported by the National Science Foundation (DEB-7824271 and DEB-8119258 to E. E. Werner and D. J. Hall and DEB-8104697 to G. G. Mittelbach). Contribution number 489 of the Kellogg Biological Station.

Literature Cited

Bagenal, T. B. 1978. Aspects of fish fecundity. Pages 75–101 in S. D. Gerking, editor. Ecology of freshwater fish production. J. Wiley and Sons, New York, New York, USA.

Caraco, T., S. Martindale, and H. R. Pulliam. 1980. Avian flocking in the presence of a predator. Nature 285:400–401.

Curio, E. 1976. The ethology of predation. Zoophysiology and ecology. Volume 7. Springer-Verlag, Berlin, Germany.

Glass, N. R. 1971. Computer analysis of predation energetics in the largemouth bass. Pages 325–363 in B. C. Patten, editor. Systems analysis and simulation in ecology. Academic Press, New York, New York, USA.

Gould, S. J., and R. C. Lewontin. 1979. The spandrels of San Marco and the Panglossian paradigm: a critique of the adaptationists programme. Proceedings of the Royal Society of London B Biological Sciences 205:581–598.

Hall, D. J., W. E. Cooper, and E. E. Werner. 1970. An experimental approach to the production dynamics and structure of freshwater animal communities. Limnology and Oceanography 15:829–928.

Hall, D. J., and E. E. Werner. 1977. Seasonal distribution and abundance of fishes in the littoral zone of a Michigan lake. Transactions of the American Fisheries Society 106:545–555.

Hassell, M. P. 1978. The dynamics of arthropod predator-prey systems. Princeton University Press, Princeton, New Jersey, USA.

Huffaker, C. B. 1958. Experimental studies on predation: dispersion factors and predator-prey oscillations. Hilgardia 27:343–383.

Jackson, P. B. N. 1961. The impact of predation especially by the tiger fish (*Hydrocynus vittatus* Cast) on African freshwater fishes. Proceedings of the Zoological Society of London 136:603–622.

Keast, A. 1977. Mechanisms expanding niche width and minimizing intraspecific competition in two centrarchid fishes. Pages 333–395 in M. K. Steere and B. Wallace, editors. Evolutionary biology. Volume 10. Plenum Press, New York, New York, USA.

Laughlin, D. R., and E. E. Werner. 1980. Resource partitioning in two coexisting sunfish: pumpkinseed (*Lepomis gibbosus*) and northern longear sunfish (*Lepomis megalotis peltastes*). Canadian Journal of Fisheries and Aquatic Sciences 37:1411–1420.

Lawrence, J. M. 1957. Estimated sizes of various forage fishes largemouth bass can swallow. Proceedings of the Southeastern Association of Game and Fish Commission 11:220–226.

Milinski, M., and R. Heller. 1978. Influence of a predator on the optimal foraging behaviour of sticklebacks (*Gasterosteus aculeatus* L.). Nature 275:642–644.

Mittelbach, G. G. 1981. Foraging efficiency and body size: a study of optimal diet and habitat use by bluegills. Ecology 62:1370–1386.

———. 1984, *in press*. Predation and resource partitioning in two sunfishes (Centrarchidae). Ecology.

Ricker, W. E. 1979. Growth rates and models. Pages 677–743 in W. S. Hoar, D. J. Randall, and J. R. Brett, editors. Fish physiology. Volume VIII. Academic Press, New York, USA.

Savino, J. F., and R. A. Stein. 1982. Predator-prey interactions between largemouth bass and bluegills as influenced by simulated, submersed vegetation. Transactions of the American Fisheries Society 111:255–266.

Sih, A. 1980. Optimal behavior: can foragers balance two conflicting demands? Science 210:1041–1043.

Stein, R. A. 1979. Behavioral response of prey to fish predators. Pages 343–353 in R. H. Stroud and H. Clepper, editors. Black bass biology and management. Sport Fishing Institute, Washington, D.C., USA.

Swingle, H. S., and E. V. Smith. 1940. Experiments on the stocking of fish ponds. Transactions of the North American Wildlife Conference 5:267–276.

Van Sickle, J. 1977. Analysis of a distributed-parameter population model based on physiological age. Journal of Theoretical Biology 64:571–586.

Wenger, A. 1972. A review of the literature concerning largemouth bass stocking techniques. Technical Series Number 13, Texas Parks and Wildlife Department, Sheldon, Texas, USA.

Werner, E. E. 1977. Species packing and niche complementarity in three sunfishes. American Naturalist 111:553–578.

———. 1982, *in press*. The mechanisms of species interactions and community organization in fish. In D. Simberloff and D. Strong, editors. Ecological communities: con-

1548   EARL E. WERNER ET AL.   Ecology, Vol. 64, No. 6

ceptual issues and evidence. Princeton University Press, Princeton, New Jersey, USA.

Werner, E. E., and D. J. Hall. 1974. Optimal foraging and the size selection of prey by the bluegill sunfish (*Lepomis macrochirus*). Ecology 55:1042–1052.

Werner, E. E., G. G. Mittelbach, and D. J. Hall. 1981. Foraging profitability and the role of experience in habitat use by the bluegill sunfish. Ecology 62:116–125.

Werner, E. E., G. G. Mittelbach, D. J. Hall, and J. F. Gilliam. 1983. Experimental tests of optimal habitat use in fish: the role of relative habitat profitability. Ecology 64: 1525–1539.

APPENDIX

The survivorship of a fish from age x_1 to age x_2 can be described by

$$l(x_1, x_2) = \exp\left[-\int_{x_1}^{x_2} \mu(x)\,dx\right], \tag{A1}$$

where x = age, and $\mu(x)$ = instantaneous mortality rate at age x (see, e.g., Hassell 1978: Appendix I). If the mortality rate is explicitly a function of size rather than age, we can rewrite $\mu(x)$ as $\mu[s(x)]$. We can then change the variable of integration from age to size to obtain an expression for survivorship from size s_1 to size s_2. Eq. A1 can thus be rewritten as:

$$l[x(s_1), x(s_2)] = \exp\left\{-\int_{x(s_1)}^{x(s_2)} \mu[s(x)]\,dx\right\}. \tag{A2}$$

Changing the variable of integration yields:

$$l(s_1, s_2) = \exp\left[-\int_{s_1}^{s_2} \mu(s)/(ds/dx)\,ds\right]. \tag{A3}$$

Since ds/dx is the growth rate, $g(s)$, we have

$$l(s_1, s_2) = \exp\left[-\int_{s_1}^{s_2} \mu(s)/g(s)\,ds\right]. \tag{A4}$$

Van Sickle (1977) has derived this equation by a different method. Intuitively, Eq. A4 represents survivorship across a size-interval because $\mu(s)/g(s)$ is the instantaneous probability of death at a particular size, since that probability is the product of $\mu(s)$ (i.e., deaths/time) and the inverse of the growth rate (a measure of the time spent at that size).

For heuristic purposes, assume that over some size-interval (s_1, s_2) the presence of the bass uniformly multiplies $\mu(s)$ by a factor c_μ and multiplies $g(s)$ by a factor c_g. Then the survivorship through a size-interval in the presence of the predator (l_p) is:

$$l_p(s_1, s_2) = \exp\left\{-\int_{s_1}^{s_2} [c_\mu \cdot \mu(s)]/[c_g \cdot g(s)]\,ds\right\}$$

$$= \exp\left[-(c_\mu/c_g)\int_{s_1}^{s_2} \mu(s)/g(s)\,ds\right]$$

$$= \left\{\exp\left[-\int_{s_1}^{s_2} \mu(s)/g(s)\,ds\right]\right\}^{(c_\mu/c_g)}. \tag{A5}$$

Since the expression in the braces is just the survivorship without the predator, $l_{np}(s_1, s_2)$, this yields:

$$l_p(s_1, s_2) = [l_{np}(s_1, s_2)]^{c_\mu/c_g}. \tag{A6}$$

This is Eq. 1. The percent reduction in survivorship is thus (the arguments of l_p and l_{np} are dropped for convenience):

$$(100)(l_{np} - l_p)/l_{np} = (100)(l_{np} - l_{np}^{c_\mu/c_g})/l_{np}$$
$$= (100)[1 - l_{np}^{(c_\mu/c_g - 1)}]. \tag{A7}$$

By estimating c_μ and c_g in this experiment, we can somewhat crudely estimate the relative intensities of the predator's "direct" effect (increased daily mortality rates) and its

"indirect" effect (reduced growth rate) on the bluegill's survivorship through a size-interval. Here we consider survivorship over the size-interval from 0.28 g (the initial mass on both pond sides) to 0.90 g (the final mass on the predator side). We take $c_g = 0.729$ (Table 2). By specifying l_p (0.28, 0.90) and l_{np} (0.28, 0.90), c_μ can be calculated from Eq. A6 and the relative effects of c_μ and c_g assessed. Below, we calculate values for l_p (0.28, 0.90) and l_{np} (0.28, 0.90) and then partition the total reduction in survivorship into components due to the direct effect (c_μ), the indirect effect (c_g), and their interaction when both occur simultaneously.

The value of l_p (0.28, 0.90) cannot be calculated exactly because fish were replaced at various times and sizes on the predator side, and we have no data on the shape of $\mu(s)$. However, we can bracket the possible values for l_p (0.28, 0.90) by calculating the survivorship as if all the replacements were added on (1) the first day of the experiment, or (2) the last day of the experiment (equivalently, not added at all). The first calculation yields a survivorship of 348/644 = 0.540 (348 fish recovered from 644 total fish). The second calculation yields 204/500 = 0.408 (348 recovered minus 144 replacements, divided by the 500 original fish). Thus we take l_p (0.28, 0.90) to lie somewhere between the extreme limits of 0.408 and 0.540. The parameter l_{np} (0.28, 0.90) also cannot be calculated exactly since we know only the survivorship to 1.13 g (the final size with no predator); l_{np} (0.28, 1.13) was 359/500 = 0.718. We can bracket the possible values of l_{np} (0.28, 0.90) by performing the calculation as if (1) no fish died after reaching 0.90 g (i.e., the mortality rate was a very strongly declining function of fish size), and (2) the mortality rates were independent of fish size. The first case yields l_{np} (0.28, 0.90) = l_{np} (0.28, 1.13) = 0.718. The second case yields l_{np} (0.28, 0.90) = 0.787. This was calculated by letting μ represent the size-independent daily mortality rate. Then $N_t = N_0 e^{-\mu t}$, where N_t = survivors at time t, N_0 = initial number of fish, t = time. Taking $t = 76$ d (the length of the experiment), $N_{76} = 359$, and $N_0 = 500$ yields $\mu = 0.004359$. The fish reached a mean size of 0.90 g after 55 d; $N_{55} = N_0 e^{-\mu(55)}$, which yields $N_{55} = 393.4$. Thus l_{np} (0.28, 0.90) = 393.4/500 = 0.787.

These survivorship ranges can now be used to estimate the relative impact of c_μ and c_g on survivorship. First, we calculate our "best estimate" of the relative effects by taking the midpoints of the survivorship ranges. Thus, we take l_{np} (0.28, 0.90) = 0.753 and l_p (0.28, 0.90) = 0.474, a reduction in survivorship of 37.1%. Application of Eq. A6 with $c_g = 0.729$ yields $c_\mu = 1.92$. If the predator had not affected the prey's growth rate (take $c_g = 1$), this model would predict survivorship to have been l_p (0.28, 0.90) = $l_{np}^{c_\mu}$ (0.28, 0.90) = $(0.753)^{(1.92)} = 0.580$, a reduction of 23.0%. Similarly, if the predator had affected only the growth rate, the model would predict l_p (0.28, 0.90) = l_{np}^{1/c_g} (0.28, 0.90) = $(0.753)^{(1/0.729)} = 0.678$, a decrease of 10.0%. Thus, we estimate that the total of 37.1% reduction in survivorship can be partitioned into 23.0% due to the "direct" effect of increased daily mortality rates acting alone, 10.0% due to the "indirect" effect of decreased growth rates acting alone, and the remainder, 4.1%, to their interaction when both factors acted simultaneously.

Finally, there are four combinations of the endpoints of the ranges of l_{np} (0.28, 0.90) and l_p (0.28, 0.90). Taking the highest estimate of l_{np} (0.28, 0.90) and the lowest estimate of l_p (0.28, 0.90) yields l_{np} (0.28, 0.90) = 0.787 and l_p (0.28, 0.90) = 0.408, a reduction of 48.2%. Taking $c_g = 0.729$ and solving Eq. A6 yields $c_\mu = 2.73$. This results in a partitioning of the 48.2% into 33.9, 8.5, and 5.8% attributable to c_μ, c_g, and their interaction, respectively. Taking the lowest estimate of l_{np} (0.28, 0.90) and the highest estimate of l_p (0.28, 0.90) yields l_{np} (0.28, 0.90) = 0.718 and l_p (0.28, 0.90) = 0.540. In this case, $c_\mu = 1.36$, and the total reduction of 24.8% can be partitioned into 11.3, 11.5, and 2.0% attributable to c_μ, c_g, and their interaction. The results of the other two combinations lie within the range of the above combinations.

Oecologia (Berlin) (1986) 70:486–494

Oecologia
© Springer-Verlag 1986

Variation in the costs and benefits of mutualism: the interaction between yuccas and yucca moths

John F. Addicott

Department of Zoology, University of Alberta, Edmonton, Alberta T6G 2E9, Canada and
Rocky Mountain Biological Laboratory, Crested Butte, CO 81224, USA

Summary. Yucca moths are both obligate pollinators and obligate seed predators of yuccas. I measured the costs and net benefits per fruit arising for eight species of yuccas from their interaction with the yucca moth *Tegeticula yuccasella*. Yucca moths decrease the production of viable seeds as a result of oviposition by adults and feeding by larvae. Oviposition through the ovary wall caused 2.3–28.6% of ovules per locule to fail to develop, leaving fruit with constrictions, and overall, 0.6–6.6% of ovules per fruit were lost to oviposition by yucca moths. Individual yucca moth larvae ate 18.0–43.6% of the ovules in a locule. However, because of the number of larvae per fruit and the proportion of viable seeds, yucca moth larvae consumed only 0.0–13.6% of potentially viable ovules per fruit. Given both oviposition and feeding effects, yucca moths decreased viable seed production by 0.6–19.5%. The ratio of costs to (gross) benefits varied from 0% to 30%, indicating that up to 30% of the benefits available to yuccas are subsequently lost to yucca moths. The costs are both lower and more variable than in a similar pollinator-seed predator mutualism involving figs and fig wasps.

There were differences between species of yuccas in the costs of associating with yucca moths. Yuccas with baccate fruit experienced lower costs than species with capsular fruit. There were also differences in costs between populations within species and high variation in costs between fruit within populations. High variability was the result of no yucca moth larvae being present in over 50% of the fruit in some populations, while other fruit produced up to 24 larvae. I present hypotheses explaining both the absence and high numbers of larvae per fruit.

Key words: Mutualism – Yucca – Yucca Moths – Gest-benefit analysis – Seed predation

An important problem in the study of dynamics and evolution of mutualism is how gross costs and benefits and net benefits vary and are regulated (Addicott 1979, 1981, 1984, 1985a, 1985b; Boucher et al. 1982; Dean 1983; Howe 1984; Keeler 1981). Study of this variation is hampered by two practical problems. First, it is difficult make direct comparisons of gross costs and benefits, because they usually arise in different biological currencies. For example, a major cost for many animal-pollinated plants is the production of nectar (Southwick 1984), while the major benefit is the transfer of pollen leading to seed development. Similar problems exist with other mutualisms, including ant-plant systems and coelenterate-algal symbioses. Second, it is difficult to

determine sources of costs and benefits in mutualisms involving a number of species, particularly where the number, kind and interdependence of mutualists may vary in time and space. In many pollination systems plants are associated with a whole suite of pollinators and pollinators with whole suites of plants. In ant-homopteran mutualisms, homopterans must interact with different species of ants, given the mosaic distribution of ant species.

There are relatively few mutualisms in which costs and benefits can be easily compared and in which there is also a small and consistent group of interacting species. Obvious examples are those systems in which a mutualist enhances seed production of a plant through one effect and diminishes seed production with another effect (e.g. Janzen 1979; Louda 1982). The interaction between yucca moths (*Tegeticula* spp., Prodoxinae, Incurvariidae) and yuccas (*Yucca* spp., Agavaceae) is one such system. Like fig wasps for figs, yucca moths are both obligate pollinators and obligate seed predators of yuccas (Powell and Mackie 1966; Riley 1892), and therefore most costs and benefits for yuccas are measurable in the same biological variable, seeds. The structure of this pollination-seed predation mutualism is also very simple. Two yuccas, *Y. whipplei* Torr. and *Y. brevifolia* Engelm., are each pollinated by their own species of moth, *T. maculata* (Riley) and *T. synthetica* (Riley), respectively, and with the exception of *Y. schottii* Engelm., each of the 20–30 other species of yuccas is apparently only pollinated by *T. yuccasella* Riley) (Davis 1967; McKelvey 1938, 1947; Webber 1953; but see below).

In this paper I describe and compare the costs and net benefits to eight species of yuccas interacting with *T. yuccasella*. The purpose of this study was to identify patterns, if any, in the interactions among yuccas and yucca moths. Are there differences between species or between populations within species in the interaction between yucca moth and yucca, and if so, what factors are associated with the differences? How variable are the number of yucca moth larvae and net seed production per fruit within populations of yuccas?

Methods

Yuccas and yucca moths

In July and August 1980, I collected a total of 690 mature fruit from eight species of yuccas in Arizona, Colorado, Montana, New Mexico and Utah, U.S.A.: *Y. schottii* Engelm., *Y. arizonica* McKelvey, *Y. baccata* Torr., *Y. glauca* Nutt., *Y. baileyi* Woot. & Standl., *Y. angustissima* Engelm., *Y. kanabensis* McKelvey, and *Y. elata* Engelm. The first three species belong to the Sarcocarpa section of the genus *Yucca*, the fruit of which are large, fleshy and indehiscent. I will refer to these as baccate species. The last five species belong to the Chaenocarpa section. Their fruit are capsular and dehiscent, and I will refer to them as capsular species. I also collected fruit from *Y. brevifolia* Engelm., but since it is pollinated by *T. synthetica* rather than *T. yuccasella*, I have excluded it from the analysis. There are five other yuccas within the region where I collected: *Y. whipplei* Torr., *Y. schidigera* Roezel, *Y. neomexicana* Woot. & Standl, *Y. gilbertiana* (Trel.), and *Y. torreyi* Shafer. Either these species did not bloom and set fruit in 1980, or I did not encounter them.

Given the intergradation between species of yuccas in many regions of the southwest (McKelvey 1938, 1947; Webber 1953; Cronquist et al. 1977), assignment of populations to a particular taxon can be difficult. Populations in central Arizona, that are apparently hybrids between *Y. angustissima* and *Y. elata*, are particularly confusing. I have assigned them to *Y. elata*, based upon fruit shape, oviposition patterns, and branching of inflorescences. The yuccas near Kanab, Utah also present a problem. Some authors consider them as a variant of *Y. angustissima* (e.g. Cronquist et al. 1977). However, based upon fruit and inflorescence characters, I place these plants close to *Y. baileyi*. In this paper I use the name *Y. kanabensis* (McKelvey 1947). Because of the confusing state of taxonomy of the genus *Yucca*, I present precise collection localities in Appendix 1. I grouped collection sites into regional populations (see Appendix 1). Until the population structure of *T. yuccasella* is studied, designations of populations will remain arbitrary.

All eight yuccas in this study are pollinated by *T. yuccasella*. However, there is some question about the homogeneity of the taxon *T. yuccasella* (Davis 1967; Miles 1983). Until the taxonomic status of these variants is clarified, it is appropriate to refer to the yucca moths in this study as *T. yuccasella*.

Procedures

Each fruit was dissected, and the numbers of viable, inviable, eaten and uneaten seeds, and the number of yucca moth larvae per locule counted. Viable seeds are dark, usually black, while inviable seeds have white seed coats and lack endosperm. For some analyses I excluded those fruit for which I could not determine whether seeds were damaged by yucca moth larvae or by other insect seed predators. Locules were classified as being constricted or unconstricted.

Data were analyzed by one-way non-parametric ANOVA with nesting, and two-way non-parametric ANOVA, using the Kruskal-Wallis test. Koch (1970) and Scheirer (1976) provide examples of the extension of the Kruskal-Wallis test to complex ANOVA designs, and they provide confirmation that tabled probability levels are appropriate. Mariscuilo and McSweeny (1977) discuss partitioning the Kruskal-Wallis test statistic, as well as procedures for planned comparisons. I made planned comparisons between baccate and capsular species and pairwise comparisons between species within each group. Other analyses involved the use of 2-way contingency tables to test for heterogeneity, and a multi-sample sign test (Mariscuilo and McSweeny 1977).

Results

Net seed production

The total number of ovules per fruit varied between ($H_7 = 291.0$, $P < 0.001$) and within ($H_{17} = 96.9$, $P < 0.001$) species of yuccas (see Tables 1 B, 2 B). Species with baccate fruit (*Y. schottii*, *Y. arizonica* and *Y. baccata*) had fewer ovules than species with capsular fruit (*Y. glauca*, *Y. baileyi*, *Y. angustissima*, *Y. kanabensis* and *Y. elata*). Species with thick-walled, capsular fruit (*Y. baileyi* and *Y. kanabensis*) had more ovules than species with thin-walled, capsular fruit, particularly *Y. angustissima*. There was also significant variation between populations within *Y. baccata*, *Y. glauca*, *Y. angustissima*, and *Y. elata*.

Net seed production in yuccas is measurable either as the absolute number or proportion of seeds that are viable and escape predation from yucca moth larvae. Absolute net seed production differed between species ($H_7 = 91.8$, $P < 0.001$) and between populations within species ($H_{17} = 154.3$, $P < 0.001$) (Tables 1 C, 2 C). The number of viable, uneaten seeds was lower in baccate species than capsular species, while *Y. kanabensis* and *Y. elata* produced more viable, uneaten seeds than did the other three capsular species. These differences existed despite significant variation between populations of *Y. baccata*, *Y. glauca*, and *Y. elata*.

A better comparative measure of the net benefits of the interaction between yuccas and yucca moths is proportional net seed production per fruit: the ratio of the number of viable, uneaten seeds to the number of ovules. Values per species ranged from 0.36 to 0.60 (Table 1 D), but were not significantly different ($H_7 = 13.5$, $P = 0.06$). However, there were differences between populations within species ($H_{17} = 142.7$, $P < 0.001$), particularly between populations of *Y. baccata*, *Y. glauca*, and *Y. elata*. Values per population ranged from 0.36 to 0.81 for *Y. glauca* and from 0.23 to 0.73 for populations of *Y. elata* (Table 2D). There was also high variation between fruit within populations. For example, the range for *Y. baccata* was from 0.051 to 0.901, and in both *Y. glauca* and *Y. elata* there were fruit in which proportional net seed production was zero. A pattern typical of this variation is shown for *Y. baccata* in Fig. 1.

Effects of oviposition on seed production

Net seed production is a function of both the proportion of ovules which do not produce viable seeds, and the proportion of viable seeds which are eaten by yucca moth larvae. There were significant differences both between species ($H_7 = 47.3$, $P < 0.001$) and between populations within species ($H_{17} = 118.1$, $P < 0.001$) in the proportion of inviable ovules (Tables 1 E, 2 E). Values were high for *Y. angustissima* compared to *Y. glauca*, *Y. kanabensis*, and *Y. elata*. However, baccate species did not differ from capsular species. There were also significant differences between populations of *Y. schottii*, *Y. glauca* and *Y. elata*.

Three factors contribute to seed inviability. First, adult yucca moths could transfer insufficient pollen for fertilization of all ovules within an ovary, or they could transfer pollen of low quality. Second, fertilized ovules may abort because of insufficient resources for seed development. Third, insertion of a yucca moth's ovipositor through the locular wall into the locular cavity may damage ovules or interfere with ovule development, leading to constriction

488

Table 1. Ovule numbers, net seed production, and components of decreased seed production caused by *T. yuccasella* adults and larvae in eight species of yuccas. Sample size (column A) applies to data in columns (B), (E), (F), and (G). Sample size in column (L) applies to columns (L), (C), (D), (I), and (K). Sample size in column (H) applies to columns (H) and (N).

Species	(A) n	(B) Mean # Ovules per Fruit	(C) Mean # Uneaten, Viable Seeds per Fruit	(D) Mean # Uneaten, Viable Seeds per Ovule per Fruit	(E) Mean # Inviable Seeds per Ovule per Fruit	(F) Mean # Constricted Locules per Fruit	(G) Proportion of Fruit Lacking Constrictions	(H) Mean Difference per Fruit between Constricted and Unconstricted Locules in the # Inviable Seeds per Ovule per Locule (n)	(I) Mean # Eaten, Viable Seeds per Viable Seed per Fruit	(J) Cost of T.y. per Ovule per Fruit	(K) Mean # T.y. Larvae per Fruit	(L) Proportion of Fruit Lacking T.y. Larvae (n)	(M) Maximum # T.y. Larvae per Fruit	(N) Proportion of Fruit with more Inviable Seeds in Constricted Locules (n)
Y. schottii	42	150.0	89.4	0.588	0.305	4.2	0.095	0.055 (26)	0.107	0.145	1.9	0.500 (42)	18	0.731 (26)
Y. arizonica	9	220.1	131.0	0.604	0.336	1.6	0.444	0.023 (5)	0.000	0.006	0.0	0.800 (5)	0	0.600 (5)
Y. baccata	108	196.9	116.1	0.589	0.391	1.9	0.444	0.184 (40)	0.006	0.064	1.5	0.697 (99)	20	0.875 (40)
Y. glauca	124	291.3	137.9	0.482	0.356	3.5	0.145	0.114 (67)	0.120	0.186	3.9	0.295 (95)	19	0.881 (67)
Y. baileyi	14	352.0	139.0	0.362	0.457	2.5	0.429	−0.06 (5)	0.136	0.109	9.2	0.200 (5)	13	0.400 (5)
Y. angustissima	28	285.8	126.0	0.425	0.457	5.0	0.100	0.058 (13)	0.147	0.195	3.3	0.364 (22)	12	0.538 (13)
Y. kanabensis	54	347.9	200.9	0.585	0.293	3.2	0.111	0.124 (42)	0.124	0.190	4.6	0.135 (52)	24	0.857 (42)
Y. elata	250	316.7	179.0	0.567	0.288	0.9	0.636	0.166 (83)	0.138	0.162	2.9	0.333 (246)	23	0.964 (83)

Table 2. Ovule numbers, net seed production, and components of decreased seed production caused by *T. yuccasella* adults and larvae in twenty five populations representing eight species of yuccas. Sample size (column A) applies to data in columns (B), (E), (F), and (G). Sample size in column (L) applies to columns (L), (C), (D), (I), and (K). Sample size in column (H) applies only to column (H)

Species	Population	(A) n	(B) Mean # Ovules per Fruit	(C) Mean # Uneaten, Viable Seeds per Fruit	(D) Mean # Uneaten, Viable Seeds per Ovule per Fruit	(E) Mean # Inviable Seeds per Ovule per Fruit	(F) Mean # Constricted Locules per Fruit	(G) Proportion of Fruit Lacking Constrictions	(H) Mean Difference per Fruit between Constricted and Unconstricted Locules in the # Inviable Seeds per Ovule per Locule (n)	(I) Mean # Eaten, Viable Seeds per Viable Seed per Fruit	(J) Cost of T.y. per Ovule per Fruit	(K) Mean # T.y. Larvae per Fruit	(L) Proportion of Fruit Lacking T.y. Larvae (n)	(M) Maximum # T.y. Larvae per Fruit
Y. schottii	Bisbee	4	138.2	87.0	0.629	0.371	5.2	0.000	0.049 (3)	0.000	0.042	0.0	1.00 (4)	0
	Nogales	8	123.1	53.1	0.451	0.547	3.3	0.125	0.037 (6)	0.018	0.032	0.1	0.875 (8)	10
	Portal	30	158.8	99.4	0.619	0.231	4.3	0.100	0.063 (17)	0.149	0.194	2.6	0.333 (30)	18
Y. arizonica	Nogales	9	220.1	131.0	0.604	0.336	1.6	0.444	0.023 (5)	0.000	0.006	0.0	0.800 (5)	0
Y. baccata	Peach Springs	24	145.9	68.0	0.462	0.531	3.6	0.125	0.100 (11)	0.000	0.060	0.1	0.958 (24)	3
	Gateway	35	205.4	132.8	0.664	0.347	0.6	0.743	0.323 (8)	0.014	0.046	2.6	0.613 (31)	20
	Kanab	28	236.8	141.2	0.615	0.370	2.2	0.321	0.184 (13)	0.008	0.075	1.8	0.536 (28)	9
	Moab	21	182.9	115.2	0.609	0.345	1.8	0.476	0.163 (8)	0.002	0.050	1.0	0.750 (16)	5
Y. glauca	Gunnison	31	260.0	93.8	0.363	0.458	5.1	0.032	0.121 (10)	0.161	0.263	4.1	0.226 (31)	11
	Poncha Springs	37	242.0	104.2	0.443	0.397	4.2	0.108	0.107 (20)	0.072	0.146	3.8	0.437 (16)	16
	Wolf Creek	40	349.0	167.7	0.487	0.325	2.1	0.300	0.098 (22)	0.197	0.231	5.6	0.187 (32)	19
	Clines Corners	16	321.5	259.6	0.813	0.161	2.1	0.063	0.140 (15)	0.025	0.074	0.7	0.500 (16)	2
Y. baileyi	Page	14	352.0	139.0	0.362	0.457	2.5	0.429	-0.06 (5)	0.136	0.109	7.4	0.200 (5)	13
Y. angustissima	Peach Springs	11	332.4	138.8	0.417	0.446	4.4	0.000	0.100 (7)	0.154	0.227	4.6	0.167 (6)	12
	Gateway	17	255.6	111.6	0.434	0.463	5.4	0.000	0.010 (6)	0.139	0.148	2.3	0.437 (16)	5
Y. kanabensis	Kanab	38	350.3	197.0	0.573	0.281	2.8	0.158	0.107 (27)	0.143	0.192	5.1	0.111 (36)	24
	Rockville	16	342.4	212.3	0.621	0.321	4.0	0.000	0.154 (15)	0.067	0.169	3.3	0.187 (16)	6
Y. elata	Cottonwood	48	312.2	170.7	0.550	0.375	1.4	0.604	0.122 (15)	0.074	0.102	1.2	0.396 (48)	4
	Sonoita	28	311.2	118.2	0.382	0.312	0.6	0.821	0.173 (3)	0.307	0.347	7.9	0.148 (27)	23
	Portal	12	370.8	194.0	0.526	0.348	0.1	0.833	-0.02 (2)	0.125	0.125	2.7	0.500 (12)	13
	Prescott	24	319.7	72.6	0.237	0.409	0.9	0.583	0.130 (9)	0.350	0.369	7.9	0.045 (22)	22
	Sedona	30	293.0	186.3	0.638	0.255	0.7	0.733	0.102 (8)	0.103	0.114	1.6	0.433 (30)	7
	Wickenburg	28	303.9	199.0	0.656	0.227	0.2	0.929	0.286 (2)	0.111	0.120	1.4	0.393 (28)	4
	Alamagordo	36	333.1	232.6	0.681	0.236	1.5	0.361	0.160 (22)	0.080	0.112	2.1	0.229 (35)	5
	Deming	44	319.7	235.0	0.738	0.198	1.0	0.500	0.245 (22)	0.062	0.102	1.2	0.455 (44)	7

490

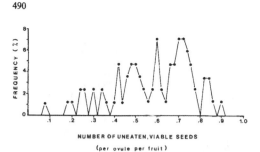

Fig. 1. Percentage frequency distribution of proportional net seed production per fruit ($n = 99$) for *Y. baccata* showing high variation between fruit

of the mature fruit at the site(s) where oviposition occurred (Riley 1892).

With the present data set I can only assess the effect of oviposition on seed production. The average number of constricted locules varied between species ($H_7 = 160.2$, $P < 0.001$) and between populations within species ($H_{17} = 71.3$, $P < 0.001$) (Tables 1 F, 2 F). These differences were due primarily to the proportion of fruit that lacked any constrictions (Table 1 G) ($X_7^2 = 147.3$, $P < 0.001$). The fruit of *Y. angustissima* were usually constricted while few fruit of *Y. elata* were constricted. Variation between populations of *Y. elata* was particularly obvious. For example, 92.5% were unconstricted in the Wickenburg population, but only 36.1% in the Alamagordo population.

Where constricted and unconstricted locules occur on the same fruit, I can assess the importance of yucca moth oviposition *per se* in the decreasing net viable seed production. To assess the qualitative effect of oviposition on seed inviability, I classified fruit into two categories: fruit in which constricted locules had either a smaller or larger proportion of inviable seeds than locules without constrictions. Using the data of Table 1 N, I conducted a multi-sample sign test, which showed that constricted locules have a greater proportion of inviable seeds per locule ($X_7^2 = 143.7$, $P < 0.001$).

To assess the quantitative effect of oviposition on seed viability, I subtracted the proportion of inviable seeds in unconstricted locules from the proportion of inviable seeds in constricted locules within individual fruit. I then subjected these values to a 1-way Kruskal-Wallis ANOVA with nesting. There were differences between species ($H_7 = 20.7$, $P < 0.01$) and between populations within species ($H_{17} = 31.1$, $P < 0.05$) (see Tables 1 G, 2 G). The effect of constrictions on seed inviability is greatest in *Y. baccata* and *Y. elata*, with over 15% of ovules per constricted locule being affected.

Seed consumption and net costs

The effects (if any) of insufficient pollen transfer, insufficient resources, and yucca moth oviposition on seed inviability will have occurred prior to yucca moth larvae consuming any significant number of seeds. Therefore, the effect of yucca moth larvae on net seed production should be measured by the number of potentially viable seeds eaten per ovule per fruit (Tables 1 I, 2 I). There were significant differences between species ($H_7 = 71.4$, $P < 0.001$) and between populations within species ($H_{17} = 98.9$, $P < 0.001$). Proportional consumption of viable seeds was lower in bac-

cate than capsular species, but there were no differences between species within either group. There was significant variation between populations within species for *Y. schottii*, *Y. glauca*, and *Y. elata*. Consumption of viable seeds was unusually high in the Sonoita and Prescott populations of *Y. elata* and low in the Clines Corners population of *Y. glauca*.

I can estimate the net cost per fruit for yuccas of associating with yucca moths by combining the effects of oviposition and seed consumption. I computed the cost of oviposition as the decrease in seed viability per locule (Tables 1 H, 2 H) times the proportion of locules constricted, which is the number of constricted locules per fruit (Tables 1 F, 2 F) divided by six. I added the result to proportional seed consumption (Tables 1 I, 2 I) to obtain net cost (Tables 1 J, 2 J). With the exception of one population of *Y. schottii*, baccate species lost less than 10% of their ovules to yucca moths. Capsular species averaged 10–20% loss, but losses were as high as 36.9% in some populations.

Seed consumption per larva

Two factors affect the proportion of viable seeds destroyed by yucca moth larvae: the total number of viable seeds eaten per larva, and the number of larvae per fruit. Data on seed consumption per yucca moth larva were obtained where feeding in one or more adjacent locules could be ascribed unequivocally to just one or two larvae. For feeding zones with a single larva feeding in a single locule, the number of seeds consumed varied from 7.2 in *Y. baccata* and *Y. schottii* to 23.6 in *Y. elata* (Table 3 B), while the proportion of the ovules (viable or inviable) consumed per larva per locule ranged from 0.18 in *Y. baccata* to 0.43 in *Y. elata* (Table 3 A). In general, small numbers and proportions of seeds of baccate species were consumed relative to capsular species. The only exception to this pattern was *Y. kanabensis*, which had relatively large seeds for a capsular species. An approximate measure of seed size was obtained by dividing the number of seeds per locule by fruit length. Regression of log of the number of seeds eaten on log of seed size was highly significant ($b = -0.892$, $r^2 = 0.453$, $n = 317$, $P < 0.001$). Thus seed size is a good predictor of seed consumption for *T. yuccasella*.

A 2-way Kruskal-Wallis ANOVA for the number of seeds consumed per larvae (Table 3 B, 3 C) shows that there were significant differences between *T. yuccasella* feeding in different species of yuccas ($H_5 = 44.9$, $P < 0.001$), and that there was a small, but consistent increase in the number of seeds eaten when a larva crosses over to feed in more than one locule ($H_1 = 5.24$, $P < 0.05$). There was no interaction between species and crossovers ($H_5 = 3.97$, NS). In general, only about 10% of all larvae fed in more than one locule. There was no increase in the number of seeds eaten per larva when 2 larvae fed simultaneously in the same locule (Table 3 B, 3 D) ($H_1 = 0.25$, NS).

Number of larvae per fruit

The second component affecting the proportion of viable seeds consumed per fruit is the number of yucca moth larvae feeding per fruit. The mean number of larvae per fruit (Tables 1 K, 2 K) differed between species ($H_7 = 58.9$, $P < 0.001$), with baccate species producing fewer yucca moth larvae, and *Y. elata* producing fewer larvae than other cap-

Table 3. Seed consumption by individual *T. yuccasella* larvae in six species of yuccas

Species	(A) Mean Proportion of Seeds Eaten per Locule, 1 Larva Present (n)	(B) Mean # Seeds Eaten by 1 Larva Feeding in 1 Locule (n)	(C) Mean # Seeds Eaten by 1 Larva Feeding in 2 Locules (n)	(D) Mean # Seeds Eaten per Larva by 2 Larvae Feeding in 1 Locule (n)
Y. schottii	0.247 (13)	7.2 (12)	19.0 (1)	
Y. baccata	0.180 (40)	7.2 (36)	11.0 (4)	6.7 (10)
Y. glauca	0.344 (53)	21.1 (40)	20.7 (13)	18.3 (7)
Y. angustissima	0.301 (13)	16.5 (9)	17.5 (4)	
Y. kanabensis	0.213 (56)	12.3 (53)	17.3 (3)	13.7 (4)
Y. elata	0.436 (143)	23.6 (125)	30.7 (38)	18.6 (17)

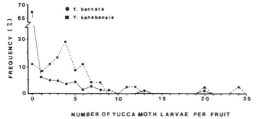

Fig. 2. Percentage frequency distribution of the number of yucca moth larvae per fruit for *Y. baccata* ($n = 99$) and *Y. kanabensis* ($n = 52$)

sular species. There was also significant variation between populations within species ($H_{17} = 80.7$, $P < 0.001$). The Clines Corners population of *Y. glauca* had few larvae, and the Sonoita and Prescott populations of *Y. elata* had many larvae.

The most striking features of the data on larvae per fruit, are that a very high proportion of fruit did not produce any yucca moth larvae (Tables 1 L, 2 L), and the high number of larvae in some fruit (Tables 1 M, 2 M). Analysis of the contingency table of number of fruit with or without yucca moth larvae against species shows highly significant heterogeneity ($X_7^2 = 66.9$, $P < 0.001$). Baccate and capsular species appear to be different, as over one half of baccate fruit produce no larvae. There were five populations of *Y. elata* in which over 39% of fruit lacked larvae. The maximum number of larvae per fruit shows the wide variation that occurred within populations, as many populations had fruit with more than 20 larvae. Representative patterns of the distribution of numbers of larvae per fruit are shown for *Y. baccata* and *Y. kanabensis* in Fig. 2.

Discussion

My results should be compared with Janzen (1979), who studied five species of *Ficus* and their agaonid wasps in Santa Rosa National Park, Costa Rica, and with Keeley et al. (1984), who studied nine species of *Yucca* from California to Texas. Janzen (1979) found that the average proportion of ovules damaged by wasps varied from 41% to 77% per population. These values are much higher than either Keeley et al. (1984) or I found in yuccas. The proportion of seeds per fruit destroyed by feeding of *T. yuccasella* larvae was 3–45% in Keeley's study and 0.6–35% in my

study. Even adding as much as 10% loss for ovules damaged by oviposition, seed loss to yuccas due to their pollinator is considerably less, on average, in *Yucca* than in *Ficus*. However, there is much greater variation in seed loss in *Yucca* than in *Ficus*. Janzen (1979) observed no less than 25% loss in any one fruit. In yuccas a high proportion of fruit showed no loss of seeds to feeding by yucca moth larvae, and in *Y. elata* there were many fruit in which there was no loss of seeds due to oviposition. Alternatively, there were also yucca fruit in which all potentially viable seeds were lost to yucca moth larvae.

The high variation in damage caused by oviposition and feeding by yucca moths is seen between species, between populations within species, and between fruit within populations. Accounting for this high variation requires an examination of the potential causes for both high and low values of seed damage. Since seed damage is most closely tied to the number of larvae per fruit, I will examine hypotheses which could explain both very low numbers and very high numbers of ovipositions and larvae per fruit.

Keeley et al. (1984) suggest three hypotheses for the absence of yucca moth larvae in individual fruit: 1) pollination by agents other than yucca moths, 2) pollination by yucca moths without oviposition, and 3) egg or larval mortality. However, there is no convincing evidence from the literature that agents other than yucca moths pollinate yuccas. Likewise, oviposition scars show that every yucca fruit has been visited at some time by yucca moths. The most obvious sign of oviposition is constriction of a locule, with a characteristic scar on both inner and outer walls of the locule. Although, there are many fruit which lack constrictions, particularly in *Y. elata* (see Tables 1 F–G, 2 F–G), lack of a constriction does not imply the absence of yucca moths at the time of flowering. There is a morph of *T. yuccasella* (or a new species losely related to *T. yuccasella*) which oviposits shallowly into the carpel wall, rather than into the locular space (Addicott 1985 b). Shallow ovipositions cause no constriction of the fruit, and instead leave a welt on the surface of the fruit. They occur in n. Arizona and s. Utah on *Y. kanabensis*, *Y. baileyi*, and *Y. angustissima*, and may occur in both the presence and absence of the normal mode of oviposition. There are also two morphs of *T. yuccasella* that oviposit in the style rather than the ovary of a yucca pistil, leaving no constriction of the fruit. This apical oviposition is common on both *Y. baccata* and *Y. elata* (Addicott unpublished work). Taking into account the different patterns and locations of oviposition, careful inspection of young fruit invariably shows evidence of inser-

492

tion of yucca moth ovipositors (Addicott unpublished work). Although ovipositor insertion could take place without oviposition, I have found by dissection that there is close to a 1:1 ratio of oviposition scars to yucca moth eggs (Addicott unpublished work). Small deviations from the 1:1 ratio are not large enough to explain the high proportion of fruit lacking larvae.

This leaves egg and/or larval mortality as the only viable explanation for the high proportion of fruit lacking yucca moth larvae. There is very little parasitism of larvae (Force and Thompson 1984), and Kingsolver (1984) estimated larval mortality of T. yuccasella in Y. glauca to be only about 9.4%. I am uncertain of sources of egg mortality, but the rates must be high. For example, in Y. kanabensis there are frequently 10–15 shallow ovipositions per locule, yet it is unusual to find more than 2 larvae per locule. Similarly, the success rate of eggs placed in the styles of Y. elata is low.

Fruit with high numbers of yucca moth larvae present a different problem. Keeley et al. (1984) observed up to 17 larvae per fruit, and I observed up to 24 larvae per fruit from Y. kanabensis in this study. I have observed populations of yuccas in other years in which every fruit had at least 30 and up to 50 larvae. Normal (deep), shallow and apical oviposition at the time of pollination do not appear to be responsible for high numbers of larvae. Kingsolver (1984) observed the behavior of ovipositing T. yuccasella and found that they avoided oviposition in locules in which an egg had already been placed. However, Aker and Udovic (1981) observed T. maculata ovipositing, not just in the ovaries of fresh flowers, but also in young fruit. I hypothesize that these secondary ovipositions are responsible for the high number of larvae per fruit (Addicott unpublished work). When there are high numbers of larvae per fruit, there are usually two cohorts of larvae, the second cohort being numerous relative to the first cohort. Eggs from secondary ovipositions are placed directly into developing seeds (Addicott in preparation), not into the locular cavity, carpal wall or style.

Although I suspect the source for the high numbers of larvae per fruit, I still do not know what causes the secondary ovipositions. Given the differences in oviposition, it could be the result of yet another morph of T. yuccasella, for example, a morph lacking maxillary tentacles (Davis 1967). Alternatively, it could simply be a facultative response of T. yuccasella when its flight season lasts longer than the flowering season of its yuccas.

The preceding has emphasized variation between fruit within populations and between populations within species, but there were still significant differences between species for all but net seed production. Differences were particularly noticeable between baccate and capsular species. Two lines of evidence suggest that these differences are real. First, the patterns I observed are similar to those found by Keeley et al. (1984) from collections made in 1979. Baccate species have relatively low numbers of yucca moth larvae per fruit and a high proportion of fruit without larvae. Second, variation between species is not just a reflection of habitat differences. For example, the consumption of viable seeds by T. yuccasella larvae was lower in Y. baccata than Y. angustissima at both Peach Springs and Gateway (Table 2I). Therefore, differences between baccate and capsular species in how they interact with yucca moths are unlikely to be artifacts of different moth densities in different habitats.

Some of the differences in losses of viable seeds due to feeding by yucca moth larvae may be due to differences in locations where feeding occurs, and a nonrandom distribution of viable and inviable seeds within locules. A high proportion of T. yuccasella larvae fed near the apex of the fruit in both Y. baccata and Y. elata. The low proportion of viable seeds consumed by T. yuccasella larvae in Y. baccata (Tables 1 H, 2 H) may be a function of the distribution of inviable seeds, there being many inviable seeds towards the apical end of Y. baccata fruit.

Proportional net seed production was not significantly different between species, and this could potentially reflect a strong regulation of the interaction between yuccas and yucca moths. First, a small number of pollination events per fruit could be associated with a small proportion of viable ovules, and a small loss due to feeding by yucca moth larvae. Higher visitation would yield more viable ovules but a greater loss of ovules due to oviposition and feeding. Oviposition behavior can change in response to the number of eggs already laid in a given ovary (Kingsolver 1984). Second, there could be selective abortion of pistils carrying large numbers of yucca moth eggs, as fruit abortion is very common in most yuccas (Addicott 1985a, Udovic and Aker 1981). However, the high variation between populations within species of yuccas and between fruit within populations does not support the hypothesis that there is strong regulation of the interaction between yuccas and yucca moths.

A detailed cost/benefit analysis of the interaction between yuccas and yucca moths is not possible with the present data set. For example, damage caused by larvae from different patterns of oviposition should be considered separately. This is particularly important, because some populations of yuccas, and even individual fruit, experience three different types of oviposition. Also, I do not yet have direct measurements of the impact of oviposition on seed inviability. This makes it difficult to accurately assess gross benefits of pollination. However, I can make a preliminary estimate of cost/benefit ratios by dividing costs of yucca moth larvae (Table 1 J) by the sum of net seed production (Table 1 D) and costs (Table 1 J). These estimates indicate that costs are about 20–30% of gross benefits for the capsular yuccas, and 0–20% for the baccate yuccas. Therefore, the obligate pollination mutualism between yuccas and yucca moths is clearly beneficial to the yuccas, but up to 30% of the potential benefits are lost to the yucca moths.

Acknowledgements. Kate Shaw and Daryl Williams assisted in collecting and/or dissecting this material. This work was partially supported by NSERC of Canada Operating Grant A9674 and a grant from the Central Research Fund of the University of Alberta.

References

Addicott JF (1979) A multispecies aphid-ant association: density dependence and species-specific effects. Can J Zool 57:558–569

Addicott JF (1981) Stability properties of 2-species models of mutualism: simulation studies. Oecologia (Berlin) 49:42–49

Addicott JF (1984) Mutualistic interactions in population and community processes. In: Price PW, Slobodchikoff CN, Gaud BS (eds) A new ecology: novel approaches to interactive systems. John Wiley & Sons, New York, pp 437–455

Addicott JF (1985a) Competition in mutualistic systems. In: Boucher D (ed) The biology of mutualism: ecology and evolution. Croom Helm, London, pp 217–247

Addicott JF (1985b) On the population consequences of mutualism. In: Case T, Diamond J (eds) Community Ecology. University of California Press, Berkeley, pp 425–436

Aker CL, Udovic D (1981) Oviposition and pollination behavior of the yucca moth, *Tegeticula maculata,* (Lepidoptera: Prodoxidae), and its relation to the reproductive biology of *Yucca whipplei* (Agavacae). Oecologia (Berlin) 49:96–101

Boucher DH, James S, Keeler KH (1982) The ecology of mutualism. Annu Rev Ecol Syst 13:315–347

Cronquist A, Holmgren AH, Holmgren NH, Reveal JL (1977) Intermountain Flora: Vascular Plants of the Intermountain West, USA The Moncotyledons, Vol. 6. Columbia University Press, New York

Davis DR (1967) A revision of the Moths of the Subfamily Prodoxinae (Lepidoptera: Incurvariidae). US Natl Mus Bull 255:1–170

Dean AM (1983) A simple model of mutualism. Am Nat 121:409–417

Force DC, Thompson ML (1984) Parasitoids of the immature stages of several southwestern yucca moths. Southwest Nat 29:45–56

Howe HF (1984) Constraints on the evolution of mutualisms. Am Nat 123:764–777

Janzen DH (1979) How many babies do figs pay for babies? Biotropica 11:48–50

Keeler K (1981) A model of selection for facultative nonsymbiotic mutualism. Am Nat 118:488–498

Keeley JE, Keeley SC, Swift CC, Lee J (1984) Seed predation due to the Yucca-moth symbiosis. Am Midl Nat 112:191–197

Kingsolver RW (1984) Population biology of a mutualistic association: *Yucca glauca* and *Tegeticula yuccasella.* Ph. D. Thesis. University of Kansas, p 125

Koch GG (1970) The use of non-parametric methods in the statistical analysis of a complex split plot experiment. Biometrics 26:105–128

Louda SM (1982) Inflorescence spiders: a cost/benefit analysis for the host plant, *Haplopappus venetus* Blake (Asteraceae). Oecologia (Berlin) 55:185–191

Mariscuilo LA, McSweeny M (1977) Nonparametric and Distribution-Free Methods for the Social Sciences. Brooks/Cole Publishing Company, Monterey, California, p 556

McKelvey SD (1938) Yuccas of the Southwestern United States, Part 1. The Arnold Arboretum of Harvard University, Jamaica Plain, Massachusetts, p 150

McKelvey SD (1947) Yuccas of the Southwestern United States, Part 2. The Arnold Arboretum of Harvard University, Jamaica Plain, Massachusetts, p 192

Miles NJ (1983) Variation and host specificity in the yucca moth, *Tegeticula yuccasella* (Incurvariidae): A morphometric approach. J Lepid Soc 37:207–216

Powell JA, Mackie RA (1966) Biological relationships of moths and *Yucca whipplei* (Lepidoptera: Gelechiidae, Blastobasidae, Prodoxidae). Univ Calif Pub Entomol 42:1–59

Riley CV (1892) The yucca moth and yucca pollination. Annu Rep Mo Bot Garden 3:99–159

Scheirer CJ (1976) The analysis of ranked data derived from completely randomized factorial designs. Biometrics 32:429–434

Southwick EE (1984) Photosynthate allocation to floral nectar: a neglected energy investment. Ecology 65:1775–1779

Udovic D, Aker C (1981) Fruit abortion and the regulation of fruit number in *Yucca whipplei.* Oecologia (Berlin) 49:245–248

Webber JM (1953) Yuccas of the southwest. US Dept Agric, Agric Monogr No 17, p 93

Received May 16, 1986

Appendix 1. Collection sites and designations of populations where yucca fruit were collected for this study

Species	Population	Location					
Y. schottii	Bisbee	6.6 km	w	Bisbee	Cochise Co.		Arizona
	Nogales	6.5 km	sw	Patagonia	Santa Cruz Co.		Arizona
		23.3 km	ne	Nogales	Santa Cruz Co.		Arizona
	Portal	4.5 km	w	Portal	Cochise Co.		Arizona
		13.8 km	w	Portal	Cochise Co.		Arizona
		3.0 km	w	Portal	Cochise Co.		Arizona
				Portal	Cochise Co.		Arizona
Y. arizonica	Nogales	3.3 km	e	Nogales	Santa Cruz Co.		Arizona
Y. baccata	Peach Springs	6.7 km	w	Peach Springs	Yavapai Co.		Arizona
	Gateway	16.0 km	e	Gateway	Mesa Co.		Colorado
		12.2 km	e	Gateway	Mesa Co.		Colorado
	Kanab	40.3 km	w	Mt. Carmel Jct.	Kane Co.		Utah
		18.7 km	w	Mt. Carmel Jct.	Kane Co.		Utah
		14.2 km	w	Mt. Carmel Jct.	Kane Co.		Utah
	Moab	2.6 km	e	LaSal	San Juan Co.		Utah
Y. glauca	Gunnison	16.5 km	e	Gunnison	Gunnison Co.		Colorado
		7.1 km	n	Gunnison	Gunnison Co.		Colorado
	Poncha Springs	21.3 km	nw	Saguache	Saguache Co.		Colorado
		3.6 km	s	Poncha Springs	Chaffee Co.		Colorado
		8.8 km	w	Poncha Springs	Chaffee Co.		Colorado
		14.2 km	w	Poncha Springs	Chaffee Co.		Colorado
	Wolf Creek	1.6 km	e	Wolf Creek	Lewis and Clark Co.		Montana
		36.0 km	e	Wolf Creek	Lewis and Clark Co.		Montana

494

Appendix 1 (continued)

Species	Population	Location				
Y. glauca	Clines Corners	22.3 km	s	Santa Fe	Santa Fe Co.	New Mexico
		7.1 km	se	Clines Corners	Torrance Co.	New Mexico
		25.7 km	se	Clines Corners	Torrance Co.	New Mexico
Y. baileyi	Page	7.1 km	w	Glen Canyon City	Kane Co.	Utah
		24.3 km	w	Black Mesa	Navajo Co.	Arizona
		24.3 km	w	Black Mesa	Navajo Co.	Arizona
		0.3 km	e	Black Mesa	Navajo Co.	Arizona
Y. angustissima	Peach Springs	27.5 km	e	Peach Springs	Yavapai Co.	Arizona
		5.0 km	w	Peach Springs	Yavapai Co.	Arizona
	Gateway	32.0 km	w	Whitewater	Mesa Co.	Colorado
		18.2 km	s	Gateway	Mesa Co.	Colorado
Y. kanabensis	Kanab	18.5 km	w	Mt. Carmel Jct.	Kane Co.	Utah
		21.5 km	n	Kanab	Kane Co.	Utah
	Rockville	11.7 km	s	Rockville	Washington Co.	Utah
Y. elata	Cottonwood	7.1 km	ne	Cottonwood	Yavapai Co.	Arizona
		5.7 km	ne	Cottonwood	Yavapai Co.	Arizona
		7.1 km	se	Cottonwood	Yavapai Co.	Arizona
	Sonoita	0.8 km	s	Huachuca City	Cochise Co.	Arizona
		15.2 km	e	Sonoita	Santa Cruz Co.	Arizona
	Portal	2.1 km	s	Rodeo	Hidalgo Co.	New Mexico
	Prescott	2.8 km	n	Dewey	Yavapai Co.	Arizona
	Sedona	25.8 km	s	Sedona	Yavapai Co.	Arizona
		19.5 km	s	Sedona	Yavapai Co.	Arizona
	Wickenburg	20.0 km	nw	Wickenburg	Yavapai Co.	Arizona
	Alamagordo	24.2 km	sw	Alamagordo	Otero Co.	New Mexico
		42.2 km	sw	Alamagordo	Otero Co.	New Mexico
		53.3 km	sw	Alamagordo	Otero Co.	New Mexico
		31.7 km	ne	Las Cruces	Dona Ana Co.	New Mexico
	Deming	38.0 km	w	Las Cruces	Dona Ana Co.	New Mexico
		44.3 km	w	Deming	Luna Co.	New Mexico
		27.5 km	n	Hachita	Grant Co.	New Mexico

Index

affirmative action, 7
age- or stage-structured models, 510–11
algal blooms, 317, 318, 319
analysis of variance, 1, 509
anthropogenic land use, and metapopulations, 402
apparency theory of plant defenses, 677–78, 680
apparent competition, 12, 13, 14, 153

BACI and BACIP designs, 4
BEF (biodiversity and ecosystem function), 325
behavioral ecology, 674–77. *See also* foraging behavior;
   territorial behavior
behavioral indirect effects, 153
behaviorally mediated effects, in trophic cascades, 322
binary matrices in biogeography, 4–5
biodiversity: BEF research, 325; limited evidence for local
   declines in, 325; mutualism as mechanism for generating, 681. *See also* species diversity
biogeographic history, and trait clusters, 674, 679
biogeography, binary matrices in, 4–5

canonical correspondence analysis, 4
carbon cycling, and HNLC regions, 318
carbon limitation, and plant defenses, 678
carrying capacity: of parasite's host population, 508;
   reflecting availability and efficient use, 154
catastrophes, and extinction risk, 511, 513
chaos, 508, 509
character associations, 674, 677–80
character displacement patterns, 155–56
climate factors, vs. evolutionary history, 679
climate regulation, and HNLC regions, 318
clutch size in birds, 673, 677
coevolution: geographic mosaic theory of, 403; indirect
   effects in, 400; trade-offs and, 680
coexistence: of coral reef fish species, 158; indirect effects
   in, 400; Lotka-Volterra theory and, 153–54; in lottery
   systems, 158; niche differences and, 151; nonequilibrial,
   17, 154; nonlinear responses to a resource and, 154;
   phylogenetic affinity and, 679; of predator and prey, 11;
   predator-mediated, 11, 12, 13–14, 16, 17, 156; spatially
   structured habitats and, 17, 401; and trade-off between
   growth and mortality risk, 682

colonization of patches: after disturbance, 16, 158; metapopulations and, 401, 402, 403
common core species, 402
community matrices, 152
community structure: competition and predation in, 12;
   competition and the niche in determining, 397; conditions limiting membership in, 151, 152; co-occurrence
   patterns of related species and, 155; equilibrium in
   niche theory and, 397; foraging and diet linking selection with, 676; indirect effects in, 12–15, 682; manipulative experiments and, 400; metacommunities and,
   402, 404; predators in regulation of, 322, 682; stability
   vs. complexity of, 397–98, 400. *See also* succession
compartments, 398, 404
competition: apparent, 12, 13, 14, 153; arguments about
   nature of, 155; character differences and, 155; co-
   occurrence of species on islands and, 155; Hutchinson/
   MacArthur paradigm and, 151–52, 155–56; Lotka-
   Volterra equations and, 152, 153, 154–55; niche theory
   and, 397; number of available resources and, 153; population regulation by, 151; predation as complementary
   force with, 12, 15; resource-ratio theory and, 154–55;
   species diversity and, 12, 14–15, 16, 151; trait combinations in plants and, 677
competitive ability vs. dispersal ability, 17, 680
competitive exclusion principle, 153, 154
conservation biology: analytical models in, 512; biodiverse
   ecosystems and, 325; commercial fishing and, 513–14;
   dialectic between basic ecology and, 514; matrix population models in, 510–11; metapopulations and, 402,
   403, 512; population sinks and, 403; reasoning from
   simple models in, 320; simulation models in, 512, 514
consumer-resource relationships, unified view of, 15
continuous traits, selection on, 680
coral reef diversity, and disturbance, 16–17, 158
core species, 402, 404
costs and benefits. *See* trade-offs

data-poor conditions, 507, 511, 512–13, 514
demographic matrix models, 510–11
demographic stochasticity, 511–13
density-dependent factors: coexistence of prey species
   and, 14; vs. density-independent factors, 507, 509–10;